Fluorescent and Luminescent Probes for Biological Activity

BIOLOGICAL TECHNIQUES

A series of Practical Guides to New Methods in Modern Biology

Series Editor
DAVID B SATTELLE

Computer Analysis of Electrophysiological Signals
J Dempster
Fluorescent and Luminescent Probes for Biological Activity, 1st edition
WT Mason (Editor)
Planar Lipid Bilayers
W Hanke & W-R Schlue
***In Situ* Hybridization Protocols for the Brain**
W Wisden and BJ Morris (Editors)
Manual of Techniques in Insect Pathology
LA Lacey (Editor)
Non-radioactive Labelling: A Practical Introduction
AJ Garman

CLASSIC TITLES IN THE SERIES

Microelectrode Methods for Intracellular Recording and Ionophoresis
RD Purves
Immunochemical Methods in Cell and Molecular Biology
RJ Mayer & JH Walker

BIOLOGICAL TECHNIQUES

Fluorescent and Luminescent Probes for Biological Activity

A Practical Guide to Technology for Quantitative Real-Time Analysis

Second Edition

Edited by

WT MASON

Life Science Resources Ltd
Cambridge, UK

ACADEMIC PRESS
San Diego · London · Boston
New York · Sydney · Tokyo · Toronto

This book is printed on acid-free paper

Academic Press
24–28 Oval Road, London NW1 7DX, UK
http://www.hbuk.co.uk/ap/

Academic Press
a division of Harcourt Brace & Company
525 B Street, Suite 1900, San Diego, California 92101-4495, USA
http://www.apnet.com

ISBN 0–12–447836–0

Library of Congress Catalogue Card Number: 99–60998
A catalogue record for this book is available from the British Library

Typeset by J&L Composition Ltd, Filey, North Yorkshire
Printed in Great Britain by The Bath Press, Bath

99 00 01 02 03 04 BP 9 8 7 6 5 4 3 2 1

Series Preface

The rate at which a particular aspect of modern biology is advancing can be gauged, to a large extent, by the range of techniques that can be applied successfully to its central questions. When a novel technique first emerges, it is only accessible to those involved in its development. As the new method starts to become more widely appreciated, and therefore adopted by scientists with a diversity of backgrounds, there is a demand for a clear, concise, authoritative volume to disseminate the essential practical details.

Biological Techniques is a series of volumes aimed at introducing to a wide audience the latest advances in methodology. The pitfalls and problems of new techniques are given due consideration, as are those small but vital details that are not always explicit in the methods sections of journal papers. The books will be of value to advanced researchers and graduate students seeking to learn and apply new techniques, and will be useful to teachers of advanced undergraduate courses, especially those involving practical and/or project work.

When the series first began under the editorship of Dr John E Treherne and Dr Philip H Rubery, many of the titles were in fields such as physiological monitoring, immunology, biochemistry and ecology. In recent years, most biological laboratories have been invaded by computers and a wealth of new DNA technology. This is reflected in the titles that will appear as the series is relaunched, with volumes covering topics such as computer analysis of electrophysiological signals, planar lipid bilayers, optical probes in cell and molecular biology, gene expression, and *it situ* hybridization. Titles will nevertheless continue to appear in more established fields as technical developments are made.

As leading authorities in their chosen field, authors are often surprised on being approached to write about topics that to them are second nature. It is fortunate for the rest of us that they have been persuaded to do so. I am pleased to have this opportunity to thank all authors in the series for their contributions and their excellent co-operation.

DAVID B SATTELLE ScD

Preface

The first edition of this book in the Biological Techniques Series was originally published in 1993. The book was strongly reviewed and welcomed as an outstanding introduction to the use of optical probes for investigating biological function in living and dead cells. The book focused on the chemical probes required for investigation, and the technologies including imaging, light measurement and computing required for their study.

This field has moved forward rapidly since 1993 on a number of fronts. The number of available probes has multiplied dramatically and some probes have become increasingly available not only as chemicals, but also as molecular vectors able to be expressed in living cells. New probes for ions, receptors and cellular components have also become available, while novel new probes for detecting gene expression in living cells have emerged which are clearly becoming much more widely accessible. Perhaps even more importantly, a number of new technologies for the detection of such probes have also emerged or advanced. CCD camera technology has moved forward dramatically, but additionally confocal technologies have been transformed, particularly in the area of multiple photon excitation. A number of new optical measurement techniques have also been developed and are rapidly coming into new use in a range of laboratories. Some such technologies are even able to probe naturally occurring reactive species in cells, and use them as endogenous probes of the molecular environment.

It is clear that the original book published in 1993 requires a second edition to address these new developments, and to update selected chapters from the original book. The second edition of the book is essentially aimed to extend the first edition in scope, and to be a valuable primer for those wishing to embark on work in the field, or develop their own methodological base to take advantage of such emerging technologies. This second edition, like the previous book, comes from a range of contributors in academia and industry.

Dr W. T. Mason

Contributors

S.R. Adams *Howard Hughes Medical Institute 0647, University of California San Diego, La Jolla, San Diega, CA 92093-0647, USA*

R. Aikens *Photometrics, 3440 East Britannia Drive, Tucson, AZ 85706, USA*

S. Antić *Dept of Cellular and Molecular Physiology, Yale University School of Medicine, PO Box 3333, 333 Cedar Street, New Haven, CT 06520, USA*

B.J. Bacskai *Howard Hughes Medical Institute 0647, University of California San Diego, La Jolla, San Diega, CA 92093-0647, USA*

M.N. Badminton *Dept of Medical Biochemistry, University of Wales College of Medicine, Heath Park, Cardiff CF4 4XN, UK*

T.C. Bakker Schut *Laboratory for Intensive Care Research and Optical Spectroscopy, Institute of General Surgery 10M, Erasmus University Rotterdam and University Hospital Rotterdam 'Dijkzigt', Dr Molewaterplein 40, 3015 GD Rotterdam, The Netherlands*

I. Bar-Am *Applied Spectral Imaging, Migdal Hatmet, Israel*

E. Blancaflor *Dept of Biology, Pennsylvania State University, 208 Mueller Laboratory, University Park, PA 16892, USA*

F. Bobanović *Life Science Resources, Abberley House, Granham's Road, Great Shelford, Cambridge, CB2 5LQ, UK.*

N. Bodsworth *Celsis Ltd, Cambridge Science Park, Milton Road, Cambridge CB4 0FX, UK*

S. Bolsover *Dept of Physiology, University College London, Gower Street, London WC1E 6BT, UK*

M.D. Bootman *Dept of Zoology, University of Cambridge, Downing Street, Cambridge CB2 3EJ, UK*

D.N. Bowser *Confocal and Fluorescence Imaging Group, Dept of Physiology, The University of Melbourne, Parkville, Victoria 3052, Australia*

H.A. Bruining *Laboratory for Intensive Care Research and Optical Spectroscopy, Institute of General Surgery 10M, Erasmus University Rotterdam and University Hospital Rotterdam 'Dijkzigt', Dr Molewaterplein 40, 3015 GD Rotterdam, The Netherlands*

H.P.J. Buschman *Laboratory for Intensive Care Research and Optical Spectroscopy, Institute of General Surgery 10M, Erasmus University Rotterdam and University Hospital Rotterdam 'Dijkzigt', Dr Molewaterplein 40, 3015 GD Rotterdam, The Netherlands, and Dept of Cardiology, Leiden University Medical Center, Leiden, The Netherlands*

A.K. Campbell *Dept of Medical Biochemistry, University of Wales College of Medicine, Heath Park, Cardiff CF4 4XN, UK*

P.J. Caspers *Laboratory for Intensive Care Research and Optical Spectroscopy, Institute of General Surgery 10M, Erasmus University Rotterdam and University Hospital Rotterdam 'Dijkzigt', Dr Molewaterplein 40, 3015 GD Rotterdam, The Netherlands*

B. Chance *Johnson Research Foundation, University of Pennsylvania, Dept of Biochemistry and Biophysics, School of Medicine, Philadelphia, PA 19104, USA*

S.H. Cody *Confocal and Fluorescence Imaging Group, Dept of Physiology, The University of Melbourne, Parkville, Victoria 3052, Australia*

L.B. Cohen *Dept of Cellular and Molecular Physiology, Yale University School of Medicine, PO Box 3333, 333 Cedar Street, New Haven, CT 06520, USA*

J.M.C.C. Coremans *Laboratory for Intensive Care Research and Optical Spectroscopy, Institute of General Surgery 10M, Erasmus University Rotterdam and University Hospital Rotterdam 'Dijkzigt', Dr Molewaterplein 40, 3015 GD Rotterdam, The Netherlands*

G.C. Cox *Australian Key Centre for Microscopy and Microanalysis, University of Sydney, New South Wales 2006, Australia*

A.A. Culbert *Dept of Biochemistry, School of Medicine, University of Bristol, Bristol BS8 1TD, UK*

R. DeBasio *Center for Light Microscope Imaging and Biotechnology, Carnegie Mellon University, 4400 Fifth Avenue, Pittsburgh, PA 15213, USA*

A.W. de Feijter *Meridian Instruments Inc., 2310 Science Parkway, Okemos, MI 48864, USA*

J. Dempster *Dept of Physiology and Pharmacology, University of Strathclyde, Glasgow G1 1XW, UK*

P.N. Dubbin *Confocal and Fluorescence Imaging Group, Dept of Physiology, The University of Melbourne, Parkville, Victoria 3052, Australia*

Z. Dubinsky *Life Sciences Dept, Bar-Ilan University, Ramat-Gan, 52900, Israel.*

C.X. Falk *Dept of Cellular and Molecular Physiology, Yale University School of Medicine, PO Box 3333, 333 Cedar Street, New Haven, CT 06520, USA*

K. Florine-Casteel *Duke University Medical Center, Box 3712, M310 Davison Building, Durham, NC 27710, USA*

N. Foote *Celsis Ltd, Cambridge Science Park, Milton Road, Cambridge CB4 0FX, UK*

M.D. Fricker *Dept of Plant Sciences, University of Oxford, South Parks Road, Oxford OX1 3RB, UK*

T.W.J. Gadella Jr *MicroSpectroscopy Center, Dept of Biomolecular Sciences, Wageningen Agricultural University, Dreijenlaan 3, NL-6703 HA, Wageningen, The Netherlands*

S. Gilroy *Dept of Biology, Pennsylvania State University, 208 Mueller Laboratory, University Park, PA 16892, USA*

E. Gratton *Laboratory of Fluorescence Dynamics, Dept of Physics, University of Illinois at Urbana, IL 61801, USA*

A.M. Gurney *Dept of Physiology and Pharmacology, University of Strathclyde, Royal College, 204 George Street, Glasgow G1 1XW, UK*

R. Haggart *Celsis Ltd, Cambridge Science Park, Milton Road, Cambridge CB4 0FX, UK*

K. Hahn *Dept of Neuropharmacology, Scripps Research Institute, 1066 Torrey Pines Road, La Jolla, CA 92037, USA*

R.P. Haugland *Molecular Probes Inc., 4849 Pitchford Avenue, Eugene, OR 97402-9165, USA*

B. Herman *Laboratories for Cell Biology, CB No 7090, Dept of Cell Biology and Anatomy, University of North Carolina School of Medicine, 108 Taylor Hall, Chapel Hill, NC 27599-7090, USA*

D. Hoekstra *Dept of Physiological Chemistry, University of Groningen, A. Deusinglaan 1, 9713 AV Groningen, The Netherlands*

M.A. Horton *Bone and Mineral Centre, Dept of Medicine, The Rayne Institute, 5 University Street, London WC1E 6KK, UK*

J. Hoyland *Life Science Resources Ltd, Abberley House, Granham's Road, Great Shelford, Cambridge CB2 5LQ, UK*

A. Ichihara *Yokogawa Research Institute Corp., 2-9-32 Nakacho, Musashino-shi, Tokyo 180-8750, Japan*

H. Ishida *Department of Physiology, School of Medicine, Takai University, Boseidai, Isehara-shi, Kanagawa Prefecture, 259-1100 Japan*

I.D. Johnson *Molecular Probes Inc., 4849 Pitchford Avenue, Eugene, OR 97402-9165, USA*

H.E. Jones *Dept of Medical Biochemistry, University of Wales College of Medicine, Heath Park, Cardiff CF4 4XN, UK*

S.R. Kain *Cell Biology Group, CLONTECH Laboratories Inc., 1020 East Meadow Circle, Palo Alto, CA 94303-4230, USA*

M.S. Kannan *Dept of Veterinary Pathobiology, University of Minnesota, St Paul MN 55108 USA*

F.H. Kasten *P.O. Box 1557, Johnson City, Tennessee 37605–1557, USA*

S. Katz *Life Sciences Dept, Bar-Ilan University, Ramat-Gan, 52900, Israel*

J.M. Kendall *Nycomed Amersham plc, Forest Farm, Whitchurch, Cardiff CF4 7YT, UK*
H. Knight *Dept of Plant Sciences, University of Oxford, South Parks Road, Oxford OX1 3RB, UK*
M.R. Knight *Dept of Plant Sciences, University of Oxford, South Parks Road, Oxford OX1 3RB, UK*
J.W. Kok *Laboratory of Physiological Chemistry, 9712 KZ, University of Groninger, Bloemsingel, The Netherlands*
J. Kolega *Dept of Anatomy and Cell Biology, State University of New York, School of Medicine and Biomedical Science, 3435 Main Street, Buffalo, NY 14214-3000, USA*
I. Kurtz *Division of Nephrology, Dept of Medicine, Center for Health Studies, 10833 Le Conte Avenue, Los Angeles, CA 90024-1689, USA*
J.J. Lemasters *Laboratories for Cell Biology, CB No 7090, Dept of Cell Biology and Anatomy, University of North Carolina School of Medicine, 108 Taylor Hall, Chapel Hill, NC 27599-7090, USA*
P. Lipp *Laboratory of Molecular Signalling, The Babraham Institute, Babraham, Cambridge CB2 4AT, UK*
D.H. Llewellyn *Dept of Medical Biochemistry, University of Wales College of Medicine, Heath Park, Cardiff CF4 4XN, UK*
L. Loew *Dept of Physiology, University of Connecticut Health Center, Farmington, CT 06030-1507, USA*
A. Lyons *Photonic Science, Millham, Mountfield, Robertsbridge, E. Sussex, TN32 5LA, UK*
C. Mackay *Institute of Astronomy, University of Cambridge, Madingley Road, Cambridge CB3 0HA, UK*
Z. Malik *Life Science Dept, Bar-Ilan University, Microscopy Unit, Ramat-Gan, 52900, Israel*
W.T. Mason *Life Science Resources Ltd, Abberley House, Granham's Road, Great Shelford, Cambridge CB2 5LQ, UK*
B.R. Masters *Laboratory of Fluorescence Dynamics, Dept of Physics, University of Illinois at Urbana, IL 61801, USA*
T.J. McCann *Life Science Resources, Abberley House, Granham's Road, Great Shelford, Cambridge CB2 5LQ, UK*
T.J. Mitchison *Dept of Pharmacology, University of California, San Francisco, CA 94143, USA*
J. Montibeller *Center for Light Microscope Imaging and Biotechnology, Carnegie Mellon University, 4400 Fifth Avenue, Pittsburgh, PA 15213, USA*
J. Myers *Center for Light Microscope Imaging and Biotechnology, Carnegie Mellon University, 4400 Fifth Avenue, Pittsburgh, PA 15213, USA*
W. O'Brien *Life Science Resources, Abberley House, Granham's Road, Great Shelford, Cambridge CB2 5LQ, UK*
C. Plieth *Dept of Plant Sciences, University of Oxford, South Parks Road, Oxford OX1 3RB, UK*
J.S. Ploem *Medical Faculty, University of Leiden, Wassenaarseweg 72, 2333 AL Leiden, The Netherlands*
P. Post *Dept of Biological Sciences, Yale University, New Haven, CT 06520, USA*
Y.S. Prakash *Assistant Professor of Anesthesiology, Mayo Clinic, 200 First Street SW, Rochester, MN 55905, USA*
G.J. Puppels *Laboratory for Intensive Care Research and Optical Spectroscopy, Institute of General Surgery 10M, Erasmus University Rotterdam and University Hospital Rotterdam 'Dijkzigt', Dr Molewaterplein 40, 3015 GD Rotterdam, The Netherlands*
T.J. Römer *Laboratory for Intensive Care Research and Optical Spectroscopy, Institute of General Surgery 10M, Erasmus University Rotterdam and University Hospital Rotterdam 'Dijkzigt', Dr Molewaterplein 40, 3015 GD Rotterdam, The Netherlands, and Dept of Cardiology, Leiden University Medical Center, Leiden, The Netherlands*
C. Rothmann *Life Sciences Dept, Bar-Ilan University, Microscopy Unit, Ramat-Gan, 52900, Israel*
G.A. Rutter *Dept of Biochemistry, School of Medicine, University of Bristol, Bristol BS8 1TD, UK*
G.B. Sala-Newby *Dept of Surgery, Bristol Heart Institute, University of Bristol, Bristol Royal Infirmary, Bristol BS2 8HW, UK*
K.E. Sawin *Dept of Biochemistry and Biophysics, University of California, San Francisco, CA 94143, USA*
G. Shankar *NPS Pharmaceuticals Inc., 420 Chipeta Way, Salt Lake City, UT 84108, USA*
C.J.R. Sheppard *Physical Optics Dept, School of Physics, University of Sydney, Sydney, New South Wales 2006, Australia*
M. Shimizu *Confocal Scanner Group, Yokogawa Electric Corporation, 2-9-32 Nakacho, Musashino-shi, Tokyo 180-8750, Japan*

S.L. Shorte *INSERM U-261, Institut Pasteur, 28 rue du Dr Roux, 75274 Paris Cedex 15, France*

G.C. Sieck *Dept of Anesthesiology, Physiology and Biophysics, Mayo Clinic, 200 First Street SW, Rochester, MN 55905, USA*

E.R. Simons *Dept of Biochemistry, Boston University School of Medicine, 80 East Concord Street, Boston, MA 02118-2394, USA*

V. Singer *Molecular Probes Inc., 4849 Pitchford Avenue, Eugene, OR 97402-9165, USA*

G. Skews *Life Science Resources, Abberley House, Granham's Road, Great Shelford, Cambridge CB2 5LQ, UK*

P.T.C. So *Dept of Mechanical Engineering, Massachusetts Institute of Technology, 77 Massachusetts Avenue, Cambridge, MA 02139, USA*

B. Somasundaram *Life Science Resources, Abberley House, Granham's Road, Great Shelford, Cambridge CB2 5LQ, UK*

T. Tanaami *Confocal Scanner Group, Yokogawa Electric Corporation, 2-9-32 Nakacho, Musashino-shi, Tokyo 180-8750, Japan*

J.M. Tavaré *Dept of Biochemistry, School of Medicine, University of Bristol, Bristol BS8 1TD, UK*

S.S. Taylor *Dept of Chemistry 0654, University of California San Diego, La Jolla, CA 92093-0654, USA*

K.M. Taylor *Tenovus Building, University of Wales College of Medicine, Heath Park, Cardiff CF4 4XX, UK*

D.L. Taylor *Carnegie Mellon University, 4400 Fifth Avenue, Pittsburgh, PA 15213-2683, USA*

J.A. Theriot *Dept of Biochemistry and Biophysics, University of California, San Francisco, CA 94143, USA*

P. Tomkins *Photonic Science, Millham, Mountfield, Robertsbridge, E. Sussex, TN32 5LA, UK*

J.E. Trosko *Dept of Pediatrics & Human Development, B240 Life Sciences, Michigan State University, East Lansing, MI 48824-1317, USA*

R.Y. Tsien *Howard Hughes Medical Institue 0647, University of California San Diego, La Jolla, San Diega, CA 92093-0647, USA*

M.H. Wade *Meridian Instruments Inc., 2310 Science Parkway, Okemos, MI 48864, USA*

N.S. White *Dept of Plant Sciences, University of Oxford, South Parks Road, Oxford OX1 3RB, UK*

D.A. Williams *Confocal and Fluorescence Imaging Group, Dept of Physiology, The University of Melbourne, Parkville, Victoria 3052, Australia*

R. Wolthuis *Laboratory for Intensive Care Research and Optical Spectroscopy, Institute of General Surgery 10M, Erasmus University Rotterdam and University Hospital Rotterdam 'Dijkzigt', Dr Molewaterplein 40, 3015 GD Rotterdam, The Netherlands*

J.-y. Wu *Georgetown Institute of Cognitive and Computational Sciences, Georgetown University, Washington DC 20007, USA*

K.-P. Yip *Dept of Molecular Pharmacology, Physiology and Biotechnology, Brown University, Providence, Rhode Island, USA*

D. Zečević *Dept of Cellular and Molecular Physiology, Yale University School of Medicine, PO Box 3333, 333 Cedar Street, New Haven, CT 06520, USA*

A.V. Zelenin *Laboratory for Functional Morphology of Chromosomes, Engelhardt Institute of Molecular Biology, Russian Academy of Sciences, Vavilov Street, Moscow 117984, Russia*

Contents

Part XI Applications of Optical Probe Imaging to Biological Problems

Colour plates appear between pages 102 and 103 (Plates 5.1 to 19.1), and 326 and 327 (Plates 23.1 to 46.2)

Available in the first edition of Fluorescent and Luminescent Probes, also edited by Bill Mason:

- **Probes for the Endoplasmic Reticulum** (Chapter 8)
  M. Terasaki
- **Probing Mitochondrial Membrane Potential in Living Cells by a J-Aggregate-forming dye** (Chapter 9)
  L.B. Chen & S.T. Smiley
- **A Small-area Cooled CCD Camera and Software for Fluorescence Ratio Imaging** (Chapter 13)
  G.J. Law & W. O'Brien
- **Multiparameter Imaging of Cellular Function** (Chapter 14)
  G.R. Bright
- **Digital Image Analysis: Software Approaches and Applications** (Chapter 15)
  G.T. Relf
- **Flow Cytometric Analysis and Sorting of Viable Cells** (Chapter 24)
  J.F. Keij & H. Herweijer

PART I

Introduction to Fluorescence Microscopy

CHAPTER ONE

Fluorescence Microscopy

JOHAN S. PLOEM
Medical Faculty, University of Leiden, The Netherlands

1.1 INTRODUCTION

1.1.1 Applications of fluorescence microscopy

As a tool in microscopy, fluorescence provides a number of possibilities in addition to absorption methods. Fluorescence probes can, for instance, be selectively excited and detected in a complex mixture of molecular species. It is also possible to observe a very small number of fluorescent molecules – approximately 50 molecules can be detected in 1 μm^3 volume of a cell (Lansing Taylor *et al.*, 1986). Furthermore, fluorescence microscopy offers excellent temporal resolution, since events that occur at a rate slower than about 10^{-8} s can be detected and measured with appropriate instrumentation. When confocal laser scanning is used in fluorescence microscopy, the theoretical limits of the spatial resolution (determined by the numerical aperture of the objective and the wavelength of the emitted fluorescence light) can be obtained in practice. In conventional microscopy, this is very difficult to obtain.

Immunofluorescence microscopy has been the most common application of fluorescence microscopy in cell biology (Coons *et al.*, 1941). The possibility of detecting multiple regions, represented by specific antigens in the same cell, by selective binding of antibodies marked with fluorophores with different fluorescence colours is often used nowadays in *in situ* hybridization studies of, for example, DNA sequences in the interphase nucleus (Nederlof *et al.*, 1990).

Fluorescence microscopy is also often used for the study of living cells (Kohen & Hirschberg, 1989). It is possible to measure, for example, the pH, free calcium and NAD(P)H concentration in the cytoplasm, as well as intercellular communications between cells. Flow cytometry as a specialized form of fluorescence microscopy (Melamed *et al.*, 1990) permits the examination of biological surfaces when cells pass a beam of excitation light from a laser. A large number of cells can be analysed in a relatively short period of time by using several fluorescent probes in this technology.

1.1.2 The nature of fluorescence

Hot bodies that are self-luminous solely because of their high temperature are said to emit incandescence.

FLUORESCENT AND LUMINESCENT PROBES, 2ND EDN
ISBN 0–12–447836–0

All other forms of light emission are called luminescence. A system emitting luminescence is losing energy. Consequently, some form of energy must be applied from elsewhere and most kinds of luminescence are classified according to the source of this energy. One speaks, therefore, of electroluminescence, radioluminescence, chemiluminescence, bioluminescence and photoluminescence. In the latter form of luminescence the energy is provided by the absorption of ultraviolet, visible or infrared light. Fluorescence is a type of luminescence in which light is emitted from molecules for a very short period of time, following the absorption of light. The emitted light is termed fluorescence if the delay between absorption and emission of photons is of the order of 10^{-8} s or less. Delayed fluorescence is the term used if the delay is about 10^{-6} s, while a delay of greater than about 10^{-6} s results in phosphorescence. All these phenomena can be seen in microscopy.

1.1.3 Fluorescent stains

Compounds exhibiting fluorescence are called fluorophores or fluorochromes. When a fluorophore absorbs light, energy is taken up for the excitation of electrons to higher energy states. The process of absorption is rapid and is immediately followed by a return to lower energy states, which can be accompanied by emission of light. The spectral characteristics of a fluorochrome are related to the special electronic configurations of a molecule. Absorption and emission of light take place at different regions of the light spectrum (Fig. 1.1). According to Stokes's law the wavelength of emission is almost always longer than the wavelength of excitation. It is this shift in wavelength that makes the observation of the emitted light in a fluorescence microscope possible. The excitation light of shorter wavelengths is prevented from entering the eyepieces by using the appropriate dichromatic (dichroic) dividing mirrors (Ploem, 1967). It should be noted that the intensity of the emitted light is weaker than that of the excitation light, as the emitted energy is much smaller than the energy needed for excitation. For different fluorochromes this may vary and is known as the quantum efficiency of the fluorophore used.

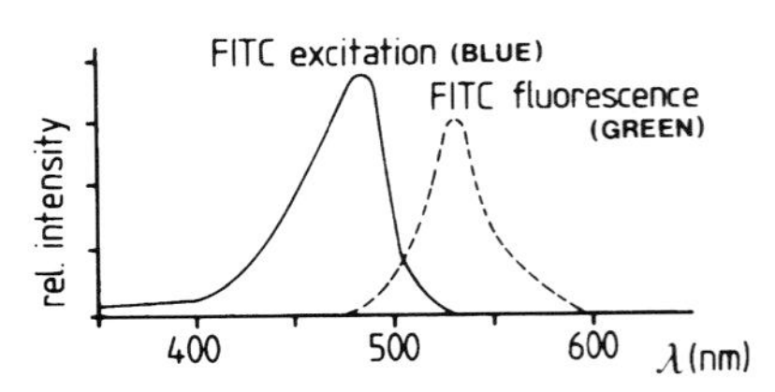

Figure 1.1 Excitation (absorption) and fluorescence (emission) spectra of fluorescein isothiocyanate (FITC).

Different fluorochromes are characterized by their absorption and emission spectra. The absorption or excitation spectrum is obtained by recording the relative fluorescence intensity at a certain wavelength when the specimen is excited with varying wavelengths. The most intense fluorescence occurs when the specimen is irradiated with wavelengths close to the peak of the excitation curve. An example of an absorption and an emission spectrum is given in Fig. 1.1. Most excitation and emission curves overlap to a certain extent.

Decrease in fluorescence during irradiation with light is called fading. The degree of fading depends on the intensity of the excitation light, the degree of absorption by the fluorophore of the exciting light and the exposure time (Patzelt, 1972). Reduction in fluorescence intensity can also be due to modification in the excited states of the fluorophore. These physicochemical changes may be caused by the presence of other fluorophores, oxidizing agents, or salts of heavy metals. This phenomenon is called quenching. Prior to microscopy a decrease in the potential to fluoresce can also occur. Preparations are therefore best stored in the dark at 4°C. To reduce fading during microscopy, agents such as DABCO (1,4-diazobicyclo-2,2,2-octane), *N*-propylgallate and *p*-phenylenediamine should be added to the mounting medium (Gilot & Sedat, 1982; Johnson & Nogueira Araujo, 1981).

1.1.4 Specialized literature on fluorescence microscopy

A number of books have been published recently on (quantitative) fluorescence microscopy and its applications. A few interesting examples are the books by Rost (1991), Kohen and Hirschberg (1989), and Lansing Taylor *et al.* (1986). Also, specialized techniques of fluorescence microscopy such as laser scanning fluorescence microscopy and confocal laser scanning microscopy have found wide applications, and consequently have been included in most recent books dealing with microscopy.

1.2 MICROSCOPE DESIGN

A fluorescence microscope is designed to provide an optimal collection of the fluorescence signal from the specimen, while minimizing the background illumination consisting of unwanted excitation light and autofluorescence. This requires rather sophisticated technology, since the specific fluorescence from the specimen can be several orders of magnitude weaker than the intensity of the exciting light. In the first place the fluorophore in the preparation must be excited with wavelengths as close as possible to the absorption peak of the fluorophore, assuming that the light source emits sufficiently in this wavelength region (Ploem, 1967). Secondly, the fluorescence emission collected by the optical system of the microscope must be maximized.

Strong excitation of the fluorophore with relatively efficient collection of the fluorescence is often not a good solution, since intense illumination may cause excessive fading of the fluorophore. Also, exciting light which does not correspond well with the excitation peak of the fluorochrome will often cause unnecessary autofluorescence of the tissue and optical parts, diminishing the image contrast. This contrast is determined by the ratio of the fluorescence emission of the specifically stained structures to the light observed in the background. For a good separation of exciting and fluorescence light the use of narrow-band excitation filters, which often have a relatively low transmission, is therefore necessary.

For easy visual observation, however, or photography with reasonably short exposure times, a sufficiently bright image is required. To that purpose a compromise between the intensity of the fluorescence and the level of background illumination must sometimes be accepted. If only a few fluorescent molecules are to be observed, not only the non-specific autofluorescence of tissue components, but also the level of autofluorescence of the glass components of the objective, immersion oil and the mounting medium can interfere with the observation of specific fluorescence. Laser scanning microscopy can provide a partial solution for these types of problems, as will be explained later in this chapter.

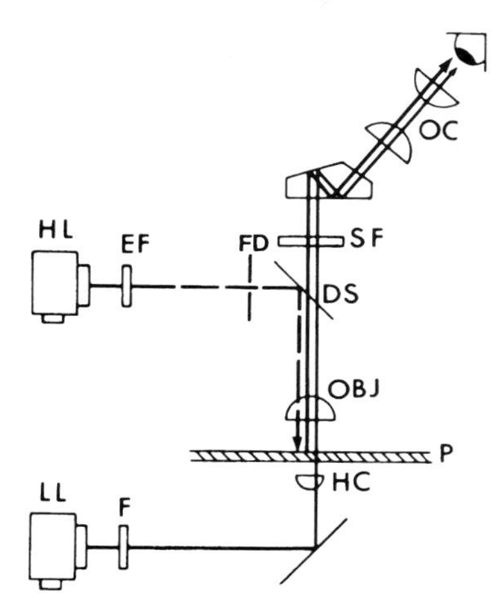

Figure 1.2 Schematic diagram of a microscope for fluorescence microscopy with transmitted (—) and incident (– – –) illumination. LL/HL = light source; EF/F = excitation filter; FD = field diaphragm; DS = dichroic mirror; SF = barrier filter; HC = condenser; P = preparation; OBJ = objective; OC = ocular.

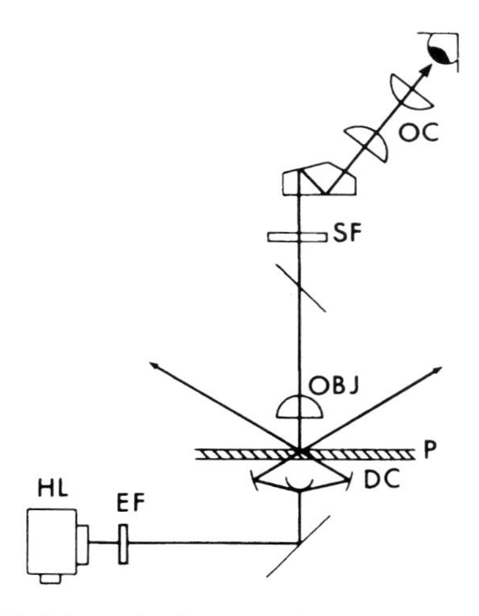

Figure 1.3 Schematic diagram of transmitted illumination with a dark-field condenser (DC). Other abbreviations as in Fig. 1.2.

1.3 TYPES OF ILLUMINATION

A fluorescence microscope is a conventional compound microscope. There are two basic types of illumination for fluorescence microscopy (FM): transmitted illumination (Young, 1961; Nairn, 1976) and incident illumination (Ploem, 1967; Kraft, 1973). The illumination pathway of transmitted light illumination is shown in Fig. 1.2. A condenser focuses the excitation light onto a microscope field. The emitted fluorescence is collected by the objective and observed through the eyepieces. In this configuration it is essential that two different lenses are used: a condenser to focus the excitation light on the specimen and an objective to collect the emitted fluorescence light. For optimal observation of fluorescing images these two lenses, which have independent optical axes, must be perfectly aligned. This is not always easy to obtain and maintain in routine use. It should also be realized that focusing of the excitation rays by the condenser onto the specimen and focusing of the objective for the observation of fluorescence are two different procedures.

For transmitted light illumination two types of condensers can be used. The excitation pathway either contains a bright-field condenser which allows all the exciting light to enter the objective or a dark-field condenser which illuminates the specimen with an oblique cone of light in such a way that no direct exciting light enters the objective (Fig. 1.3). The latter type of condenser, of course, facilitates separation of

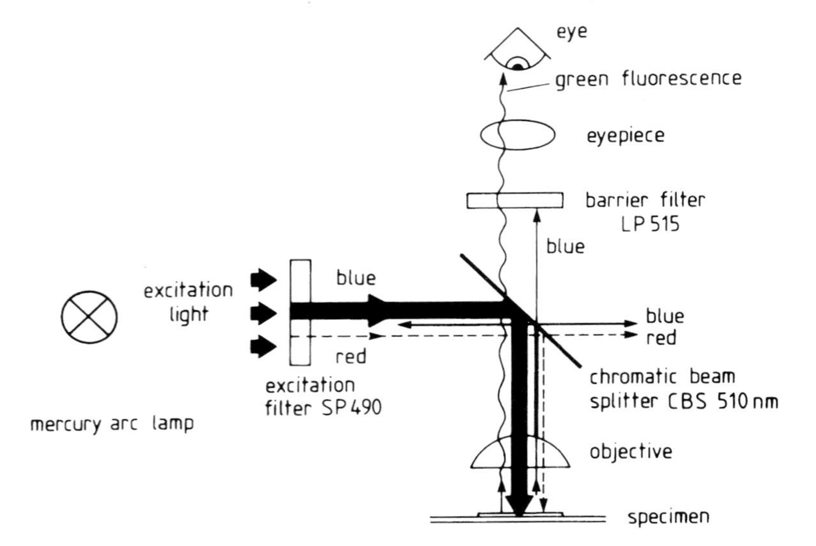

Figure 1.4 Light path in incident illumination. The vertical illuminator equipped with a chromatic beam-splitter has a high reflectance for blue excitation light and a high transmittance for green fluorescence light.

fluorescence from excitation light. Due to the fact that high-performance interference filters only became available after 1970, dark-field illumination using coloured glass filters was the best method to remove unwanted excitation light from the observing light path until 1970. In dark-field condensers part of the aperture is obscured to prevent the light from entering the objective, which must be used at a limited numerical aperture (NA) in order to avoid entrance of unwanted excitation light. Often a working aperture of less than about NA=0.7 is used, whereas good-quality objectives might have apertures of NA=1.4.

As mentioned above, originally, coloured glass filters of the Schott UG1, BG12, etc. type were used to select the excitation light. With these filters it was not possible to absorb all the excitation light with a barrier filter when bright-field illumination was used. Hence, the popularity of dark-field illumination, which did not put such high requirements on excitation and fluorescence filters. With modern high-performance interference filters it is now much easier to use full aperture transmitted bright-field illumination. Furthermore, dark-field condensers do not allow a combination of transmitted fluorescence with phase-contrast or differential interference contrast.

Incident or epi-illumination fluorescence microscopy is shown in Figs 1.2, 1.4 and 1.5. To focus the excitation light onto the specimen and to collect the emitted light from the fluorescing specimen only the objective lens is used. The advantage of epi-illumination is that the same lens system acts as objective and condenser. Focusing the objective onto the specimen results in proper alignment of the microscope, with the same alignment for excitation light and the observed fluorescence light. The illuminated field is the field of view.

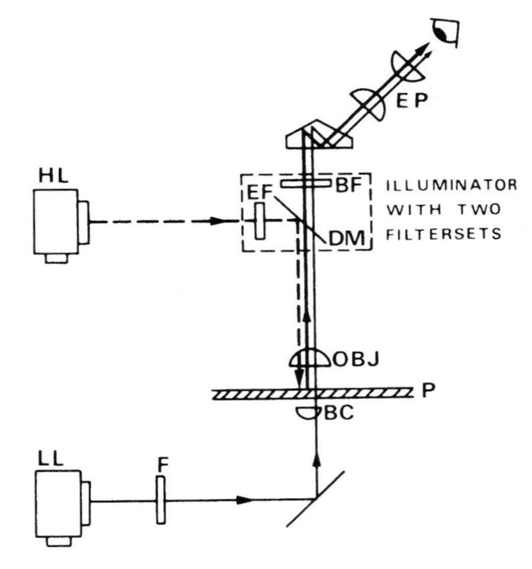

Figure 1.5 Excitation filters (EF), dichroic mirrors (DM) and barrier filters (BF) are mounted in one filter block which may contain up to four of such filter sets for different applications. HL/LL = light source; F = filter; BC = bright-field condenser; P = preparation; OBJ = objective; EP = eye pieces.

To direct the excitation light onto the specimen, a special type of mirror – a chromatic beam-splitter (CBS), also known as a dichroic mirror – is positioned above the objective. These mirrors have a special interference coating, which reflects light shorter than a certain wavelength and transmits light of longer wavelengths. Thus these mirrors effectively reflect the shorter wavelengths of the exciting light onto the specimen and transmit the longer wavelengths of the emitted fluorescence towards the eyepieces.

Also in incident illumination, a relatively small

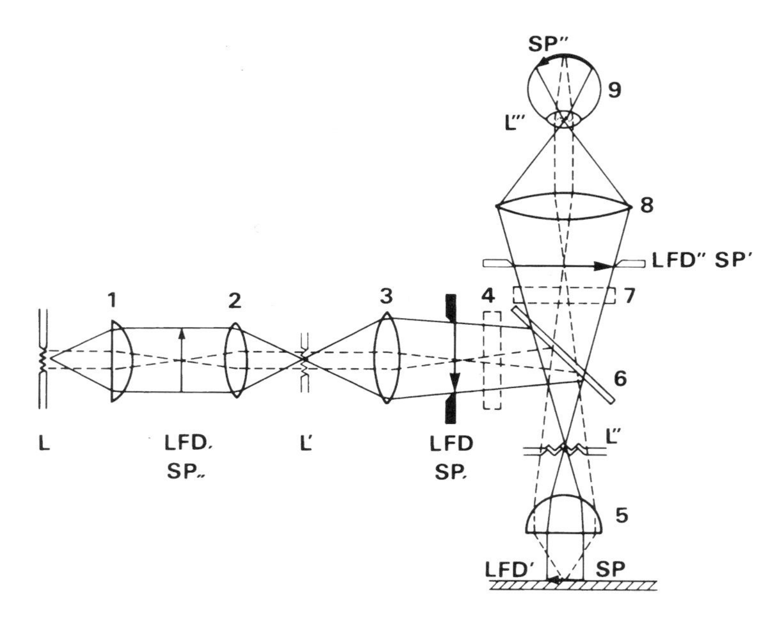

Figure 1.6 Images in Koehler illumination. —— = imaging light path; – – – – – – = illuminating light path. L = light source; LFD = luminous field diaphragm; SP = specimen; 1 = collector; 2 and 3 = auxiliary lenses; 4 = excitation filter; 5 = objective/condenser in epi-illumination; 6 = chromatic beam-splitter; 7 = barrier filter; 8 = eyepiece; 9 = eye. L′, L″, L‴, LFD′, LFD″, SP′ and SP″ are forward images; LFD, , SP, and SP,, are backward images.

amount of exciting light may be reflected by the specimen or optical parts in the direction of the eyepieces. This unwanted excitation light is effectively deflected out of the observation light path by the same chromatic beam-splitter, blocking this light from reaching the eyepieces. In principle, the CBS acts thus as both excitation and barrier filter. In practice, an additional barrier filter is, however, still needed to eliminate any residual unwanted excitation light. Figure 1.4 gives an example of the use of a chromatic beam-splitter. Chromatic beam-splitters exist for separation of all regions of the light spectrum, from the UV (300 nm) to the far red (700 nm). They are mounted in units together with an excitation and barrier filter, especially selected for each separate wavelength range (Fig. 1.5). Units for UV, violet, blue, green and red excitation are provided by several microscope manufacturers.

Epi-illumination often makes use of a vertical illuminator allowing an illuminating light path according to the Koehler principle (Fig. 1.6). The image of the light source is focused onto an iris diaphragm which is conjugate with the entrance pupil or back aperture of the objective lens. This iris diaphragm therefore determines the illuminated aperture. Opening and closing of this aperture diaphragm results in an increase or decrease of the intensity of the illumination, without changing the size of the illuminated field. In addition, a field diaphragm is present which is brought into focus on the specimen plane. This diaphragm controls the size of the illuminated area of the specimen without affecting the intensity of the illuminated area. Closing the field diaphragm as much as the specimen observed allows, generally increases the image contrast of the specimen due to the decrease in autofluorescence of optical parts and a further elimination of still remaining unwanted excitation light.

Moreover, epi-illumination permits an easy changeover or combination between fluorescence and transillumination microscopy, since the substage illumination remains available. Combinations of fluorescence with, for example, phase-contrast microscopy, differential interference contrast and polarization microscopy make it possible to compare the distribution of fluorescence in a specimen, while these transmitted light contrast methods give insight in the structure of the specimen.

In general, epi-illumination is also used in confocal fluorescence scanning microscopy, and in inverted microscopes used for the study of living cells (Ploem *et al.*, 1978). In the latter instrument the epi-illuminator is mounted underneath the stage supporting the dishes or trays in which cells are grown or collected. Objectives for inverted microscopy should be selected such that they have a sufficient working distance to enable focusing on the cells on the plastic bottom of the tray. A disadvantage is that some plastics show a considerable autofluorescence when excited with short wavelengths. Preferably narrow-band long-wavelength blue or green excitation light should be used in combination with fluorochromes having absorption peaks in this wavelength area.

1.4 LIGHT SOURCES

Four major characteristics of light sources must be considered: (1) the spectral distribution of the emitted

Table 1.1 Lamps for fluorescence microscopy in order of intrinsic brilliancy of the arc.

Lamp	*Mean luminous density (cd cm^{-2})*	*Wavelength region*
Hg 100 W	170 000	Main peaks at 366, 405, 436, 546 and 578 nm
Hg 200 W	33 000	
Hg 50 W	30 000	
Xe 75 W	40 000	Continuum and peaks > 800 nm
Xe 450 W	35 000	
Xe 150 W	15 000	
Halogen		Little UV & violet emission; higher intensity towards longer wavelengths
Various lasers		Specific lines

wavelengths; (2) the spectral density of the radiance of the arc or filament representing the radiant intensity per unit area; (3) the uniformity of the illumination in the microscope field; and (4) the stability of the light output over time and the spatial stability of the arc in high-pressure lamps.

The choice of light source is determined by the excitation spectrum of the fluorochrome and its quantum efficiency, the number of fluorochrome molecules that one wants to detect and the sensitivity of the detector used: human eye, film, photomultiplier or TV (CCD) camera. Halogen, mercury and xenon high-pressure arc lamps, and various laser light sources are available. Halogen and xenon lamps have more or less continuous emission spectra; mercury arc lamps have strong emission peaks, and laser light sources emit their energy in multiple lines. The choice of the light source depends also on the mode of illumination. With full field illumination halogen or arc lamps are suitable. In scanning illumination, as is mostly used in confocal fluorescence microscopy, multiple small spots in the specimen must be illuminated sequentially, and only an intense small light beam from a laser can provide sufficient photons to allow a relatively fast scanning of a microscope field. Laser light is coherent and can cause interference phenomena in the imaging of the microscope. With non-perfect excitation and barrier filter systems, leaking (unwanted) exciting laser light can cause interference images. Optical systems are therefore adapted to make laser light non-coherent for use in a microscope set-up.

Weak fluorochromes with low quantum efficiency (low Q) or low numbers of fluorochrome molecules require more excitation light for viewing than strong fluorochromes. Often strong light from an entire laser light source is concentrated on a small (0.5–1 μm) spot in the specimen, as in (confocal) laser scanning fluorescence microscopy.

Tungsten halogen (12 V, 50 and 100 W) lamps are suitable and inexpensive light sources for routine investigations, provided that the specimen emits fluorescence of sufficiently high intensity. These lamps can be used for both transmitted and incident light illumination, and can be switched on and off easily and frequently without damage to the lamp (Tomlinson, 1971).

A mercury lamp has peak emissions at, for example, 366, 405, 436, 546 and 578 nm, but also a strong background continuum. In the blue region, for instance, this continuum is still stronger than that given by a tungsten halogen lamp. If UV or violet, or green light is required, the mercury peaks at these wavelength ranges are preferred (Thomson & Hageage, 1975). Mercury lamps are available at 50, 100 and 200 W. It should be noted that the 100 W lamp has a smaller arc than the 50 and 200 W lamps. Ideally, the collecting lens of the lamphouse should provide an image of the arc onto the entrance pupil of the objective used for epi-illumination. It is clear that a collecting lens of fixed focal length cannot project different sizes of arcs in such a way that the entrance pupil of an objective is always homogeneously filled with an image of the arc for Koehler illumination. A zoom collecting lens should be constructed by the optical industry to solve this problem. Inhomogeneous illumination can thus not always be avoided. For homogeneous illumination of an entire microscopic field, which is desirable in fluorescence image analysis, the very large arc of a xenon 450 high-pressure lamp is sometimes used.

The mercury lamps have a limited lifetime (about 200 burning hours). They are mostly operated on AC current supply. The HBO 100 W can be operated on DC supply for increased stability in microfluorometry. The filter sets developed for fluorescence microscopy are mostly chosen in relation to the location of the major mercury emission peaks in the emission spectrum.

Xenon lamps emit a wide and flat spectrum of rather constant energy from UV to red, without strong peaks (Tomlinson, 1971). They are available as 75, 150 and 450 W with lifetimes of 400, 1200 and 2000 burning hours respectively. Xenon lamps should be handled with care, because even cold lamps are under relatively high pressure, and safety eyeglasses should be used during removal and replacement. The lamps are operated on DC current supply. Unfortunately the xenon 450 W DC operated lamp, which has a relatively long lifetime, needs a rather expensive power supply. For their use in microfluorometry the lamps should be burnt in, under conditions of low mechanical vibrations, e.g. during the night and with a voltage stabilizer to overcome large voltage fluctuations. This creates fewer and more stable burning points, resulting in greater stability of the arc.

Laser light sources emit strong lines which provide monochromatic radiation of very high energy. As such they provide, therefore, potential light sources for special purposes fluorescence microscopy applications that need such types of excitation light (Bergquist & Nilsson, 1975; Wick *et al.*, 1975). Lasers can provide continuous output of energy or operate in a pulsed mode. With the use of short pulses of excitation energy (1 μs to 1 ns), delayed fluorescence phenomena can be studied (Jovin & Vaz, 1989; Beverloo *et al.*, 1990). Lasers are also used in fluorescence scanning confocal microscopy (Wilke, 1983; see also Chapter 17 of 1st edition).

Without aiming at confocal microscopy, it is possible to use laser scanning microscopy only for illumination of the field (Ploem, 1987). In this set-up, a vibrating mirror system is used to generate a meander of a few hundred thousand laser illuminated spots (0.5 μm) over the entire microscopic field in less than a second by using epi-illumination fluorescence microscopy and a photomultiplier for the recording of the fluorescence of each single spot. Since the energy of the entire laser output is concentrated on each 0.5 μm spot, an extremely high excitation energy is obtained. Since only a small pencil of light passes the objective lens at any moment for the illumination of one spot, the autofluorescence of glass in the objective contributing to the background light is low. It is especially low in relation to conventional microscopy, where the entire objective is filled with a massive excitation light beam needed to illuminate the entire microscope field simultaneously.

Modern fluorescence microscopy requires a range of light sources to meet the varying demands of the various applications. Very low irradiation may be required in combination with a very sensitive camera system, in order to avoid photo damage; extremely strong laser excitation may be wanted to kill living cells; and the wavelengths of the illumination will vary from deep UV (250 nm) to infrared. Since these types of illumination cannot be provided by a single light source, several lamp housings may be attached to one fluorescence microscope for an easy interchange of illumination.

1.4.1 Lamp housings

Correct alignment of the arc or high-pressure lamps is extremely important for the fluorescence yield. Therefore, the quality of a lamp housing can almost be judged by the stability of a correct alignment of the arc made in the factory, or by the efficiency of user accessible knobs for two directional arc alignment. It should be feasible to obtain a homogeneous illumination of the microscopic field and it should be possible to focus the lamp collector to project an image of the arc on the entrance pupil of the substage condenser with transmitted illumination or on the entrance pupil of the objective in epi-illumination.

Lamp housings usually have filter holders for inserting filters for infrared elimination and colour filters. Heat and infrared filters should be of the reflecting type rather than of the absorbing type, since these crack less frequently. These heat-reflecting filters should always be placed closer to the lamp than the coloured filters to prevent excessive infrared absorption by the latter.

1.5 FILTERS

Filters are very important components in the fluorescence microscope. Filter choice depends on the light source, and on the spectral characteristics, quantum efficiency and distance in wavelength between excitation and emission peak of the fluorochromes used. The main types of filters in fluorescence microscopy are: colour glass filters, interference filters, and a special type of interference filter placed at 45° to the light beam, known as dichroic mirrors or chromatic beam-splitters (Fig. 1.4). Colour glass filters are mostly made by adding certain oxides of various heavy metals to the glass. Although to the naked eye a colour filter transmits only light from one colour, the transmission curve has in fact a fairly broad base. Thus while there will be a peak transmission of one colour, some light from the neighbouring regions of the spectrum will also be transmitted. The concentration of the added oxides and the thickness of the glass determine how much of the light is absorbed. The remaining light is transmitted. If the absorption extends into the infrared regions of the spectrum, it will cause a considerable heating and may lead to cracking if the filters are used in combination with a powerful light source such as high-pressure arc lamps. For this reason it is desirable to place a heat-reflecting filter between the colour filter and the lamp (e.g. Calflex filter from Schott). Colour filters transmit rather broad wavelength ranges and are therefore known as broad-band filters.

Interference filters consist of many layers of thin film with different refractive indices, sequentially deposited upon a flat glass surface. Interference filters transmit light of well-defined wavelengths resulting from the passage of light through layers of different refractive indices and from reflection by the surface of these layers. As the spectral characteristics of these filters depend on very precise maintenance of the gap between the semitransparent coatings, interference filters are made for very narrow tolerances and are accordingly much more expensive than glass filters. If the filters are tilted along the optical axis the spectral properties will change. Due to the construction of interference filters a shift towards shorter wavelengths occurs when the angle of incidence increases. Sometimes this shift is used in the fine-tuning of a filter to obtain a precise peak wavelength by introducing a small angle of the filter in relation to the optical path of the microscope.

A filter can be described according to its half bandwidth (HB) indicating the transmission width at 50% on either side of the transmission peak. The interference filters are defined into narrow-band and wide-band filters according to the wavelength band they transmit. Some interference filters do not have a symmetrical (bell-shaped) transmission curve but a sharp slope. When such a filter transmits light of longer wavelengths and blocks short wavelengths it is known as a long-pass (LP) filter. A filter which transmits short wavelengths and blocks long wavelengths is defined as a short-pass (SP) filter.

Recently, interference filters with very complex transmission characteristics have been developed for flow cytometry and fluorescence microscopy (Omega Optical Inc., USA) that enable the simultaneous excitation of two or three fluorochromes.

Filters can also be characterized by their position in the microscope (excitation or emission side). Consequently the terminology used by different manufacturers is quite confusing.

1.5.1 Excitation filters

Excitation filters are used to isolate a limited region of the light spectrum in correspondence with the absorption peak of the fluorochrome. In addition, almost all the light in the wavelength range of the fluorescence emission of the fluorochrome must be removed from the illumination light beam, since the barrier filters (used above the objective to block unwanted excitation light that otherwise would reach the eyepieces) are usually not perfect and will still transmit a very small amount of excitation light. Due to the fact that in many applications also very weakly fluorescing objects are to be observed, the amount of unwanted excitation light still passing the barrier filter must be minimized. The problem of eliminating unwanted excitation light from the observed microscope field becomes even more pronounced if the fluorochrome has excitation and fluorescent peaks which are rather close to each other, like FITC. Filters with a high transmission close to the excitation peak of a fluorochrome and which also strongly block unwanted excitation light in the fluorescence wavelength range of such a fluorochrome are relatively difficult to manufacture and expensive.

The choice of an excitation filter must be made on the image contrast finally required for the intended application. Glass filters like the BG (blue glass) and UG (ultraviolet glass) filters are still in use. They have rather broad transmission characteristics. Interference filters are more selective. A disadvantage of these filters in the past was their low transmission value (30–60%). Modern technology has enabled the development of band-pass filters with high-transmission (90%) narrow-band characteristics and very good suppression of unwanted excitation light in the wavelength range of the expected fluorescence. Short-pass filters (SP) transmit shorter wavelengths and effectively block longer wavelengths (Rygaard & Olson, 1969; Ploem, 1971; Lea & Ward, 1974).

The recently developed filters for the simultaneous excitation of 2 or 3 fluorochromes should be used in combination with corresponding dichroic mirrors and barrier filters to allow observation of 2 or 3 fluorescence colours (Fig. 1.7).

1.5.2 Barrier filters

Barrier filters are used to block the unwanted excitation light in the wavelength range of the fluorescence emission. Mostly colour glasses are used with a high transmission for the longer wavelengths (90% or higher) and a very effective blocking of shorter wavelengths. Colour glass barrier filters absorbing short wavelength excitation light may fluoresce which may lead to a decrease in the image contrast. Barrier filters for some applications requiring an extremely dark background are therefore coated with an interference filter layer that will reflect most excitation light and prevent autofluorescence of the barrier filter.

In some applications not all the fluorescence light longer than a certain wavelength is wanted for observation, but only the fluorescence in a limited wavelength range (e.g. the narrow emission peak of FITC). This is achieved by adding an extra band or a short-pass interference filter to the barrier filter or by coating the colour glass barrier filter with an interference coating, selecting a narrow wavelength band. Such filter combinations can be defined as fluorescence selection filters.

Recently, barrier filters of the interference type have been manufactured which permit the observation of 2 or 3 fluorescence colours simultaneously (Fig. 1.7). Such filters have a complex transmittance curve with several wavelength bands of high transmission for fluorescence and several wavelength regions for strong blocking of unwanted excitation light. Such filters

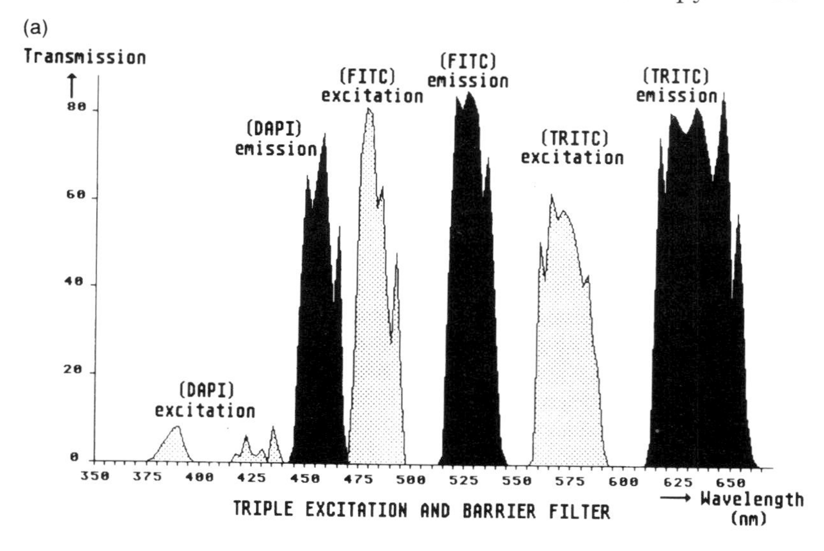

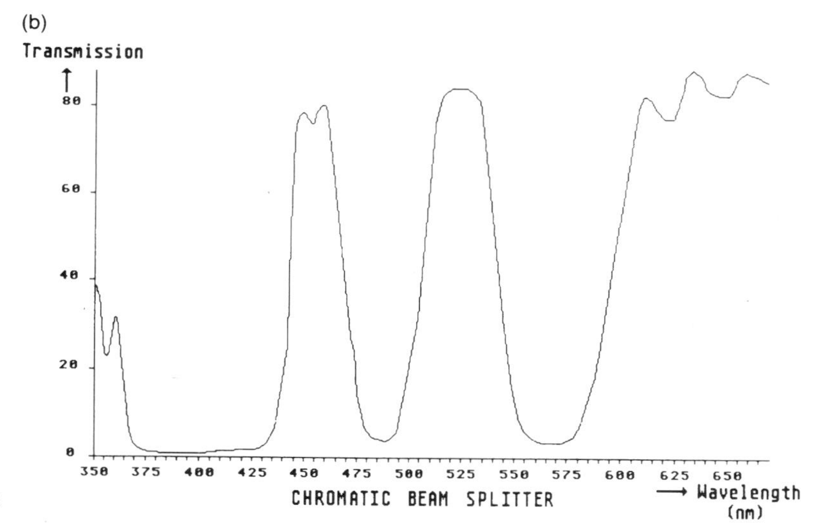

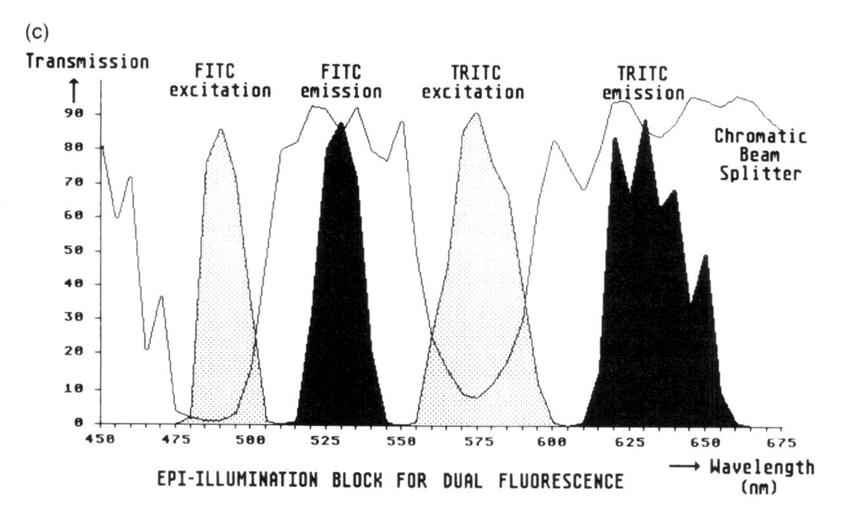

Figure 1.7 (a) Spectral characteristics of an excitation filter (lightly shaded area) and a barrier filter (darkly shaded area) that enable the simultaneous excitation of three fluorochromes (DAPI, FITC and TRITC) and the observation through the barrier filter of blue, green and red fluorescence (interference filters manufactured by Omega Optical Inc., USA).
(b) Spectral characteristics of a chromatic beam-splitter (dichroic mirror) that must be used in combination with the excitation filter in (a) to simultaneously excite three fluorochromes in epi-illumination fluorescence microscopy (dichroic mirror manufactured by Omega Optical Inc., USA). (c) Spectral characteristics of an excitation filter (lightly shaded area), a chromatic beam-splitter (dichroic mirror) and a barrier filter (darkly shaded area) for the excitation of FITC and TRITC (interference filters manufactured by Omega Optical Inc., USA).

must be used in combination with special excitation interference filters, exactly matching the transmission of the barrier filter.

1.5.3 Chromatic beam-splitters (CBS)

Chromatic beam-splitters (CBS), also known as dichroic mirrors, reflect light of wavelengths shorter than the specified wavelength and transmit light of longer wavelengths. They are placed at an angle of 45° to the optical axis and reflect excitation rays into the objective in epi-illumination, where the objective also serves as a condenser (Figs 1.2, 1.4 and 1.5). Recently, chromatic beam-splitters have been made for simultaneous fluorescence of 2 or 3 fluorochromes (Fig. 1.7) in epi-illumination. They should be used in combination with the appropriate excitation and barrier filters for dual or triple fluorescence excitation and observation.

1.5.4 Multi-wavelength epi-illuminators

Effective epi-illumination can only be achieved by combining a closely matched excitation filter, a chromatic beam-splitter and a barrier filter for each main fluorescence colour. They are usually mounted together in a filter block which can be inserted in an epi-illuminator (Fig. 1.5). Multi-wavelength vertical illuminators are available with sliding or revolving filter blocks permitting epi-illumination in several wavelength bands. Various filter combinations for different wavelengths are given in Table 1.2.

The newly developed combinations of an excitation filter, a chromatic beam-splitter and barrier filter, when mounted in one block, permit the excitation and observation of two fluorochromes with two different fluorescent colours (Fig. 1.7).

1.6 OBJECTIVES AND EYEPIECES

In epi-illumination the microscope objective also serves as a condenser. The obtained result therefore strongly depends on the choice of the objective. Not all objectives are suited for fluorescence microscopy. The glasses used for such objectives must show very little autofluorescence. This is especially important with very weak fluorescence signals. In testing an objective it is necessary to distinguish still remaining,

Table 1.2 Typical combinations of excitation filters, chromatic beam-splitters and barrier fluorescence emission filters (often combined in sets or blocks, which can be inserted in multi-wavelength epi-illuminators). Most filters, except the ones marked, can be obtained from all major fluorescence microscope manufacturers.

Excitation light	*Excitation filter (nm)*	*CBS*	*Barrier filter*	*Fluorescence colours*
UV (365 nm)	340–380 or 350–410	400 or 410	LP 430	Violet, blue-green, yellow, orange, red and infrared
Violet (405 nm)	350–460 or 420–490	455 or 460	LP 470	Blue, green, yellow, orange-red, infrared
Blue (470 nm)	450–490 or 470–490	500 or 510	LP 520	Green, yellow-orange, red and infrared
Green (546 nm)	515–560 or 530–560	580	LP 580	Yellow, orange-red and infrared
Yellow (560 nm)	550–570[a]	595[a]	LP 635[a]	Orange, red and infrared
Orange (590 nm)	580–600[a]	620[a]	LP 660[a]	Red and infrared
Red (630 nm)	610–650[b]	660[b]	LP 670[b]	Infrared
UV, blue and green	Multiple transmittance and reflection bands[a,c]			UV, blue and green

[a] Glen Spectra Limited, UK.
[b] Chroma Technology Group, USA.
[c] Omega Optical Inc, USA.

unwanted excitation light, and autofluorescence of the mounting medium, the specimen and immersion oils from the autofluorescence of the objective itself. Objectives for fluorescence microscopy should have a relatively high numerical aperture in combination with a relatively low magnification. Examples of such objective lenses are oil-immersion objectives 10× with a numerical aperture (NA) of 0.45 and 40× with a NA of 1.30 (Leica, Germany). Also, dry objectives with relatively high NA, considering their magnification, are now manufactured. Water-immersion objectives made for fluorescence microscopy offer the advantage of avoiding the autofluorescence of some immersion oils and in addition permit the study of live cells by dipping the objective directly into the cultivating medium.

The fluorescence intensity obtained is proportional to the square power of the numerical aperture (NA) of both condenser and objective in transmitted and to the fourth power of the objective in epi-illumination. The brightness is inversely related to the magnification of the objective. Fluorescence microscopy thus preferably has to be carried out with objectives of high NA in combination with low-power eyepieces.

REFERENCES

Bergquist N.R. & Nilsson P. (1975) *Ann. NY Acad. Sci.* **254**, 157–162.
Beverloo H.B., Schadewijk A. van, Gelderen-Boele S. & Tanke H.J. (1990) *Cytometry* **11**, 784–792.
Coons A.H., Creech H.J. & Jones R.N. (1941) *Proc. Soc. Exp. Biol. Med.* **47**, 200–202.
Giloh H. & Sedat J.W. (1982) *Science* **217**, 1252–1255.
Johnson G.D. & Nogueira Araujo G.M. (1981) *J. Immunol. Methods* **43**, 349.
Jovin T.M. & Vaz W.L.C. (1989) *Methods Enzymol.* **172**, 471–513.
Kohen E. & Hirschberg J.G. (1989) *Cell Structure and Function by Microspectrofluorometry.* Academic Press, San Diego.
Kraft W. (1973) Fluorescence microscopy and instrument requirements. Leitz Technical Information 2, pp. 97–109.
Lansing Taylor D., Waggoner A.S., Murphy R.F., Lanni R. & Birge R.R. (1986) *Applications of Fluorescence in the Biomedical Sciences.* Alan Liss, New York.
Lea D.J. & Ward D.J. (1974) *J. Immunol. Methods* **5**, 213–215.
Melamed M.R., Lindmo T. & Mendelsohn M.L. (1990) *Flow Cytometry and Sorting.* John Wiley & Sons, New York.
Nairn R.C. (1976) *Fluorescent Protein Tracing.* E. & S. Livingstone, Edinburgh.
Nederlof P.M., Flier S. van der, Wiegant J., Raap A.K., Tanke H.J., Ploem J.S. & Ploeg M. van der (1990) *Cytometry* **11**, 126–131.
Patzelt W. (1972) *Leitz-Mitt. Wiss. u. Techn.* **V/7**, 226–228.
Ploem J.S. (1967) *Z. wiss. Mikrosk. u. mikrosk. Techn.* **68**, 129–142.
Ploem J.S. (1971) *Ann. NY Acad. Sci.* **177**, 414–429.
Ploem J.S. (1987) *Appl. Optics* **26**, 3226–3231.
Ploem J.S., Tanke H.J., Al I. & Deelder A.M. (1978) In *Immunofluorescence and Related Staining Techniques,* W. Knapp, K. Holubar & G. Wick (eds). Elsevier, Amsterdam, pp. 3–10.
Rost F.W.D. (1991) *Quantitative Fluorescence Microscopy.* Cambridge University Press, Cambridge.
Rygaard J. & Olson W. (1969) *Acta Pathol. Microbiol. Scand.* **76**, 146–148.
Thomson L.A. & Hageage G.J. (1975) *Appl. Microbiol.* **30**, 616–624.
Tomlinson A.H. (1971) *Proc. Microsc. Soc.* **7**, 27–37.
Wick G., Schauenstein K., Herzog F. & Steinbatz A. (1975) *Ann. NY Acad. Sci.* **254**, 172–174.
Wilke V. (1983) *Proc. of SPIE* **396**, 164–172.
Young M.R. (1961) *Q. J. Microsc. Sci.* **102**, 419–449.

PART II

Optical Probes and Their Applications

CHAPTER TWO

Introduction to Fluorescent Probes: Properties, History and Applications

FREDERICK H. KASTEN
Department of Anatomy, Louisiana State University Medical Center, New Orleans, LA, USA

2.1 INTRODUCTION

The availability of sensitive and selective fluorescent probes for living cells has opened new horizons in cell biology. With the aid of the modern epifluorescence microscope and video intensification microscopy, in combination with fluorescent probes, fluorescent-labelled organelles and molecules can be visualized, measured, and the information stored. The fluorescence signal superimposed against a dark background permits sharper cytologic detail to be observed than with a comparably stained specimen in the ordinary light microscope. This enables cell organization and function to be analysed with a precision and clarity not previously possible (Rost, 1980; Willingham & Pastan, 1982; Sisken *et al.*, 1986; Spring & Smith, 1987; Taylor & Salmon, 1989). Single fluorescent microtubules have been detected (Sammak & Borisy, 1988). With the addition of the confocal principle applied to imaging in fluorescence and integration with computers, precise optical sectioning and analyses of living and fixed cells are possible. In the confocal system, the fluorescent contributions of out-of-focus areas are eliminated during laser scanning and the limits of resolution are extended. Enhanced imaging at high numerical apertures is realized. Also, three-dimensional reconstructions and measurements are obtained based on the accumulated optical sections (Stelzer & Wijnaendts-van-Resandt, 1989; Wilson, 1990; Herman & Jacobson, 1990; Kohen *et al.*, 1991).

It is the purpose of the present chapter to summarize the nature of fluorescence, the properties of fluorescent probes, the historical developments leading from early use of fluorochromes to modern fluorescent probes, and to summarize their applications in living cells. Further details of probes employed for specific applications are given in chapters elsewhere in this volume. Additional information is to be found in other reviews (Kasten, 1967, 1981, 1983a, 1989; Waggoner, 1986; Wang & Taylor, 1989; Taylor & Wang, 1989; Haugland, 1992; Kapuscinski & Darzynkiewicz, 1990; Darzynkiewicz & Crissman, 1990; Herman & Jacobson, 1990).

FLUORESCENT AND LUMINESCENT PROBES, 2ND EDN
ISBN 0-12-447836-0

2.2 NATURE OF FLUORESCENCE AND PROPERTIES OF FLUORESCENT PROBES

Fluorescence is a form of luminescence which occurs after photons of light are absorbed by a molecule known as a fluorophore, fluorochrome, or fluorescent probe at the ground electronic state. The molecule is raised to an excited state as a result of electron transfer to a higher energy orbit. This excess energy is dissipated when the electron returns to the original ground state, releasing a quantum of light. The time required for absorption is immediate, about 10^{-15} s, whereas the fluorescence lifetime is approximately 10^{-8} s. The fluorescence phenomenon was first described by Brewster in 1838. The term fluorescence was coined by Stokes in 1852. Phosphorescence is a type of luminescence that persists after the exciting light is turned off. It has a lifetime of several seconds or longer because the excited electron first stops at an intermediate triplet state before reaching the ground state. According to Stokes' law, the fluorescent light is of a longer wavelength than the absorbed light. The law was extended in 1875 by Lommel, who stated that the molecule must first absorb radiation in order to exhibit fluorescence.

The difference in energy levels associated with absorption and fluorescence characterizes the absorption and emission wavelength maxima. The absorption intensity or extinction coefficient, ε, reflects the probability of absorption. Fluorescein (FITC) has an extinction coefficient maximum of 75 000 cm^{-1} M^{-1}. Unusually high extinction coefficients are given by the algal-derived phycobiliproteins, which have multichromophore complexes. For instance, phycoerythrin has an extinction coefficient greater than 10^6 cm^{-1} M^{-1}.

The emission intensity relates directly to the quantum yield, ϕ, which is the ratio of quanta released to quanta absorbed. Fluorochromes have characteristic quantum yields of efficiencies that range from 0.1 to almost 1. For practical purposes, the quantum yield should be close to 0.4 or greater when the fluorochrome is bound to the cell structure or molecule. The fluorescence intensity of a probe is determined by the product of ε and ϕ.

The third important characteristic of a fluorochrome or fluorescent probe is the fluorescence lifetime or excited state lifetime, τ, which is the average time that a molecule remains in the excited state. Short fluorescence lifetimes permit the greatest sensitivity to be achieved since multiple excitations can be achieved if the molecule is quickly relaxed after a prior excitation event. Most fluorochromes have emission lifetimes on the order of nanoseconds. Fluorescein has a τ of about 4 ns. It has been pointed out that unusually long lifetimes can be valuable in high-sensitivity detection (Waggoner, 1986). In cases where scattered light and autofluorescence of short lifetimes create interference with the desired fluorescence signal, it is desirable to use long-lived fluorochromes in combination with an appropriate photomultiplier tube.

Another property of fluorescent probes that needs to be considered in selecting a suitable dye is the wavelength of maximum absorption or excitation. Vital probes of cell vitality (SITS), membranes (ANS, DPH, 'Long Name', NPN), and ions (fura-2, indo-1, quin-2, SBQ) all require excitation in the long-wave UV to produce fluorescence in the visible range. This requires suitable UV-emitting light sources with attending protection for personnel. Also, there may be interfering autofluorescence from native cytoplasmic flavins, flavoproteins and NADPH. In the case of fixed cells, these metabolites are unlikely to be a problem and blue-fluorescing DNA-binding probes, like DAPI and Hoechst 33258, are useful.

The photobleaching of some probes is a serious problem. This is commonly observed with fluorescein-labelled cells in the fluorescence microscope, especially during photographic exposures. It is not usually a problem in flow cytometry because the individual cells are in the laser beam only a short time. In the confocal microscope, laser photobleaching is reduced by cutting down the number of optical scans. However, when line-averaging is necessary to reduce background noise, bleaching can be observed with FITC and Nile red. Some chemical agents like phenylenediamine (Johnson *et al.*, 1982) and propyl gallate (Giloh and Sedet, 1982) in the glycerol mounting medium help to reduce fading, but this is not possible in studies of living cells. To counter this photobleaching effect, the light intensity may be reduced, sensitive video cameras can be used, and photographic film with high sensitivity can be employed to reduce exposure time. Colour film with an ASA rating of 3200 is available without pushing. High-sensitivity black-and-white film is also on the market.

Fluorescein substitutes have been sought and the Bodipy fluorochrome is now recommended (Haugland, 1990). The Bodipy fluorophore, boron dipyrromethene difluoride, is said to have high photostability (Wories *et al.*, 1985) and other desirable features. The absorption peak is similar to that of fluorescein (505 nm compared with 490 nm), the emission peak is at almost the same wavelength (520 nm compared with 519 nm), the extinction coefficients are almost identical (about 75 000 cm^{-1} M^{-1}) and the quantum yields are similar. In other ways, Bodipy overcomes certain deficiencies of fluorescein and seems too good to be true. Fluorescein is pH-sensitive in the physiological range, which limits its application in living cells. The emission curve of fluorescein exhibits a broadness on the long wavelength side, which causes some overlap with other dyes used in two-colour fluorescence. Also, fluorescein conjugates have negative charges, which limit their use in examining surface membranes and receptors. According to Haugland (1990), Bodipy offers advantages over fluorescein in addition to the improved photostability. Bodipy has a narrow emission spectrum with a large Stokes' shift and gives less overlap with certain red fluorochromes, like Texas Red. Bodipy is relatively lipophilic and can be bound to certain compounds for receptor studies that cannot be done with fluorescein.

For additional details on the nature of fluorescence and on the properties of fluorescent probes, the articles by Waggoner (1986) and Taylor and Salmon (1989) should be consulted.

2.3 HISTORICAL DEVELOPMENTS

2.3.1 Fluorescence microscope

The first fluorescence microscope was developed over 80 years ago by Heimstädt (1911) and Lehmann (1913) as an outgrowth of the UV microscope. The instrument used a high-powered arc lamp to generate UV light, a modified Wood's filter (nitrosodimethylaniline solution with copper sulphate) as a primary filter, a dark-field condenser, Uviol secondary filter, and quartz optics. The microscope was used to investigate the autofluorescence of bacteria, protozoa, plant and animal tissues, and bioorganic substances, such as albumin, elastin and keratin (Stübel, 1911; Tswett, 1912; Wasicky, 1913; Provazek, 1914). A history of these developments and subsequent technological advances are presented elsewhere (Kasten, 1983a, 1989).

2.3.2 Synthesis of coal-tar dyes and early uses

The first synthetic coal-tar dye, mauve or aniline purple (CI 50245),* was made accidentally by William Perkin in 1856 (cf. Perkin, 1906). This breakthrough was followed by feverish attempts on the part of many chemists to synthesize other dyes. Using oxidized aniline and the approaches suggested by Perkin, numerous dyes were produced, the first of which was magenta in 1859, also known as rosaniline or fuchsin (CI 42510). The great need for textile dyes with wide-ranging colours and resistance to bleaching by light stimulated further commercial interest. Between the time of Perkin's discovery and the invention of the fluorescence microscope, about 55 years, scores of new dyes became available. Among this group of dyes synthesized in the late nineteenth century were pararosaniline (CI 42500), methyl violet or gentian violet (CI 42535), crystal violet (CI 42555), methyl green (CI 42585), malachite green (CI 42000), brilliant green (CI 42040), safranin O (CI 50240), methylene blue (CI 52015), gallocyanin (CI 51030), and numerous azo dyes like Bismarck brown R (CI 21000). A small German firm known as Dr G. Grüblers Chemisches Laboratoriums first opened in 1880. Grübler tested and packaged the most desirable dyes for biologists and medical researchers.* This quality assurance of selected, high-quality German dyes was of great value to laboratory workers, who took advantage of the newly available dyes to stain histological, haematological and bacteriological material and to develop new staining methods (Kasten, 1983b).

Other dyes produced during this period included xanthene and acridine derivatives, which were highly fluorescent. Some of the well-known xanthenes were pyronin Y (G) (CI 45005), rhodamine B (CI 45170), fluorescein (CI 45350), eosin Y (CI 45380), and erythrosin (CI 45430). Some of the early acridines were phosphine (CI 46045), acridine yellow (CI 46025), acridine orange (CI 46005), acriflavine (CI 46000), and coriphosphine O (CI 46020). A few fluorescent dyes were derived from other chemical groups, such as auramine O (diphenylmethane, CI 41000), Calcofluor white (stilbene, CI 40621), brilliant sulphoflavine (amino ketone, CI 56205), neutral red (azin, CI 50040), and pararosaniline (CI 42500). Table 2.1 lists some of the common fluorochromes and the year when each was synthesized. Common acridine dyes used in histology and histochemistry are described by Kasten (1973). General properties of dyes are given elsewhere (Harms, 1965; Lillie, 1977; Green, 1990).

In spite of the fact that many fluorescent dyes were available to microscopists by the beginning of this century, few were actually used. Histologists, cytologists and bacteriologists favoured strong-staining red, violet, blue and green dyes, which were largely non-fluorescent or only weakly fluorescent in solution. Pyronin Y and eosin Y, which are red dyes, were exceptional cases. Dyes of the acridine group, which usually stain cells yellow, were less commonly used. To illustrate with several examples, basic fuchsin, a mixture containing red-staining pararosaniline and rosaniline dyes, became an important nuclear stain in histology following its introduction by Waldeyer in 1863. It had a great impact in bacteriology, particularly in the Ziehl-Neelsen method for

* CI stands for 'Colour Index' and the number following is that assigned in the 3rd edition (*Colour Index*, 1971).

* In 1897, the Grübler firm became known as Dr K. Hollborn & Söhne.

Table 2.1 Common biological fluorochromes.

Fluorochrome	*Year of synthesis*	*CI No.*[a]
Acridine orange	1889	46005
Acridine red 3B	1891	45000
Acridine yellow	1889	46025
Acriflavine	1910	46000
Auramine O	1883	41000
Brilliant sulphoflavine	1927	56205
Calcein	1956	—
Chrysophosphine 2G	1922	46040
Congo red	1884	22120
Coriphosphine O	1900	46020
Eosin B	1875	45400
Eosin Y	1871	45380
Erythrosin B	1876	45430
Euchrysin	1922	46040
Flavophosphine N	1887	46065
Fluorescein	1871	45350
Neutral red	1879	50040
Nile blue A	1888	51180
Oxytetracycline	1950	—
Pararosaniline	1878	42500
Phosphine 5G	1900	46035
Phosphine GN	1862	46045
Primulin	1887	49000
Proflavine	1910	—
Prontonsil	1932	—
Pyronin Y (G)	1889	45005
Quinacrine (Atabrine)	1934	—
Rheonine	1894	46075
Rhodamine 3GO	1895	45210
Rhodamine 5G	1902	45105
Rhodamine 6G	1892	45160
Rhodamine B	1887	45170
Rhodamine G	1891	45150
Rhodamine S	1888	45050
Sulphorhodamine B	1906	45110
Thiazole yellow G	1893	19540
Thioflavine S	1888	49010

[a] The CI no. refers to the *Colour Index* no., a specific designation for the chemical structure of a dye as listed in the *Colour Index* (1971).

demonstrating acid-fast microorganisms like the tubercle bacillus. Gentian violet was introduced into microtechnique by Weigert and by Ehrlich in 1881 and 1882 and became an essential component of the Gram stain in 1884. Methylene blue was employed by Ehrlich as the first important vital stain in 1885; he demonstrated its affinity for nerve tissue. Ehrlich introduced many other dyes into the field, like his famous triacid mixture (methyl green, acid fuchsin, orange G).

The acridines and other fluorescent dyes were not used on fixed cells and tissues in fluorescence microscopy until the early 1930s, more than 20 years after the fluorescence microscope was developed. Krause's three-volume *Enzklopädie der Mikroscopischen Technik* (Krause, 1926a, b, 1927) did not mention any uses for acridine dyes and failed to include a discussion of the acridines, although other dye groups were included. The section on the fluorescence microscope in this otherwise valuable reference referred only to its application in detecting autofluorescence in tissue sections. Microscopists failed to appreciate the fact that fluorochromes could impart added sensitivity and clarity to stained tissues when viewed by fluorescence microscopy. It was mistakenly felt that microscopic observations by induced fluorescence via fluorochromes would introduce artifacts and misinterpretations.

Although fluorescent dyes were not utilized to stain fixed tissues and cells for many years, researchers made use of the dyes in other ways. Fluorescein was known to produce an intense yellow-green fluorescence in aqueous solution. Its sodium salt (uranin) produced a pale green fluorescence even when diluted 1 part to 16×10^6 parts of water (Fay, 1911). The dye was employed by Ehrlich (1882) to track the pathway of aqueous humour in the eye. In 1906, fluorescein was the first fluorescent dye to be used for tracing underground waters in the United States.

With the beginning of the First World War and the need to treat infected wounds, the efficacy of acriflavine as an antiseptic became established (Browning *et al.*, 1917). The dye was referred to as 'flavine' by British researchers. Proflavine, a close relative of acriflavine, was another useful antiseptic. Both of these dyes would later prove to be valuable fluorescent probes of nucleic acids. Many other diaminoacridine compounds were prepared for experimental and clinical trials, none of which proved to be superior to acriflavine (Browning, 1922; Albert, 1951). Tissue cultures were tested for their response to acriflavine and proflavine to determine cell toxicity levels and effects on bacterial-infected cultures (Mueller, 1918; Hata, 1932; Jacoby *et al.*, 1941). However, the culture system failed to aid in predicting the value of aminoacridine antiseptics as local chemotherapeutic agents (Browning, 1964).

2.3.3 First usage of fluorochromes in living cells

By the early 1900s, pharmacologists and experimental therapeuticists showed a great interest in the action of fluorescent dyes in sensitizing microorganisms to light. This dye-enhanced light inactivation became known as photodynamic inactivation. Research on this subject was stimulated by the appearance of an important volume by Tappeiner and Jodlbauer (1907). Acridine dyes were shown to be effective agents in treating trypanosomes (Werbitzki, 1909). Ehrlich's use of acriflavine for combating this protozoan in infected mice gave dramatic results (Ehrlich & Benda, 1913). The dye was referred to as 'trypaflavin' because of its influence on trypanosomes. Microscopists used the bright-field microscope to observe the binding of such dyes to microorganisms.

The protozoologist Provazek (1914) was apparently the first person to employ the fluorescence microscope to study dye binding to living cells. He added various fluorochromes and drugs (fluorescein, eosin, neutral red, quinine) to cultures of the ciliate *Colpidia* and viewed the induced fluorescence of the cells. He grasped the fundamental significance of this new experimental approach and stated that the object was:

> To introduce into the cell certain substances of different types, without regard as to whether they are stains or colorless drugs, on the assumption that they follow definite distribution laws and collect under certain circumstances in particular functional elements inside the cell so that they effectively illuminate the partial functions of the cell in the dark field of the fluorescence microscope.

2.3.4 Developments in vital and supravital fluorochroming*

The introduction of fluorochromes into fluorescence microscopy in 1914 marked a giant step forward in experimental cytology. The report by Provazek was the first to demonstrate vital fluorochroming. Previously, vital dyes, including fluorochromes, were tested on protozoa but observed only with the bright-field microscope. Supravital fluorochroming had its start in 1932 when Jancso, a Hungarian pharmacologist, injected several different fluorochromes into rodents previously infected with trypanosomes. Examination of blood smears in the fluorescence microscope revealed specific binding of the dyes to nuclei and basal bodies of the blood-borne trypanosomes. Acriflavine displayed especially strong binding to these structures. Because of the unusual interest in chemotherapy in the 1930s, attempts were made to locate fluorochromes that would bind to the malaria organism in infected animals and humans. Of the various fluorochromes tested, quinacrine (atabrine) was found by fluorescence microscopy to be selectively taken up by circulating plasmodia within 10 min after dye injection (Fischl & Singer, 1935; Bock & Oesterlin, 1939; Patton & Metcalf, 1943). Acriflavine was added to fibroblast cultures and shown by fluorescence microscopy to inhibit cell division at the concentrations tested (Bucher, 1939). Additional details of the early investigations into vital fluorochroming are found in the volume by Drawert (1968).

From these studies, it became clear that certain aminoacridines bound preferentially to components in nuclei. Attempts to determine the mode of action of such binding led to the discovery that the dyes had an affinity for nucleic acids. This strong interaction was demonstrated with purified nucleic acids *in vitro* (DeBruyn *et al.*, 1953; Peacocke & Skerrett, 1956) and in fixed cells at a low pH range (Armstrong, 1956; Bertalanffy & Bickis, 1956; Schümmelfeder *et al.*, 1957). This is discussed further in Section 2.5.

2.3.5 *In vivo* fluorochroming*

After Ehrlich demonstrated that fluorescein could be used to follow the path of aqueous humour in the eye, the slit-lamp microscope was adapted by Thiel to observe the dye's fluorescence within the eye. The fluorescence microscope, which uses light transmitted through a condenser, could not be employed to examine opaque specimens from most living organs. In 1929, the fluorescence microscope was modified markedly by Philipp Ellinger, a pharmacologist at Heidelberg University, in collaboration with a young anatomist, August Hirt (Ellinger & Hirt, 1929a). Ellinger was interested to examine the microcirculation in the kidney of the exteriorized organ with the aid of fluorochromes previously injected into the animal. This new fluorescence microscope utilized vertical illumination that was directed into the microscope tube laterally and then passed through the objective to the specimen. The emitted fluorescence was transmitted back up the tube to the eye.

* According to classical usage, vital fluorochroming or staining is the non-toxic staining of living cells or tissues in the organism. An example is the intracellular uptake of colloidal azo dyes like trypan blue by macrophages and Kupffer cells *in vitro*. The term intravital staining is sometimes used synonymously with vital staining. Supravital fluorochroming or staining means the addition of dyes to an *in vitro* solution containing cells previously removed from an organism. It may also refer to the staining of living cells within a recently killed animal. The stain binds to cytoplasmic organelles, like the binding of Janus green to mitochondria or of neutral red to cytoplasmic granules. Cell biologists today commonly apply the term vital stain or fluorochrome to dyes added to cultured cells.

* *In vivo* or intravital fluorochroming refers to the non-toxic staining of living cells and tissues in the organism with observations by microscopy.

Other essential components included appropriate filters and a water-immersion objective. The instrument was called an 'intravital microscope' and may be considered as the first epifluorescence microscope. Dilute fluorescein and acriflavine solutions were used to study the physiology of urine formation (Ellinger & Hirt, 1929b). During the Second World War, Hirt carried out unethical experiments on humans in which he hoped to examine human tissues *in vivo* with the intravital fluorescence microscope (Kasten, 1991).

The intravital fluorescence microscope attracted the attention of other researchers and a variety of fluorochromes were employed by them; brilliant phosphine G (probably CI 46045), germanin S, primulin yellow (CI 49000), rheonin A (CI 46075), thiazole yellow (CI 19540), and thioflavine (probably CI 49010). During the 1930s, with this new technique, fluorescence observations were made of the microcirculation in living skin, liver, kidney, conjunctiva and adrenal gland (Singer, 1932; Franke & Sylla, 1933; Pick, 1935; Heuven, 1936; Grafflin, 1938; Schmidt-LaBaume & Jäger, 1939). For physiological studies, vital fluorochromes needed to be used under conditions of isotonicity, non-toxicity, and non-quenching. A review of these early investigations is given by Ellinger (1940), Price and Schwartz (1956), and Kasten (1983a). In recent years, intravital microscopy of the microcirculation has enjoyed renewed popularity as a result of access to modern epifluorescence instruments, scanning microfluorometry, sensitive charge-coupled video cameras, and time-frame generators for data evaluation (cf. Bollinger *et al.*, 1983; Witte, 1989). As an example, such instrumentation was utilized with acridine orange to quantify the hepatic microcirculation in rodents (Menger *et al.*, 1991).

2.4 APPLICATION OF FLUOROCHROMES IN HISTOLOGY AND MICROBIOLOGY

2.4.1 Histology

Until 1929, microscopic work with the fluorescence microscope was confined to observations of tissue and cellular autofluorescence (porphyrins, native cytoplasmic proteins, chlorophyll) and to the detection of living protozoa with fluorochromes. The first report of the use of a fluorochrome on fixed tissue sections was by the dermatologist Sigwald Bommer (1929). He employed a dilute solution of acriflavine on skin sections and observed a selective green-yellow fluorescence of cell nuclei. Bommer suggested the possibility of a fundamental cytochemical basis for the nuclear fluorescence he observed.* He raised the question: 'To what extent is it possible to establish definite affinities between certain stains and tissue constituents when used in a certain dilution and with a definite technique?'

The next milestone in this field was achieved in Vienna by the young pathologist Herwig Hamperl, in collaboration with Max Haitinger, an expert in fluorescence microscopy (Haitinger & Hamperl, 1933). They examined the staining properties of more than 65 different fluorochromes on formalin-fixed frozen sections of animal tissues to see which ones would produce differential fluorescence. Out of this empirical survey, they recommended 35 fluorochromes worthy of use in normal and pathological histology. Hamperl (1934) extended the study to paraffin-embedded tissues and described the differential binding affinities of many fluorochromes and recommended staining methods. At this time, the word fluorochrome was created to mean fluorescent compounds which are bound selectively to individual tissue structures without disturbing the autofluorescence of other tissue elements (Haitinger, 1934). To distinguish these applied dyes from natural autofluorescing substances in cells, fluorochromes were also termed secondary fluorochromes and the process of staining as secondary fluorochroming. Autofluorescing substances like the porphyrins were called primary fluorochromes. In modern usage, the word fluorochrome has come to mean any fluorescent dye, regardless of its staining properties and effect on native autofluorescence.

The studies of Hamperl and Haitinger served as a foundation for later investigators, who applied the methods to their own fields and added modifications in staining protocols. It was clear that secondary fluorochroming produced brilliant colours in tissues and cells with striking contrast and sensitivity. The impact of Hamperl and Haitinger's work was felt first at the University of Vienna. Here, various investigators collaborated with Haitinger, who maintained a fluorescence microscopy facility and popularized the technique (see dedication volume to Haitinger by Braütigam and Grabner, 1949). The initial histomorphological applications of the new fluorochroming methods were in the fields of botany (Haitinger & Linsbauer, 1933), the nervous system (Exner & Haitinger, 1936), pathology (Haitinger & Geiser, 1944; Eppinger, 1949), and cytology (Bukatsch, 1940).

When fluorochromes were used with plant material,

* Bommer was apparently unaware of the discovery of DNA in animal and plant cell nuclei by Feulgen and Rossenbeck in 1924 (Kasten, 1964) although Bommer and Feulgen were fellow faculty members at Justus-Liebig University in Giessen.

some unusual colour combinations were produced; autofluorescence often persisted together with the induced fluorescence. For instance, when a section of wood was stained with Magdala red (CI 50375) and examined in the fluorescence microscope, there was revealed red-fluorescing cuticle, orange-red primary and secondary phloem, blue cork cambium, and intense red xylem (Haitinger & Linsbauer, 1935). With the use of certain fluorochromes, like coriphosphine O, two different colours were induced in the same cell, revealing orange-fluorescing cytoplasm and yellow-fluorescing nuclei (Bukatsch & Haitinger, 1940). This multicoloured fluorescence, known as metachromasia, was seen as well in mixed populations of live and dead cells when the mixture was treated with acridine orange (Strugger, 1940). For instance, living epidermal plant cells fluoresced green at a pH of 5.7–8.0 and dead cells fluoresced red. Further details of Strugger's pioneering work with acridine orange will be discussed later.

2.4.2 Microbiology

Tubercle bacilli had been examined in the fluorescence microscope as early as 1917 by virtue of their autofluorescing properties (Kaiserling, 1917). The first report of the fluorochroming of microorganisms came 20 years later by Hagemann at the Hygienic Institute of the University of Cologne, who used berberine (CI 75160) to fluorochrome lepra bacilli (Hagemann, 1937a) and primulin (CI 49000) to visualize viruses (Hagemann, 1937b). Other fluorochromes that proved useful were thioflavine S (CI 49005, Hagemann, 1939; Levaditi & Reinie, 1939) and mordanted morin (CI 756609) and thioflavine (Hagemann, 1937c). Attempts were made to find a more sensitive staining method than the Ziehl-Neelsen procedure to detect tubercle bacilli. Eventually, auramine O (CI 41000) was found satisfactory in combination with an acid–alcohol treatment of the smears (Hagemann, 1938). Auramine O produced brilliant yellow fluorescence from stained organisms and the technique became popular in the United States through the work of Richards (1941). He examined the mechanism of staining and concluded that the specificity for the tuberculosis organism was due to dye binding to mycolic acid, the acid-fast component in the bacterial cell wall (Richards, 1955). With tuberculosis still a leading cause of death in the world among communicable diseases, laboratory detection still relies in part on the carbol–auramine O fluorescence method, often with a rhodamine or acridine orange counterstain. For reviews of the literature on fluorochroming of microorganisms, see Ellinger (1940), Strugger (1949), Duijn (1955), Price and Schwartz (1956), and Kasten (1983a, 1989).

2.5 INTRODUCTION OF ACRIDINE ORANGE INTO CELL PHYSIOLOGY, CYTOLOGY AND CYTOCHEMISTRY

Acridine orange (AO), a basic dye, was synthesized by Benda in 1889 and was produced by Badische Anilin & Soda Fabrik. Although AO was available from Dr K. Hollborn & Söhne, according to their catalogue of 1932, the dye was overlooked by Hamperl and Haitinger in their extensive survey of fluorochromes for possible value in fluorescence microscopy (Haitinger & Hamperl, 1933; Hamperl, 1934).

AO was first introduced as a fluorochrome into fluorescence microscopy, independently by Bukatsch and Haitinger (1940) and by Strugger (1940). Bukatsch and Haitinger found that AO was suitable as a vital fluorochrome in living plant cells, staining cell nuclei. Mitotic chromosomes were also fluorochromed (Bukatsch, 1940). These workers did not notice any unique fluorescent properties of AO.

Siegfried Strugger, a plant cell physiologist at the University of Münster, discovered the extraordinary ability of AO to fluorochrome live and dead cells in different colours (Strugger, 1940). This finding had its basis in a long series of papers published between 1931 and 1940 in which Strugger investigated the vital staining of cells with other dyes by bright-field microscopy. Most workers had not considered the influence of pH of the staining solution when examining fluorescence of tissues and cells; dyes were simply prepared in dilute solutions. However, Hercik (1939) reported that intravital staining of onion epidermal cells with fluorescein at pH 1.5 produced different fluorescence patterns, according to whether the cells were viable or not. As a cell physiologist, Strugger was acutely aware of the importance of pH in the binding between charged fluorochrome ions and intracellular constituents. He also recognized the influence that dye concentration might have on the presence of dissociated and undissociated forms of the dye in solution. His findings were made and reported during the war years. There were about 20 research papers published between 1940 and 1944, based on fluorescence microscopic investigations with AO and a few other fluorochromes. Immediately after the war, he was commissioned by the occupational authorities to summarize the German wartime research on cell physiology and protoplasm of plant cells (Strugger, 1946). He was recruited by the US government under Project Paperclip and did research in the United States. After returning to Germany, Strugger wrote two books, one of which dealt with fluorescence microscopy (Strugger, 1949). Strugger's seminal contributions led directly to the modern use of AO as a fluorescent probe for nucleic acids in fluorescence microscopy and flow cytometry. Because of the significance of Strugger's work, this will be covered in more detail here.

In his first and most important research work in fluorescence microscopy. Strugger systematically examined the uptake and storage of AO by living plant cells (Strugger, 1940). The fluorescence colour within cells was shown to depend in part on dye concentration, called the 'concentration effect'. Others later referred to this as the 'Strugger effect'. At low concentrations (1:5 000–1:100 000), the fluorescent colour was green whereas at high concentration (1:100), the colour was red. Intermediate dye concentrations produced a yellow colour. Also, Strugger showed that AO could discriminate between live and dead plant cells by fluorescence microscopy, according to the pH of the staining solution. At a pH of 5.7–8.0, living cells fluoresced green and dead cells appeared red. With dead cells, AO produced a red cytoplasm above pH 4.7. The nuclei were red at low pH and shifted to yellow-green, beginning at a pH of 6.8. Strugger's observations on bicolour fluorescence from populations of live and dead cells were confirmed (Bukatsch, 1941; Bucherer, 1943). Strugger and associates extended these vitality experiments to yeast, slime moulds, bacteria and sperm (Strugger, 1940/41; Strugger & Hilbrich, 1942; Strugger & Rosenberger, 1944). The AO method was important since previous techniques for distinguishing live and dead cells failed to give such striking and clear-cut results. Also, the observation of bicolour fluorescence with AO staining, a type of metachromasia, attracted attention to this phenomenon. Strugger pioneered in the use of fluorescent pH indicators in cell physiology (Strugger, 1941).

According to Strugger, the basis for bicolour fluorescence after vital fluorochroming depended on the relative binding of AO cations by cell proteins. In live cells, the concentration of bound dye cations was low, producing green fluorescence, whereas in damaged (yellow fluorescence) and dead (red fluorescence) cells, there would be a progressive increase in AO cation binding by proteins through electrostatic means. In viable cells, there would presumably be few accessible electronegative charges present on proteins. Following injury or cell death, Strugger postulated that a disturbance occurs in the submicroscopic protein scaffold, making accessible more negative charges. The copper-red fluorescence seen in dead cytoplasm would be the visible manifestation of the fine structural alterations. Strugger's AO method attracted the attention of Adolph Krebs, who systematically examined alpha particle radiation damage to cells with the aid of AO and fluorescence microscopy (Krebs, 1944). After the war, Strugger and Krebs worked together with Gierlach at Fort Knox, Kentucky, where additional radiation studies were done on onion cells using AO as a new experimental tool in radiation biology (Krebs & Gierlach, 1951; Strugger *et al.*, 1953).

AO was also shown by Strugger to be favourable for determining the isoelectric point (IEP) of cellular proteins. In alcohol-fixed material, protein-containing structures emitted a green fluorescence below the IEP and a copper-red fluorescence above the IEP.

The interaction of AO with living cells was examined further by many workers. Strugger's interpretation of the differential fluorescence of living and dead cells has been questioned. The bicolour fluorescence was suggested to be related to cellular metabolic activity (Schümmelfeder, 1950), binding to DNA, mononucleotides in mitochondria, and polysaccharides (Austin & Bishop, 1959), lysosomes (Robbins & Marcus, 1963; Robbins *et al.*, 1964), and nucleoprotein complexes (Wolf & Aronson, 1961). Dye solutions of AO exhibit metachromasia due most likely to the formation of species of dye monomers, dimers and polymers (Zanker, 1952; Steiner & Beers, 1961). AO–nucleic acid solutions form different complexes, including dye intercalation between base layers of DNA (green) and dye interaction with phosphate groups on nucleic acid surfaces, referred to as stacking (Loeser *et al.*, 1960; Bradley, 1961; Steiner & Beers, 1961).

Binding of AO to fixed cells was studied in detail by Schümmelfeder (1948, 1956), who emphasized the importance of pH in the dynamics of AO binding to intracellular constituents. The application of the pH principle to determine the IEP of tissue proteins was verified (Schümmelfeder & Stock, 1956; Schümmelfeder, 1956). Under acidic staining conditions, the AO dye cation was shown by various workers to stain acid components, like the acidic mucopolysaccharides found in cartilage and mast cell granules, and nucleic acids of cells. Independently, three groups of investigators discovered that under controlled conditions of staining with AO, DNA of fixed interphase nuclei and chromosomes fluoresced yellow-green to green whereas regions rich in RNA (nucleolus, basophilic regions of cytoplasm) fluoresced orange to red (Armstrong, 1956; Bertalanffy & Bickis, 1956; Schümmelfeder *et al.*, 1957). It was suggested that the colour differences were due to molecular size variations and configuration of the two nucleic acids (polymerization, denaturation) and not to intrinsic chemical differences between RNA and DNA (Schümmelfeder, 1958; Aldridge & Watson, 1963). An impressive microspectrofluorometric investigation of the mechanism of AO binding to purified and intracellular nucleic acids was done by Rigler (1966). He confirmed that the orderliness of the secondary structure of DNA (accessibility of DNA-phosphates) had a profound influence on AO binding.

It became apparent with AO fluorochroming that

Table 2.2 Fluorochromes used in biological microscopy.[a]

Fluorochrome	*CI no.*	*Chemical group*	*Acidic or basic*	*Biological applications*
Acid fuchsin	42685	Arylmethane	A	Counterstain, elastic fibres and other connective tissues supravital fluorochrome (plant cells), pH indicator
Acridine red 3B	45000	Xanthene	B	Histology
Acridine orange	46005	Acridine	B	Histology, cell viability, DNA intercalator, nucleic acids (fluorochromasia), cytodiagnosis, isoelectric point of proteins, bacteria, viruses, mast cells, supravital fluorochrome (plant and animal cells), protozoa, tumour localization, sperm, lysosomes, acid mucopolysaccharides, plant tissues, pH indicator, amyloid
Acridine yellow	46025	Acridine	B	Histology, insect tissues, cytodiagnosis, protozoa, viruses, tubercle bacilli, Schiff-type reagent
Acriflavine (trypaflavine, mixture of 3,6-diamino-10-methyl acridinium chloride and 3,5-acridinediamine or proflavine)	46000	Acridine	B	Histology, plant tissues, intercalating dye for nucleic acids (fluorochromasia), Schiff-type reagent, protozoa, viruses, vital stain, bacteriostatic agent, intravital fluorochrome, inhibits mitochondriogenesis
Alizarin red S	58005	Anthraquinone	A	Bone and bone growth
Alizarin cyanine BBS	58610	Anthraquinone	A	Histology
Aniline blue	42755	Arylmethane	A	Plant cell walls (callose), β-glucans, eosinophils, glycogen
Atabrine (see quinacrine dihydrochloride)				
Auramine O	41000	Arylmethane	B	Tubercle bacilli (acid-fast bacteria), bacterial counting, viability (plant tissue), blood cells, Schiff-type reagent; fluorescent complex with horse-liver alcohol dehydrogenase
Aurophosphine (see phosphine 5G)				
Basic fuchsin (see pararosaniline)				
Benzoflavine (see flavophosphine N)				
3,4-Benzopyrene				Lipids (suspect cancer agent)
Brilliant cresyl blue	51010	Oxazine	B	Lipids, vital staining of blood, reticulocytes in blood smears, protozoa, chromosomes
Berberine sulphate	75160	Natural plant alkaloid	B	Histology, wood tissue, vital fluorochrome (botany, protozoa), insect histology, bacteria, mitochondria, viruses, nucleic acids (nuclei), heparin, antibacterial and antimalarial agent, chromosome banding, mast-cell granules
Brilliant sulphoflavine	56205	Aminoketone	A	Protein stain, bacterial spores
Calcein (active part is DCAF, 3,6-hydroxy 24-bis-[*N*,*N'*-di-(carboxy methyl)-aminomethyl] fluoran (Fluorexone)	—	Xanthene	A	Bone growth, eosinophils
Calcein blue	—	Xanthene	A	Bone growth
Calcofluor white M2R (Cellufluor)	40622	Stilbene	A	Plant cell walls, β-glucans, microorganisms, vital fluorochrome, fungi in tissue sections
Chelidonium	—	Natural plant extract		Fat, nuclei
Chlortetracycline (Aureomycin)	—	Natural	A	Bone growth, antibiotic, membrane-bound Ca^{2+}

Table 2.2 Continued

Fluorochrome	*CI no.*	*Chemical group*	*Acidic or basic*	*Biological applications*
Chrysophosphine 2G	46040	Acridine	B	Nuclei, mast cells, wood, acid mucopolysaccharides, Schiff-type reagent, amyloid
Congo red	22120	Disazo	A	Counterstain, amyloid, probe for conformation of nucleotide-binding enzymes
Coriphosphine O	46020	Acridine	B	Histology, nucleic acids (fluorochromasia), mast cells, haematology, wood, bacteria, plant cell walls, juxtaglomerular granules, fat, acid mucopolysaccharides, Schiff-type reagent, red fluorescence of diffuse neuroendocrine cells by 'masked basophilia'
Eosin B	45400	Xanthene	A	Counterstain, muscle, haemoglobin
Eosin Y	45380	Xanthene	A	Counterstain, muscle, plant tissue, immunofluorescence label, constituent of blood stains, histones
Erythrosin B	45430	Xanthene	A	Counterstain, supravital, dental disclosing agent for plaque, immunofluorescence label
Esculin (6,7-dihydroxycoumarin 6-glucoside)	—	Natural plant glucoside		Vital fluorochrome (protozoa), antimalarial agent
Ethyl eosin	45386	Xanthene	A	Counterstain
Euchrysine 2GNX	(see chrysophosphine 2G)			
Evans blue	23860	Disazo	A	Counterstain, fluoresces red when bound to protein, diagnostic aid (blood volume determinations), retrograde procedure for axonal branching with DAPI and primuline, teratogen (suspect cancer agent)
Flavophosphine N (benzoflavine)	46065	Acridine	B	Histology, Schiff-type reagent, supravital fluorochrome (plant physiology), pH indicator
Fluorescein (uranin, Na salt)	45350	Xanthene	A	pH indicator, intravital fluorochrome (microcirculation), insect histology, viruses, diagnostic aid (ophthalmology and central nervous system tumours), immunofluorescence label, vital (plants), dental disclosing agent for plaque, circulation time
Gernaine B	14930	Monoazo	A	Fat, cell nuclei, elastic fibres
Isamine blue (brilliant dianyl blue)	42700	Arylmethane	A	Connective tissue, nervous system, supravital
Magdala red	50375b	Azine	B	Fat, mucus, plant cell walls
Mercurochrome	—	Xanthene	A	Counterstain, leukocytes, protein-bound SH and S-S groups
Methylene blue	52015	Thiazine	B	Fat, histology, blood stain constituent, bacteriological stain, antidote to cyanide poisoning
Methyl green	42585	Arylmethane	B	Cell nuclei, polymerized nucleic acid (DNA), histology, gonococci, mast cells, (irritant)
Morin (3,5,7,2′,4′-pentahydroxyflavanol	75600	Natural plant flavone	A	Cell nuclei (nucleic acids when complexed with aluminium ammonium sulphate), dye complexes with spirochaetes, trypanosomes, metal detection (Al)
Neutral acriflavine (see acriflavine)				
Neutral red	50040	Azine	B	Lipids, mast cells, bacteria, supravital fluorochrome, Schiff-type reagent, histology
Nile blue A	51180	Oxazine	B	Fat, differentiating melanins and lipofuscins, Schiff-type reagent

Table 2.2 Continued

Fluorochrome	*CI no.*	*Chemical group*	*Acidic or basic*	*Biological applications*
Nile red (Nile blue A oxazone)	51180	Oxazine	B	Lipid droplets, plant cell microsomes
Oxytetracycline	—	Natural	A	Bone growth, antibiotic
Pararosaniline (main component of fuchsin)	42500	Arylmethane	B	Cell nuclei, elastic tissues, Schiff reagent, bacilli, anti-schistosomal and topical anti-fungal agnet, caries stain discloser (suspect cancer agent)
Phenosafranin	50200	Azine	B	Histology, Schiff-type reagent
Phloxine B	45405	Xanthene	A	Counterstain
Phosphine 5G (Aurophosphine)	46035	Acridine	B	Mast cells, acid mucopolysaccharides, Schiff-type reagent
Phosphine GN (phosphine 3R)	46045	Acridine	B	Lipids, cell nuclei, insect histology, nerve tissue, Schiff-type reagent
Primulin	49000	Thiazole	A	Intravital fluorochrome, plant cell walls, cell nuclei, protozoa, viruses, proteins
Procion red	18159	Monoazo	A	Bone growth
Procion yellow M4RS (MX-4R)	Reactive Orange 14	Monoazo	A	Vital fluorochroming of neurons and functional connections after introduction into cells by electrophoresis, label for newly forming bone
Proflavine (similar to acriflavine)				
Prontosil (sulphamido-chrysoidine)	—	Monoazo	B	Vital fluorochrome (insects and plants), connective tissue fibres
Pseudoisocyanin	—	Quinolin	B	Neurosecretion, cysteic acid groups in proteins
Pyronin Y (G)	45005	Xanthene	B	RNA preferentially (usually in combination with methyl green), single-stranded nucleic acids, bacteria, supravital fluorochrome (plant tissues), pH indicator, plasma cells
Quinacrine dihydrochloride (Atabrine)	—	Acridine	B	DNA, chromosome Q banding, vital fluorochrome (protozoa), nerve fibres, tumour localization, antimalarial and antihelminthic agent, (light-sensitive)
Quinine	—	Natural plant alkaloid	A	Vital fluorochrome (protozoa), insect histology, antimalarial agent, (light-sensitive)
Rheonin A	46075	Acridine	B	Histology, fungi, Schiff-type reagent
Rhodamine B	45170	Xanthene	Neut.	Vital fluorochrome (plant cell sap, mitochondria, bacteria, fat, viruses, metal detection, immuno-fluorescence label (suspect cancer agent)
Rhodamine G	45150	Xanthene	B	Histology, vital fluorochrome, viruses
Rhodamine 3GO	45210	Xanthene	B	Histology, Schiff-type reagent
Rhodamine 6G	45160	Xanthene	B	Vital fluorochrome (mitochondria), wood (suspect cancer agent)
Rhodamine S	45050	Xanthene	B	Histology
Rhodanile blue (complex of Nile blue A and rhodamine B)	—	Oxazine-xanthene	B	Histological differentiation
Rhodindine (related to Magdala red)	50375a	Azine	B	Schiff-type reagent
Rhubarb	—	Natural plant extract		Histology
Rivanol	—	Acridine	B	Supravital (protozoa), bacteria, leukocytes, cell nuclei, Schiff-type reagent, mast cells
Rose bengal	45400	Xanthene	A	Fat, liver, bacteria, hepatic function determination

Table 2.2 Continued

Fluorochrome	*CI no.*	*Chemical group*	*Acidic or basic*	*Biological applications*
Safranin O	50240	Azine	B	Histology, nuclei and chromosomes, Schiff-type reagent, plant tissues, starch granules
Sanguinarine	—	Natural plant alkaloid		Insect tissues
Stilbene	40000	Stilbene	A	Proteins, insect tissues
Sulphorhodamine B (Lissamine rhodamine B 200)	45100	Xanthene	A	Immunofluorescence label
Tetracycline	—	Natural	A	Bone growth, mitochondria, cancer localization, antibiotic
Thiazin red R	14780	Monoazo	A	Fat, amyloid, proteins
Thiazole yellow G (Titan yellow)	19540	Monoazo	A	Histology, vital fluorochrome, Mg detection
Thioflavine S	49010	Thiazole	A	Histology (fluorochromasia), intravital fluorochrome (blood vessels), leukocytes, bacteria, amyloid, protozoa, myelin
Thioflavine T	49005	Thiazole	B	Histology, tubercle bacilli, phospholipids, mast cells, insect histology, amyloid
Titan yellow (see thiazole yellow G)				
TMPP (meso-tetra (4-*N*-methylpyridyl) porphine		Porphyrin	B	Chromatin (DNA)
TPPS (tetraphenylporphin sulphonate)		Porphyrin	A	Elastic fibres
Trypaflavine (see acriflavine)				
Trypan blue		Disazo	A	Dye exclusion test for cell vitality, teratogen, fluoresces red when bound to protein
Uranin (see fluorescein)				
Uvitex 2B		Stilbene	A	Fungi in tissue sections
Vasoflavine (see thioflavine S)				
Xylenol orange	—		A	Sites of calcification (bone growth)

[a] The word *fluorochrome* was coined by Haitinger in 1934 to denote fluorescent dyes used in biological staining to induce secondary fluorescence in tissues. Data listed above were derived from many sources. In addition to obtaining information from published research and review articles, other material was assembled from the *Colour Index* (1971), *Reichert's Fluorescence Microscopy with Fluorochromes* (1952), *Conn's Biological Stains* (1977), *Handbuch der Farbstoffe für die Mikroskopie* (Harms, 1965), and catalogues of the Aldrich Chemical Co., Eastman Kodak Co., Polysciences, Inc., and Sigma Chemical Co.

brilliant colour differences could be easily seen between cancer cells, with their hyperchromatic nuclei and high RNA content, and normal cells. The AO technique was incorporated into exfoliative cytology as a rapid screening test for cervical cancer and other malignancies (Bertalanffy & Bickis, 1956; Dart & Turner, 1959). AO proved to be a sensitive cytochemical fluorochrome for the detection and identification of nucleic acids in purified and viral-infected cells (Armstrong & Niven, 1957; Mayor, 1963). The organization of DNA in chromosomes was investigated using polarized fluorescence microscopy (MacInnes & Uretz, 1966). As a stain for DNA in chromosomes, AO ordinarily gives uniform fluorescence along the length of chromosome arms. However, the dye produces reverse banding (R-bands) when it is used after pretreatment with the antibiotics distamycin (AT-specific) or actinomycin D (GC-specific) or hot phosphate buffer (Comings, 1978; Gustashaw, 1991). Microfluorometry was employed to obtain quantitative information about the content of RNA and DNA in single cells and DNA molecular alterations (Rigler, 1966).

The metachromatic fluorochrome AO has proved to be a valuable nucleic acid probe in modern flow cytometry when conditions of dye binding are well-controlled. For reviews, see Melamed and Darzynkiewicz (1981) and Darzynkiewicz (1991). A review of the older AO literature is given by Kasten (1967). The chapter by Zelenin (Chapter 9) in the present volume gives additional details about AO as a fluorescent probe.

Table 2.3 Fluorescent probes used in modern cell and molecular biology.[a]

Name	*Wavelength (nm)* Excitation	*Wavelength (nm)* Emission	*Applications*
Immunocytochemical fluorophores, conjugates and lectins			Tracers for various proteins, receptors, and mono- and polysaccharides
Allophycocyanin (AP)	620	660	
Allophycocyanin cross-linked (AP-XL)	650	660	
AMCA (7-amino-4-methylcoumarin-3-acetic acid	350	450	
Bodipy (borondipyromethene difluoride dye)	505	512	
CT-120 (coumarin 120 thiolactone)	345	410	
CT-339 (coumarin 330 thiolactone)	350	420	
Coumarin 138	365	460	
Dansyl chloride	340	578	
2,7′-Dichlorofluorescein	513	532	
4′,5′-Dimethylfluorescein	510	535	
5-DTAF (5-(4,6-dichlorotriazinyl) aminofluorescein)	495	530	
Eosin-5-isothiocyanate	524	548	
Erythrosin-5-isothiocyanate	535	558	
FITC (fluorescein-5-isothiocyanate)	490	520	
Fluorescein anhydride (FA)	490	520	
3-HFT (3-hydroxyflavone thiolactone)	350	415	
Lucifer yellow CH	435	530	
N-Methylanthranyloyl	350	440	
NBD (nitrobenzoxadiazole)	468	520	
Phycocyanin (PC)	620	650	
Phycoerythrin B (PE-B)	545	576	
Phycoerythrin R (PE-R)	495, 545	578	
Princeton red anhydride (PRA)	490	580	
RITC (rhodamine B isothiocyanate)	570	595	
Sulphorhodamine B sulphonyl chloride (lissamine rhodamine B sulphonyl chloride)	570	590	
Sulphorhodamine 101 (sulphonyl chloride, Texas Red)	596	620	
TMRA (tetramethylrhodamine anhydride)	550	570	
TRITC (tetramethylrhodamine-5-isothiocyanate)	541	572	
XL-Allophycocyanin (XL-AP)	620	660	
XRITC (tetra-*N*-cyclopropyl-rhodamine isothiocyanate)	578	604	
Site-selective probes (vital)			
Acridine orange	490	590	Lysosomes, nuclei, fluorescent counterstain for retrogradely labelled neuronal tracers
Acridine orange-10-dodecyl bromide	493	520	Mitochondria
Alizarin complexone	580	645	Sites of calcification (bone growth)
AMA (3-amino-6-methocyacridine)	UV	Y,G	Lysosomes, nuclei (fluorochromasia)
Bis-ANS	385	500	Inhibitor of microtubule assembly and RNA polymerase
Bisbenzimide	360	530	Retrograde labelling of neuronal nuclei
Calcein	495	520	Sites of calcification (bone growth)
Calcein blue	375	435	Sites of calcification (bone growth)
Cascade Blue hydrazide	376, 389	423	Covalent labelling of microinjected cells
Colcemid-NBD	468	520	Tubulin polymerization detection

Table 2.3 Continued

Name	Wavelength (nm)		Applications
	Excitation	*Emission*	
DACK (dansylalanyllysyl chloromethyl ketone)	350	450	Inhibitor of acrosin (mammalian sperm)
DAPI (4′,6-diamidino-2-phenylindole)	372	456	Tubulin polymerization detection without interfering with microtuble assembly, retrograde labelling of neurons
DASPEI (2-(4-dimethyl aminostyryl)-*N*-ethyl-pyridinium iodide)	429	557	Mitochondria (metabolic state affects fluorescence response)
DASPMI	429	557	Mitochondria (yellow), membranes (green), nucleus (red-orange)
DEQTC (1,3′-diethyl-4,2′-quinolylthiacyanine iodide)	502	529	Reticulocytes
5,7-DHT (5,7-dihydroxytryptamine)	UV-B	G	Injection induces fluorescence in distant populations of amacrine cells of retina
$DiIC_{18}$ (3) (1,1′-dioctadecyl-3,3,3′,3′-tetramethylindocarbocyanine perchlorate)	547	571	Long-term cell tracing *in vitro*
Dihydroethidium (*see* Nucleic acid probes and Cell viability probes)			
Dihydrorhodamine	UV (510)	320 (534)	Colourless but oxidizes inside cells to rhodamine 123 that vitally stains mitochondria
$DiOC_1(3)$	482	510	Reticulocytes
$DiOC_6(3)$ (3,3′-dihexyloxacarbocyanine iodide)	478	496	Endoplasmic reticulum (also in fixed cells), mitochondria
$DiOC_7(3)$ (3,3′-diheptyloxacarbocyanine iodide)	488	540	Penetration probe into spheroids or tumour cords, mitochondria of plant cells, distinguish cycling from non-cycling fibroblasts
$DiSC_1(3)$ (3,3′-dimethyl-thiacarbocyanine iodide)	551	568	Endoplasmic reticulum
Dopamine	UV-B	G	Injection induces fluorescence in distinct populations of amacrine cells of retina
DPPAO (lecithin analogue of acridine orange)	UV	G	Mitochondria (also in fixed cells)
Ethidium bromide	545	610	Usefule Nissl fluorochrome after FITC labelling of nervous system
Evans blue	550	610	Retrograde labelling of neuronal cytoplasm
Fast blue (*trans*-1-(5-amidino-2-benzofuranyl)-2-(6-amidino-2-indolyl) ethylene dihydrochloride	360	410	Retrograde labelling of neuronal cytoplasm
FITC-dextran	490	520	Fluid phase pinocytosis, loading cells with macromolecules
FluoroBora 1 (3-(dansylamido) phenylboronic acid)	UV-B	Y-G	Lysosomes
FluoroBora 2 (3-(darpsylamidyl)-1-phenylboronic acid)	UV	B-W	Golgi apparatus
FluoroBora T-acriflavine (3-amino, 6-7′(7′,8′,8′-tri) cyanoquinodimethane phenyl-boronic acid)	470	530, 590	Lipid- and water-soluble FluoroBora with trace of acriflavine penetrates cells and produces yellow-green chromatin and orange cytoplasm
Fluoro-Gold	323	408	Retrograde labelling of neuronal cytoplasm, fluoresces gold at neutral pH and blue at acid pH

Table 2.3 Continued

Name	Wavelength (nm)		Applications
	Excitation	Emission	
Granular blue (2-(4-(4-amidino-phenoxy) phenyl) indol-6-carbox-amidin-dihydrochloride)	375	410	Retrograde labelling of neuronal cytoplasm
Hexanoic ceramide-NBD (C_6-NBD-ceramide)	475	525	Golgi apparatus and lipid transport pathways in cytoplasm
Hoechst 33258 (bisbenzimide trihydrochloride)	365	465	Mycoplasma detection, chromosomal bands and interbands
Hydroethidine (*see* dihydroethidium under Nucleic acid probes and Cell vitality probes)			
Lucifer yellow CH (LY)	430	535	Neurons and functional connections after microinjection into cells or fluid phase pinocytosis
Merocyanine 540	500	572	Mitochondria, binds to leukaemic cells, axons stain
MUA (4-methyl-umbelliferone derivatives)	340	430	Lysosomal enzymes, used to detect lysosomal storage diseases
NAO (10-*N*-nonyl-acridine orange chloride)	492	522	Mitochondria (also in fixed cells)
Nile red (Nile blue A oxazone)	450–500 515–560	530 605	Neutral lipids, cholesterol, phospholipids in cellular cytoplasmic droplets and lysosomes, foam cells and lipid-loaded macrophages, (excitation and emission spectra vary greatly according to hydrophobicity of environment)
NPN (*N*-phenyl-1-naphthylamine)	340	420	Detects early lymphocyte activation
Nuclear yellow (Hoechst S-769121)	360	460	Retrograde labelling of neuronal nuclei
Phallacidin-Bodipy	505	512	F-actin
Phallacidin-NBD	468	520	F-actin
Phalloidin-fluorescein	490	520	F-actin
Phalloidin-rhodamine	540	580	F-actin
Phallotoxin-phenylcoumarin	387	465	F-actin
Procion yellow M4RS (MX-4R)	488	530	Neurons and functional connections after electrophoretic injection into cells
p-Bis-(2-chloroethyl)-amino-benzilidene-cinnamonitrile fluorochromes (nitrogen mustard derivatives with stilbene-like structures)	360–400 450–480	520–550 550–590	Cell fluorochromes useful in combination with autofluorescing coenzymes
Pyronin Y	545	580	Mitochondria, arrests cells in G1 phase
Rhodamine 123	510	534	Mitochondria (increased accumulation and retention in mitochondria of carcinoma cells), also fixed cells
Rhodamine B hexyl ester, chloride	555	579	Endoplasmic reticulum
Rhodamine-dextran	570	595	Fluid phase pinocytosis, loading cells with macromolecules
Rhodamine 6G	530	590	Mitochondria
Texas Red-ovalbumin	596	620	Absorptive pinocytosis
Thiazole orange (TO)	509	533	Reticulocytes and malaria parasites (haematology and flow cytometry)
Thioflavin T	370	418	Amyloid plaque core protein (APCP), reticulocytes (flow cytometry)
'True Blue' (*trans*-1,2,-bis (5-amido-2-benzofuranyl) ethylene-dihydrochloride	373	404	Retrograde labelling of neuronal cytoplasm
Tubulin-DTAF	495	530	Microtubules
Tubulin-NBD	468	520	Microtubules

Table 2.3 Continued

Name	Wavelength (nm)		Applications
	Excitation	*Emission*	
Xylenol orange	377	610	Sites of calcification (bone growth)
Nucleic acid probes			
ACMA (9-amino-6-chloro-2-methoxyacridine)	430	474	DNA, chromosome Q-banding, AT-specific DNA
Acridine ethidium heterodimer	492 528	627 634	AT-rich DNA and total DNA according to excitation used
Acridine orange	490	530, 640	Distinguish and measure single- and double-stranded nucleic acids by fluorochromasia (intercalates into double-stranded nucleic acids, binds to phosphate groups), chromosome banding
Acriflavine-Feulgen	455	515	DNA
Adriamycin	480	555	Intercalates in GC-specific DNA, chromosome D banding, Y chromosome, antineoplastic anthracyline (mostly nuclear fluorescence)
7-Amino-AMD (7-amino-actinomycin D)	555	655	Intercalates in GC-specific DNA
Auramine O-Feulgen	460	550	DNA
BAO-Feulgen (bisamino-phenyloxadizole)	380	470	DNA
Bis-ANS	385	500	Inhibits RNA polymerase
Carminomycin	470	550	Antineoplastic anthracycline (cytoplasmic fluorescence)
Chromomycin A3	450	570	GC-specific DNA in presence of Mg^{2+}, antineoplastic antibiotic
DAMA (3-dimethylamino-6-methoxyacridine)	467	549, 617	RNA and DNA (fluorochromasia)
DAPI (4′,6-diamidino-2-phenylindole HCl)	372	456	AT-specific double-stranded DNA, chromosome Q banding, distinguish between yeast mitochondrial and nuclear DNA, viral and mycoplasma DNA infection in cells
Daunomycin (daunorubicin HCl)	475	550	DNA, chromosome D banding, Y chromosome, antineoplastic anthracycline (nuclear fluorescence)
Dihydroethidium (hydroethidine, reduced ethidium bromide)	370, 525	420, 605	DNA (hydroethidine is enzymatically oxidized in living cells to form ethidium bromide, which intercalates into nuclear chromatin (red), cytoplasm fluoresces blue-white in lipoidal pockets)
$DiOC_1(3)$	482	510	Nucleic acids
DIPI (4′,6-bis(2′-imidazolinyl-4H,5H)-2 phenylindole)	355	450	AT-specific DNA, chromosome Q-banding
Ellipticine	450	525	RNA and DNA (intercalates into double-stranded nucleic acids)
Ethidium bromide (homidium bromide)	545	610	RNA and DNA (intercalates into double-stranded nucleic acids of fixed cells), viability assay
Ethidium monoazide	510	600	RNA and DNA (intercalates into double-stranded nucleic acids), viability assay

Table 2.3 Continued

Name	Wavelength (nm) Excitation	Emission	Applications
Hoechst 33258 (bisbenzimide trihydrochloride)	365	465	AT-specific DNA (vital and fixed cells), chromosome Q-banding, DNA synthesis quenching of fluorescence detects incorporation of 5-BrdU into DNA), viral and mycoplasma DNA infection in cells
Hoechst 33342 (bisbenzimidazole derivative)	355	465	AT-specific DNA (vital and fixed cells), may be effluxed rapidly from certain cells so as to prevent DNA binding
Homoethidium (*see* dihydroethidium)			
Hydroxystilbamidine	360	450, 600	AT-specific DNA (both peaks appear) and RNA (only 450 nm emission seen)
Mithramycin (aureolic acid)	395	570	GC-selective DNA, antineoplastic antibiotic
Nogalomycin	480	560	DNA, antineoplastic anthracycline
Olivomycin	430	545	DNA, chromosome R banding, antineoplastic antibiotic
Oxazine 750 (OX750)	690	699	DNA (excitable by helium-neon laser)
Proflavine	455	515	Nucleic acids, AT-specific DNA
Propidium iodide	530	615	RNA and DNA (intercalates into double-stranded nucleic acids of fixed cells), viability assay
Pyronin Y	540	570	RNA (preferentially in presence of methyl green or Hoechst 33342)
Quinacrine dihydrochloride (atabrine)	436	525	AT-specific DNA, chromosome Q-banding, Y chromosome
Quinacrine mustard	385	525	AT-specific DNA, chromosome Q-banding
Rhodamine 700 (LD 700)	659	669	DNA (excitable by helium-neon laser)
Rhodamine 800 (R 800)	700	715	DNA (excitable by helium-neon laser)
Rubidazone	480	560	Antineoplastic anthracycline (cytoplasmic fluorescence)
RuDIP (tris-(4,7-diphenyl-phenanthroline) ruthenium (III)	453 470	480 630	Metal complex to distinguish handedness of DNA helices
Thiazole orange (TO)	509	533	Nucleic acids (*see* Site-selective probes)
Thioflavine T	422	487	Nucleic acids (*see* Site-selective probes)
TMPP (Meso-tetra (4-*N*-methylpyridyl porphine)	436	655	DNA
4,5′,8-Trimethylpsoralen (trioxsalen)	338	420	Intercalates into double-stranded DNA and covalently adds to pyrimidines upon UV illumination
'True Blue' (*trans*-1,2-bis (5-amido-2-benzofuranyl) ethylene-dihydrochlorine)	373	404	AT-specific DNA
Protein probes and functional groups			
Acridine orange	490	530	Acidic groups of proteins after hot TCA extraction
Ammonium 7-fluoro-2-oxa-1,2-diazole-4-sulphonate	380	515	Thiol groups
ANS (8-anilino-1-naphthalene sulphonic acid)	385	485	Proteins (hydrophobic probe)
Anthracene-9-carboxaldehyde carbohydrazone	393	456	Aldehyde probe
Bis-ANS (1,1′-bi(4-anilino) naphthalene-5,5′-disulphonic acid, dipotassium salt)	385	500	Proteins (dimer of ANS, binds at multiple sites)
Brilliant sulphoflavine (BSF)	420	520	Histone proteins at pH 8.0 after DNA extraction, total proteins at pH 2.8

Table 2.3 Continued

Name	Wavelength (nm) Excitation	Emission	Applications
CPM (*N*-(4-(7-diethylamino-4-methylcoumarin-3-yl) maleimide	385–390	465	Thiol groups
DAB-ITC (4-*N*, *N*-dimethylamino-benzene-4′-isothiocyanate)	430		Amino acid probe
Dansyl chloride (5-dimethylamino-1-naphthalene-sulphonyl chloride)	335	500	Histone proteins and protamines
Eosin	522	551	Histones at pH 10.0 or higher
FITC (fluorescein-5-isothiocyanate)	490	520	Proteins
Fluoral-P (4-amino-3-pentene-2-one)	410	510	Aldehyde probe
Fluorescamine	390	460	Proteins at cell surface (primary amine binding creates fluorescent complex)
FDA (fluorescein diacetate)	490	520	Esterases
Fluorescein mercuric acetate	B	Y	Thiol groups of proteins (nuclear non-histone proteins after SH reduction to disulphides)
Formaldehyde-induced monoamine fluorophores	410–415	475, 525	Biogenic monamines (dopamine, noradrenaline, 5-hydroxytryptamine)
9-Hydrazine acridine	420	500	Aldehyde probe
'Long Name' (stilbene disulphonic acid derivative, optical brightener)	350	460	Proteins
Mercurochrome	B	Y	Thiol groups of proteins (nuclear nonhistone proteins after SH reduction to disulphides)
Mercury orange (1(4-chloro-mercury-phenyl-azo-2-naphthol))	450–490	600	Glutathione (non-aqueous solvents), thiol groups of proteins (nuclear nonhistone proteins after SH reduction to disulphides)
MUA (4-methylumbelliferone acetate)	540	430	Esterases
OPT (O-phthaldehyde)	UV	B	Polyamines at pH 7–9 (spermidine and spermine)
Primuline	365	455	Proteins
RITC (rhodamine B-isothiocyanate)	570	595	Proteins
Salicoyl hydrazine	320	400	Aldehyde probe
SBD-Cl (4-chloro-7-sulphobenzo-furan, ammonium salt)	380	510	Thiol groups
SITS (4-acetamido-4-isothiocyanatostilbenene-2, 2′-disulphonic acid, disodium salt)			
Sulphorhodamine 101	576	602	Proteins
XRITC (tetra-*N*-cyclopropyl-rhodamine isothiocyanate)	578	604	Proteins
Cell viability probes			
Acridine orange	490	530, 640	Vital fluorochrome (monomeric dye form is green in living cells, aggregated dye form is red in dead cells)
Calcein AM (acetoxymethyl ester)	495	520	Vital fluorochrome
Calcofluor white M2R (CFW)	UV	B	Stains non-viable animal cells, walls of live plant cells
CFDA (5(6)-carboxyfluorescein diacetate)	495	520	Cell vitality probe based on production of intracellular carboxyfluorescein by esterases, detects permeable channels between cells
Chrysophosphine 2G (euchrysine 2GNS)	435	515	Vital fluorochrome (similar to acridine orange)

Table 2.3 Continued

Name	*Wavelength (nm)*		*Applications*
	Excitation	*Emission*	
Dihydroethidium (Hydroethidine, reduced ethidium bromide)	370, 535	420, 585	Nuclei fluoresce red in living cells due to production of ethidium bromide by enzymatic oxidation of dihydroethidine, cytoplasm blue to blue-white
Dihydrorhodamine 123	320	—	Colourless but oxidizes inside cells to rhodamine 123 that vitally stains mitochondria
Ethidium monoazide	460	600	Fluorochromes dead cells, usable after cell fixation
FDA (fluorescein diacetate)	490	520	Cell vitality probe based on production of intracellular fluorescein by esterases, may be used in combination with propidium iodide
Fluorescein	490	520	Detection of fluorescence depolarization (protein-bound dye) in living cells using fluorescence anisotropy measurements
Fluorescein digalactoside	490	520	Monitoring galactosidase gene activity by formation of intracellular fluorescein
FluoroBora T	470	590	Vital fluorochrome (*see* other FluoroBora derivatives under Site-selective probes and Membrane probes and receptors)
Hydroethidine (*see* dihydroethidium)			
Propidium iodide	470	615	Stains dead cells with membrane damage (used with green fluorescing vital dyes like fluorescein)
Pyronin Y	540	570	Vital fluorochrome (mitochondria)
Rhodamine 123	510	534	Vital fluorochrome (mitochondria), accumulates in carcinoma cells, sperm, distinguish cycling from non-cycling cells
SITS (4-acetamido-4-isothiocyanatostilbene-2,2′-disulphonic acid, disodium salt)	350	420	Vital fluorochrome
Vita blue dibutyrate-14 (VBDB-14)	524	570	Vital fluorochrome based on production of fluorescent derivative of fluorescein by esterases
Membrane probes and receptors (vital)			
Acridine orange	490	530, 640	Multilamellar liposomes
Acridine orange-10-dodecyl bromide	500	534, 568	Surfactant micelles (monomer-dimer spectral change)
9-Amino-acridyl propranolol	B	Y	β-Adrenergic receptors
ANS (8-anilino-1-naphthalene sulphonic acid)	385	485	Localizes at interfaces of hydrophilic and hydrophobic membrane regions
Anthroyl ouabain	362, 381	485	Cardiac glycoside receptors
α-Bungerotoxin-tetramethylrhodamine	518, 551	575	Cholinergic receptors
Dansyl lysine	340	515	Membranes with low cholesterol content, leukaemic cells
Dansyl phorbol acetate	341	496	Tumour promoter analogue binds to receptors
Dexamethasone rhodamine	542	566	Glucocorticoid receptors
$DiIC_{18}$ (3) (1,1′-dioctadecyl-3,3,3′,3′-tetramethylindocarbocyanine perchlorate)	549	568	Cationic lipophilic probe used to study fusion and lateral diffusion in membranes, retained for long periods within neurons
DPH (diphenylhexatriene)	351	430	Hydrophobic probe, fluorescence polarization probe of membrane fluidity
Fluorescamine	390	460	Cell surface proteins (non-fluorescent until bound to NH_2-groups)

Table 2.3 Continued

Name	Wavelength (nm) Excitation	Emission	Applications
FluoroBora P (3-(pyrenesulphamido-phenyl-boronic acid))	UV	W-Y, V	Hydrophilic areas (white-yellow), hydrophobic areas (violet)
'Long Name' (stilbene disulphonic acid derivative, optical brightener)	350	460	Intracellular membranes of granulocytes
Naloxone fluorescein	490	520	Opioid receptors
NPN (*N*-phenyl-1-naphthylamine)	340	420	Hydrophobic probe (non-fluorescent until bound to cell membranes or lipids)
1-Pyrene butyryl choline bromide	342	378–420	Synaptic localization
Rhodamine B octadecyl ester	560	590	Hydrophobic probe, membrane fusion assay
TMA-DPH	351	430	Outer plasma membrane, fluidity
Cell membrane potentials and pH (vital)			
ADB 1,4 diacetoxy-2,3-dicyano-benzene	351	450–476	Intracellular pH (emission peak shifts in alkaline state)
BCECF (2′7′biscarboxyethyl-5,6-carboxyfluorescein	500	530, 620	Intracellular pH
BCECF-AM (2′,7′bis-carboxy-ethyl-5,6-carboxyfluorescein tetraacetoxymethyl ester)	500	530, 620	Intracellular pH (transmitted into cells and enzymatically hydrolysed by esterases to BCECF)
Bis-oxonol (DiSBa-C_2 (3))	540	580	Membrane potentials
Carboxy SNAFL-2 (semi-naphthofluorescein)	485, 514 (acid) 547 (base)	546 (acid) 630 (base)	Monoexcitation–dual-emission pH indicator
Carboxy SNARF-1 (semi-naphthorhodofluor)	518, 548 (acid) 574 (base)	587 (acid) 630 (base)	Monoexcitation–dual-emission pH indicator
1,4-Diacetoxyphthalonitrile	350	420–440 500–580	Intracellular pH
DIDS (4,4′-diisothiocyano-2,2′-stilbenedisulphonic acid)	340	430	Anion transport inhibitor
$DiOC_7(3)$ (3,3′-diheptyloxacarbocyanine iodide)	482	511	Membrane potentials
$DiOC_2(3)$	482	500	Membrane potentials
$DiOC_2(5)$	579	603	Membrane potentials
9-(*N*-Dodecyl) aminoacridine	430	475	pH gradients across membranes
Merocyanine 540	500	572	Membrane potentials, differentiation marker in cancer research
Oxonol V	609	645	Membrane potentials
Rhodamine 123	510	534	Mitochondrial and plasma membrane potentials
Vita blue dibutyrate-14 (VBDB-14)	609 (base) 524 (acid)	665 (base) 570 (acid)	Intracellular pH (dual fluorescence)
WW 781	603	635–645	Membrane potentials
Ionic probes (vital)			
Aequorin	—	469	Luminescent protein that emits light in presence of Ca^{2+}
9-Anthronyl choline iodide	320	420	Ca^{2+} binding to calmodulin
Calcein (active part is DCAF, 3,6-dihydroxy-24-bis-(*N,N′*-di (carbomethyl)-aminomethyl) fluoran (Fluorexone)	305, 490	520	Ca^{2+} and Mg^{2+}

Table 2.3 Continued

Name	*Wavelength (nm)* Excitation	Emission	Applications
CTC (7-chlortetracycline, Aureomycin)	345, 400	430, 520	Free Ca^{2+} near membranes
Fluo-3	506	526	Ca^{2+}
Fura-2	335, 362	505	Ca^{2+} (bound Ca^{2+} excites maximally at 335 nm while free dye excites at 362 nm ratio imaging)
Furaptra (mag-fur-2)	376 (low ion)	506	Mg^{2+} (microscopy and ratio imaging)
	344 (high ion)	492	
Indo-1	331	410	Ca^{2+}
Mag-indo-1	354 (low ion)	475	Mg^{2+} (flow cytometry)
	349 (high ion)	419	
PBF1	346 (low ion)	551	K^+
	334 (high ion)	525	
Quin-2	339	492	CA^{2+}
Rhod-2	553	576	Ca^{2+}
SBFI	340/380	505	Na^+ microinjected into cells or by use of AM ester)
SPQ	344	450	Cl^-
TnC_{DABZ} (troponin C dansyl-aziridine)	340	514	Ca^{2+} binding to Ca^{2+}-specific regulatory sites of troponin C
TSQ	335	376	Zn^{2+} in presynaptic boutons

Information compiled from numerous sources, including published research articles and catalogues of companies that specialize in supplying fluorescent probes to immunologists and cell and molecular biologists. Many of the probes listed above are employed in flow cytometry. I have utilized the *Handbook of Fluorescent Probes and Research Chemicals* (Haugland, 1992) and published material supplied by Biomeda Corporation, Eastman Kodak, Sigma Chemical Co., and Polysciences, Inc. I am particularly grateful to Dr Paul Gallop (Massachusetts General Hospital), Dr Natalie S. Rudolph (Viomedics), Dr Richard P. Haugland (Molecular Probes), and Ms Larissa Korytko (Eastman Kodak Co.), who provided valuable information and gave generously of their time in discussing particular fluorescent probes. The spectral data shown represent published excitation and emission peaks. However, slight spectral shifts can be expected according to the solvent used, pH of the system, and whether or not the probe is bound to its substrate. In a few cases, colours are given (B = blue, G = green, UV = ultraviolet, W = white, Y = yellow). Trademarks are assigned for Bodipy (Molecular Probes), Cascade Blue (Molecular Probes), Cellufluor (Polysciences), FluoroBora (Childrens Hospital), Fluoro-Gold (Fluorochrome, Inc.), Hydroethidine (Prescott Labs.), Lissamine (Imperial Chemical Industries), and SNAFL, SNARF, and Texas Red (Molecular Probes).

2.6 GENERAL APPLICATIONS OF FLUORESCENT PROBES

Since the employment of fluorochromes in microscopy by the early pioneers, Bommer, Ellinger and Hirt, Hamperl and Haitinger, Strugger, and Schümmelfeder, dozens of fluorescent dyes have been tested for histological and histochemical specificity. Some have been used as vital fluorochromes. Table 2.2 gives a general listing of the dyes and their applications.

In modern cytochemistry and cell biology, fluorescent probes have been designed and employed as fluorescent tags. Some, like fluorescein, TRITC, phycoerythrin, and Bodipy, are covalently bound to isothiocyanates or linked to chlorotriazinyl derivatives and hydroxysuccinimido esters. They are water-soluble and have been attached to antibodies, lectins, hormones and other macromolecules. Other fluorescent probes are non-covalently linked to macromolecules, ions and organelles within cells. Some probes have been used to evaluate cell viability, pH and membrane potentials. A summary of many of these fluorescent probes is given in Table 2.3. Other chapters in this volume give further information about specific groups of probes.

ACKNOWLEDGEMENTS

It is a pleasure to recognize the contribution of Ms Paula Porter, who typed the manuscript. My research is supported in part by the Biological Stain Commission.

REFERENCES

Albert A. (1951) *The Acridines, Their Preparation, Properties and Uses.* Arnold, London.
Aldridge W.G. & Watson M. (1963) *J. Histochem. Cytochem.* **11**, 773–781.
Armstrong J.A. (1956) *Exp. Cell Res.* **11**, 640–643.
Armstrong J.A. & Niven J.S.F. (1957) *Nature* **180**, 1335–1336.
Austin C.R. & Bishop M.W.H. (1959) *Exp. Cell Res.* **17**, 33–43.
Barch M.J. (ed.) (1991) *The ACT Cytogenetics Laboratory Manual* (2nd edition). Raven Press, New York.
Bertalanffy L.v. & Bickis I. (1956) *J. Histochem. Cytochem.* **4**, 481–493.
Bock E. & Oesterlin M. (1939) *Zbl. Bakteriol.* **143**, 306–318.
Bollinger A., Franzeck U.K. & Jäger K. (1983) *Prog. Appl. Microcirc.* **3**, 97–118.
Bommer S. (1929) *Acta Derm. Venereol.* **10**, 253–315.
Bradley D.F. (1961) *Trans. NY Acad. Sci.* **24**, 64–74.
Braütigam F. & Grabner A. (eds) (1949) *Beiträge zur Fluoreszenzmikroskopie.* Verlag Georg Fromme & Co., Vienna.
Browning C.H. (1922) *Nature* **109**, 750–751.
Browning C.H. (1964) In *Experimental Chemotherapy*, Vol. 2, Pt 1, *Chemotherapy of Bacterial Infections*, R.J. Schnitzer & Frank Hawking (eds). Academic Press, New York, pp. 2–36.
Browning C.H., Gulbransen R., Kennaway E.L. & Thorton L.H.D. (1917) *Br. Med. J.* **1**, 73–78.
Bucher O. (1939) *Z. Zellforsch.* **29**, 283–322.
Bucherer H. (1943) *Zbl. Bakteriol.* **106**, 81–88.
Bukatsch F. (1940) *Z. Gesamte Naturwissen.* p. 90.
Bukatsch F. (1941) *Zts. ges. Naturwiss.* **7**, 288 (cited Strugger, 1946, p. 4).
Bukatsch F. & Haitinger M. (1940) *Protoplasma* **34**, 515–523.
Colour Index (1971) (3rd edition). Society of Dyers and Colourists, Bradford, Yorkshire, UK.
Comings D.E. (1978) *Ann. Rev. Genet.* **12**, 25–46.
Dart L.H. Jr. & Turner T.R. (1959) *Lab. Invest.* **8**, 1513–1522.
Darzynkiewicz Z. (1979) In *Flow Cytometry and Sorting*, M.R. Melamed, P.F. Mullaney & M.L. Mendelsohn (eds). John Wiley, New York, pp. 283–316.
Darzynkiewicz Z. (1991) In *Flow Cytometry*, Z. Darzynkiewicz & H.A. Crissman (eds). Academic Press, San Diego, pp. 285–298.
Darzynkiewicz Z. & Crissman H.A. (eds) (1990) *Flow Cytometry.* Academic Press, San Diego.
DeBruyn P.P.H., Farr R.S., Banks H. & Morthland F.W. (1953) *Exp. Cell Res.* **4**, 174–180.
Drawert H. (1968) *Vitalfärbung und vitalfluorochromierung Pflanzlicher Zellen und Gewebe.* Protoplasmatologia Bd. 2. Teil D. Heft 3, Springer-Verlag, Vienna.
Duijn C. van Jr. (1955) *Microscope* **10**, 122–128.
Ehrlich P. (1882) *Dtsch. Med. Wschr.* **8**, 21–22, 36–37, 54–55.
Ehrlich P. & Benda L. (1913) *Ber. Dtsch. Chem. Gesel.* **46**, 1931–1951.
Ellinger P. (1940) *Biol. Rev. Cambridge Philos. Soc.* **15**, 323–350.
Ellinger P. & Hirt A. (1929a) *Z. Anat. Entwcklgeschl.* **90**, 701–802.
Ellinger P. & Hirt A. (1929b) *Arch. Exp. Pathol. Pharmakol.* **149**, 285–297.
Eppinger H. (1949) In *Beiträge zur Fluoreszensmikroskopie*, F. Braütigam & A. Grabner (eds). Verlag Georg Fromme & Co., Vienna, pp. 37–45.
Exner R. & Haitinger M. (1936) *Psychiatr. Neurol. Wschr.* **38**, 183–187.
Fay I.W. (1911) *Chemistry of the Coal-Tar Dyes.* D. Van Nostrand, New York.
Fischl V. & Singer E. (1935) *Z. Hyg. u. Infekt.* **116**, 348–355.
Franke F. & Sylla A. (1933) *Z. Exp. Med.* **89**, 141–158.
Giloh H. & Sedet J.W. (1982) *Science* **217**, 1252–1255.
Grafflin A.L. (1938) *J. Cell. Comp. Physiol.* **12**, 167–170.
Green F.J. (1990) *The Sigma-Aldrich Handbook of Stains, Dyes and Indicators.* Aldrich Chemical Co., Inc., Milwaukee.
Gustashaw K.M. (1991) In *The ACT Cytogenetics Laboratory Manual* (2nd edition), M.J. Barch (ed.). Raven Press, New York, pp. 205–269.
Hagemann P.K.H. (1937a) *Dtsch. Med. Wschr.* **63**, 514–518.
Hagemann P.K.H. (1937b) *Münch. Med. Wschr.* **84**, 761–765.
Hagemann P.K.H. (1937c) *Münch. Med. Wschr.* **20**, 761–765.
Hagemann P.K.H. (1938) *Münch. Med. Wschr.* **85**, 1066–1068.
Hagemann P.K.H. (1939) *Arch. Exp. Zellforsch.* **22**, 459–462.
Haitinger M. (1934) In *Abderhalden's Handbuch der Biologischen Arbeitsmethoden*, Abt. 11, Teil 3. Urban & Schwartzenberg, Berlin, pp. 3307–3337.
Haitinger M. & Geiser P. (1944) *Virch. Arch.* **312**, 116–137.
Haitinger M. & Hamperl H. (1933) *Z. Mikrosk. Anat. Forsch.* **33**, 193–221.
Haitinger M. & Linsbauer L. (1933) *Beih. Bot. Zbl.* **50**, 432–444.
Haitinger M. & Linsbauer L. (1935) *Bot. Zbl.* **53**, 387–397.
Hamperl H. (1934) *Virch. Pathol. Anat. Arch.* **292**, 1–51.
Harms, H. (1965) *Handbuch der Farbstoffe für die Mikroskopie.* Staufen Verlag, Kamp-Lintford.
Hata S. (1932) *Kitasato Arch. Exp. Med.* **9**, 1–71.
Haugland R.P. (1992) *Handbook of Fluorescent Probes and Research Chemicals.* Molecular Probes, Eugene, Oregon.
Haugland R.P. (1990) In *Optical Microscopy for Biology*, B. Herman & K. Jacobson (eds). Wiley-Liss, New York, pp. 143–157.
Heimstädt O. (1911) *Z. Wiss. Mikrosk.* **28**, 330–337.
Hercik F. (1939) *Protoplasma* **32**, 527–535.
Herman B. & Jacobson K. (1990) *Optical Microscopy for Biology.* Wiley-Liss, New York.
Heuven J.A. van (1936) *Ned. Tijdscher. Geneesk.* **80**, 1728–1732.
Jacoby F., Medawar P.G. & Willmer E.N. (1941) *Br. Med. J.* **2**, 149–153.
Jancśo N. von (1932) *Klin. Wschr.* **11**, 689.
Johnson G.D., Davidson R.S., McNamee K.C., Russel G., Goodwin D. & Holborow E.J. (1982) *J. Immunol. Meth.* **55**, 231–242.
Kaiserling C. (1917) *Z. Tuberc.* **27**, 156–161.
Kapuscinski J. & Darzynkiewicz Z. (1990) In *Flow Cytometry*, Z. Darzynkiewicz & H.A. Crissman (eds). Academic Press, San Diego, pp. 655–669.
Kasten F.H. (1964) In *100 Years of Histochemistry in Germany*, W. Sandritter & F.H. Kasten (eds) F.K. Schattauer-Verlag, Stuttgart, pp. 97–101.
Kasten F.H. (1967) *Int. Rev. Cytol.* **21**, 141–202.
Kasten F.H. (1973) In *Encyclopedia of Microscopy and Microtechnique*, P. Gray (ed.). Reinhold, New York, pp. 4–7.
Kasten F.H. (1981) In *Staining Procedures* (4th edition), G. Clark (ed.). Williams & Wilkins, Baltimore, pp. 39–103.
Kasten F.H. (1983a) In *History of Staining* (3rd edition), G. Clark & F.H. Kasten (eds). Williams & Wilkins, Baltimore, pp. 147–185.

Kasten F.H. (1983b) In *History of Staining* (3rd edition), G. Clark & F.H. Kasten (eds). Williams & Wilkins, Baltimore, pp. 186–252.
Kasten F.H. (1989) In *Cell Structure and Function by Microspectrofluorometry*, E. Kohen & J.G. Hirschberg (eds). Academic Press, San Diego, pp. 3–50.
Kasten F.H. (1991) In *Historians and Archivists: Essays in Modern German History and Archival Policy*, G.O. Kent (ed.). George Mason Press, Fairfax, VA, pp. 173–208.
Kohen E., Kohen C., Hirschberg J.G., Santus R., Morlière P., Kasten F.H. & Ghadially F.N. (1991) In *Encyclopedia of Human Biology*, Vol. 5, R. Dulbecco (ed.). Academic Press, San Diego, pp. 561–585.
Krause R. (1926a) *Enzyklopädie der Mikroskopischen Technik* (3rd edition), Vol. 1. Urban & Schwartzenberg, Berlin.
Krause R. (1926b) *Enzyklopädie der Mikroskopischen Technik* (3rd edition), Vol. 2. Urban & Schwartzenberg, Berlin.
Krause R. (1927) *Enzyklopädie der Mikroskopischen Technik* (3rd edition), Vol. 3. Urban & Schwartzenberg, Berlin.
Krebs A. (1944) *Strahlentherap.* **75**, 346 (cited in Strugger, 1946, p. 43).
Krebs A.T. & Gierlach Z.S. (1951) *Am. J. Roentgenol.* **65**, 93–97.
Lehmann H. (1913) *Z. Wiss. Mikrosk.* **30**, 417–470.
Levaditi R.O. & Reinie L. (1939) *C.R. Soc. Biol.* **131**, 916–919.
Lillie R.D. (1977) *H.J. Conn's Biological Stains* (9th edition). Williams & Wilkins, Baltimore.
Loeser C.N., West S.S. & Schoenberg M.D. (1960) *Anat. Rec.* **138**, 163–178.
MacInnes J.W. & Uretz R.B. (1966) *Science* **151**, 689–691.
Mayor H.D. (1963) *Int. Rev. Exp. Pathol.* **2**, 1–45.
Melamed M.R. & Darzynkiewicz Z. (1981) In *Histochemistry: The Widening Horizons of its Applications in the Biomedical Sciences*, P.J. Stoward & J.M. Polak (eds). John Wiley & Sons, Chichester, pp. 237–261.
Menger M.D., Marzi I. & Messmer K. (1991) *Eur. Surg. Res.* **23**, 158–169.
Mueller J.H. (1918) *J. Pathol. Bacteriol.* **22**, 308–318.
Patton R.L. & Metcalf R.L. (1943) *Science* **98**, 184.
Peacocke A.R. & Skerrett J.N.H. (1956) *Trans. Faraday Soc.* **52**, 261–279.
Perkin W.H. (1906) *Science* **24**, 488–493.
Pick J. (1935) *Z. Wiss. Mikrosk.* **51**, 338–351.
Price G.R. & Schwartz S. (1956) In *Physical Techniques in Biological Research*, Vol. 3, G. Oster & A.W. Pollister (eds). Academic Press, New York, pp. 91–148.
Provazek S. von. (1914) *KlěLnwelt.* **6**, 37.
Reichert's Fluorescence Microscopy with Fluorochromes (1952) (2nd edition). Optische Werke C. Reichert, Wien.
Richards O.W. (1941) *Science* **93**, 190.
Richards O.W. (1955) In *Analytical Cytology*, R.C. Mellors (ed.) McGraw-Hill, New York, pp. 5/1–5/37.
Rigler R. Jr. (1966) *Acta Physiol. Scand.* **67** (Suppl. 267), 1–122.
Robbins E. & Marcus P.I. (1963) *J. Cell Biol.* **18**, 237–250.
Robbins E., Marcus P.I. & Ganatas, N.K. (1964) *J. Cell Biol.* **21**, 49–62.
Rost F.W.D. (1980) In *Histochemistry. Theoretical and Applied* (4th edition), Vol. 1, *Preparative and Optical Technology*, A.G.E. Pearse (ed.). Churchill Livingstone, Edinburgh, pp. 346–378.
Sammak P.J. & Borisy G.G. (1988) *Cell Motil. Cytoskel.* **10**, 237–245.
Schmidt-LaBaume F. & Jäger R. (1939) *Arch. Dermatol. Syphilas.* **179**, 531–542.
Schümmelfeder N. (1948) *Naturwissenschaften* **35**, 346.
Schümmelfeder N. (1950) *Virch. Arch.* **318**, 119–154.
Schümmelfeder N. (1956) *Z. Zellforsch.* **44**, 488–494.
Schümmelfeder N. (1958) *Acta Histochem.* Suppl. Bd. **1**, 148–151.
Schümmelfeder N. & Stock K.-F. (1956) *Z. Zellforsch.* **44**, 327–338.
Schümmelfeder N., Ebschner K.J. & Krogh E. (1957) *Naturwissenschaften* **44**, 467–468.
Singer E. (1932) *Science* **75**, 289.
Sisken J.E., Barrows G.H. & Grasch S.D. (1986) *J. Histochem. Cytochem.* **34**, 61–66.
Spring K.R. & Smith P.D. (1987) *J. Microsc.* **147**, 265–278.
Steiner R.F. & Beers R.F. Jr. (1961) *Polynucleotides*. Elsevier, Amsterdam.
Stelzer E.H.K. & Wijnaendts-van-Resandt R.W. (1989) In *Cell Structure and Function by Microspectrofluorometry*, E. Kohen & J.G. Hirschberg (eds). Academic Press, San Diego, pp. 131–143.
Strugger S. (1940) *Jena Z. Naturwiss.* **73**, 97–134.
Strugger S. (1940/41) *Z. Wiss. Mikrosk. u. Mikrosk. Tech.* **57**, 415–419.
Strugger S. (1941) *Flora* **35**, 101–134.
Strugger S. (1946) *Fiat Review of German Science* **52**, Biologie Teil 1, 1–50.
Strugger S. (1949) *Fluoreszenzmikroskopie und Mikrobiologie*. Verlag M. & H. Schaper, Hannover.
Strugger S. & Hilbrich P. (1942) *Dtsch. tierärztl. Wschr.* **50**, 121–130.
Strugger S. & Rosenberger G. (1944) *Dtsch. tierärztl. Wschr.* (cited in Strugger, 1946, p. 5).
Strugger S. Krebs A.T. & Gierlach Z.S. (1953) *Am. J. Roentgenol.* **70**, 365–375.
Stübel H. (1911) *Pflügers Arch.* **142**, 1–14.
Tappeiner H. von & Jodlbauer A. (1907) *Die Sensibilisierende Wirkung Fluoreszierender Substanzen Gesammelte Untersuchungen über die Photodynamischer Erscheinung*. FCW Vogel Verlag, Leipzig.
Taylor D.L. & Salmon E.D. (1989) In *Fluorescence Microscopy of Living Cells in Culture Part A. Fluorescent Analogs, Labeling Cells, and Basic Microscopy*, Y.-I. Wang & D.L. Taylor (eds). Academic Press, San Diego, pp. 207–237.
Taylor D.L. & Wang Y-I. (eds) (1989) *Fluorescence Microscopy of Living Cells in Culture. Part B. Quantitative Fluorescence Microscopy – Imaging and Spectroscopy.* Academic Press, San Diego.
Tswett M. (1912) *Ber. Dtsch. Bot. Ges.* **29**, 744–746.
Venkataraman K. (1952) *The Chemistry of Synthetic Dyes*, Vol. 2. Academic Press, New York.
Waggoner A.S. (1986) In *Applications of Fluorescence in the Biomedical Sciences*, D.L. Taylor (ed.). Alan R. Liss, New York, pp. 3–28.
Wang Y-I. & Taylor D.L. (eds) (1989) *Fluorescence Microscopy of Living Cells in Culture. Part A. Fluorescent Analysis, Labeling Cells, and Basic Microscopy.* Academic Press, San Diego.
Wasicky R. (1913) *Pharm. Post.* **46**, 877–878.
Werbitzki F.W. (1909) *Zbl. Bakteriol.* **53**, 303–315.
Willingham M.C. & Pastan I.H. (1982) *Methods Enzymol.* **98**, 266–283.
Wilson T. (1990) *Confocal Microscopy.* Academic Press, London.
Witte S. (1989) *Res. Exp. Med.* **189**, 229–239.
Wolf M.K. & Aronson S.B. (1961) *J. Histochem. Cytochem.* **9**, 22–29.
Wories H.J., Kopec, J.H., Lodder G & Lugtenberg J. (1985) *Recl. Trav. Chim. Pays-Bas.* **104**, 288–291 (cited by Haugland, 1990).
Zanker V. (1952) *Z. Phys. Chem.* **199**, 225–258.

CHAPTER THREE

Intracellular Ion Indicators

R.P. HAUGLAND & I.D. JOHNSON
Molecular Probes, Inc., Eugene, OR, USA

3.1 INTRODUCTION

Fluorescent indicators that selectively respond to ions in living cells have provided a wealth of new information on cellular physiology over the last 15 years. Intracellular ion fluxes can be spatially and temporally mapped by fluorescence imaging, and variations in their magnitude among cell populations can be assessed by flow cytometry. These types of information are not readily accessible using other measurement techniques.

Fluorescent indicators have been developed for most biologically important ions including calcium, hydrogen (pH), magnesium, sodium, potassium and chloride. Certainly the most successful and widely used indicators have been for intracellular calcium and pH. Within each class, specialization of ion-binding and spectroscopic properties confers suitability for particular applications. This chapter reviews and evaluates properties of indicators in current use. In so doing, it is also necessary to consider various practical limitations and measurement artifacts affecting the use of fluorescent indicators, many of which originate from the ways in which these exogenous molecular probes are processed and accommodated by living cells.

3.2 GENERAL PROPERTIES OF INTRACELLULAR ION INDICATORS

In the following discussion, ion-binding and spectroscopic properties that are required in a functional fluorescent indicator are divided into seven categories. More specialized properties of indicators for calcium, magnesium, sodium, potassium, chloride and pH are discussed in turn in Section 3.3.

3.2.1 Stoichiometry

The indicator binding stoichiometry is the number of target ions bound per indicator at saturation. A 1:1 stoichiometry presents a minimum of complexity in quantitative data analysis and is common to most of the fluorescent indicators described in this chapter. There are, however, exceptions among indicators used

FLUORESCENT AND LUMINESCENT PROBES, 2ND EDN
ISBN 0-12-447836-0

with other detection techniques. The stoichiometry of calcium binding is 3:1 for the bioluminescent indicator aequorin (Blinks *et al.*, 1978) and 1:2 for the metallochromic indicator arsenazo III (Scarpa *et al.*, 1978).

3.2.2 Specificity

The relationship between indicator fluorescence and target ion concentration may be affected by competing binding equilibria, particularly in the complex ionic environments within cells. Specificity for the target ion is therefore a primary objective in the design of fluorescent indicators. Where competing equilibria cannot be eliminated by design, their effects must be considered in the calculation of indicator dissociation constants. For instance, intracellular magnesium ion concentrations are generally about 1 mM, whereas the concentration of intracellular calcium is usually submicromolar. The BAPTA chelator, the basis of most current fluorescent calcium indicators (Fig. 3.1), was designed to provide the requisite binding selectivity (>10^4-fold) to enable changes in low intracellular Ca^{2+} levels to be detected against a much larger Mg^{2+} background (Tsien, 1980; Grynkiewicz *et al.*, 1985). On the other hand, BAPTA-based indicators bind various heavy metal cations (Mn^{2+}, Zn^{2+}, Pb^{2+}) with substantially higher affinity than Ca^{2+}(Hinkle *et al.*, 1992; Atar *et al.*, 1995). Endogenous levels of these ions may be sufficient to cause perturbations to calcium measurements (Snitsarev *et al.*, 1996). The Ca^{2+} dissociation constant of BAPTA-based indicators increases as pH decreases (Tsien, 1980; Lattanzio & Bartschat, 1991), requiring corrections to the magnitude of apparent calcium fluxes in some cases (Beatty *et al.*, 1993). The selectivity of crown ether-based sodium and potassium indicators (Fig. 3.1) is much lower than that of indicators for divalent cations (Minta and Tsien, 1989). Chloride indicators (Section 3.3.4) do not have any selective affinity for their target ion, and are in fact more sensitive to bromide and iodide ions than they are to chloride. Their efficacy as intracellular chloride indicators is dependent on the absence of other halide ions.

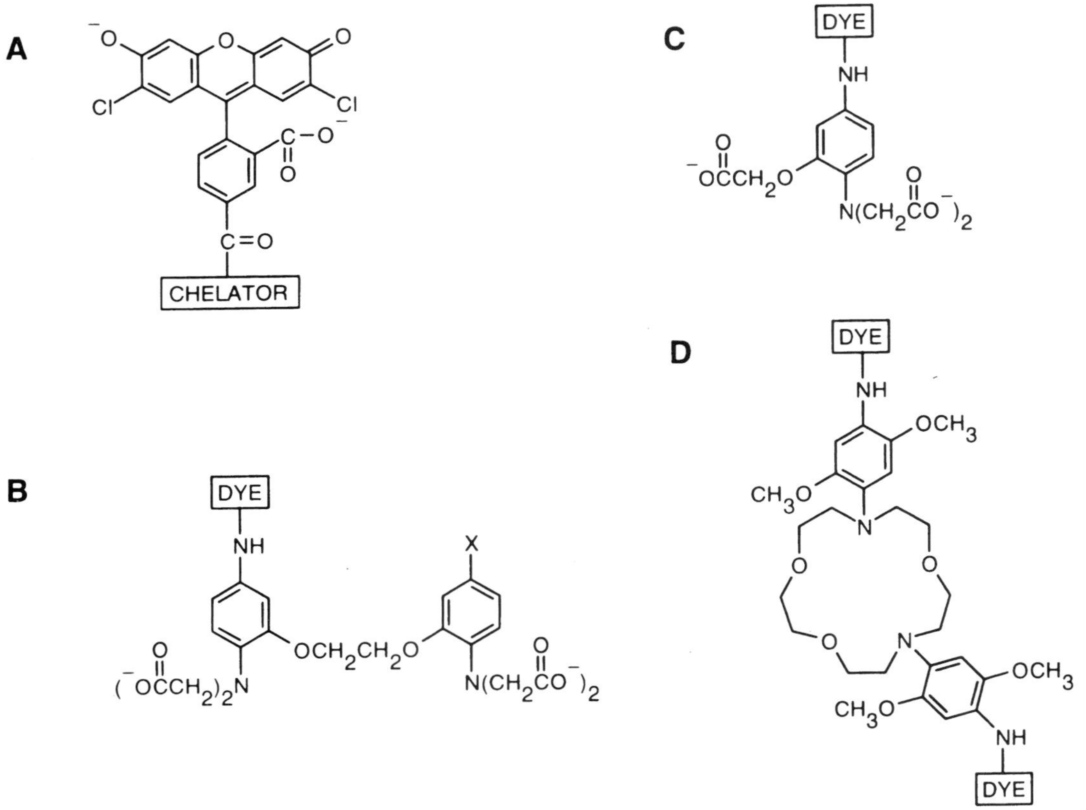

Figure 3.1 Ion indicators derived from the same fluorescent dye (5-carboxy-2′,7′,-dichlorofluorescein, (A) by coupling to chelators with different ion-binding specificities. (B) Calcium Green-1 (BAPTA chelator, X = H) or Calcium Green-5N (5-nitro BAPTA chelator; X = NO_2). (C) Magnesium Green (APTRA chelator). (D) Sodium Green (diaza crown ether chelator).

3.2.3 Dissociation constant

Fluorescent indicators operate by generating an optical response (see below, Section 3.2.4) that is coupled to the binding of the target ion. Under normal conditions of use, the response is sigmoidal (see Fig. 3.4) and is approximately maximal at the ion concentration corresponding to the indicator dissociation constant (K_d). The operational concentration range of the indicator typically extends over about two orders of magnitude, centred around the K_d. Clearly, the indicator dissociation constant must be matched to the physiological concentration range of the target ion if a detectable response is to be obtained. Unfortunately, several environmental factors – including pH, temperature, ionic strength, viscosity and protein binding – modify the effective intracellular dissociation constant relative to values determined in controlled aqueous solutions. The data in Table 3.1 illustrate the scope of this problem.

A major reason for the successful application of fluorescent calcium indicators is the exceptionally large increases in intracellular Ca^{2+} concentrations that occur when cells are stimulated. The concentration of Ca^{2+} may change from a resting level as low as perhaps 10 nM to as high as 10 μM or even higher. On the other hand, the changes in intracellular pH or Mg^{2+} concentration that occur on stimulation are commonly less than a few tenths of a log unit. A calcium indicator such as fura-2, which has a dissociation constant of 145 nM in 10 mM MOPS buffer at 22°C (Table 3.2), is nearly saturated with Ca^{2+} above ~1 μM free Ca^{2+} (Fig. 3.2). Fortunately, it is possible to 'tune' the dissociation constant – and therefore the optical response – of most indicators by choosing appropriate chemical substituents and chelating or sensing moieties (Fig. 3.1).

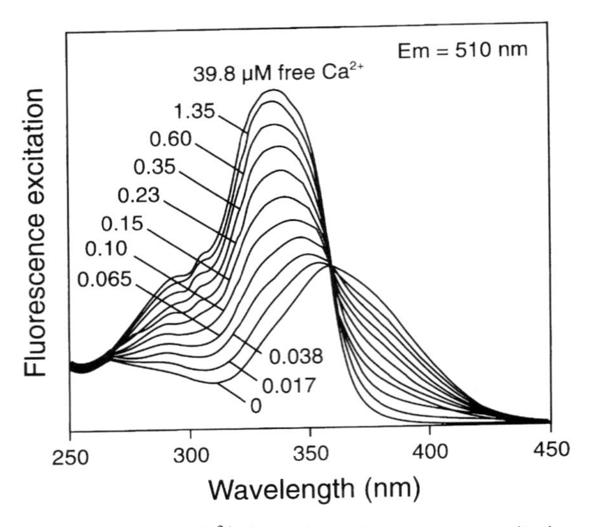

Figure 3.2 The Ca^{2+}-dependent fluorescence excitation spectra of fura-2.

3.2.4 Optical response

Changes in spectroscopic properties in response to binding of the target ion are fundamental to all optical indicators. In many cases this change is a shift in the absorption wavelengths. Changes in indicator fluorescence on ion binding typically can take several forms: (1) an increase or decrease in the fluorescence quantum yield of the indicator with little change in either the absorption or fluorescence spectra (e.g. fluo-3, Calcium Green, mag-fluo-4 and SPQ); (2) a shift of the absorption, and therefore the fluorescence excitation spectrum, to shorter wavelengths with little shift in the emission maximum (e.g. fura-2, Fura Red, mag-fura-2, SBFI, PBFI and BCECF); or (3) a shift in both the absorption (excitation) and

Table 3.1 Comparison of solution and intracellular K_d values.

Indicator (ion)	*Solution K_d*	*Intracellular K_d* [a]	*Cell/tissue type*
Fluo-3 (Ca^{2+})	390 nM [b]	2570 nM	Frog skeletal muscle [*Biophys. J.* **65**, 865 (1993)]
Fura-2 (Ca^{2+})	145 nM [b]	371 nM	U373-MG astrocytoma [*Cell Calcium* **21**, 233 (1997)]
Indo-1 (Ca^{2+})	230 nM [b]	844 nM	Rabbit cardiac myocyte [*Biophys J.* **68**, 1453 (1995)]
Mag-fura-2 (Mg^{2+})	1.9 mM [c]	5.4 mM	Guinea pig smooth muscle [*Biophys. J.* **73**, 3358 (1997)]
SBFI (Na^+)	3.8 mM [d]	26.6 mM	Porcine adrenal chromaffin cells [*J. Neurosci. Meth.* **75**, 21 (1997)]

[a] Values determined in cellular environments listed in the next column.
[b] Values determined at 22°C in 100 mM KCl, 10 mM MOPS pH 7.2, 0 to 10 mM CaEGTA.
[c] Values determined at 22°C in 115 mM KCl, 20 mM NaCl, 10 mM TRIS pH 7.05, 0 to 35 mM Mg^{2+}.
[d] Value determined at 22°C in 10 mM MOPS, pH 7.0, 0 to 135 mM Na^+.

Table 3.2 Fluorescent indicators for divalent cations.

Name	*EX/EM ion-free (nm)*[a]	*EX/EM ion-bound (nm)*[a]	*Measurement*[b]	K_d *(*Ca^{2+}*) (nM)*[c]	K_d *(*Mg^{2+}*) (mM)*[c]
Ca^{2+} indicators (BAPTA chelator)					
Indo-1	346/475	330/401	1	230	
Fura-2	363/512	335/505	2	145	9.8
BTC	464/533	401/529	2	7000	
Fura Red	472/657	436/637	2	140	
Fluo-3	503/none	506/526	3	390	9.0
Fluo-4	491/none	494/518	3	345	
Fluo-5N	491/none	494/516	3	90 000	
Calcium Green-1[d]	506/531	506/531	3	190	
Calcium Green-5N[d]	506/532	506/532	3	14 000	
Rhod-2	549/none	552/581	3	570	
Calcium Orange	549/575	549/575	3	185	
X-rhod-1	581/none	584/600	3	700	
X-rhod-5N	581/none	584/600	3	350 000	
Calcium Crimson	590/615	590/615	3	185	
Mg^{2+} indicators (APTRA chelator)					
Mag-indo-1	349/480	330/417	1	35 000	2.7
Mag-fura-2	369/511	330/491	2	25 000	1.9
Mag-fluo-4	492/none	494/516	3	22 000	4.7
Magnesium Green	506/531	506/531	3	6000	1.0

[a] Fluorescence EXcitation and EMission maxima.
[b] 1 = dual-wavelength emission ratio, 2 = dual-wavelength excitation ratio, 3 = single wavelength emission intensity.
[c] Dissociation constants were determined in aqueous buffers and are quoted from *Handbook of Fluorescent Probes and Research Chemicals*, 6th edn (Haugland, 1996) and other literature sources. These values vary considerably depending on a number of environmental and instrumental factors as discussed in the text.
[d] Oregon Green 488 BAPTA-1 and Oregon Green 488 BAPTA-5N are fluorinated derivatives of Calcium Green-1 and Calcium Green-5N, respectively, with excitation maxima closer to the 488-nm argon-ion laser line (EX = 494 nm).

emission spectra to shorter wavelengths (e.g. indo-1, mag-indo-1 and the SNARF and SNAFL series of pH indicators). Changes in other optical properties of indicators, such as the excited state decay lifetime, can also be used (Lakowicz *et al.*, 1992).

A feature common to the spectral response of indicators that undergo an excitation or emission shift is the presence of an 'isosbestic point' in the absorption, excitation and/or emission spectra, defined as a wavelength point at which the indicator fluorescence is insensitive to ion binding. Isosbestic points are found when two (and only two) spectroscopically detectable species, such as the ion-free and ion-bound forms of an indicator, are in chemical equilibrium. Absence of an isosbestic point in an indicator response curve may be evidence of additional equilibria or contamination by an ion-insensitive species. Examples of isosbestic points are seen in both the excitation spectra of fura-2 (Fig. 3.2) and the emission spectra of carboxy-SNARF-1 (see Fig. 3.7). Differences in the fluorescence quantum yields for the ion-free and ion-bound indicator may result in the absorption and excitation isosbestic points being different. For instance, the isosbestic point of BCECF is at 465 nm in absorption measurements and at 439 nm in excitation measurements (Paradiso *et al.*, 1987).

For indicators such as fura-2 that undergo a spectral shift on ion binding, the fluorescence intensity resulting from excitation at the isosbestic point is directly proportional to the amount of dye in the sample and *independent* of the ion concentration. The intensity excited at any other wavelength depends both on the indicator concentration and the ion concentration. Because the intensities at any two wavelengths are expressed in arbitrary fluorescence units per mole of indicator, forming a ratio of the intensities removes the indicator concentration dependence. Thus the *ratio* of intensities measured at any two wavelengths depends on the ion concentration, but is *independent* of the indicator concentration.

Indicators that undergo an absorption or emission wavelength shift upon ion binding have particular advantages for measurements in living cells. As long as the intensities are significantly above background, intensity ratioing can reduce problems in quantitation resulting from unequal loading or distribution of the probe, variations in cell thickness, photobleaching and leakage of the indicator from intact cells. It is more difficult to obtain quantitative information on ion concentrations using indicators such as fluo-3 and Calcium Green-1 that do not exhibit ion-dependent spectral shifts and are therefore not directly amenable to ratiometric detection. In these cases, ion-dependent ratio signals can be obtained by simultaneous use of a pair of spectrally contrasted indicators such as fluo-3 and Fura Red (Lipp and Niggli, 1993; June & Rabinovitch, 1994). The validity of this method is dependent on the consistency of the relative intracellular concentrations of the two indicators from cell to cell, a requirement that is not always fulfilled (Floto *et al.*, 1996). Other preconditions such as equivalent intracellular distributions and Ca^{2+} binding kinetics for the two indicators appear to be more generally valid (Floto *et al.*, 1996; Lipp and Niggli, 1993).

3.2.5 Cell loading

It is obviously essential to deliver fluorescent indicators to the intracellular region of interest (typically the cytoplasm) for ion concentration measurements. A variety of disruptive or passive cell loading techniques have been developed. Disruptive techniques include microinjection, hypotonic shock permeabilization, perfusion via patch pipette and electroporation. Passive techniques are based on membrane-permeant precursors from which the ion-sensitive indicator is generated inside the cell via chemical or enzymatic conversion. By far the most widely applied passive loading method has been the use of cell permeant acetoxymethyl (AM) or acetate ester derivatives that are hydrolysed by non-specific esterases to yield anionic Ca^{2+}, Mg^{2+}, Na^{+}, K^{+} and pH indicators. Other non-polar precursors used for passive loading include the reduced derivative of the chloride indicator MEQ (see Section 3.3.4) and the free acid forms of polycarboxylate Ca^{2+} indicators (Bush & Jones, 1987).

The lipophilic nature of AM esters gives them low solubility in aqueous media. In most cases they are added to the culture medium from stock solutions in dimethylsulfoxide (DMSO) to give a final concentration of about 1–10 μM. Additives such as the non-ionic detergent Pluronic F-127 are used to aid dispersion in some cases. Incubation at room temperature to 37°C for 10 minutes to 1 hour is usually sufficient for passive uptake through the membrane and subsequent intracellular hydrolysis. This basic protocol requires adaptation for different cell types and different indicators. Experiments in frog muscle fibres have revealed an inverse relationship between the loading rates of AM esters and their molecular weights (Zhao *et al.*, 1997). The chemical by-products of AM ester hydrolysis – acetate, protons and formaldehyde – all have potential deleterious effects on living cells (Tiffert *et al.*, 1984), but do not appear to cause serious problems at typically used loading levels. Final intracellular indicator concentrations obtained are 5–100 μM or even higher (Tran *et al.*, 1995; Berlin *et al.*, 1994; Negulescu & Machen, 1990). At such concentrations, the indicator may constitute a significant fraction of the total intracellular Ca^{2+} buffering capacity (Berlin *et al.*, 1994), resulting in damping of high-amplitude ion-concentration transients.

The efficiency of enzymatic conversion of AM esters to free indicators may vary considerably from one cell type to another (Oakes *et al.*, 1988). Failure to remove even one AM ester group from the calcium ion-chelating carboxylic acids of indicators such as fura-2 results in a highly fluorescent – but calcium-insensitive – probe (Scanlon *et al.*, 1987). Detection of partially hydrolysed AM esters will result in underestimation of intracellular calcium levels. Fluorescence quenching by Mn^{2+}, which binds only to completely de-esterified indicators, can be used to quantitate these effects (Tran *et al.*, 1995).

3.2.6 Intracellular processing

The environment and distribution of fluorescent indicators within cells can significantly impact the execution and interpretation of experiments. Compartmentalization, the accumulation of indicators in membrane-enclosed organelles (e.g. mitochondria, endoplasmic reticulum, nucleus) rather than their intended destination in the cytoplasm, is an inherent problem with AM ester loading. Loading at room temperature rather than at 37°C reportedly reduces compartmentalization (Di Virgilio *et al.*, 1990). Vacuolar compartmentalization can be particularly problematic in plant and fungal cells (Read *et al.*, 1992; Slayman *et al.*, 1994). Although compartmentalization is usually an undesirable side effect, it can also be exploited to measure Ca^{2+} at the subcellular level. Examples include measuring Ca^{2+} levels in intracellular stores (Hofer & Machen, 1993; Hofer & Schulz, 1996) using mag-fura-2 AM, and in mitochondria using rhod-2 AM (Jou *et al.*, 1996). Several indicators have been deliberately designed for targeting to particular locations within cells, including the plasma membrane (Etter *et al.*, 1994), intracellular membranes (Horne & Meyer, 1997) and the nucleus (Allbritton *et al.*, 1994).

Binding of indicators to intracellular proteins causes elevated K_d values and altered spectral properties (Hove-Madsen & Bers, 1992; Baker *et al.*, 1994).

The fraction of protein-bound indicator in cells appears to be less for APTRA derivatives such as mag-fura-2 than for BAPTA derivatives such as fura-2 (Konishi *et al.*, 1991; Zhao *et al.*, 1996). For cells that can be loaded by disruptive techniques, such as electroporation (Teruel & Meyer, 1997) and microinjection, the best means for avoiding both protein binding and compartmentalization may be to use dextran-coupled indicators (Read *et al.*, 1992). Conjugation of the pH indicator BCECF to a dextran was shown to prevent binding to intracellular proteins (Bright *et al.*, 1989). Dextran conjugates of all the major classes of ion indicator discussed in this chapter are commercially available (Haugland, 1996). An important recent technical development has been the use of calcium indicator–dextran conjugates to label neurons via retrograde or anterograde transport, allowing real-time imaging of neuronal activity (O'Donovan *et al.*, 1993).

Once hydrolysed intracellularly to the free indicator, most calcium- and magnesium-ion indicators are well retained in viable cells because of their high ionic charges. Certain cells, such as macrophages, however, actively secrete the indicator (Di Virgilio *et al.*, 1990). Secretion can sometimes be blocked by the addition of drugs such as probenecid and sulfinpyrazone. Intracellular retention of pH indicators is primarily determined by their ionic charge. BCECF, with 4–5 negative charges at physiological pH, is the best retained pH indicator that is not polymer-conjugated, whereas fluorescein, with only 1–2 negative charges, usually leaks from cells within minutes. The impact of indicator leakage on experimental data varies between different detection techniques. In microscopy of continuously perfused cells or tissues, and in flow cytometry, extracellular indicator is washed away in the flowing medium and is not detected. But for measurements made on cell suspensions in cuvettes or microplate wells, extracellular fluorescence remains within the field of view of the detection system and cannot be discriminated from the intracellular component without an additional physical probe such as Mn^{2+} quenching (McDonough and Button, 1989).

3.2.7 Fluorescence output

It is essential that the fluorescence can be detected above any background due to the sample or other probes that may be present. As with fluorescent dyes in general, detection sensitivity is highest for indicators with strong absorption (i.e. large extinction coefficient at the excitation wavelength) and high fluorescence quantum yields. High sensitivity detection allows loading concentrations to be reduced, mitigating some of the practical problems described above (intracellular buffering, incomplete AM ester hydrolysis). It must be possible to excite the probe using illumination that is minimally absorbed or scattered by cells and tissue samples. The development of multiphoton excitation microscopy, in which infrared light is used to excite indicators that have single photon absorption spectra in the UV-visible range, has provided a tremendous advance in capability in this respect (Denk *et al.*, 1995). With conventional excitation sources, longer wavelength indicators such as X-rhod-1 (Fig. 3.3) are sometimes preferred because reduced cellular autofluorescence usually results (Brooke *et al.*, 1996). In some cases selection of the probe depends on the instrumentation available. This is particularly true of equipment that uses 488-nm argon-ion laser excitation, including flow cytometers and laser scanning confocal microscopes. The ultraviolet argon-ion laser lines at 351 nm and 364 nm can be used to excite indo-1 and fura-2 for flow cytometry or confocal microscopy (June & Rabinovitch, 1994; Nitschke *et al.*, 1997).

Resistance to photobleaching is important in imaging applications, particularly in the absence of a ratiometric ion-binding response. Reagents that are typically used to reduce dye bleaching in immunofluorescence microsopy cannot be tolerated by living cells. A biocompatible analogue of vitamin E ('Trolox') has been found to inhibit the photodegradation of indo-1 to a fluorescent but Ca^{2+}-insensitive species (Scheenen *et al.*, 1996).

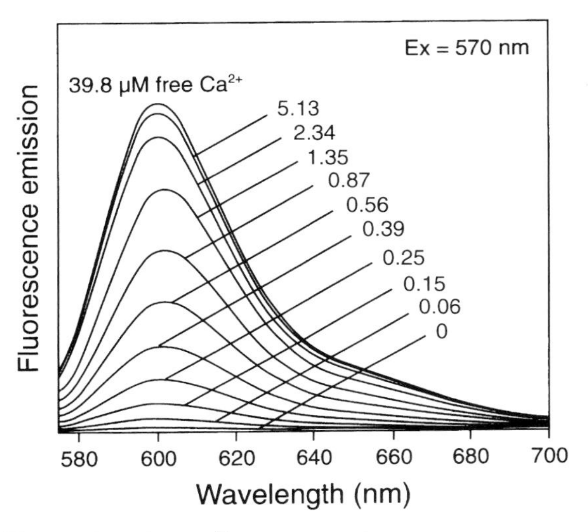

Figure 3.3 The Ca^{2+}-dependent fluorescence emission spectra of X-rhod-1.

3.3 EXAMPLES OF INTRACELLULAR ION INDICATORS

3.3.1 Calcium indicators

By far the most significant group of ion indicators has been those for intracellular calcium. Almost all of these are fluorescent derivatives of the chelator BAPTA (Fig. 3.1), which is an aromatic analogue of the calcium-selective chelator ethyleneglycol-bis(β-aminoethylether)-*N,N,N'*,N'-tetraacetic acid (EGTA). BAPTA is described along with the prototype calcium indicator, quin-2, in the classic paper by Dr Roger Tsien (1980). The major variations among indicators in current use (Table 3.2) are in excitation/emission wavelength range, dissociation constant (K_d) and spectroscopic response to ion binding (fluorescence intensity change or spectral shift). Spectral response curves for most of the indicators are given in the *Handbook of Fluorescent Probes and Research Chemicals* published by Molecular Probes Inc. (Haugland, 1996). Although they are predominantly used to examine suspensions or adherent monolayers of mammalian cells, these indicators have also been successfully applied in yeast, bacteria, plant cells, intact tissues and whole organs. The ultraviolet-excited ratiometric indicators are preferred for quantitative imaging (fura-2) and flow cytometry (indo-1). Fluo-3 (Minta *et al.*, 1989), Calcium Green-1, and other indicators excited in the visible wavelength range are most suitable for imaging Ca^{2+} dynamics, and for experiments involving photoactivation of caged chelators, second messengers and neurotransmitters (Wang & Augustine, 1995; Parker *et al.*, 1996). A distinguishing feature of these indicators is the dynamic range of the fluorescence intensity increase upon binding Ca^{2+} (Fig. 3.4) Calcium Green-1, Calcium Orange and Calcium Crimson have the advantage of being more fluorescent in both the Ca^{2+}-bound and Ca^{2+}-free states than fluo-3, rhod-2 and X-rhod-1, respectively (Fig. 3.4). This makes both stimulated and resting cells more readily detectable. Consequently, excitation intensity can be attenuated to minimize phototoxicity, and detection of small groups of loaded cells in tissues is more practicable. On the other hand, indicators of the fluo-3 type, which produce fluorescence intensity increases of 100-fold or more on ion binding, are better suited for detection of small incremental changes in Ca^{2+} levels.

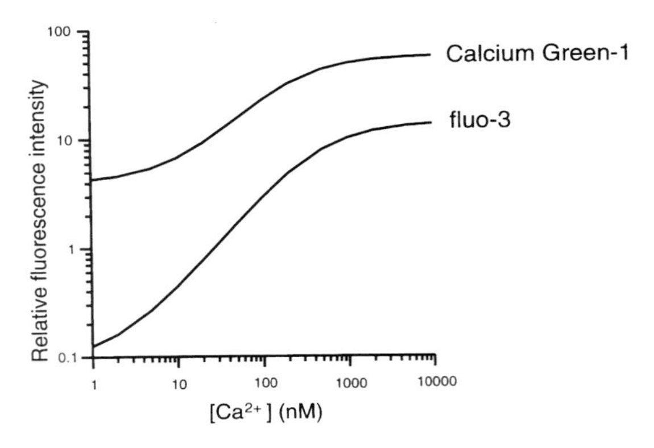

Figure 3.4 Comparison of fluorescence intensity responses to Ca^{2+} for fluo-3 and Calcium Green-1. Responses were calculated from the Ca^{2+} dissociation constants for the two indicators (Table 3.2) and the extinction coefficients and fluorescence quantum yields of their ion-free and ion-bound forms. They therefore represent the relative fluorescence intensities that would be obtained from equal concentrations of the two indicators excited and detected at their peak wavelengths.

The 5-nitro BAPTA ('5N') indicators such as Calcium Green-5N (Fig. 3.5), and APTRA-based magnesium indicators such as mag-fura-2 and mag-fluo-4, have Ca^{2+} dissociation constants above ~1 μM and are therefore useful for detecting intracellular calcium levels in the micromolar range that would saturate the response of indicators such as fura-2. Such elevated calcium levels are associated with activation of smooth muscle (Konishi *et al.*, 1991; Zhao *et al.*, 1996), neurons (Eilers *et al.*, 1995, Rajdev & Reynolds, 1993) and intracellular calcium stores (Parker *et al.*, 1996). Because the ion dissociation rates of these indicators are faster, they are more suitable for tracking rapid ion flux kinetics (Konishi *et al.*, 1991; Regehr & Atluri, 1995; Zhao *et al.*, 1996; Escobar *et al.*, 1997) than high-affinity indicators with K_d <1 μM.

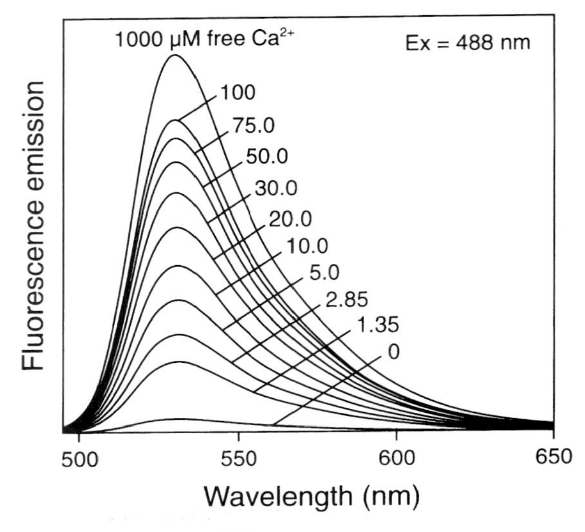

Figure 3.5 The Ca^{2+}-dependent fluorescence emission spectra of Calcium Green-5N.

3.3.2 Magnesium indicators

All of the current magnesium indicators in Table 3.2 are variants of the calcium indicators in which the BAPTA chelator is replaced by a triacetic acid analogue referred to as APTRA (Raju *et al.*, 1989; Fig. 3.1). The two groups of indicators exhibit parallel spectroscopic responses to ion binding (mag-fura-2 is like fura-2, Magnesium Green is like Calcium Green-1, mag-fluo-4 is like fluo-4, etc.). APTRA-based indicators respond to Mg^{2+} concentrations in the range of about 0.1–10 mM, making them useful for biochemical analysis of the role of Mg^{2+} as a cofactor in enzymatic catalysis (Srivastava *et al.*, 1995; Sun & Budde, 1997) as well as for intracellular measurements. Physiological changes in concentrations of intracellular magnesium are smaller and slower than calcium fluxes (Konishi *et al.*, 1993; Koss *et al.*, 1993; Zhang & Melvin, 1996) and are consequently more difficult to measure accurately. High-affinity binding of Ca^{2+} (Table 3.2) with a spectral response that is practically indistinguishable from that of Mg^{2+} presents a particular problem for APTRA-based indicators. In the case of mag-fura-2, interference with Mg^{2+} measurements becomes significant when Ca^{2+}concentrations exceed about 1 μM (Hurley *et al.*, 1992; Koss *et al.*, 1993).

3.3.3 Sodium and potassium indicators

The intracellular sodium indicators SBFI (Minta and Tsien, 1989) and Sodium Green incorporate a diaza crown ether chelator (Fig. 3.1), yielding 20 to 40-fold selectivity for sodium versus potassium. Their dissociation constants for Na^+ are substantially modified in the presence of K^+ (Minta and Tsien, 1989; Haugland, 1996). The fluorophore portion of SBFI is similar in structure and spectral properties to that of fura-2 (although the fluorescence quantum yield of SBFI is considerably lower), allowing use of the same ratiometric measurement configuration. Sodium Green can be excited at 488 nm for flow cytometry (Amorino & Fox, 1995) and confocal microscopy (Friedman & Haddad, 1994), and undergoes an 8-fold increase in fluorescence intensity on binding Na^+, without a spectral shift. In some cell types, the Na^+-dependent fluorescence increase of Sodium Green is severely attenuated, probably as a result of binding to proteins. Although AM ester loading of SBFI is sometimes difficult (Negulescu and Machen, 1990), its utility is well established, with at least 100 published applications to date. PBFI, an analogue of SBFI with an enlarged cryptand cavity to accommodate potassium ions, has only about 2-fold selectivity for K^+ over Na^+ (Minta and Tsien, 1989); published applications have been relatively few.

3.3.4 Chloride indicators

All of the intracellular chloride indicators in current use are 6-methoxyquinolinium derivatives, the prototype of which is 6-methoxy-*N*-(3-sulfopropyl)quinolinium (SPQ) (Wolfbeis & Urbano, 1982; Illsley & Verkman, 1987). Chloride detection sensitivity has been improved by modifications of the quinolinium *N* substituent (Verkman *et al.*, 1989; Biwersi *et al.*, 1992). All these indicators detect chloride via diffusion-limited collisional quenching. This detection mechanism is different from that of most other fluorescent ion indicators. It involves a transient interaction between the excited fluorophore and a halide ion; no ground state complex is formed. The efficiency of this process is characterized by the Stern-Volmer quenching constant (K_{SV}), the reciprocal of the ion concentration that produces 50% of maximum quenching. Quenching is not accompanied by spectral shifts and consequently ratio measurements are not feasible. Several properties of these indicators are less than optimal and give rise to technical problems:

(1) Low fluorescence output that results primarily from low absorptivity. This necessitates use of a large amounts of indicator and results in low detection sensitivity. The excitation wavelength is in the ultraviolet region, which may result in high autofluorescence and photodamage to cells. However excitation by the ultraviolet output of argon-ion lasers (351 nm and 364 nm) is feasible for confocal microscopy and flow cytometry (Pilas and Durack, 1997; Inglefield & Schwartz-Bloom, 1997).
(2) Rapid leakage of the dye from cells. With the exception of diH-MEQ (see below), chloride indicators must be loaded into cells by long-term incubation (up to 8 hours) in the presence of a large excess of dye or by brief hypotonic permeabilization. Since membranes are slightly permeable to the indicator, rapid leakage may occur. Experimentally determined estimates of leakage vary quite widely (Pilas and Durack, 1997; West & Molloy, 1996; Koncz & Daugirdas, 1994).
(3) Variation of the quenching constant with viscosity. Since the quenching is a diffusional process, it shows a high sensitivity to the viscosity of the medium. For SPQ, K_{SV} is reported to be 118 M^{-1} in aqueous solution and 12 M^{-1} inside cells (Krapf *et al.*, 1988). For MQAE, currently the most widely used intracellular chloride indicator, K_{SV} values of 25–28 M^{-1} have been determined in various cell types (West & Molloy, 1996; Koncz & Daugirdas, 1994), compared with the solution value of 200 M^{-1}.

Verkman and his colleagues have attempted to overcome some of the limitations of SPQ and MQAE by attaching 6-methoxyquinolinium dyes to dextrans (Biwersi *et al.*, 1992). Biwersi and Verkman have also described a freely membrane-permeant, chemically reduced form of 6-methoxy-*N*-ethylquinolinium (Biwersi & Verkman, 1991). Called diH-MEQ, this colourless, nonfluorescent probe is spontaneously oxidized inside cells to generate the indicator MEQ. The susceptibility of diH-MEQ to air oxidation makes it necessary to prepare the probe shortly before use by a straightforward chemical reduction procedure. DiH-MEQ has been successfully used to label living brain tissue, allowing neurotransmitter-mediated changes in intracellular chloride to be detected by confocal microscopy (Inglefield & Schwartz-Bloom, 1997).

3.3.5 pH indicators

Sensitivity of the absorption and emission of some fluorescent dyes to pH has been known for many years. In most cases, the pH sensitivity results from ionization of a phenolic moiety on the dye; fluorescein and 7-hydroxy-4-methylcoumarin (β-methylumbelliferone) are well-known examples. The pK_a of pH-sensitive dyes can be adjusted by making structural modifications. For example, addition of electron-withdrawing fluorine substituents lowers the pK_a of fluorescein from 6.4 to 4.7 (Haugland, 1996). As with the other ion indicators, a pK_a near the pH of the compartment whose pH is to be measured is necessary. In most cases this is a near-neutral pH; however, measurements of pH in endosomes may require the use of a pH indicator whose pK_a is in the range of 4–6.

Intracellular delivery of fluorescent pH indicators in the form of membrane-permeant, esterase-activated precursors was first implemented with carboxyfluorescein diacetate (Thomas *et al.*, 1979). Problems with intracellular retention and the lower-than-optimal pK_a of carboxyfluorescein were quickly rectified by the introduction of 2′,7′-bis-(2-carboxyethyl)-5(6)-carboxyfluorescein (BCECF) in 1982 (Rink *et al.*, 1982). Compared to carboxyfluorescein, BCECF has a higher pK_a (7.0 versus 6.4) and a higher net charge (−5 versus −3 when fully ionized). Protonation of BCECF results in a shift of the absorption maximum from 503 nm to 482 nm, accompanied by decrease of the molar extinction coefficient. These changes are replicated in the fluorescence excitation spectrum (Fig. 3.6), although the replication is not exact because the fluorescence quantum yield of the acid form is much lower than that of the base form. This response enables BCECF to be used for dual-excitation ratiometric measurements (Bright *et al.*, 1987). Although its utility is well established, the

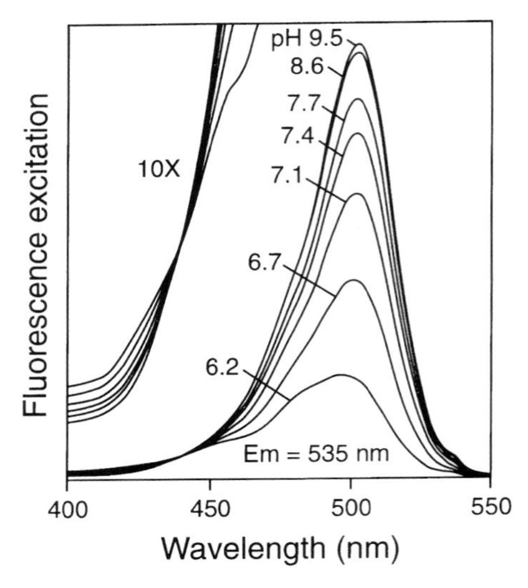

Figure 3.6 The pH-dependent fluorescence excitation spectra of BCECF. A 10-fold enlargement of the short-wavelength region is necessary to clearly display the isosbestic point at 439 nm.

pH-dependent spectral properties of BCECF are less than optimal in two respects. Firstly, the fluorescence excitation isosbestic point is quite far from the excitation maximum, giving poor signal-to-noise characteristics in ratio imaging microscopy (Bright *et al.*, 1987). Secondly, pH-dependent changes in the fluorescence emission spectral profile are very small, so that although dual emission ratio measurements are possible (Boyer & Hedley, 1994), they are not often performed. A minor structural modification, replacing the carboxyethyl substituents of BCECF with carboxypropyl groups, produces improvements in both properties without affecting the pK_a. The new derivative, BCPCF has an isosbestic point in its excitation spectrum at 454 nm (compared to 439 nm for BCECF), and the pH-modulated emission spectra exhibit an isosbestic point at 504 nm (Liu *et al.*, 1997).

The SNARF and SNAFL pH indicators undergo pH-dependent shifts in both their excitation and emission spectra. The pK_a values of these dyes are in the range 7.4–7.8 (Whitaker *et al.*, 1991). The quantum yield of the basic form of carboxy SNARF-1 is greater than that of its acidic form. This compensates for the lower absorption of its basic form at 480–520 nm and results in a particularly useful dual-emission optical response (Fig. 3.7). Because carboxy SNARF-1 can be excited by the argon-ion laser at 488 nm or 514 nm and also by the argon-krypton laser at 568 nm, it is usually the preferred pH indicator for flow cytometry and laser scanning microscopy (Boyer & Hedley, 1994; Chacon *et al.*, 1994). Carboxy SNARF-1 is also well suited for use in combination with fura-2 or indo-1 for simultaneous measurements of pH and

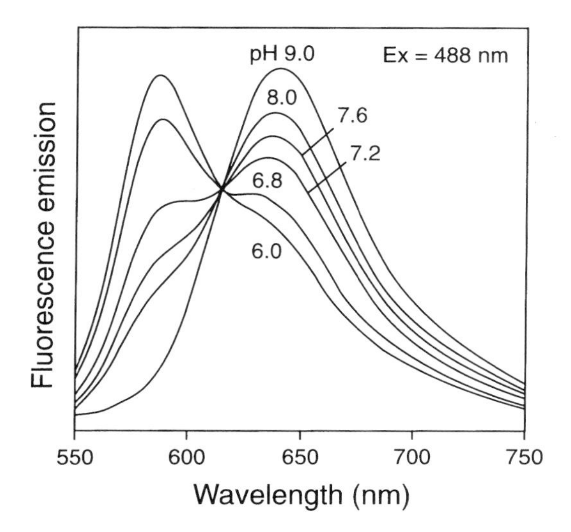

Figure 3.7 The pH-dependent fluorescence emission spectra of carboxy SNARF-1 obtained using excitation at 488 nm.

Ca^{2+} (Beatty *et al.*, 1993; Martínez-Zaguilán *et al.*, 1996). The SNAFL dyes have stronger emission from their acidic forms, which results in this class of dyes being more useful as dual-excitation indicators (Whitaker *et al.*, 1991).

Unlike the cytoplasmic pH indicators, which are loaded into cells as permeant esters, indicators for acidic organelles are often introduced by attaching them to ligands such as transferrin or chemotactic peptides that are internalized via receptor-mediated endocytosis (Murphy *et al.*, 1984; Yamashiro *et al.*, 1984; Fay *et al.*, 1994). The pH-dependent fluorescence of these probes can be used to track the internalization process itself, or to detect physiological or pathological variations in endosomal pH (Barasch *et al.*, 1991). Fluorescein (pK_a=6.4) has little response below pH 5; the range of sensitivity can be extended by using Oregon Green 488, a fluorinated derivative of fluorescein with a pK_a of 4.7 (Vergne *et al.*, 1998). LysoSensor dyes are a new series of pH indicators that selectively accumulate in acid organelles owing to protonation of a weakly basic secondary amine substituent (Cousin & Nicholls, 1997; Gu *et al.*, 1997). Unlike phenolic pH indicators, LysoSensor dyes are usually more fluorescent in acidic environments than they are at neutral or basic pH. The various LysoSensor dyes have pK_a values ranging between 4.2 and 7.5 (Haugland, 1996).

3.4 CONCLUSIONS

The numerous applications of the ion indicators described above are beyond the scope of this chapter. New applications, such as high-throughput pharmacological screening (Schroeder and Neagle, 1996) continue to emerge, stretching the capabilities of existing indicators and motivating the development of new ones. All existing indicators have practical limitations to some degree. As these limitations become increasingly well characterized, they can be rectified, circumvented, or at least properly accounted for. The efforts that are being made in this direction are very well compensated by the unique insights into cellular physiology that are obtainable as a result. The design and synthesis of new indicators continues, following the objectives outlined in Section 3.2. Calcium indicators can now be selected from an extensive array to meet diverse experimental requirements. However the 'ultimate' 488 nm-excitable dual-emission indicator has so far eluded development efforts. Development of optimal fluorescent indicators for intracellular pH is probably close to completion. The current sodium, potassium and chloride indicators are more limited, both in terms of their properties and in the number of available variants. Improvements should be attainable through application of the experience accumulated in the development of the more advanced Ca^{2+} and pH indicators.

ACKNOWLEDGEMENTS

Thanks to Kyle Gee and Cailan Zhang of Molecular Probes for providing data for Table 3.2 and Fig. 3.3.

REFERENCES

Allbritton N.L., Oancea E., Kuhn M.A. & Meyer T. (1994) *Proc. Natl Acad. Sci. USA* **91**, 12458.
Amorino G.P. & Fox M.H. (1995) *Cytometry* **21**, 248.
Atar D., Backx P.H., Appel M.M., Gao W.D. & Marban E. (1995) *J. Biol. Chem.* **270**, 2473.
Baker A.J., Brandes R., Schreur J.H.M., Camacho S.A. & Weiner M.W. (1994) *Biophys J.* **67**, 1646.
Barasch J., Kiss B., Prince A., Saiman L., Gruenert D. & Al-Awqati Q. (1991) *Nature* **352**, 70.
Beatty D.M., Chronwall B.M., Howard D.E., Wiegmann T.B. & Morris S.J. (1993) *Endocrinology* **133**, 972.
Berlin J.R., Bassani J.W.M. & Bers D.M. (1994) *Biophys. J.* **67**, 1775.
Biwersi J. & Verkman A.S. (1991) *Biochemistry* **30**, 7879.
Biwersi J., Farah N., Wang Y-X., Ketcham R. & Verkman A.S. (1992) *Am. J. Physiol.* **262**, C243.
Blinks J.R., Mattingly P.H., Jewell B.R., Van Leeuwen M., Harrer G.L. & Allen D.G. (1978) *Methods Enzymol.* **57**, 292.
Boyer M.J. & Hedley D.W. (1994) *Methods Cell Biol.* **41**, 135.
Bright G.R., Rogowska J., Fisher G.W. & Taylor D.L. (1987) *J. Cell Biol.* **104**, 1019.
Bright G.R., Whitaker J.E., Haugland R.P. & Taylor D.L. (1989) *J. Cell Physiol.* **141**, 410.

Brooke S.M., Trafton J.A. & Sapolsky R.M. (1996) *Brain Res.* **706**, 283.
Bush D.S. & Jones R.L. (1987) *Cell Calcium* **8**, 455.
Chacon E., Reece J.M., Nieminen A.-L., Zahrebelski G., Herman B. & Lemasters J.J. (1994) *Biophys. J.* **66**, 942.
Cousin M.A. & Nicholls D.G. (1997) *J. Neurochem.* **69**, 1927.
Denk W., Piston D.W. & Webb W.W. (1995) In *Handbook of Biological Confocal Microscopy*, 2nd edn, J.B. Pawley (ed). Plenum Press, New York, pp. 445–458.
Di Virgilio F., Steinberg T.H. & Silverstein S.C. (1990) *Cell Calcium* **11**, 57.
Eilers J., Callewaert G., Armstrong C. & Konnerth A. (1995) *Proc. Natl. Acad. Sci. USA* **92**, 10272.
Escobar A.L., Velez P., Kim A.M., Cifuentes F., Fill M. & Vergara J.L. (1997) *Pflügers Arch. Eur. J. Physiol.* **434**, 615.
Etter E.F., Kuhn M.A. & Fay F.S. (1994) *J. Biol. Chem.* **269**, 10 141.
Fay S.P., Habbersett R., Domalewski M.D., Posner R.G., Houghton T.G., Pierson E., Muthukumaraswamy N., Whitaker J., Haugland R.P., Freer R.J. & Sklar L.A. (1994) *Cytometry* **15**, 148.
Floto R.A., Mahaut-Smith M.P., Somasundaram B. and Allen J.M. (1996) *Cell Calcium* **18**, 377.
Friedman J.E. & Haddad G.G. (1994) *Brain Res.* **663**, 329.
Grynkiewicz G., Pozzan T. & Tsien R.Y. (1985) *J. Biol. Chem.* **260**, 3440.
Gu F., Aniento F., Parton R.G. & Gruenberg J. (1997) *J. Cell Biol.* **139**, 1183.
Haugland R.P. (1996) *Handbook of Fluorescent Probes and Research Chemicals*, 6th edn, Molecular Probes, Inc., Eugene, OR. A continuously updated HTML edition of this publication is available on the internet at <http://www.probes.com/handbook/toc.html>.
Hinkle P.M., Shanshala E.D. & Nelson E.J. (1992) *J. Biol. Chem.* **267**, 25553.
Hofer A.M. & Machen T.E. (1993) *Proc. Natl. Acad. Sci. USA* **90**, 2598.
Hofer A.M. & Schulz I. (1996) *Cell Calcium* **20**, 235.
Horne J.H. & Meyer T. (1997) *Science* **276**, 1690.
Hove-Madsen L. & Bers D.M. (1992) *Biophys J.* **63**, 89.
Hurley T.W., Ryan M.P. & Brinck R.W. (1992) *Am. J. Physiol.* **263**, C300.
Illsley N.P. & Verkman A.S. (1987) *Biochemistry* **26**, 1219.
Inglefield J.R. & Schwartz-Bloom R.D. (1997) *J. Neurosci. Methods* **75**, 127.
Jou M.-J., Peng T.-I. & Sheu S.-S. (1996) *J. Physiol.* **497**, 299.
June C.H. & Rabinovitch P.S. (1994) *Methods Cell Biol.* **41**, 149.
Koncz C. & Daugirdas J.T. (1994) *Am. J. Physiol.* **267**, H2114.
Konishi M., Hollingworth S., Harkins A.B. & Baylor S.M. (1991) *J. Gen. Physiol.* **97**, 271.
Konishi M., Suda N. & Kurihara S. (1993) *Biophys. J.* **64**, 223.
Koss K.L., Putnam R.W. & Grubbs R.D. (1993) *Am. J. Physiol.* **264**, C1259.
Krapf R., Berry C.A. & Verkman A.S. (1988) *Biophys. J.* **53**, 955.
Lakowicz J.R., Szmacinski H. & Johnson M.L. (1992) *J. Fluorescence* **2**, 47.
Lattanzio F.A. & Bartschat D.K. (1991) *Biochem. Biophys. Res. Commun.* **177**, 184.
Liu J., Diwu Z. & Klaubert D.H. (1997) *Bioorg. Med. Chem. Lett.* **7**, 3069.
Lipp P. and Niggli E. (1993) *Cell Calcium* **14**, 359.
Martínez-Zaguilán R., Parnami G. & Lynch R.M. (1996) *Cell Calcium* **19**, 337.
McDonough P.M. & Button D.C. (1989) *Cell Calcium* **10**, 171.
Minta A. & Tsien R.Y. (1989) *J. Biol. Chem.* **264**, 19449.
Minta A., Kao J. & Tsien R.Y. (1989) *J. Biol. Chem.* **264**, 8171.
Murphy R.F., Powers S. & Cantor C.R. (1984) *J. Cell Biol.* **98**, 1757.
Negulescu P.A. & Machen T.E. (1990) *Methods Enzymol.* **192**, 38.
Nitschke R., Wilhelm S., Borlinghaus R., Leipziger J., Bindels R. & Greger R. (1997) *Pflügers Arch. Eur. J. Physiol.* **433**, 653.
Oakes S.G., Martin W.J., Lisek C.A. & Powis G. (1988) *Anal. Biochem.* **169**, 159.
O'Donovan M.J., Ho S., Sholomenko G. & Yee W. (1993) *J. Neurosci. Meth.* **46**, 91.
Paradiso A.M., Tsien R.Y., Demarest J.R. & Machen T.E. (1987) *Am. J. Physiol.* **253**, C30.
Parker I., Choi J. & Yao Y. (1996) *Cell Calcium* **20**, 105.
Pilas B. & Durack G. (1997) *Cytometry* **28**, 316.
Rajdev S. & Reynolds I.J. (1993) *Neurosci. Lett.* **162**, 149.
Raju B., Murphy E., Levy L.A., Hall R.D. & London R.E. (1989) *Am. J. Physiol.* **256**, C540.
Read N.D., Allan W.T.G., Knight H., Knight M.R., Malho R., Russell A., Shacklock P.S. & Trewavas A.J. (1992) *J. Microscopy* **166**, 57.
Regehr W.G. & Atluri P.P. (1995) *Biophys. J.* **68**, 2156.
Rink T.J., Tsien R.Y. & Pozzan T. (1982) *J. Cell Biol.* **95**, 189.
Scanlon, M., Williams, D.A. & Fay, F.S. (1987) *J. Biol. Chem.* **262**, 6308.
Scarpa A., Brinley F.J., Tiffert T. & Dubyak G.R. (1978) *Ann. NY Acad. Sci.* **307**, 86.
Scheenen W.J.J.M., Makings L.R., Gross L.R., Pozzan T. & Tsien R.Y. (1996) *Chem. Biol.* **3**, 765.
Schroeder K.S. & Neagle B.D. (1996) *J. Biomol. Screening* **1**, 75.
Slayman C.L., Moussatos V.V. & Webb W.W. (1994) *J. Exp. Biol.* **196**, 419.
Snitsarev V.A., McNulty T.J. & Taylor C.W. (1996) *Biophys. J.* **71**, 1048.
Srivastava D., Fox D.A. & Hurwitz R.L. (1995) *Biochem. J.* **308**, 653.
Sun G. & Budde R.J.A. (1997) *Biochemistry* **36**, 2139.
Teruel M.N. & Meyer T. (1997) *Biophys. J.* **73**, 1785.
Thomas J.A., Buchsbaum R.N., Zimniak A. & Racker E. (1979) *Biochemistry* **18**, 2210.
Tiffert T, Garcia-Sancho J. & Lew V.L. (1984) *Biochim. Biophys. Acta* **773**, 143.
Tran N.N.P., Leroy P., Bellucci L., Robert A., Nicolas A., Atkinson J. & Capdeville-Atkinson C. (1995) *Cell Calcium* **18**, 420.
Tsien R.Y. (1980) *Biochemistry* **19**, 2396.
Vergne I., Constant P. & Lanéelle G. (1998) *Anal. Biochem.* **255**, 127.
Verkman A.S., Sellers M.C., Chao A.C., Leung T. & Ketcham R. (1989) *Anal. Biochem.* **178**, 355.
Wang S.H. & Augustine G.J. (1995) *Neuron* **15**, 755.
West M.R. & Molloy C.R. (1996) *Anal. Biochem.* **241**, 51.
Whitaker J.E., Haugland R.P. & Prendergast F.G. (1991) *Anal. Biochem.* **194**, 330.
Wolfbeis O.S. & Urbano E. (1982) *J. Heterocyclic Chem.* **19**, 841.
Yamashiro D.J., Tyeko B., Fluss S.R. & Maxfield F.R. (1984) *Cell* **37**, 789.
Zhao M., Hollingworth S. & Baylor S.M. (1996) *Biophys. J.* **70**, 896.
Zhao M., Hollingworth S. & Baylor S.M. (1997) *Biophys. J.* **72**, 2736.
Zhang G.H. & Melvin J.E. (1996) *J. Biol. Chem.* **271**, 29067.

CHAPTER 4

Fluorescent Imaging of Nucleic Acids and Proteins in Gels

VICTORIA L. SINGER & RICHARD P. HAUGLAND
Molecular Probes, Inc., Eugene, OR, USA

4.1 INTRODUCTION

Fluorescent stains have been commonly used to visualize nucleic acids in electrophoretic gels ever since the landmark paper by Sharp, Sugden and Sambrook in 1973 (Sharp *et al.*, 1973) using ethidium bromide and SV40 DNA to detect restriction endonuclease activity in extracts of *Haemophilus parainfluenzae*. At first, available fluorescence-based methods for detecting and documenting the presence of nucleic acids in stained gels were limited to black and white Polaroid® photography with ultraviolet light transillumination. However, in recent years, an increasing number of sophisticated imaging systems have been developed specifically for this application. These systems include automated sequencers, laser scanners, charge-coupled device (CCD) and video camera systems used in combination with ultraviolet transilluminators or visible light transilluminators, and a xenon lamp-based epi-illuminator. The new imaging instruments have made it possible to carry out quantitative analysis of the abundance of electrophoretic species, in addition to qualitative band identification and size analysis. They have also made it possible to do multiparameter analysis, which is particularly useful in DNA sequencing and DNA typing. The proliferation of these new imaging instruments has also in turn driven the development of new fluorescent gel stains and reactive fluorophore dye labels for nucleic acids and for proteins. The use of fluorescent stains and fluorophore labels to detect proteins in electrophoretic gels lagged considerably behind their use in nucleic acid detection. This lag was partly due to the relative paucity of fluorescent stains available for this application and also to the ready availability of sensitive colorimetric detection methods. However, some very sensitive fluorescent protein stains are now available.

This chapter reviews the fluorescent stains and fluorophore labels in common use with imaging instrumentation for detecting nucleic acids and proteins in gels. In addition, bioanalytical applications of these dyes are described.

FLUORESCENT AND LUMINESCENT PROBES, 2ND EDN
ISBN 0–12–447836–0

4.2 GENERAL PROPERTIES OF FLUORESCENT NUCLEIC ACID STAINS

Fluorescent nucleic acid gel stains are dyes that have low fluorescence when free in solution, but exhibit substantial increases in fluorescence (fluorescence enhancement) upon binding non-covalently to nucleic acids in gels. Such dyes belong mainly to two distinct chemical classes: phenanthridines and cyanine dyes. Indoles, imidazoles and acridines, although well-known nucleic acid stains for cells, are generally not used in combination with gel electrophoresis. This is primarily because of the difficulty of photographing gels stained with blue fluorescent dyes, the poor match of excitation wavelengths to commonly available light sources, and their relatively poor fluorescence enhancement.

Binding of nucleic acid stains occurs via a variety of different modes, including intercalation between adjacent base pairs like ethidium bromide (Reinhardt & Krugh, 1978), bis-intercalation like YOYO-1 (Hansen *et al.*, 1996) and ethidium homodimer, or binding in the minor groove of the helix like DAPI (Kapuscinski, 1995), Hoechst 33258 and Hoechst 33432. Many nucleic acid stains are cationic, although some, such as the acridines, are essentially neutral. In each case, the fluorescence enhancement is thought to mainly result from two factors: immobilization of the dye molecule in a planar conformation (allowing extensive electron delocalization to occur) and exclusion of solvent (water) from interactions with the dye that might otherwise quench fluorescence. Molecules exhibiting the largest fluorescence enhancements are those that have the greatest molecular flexibility (thus the lowest fluorescence) when free in solution and the greatest degree of immobilization (thus the greatest rigidity) when bound by nucleic acids. Dyes such as the phenanthridines, which have significant fluorescence in the unbound state, are generally fairly rigid molecules. Many of the unsymmetrical cyanine dyes tend to be relatively flexible in comparison and thus have virtually no fluorescence when free in solution and large enhancements upon binding to nucleic acids (Haugland *et al.*, 1995; Yue *et al.*, 1997).

Interestingly, although the protein stains described below exhibit essentially no sensitivity for detection of nucleic acids, many nucleic acid stains can detect proteins. Detection is presumably due to exclusion of the dye from interactions with water and may also result from the molecule being held in a somewhat rigid conformation in a detergent micelle surrounding the protein or when interacting with hydrophobic portions of proteins. The SYBR dyes described below exhibit reasonably good sensitivity for detecting proteins in SDS gels; however, their sensitivity is not as good as that of the SYPRO dyes and their fluorescence intensity upon binding is much lower. The sensitivity of the SYBR dyes for detecting proteins is generally at least 100-fold poorer on a mass basis than it is for detecting nucleic acids. In our hands, the most sensitive of the SYBR dyes for protein detection seems to be SYBR Green I stain (data not shown).

4.3 EXAMPLES OF FLUORESCENT NUCLEIC ACID GEL STAINS

The properties of selected fluorescent nucleic acid stains, are listed in Table 4.1.

4.3.1 Ethidium bromide

Ethidium bromide is a phenanthridine monomer dye (Fig. 4.1), which exhibits a 20- to 25-fold fluorescence enhancement upon binding to double-stranded DNA (LePecq & Paoletti, 1967). The dye intercalates between adjacent base pairs in the double-stranded DNA molecule (Reinhardt & Krugh, 1978). The fluorescence enhancement observed upon ultraviolet light excitation of intercalated ethidium bromide is also thought to be due in part to energy transfer from the bases of the DNA to the dye (LePecq & Paoletti, 1967). The primary sequence of the DNA has relatively little effect on nucleic acid binding, making this dye a good general-use stain. The dye penetrates electrophoretic gels rapidly and efficiently. Ethidium bromide can be precast in gels, used as a poststain, and is also occasionally used as an RNA prestain in formaldehyde–agarose gel electrophoresis.

4.3.2 SYBR dyes

The SYBR family of nucleic acid stains are substituted unsymmetrical cyanine dyes that are virtually nonfluorescent when free in aqueous solution, but exhibit extremely large fluorescence enhancements upon binding to nucleic acids (Haugland *et al.*, 1995; Yue *et al.*, 1997). The binding modes of these dyes are not known and may be complex. Binding can be efficiently inhibited by the presence of cations, particularly divalent cations, and the dyes can be removed by ethanol precipitation or interactions with ionic detergents.

SYBR Green I stain has a quantum yield of ~0.8 upon binding to double-stranded DNA (dsDNA; Singer *et al.*, 1994), making it one of the brightest

Table 4.1 Properties of selected fluorescent nucleic acid gel stains.

Stain	*Excitation maxima*[a] *(nm)*	*Emission maximum*[a] *(nm)*	*Major applications*
Ethidium bromide	300, 520	605	General use stain for all nucleic acids
SYBR Green I stain	290, 380, 495	520	High-sensitivity double-stranded DNA poststain
SYBR Green II stain	290, 345, 490	510	High-sensitivity single-stranded DNA and RNA poststain
SYBR Gold stain	300, 500	535	General use high-sensitivity gel poststain
Vistra Green stain	495	520	High-sensitivity gel stain for use with laser scanners
GelStar stain	490	530	High-sensitivity stain for casting in gels
Radiant Red stain	300, 525	615	Moderately sensitive poststain for formaldehyde agarose RNA gels

[a] Spectra were obtained bound to double-stranded DNA in buffer; wavelengths are rounded to the nearest 5 nm.

Figure 4.1 Structure of ethidium bromide.

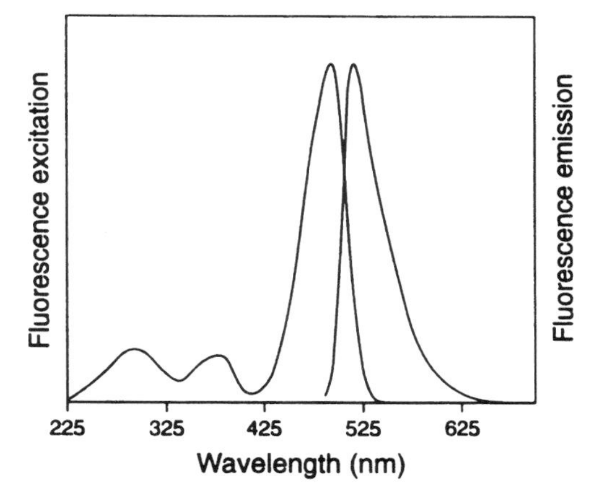

Figure 4.2 Excitation and emission spectra of SYBR Green I stain bound to double-stranded DNA.

stains available for nucleic acid detection. Because it has a high fluorescence enhancement upon binding to dsDNA (~800- to 1000-fold), it is especially useful for applications that require high signal-to-background ratios. Examples of such assays include reverse transcription PCR (Schneeberger *et al.*, 1995; Su *et al.*, 1997), telomerase assays (Fong *et al.*, 1997), apoptosis ladder detection, forensic DNA typing (Morin & Smith, 1995; Kishida *et al.*, 1995; Worley *et al.*, 1997), kinetic or real-time PCR analysis (Ririe *et al.*, 1997; Wittwer *et al.*, 1997), comet assays, nuclease detection (Yasuda *et al.*, 1998), detecting DNA damage in pulsed field gels (Kiltie and Ryan, 1997), dsDNA capillary electrophoresis (Skeidsvoll & Ueland, 1995; Takai *et al.*, 1997) and studying protein–nucleic acid interactions via bandshift (electrophoretic mobility shift) analysis. It is significantly less sensitive for detecting single-stranded DNA (ssDNA) compared to dsDNA (Stothard *et al.*, 1997). SYBR Green I stain has such a high affinity for dsDNA that it can also be used to prestain DNA prior to sample loading. The dye remains bound during electrophoresis and can be used to visualize individual DNA bands directly as they migrate. SYBR Green I stain has two excitation peaks in the ultraviolet, one of which is well-matched to the 254 nm light produced by some side-arm epi-illuminators and short wavelength transilluminators (Fig. 4.2). The greatest sensitivity for staining with this dye is achieved when using these short wavelength light sources (Fig. 4.3). SYBR Green I stain also has a visible excitation peak at ~500 nm that is spectrally well matched to several commercially available laser scanners.

SYBR Green II stain has the unusual characteristic of having a higher quantum yield upon binding to RNA than to dsDNA. It is spectrally well matched to the green emission filters available with many CCD-based imaging systems and has been

(A)

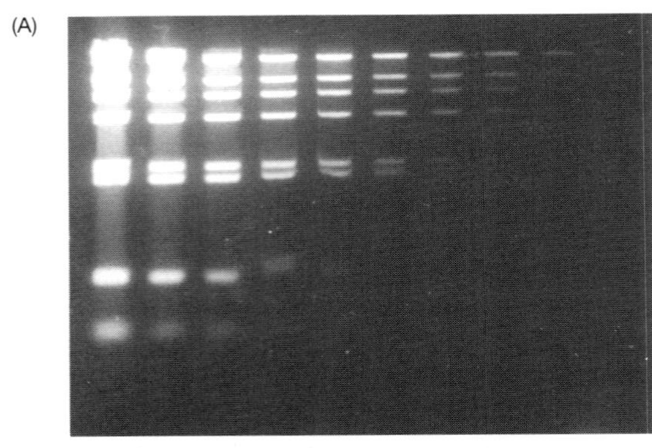

(B)

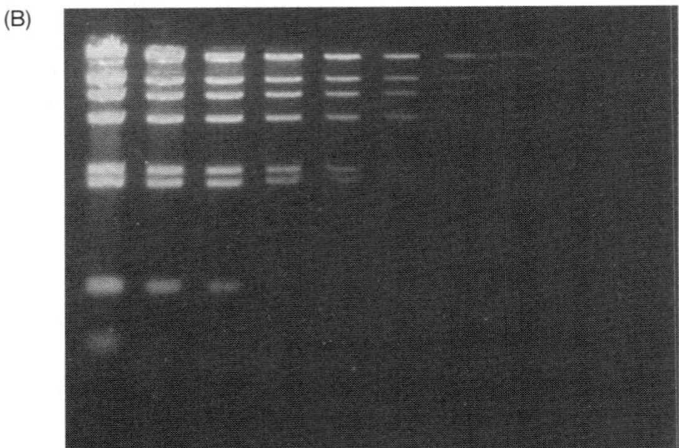

(C)

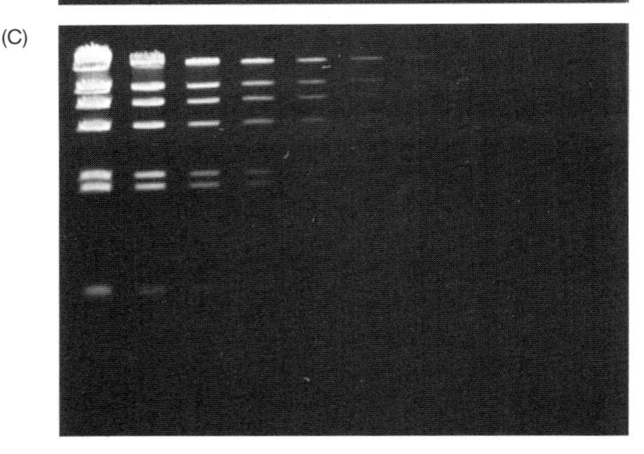

Figure 4.3 Comparison of SYBR Green I stain and ethidium bromide sensitivities for detecting double-stranded DNA. Identical 3-fold dilution series of double-stranded DNA were separated on electrophoretic gels and stained with SYBR Green I stain (A, B) or ethidium bromide (C). The gel stained with ethidium bromide was then destained for 30 minutes. Gels were photographed using 254 nm epi-illumination (A) or 300 nm transillumination (B, C) through appropriate photographic filters, using Polaroid black-and-white print film.

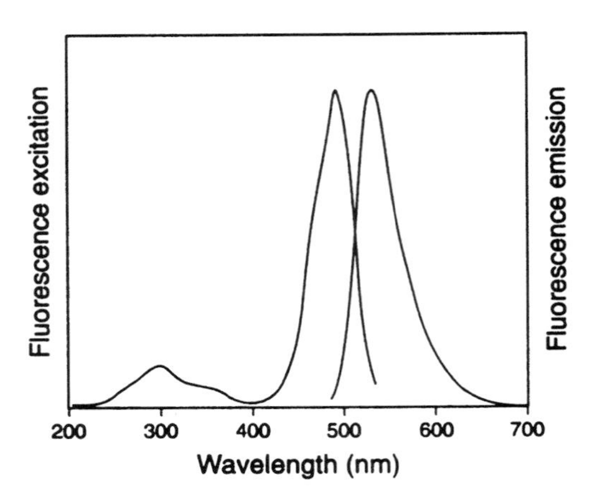

Figure 4.4 Excitation and emission spectra of SYBR Gold stain bound to double-stranded DNA.

extensively used for detecting RNA in denaturing gels (Chen & Boynton, 1995), single-strand conformation polymorphism analysis (Emanuel *et al.*, 1996; Law *et al.*, 1996), denaturating gradient gel electrophoresis (Stoner *et al.*, 1996) and reverse transcription (cDNA synthesis). SYBR Green II stain has very similar spectra to SYBR Green I stain. It is very sensitive for detecting ssDNA as well as being reasonably sensitive for dsDNA (Stothard *et al.*, 1997).

SYBR Gold stain penetrates electrophoretic gels much more rapidly and with greater efficiency than either of the SYBR Green gel stains. This is particularly true for thick and high-percentage gels. The quantum yield of SYBR Gold stain is similar (~0.6) when bound to dsDNA, ssDNA or RNA, making it especially useful for staining nucleic acids in denaturing gels. It is particularly sensitive for detecting glyoxylated RNA and RNA separated in formaldehyde–agarose gels. SYBR Gold stain is also extremely photostable, making it the best dye to choose for excising bands from gels for DNA or RNA fragment purification. The ultraviolet excitation peak for SYBR Gold stain is much better matched to standard 300 nm transilluminators than the ultraviolet excitation peaks for the SYBR Green stains (Fig. 4.4). Thus it is a better dye choice for use with those instruments.

GelStar stain is the best choice of the unsymmetrical cyanine dyes for precasting in gels because it has relatively little effect on electrophoretic mobility and has relatively high sensitivity (approximately 6-fold greater sensitivity than ethidium bromide in most applications). GelStar stain can be used with laser scanners or ultraviolet transilluminators. It has spectra that are similar to those of the SYBR Green gel stains.

Vistra Green stain has an extremely low background fluorescence, making it very well suited for use with laser scanners. Its background fluorescence in these instruments is as much as several-fold lower than that of SYBR Green I stain. It also causes less

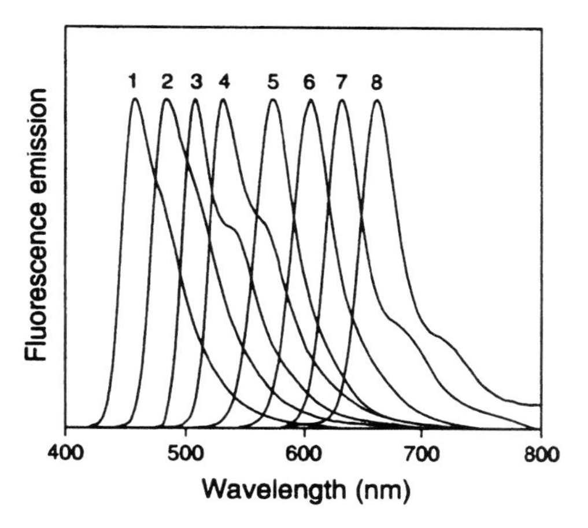

Figure 4.5 Emission spectra of the unsymmetrical cyanine dimer dyes bound to double-stranded DNA. 1, POPO-1; 2, BOBO-1; 3, YOYO-1; 4, TOTO-1; 5, POPO-3; 6, BOBO-3; 7, YOYO-3; 8, TOTO-3.

mobility perturbation than SYBR Green I stain when precast in gels.

4.3.3 Other Unsymmetrical Cyanine Dyes

Dyes such as YO-PRO-1 have also been used as gel stains and have sensitivities between those of ethidium bromide and the SYBR Green gel stains. Because the fluorescence excitations and emissions of these dyes span much of the visible spectrum (Fig. 4.5), it is possible to choose a dye from among this dye set that is spectrally ideal for any instrument choice. The dimer versions of these dyes, YOYO-1 etc., have much higher affinities (Haugland, 1996) and can be used as electrophoretic prestains (Rye *et al.*, 1991, 1992, 1993; Glazer & Rye, 1992). The red fluorescent SYTO dyes, SYTO 59 and SYTO 61, have moderately high affinities for nucleic acids and can be used as sensitive electrophoretic poststains in combination with gel scanners containing 633 nm red HeNe laser excitation sources. Many of these dyes have also been used to detect DNA via capillary electrophoresis (Srinivasan *et al.*, 1993a, b; Butler *et al.*, 1994; Zhu *et al.*, 1994; Oda *et al.*, 1996; Clark & Mathies, 1997; Madabhushi *et al.*, 1997; see also BioProbes 25, p. 22 (1997)).

4.3.4 Other Phenanthridine Dyes

Ethidium homodimer has an extremely high affinity for dsDNA and can be used as an electrophoretic prestain, like the cyanine dimer dyes discussed above (Glazer *et al.*, 1990; Rye *et al.*, 1991; Glazer & Rye, 1992). Radiant Red stain, another member of this chemical class, is primarily used as a poststain for visualizing RNA in formaldehyde–agarose gels. In this application, its sensitivity is somewhat lower than that of SYBR Green II stain or SYBR Gold stain. It is visualized using 300 nm excitation light.

4.4. GENERAL PROPERTIES OF FLUOROPHORE LABELS USED TO DETECT NUCLEIC ACIDS

There are a wide variety of fluorophore labels that have been used to detect nucleic acids in gels. These dyes are mainly available as amine-reactive succinimidyl esters or isothiocyanates, which are reacted with primary amines incorporated primarily at the 5′ end during synthesis of a synthetic oligonucleotide. Nucleic acids can also be labelled using maleimides or iodoacetamides to modify synthetically incorporated thiol groups. Many fluorophores are also available as nucleotide conjugates, which can be incorporated enzymatically into nucleic acids prior to electrophoresis. Some examples of available labels and their excitation and emission maxima are shown in Table 4.2, along with examples of the types of instruments suitable for detecting the labelled products.

4.4.1 Amine reactive dyes

Amine reactive dyes can be used to label amine-modified oligonucleotides after synthesis. There are a wide variety of such labels available and thus this is probably the most flexible labelling method in current use, in terms of fluorophore selection. However, this method cannot be applied to long RNA or DNA molecules. Such labelled oligonucleotides are useful as sequencing primers, PCR primers (particularly for DNA typing), and as hybridization probes, particularly for RNA *in situ* hybridization, but also for blot detection.

4.4.2 Fluorophore-labelled phosphoramidites

Labelled phosphoramidites allow the construction of a fluorophore-labelled oligonucleotide primer on the column, during synthesis of the oligonucleotide. Labelling in this way is probably the most convenient method available for most researchers. However, there are at present relatively few choices of fluorophore-labelled phosphoramidites. Commercially available fluorophore choices include fluorescein (FAM), tetramethylrhodamine (TAMRA), JOE, Cy3 and Cy5. The

Table 4.2 Fluorophore labels in common use with gel imaging systems.

Fluorophore label	*Excitation/emission maxima (nm)*	*Commercially available reactive forms*[a]	*Commercially available nucleotides*	*Suitable gel imaging instruments*
Pyrene	340/375	SE, ITC, iodoacetamide, maleimide	UTP, CTP, dUTP, dCTP, dATP	UV light sources
AMCA	350/440	SE	dUTP, dCTP, dATP, dGTP, UTP, CTP, ATP	UV light sources and Polaroid photography
Cascade Blue	400/420	Acetyl azide	dUTP	UV light sources and Polaroid photography
Diethylaminocoumarin	410/475	ITC, SE	dUTP, dCTP, dATP	UV lights sources and Polaroid photography
Fluorescein (FAM)	495/520	ITC, SE, phosphoramidite	UTP, ATP, CTP, dUTP, dATP, dCTP, ddATP[b], ddUTP[b], ddCTP[b], ddGTP[b]	Sequencers, laser scanners, CCD-camera systems, xenon-lamp epi-illuminator
BODIPY FL	505/515	SE	UTP, dUTP	Sequencers, laser scanners, CCD-camera systems, xenon-lamp epi-illuminator
Rhodamine 110	495/520		ddUTP[b], ddCTP[b], ddATP[b], ddGTP[b]	Sequencers, laser scanners, CCD-camera systems, xenon-lamp epi-illuminator
Oregon Green 488	495/520	SE	dUTP	Sequencers, laser scanners, CCD-camera systems, xenon-lamp epi-illuminator
Alexa 488	490/520	SE, maleimide	dUTP	Sequencers, laser scanners, CCD-camera systems, xenon-lamp epi-illuminator
Rhodamine Green	505/530	SE	UTP, dUTP	Sequencers, laser scanners, CCD-camera systems, xenon-lamp epi-illuminator
Eosin	520/545	ITC, maleimide, iodoacetamide	ddCTP	Sequencers, laser scanners, CCD-camera systems, xenon-lamp epi-illuminator
Alexa 532	525/550	SE	dUTP	Sequencers, laser scanners, CCD-camera systems, xenon-lamp epi-illuminator
2′,7′-Dimethoxy-4′, 5’-dichloro-6-carboxyfluorescein (JOE)	525/550	SE, phosphoramidite	ddUTP[b], ddCTP[b], ddATP[b], ddGTP[b]	Sequencers, laser scanners, CCD-camera systems, xenon-lamp epi-illuminator
Naphthofluorescein	510/560		dUTP, dCTP, dATP, UTP, CTP	Sequencers, laser scanners, CCD-camera systems, xenon-lamp epi-illuminator
Alexa 546	555/570	SE	dUTP	Sequencers, laser scanners, CCD-camera systems, xenon-lamp epi-illuminator

Table 4.2 Continued

Fluorophore label	*Excitation/ emission maxima (nm)*	*Commercially available reactive forms*[a]	*Commercially available nucleotides*	*Suitable gel imaging instruments*
Cy3	550/570	SE, phosphoramidite	UTP, dUTP, dCTP	Sequencers, laser scanners, CCD-camera systems, xenon-lamp epi-illuminator
Tetramethylrhodamine	550/570	ITC, SE, phosphoramidite	ddUTP[b], ddCTP[b], ddATP[b], ddGTP[b], UTP, dUTP	Sequencers, laser scanners, CCD-camera systems, xenon-lamp epi-illuminator
Rhodamine 6G	530/550		ddUTP[b], ddCTP[b], ddATP[b], ddGTP[b]	Sequencers, laser scanners, CCD-camera systems, xenon-lamp epi-illuminator
Alexa 568	575/600	SE	dUTP	Sequencers, laser scanners, CCD-camera systems, xenon-lamp epi-illuminator
Lissamine Rhodamine, Rhodamine Red	570/590	SC, SE	ddGTP, dUTP, dCTP, dATP, dGTP, UTP, CTP, ATP	Sequencers, laser scanners, CCD-camera systems, xenon-lamp epi-illuminator
Carboxy-X-rhodamine (ROX)	585/610	SE	ddUTP[b], ddCTP[b], ddATP[b], ddGTP[b]	Sequencers, laser scanners, CCD-camera systems, xenon-lamp epi-illuminator
Texas Red	595/615	SE, SC	UTP, dUTP	Sequencers, laser scanners, CCD-camera systems, xenon-lamp epi-illuminator
BODIPY TR	595/625	SE	UTP, dUTP	Sequencers, laser scanners, CCD-camera systems, xenon-lamp epi-illuminator
BODIPY 630/650	630/650	SE	dUTP	Laser scanners, CCD-camera systems, xenon-lamp epi-illuminator
BODIPY 650/665	650/670	SE	dUTP	Laser scanners, CCD-camera systems, xenon-lamp epi-illuminator
Cy5	650/670	SE, phosphoramidite	dUTP, dCTP	Laser scanners, CCD-camera systems, xenon-lamp epi-illuminator

[a] ITC, isothiocyanate; SE, succinimidyl ester; SC, sulfonyl chloride.
[b] Single isomer conjugates are available.

applications of these labelled oligonucleotides are essentially identical to those described above for amine reactive dye labelling.

4.4.3 Energy transfer dyes

Because many of the high-sensitivity gel imaging systems use laser light sources, it would be useful to have a set of dyes with spectrally distinct emissions, which all excite efficiently with the same laser source. A series of energy transfer dyes have been constructed for this purpose (Ju *et al.*, 1995, 1996; Hung *et al.*, 1996). These dyes all use a fluorescein donor and classical sequencing dye acceptors and are efficiently excited with the argon-ion laser used in automated sequencing. The advantage that sequencing primers labelled with these dyes provide over primers labelled with conventional dyes is the higher signals of the longer wavelength emission dyes.

4.4.4 Fluorophore-labelled nucleotides

Fluorophore-labelled nucleotides are also in common use for labelling DNA hybridization probes. With the exception of labelled dideoxy nucleotides, labelled nucleotides are mainly used to produce hybridization probes for fluorescence *in situ* hybridization and for chip-based array analysis. One of the main reasons this labelling method is rarely used for direct visualization of nucleic acids in gels is that the addition of the fluorophore label alters the electrophoretic mobility of the nucleic acid being analysed. While it is trivial to use software to compensate for a single mobility change due to the presence of a single 5′ or 3′ label, it is difficult to compensate for unknown numbers of incorporated dye labels. Dideoxy nucleotides do not have this problem, since a single label is incorporated per DNA strand, at the 3′ end, allowing precise determination of the size of the labelled electrophoretic band. Thus, these nucleotides are commonly used in sequencing, particularly with laser-excited automated sequencers.

4.5 GENERAL PROPERTIES OF FLUORESCENT PROTEIN GEL STAINS

Fluorescent protein gel stains are of two types: (1) dyes that primarily bind non-covalently to detergent-coated proteins and exhibit large fluorescence enhancements upon doing so, and (2) dyes that react with protein amines, forming covalent adducts. Dyes in the former category, such as Nile red and the SYPRO dyes, have low fluorescence in aqueous solution and substantially increased fluorescence emission when they are in hydrophobic environments. It is not known if immobilization of the dyes plays a role in the fluorescence enhancement. Dyes in the latter category are of two types: those that are not fluorescent prior to adduct formation and those that are fluorescent in the reactive state as well as when conjugated.

4.6 EXAMPLES OF FLUORESCENT PROTEIN GEL STAINS

There are a wide variety of protein gel stains available, many of which yield a coloured product upon reaction with amine side groups of polypeptides. The most commonly used fluorescent gel stains, on the other hand, are environmentally sensitive dyes that have minimal fluorescence when free in aqueous solution and exhibit a large fluorescence enhancement upon binding to detergent-coated proteins. In addition, there are some amine-reactive fluorescent and luminescent stains available.

4.6.1 SYPRO Dyes

The SYPRO dyes are merocyanine dyes that are essentially non-fluorescent when free in solution, but become intensely fluorescent in hydrophobic environments (Steinberg *et al.*, 1996a; Haugland *et al.*, 1997). Thus staining is rapid and does not require either a prefixation step or a destaining step. These dyes appear to act by binding to the detergent coat surrounding SDS-coated proteins and by binding to hydrophobic regions of electrophoretically separated proteins. Binding of the SYPRO dyes is readily reversible. Dye can be removed by soaking gels in detergent or methanol, or soaking in water or acetic acid solution. Thus staining will not interfere with antibody generation. Staining is performed in acetic acid solution or transfer buffer and does not interfere with Western blot analysis (Steinberg *et al.*, 1996b; Hamby, 1996), microsequencing (Hamby, 1996) or mass spectrometry (Lauber *et al.*, 1998). The acetic acid staining solution is relatively stable and can be reused three to four times. The SYPRO dyes do not appear to interact with nucleic acids or amino acids and have minimal interactions with lipopolysaccharides (Steinberg *et al.*, 1996b). Staining has little dependence on amino acid content (Steinberg *et al.*, 1996a), since most proteins bind an approximately equal amount of SDS on a mass basis (Reynolds and Tanford, 1970a,b). Stained gels can be visualized using 300 nm excitation and Polaroid (Steinberg *et al.*, 1997) or CCD-camera based image acquisition, or using commercially available laser scanners (Steinberg *et al.*, 1996a). Detection sensitivity in many cases rivals that of silver staining (Fig. 4.6), with single nanogram per band sensitivity levels in minigels (Steinberg *et al.*, 1996a). Gels stained with the SYPRO dyes can also be subsequently stained using colorimetric stains such as Coomassie^R Brilliant Blue or silver. The SYPRO dyes have also been used for capillary electrophoresis (Harvey *et al.*, in press).

4.6.2 Nile red

The lipid stain Nile red (9-ethylamino-5H-benzo[α]-phenoxazine-5-one) has also been used to detect proteins in SDS polyacrylamide gels (Daban *et al.*, 1991). The mode of binding of Nile red is presumably

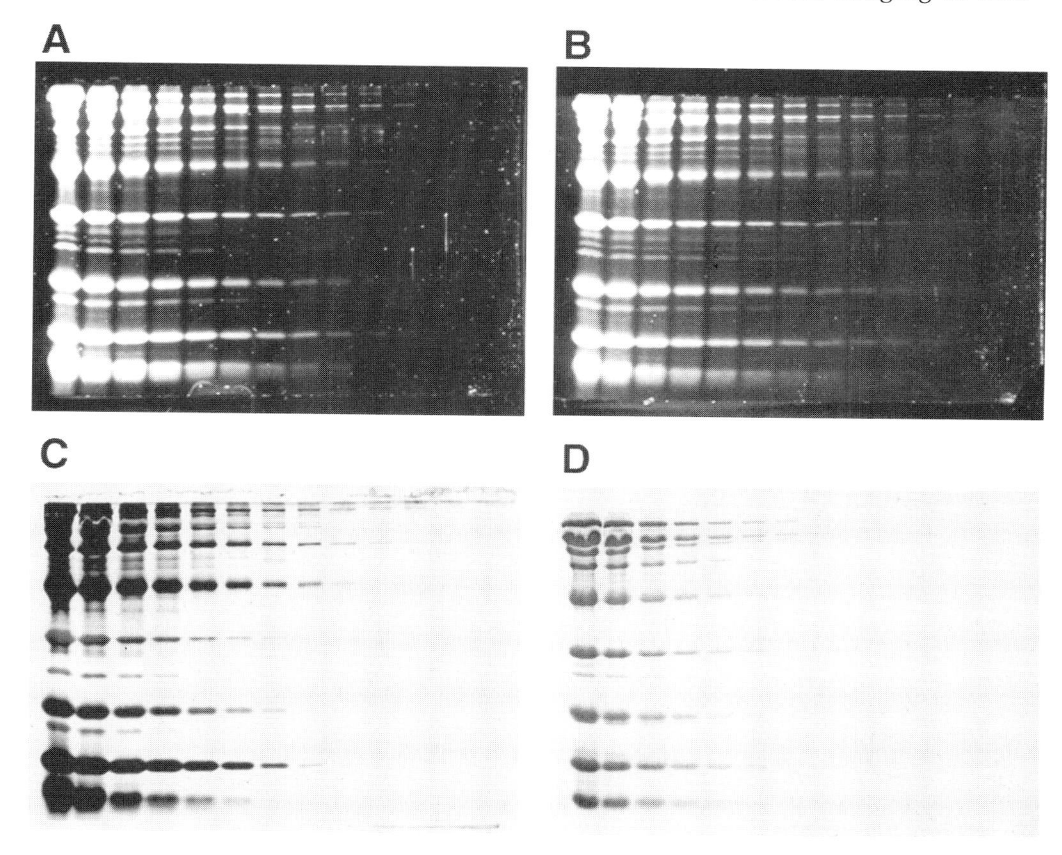

Figure 4.6 Sensitivity of SYPRO Orange stain (A), SYPRO Red stain (B), silver staining (C) and Coomassie Brilliant Blue staining (D) for detecting molecular weight marker proteins in SDS polyacrylamide electrophoresis.

similar to that of the SYPRO dyes, in that the dye appears to associate with hydrophobic regions of proteins and with the SDS coat surrounding proteins in denaturing gels. Nile red is relatively insoluble in aqueous solutions and is first dissolved in acetone, then mixed rapidly with water immediately prior to application to the gel. The gel is then agitated vigorously to allow the dye to penetrate the gel matrix before it precipitates in the staining container and on the gel surface. Gels are briefly rinsed to remove surface precipitates prior to imaging. Nile red–detergent–protein complexes are relatively photolabile, thus gel photography and imaging must be done efficiently or else signal degradation occurs. Stained gels can be imaged using 300 nm transillumination and Polaroid film or CCD-camera based systems. The sensitivity obtained with Nile red staining is between that obtained with Coomassie Brilliant Blue and that obtained with the SYPRO dyes.

4.6.3 ANS

ANS (1-anilinonaphthalene-8-sulfonate) has also been used as a fluorescent protein stain for SDS gels, with sensitivity near that of Coomassie Brilliant Blue (Aragay *et al.*, 1985). It has an apparent binding mode similar to that of Nile red. Dye–protein complexes have an excitation maximum ~370 nm and emission maximum ~480 nm.

4.6.4 Fluorescamine

Fluorescamine is an amine-reactive dye that is nonfluorescent when free in aqueous solution, but forms intensely fluorescent adducts upon reaction with protein primary amines. It has been used to stain proteins prior to SDS tube gel electrophoresis, with detection limits ~0.5 μg per band for most proteins and a linear

Table 4.3 Examples of gel imaging systems that can be used with some protein and nucleic acid stains.

	Laser scanners					*UV illumination, CCD or video cameras*					
Strain	*FMBIO*[a]	*STORM*[b]	*FluorImager*[b]	*LAS-1000*[c]	*FLA-2000*[c]	*Alpha-Imager*[d]	*Eagle Eye*[e]	*Arthur*[f]	*Fluor S*[g]	*EDAS*[h]	*FOTO/Analyst*[i]
Nucleic acid stains											
SYBR Green I stain	+	+	+	+	+	+	+	+	+	+	+
SYBR Green II stain	+	+	+	+	+	+	+	+	+	+	+
SYBR Gold stain	+	+	+	+	+	+	+	+	+	+	+
Ethidium bromide	+	+/−	+	+	+	+	+	+	+	+	+
Protein stains											
SYPRO Orange stain	−	+	+	+	+	+	+	+	+	+	?
SYPRO Red stain	+	+	−	?	?	+/−	+	+	+/−	+/−	?
Nile red	+	+/−	−	?	?	+	+	+	+	?	?

[a] Hitachi Software Engineering America, Ltd., South San Francisco, CA.
[b] Molecular Dynamics, Inc., Sunnyvale, CA.
[c] Fuji Photo Film Co., Ltd, Tokyo, Japan.
[d] Alpha Innotech Corp., San Leandro, CA.
[e] Stratagene Cloning Systems, Inc., La Jolla, CA.
[f] EG & G Wallac, Turku, Finland.
[g] Bio-Rad Laboratories, Hercules, CA.
[h] Eastman Kodak Co., Rochester, NY.
[i] FotoDyne, Inc., New Berlin, WI.

quantitation range extending up to 7–12.5 μg protein per band (Ragland *et al.*, 1974). Sensitivity is dependent mainly on the lysine (or free primary amine) content of the proteins. Bound fluorescamine has an excitation maximum at ~390 nm and emission maximum at ~475 nm, making it more useful with ultraviolet illuminators than laser scanners. However, the blue fluorescence is difficult to detect using most commercially available gel imaging systems.

4.7 PROTEIN LABELLING

Electrophoretically separated proteins can also be visualized using a reactive fluorophore label. Unlike nucleic acids, proteins contain abundant, highly reactive primary amines and thiols that can be coupled to a wide variety of fluorophores. Covalent labelling can be performed prior to electrophoresis, but in this case there is often extensive distortion of electrophoretic mobility (particularly if an amine-reactive reagent is used to label proteins prior to isoelectric focusing). This type of procedure has been used in capillary electrophoresis, for example, to label selectively phosphoserine moieties (Fadden & Haystead, 1995). Alternatively, samples in the gel itself can be directly labelled or samples can be labelled after the first dimension gel is run. Although these techniques are not presently in common use, because of the interest engendered by the proteomics field, it is likely such methods will be more commonly used in the future. The advantage of labelling before electrophoretic separation is that two biologically different samples labelled with two different fluorophores can be simultaneously electrophoresed, so that the relative abundance and distribution of a wide variety of proteins can be studied in 2-dimensional gels. The goal in such experiments is to select a pair of dyes that are spectrally distinct from one another, label proteins with approximately the same efficiency, cause approximately the same electrophoretic distortion to the modified protein, and can be visualized with a commercially available imaging system.

4.8 CONCLUSIONS

There are a wide variety of nucleic acid and protein gel stains available, with spectral properties matched to a large number of commercially available detection systems (Table 4.3). The choice of the best stain for a particular application depends in large part on the instrumentation available for visualizing the dye and on the results desired.

ACKNOWLEDGEMENTS

Thanks to Rabiya Tuma, Matthew Beaudet, Laurie Jones, Thomas Steinberg, and Xiaokui Jin for providing data for this chapter.

REFERENCES

Aragay A.M., Diaz P. & Daban J.-R. (1985) *Electrophoresis* **6**, 527–531.

Butler J.M., McCord B.R., Jung J.M., Wilson M.R., Budowle B. & Allen R.O. (1994) *J. Chromatogr. B* **658**, 271–280.

Chen S.-C. & Boynton A.L. (1995) *J. Chinese Biochem. Society* **32**, 1–7.

Clark S.M. & Mathies R.A. (1997) *Anal. Chem.* **68**, 1355–1363.

Daban J.R., Bartolomé S. & Samsó M. (1991) *Anal. Biochem.* **199**, 169–174.

Emanuel J.R., Damico C., Ahn S., Bautista D. & Costa J. (1996) *Diagn. Mol. Pathol.* **5**, 260–264.

Fadden P. & Haystead T.A.J. (1995) *Anal. Biochem.* **225**, 81–88.

Fong D., Burke J.P. & Chan M. M.-Y. (1997) *Biotechniques* **23**, 1029–1032.

Glazer A.N., Peck K. & Mathies R.A. (1990) *Proc. Natl. Acad. Sci. USA* **87**, 3851–3855.

Glazer A.N. & Rye H.S. (1992) *Nature* **359**, 859–861.

Hamby R.K. (1996) *Am. Biotechnol. Lab.* **14**, 12.

Hansen L.F., Jensen L.K. & Jacobsen J.P. (1996) *Nuc. Acids Res.* **24**, 859–867.

Harvey M.D., Bandillar D. & Banks P.R. (1998) *Electrophoresis*, in press.

Haugland R.P., Yue S., Millard P. & Roth B. (1995) US Patent No 5, 436,134.

Haugland R.P. (1996) *Handbook of Fluorescent Probes and Research Chemicals*, 6th edn, M.T.Z. Spence (ed). Molecular Probes, Oregon.

Haugland R.P., Singer V.L., Jones L.J. & Steinberg T.H. (1997) US Patent No. 5,616,502.

Hung S.-C., Ju J., Mathies R.A. & Glazer A.N. (1996) *Anal. Biochem.* **238**, 65–170.

Ju J., Ruan C., Fuller C.W., Glazer A.N. & Mathies R.A. (1995) *Proc. Natl. Acad. Sci. USA* **92**, 4347–4351.

Ju J., Glazer A.N. & Mathies R.A. (1996) *Nat. Med.* **2**, 246–249.

Kapuscinski J. (1995) *Biotech. Histochem.* **70**, 220–233.

Kiltie A.E. & Ryan A.J. (1997) *Nucleic Acids Res.* **25**, 2945–2946.

Kishida T., Tamaki Y., Kuroki K., Fukuda M. & Wang W. (1995) *Nippon Hoigaku Zasshi* **49**, 299–303.

Lauber W., Carroll J. & Duffin K. (1998) *Proceedings of the 2nd International Business Communications International Conference on Proteomics*. International Business Communications, Southborough, MA.

Law J.C., Facher E.A. & Deka A. (1996) *Anal. Biochem* **236**, 373–375.

LePecq J.-B. & Paoletti C. (1967) *J. Mol. Biol.* **27**, 87–106.

Madabhushi R.S., Vanier M., Dolnik V., Enad S., Barker D.L., Harris D.W. & Mansfield E.S. (1997) *Electrophoresis* **18**, 104–111.

Morin P.A. & Smith D.G. (1995) *Biotechniques* **19**, 223–228.

Oda R.P., Wick M.J., Rueckert L.-M., Lust J.A. & Landers J.P. (1996) *Electrophoresis* **17**, 1491–1498.

Ragland W.L., Pace J.L. & Kemper D.L. (1974) *Anal. Biochem.* **59**, 24–33.

Reinhardt C.G. & Krugh T.R. (1978) *Biochemistry* **17**, 4845–4854.

Reynolds J.A. & Tanford C. (1970a) *Proc. Natl. Acad. Sci. USA* **66**, 1002–1007.

Reynolds J.A. & Tanford C. (1970b) *J. Biol. Chem.* **245**, 5161–5165.

Ririe K.M., Rasmussen R.P. & Wittwer C.T. (1997) *Anal. Biochem.* **245**, 154–160.

Rye H.S., Quesada M.A., Peck K., Mathies R.A. & Glazer A.N. (1991) *Nucleic Acid Res.* **19**, 327–333.

Rye H.S., Yue S., Wemmer D.E., Quesada M.A., Haugland R.P., Mathies R.A. & Glazer A.N. (1992) *Nucl. Acids Res.* **20**, 2803–2812.

Rye H.S., Yue S., Quesada M.A., Haugland R.P., Mathies R.A. & Glazer A.N. (1993) *Methods Enzymol.* **217**, 414–431.

Schneeberger C., Speiser P., Kury F. & Zeillinger R. (1995) *PCR Methods Appl.* **4**, 234–238.

Sharp P.A., Sugden B. & Sambrook J. (1973) *Biochemistry* **12**, 3055.

Singer V.L., Jin X., Ryan D. & Yue, S. (1994) *Biomed. Products* **19**, 68–69.

Skeidsvoll J. & Ueland P.M. (1995) *Anal. Biochem.* **231**, 359–365.

Srinivasan K., Morris S.C., Girard J.E., Kline M.C. & Reeder D.J. (1993a) *Appl. Theor. Electrophor.* **3**, 235–239.

Srinivasan K., Girnard J.E., Williams P., Roby R.K., Weedn V.W., Morris S.C., Kline M.C. & Reeder D.J. (1993b) *J. Chromatogr. A* **652**, 83–91.

Steinberg T.H., White H.M. & Singer V.L. (1997) *Anal. Biochem.* **248**, 168–172.

Steinberg T.H., Jones L.J., Haugland R.P. & Singer V.L. (1996a) *Anal. Biochem.* **239**, 223–237.

Steinberg T.H., Haugland R.P. & Singer V.L. (1996b) *Anal. Biochem.* **239**, 238–245.

Stoner D.L., Browning C.K., Bulmer D.K., Ward T.E. & MacDonell M.T. (1996) *Appl. Environ. Microbiol.* **62**, 1969–1976.

Stothard J.R., Frame I.A. & Miles M.A. (1997) *Anal. Biochem.* **253**, 262–264.

Su S., Vivier R.G., Dickson M.C., Thomas N., Kendrick M.K., Williamson N.M., Anson J.G., Houston J.G. & Craig F.F. (1997) *Biotechniques* **22**, 1107–1113.

Takai H., Yamakawa H., Ohara O. & Sakaguicchi-Inoue J. (1997) *Biotechniques* **23**, 58–60.

Wittwer C.T., Herrmann M.G., Moss A.A. & Rasmussen R.P. (1997) *Biotechniques* **22**, 130–138.

Worley J., Lee S., Ma M.S., Eisenberg A., Chen H.-Y. & Mansfield E. (1997) *Biotechniques* **23**, 148–153.

Yasuda T., Takeshita H., Nakazato E., Nakajima T., Hosomi O., Nakashima Y. & Kishi K. (1998) *Anal. Biochem.* **255**, 274–276.

Yue S., Singer V.L., Roth B., Mozer T., Millard P., Jones L., Jin X. & Haugland R.P. (1997) U.S. Patent No. 5,658,751.

Zhu H., Clark S.M., Benson S.C., Rye H.S., Glazer A.N. & Mathies R.A. (1994) *Anal. Chem.* **66**, 1941–1948.

PART III

Using Optical Probes in Cells – Practicalities, Problems and Pitfalls

CHAPTER 5

Introducing and Calibrating Fluorescent Probes in Cells and Organelles

DAVID N. BOWSER, STEPHEN H. CODY, PHILIP N. DUBBIN & DAVID A. WILLIAMS

Confocal & Fluorescence Imaging Group, Department of Physiology, The University of Melbourne, Parkville, Victoria, Australia

5.1 INTRODUCTION

The last decade has seen the rapid escalation in interest in the regulation of intracellular processes by changes in both the levels and distribution of a number of physiologically important cations such as Ca^{2+}, H^+, Na^+, K^+ and Mg^{2+}. Much of this interest is directly related to improvements in both the methodologies of digital imaging microscopy (for review see Moore *et al.*, 1990; Tsien & Harootunian, 1990; Coppey Moisan *et al.*, 1994; Carrington *et al.*, 1995), and in the design and chemical synthesis of fluorescent probes for detection of specific intracellular cations (Grynkiewicz *et al.*, 1985; Minta *et al.*, 1989; Tsien 1992; Gonzalez & Tsien, 1997).

The ultimate aim in the use of an ion-sensitive fluorophore is to obtain a measurement of the average concentration or intracellular distribution of the ion in question, under conditions where the result itself is not influenced by the methodology or through the presence of the fluorophore. General problems encountered in minimizing the error in such measurements include: (1) cell damage or by-product liberation while introducing readily detected levels of the fluorophore, (2) high intracellular levels of fluorophore which buffer the concentration of the ion in question, (3) limiting the distribution of the probe to the intracellular sites of interest (e.g. cytosol or individual organelles) and (4) determining the effects of the intracellular environment (e.g. viscosity, binding) on the behaviour of the fluorophore.

Innumerable research articles detailing the use of ion sensitive fluorophores have been published. In 1990, a special issue of the journal *Cell Calcium* (Feb/Mar 1990) brought together a number of articles which emphasized the intricacies in making accurate intracellular ion measurements at high spatial and/or temporal resolution with calcium indicative fluorophores. Similar research articles have been published on the calibration of fluorophores for Na^+ (Negulescu *et al.*, 1990), Mg^{2+} (Jung *et al.*, 1996) and pH (Buckler & Vaughan-Jones, 1990; Bassnett *et al.*, 1990). Such papers use numerous cell types to describe a number of useful experimental paradigms for the calibration of fluorescent signals. It is not the aim of the present chapter to summarize all of this

FLUORESCENT AND LUMINESCENT PROBES, 2ND EDN
ISBN 0-12-447836-0

information, but rather to emphasize a number of general principles of loading and calibration.

An established area of optical methodology in physiological research is that of laser scanning confocal microscopy (for review see Pawley, 1995). This methodology removes most of the out-of-focus information from 2-dimensional images, and has allowed for improvements in spatial resolution and production of optical sections of living cells in isolation or within tissues. Coupled with specific fluorescent probes, confocal microscopy may provide for highly accurate spatial 2-D and 3-D distribution maps of ion concentrations in cells. Many of our projects have involved this combination and a number of the calibration issues that we will deal with are a result of the specifications of the hardware of the present laser scanning confocal microscopes. The utilization of argon- or argon/krypton-ion lasers in most commercially available systems limits excitation wavelengths to the major bands emitted by such lasers, generally 488 and 514 nm in argon-ion lasers and 488, 568 and 647 nm in argon/krypton-ion lasers. As such there has been frequent use of ion-specific fluorophores, which are structurally related to fluorescein or rhodamine and are excited in the visible wavelength spectrum. In particular the Ca^{2+} indicators fluo-3, Calcium Green and fura-red, and the SNARF and SNAFL series of pH indicators (Molecular Probes™) are being widely employed for quantitative measurements of cation concentrations with confocal microscopy.

Laser-scanning confocal microscopy has offered much in the study of cellular physiology and pathology. In motile cell systems, such as isolated muscle cells, the confounding effects of cell contraction (or movement) and out-of-focus information on fluorescence levels can be minimized through the ability to confine data acquisition to restricted and well-defined volumes within the cell. These volumes may be represented by complete 2-dimensional slices of the cell or by specific areas of interest within the cell, both of which have been demonstrated in some of the studies we will describe.

5.2 GENERAL PRINCIPLES OF THE LOADING PROCESS

The loading protocol used to introduce single or numerous fluorophores into cells should be specific for cell type and intracellular compartment to be examined. When examining the cytosolic concentration of a particular ion, all internalized indicator should be normally confined to the cytosol rather than allowed to accumulate in intracellular organelles such as mitochondria and nucleus. Of course there are experimental situations where the aim is to study the ion fluxes within an organelle system, in which case the loading protocol can be altered to optimize this possibility. Fluorophore concentrations should be restricted to levels which, although allowing for acceptable signal-to-noise and signal-to-autofluorescence ratio, minimize the potential for buffering of the target ion. A number of methods have been devised to introduce fluorophores into the internal environment of cells or tissues. The most widely used of these include: (1) incubation of cells (tissues) with esterified derivatives of fluorophores (the most commonly used technique), (2) pressure injection or ionophoresis of cells with microelectrodes containing acidic (charged) dyes, (3) transfection of cells with a gene for a fluorescent molecule, and (4) bathing cell suspensions with fluorophores and a chemical agent which transiently and reversibly increases the permeability of the cell membrane to low molecular weight substances.

5.2.1 Esterified Fluorophores

The factors which influence the dye levels and distribution that result from incubation with esterified dye derivatives can be divided into two main types: fixed (cell-specific) and variable (experimentally modifiable). Cell-specific factors include the surface area of the cell membrane and the location and amount of cellular esterases that carry out the de-esterification of internalized dye. It is these factors which may be the major determinant of the inefficient loading which has been long reported to occur in certain cell types such as polymorphonuclear leukocytes (neutrophils) (Scanlon *et al.*, 1987).

Incubation time and temperature, agitation of the incubation mixture, and initial dye concentration (external) and cell density are the experimental variables which influence the success of the loading process and should be modified to suit the cell type in use and the experimental requirements. In general, experiments which require kinetic measurements of changes in cytosolic ion concentrations in single cell preparations require short loading times (10–30 minutes), high (>30°C) temperatures, low dye concentrations and high cell densities. The latter two variables have a complex influence on internal dye concentrations, as has previously been described in detail (Moore *et al.*, 1990). In short, the combination of these factors controls, in a probabilistic fashion, the maximum potential number of dye molecules that associate with (and become internalized by) an individual cell. In some situations the ion concentrations of organelles such as the mitochondria, nucleus and endoplasmic reticulum are also of experimental interest, as may be deep lying cell layers in a multicellular preparation. These cases necessitate long loading times (>60 minutes), lower temperatures (<30°C), high dye concentrations and/or low cell densities.

Temperature has an obvious influence on the loading process and the distribution of fluorophore, particularly in tissue samples and intact organs such as whole muscles. High temperatures within the physiological range (30–37°C) induce elevated esterase enzyme activity. Internalized dye is likely to be rapidly cleaved in the first compartment (usually the cytosol) or cell it enters (if in a tissue preparation) and as a result the distribution of the dye is restricted. Lowering the temperature of the incubation mixture below this range has dramatic effects on enzymatic processes such as de-esterification (i.e. high Q10) but has much less influence on a physical process (low Q10) such as diffusion. Fluorophores are able to diffuse extensively before cleavage restricts their distribution. In practice we have found that it is essential that low-temperature incubation (5–10°C) should be followed by a period of incubation at higher temperature. This ensures that partially cleaved dye does not accumulate in cells or organelles as these moieties have been shown to introduce significant error into determination of intracellular Ca^{2+} concentrations (Luckoff, 1986; Highsmith *et al.*, 1986; Scanlon *et al.*, 1987). This loading paradigm has proved to be particularly useful for large multicellular structures such as tumour cell spheroids (Cody *et al.*, 1993), intact skeletal muscles (Fig. 5.1), whole-heart perfusions (Minamikawa *et al.*, 1997) and also in tissues which possess high levels of extracellular esterase activity (Williams and Fay, 1990; Gehring *et al.*, 1990). In addition, some recent studies have used a similar strategy specifically to enhance the loading of some indicators into intracellular compartments such as mitochondria (Trollinger *et al.*, 1997).

5.2.2 Use of non-esterified fluorophores

The ease of use of esterified derivatives for introducing fluorophores into the majority of cell types would seem to make alternative methods unnecessary. However, the often significant fluorescence contribution of both dye within intracellular organelles (Allen *et al.*, 1992; Griffiths *et al.*, 1997a) and uncleaved dye (Luckoff, 1986; Highsmith *et al.*, 1986; Scanlon *et al.*, 1987) to the measured fluorescent signals, and the possibility of dye binding to intracellular sites within cells (Baylor & Hollingworth, 1988), have been the impetus for the formulation of alternatives which allow introduction of cell impermeant, charged forms directly into cells.

Several of these methods, such as pressure injection and ionophoresis, require the use of fine-tipped (<1–2 μm diameter) microelectrodes to pierce the plasmalemma of the target cell. Positive pressure or the passage of electric current (generally negative) is

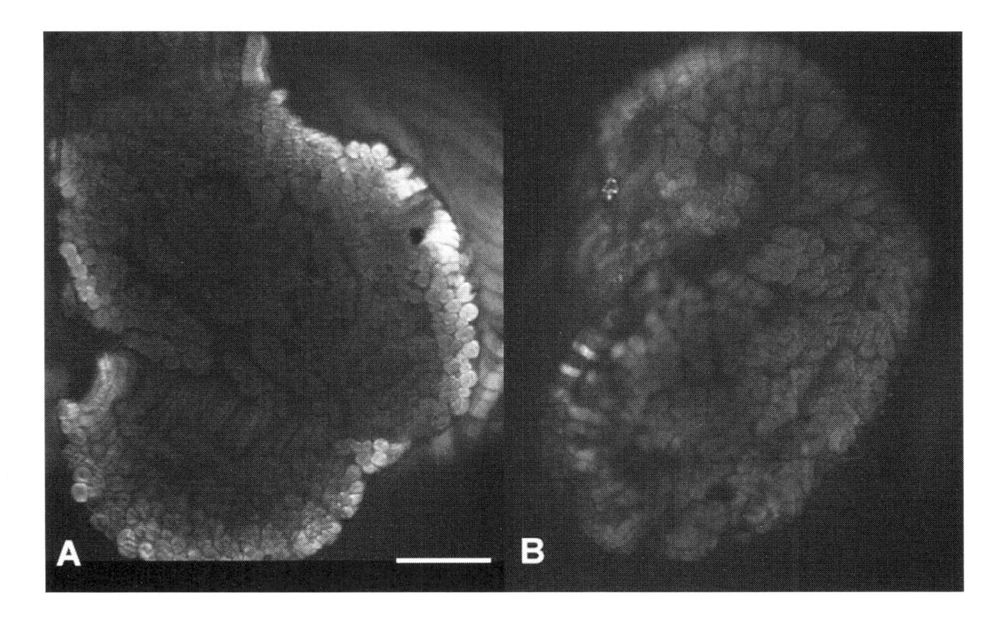

Figure 5.1 A comparison of the loading of the pH-sensitive fluorophore SNARF at 22°C (panel A) and 5–10°C (panel B) in an intact skeletal muscle. In the high-temperature-loaded skeletal muscle, fluorescence is only imaged in the outer fibres. Incubation at lower temperatures followed by a period of higher temperature (for complete cleavage of the AM group) enabled the detection of fluorescence in the inner muscle fibres. Scale bar: 250 μm.

then used to expel fluorophore into the cytosol. This process necessarily results in a small degree of cell damage which is often reversible, and where multicellular tissues are used may necessitate multiple impalements to produce enough loaded cells for fluorescence measurements. However, the use of ionophoresis electrodes for cell loading provides a method for accurately calculating the level of fluorophore introduced into target cells. This method, described by Williams *et al.* (1990b), requires knowledge of the pulse amplitude and duration of the ionophoresis protocol, and the volume of the injected cell. The tip diameter of glass microelectrodes varies greatly and as such it was essential to eject fura-2 from a standard microelectrode (e.g. one of a batch from a program-controlled microelectrode puller) into a solution droplet mimicking the ionic composition of the cytosol to calibrate the expulsion rate of electrodes. The fura-2 concentration of the droplets was determined from a standard curve of solution absorption measurements made spectrophotometrically at 360 nm. The fura-2 concentration could then be determined for a range of ionophoresis protocols and allowed formulation of a calibration curve relating total charge passed by an electrode (pulse amplitude × duration) to number of moles of ejected fura-2. Values were then expressed in molar terms by using an estimation of the volume of the target cell, or even more accurately, by using the cytosolic volume fraction of the cell (which was usually the target of the electrode).

Many electrophysiologists use patch microelectrodes to make recordings of whole cell currents while perfusing the intracellular environment with fluorophores contained in the electrode. This coupling of methods has resulted in elegant studies of ionic currents and ion fluxes in many different cell types (Neher & Almers, 1986; Barcenas-Ruiz *et al.*, 1987; Cannell *et al.*, 1987). The major concern that is raised about measurements of intracellular ion concentrations made under these circumstances is that the cytosolic contents may be modified through dilution by the significantly larger volume of the patch electrode.

5.2.3 Chemical methods for reversible permeability

Chemical agents have also been used to increase the permeability of cell membranes to small molecular weight substances such as the cation-sensitive fluorophores. These methods have similarities in that they induce a state of enhanced cell membrane permeability through incubation of cells (or tissues) with certain chemical agents. These treatments include low pH solutions (acid loading; Bush & Jones, 1988), and solutions with high ATP (Rozengurt & Heppel, 1975).

The use of the detergent digitonin to enhance the uptake of dyes by plant cells has been described (Timmers *et al.*, 1991). Incubation of carrot somatic embryos with fluo-3/AM alone, or with fluo-3 (acidic form) in the presence of either low pH (acid loading) or the dispersant Pluronic F-127 (Molecular Probes), resulted in little appreciable dye internalization. However, if a low concentration (0.025–0.1%, w/v) of digitonin was added to the incubation medium containing fluo-3, significant levels of intracellular fluorescence were recorded in the cytosol of embryo cells. Importantly, this treatment did not interfere with the normal development of the embryos, indicating that the degree of membrane damage was low or was transient.

Perhaps the least utilized of the chemical methods for introducing low molecular weight substances into cells is a method of reversible hyperpermeabilization (Sutherland *et al.*, 1980). This technique utilizes the transient removal of divalent cations to make the external plasmalemma leaky during which time constituents of the external medium (including introduced species such as fluorophores) become internalized in the treated tissue. Although this procedure was adopted to introduce the photoprotein aequorin into smooth muscle cells and tissue (DeFeo & Morgan, 1985), there are no reports of its use with fluorescent probes. In our preliminary studies the technique was found to be well suited to loading tissue samples with the de-esterified (acidic) forms of Ca^{2+}-sensitive indicators.

5.2.4 Transfection of cDNA for fluorescent proteins

A large proportion of the future of fluorescence technology lies in the selective expression of fluorescent proteins, engineered to target specific intracellular organelles, ion channels or simply to indicate gene expression. The best characterized is the green fluorescent protein (GFP) isolated from the jellyfish *Aequorea victoria*, and cloned by Prasher *et al.* (1992). Excitation of wild-type GFP at 395 nm produces green emission, while selected point mutations of the complementary DNA (cDNA) result in better spectral properties. S65T is a particularly useful mutation; it has spectral characteristics that are particularly useful for confocal microscopy and fluorescence-activated cell sorting (FACS), with excitation and emission maxima of approximately 490 and 510 nm, respectively (Cubitt *et al.*, 1995). The cDNA sequence for GFP can be readily concatenated with a targeting signal for an organelle system such as the mitochondria, endoplasmic reticulum (ER) or nuclei, and included in a viral vector (Rizzuto *et al.*, 1995, 1996). Alternatively, the cDNA sequence can be concatenated with sequences for ion channels (Trouet *et al.*, 1997) or receptors (Htun *et al.*, 1996).

While expression of GFP enables visualization of cellular structures, it does not provide us with information regarding ion concentrations. Such measurements can be made with the Ca^{2+} sensitive photoprotein aequorin targeted in a similar fashion to GFP. Thus, measurements of mitochondrial, ER or nuclear Ca^{2+} can be made with targeted recombinant aequorin (Rizzuto *et al.*, 1992, 1994). Unfortunately, it requires coelenterazine as a cofactor, and is irreversibly consumed upon binding Ca^{2+}. These limitations have been largely overcome by a new generation of fluorescent indicator that combines the ion sensitivity of fluorescent chelators (e.g. fluo-3, fura-2) with the targetability of aequorin. These so-called 'chameleons' are an engineered protein sequence consisting of a cyan-emitting GFP mutant, calmodulin, the calmodulin-binding protein, and a yellow-emitting GFP mutant; forming a 'dumbbell' arrangement, with GFP mutants at either end. Upon binding calcium, the conformational change in calmodulin brings the GFP mutants closer together, inducing fluorescence resonance energy transfer (FRET) from the cyan- to the yellow-emitting GFP (Heim & Tsien, 1996; Miyawaki *et al.*, 1997). Such technology has obvious applications for the measurement of ionic concentrations within organelles, but may also allow visualization of local Ca^{2+} gradients in proximity to ion channels.

5.3 GENERAL PRINCIPLES OF THE CALIBRATION PROCESS

Calibration of intracellular signals of ion-sensitive fluorophores has been the subject of a number of research articles (Williams & Fay, 1990; Groden *et al.*, 1991; Owen, 1991; Uto *et al.*, 1991; Lipp & Niggli, 1993; Szmacinski & Lakowicz, 1995; Henke *et al.*, 1996). Most of the problems encountered in the calibration process result from dye that is internalized by cells but which exhibits properties that differ from those of the acidic (ionized) form under defined ionic conditions. The major conditions which have been proposed to alter the intracellular behaviour of fluorescent dyes include viscosity (Poenie, 1990) and intracellular binding of internalized dye (Baylor & Hollingworth, 1988; Blatter & Wier, 1990).

Techniques which attempt to define the intracellular fluorescence of an indicator directly have the greatest potential for producing accurate measurements of ionic changes in experimental situations. These techniques are characteristically difficult and require well-defined buffering systems for the specific ion, and solutions containing mixtures of ionophores which allow equilibration of the intracellular and extracellular pools of the required ions. Ionomycin (Calbiochem), a cation ionophore with a high specificity for Ca^{2+} over other ions (i.e. Mg^{2+}, K^{+} and Na^{+}), is often employed for equilibration of Ca^{2+} pools, but the Ca^{2+}-transporting ability exhibits strong pH sensitivity (Lui and Hermann, 1978). As such it is essential to elevate significantly the pH of the medium bathing the cells or tissue (pH 8–10) to facilitate Ca^{2+}-equilibration, but this is not always possible given the constraints of many experiments. Alternative pH-insensitive ionophores include A23187 (Calbiochem), which exhibits autofluorescence at the wavelength combinations employed with fura-2, and the non-fluorescent, brominated derivative Br-A23187. This latter ionophore has found increased usage with the UV-excitable fluorophores fura-2 and Indo-1, whereas A23187 is suitable for Ca^{2+}-equilibration of cells loaded with the dyes excitable in the visible spectrum (e.g. fluo-3, Calcium Green and fura-red).

A useful method for the *in vivo* calibration of fura-2, involving the construction of a calibration curve from intracellular fluorescence measurements, has been described for sections of isolated mammalian skeletal muscle fibres (Bakker *et al.*, 1993). Fibres were maintained under paraffin oil and exposed to a range of mixtures of strongly Ca^{2+}-buffered solutions of a known free $[Ca^{2+}]$, and concentrations of Mg^{2+}, K^{+}, H^{+} and organic anions that approximated those of the intact fibre. Calcium concentration was buffered with either of the chelators EGTA and BAPTA, the latter constituent being used because of its insensitivity to pH in the physiological range.

Small droplets of the Ca^{2+}-buffered solution along with fura-2 free acid (100 μM) and saponin (100 mg ml^{-1}) were applied (pressure ejection from a patch pipette) under oil directly to the fibre bundle. Saponin permeabilized the sarcolemma, allowing equilibration of the bathing solution with the cytosol of the fibres. The bundles were equilibrated in test solutions until the fibres fluoresced uniformly at a stable intensity level (usually 15 minutes). Paraffin oil restricted the leakage of fura-2 from the bundle, even when fibres went into supercontracture at high Ca^{2+} concentrations. The fluorescence was recorded following excitation at both 350 and 380 nm, and calibration curves of the fluorescence ratio (350/380 nm) and Ca^{2+} concentration were derived with a non-linear curve-fitting routine employing a least sum of squares procedure. Values for the K_d of fura-2 resulting from the *in vivo* technique were identical to those found *in vitro* in the same Ca^{2+}-buffered solutions. However, the constants that are normally employed in the calibration of fura-2 were all found to be lower in the muscle fibres than the *in vitro* values.

R_{max} (14.6 0.05%) and R_{min} (9.4 0.15%) were both lowered by a similar amount as shown previously in smooth muscle cells (Williams *et al.*, 1985); this emphasizes the importance of employing the most accurate constants for the calibration procedure.

Where it is not possible to implement techniques such as this, a comprehensive *in vitro* calibration process is still desirable as this will allow determination of the dissociation constant of the dye–ion complex, and calibration parameters (R_{max} and R_{min}), under the conditions that will be employed experimentally with cell or tissue preparations. Instead, many research groups simply adopt the values presented in the initial description of fura-2 characteristics (Grynkiewicz *et al.*, 1985). This represents poor experimental practice as these values were not meant to be used under all experimental conditions in any cell or tissue type.

5.4 PUTTING PRINCIPLES INTO PRACTICE

To cover the essential issues relevant to the introduction and calibration of all the fluorescent probes in living cells and tissues is an onerous task. The list of available fluorophores is immense and is growing all the time (consider the progressive growth in size of the Molecular Probes™ Catalogue). In an effort to cover the most important issues, we will remain within our direct area of expertise and summarize the major projects completed over the last decade, while discussing the major difficulties encountered with the use of fluorescent probes and detailing the solutions devised to alleviate or minimize these problems.

5.4.1 Spontaneous calcium changes in isolated cardiac myocytes

We have sought to investigate the properties of propagated spontaneous calcium release (SCR) with laser scanning confocal microscopy by visualizing the intracellular fluorescence fluctuations of a visible wavelength Ca^{2+}-indicator, fluo-3 (Minta *et al.*, 1989). The potential advantages of this combination include quantification of Ca^{2+} wave propagation rates in single cardiac myocytes and the potential for lengthy observations of spontaneous calcium release (SCR), propagation, and interaction in contracting cardiomyocytes.

Suspensions of cardiomyocytes prepared by a standard enzymatic technique from rat left ventricles were loaded with fluo-3 by incubation with the esterified derivative (fluo-3/AM) at 30°C for 30 minutes. Individual cells were viewed with a laser scanning confocal microscope (BioRad MRC-500, -600 or -1000, Oxfordshire, UK) coupled to an inverted Olympus IMT-2 microscope.

One major advantage of working with contractile cells is that there is a readily measured physiological parameter – contraction (time and magnitude) – that allows direct determination of any deleterious effects that may result from the presence of the fluorophore or its by-products within cells. Analysis of cell shortening following electrical stimulation of fluo-3-containing cells indicated that the isotonic contractile kinetics (maximum cell shortening, contraction time, maximum rate of shortening and lengthening) were indistinguishable from those measured in similar cells which did not contain the fluorophore. This strongly suggests that the levels of internalized fluorophore employed in this type of study had little effect, through direct Ca^{2+}-buffering or by-product generation, on the physiological parameters (contraction or Ca^{2+} levels) investigated in these cells.

Spontaneously contracting cells exhibited areas of high intensity fluorescence, which propagated throughout the cell in unison with the localized bands of contraction. An example of this type of activity can be seen in the sequence of images displayed in Fig. 5.2. Calcium-dependent fluorescence originated at a localized site (upper left) of the cell and gradually propagated throughout the cell. The circular front of the fluorescence wave indicates that the release and diffusion processes were equal in all directions and generally not affected by the different intracellular structures encountered along and across the cell. As we have previously described (Williams, 1990; Williams *et al.*, 1992), the fluorescence bands are of limited propagation velocity (50–150 µm s^{-1}), are not subject to contraction or movement artifacts, and are exclusively due to localized Ca^{2+} changes which can be attributed to a spontaneous and propagated Ca^{2+}-release process.

The result of one of the control experiments performed to determine how changes in cell geometry may affect fluorescence intensity is shown in Fig. 5.3. This series shows the fluorescence of a spontaneously contracting cardiomyocyte loaded with a recently released Ca^{2+}-sensitive fluorophore fura-red (Molecular Probes). When excited at 488 nm this fluorophore responds to Ca^{2+} increases by exhibiting decreased fluorescence. As a result we would predict that Ca^{2+} waves in contracting cells should be accompanied by bands of reduced fluorescence propagating along the cell. This is unless the cell contraction resulting from the contraction band is sufficient to cause a large local volume increase and, as a result, a localized increase in fluorescence intensity. As shown in Fig. 5.3 a dark band, again with a circular wave-front, originates locally and passes along this cell. It is evident from this behaviour that

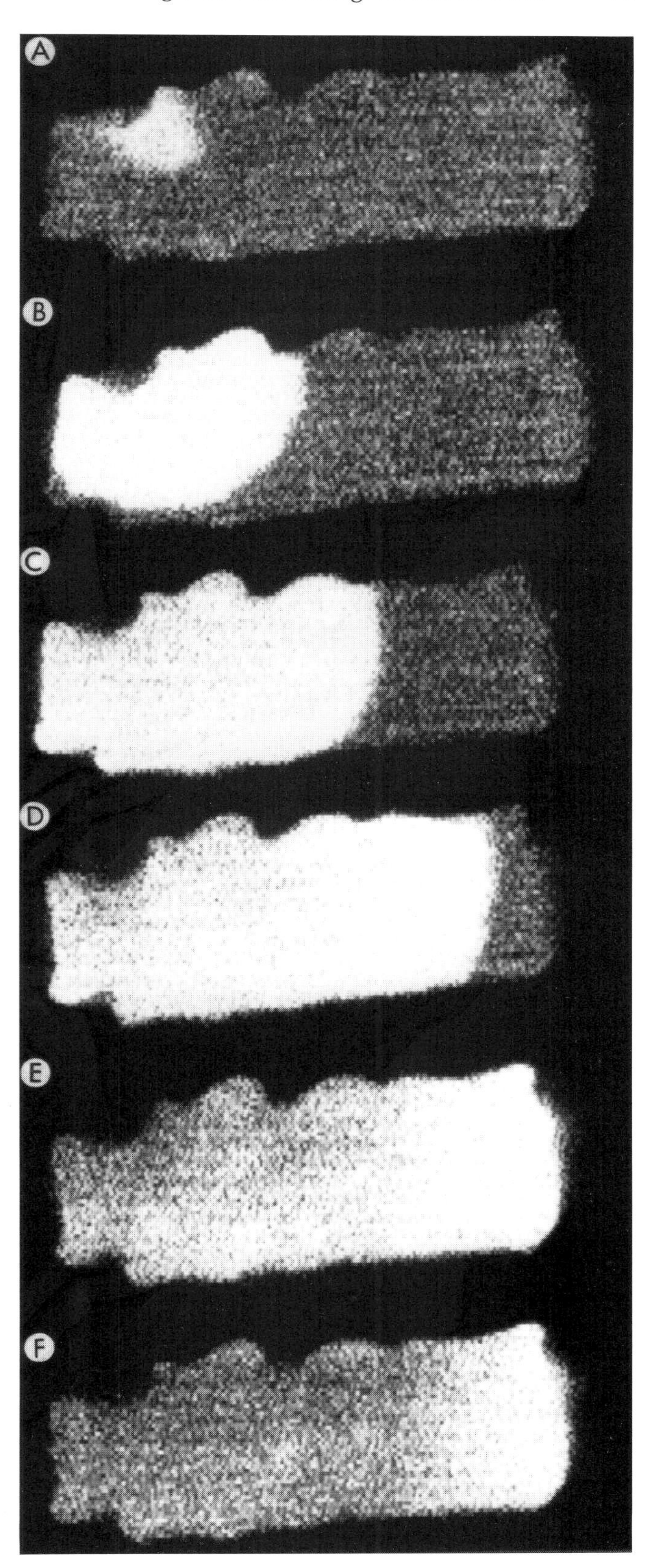

Figure 5.2 Changes in fluorescence intensity in a single spontaneously contracting cardiomyocyte loaded with fluo-3. A broad band of high fluorescence intensity propagated along the cell from upper left to right. Initial cell length (panel A): 100 μm.

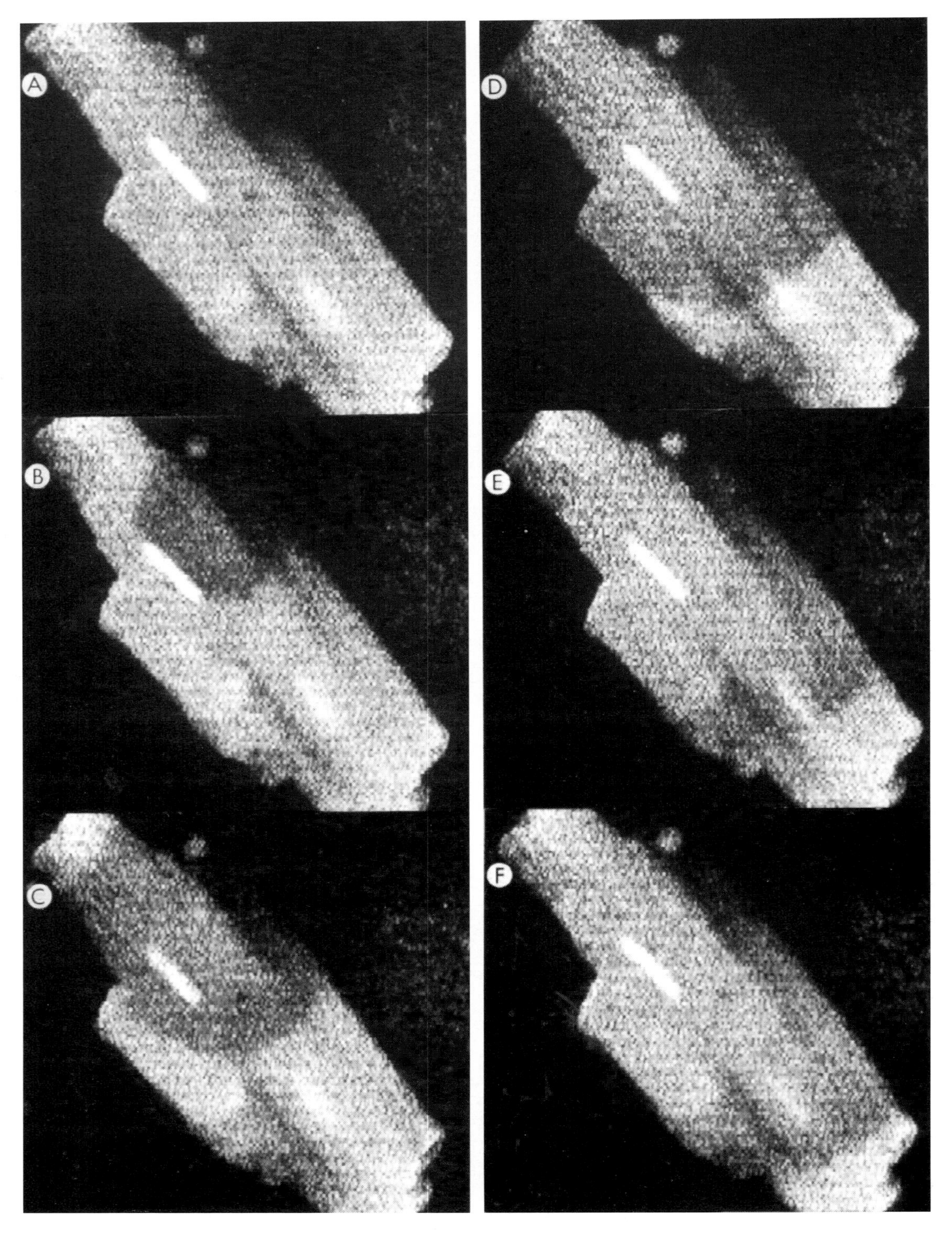

Figure 5.3 The change in intensity of fura-red fluorescence in an isolated spontaneously contracting cardiac cell. With this fluorophore areas of diminished intensity reflect local increases in Ca^{2+} level, which clearly propagate throughout the cell. Initial cell length (panel A): 105 μm.

the increases in intensity recorded under similar conditions with fluo-3 (see Fig. 5.2) exclusively represent Ca^{2+} fluctuations.

Accurate calibration of the fluctuations in fluorescence intensity of internalized fluo-3 in terms of ionized Ca^{2+} levels presented a number of challenges that are still common in the use of a fluorophore that does not undergo a significant Ca^{2+}-dependent change in spectral characteristics. Variation of the absolute concentration of intracellular fluorophore within cells or organelles, or as a function of time, could produce intensity variations that could be erroneously interpreted as ion distribution differences. This was particularly evident in cells which exhibited large regional variations in volume (e.g. flattened cells which taper at the edges), cells with large volumes occupied by intracellular organelles, or cells which changed shape during the activity of interest (e.g. contractile cells).

Calibration of fluo-3 fluorescence intensity in cardiac myocytes was performed in several ways. The simplest calibration resulted from the determination of the ratio of maximum (F_{max}) and minimum (F_{min}) fluorescence intensity levels for the fluorophore within individual cells. Experimental fluorescence levels could be scaled relative to these fluorescence limits to allow for the determination of [Ca^{2+}] at any point in time, as has previously been described in detail for quin2 (Rink & Pozzan, 1985). The resulting value for F_{max}/F_{min} of 4.95 ± 0.04 (n=14 individual cells) for cardiac cells varied little from that obtained with fluo-3 acid in Ca^{2+}-containing solutions (5.10 ± 0.17, n=3) under the same experimental conditions. Therefore, these data indicated that internalized fluo-3 exhibited similar properties to those of the indicator *in vitro*.

Alternatively, cells were loaded with both fluo-3 and a second Ca^{2+}-sensitive fluorophore fura-2 to allow for implementation of a cross-calibration technique described by Williams (1990). The direct Ca^{2+}-calibration of fura-2 ratio images, as has been described in detail elsewhere (Moore *et al.*, 1990), provided a baseline average [Ca^{2+}] for individual non-stimulated cells upon which changes in fluo-3 fluorescence intensity could be expressed as [Ca^{2+}] changes. This technique also requires knowledge of the fluorescence enhancement of fluo-3 upon Ca^{2+} binding (F_{max}/F_{min}) for the experimental system in use.

More recently, we have demonstrated the use of fura-red to monitor ratiometrically cytosolic [Ca^{2+}]. The unique excitation spectra for fura-red allows dual-excitation at 458 nm (close to the isosbestic point) and 488 nm, while collecting the emission fluorescence above 585 nm. Calculation of the 458/488 nm emission ratio was conducted in cells exposed to ionophore (Br-A23187, Calbiochem) in solutions of known [Ca^{2+}], and appropriate calibration curves were constructed in a similar method to the calibration of fura-2. The advantage of fura-red is that its excitation frequencies are in the visible spectrum, allowing it to be imaged with laser scanning confocal microscopy using argon-ion lasers. This has the added benefit of allowing the mapping of intracellular variations in [Ca^{2+}].

These calibration methods have allowed us to make accurate determinations of Ca^{2+} levels in many spontaneously contracting cardiac cells, as described in detail in previous reports (Williams, 1990; Williams *et al.*, 1992; Hayes *et al.*, 1996). Localized changes in Ca^{2+} concentrations from resting levels approximately 200 nM to peak values of between 800 nM and 1.5 μM were responsible for the contraction bands evident in the cells we investigated. These procedures have increased the certainty with which quantitative information can be derived from single wavelength indicators such as fluo-3 and Calcium Green, and other indicators (e.g. BCECF) which may be confined to a single wavelength mode of data acquisition by the constraints of the standard hardware configurations.

Ca^{2+} oscillations occur in unstimulated cardiac cells from many species, with physiological Ca^{2+} levels in the bathing solutions, and with minimal experimental perturbations of the Ca^{2+} loading state of the cell. Such a powerful and fundamental phenomenon occurring under relatively physiological conditions is clearly relevant to an understanding of the generation of cardiac arrhythmias. The techniques described may allow the necessary investigation.

5.4.2 Ionized calcium in living plant cells

With the realization of the importance of calcium ions in mediation of plant cellular responses (for review, see Hepler & Wayne, 1985), there is impetus to improve existing technologies to enable monitoring of intracellular calcium dynamics, especially in response to growth hormones and other physiological stimuli such as gravity and light. However, plant cells are difficult to load with fluorescent calcium indicators, most likely as a result of extracellular hydrolysis of the esterified derivatives of the fluorophores (Cork, 1985), or the incomplete hydrolysis of the internalized dyes (Brownlee & Wood, 1986). Problems of a different kind arise with both intact plant (and animal) tissues, involving uncertainty in analysing signals emanating from individual cells within a given image plane because of contaminating out-of-focus information. To circumvent some of the technical difficulties evident with plant cells we have utilized fluo-3 and a laser scanning confocal microscope to study the role of cations (Ca^{2+} and H^{+}) in plant cell physiology. This combination enables these processes to be studied at the cellular level within intact plant tissue preparations, which is an important consideration given that few functional single cell preparations have been isolated from plant tissues.

To maximize the access of fluorescent esters to cells within plant tissues we found it necessary to remove the barrier provided by the waxy outer cuticle found in intact tissue samples such as coleoptile tips. This was most efficiently achieved by scraping the tissue surface with sharp blades or by applying adhesive tape to, and removing it from, the surface of the coleoptile. The latter alternative enhanced loading in larger areas of the tissue, with scraping alone resulting in significant internalization of dyes in the localized areas of cuticle removal. Rinsing tissue samples in fresh tap water (40 μM [Ca^{2+}]) terminated dye access. When internalized dye was isolated from plant tissues by mechanical homogenization it was invariably found to be fully Ca^{2+}-sensitive (Scanlon *et al.*, 1987), indicating that plant cells, once given access to the esterified derivatives of ion-sensitive indicators, are capable of complete intracellular cleavage of the molecules.

A low magnification image (Fig. 5.4(A)) of a coleoptile slice following illumination by conventional epifluorescence microscopy shows the presence of diffuse fluorescence with little indication of cellular detail. The autofluorescence of the two major vascular bundles is the major discernible feature. This image illustrates one of the optical problems that make quantitative evaluation of cellular fluorescence within plant cells difficult, even when successful dye loading has been achieved. Contaminating fluorescence, which emanates from focal planes above and below the plane of interest, impinges upon the image plane and is a problem in all conventional wide-field images, particularly in large 3-D structures such as tissue samples.

An image of the same coleoptile slice, acquired with the laser scanning confocal microscope employing the same objective (see Fig. 5.4(B)), shows a marked improvement in resolution of individual cells. Epidermal layers and the cortical cell mass are clearly visible. Spatial heterogeneity in intracellular fluorescence is clearly evident in higher magnification images (see Fig. 5.5). Vacuoles are discernible but contained little fluorescence, suggesting that the majority of internalized fluo-3 was cleaved and retained by the small volume of cytosol surrounding the large vacuoles before it was able to diffuse into the vacuolar compartment. The slow leakage, or transport of dye into vacuoles via non-specific anion transport mechanisms (Malgaroli *et al.*, 1987), ensures that changes in fluorescence recorded within 60 minutes of loading exclusively reflect changes in cytosolic [Ca^{2+}].

It is clear that the many problems associated with the imaging of fluorescent probes to make accurate measurements of intracellular ions in plant tissues can be effectively eliminated by: (1) removing the waxy cuticle that presents the major barrier for dye internalization in plant tissues; and (2) employing confocal microscopy to reduce the 3-D spread of contaminating out-of-focus information in image planes. As a result we were able to show for the first time with high spatial resolution the elevations in cytosolic Ca^{2+} of maize coleoptiles in response to application of ionophores and plant growth hormones (Williams *et al.*, 1990a). Ca^{2+} changes were calibrated from the fluorescence changes using the same techniques that we have described for cardiomyocytes. Using the techniques and paradigms described here we were also able to visualize changes in [Ca^{2+}] and pH in epidermal and cortical cells as a result of

(A)

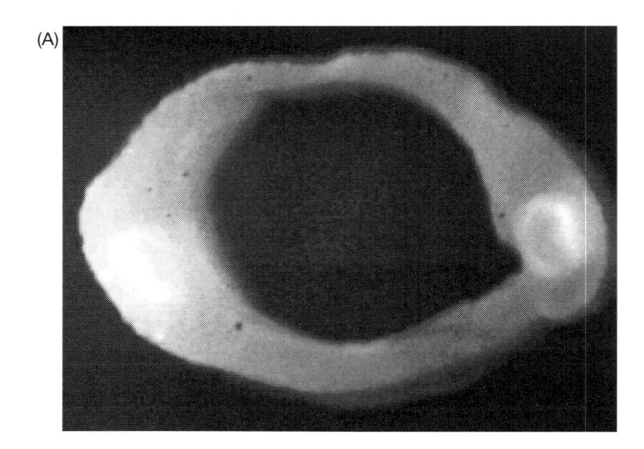

(B)

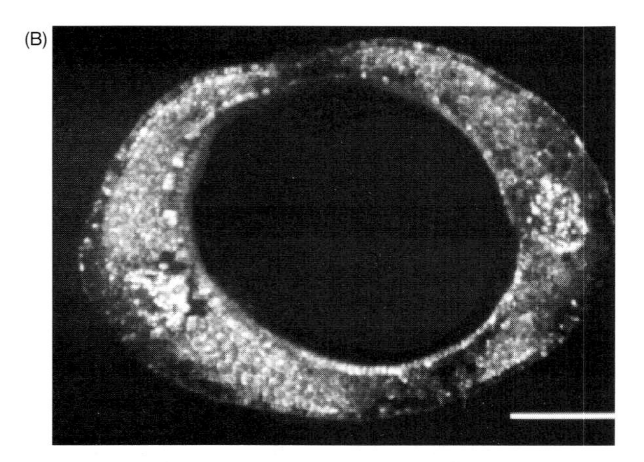

Figure 5.4 (A) Image of the fluorescence of a fluo-3-loaded coleoptile tip as captured by a silicon-intensified-target (SIT) camera following standard epifluorescence illumination.
(B) The same coleoptile tip viewed immediately after with laser scanning confocal microscopy. Both images are the average of eight consecutive video frames. Scale bar: 250 μm. Reproduced by permission of *Cell Calcium*.

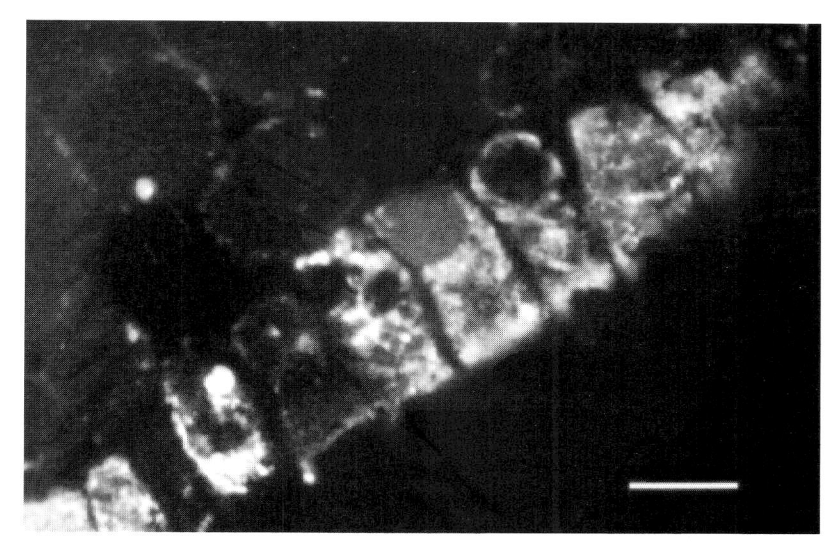

Figure 5.5 High magnification confocal image of a small segment of a fluo-3-loaded coleoptile tip. The intra- and extracellular fluorescence heterogeneity is readily apparent. Scale bar: 25 μm. Reproduced by permission of *Cell Calcium*.

physiological stimuli such as gravitropism or phototropism (Gehring *et al.*, 1990).

5.4.3 Localization of cell nuclei

Fluorescent molecules are commonly used to elucidate specific features of cell structure. Apart from their use as labels for numerous monoclonal and polyclonal antibodies, there are fluorophores which associate preferentially with specific classes of molecules within cells. We have frequently used ethidium bromide, which intercalates with the double-stranded conformation of DNA and also associates with single-stranded RNA, to investigate the structure, number and position of cell nuclei in both isolated cardiac cells, and single muscle fibres from normal and diseased (dystrophic) skeletal muscles (Williams *et al.*, 1993). This fluorophore, although a polar molecule, has significant cell membrane permeability and, following a short (10 minutes) incubation period, it quickly enters both cell types readily associating with intracellular nucleic acids.

The images shown in Fig. 5.6 give examples of the two most frequently occurring patterns of nuclear distribution in rat cardiomyocytes. A small number of cells possessed a single centrally placed nucleus (Fig. 5.6(A)), while the majority of individual cells (>90%) had two nuclei placed one at each end of the cell (Fig. 5.6(B)). The numerous strands of mitochondria were also evident in both cell images and are due to the fluorescent staining of the mitochondrial RNA content.

Whereas nuclei are generally distributed in a peripheral spiral in adult skeletal fibres, the presence of centrally located nuclei is thought to be indicative of a muscle fibre that has regenerated following a damage/degeneration sequence (Harris & Johnson, 1978). In muscular dystrophy there is a large turnover of skeletal muscle with continual degeneration and regeneration of individual fibres. In studies of single muscle fibres from the dystrophic *mdx* mouse we have been able to identify the nuclear distribution pattern in the same fibres from which contractile and morphological properties were measured (Williams *et al.*, 1992; Head *et al.*, 1992). By doing so we sought to correlate function and the regenerative status (i.e. original or regenerated) of the fibres.

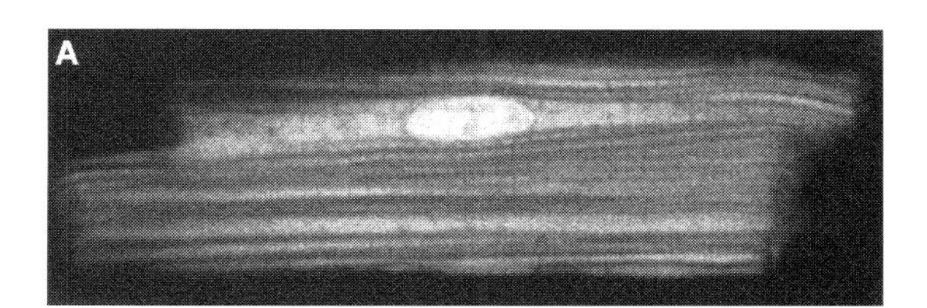

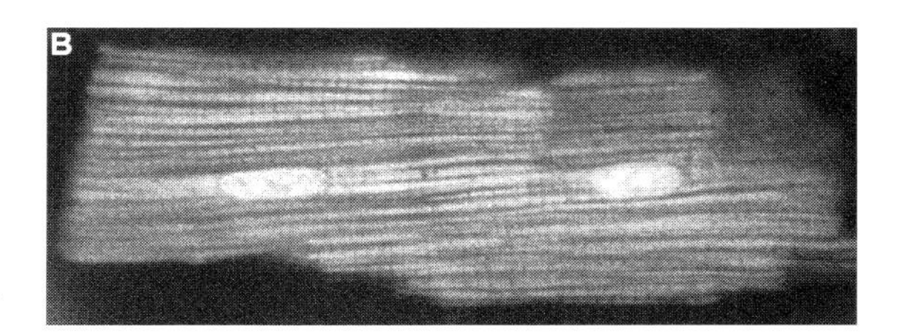

Figure 5.6 Laser scanning confocal microscopic images of individual cardiomyocytes following a 10 minute incubation with ethidium bromide. A small percentage (<10%) of all cells possessed a single, centrally placed nucleus (A), while the majority (90%) of cells were binucleate (B). Cell lengths: (A) 95 μm; (B) 150 μm.

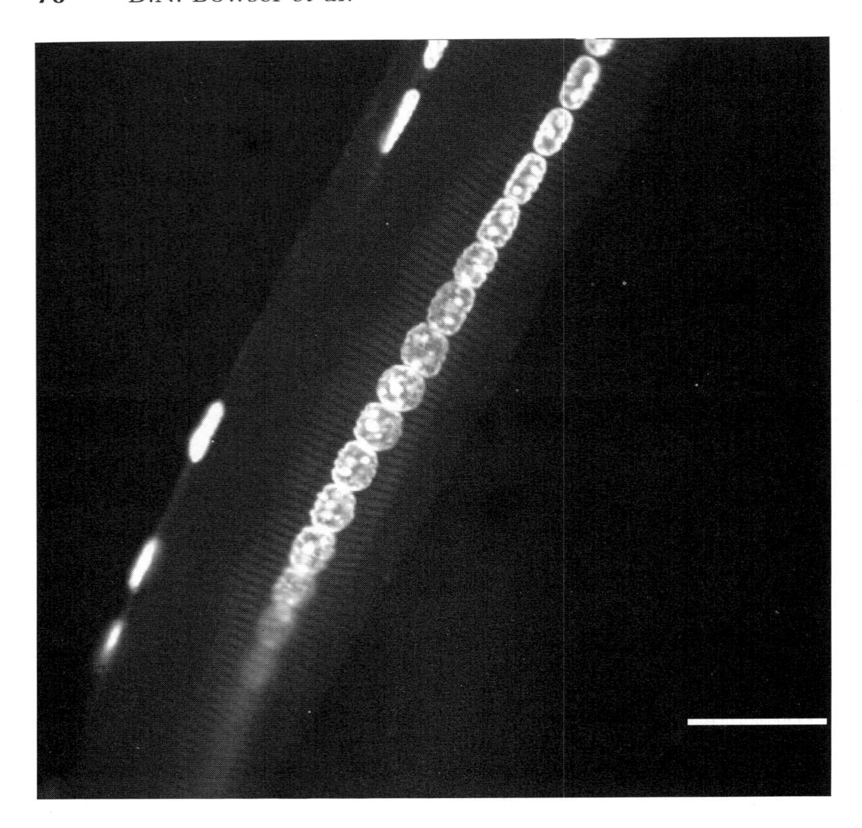

Figure 5.7 A single, enzymatically isolated skeletal muscle fibre from a dystrophic mdx mouse as stained with ethidium bromide. A long chain of centrally located nuclei is evident, as are a few peripheral nuclei. Scale bar: 25 μm.

Figure 5.7 illustrates the nuclear distribution pattern of one such fibre and shows a distinctive strand of nuclei placed centrally within the fibre. Rather than finding just the two distribution patterns for nuclei – central strands (as illustrated) and peripheral spirals (not shown) – we found a myriad of combinations of these two patterns. These observations suggest that dystrophic fibres do not necessarily undergo complete degeneration following damage but instead are capable of repeated local repair processes, leading to local variations in the nuclear distribution within an individual fibre. This has important implications for interpretation of the disease aetiology and is the subject of further investigation.

5.4.4 Localization and function of cellular mitochondria

Fluorescence techniques have been applied to investigations of cellular mitochondria, particularly in examining mitochondrial membrane potential, distribution and [Ca^{2+}]. These techniques have been used in our investigations into the role cardiac mitochondria play in the development and propagation of spontaneous calcium waves, as described earlier.

Most *in vivo* investigations of mitochondrial calcium have used fura-2 (Steinberg *et al.*, 1987; Gunter *et al.*, 1988) or indo-1 (Miyata *et al.*, 1991; Di Lisa *et al.*, 1993; Hehl *et al.*, 1996; Griffiths *et al.*, 1997a,b), preferentially quenched with $MnCl_2$ to remove cytosolic fluorescence. The use of these probes is based on evidence of mitochondrial loading and removal of the acetoxymethyl ester group by mitochondrial esterases. Further incubation of these cells with $MnCl_2$ quenched the cytosolic fluorescence of the particular probe. In our investigations of mitochondrial Ca^{2+}, we have used the single emission, calcium-sensitive fluorophore rhod-2 (Molecular Probes, OR, USA), which has a net positive charge favouring its accumulation into normally polarized mitochondria. Additionally, rhod-2 can be reduced so that only mitochondrial rhod-2 is oxidized to the active form (Mix *et al.*, 1994). Briefly, a small amount of sodium borohydride is added to a 50 μg vial of rhod-2/AM dissolved in DMSO. The colourless product, dihydrorhod-2/AM (DHRhod2), is then used for cell loading. Upon entering mitochondria, DHRhod2 is rapidly oxidized to the active form, and mitochondrial esterases cleave the acetoxymethyl ester group. Rat ventricular myocyte suspensions are incubated for 30 minutes at 30°C in the presence of dihydrorhod-2. The cells are then washed with additional buffer, and left to equilibrate for 10 minutes at room tem-

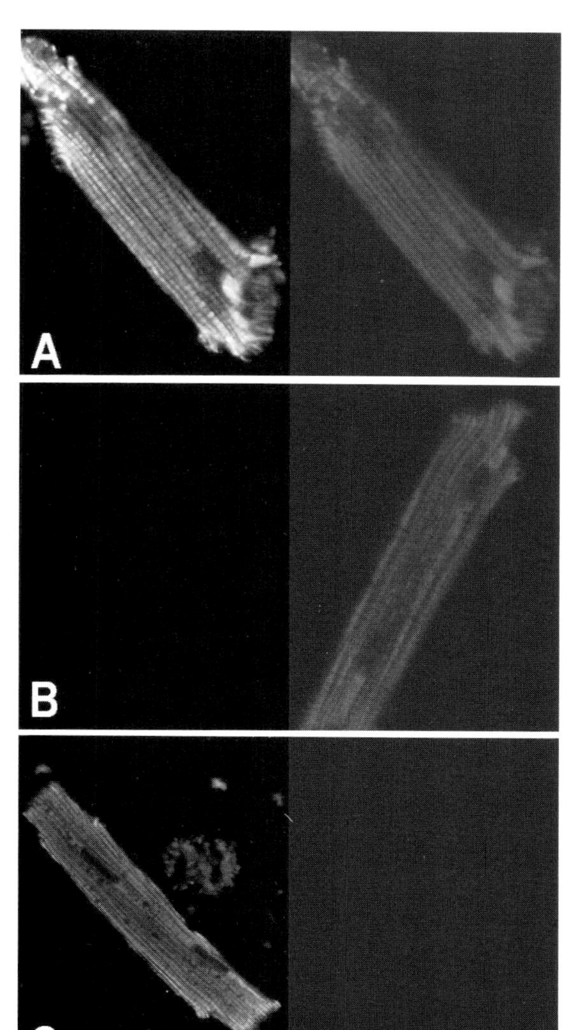

Figure 5.8 Rat ventricular myocytes loaded with dihydrorhod-2 and/or MitoTracker™ Green. (A) Dual-loaded myocyte demonstrating co-localized fluorescence of dihydrorhod-2 (right) and MitoTracker™ Green (left). (B) Image of a dihydrorhod-2-loaded cardiac myocyte. (C) Image of a MitoTracker™ Green-loaded cardiac myocyte. All cells were treated identically regarding concentration of fluorophore, incubation time and temperature, as well as confocal settings (aperture, gain, black level) during image acquisition.

perature (20–22°C). Myocytes were imaged with a BioRad MRC-1000 laser scanning confocal microscope, allowing excitation of the rhod-2 at 488 nm, and collection of the fluorescence through a 580±16 nm bandpass filter. The loading pattern seen is typical of mitochondrial distribution in cardiac myocytes (Fig. 5.8).

Mitochondrial localization of fluorescence was confirmed through the co-loading of DHRhod2 with the mitochondrial-specific probe MitoTracker™ Green (Molecular Probes, OR, USA). MitoTracker™ Green covalently binds to the inner mitochondrial membrane, and is unresponsive to alterations in membrane potential and mitochondrial calcium concentration (Poot *et al.*, 1996). Low concentrations of MitoTracker™ Green were used owing to the probe's intense fluorescence compared with rhod-2, otherwise incubation times and temperature were identical to those used to load rhod-2 alone. Both probes could be simultaneously imaged using the BioRad MRC-1000 confocal imaging system (Fig. 5.8). Bleed-through of the fluorescence of MitoTracker™ Green, into the DHRhod2 detector, was examined in single cells loaded with only one of the probes. These cells were imaged using identical settings (gain, iris, and black level), concentrations and incubation times (Figs 5.8(A), (B)).

Similar co-loading protocols can be used to image calcium simultaneously in both cytosolic (fluo-3/AM) and mitochondrial (rhod-2/AM) compartments. These probes were both excited with the 488 nm line of an argon-ion laser and appropriate filters were used to separate the fluorescence from each probe to a different detector. Figure 5.9 shows a cardiac myocyte, co-loaded with fluo-3 and DHRhod2, with a Ca^{2+} wave propagating through the cytosol. This method was used to investigate simultaneously the potential influence of mitochondrial calcium on cytosolic calcium levels and the characteristics of propagating calcium waves (Bowser *et al.*, 1998).

5.4.5 Mapping of intracellular pH in isolated cells

The measurement of pH in a variety of different cell types utilizing the unique emission spectra of SNARF-1 in combination with microspectrofluorimetry has been reported by a number of groups (Buckler & Vaughan-Jones, 1990; Bassnett, 1990; Seksek *et al.*, 1991; Mariot *et al.*, 1991; Martinez-Zaguilan *et al.*, 1991; Opitz *et al.*, 1994). The final fluorescence measurement obtained in these studies represented the average intracellular pH with little distinction of the pH of individual intracellular organelles. The confocal microscope can be readily configured for dual-channel emission detection and the emission spectrum of SNARF-1 can be divided at the isosbestic point with the separate emission bands sent to different detectors (Cody *et al.*, 1993; Dubbin *et al.*, 1993). Although the gain and black level of each channel are usually adjusted individually, if quantitative measurements are desired then it is essential to select settings which will cope with the range of fluorescence intensity fluctuations that are expected to occur, and to maintain these settings throughout

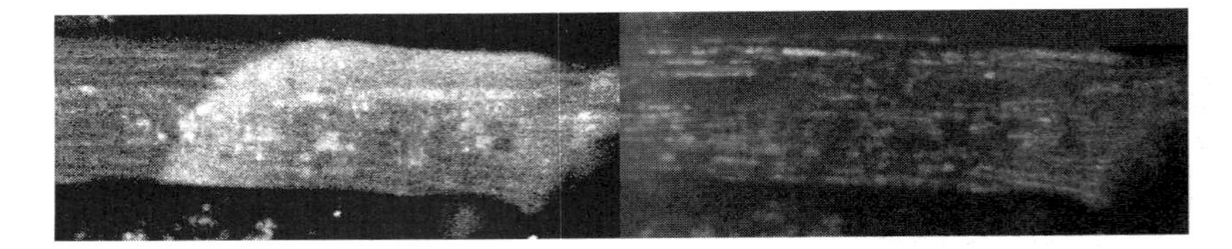

Figure 5.9 Rat ventricular myocyte co-loaded with dihydrorhod-2 (right) and fluo-3 (left). A Ca^{2+} wave (fluo-3) can be seen progressing through the cytosol, with minimal fluorescence change in the dihydrorhod-2 channel.

the experiment and calibration process. The thickness of the optical sections that contributed to the fluorescence emission recorded by each detector was also matched by using equivalent confocal apertures in each emission pathway.

As we have already suggested, experimentation with concentration and incubation times is necessary for optimization of loading for each cell type used. As a result there have been a wide variation of conditions that have resulted in successful loading of SNARF-1. We have used cultured rat aortic smooth muscle and rabbit proximal tubular cells, which were isolated and grown on glass coverslips. The cells were placed in a temperature-controlled perfusion bath and perfused with a HEPES-buffered Krebs solution for incubation with 20 μM carboxy-SNARF-1/AM (esterified derivative) for 20 minutes at 37°C followed by a 10 minute period where cells were perfused in fresh, dye-free solutions.

In agreement with previous studies we have found that the media used for most cell culture procedures often present problems for the accurate measurement of intracellular ion concentrations. Cultured cells should be grown in a medium which is free of pH indicator (the most common of which is Phenol red), as these indicators and their by-products do accumulate in cells during growth, and often contribute to the fluorescence output of the loaded cells. We have found that several cell passages in a medium free of pH indicator is generally sufficient for removal of the contaminating species. In addition, the aliphatic amines, which are also found in cell culture media, are capable of extracellular cleavage of ester groups, and it is essential to replace the culture medium before loading cells.

A typical confocal image of a culture of rabbit proximal tubule cells loaded with SNARF-1 is shown in Fig. 5.10. The fluorescence intensity of each individual image pair indicated that the uptake and cleavage of the SNARF-1/AM was consistent between cells and resulted in relatively uniform distribution of fluorophore within each cell. In contrast, there are reports that the majority of the dye becomes associated preferentially with the cytosol (Blank *et al.*, 1990).

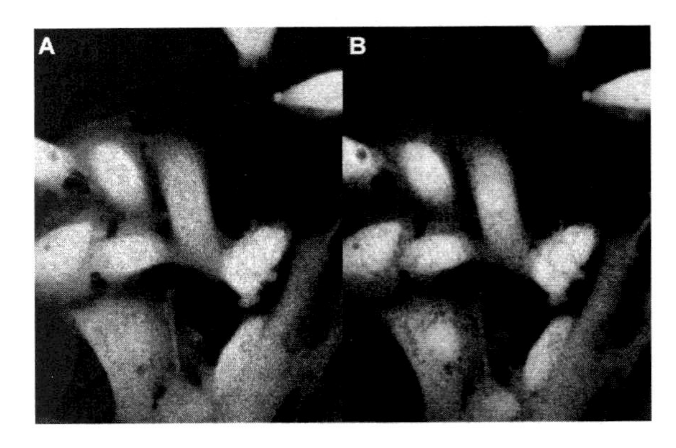

Figure 5.10 Fluorescence images of a field of rabbit proximal tubule cells loaded with the pH-sensitive fluorophore SNARF-1. Images represent wavelengths of the emission spectrum above (A) and below (B) 610 nm. Scale bar: 25 μm.

A ratio image of the SNARF-1 emission is displayed in Fig. 5.11(A) and represents the pixel-by-pixel division of the image pair (Fig. 5.10) for these cells. The majority of cells showed a relatively uniform cytosolic pH level. However, in most cases there were pH gradients evident between the cytosol and nucleus of individual cells, with the nuclear regions displaying predominantly higher pH than the cytosol. At the end of each experiment the absolute intracellular pH levels were calibrated *in vivo* by treatment of each cell with nigericin (20 mM) in solutions of high $[K^+]$ with a range of defined pH levels. Calibration curves constructed from these known pH standards were then used to calculate pH for images acquired during the experiment. We noted that the calibration curves of each individual cell within the same field of view were rarely identical. The fact that almost all cells showed a unique calibration curve may be explained by slight differences in the behaviour of the probe within each cell.

The basal cytosolic and nuclear pH levels within the cells that we examined were 7.35±0.08 and 7.44±0.02 (n=9) for rabbit proximal tubule cells, and 7.38±0.01 and 7.54±0.08 (n=5) for rat aortic smooth muscle cells, respectively (Dubbin *et al.*, 1993). There was a large variation in the basal intracellular pH of several

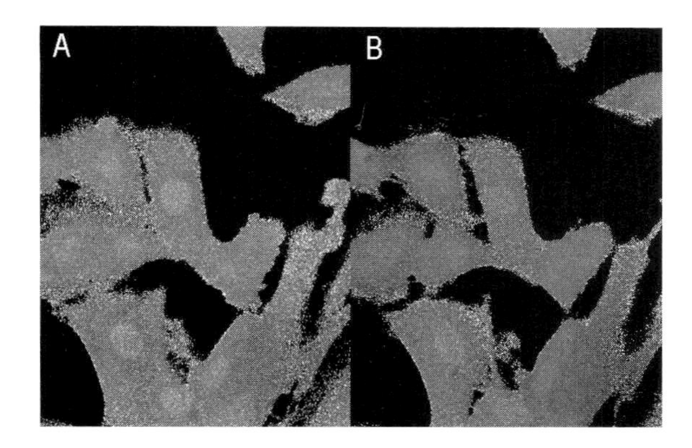

Figure 5.11 Ratio (Fig. 5.10(A)/Fig. 5.10(B)) images of SNARF-1 emission in the same field of cells, prior to (A) and following (B) the addition of the ionophore nigericin. The distinct differences in pH levels of the nucleus and cytosol of individual cells were virtually abolished by the addition of ionophore. Scale bar: 25 μm.

different cell cultured types. As also shown by several other groups (Bassnett *et al.*, 1990; Mariot *et al.*, 1991; Seksek *et al.*, 1991), the intracellular pH in a single population of cultured cells varied by as much as one unit of pH, so the importance of making measurements in large numbers of cells within a group becomes apparent. This variation makes it essential that investigation of the effects of physiological agonists be conducted at the level of individual cells.

In order to determine the ability of SNARF-1 to reflect accurately changes of pH within living cells we manipulated the internal pH of the cells experimentally without damaging their integrity. To do this we utilized a short perfusion with a small quantity of ammonium chloride (NH_4Cl). The cells responded with a typical transient rise in pH on addition and a fall after removal of NH_4Cl. By altering the concentration of NH_4Cl applied (0.1 mM to 10 mM) we were able to control the initial rise in pH. We found that we could accurately resolve changes as small as 0.03–0.05 pH units. It was also interesting to note that when NH_4Cl was added the pH of the nucleus and cytosol both increased and the pH gradient between these compartments was still evident for all concentrations. Using high concentrations of nigericin (10–20 mM) we were able to remove this gradient during control experimental conditions (see Fig. 5.8(B)) which indicated that the nuclear membrane was actively involved in controlling the intranuclear pH. More importantly this also showed that the difference was not due to an artifact of differential uptake or behaviour of the probe in different cellular compartments. The technique of coupling SNARF-1 with a laser scanning confocal microscope system can provide high-resolution spatial information on intracellular pH levels within single living cells during physiological manipulations.

5.4.6 Intracellular pH gradients in mammalian tumour spheroids

Cancer research has employed cell spheroids as models of tumour tissue because they are poorly vascularized, have radial proliferation, and show distinct pO_2 gradients with the tissue. It is also thought that such tissues should exhibit pH gradients, and measurements made with pH electrodes have shown a pH of 7.3 at the surface decreasing to 7.0 at a depth of 400 μm within the spheroid, thus supporting this contention (Acker *et al.*, 1987; Carlsson & Acker, 1988). However, the authors of these studies expressed reservations as to whether the microelectrodes used were exclusively reporting levels of intracellular pH because of uncertainty in the location of the tip within the spheroid and because of the tip size of the electrode (2–3 μm). Measurements were therefore conservatively interpreted as representing the pH of the extracellular space within the spheroid. However, this was still an important observation given the known involvement of pH changes in mediating growth response in tumour cells (Carlsson & Acker, 1988).

In a collaborative study with the Department of Surgery, Royal Melbourne Hospital (Cody *et al.*, 1993), we combined confocal microscopy with pH-sensitive fluorophores to determine unequivocally whether these extracellular pH gradients were mimicked in the intracellular environment of the spheroid cells. Rat C6 glioblastoma spheroids (Benda *et al.*, 1971) were cultured with well-established techniques (Yuhas *et al.*, 1977; Sutherland *et al.*, 1971). Loading of spheroids with SNARF-1/AM (ester derivative) was performed in the dark at 10°C for 90 minutes under constant agitation, to allow for penetration of dye into the spheroids used in these experiments which ranged from 600 to 1500 μm in diameter. However, the limitation to observations of the pH of cells deep within the spheroid was not one of dye distribution, but rather stemmed from the inability to focus on cells throughout the full depth of the spheroid with the low magnification (×10) and therefore low numerical aperture (NA=0.5) objectives required to observe the entire spheroid with the confocal microscope. Therefore, following incubation with SNARF-1/AM spheroids were embedded in agar (2% Sea Plaque; Bioproducts, ME, USA) at 37°C for 15 minutes. Agar was solidified with brief (2 minutes) immersions in crushed ice. The embedding step at 37°C also facilitated de-esterification of dye internalized throughout the spheroid. Embedded spheroids were then carefully bisected with a disposable microtome blade (featherblade) to avoid tissue distortion and reveal the necrotic core of the tissue.

A ratio image of the SNARF-1 emission of a spheroid is shown in Plate 5.1(A). There is evidence of a striking zonation in intracellular pH within the spheroid with a band of cells of uniform pH surrounding a distinctly more acidic in core of cells. Analysis of pH in a transect across the tissue (Plate 5.1 (B)) shows that the cells in the outer mass maintained a pH of 7.5–7.6 while the pH of cells in the central cell zone fell progressively to 7.1–7.2 at the heart of the spheroid. These results were the first clear measurements of cytosolic pH in the cell layers of tumour spheroids. The values were in the same range as extracellular values recorded with microelectrodes and confirmed a previous hypothesis that both extracellular and intracellular environments will exhibit similar pH gradients (Acker *et al.*, 1987).

ACKNOWLEDGEMENTS

We are grateful for the technical assistance of Christine Goulter in many of these studies, which were supported by the National Heart Foundation (NHF) of Australia, the National Health and Medical Research Council (NH&MRC) of Australia and the Australian Research Council (ARC).

REFERENCES

Acker H., Carlsson J., Holtermann G., Nederman T. & Nylen T. (1987) *Cancer Res.* **47**, 3504–3508.
Allen S.P., Stone D. & McCormack J.G. (1992) *J. Mol. Cell. Cardiol.* **24**, 765–774.
Bakker A.J., Head S.I., Williams D.A. & Stephenson D.G. (1993) *J. Physiol.* **460**, 1–13.
Barcenas-Ruiz L., Beuckelmann D.J. & Wier W.G. (1987) *Science* **238**, 1720–1722.
Bassnett S. (1990) *J. Physiol.* **431**, 445–464.
Bassnett S., Reinisch L. & Beebe D.C. (1990) *Am. J. Physiol.* **258**, C171–178
Baylor S.M. & Hollingworth S. (1988) *J. Physiol.* **403**, 151–192.
Benda P., Someda K., Messer J. & Sweet W.H. (1971) *J. Neurosurg.* **34**, 310–323.
Blank P.S., Silverman H.S., Chung O.Y., Stern M.D., Hansford R.G., Lakatta E.G. & Capogrossi M.C. (1990) *Biophys. J.* **57**, 137a.
Blatter L.A. & Wier W.G. (1990) *Biophys. J.* **58**, 1491–1499.
Bowser D.N., Minamikawa T., Nagley P. & Williams D.A. (1998) *Biophys. J.* **75**, 2004–2014.
Brownlee C. & Wood, J.W. (1986) *Nature (London)* **320**, 624–626.
Buckler K.J. & Vaughan-Jones R.D. (1990) *Eur. J. Physiol.* **417**, 234–239.
Bush D.S. & Jones R.L. (1988) *Eur. J. Cell Biol.* **46**, 466–469.
Cannell M.B., Berlin J.R. & Lederer W.J. (1987) *Science* **238**, 1419–1423.
Carlsson J. & Acker H. (1988) *Int. J. Cancer,* **42**, 715–720.
Carrington W.A., Lynch R.M., Moore E.D., Isenberg G., Fogarty K.E. & Fay F.S. (1995) *Science* **268**, 1483–1487.
Cody S.H., Dubbin P.N., Beischer A., Duncan N.D., Hill J.S., Kaye A. & Williams D.A. (1993) *Micron* **24**, 573–580.
Coppey Moisan M., Delic J., Magdelenat H., Coppey J. (1994) *Methods Mol. Biol.* **33**, 359–393.
Cork R.J. (1985) *Plant Cell Environ.* **9**, 157–160.
Cubitt A.B., Heim R., Adams S.R., Boyd A.E., Gross L.A. & Tsien R.Y. (1995) *Trends. Biochem. Sci.* **20**, 448–455.
DeFeo T.T. & Morgan K.G. (1985) *J. Physiol.* **369**, 269–282.
Di Lisa F., Gambasi G., Spurgeon H. & Hansford R.G. (1993) *Cardiovasc. Res.* **27**, 1840–1844.
Dubbin P.N., Cody S.H. & Williams D.A. (1993) *Micron* **24**, 581–586.
Gehring C.A., Williams D.A., Cody S.H. & Parish R.W. (1990) *Nature (London)*, **345**, 528–530.
Gonzalez J.E. & Tsien R.Y. (1997) *Chem. Biol.* **4**, 269–277
Griffiths E.J., Stern M.D. & Silverman H.S. (1997a) *Am. J. Physiol.* **273**, C37–C44.
Griffiths E.J., Wei S-K., Haigney M.C.P., Ocampo C.J., Stern M.D. & Silverman H.S. (1997b) *Cell Calcium* **24**, 321–329.
Groden D.L., Guan Z. & Stokes B.T. (1991) *Cell Calcium* **12**, 279–288.
Grynkiewicz G., Poenie M. & Tsien R.Y. (1985) *J. Biol. Chem.* **260**, 3440–3450.
Gunter T.E., Restrepo D. & Gunter K.K. (1988) *J. Physiol.* **255**, C304–C310.
Harris J.B. & Johnson M.A. (1978) *Clin. Exp. Pharmacol. Physiol.* **5**, 587–600.
Hayes A., Cody S.H., Williams D.A. (1996) *Int. Rev. Exp. Pathol.* **36**, 197–212.
Head S.I., Williams D.A. & Stephenson D.G. (1992) *Proc. Roy. Soc. Lond.* **248**, 163–169.
Hehl S., Golard A. & Hille B. (1996) *Cell Calcium* **20**, 515–524.
Heim R. & Tsien R.Y. (1996) *Curr. Biol.* **6**, 178–182.
Henke W., Cetinsoy C., Jung K., Loening S. (1996) *Cell Calcium.* **20**, 287–292.
Hepler P.K. & Wayne R.O. (1985) *Annu. Rev. Plant Physiol.* **36**, 397–439.
Highsmith S., Bloebaum P. & Snowdowne K.W. (1986) *Biochem. Biophys. Res. Commun.* **138**, 1153–1162.
Htun H., Barsony J., Renyi I., Gould D.L. & Hager G.L. (1996) *Proc. Natl. Acad. Sci. USA* **93**, 4845–4850.
Jung D.W., Chapman C.J., Baysal K., Pfeiffer D.R. & Brierley G.P. (1996) *Arch. Biochem. Biophys.* **332**, 19–29.
Lipp P. & Niggli E. (1993) *Cell Calcium.* **14**, 359–372.
Luckoff A. (1986) *Cell Calcium* **7**, 233–248.
Lui C. & Hermann T.E. (1978) *J. Biol. Chem.* **253**, 5892.
Malgaroli A., Milani D., Meldelosi J. & Pozzan T. (1987) *J. Cell Biol.* **105**, 2145–2155.
Mariot P., Sartor P., Audin J. & Dufy B. (1991) *Life Sci.* **48**, 245–252.
Martinez-Zaguilan R., Martinez G.M., Lattanzio F., & Gillies R.J. (1991) *Am. J. Physiol.,* **260**, C297–C307.
Minamikawa T., Cody S.H. & Williams D.A. (1997) *Am. J. Physiol.* **272**, H236–243.
Minta A., Kao J. & Tsien R.Y. (1989) *J. Biol. Chem.* **264**, 8171–8178.
Mix T.C.H., Drummond R.M., Tuft R.A. & Fay F.S. (1994) *Biophys. J.* **66**, A97.
Miyata H., Silverman H.S., Sollott S.J., Lakatta E.G., Stern M.D. & Hansford R.G. (1991) *Am. J. Physiol,* **261**, H1123–H1134.

Miyawaki A., Llopla J., Heim R., McCaffery J.M., Adams J.A., Ikura M. & Tsien R.Y. (1997) *Nature* **388**, 882–887.
Moore E., Becker P.L., Fogarty F.S., Williams D.A. & Fay F.S. (1990) *Cell Calcium,* **11**, 157–179.
Negulescu P.A., Harootunian A., Tsien R.Y. & Machen T.E. (1990) *Cell Regul.* **1**, 259–268.
Neher E. & Almers W. (1986) *EMBO Journal,* **5**, 51–53.
Opitz N., Merten E. & Acker H. (1994) *Pflugers Arch.* **427**, 332–342.
Owen C.S. (1991) *Cell Calcium* **12**, 385–393.
Pawley J.B. (1995) *Handbook of Biological Confocal Microscopy*, 2nd edn. Plenum Press, New York.
Poenie M. (1990) *Cell Calcium* **11**, 85–91.
Poot M., Zhang Y.-Z., Krämer J.A., Wells K.S., Jones L.J., Hanzel D.K., Lugade A.G., Singer V.L. & Haugland R.P. (1996) *J. Histochem. Cytochem.* **44**, 1363–1372.
Prasher D.C., Eckenrode V.K., Ward W.W., Prendergast F.G. & Cormier M.J. (1992) *Gene* **111**, 229–233.
Rink T.J. & Pozzan T. (1985) *Cell Calcium* **6**, 133–144.
Rizzuto R., Simpson A.W.M., Brini M. & Pozzan T. (1992) *Nature* **358**, 325–327.
Rizzuto R., Brini M. & Pozzan T. (1994) *Methods Cell Biol.* **40**, 339–357.
Rizzuto R., Brini M., Pizzo P., Murgia M. & Pozzan T. (1995) *Curr. Biol.* **5**, 635–642.
Rizzuto R., Brini M., De Giorgi F., Rossi R., Heim R., Tsien R.Y. & Pozzan T. (1996) *Curr. Biol.* **6**, 183–188.
Rozengurt E. & Heppel L.A. (1975) *Biochem. Biophys. Res. Commun.* **67**, 1581–1585.
Scanlon M., Williams D.A. & Fay F.S. (1987) *J. Biol. Chem.* **262**, 6308–6312.
Seksek O., Henry-Toulme N., Sureau F. & Bolard J. (1991) *Anal. Biochem.* **193**, 49–54.
Steinberg S.F., Bilezikian J.P. & Al-Awquati Q. (1987) *Am. J. Physiol.* **253**, C744–C747.
Sutherland P.J., Wendt I.R. & Stephenson D.G. (1980) *Proc. Aust. Physiol. Pharmacol. Soc.* **11**, 160P.
Sutherland R.M., McCredie J.A. & Inch W.R. (1971) *J. Natl. Cancer Inst.* **46**, 113–117.
Szmacinski H. & Lakowicz J.R. (1995) *Cell Calcium* **18**, 64–75.
Timmers A.C.J., Reiss H.D. & Schel J.H.N. (1991) *Cell Calcium* **12**, 515–521.
Trollinger D.R., Cascio W.E. & Lemasters J.J. (1997) *Biochem. Biophys. Res. Commun.* **236**: 738–742.
Trouet D., Nilius B., Voets T., Droogmans G. & Eggermont J. (1997) *Eur. J. Physiol.* **434**, 632–638.
Tsien R.Y. (1992) *Am. J. Physiol.* **263**, C723–C728.
Tsien R.Y. & Harootunian A.T. (1990) *Cell Calcium* **11**, 93–109.
Uto A., Arai H. & Ogawa Y. (1991) *Cell Calcium,* **12**, 29–38.
Williams D.A. (1990) *Cell Calcium* **11**, 589–597.
Williams D.A. & Fay F.S. (1990) *Cell Calcium* **11**, 75–90.
Williams D.A., Fogarty K.E., Tsien R.Y. & Fay F.S. (1985) *Nature* **318**, 558–561.
Williams D.A., Cody S.H., Gehring C.G., Parish R.W. & Harris P.J. (1990a) *Cell Calcium* **11**, 291–298.
Williams D.A., Head S.I., Bakker A.J. & Stephenson D.G. (1990b) *J. Physiol.* **428**, 243–256.
Williams D.A., Delbridge L.M., Cody S.H., Harris P.J. & Morgan T.O. (1992) *Am. J. Physiol.* **262**, C731–C742.
Williams D.A., Head S.I., Lynch G.S. & Stephenson D.G. (1993) *J. Physiol. (Lond.)* **460**, 51–67
Yuhas J.M., Li A.P. & Martinez A.O. (1977) *Cancer Res.* **37**, 3639–3643.

CHAPTER SIX

Electroporation: A Method for Introduction of Non-permeable Molecular Probes

FEDJA BOBANOVIĆ
Life Science Resources Ltd, Cambridge, UK, and Laboratory of Biocybernetics, Faculty of Electrical Engineering, Ljubljana, Slovenia

6.1 INTRODUCTION

The ability to probe intracellular compartments of living cells greatly facilitates the study of the intracellular mechanisms that regulate cell physiology. Fluorescent probes offer us unparalleled scope to visualize complex subcellular structures and follow dynamic events within living cells with minimum physiological perturbations. Entry of the probes into the cells is dependent on the physiochemical properties of the probe, i.e. hydrophilicity–lipophilicity, electric charge, acid or base strength (p*K*), as well as the overall size of the probe.

In many cases, passive diffusion through the plasma membrane does not allow accumulation of the probe within the cell. To achieve a level of loading which generates sufficient optical information to be followed by sensitive cameras or photomultiplier tubes, different methods can be used. One method is based on changing the physiochemical properties of the fluorescent probes, e.g. by masking the charge of the probe. This can be achieved by either esterifying the carboxy groups with a chemical, for example one with acetoxymethyl groups (AM esters), or by applying a low pH (methods addressed more detailed in other chapters).

Another method used for intracellular introduction of probes creates alternative membrane pathways through which probes can move following concentration and electrical gradients or by application of a pressure force. These alternative pathways can be chemically created by permeabilizing reagents such as digitonin and Triton® X-100, or by physical means using single cell procedures such as microinjection and patch pipette perfusion, or by bulk loading procedures such as scrape loading and electroporation.

Although convenient to use, chemical permeabilization cannot be regarded as a general tool for intracellular introduction since it has a considerable effect on normal cell physiology. Most often, an alternative membrane pathway is established by introduction of a very thin glass pipette into the cell interior, i.e. microinjection, which allows loading of only one cell at a time.

Electroporation, which is also regarded as mass microinjection, offers advantages associated with both chemically and physically originated methods. Electroporation creates many small pores spread

FLUORESCENT AND LUMINESCENT PROBES, 2ND EDN
ISBN 0–12–447836–0

over the membrane (Chang, 1992), in contrast to the single large pathway associated with microinjection. Unlike microinjection, where the probe can only be introduced into one cell at a time, using electroporation the electroloading can be assayed at the cell population level. In this respect it is similar to chemical permeabilization. Another advantage of this method is purely physical and is based on the fact that electroporation is a controllable ON–OFF phenomenon. It acts almost instantaneously (less than a μs), but even more important is the fact that electroporated cells can be permeabilized for only a short period of time. Further advantages of electroporation derive from its ability to manipulate pore dimension and lifetime, which gives the experimenter additional control over transmembrane transport during permeabilization.

Although electroporation is regularly mentioned as a general tool for introduction of fluorescent probes, the method is not widely used in everyday practice. We can identify at least three reasons why electroloading of fluorescent probes is still used only rarely: (1) AM ester loading, which is by far the most popular method for loading fluorescent ion indicators (good candidates for electroloading), is non-invasive and technically straightforward; (2) although the electroporation technique is easy to employ, the phenomenon of electrically induced pore formation is a complex process which has to be extensively tested for optimized conditions/parameters separately for each particular purpose and cell type; (3) there is a lack of commercially available electroporation systems designed specifically for the purpose of loading fluorescent indicators.

Electroporation can be modified to fit specific requirements important for loading fluorescence probes, e.g. it can be altered to enable loading of cells which are adherent on coverslips. The process can also be undertaken in very small volumes, which increases cost-efficiency of the method. Multipolar electroporation can be employed to enable loading of cells which are of different shapes and relative position to the exposure, and fast temperature control of the electroporated sample can be achieved. However, modification of equipment usually requires time, substantial effort and comprehensive knowledge of the problem. Nevertheless, electroporation is considered to be potentially a good alternative to microinjection in cases where a larger number of cells is to be studied or when the cells are of rather small and/or of flat shape, which makes microinjection even more difficult.

The many reviews of electroporation provide excellent sources of information. The intention of this chapter is not to challenge such comprehensive texts, but rather to serve as a starter which should allow easier reading of such detailed literature. Unfortunately, well-tested protocols for electroloading of a specific probe into a particular cell type will only become available once greater interest has been shown in the method and protocols have been published in the literature, as is the case with protocols for genetic manipulation by electroporation of DNA and RNA into cells and microorganisms.

6.2 BASIC CONCEPT OF ELECTROPORATION

Electroporation is defined as the phenomenon in which a normally selective biological membrane suddenly becomes permeable to a wide variety of ions and molecules (Zimmerman, 1986; Mir *et al.*, 1988; Neumann & Boldt, 1990; Tsong, 1991; Weaver, 1993; Weaver & Chizmadzhev, 1996). The phenomenon is based on the exposure of cells to an intense electric field which causes transient localized disruptions, or ruptures, of the outer cellular membrane through which molecules can pass relatively unhindered. Electroporation can be described as a sequence of the following events:

(1) electric field generation,
(2) polarization of the outer membrane,
(3) electropore formation,
(4) transmembrane transport, and
(5) electropore resealing.

Formation and resealing of the electropores appear to be complex events and are still not fully understood. However, it is known that pore development is an indirect effect of the application of an extracellular electric field and its electrostatic interaction with an outer membrane of the cell. Due to the specific dielectric characteristics of the extracellular solution and exposed cell, an extracellular electric field is strongly amplified across the plasma membrane. This changes membrane polarization and concurrently electric field energy within the membrane. In the parts of the membrane where polarization exceeds several hundred mV, lipid molecules of the bilayer start to reorientate to form pores. The number of electropores created, the size of the pores and the pore lifetime, along with the driving forces that cause directional movement of the molecules, are parameters that determine which and how many of the normally non-permeant molecules will be transported through the membrane.

Large, long-lasting pores have a greater capacity to transport molecules than small pores which reseal within milliseconds. However, such pores also cause greater loss of intracellular components and more disturbance to the cell. The ultimate goal of electroporation is to enable transmembrane transport of a certain amount of non-permeant molecules with no long-term effect on membrane permeability and consequently cell physiology.

The overall efficiency of electroporation can be schematically represented through the sequence of electroporative events mentioned above (Fig. 6.1). Although electroporation is a bulk loading procedure, each cell in the population is affected separately and shows its own characteristic response to the external electric field. Such individuality in response depends on the dimension of the cell, its shape, its relative position to the direction of the electric field, and the structure of particular parts of the membrane.

One can assume that each cell in a population can be affected by the field in three different ways (Fig. 6.1(A)): underexposed (cells X), overexposed (cells Z), or optimally exposed (cells Y). In underexposed cells, the electric field is insufficient to create electropores whose number, size and lifetime would enable loading of the desired quantity of the molecules present in extracellular space. In optimally and overexposed cells, electroporation provides sufficient loading, but an overexposed cell remains permeabilized for a longer time and will eventually die. Whether or not the cell survives also depends on the cell type, the extracellular medium composition and the overall change of intracellular composition.

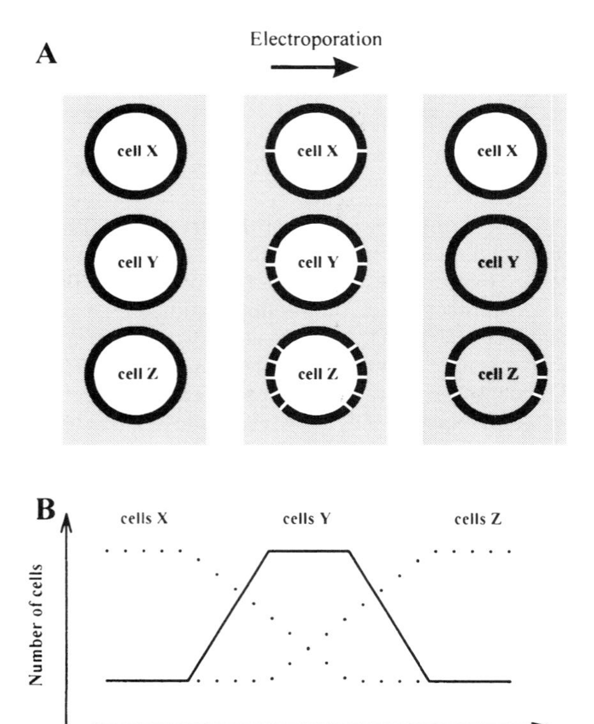

Figure 6.1 (A) Schematic representation of the efficiency of electroporation with respect to the individuality of responses of the cells in a population. The efficiency is discussed on the assumption that each cell in a population can be underexposed (cell X), overexposed (cell Z) or optimally exposed (cell Y). (B) The hypothetical relationship between exposure parameters, i.e. E and t, and number of under-, over-, and optimally exposed cells.

In general, with increasing intensity and duration of the electric field exposure, the number of underexposed cells will decrease, while the number of overexposed cells will tend to increase (Fig. 6.1(B)). It is plausible to say that efficiency of electroporation is determined by the number of sufficiently permeabilized but undamaged cells. The number of such cells is dependent on electric field parameters in a 'window' fashion (Fig. 6.1(B)). For smaller molecules, such as free salt forms of the fluorescent probes and other molecules whose molecular weight does not exceed several kDa, such a window is usually wider and even modest optimization can bring satisfactory results. However, if molecules of the smallest size are to be loaded to high intracellular concentrations, the permeabilization parameters have to be increased and the window in which electroporation provides efficient loading will be narrow. For larger molecules such as dextran-conjugated probes, a careful optimization is even more important since introduction of such molecules already requires more intense exposure.

In contrast with chemical agents, where the membranes of the cells are equally affected and only a few parameters can be altered, electroporation is a method which can be optimized by changing many parameters of an exposure field as well as the conditions in which permeabilization is performed. As well as the field strength (intensity), which is usually considered to be the major parameter, it is important to understand that the electric field can be generated as a signal of various shapes, duration and directions. Moreover, such a signal can be applied as a single-shot treatment or repeated a number of times with chosen frequency.

It is also important to mention that the conditions in which electroporation is performed, such as temperature, ion composition and osmolarity, offer substantial control over the efficiency of the method. Both exposure and conditional parameters can be used to find the window of satisfactory loading. In theory, changing exposure parameters can also widen, or even create, the window for an application that leads to optimal loading. For example, if one was to test the efficiency of electroloading just by varying the intensity of a single exponential pulse in a case where X and Z curves (Fig. 6.1(B)) already show considerable overlap, successful loading could not be achieved. In such a case, changing the duration, shape, number, frequency or field direction and temperature of the sample could markedly change the balance between under- and overexposed cells and consequently the efficiency of electroporation. However, such a wide choice of parameters can also cause confusion, since it is impossible to test every combination of parameters for each application. For this reason, it may be more cost-effective to understand the underlying mechanisms of electroporation and how certain parameters correspond to the phenomenon instead of randomly testing different combinations.

6.3 ELECTRIC FIELD GENERATION AND MONITORING

Electroporation systems usually consist of three major parts: (1) a generator to provide the electrical signals; (2) a chamber containing electrodes in which the cells are subject to the exposure; and (iii) a monitoring system, which is an optional but highly recommended component. The generator used for electroporation needs to be a high-voltage, high-power device because the electric field strength necessary for electroporation is typically between 0.2 kV cm^{-1} and 10 kV cm^{-1}, with momentary power during the exposure exceeding 10 kW. In practice, capacitors in the generator which are used to store electrical energy are charged slowly with low power and then discharged very rapidly, delivering the required high-power pulse (Hofmann and Evans, 1986; Potter, 1988). Electroporating chambers can be designed in different geometrical shapes, depending on their specific purpose. The metallic electrodes are usually mounted on opposite sides of the chamber, which is filled with bath solution or cell suspension.

To generate the electric field in the chamber, an electric circuit consisting of a voltage generator, electrodes and bath solution is established. When the electric circuit is closed, voltage from the generator charges the electrodes and forces positive ions and other molecules with positive net charge in the solution to move towards the cathode, while negatively charged ions and molecules move towards the anode. Such directed movement constitutes electric (mainly ionic) current in the solution. The electric field in the solution is proportional to the specific resistivity (ρ) of the medium and the ionic current density (J):

$$E = \rho J \quad (\text{V cm}^{-1}) \qquad [6.1]$$

In general, the electric field E can be position-dependent ($E = E(x,y,z)$), where x, y and z represent spatial coordinates, and/or time-dependent ($E = E(t)$). If the intensity of the field does not change with position in the chamber ($E(x,y,z)$ = constant), then a homogeneous field has been generated. An electric field which does not change in time ($E(t)$ = constant) is called a static electric field. Since electroporation is based on a brief, high-voltage exposure, application of a static electric field does not have any practical implementation. However, a square pulse of fixed duration can be considered as a quasi-static exposure. This simplifies calculation of the induced membrane polarization and prediction of electroporated sites on the membrane.

Although closely linked, it is important to distinguish between the electrical signal from the generator ($U(t)$) and the electric field which the cell actually experiences. Generally, $U(t)$ can be of various kinds of voltage functions ranging from a pulse of simple exponential decay to multipolar bursts of high-frequency pulses of complex shapes. In each case $E(x,y,z,t)$ in the chamber will tend to follow the voltage signal from the electrodes with respect to the specific geometry of the chamber and the electrodes. For a rectangular chamber with parallel-plate electrodes located at a distance d the $E(t)$ will be homogeneous between the electrodes and is estimated from $U(t)$ as

$$E(t) = U(t)/d \quad (\text{V cm}^{-1}) \qquad [6.2]$$

This means that voltage applied from electrodes 1 mm apart will produce an electric field ten times stronger than the same voltage applied from electrodes 1 cm apart. If the shape of the chamber is not rectangular and/or the electrodes are not parallel plates, the generated electric field is inhomogeneous and must be calculated for each particular position in the chamber.

It is important to mention that the actual voltage delivered to the solution can be substantially lower than what is normally assumed to be the generator output voltage. This causes an overestimation of the electric field intensity in the chamber. The effect is caused by internal resistance of the generator and the resistance of the electrodes which compete with the resistance of the solution in the absorption of the voltage drop. For this purpose it is important that the resistance of the solution is high enough in comparison with the resistance of the rest of the electric circuit.

The resistance of the chamber can be calculated from the specific resistivity of the medium and the geometry of the chamber. If the chamber is of rectangular shape, then the resistance (R_{CH}) is calculated as:

$$R_{CH} = \rho L/A \qquad [6.3]$$

where ρ is the specific resistivity of the medium, A is the cross-sectional area of the chamber filled with solution, and L is the distance between the electrodes. To modify the resistivity of the sample, either the geometry of the chamber (A, L) or the ion composition of the solution (ρ) can be changed. However, changing the resistivity of the sample may also affect the actual electric field exposure. This is particularly the case if a capacitor discharge is used to generate the electric field. With such a device the exposure electric field is an exponentially declining pulse in which time constant (RC) is defined by the capacitance of the generator (C) and resistivity of the electric circuit (R):

$$E(t) = E \exp(-t/RC) \quad [6.4]$$

Changing the resistivity of the chamber (R_{CH}) changes the resistivity of electric circuit (R), and consequently changes the shape and duration of the electric field exposure. If a square pulse generator is used, the pulse shape should be much less sensitive to modulation of R_{CH}. This is because such generators use a partial discharge of the capacitors, providing almost constant exposure field during the pulse. However, bearing in mind that square pulse generators store electrical energy in the same way as exponential pulse generators, the wrongly designed chamber (lower R_{CH} than a designed generator requires) may substantially increase the discharge of the capacitors during the exposure, which will cause distortion of the pulse shape.

Since modifications of the electroporation systems may lead to serious misjudgement of the exposure parameters, it is important to measure the actual electric field in the chamber. The best way to do this is to measure voltage drop in the solution using a separate set of electrodes, a high-voltage probe and an oscilloscope. Although different electrodes can be used to measure such a relatively high voltage, non-polarizable silver–silver chloride or platinum–platinum black electrodes would provide the most accurate results. The electrodes should be immersed into the solution at an accurately defined distance which will allow the calculation of the electric field from the measured voltage (equation [6.2]). There are several different designs of electric field probes containing two or more electrodes which allow the measurement of intensity and direction of the field in volume conductors. If the chamber is of an irregular shape, voltage should be measured in different parts of the chamber and the actual level of the field inhomogeneity estimated. The use of an oscilloscope which provides information on how voltage is changing with time allows measurement of not only the intensity, but also the duration and shape of the electric field. Such information about exposure parameters is crucial for the successful design and use of electroporation.

6.4 POLARIZATION OF THE OUTER MEMBRANE

When an extracellular electric field is established in the chamber, the outer membranes of the cells start to experience an electrostatic interaction which results in a change in membrane polarization. To model this effect one has to solve Laplace's equation, the solution of which shows the distribution of electric potentials for certain dielectric and geometric structure and stimulation conditions.

In a simplified example, an electrically stimulated cell may be considered as a thin-walled sphere made of pure dielectric. A homogeneous, static electric field applied to the surrounding cell environment then induces a change in membrane potential (ΔV) (polarization) which varies with position (point P on Fig. 6.2) on the cell surface according to the following equation:

$$\Delta V = -(3\ Er \cos \varphi)/2 \quad [6.5]$$

where E is the intensity of applied homogeneous static electric field, r is the radius of the cell, and φ is the angle between the field direction and the radial vector from the cell centre to the cell surface at the position where the membrane potential is considered (the cathode lies at $\varphi = 0°$). The total membrane potential V consists of resting potential V_{rest} (which exists when no external field is applied) and the superimposed field-induced ΔV, as expressed by the equation:

$$V = V_{rest} + \Delta V \quad [6.6]$$

In practical terms, the external field E causes a hyperpolarization (V_{hyp}) at the cell hemisphere nearest the anode and a depolarization (V_{dep}) at the opposite cell hemisphere, nearest the cathode. The most pronounced potential change occurs at those membrane portions that directly face the electrodes, where φ is 0° and 180°, respectively, and cos φ reaches its maximum values ± 1. Little or no change is induced by the electric field in the cell wall along the circumference between the hemispheres described above. These portions of the membrane are parallel to the electric field ($\varphi = 90°$ and $\varphi = 270°$), and so the membrane potential remains unchanged ($\cos \varphi = 0°$), i.e. $V(\varphi)$ equals V_{rest}. Figure 6.2 shows the membrane potential $V(\varphi)$ as a function of the angle φ in polar coordinates in unexposed (A) and exposed (B) spherical cells. Each figure illustrates the voltage in the cell membrane around a section through an ideal, spherical cell. The cell is shown schematically at the centre of each drawing. The origin of the coordinates

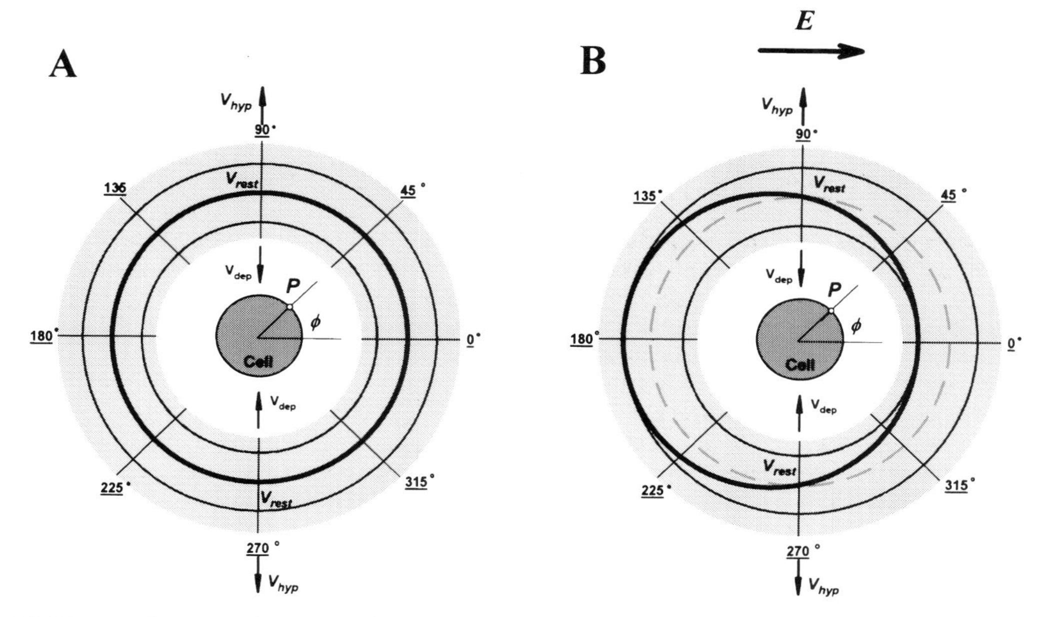

Figure 6.2 Diagram of transmembrane potential versus angle, in polar coordinates, in a theoretical model of a spherical cell with no applied electrical field (A) and with an applied homogeneous, static electric field (B). In unexposed cells ($E = 0$; $\Delta V = 0$) the membrane potential is constant and equal to the resting potential all over the cell membrane ($V(\varphi) = V_{\text{rest}}$). This is illustrated by means of the central circle concentric to the origin (A). When an external electric field E is applied, $V(\varphi)$ becomes an eccentric circle shifted towards the anode, as shown in (B). This indicates that one side of the cell is hyperpolarized (the left side in (B)) while the other side is depolarized (the right side in (B)). $V(\varphi)$ is rotationally symmetric about $\varphi = 0°$, which is the direction of the electric field E.

is the centre of the cell. For easier understanding, V_{rest} is assumed to be 0 V, while hyperpolarizing voltages radiate outwards and depolarizing voltages radiate inwards. This theoretical consideration of field-induced polarization of the cell membrane has also been confirmed experimentally using a microscopic video imaging technique and fast voltage-sensitive dyes (Kinosita *et al.*, 1988, 1992; Hibino *et al.*, 1993).

If the cell is not perfectly spherical, then equation [6.5] does not provide an accurate estimation of the induced membrane polarization. For the majority of common cells, whose geometry cannot compare to a hollow sphere, Laplace's equation must be solved numerically. For this purpose computer calculation implementing the finite-difference or finite-element method is used (Miklavčič *et al.*, 1998). The same numerical approach has to be used if the effect of an inhomogeneous field is to be calculated.

If the electric field is considered as a time-dependent function ($E(t)$), then the induced polarization will be a time-dependent function $\Delta V(t)$ that will tend to follow the amplitude of the field E with respect to a 'relaxation time' of the membrane (τ), which is defined by the specific resistivity and specific capacitance of the membrane (Kotnik *et al.*, 1997):

$$\Delta V(\varphi, t) = 1.5\, E\, r \cos\varphi\, [1 - \exp(-t/\tau)] \qquad [6.7]$$

When an electric field is applied, the membrane does not experience the applied pulse immediately because the membrane capacitance has initially to charge through the resistance of the membrane and the extracellular solution. For a square pulse, a duration of 1 μs is required to build the polarization of the membrane near to the induction predicted for static field stimulation. Generally, with increasing frequency of stimulating electric field, the induced membrane polarization will decrease. The lower the conductance of the solution and the larger the cell size, the more this effect is emphasized. The reasonable limit of high-frequency electroporation is about 1 MHz (Kotnik *et al.*, 1998).

6.5 ELECTROPORE FORMATION AND RESEALING

Whilst field-induced polarization has been theoretically understood and experimentally shown, the consequent pore formation and resealing still remain unclear. However, it is known that when polarization of the cell membrane increases to the threshold level (reported to be between 0.2 and 1.0 V), permeabilization of the lipid structure begins to occur (Tekle *et al.*, 1990; Teissié & Rols, 1993). In Fig. 6.2 these threshold levels are shown as outer and inner circles, respectively. As expected, in regions where the membrane faces the electrodes, $V(\varphi)$ reaches the threshold level first and consequently causes permeabilization of these regions. Since V_{rest} is negative for most living cells, the total membrane potential during electric field exposure has the highest absolute values at the hyperpolarization side of the cell. In practice this results in a highly uneven and asymmetrical permeabilization of the membrane (Tekle *et al.*, 1994).

Since there is very little direct experimental evidence regarding electropore creation and resealing, progress toward a mechanistic understanding has been based mainly on theoretical models involving transient aqueous pores (Neumann *et al.*, 1992). The present view is that both the transient pore population, and possibly a small number of metastable pores, may contribute to the electrically induced permeabilization (Weaver and Chizmadzhev, 1996). Heterogeneous distribution of pore sizes can be expected for at least two reasons: (1) uneven polarization of the membrane, which is a primary event of underlying pore formation, and (2) participation of the thermal fluctuation together with polarization energy within the membrane in making pores. However, there are also indications that larger, long lasting pores which are believed to be of hydrophilic character may develop from the smaller precursors – hydrophobic pores. Although the existence of hydrophobic pores is plausible, it is unlikely that they can be visualized by any present form of microscopy because of their small size and short lifetime. According to Neumann *et al.* (1992), the transition from hydrophobic to hydrophilic pores occurs when pore diameter reaches a critical value which is equal to the thickness of the membrane (Fig. 6.3). Present knowledge also suggests that the minimum pore size is about 1 nm and that the

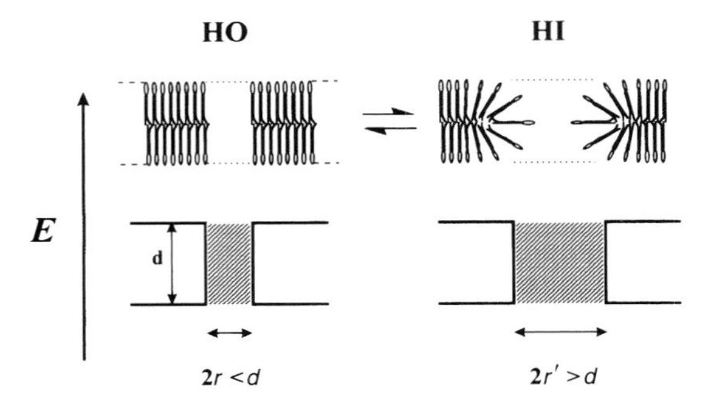

Figure 6.3 Schematic representation of aqueous pores formed as a consequence of a strong electric field (E) in the lipid bilayer. On the left of the figure is a hydrophobic pore (HO) which is smaller in size and of shorter lifetime. A larger, hydrophilic pore (HI) is believed to develop from a hydrophobic pore if the size reaches the critical dimension ($2r=d$). The figure is adapted in part from Neumann *et al.* (1992).

fractional aqueous area of the membrane caused by electroporation is less than 0.1% of the total membrane area, even with the strongest exposure.

Shortly after pore formation, spontaneous recovery of the membrane begins to return the membrane to its original state in which only a few pores existed. Both artificial planar bilayer membranes and membranes of the living cells are capable of experiencing reversible electroporation. Small hydrophobic pores reseal in the timescale of milliseconds, while metastable pores can be open for hours and sometimes do not reseal at all. Irreversible electroporation is not fully understood, but it is thought that the membrane integrity may be destroyed by expansion of one or more supracritical pores.

6.6 TRANSMEMBRANE TRANSPORT

The most important result of electroporation in terms of biological research is the massive increase in molecular transport that it causes. The transport is dependent on not only the number and characteristics of the created pores, but also on underlying driving forces that cause movement of molecules and ions in the preferential direction (Weaver, 1995).

The main characteristics of the electropores which influence transmembrane transport are their size and lifetime. Although electroporation results in the creation of a heterogeneous pore population, one can still determine an upper limit for pore size and lifetime for a particular set of exposure parameters. This means that electroporation offers a unique ability to exert size-exclusion in the molecules that are transported via its newly created pores. This feature makes it possible to control the size of molecules that can move across the membrane.

The three main mechanisms which contribute to molecular movement through the created pores are drift, diffusion and convection. Drift is defined as the velocity in response to local physical fields. In the case of electroporation the local electrical field is represented by the polarization of the plasma membrane. This means that induced polarization of the membrane has a dual role: (1) it causes the formation of the pores, and (2) it provides a local driving force for charged molecules and ions. Whether the electrical force is in favour of electroloading depends on the polarization of the molecules and the membrane. Diffusion is a random microscopic process which depends on local concentration gradients. The difference between the intracellular and extracellular concentration generates the driving force which, in addition to the electric field gradient, defines transmembrane movement of the molecules. Convection, the last mechanism for transport through the pores, is associated with fluid flow carrying dissolved molecules. However, it probably does not contribute to a major extent in electroloading.

It is important to note that although electroporation is commonly used for an introduction of desired molecules into the cells, such created pores enable unhindered transport of all molecules as long as the pore is large enough to accommodate the molecule. This means that as electroloading is occurring, porated cells are undergoing the 'electroemptying' of intracellular components which follow the same physiochemical principles mentioned above.

As well as its ability to transport molecules via electrically created pores, it has been reported that electroporation can induce rapid uptake of exogenous molecules by endocytosis (Glogauer *et al.*, 1993). This observation is not general and probably relies on the triggering of a signalling cascade, either by an alteration in the intracellular concentration of signalling molecules as a result of permeabilization, or by effects associated with possible conformational changes directly caused by the electric field.

6.7 PRACTICAL CONSIDERATIONS OF ELECTROPORATION

6.7.1 Electrical and environmental parameters

The knowledge that electroporation is a direct consequence of a field-induced polarization, and that such polarization can be estimated by Eq. 6.5, makes it possible to identify the main parameters that influence the level of permeabilization as: (1) electric field strength E; (2) electric field duration t, and (3) the dimension of the cell (for spherical cell r).

Both exposure parameters positively influence electroporation, i.e. the stronger the field and longer the exposure is applied the greater the amount of permeabilization will occur. However, such simplification of electroporation parameters can be misleading. This is because polarization is a position-dependent function ($\cos \varphi$), which means that the permeabilized area is dependent on E in a complex non-linear manner. The result of this is that if the applied field is increased from an initial value below the electroporation threshold, then as the applied field starts to exceed the threshold the permeabilized area increases rapidly, but as the field continues to rise the permeabilized area increases much more slowly. Thus increasing the intensity of a field is not always an effective way to increase the area of electroporation. A further problem that arises from simply increasing the field strength is that the pores at the centre of the electroporated area are effectively created by a membrane field which is much stronger than the field which creates pores at the edge of the electroporated area. Such uneven permeabilization can diminish the size-exclusion feature of electroporation, creating a very heterogeneous pore population. Pores created with very strong fields are much more likely to prejudice cell viability.

As mentioned above, the level of electroporation can also be increased by prolongation of the field exposure. For square pulses, the duration of high-intensity exposure (electric fields of 1 kV cm^{-1} and above) is usually in the microsecond region, while pulse duration of a low-intensity field (several hundred V cm^{-1}) can be up to a few hundred milliseconds. Since there are different shapes of pulses in use, it is difficult directly to compare and standardize electroporation parameters. For instance, the meaning of the field intensity and duration is completely different for a square pulse than it is for an exponential exposure. While the square pulse retains the maximal field intensity for an entire exposure period, an exponential exposure delivers a very short peak of maximal field followed by an exponential decay of intensity. To somehow compare these exposures a method of delivering the same energy to a sample can be implemented, but it is a question whether the energy factor is directly associated with a level of electroporation. As well as square and exponential pulses, in some studies sinusoidal, DC-shifted sinusoidal, square-bipolar and triangular shape of exposure have been used. So far, there are no general recommendations as to which shape of field should be used for which purpose, but for electroporation of mammalian cells a square pulse is used more often and an exponential pulse is associated more with electroporation of microorganisms.

Total exposure time can also be increased by applying a succession of pulses (n), each of duration t. Either increasing duration t or the number n will increase induced membrane permeabilization, but the threshold level of electroporation will remain unchanged. However, as long as the total duration nt is kept constant, the permeability induced by electric field above the threshold value increases with increasing number of pulses (Kinosita & Tsong, 1977a; Schwister & Deuticke, 1985; Serpersu *et al.*, 1985; Rols & Teissié, 1990a).

Equation (6.5) also indicates that induced polarization is proportional to the size of the cell. This is the reason why microorganisms have to be exposed to much higher fields than animal cells, which are larger. A typical field intensity for electroporation of animal cells is about 1 kV cm^{-1}, while for electroporation of microorganisms the field intensity used is usually ten or more times higher. The size-effect of electroporation also means that if cells in the population are not of the same size, different levels of permeabilization can be expected. This effect can be partially overcome by electroporation with a pulsed electric field oscillating at radio frequency (RF). The main advantage of using such an oscillating electric field is to counterbalance the cell size-effect with an opposite effect of cell relaxation (Chang *et al.*, 1992). The further advantage of using oscillating fields relies on the induced vibration motion of molecules in the cell membrane it creates due to an electromechanical coupling effect. This motion is equivalent to a localized sonication, which can produce mechanical fatigue in the membrane and thus enhance the formation of the pores.

The most significant drawback of conventional electroporation is that the electric field causes permeabilization primarily restricted to the areas of the cell facing the electrodes. This means that cell shape is a very important determinant of electroporation efficiency. Only perfectly spherical cells undergoing spontaneous movement allow a homogeneous electropermeabilization of populations of cells. Unfortunately not all cells can be electroporated under conditions in which they retain spherical shape. Instead, many situations exist where it would be advantageous to be able to introduce molecules into adherent cells still attached to their substrate. This is especially the case for the cells which are to be used for microscopic visualization. It is also worth noting that some cells, such as neurons and endothelial cells, need to adhere to a basal surface to maintain their characteristic morphology and physiology. Such adherent cells usually flatten considerably and occupy unpredictable irregular shapes. A common finding with adherent cells is that the sensitivity of various regions of the membrane to electroporation is not uniform. This means that the chances of poration depend upon the orientation of the cells relative to the direction of the electric field. Therefore, a significant proportion of cells may not be permeabilized at all if they are in an awkward orientation.

The increase in density of the treated cells has a negative effect on the efficiency of electroporation (Susil *et al.*, 1998). This is again very important if confluent layers of the cells or tissue slices are to be electroporated, which means that stronger exposure conditions have to be implemented.

To increase efficiency of electroloading of cells with complex geometry we developed a multipolar high-frequency electroporation system (patent no. GB9804246.8) which was designed specifically for electroporation of the cells adherent on coverslips. The system consists of an electroporation chamber containing six electrodes and associated electronic circuitory. The electrodes are positioned at 60° intervals around a central portion of the chamber in which the coverslip containing cells is placed. During the exposure, the direction of the electric field is switched through a succession of six directions with frequency up to 100 kHz. In each orientation of the field, a sufficient field strength is applied to cause permeabilization in the cell surface nearest the electrodes. Thus, each time the field switches to a new direction, a new area of permeabilization is correspondingly induced. Such a method allows permeabilization of a much larger area of the cell membrane than normally occurs with a monopolar field with substantially no increase in the strength of the field. The importance of multidirectional exposure is particularly emphasized if confluent cells are to be electroloaded (data not shown).

Another plausible but often forgotten tip for electroloading is that electroporation, if carefully assayed, is a fully reversible process. The practical implementation of this is that for some applications it could be more appropriate to try to load the cells during a series of consecutive exposures, each delivered after the period of time needed to recover the cells, than to try to load the cells in 'one go' by increasing the parameters of exposure. We have managed to induce repetitive loading in different cell types for ten and more times, over a period of 2 hours. Certainly, in such a case more attention should be paid to environmental parameters to keep the cells in good condition for a longer time.

As well as electrical parameters of the exposure and geometrical parameters of the biological sample which relate directly to the nature of electric field action, there are also environmental parameters which can substantially affect the level of electroloading. The most important is the temperature of the sample. Electroporation can be performed in a broad range of temperatures varying from 4°C (on ice) to 37°C, but the kinetics of pore resealing is strongly dependent on the temperature of the medium: the higher the temperature, the more rapid resealing. There is a difference in the kinetics of resealing by about one order of magnitude between 37°C and 20°C, and another between 20°C and 4°C (Kinosita & Tsong, 1977b; Serpersu *et al.*, 1985; Escande-Géraud *et al.*, 1988; Michel *et al.*, 1988). The resealing process also appears to be dependent on ionic strength (Rols & Teissié, 1989), osmotic pressure (Rols & Teissié, 1990b) and the presence of membrane perturbing agents (Rols & Teissié, 1990b).

It is important to mention that electric field exposure generates a temperature rise in the chamber. This temperature rise is dependent on the amount of electrical energy delivered to the samples, as the electrical energy is transformed to heat. Heating of the sample (ΔT) rises linearly with increasing exposure time (t) of electroporation, but it rises to the square of the electric field intensity (E). For a square pulse, the temperature rise (ΔT) in a medium of specific resistivity ρ can be calculated as:

$$\Delta T = E^2 \, t/4.2r \qquad [6.8]$$

This means that if high-intensity electric fields are implemented, careful calculation or measurement of the temperature rise should be carried out. The increase in temperature can be almost abolished if a low-conducting, basically non-ionic, solution is used, but such an approach reduces the ability to manipulate the composition of the extracellular solution in order to achieve optimal parameters. Also with regard to the ion composition of the extracellular solution, one should pay particular attention to the concentration of divalent cations, especially Ca^{2+} and Mg^{2+}, since both are important in stabilizing the lipid structure. However, a prolonged increase in intracellular Ca^{2+} concentration can lead to cell death.

6.7.2 Chamber design

Practical considerations regarding the design of an electroporation chamber are mostly concerned with the correct geometrical construction for the particular purpose. For a successful chamber design one should take particular care that the electric field in the chamber is homogeneous, even if more than one pair of electrodes is being used. This can be achieved by the symmetrical placement of electrodes and coupling of several electrodes together during the exposure. Since the current density on the electrodes may be very high during the pulse, a substantial amount of toxic electrolytic product may be generated. This can be particularly important if electroporated cells are to be used for studies of normal cell physiology. To avoid the harmful effects of these products, electrodes can be kept separate from the cells, and cells constantly perfused with fresh solution. The use of electrodes made of inert materials such as platinum should further diminish possible toxicity. Prevention of toxic effects can be assured by the application of charge-balanced exposure which generates substantially less toxic product than in a monopolar electric field. The last way of decreasing production of electrolysis is by electroporating in low-ionic medium which decreases the total current flow through the chamber.

It is also worth considering that it is better to design the chamber in such a way to allow electroporation in a small volume of extracellular solution. This recommendation deals more with the cost-effectiveness of the method rather than any technical advantage, although smaller volumes are easier to heat and cool if temperature is to be manipulated. Cost-effectiveness is important when dealing with the fact that the electroporation usually does not allow loading of more than a few per cent of extracellular molecules. This means that a reasonable starting point for a chosen concentration of extracellular molecules to be introduced is about 100 times higher than the final intracellular concentration. When the electroporation parameters are optimized this concentration can be decreased, but the simple fact is that the greater the concentration gradient you have, the less severe the parameters of electroporation you can use.

In the case of loading fluorescent probes, especially ones that are of heavier molecular weight, for example dextran-conjugated forms, the price of the loading in bigger volumes may soon be unacceptable unless a better price is negotiated with a supplier. Our chambers are designed to carry the electroloading procedure in 25–100 μl of extracellular solution, which leads to an acceptable price per experiment ratio for most applications.

6.8 EXPERIMENTAL EVIDENCE

Although there are numerous reports concerning electroporation that can be very helpful when attempting to optimize the electroporation procedure, very few of them can be directly implemented for electroloading of fluorescent probes. In the following paragraphs several reports which show that fluorescent probes can be successfully loaded with minimal cell damage are discussed.

An interesting study that implemented electroloading to examine intracellular Ca^{2+} signalling has been published recently by Horne and Meyer (1997). This report is important because it shows that several different molecules, which are good candidates to be introduced in a range of studies, can be loaded to the operational level without affecting cell viability or physiology. The authors used a custom-made system to introduce: (1) dextran-based calcium indicator CAAX green, an immobilized indicator that binds non-specifically to internal membranes, which was used to provide information on highly localized Ca^{2+} events that occur near the releasing sites; (2) calcium chelator EGTA, which was used to restrict Ca^{2+} signals to their initiation sites and prevent global response; and (3) caged InsP3 which was used to trigger Ca^{2+} from $InsP_3$-sensitive stores. As well as providing evidence that electroporation can be successfully used for the molecules mentioned, a further important observation can be made. Since the various Ca^{2+} indicators are variations of the non-fluorescent calcium chelators EGTA and BAPTA, the ability to load cells to a very high intracellular concentration of EGTA, estimated in the study to be up to 1 mM, is a good indication that the method could be successfully implemented for loading free-salt forms of Fura-2, Fluo-3, Indo-1 and other Ca^{2+} indicators including Bis-fura-2, which at present is not available in AM form.

Another study of particular interest for the purpose of loading fluorescent probes investigated the effect of electric field pulse shape, intensity and duration and loading of fluorescence-labelled dextrans of different weights (Liang *et al.*, 1988). It was shown that dextrans of less than 41 kDa can be loaded with fields of lower pulse strength and shorter pulse duration. There was no detectable advantage in using a rectangular pulse over using an exponential decay pulse of similar power.

Besides the system mentioned above (Horne & Meyer, 1997), there are also further designs of electropermeabilization systems suitable for loading adherent cells which can then be used for microscopic visualization. One of them uses cells which are grown on electrically conductive, optically transparent indium tin oxide, which serve as an electrode (Raptis *et al.*, 1994, 1995). Prevention of toxic effects can be particularly important in such a chamber since cells are very close to the electrodes. If the method is to be adopted one should also be aware that transparency of indium tin oxide decreases as UV wavelengths are approached.

Another system for electroporation adherent cells comprises a chamber with electrodes that can be positioned at a defined distance above the cell layer on the coverslip (Bright *et al.*, 1996). Unfortunately, parameters of electrical exposures in that report are given as generator voltage output and capacitance instead of electric field intensity, shape and exposure duration. This prevents exact estimation of the electric field that the cells actually experienced in the chamber.

We have also developed an electroporation system specifically for loading of adherent cells. It is based on high-frequency multipolar exposure which can be performed in a well-defined, near-homogeneous field in a charge-balanced manner. The chambers are of small volumes and non-toxic platinum electrodes are used. There is a possibility of fast perfusion as well as heating and cooling of the sample. Although the system is designed for attached cells it can also be used for free-floating cells.

One of the areas where electroloading of fluorescent probes may have important implementation is in plant science, where the AM form of the probes does not provide a sufficient loading. Authors reported partial or complete success in loading fluorescent probes in protoplasts from the root tip of mung beans (Gilroy *et al.*, 1986), *Dryopteris* spores (Scheuerlein *et al.*, 1991) and pollen grains (Obermeyer and Weisenseel, 1995).

6.9 SUMMARY

Although it is accepted that electroporation potentially provides a simple tool for cell loading, the method still has not reached the level of broad application in cell science. Scepticism about the wider use of electroporation has probably been due to the lack of interdisciplinary approaches to understanding the underlying mechanisms of the phenomenon. This has caused some unjustified concern that electroporation must affect cell viability and physiology. In recent years we have witnessed considerable effort in pursuing basic knowledge about the phenomenon and in showing that electroporation could be an effective tool, yet gentle on the cells.

Formation of pores which are of smaller size and reseal quickly enable selective loading with controllable loss of intracellular components. Many molecules which are targets for introduction into living cells, such as various fluorescent probes, are of small size and so there is no need to create pores larger than required as the same effect can be achieved with less membrane and cell disruption. Such treatment can then be repeated over a period of time until the desired level of loading is achieved.

Electroporation has a number of advantages: (1) the method potentially has a wide application in cell science where ester loading of cells is not practical and where such work has to be undertaken by microinjection; (2) for rapid loading of large populations of cells; (3) for *in situ* loading of cells in tissue slices; and (4) in other areas where it is required to preserve cell viability during introduction of novel species such as antisense oligodeoxynucleotides.

However, is should be emphasized that for sophisticated manipulation of intracellular contents, electroporation must be used with particular care and understanding and specialized systems must be developed.

ACKNOWLEDGEMENTS

Development of the multipolar electroporation system was financially supported by Babraham's Institute DEBS Initiative. The author would also like to acknowledge the generous financial support of the EMF Biological Trust, the Royal Society Cooperation Grant Scheme and the continuous support and enthusiasm of Dr W.T. Mason, Dr P. Lipp, Dr M.D. Bootman, Professor L. Vodovnik and Professor D. Miklavčič. All credit for electronic and mechanical design goes to Mr J.

Osborne and Mr F. Head, and credit for experimental work goes to Dr N.A. Parkinson and Miss A.O. Ige.

REFERENCES

Bright G.R., Kuo N.-T., Chow D., Burden S., Dowe C. & Przybylski R.J. (1996) *Cytometry* **24**, 226–233.

Chang D.C. (1992) In *Guide to Electroporation and Electrofusion*, D.C. Chang, B.M. Chassy, J.A. Saunders & A.E. Sowers (eds). Academic Press, London, pp. 9–27.

Chang D.C., Hunt J.R., Zheng Q. & Gao P.-Q. (1992) In *Guide to Electroporation and Electrofusion*, D.C. Chang, B.M. Chassy, J.A. Saunders & A.E. Sowers (eds). Academic Press, London, pp. 303–326.

Escande-Géraud M.L., Rols M.-P., Dupont M.A., Gas N. & Teissié J. (1988) *Biochim. Biophys. Acta* **939**, 247–259.

Gilroy S., Hughes W.A. & Trewavas A.J. (1986) *FEBS Lett.* **199**, 217–221.

Glogauer M., Lee W. & McCulloch C.A.G. (1993) *Exp. Cell Res.* **208**, 232–240.

Hibino M., Itoh H. & Kinosita K. Jr (1993) *Biophys. J.* **64**, 1789–1800.

Hofmann G.A. & Evans G.A. (1986) *IEEE Eng. Med. Biol.* **5**, 6–25.

Horne J.H. & Meyer T. (1997) *Science* **276**, 1690–1693.

Kinosita K. Jr & Tsong T.Y. (1977a) *Biochim. Biophys. Acta* **471**, 227–242.

Kinosita K. Jr & Tsong T.Y. (1977b) *Nature* **268**, 438–441.

Kinosita K. Jr, Ashikawa I., Saita N., Yoshimura H., Itoh H., Nagayama K. & Ikegami A. (1988) *Biophys. J.* **53**, 1015–1019.

Kinosita K. Jr, Hibino M., Itoh H., Shigemori M., Hirano K., Kirino Y. & Hayakawa T. (1992) In *Guide to Electroporation and Electrofusion*, D.C. Chang, B.M. Chassy, J.A. Saunders & A.E. Sowers (eds). Academic Press, London, pp. 29–47.

Kotnik T., Bobanović F. & Miklavčič D. (1997) *Bioelectrochem. Bioenerg.* **43**, 285–291.

Kotnik T., Miklavčič D. & Slinik T. (1998) *Bioelectrochem. Bioenerg.* **45**, 3–16.

Liang H., Purucker W.J., Stenger D.A., Kubiniec R.T. & Hui S.W. (1988) *Biotechniques* **6**, 550–558.

Michel M.R., Elgizoli M., Koblet H. & Kempf, C. (1988). *Experientia* **44**, 199–203.

Miklavčič D., Beravs K., Šemrov D., Čemažar M., Demšar F. & Serša G. (1998) *Biophys.* J. **74**, 2152–2158.

Mir L.M., Banoun H. & Paoletti C. (1988) *Exp. Cell Res.* **175**, 15–25.

Neumann E. & Boldt E. (1990) *Clin. Biol. Res.* **343**, 69–83.

Neumann E., Sprafke A., Boldt E. & Wolf H. (1992) In *Guide to Electroporation and Electrofusion*, D.C. Chang, B.M. Chassy, J.A. Saunders & A.E. Sowers (eds). Academic Press, London, pp. 77–90.

Obermeyer G. & Weisenseel M.H. (1995) *Protoplasma* **187**, 132–137.

Potter H. (1988) *Anal. Biochem.* **174**, 361–373.

Raptis L.H., Brownell H.L., Firth K.L. & MacKenzie L.W. (1994) *DNA Cell Biol.* **13**, 963–975.

Raptis L.H., Liu S.K.-W., Firth K.L., Stiles C.D. & Alberta J.A. (1995) *Biotechniques* **18**, 104–114.

Rols M.-P. & Teissié J. (1989) *Eur. J. Biochem.* **179**, 109–115.

Rols M.-P. & Teissié J. (1990a) *Biophys. J.* **58**, 1089–1098.

Rols M.-P. & Teissié J. (1990b) *Biochemistry* **29**, 4561–4567.

Scheuerlein R., Schmidt K., Poenie M. & Roux S.J. (1991) *Planta* **184**, 166–174.

Schwister K. & Deuticke B. (1985) *Biochim. Biophys. Acta* **816**, 332–348.

Serpersu E.H., Kinosita K. Jr & Tsong T.Y. (1985) *Biochim. Biophys. Acta* **812**, 779–785.

Susil R., Semrov D. & Miklavčič D. (1998) *Electro Magnetobiology* **17**, 391–399.

Teissié J. & Rols M.-P. (1993) *Biophys. J.* **65**, 409–413.

Tekle E., Astumian R.D. & Chock P.B. (1990) *Biochem. Biophys. Res. Commun.* **172**, 282–287.

Tekle E., Astumian R.D. & Chock P.B. (1994) *Proc. Natl. Acad. Sci. USA* **91**, 11512–11516.

Tsong T.Y. (1991) *Biophys. J.* **60**, 297–306.

Weaver J.C. (1993) *J. Cell. Biochem.* **51**, 426–435.

Weaver J.C. (1995) In *Animal Cell Electroporation & Electrofusion Protocols*, J.A. Nickoloff (ed). Humana Press, Totowa, New Jersey, pp. 3–28.

Weaver J.C. & Chizmadzhev Y.A. (1996). *Bioelectrochem. Bioenerg.* **41**, 135–160.

Zimmerman U. (1986) *Rev. Physiol. Biochem. Pharmacol.* **105**, 175–256.

CHAPTER SEVEN

Imaging Reality: Understanding Maps of Physiological Cell Signals Measured by Fluorescence Microscopy and Digital Imaging

SPENCER L. SHORTE[1] & STEPHEN BOLSOVER[2]
[1] INSERM Unité 261 – Neurobiologie Cellulaire et Moleculaire, Institut Pasteur, Paris, France
[2] Department of Physiology, University College London, UK

7.1 INTRODUCTION

Fluorescent probes can yield quantitative and qualitative information about physiological and biochemical properties in biological samples. Various electronic imaging techniques allow cell-associated fluorescence to be visualized and captured from the microscope as digitized electronic images. Thus, the magnitude of putative biological signals can be estimated and mapped spatially and temporally as they change inside a living cell. However, realizing the biological significance of information collected this way requires data interpretation based on an understanding of the underlying principles of the techniques utilized.

A digital image on a computer screen is, in principle, no different from the line on an oscilloscope, or the banding pattern on a gel. It is data, on the basis of which we form experimentally testable hypotheses and extrapolate biological meaning. However, fluorescence imaging techniques are sophisticated, and the facile representation of an image on a computer screen can be misleading. The information comprising the matrix of a digital image is complex – it is a convoluted two-dimensional numerical map of a three-dimensional reality that changes with time. This must be understood in order to realize the analysis of data from fluorescence imaging experiments. The experimenter must be aware of the possible problems that can affect the particular approach used. True *artifacts* arise from perturbation of the measured signal due to the experimental manipulation or the equipment. Many of the issues discussed below are not artifacts in the strict sense. Usually, the data output from a fluorescence imaging system does indeed have its origin in the interaction of the fluorophore with the biology, but the relationship is not always simple or obvious. Thus, the digital image may reflect the specific interaction of a fluorescent probe with some biological aspect, but it can be dangerous to assume that the information is entirely free from distortions. The experimenter must also, therefore, be able to interpret the data correctly.

FLUORESCENT AND LUMINESCENT PROBES, 2ND EDN
ISBN 0-12-447836-0

This is the key to reaping maximum benefit from what is becoming a powerful and highly flexible collected armoury of techniques for single cell physiology. Here we detail a general discussion of the underlying principles and the current state of the art, intended as a practical overview.

7.2 GENERIC CONSIDERATIONS FOR THE USE OF FLUORESCENT INDICATORS

7.2.1 Fluorescent probes of biological changes

There are a whole variety of fluorescent indicator dyes and conjugates available, enabling relatively specific monitoring of a variety of physiological parameters in living cells (see Haugland, 1996). Specific ion indicators are historically the most popular and are exemplified by the ever growing family of calcium binding dyes that began with quin-2 (Tsien *et al.*, 1982). These, like all fluorescent reporter molecules, ideally exhibit fluorescent emission spectra that are altered proportionally to some specific biological activity, in this case a specific ion species. More recently, new fluorescent probes have become available to measure other physiological parameters. For example, voltage-sensitive dyes have been put to good use, answering questions which up until now were difficult to address using single cell electrophysiology patch-clamp techniques (Zecevic, 1996). Novel lipophilic membrane indicators have been used to measure neurotransmitter release with resolution sufficient to reveal release patterns from single boutons (Ryan *et al.*, 1993), and membrane reuptake following exocytosis in endocrine cells (Shorte *et al.*, 1995; Smith & Betz, 1996). Exocytosis has also been measured using synaptolucins (protein chimaeras comprising specific neurotransmitter vesicle protein coupled to bioluminescent luciferase): when expressed in PC12 cells these enabled neurotransmitter release to be monitored directly by fluorescence (Miesenböck & Rothman, 1997). Another protein chimaera comprising calcium binding protein calmodulin attached to mutated spectral variants of Green Fluorescent Protein (GFP) has allowed a novel means for measurement of cytosolic calcium ion concentration (Miyawaki *et al.*, 1997). The latter approach used Fluorescence Resonance Energy Transfer (FRET; Stryer, 1959; Fairclough & Cantor, 1978) to monitor intramolecular energy transfer over GFP2-calmodulin. FRET measurements have also been used, in living cells, to measure membrane voltage changes (Gonzalez & Tsien, 1995) and intracellular cAMP concentration (Adams *et al.*, 1991). Such novel approaches hold much promise for the future and are likely to provide the means to measure a whole host of biological parameters (Tsien *et al.*, 1993; Heim & Tsien, 1996; Pozzan, 1997; see also other chapters). Nonetheless, in common with more established techniques, the same general assumptions need to be addressed with adaptive provisos to allow meaningful data interpretation.

7.2.2 Ideal relationships between fluorescence and biological changes

The change in the spectral characteristics of a fluorescent indicator gives us information about the qualitative and quantitative progression of a biological signal in labelled intracellular compartments of a living cell. The simplest measurement of dye fluorescence tells us how brightly the dye is fluorescing at the moment it is excited at a particular wavelength (given a selected bandwidth for the emission light) – no more and no less. More sophisticated techniques allow other fluorescence parameters to be measured too, for example FRAP (Fluorescence Recovery After Photobleaching), dichroism and fluorescence polarization. However, for this discussion it is sufficient to consider only fluorescence intensity measurements, since the assertions hold true for other techniques too.

Ideally, the magnitude of fluorescence intensity has a simple calibratable relationship to the magnitude of the physiological signal being measured, over a wide dynamic range. Intensity changes, given a number of hypotheses, serve to allow the experimenter to extrapolate conclusions concerning the characteristics of the biological signal inside the intracellular, dye-containing compartment. For the cell physiologist, this is especially interesting when quantitative estimates of the magnitude of a biological signal can be made. For example, cytosolic ion concentrations where for an indicator dye I which binds a specific ion species X, ideal fluorescence measurements give information about the spatial distribution of changing ion concentration over time. However, in order to make statements about ion concentrations on the basis of fluorescence measurements, we must make three hypotheses concerning (1) the behaviour of the dye, (2) the location of the dye and (3) the effect of the measurement protocol on the biological phenomenon being studied. We will now consider these three hypotheses, and the extent to which they are likely to be valid under particular circumstances, in turn.

7.2.3 The link between measured fluorescence and estimated ion concentration consists of three hypotheses

7.2.3.1 Hypothesis 1: The measured fluorescence intensity change has a simple, quantifiable relationship to the ion concentration

For example, fluo-3 is one of a family of visible-wavelength-excited calcium binding dyes, whose fluorescence emission is altered by changes in calcium concentration (Minta *et al.*, 1989). Thus, one explanation for an increase in fluorescence is an increase in the amount of dye complexed with calcium. However, this is not the only possible explanation. For many fluorescent probes including fluo-3, a change in the fluorescence signal may instead be a consequence of a change in the chemical environment of the dye; notably the viscosity and ionic strength can alter fluorescence emission non-specifically (Poenie 1990; Perez-Terzic *et al.*, 1997). Other sources of change that might lead to fluorescence changes must also be considered. For example, contaminant heavy metal ion species are present in some cell types (and laboratory water sources), and have a particularly high affinity for some ion indicator dyes whereby they can displace the ion species of interest at very low concentrations (Grynkiewicz *et al.*, 1985; Kao *et al.*, 1989; Haugland, 1996). Suitable control experiments use intracellular (cell-permeable) heavy metal ion chelators such as TPEN [*N*,*N*,*N*′,*N*′-tetrakis-(2-pyridylmethyl)-ethylene-diamine] to 'mop' up such contamination. Where contamination exists addition of TPEN leads to a change in the primary fluorescence signal where no change in the concentration of the ion of interest has occurred.

Fluo-3 is excited at a single wavelength at which a single emission wavelength reports fluorescence changes optimally. As such, dye concentration and cell thickness can alter the reported signal, and the ion concentration cannot be calculated from the fluorescence signal alone. In this respect, ratiometric indicator dyes are an advantageous alternative since they allow measurement of ion concentrations independently from dye concentration, sample thickness etc. The underlying assumption of ratiometric imaging is an extension of Hypothesis 1, thus, the fraction of dye I that has bound ion X ([X:dye]/[dye]) with a dissociation constant (K_d) depends on the local free ion concentration [X] according to the equation [X:dye]/[dye] = [X]/K_d. Ratiometric dyes allow two independent fluorescence signals, whose relative intensities depend on the fraction of dye I complexed with ion X, to be measured using either different excitation or emission wavelengths. Combined as a ratio and calibrated, the measurement allows an estimate of [X] (Grynkiewicz *et al.*, 1985). Of course, the provisos noted above also hold true here. Thus, any change in the chemical environment of the dye-containing compartment may alter the affinity of the dye for X (Perez-Terzic *et al.*, 1997). Also, if spatial and/or temporal gradients of X exist, the finite rates of formation and dissociation of the X:dye complex will mean that [X:dye]/[dye] is not a simple function of [X]. In particular, changes of X that are brief in time or restricted in space may be heavily filtered in the resulting dye signal (Hollingworth & Baylor, 1986; Nowycky & Pinter, 1993). The most popular ratiometric dye for measuring intracellular calcium concentration is fura-2 (Grynkiewicz *et al.*, 1985), however, this dye is excited in the UV range and has led some groups to find alternatives. Notably the use of dye mixtures such as fluo-3 plus Fura Red to allow a ratiometric approach to be utilized with visible wavelength excitation (see Haugland, 1996). However, dye combination gives rise to a whole set of other problems and it is well advised that care should be exercised when considering such an approach (Floto *et al.*, 1995).

7.2.3.2 Hypothesis 2: Cellular fluorescence reports from a homogeneous subcellular compartment, and spatial gradients reflect biological phenomenon

It is not uncommon to observe subcellular anisotropy of measured fluorescent signal. For example, cells loaded with fluo-3 or fura-2, and observed at a single excitation wavelength can appear to have two major compartments – cytosolic, and nuclear. However, it is not necessarily true that all the fluorescence observed is derived from these subcellular compartments. Careful consideration is necessary before interpreting a spatial gradient as biologically meaningful. Implicit conclusions are common in the literature, as is confusion in nomenclature. We emphasize that nuclear and nucleoplasmic are distinct cellular compartments. The nucleoplasm is the aqueous medium within the inner nuclear membrane, in which chromatin is suspended. The nucleus comprises the nucleoplasm and chromatin, the inner and outer nuclear membranes, and a second aqueous space, the intermembrane space, continuous with the lumen of the endoplasmic reticulum. Depending on the techniques used to load the dye, one or other or both of these compartments may contain the dye (Gerasimenko *et al.*, 1995). Next, the cytoplasm describes everything (including organelles) that lies between the outer nuclear membrane and the plasma membrane. The aqueous medium of the cytoplasm, in which the cell's organelle constituents are suspended (e.g. cytoskeleton, mitochondria, Golgi, etc.), is the cytosol. Again, depending on the techniques used to load the dye, it may be confined to the cytosol. However, many loading procedures also result in the dye becoming concentrated in one or more of the membrane-limited organelle compartments. This differential cytoplasmic loading can be a source of apparent cytosolic spatial gradients (Plate 7.1) – when the experimenter is unaware of such dye localization, and assumes that all dye is in the cytosolic and/or nucleoplasmic compartments, severe misinterpretation can result (Connor, 1993; Al-Mohanna *et al.*, 1994). Unwanted organelle dye loading can be detected by lysing the plasma membrane chemically (e.g. with digitonin; Al-Mohanna *et al.*, 1994). The free cytosolic dye then leaks from the cell, and the remaining signal can be attributed to dye that is trapped. On the other hand, it is worth noting that dye localization can also yield extremely interesting physiological information, particularly with regard to ion buffering by organelles. Dye-loading techniques intended to partition the dye into specific subcellular organelle compartments have proved useful for a number of experimental paradigms (Connor, 1993; Hajnoczky *et al.*, 1995; van de Put & Elliot, 1996; Korkotian & Segal, 1997). Indeed, some indicators are now largely used to target organelles, for example the calcium indicator rhod-2 is useful for its preferential partitioning to mitochondrial compartments (Rutter *et al.*, 1996) and C18-fura-2 (a hydrophobic variant of fura-2) is reported to partition to plasma membranes, thereby reducing the contribution of signal measured from the bulk cytosol (Etter *et al.*, 1994).

Fluorescent probes must always be introduced to the cell of interest. In the case of fluorescent ion indicators, most are available as acetoxy methyl (AM) esters. In this form the dyes are membrane permeable since the AM residues are uncharged. Incubation of cells in solutions containing the AM form allows the dye to enter the cytosol and, from there, other cellular compartments until it is cleaved by endogenous non-specific esterases, resulting in the formation of the free acid (FA) dye species. The FA dye form is negatively charged by virtue of the ester-to-carboxylic group transformation, and is no longer permeable to membranes. It thus remains trapped in subcellular compartments, from which it reports fluorescence. In many cases, we want to know the ion concentration in the cytosol and nucleoplasm, and do not want dye to enter other compartments, but three different processes can operate to cause dye to load into organelles. These are described below.

Movement of the free, membrane impermeant dye from the cytosol to organelles

The term membrane impermeant refers to restriction of dye movement by simple diffusion through the lipid bilayer, which is close to impossible for most indicators since they are multiply charged. However, dye molecules can be moved across membranes by an anion transporter. Since this is a specific protein, one would expect it to be present on a discrete subset of organelles. In plants, the effect is particularly prominent for the vacuole, which rapidly clears dyes such as fura-2 from the cytosol. In animal cells, the organelles that take up anionic dye are less well characterized. Uptake can be prevented by use of dextran-conjugated dyes, or by blocking the transporter with drugs such as sulfinpyrazone and probenicid (DiVirgilio *et al.*, 1990).

Pinocytosis/endocytosis of dye micelles

Acetoxymethyl esters are prone to micelle formation in saline solutions (at biological ionic strength) since they behave as weak detergents. The AM dye micelles, and free AM dye molecules are easily adsorbed into pinocytotic vesicles that are the major membrane recycling path for many cell types. AM dye trapped in this way can lead to the formation of an internal dye pool partitioned from the cytosol and therefore responsible for a dye loading problem (Plate 7.1 and Fig. 7.1). Such problems are amplified particularly for exocrine, endocrine and neuronal cells where, in addition to pinocytotic/endocytotic uptake, membrane retrieval following regulated exocytosis (hormone or neurotransmitter release) can lead to a large percentage of AM dye becoming trapped inside intracellular compartments including the Golgi, the endoplasmic reticulum, lysosomal and endosomal pools, and ultimately newly synthesized hormone storage granules. It is sometimes possible to detect micelle uptake by the appearance of punctate fluorescence spots inside the cell. Equally, they can be seen in the cell-free, dye-containing loading solution too (Karaki, 1989). Pinocytotic and endocytotic micelle uptake can be reduced by carrying out AM loading below about 20°C. In most systems, this temperature slows the rate of pinocytosis and endocytosis, but it increases the time necessary to load the cytosol sufficiently. To compensate, dye micelle dispersion into aqueous solution can be improved by making the dye up in physiological solutions containing low concentrations of other weak detergents. This increases the rate and efficacy of loading. Pluronic detergent F-127 is a popular choice and has been used at final concentrations between 0.00025% and 0.001% (w/v) to improve loading with fura-2 (Poenie *et al.*, 1986). Another compound acting analogously is Cremophor EL, a mixture of castor oil and ethylene glycol. This is thought to be less toxic for cells than weak detergents, so can be used at between 0.005% and 0.01% (v/v) final concentration, but dispersion must be aided by light sonication. These suggestions, however, are not 'cure-alls' – they simply

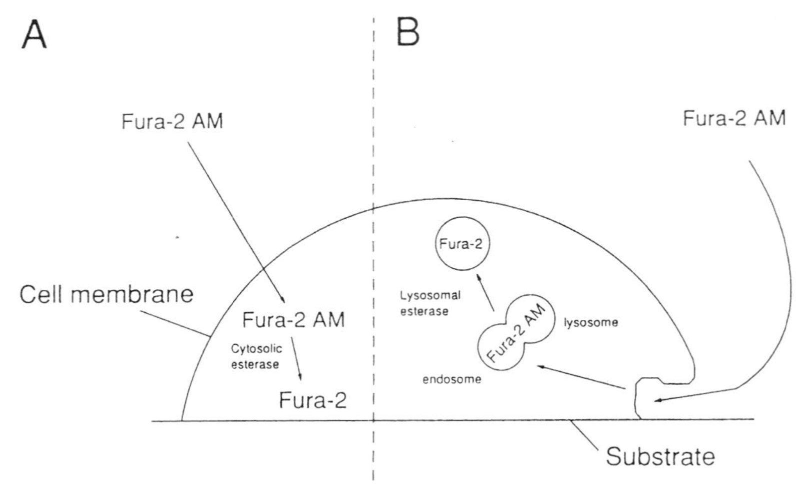

Figure 7.1 Schematic diagram of AM ester loading into living cell. In this scheme the calcium indicator fura-2 is depicted; however, it applies to all AM ester loading protocols. (A) Fura-2 AM traverses the plasma membrane and is converted to the free acid form by endogenous cellular esterases present in the cytosol. This process leads to veritable cytosolic calcium being reported by the dye's fluorescence. However, uptake of the dye can also occur by endosomal uptake (B), especially at the cell–substrate interface. This mode of uptake partitions the dye from the cytosol because it transports the dye through intracellular membrane sorting pathways. This can easily result in erroneous signal being detected (see Plate 7.1).

widen the window within which optimal loading can be achieved routinely for a given cell type.

Movement of the AM ester from the cytosol to the organelle

Since the AM ester is membrane permeant, it can also pass from the cytosol to the interior of membranous organelles. The amount of dye that ends up trapped in particular organelles, versus the cytosol/nucleoplasm, therefore depends largely on the relative activities of esterase in the various compartments and varies greatly between cell types. We are unaware of any protocols that can reduce this component of organelle loading other than eschewing AM loading in favour of direct injection of membrane-impermeable dye (Plate 7.1). This can be achieved using a sharp micropipette, or in the case of electrophysiological experiments the cell can be dialysed through a dye-containing patch pipette so that dye enters the cell cytosol by dialysis after the whole cell configuration is established. Such an approach has been used to maximum benefit in neuronal cell types where very high concentrations of dye are introduced, allowing fluorescence to be recorded from tiny subcellular dendritic processes that would normally take many hours to load by AM incubation (Schiller *et al.*, 1995; Spruston *et al.*, 1995; Helmchen *et al.*, 1996). If this approach is chosen, then we recommend the use of dextran-conjugated dyes rather than the free acids since in this way all three modes of organelle loading are absent.

7.2.3.3 *Hypothesis 3: The kinetics of a biological signal measured by fluorescence in an observed cell reflects the situation in unmanipulated cells*

This assertion need not be true, and is often the most difficult to design controls for. Measurement of dye fluorescence introduces a number of factors unique to the observed cell and these are, therefore, artifacts sensu strictu. The dye itself can be considered more or less toxic, affecting the cell's physiology (Bolsover *et al.*, 1986), and the method used to load the dye into the cell may also cause damage (Rand *et al.*, 1994). In this context it is worth reiterating that many of the popular ion indicator dyes are weak detergents themselves and at high concentrations can alter the permeability of biological membranes. Another primary issue for concern to the cell physiologist is that the fluorescent ion indicators increase unwanted buffering of the ion X inside dye-containing compartments. Buffering and mobility are particularly important considerations when measuring fast changes from very small subcellular compartments, for example postsynaptic dendritic spines in neurons (Helmchen *et al.*, 1996). Under certain conditions, the dye may behave as a significant buffer, and where the dye is diffusionally mobile inside the cell it will also contribute to and alter the mobility of the ion of interest (Gabso *et al.*, 1997). The net result is a dynamic smearing of ion concentration kinetics, creating false spatial and temporal gradients in [X]. This is especially profound in systems where [X] kinetics are driven by a feedback component. For example, calcium-induced calcium release in some cell types drives repeating cytosolic calcium spikes. These depend on the ability of the calcium release organelle to detect changes in the cytosolic calcium levels, and are therefore sensitive to relatively small changes in the buffering capacity of the cytosol (Bootman & Berridge, 1995). In the extreme they can be abolished by dye loading (Bolsover & Silver, 1991).

There are, in general, two excitation illumination sources used for microscope imaging – conventional and laser light. In either case specific wavelengths are selected in order to give optimal excitation of the chosen indicator dye. The excitation wavelengths used most commonly range from near ultraviolet (around 340 nm), through to visible wavelengths ranging from blue through green to red (400 nm to 700 nm). In some special cases infrared wavelengths are also used to excite fluorescent indicator dyes, for example in the case of multiphoton excitation microscopes (Denk *et al.*, 1990; Xu *et al.*, 1996; Denk & Svoboda, 1997). However, excitation light can cause unwanted photochemical effects to occur in cells. Excitation light sources are capable of delivering high energy that can be deleterious for the observed cell, depending on the power intensity, area of illumination and wavelength of excitation light. Short wavelengths (e.g. UV light), high power intensity illumination (e.g. lasers), combined with high numerical aperture objectives (see final section) can result in high rates of energy delivery that can damage, or alter, the biology being observed (Byrne & Duncan, 1990). There are three primary considerations – temperature, free radical generation, and photochemical effects.

High-energy light can cause microdomain temperature increases. Incident excitation light dissipates its energy over the sample as heat. For example, it was recently concluded that microdomain temperature increases were responsible for an excitation light-induced increase in fluo-3 signal measured from cultured fibroblasts using a single photon laser confocal microscope (Lui *et al.*, 1997). These workers suggested that the observed erroneous, laser induced rise in fluo-3 fluorescence was due to a residual intracellular pool of non-fluorescent AM dye being more rapidly converted to the fluorescent free acid form during microdomain heating induced by the laser's power. Temperature increases during illumination by relatively long wavelengths used for optical trapping microscopy (1000 nm), and two-photon excitation (700–900 nm), have been carefully assessed. Increases reported were 0.1 $\pm$0.3°C per 100 mW of laser power (Liu *et al.*, 1996), and this is significant given that continuous wave lasers typically operate around 400 mW power. These workers assessed the extent of other photochemical effects including changes in DNA structure, cell viability and intracellular pH. They concluded that the long wavelength laser light sources interfered significantly with the physiology of living samples tested. Another report showed that cloning efficiency was reduced, membrane permeability changed, and UVA-like stress was induced by irradiation of CHO cells at 760 nm. It was concluded that the effects arose in part from photochemical activation of cellular multiphoton absorption (König *et al.*, 1995). Thus, exposure to long wavelength (low-energy) light can cause unexpected changes in the biology of living samples. This suggests that visible and UV sources (high-energy light) are also likely to exert phototoxic effects.

Chemical phototoxicity arises from increased generation of free radicals and direct cellular damage. Free radicals are generated from two sources. Firstly, the fluorescent indicator itself can be bleached by light and in this state can spontaneously degrade yielding exogenous free radicals. Secondly, biological molecules are vulnerable to light-induced damage which can result in free radical generation. For example, light-induced cellular stress (e.g. light damage to macromolecules) leads to upregulation of lysosomal activity, an endogenous source of free radical generation. Further to this point, macromolecules necessary for essential cell processes can be insidiously destroyed during illumination and this can cause changes in the observed biology. For studies of developmental processes this may be very important. For example, during early development of fertilized zebra fish embryos, the microtubules of the cytoskeleton are crucial to correct blastula formation. They are exquisitely sensitive to damage by UV light, as found by Strähle & Jesuthasan (1993) using 245 nm illumination, a wavelength that finds little application for commonly used fluorophores. However, such UV-induced cellular damage could also result from accumulated exposure to near UV wavelengths like those used to excite fura-2 fluorescence (340–380 nm). It is noteworthy that a cell's ability to compensate metabolically for free radicals is finite. Thus, accumulative effects can occur when cells are subjected to repeated cycles of illumination like those used for studies employing caged compounds (where high-energy UV flash exposures are repeated; Wang & Augustine, 1995), or time lapse ion measurements. The only way to reduce these problems is to minimize the intensity of excitation light as far as is possible. This is normally achieved using neutral density filters in the excitation light path, commonly reducing output light intensity to between 1% and 5% of that available.

7.3 OPTIMIZATION OF FLUORESCENT LIGHT DETECTION AND BACKGROUND LIGHT CORRECTION

7.3.1 Light Detectors

There are two ways in which fluorescent light emission can be detected electronically – using photomultipliers (and photodiode arrays) or using low level light video imagers, e.g. intensified charge coupled device (CCD) cameras (Tomkins & Lyons, 1993; Art, 1995). Confocal microscopes use photomultipliers for detection, whereas conventional microscopes, and multiphoton excitation microscopes use either photomultipliers or CCD cameras. Photodiode arrays are less popular, finding use for applications that necessitate very high temporal resolution (<0.2 ms per frame, see Senseman, 1996). Photodiode arrays, photomultipliers and electronic cameras are all analogue devices – detection of incident light photons results in an analogue voltage signal output from the detector. This voltage signal must then be digitized before it can be output to a computer for storage, analysis or rendered as an image on a computer screen. As such, regardless of the light detector used in a particular system, there are several guiding principles that hold true for all analogue-to-digital optical detection systems used to measure fluorescence signals from living cells.

7.3.2 Minimizing background signal

Electronically captured images comprise, in addition to the veritable signal of the fluorophore, electronic and optical background signals which combine to produce a background signal that can be defined experimentally as detected signal that does not arise specifically from intracellular dye fluorescence. Background light has two components, acellular and cellular. The importance of correcting for each cannot be overemphasized, since a failure to compensate appropriately for background can lead to considerable data misinterpretation (see Plate 7.2).

7.3.2.1 Acellular background

Acellular background is that which is still present when the cells are removed from the microscope. It has a number of components whose relative amplitude can vary considerably between different set-ups and from moment to moment. Stray light, i.e. from ambient sources such as room lights and windows, is an obvious primary source, and can be considerable when the light detector is very sensitive (which is normally the case). Electronic noise arising from the detector, and the A/D converter electronics (Tomkins & Lyons, 1993) also contributes. Another source of acellular background is incompletely blocked excitation light. This, too, can be considerable, especially when the excitation light wavelength is close to the emission wavelength being detected. It is a result of light leakage through dichroic mirrors and excitation/emission barrier filters which, even when new, are never 100% efficient. For conventional illumination systems, the problem can be made worse still where excitation filters are old – the many hundreds of hours of use that these filters can endure leads to light leakage developing due to accumulated heat damage. Most manufacturers of customized filter sets offer various qualities of glass and coating, which are more or less expensive depending on the additional treatments implemented to avoid this deterioration of optical efficiency. A cheap alternative is to introduce heat block filters into the excitation light path, close to the illumination source.

Technical approaches that utilize fluorescent dye incubation in the observation chamber itself can result in residual dye signal remaining extracellularly. For example, AM dye loading in tissue slice preparations leads to considerable background signals (Leinekugel *et al.*, 1995) because removal of extracellular dye is slow due to its being non-specifically bound to the extracellular matrix. Also, the use of lipophilic membrane dyes such as FM1-43 to measure membrane turnover necessitates their continuous presence in the perfusion medium during the experiment, and although they are not fluorescent in aqueous medium they do weakly bind and fluoresce in the presence of low concentrations of globular protein additives, e.g. albumin. Alternatively, acellular background light can come from the sample incubation chamber (depending on the materials used for its manufacture), and/or the incubation media – both may yield fluorescence/phosphorescence in their own right. Fortunately, objective immersion oils are now available that show little fluorescence across the range of commonly used excitation wavelengths.

Acellular background is relatively easy to correct for, although one should never assume that it is uniform either in space or time, and allowing for one type of variation usually involves ignoring the other. Thus many software packages provide facilities that allow the experimenter to capture a control background image from a field containing no cells, which can then be subtracted on a pixel by pixel basis from the experimental data. This approach accounts for uneven field illumination, which gives rise to spatially non-uniform acellular background, but will miss changes of background occurring during the experiment.

Alternatively, acellular background can be estimated independently for each image of a time series from areas of the field without cells. This approach accounts for variation of background with time, such as can occur due to build up of phosphorescence in the immersion oil, or fluorescence of an added drug, but cannot precisely correct for spatial variation.

7.3.2.2 Cellular background

Cellular background arises from autofluorescence of endogenous cell constituents such as NADH, riboflavin and flavin coenzymes (Taylor & Salmon, 1989). Autofluorescence is never uniform, in either space or time. For example, the cytoplasm of cells fluoresces more brightly than does the nucleus (O'Malley, 1994). Clearly, if not compensated for this can produce apparent spatial gradients of ion concentration where really none exists, and it is essential to design control experiments to confirm that the perceived gradients are real (O'Malley, 1994). One approach used with fluorescent ion indicator dyes is to add a high concentration of 'quenching' ion (e.g. manganese for fura-2). The ion enters the cells and quenches the fluorescence of the indicator, allowing a background image to be sampled from the cell-containing field. Autofluorescence forms a part of that field and can therefore be subtracted (Kao *et al.*, 1989). However, the intensity of autofluorescence signal can also change with cellular redox state (Duchen, 1992; Duchen *et al.*, 1993), which means that the background may change during an experiment. It is impossible to measure background at each time point during the course of an experiment, so autofluorescence is notoriously difficult to correct for. Particular solutions must be found for particular experimental situations. No general solution applies.

7.3.3 Optimizing the input–output relationship of the light detector

Figure 7.2 represents the input–output (I/O) relation of an optimally adjusted detector. On the horizontal axis is the light incident on the detector, that is, the fluorescence from the object plus acellular background. For simplicity, this is represented on an arbitrary scale from 0 to 255. The vertical axis represents the digital signal output from the light detector to the computer. The dynamic resolution of optical detection devices currently available ranges between 8 and 32 bits, so that the digital output after A/D conversion gives integers in a fixed range. For our example we have an 8-bit device, and integers are therefore in the range 0 to 255 (i.e. 2^8 increments), as shown. Thus, the basic property of this optimally adjusted I/O relation-

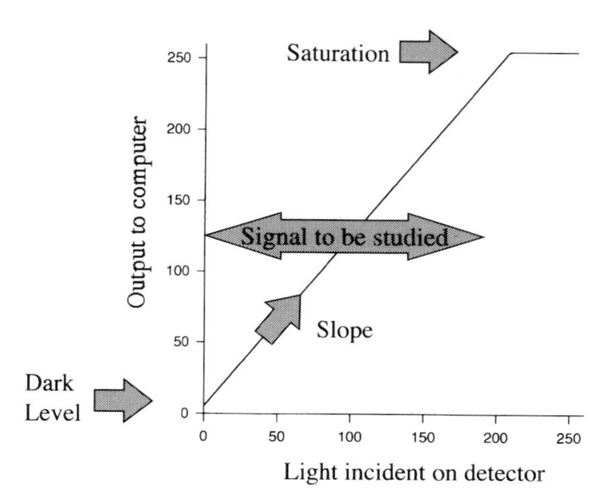

Figure 7.2 The relationship between digital output and the intensity of incident fluorescent light measured using an analogue-to-digital light detection system. The x-axis shows the intensity of incident fluorescent light (arbitrary units), and the y-axis shows the output of a generic 8-bit A/D light detection system. The dynamic range of the system is represented by the values over which the slope increases linearly between the dark level and saturation values indicated. Thus, in this example, light intensities above a value of 200 on the arbitrary scale of the x-axis will not be detected. The signal to be studied must therefore be restricted to values in this range.

ship is that the entire range of the signal to be studied is coded in a linear fashion by the output to the computer. Such a linear relationship is, of course, a prerequisite for further quantitative analysis of the data.

The dark level (often called the offset, or the black level) is the signal received by the computer when there is no light incident upon the detector. In many imaging systems it is possible to adjust the dark level to be negative, so that only light intensities above a threshold level are reported to the computer. This is an advantage to cytologists who want to use the entire (0 to 255) dynamic range of the output to record the structure they are interested in. However, dynamic signal imaging (e.g. ion imaging) requires that the dark level *must not* be used in this way. Rather, it is essential that the dark level be just slightly positive, to ensure that the output sent to the computer is a linear function of the incident light, even at the dimmest regions of the image.

Figure 7.3 shows a field of cells imaged by a Noran confocal microscope. In such a field, one would adjust the dark level such that the pixel values for regions outside the cells, such as point a, are 1 or above. Many instruments provide a software LUT (Look-Up Table) in which values of zero are represented by a colour that stands out dramatically against a grey-scale image on the computer screen. This greatly aids adjustment of the dark level since one simply turns up the dark level until the colour disappears. The slope (Fig. 7.2) reflects what is often called the gain, and determines the range of light intensities that are linearly coded within the dynamic output range. Clearly, if the gain is set too low, the slope will be shallow, and biologically interesting fluorescence changes may not be seen simply because they are not coded for in the output. If the gain is too high, the slope will be steep, and the brightest parts of the image will saturate (i.e. will be reported as 255). Figure 7.3 is saturating in region b. Here the operator has chosen to allow the detector to saturate for one cell, which will therefore be ignored in post-experiment analysis, in order to code the fluorescence changes optimally in the other cells. Dyes such as fluo-3 that dramatically increase their fluorescence upon binding calcium pose problems, since an unstimulated cell that lies well within the linear region of the I/O relation may still saturate the detector when the calcium concentration is high. Figure 7.3 shows an example where the operator got the slope just right: cell c, which appears dim at rest, became bright when cytosolic calcium concentration rose, but nowhere did the pixel values exceed 255. Regions of the image where the output of the light detector saturates must be excluded from quantitative analysis. However, only those pixels and those time points for which saturation occurred need to be excluded (unless the signal is noisy, so that signal clipping is a problem – see Silver *et al.*, 1992). In this respect saturation (see Plate 7.2E) is a less critical problem than inappropriate adjustment of the dark level, which can render the entire data set useless.

7.4 3-D SPATIAL MAPS OF FLUORESCENT SIGNALS

This final section describes some inherent consequences of conventional fluorescence microscope digital imaging of three-dimensional fluorescent light signals. During the last five years some very sophisticated developments have emerged as the current state of the art, and are promising a way to quantifiably map such spatial light signals. For detailed discussions the reader is urged to refer to the cited articles.

7.4.1 The practical contribution of out-of-focus light

The most important optical characteristics for fluorescence microscopy are the refractive index of the

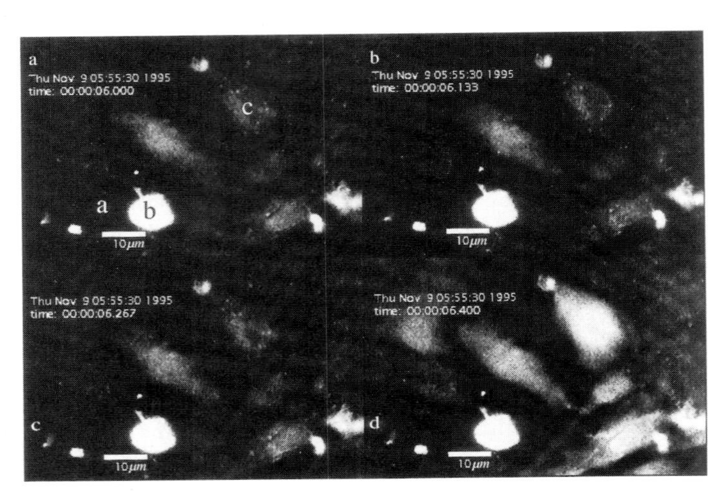

Figure 7.3 Evoked intracellular calcium increase recorded from cells loaded with fluo-3, and detected using a high-speed confocal microscope. Rat cardiac myocytes were loaded with two calcium indicators: fluo-3 (seen in the nucleus and cytosol) and rhod-2 (at least partially sequestered in the mitochondria). The images were acquired using the Odyssey XL confocal laser scanning microscope (Noran Instruments Inc., Middleton, Wisconsin). Images were acquired at a rate of 120 images per second. Averaging every successive 16 frames yielded noise-free images at a final rate of 7.5 images per second. These images show the propagation of calcium waves within rat cardiocytes during spontaneous contractions. The area labelled a is a region of the image that is outside the cells. While b is a region that is saturating the detector. Fluorescence of the cell indicated c is always within the linear range of the detector; this cell shows a dramatic fluorescence increase during the recording sequence. Images courtesy of Dr John Barrett, Department of Physiology and Biophysics, University of Miami (see text).

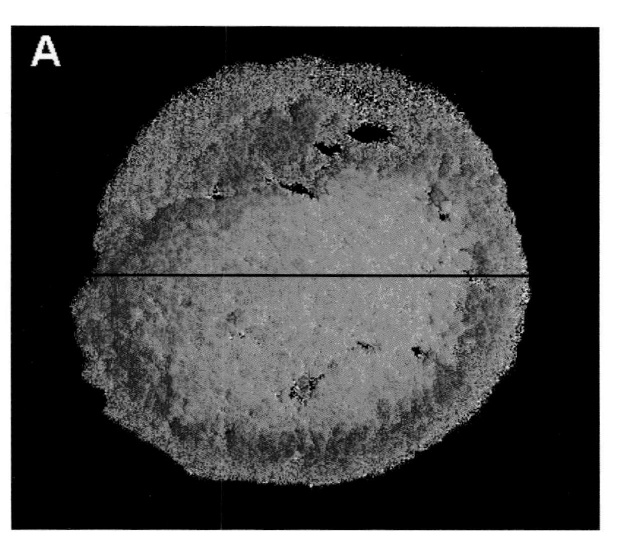

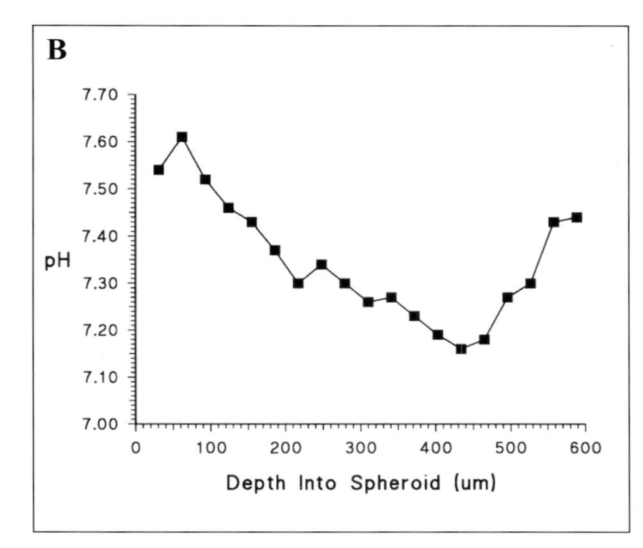

Plate 5.1 (A) A ratio image of the intracellular fluorescence of a SNARF-1-loaded rat C6 glioblastoma spheroid following mechanical bisection. The pH levels of distinct areas of a transect (indicated by the line) through the spheroid are shown graphically in panel (B).

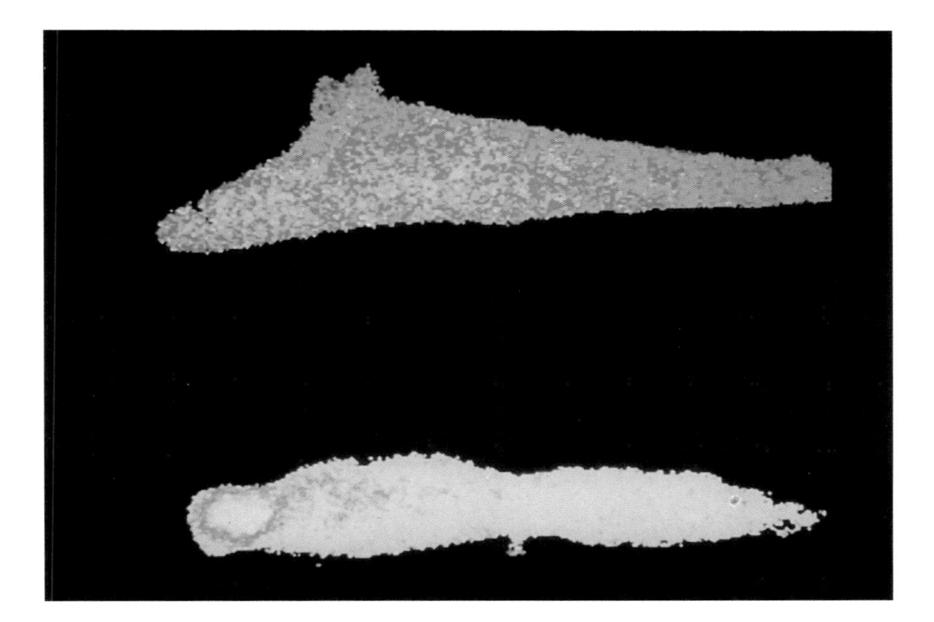

Plate 7.1 The presence of calcium indicator dye inside organelles can cause large apparent signal gradients. The two ratiometric images are pseudo-coloured maps of cytosolic calcium measured using fura-2 in the leading edge of nerve cell axons (the growth cone). Warmer colours indicate higher calcium levels. For each image, the cell body is out of the field of view to the right. The uppermost image shows the 'true' cytosolic calcium signal, as reported by fura-2 free acid injected directly into the cytosol. The signal is relatively uniform and calcium levels are low. For comparison, the lower image shows the signal reported after fura-2 AM loading of the cell at 37°C. Here, the map of calcium shows distinct gradients, notably at the tip of the growing axon. However, this is not a true cytosolic calcium signal: the gradient is due to accumulation of dye into endcytotic vesicles.

Plate 7.2 Artifactual calcium gradients reported by fura-2, caused by inappropriate adjustment of light detection system and background subtraction errors. Images are colour-coded ratiometric maps of calcium reported from a computer-generated model of a fura-2-loaded cell. The ratio levels are calibrated to the colours shown on the right-hand scale. (A) A raw fluorescence image of the dye-loaded cell. The image is brightest at the centre and grades radially to a uniform background. (B) An error-free ratiometric image of the unstimulated cell reports uniformity of cytosolic calcium signal. (C) Ratiometric image of unstimulated cell when the background subtracted from the raw image was overestimated by 20%. The image is not significantly different to the error-free image in (B). (D) Error-free ratiometric image of cell with high cytosolic calcium following stimulation. The signal is still uniform. (E) 'Bull's Eye' artifact in ratiometric image of stimulated cell. The camera detection system has saturated at the centre of the cell; the calcium is reported correctly outside the central region. (F) Another bull's eye artifact, but caused this time by inappropriate background subtraction. Background light was overestimated by 20%. The calcium rise reported is no longer uniform; instead a distinct and artifactual gradient is now apparent. In direct contrast to (E), the error is greatest at the cell edge. Note that the mistake – an overestimation of background by 20% – is the same here as in (C), but the effect on the calcium map produced is dramatically different.

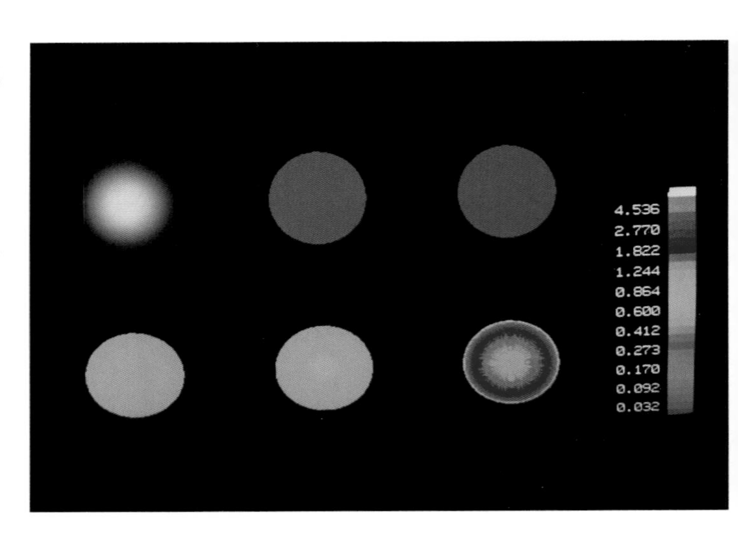

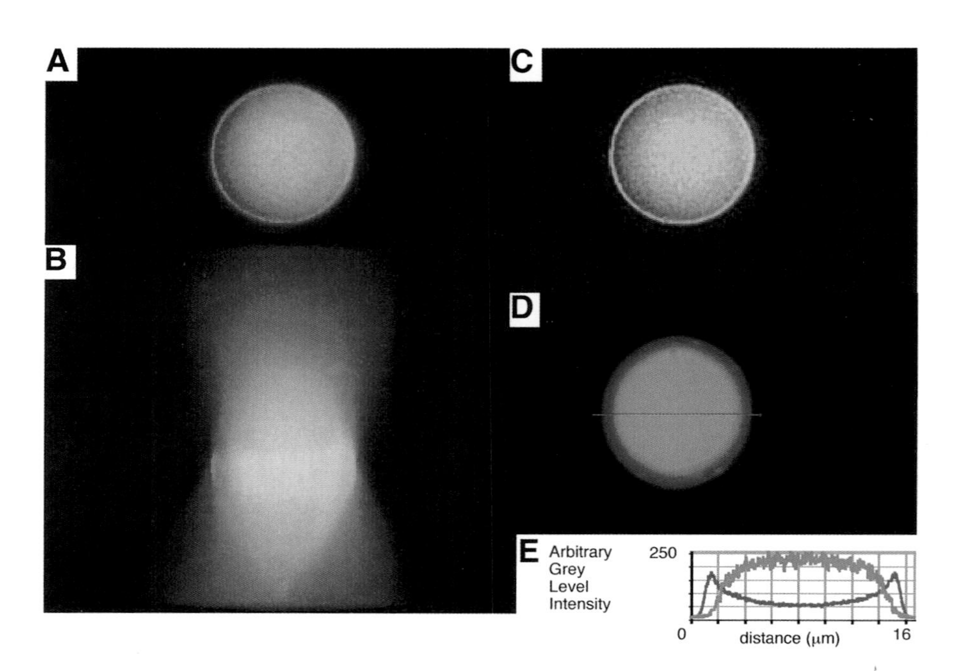

Plate 7.3 The spread of out-of-focus fluorescent light and image distortion as seen by a conventional microscope, and enhanced images using deconvolution, or confocal microscopy. (A) A 15 μm diameter red shell, green interior bead (FocalCheck™, Molecular Probes, OR, USA) was first visualized through its central plane, using two different excitation wavelengths and corresponding emission wavelengths for green (515 nm) and red (680 nm) fluorescent light. The raw black and white images were colour-coded with red or green LUT tables, respectively, and then merged to an RGB composite average as shown. (B) Images were collected as described for (A), except a stack of 100 was collected at each wavelength, representing 100 0.45 μm steps along the *z*-axis to the focal plane. The red and green image stacks were then rendered as a three-dimensional projection, coloured using corresponding red and green LUT tables, and merged. (B) shows the *x,z* projection of the rendered series (see text for discussion). The green signal is greatest at the centre of the bead, where the microscope is looking through the maximum thickness of fluorescent material. (C) shows the same image as (A) except in this case, before colorization and merging, the image pair was treated with the appropriate (wavelength-dependent) 'nearest neighbour' deconvolution algorithm (OpenLab software) to remove out-of-focus haze. Deconvolution has increased the intensity and sharpness of the red shell. However, the green signal is still markedly greater at the centre than at the edge. For comparison, (D) shows the same bead type imaged on a confocal microscope. Resolution of the red shell is not as sharp as in the deconvolved image (C), but the confocal signal more closely approximates the true pattern of fluorescence density in that the green signal is relatively uniform across the bead. (E) shows these phenomena graphically and plots the red and green intensity profiles measured from pixels indicated by the horizontal red line in image (D). In (A) – (C), the objective used was an Olympus 60 ×, 0.9 NA (water immersion/IR). The CCD light detection system was the Princeton Instruments (USA) PentaMax (12-bit) equipped with GENIV intensifier, and the acquisition/processing software was OpenLab 1.7.5 (Improvision, UK). (D) and (E) were acquired using a Zeiss 510 CLSM confocal microscope system, equipped with a 63 × objective (NA 1.2, oil immersion), with pinholes set at 1.0 Airey unit.

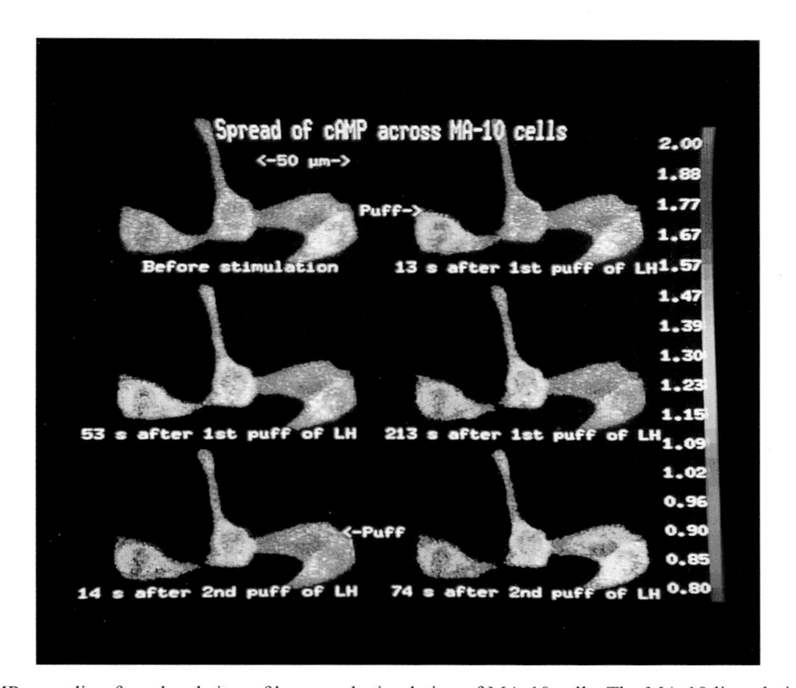

Plate 11.1 Imaging of cAMP spreading from local sites of hormonal stimulation of MA-10 cells. The MA-10 line, derived from a Leydig cell tumour (Ascoli, 1981; Podesta *et al.*, 1991) was grown on glass coverslips and maintained in cell culture in a humidified incubator at 37°C, supplied with 5% CO_2 in Waymouth's medium supplemented with 15% horse serum. These cells are stimulated by gonadotrophic hormones such as human chorionic gonadotrophin (hCG) and luteinizing hormone (LH) to elevate cAMP and increase both biosynthesis and secretion of a mixture of steroids, predominantly progesterone (Ascoli, 1981). Cells were washed with HEPES-buffered Hank's balanced salt solution (HBSS) and placed in a recording chamber maintained at 33°C. Four individual cells were microinjected with FlCRhR[1] and imaged with a high-speed, dual-emission confocal microscope (Tsien, 1990). Excitation at 488 nm was accomplished with an argon laser and simultaneous single wavelength emission images (500 –535 and >560 nm, respectively) were acquired and stored on an optical memory disk. The single wavelength images were background corrected, and a log ratio image was calculated (short/long wavelength). The log ratio image was corrected for any shading errors by subtraction of the log ratio image obtained from a uniform field of a 1:1 (v/v) mixture of 1mM fluorescein in aqueous pH 7.6 potassium phosphate with 1mM rhodamine B in propylene glycol. The corrected image was then pseudocoloured to represent changes in the ratio value. The scale calibrates the colour scale in terms of ratios from 0.8 to 2.0, where 1.0 is defined as the emission ratio obtained from the reference fluorescein–rhodamine B mixture, and low ratios represent low cAMP values and high ratios represent higher levels of cAMP. The resting ratios begin at or slightly below 1.0, and probably correspond to effectively zero free cAMP. A micropipette filled with 20 ng ml^{-1} LH was then brought within 5 – 10 μm of the left-most cell, and at a time = 0, a brief pressure pulse (puff) was manually applied with a syringe to release the hormone locally. The second panel shows the same field of cells 13 s after the puff. In this panel, the left-most cell displays a gradient from orange (ratio about 1.3) near the site of stimulation to blue-green further away. In panel 3, 53 s after the puff, the left-most cell has nearly reached a maximum and its neighbour is beginning to respond. In panel 4, 213 s after the puff, the second cell has increased its cAMP further, either by bath diffusion of the hormone or by diffusion of cAMP through gap-junctions. However, the ratio has still not increased in the two right-most cells, showing that propagation is spatially limited. In panel 5, the pipette was moved to the other side of the field and the two right-most cells were puffed with LH; within 14 s, a small beachhead of cAMP elevation was observable. In panel 6, 74 s after the second puff, the cAMP increase had spread throughout both cells. The dark oval zones within the first and third cells are their nuclei, which partially exclude the FlCRhR. This experiment was performed by Brian Bacskai and Clotilde Randriamampita in collaboration with José Lemos, Ernesto Podesta and Mario Ascoli.

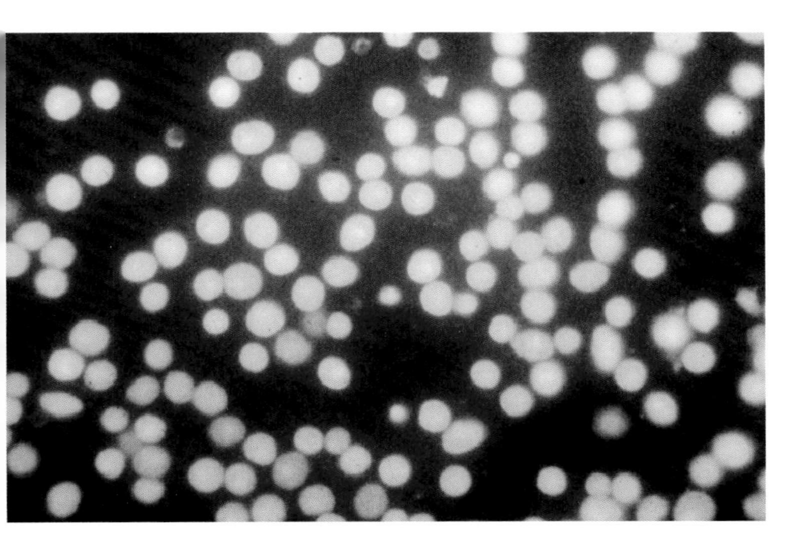

Plate 12.1 Fluorescence micrograph of fura-2/AM-loaded pituitary cells.

Plate 12.2 Selection of charge-coupled devices (CCDs) from a range of manufacturers used in digital cameras employed for ion imaging.

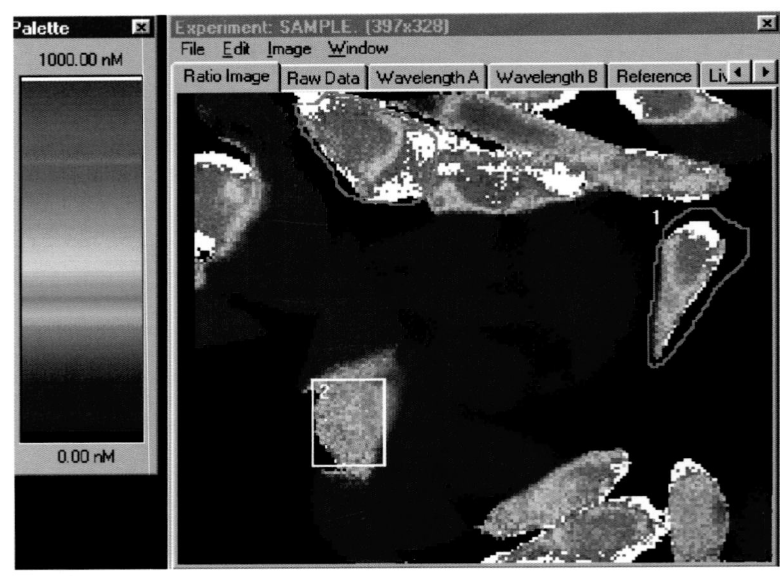

Plate 12.3 Ratio image of two images acquired at 340 and 380 nm excitation wavelengths using fura-2-loaded rat osteoblast preparation.

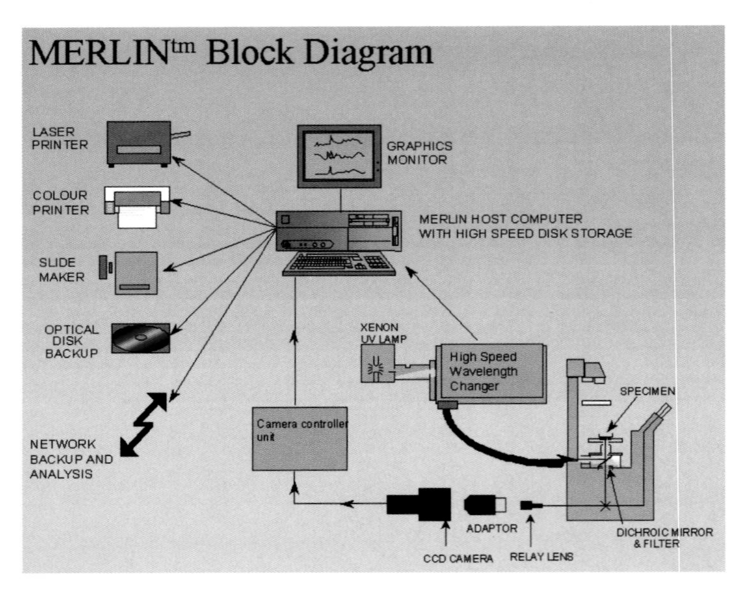

Plate 12.4 Schematic diagram of the MERLIN ion imaging system.

Plate 12.5 User interface of MERLIN ion imaging system showing a time sequence of cardiac myocytes during a spontaneous contraction, using 12-bit data collection. The calcium wave of increasing concentration can be seen to move from the bottom right of the cell up to the top left. The calcium concentration versus time plot can be displayed live on-line simultaneously, as shown on the right of the figure.

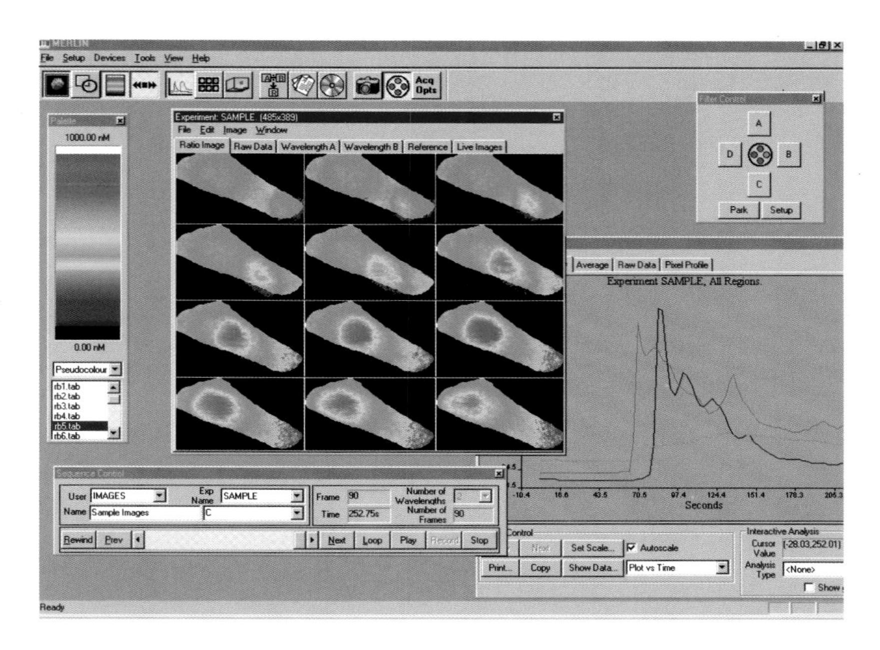

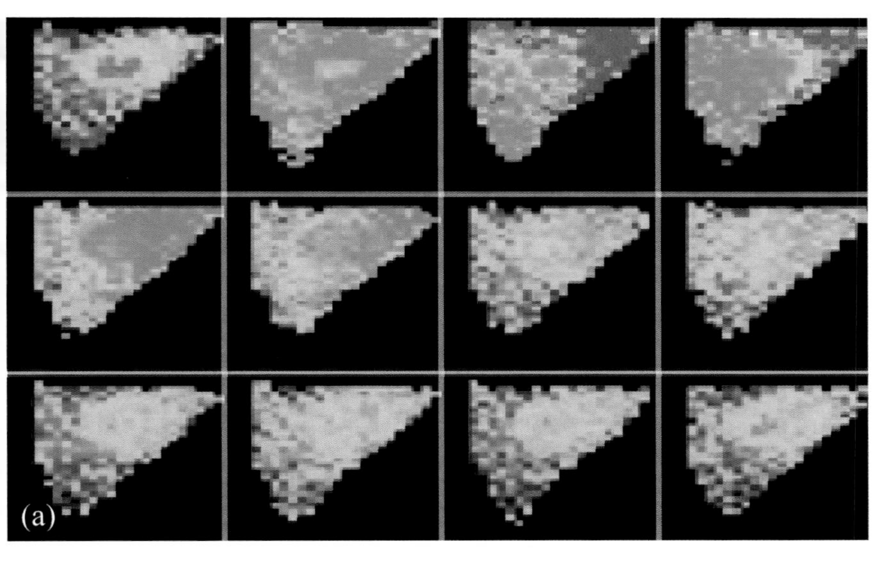

Plate 12.6 (a) A cardiac myocyte imaged using binning of the pixels on the CCD chip, which in this case was a high-speed 12-bit EEV37 CCD in an LSR UltraPix camera interfaced to MERLIN. (b) Time plot of the cardiac myocyte imaged in part (a) showing high-speed calcium transients arising from spontaneous contractions of the single muscle cell.

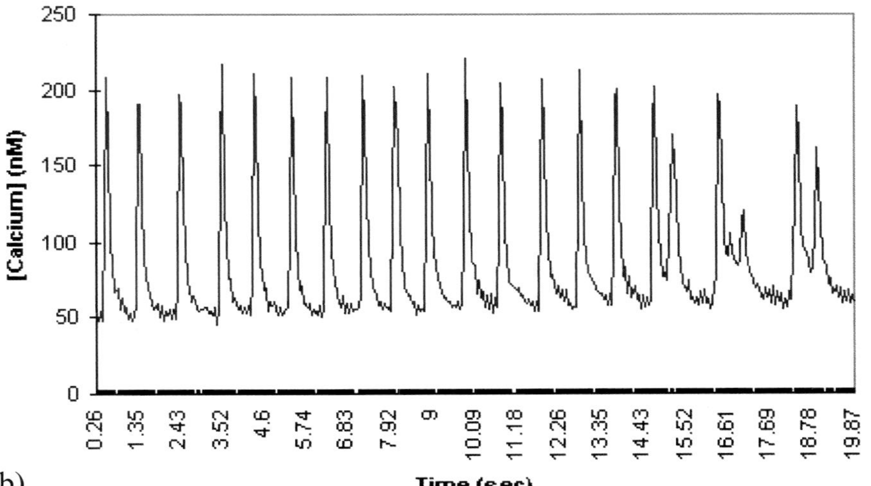

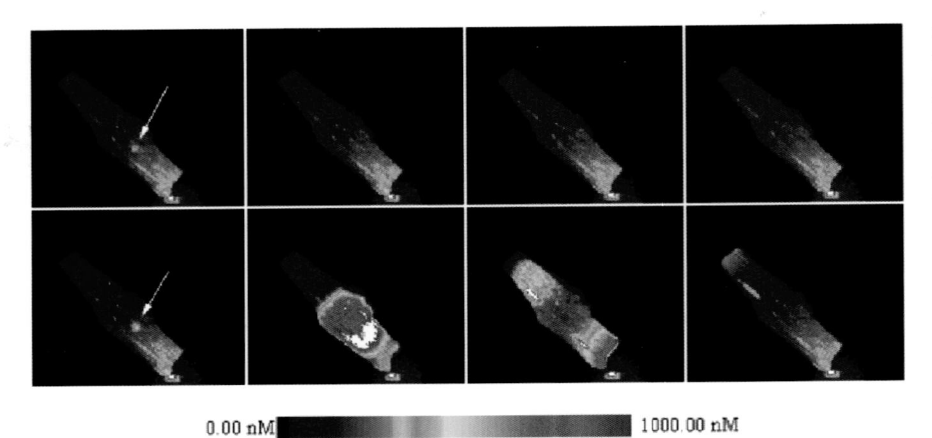

Plate 12.7 Calcium sparks and a calcium wave imaged with the MERLIN 12-bit ion imaging system, with the subthreshold sparks being shown by arrows. These images also utilized the UltraVIEW confocal scanning attachment to improve optical resolution, and this enabled the sparks to be visualized.

Plate 12.8 Time-dependent calcium ion oscillations imaged using the 12-bit MERLIN system, with four cells being plotted simultaneously, allowing inter-cell comparison.

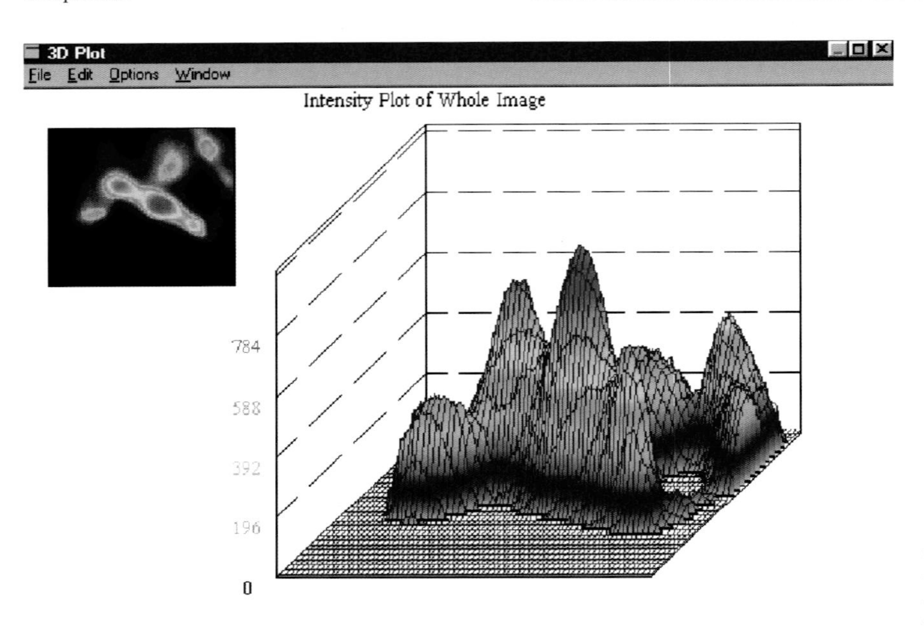

Plate 12.9 Digital 3-D plot of cardiac myocyte shown in the upper left-hand panel of the figure, allowing representation of the calcium concentration profile throughout the volume of the cell.

Plate 12.10 User interface of the PhoClamp system, which is here shown with a record of a voltage clamped pituitary cell loaded with indo-1. Top trace shows the calcium transient evoked by a voltage-activated calcium current. Bottom trace shows the short duration calcium current. The upper trace is displayed at a time resolution which is ten times slower than the voltage/current record. This system also allows current–voltage plots and a wide range of other analyses to be performed.

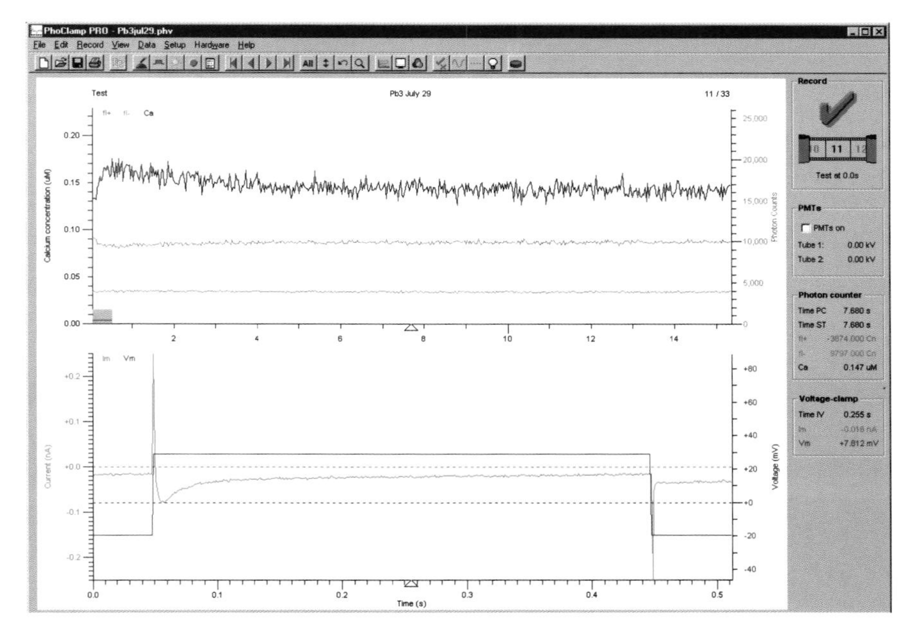

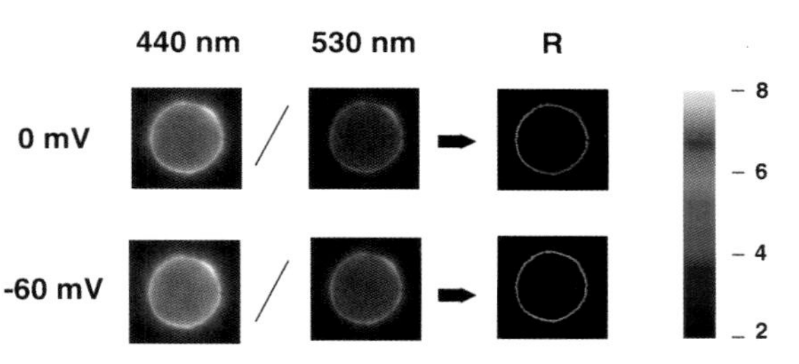

Plate 14.1 Calibration of di-8-ANEPPS in neuroblastoma N1E-115 cells. Fluorescence ratio (*R*) of a neuroblastoma N1E-115 cell clamped at 0 mV and –60 mV. This undifferentiated cell was stained with 1 μM di-8-ANEPPS in buffer solution containing 0.05% Pluronic F127. The left and the middle images at each voltage were the raw images taken at the excitation wavelengths of 440 nm and 530 nm, respectively. After background subtraction and flat-field correction, the pseudocolour ratio image (right) was obtained by dividing the 440-nm image by the 530-nm image. *R* was taken as the average value of the pixels corresponding to the contour of the cell membrane, using the ring stain pattern of fluorescence to segment the membrane contour. (Reproduced, with permission, from Zhang *et al.*, 1998).

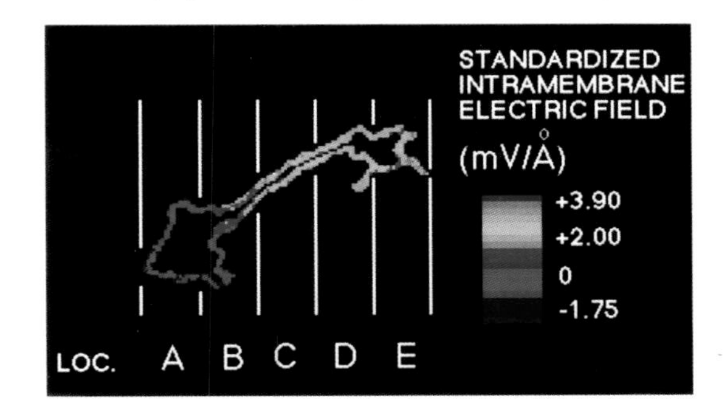

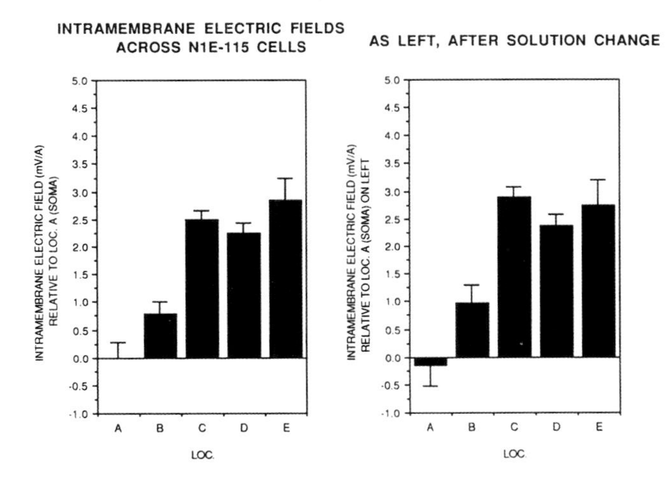

Plate 14.2 Measurement of intramembrane electric fields along neurons reveals regional variations. The colour image at the top is a map of the intramembrane electric field along the surface of a single N1E-115 neuroblastoma cell. The colours are calibrated in units of mV per Å as indicated on the scale at the right. Each cell was divided into five equally sized locations. Location 'A' always encompassed only membrane from the soma; 'B' included membrane from the soma, the axon hillock and proximal axon; 'C' included only axonal membrane; 'D' included membrane from the distal axon and proximal growth cone, and location 'E' included only distal growth cone membrane. Intramembrane electric field variations were referenced to the average in location 'A', taken as 0. The bar graph at the bottom left shows the average intramembrane electric field in each location as a mean of 20 different N1E-115 cells (error bars indicate SE). Note that the locations appear to fall into two distinct populations, one from the soma and a separate one from the neurite; statistical analysis confirmed this (Bedlack *et al.*, 1994). A control exchange of salt solution (bottom right) did not change the absolute intramembrane electric field in any location. (Reproduced, with permission, from Bedlack *et al.*, 1994.)

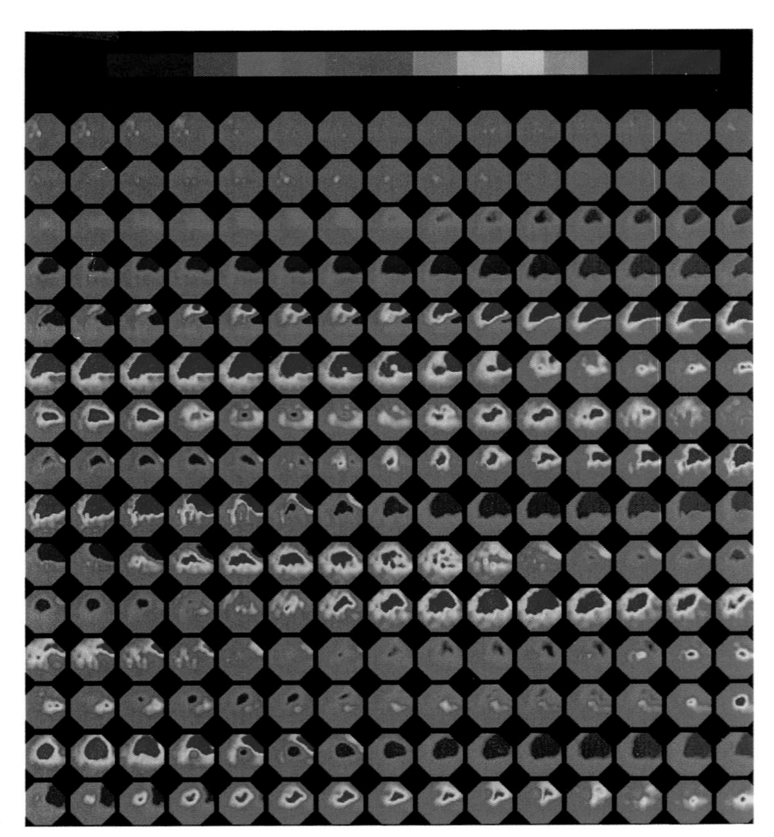

Plate 15.1 Pseudocolour images of the turtle visual cortex in response to a looming stimulus. The depolarizing events are both complex and variable. The first 1350 ms following the stimulus are shown. The data were filtered with a bandpass of 2–50 Hz (similar to Fig. 15.9 (B)) (gaussian high-pass and Butterworth low-pass) before making the display. Frames were recorded every millisecond but only one frame every 6 ms is shown. The first frame (upper left corner) occurs at the onset of the stimulus. The last frame (lower right) is 1350 ms later. The pseudocolour pictures were made by assigning red to the largest depolarization on a pixel in the middle of the frame and purple to the largest hyperpolarization on that pixel. Modified from Prechtl *et al.* (1997).

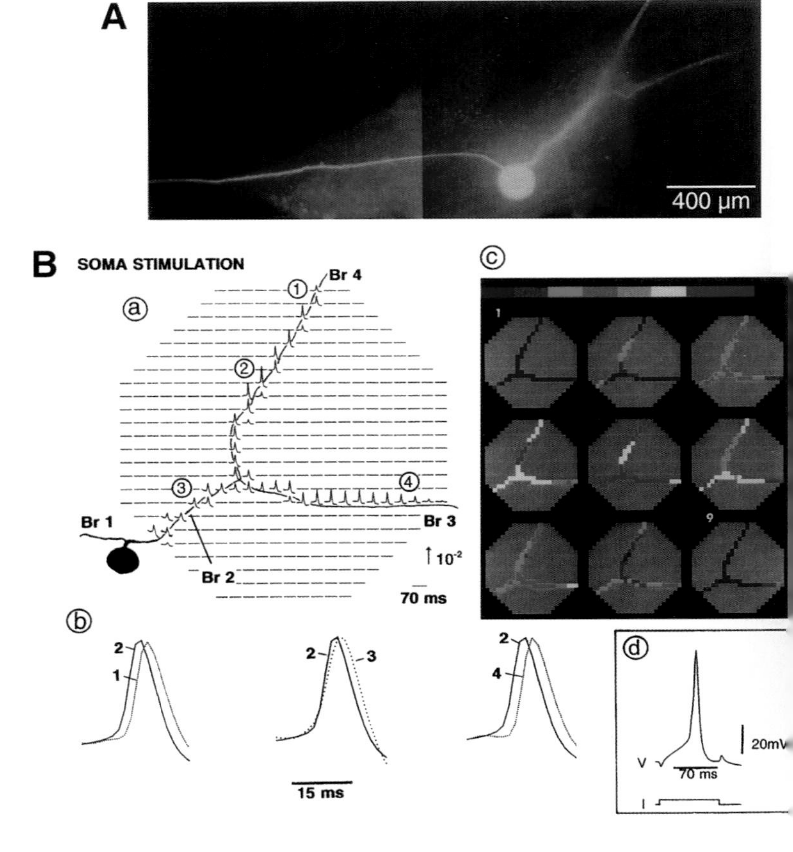

Plate 16.1 (A) the giant metacerebral neuron from the left cerebral ganglion 12 hours after injection with the fluorescent voltage-sensitive dye, JPW1114. The cell body and main processes are clearly visible in the unfixed preparation. Excitation wavelength, 540±30 nm; dichroic mirror, 570 nm, barrier filter, 610 nm. (B)(a) Voltage-sensitive dye recording of action potential signals from elements of the photodiode array positioned over the image of axonal arborization of a metacerebral cell in the left cerebral ganglion. Axonal branches are marked Br1–4. Spikes were evoked by transmembrane current steps, as shown in (d), delivered through the recording microelectrode in the soma. Each optical trace in (a) represents 70 ms of recording centred around the peak of the spike as indicated by the time bar in (d). Each diode received light from 50 x 50 μm area in the object plane. (b) Recordings from four different locations indicated in panel (a), scaled to the same height, are compared to determine the site of the origin of the action potential and the direction of propagation. (c) Colour-coded representation of the data shown in (a) indicating the size and location of the primary spike trigger zone and the pattern of spike propagation. Consecutive frames represent data points that are 1.6 ms apart. The colour scale is relative with the peak of the action potential for each detector shown in red (modified from Zečević, 1996).

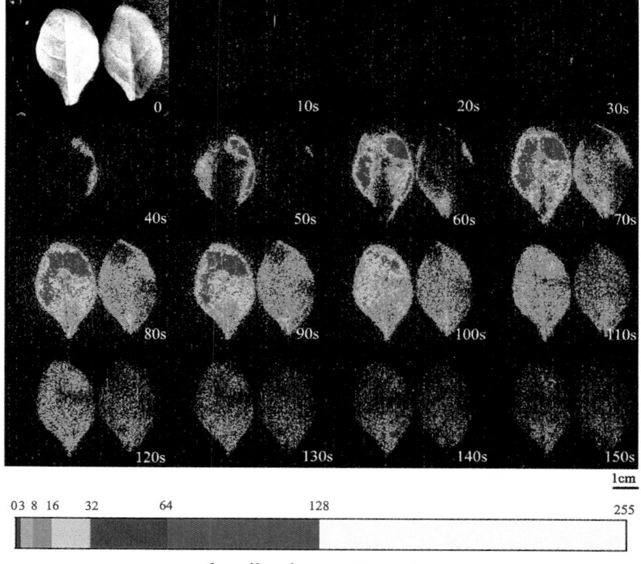

Plate 17.1 Effect of gradual cooling on the Ca^{2+}-dependent uminescence in tobacco leaves. Aequorin was reconstituted by nfiltration with coelenterazine and placed on a Peltier element that educed the temperature from 25°C to 2°C. Luminescence was imaged ısing the Photek camera. The Ca^{2+}-dependent luminescence was first letected between 30 and 50 s and increased as the temperature lecreased.

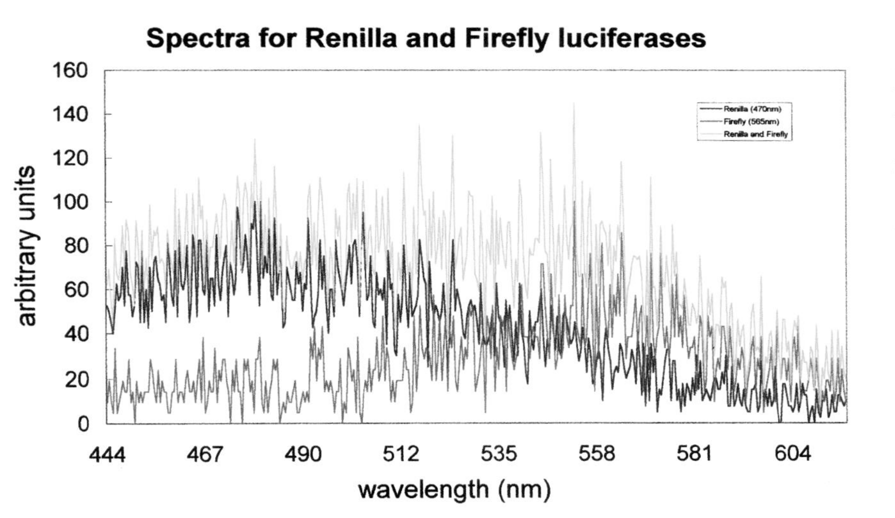

Plate 17.2 Simultaneous spectra of *Renilla* and *Photinus* luciferases. The measurements were performed by combining a fibreoptic bundle with a monocromator and spraying the spectra over the photocathode of the Photek camera. The calibration was performed using narrow bandpass (10 nm) interference filters. Blue trace, *Renilla* luminescence; red trace, firefly luminescence; yellow trace, simultaneous measurement.

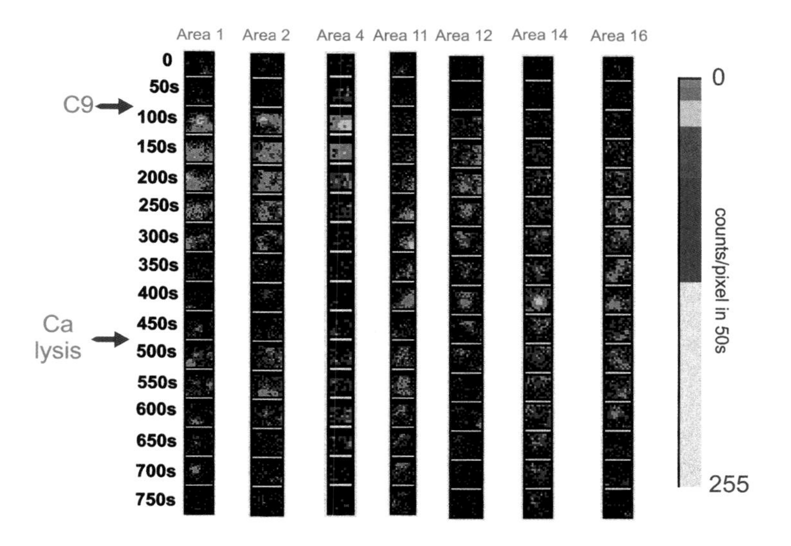

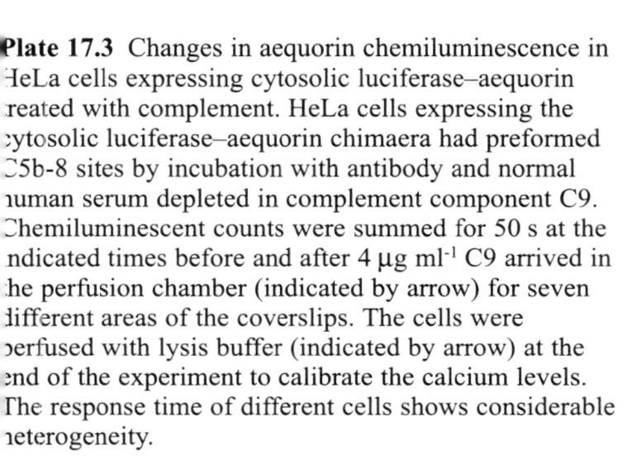

Plate 17.3 Changes in aequorin chemiluminescence in HeLa cells expressing cytosolic luciferase–aequorin reated with complement. HeLa cells expressing the ytosolic luciferase–aequorin chimaera had preformed C5b-8 sites by incubation with antibody and normal uman serum depleted in complement component C9. Chemiluminescent counts were summed for 50 s at the ndicated times before and after 4 μg ml^{-1} C9 arrived in he perfusion chamber (indicated by arrow) for seven lifferent areas of the coverslips. The cells were perfused with lysis buffer (indicated by arrow) at the end of the experiment to calibrate the calcium levels. The response time of different cells shows considerable heterogeneity.

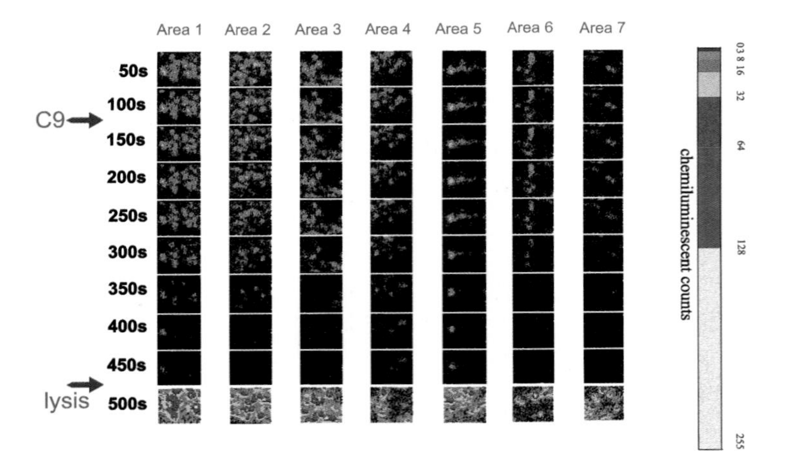

Plate 17.4 Images of firefly luciferase chemiluminescence as an indicator of ATP levels in HeLa cells expressing cytosolic luciferase–aequorin and treated with complement. HeLa cells expressing the cytosolic luciferase–aequorin chimaera had preformed C5b-8 sites by incubation with antibody and normal human serum depleted in complement component C9. Chemiluminescent counts were summed for 50 s at the indicated times before and after 4 µg ml^{-1} C9 arrived in the perfusion chamber (indicated by arrow) for seven different areas of the coverslips. The cells were perfused with lysis buffer (indicated by arrow) at the end of the experiment to assess cell integrity. The response time of different cells shows considerable heterogeneity.

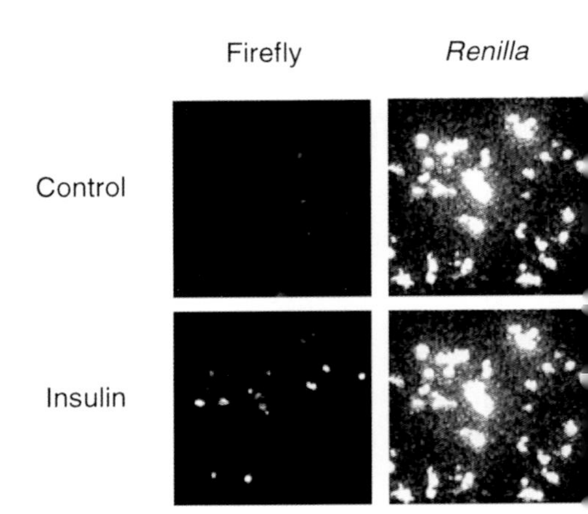

Plate 18.1 Expression of luciferase and *Renilla* activities in 3T3 L1 adipocytes. 3T3 L1 adipocytes were microinjected with a mixture of three plasmids: (1) pSG424.Elk1 (83-428) (0.05 mg ml^{-1}), which drives the expression of a fusion protein consisting of the Elk-1 transactivation domain and the yeast GAL4 DNA binding domain; (2) the reporter plasmid, pGL3.G5E4Δ38 (0.1 mg ml^{-1}), which contains a cytosolically targeted firefly luciferase controlled by a minimal promoter containing 5 GAL4 binding sites (Promega Corp.); and (3) *Renilla* luciferase under CMV promoter control (pRL-CMV, 0.05 mg ml^{-1}; Promega). The cells were treated with or without insulin. Firefly luciferase was imaged followed by *Renilla* luciferase 24 hours later, by photon counting for 15 min each (see text).

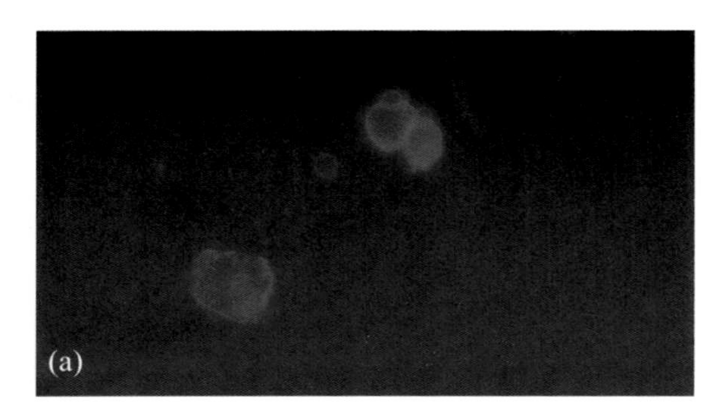

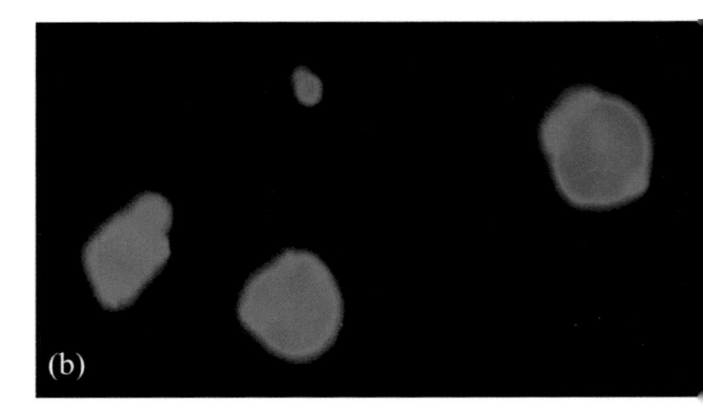

Plate 19.1 Detection of annexin V binding by fluorescence microscopy. Apoptosis was induced in Jurkat cells by incubation with 200 ng ml^{-1} anti-Fas monoclonal antibody (clone CH-11) for 14 hours. Cells were collected by centrifugation, and incubated with 0.5 µg ml^{-1} annexin V conjugate for 10 minutes in the dark. (A) Annexin V-FITC; (B) Annexin V-EGFP.

medium through which the fluorescent light must travel to reach the lens, and the numerical aperture (NA) and magnification (m) of the objective used. The primary importance of these parameters arises because they define the intensity (I) of excitation and emission light, the two-dimensional x,y-axis point resolution (γ), and the focal depth (D) of the microscope. The NA value of an objective is a measure of the light-gathering ability of an objective lens, and is a function of the refractive index (n) and the half angle (θ) of the cone of light rays entering the objective. Thus:

$$NA = n \sin \theta$$

Since, in fluorescence microscopy, the objective also serves as the light condenser, the intensity (I) of light passing through the objective varies proportionally to magnification, m, and the numerical aperture (see Taylor & Salmon, 1989; Young, 1989) according to:

$$I \propto NA^4/m^2$$

The theoretical x,y-axis point resolution is the minimum distance that can be resolved by the objective between two points of fluorescent emission. Using Raleigh's criterion for resolving power, the resolving power of the objective (γ) is given by:

$$\gamma = 0.61\lambda/NA$$

Where λ is the wavelength of measured fluorescence. However, the depth of field is the most important parameter determining z-axis resolution, which is in effect the distance that a single point object can be moved in the z-axis before the image changes. The depth of field, D is given by:

$$D = \lambda/[4\ n \sin^2 (\theta/2)]$$

θ can be derived by solving the equation for NA (see Taylor & Salmon, 1989). Thus, for a 63×, oil immersion objective (NA=1.2); where λ = 500 nm, the calculated values are:

$$\gamma = 0.25\ \mu m \quad \text{and} \quad D = 0.42\ \mu m \quad (n = 1.515)$$

Whereas, for a 60× water immersion objective designed for long-distance work (NA=0.9), the calculated values were:

$$\gamma = 0.34\ \mu m \quad \text{and} \quad D = 0.71\ \mu m \quad (n = 1.33)$$

As mentioned at the beginning of this chapter, a digital image is a two-dimensional representation of three-dimensional information. Thus, the individual numerical pixels which comprise the image are distorted since they include not only the true signal emerging from a single point in the focal plane, but also varying degrees of additional signal accumulated from fluorescent out-of-focus volumes, above and below the focal plane, but incident within the focal depth. To illustrate this point, Plate 7.3(A) shows the two-dimensional image of a red shell, green interior fluorescent bead (15 μm) on a conventional fluorescence microscope. The bead was imaged through its central plane, at two different excitation wavelengths, and the images were superimposed. The image is relatively clean, and compares well to an image of the same bead on a confocal microscope (Plate 7.3(D)). Next, the bead was imaged using a fast piezo-electric z-axis objective drive, and 100 images were collected at 0.45 μm steps so that the entire volume of the bead was imaged. These images were then rendered on a computer to produce the three-dimensional x,z view projection shown in Plate 7.3(B). From this projection, the true optical consequences of using a conventional microscope are apparent. Firstly, the image is blurred primarily because of the interference of out-of-focus light inside the focal depth of each sample volume. Out-of-focus light is bright and extensive, forming two asymmetrical cones above and below the central (focal) plane of the sample, whose spread diameters are nearly twice that of the bead itself. As such, one can expect extreme interference of out-of-focus light throughout the sample volume. Thus, although the two-dimensional image through the focal plane of the bead (Plate 7.3(A)) looks undistorted, the measured intensity of each pixel in the image must be corrupted to varying degrees by the out-of-focus light as mapped in Plate 7.3(B). This is why the signal intensity across the green interior of the bead in Plate 7.3 (which is designed to be homogeneous), is observed to be heterogenous using the conventional microscope, particularly compared with the confocal microscope image (Plate 7.3(D)). A second point is that the 'waistband' of red (Plate 7.3(B)) is a result of the wide (0.71 μm) focal depth of the objective used. In the focal plane of the bead the entire focal depth is filled with red shell due to the fact that the bead is a sphere, and the incident angle of the thin red shell to the focal plane is at its highest. This resulted in the red signal being integrated onto fewer adjacent camera pixels. However, in subsequent planes the incident angle of the shell to the focal plane is reduced, so that the red shell signal was integrated across more pixels in each two-dimensional image of the series. Thus, the intensity of red signal on any given pixel fell as the focal plane was moved away from the centre of the sample. In the extreme, no clear red signal was detected near the top of the bead. On the other hand, at the interface between the bead and the mounting substrate (a glass slide) a strong red signal was recorded. This is due to glare and backscattering (Young, 1989) of the coincident out-of-focus light emitted from the red shell, acting to reinforce the red signal at the substrate interface. Finally, we noted the expected elongation effect which makes the sphere of the bead look more like an egg, or ellipse. This results from inherent *spherical aberration* inherent in the microscope's optics and the maximum separation estimate used to render the images (see Moss, 1992).

7.4.2 Removing out-of-focus distortion

To avoid the contribution of out-of-focus light inherent to digital images acquired using a conventional fluorescence microscope, confocal microscopes have become a popular alternative. Confocal microscopes capture light only from the focal plane, whose limits can be set to approach the focal depth of the objective (Shotton, 1989; Shotton & White, 1989), as compared to conventional fluorescence microscopes which capture light from the entire thickness of the tissue sample (see Silver *et al.*, 1992).

However, confocality comes with its own problems. The implication is that vertical resolution is better than from a conventional microscope, but this is only true in as much as out-of-focus light does not corrupt the image within the focal depth. Another practical limitation is that because they work by rejecting light emergent from outside a confocal depth, there is an inevitable loss of information, which translates into optical inefficiency. As described earlier, optical inefficiency can be expensive when working with fragile biological samples sensitive to phototoxicity. For confocals developed to capture images at very high rates approaching video rate, this can require extremely high-energy light input that results in sample lifetimes being reduced to seconds from the beginning of an experimental acquisition series. Finally, it is important to understand that the confocal depth is itself a finite volume, and as such does contain a finite amount of out of focus information (Wijnaends van Resandt, *et al.*, 1985; Kempen *et al.*, 1997).

These disadvantages may be overcome through two major advances. Firstly, the development of multiphoton excitation which limits excitation of the fluorophore to the focal plane only (Denk *et al.*, 1990; Xu *et al.*, 1996; Denk & Svoboda, 1997). In principle, therefore, bleaching problems are considerably reduced and phototoxicity is minimized. Along these lines, a detailed technical comparison of single photon versus multiphoton three-dimensional ion imaging is described well elsewhere (Sako *et al.*, 1997). Secondly, digital image restoration (DIR) has found wide application, improving conventional microscope images to yield superior resolution to raw confocal images (Carter *et al.*, 1993; Carrington *et al.*, 1995). DIR uses postacquisition image processing algorithms intended to remap primary digital images containing out-of-focus fluorescent light. The result is a deconvolved image more closely representative of the true signal in the focal plane. This calculation-intensive approach in effect 'decodes' the contribution of out-of-focus light. Many different DIR algorithms are available of which the Carrington and Richardson–Lucy types are popular and exemplify different approaches (Kempen *et al.*, 1997). The study by Kempden *et al.* (1997) demonstrates how confocal microscopy can be combined with DIR to remove the finite out-of-focus light in confocal images. Indeed, DIR can even provide greatly enhanced images from very sophisticated microscope acquisition systems. A good example of this used 4Pi microscopy (a variation of multiphoton excitation microscopy, with improved axial resolution), and DIR to visualize actin filaments with a vertical resolution of 100 nm (Hell *et al.*, 1997). This resolution approaches that previously considered purely hypothetical for optical microscopes.

DIR algorithms are now widely available from a variety of commercial manufacturers. The popular nearest-neighbour (*nn*) algorithms, remove out-of-focus light from single image planes using an ideal Point Spread Function (PSF). The PSF is a function describing how three-dimensional light spreads from a sub-resolution volume, and defines the results of DIR algorithms. It has been argued that *nn* algorithms preclude true quantitative measurements because they utilize idealized PSFs which cannot account for the *in situ* conditions under which the experimental images are acquired. In particular, three-dimensional information like the intensity of two points in different vertical planes, and the distance between them, may be altered using this sort of approach (see Holmes & Liu, 1992). However, we find *nn* algorithms do allow good qualitative image enhancement in two dimensions, but only under certain conditions. If used appropriately, they can reveal visual information that is not otherwise resolved. For example, a zebra fish embryo was injected deep inside the yolk sac with fluorescein-conjugated tubulin molecules (Fig 7.4), and the distribution of fluorescence was monitored over time. Thirty minutes following the injection, striations of fluorescence were just about visible against a large fluorescent background in the raw images (Fig. 7.4(A) and (B)). Subsequently the raw images were processed using a *nn* algorithm, and this revealed distinct fluorescent microtubule filaments (Fig. 7.4(C) and (D)). The striking enhancement of the images in Fig. 7.4 is interesting compared with the lack of obvious enhancement noted for the bead

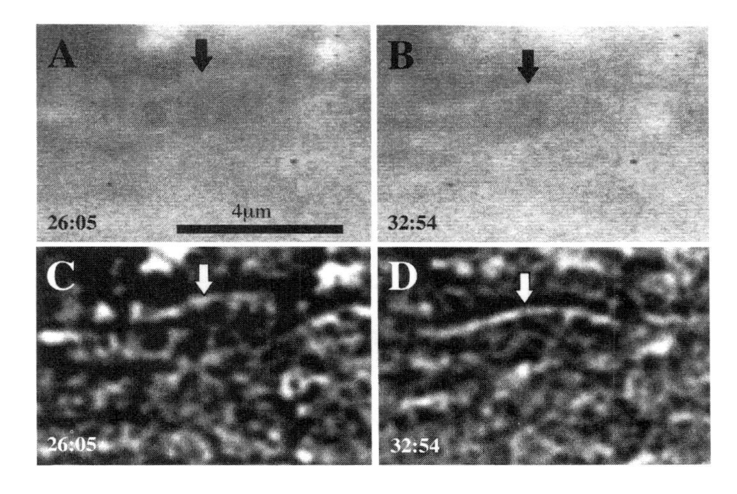

Figure 7.4 Digital image reconstruction of raw fluorescence images recorded from fluorescein-labelled tubulin molecules inside living zebra fish embryo yolk sac. A developing zebra fish embryo was injected inside its yolk sac using a micropipette containing fluorescein-conjugated tubulin. The distribution of fluorescence was then monitored using the same conventional microscope as described for Plate 7.3, except the light detector was an intensified BioLumi2 (Photonic Sciences, UK). Raw fluorescence images (panels A and B) were recorded approximately 26 and 33 minutes, respectively, after the injection. These images were then processed using two gaussian filters, and the appropriate nearest neighbour DIR algorithm based on an ideal PSF (panels C and D). The qualitatively enhanced images revealed a microtubule in the field of view. The microtubule appeared to accumulate fluorescent label with time, probably through incorporation of free fluorescein-labelled tubulin by subunit exchange. Scale bar: 4 μm.

sample (cf. Plate 7.3(A) and (B)). This difference is probably due to the validity of the ideal PSF used under each condition. As noted for the bead in Plate 7.3, fluorescent signals near the bottom of the mounting chamber can be markedly altered due to backscattered light, and glare. This factor can also greatly affect the accuracy and effectiveness of DIR image treatments because backscattered light ensures that a PSF measured from a fluorescent bead on the very bottom of the mounting chamber will be different from the PSF of a bead suspended a few tens of micrometres above. The asymmetric distortion of the fluorescent bead imaged in Plate 7.3(B) illustrates this point. On the other hand, the fluorescent microtubule imaged in Fig. 7.4, was suspended deep inside an embryo, more than 100 μm above the mounting chamber floor. In this case, the ideal PSF used in the DIR algorithm was probably a very much better approximation of the real PSF of the microscope in this particular spatial plane. This is an extremely important point to bear in mind with regard to using DIR on thin samples, such as cell cultures. For example, DIR imaging of a cell attached to growth substrate, with a diameter approaching that of the bead shown in Plate 7.3, would not necessarily be improved by an ideal PSF-based algorithm treatment. Indeed, it is likely that this will corrupt the images. Under such circumstances, the very best approach would utilize a direct measurement of the 'real' PSF from the particular microscope that one is using. Since distortion is systematic, using a measured PSF would account for the non-ideal behaviour of light near the mounting substrate. A PSF calibration measurement is easily achieved using sub-resolution (0.1–0.4 μm) fluorescent beads, which are imaged at the appropriate wavelengths in three dimensions. The best DIR software packages allow the three-dimensional map of the light spread from the microbead to be used to calculate a real PSF, to be used with the DIR algorithm of choice. This is the only approach that could work well for DIR enhancement of relatively thin samples, and it is technically difficult to achieve good results in living cells. For this reason, the literature at this time contains few examples of DIR-based studies on living cells. However, DIR approaches to quantitatively improving image verity are beginning to find application for fluorescent imaging in living cells, and do promise a future of routine four-dimensional image mapping of biological signals.

ACKNOWLEDGEMENTS

S.L. Shorte thanks Dr P. Bregestovski for collaboration in the experiment illustrated in Fig. 7.4, and INSERM, Le Ministère de l'Education Nationale de L'Enseignement Supérieur de La Recherche et de L'Insertion Professionnelle, the European Commission, and the Fondation pour le Recherche Medicale for support. Stephen Bolsover's work is supported by Action Research and the Wellcome Trust. The experiment of Plate 7.1 was performed by Dr R. Angus Silver.

REFERENCES

Adams S.R., Harootunian A.T., Buechler Y.J., Taylor S.S. & Tsien R.Y. (1991) *Nature* **349**, 694–697.
Al-Mohanna F.A., Caddy K.W.T. & Bolsover S.R. (1994) *Nature* **367**, 745–750.
Art J. (1995) In *Handbook of Biological Confocal Microscopy*, J.B. Pawley (ed). Plenum Press, New York, pp. 183–196.
Bolsover S.R. & Silver R.A. (1991) *Trends Cell Biol.* **1**, 71–74.
Bolsover S.R., Brown J.E. & Goldsmith T.H. (1986) *Soc. Gen. Physiol. Ser.* **40**, 285–310.
Bootman M.D. & Berridge M.J. (1995) *Cell* **83**, 675–678.
Byrne J. & Duncan C.J. (1990) *Biochem. Soc. Trans.* **18**, 609.
Carrington W.A., Lynch R.M., Moore E.D.W., Isenberg G., Fogarty K.E. & Fay F.S. (1995) *Science* **268**, 1483–1487.
Carter K.C., Bowman D., Carrington W., Fogarty K., McNeil J.A., Fay F.S. & Lawrence J.B. (1993) *Science* **259**, 1330–1335.
Connor J.A. (1993) *Cell Calcium* **14**, 185–200.
Denk W. & Svoboda K. (1997) *Neuron* **18**, 351–357.
Denk W., Strickler J.H. & Webb W.W. (1990) *Science* **248**, 73–76.
DiVirgilio F., Steinberg T.H. & Silverstein S.C. (1990) *Cell Calcium* **11**, 57–62.
Duchen M.R. (1992) *Biochem. J.* **283**, 41–50.
Duchen M.R., Smith P.A. & Ashcroft F.M. (1993) *Biochem. J.* **294**, 35–42.
Etter E.F., Kuhn M.A. & Fay F.S. (1994) *J. Biol. Chem.* **269**, 10141–10149.
Fairclough R.H. & Cantor C.R. (1978) *Methods Enzmol.* **48**, 347–379.
Floto R.A., Mahaut-Smith M.P., Somasundaram B. & Allen J.M. (1995) *Cell Calcium* **18**, 377–389.
Gabso M., Neher E. & Spira M.E. (1997) *Neuron* **18**, 473–481.
Gerasimenko O.V., Gerasimenko J.V., Tepikin A.V. & Peterson O.H. (1995) *Cell* **80**, 439–444.
Gonzalez J.E. & Tsien R.Y. (1995) *Biophys. J.* **69**, 1272–1280.
Grynkiewicz G., Poenie M. & Tsien R.Y. (1985) *J. Biol. Chem.* **260**, 3440–3450.
Hajnoczky G., Robb Gaspers L.D., Seitz M.B. & Thomas, A.P. (1995) *Cell* **82**, 415–424.
Haugland R.P. (1996) In *Handbook of Fluorescent Probes and Research Chemicals*, 6th edn, M.T.Z. Spence (ed). Molecular Probes, OR, USA.
Heim R. & Tsien R.Y. (1996) *Curr. Biol.* **6**, 178–182.
Hell S.W., Schrader M. & Van der Voort H.T.M. (1997) *J. Microsc.* **187**, 1–7.
Helmchen F., Imoto K. & Sakmann B. (1996) *Biophys. J.* **70**, 1069–1081.
Hollingworth S. & Baylor S.M. (1986) *Soc. Gen. Physiol. Ser.* **40**, 261–283.
Holmes T.J. & Liu Y.-H. (1992) In *Visualization in Biomedical Microscopies*, A. Kriete (ed). VCH Verlagsgesellschaft mbH, Weinheim, pp. 283–327.
Karaki H. (1989) *J. Exp. Med. (Japan)* **7**, 626–631.
Kao J.P.Y., Harootunian A.T. & Tsien R.Y. (1989) *J. Biol. Chem.* **264**, 8179–8184.
Kempen G.M.P., Van Vliet L.J., Verveer P.J. & Van der Voort H.T.M. (1997) *J. Microsc.* **185**, 354–365.
König K., Liang H., Berns M.W. & Tromberg B.J. (1995) *Nature* **377**, 20–21.
Korkotian E. & Segal M. (1997) *J. Neurosc.* **17**, 1670–1682.
Leinekugel X., Tseeb V., Ben-Ari Y. & Bregestovski P. (1995) *J. Physiol. (Lond.)* **487**, 319–329.
Liu Y., Sonek G.J., Berns M.W. & Tromberg B.J. (1996) *Biophys. J.* **71**, 2158–2167.
Lui P.P.Y., Lee M.M.F., Ko S., Lee C.Y. & Kong S.K. (1997) *Biol. Signals* **6**, 45–51.
Miesenböck G. & Rothman J.E. (1997) *Proc. Natl Acad. Sci. (USA)* **94**, 3402–3407.
Minta A., Kao J.P.Y. & Tsien R.Y. (1989) *J. Biol. Chem.* **264**, 8171–8178.
Miyawaki A., Llopis J., Heim R., McCaffery J.M., Adams J.A., Ikura M. & Tsien R.Y. (1997) *Nature* **388**, 882–887.
Moss V.A. (1992) In *Visualization in Biomedical Microscopies*, A. Kriete (ed). VCH Verlagsgesellschaft mbH, Weinheim, pp. 19–43.
Nowycky M.C. & Pinter M.J. (1993) *Biophys. J.* **64**, 77–91.
O'Malley D.M. (1994) *J. Neurosc.* **14**, 5741–5758.
Perez-Terzic C., Stehno Bittel L. & Clapham D.E. (1997) *Cell Calcium* **21**, 275–282.
Poenie M. (1990) *Cell Calcium* **11**, 85–91.
Poenie M., Alderton J., Steinhardt R. & Tsien R.Y. (1986) *Science* **233**, 886–889.
Pozzan T. (1997) *Nature* **388**, 834–835.
van de Put F.H.M.M. & Elliot A.C. (1996) *J. Biol. Chem.* **271**, 4999–5006.
Rand M.N., Leinders-Zufall T., Agulian S. & Kocsis J.D. (1994) *Nature* **371**, 291–292.
Rutter G.A., Burnett P., Rizzuto R., Brini M., Murgia M., Pozzan T., Tavare J.M., Denton R.M. (1996) *Proc. Natl. Acad. Sci.* **93**, 5489–5494.
Ryan T.A., Reuter H., Wendland B., Schweizer F.E., Tsien R.W. & Smith S.J. (1993) *Neuron* **11**, 713–724.
Sako Y., Sekihata A., Yanagisawa Y., Yamamoto M., Shmida Y., Ozaki K. & Kusumi A. (1997) *J. Microsc.* **185**, 9–20.
Schiller J., Helmchen F. & Sakmann B. (1995) *J. Physiol. (Lond.)* **487**, 583–600.
Senseman D.M. (1996) *J. Neurosci.* **16**, 313–324.
Shorte S.L., Stafford S.J.V., Collett V.J. & Schofield J.G. (1995) *Cell Calcium* **18**, 440–454.
Shotton D.M. (1989) *J. Cell. Sci.* **94**, 175–206.
Shotton D.M. & White N. (1989) *Trends Biochem. Sci.* **14**, 435–439.
Silver R.A., Whitaker M.J. & Bolsover S.R. (1992) *Pflugers Arch.* **420**, 595–602.
Smith C.B. & Betz W.J. (1996) *Nature* **380**, 531–534.
Spruston N., Schiller Y., Stuart G. & Sakmann B. (1995) *Science* **268**, 297–300.
Strähle U. & Jesuthasan S. (1993) *Development* **119**, 909–919.
Stryer L. (1959) *Biochim. Biophys. Acta* **35**, 243–244.
Taylor D.L. & Salmon E.D. (1989) In *Methods in Cell Biology, vol. 29: Fluorescence Microscopy of Living Cells in Culture*, Y.-L. Wang & D.L. Taylor (eds). Academic Press, London, pp. 207–237.

Tomkins P. & Lyons A. (1993) In *Fluorescent and Luminescent Probes for Biological Activity*, 1st edn, W.T. Mason (ed). pp 264–275. Academic Press, London, pp. 264–275.
Tsien R.Y., Pozzan T. & Rink T.J. (1982) *J. Cell Biol.* **94**, 325–334.
Tsien R.Y., Bacskai B.J. & Adams S.R. (1993) *Trends Cell Biol.* **3**, 242–245.
Wang S.S.-H. & Augustine G.J. (1995) *Neuron* **15**, 755–760.
Wijnaends van Resandt R., Marsman H.B.J., Kaplan R., Davoust J., Stezler E.K.H. & Stricker R. (1985) *J. Microsc.* **138**, 29–34.
Xu C., Zipfel W., Shear J.B., Williams R.M. & Webb W.W. (1996) *Proc. Natl Acad. Sci.* (*USA*) **93**, 10763–10768.
Young I.T. (1989) In *Methods in Cell Biology, vol. 29: Fluorescence Microscopy of Living Cells in Culture*, Y.-L. Wang & D.L. Taylor (eds). Academic Press, London, pp. 1–43.
Zecevic D. (1996) *Nature* **381**, 322–325.

CHAPTER EIGHT

Fluorescent Probes in Practice – Potential Artifacts

J. HOYLAND
Life Science Resources, Cambridge, UK

8.1 INTRODUCTION

In reviewing the potential artifacts and pitfalls in the use of fluorescent probes which may befall the unwary, it is important to look at the general application modes in which fluorescent probes may be used. For example, fura-2, the calcium-sensitive probe, may be used either as a single-wavelength qualitative probe simply to image the spatial distribution of calcium. Alternatively, it may be used in dual-wavelength ratio mode in a calibrated system to quantify calcium concentrations and fluxes. The potential artifacts will be very different in the two systems, indeed, simply utilizing fura-2 in its ratio mode corrects most of the problems associated with single-wavelength measurements.

Most of the following chapter refers to the potential artifacts which may affect the *measurement* of ion concentrations. It therefore generally relates to imaging and photometry for *ratio* measurements rather than simple fluorescence. The variety of difficulties experienced in dynamic video imaging are as diverse as the measurements we make. This contribution, therefore, partly re-states the obvious, establishing an understanding of the basics, but more importantly hopes to stimulate constructive thought on recognizing and solving the inevitable problems which arise when systems and probes are pushed toward their limits.

8.2 PHOTOBLEACHING

All fluorescent probes will photobleach to a greater or lesser extent when excited with a suitable wavelength, at a rate proportional to the intensity of the incident light. While this may not present a problem for some applications such as the simple spatial mapping of hormone distribution, it does seriously affect any attempt to quantify ion concentrations using single-wavelength probes.

The most obvious practical way to reduce photobleaching is to minimize the light reaching the probe. Unfortunately, reducing the incident light will also reduce the emitted light intensity, so optimum conditions for image analysis must include the following.

FLUORESCENT AND LUMINESCENT PROBES, 2ND EDN
ISBN 0–12–447836–0

8.2.1 Optimal loading

Optimal loading of a particular probe in a particular cell type should be assessed by experiment. A maximal signal will only result from maximal loading, but care must be taken to ensure the probe is not significantly buffering the ion of interest. A compromise must therefore be sought between maximizing the signal and minimizing probe concentration. In practice, intracellular probe concentrations of 30–100 μM are usually suitable. Loading may be achieved either by direct microinjection or by acetoxymethyl ester loading across the cell membrane (Tsien, 1981).

8.2.2 Maximum sensitivity of data collection

The availability of suitably sensitive detectors has played a crucial role in the development of technology to measure dynamic ion changes in living cells. While adequately sensitive photomultiplier tubes have been available for some years, more recent developments in intensified charge-coupled device (CCD) video cameras have allowed direct imaging of probes for measurement of ion concentrations and fluxes. In addition, fast computer access and processing now allows ratiometric techniques to be applied to video imaging, eliminating many of the possible artifacts and errors associated with single-wavelength techniques. Other factors which will improve sensitivity are the use of excitation and emission filters which are well-matched to the probe. The emission filter should have a bandwidth covering at least 90% of the emission spectrum. The use of an objective of the highest numerical aperture available will also improve sensitivity.

8.2.3 Ratiometric measurements

Utilizing ratiometric measurements will correct for uneven probe loading between cells and across each cell. It will correct for differences in cell membrane thickness, where the membrane unevenly absorbs some of the emitted light, and it will correct for part of the photobleaching problem. Correction of these artifacts is, however, limited to the dynamic range of the detection system and any combination which takes either of the images used for the ratio outside its dynamic range will cause problems as described in the section below. Furthermore, significant photobleaching has been shown in fura-2 to not only reduce the light output of the probe, but to also shift the emission spectrum (Becker & Fay, 1987). At just 8% photobleaching, using a standard ratio calibration, this results in an underestimation of up to 20% of 200 nM $[Ca^{2+}]_i$.

8.2.4 Minimum illumination time

Minimize the time actually illuminating the probe by exciting it only while capturing data. Extraneous room light must be kept to a minimum as it will contribute to photobleaching not only during an experiment but also while loading and storing loaded cells before an experiment.

8.2.5 Minimum oxygen

Minimize the oxygen concentration as it has been shown to play a major role in photobleaching of the fura-2/calcium system (Becker & Fay, 1987). The lowest concentration concomitant with good cell viability will greatly reduce the rate of bleaching.

8.3 DYNAMIC RANGE

Whatever type of system is employed to detect the emission from fluorescent probes, it will be limited to a specific dynamic range. That is, there will be a lower level at which detection is not possible and there will be an upper level at which the system saturates. The value at the maximum divided by that at the minimum is known as the dynamic range of the system. In addition to the values for the system as a whole, due regard must be paid to saturation in any single part of the system.

For example, an imaging system will have an overall dynamic range which is governed by the range of the probe, the range of the camera (which may have two intensifier stages, both capable of saturation and a CCD detector whch may saturate) and the range of the analogue-to-digital converter (ADC) used to digitize the image. There will also be a maximum ratio set by the ratioing range in the calibration table. In an ideal system the effective 'gain' of each component is set so they all saturate at the same level. This, however, is not always possible and great care must be taken to ensure that no part of the system will either saturate or fall below its minimum detection threshold during an experiment. (Fig. 8.1)

The most usual saturation artifact in ratio imaging is perhaps concerned with the setting of the maximum ratio level when ratioing. This is necessary to optimize the display of changes in ion concentration. If, for example, a cell undergoes a change of 200 nM during an experiment, it would be unwise to display it on an

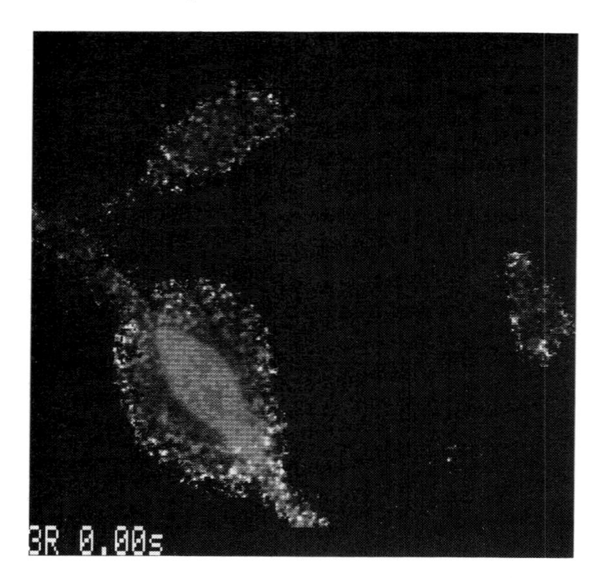

Figure 8.1 A typical ratio image of a cell which is saturating at one or both excitation wavelengths. The edges of the cell are usually ill-defined and it may appear larger than its real size due to scattered light. The ratio image from a properly illuminated cell will usually appear relatively uniform across its surface but show considerable pixel-to-pixel noise consistent with the use of image intensifiers. The saturated cell exhibits a 'flat' centre containing unusually low pixel-to-pixel noise.

axis with a span of 5000 nM. In this case the maximum level set when calculating the ratios would be set to about 500 nM to make the changes more obvious. If, however, this level is set, and ion concentrations exceed this value, then the system will simply show the maximum value for all higher ratios, resulting in a serious under-reading error. The better image-analysis systems software do, however, allow 'test ratioing' and inform the user when this condition is approached. Care must be taken, however, to 'test ratio' the *highest* ratio images of a sequence.

Another frequent situation resulting in under-reading of ion concentration occurs when a component of the image-collection system saturates; this may be the ADC, the CCD chip or the intensifier stages. The safest way to eliminate this type of potential artifact is to collect and inspect the original images before they are ratioed, as the changes in fluorescence during an experiment cannot always be predicted. Some systems which perform the ratioing operation 'on line', and do not allow access to the original images, leave themselves open to this artifact.

The problems that occur when images fall below the minimum detectable value are less serious and more easily overcome. With the better software, a minimum threshold level is set, below which all values are treated as zero and therefore do not show up on the image. In addition, any pixels which then have a zero value automatically result in a zero value ratio and are therefore not displayed on the ratio image. This problem may develop during an experiment due to a combination of changing ion concentration, photobleaching and probe leakage or exocytosis. It causes a deterioration in the ratio image quality as pixels disappear but does not indicate erroneous values.

8.4 PROBE LOADING

Probes may be loaded into the cytoplasm either by direct microinjection of the free acid probe through the plasma membrane or by a membrane permeable chemical form which is hydrolysed and trapped inside the cell. Microinjection may be regarded, from a chemical point of view, as the simplest option but requires delicate and expensive facilities. The most usual method of loading in imaging laboratories is therefore by the chemical method.

Most probes in general use are now available in an acetoxymethyl ester (AM) form which passes readily across most cell membranes. Once inside, non-specific esterases hydrolyse it to its free acid form. As the cell membrane is not permeable to this form of the probe it is trapped and concentrated inside. For example, loading with a solution of only 4 µM outside the cell for 30 min at 37°C may result in intracellular concentrations of fura-2 free acid of 50–100 µM.

As with all experimental systems, care must be taken to ensure that the probe is not significantly affecting the measurement of the parameter of interest. In this case there is a risk of the probe buffering the ion of interest and experimental controls must be devised to ensure this is not the case. A simple 'dose–response' curve for probe loading should reveal the maximum safe loading concentration for a particular probe in a cell type. In addition, it should also be noted that probes and/or the carrier (often dimethylsulphoxide, DMSO) are generally toxic to many cell types. They should, therefore, be used within 2–3 h of loading, preferably as soon as hydrolysation is judged to be complete, which may be 10–20 min after washing the extracellular probe away. A further wash at this point is advisable to remove any unhydrolysed probe which has passed out of the cells across the plasma membrane.

It should also be noted that the use of AM probes for loading can result in loading of organelles within the cells (Steinberg *et al*, 1987; DiVirgilio *et al.*, 1990). While these may give a good imaging signal, the probe inside may not be available to cytosolic free ions. Muted responses in some cell types may therefore be caused by the presence of organelles rather than small changes in ion concentration as indicated. Loading the cells with free acid probe by microinjection would not load the organelles and would reveal the real extent of the cytosolic free ion concentration.

8.5 ION CALIBRATION

Calibration of a ratiometric imaging system may be performed either by measuring the ratios of solutions of known ion concentration or by loading the cells with probe and forcing maximum and minimum ratios. First the loaded cells are made permeable with a suitable ionophore, allowing free ions from the extracellular medium to saturate the probe. This will achieve the maximum ratio possible. A chelator of the ion of interest is then added to preferentially bind all the free ions to result in the minimum ratio obtainable. Maximum ratios obtainable by saturating the probe in *solutions* can be up to double that of saturated probe inside *cells*. It is therefore most desirable to calibrate a system using the cells to be used in experiments rather than any other method. The reasons for this difference are not entirely clear but evidence suggests that intracellular viscosity (Poenie, 1990) and compartmentalization of the probe are the main causes. The choice of ionophore will be governed by the ion of interest but it should be noted that some have been shown to be fluorescent at the wavelengths used to excite some probes.

Difficulties in calibration may be encountered if the cells of interest contain secretory granules. Permeabilizing the cells with ionophore has been shown also to permeabilize secretory granules, especially to H^+ ions. (Almers & Neher, 1985). Internal compartments of secretory granules may be quite acidic so this can result in a significant acidification in intracellular pH during calibration. A large variation in the values obtained for maximum (R_{max}) and minimum (R_{min}) ratios have been reported even in supposedly homogeneous cell populations. While the ion readily saturates the probe for the R_{max} value, chelating the ion to measure R_{min} may be more difficult and can take 15–30 min to reach a plateau. An alternative method to acquire the (R_{min}) value is to use $MnCl_2$ to quench the fura-2 signal (see Hesketh *et al.*, 1983).

8.6 CELL MOVEMENT AND FAST ION FLUXES

Cells used for image analysis of fluorescent probes are usually plated on thin glass coverslips to allow transmission of ultraviolet wavelengths and enable viewing on an inverted stage microscope. Cells used in ratiometric image analysis must remain stationary during the experiment. Usually, experimental protocol requires changing the extracellular medium either for the addition of an agonist or perfusion of the cells. It is therefore imperative that they adhere firmly to the coverslip. Even a small movement of the cells while adding an agonist is likely to result in a crescent-shaped artifact of apparently high ion concentration caused by shifting of the denominator image when ratioing (Fig. 8.2). In the worst case cells are completely lost from the field of view. Some cell types in culture on the glass coverslip adhere quite well whereas others require it to be coated with a substrate such as poly-L-lysine to aid adhesion. The choice of substrate will depend on the cell type and application; for example, fibrinogen has been used to bind human platelets (Heemskert *et al.*, 1992) but care must be taken to ensure that it does not interfere with the normal function of the cell.

In a dual-emission ratiometric system, images are captured sequentially and then ratioed. The resultant images are therefore not true ratios as the individual images are not captured at the same time. The shorter the time between capturing image pairs, the better the approximation to a true ratio measurement. The better image analysis systems have capture rates up to 25 frames per second or 40 ms between images used to calculate a ratio. While the time between pairs may be

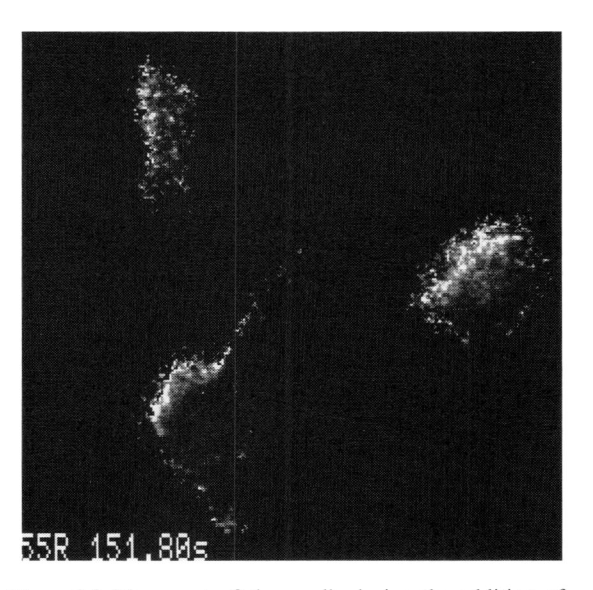

Figure 8.2 Movement of these cells during the addition of agonist results in a high ratio on the top left edge which may be interpreted as a fast polar response. This example of the GH3 cell line loaded with fura-2 is relatively obvious as all the cells in the field exhibit the effect. If, however, just one cell from a field of 20 moves, the artifact may be much more believable.

set to be very much longer, thus enabling slower, longer experiments to be performed while ensuring the ratio image is a close approximation to a true ratio.

Very fast changes in ion concentration will also be subject to a similar error. This will result in under-read values during fast transient increases in ion concentration and over-read values as the concentration decreases. The only solution in this case is to increase the data capture rate by moving to a faster system. While they have other limitations, ratio photometric systems may be more appropriate with capture rates of up to 200 Hz.

8.7 AUTOFLUORESCENCE

Many of the probes employed to monitor intracellular ion concentrations require excitation wavelengths well into the ultraviolet region. Unfortunately, a number of natural peptides are also fluorescent in this region. This problem is especially prevalent in plant tissue and mammalian pancreatic cells. If the component from autofluorescence is small compared to the contribution from the probe it may be disregarded. Otherwise, levels of autofluorescence must be recorded before loading so they may be subtracted from the loaded images. In practice, this is rather difficult as the excitation intensity must be set, to measure autofluorescence, before loading with probe. Preliminary experiments must, therefore, establish a loading protocol which gives a consistent level of fluorescence. Plated cells must also be either kept on the microscope stage while loading or accurately repositioned afterwards.

8.8 INTERACTIONS BETWEEN MULTIPLE PROBES

Currently, the better imaging systems allow simultaneous monitoring of multiple probes. Intracellular pH and calcium, for example, are known to interact and play pivotal roles in cell signalling and secretion. BCECF and fura-2 may be used to monitor pH and calcium, their excitation wavelengths are 440, 490 nm and 340, 380 nm respectively. Emission spectra from both probes overlap at about 520 nm so they may both be monitored using a single emission filter set. There will, however, be significant optical cross-talk between the probes both in the combined emission spectrum and as emission from the lower wavelength probe exciting the other. Fura-2 in this case emits a broad spectrum, peaking at 510 nm, so there is a significant proportion at 490 nm, the upper excitation wavelength of BCECF. Experimental protocols must, therefore, be devised to measure both types of cross-talk in the cells of interest so corrections may be made (Fig. 8.3). It is also advantageous to keep the lower wavelength probe concentration as low as possible consistent with adequate signal for good signal-to-noise. This may usually be achieved by weighting the loading concentration of each probe (Zorec *et al.*, 1993).

Care must also be taken when loading the ester form of multiple probes. It has been found essential to load probes simultaneously as the probes appear to

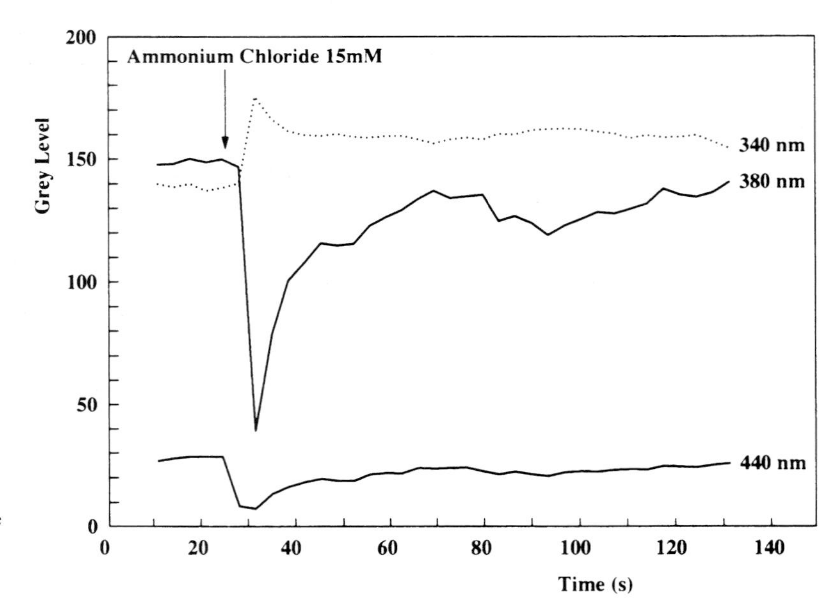

Figure 8.3 The optical cross-talk of fura-2 loaded bovine lactotrophs from excitation at 380 nm and 440 nm. Analysis here reveals cross-talk of 14.4 ± 3.6% which must be corrected for when using multiple probes.

compete for the intracellular esterase activity. Great difficulty has been experienced with sequential loading of multiple probes.

8.9 AVERAGING AND INTENSIFIER NOISE

Some applications, such as monitoring fast ion fluxes, call for maximal capture rates. Others, such as mapping the spatial distribution of ion concentrations within cells, may call for slower rates with higher spatial resolution. For the reasons mentioned above, most video imaging systems currently use intensified CCD cameras to capture images. They do, however, have an inherent problem in that microchannel plate intensifiers introduce a degree of electronic noise into the final image. The most effective way to improve the signal-to-noise ratio is to average a number of video frames (Mason *et al.*, 1990). This may either be achieved in software after storing the frames or, in the better systems, on-line prior to storage and subsequent analysis. There is, of course, a trade-off between temporal resolution and signal-to-noise ratio so the experimental objectives must dictate the balance. Generally, averaging four or eight video frames gives adequate signal-to-noise ratio for many purposes which gives a temporal resolution of 0.3 to 0.6 s between ratio image pairs.

8.10 PROBE LEAKAGE AND EXOCYTOSIS

The cell membrane is not completely impermeable to the free acid form of fluorescent probes. There is a small leakage of probe to the extracellular medium (DiVirgilio *et al.*, 1990). In a constantly perfused system this causes no problems other than a very small drift of intensity which is easily accommodated by ratiometric image analysis. In static systems with no perfusion, long-term leakage may increase the background fluorescence slightly but the effect is minimal.

A potentially serious effect has, however, been reported in mast cells (Almers & Neher, 1985). Acetoxymethyl ester loading here appears to load secretory granules with probe. On stimulation, the granules are released from the cell membrane to deposit their high concentrations of probe to the immediately surrounding area. This appears as a ring of high ion concentration around the edge of the cell and may be misinterpreted as an ingress of ions around the periphery. Examination of the images prior to ratioing would reveal the difference, a loss of probe would result in a sharp reduction of emission at both ratio wavelengths.

8.11 PROBE KINETICS

Most of the work on probe kinetics has been performed in solutions rather than cells. Measurements of association and dissociation constants have been performed for fura-2 and azo-1 (Kao & Tsien, 1988) by the temperature jump relaxation method and for fura-2 and indo-1 (Jackson *et al.*, 1987) by stopped flow measurements. The practical implications of this work are that fura-2 requires 5–10 ms to reach equilibrium at 20°C in solution of ionic strength 140 mM. While the response will be faster at 37°C, other factors such as viscosity and spatial microheterogeneity may slow intracellular measurements. Some fast calcium fluxes may, therefore, be misinterpreted as the probe kinetics may be the limiting factor. However, the main rate-limiting factor for a system will be the maximum capture rate of the system. For imaging systems this will generally be video frame-rate or 40 ms per image. Photometric systems can run much faster and, in some cases, exceed the response time of the probe.

REFERENCES

Almers W. & Neher E. (1985) *FEBS Lett.* **192**, 13–18.

Becker P.L. & Fay F.S. (1987) *Am. Phys. Soc. Special Comm.* C613–C618.

DiVirgilio F., Steinberg T.H & Silverstein S.C. (1990) *Cell Calcium* **11**, 57–62.

Grynkiewicz G., Poenie M. & Tsien R.Y. (1985) *J. Biol. Chem.* **260**, 3440–3450.

Heemskerk J.W.M., Hoyland J., Mason W.T. & Sage S.O. (1992) *Biochem. J.* **283**, 379–383.

Hesketh T.R., Smith G.A., Moore J.P., Taylor M.V. & Metcalfe J.C. (1983) *J. Biol. Chem.* **258**, 4876–4882.

Jackson A.P., Timmereman M.P., Bagshaw C.R. & Ashley C.C. (1987) *FEBS Lett.* **216**, 35–39.

Kao J.P.Y. & Tsien R.Y. (1988) *Biophys. J.* **53**, 635–639.

Mason W.T., Hoyland J., Rawlings S.R. & Relf G.T. (1990) *Methods Neurosci.* **3**, 109–135.

Poenie M. (1990) *Cell Calcium* **11**, 85–91.

Steinberg S.F., Bilezikian J.P. & Al-Awquti Q. (1987) *Am. Phys. Soc. Special Comm.* C744–C747.

Tsien R.Y. (1981) *Nature* **290**, 527–528.

Zorec R., Hoyland J. & Mason W.T. (1993) *Eur. J. Physiol.* (in press).

PART IV

Optical Probes for Specific Molecules, Organelles and Cells

CHAPTER NINE

Acridine Orange as a Probe for Cell and Molecular Biology*

ALEXANDER V. ZELENIN
Engelhardt Institute of Molecular Biology, Russian Academy of Sciences, Moscow, Russia

9.1 INTRODUCTION

A vast number of stains and dyes are used in biology and medicine. In the specialized field of fluorescent probes, the basic fluorescent dye acridine orange (AO) emerges among the most popular ones.

The use of AO in different areas of cytochemistry and cell and molecular biology has been reviewed many times (Bertalanffy, 1963; Kasten, 1967; Zelenin, 1967, 1971; Meissel & Zelenin, 1973). The more recent reviews (Darzynkiewicz, 1990a; Darzynkiewicz & Kapuscinski, 1990) have, however, mostly dealt with the use of the stain in flow cytometry. To avoid large overlaps and to make this chapter complementary to the literature mentioned above, I have tried to cover different fields of AO application and to give more attention to some uses other than flow cytometry. Particular attention is paid to the most important works published in Russian and, therefore, largely inaccessible to Western readers.

* To the memory of my teacher Professor Maxim N. Meissel, the founder of fluorescence microscopy in Russia.

9. 2 HISTORICAL REMARKS

Acridine orange (AO) was synthesized in 1889 (for review see Albert, 1966; Kasten, 1967; Zelenin, 1967; 1971; Darzynkiewicz & Kapuscinski, 1990), but its first biological applications were reported only in 1940 by Bukatsch & Haitinger (1940) and Strugger (1940a, b). Strugger carried out a fundamental investigation on the interaction of AO with plant cells and found that living and dead cells differed in their stainability by AO. Among those who made the most substantial contribution to the introduction of AO into biology, M. Meissel, J. Armstrong, L. von Bertalanffy and N. Schummelfeder in the 1950s, and R. Rigler and Z. Darzynkiewicz, at a later period, deserve particular mention.

The work of Maxim Meissel and his co-workers, carried out in Moscow and published mostly in Russian, was of great importance. Together with his colleague Dr. V. Korchagin, Meissel undertook the first investigation of AO binding to nucleic acids *in vitro*. It was shown (Meissel & Korchagin, 1952) that addition of AO to purified DNA and RNA caused

FLUORESCENT AND LUMINESCENT PROBES, 2ND EDN
ISBN 0–12–447836–0

their green and red fluorescence, respectively. These investigations stimulated intensive work on fluorescence microscopy in Russia. AO fluorescence microscopy began to be used on a large scale for the study of cancer (Meissel & Gutkina, 1953) and X-ray damaged cells (Meissel & Sondak, 1956; for review, see Meissel & Zelenin, 1973). Later, AO was successfully applied to the investigation of nuclei acid secondary structure *in vitro* (Borisova & Tumerman, 1964, 1965).

The rapid and successful development of fluorescence microscopy in Russia was closely related to the invention and introduction of appropriate fluorescence instruments by Dr. Eugeny M. Brumberg in Leningrad. In 1948 he suggested the use of a semi-reflective dichroic mirror in fluorescence microscopy (Brumberg & Girshgorin, 1947) and later developed a special opaque illuminator which allowed the easy conversion of a conventional light microscope into a fluorescence one (Brumberg & Krilova, 1953). In the mid-1950s Brumberg's vertical fluorescence opaque illuminators with dichroic mirrors began to be produced and sold by the Leningrad Optical Factory (LOMO). Shortly after that a special fluorescence microscope with easily changeable transmitted and opaque illumination was constructed and, in 1959, LOMO began to offer such microscopes (ML-1) for sale.

Brumberg's illumination system, slightly modified and improved, was later introduced into the Western fluorescence microscope market under the name of 'Ploem opack'. It is only recently that the fundamental role of Brumberg in the creation of modern fluorescence microscopy instrumentation has begun to be appreciated by the international scientific community (Reichman, 1994).

9.3 AO AS A FLUORESCENT DYE

Acridine orange (AO), 3,6-dimethylaminoacridine (Fig. 9.1), is a weak base which is readily soluble in water. It is usually prepared and sold as a zinc complex of a hydrochloride salt.

The dye has been shown to exist in aqueous solution in two main forms (Zanker, 1952a,b; Morosov, 1963a,b; Borisova & Tumerman, 1964). The first form is characterized by (1) an absorption maximum at 494 nm, (2) a green fluorescence with a maximum at 530 nm, and (3) a fluorescence lifetime of 2×10^{-9} s. This form is characteristic of AO in highly diluted solutions (10^{-7} M) in which the stain molecules are presented as single monomers.

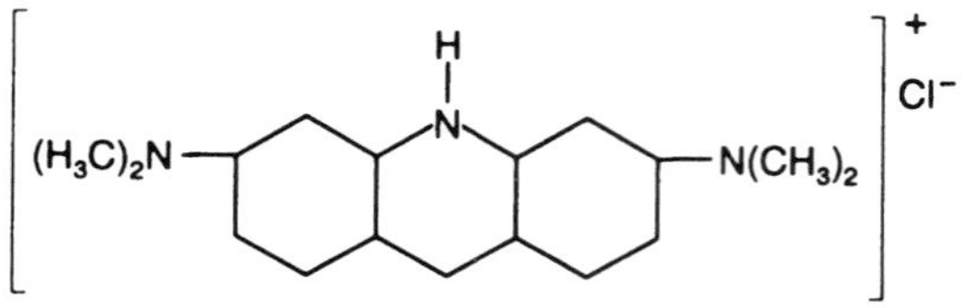

Figure 9.1 The structure of acridine orange.

The second form is characterized by (1) an absorption maximum at 465 nm, (2) a red fluorescence with a maximum at 640 nm, and (3) by a fluorescence lifetime of 20×10^{-9} s. This form is characteristic of AO in concentrated solutions ($>10^{-2}$ M) in which the stain molecules are presented as dimers.

AO solution of intermediate molarity has mixed spectral properties.

As well as fluorescence, AO in both monomeric and dimeric form has phosphorescence which is quite strong at low temperatures (Zanker *et al.*, 1959; Bradley, 1961; Morosov, 1963a,b; Morosov & Savenko, 1977). This is true for red AO emission, both in solutions and in complexes with single-stranded nucleic acids (for details, see Darzynkiewicz & Kapuscinski, 1990). Thus, strictly speaking, general AO emission should be called luminescence and not fluorescence. Nevertheless, in most current publications on AO, for the sake of convenience the term fluorescence is used. We also prefer to use the latter term for all types of AO emission. It must be taken into consideration that the equipment usually applied to the investigation of AO properties, such as fluorometers (Borisov & Tumerman, 1959) or flow cytometers, register only very short signals (shorter than 10^{-6} s), i.e. only fluorescence.

9.4 SPECTRAL PROPERTIES OF AO IN COMPLEXES WITH NUCLEIC ACIDS AND OTHER BIOPOLYMERS

The first work on the AO interaction with nucleic acids *in vitro* was performed by Meissel & Korchagin (1952). In this investigation the data on the fluorescence properties of AO complexes with DNA and RNA were obtained by means of visual fluorescence microscopy. Later, spectral investigations gave detailed characteristics of these complexes.

Two types of AO complexes with nucleic acids have been described. The A-type complex (Steiner & Beers, 1958, 1959, 1961) arises on interaction of AO with double-stranded DNA or double-stranded regions of RNA with a ratio of the number of nucleotide molecules to number of AO molecules of 4:1 or less. This type of complex is characterized by (1) an absorption maximum at 502–504 nm, (2) green fluorescence with a maximum at 530 nm, and (3) a fluorescence yield and lifetime 2–2.5 times higher than that of AO in a diluted solution when the dye is in the monomeric form.

The AO properties in the A-type complex are quite close to those of AO in highly diluted solutions, the only differences being a shift of the absorption maximum from 594 to 504 nm and in some instances an increase in fluorescence yield and lifetime (Borisova *et al.*, 1963; Borisova & Tumerman, 1964). This means that in the A-type complex AO molecules are located on nucleic acid molecules at such a distance that no dimeric complexes can be formed.

It was suggested more than 30 years ago (Lerman, 1961, 1963, 1964) that a complex of this type is formed as a result of intercalation of an amino acridine molecule between the nucleic acid base pairs (Fig. 9.2). This model was confirmed later by X-ray analysis (Wang, 1974; Berman & Young, 1981; Waring, 1981). For a more recent review, see Darzynkiewicz & Kapuscinski (1990). The B-type complex (Steiner & Beers, 1958, 1959, 1961) differs considerably from the A-type. It is characterized by the following properties: (1) an absorption maximum at 475 nm, (2) a red fluorescence with the maximum at 640 nm, and (3) a decrease in the fluorescence yield and sharp (10-fold) increase in the fluorescence lifetime of the AO molecule (Borisova *et al.*, 1963; Borisova & Tumerman, 1964). A detailed investigation of crystalline acridine complexes with the single-stranded regions of nucleic acids was performed by Neidle *et al.* (1978).

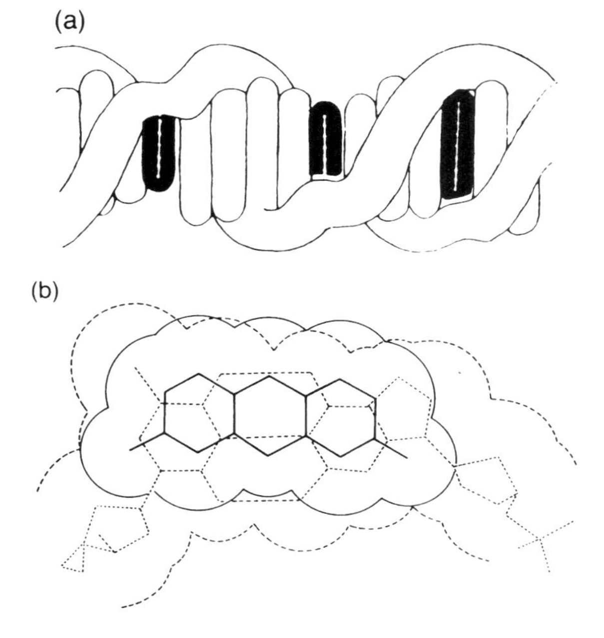

Figure 9.2 The structure of DNA–acridine complexes. (a) Acridine molecules intercalated into DNA. (b) A molecule of acridine superposed over a nucleotide pair with the deoxyribose phosphate chain in an extended configuration. (From Lerman, 1964.)

The properties of this complex are similar to those of AO in dimer form. The B-type complex is formed when AO is combined with single-stranded regions of RNA and denatured DNA with the creation of AO dimers.

AO was shown to form salt-like complexes with mucopolysaccharides with red fluorescence (Kuyper, 1957; Appel & Zanker, 1958; Saunders, 1964). At high pH (>7), when protein carbohydrates are protonized, AO binds to proteins to form the red fluorescent complexes.

9.5 AO IN THE STUDY OF NUCLEIC ACIDS *IN VITRO*

The key work in this area was performed by Borisova and Tumerman (Borisova *et al.*, 1963; Borisova & Tumerman, 1964, 1965). It has been established that investigation of AO properties allows the quantification of the strandedness of nucleic acid in solution (Figs 9.3 and 9.4). It has been shown (Borisova *et al.*, 1963) that tRNA in solution contains no more than 28–30% single-stranded regions. These data coincide well with the results of direct X-ray analysis of tRNA crystals (Kim *et al.*, 1974).

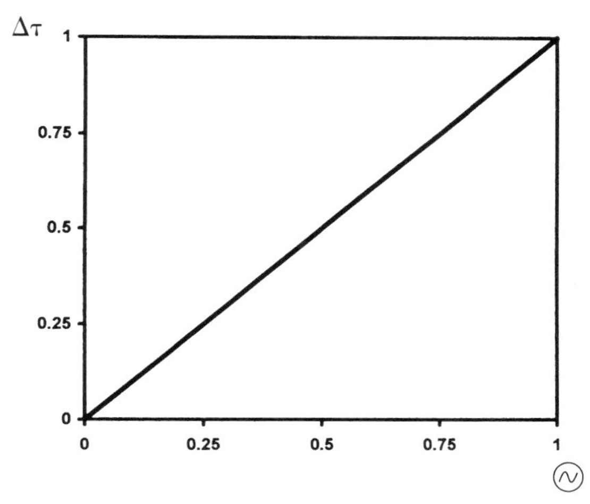

Figure 9.3 Relative dependence of AO fluorescence lifetime ($\Delta\tau$) on the content of the denatured DNA in mixture.

Abscissa – content of the denatured DNA in mixture
Ordinate:

$$\Delta\tau = \frac{(\tau \text{ red} - \tau \text{ green})t}{(\tau \text{ red} - \tau \text{ green})\ 90^\circ}$$

where τ is life-time of AO fluorescence in green (τ green) and red (τ red) spectrum parts. τ is taken as a measure of AO monomers and dimers binding to DNA; ⊚ = degree of DNA despiralization. (From Borisova & Tumerman, 1964.)

Later, the properties of pre-mRNA in nuclear particles and in free state in solution were studied. It was found that pre-mRNA in nuclear particles was almost completely single-stranded whereas in solution about 70% of the same mRNA was double-stranded (Borisova *et al.*, 1981). Similar characteristics (70% double-strandedness) were found for the MS2-phage mRNA in solution (Borisova & Grechko, 1984).

Among more recent applications of the AO method, the investigation of some unusual DNA structures of interest with parallel-stranded orientation should be mentioned. It has been shown that parallel DNA helices may contain both AT and GC base pairs (Borisova *et al.*, 1993). The AO fluorescence lifetime measurement revealed non-paired thymidine residues in four-stranded nucleic acid structures with parallel orientation of (GT)*n* strands (Borisova *et al.*, 1992).

AO itself appears to be also very useful for studying structural organization of DNA–RNA triplexes (for review, see Plum *et al.*, 1995). It has been shown that addition of AO to nucleic acid triple complexes increases their stability by stimulating RecA protein binding to double-stranded DNA. The stabilizing effect depends on AO concentration. One to two AO molecules per 10 triplets stabilize the complex whereas intercalation of the third molecule results in an essential complex destabilization (Shchyolkina and Borisova, 1997).

9.6 AO IN NUCLEIC ACID CYTOCHEMISTRY

9.6.1 AO as a direct probe for nucleic acids

When used under appropriate conditions (fixation, pH 4.2–4.6, AO concentration 5.10^{-4} M, sufficient washing-off of the unbound stain excess), AO acts as an excellent cytochemical probe for nucleic acids. It gives bright green fluorescence to DNA (nucleus) and deep red fluorescence to RNA (cytoplasm and nucleoli). An important condition for obtaining such a picture is proper fixation. The fixative should preserve nucleic acids and should not contain any denaturing DNA components (formaldehyde, acids, etc). The AO concentration must be high enough for dimer formation on single-stranded regions of nucleic acids.

Over 40 years' experience allows me to say with complete confidence that the AO staining technique is the best cytochemical approach to nucleic acid investigation in routine work. In our laboratory it is used in all experiments in which quick and reliable screening of the cells for the presence of nucleic acids is required.

The cytochemical interpretation of the fluorescence microscopical picture obtained with AO is based on experiments with cell pretreatments with nucleases (Armstrong, 1956; Bertalanffy & Bickis, 1956; Schummelfeder *et al.*, 1957) as well as on the results described above on AO binding to nucleic acids *in vitro*. However, some particular problems of the cytochemical specificity of the obtained pictures should be discussed.

No doubt usually arises as to the nature of green fluorescence induced in the cell structures by the AO treatment. A strong A-type complex is specific for AO binding to double-stranded nucleic acids. It should be noted, however, that the sensitivity of the method, at least under visual observation, is not sufficient to detect the green fluorescence of double-stranded mitochondrial DNA in the cytoplasm. At the same time, double-stranded RNAs of some viruses (reovirus) acquire green fluorescence after AO staining which is easily detected microscopically (Gomatos *et al.*, 1962; Gomatos & Tamm, 1963).

In most cases the red fluorescence in AO-treated preparations indicates the presence of RNA. However, red fluorescence is also characteristic of single-stranded DNA of certain phages (Mayor & Hill, 1961) and viruses (Jamison & Mayor, 1965) and is also easily detected cytochemically.

It should also be noted that the AO dimers responsible for the red fluorescence can be formed when AO is bound to any polyanions (acid polysaccharides, protonized proteins, etc.). Of these, acid polysaccharides are the most important (for details, see Section 9.12).

These facts should be borne in mind when any new type of cell is taken for investigation.

Although RNA has been shown to contain large double-stranded regions *in vitro*, the whole cytoplasm expresses only red fluorescence, thus indicating that its RNA is completely in single-stranded form, most likely due to its denaturation under the action of fixative and AO in the course of staining.

When nucleic acids are investigated visually after AO treatment, this RNA denaturation is usually quite sufficient. For quantitative investigation, additional RNA denaturation is necessary.

The possibility of nucleic acid quantitation with AO requires special discussion. Proteins in nucleoprotein complexes restrict ligand binding to nucleic acids; this is particularly true for the nuclear chromatin. DNA in chromatin is only partially accessible to AO; this accessibility is highly dependent on the DNA–histone interaction and reflects the chromatin functional state. These facts open up the possibility of using AO for the investigation of chromatin functional state (for details, see Section 9.9). At the same time, they show that for nucleic acid quantitation the proteins that restrict AO binding to nucleic acids should be removed, or at least their action should be minimized. For that purpose a special technique has been suggested (Darzynkiewicz *et al.*, 1975; Darzynkiewicz, 1990a). The method consists of permeabilization of cells with a detergent in the presence of acid and subsequent staining with AO. Acid treatment results in dissociation and extraction of histones, which probably makes subsequent DNA assay quantitative (but this is not proven). At the same time AO staining under the suggested conditions causes a complete denaturation of double-stranded RNA regions and the whole RNA thus acquires the red fluorescence.

These modifications of the method have been successfully used for simultaneous DNA/RNA estimation by flow cytometry in different cell systems.

A rapid increase in red fluorescence intensity has been detected in quiescent (non-dividing) cells stimulated to proliferate (Darzynkiewicz *et al.*, 1979a; Darzynkiewicz, 1990a). DNA/RNA analysis by staining the cell with AO has been found to be helpful in differential diagnosis of bone tumours and the evaluation of treatment response (El-Naggar *et al.*, 1995).

A special technique for evaluation of the cytoplasmic ribosome condition was developed recently by Gordon *et al.* (1997). These workers measured the ratio of red to green fluorescence intensity in fixed brain slices of hippocampal neurons from ground squirrels. This revealed different accessibility of active and inactive ribosomes for AO binding.

9.6.2 Quenching of the fluorescence of AO bound to DNA

9.6.2.1 Actinomycin D as an energy acceptor

When AO is bound to DNA *in vitro* simultaneously with an energy acceptor, its fluorescence is quenched, partially or completely, owing to the energy transfer. Different substances may be, and are, used for that purpose.

Borisova *et al.* (1968) showed in experiments *in vitro* that the antibiotic actinomycin D quenched the fluorescence of AO bound to pure DNA and DNA in chromatin. A method for the estimation of the length of chromatin DNA regions not accessible to ligands was proposed. Approximately half of the DNA in native chromatin was shown to be covered with proteins, and thus not accessible to AO, whereas the other half was easily ligand-accessible and therefore protein-free. The lengths of the DNA regions of both types were roughly estimated to be more than 100 base pairs. These early findings coincide closely with the modern concepts of chromatin subdivision into nucleosome and linker parts.

Actinomycin D quenching of the AO fluorescence was later applied to cytochemical investigation of nucleic acids. It was found (Zelenin *et al.*, 1976) that simultaneous treatment of fixed cytochemical preparations with AO and actinomycin D resulted in suppression of fluorescence of AO bound to the DNA-containing cell structures (nucleoplasm). At the same time fluorescence of RNA-containing structures (cytoplasm and nucleoli) remained unchanged. These observations are consistent with the selective binding of actinomycin D to the DNA guanines.

Distinguishing between double-stranded DNA and RNA is a possible application of this approach. As has been shown above, the double-stranded RNA of certain viruses forms green fluorescing complexes with AO. Cytochemical identification of such RNA by conventional methods is rarely possible because it is usually RNAse resistant (Gomatos & Tamm, 1963). But the fluorescence of these AO–RNA complexes should not be quenched by actinomycin D. The actinomycin D specificity for guanines also opens up a possibility for application of this approach to the cytochemical detection of long AT-rich chromosome regions.

9.6.2.2 5-Bromodeoxyuridine as an energy acceptor

5-Bromodeoxyuridine (BrdUrd), incorporated into cellular DNA as a thymidine analogue, quenches fluorescence of different fluorochromes, including AO, applied to these cells. BrdUrd incorporation into DNA reduces both fluorescence intensity (Latt, 1976) and phosphorescence of AO (Galley & Purkey, 1972). Labelling of cells with BrdUrd followed by the AO staining may thus be used for the investigation of the cell cycle. According to Darzynkiewicz *et al.* (1983), all BrdUrd-treated cells, regardless of the cell cycle phase, have shown a decreased fluorescence. The BrdUrd-attributed suppression of the cell fluorescence has been found to be large enough to separate totally BrdUrd-labelled from unlabelled mitotic cell populations. It has been concluded that the BrdUrd–AO flow cytometry method provides an alternative to the autoradiography technique for obtaining a fraction of labelled mitosis. Elsewhere, BrdUrd–AO flow cytometry has been successfully used for discrimination between cycling and non -cycling cells (Darzynkiewicz *et al.*, 1978). In more recent investigations, however, AO has been mostly replaced by different, more DNA-specific stains (ethidium, bis-benzimides, mithromycin, etc); for reviews see Bohmer (1990), Crissman & Steinkamp (1990), Poot *et al.* (1990).

9.7 AO DNA STAINING AFTER ACID PRETREATMENTS

Acid treatment, even if mild, brings about at least partial apurinization of DNA and thus changes its stainability by AO. As a result, AO forms dimers at depurinized regions, thus causing the DNA red fluorescence. This makes it necessary to avoid the presence of acids in fixatives used for AO conventional fluorescence microscopy of nucleic acids. At the same time it opens up important possibilities for the investigation of the DNA state *in situ*.

It has been shown in several investigations (Schummelfeder *et al.*, 1957; Nash & Plaut, 1964; Roschlau, 1965) that, when preparations containing double-stranded DNA are exposed to a short acid hydrolysis, subsequent staining with AO results in the red or yellow-red fluorescence of DNA-containing structures instead of the standard green fluorescence.

These observations have been expanded in two directions. Firstly, Darzynkiewicz *et al.* (1975, 1979b) suggested a flow cytometry technique for measurement of the sensitivity of DNA *in situ* to acid denaturation. The ratio of the red to green fluorescence was taken as a measure of DNA denaturation. Some aspects of the procedure (proper fixation, staining at the equilibrium conditions, etc.) are critical for success (for details see Darzynkiewicz, 1990b).

This method was successfully applied to the investigation of exponentially growing cells, to the study of quiescent versus cycling cells and non-differentiated versus differentiated cells (Darzynkiewicz, 1990b), as well as to the detection of structural changes in sperm chromatin during maturation and in the presence of damaging factors (Karabinus *et al.*, 1997; Avarindan & Mongdal, 1998).

An interesting application of the method is based on the higher accessibility of mitotic chromatin to the acid treatment compared with that in the interphase nuclei (Darzynkiewicz *et al.*, 1975).

Secondly, Zelenin *et al.* (1977) used AO staining after acid pretreatment for the investigation of chromatin in *Chironomus* giant chromosomes in cytological preparations. After very mild hydrolysis, as well as in control preparations, all chromosome regions showed green fluorescence. After long hydrolysis all chromosome bands fluoresced red. Intermediate hydrolysis gave red fluorescence in all transcriptionally inactive bands, including the centomeric ones, whereas transcriptionally active puffing regions were green. The most likely explanation of these observations is that transcriptionally active chromosome regions are less susceptible to acid owing to the protective action of the non-histone proteins present there. The method has some advantages over the method currently used for the study of chromatin properties (see Section 9.9), as it does not require cytofluorometry and therefore can be applied to the investigation of cells and their components in conventional cytological preparations.

Some additional aspects of the application of acid–AO technique for the evaluation of the chromatin functional state are discussed in Section 9.9.

9.8 AO IN THE STUDY OF DNA THERMAL DENATURATION

Experiments *in vitro* (Borisova & Tumerman, 1964) demonstrated the applicability of AO spectrofluorometry for distinguishing between single- and double-stranded nucleic acids and, in particular, for the study of the process of thermal DNA denaturation–renaturation (Fig. 9.4). A detailed technique for the investigation of DNA melting *in situ*, i.e. inside intact cells, was later developed (Rigler & Killander, 1969; Rigler *et al.*, 1969). This technique involves heating and cooling the preparations in the presence of formaldehyde, to prevent DNA renaturation, and subsequent estimation of the degree of DNA denaturation based on measurement of the ratio of red to green fluorescence by AO staining (coefficient α). The cell treatment in the presence of formaldehyde is the crucial step in this method. DNA *in situ* is in such a compact form that in the absence of formaldehyde denatured DNA undergoes immediate and complete renaturation.

The curves obtained for DNA melting *in situ*

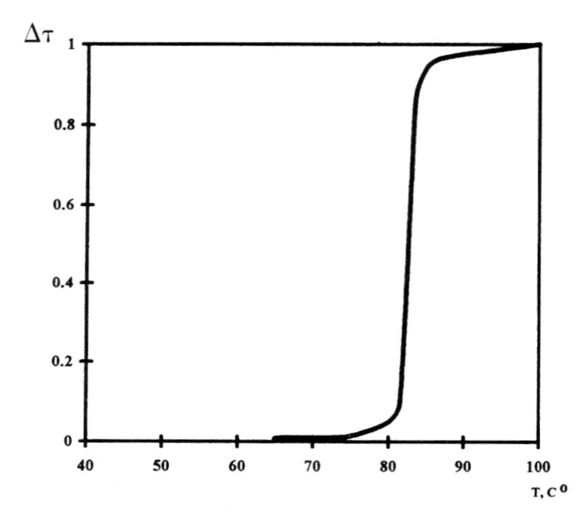

Figure 9.4 Fage T2 DNA melting curve (citrate buffer, ion strength 0.08). Abscissa: temperature. Ordinate: for designation see legend to Figure 9.3. (From Borisova & Tumerman, 1965.)

proved to be similar to those for DNA melting *in vitro*. It was found that the susceptibility of DNA to the action of heat varied in cells with different proliferative activity. This was thought to be due to the protective action of histones in chromatin. The biological results received by this technique are discussed in Section 9.9.

Some improvements of the method were later suggested, including changing of the AO staining conditions (Liedeman & Bolund, 1976b) and the development of a technique for investigation of DNA thermal stability by flow cytometry (Darzynkiewicz *et al.*, 1974, 1975). The latter method, which includes use of alternative apparatuses as well as different techniques for preparation of cell specimens, quickly became widely used for the study of large cell populations, primarily the peripheral blood cells and spermatozoa. The application of these approaches to the study of the chromatin functional state is discussed in Section 9.9.

The use of the DNA denaturation–renaturation approach for chromosome banding is described in Section 9.11.

9.9 AO IN THE STUDY OF THE CHROMATIN FUNCTIONAL STATE

9.9.1 Introduction

The chromatin functional state, i.e. the ability of its DNA to serve as a primer in RNA synthesis, is usually studied by biochemical (*in vivo*) and molecular biological (*in vitro*) techniques. However, the chromatin *in situ*, i.e. inside the cells in cytological smears, suspension and, sometimes, in histological preparations, retains most characteristics of the chromatin *in vivo*, at least not less than the chromatin *in vitro* (Table 9.1).

A large number of different techniques have been suggested for the investigation of chromatin properties *in situ*. In my review on the subject published 15 years ago more than 20 such techniques were listed (Zelenin, 1982). During recent years this number has more than doubled. These techniques are based on different approaches, most of which fall outside of the scope of this chapter and are discussed in detail elsewhere (Ringertz, 1969; Zelenin, 1977, 1982; Zelenin & Kushch, 1985; Darzynkiewicz, 1990c). Here I will confine myself to the investigation of the chromatin functional state using AO.

Such investigation is based on two different approaches: (1) estimation of direct AO binding to cellular chromatin, and (2) study of spectral properties of cells stained with AO after different pretreatments.

Table 9.1 Principal properties of chromatin *in situ*.

- Chromatin in preparations *in situ* is able to act as a primer in the RNA and DNA polymerase reactions catalysed by exogenous polymerases.
- Chromatin *in situ* contains endogenous RNA and DNA polymerases in amounts sufficient to catalyse RNA and DNA synthesis.
- The properties of chromatin *in situ* can be modified in a similar way to those of chromatin *in vitro*.
- Changes in physicochemical properties of chromatin *in situ* reflect well the chromatin changes *in vitro* and *in vivo*.

9.9.2 Direct AO binding to the cellular chromatin

The estimation of AO binding to DNA inside chromatin (Rigler, 1966) was the first and, for a certain period, the most popular approach to chromatin study *in situ* (for reviews see Rigler, 1969; Ringertz, 1969; Auer, 1972; Ringertz & Bolund, 1974; Zelenin, 1977).

The method is based on the assumption that in chromatin AO binds only to DNA regions not covered with histones. An increase in the cell genome activity accompanied by weakening of the DNA–histone interaction causes an increase in the number of AO molecules attached to DNA which can be detected cytofluorometrically. This result was first revealed in experiments with peripheral blood lymphocytes exposed to phytohaemagglutinin (Killander & Rigler, 1965, 1969; Rigler, 1966; Rigler & Killander, 1969) and then extended to a number of different quiescent cells stimulated to proliferate by different agents: the peripheral blood and lymph node lymphocytes in 'crowded' cultures (Auer *et al.*, 1970; Zelenin & Vinogradova, 1973); hepatocytes of regenerating liver (Kolesnikov *et al.*, 1973; Alvarez, 1974; Kushch *et al.*, 1974, 1978); kidney cells induced to grow in culture (Auer *et al.*, 1973); quiescent cultured cells after addition of fresh serum to the culture medium (Terskikh *et al.*, 1976); hen erythrocytes fused with dividing culture cells (Bolund *et al.*, 1969); etc.

It was shown that removal of histone H1 from the chromatin *in situ* resulted in a substantial increase in AO binding to chromatin (Zelenin & Vinigradova, 1973), whereas cell treatment with 0.3–0.35 M sodium chloride, which presumably removed non-histone proteins, caused a decrease in the AO binding to chromatin of activated cells (Kushch *et al.*, 1974, 1980) and had no influence on AO binding to non-activated cells. In both cases the difference in AO binding to quiescent and activated cells disappeared.

In most of these experiments the results obtained by AO binding were directly or indirectly confirmed by alternative techniques such as binding of other ligands (actinomycin D, ethidium, etc.), investigation of DNA accessibility to the action of damaging factors (heat, acids) and histone stainability.

A particularly good correlation was found in many cases between AO cytofluorometry data and the results of the *in vivo* DNA template activity estimation (Fig. 9.5); for review see Zelenin (1977, 1982), Zelenin and Kushch (1985). In our experiments with liver cells activated by partial hepatectomy (Kolesnikov *et al.*, 1973; Kushch *et al.*, 1974; Sondore *et al.*, 1978), almost complete coincidence was found between the results *in situ* and *in vitro* (on isolated chromatin) (Fig. 9.6).

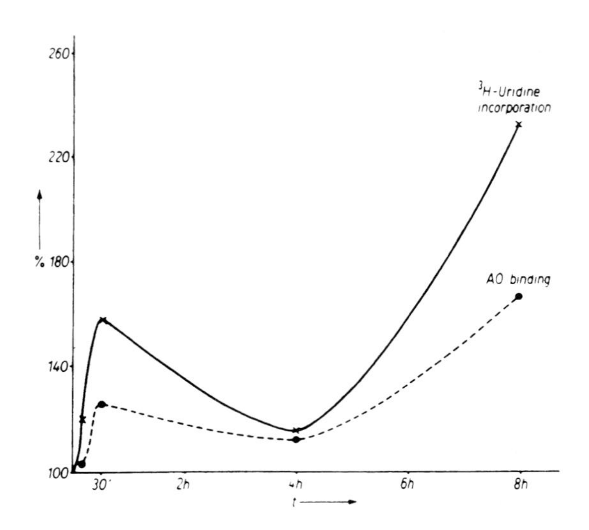

Figure 9.5 AO binding by chromatin and labelled uridine incorporation in RNA of quiescent Chinese hamster cells after growth stimulation. Growth stimulation was achieved in the stationary culture by replacement of nutritive medium. Abscissa: time after stimulation (min). Ordinate: AO fluorescence intensity and radioactive label incorporation (% to the control). A good correlation between both curves is seen. (From Terskikh *et al.*, 1976.)

The concept of chromatin activation was conceived on the basis of these observations (Killander & Rigler, 1965; Rigler, 1966). It was shown later that at least some chromatin properties varied at different stages of this process (Sondore *et al.*, 1978) and three steps of chromatin activation were described in dormant cells stimulated to proliferate (Zelenin & Kushch, 1985).

In parallel experiments, a marked decrease of AO binding to chromatin in the course of genome inactivation was described on four cell systems: spermatogenesis (Gledhill *et al.*, 1966), red blood cell (Kernell *et al.*, 1971) and lymphocyte maturation (Fig. 9.7, Manteifel *et al.*, 1973; Zelenin, 1977; Kushch *et al.*, 1980) and contact cell inhibition in culture cells (Zetterberg & Auer, 1970). A similar finding was recently obtained from flow cytometry of maturing human sperms (Weissenberg *et al.*, 1995; Golan *et al.*, 1996).

Soon after the development of Rigler's method there began to appear in the literature publications claiming that the results obtained by Rigler's original technique might depend not only on the chromatin functional state, but also on the local cell density in a preparation (Bolund *et al.*, 1970; Ringertz & Bolund, 1970; Liedeman *et al.*, 1975).

Attempts to improve Rigler's technique were reported (Kolesnikov *et al.*, 1973; Smets, 1973; Liedeman & Bolund, 1976a,b). Thus in all investigations carried out in our group since the beginning of the 1970s (Zelenin *et al.*, 1974; for review see Zelenin, 1977, 1982; Zelenin & Kushch, 1985) a modification of Rigler's original technique has been used. In this modification (Kolesnikov *et al.*, 1973) the acetylation step is omitted and, after staining in a high concentration of AO (10^{-4} M) the preparations are washed in

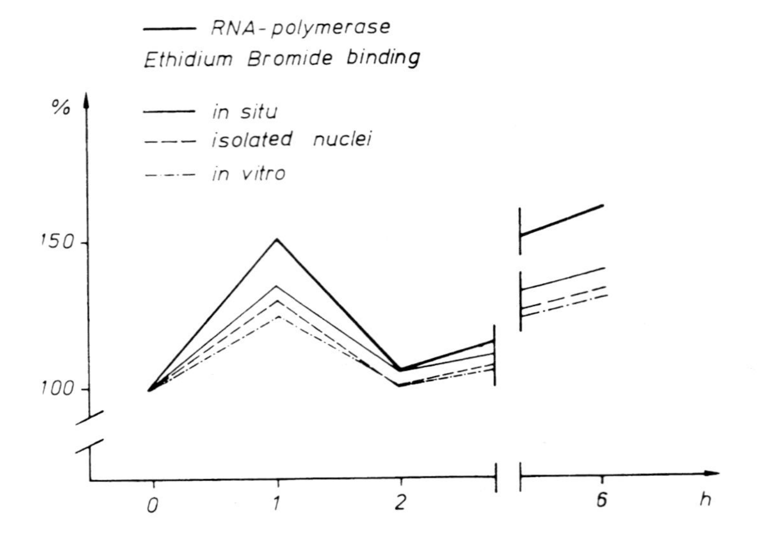

Figure 9.6 Changes in the rat liver chromatin properties after partial hepatectomy. Abscissa: time after operation (h). Ordinate: AO fluorescence intensity and radioactive label incorporation (% of control). The figure is based on the data from the work of Sondore *et al.* (1978).

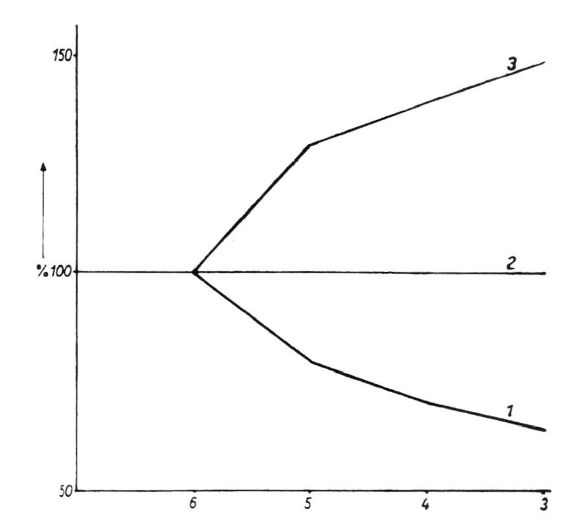

Figure 9.7 Changes in the cytochemical properties of the lymph node lymphocytes during their maturation. Abscissa: cell diameter in arbitrary units. Ordinate: (1) mean values of AO binding; (2) Feulgen DNA staining; (3) histone stainability with an acid dye (primulin). Decrease in the cell diameter is taken as a measure of lymphocyte maturation. Cytochemical parameters for the lymphocytes with the maximal diameter were taken as 100%. (From Manteifel *et al.*, 1973.)

a low concentration of AO (10^{-6} M). For the subsequent cytofluorometry the cell preparations are placed in the latter solution. This modification, known as semi-equilibrium staining, is similar to the cell staining in equilibrium conditions proposed by Liedeman (Liedeman *et al.*, 1975; Liedeman & Bolund, 1976a).

At the same time flow cytometric techniques for estimation of the chromatin functional state began to be developed. These techniques were mainly based on indirect approaches such as the study of chromatin susceptibility to the action of heat or acids with subsequent AO staining (for details see Section 9.9.3) and enabled quick investigation of vast cell populations. Simultaneously new flow cytofluorometers were developed and became more widely available. As a result, from the early 1980s in many laboratories the estimation of the amount of AO directly bound to DNA for the study of chromatin properties was gradually replaced by indirect flow cytometrical methods (Darzynkiewicz, 1990c; Darzynkiewicz & Kapuscinski, 1990).

Our long experience has proved the direct estimation of AO binding to the nuclear chromatin to be a reliable approach to the estimation of its functional state (i.e. the total chromatin template activity).

This conclusion is based on results obtained from different cell systems for which the local cell density in a preparation is irrelevant, and on a correlation of AO binding with chromatin template activity. In addition to the data presented above, other results from our group (Smol'yaninova *et al.*, 1991) may be cited (Fig. 9.8). The data discussed above, taken from many investigations, including ours, show a good coincidence between results obtained by AO cytofluorometry and those obtained using alternative techniques such as chromatin susceptibility to damaging factors, binding of different ligands including non-intercalating ones, histone stainability, etc.

It must be remembered, however, that only qualitative or semiquantitative conclusions may be drawn from the results obtained by these techniques. Thus, from the data presented in Fig. 9.8 one may certainly conclude that chromatin is more active 45 minutes

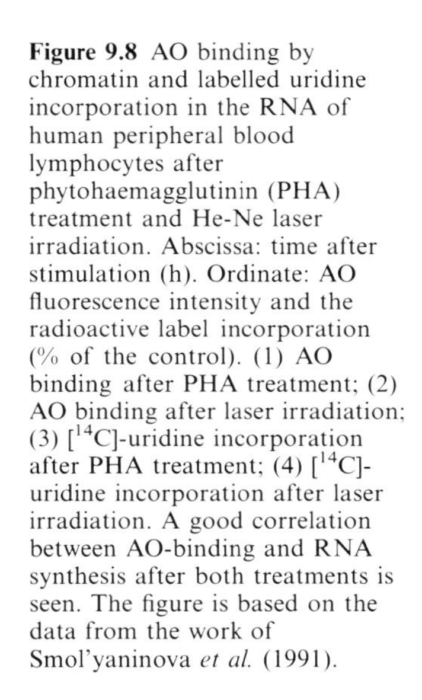
Figure 9.8 AO binding by chromatin and labelled uridine incorporation in the RNA of human peripheral blood lymphocytes after phytohaemagglutinin (PHA) treatment and He-Ne laser irradiation. Abscissa: time after stimulation (h). Ordinate: AO fluorescence intensity and the radioactive label incorporation (% of the control). (1) AO binding after PHA treatment; (2) AO binding after laser irradiation; (3) [^{14}C]-uridine incorporation after PHA treatment; (4) [^{14}C]-uridine incorporation after laser irradiation. A good correlation between AO-binding and RNA synthesis after both treatments is seen. The figure is based on the data from the work of Smol'yaninova *et al.* (1991).

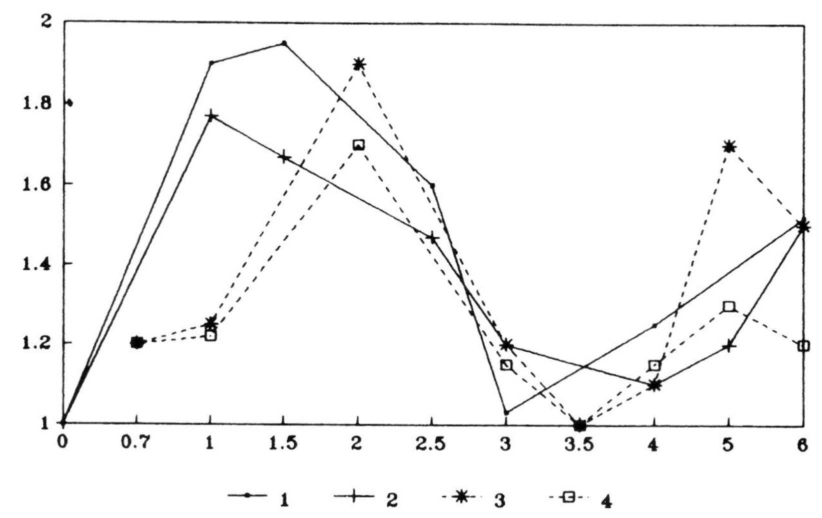

after laser irradiation than before it, but it is impossible to give a numerical value to the activity increase.

9.9.3 Investigation of spectral properties of chromatin stained with AO after different pretreatments

These methods are based on the assumption that both DNA–protein interactions and the chromatin superstructure affect the susceptibility of DNA to heat or acid, and that DNA denaturation in chromatin is different in cells with different genome activity. Subsequent AO staining allows the degree of DNA denaturation to be determined (for details see Sections 9.7 and 9.8). The advantage of this approach, compared with the method based on direct AO binding, is that it is possible to obtain the characteristics for a single cell expressed as the ratio of red to green fluorescence (coefficient α) without the need to subtract the background. Specialized techniques based on static (Rigler *et al.*, 1969; Liedeman & Bolund, 1976b) and flow (Darzynkiewicz *et al.*, 1974, 1975, 1979b; Darzynkiewicz, 1990c) fluorometry were developed and became widely used for the investigation of chromatin properties in single cells and large cell populations.

There are still many contradictions connected with these methods. In most cases flow cytometry techniques reveal a decreased susceptibility of active chromatin to damaging factors compared with inactive chromatin. This phenomenon was assumed to be unrelated to the histone H1, to modifications (phosphorylation, acetylation, etc.) of core histones, or to the presence of HMG proteins, but was thought to be dependent on the higher order of the chromatin structure (Darzynkiewicz, 1990c).

At the same time, in work performed on cytological preparations, the opposite trend – increased susceptibility of active chromatin to the action of damaging factors – was described (Rigler *et al.*, 1969; Ringertz, 1969; Zelenin, 1977, 1982). Experiments in which salt treatment was used to remove H1 and non-histone proteins from the preparations, demonstrated the involvement of these proteins in the thermal stability of DNA in chromatin (Zelenin & Vinogradova, 1973; Kushch *et al.*, 1974). For an attempt to explain these divergences, see Darzynkiewicz (1990c) and Darzyinkiewicz & Kapuscinski (1990).

It may be concluded that when used under strict conditions these techniques (both static and flow) give us reliable qualitative and semiquantitative information about the functional state of chromatin *in situ*. Quantitative conclusions on the subject should, however, be made with great caution (Liedeman & Bolund, 1976b).

9.9.4 Concluding remarks: practical uses of chromatin cytochemistry

There are two sets of AO methods which allow us to judge the functional state of chromatin *in situ* based on different approaches to the preparation of the cell specimens (on slides and in suspension) and the determination of fluorescence properties (static and flow cytometry); these probably reflect different features of chromatin structure (the DNA–histone interactions and the higher order of chromatin structure). These factors should be taken into account when results obtained by alternative techniques are compared or the choice of a particular technique is being made.

Flow cytometry is more convenient and more applicable for chromatin investigation when large cell populations are being studied and suitable material is available. At the same time chromatin smear cytochemistry is useful in cases when investigation of different cells in cytological preparations is required, for example in cell hybrids. These methods should, therefore, be regarded as complementary, with their own particular fields of application.

AO chromatin cytochemistry has been successfully applied to many problems of medicine and agriculture including cancer diagnosis (Kunicka *et al.*, 1987), diagnosis and prognosis of septicaemia in children (Karachunski *et al.*, 1987), as well as to the investigation of pea root meristematic cell germination (Troyan *et al.*, 1984); for reviews see Zelenin and Kushch (1985) and Darzynkiewicz (1990c).

9.10 FLUORESCENCE POLARIZATION OF AO IN STUDIES OF BIOPOLYMERS

9.10.1 Fluorescence polarization of AO bound to DNA

This line of investigation was started by MacInnes & Uretz (1966), who studied the degree of polarization of AO bound to the polythene chromosomes of *Drosophila virilis*. As a result of their investigation, these authors suggested that the interband DNA was not supercoiled but lay parallel to the chromosome axis. Such an organization of the interband DNA was later demonstrated by direct electron microscopical observations (Ananiev & Barsky, 1985).

In other works polarized fluorescence microscopy was applied to the study of DNA organization in the sperm of different organisms (MacInnes & Uretz, 1968; Vinogradov *et al.*, 1980). Using a two-channel polarization microfluorometer Zotikov (1982) observed changes in the polarization of AO bound to *Tetrachimena periformis* DNA during the cell division cycle. Developing this approach, Zotikov & Zelenin

(1987) found a marked decrease in degree of AO polarization after histone H1 removal from the chromatin by fixed cell treatment with 0.6 M sodium chloride.

The results cited above suggest that measurement of the degree of polarization of the AO fluorescence provides useful information about the structural organization of DNA *in situ*.

Shurdov & Gruzdev (1984) applied the fluorescence polarization measurement for the investigation of the tertiary structure of bacteriophage DNA and proposed a new model of such a structure.

9.10.2 Fluorescence polarization of AO bound to proteins and lipids

Gamaley & Kaulin (1988) studied the fluorescence anisotropy of AO bound to thick protofibrils of muscle fibre. The AO orientation was found to decrease as the temperature increased. The authors concluded that hydrophobic interactions contributed greatly to the stabilization of the myosin structure.

AO fluorescence polarization assay has also been successfully used for the investigation of myelin membrane structure (Vorobyev *et al.*, 1985).

9.11 AO IN CHROMOSOME BANDING

At least three methods of chromosome banding based on the use of AO have been suggested:

(1) The R-banding technique (Bobrov & Modan, 1973; Comings *et al.*, 1973; Verma & Lubs, 1975) includes chromosome preparation pretreatment in DNA denaturing conditions (heat denaturation, formaldehyde or trypsin treatment, etc.) with subsequent AO staining. The regions between Q(G)-bands acquire the orange-red fluorescence, while the other parts of the chromosome are weakly green. The picture is the reverse of the conventional Q/G-banding pattern, thus explaining the name of the technique ('R' for 'reverse' banding). Some minor differences may appear in the AO picture, depending on the type of pretreatment (Bobrov & Modan, 1973). The cytochemical nature of the method is rather obscure but the technique certainly enhances the possibilities of the chromosome analysis.
(2) A technique which reveals fast and slowly reassociating chromosome regions was suggested by Stockert & Lisanti (1972). The chromosomes are denatured by heat, then treated in conditions which allow their partial renaturation (a short time of reassociation, cooling in the presence of formaldehyde). They are then stained with AO. As a result, the centromeric regions containing highly repetitive DNA acquire green fluorescence while the other parts of the chromosomes acquire red fluorescence. The banding pattern resembles that of conventional C-banding.
(3) The T-banding technique (Dutrillaux, 1973) is also based on chromosome thermal denaturation and AO treatment but under different conditions (heating in a diluted phosphate buffer at 87°C, staining in a highly concentrated AO solution). As a result, distinct spots with a very bright green fluorescence appear in the telomeric parts of chromosomes – hence the name of the technique. It reveals highly repetitive sequences, for example the TTAGGG telomeric repeat. Application of this technique allowed the author to localize precisely the juxta-telomeric break points barely detectable by other banding methods.

Although all three techniques are rather similar, the banding patterns obtained by them are different, most likely due to some variations in experimental conditions.

Another use of AO in chromosome analysis is based on its interaction with DNA *in vivo* and the resulting changes in chromatin structure (Frenster, 1971), as well as its suppression of mitotic chromosome condensation (Matsubara & Nakagome, 1983). The latter property allows the use of AO to increase the resolution of chromosome banding, similar to ethidium (Ikeuchi, 1984; Chiryaeva *et al.*, 1989). It should be noted that the use of AO is appropriate whenever very mild chromosome elongation is required. When more strong decondensation is necessary, the utilization of another acridine amino derivative – 9-aminoacridine – may be recommended (Muravenko *et al.*, 1998).

9.12 AO IN ACID POLYSACCHARIDE HISTOCHEMISTRY

Acridine diamino derivatives, including AO as a bisdiamine, form salt-like complexes with the acid polysaccharides, characterized by bright red fluorescence (Kuyper, 1957; Appel & Zanker, 1958) and a shift in absorption typical for the AO dimer form (Saunders, 1962).

Stone & Bradley (1967) demonstrated the applicability of AO for quantitative analysis of acid polysaccharide of four groups: polycarboxylates, polysulfates, mucopolysaccharides and heparinoids.

The ability of AO to induce red fluorescence in cell structures containing acid mucopolysaccharides was described in all the early publications on the use of AO in nucleic acid cytochemistry (Armstrong, 1956; Bertalanffy & Bickis, 1956; Schummelfeder *et al.*, 1957), but mostly with the aim of distinguishing between these two types of biopolymers. Details of the fluorescence cytochemistry of acid mucopolysaccharides were developed by Saunders (1962, 1964) and Zelenin & Stepanova (1968).

The method suggested by Saunders (1962, 1964) is based on two approaches:

(1) prestaining of preparations with AO (0.1%), followed by washing in solutions of progressively increasing concentration of sodium chloride; and
(2) pretreatment of preparations with cetyltrimethylammonium chloride, followed by AO staining.

A combination of these methods allowed the author to distinguish between different types of acid mucopolysaccharides: washing with 0.3 M NaCl removed the hyaluronic acid staining, washing with 0.6 M NaCl eliminated staining of hyaluronic acid and chondroitin sulfuric acid, and washing with 0.8 M NaCl removed the staining of both these mucopolysaccharides as well as heparin. After the cetyltrimethylammonium treatment hyaluronic acid acquired red fluorescence.

Zelenin & Stepanova (1968) suggested staining with AO at different pH to distinguish between different types of acid mucopolysaccharides. At pH 4.5 all acid mucopolysaccharides, including hyaluronic acid, acquired the red fluorescence, whereas at pH 2.0 this fluorescence was acquired only by the sulfonated ones (heparin and chondroitin sulfates). If necessary, RNAse or different hyaluronidases could be applied.

Our experience (Zelenin, 1967) allows us to recommend a combinations of both Saunders' and our approaches. To eliminate a false positive staining which may be present in epithelial cells, highly diluted AO solutions can be used since chondroitin sulfuric acid and heparin still fluoresce red under these conditions (Saunders, 1962).

The approaches discussed above are all suggested for fixed preparations and there have been few reports on the applicability of AO for the investigation of acid polysaccharides *in vivo*. However, when a preparation of subcutaneous connective tissue is treated with AO in vital or supravital staining (see Section 9.14.1) a large number of big, bright red fluorescent heparin granules can be observed in the mast cells. Other cells in the preparation (fibroblasts, leukocytes, etc.) give the typical appearance of living cells.

These observations are consistent with the data of Nakamura *et al.* (1980), who demonstrated the formation of a complex between AO and heparin, both in solution and inside neoplastic mast cells. It may thus be concluded that, at least in some cases (mast cells), AO reveals acid mucopolysaccharides *in vivo*.

9.13 AO BINDING TO PROTEINS

As was noted above (see Section 9.4), at high pH (>7), when the protein carbohydrates are protonized, AO binds to proteins with the formation of red fluorescing complexes. Filatova *et al.* (1973) used AO as a probe to investigate the structural state of actin in solution and found that its viscosity increased with increase of the ionic strength. The effect was characteristic of actin only and was not observed with such proteins as myosin, tropomyosin and serum albumin.

For works on fluorescence polarization of AO bound to proteins, see Section 9.10.2

9.14 AO BINDING TO A LIVING CELL

9.14.1 General trends: comparison of pictures of fixed and living cells

AO was applied to the investigation of living cells in the very first work on the biological use of this fluorescent stain (Strugger, 1940a,b). Among the long list of authors who have since used AO for this purpose, Meissel in Moscow deserves particular mention. Reviews on *in vivo* use of AO can be found in Kasten (1967), Zelenin (1967, 1971), Meissel & Zelenin (1973) and Darzynkiewicz & Kapuchinsky (1990).

The appearance of a living cell stained with AO differs markedly from that of a fixed one. A fixed cell stained with AO has a bright green nucleus with orange-red nucleoli inside it. The cytoplasm has intense deep red fluorescence.

In a living cell the weakly fluorescing green nucleus can usually be distinguished; some authors claim, however, that when AO is used in very low concentration the nucleoplasm is indistinguishable from the background (Delic *et al.*, 1991). More bright green fluorescence of nucleoli is seen in all cases. The fluorescence of cytoplasm is so weak that it is almost indistinguishable from the background. At the same time there are a number of small granules in the cytoplasm with bright red fluorescence. These granules present the most striking feature of a living cell stained with AO.

There are a number of different approaches to staining living cells with AO: a preparation of cultured cells on a coverslip or a smear of different animal and plant cells can be placed into the AO solution; AO can be added to the cell suspension; or AO can be injected intravenously or intraperitoneally into a living animal.

The fluorescence-microscopical picture is highly dependent on the staining conditions, especially on the AO concentration. The typical picture described above is usually obtained when AO is used at a concentration of about 5 × 10^{-6} M. This concentration varies, however, depending on the ratio of cells to the volume of the stain solution. The pH of the staining solution is also very important; it should be about neutral or a bit higher (7.2–7.4). The staining medium should not contain any other stains or chemicals which may fluoresce themselves or quench the AO fluorescence. At the same time, addition of glucose to the staining solution is desirable (Zelenin, 1971).

Staining under these conditions is usually called vital. Strictly speaking, the term should mean that the studied cell retains all the properties of an unstained one. This is, however, not possible in the case of any 'vital' stain, including AO. AO in a living cell binds with nucleic acids, thus interfering with their synthesis and inhibiting protein synthesis and mitotic activity (Goldberg *et al.*, 1963; Zelenin & Liapunova, 1964a, 1966; Zelenin, 1971). If applied for a long time, AO reduces the size of nucleoli (Zelenin, 1971) and alters the morphology of red fluorescent granules (for details, see Section 9.14.3). Still, the main features of a living cell are preserved under these conditions. Therefore such staining is often called 'supravital'.

If AO is used in very high concentrations (10^{-2} M–10^{-3} M) or for too long, the cell can be damaged or even killed by the action of the dye. The first feature of such damage is the appearance of orange-red nucleoli. Later, the nucleoplasm acquires bright green fluorescence and, in the last stage of such 'mortal' cell treatment, the cytoplasm appears diffuse red. The fluorescence appearance of such a cell is quite similar to that of a fixed one. A typical example of mortal staining is provided by the work of Bertalanffy & Bickis (1956). Sometimes such staining is not quite 'mortal' and a cell can be revived at least into a 'supravital' state if placed into a fresh culture medium, in particular one containing an excess of glucose.

9.14.2 Cytochemical interpretation of microscope images

Reliable interpretation of microscope images is difficult for a living cell because standard cytochemical approaches such as enzyme pretreatments are incompatible with the living state. Circumstantial evidence is therefore mostly used for that purpose.

Important information has been obtained through spectroscopical investigations. It has been shown that the absorption spectrum of AO shifts after staining of a living cell in the same way as when AO is bound to the nucleic acids *in vitro* (Loeser *et al.*, 1960). Instead of one peak of absorption characteristic of AO in highly diluted solutions (λ_{abs} 494 nm), two new peaks appear, with maxima at 502 nm and 470 nm, respectively. As was mentioned previously, the 502 nm peak is characteristic of AO bound to the double-stranded nucleic acids. Thus it can be assumed that the green fluorescence of a living cell stained with AO reflects its binding to the nucleic acids.

This assumption in connection to the nuclear fluorescence is confirmed by the following observations:

(1) Like other acridine amino derivatives, AO inhibits RNA synthesis *in vivo* (Goldberg *et al.*, 1963; Zelenin & Liapunova, 1966; Zelenin, 1971). These data may be assumed to indicate AO binding to DNA.
(2) It is well known that cell treatment with low doses of actinomycin D results in a gradual reduction of the RNA component of nucleoli. AO vital staining reveals in such cells a marked reduction of green fluorescent nucleoli (Zelenin, 1971).

In contrast to the data described above, Delic et al. (1991) concluded that AO does not intercalate into the nuclear DNA of a living cell. Intensified fluorescence microscopy, coupled with a digital imaging system, was used in this work. It allowed the authors to obtain good fluorescence pictures at very low AO concentration (10^{-6} M) and low levels of excitation. Under the conditions of these experiments the only fluorescence registered in the cell was green fluorescence of nucleoli and red fluorescence of cytoplasmic granules. It is quite possible that under these conditions AO does not bind to DNA and thus causes less damage to the cell. Such staining is possibly more nearly truly 'vital'.

Interpretation of the cytoplasmic pictures is more difficult. It has been shown that AO specifically inhibits protein synthesis in the cell, this effect being due to its binding to cytoplasmic RNAs, tRNA in particular (Zelenin & Liapunova, 1964a; Zelenin, 1971). It can be assumed that AO binds to the cytoplasmic RNAs and is responsible for the weak green fluorescence of the cytoplasm usually seen in AO-treated cells.

But the most interesting effect is connected with the red cytoplasmic granules, which were shown to be lysosomes which accumulated AO.

9.14.3 *In vivo* lysosome investigation

The presence of red cytoplasmic granules is the most striking feature of a living cell exposed to the action of AO. These granules were first found by Vonkennel & Wiedemann in 1944 and later described in many animal cells (see Bertalanffy & Bickis, 1956; Meissel & Zelenin, 1973). Their cytochemical and morphological nature was for many years a subject of some controversy. At first they were regarded as a result of a complex formation between AO and the cytoplasmic RNA. This assumption was based on the *in vitro* data on the red fluorescence of AO in complex with RNA (Meissel & Korchagin, 1952). Some authors regarded the granules as complexes between AO and acid mucopolysaccharides or proteins, or as AO-stained mitochondria (for review, see Zelenin, 1967, 1971; Meissel & Zelenin, 1973).

Later, however, these granules were proved to be lysosomes and lysosome-related structures that had accumulated AO (Koenig, 1963; Robbins *et al.*, 1964; Zelenin & Liapunova, 1964b; Zelenin *et al.*, 1965; Zelenin, 1966; Dingle & Barrett, 1967). It was shown that even isolated lysosomes retained the capacity to concentrate AO inside them (Dingle & Barrett, 1967; Cononico & Bird, 1969). AO was therefore named as an excellent non-enzymatic lysosome marker (Cononico & Bird, 1969).

Almost immediately AO* became to be widely used for *in vivo* lysosome investigations of primary and secondary lysosomes (Allison & Young, 1964, 1969; Blume *et al.*, 1969) and in special lysosome analogues such as acrosomes (Allison & Hartee, 1970).

The bright red fluorescence of AO lysosomes shows that AO is present in the dimer form, whereas in other cellular structures which fluoresce green it is in a monomer form. This means that the AO concentration inside the lysosomes is at least 1000 times higher than that in the cytosole. Preferential accumulation of AO in lysosomes was demonstrated directly by Dingle & Barrett (1968) in experiments with tritium-labelled AO.

The cytophysiological mechanism of AO accumulation in lysosomes merits special discussion.

It was shown that the capacity of lysosomes to concentrate AO inside them is connected with the presence on their membrane of a proton pump responsible for the maintenance of low pH inside lysosomes (De Duve *et al.*, 1974; Yamashiro *et al.*, 1983; Moriyama *et al.*, 1982, 1984). Low pH causes accumulation inside lysosomes of different cationic compounds including AO. It is, however, possible that red lysosome fluorescence is due not only to the high concentration of AO, but also to its binding to some acid substrate (polymer) which is capable of binding AO and thus facilitates formation of dimers.

* In some early papers (Dingle & Barrett, 1967; Blume *et al.*, 1969; Unanue *et al.*, 1969; Allison & Hartee, 1970) a fluorochrome called euchrysine 3R was reported to be used for that purpose. As we have noted later (Zelenin, 1971), euchrysine 3R is just one of numerous trademarks of AO (*Colour Index*, 1956). Allison & Young (1969) themselves showed the chromotographic identity of euchrysine 3R with AO.

One candidate for such a substrate is a strongly acidic component capable of binding cations, which was described in lysosomes by Barret & Dingle (1967).

AO vital and supravital staining may be used for detection of lysosomes, investigation of their distribution, localization and morphology. It is very useful for investigation of secondary lysosomes and for the study of the process of lysosome fusion, for example with phagosomes. It should be noted that only acified lysosome-like vesicles acquire red fluorescence after AO staining. If applied in low concentration for a long time (days), AO staining reveals the general accumulative capacity of the cellular lysosomes. Under such conditions, the lysosomes markedly enlarge in size and are concentrated around the nucleus (Zelenin, 1966, 1971).

Recent data (Mpoke & Wolfe, 1997) indicate that such perinuclear lysosome location may be connected with apoptotic changes in the cells (for details, see Section 9.16).

In parallel, lysosome staining with AO opens up important possibilities in the study of cell physiology. The capacity of a cell to concentrate AO in its lysosomes indicates that their proton pumps functions normally. This requires an adequate energy supply. It was shown that the AO accumulation inside lysosomes is an energy-dependent process (Weissmann & Gilgen, 1956; Kirianova & Zelenin, 1970; Zelenin, 1971). Thus the presence of the red cytoplasmic granules in the AO-treated cells indicates that the energy-supplying mechanisms function normally in these cells.

Although the AO method of lysosome investigation was suggested and developed rather a long time ago, it is still widely used. This approach was recently successfully applied by Del Bing *et al.* (1991) to the investigation of the cytotoxic action of the DNA topoisomerase I inhibitor camptophecin. The inhibitor had no effect on AO uptake into lysosomes. This allowed the authors to conclude that the cytotoxic effect of this substance was not connected with its action on the maintenance of the proton pump.

In the works analysed above the AO-stained lysosomes were studied by conventional (static) fluorescence microscopy. In parallel to these, AO has also been widely used in flow cytometry. Lysosome staining with AO has been successfully used in the flow cytometric investigation of living blood cells (Melamed *et al.*, 1972, 1974) as well as in distinguishing between different types of lung cells (Wilson *et al.*, 1986).

At present AO lysosome staining is intensively used for the detection of cell viability (see Section 9.15) and the investigation of apoptosis (see Section 9.16).

9.15 AO IN THE STUDY OF CELL VIABILITY

In the very first publication on the biological application of AO, Strugger (1940b, 1942) demonstrated the possibility of its use for distinguishing between dead and living cells. In his experiments dead cells fluoresced red, whereas living cells fluoresced green. These findings were carefully studied in many works. Some authors confirmed these observations but others were unable to do so (see reviews by Bertalanffy & Bickis, 1956; Zelenin, 1967). It is difficult to give a complete explanation of Strugger's observations. It is possible, however, that in successful experiments when the 'Strugger effect' was observed, the staining had been carried out with such AO concentrations and under such conditions as to cause the red fluorescence of the ribosomal RNA of the dead or damaged cells. At the same time, those cells had very few or no lysosomes (this is typical for plant, yeast and bacterial cell mostly used in those early experiments), and thus acquired only green fluorescence after staining *in vivo*.

Recently, Mason & Lloyd (1997) applied Strugger's approach to the evaluation of bacterial susceptibility to the action of the antibiotic gentamicin. Using the flow cytometric detection of green/red AO fluorescence of *Escherichia coli* cells, the authors found a good correlation between the numbers of green fluorescing cells and the colony-forming units.

After the discovery that AO could be actively concentrated in lysosomes of living cells, it began to be used for the study of cell viability. As pointed out above (Section 9.14.3), the presence of red fluorescing granules in the cytoplasm of vitally (supravitally) stained cells is a reliable indication that their lysosomes are normal and the cell energy-supply mechanisms are functioning. In our laboratory we use AO staining as a primary conventional test of the viability of culture cells.

More sophisticated applications of AO for that purpose have been described. Singh & Stephens (1986) developed a two-stain method for the determination of cell viability. They applied two fluorescent dyes for that purpose – AO and Hoechst 33258. Under the conditions used AO penetrated only into the living cells whereas H 33258 penetrated only into the dead ones. When these dyes were used in combination, the dead cells fluoresced brilliant blue and the living ones green.

Singh & Stephens (1986) used a conventional fluorescence microscope. Other authors applied flow cytometry for that purpose. This approach is now widely used in such investigations (Darzynkiewicz & Kapuchinscki, 1990; Pollack & Ciancio, 1990). Bohmer (1984, 1985) combined AO with ethidium to detect living and dead cells. Three groups of cells treated with agents which caused complete cell cycle arrest were found: vital cells, stained only with AO; dead cells, in which ethidium stained only the cytoplasm; and dead cells in which ethidium stained both cytoplasm and nucleus. The method proved itself to be more sensitive than the conventional trypan blue exclusion test. Two steps of the cell death process were revealed.

More recently AO-ethidium flow cytometry has been used to distinguish between apoptotic and necrotic cells (Echaniz *et al.*, 1995).

9.16 AO IN THE STUDY OF APOPTOSIS

Owing to its versatile properties (Darzynkiewicz, 1990c), AO has proved to be particularly useful for the investigation of both nuclear and cytoplasmic changes in apoptosis.

When applied to fixed preparations, AO readily reveals general nucleus morphology, the degree of chromatin condensation, the presence or absence of chromatin fragmentation, clumping, and the appearance of perinuclear granules of distracted chromatin. Apoptotic cells are identified by characteristically bright, condensed and fragmented nuclei with an intact cell membrane (Abrams *et al.*, 1993; Padayatty *et al.*, 1997).

More interesting results are obtained by supravital staining with AO. As described above (see Section 9.14.3), in a living 'normal' cell treated with AO under vital (supravital) conditions, one can observe the weakly fluorescing green nucleus with more brightly green fluorescing nucleoli; in the cytoplasm bright red lysosomes are usually distinguished.

The fluorescence image after AO supravital staining of apoptotic cells differs drastically: the nucleus is intensively stained by AO, chromatin granules are easily detected, and all features of nucleic changes found in the AO-stained fixed cells (chromatin fragmentation, condensation, etc.) are clearly revealed. In addition, AO staining allows the investigator to detect the state of the cytoplasm; at early stages of apoptosis a perinuclear concentration of lysosomes is revealed, whereas at later stages red fluorescing lysosomes disappear.

This change in the fluorescence microscopical appearance received the name of the AO redistribution (relocation) test (Yuang *et al.*, 1997, 1998).

Even more striking results can be obtained when AO is used in combination with the DNA-specific fluorochrome Hoechst 33342. This dye combination was successfully applied to the study of macronuclear elimination in *Tetrahymena periformis* (Mpoke &

Wolfe, 1997). The lysosomes were found to be clustered around the degenerating nuclei and the nuclei themselves were stained orange-red. The results obtained are consistent with the suggestion that apoptotic nuclei become acidified, possibly due to fusion with lysosomes.

The AO-ethidium vitality test has been successfully applied to the study of the blood lymphocyte necrosis and apoptosis of patients with HIV infection (Echaniz *et al.*, 1995). The technique revealed changes in the range from the typical features of cells undergoing programmed cell death (apoptosis) to those observed in necrotic cells.

Another application of AO to the apoptosis investigation is its use as an apoptosis inducer (Brunk *et al.*, 1997; Hellquist *et al.*, 1997). Cultured human fibroblasts which had accumulated AO in lysosomes were irradiated with blue light to bring about the photo-oxidative disruption of the lysosome membrane. The procedure resulted in apoptotic death of the irradiated cells. The authors concluded that the release of lysosome proteases and endonucleases may induce apoptosis.

9.17 AO IN FLOW CYTOMETRY

The use of AO in flow cytometry has been reviewed in detail (Darzynkiewicz, 1990a; Darzynkiewicz & Kapuchinscki, 1990). Therefore, in this section only the most important directions of these investigations will be mentioned.

(1) Determination of the RNA/DNA ratio in the cell cycle study, study of the activation of proliferation of quiescent cells, and investigation of the changes in cell transcriptional activity, etc.
(2) Investigation of chromatin properties after acid or thermal denaturation in cells with different genome activity (stimulation of proliferation in quiescent cells, determination of mitotic activity, etc.).
(3) Cell cycle analysis by BrdUrd-suppressed AO fluorescence, including discrimination between cycling and non-cycling cells.
(4) Investigation of cell viability.
(5) Study of apoptosis (Echaniz *et al.*, 1995; Gorman *et al.*, 1997; Darzynkiewicz *et al.*, 1997).

9.18 OTHER APPLICATIONS OF AO

(1) When applied to a living cell, AO acts as a biologically active substance which inhibits nucleic acid (Goldberg *et al.*, 1963; Zelenin & Liapunova, 1966) and protein (Zelenin & Liapunova, 1964a) synthesis and changes chromatin structure (Frenster, 1971).
(2) In some cases it is not AO but its different derivatives that are used as fluorescent probes; for example, a hydrophobic acridine orange derivative was used to study amphiphilic peptide structure (Morii *et al.*, 1991). The AO derivative 10-*N*-nonyl acridine orange interacts with acid phospholipids, especially cardiolipin (Petit *et al.*, 1992), inducing red fluorescence; this method has been applied successfully to cardiolipin quantification (Gallet *et al.*, 1995).
(3) Among the new and very promising intercalators related to AO, bis- and tris-acridines should be mentioned (Chen *et al.*, 1978; Denny *et al.*, 1985; Gaugain *et al.*, 1984).
(4) A number of other, more specialized, applications of AO may be found in the current literature via the Internet.

REFERENCES

Abrams J.M., White K., Fessler L.I. & Steller H. (1993) *Development* **117**, 29–43.
Albert A. (1966) *The Acridines.* London, Edward Arnold.
Allison A.G. & Hartee E.F. (1970) *J. Reprod. Fertil.* **21**, 501–515.
Allison A.C. & Young M.R. (1964) *Life Sci.* **3**, 1407–1414.
Allison A.C. & Young M.R. (1969) In *Lysosomes in Biology and Pathology*, Part 2, J.T. Dingle & H.B. Fell (eds). North-Holland, Amsterdam and London, pp. 603–628.
Alvarez M.R. (1974) *Exp. Cell Res.* **83**, 225–230.
Ananiev E.V. & Barsky V.E. (1985) *Chromosoma* **19**, 104–112.
Appel W. & Zanker V. (1958) *Z. Naturforsch.* **13b**, 126–134.
Armstrong J.A. (1956) *Exp. Cell Res.* **11**, 640–643.
Auer G. (1972) *Cytochemical Properties of Chromatin Related to Changes in Cellular Growth Activity.* Almqvist and Wikseli Informationindustri AB, Uppsala, pp. 1–78.
Auer G., Zetterberg A. & Killander D. (1970) *Exp. Cell Res.* **62**, 32–38.
Auer G., Moore G.P.M., Ringertz N.R. & Zetterberg A. (1973) *Exp. Cell Res.* **76**, 229–233.
Avarindan G.R. & Mongdal N.R. (1998) *Arch. Androl.* **40**, 29–41.
Barret R.E. & Dingle J.T. (1967) *Biochem. J.* **105**, p. 20.
Bertalanffy L. von (1963) *Protoplasma* **57**, 51–83.
Bertalanffy L. von & Bickis I. (1956) *J. Histochem. Cytochem.* **4**, 481–493.
Berman H.M. & Young, P.R. (1981) *Ann. Rev. Biophys. Bioeng.* **10**, 87–14.
Blume R.S., Clade P.R. & Chessin L.N. (1969) *Blood* **33**, 87–99.
Bobrov M. and Modan K. (1973) *Cytogenet. Cell Genet.* **12**, 145–156.
Bohmer R.M. (1984) *Cell Tissue Kinet.* **17**, 593–600.
Bohmer R.M. (1985) *Cytometry* **6**, 215–218.
Bohmer R.M. (1990) In *Methods in Cell Biology*, vol. 33: Flow Cytometry, Z. Darzynkiewicz & H.A. Crissman (eds). Academic Press, London, pp. 173–184.

Bolund L., Ringertz N.R. & Harris H. (1969) *J. Cell Sci.* **4**, 71–87.
Bolund L., Darzynkiewicz Z. & Ringertz N.R. (1970) *Exp. Cell Res.* **62**, 76–81.
Borisov A. Yu. & Tumerman L.A. (1959) *Izv. Acad. Nauk. Ser. Phys.* **23**, 97–101.
Borisova O.F. & Grechko V.V. (1984) *Studia Biophys.* **101**, 133–134.
Borisova O.F. & Tumerman L.A. (1964) *Biofizika* **9**, 537–544.
Borisova O.F. & Tumerman L.A. (1965) *Biofizika* **10**, 32–36.
Borisova O.F., Kisselev L.L. & Tumerman L.A. (1963) *Dokl. Acad. Nauk.* **152**, 1001–1004 .
Borisova O.F., Horachek P., Gursky G.V., Minyat E.E. & Tumanyan V.G. (1968) *Izv. Acad. Nauk Ser. Phys.* **32**, 1317–1324.
Borisova O.F., Krichevskaya A.A. & Samarina O.P. (1981) *Nucleic Acids Res.* **9**, 663–681.
Borisova O.F., Shchyolkina A.K., Timofeev E.N., Florentiev V.L. & Churikov N.A. (1992) *FEBS Lett.* **306**, 140–142.
Borisova O.F., Shchyolkina A.K., Chernov B.K. & Churikov N.A. (1993) *FEBS Lett.* **322**, 304–306.
Bradley D.F. (1961) *Trans. NY Acad. Sci. Ser II* **24**, 64–74.
Brunk U.T., Dalon H., Roberg K. & Helmquist H.B. (1997) *Free Radic. Biol. Med.* **23**, 616–626.
Brumberg E.M. & Girshgorin, A.S. (1947) An opack-illuminator, primarily for luminescence microscopy. USSR Patent, Klass 42, 14/05, No. 78635.
Brumberg E.M. & Krilova G.H. (1953) *Jurnal obshchey biologii (Moscow)* **14**, 461–464.
Bukatsch F. & Haitinger M. (1940) *Protoplasma* **34**, 515–523.
Chen T.K., Fico R. & Canellakis E.S. (1978) *J. Med. Chem.* **21**, 868–874.
Chiryaeva O.G., Amosova A.V., Efimov A.M., Smirnov A.F., Kaminir L.B., Yakovlev A.F. & Zelenin A.V. (1989) In *Cytogenetics of Animals*, C.R.E. Halnan (ed). C.A.B. International., Wallingford, U.K., pp. 211–219.
Colour Index (1956) In 4 volumes. Bradford Society of Dyers and Colourists, Lowell.
Comings D.E., Aveling E., Okada T.A. & Wyandt, H.E. (1973) *Exp. Cell Res.* **77**, 469–493.
Cononico P.G. & Bird J.W.C. (1969) *J. Cell Biol.* **43**, 367–371.
Crissman H.A. & Steinkamp J.A. (1990) In *Methods in Cell Biology*, vol. 33: Flow Cytometry, Z. Darzynkiewicz & H.A. Crissman (eds). Academic Press, London, pp. 199–206.
Darzynkiewicz Z. (1990a) In *Methods in Cell Biology*, vol. 33: Flow Cytometry, Z. Darzynkiewicz & H.A. Crissman (eds). Academic Press, London, pp. 285–298.
Darzynkiewicz Z. (1990b) In *Methods in Cell Biology*, vol. 33: Flow Cytometry, Z. Darzynkiewicz & H.A. Crissman (eds). Academic Press, London, pp. 337–352.
Darzynkiewicz Z. (1990c) In *Flow Cytometry and Sorting*, M.R. Melamed, T. Lindmo & M.L. Mendelson (eds). Wiley-Liss, New York, pp. 515–340.
Darzynkiewicz Z. & Kapuscinski J. (1990) In *Flow Cytometry and Sorting*, M.R. Melamed, T. Lindmo & M.L. Mendelson (eds). Wiley-Liss, New York, pp. 291–314.
Darzynkiewicz Z., Troganos F., Sharpless T. & Melamed M.R. (1974) *Biochem. Biophys. Res. Commun.* **59**, 392–399.
Darzynkiewicz Z., Troganos F., Sharpless T. & Melamed M.R. (1975) *Exp. Cell Res.* **90**, 411–428.
Darzynkiewicz Z., Andreeff M., Troganos F., Sharpless T. & Melamed M.R. (1978) *Exp. Cell Res.* **115**, 31–35.
Darzynkiewicz Z., Evenson D., Staiano-Coico L., Sharpless T. & Melamed M.R. (1979a) *J. Cell Physiol.* **100**, 425–438.
Darzynkiewicz Z., Troganos F., Andreeff M., Sharpless T. & Melamed M.R. (1979b) *J. Histochem. Cytochem.* **27**, 478–485.
Darzynkiewicz Z., Troganos F. & Melamed, M.R. (1983) *Cytometry* **3**, 345–348.
Darzynkiewicz Z., Juan G., Li X., Gorczyca W., Mukarami T. & Troganos F. (1997) *Cytometry* **27**, 1–20.
De Duve C., de Barsy T., Poole B., Trouet A., Tulkens P. & Van Hoof, F. (1974) *Biochem. Pharmacol.* **23**, 2495–2531.
Del Bing G., Lassota P. & Darzynkiewicz Z. (1991) *Exp. Cell Res.* **193**, 27–35.
Delic J., Coppey J., Magdelenat H. & Coppey-Moisan M. (1991) *Exp. Cell Res.* **194**, 147–153.
Denny W.A., Atwell G.J., Baquley B.C. & Walelin L.B.G. (1985) *J. Med. Chem.* **28**, 1568–1574.
Dingle J.T. & Barrett A.J. (1967) *Biochem. J.* **105**, 19–20.
Dingle J.T. & Barrett, A.J. (1968) *Biochem. J.* **109**, 19.
Dutrillaux B. (1973) *Chromosoma* **41**, 395–402.
El-Naggar A.K., Hurr K., Tu N.Z., Teage K., Raymond K.A., Ayala A.G. & Murray J. (1995) *Cytometry* **19**, 256–262.
Echaniz P., de Juan M.D. & Cuadraro E. (1995) *Cytometry* **19**, 164–170.
Filatova L.G., Strankfeld I.G. & Moskalenko J.E. (1973) *Biophysica* (Moscow) **18**, 760–762.
Frenster, J.H. (1971) *Cancer Res.* **31**, 1128–1133.
Gallet P.F., Maftah A., Petit J.M., Denis-Gay M. & Jullien R. (1995) *Eur. J. Biochem.* **228**, 113–119.
Galley W.C. & Purkey R.M. (1972) *Proc. Natl. Acad. Sci.* **69**, 2198–2202.
Gamaley J.A. & Kaulin A.B. (1988) *Tsitologia* (Leningrad) **30**, 49–51.
Gaugain B., Markovits J., Le Pecq J.-B. & Roques B.R. (1984) *FEBS Lett.* **169**, 123–126.
Gledhill B.L., Gledhill M.P., Rigler R. & Ringertz N.R. (1966) *Exp. Cell Res.* **41**, 652–665.
Golan R., Cooper T.G., Oschry Y., Obwerpenning F., Schulze H., Chochat L. & Lewin L.M. (1996). *Hum. Reprod* **11**, 1457–1462.
Goldberg I.H., Reich E. & Rabinowitz M.M. (1963) *Nature* **199**, 44–46.
Gomatos P.J. & Tamm J. (1963) *Proc. Natl. Acad. Sci. USA* **49**, 707–714.
Gomatos P.J., Tamm I., Dales S. & Franklin R.M. (1962) *Virology* **17**, 441–454.
Gordon R. Ya, Bocharova L.S., Kruman I.I., Popov V.I., Kazantsev A.P., Khutzian S.S. & Karnaukov V.N. (1997) *Cytometry* **29**, 215–221.
Gorman A.M., Samali A., McGowan A.J. & Cotter T.G. (1997) *Cytometry* **29**, 97–195.
Hellquist H.B., Svensson H. & Brunk U.T. (1997) *Redox Rep.* **3**, 65–70.
Ikeuchi T. (1984) *Cytogen. Cell Genet.* **38**, 56–63.
Jamison R.M. & Mayor H.D. (1965) *J. Bacteriol.* **90**, 1486–1488.
Karabinus D.S., Vogler C.J., Saacke R.G. & Evenson D.P. (1997) *J. Androl.* **18**, 549–555.
Karachunski A.I., Volodin N.N. & Zelenin, A.V. (1987) *Pediartya* (Moscow) **2**, 67–72.
Kasten, F.H. (1967) *Int. Rev. Cytol.* **21**, 141–202.
Kernell A.M., Bolund L. & Ringerz, N.R. (1971) *Exp. Cell Res.* **65**, 1–6.
Killander D. & Rigler R. (1965) *Exp. Cell Res.* **39**, 701–712.
Killander D. and Rigler R. (1969) *Exp. Cell Res.* **54**, 163–170.
Kim S.H., Suddath F.L., Quegley G.J., McPerson A., Susman J.L., Wang A.H.J., Seeman, N.C. & Rich, A. (1974) *Science* **185**, 435–440.

Kirianova E.A. & Zelenin A.V. (1970) *Dokl. Acad. Nauk* **190**, 451–454.

Koenig H. (1963) *J. Cell. Biol.* **19**, 87A.

Kolesnikov V.A., Kushch A.A. & Zelenin A.V. (1973) *Dokl. Akad. Nauk* **212**, 489–501.

Kunicka J.E., Darzynkiewicz Z. & Melamed, M.R. (1987) *Cancer Res.* **47**, 3942–3947.

Kushch A.A., Kolesnikov V.A. & Zelenin A.V. (1974) *Exp. Cell Res.* **86**, 419–422.

Kushch A.A., Terskikh V.V., Kolesnikov V.A., Niyazmatov A.A., Koslov Yu. V. & Zelenin A.V. (1978) *Cytobiologie* **16**, 161–170.

Kushch A.A., Niyazmatov A.A. & Zelenin A.V. (1980) *Cell Differ.* **9**, 291–304.

Kuyper Ch.M.A. (1957) *Exp. Cell Res.* **13**, 198–200.

Latt, S.A. (1976) In *Chromosomes To-day*, P. Pearson & K. Lewis (eds). vol. 5, Wiley, New York, pp. 367–372.

Lerman L.S. (1961) *J. Mol. Biol.* **3**, 18–31.

Lerman L.S. (1963) *Proc. Natl. Acad. Sci. USA* **49**, 94–102.

Lerman L.S. (1964) *J. Cell. Compar. Physiol* **64**, 1–18.

Liedeman R.R. & Bolund L. (1976a) *Exp. Cell Res.* **101**, 164–174.

Liedeman R.R. & Bolund, L. (1976b) *Exp. Cell Res.* **101**, 175–183.

Liedeman R.R., Matveyeva N.P., Vosrticova S.A. & Prilipko L.L. (1975) *Exp. Cell Res.* **90**, 105–110.

Loeser C.N., West S.S. & Schoenberg M.D. (1960) *Anat. Rec.* **138**, 163–187.

MacInnes J.W. & Uretz R.B. (1966) *Science* **151**, 689–691.

MacInnes J.W. & Uretz R.B. (1968) *J. Cell Biol.* **38**, 426–436.

Manteifel V.M., Vinogradova N.G. & Zelenin A.V. (1973) *Izv. Acad. Nauk Ser. Biol.***6**, 740–743.

Mason D.J. & Lloyd D. (1997) *FEMS Microbiol. Lett.* **153**, 199–204.

Matsubara T. & Nakagome S. (1983) *Cytogenet. Cell Genet.* **35**, 148–151.

Mayor H.D. & Hill N.O. (1961) *Virology* **14**, 264–266.

Meissel M.N. & Gutkina A.V. (1953) *Dokl. Akad. Nauk* **91**, 647–650.

Meissel M.N. & Korchagin V.A. (1952) *Bull. Exp. Biol. Mediz.* **5**, 49–51.

Meissel M.N. & Sondak V.A. (1956) *Biofizika* **1**, 262–273.

Meissel M.N. & Zelenin A.V. (1973) In *Unity Through Diversity*, vol. 9, part II: A festschrift for Ludwig von Bertalanffy, W. Gray & N.D. Rizzo (eds). Gordon & Breach, New York, London and Paris, pp. 701–738.

Melamed M.R., Adams L.R., Traganos F., Zimring A. & Kamentsky L.A. (1972) *Am. J. Clin. Pathol.* **57**, 95–102.

Melamed M.R., Adams L.R., Traganos F. & Kamentsky L.A. (1974) *J. Histochem. Cytochem.* **22**, 526–530.

Morii H., Ishimura K. & Uedaira (1991) *Proteins* **11**, 133–141.

Moriyama Y., Takano T. & Ohkuma S. (1982) *J. Biochem. (Tokyo)* **92**, 1333–1336.

Moriyama Y., Takano T., Ohkuma S. (1984) *J. Biochem. (Tokyo)* **93**, 927–930.

Morosov Yu.V. (1963a) *Biofizika* **8**, 167–171.

Morosov Yu.V. (1963b) *Biofizika (Moscow)* **8**, 331–334.

Morosov Yu.V. & Savenko A.K. (1977) *Mol. Photochem.* **8**, 1–43.

Mpoke S.S. & Wolfe J. (1997) *J. Histochem. Cytochem.* **45**, 675–683.

Muravenko O.V., Samatadze T.E., Fedotov A.R. & Zelenin A.V. (1998) *Cytometry, Suppl.* **9**, p. 56.

Nakamura N., Hurst R.E., West S.S., Menter J.M., Golden J.F., Corlis D.A. & Jones D.D. (1980) *J. Histochem. Cytochem.* **28**, 223–230.

Nash D. & Plaut W. (1964) *Proc. Natl. Acad. Sci. USA* **51**, 731–735.

Neidle S., Taylor G. & Saunderson M. (1978) *Nucleic Acids Res.* **5**, 4417–4422.

Padayatty S.J., Marcelli M., Shao T.C. & Cunningam G.R. (1997) *J. Clin. Endocrinol. Metabol.* **82**, 1434–1439.

Petit J.M., Maftah A., Ratinaud M.H. & Jullien R. (1992) *Eur. J. Biochem.* **209**, 267–273.

Plum G.E., Pilch D.S., Singleton S.F. & Breslauer K.J. (1995) *Ann. Rev. Biophys. Biomol. Struct.* **24**, 319–350.

Pollack A. & Ciancio, G. (1990) In *Methods in Cell Biology*, vol. 33: Flow Cytometry, Z. Darzynkiewicz & H.A. Crissman (eds). Academic Press, London, pp. 19–24.

Poot M., Kubbies M., Hoehn H., Grossmann A., Chen Y. & Rabinovich P. (1990) In *Methods in Cell Biology*, vol. 33: Flow Cytometry, Z. Darzynkiewicz & H.A. Crissman (eds). Academic Press, London, pp. 185–198.

Reichman J. (1994) In *Handbook of Optical Filters for Fluorescence Microscopy*. Chroma Technology Corp., USA, pp. 1–65.

Rigler R. (1966) *Acta Physiol. Scand.* **67** (suppl. 267), 1– 122.

Rigler R. (1969) *Ann. NY Acad. Sci.* **157**, 211–224.

Rigler R. & Killander D. (1969) *Exp. Cell Res.* **54**, 171–179.

Rigler R., Killander D., Bolund L. & Ringertz N.R. (1969) *Exp. Cell Res.* **55**, 215–224.

Ringertz N.R. (1969) In *Handbook of Molecular Cytology*, A. Lima-de-Faria (ed). North-Holland, Amsterdam, pp. 656–687.

Ringertz N.R. & Bolund L. (1970) *Exp. Cell Res.* **55**, 205–214.

Ringertz N.R. & Bolund L. (1974) *Int. Rev. Exp. Pathol.* **8**, 83–116.

Robbins E., Marcos P.J. & Ganatos N.K. (1964) *J. Cell Biol.* **21**, 49–62.

Roschlau G. (1965) *Histochemie* **5**, 396–406.

Saunders A.M. (1962) *J. Histochem. Cytochem.* **10**, 683–684.

Saunders A.M. (1964) *J. Histochem. Cytochem.* **12**, 164–170.

Schummelfeder N., Ebschner K.-J. & Krogh E. (1957) *Naturwissenschaften* **44**, 467–468.

Shchyolkina A.K. & Borisova O.F. (1997). *FEBS Lett.* **419**, 27–31.

Shurdov M.A. & Gruzdev A.D (1984) *FEBS Lett.* **165**, 238–242.

Singh N.P. & Stephens R.I. (1986) *Stain Technology* **61**, 315–318.

Smets L.A. (1973) *Exp. Cell Res.* **79**, 239–243.

Smol'yaninova N.K., Karu T.I., Fedoseeva G.E. & Zelenin A.V. (1991) *Biomed. Science* **2**, 121–126.

Sondore O.Yu., Fedoseeva G.E., Kadykov V.A. & Zelenin, A.V. (1978) *Mol. Biol. Rep.* **4**, 137–141.

Steiner R.F. & Bears R.F. (1958) *Science* **127**, 335–339.

Steiner R.F. & Bears R.F. (1959) *Arch. Biochem. Biophys.* **18**, 75–92.

Steiner R.F. & Bears R.F. (1961) In *Polynucleotides, Natural and Synthetic Nucleic Acids.* Elsevier, Amsterdam and London, pp. 301–323.

Stockert J.C. & Lisanti J.A. (1972) *Chromosoma* **37**, 117–130.

Stone A.L. & Bradley D.F. (1967) *Biochim. Biophys. Acta* **148**, 172–192.

Strugger S. (1940a) *Dtsch. tiearartztl. Wochenschr.* **48**, 645–692.

Strugger S. (1940b) *Jenaische Z. Naturforsch.* **73**, 97–134.

Strugger S. (1942) *Dtsch. tiearartztl. Wochenschr.* **50**(5/6), 51–61.

Terskikh V.V., Kirianova E.A., Zosimovskaya A.I. & Zelenin A.V. (1976) *Tsitologia (Leningrad)* **18**, 1085–1089.

Troyan V.M., Kolesnikov V.A., Kalinin F.L. & Zelenin A.V. (1984) *Plant Sci. Lett.* **33**, 213–219.

Traganos F., Darzynkiewicz Z., Sharpless T. & Melamed M. (1977) *J. Histochem. Cytochem.* **25**, 46–56.
Unanue E.R, Askonas B.A. & Allison A.C. (1969) *J. Immunol.* **103**, 71–78.
Verma R.S. & Lubs H.A. (1975) *Am. J. Hum. Genet.* **27**, 110–117.
Vinogradov A.E., Rosanov Ju.M. & Barsky I.Ya. (1980) *Tsitologia (Leningrad)* **22**, 1314–1322.
Vonkennel and Wiedemann. (1944) *Dtsch. med. Wochenschr.* **70**, 27–28, 529–532.
Vorobyev M.V., Gamaley J.A. & Kaulin A.B. (1985). *Tsitologia (Leningrad)* **27**, 422–426.
Wang J.C. (1974) *J. Mol. Biol.* **89**, 783–801.
Waring M.J. (1981) *Ann. Rev. Biochem.* **50**, 159–192.
Weissenberg R., Bella R., Vossefi S. & Lewin I.M. (1995) *Andrologia* **27**, 341–344.
Weissmann G. & Gilgen A. (1956) *Z. Zellforsch.* **44**, 292–299.
Wilson J.S., Steinkamp J.A. & Lehnert B.E. (1986) *Cytometry* **7**, 157–162.
Yuang X.M., Li W., Olsson A.G. & Brunk U.T. (1997) *Atherosclerosis* **133**, 153–161.
Yuang X.M., Li W., Olsson A.G. & Brunk U.T. (1998) *Arteriocler. Thromb. Vasc. Biol.* **18**, 177–184.
Yamashiro D.J., Fluss S.R. & Maxfield F.R. (1983) *J. Cell Biol.* **97**, 929–934.
Zanker V. (1952a) *Z. Phys. Chem.* **199**, 255–258.
Zanker V. (1952b) *Z. Phys. Chem.* **200**, 250–292.
Zanker V., Held M. & Rammensee H. (1959) *Z. Naturforsch.* **146**, 789–801.
Zelenin A.V. (1966) *Nature* **212**, 425–426.
Zelenin A.V. (1967) *Luminescence Cytochemistry of Nucleic Acids*. Nauka, Moscow, pp. 1–134.
Zelenin A.V. (1971) *Interaction of Aminoacridines with a Cell*. Nauka, Moscow, pp1–231.
Zelenin A.V. (1977) *Biol. Zentralblad.* **86**, 407–422.
Zelenin A.V. (1982) *Acta Histochem.* Suppl. Band **26**, 179–187.
Zelenin A.V. & Kushch A.A. (1985) *Mol. Biol.* **19**, 242–250.
Zelenin A.V. & Liapunova E.A. (1964a) *Nature* **204**, 45–46.
Zelenin A.V. & Liapunova E.A. (1964b) In *Second International Congress on Histo- and Cytochemistry*. Abstracts, Frankfurt/Main, p. 217.
Zelenin A.V. & Liapunova E.A. (1966) *Farmakol. toksikol. (Moscow)* **4**, 481–483.
Zelenin A.V. & Stepanova N.G. (1968) *Arkchiv anatomii i gistologii (Leningrad)* **54**, 82–88.
Zelenin A.V. & Vinogradova N.G. (1973) *Exp. Cell Res.* **82**, 411–414.
Zelenin A.V., Biriuzova V.I., Vorotnitskaya, N.E. & Liapunova, E.A. (1965) *Dokl. Acad. Nauk* **162**, 925–927.
Zelenin A.V., Shapiro I.M., Kolesnikov V.A. & Senin V.M. (1974) *Cell Differ.* **3**, 95–101.
Zelenin A.V., Kirianova E.A., Kolesnikov V.A. & Stepanova N.G. (1976) *J. Histochem. Cytochem.* **24**, 1169–1172.
Zelenin A.V., Stepanova N.G. & Kiknadze I.I. (1977) *Chromosoma* **64**, 327–335.
Zetterberg A. & Auer G. (1970) *Exp. Cell Res.* **62**, 262–270.
Zotikov A.A. (1982) *Acta Histochem.* Suppl. Band **26**, 219–221.
Zotikov A.A. & Zelenin A.V. (1987) *Tsitologia (Leningrad)* **29**, 1398–1401.

CHAPTER TEN

Fluorescent Lipid Analogues: Applications in Cell and Membrane Biology

JAN WILLEM KOK & DICK HOEKSTRA
Laboratory of Physiological Chemistry, University of Groningen, The Netherlands

10.1 INTRODUCTION

Recent advances in studies involving the structure and dynamics of membranes have shown that fluorescent lipid probes have become important and, frequently, indispensable tools in this area of research. These probes are applied in investigations as diverse as those dealing with biophysical aspects of membranes, including lateral mobility, phase transitions and phase separations, but also in studies of the cell biology of membranes, including membrane flow and lipid trafficking.

The type of probe to be used for a particular study depends very much on the sort of research involved. For example, to characterize overall biophysical properties of membranes, the probe to be used need not necessarily resemble that of a lipid. Essential is usually its lipid-like properties and its incorporation in the hydrophobic core or lipid phase of the membrane. A typical and classical probe in this respect is 1,6-diphenyl-1,3,5-hexatriene (DPH) and its derivatives (Shinitzky & Barenholz, 1978; Loew, 1988). By contrast, in studies aimed at revealing the cell biology of lipids, including intracellular processing and metabolism, fluorescent lipid probes are desirable that much more closely resemble the structure of the natural lipids. Therefore, in this case a particular lipid analogue can be studied, which is derived from the natural lipid by derivatization with a fluorescent tag (Pagano, 1989; Pownall & Smith, 1989). Especially these latter studies have added another element to the application of a fluorescent probe, other than just as an alternative for radiochemical or spin probes. This element involves fluorescence microscopy, which in studies of intracellular lipid trafficking, allows direct visualization of the probe without perturbation of cellular integrity.

In the following sections we will describe these various aspects of the application of fluorescent lipid probes. The properties of some frequently employed lipid probes, in particular nitrobenzoxadiazole-derivatized-lipids (NBD-lipids), will be summarized. Some attention will also be paid to *'lipid-like'* probes and in this case we will limit the discussion to the application of fluorescent compounds that are attached to an acyl or alkyl chain ('fatty acid' probes). Practical aspects and various examples will be given as

FLUORESCENT AND LUMINESCENT PROBES, 2ND EDN
ISBN 0–12–447836–0

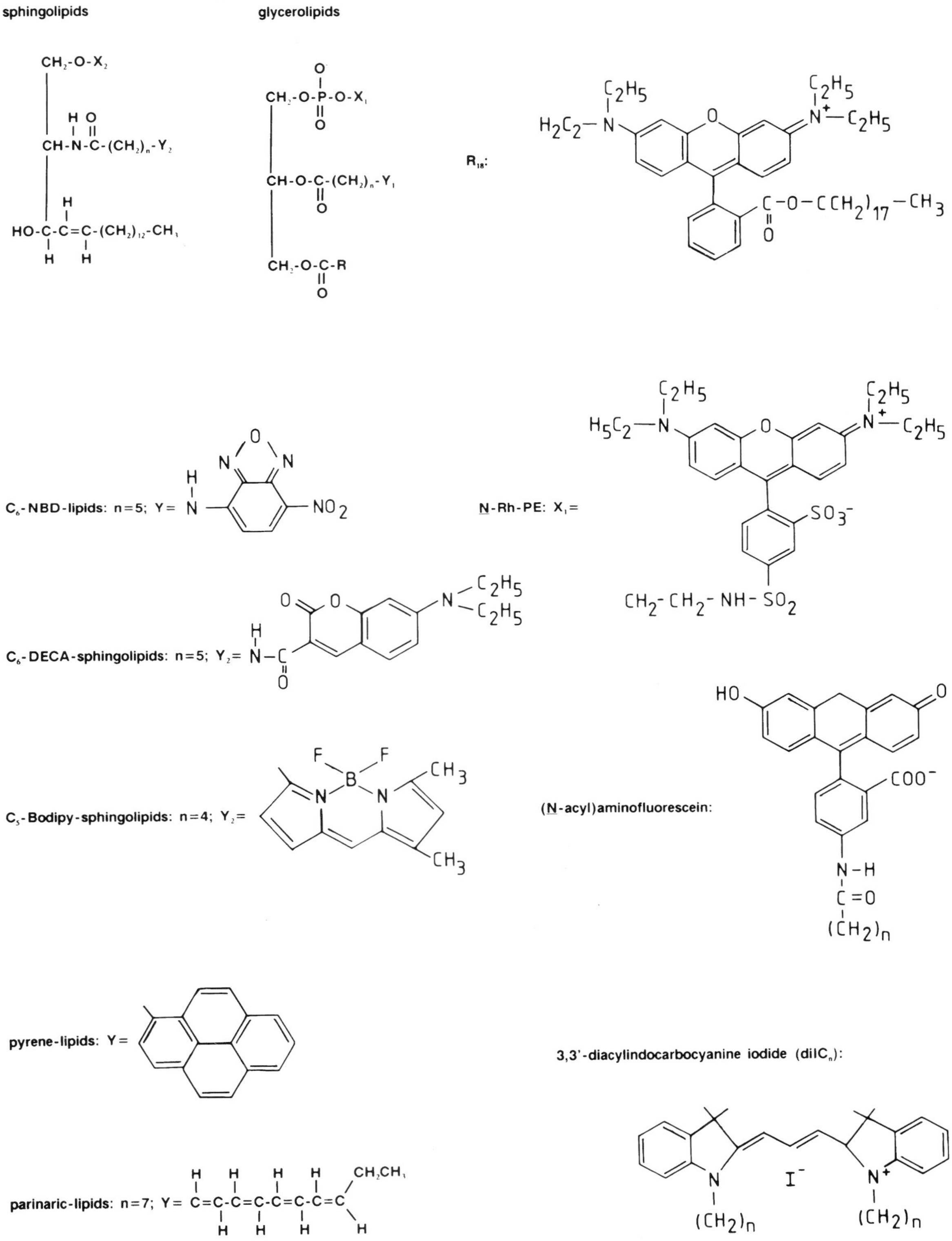

sphingolipids

glycerolipids

R₁₈:

C_6-NBD-lipids: n=5; Y=

N-Rh-PE: X_1=

C_6-DECA-sphingolipids: n=5; Y_2=

C_5-Bodipy-sphingolipids: n=4; Y_2=

(N-acyl)aminofluorescein:

pyrene-lipids: Y=

3,3'-diacylindocarbocyanine iodide ($diIC_n$):

parinaric-lipids: n=7; Y=

Figure 10.1 Structures of fluorescent lipid analogues.

to the fruitful application of these fluorescent lipid and lipid-like probes in areas such as lipid trafficking, lipid metabolism, membrane fusion, lipid transfer and in issues dealing with the polarized and asymmetric distribution of lipids.

10.2 FLUORESCENT LIPID ANALOGUES

10.2.1 Structure, availability and synthesis

Typical fluorescent lipid analogues that have found their application in studies as diverse as membrane flow, intracellular lipid trafficking, membrane fusion, lipid phase separations and transitions are the so-called NBD-lipids. These lipid analogues contain a 7-nitrobenz-2-oxa-1,3-diazol-4-yl (NBD) group. The fluorophore can be attached via a short chain (C_6) or longer spacer (C_{12}) to the lipid backbone, which can be mono-acylglycerol (in NBD-phospholipids) or sphingosine (in NBD-sphingolipids; Fig. 10.1). Alternatively, NBD can be attached to the headgroup of phosphatidylethanolamine (PE) via an amide bond. Thus, in this case a typical headgroup-labelled lipid analogue is obtained.

Several NBD-phospholipids are commercially available (Avanti; Molecular Probes). These include the fluorescent derivatives of phosphatidylcholine (PC), PE, phosphatidylglycerol (PG) and phosphatidic acid (PA), all of which are available with C_6 and C_{12} spacers. C_6-NBD-PS is not available but can be synthesized from C_6-NBD-PC by a base-exchange reaction (Comfurius *et al.*, 1990). Of the C_6-NBD-sphingolipids C_6-NBD-ceramide and C_6-NBD-sphingomyelin are now available (Molecular Probes), but all the glycosphingolipids must be synthesized in one's laboratory.

Glycosphingolipids consist of a long-chain sphingosine backbone, to which a fatty acid is attached via an amide linkage. This makes up the hydrophobic moiety, called ceramide. The carbohydrate moiety, which can vary from a single hexose (usually glucose or galactose) to a complex structure of many linked hexose units, is attached at the primary hydroxy group of the sphingoid base (Wiegandt, 1985).

To synthesize a C_6-NBD-(glyco)sphingolipid, the fatty acid is to be replaced by C_6-NBD-hexanoic acid. Thus one needs the deacylated (lyso-) form of the parent lipid. Several deacylated (glyco)sphingolipids are commercially available (Sigma). These include sphingosine (deacylated ceramide), sphingosylphosphorylcholine (deacylated sphingomyelin), psychosine (deacylated galactosylceramide), glucopsychosine (deacylated glucosylceramide) and lysosulphatide. When a deacylated sphingolipid is not available (or too expensive), as in the case of lactosylceramide, the parent lipid can be deacylated according to a modified procedure of Goda *et al.* (1987). In our laboratory we routinely apply the following procedure: 10 mg of lactosylceramide (Sigma) is added to 9 ml *n*-butanol/1 ml 10 N KOH in a round-bottom flask. The mixture is stirred and heated in a heating mantle for 6 h at 117°C (boiling point of *n*-butanol), while refluxing. Thereafter the solvent is evaporated under a nitrogen flow, while keeping the temperature at about 40°C to speed up the evaporation. The dried material is then dissolved in 16 ml of $CHCl_3/CH_3OH/H_2O$ (2:2:1), shaken vigorously and centrifuged to separate phases. The lower chloroform phase is washed with 4 ml of water and then evaporated. The dried material is separated on preparative TLC using $CHCl_3/CH_3OH/CaCl_2$ (0.2 %) in H_2O (50:42:11) as the solvent system. Lactosylsphingosine (R_f 0.39) is well-separated from lactosylceramide (R_f 0.68) and also from the degradation products glucosylsphingosine (R_f 0.44) and glucosylceramide (R_f 0.81). The lactosylsphingosine band is scraped from the TLC plate and washed with 20 ml $CHCl_3/CH_3OH$ (1:1) followed by 20 ml methanol to remove the lipid from the silica. After evaporation the product is tested for purity on TLC, using orcinol staining (Kundu, 1981) and, if necessary, subjected to another preparative TLC procedure.

The reacylation with C_6-NBD-hexanoic acid (Sigma) is performed as described by Kishimoto (1975) with slight modifications. The C_6-NBD-sphingolipid is synthesized from free fatty acid and deacylated sphingolipid by an oxidation–reduction condensation with triphenylphosphine and 2,2′-dipyridyldisulphide, a method originally devised for peptide synthesis (Mukaiyama *et al.*, 1970). Typically, 10 μmol (6.2 mg in case of lactosylsphingosine) of the deacylated sphingolipid is mixed with 10 μmol (2.9 mg) of C_6-NBD, 20 μmol (5.2 mg) of triphenylphosphine and 20 μmol (4.4 mg) of 2,2′-dipyridyldisulphide and dissolved in 200 μl of chloroform. The mixture is stirred overnight at room temperature under an atmosphere of nitrogen. The material is then applied on preparative TLC plates, which are run in a solvent system consisting of $CHCl_3/CH_3OH$/20% (wt/vol.) NH_4OH (70:30:5) for C_6-NBD-(glyco)sphingolipids, or $CHCl_3/CH_3OH$/HAc (90:2:8) for C_6-NBD-ceramide. C_6-NBD-lactosylceramide (R_f 0.16) is well-separated from C_6-NBD (R_f 0.30) and from high-R_f-value side-products, as well as from the other reactants (Fig. 10.2) The NBD-band of interest is scraped from the plate and washed with 20 ml $CHCl_3/CH_3OH$ (1:1) followed by 20 ml of methanol to remove the lipid from the silica. Purity of the product is tested in several TLC solvent systems (basic, neutral, acidic) and repurified on TLC if necessary to finally yield a single spot. The final yield of the deacylation/reacylation procedure varies from

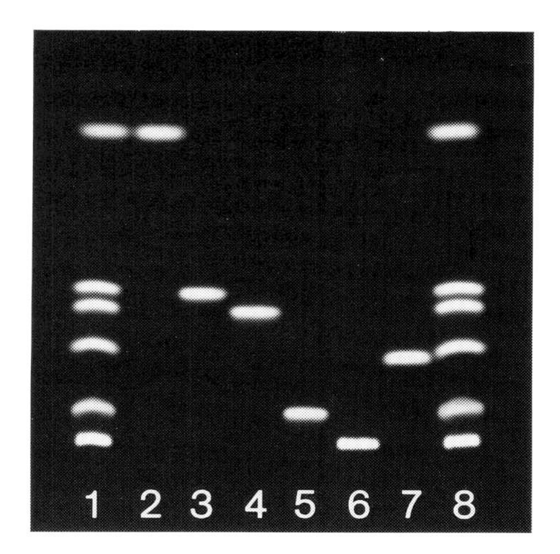

Figure 10.2 Separation of C_6-NBD-sphingolipids by thin-layer chromatography. Five different C_6-NBD-sphingolipids were synthesized from the corresponding deacylated parent lipids and C_6-NBD-hexanoic acid. The NBD-lipids were separated on a TLC plate employing $CHCl_3/CH_3OH/20$ % (wt/vol.) NH_4OH (70:30:5) as the running solvent system. (Individual C_6-NBD-sphingolipids from left to right are: C_6-NBD-ceramide (lane 2); C_6-NBD-glucosylceramide (lane 3); C_6-NBD-galactosylceramide (lane 4); C_6-NBD-lactosylceramide (lane 5) and C_6-NBD-sphingomyelin (lane 6); in lane 7 C_6-NBD-hexanoic acid was spotted.

10 to 25%, depending on the type of lipid. The above described procedure has the advantage that an extra (preceding) step of derivatization of the fatty acid to, for instance, an *N*-hydroxysuccinimide ester is not necessary. In the case of C_6-NBD-hexanoic acid, however, the succinimidyl ester is available (Molecular Probes), which renders this reaction a good alternative to the above described condensation reaction, with similar yields. Typically, 10 μmol of the deacylated sphingolipid and 15 μmol (5.9 mg) of succinimidyl-6-(7-nitrobenz-2-oxa-1,3-diazol-4-yl)aminohexanoate (Molecular Probes) are dissolved in 200 μl of $CHCl_3/CH_3OH/(C_2H_5)_3N$ (8:2:0.1) and allowed to react overnight at room temperature and under nitrogen (cf. Gardam *et al.*, 1989).

In the case of gangliosides, which are acidic glycosphingolipids containing one or more *N*-acetylneuraminic acid (NANA) residues in the carbohydrate moiety, the NBD labelling procedure becomes more complex. This is due to a concomitant deacetylation of the NANA during the deacylation of the lipid. After the labelling procedure the ganglioside has to be subjected to re-*N*-acetylation with acetic anhydride. Furthermore, the deacylation is performed with tetramethylammonium hydroxide instead of KOH and the reacylation with either the succinimide ester (Schwarzmann & Sandhoff, 1987) or the mixed anhydride of the fatty acid and chlorocarbonate, thus involving a fatty acid activation step (for details see Sonnino *et al.*, 1986; Acquotti *et al.*, 1986).

Fluorescently labelled lipids offer the great advantage that their intracellular trafficking can be monitored directly in living cells with fluorescence microscopy. When combined with the use of fluorescently tagged protein (Dunn & Maxfield, 1990), this offers the possibility to study lipid trafficking in relation to the routes by which protein ligands like transferrin are processed, which are well-established (Van Renswoude *et al.*, 1982; Chiechanover *et al.*, 1983). Thus one can perform co-internalization studies of C_6-NBD-lipids and proteins tagged with fluorophores such as those of, for example, the rhodamine group (lissamine rhodamine B (LR), (tetramethyl)rhodamine (T)RITC, Texas Red (TR)) (Kok *et al.*, 1989, 1990; Koval & Pagano,1989).

The labelling of lipids with fluorophores other than NBD, displaying different spectral properties, makes it also possible to perform double-labelling studies with different types of lipids (Uster & Pagano, 1986; Kok *et al.*, 1991). Using different combinations of fluorophores, even triple-labelling studies can be carried out. One has to take into account, however, that the overall physico-chemical properties of the lipid analogues can be influenced by the (type of) fluorescent tag, as discussed below. An example of the double-labelling strategy involving lipids is the use of 7-diethylaminocoumarin-3-carboxylic acid (DECA) as an alternative for (C_6-)NBD (Gardam *et al.*, 1989), which emits blue fluorescence instead of the green fluorescence of NBD. Microscopically, DECA can be separated from NBD by using appropriate filter sets. C_6-DECA-sphingolipids can be synthesized by first coupling an activated DECA fluorophore to hexanoic acid followed by linking this C_6-DECA to the deacylated sphingolipid with a condensation reaction (Kishimoto, 1975) as described above for the C_6-NBD-sphingolipids. The C_6-DECA can be synthesized as follows: 6-amino-*n*-hexanoic acid (Sigma) is dissolved in methanol at a concentration of 20 mg ml^{-1}. Then 100 μl (2 mg or 15 μmol) of this solution is added to 200 μl $CHCl_3/CH_3OH/(C_2H_5)_3N$ (8:2:0.1) containing 8.8 mg (20 μmol) of DECA-succinimidyl ester and incubated overnight, under nitrogen and at room temperature. Thereafter the C_6-DECA is purified by preparative TLC employing a solvent system consisting of $CHCl_3/CH_3OH/20\%$ (wt/vol.) NH_4OH (70:30:5), followed by scraping of the C_6-DECA band (R_f is 0.46, i.e. well-separated from DECA-succinimidyl ester (R_f 0.84) and washing as described above.

A disadvantage of the NBD-fluorophore is that it rapidly bleaches during observations with the fluorescence microscope. Therefore, photography is sometimes difficult and often tedious, while long-time or repeated observation of single cells at high intensity is virtually impossible. Recently, a new fluorophore, boron dipyrromethane difluoride (Bodipy, Molecular Probes), has become available, which has a higher quantum yield, an improved photostability and which can be linked to (sphingo)lipids, analogous to NBD. A highly interesting and exciting property of the probe is that it shows a density-dependent fluorescence emission shift from 515 to 620 nm. This property makes it possible to monitor the generation or elimination of high local concentrations of probe-carrying lipids in membranes of different cellular compartments (Pagano *et al.*, 1991). A disadvantage of the use of Bodipy-labelled lipids is that due to the less polar character of the fluorophore (in comparison to NBD), the lipid cannot be fully recovered from the plasma membrane by a so-called 'back-exchange', as discussed below.

Other fluorescent lipid analogues that have frequently been used in studies of the dynamics of membranes are analogues, labelled synthetically or biosynthetically with pyrene or parinaric acids. These probes are primarily used in experiments involving fluorometric measurements in the fluorometer (for a review see Pownall & Smith, 1989), while their spectral properties preclude their fruitful application in fluorescence microscopy. Although parinaric acids are most similar to natural fatty acids, these probes are particularly difficult to work with because of their rapid decomposition, due to their air-sensitivity. A spectral inert gas atmosphere is therefore necessary. The reader is referred to several excellent reviews on these lipid analogues, that were recently published elsewhere (Loew, 1988; Pownall & Smith, 1989).

10.2.2 Some properties and practical implications

C_6-NBD-lipids can easily be inserted into the plasma membrane of cells and, inversely, also recovered from the plasma membrane by a 'back-exchange'. This is due to the fact that these lipid analogues are less hydrophobic than their natural counterparts and are therefore able to transfer spontaneously from a suitable donor system to a cellular membrane. C_6-NBD-PC has been shown to transfer between vesicle populations (Pagano *et al.*, 1981), between vesicles and cells (Struck & Pagano, 1980) and between biological membranes (Kok *et al.*, 1990). This is not a typical feature of C_6-NBD-labelled lipids. In general, the transfer rate of fluorescent lipid analogues is dependent on the length of the acyl chain to which the probe is attached. For pyrene-labelled lipids, it has been shown that when the number of methylene units decreases, the transfer rate increases. This appears to be directly related to the increased solubility of the analogue in water (Pownall & Smith, 1989; Chattopadhyay, 1990). In contrast, the tendency of exogenously incorporated acyl or alkyl derivatives to leave the bilayer decreases with increasing chain length (Frank *et al.*, 1983; Nichols & Pagano, 1983). Finally, headgroup-labelled lipid derivatives, such as *N*-NBD-PE and *N*-(lissamine rhodamine B sulphonyl) phosphatidylethanolamine (*N*-Rh-PE) or lipid-like derivatives like octadecyl rhodamine B chloride (R_{18}) behave *in this respect* like natural lipids, showing monomeric transfer kinetics in the order of days (Keller *et al.*, 1977; Struck *et al.*, 1981; Hoekstra, 1982a; Nichols & Pagano, 1983; Hoekstra *et al.*, 1984).

The proper intercalation of a fluorescent lipid analogue in the lipid phase of the plasma membrane of the cells can be shown (Struck and Pagano, 1980; Kok *et al.*, 1990) by measuring lateral diffusion rates of the lipid with a technique based on fluorescence recovery after photobleaching (FPR; Axelrod *et al.*, 1976; Jacobsen *et al.*, 1976). When properly inserted, fluorescent lipid analogues display diffusion rate constants of approximately 10^{-8} cm^2 s^{-1} (Edidin, 1992).

Apart from using vesicles as donor membranes for lipid probe insertion into acceptor membranes via monomeric transfer through the aqueous phase, other procedures have also been developed. In the case of C_6-NBD-ceramide a method for introduction into cellular membranes has been described (Pagano & Martin, 1988), that makes use of NBD-lipid/BSA complexes. The lipid is dissolved in ethanol and injected under vortexing into a defatted BSA-containing buffer, followed by dialysis. The complex is then added to the cells resulting in transfer of the lipid from the BSA complex to the cell membrane.

A generally applicable method for the insertion of C_6-NBD-lipids into the plasma membrane of cells involves the use of ethanol micelles (Kok *et al.*, 1989, 1990). Appropriate amounts of C_6-NBD-lipid are dried under nitrogen and subsequently solubilized in absolute ethanol. An aliquot of the ethanolic solution (0.5% (v/v), final concentration) is injected into the appropriate buffer under vigorous vortexing. This solution (usually containing 4–5 μM NBD-lipid) is then added to the cells. It turns out that with this method it is also possible to insert *N*-Rh-PE into the plasma membrane of cells (Kok *et al.*, 1990). This analogue is a fluorescent PE derivative (but no longer has specific PE properties, see below) which has the fluorophore (rhodamine) attached to the amino group of the headgroup of the lipid. *N*-Rh-PE is defined as a non-exchangeable lipid analogue, based on observations that the probe, when incorporated into liposomal bilayers, does not spontaneously transfer between the labelled liposomes and non-labelled artificial or biological membranes (Pagano *et al.*, 1981; Struck *et al.*, 1981; Hoekstra *et*

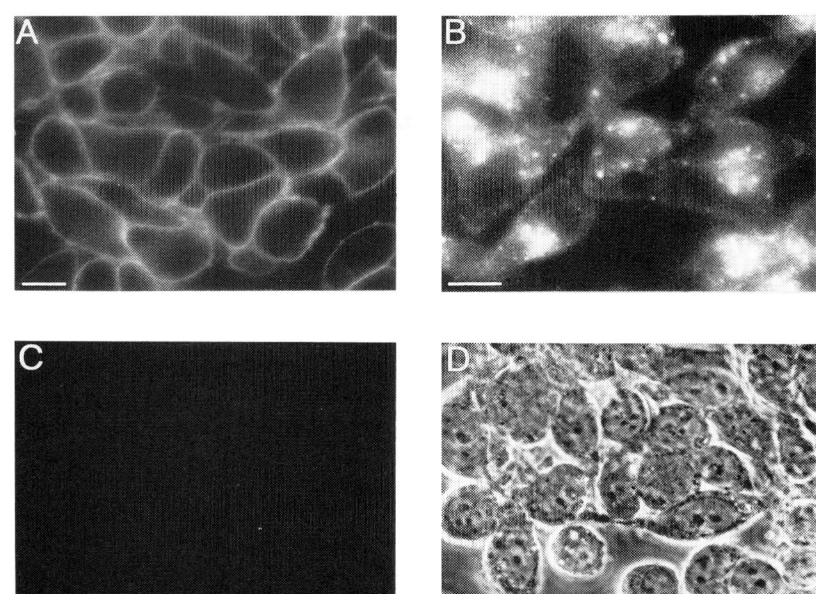

Figure 10.3 Plasma membrane insertion, endocytosis, and back-exchange of a C_6-NBD-lipid. Baby hamster kidney cells were labelled at 2°C with C_6-NBD-glucosylceramide (5 mol %)-containing DOPC liposomes for 30 min (A). Subsequently either a back-exchange was performed with 5% BSA/Hank's for 30 min at 2°C (C and D), or the labelled cells were first incubated at 37°C for 30 min, followed by a back-exchange (B). Note that the back-exchange removes all plasma membrane-inserted C_6-NBD-lipid. Similar results were obtained when the cells were labelled with the ethanol-injection method or when the back-exchange was performed with DOPC SUV (bars = 10 μm). (For details, see Kok *et al.*, 1989.)

al., 1988). However, when baby hamster kidney (BHK) cells are incubated at 2°C with *N*-Rh-PE (dispersed in ethanol), the lipid is inserted into the plasma membrane, as indicated by several criteria including FPR measurements (Kok *et al.*, 1990).

The exchangeable NBD-lipids can be removed from the outer leaflet of the plasma membrane ('back-exchange') by incubating the cells with either dioleoylphosphatidylcholine (DOPC) small unilamellar vesicles (SUV) (Struck & Pagano, 1980) or (bovine) serum albumin (Van Meer *et al.*, 1987; Kok *et al.*, 1989). The vesicles or the albumin function as acceptors for the NBD-lipid molecules. If a sufficient amount of these acceptors is added to the cells (usually two consecutive washes), the NBD-lipid can be quantitatively removed from the outer leaflet of the plasma membrane (Fig. 10.3). As anticipated, this back-exchange procedure does not work for the non-exchangeable *N*-Rh-PE (Kok *et al.*, 1990).

When the C_6-NBD-lipids are compared to their natural counterparts in cells, several arguments exist in favour of a good analogy between the probe and the natural lipid. First, C_6-NBD-PA is metabolized in cells initially to C_6-NBD-DG, followed by synthesis of C_6-NBD-TG and C_6-NBD-PC. Furthermore, the cell recognizes these products and sorts them to different intracellular compartments, since only C_6-NBD-TG becomes associated with lipid droplets (Pagano & Sleight, 1985; Pagano & Longmuir, 1985). Secondly, analogous to radioactively labelled PE, C_6-NBD-PE undergoes transbilayer movement at the plasma membrane (Pagano & Sleight, 1985; Sleight & Pagano, 1985). Similarly, when C_6-NBD-PS is supplied to erythrocytes, it inserts into the *outer* leaflet of the membrane followed by translocation to the *inner* leaflet, resulting in an asymmetric distribution among the leaflets. By contrast, C_6-NBD-PC does not translocate (Connor & Schroit, 1987). These results are quite compatible with the naturally occur ring situation, where aminosphospholipids preferentially reside in the *inner* leaflet, while the choline phospholipids are predominantly localized in the *outer* leaflet of the red blood cell membrane (Verkleij *et al.*, 1973; Gordesky *et al.*, 1975). Thirdly, the metabolism and intracellular transport of C_6-NBD-ceramide seem to reflect the behaviour of endogenous ceramide in the following ways.

(1) C_6-NBD-ceramide is metabolized to C_6-NBD-sphingomyelin and C_6-NBD-glucosylceramide (Lipsky & Pagano, 1983, 1985a; Van Meer *et al.*, 1987) and to other glycosphingolipids (Kok *et al.*, 1991), indicating that the respective biosynthetic enzymes are capable of recognizing the NBD-labelled substrate. Furthermore, when radiolabelled ceramide or C_6-NBD-ceramide are incubated with isolated subcellular fractions prepared from cultured fibroblasts, no preferential metabolism of the different ceramides is seen (Lipsky & Pagano, 1985a).
(2) The half-life for transport of C_6-NBD-sphingomyelin and C_6-NBD-glucosylceramide from the Golgi apparatus to the plasma membrane (Lipsky & Pagano, 1983, 1985a) is consistent with the half-life for the appearance of newly synthesized, radiolabelled neuronal gangliosides at the plasma

membrane (Miller-Podraza & Fishman, 1982). Furthermore, monensin inhibits the appearance of both isotopically labelled glycosphingolipids (Saito *et al.*, 1984) and C_6-NBD-glucosylceramide (Lipsky & Pagano, 1985a and our own unpublished results) at the cell surface.

(3) In polarized Madin-Darby canine kidney (MDCK) cells C_6-NBD-glucosylceramide is sorted from C_6-NBD-sphingomyelin and preferentially delivered to the apical surface (Van Meer *et al.*, 1987; Van't Hof & Van Meer, 1990). This polarized delivery is consistent with the known enrichment of glycosphingolipids in the apical membrane, thus indicating that the C_6-NBD-sphingolipids are sorted and transported similarly as compared to their natural counterparts.

Finally, in the endocytic uptake pathway C_6-NBD-sphingolipids can be recycled to the plasma membrane (Kok *et al.*, 1989; Koval & Pagano, 1989) and sorted from each other (Kok *et al.*, 1991). The recycling of intact C_6-NBD-glucosylceramide molecules to the cell surface after their initial uptake by endocytosis (Kok *et al.*, 1989) is consistent with the notion that glycolipids are thought to reside mainly in the outer leaflet of the plasma membrane, indicating that this class of lipids in particular is subject to sorting (Hakomori, 1981). In liver, gangliosides can be resynthesized from recycled glucosylceramide (Trinchera *et al.*, 1990), indicating that the lipid is sorted during inbound cellular traffic, thereby avoiding degradation in the lysosomal compartment. The sorting of C_6-NBD-glucosylceramide from other (glyco)sphingolipids, like C_6-NBD-sphingomyelin during endocytosis in HT29 cells, indicates that also in the endocytic pathway C_6-NBD-sphingolipids are recognized by the cellular sorting and transport machinery (cf. (3) above).

Taken together, valuable insight into the cell biology of lipids has been obtained from the use of C_6-NBD analogues of lipids. However, there are some indications that the use of these analogues is restricted and should always be critically evaluated.

(1) Studies aimed at investigating the role of fatty acid heterogeneity are, by definition, impossible, especially in the case of the C_6-NBD-sphingolipids. Reacylation of lipids cannot be monitored, since the marker (C_6-NBD) is lost after deacylation and is not reused for synthesis. On the one hand, this property offers the advantage of a relatively simple interpretation since the fate of only one labelled lipid derivative is usually monitored. On the other hand, however, it clearly distinguishes NBD-lipid analogues from pyrene and parinaric fatty acid-labelled lipids. The latter derivatives even more closely resemble natural lipids as both fatty acid probes can be biosynthetically incorporated into cellular lipids (Loew, 1988; Pownall & Smith, 1989). In contrast to C_6-NBD-PC and other NBD-phospholipids, which are readily deacylated (Sleight & Pagano, 1984; Moreau, 1989; Kok *et al.*, 1990), the C_6-NBD-sphingolipids do not seem to be subject to this type of catabolism (Kok *et al.*, 1989; Koval & Pagano, 1989), although it is known that their natural counterparts are (Trinchera *et al.*, 1990).

Another reason for caution in studies involving acyl chain structure stems from studies by Chattopadhyay and co-workers (1988, 1990). They show that the NBD group of C_6-NBD-lipids loops back to the polar region of the membrane, instead of being fully embedded in the hydrophobic core of the membrane. In a recent review (Chattopadhyay, 1990) the use of NBD-labelled lipids as analogues of natural lipids in membrane and cellular processes is critically evaluated. The conclusion is reached that in spite of the perturbed acyl chain conformation the NBD-labelled lipids have served as reasonably good analogues for these studies.

(2) With regard to lipid anabolism, the more complex glycolipids and gangliosides are not synthesized from the C_6-NBD-ceramide precursor. Apparently the biosynthetic enzymes for glucosylceramide and galactosylceramide are able to recognize the NBD-labelled ceramide substrate. Furthermore, C_6-NBD-glucosylceramide can also serve as a substrate for lactosylceramide synthesis. However, C_6-NBD-GM3 synthesis is not observed in BHK cells, although these cells do contain the natural counterpart of this lipid (Kok *et al.*, unpublished observations).

(3) The ratio of (glyco)sphingolipids synthesized from fluorophore-labelled ceramide may depend on the type of fluorescent probe (C_6-NBD vs. C_6-Bodipy and C_6-DECA) (Pagano *et al.*, 1991; Kok *et al.*, unpublished observations).

In conclusion, although NBD-labelled lipid analogues have shown their value in numerous cell biological applications, one should be careful in interpreting results from metabolic studies employing these fluorescent lipids, especially regarding absolute amounts of synthesis. Also, the suitability of the C_6-NBD label for the study of complex glycolipids (gangliosides) is questionable.

To study dynamic membrane properties such as membrane fusion, lipid translocation, lipid phase transitions and separations, advantage can also be taken of the specific fluorescent properties of certain lipid probes. These properties include concentration-dependent self-quenching of the probes and their ability to engage in resonance energy transfer (RET). In

Section 6.2 several examples of experiments that rely on these principles will be discussed. Given the versatile use of approaches that rely on resonance energy transfer, it is useful to outline briefly here the basic features of such an approach. In essence, RET involves the interactions that may occur between two different fluorophores, provided that the emission band of one fluorophore, the energy donor, overlaps with the excitation band of the second fluorophore, the energy acceptor. Energy transfer takes place when the probes are in close physical proximity. It involves the transfer of excited state energy from the donor to the acceptor. This energy is derived from a photon absorbed by the energy donor upon excitation. The acceptor will then fluoresce as though it had been excited directly. The energy transfer efficiency depends on the extent of overlap of the donor's emission spectrum and the acceptor's absorption spectrum and the distance between the probes. The energy transfer efficiency, E, is given by the equation $E = (1 - F/F_0) \times 100\%$, where F is the fluorescence measured in the presence of the acceptor and F_0 represents the fluorescence read in the absence of acceptor (usually obtained by reading fluorescence after addition of detergent, i.e. at infinite dilution, see below). With regard to distance, the efficiency of energy transfer is proportional to the inverse of the sixth power of the distance between the donor and acceptor (Fung & Stryer, 1978). The distance between two fluorophores can then be calculated, using the equation:

$$R = R_0[(0.5/E)^{1/6}]$$

in which R_0 is the so-called Förster distance, i.e. the distance between donor and acceptor at which E equals 50%.

A convenient energy transfer couple that has found wide application is that involving NBD-labelled lipid as energy donor, and *N*-Rh-PE as energy acceptor. Studies of membrane fusion (Struck *et al.*, 1981; Hoekstra, 1982a), protein-mediated and spontaneous exchange and transfer of lipids (Nichols & Pagano, 1983) and (transbilayer) flip-flop studies (Connor & Schroit, 1987; Pagano, 1989) have all benefited from employing this particular couple. In this system, transfer efficiency can be translated in the ability of rhodamine to quench NBD fluorescence. (Note that the transfer efficiency also depends on the concentration of the energy acceptor.) In practice, this means that the reaction (lipid transfer, flip-flop or fusion) is monitored at the NBD emission maximum of approximately 530 nm, while exciting the donor at 475 nm. It should be noted that energy transfer can also be measured between suitably labelled protein molecules and a lipid probe. In the following section, various examples will be discussed of the use of fluorescent lipid probes in cell and membrane biology.

10.3 APPLICATIONS

10.3.1 Applications in cell biology

C_6-NBD-lipids are very useful tools in studying the trafficking of individual types of lipids (for reviews see Pagano & Sleight, 1987; Simons & Van Meer, 1988; Van Meer, 1989; Hoekstra *et al.*, 1989; Schwarzmann & Sandhoff, 1990; Koval & Pagano, 1991). The fate of the lipid can be followed directly with fluorescence microscopy. Furthermore, these observations can be combined with TLC analysis and sensitive fluorometric quantitation, providing information about the localization of the lipid in relation to its metabolism.

For microscopical observation of endocytic trafficking pathways, cells are grown on glass coverslips (coated with, for instance, collagen if necessary, depending on the cell type) in Petri dishes, till confluency. Before labelling with the desired lipid probe, the cells are cooled to 2°C. At this temperature all trafficking processes cease. This allows the fluorescent lipid to be inserted into the plasma membrane of the cell (see Section 10.2.2) and 'synchronized' endocytosis to be triggered, which is done by elevating the temperature to 37°C (Fig. 10.3). When a comparison is to be made to the trafficking of a protein ligand, for example transferrin (Tf), the cells are also incubated with LR-Tf at 2°C, allowing the protein to bind to its receptor. After washing away free ligand, the cells are warmed instantly to 37°C by adding prewarmed (Hank's) buffer, allowing the cells to internalize the lipid and the protein for a defined period of time. Thereafter the cells are cooled again to 2°C to 'fix' the trafficking processes. Subsequently, a back-exchange is carried out to remove the residual plasma membrane NBD-lipid pool. Since only a small (10–20%) fraction of the original plasma membrane (NBD-)lipid pool is internalized, visualization of intracellular fluorescence is difficult without such a back-exchange. Employing the foregoing protocol we have shown that only 2 min of incubation at 37°C is necessary for C_6-NBD-glucosylceramide to reach early endosomes (Fig. 10.4), as indicated by colocalization with LR-Tf, which by definition labels early endosomes, seen as peripheral vesicles, after 2 min of receptor-mediated endocytosis (Kok *et al.*, 1989). When studying such short time (2 min) trafficking with microscopy, it is necessary to somehow fix the lipid inside the intracellular compartment. This can be achieved by cooling the microscope stage to 2°C or by using chemical fixatives. A fixation method has been

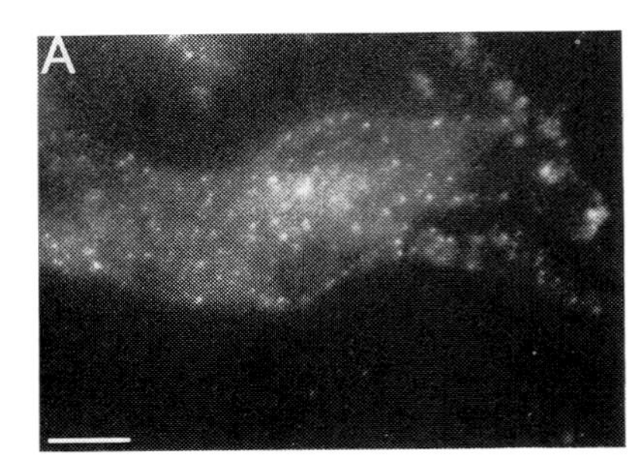

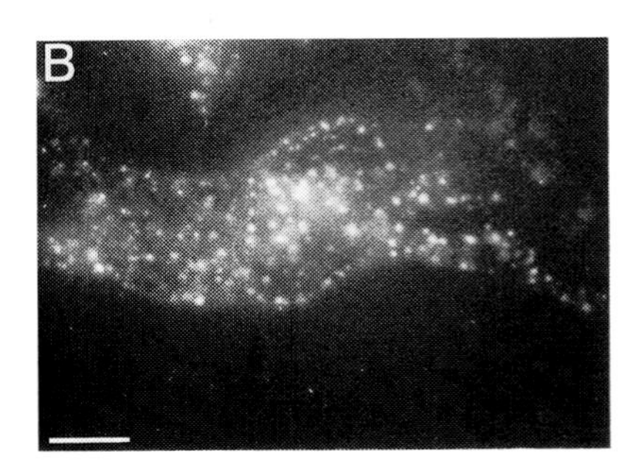

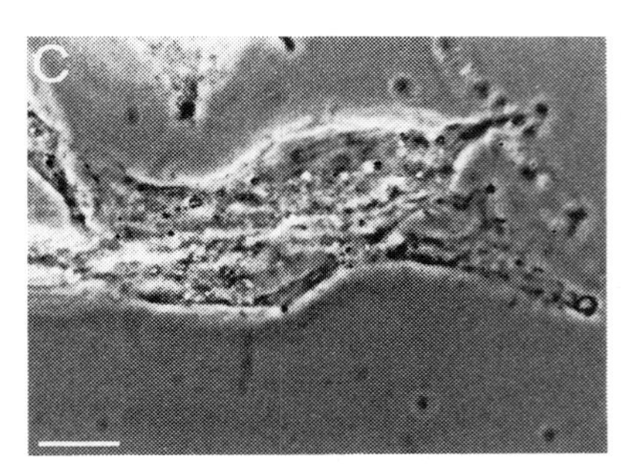

Figure 10.4 Co-internalization of C_6-NBD-glucosylceramide and LR-Tf. Baby hamster kidney cells were labelled at 2°C with C_6-NBD-glucosylceramide (5 mol %)-containing liposomes for 30 min, washed, and labelled with LR-Tf (30 min at 2°C; 0.2 mg ml^{-1}). Subsequently, the cells were incubated at 37°C for 2 min, immediately cooled to 2°C, and subjected to a back-exchange with a 5% BSA/Hank's solution for 30 min at 2°C. The cells were fixed at room temperature (see Section 10.2.1) and examined by fluorescence microscopy, using appropriate filter sets for visualizing either NBD or rhodamine fluorescence. (A) NBD(-lipid); (B) LR(-Tf); (C) corresponding phase-contrast image. Note that the lipid and the protein colocalize in early endosomes (bars = 10 μm). (For further details, see Kok *et al.*, 1989.)

developed for glycolipids (Kok *et al.*, 1989). The fixative, containing periodate, lysine and paraformaldehyde, stabilizes carbohydrate moieties. Stabilization involves the carbohydrates being oxidized by periodate and cross-linked by lysine (McLean & Nakane, 1974). In practice, cells are fixed for 30 min at room temperature with 2% (wt/vol.) formaldehyde in a buffer, containing 20 mM sodium phosphate (pH 10.4), 100 mM lysine, 60 mM sucrose, and 100 mM sodium periodate, and are post-fixed sequentially with 4% formaldehyde and 6% formaldehyde, both in the same buffer, for 5 and 10 min respectively.

The intracellular endocytic pathways of sphingolipids have been carefully studied in fibroblasts (Kok *et al.*, 1989; Koval & Pagano, 1989, 1990). One aspect of lipid trafficking that has become evident is that, like proteins, lipids can also be subject to recycling, i.e. after their initial uptake into the cell by endocytosis, original molecules return to the cell surface. In this respect, it is important to note the intramembrane topology of the NBD-lipids, that are not subject to transbilayer movement (Fig. 10.5). The NBD-lipid molecules are inserted into the outer leaflet of the plasma membrane, which means that they will be trapped inside an endocytic vesicle that buds from the membrane. During its consecutive trafficking inside the cell the lipid will remain locked up inside organelles and when recycling occurs it will reappear on the outer leaflet of the plasma membrane. Thus, after recycling, the lipid is susceptible to back-exchange, allowing quantification of the extent of recycling (as described below). Recycling of C_6-NBD-glucosylceramide in BHK cells occurs from both early and late endosomes, while another fraction of the lipid cycles via the Golgi area, possibly involving the trans-Golgi network (TGN) (Kok *et al.*, 1989).

The ability to undergo transbilayer movement ('flip-flop') depends on the type of lipid (see also Section 10.3.2). Among the (NBD-)lipids that are not subject to flip-flop are the choline phospholipids (PC, SM) (Sleight & Pagano, 1984; Koval & Pagano, 1989) and the glycolipids (Kok *et al.*, 1989, 1991). However, the amino-phospholipids PE and PS are involved in rapid transmembrane movement (Pagano & Longmuir, 1985; Connor & Schroit, 1987), probably involving a protein translocator (Connor & Schroit, 1988). This translocation occurs at temperatures around 7°C and above, and may lead to subsequent labelling of intracellular organelles by non-vesicular movement. The lipid precursors C_6-NBD-DG and C_6-NBD-ceramide already translocate at 2°C and label all intracellular membranes. At 37°C coordinated transport to specific organelles and metabolism occur. In the case of C_6-NBD-ceramide accumulation takes place in the Golgi apparatus. This is accompanied by metabolism to C_6-NBD-glucosylceramide, C_6-NBD-sphingomyelin and other glycosphingolipids (see Section 10.2.2). Because of this accumulation the ceramide can be used as a marker for the Golgi apparatus, both for fluorescence and electron microscopy (Lipsky & Pagano, 1985b; Pagano *et al.*, 1989, 1991). The metabolic products

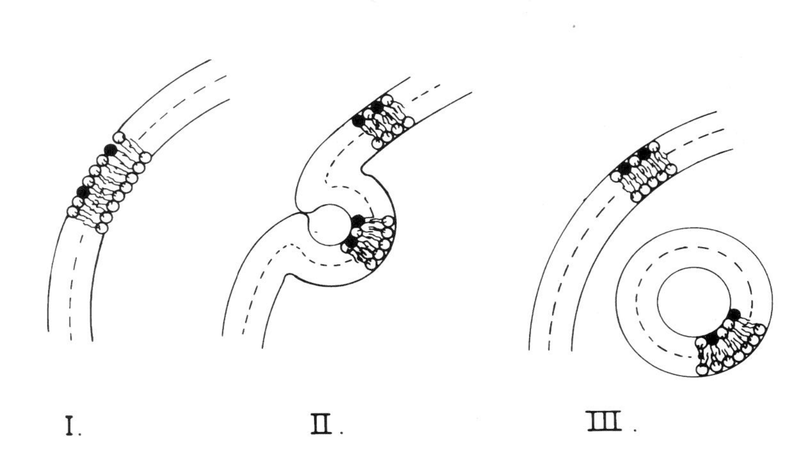

Figure 10.5 Membrane topology of C_6-NBD-lipids that do not flip-flop during trafficking. When cells are labelled with C_6-NBD-glucosylceramide (or any other NBD-lipid that is not subject to flip-flop), the lipid resides in the *outer* leaflet of the plasma membrane (I). Upon internalization the NBD-lipid is 'locked up' inside the endocytic vesicle (III), which is formed after budding (II) from the plasma membrane. In the absence of flip-flop, the lipid will remain in the *inner* leaflet of all intracellular organelles, reached by vesicular trafficking and fusion of the vesicles. Eventually the lipid analogue may reappear in the *outer* leaflet of the plasma membrane after recycling.

are subsequently transported to the plasma membrane by vesicular carriers. That transport is indeed vesicular – although the carrier vesicles have never been visualized – is supported by the kinetics of transfer, the inhibition by monensin (Lipsky & Pagano, 1985a), and the absence of transport in mitotic cells. By contrast, the transport of PE is thought not to be mediated by vesicular carriers (Kobayashi & Pagano, 1989).

In conclusion, vesicular trafficking of sphingolipids can be studied both in the endocytic pathway (uptake of plasma membrane-localized (glyco)sphingolipids) and the biosynthetic pathway (transport of newly synthesized (glyco)sphingolipids). However, the outbound traffic to the plasma membrane is more difficult to compare directly to that of proteins, since methods are not available to introduce fluorophores into proteins during their biosynthesis.

The studies on (sphingo)lipid trafficking in fibroblasts have been further extended to other cell types, including polarized cells (Van Meer *et al.*, 1987; Van't Hoff & Van Meer, 1990; Crawford *et al.*, 1991; Kok *et al.*, 1991). In both MDCK and Caco-2 polarized epithelial cells sorting of the sphingolipids C_6-NBD-glucosylceramide and C_6-NBD-sphingomyelin after their synthesis in the Golgi apparatus occurs, glucosylceramide being preferentially delivered to the apical membrane of the cell. The cells are grown as tight monolayers on filter supports. The basolateral domain of the cells, which faces the filter, is separated from the apical domain by tight junctions. These junctional complexes block the lateral diffusion of lipids in the outer leaflet of the plasma membrane (see Section 10.3.2). Since BSA can pass the pores in the filter the apical and basolateral membrane domains can thus be analysed separately for NBD-lipid content by a back-exchange. In this way, one can separate not only the plasma membrane NBD-lipid pool from the intracellular pool, but, in addition, pools of different plasma membrane domains (Van Meer *et al.*, 1987). In HT29 cells C_6-NBD-glucosylceramide is sorted from other sphingolipids (including sphingomyelin) in the endocytic pathway, being preferentially transported to the Golgi apparatus. Furthermore, this sorting phenomenon is only observed in the undifferentiated HT29 cell type and not in a differentiated line that was obtained by clonal selection (Kok *et al.*, 1991). Thus sphingolipid trafficking has been studied in relation to cell polarity and differentiation. The relevance of these kind of studies is clear when considering the involvement of (glyco)sphingolipids in processes such as cellular interaction, differentiation and oncogenesis, as reviewed by Hakomori (1981, 1984).

The observed sorting of sphingolipids both in the endocytic and biosynthetic pathways suggests that mechanisms are operative in cells that recognize, select and target specific sphingolipids. It is tempting to assume that proteins are involved in these kinds of processes. A promising approach to identify such putative sorting proteins is the use of 5-[^{125}I]iodonaphthyl-1-azide (INA). This is a photoactivatable probe which, upon activation with UV light, covalently couples to adjacent molecules, which thus become labelled with ^{125}I. Interestingly, the INA can also be indirectly activated through excitation of adjacent fluorophores (photosensitizers) (Raviv *et al.*, 1987). Conveniently also the NBD-labelled lipids can serve as appropriate photosensitizers for the INA (Rosenwald *et al.*, 1991; and our own unpublished observations). Thus one can, in principle, identify proteins that are in the vicinity of the NBD-lipids and therefore may be functionally related to these lipids. This approach offers the possibility of combining biochemical data on proteins involved in lipid trafficking with the (known) intracellular fate of the lipid, as revealed by fluorescence microscopy. This approach is currently employed in our laboratory and is compared to an approach involving the use of photoaffinity labelled (glyco)sphingolipids, that have a photoactivatable group in the fatty acid moiety (cf. Sonnino *et al.*, 1989).

A step nearer to the *in situ* situation in the organism is the use of primary cultured cells, like hepatocytes. Hepatocytes are polarized cells with bile canalicular membrane domains that are structurally and functionally separate from sinusoidal membrane domains. From liver so-called hepatocyte couplets can be isolated. A couplet is the smallest functioning hepatobiliary unit (Gautam *et al.*, 1987), that consists of two hepatocytes that enclose a bile canaliculus in between them. In this system biliary secretion of lipids can be studied in relation to intracellular trafficking of these lipids. Recently C_6-NBD-ceramide was used to study the intracellular effects of bile salts connected to their capacity to induce biliary lipid excretion (Crawford *et al.*, 1991). Taurocholate induces translocation of the Golgi apparatus to a peri-canalicular location after which excretion occurs of the metabolic products C_6-NBD-glucosylceramide and C_6-NBD-sphingomyelin in bile. One practical disadvantage of the couplet system in microscopical studies is the fact that these cells have a spherical shape, which makes focusing of intracellular structures containing fluorescent lipids difficult. However, this problem can be overcome by using confocal laser scanning microscopy to optically section the cells, which drastically improves the focus in the *z*-dimension. We are currently using this technique to study the exact localization of intracellular endocytic organelles in relation to the canaliculus between the couplet-forming cells.

Cellular processing of NBD-lipids can be easily analysed and quantified. In order to quantify trafficking or analyse metabolism, 1–5 $\times$ 10^7 cells are incubated with the lipid (precursor) at 37°C for the desired time. A back-exchange can then be performed to separate the plasma membrane NBD-lipid pool from the intracellular pool. Thereafter, both the back-exchange fraction and the cellular fraction are subjected to lipid extraction according to Bligh and Dyer (1959). Briefly, the volume to be extracted is mixed with an equal volume of methanol and two volumes of chloroform. The mixture is vortexed vigorously and centrifuged to separate phases. The lower chloroform phase is collected and the upper methanol/water phase re-extracted with an equal volume of chloroform. The two chloroform phases are combined and dried. The dried lipid is either directly measured (as described below) after dissolving it in 1% (vol./vol.) Triton X-100, or run on TLC for analysis of metabolism. For TLC analysis the dried lipid is taken up in a small (200 µl) volume of $CHCl_3/CH_3OH$ (1:1) and applied on high-performance (HP) TLC plates (Merck). NBD-lipids can be run in one dimension, or if this does not allow sufficient separation, in two dimensions. A system employing the solvent $CHCl_3$/CH_3OH/20% (wt/vol.) NH_4OH (70:30:5) in the first and $CHCl_3/CH_3OH/HAc/H_2O$ (90:40:12:2) in the second dimension separates most of the neutral NBD-(glyco)sphingolipids, which can be identified by comparison to synthesized standards. The individual spots are quantified as follows. The spots are scraped from the plate and the silica is suspended in a 1% (vol./vol.) Triton X-100 solution. This solution is then shaken vigorously at 37°C for 1 h, followed by spinning down the silica. The supernatant is measured in a fluorometer against a standard curve of C_6-NBD-PC in 1% Triton X-100. Corrections should be made for differences in efficiency of removal of the NBD-lipids from the silica. (In the case of sphingomyelin the efficiency is 10% lower than for the other sphingolipids, independent of the absolute amounts.)

In the endocytic pathway (NBD-)lipids are catabolized. Examples are the breakdown of C_6-NBD-PA to C_6-NBD-DG (Pagano & Longmuir, 1985) and the deacylation of C_6-NBD-PC (Sleight & Pagano, 1984), both occurring in the plasma membrane. NBD-(glyco)sphingolipids are not deacylated during their intracellular trafficking. De/reacylation with C_6-NBD can be excluded since C_6-NBD, once liberated, is not reused for synthesis. Although deacylation does occur with radiolabelled sphingolipids (Trinchera *et al.*, 1990; and our own observations), it is convenient that trafficking and processing of NBD-(glyco)sphingolipids can be followed without losing the NBD label. The NBD-sphingolipids are degraded to NBD-ceramide, probably in the lysosomes (Koval & Pagano, 1989, 1990). However, during every cycle of the lipid from the plasma membrane into the cell and back to the plasma membrane (recycling) only a small fraction is degraded (Koval & Pagano, 1990; Kok *et al.*, 1989). The formed NBD-ceramide can move from the lysosomes to the Golgi apparatus, where it can be reused for the synthesis of sphingolipids (see Section 10.2.2). Thus, these lipids can be remodelled during their trafficking inside the cell according to the demands of the cell. Extensive remodelling is a long-term process (hours to days) compared to the time-scale of trafficking, where an equilibrium situation is usually reached after *ca.* 30 min of incubation at 37°C. In HT29 cells equilibrium levels of sphingolipids synthesized from C_6-NBD-ceramide are reached after *ca.* 24 h of incubation at 37°C. A similar composition is reached when the cells are incubated with C_6-NBD-glucosylceramide or C_6-NBD-sphingomyelin, indicating that in the long term, complete remodelling occurs. In HeLa cells the more complex glycolipid GM1, when labelled with pyrene in the fatty acid moiety, gives rise to catabolic (GM2 and GM3) and anabolic (GD1a) derivatives after a 24 h incubation at 37°C, whereas after 2 h metabolic processing has not yet occurred (Masserini *et al.*, 1990).

In HT29 cells the composition of NBD-(glyco)-sphingolipids is changing upon differentiation of the cells (Babia *et al.*, manuscript submitted). These long-term metabolic differences related to differentiation are reflected in short-term differences in trafficking (see above). Since it is known that the ratio of the various sphingolipids synthesized from the precursor ceramide depends on the type of fluorescent probe used (as discussed in Section 10.2.2), one should not base firm conclusions on absolute amounts of synthesis. However, in the HT29 system, when studying metabolism related to differentiation, this problem is overcome, since only relative differences in metabolism are considered. Therefore, these studies do provide insight into the changing sphingolipid demands of the cell during its differentiation.

So far we have discussed the specific trafficking of individual lipid molecules, probed with fluorophores in the fatty acid moiety. These lipids do not monitor the overall membrane flow during, for example, endocytosis. However, the PE derivative described above, labelled with rhodamine in the headgroup (*N*-Rh-PE), can serve this purpose. Since the lipid in this case has a modified headgroup, it is no longer comparable to the natural PE, and is not recognized as such by the cell. This implies, for example, that the lipid derivative is not translocated across the membrane, as observed for acyl chain labelled PE (Section 10.3.2). It appears, however, that *N*-Rh-PE follows the general flow of membranes during endocytic membrane uptake, after its insertion into the plasma membrane (see Section 10.2.2). In fact, it turns out that *N*-Rh-PE follows the membrane flow in the fluid-phase endocytic pathway (Kok *et al.*, 1990), since (1) it is delivered to the lysosomes, as shown by microscopy and cellular fractionation; (2) it is not found in early endosomes that are labelled by FITC-Tf, a marker for the receptor-mediated endocytic pathway (unpublished observation); (3) it colocalizes with Lucifer yellow (a fluid-phase marker) throughout its intracellular trafficking pathway. Thus, *N*-Rh-PE can be used to monitor this pathway and as a marker for lysosomes, where it has accumulated after a 1 h incubation at 37°C.

The most commonly used fluorescent markers for fluid-phase endocytosis are Lucifer yellow and FITC-dextran (for a review see Swanson, 1989). However, Lucifer yellow appears to cross intracellular membranes in some cell types (Swanson, 1989), whereas the use of FITC-dextran of commercial sources may be frustrated by the presence of low molecular weight impurities that easily penetrate the cell (Preston *et al.*, 1987) (purification protocols now exist to remove these impurities (Cole *et al.*, 1990)). Thus, the intracellular labelling observed with these fluid-phase markers may, in part, be a consequence of aspecific exchange instead of vesicular trafficking. Using *N*-Rh-PE, which is a non-exchangeable membrane-associated probe, this problem cannot arise. Furthermore, fluid-phase uptake also occurs as a consequence of receptor-mediated endocytosis, so fluid-phase probes do not necessarily discriminate between the pathways of pinocytosis and receptor-mediated endocytosis. The fluorescent phospholipid analogue, however, seems to do so, since it is not found to colocalize with ligands (Tf) probing the receptor-mediated pathway. Finally, because of its rhodamine fluorophore *N*-Rh-PE is a convenient probe for performing collabelling studies with NBD-lipids, to monitor their possible trafficking along the fluid-phase endocytic pathway.

10.3.2 Applications in membrane biology

Apart from following the intracellular fate of a defined species of a phospho- or glycosphingolipid analogue, there are numerous applications of fluorescent lipid and lipid-like derivatives, aimed at investigating the dynamics of membranes in general, including overall biophysical properties. It is virtually impossible to present even a far-from-complete overview. To be explicit, therefore, only a few typical examples are mentioned here. Other applications and further details can be found in several recent fluorescence textbooks (Laskowicz, 1983; Loew, 1988).

The occurrence of phase separations, phase transitions and the relative 'fluidity' of membranes has been determined with fluorescent probes such as diphenylhexatriene (DPH). The probe can also be coupled to phospholipids (Parente & Lentz, 1985) and the trimethylammonium salt derivative (TMA-DPH) has been employed in measuring membrane fusion during exocytosis (Bonner *et al.*, 1986; see below). DPH displays fluidity-sensitive fluorescence polarization characteristics (Shinitzky & Barenholz, 1978). In aqueous solutions, its fluorescence is negligible. Membranes are labelled by adding the probe, solubilized in tetrahydrofuran, to the system of interest at a relatively low concentration (micromolar). DPH fluorescence polarization, p, is calculated after measuring fluorescence (excitation and emission wavelengths for DPH are 360 and 428 nm respectively) with polarizers in crossed and parallel positions, using the equation:

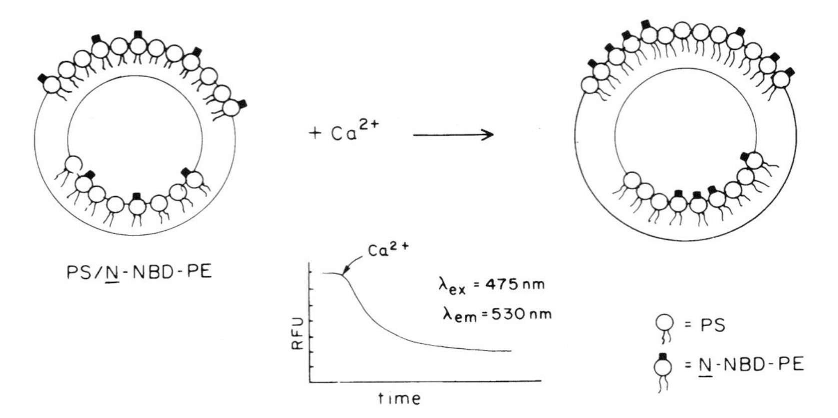

Figure 10.6 Phase separation in membranes, monitored by fluorescence self-quenching. When Ca^{2+} is added to *N*-NBD-PE (5 mol %)-containing PS liposomes, phase separation occurs, leading to a local increase of the concentration of the fluorescent lipid analogue, with a concomitant fluorescence self-quenching. This self-quenching can be continuously monitored in a fluorometer as a decrease in the NBD fluorescence signal.

$$P = \frac{I_{II} - I_{\perp}}{I_{II} + I_{\perp}}$$

where I_{II} and $I_{\perp}$ represent fluorescence intensities detected with the polarizers oriented in parallel and perpendicular positions, respectively, to the direction of polarization of the excitation light. $I_{\perp}$ is corrected for the intrinsic polarization of the instrument.

A typical feature of many fluorescent probes is their concentration-dependent self-quenching. NBD-labelled lipid analogues also show this prominent behaviour which has been exploited, among others, to monitor the occurrence and kinetics of lipid phase separations in membranes (Hoekstra, 1982a; Silvius & Gagne, 1984). The principle of the assay is shown in Fig. 10.6. It involves the incorporation of an acyl (C_6-NBD) or headgroup-labelled (*N*-NBD-PE) phospholipid analogue in lipid bilayers at a concentration of approximately 5 mol % with respect to total lipid. The occurrence of phase separation leads to an increase in the apparent concentration of the fluorescent lipid analogue as it is 'squeezed out' of the separating phases. The phase separation is thus revealed as a quenching of fluorescence, the kinetics of which can be monitored continuously in a fluorometer. DPH- and pyrene-labelled phospholipid analogues can similarly be used for this purpose. Relative clustering of monomers will enhance excimer fluorescence (Kido *et al.*, 1980; Parente & Lentz, 1986). Among others, these types of experiments have been of relevance in analysing and interpreting the role of lipid-phase separations in mechanisms involved in membrane fusion of artificial membrane systems (Hoekstra, 1982b; Silvius & Gagne, 1984).

NBD-labelled lipid analogues have also been employed in studies involving polymorphic transitions of certain phospholipids (Hong *et al.*, 1988; Stubbs *et al.*, 1989). In general, the fluorescent properties of a probe are quite sensitive to the environment of the probe. For example, when *N*-NBD-PE is incorporated in a bilayer-forming phospholipid system, the headgroup protrudes into the aqueous bilayer/water interface, sensing a polar environment (see, for example, Chattopadhyay & London, 1988; Chattopadhyay, 1990). In an apolar environment – which can be simulated by determining spectral properties of the fluorescent probe in an apolar environment using defined organic solvents of distinct polarity (see, for example, Hoekstra *et al.*, 1984) – the fluorescence quantum yield commonly increases while there is a blue shift in the emission maximum (Edidin, 1981, 1992). The latter phenomenon is seen when the *N*-NBD-PE analogue is incorporated at relatively low concentrations (<1 mol %) into lipid bilayers that contain lipids which can adopt non-bilayer structures, such as a hexagonal H_{II}-phase. Thus a transition from bilayer to non-bilayer structure can be reported by such labelled lipid analogues as an increase in fluorescence, occurring around the temperature at which such a transition can take place. The underlying mechanism of these fluorescence changes involves, however, a dehydration of the lipid phase (Stubbs *et al.*, 1989). Thus, this feature is not 'typical' for an H_{II} transition and it is evident, therefore, that one cannot solely rely on shifts in quantum yield and emission wavelength maxima when determining non-bilayer lipid transitions. Thus far, the application has only been reported in artificial membrane systems.

Lipid exchange or transfer is a prominent aspect of studies involving the dynamics of lipids and membranes (Fig. 10.7). In studies of membrane fusion, such events might interfere with assays that rely on lipid mixing and, hence, should be excluded in those particular cases (see below). However, biologically relevant lipid transfer processes have been claimed, catalysed by specific and non-specific lipid exchange proteins. Fluorescent lipid probes have been elegantly applied to measure the kinetics of protein-mediated phospholipid exchange between artificial bilayers (Nichols & Pagano, 1983). For this purpose, the

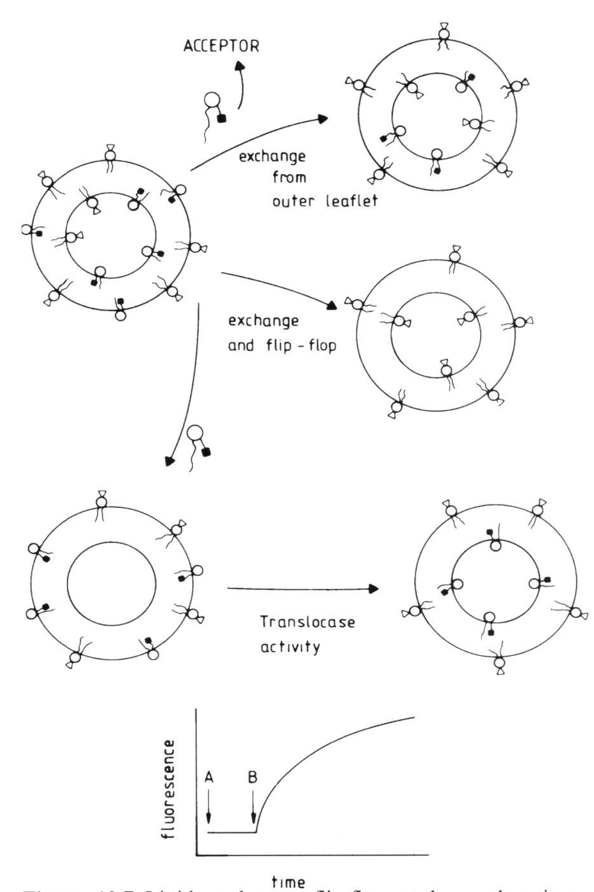

Figure 10.7 Lipid exchange, flip-flop and translocation. Starting from the upper left vesicle, which contains both a fatty-acyl labelled NBD(■)-lipid and the headgroup labelled *N*-rhodamine(△)-PE in both leaflets, the NBD-lipid can be exchanged from the outer leaflet, either spontaneously (C_6-NBD-analogues) or by an exchange protein (C_{12}-NBD-analogues). This results in the upper right situation. When the NBD-lipid can also flip-flop (for instance, C_6-NBD-ceramide) from the inner to the outer leaflet, the probe can be completely removed from the starting vesicle, provided that a sufficient amount of acceptor (membrane) is present.

In the lower part of the figure the insertion of a NBD-lipid in the outer leaflet (already containing *N*-Rh-PE) of a vesicle is depicted, followed by its translocation to the inner leaflet by a translocase activity. Insertion leads to efficient energy transfer (level A in kinetic curve).

In the case of both exchange and translocation a decrease of RET between the NBD-lipid and the *N*-Rh-PE occurs (B), that can be measured in the fluorometer as an increase of the NBD fluorescence signal.

approach was based on the principle of resonance energy transfer using *N*-Rh-PE as the energy acceptor and various NBD-labelled phospholipid species. Donor vesicles were labelled with 1 mol % each of *N*-Rh-PE and the NBD-labelled lipid of interest. In this case C_{12}-NBD lipid derivatives were used which, compared to C_6-NBD derivatives (see previous section), show a relatively low spontaneous intermembrane transfer rate, i.e. negligible on the time-scale of the exchange experiments. Upon addition of unlabelled acceptor vesicles and the specific bovine liver PC-transfer protein or the bovine liver non-specific lipid transfer protein, the occurrence and lipid specificity of transfer were monitored. This was revealed by the relief of NBD fluorescence quenching, when energy transfer with *N*-Ph-PE, occurring in the original doubly labelled bilayers, is eliminated. The latter occurred when NBD-lipid, but not *N*-Rh-PE, transferred to unlabelled acceptor membranes, catalysed by the transfer protein. With the PC-specific exchange protein, the headgroup specificity of the various NBD analogues was maintained, while the acceptor membrane was found to require PC. Furthermore, with the non-specific transfer protein the same order of lipid specificity was seen as that of natural lipids (PA>PC>PE). The authors concluded, therefore, that the data obtained with NBD-lipids are fully consistent with data obtained by other techniques (see Nichols & Pagano, 1983). Evidently, in this case applications of fluorescent lipid analogues offers the advantage over radiolabelled assays that separation of reactants and products is not required. Pyrene-labelled phospholipids have also been used in protein-mediated lipid transfer studies. Density-dependent changes in excimer fluorescence were used as a measure for lipid transfer (Massey *et al.*, 1985; Somerharju *et al.*, 1987). This procedure is less sensitive than the RET approach and requires, therefore, a higher lipid probe concentration.

In solving issues related to protein structure and function, resonance energy transfer measurements are frequently applied to determine intra- and intermolecular distances. A similar approach was recently applied in studies aimed at understanding the functioning of specific translocases in biological membranes that mediate (transbilayer) flip-flop of certain lipids (Connor & Schroit, 1987). It is assumed that phospholipids are asymmetrically distributed in biological membranes, aminophospholipids such as phosphatidylserine (PS) and phosphatidylethanolamine (PE) residing in the inner leaflets of natural membranes, while choline-containing lipids such as PC and SM are found mainly in the outer leaflet (Op den Kamp, 1979). Various experimental approaches have indeed shown that exogenously added PS and PS analogues, including spin-labelled or fluorescently tagged derivatives, translocate in an ATP-dependent and protein-mediated manner from initial site of insertion in the outer leaflet to the inner leaflet, thereby adopting an asymmetric distribution (Seigneuret & Deveaux, 1984; Daleke & Huestis, 1985; Tilley *et al.*, 1986; Zachowski *et al.*, 1987; Martin & Pagano, 1987;

Connor & Schroit, 1987). When adding C_6-NBD derivatives of PE and PS to cultured cells (fibroblasts), it can be seen that in contrast to the PC analogue, the former lipids rapidly internalize at a threshold temperature of about 7°C, i.e. at conditions where commonly operating endocytic mechanisms are not yet active (Martin & Pagano, 1987). The transbilayer movement was found to be stereospecific. These studies indicated that intracellular membranes became labelled as a result of ATP-dependent flip-flop of the exogenously supplied lipids, followed by diffusion of the lipid analogues through the cytosol between inner leaflet and other intracellular membranes. In erythrocyte membranes – in which the translocase activity has been most extensively characterized thus far (see for example, Connor & Schroit, 1988), a resonance energy transfer technique has been applied showing the occurrence of the selective translocation of exogenously supplied C_6-NBD-PS to the inner leaflet, whereas C_6-NBD-PC remained in the outer leaflet (Connor & Schroit, 1987). As energy acceptor *N*-Rh-PE was used. The principle of the approach and the interpretation of the results were based upon the rationale that a symmetric distribution of the probes should give rise to a strong energy transfer efficiency whereas relatively weak energy transfer should be seen when the probes are in opposing leaflets. These qualitative distinctions can be readily revealed by this approach, but quantitative calculations as to the exact distances between donor and acceptor lipid are, in all probability, hampered by the fact that acyl chain-labelled NBD, due to its polar character, may have a tendency to loop back to the bilayer/water interface (Chattopadhyay & London, 1988), while exogenous insertion of *N*-Rh-PE (as done in this study) may lead to a non-random and clustered distribution of the probe (Kok *et al.*, 1990). Since energy transfer depends on the concentration of the energy acceptor this may lead to an erroneous outcome. This may be particularly the case when the calculations are also based upon and compared with model systems, where probe randomization occurs when preparing the bilayer. In spite of this quantitative limitation the RET approach may give qualitative insight as further supported by several independent control experiments, described in the study of Connor and Schroit (1987). It was demonstrated that the exogenously inserted NBD-PS was inaccessible to back-exchange. Also, the lipid analogue is not derivatized by TNBS. TNBS derivatization was seen when the PS transporter activity was inhibited by using inhibitors such as sulphydryl-reactive compounds.

Finally, the occurrence of *spontaneous* flip-flop can also be demonstrated to occur, using RET. As noted above (previous section), NBD-ceramide shows a remarkable tendency to flip-flop, i.e. the analogue rapidly gains intracellular access when endocytic internalization is not operative. In model membranes, ceramide and *N*-Rh-PE were symmetrically incorporated under conditions of efficient energy transfer (usually 1 mol % each). Subsequently, a large excess of unlabelled membranes was added. A reduction in energy transfer efficiency of about 50% would be anticipated when only the outer leaflet lipid would be accessible to transfer/exchange. On the other hand, complete removal and hence flip-flop ability of the NBD-ceramide should be reflected by a complete elimination of resonance energy transfer. Indeed, the latter was observed (Pagano, 1989), indicating and confirming the high susceptibility of NBD-ceramide to engage in spontaneous flip-flop. In this regard it should be noted that spontaneous flip-flop of natural lipids is usually in the order of days. The kinetics of exchange/transfer are probably codetermined by the chemical and physical nature of the derivative, including polarity properties. In fact, the probe-dependent metabolic fate of the analogues further exemplifies the point (see, for example, Pagano & Martin, 1988).

The artificial nature of certain lipid(-like) probes and the ability to undergo flip-flop as a result of that nature, has been of relevance in defining the role of tight junctions in maintaining distinct lipid domains in polarized cells. In such cells, the lipid compositions of the apical and basolateral membranes are different, the apical membrane being particularly enriched in (glyco)sphingolipids. By adding various fluorescent lipid analogues to the apical or basolateral plasma membrane domains of polarized epithelial cells it has been shown that tight junctions prevent randomization of lipids between the two membrane domains, provided that the lipids are restricted to the outer leaflet of the plasma membrane bilayer (Dragsten *et al.*, 1981; Spiegel *et al.*, 1985; Van Meer & Simons, 1986). In contrast, lipids present in the inner leaflet of the plasma membrane are free to mix by lateral diffusion.

The experiments were carried out with confluent monolayers of epithelial cells, grown in such a way (among others, by growing the cells on filters) that it enabled selective labelling of apical membranes. Dragsten *et al.* (1981) selected four lipid(-like) probes (Fig. 10.1) that were inserted into the apical membrane by adding them to medium bathing the apical surface of a cell monolayer. Fluorescence microscopy was employed to visualize the distribution of the probes. Restriction of the probe to the apical membrane can be seen as incomplete rings or arcs of fluorescence obtained by focusing the microscopy part way up a dome of cells so that apical and basolateral surfaces of individual cells are in focus simultaneously. Thus labelling the cells with C_6-NBD-PC, *N*-hexadecanoyl amino fluorescein or rhodamine-labelled gangliosides (Spiegel *et al.*, 1985) showed a fluorescence distribution typical of probes exclusively restricted to the apical surface displaying lateral diffusion coefficients of *ca.* 10^{-8} cm^2 s^{-1}, similar to other lipid probes (Edidin, 1992).

However, both $diIC_{16}$ (a hexadecyl carbocyanine) and dodecanoyl amino fluorescein labelled the entire cell membrane, in spite of their initial insertion in the outer leaflet of the apical membrane only. This suggests their passage over the tight junction. By using agents capable of quenching the fluorescence of the lipid probe specifically it could be shown that $diIC_{16}$ and dodecanoyl amino fluorescein were capable of flipping across the apical membrane, thus gaining access to the inner leaflet of that domain. It was suggested, therefore, that the tight junction apparently presents an impassable barrier only to lipid components that are exclusively present in the outer leaflet of the plasma membrane. Lipids that have gained access to the inner leaflet diffuse freely, leading to intermixing of inner leaflet apical and basolateral lipids.

Evidently, proof for such a hypothesis would be obtained by directly inserting one and the same lipid either exclusively in the outer leaflet of the apical membrane or alternatively in either the inner or both leaflets, which would reveal the distinct differences in subsequent lipid distribution. Only in the latter case should randomization over all membrane domains occur. Such an experiment became possible when it was recognized that certain viruses can bud from specific epithelial membrane domains, either apical or basolateral (Hoekstra, 1990b; Hoekstra & Nir, 1991). The experiment (Van Meer & Simons, 1986) involved the use of cells specifically expressing the influenza fusion glycoprotein HA at the apical surface obtained by infecting the cells with influenza virus. Subsequently, lipid vesicles were prepared with a symmetric or asymmetric distribution of *N*-Rh-PE. The asymmetric localization of *N*-Rh-PE (in the outer leaflet of the liposome) was obtained when preparing the vesicles by a reconstitution procedure using octylglucoside. In addition, the vesicles contained a ganglioside acting as an anchorage site for the viral HA protein, which will also cause the vesicles to fuse with the cell surface. By this procedure *N*-Rh-PE could be inserted almost exclusively in the outer leaflet of the apical membrane (with asymmetrically labelled vesicles) or in both leaflets using the symmetrically labelled vesicles. In the former case, the lipid probe did not pass the tight junction whereas in the latter case rapid diffusion to the basolateral plasma membranes also occurred.

These results thus elegantly confirmed the results by Dragsten *et al.* (1981) that the tight junction forms an exclusive diffusion barrier in the *exoplasmic* leaflet of the plasma membrane, while lipids in the cytoplasmic leaflet appear to be capable of freely diffusing between both domains. With regard to the application of fluorescent lipid probes it should be emphasized that in this particular case physico-chemical properties determine localization of the probe rather than specific sorting and/or recognition phenomena (see previous section). When inserted in the apical membrane, any fluorescent lipid incapable of flip-flop will remain in the apical domain. Thus, the rhodamine-labelled ganglioside (Spiegel *et al.*, 1985) being restricted to the apical surface is, by itself, no proof for enrichment of glycosphingolipids in the exoplasmic apical membrane.

In the past decade, considerable progress has been made in understanding the mechanism(s) of membrane fusion, an event crucial to numerous cell biological phenomena such as endocytic and biosynthetic transport processes, the infectious entry of viruses into cells and fertilization (Ohki *et al.*, 1988; Wilschut & Hoekstra, 1990). An important aspect of these studies has been the ability to reveal the occurrence of fusion in a simple and reliable manner and to monitor this process in a continuous fashion so that parameters affecting the fusion event (thus providing insight as to how fusion is accomplished) can be readily revealed. The application of appropriate fluorescent lipid(-like) probes has played a prominent role in this respect (Düzgüneş & Bentz, 1988; Hoekstra, 1990a; Hoekstra & Klappe, 1993). The type of lipid probes that are crucial to fusion assays are those that remain associated with the membranes with which they were first associated. Thus the lipid probes of choice should be essentially non-exchangeable. This implies, therefore, that those probes that contain short-chain fatty acids (such as C_6-NBD-lipids) and which show a relatively enhanced water-solubility, are not suitable for application in assays reporting membrane fusion. Rather, typical probes for a lipid mixing assay are those that have the fluorophore coupled to lipid headgroups, such as the *N*-NBD-PE/*N*-Rh-PE couple (Struck *et al.*, 1981) or to relatively long acyl or alkyl chains (as in the lipid-like probe, octadecyl rhodamine B chloride, R_{18} (Hoekstra *et al.*, 1984) or in case of the RET couple (12-CPS)-18PC/(12-DABS)-18-PC, in which CPS[[*N*-[4-[7-(diethylamino)-4-methylcoumarin-3-yl]phenyl]carbamoyl]methyl]thio and DABS (4-(4-(dimethylamino)phenyl)sulphonyl)methyl amino are coupled to the glycerol backbone via C_{12} spacers (Silvius *et al.*, 1987). As noted above, the effective length of the acyl chain will ensure the probe's proper intercalation in and association with the membrane, as observed for natural lipids.

Application of fluorescent lipid probes in studies of membrane fusion has generally followed the principle

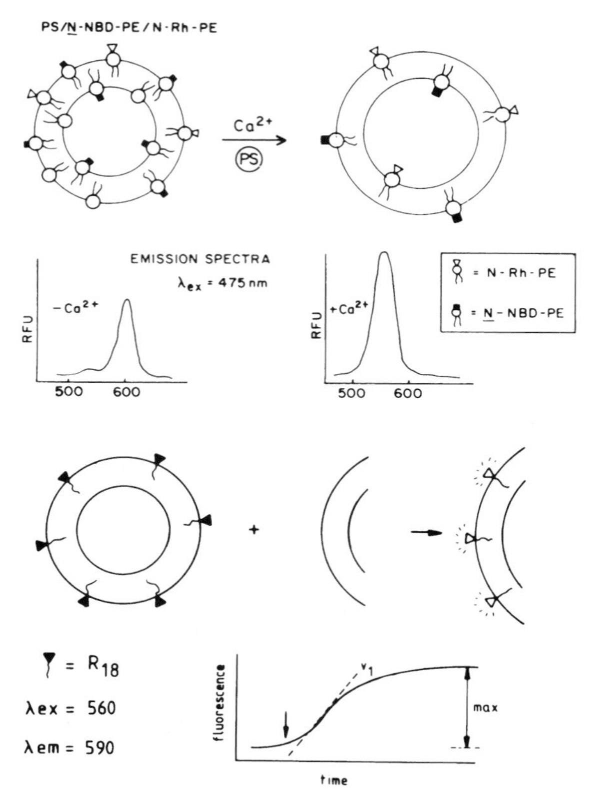

Figure 10.8 Lipid-mixing assays to monitor membrane fusion. Membrane fusion can be measured by employing the *N*-NBD-PE/*N*-Rh-PE couple. Fusion is monitored by measuring the increase in fluorescence emission of the NBD fluorophore (see emission spectra). This increase is a result of the decrease in quenching of the NBD by the rhodamine (due to resonance energy transfer) upon dilution of both probes after mixing of the lipids with those of the unlabelled fusion target membrane.

In the lower part of the figure, the principle of relief of self-quenching is depicted, which can be employed to measure fusion using R_{18} as a probe. When a vesicle containing a high (self-quenching) concentration of R_{18} fuses with an unlabelled vesicle, dilution of the R_{18} probe occurs, resulting in an increase of rhodamine fluorescence (see kinetic curve). With both assays fusion can be continuously monitored, allowing determination of parameters such as the initial fusion rate (V_1) and the extent of fusion, which are frequently used to describe the fusion process.

of monitoring the relief of fluorescence 'quenching', either as a result of resonance energy transfer or self-quenching, occurring when labelled membranes fuse with non-labelled membranes (Fig. 10.8). Over other assays, the use of fluorescence assays to monitor fusion offers a number of advantages. These include, among others, its relatively high sensitivity, its versatility and convenience, its ability to detect and monitor the onset and kinetics of the fusion process, and the relative ease with which quantitative data are required. In addition, in cases in which fusion of biological membranes is involved, it is usually possible to examine the fusion event by fluorescence microscopy.

Pyrene-labelled lipid probes have been widely used to study various aspects of membrane dynamics, including membrane fusion (Schenkman *et al.*, 1981; Pal *et al.*, 1988). The probe is usually attached to the acyl chain (of at least 16 carbons in length, to avoid spontaneous transfer), thus substituting for one of the lipid's fatty acids. Since insertion as such in a biological membrane is difficult to accomplish, the derivatized lipid is commonly used in a semi-artificial system, involving the interaction of artificial bilayers with biological membranes or in a fusion model system consisting of reconstituted biological membranes, as, for example, in the case of reconstituted viral envelopes (Anselem *et al.*, 1986; Hoekstra, 1990b). A particular advantage of pyrene-fatty acid probes is, however, that they can become metabolically incorporated into 'natural' phospholipids. This approach has been taken to study the fusion properties of an enveloped virus (vesicular stomatitis virus), which infects a cell by fusion with intracellular membranes after receptor-mediated endocytic internalization (Pal *et al.*, 1988). The principle of measuring fusion with pyrene-labelled lipids involves the (discontinuous) determination of changes in the so-called excimer/monomer fluorescence intensity ratio, occurring when pyrene-labelled membranes fuse, and thus dilute, with non-labelled membranes (cf. Hoekstra & Klappe, 1993). The fluorescence emission spectrum of pyrene is sensitive to the concentration of the probe. At relatively high (5–10 mol %) concentrations, the emission fluorescence shows a typical 'excimer' emission at 470 nm (excitation wavelength is 330 nm). The excimeric state of excited pyrene dimers reduces in number when the concentration decreases (Pownall & Smith, 1989), occurring when labelled membranes fuse with non-labelled membranes. This reduction is correlated to the emitted fluorescence derived from monomers, which is detected at an emission wavelength of 385 nm. Since the *E*/*M* ratio is linearly dependent on the concentration of probe up to *ca.* 7 mol %, the change in ratio can be directly correlated to the extent of fusion.

The fluorescent lipid-like probe octadecyl rhodamine B chloride (R_{18}) is extensively used to study fusion between intact biological membranes, in particular the fusion between viruses and cells (Fig. 10.8). The principle relies on the surface density-dependent self-quenching properties of the dye, implying that the fluorescence intensity increases when the density of the probe decreases and vice versa (Hoekstra *et al.*, 1984). The approach involves the labelling of one membrane preparation at an R_{18} concentration such that fluorescence self-quenching occurs. Fusion is then monitored continuously, at excitation and emission wavelengths of 560 and 590 nm, respectively, in a fluorometer. Essentially, the increase in rhodamine fluorescence is followed, occurring when the labelled membrane merges with the unlabelled membrane.

In practice, the procedure can be carried out as follows. Membrane labelling is done by rapidly injecting an ethanolic solution (final concentration <1% (v/v)) of R_{18} in the membrane mixture, while vigorously vortexing. The suspension is left in the dark for approximately 20–30 min, although shorter incubation periods suffice. Non-incorporated probe – upon labelling of Sendai virus some 20–30% – is removed by gel filtration (Sephadex G75, 1×15 cm), and adsorbs on top of a column. The final concentration of the probe in the membrane that is labelled should be 6–7 mol %, so as to obtain a self-quenching of approximately 70–80%. This is determined by adding Triton X-100 (1% (v/v)) which dilutes the probe infinitely and does not affect its quantum yield. The detergent is also used to calibrate the fluorescence scale by setting the fluorescence of the labelled membrane preparation at the 'zero' level (no fusion) while 100% is obtained after addition of Triton X-100, and a correction for sample dilution. Fluorescence dequenching is thus used as a reflection of the occurrence of fusion, while the rate of dequenching corresponds to the rate of fusion. In this respect, it is important to establish that probe redistribution is not rate-limiting, i.e. the kinetics of dequenching should accurately reflect the kinetics of the fusion reaction. In spite of the *assumption* that probe mixing is very fast and not rate-limiting in the overall fusion event, the reasonability of this assumption has only been proven in a few cases (Rubin & Chen, 1990). Thus, redistribution of R_{18} is not rate-limiting in virus–cell fusion but is close to rate-limiting in viral spike–protein-mediated cell–cell fusion.

Apart from fluorometric measurements, appropriate fluorescent probes also offer the possibility to study fusion events by advanced fluorescence microscopic techniques. Recently, a technique has been developed that employs digital fluorescence imaging to detect and quantify fusion of viruses with cells by sequential imaging obtained at low virus/cell ratios (Georghiu *et al.*, 1989). In this case, single cells can be analysed in contrast to an average response of an entire cell population, as obtained by fluorometric measurements. Single cell analysis has also been carried out by application of a technique involving fluorescence recovery after photobleaching (Aroeti & Henis, 1986). In the latter case the lateral mobility of a fluorescent derivative such as *N*-NBD-PE has been monitored after interaction of reconstituted viral envelopes, containing the probe, with cells. The occurrence of fusion is then determined by measuring the mobile fraction of the cell-associated fluorophore. This is determined by following the recovery of fluorescence after bleaching of a relatively small spot on the cell surface with a high-energy laser beam. Note that this approach is similar to that used to measure diffusion rate constants for membrane-inserted lipid probes (see Section 10.2.2).

In principle, the *N*-NBD-PE, as an energy donor, can also be employed in conjunction with an energy acceptor, such as *N*-Rh-PE to measure fusion by resonance energy transfer, at probe concentrations low enough (usually less than 1 mol % with respect to total lipid) to avoid self-quenching (Fig. 10.8). The principle of resonance energy transfer has been described above. Numerous combinations of probes fulfilling the typical features of an overlap between donor emission spectrum and excitation spectrum of the acceptor have been employed for the purpose of determining and monitoring fusion (see, for example, Gibson & Loew, 1979; Vanderwerf & Ullman, 1980; Uster & Deamer, 1981; Loew, 1988; Düzgüneş & Bentz, 1988). Usually, the relief of energy transfer is followed by measuring the increase in donor fluorescence as a function of time (Fig. 10.8), occurring when labelled membranes fuse with unlabelled target membranes.

Assays based on resonance energy transfer have been applied primarily in fusion involving artificial membranes. This is due to the fact that spontaneous insertion of the RET probes in a manner suitable for efficient energy transfer is difficult to accomplish by exogenous addition of the probes, as would be required for incorporation in biological membranes. However, liposomes or reconstituted biological membranes allow rapid incorporation, since the fluorescent lipid probes are readily incorporated during the procedure of preparing these membrane preparations (see, for example, Hoekstra & Klappe, 1993).

Finally, as already noted above, particularly in studies involving the use of fluorescent lipid probes as reporters of membrane fusion, it is essential to exclude that lipid mixing has been accomplished by spontaneous transfer or flip-flop of the *probes* between membranes. It is therefore crucial that with any probe or combination of lipid probes used for this purpose and in the system in which fusion is studied, appropriate control experiments are carried out. This can be done, for example, by comparing different assays in the same system or by using a particular assay in a system at conditions where fusion is known not to occur as by using specific inhibitors or by proteolytically cleaving proteins as in the case of viruses. Provided that such controls are carried out properly, these assays can be valuable tools for analysing mechanisms of membrane fusion, adding another versatile aspect to employment of fluorescent lipid probes in the (cell) biology of membranes.

ACKNOWLEDGEMENTS

Work cited in this paper and carried out in the authors' laboratory was supported by NIH Grant AI 255534 and by The Netherlands Organization for Scientific Research (NWO/SON). The secretarial assistance of Mrs Rinske Kuperus is gratefully acknowledged.

REFERENCES

Acquotti D., Sonnino S., Masserini M., Casella L., Fronza G. & Tettamanti G. (1986) *Chem. Phys. Lip.* **40**, 71–86.
Amselem S., Barenholz Y., Loyter A., Nir S. & Lichtenberg D. (1986) *Biochim. Biophys. Acta* **860**, 301–313.
Aroeti B. & Henis Y.I. (1986) *Biochemistry* **25**, 4588–4596.
Axelrod D., Koppel D.E., Schlessinger J., Elson E. & Webb W.W. (1976) *Biophys. J.* **16**, 1055–1069.
Bligh E.G. & Dyer W.J. (1959) *Can. J. Biochem. Physiol.* **37**, 911–917.
Bronner C., Landry Y., Fonteneau P. & Kuhry J.G. (1986) *Biochemistry* **25**, 2149–2154.
Chattopadhyay A. (1990) *Chem. Phys. Lip.* **53**, 1–15.
Chattopadhyay A. & London E. (1988) *Biochim. Biophys. Acta* **938**, 24–34.
Chiechanover A., Schwartz A.L., Dautry-Varsat A. & Lodish H.F. (1983) *J. Biol. Chem.* **258**, 9681–9689.
Cole L., Coleman J., Evans D. & Hawes C. (1990) *J. Cell. Sci.* **96**, 721–730.
Comfurius P., Bevers E.M. & Zwaal F.A. (1990) *J. Lipid Res.* **3**, 1719–1721.
Connor J. & Schroit A.J. (1987) *Biochemistry* **26**, 5099–5105.
Connor J. & Schroit A.J. (1988) *Bochemistry* **27**, 848–851.
Crawford J.M., Vinter D.W. & Gollan J.L. (1991) *Am. J. Physiol.* **260**, G119–G132.
Daleke D.L. & Huestis W.H. (1985) *Biochemistry* **24**, 5406–5416.
Dragsten P.R., Blumenthal R. & Handler J.S. (1981) *Nature* **294**, 718–722.
Dunn K.W. & Maxfield F.R. (1990) In *Noninvasive Techniques in Cell Biology*, J.K. Foskett & S. Grinstein (eds). Wiley-Liss, New York, pp. 153–176.
Düzgüneş N. & Bentz J. (1988) In *Spectroscopic Membrane Probes*, L.D. Loew (ed.). CRC Press, Boca Raton, pp. 117–159.
Edidin M. (1981) In *New Comprehensive Biochemistry*, Vol. I, A. Neuberger & L.L.M. van Deenen (eds). Elsevier, Amsterdam, pp. 37–82.
Edidin M. (1992) In *The Structure of Biological Membranes*, P.L. Yeagle (ed.). CRC Press, Boca Raton, pp. 539–572.
Foster T. (1948) *Ann. Phys. (Leipzig)* **2**, 55–75.
Frank A., Bazenholz Y., Lichtenberg D. & Thompson T.E. (1983) *Biochemistry* **22**, 5647–5651.
Fung B.K.-K. & Stryer L. (1978) *Biochemistry* **17**, 5241–5248.
Gardam M.A., Itovitch J.J. & Silvius J.R. (1989) *Biochemistry* **28**, 884–893.
Gautam A., Ng O.-C. & Boyer J.L. (1987) *Hepatology* **7**, 216–223.
Georghiu G., Morrison I.E.G. & Cherry R.J. (1989) *FEBS Lett.* **250**, 487–492.
Gibson G.A. & Loew L.M. (1979) *Biochem. Biophys. Res. Commun.* **88**, 135–140.
Goda S., Kobayashi T. & Goto I. (1987) *Biochim. Biophys. Acta* **920**, 259–264.
Gordesky S.E., Marinetti G.V. & Love R. (1975) *J. Membr. Biol.* **20**, 111–132.
Hakomori S.-I. (1981) *Ann. Rev. Biochem.* **50**, 733–764.
Hakomori S.-I. (1984) *Trends Biochem. Sci.* **9**, 453–458.
Hoekstra D. (1982a) *Biochemistry* **21**, 1055–1061.
Hoekstra D. (1982b) *Biochemistry* **21**, 2833–2840.
Hoekstra D. (1990a) *Hepatology* **12**, 615–665.
Hoekstra D. (1990b) *J. Bioenerg. Biomembr.* **22**, 121–155.
Hoekstra D. & Klappe K. (1993) *Methods Enzymol.* (in press).
Hoekstra D. & Nir S. (1991) In *The Structure of Biological Membranes*, P.L. Yeagle (ed.), CRC Press, Boca Raton, pp. 949–996.
Hoekstra D., De Boer T., Klappe K. & Wilschut J. (1984) *Biochemistry* **23**, 5675–5681.
Hoekstra D., Klappe K., Stegmann T. & Nir S. (1988) In *Molecular Mechanisms of Membrane Fusion*, S. Ohki, D. Doyle, S.W. Hui & E. Mayhew (eds). Plenum Press, New York, pp. 399–412.
Hoekstra D., Eskelinen S. & Kok, J.W. (1989) In *Organelles in Eukaryotic Cells. Molecular Structure and Interactions*, J.M. Tager, A. Azzi, S. Papa & F. Guerreri (eds). Plenum Press, New York, pp. 59–83.
Hong K., Baldwin P.A., Allen T.M. & Papahadjopoulos D. (1988) *Biochemistry* **27**, 3947–3955.
Jacobsen K., Derzko Z., Wu E.S., Hou Y. & Poste G. (1976) *J. Supramol. Struct.* **5**, 565–576.
Keller P.M., Person S. & Snipes W. (1977) *J. Cell Sci.* **28**, 167–177.
Kido N., Tanaka F., Kaneda N. & Yagi K. (1980) *Biochim. Biophys. Acta* **603**, 255–265.
Kishimoto Y. (1975) *Chem. Phys. Lip.* **15**, 33–36.
Kobayashi T. & Pagano R.E. (1989) *J. Biol. Chem.* **264**, 5966–5973.
Kok J.W., Eskelinen S., Hoekstra K. & Hoekstra D. (1989) *Proc. Natl. Acad. Sci. USA* **86**, 9896–9900.
Kok J.W., ter Beest M., Scherphof G. & Hoekstra D. (1990) *Eur. J. Cell Biol.* **53**, 173–184.
Kok J.W., Babia T. & Hoekstra D. (1991) *J. Cell Biol.* **114**, 231–239.
Koval M. & Pagano R.E. (1989) *J. Cell Biol.* **108**, 2169–2181.
Koval M. & Pagano R.E. (1990) *J. Cell Biol.* **111**, 429–442.
Koval M. & Pagano R.E. (1991) *Biochim. Biophys. Acta* **1082**, 113–125.
Kundu S.K. (1981) *Meth. Enzymol.* **72**, 185–204.
Laskowicz J.R. (ed.) (1983) *Principles of Fluorescence Spectroscopy.* Plenum Press, New York.
Lipsky N.G & Pagano R.E. (1982) *Proc. Natl. Acad. Sci. USA* **80**, 2608–2612.
Lipsky N.G. & Pagano R.E. (1985a) *J. Cell Biol.* **100**, 27–34.
Lipsky N.G. & Pagano R.E. (1985b) *Science* **228**, 745–747.
Loew L.M. (ed.) (1988) *Spectroscopic Membrane Probes*, Vol. 1-III, CRC Press, Boca Raton.
McLean I.W. & Nakane P.K. (1974) *J. Histochem. Cytochem.* **22**, 1077–1083.
Martin O.C. & Pagano R.E. (1987) *J. Biol. Chem.* **262**, 5890–5898.
Masserini M., Giuliani A., Palestini P., Acquotti D., Pitto M., Chigorno V. && Tettamanti G. (1990) *Biochemistry* **29**, 697–701.
Massey J.B., Hickson-Bick D., Via D.P., Gotto Jr. A.M. & Pownall H.J. (1985) *Biochim. Biophys. Acta* **835**, 124–131.

Miller-Podraza H. & Fishman P.H. (1982) *Biochemistry* **21**, 3265–3270.
Moreau R.A. (1989) *Lipids* **24**, 691–699.
Mukaiyama T., Matsueda R. & Suzuki M. (1970) *Tetrahedron Lett.* **22**, 1901–1904.
Nichols J.W. & Pagano R.E. (1983) *J. Biol. Chem.* **258**, 5368–5371.
Ohki S., Doyle D., Flanagan T., Hui S. & Mayhew E. (eds) (1988) *Molecular Mechanisms of Membrane Fusion.* Plenum Press, New York.
Op den Kamp J.A.F. (1979) *Ann. Rev. Biochem.* **48**, 47–71.
Pagano R.E. (1989) *Methods Cell Biol.* **29**, 75–85.
Pagano R.E. & Longmuir K.J. (1985) *J. Biol. Chem.* **260**, 1909–1916.
Pagano R.E. & Martin O.C. (1988) *Biochemistry* **29**, 4439–4445.
Pagano R.E. & Sleight R.G. (1985) *Science* **229**, 1051–1057.
Pagano R.E., Martin O.C., Schroit A.J. & Struck D.K. (1981) *Biochemistry* **20**, 4920–4927.
Pagano R.E., Longmuir K.J. & Martin O.C. (1983) *J. Biol. Chem.* **258**, 2034–2040.
Pagano R.E., Sepanski M.A. & Martin O.C. (1989) *J. Cell Biol.* **109**, 2067–2079.
Pagano R.E., Martin O.C., Kang H.C. & Haughland R.P. (1991) *J. Cell Biol.* **113**, 1267–1279.
Pal R., Barenholz Y. & Wagner R.R. (1988) *Biochemistry* **27**, 30–36.
Parente R.A. & Lentz B.R. (1985) *Biochemistry* **24**, 6178–6185.
Parente R.A. & Lentz B.R. (1986) *Biochemistry* **25**, 1021–1026.
Pownall H.J. & Smith L.C. (1989) *Chem. Phys. Lip.* **50**, 191–211.
Prendergast F.G., Haughland R.P. & Callahan P.J. (1981) *Biochemistry* **20**, 7333–7338.
Preston R.A., Murphy R.F. & Jones E.W. (1987) *J. Cell Biol.* **105**, 1981–1987.
Raviv Y., Salomon Y., Gitler C. & Bercovici T. (1987) *Proc. Natl. Acad. Sci. USA* **84**, 6103–6107.
Rosenwald A.G., Pagano R.E. & Raviv Y. (1991) *J. Biol. Chem.* **266**, 9814–9821.
Rubin R.J. & Chen Y. (1990) *Biophys. J.* **58**, 1157–1167.
Saito M., Saito M. & Rosenberg A. (1984) *Biochemistry* **23**, 1043–1046.
Schenkman S., Aranjo P.S., Dijkman R., Quina F.H. & Chaimovich H. (1981) *Biochim. Biophys. Acta* **649**, 633–641.
Schwarzmann G. & Sandhoff K. (1987) *Methods Enzymol.* **138**, 319–341.
Schwarzmann G. & Sandhoff K. (1990) *Biochemistry* **29**, 10865–10871.
Seigneuret M. & Devaux P.F. (1984) *Proc. Natl. Acad. Sci. USA* **81**, 3751–3755.
Shinitzky M. & Barenholz Y. (1978) *Biochim. Biophys. Acta* **515**, 367–394.
Silvius J.R. & Gagne J. (1984) *Biochemistry* **23**, 3241–3247.
Silvius J.R., Leventis R., Brown P.M. & Zuckermann M. (1987) *Biochemistry* **26**, 4279–4287.
Simons K. & Van Meer G. (1988) *Biochemistry* **27**, 6197–6202.
Sleight R.G. (1987) *Ann. Rev. Physiol.* **49**, 193–208.
Sleight R.G. & Pagano R.E. (1984) *J. Cell Biol.* **99**, 742–751.
Sleight R.G. & Pagano R.E. (1985) *J. Biol. Chem.* **260**, 1146–1154.
Somerharju P.J., Van Loon D. & Wirtz K.W.A. (1987) *Biochemistry* **26**, 7193–7199.
Sonnino S., Acquotti D., Riboni L., Giuliani A., Kirschner G. & Tettamanti G. (1986) *Chem. Phys. Lip.* **42**, 3–26.
Sonnino S., Chigorno V., Acquotti D., Pitto M., Kirschner G. & Tettamanti G. (1989) *Biochemistry* **28**, 77–84.
Spiegel S., Blumenthal R., Fishman P.H. & Handler J.S. (1985) *Biochim. Biophys. Acta* **821**, 310–318.
Struck D.K. & Pagano R.E. (1980) *J. Biol. Chem.* **255**, 5405–5410.
Struck D.K., Hoekstra D. & Pagano R.E. (1981) *Biochemistry* **20**, 4093–4099.
Stryer L. (1978) *Ann. Rev. Biochem.* **47**, 819–832.
Stubbs C.D., Williams B.W., Boni L.T., Hoek J.B., Taraschi T.F. & Rubin E. (1989) *Biochim. Biophys. Acta* **986**, 89–96.
Swanson J. (1989) *Methods Cell Biol.* **29**, 137–151.
Tilley L., Cribier S., Roelofsen B., Op den Kamp J.A.F. & Van Deenen L.L.M. (1986) *FEBS Lett.* **194**, 21–27.
Trinchera M., Ghidoni R., Sonnino S. & Tettamanti G. (1990) *Biochem. J.* **270**, 815–820.
Uster P.S. & Deamer D.W. (1981) *Arch. Biochem. Biophys.* **209**, 385–395.
Uster P.S. & Pagano R.E. (1986) *J. Cell Biol.* **103**, 1221–1234.
Vanderwerf P. & Ullman E.F. (1980) *Biochim. Biophys. Acta* **596**, 302–314.
Van Meer G. (1989) *Ann. Rev. Cell Biol.* **5**, 247–275.
Van Meer G. & Simons K. (1986) *EMBO J.* **5**, 1455–1464.
Van Meer G., Stelzer E.H.K., Wijnaendts-Van-Resandt R.W. & Simons K. (1987) *J. Cell Biol.* **105**, 1623–1635.
Van Renswoude J., Bridges K.R., Hartford J.B. & Klausner R.D. (1982) *Proc. Natl. Acad. Sci. USA* **79**, 6186–6190.
Van't Hof W. & Van Meer G. (1990) *J. Cell Biol.* **111**, 977–986.
Verkleij A.J., Zaal R.F.A., Roelofsen B., Comfurius P., Kastelijn D. & Van Deenen L.L.M. (1973) *Biochim. Biophys. Acta* **323**, 178–193.
Wiegandt H. (1985) *New Comprehensive Biochemistry*, Vol. 10. Elsevier, Amsterdam.
Wilschut J. & Hoekstra D. (eds) (1990) *Membrane Fusion.* Marcel Dekker, New York.
Zachowski A., Herrmann A., Paraf A. & Devaux P.F. (1987) *Biochim. Biophys. Acta* **897**, 197–200.

CHAPTER ELEVEN

Optical Probes for Cyclic AMP

STEPHEN R. ADAMS,[1] BRIAN J. BACSKAI,[1] SUSAN S. TAYLOR[2] AND ROGER Y. TSIEN[1]

[1] Howard Hughes Medical Institute 0647, University of California San Diego, La Jolla, CA, USA

[2] Department of Chemistry 0654, University of California San Diego, La Jolla, CA, USA

11.1 RATIONALE FOR CREATING OPTICAL PROBES FOR CYCLIC AMP

Cyclic 3′,5′-adenosine monophosphate (usually abbreviated to cyclic AMP or cAMP) was the first molecule to be explicitly recognized as a 'second messenger', or mediator between extracellular stimuli ('first messengers') and intracellular biochemical responses (Rall & Sutherland, 1961; Robison *et al.*, 1971). For several years after the initial delineation of its role in transducing hormone signals, the involvement of cAMP in nearly every imaginable example of signal transduction was hypothesized and tested. Eventually it became clear that cAMP is only one member, albeit one of the most important, of a variety of second messengers and transduction pathways.

The longest-established roles of cAMP, mediating the immediate intracellular actions of numerous peptide hormones and transmitter substances, are still of tremendous clinical and pharmacological importance. In addition, new roles for cAMP under intense investigation include vertebrate olfaction (Nakamura & Gold, 1987), invertebrate neuronal plasticity (Schacher *et al.*, 1988), signalling in yeast (*Saccharomyces*; Broach & Deschenes, 1990) and slime moulds (*Dictyostelium*; Van Haastert, 1991), and the control of gene expression (Karin, 1989). Also, rapid progress continues to be made in the biochemistry and molecular biology of the proteins involved in the cAMP signalling cascade, e.g. β-adrenergic receptors (O'Dowd *et al.*, 1989), G-proteins (Bourne *et al.*, 1991), adenylyl cyclase (Gilman, 1990), cyclic-AMP-dependent protein kinase (A-kinase; Taylor *et al.*, 1990a,b), cAMP phosphodiesterases (Beavo & Reifsnyder, 1990), and kinase substrates such as the cAMP-response-element binding protein (CREB) mediating gene activation (Montminy & Bilezikjian, 1987; Meinkoth *et al.*, 1991).

However, one area where cAMP research has still been lagging is the detailed exploration of the spatial and temporal dynamics of cAMP signalling. This gap is particularly noticeable in comparison with the study of intracellular Ca^{2+}, the only other second messenger known to be of comparable ubiquity and importance to cAMP. Optical probes for Ca^{2+} have revealed that Ca^{2+} signalling is often remarkably complex in fine structures such as heterogeneity of nominally identical neighbouring cells, spatial gradients of cytosolic $[Ca^{2+}]_i$

FLUORESCENT AND LUMINESCENT PROBES, 2ND EDN
ISBN 0–12–447836–0

in individual cells, and temporal oscillations in response to steady stimuli (Tsien & Poenie, 1986; Berridge *et al.*, 1988; Tsien & Tsien, 1990; Meyer & Stryer, 1991). Their probable function is to enable ensembles of cells to generate multivariate or vectorial responses, which are too complex to be coded by a single, slowly varying, spatially uniform scalar concentration of a messenger substance. Somewhat analogous intracellular compartmentation of cAMP has long been suspected on indirect grounds such as discrepancies between the total cAMP, measured after destruction of the tissue, and physiological functions believed to be controlled by cAMP (Terasaki & Brooker, 1977; Hayes & Brunton, 1982; Greenberg *et al.*, 1987; Bode & Brunton, 1988; Aass *et al.*, 1988; Akil & Fisher, 1989; Murray *et al.*, 1989). The development of non-destructive optical imaging of cAMP in living cells would at long last enable direct exploration of the spatio-temporal intricacies of cAMP signalling. A further benefit would be the ability to measure cAMP in single cells, particularly valuable when the cells are scarce or embedded amongst other cell types or when the cell's electrophysiological behaviour is being simultaneously recorded.

11.2 PREVIOUS METHODS FOR MEASURING cAMP OR IMAGING RELATED MOLECULES

11.2.1 Assay in cell lysates

By far the most common existing method for measuring cAMP is to lyse the tissue by acid quenching, freezing, or other destructive methods, then to measure cAMP in the soluble supernatant by various binding or enzymatic assays, the most popular of which is radioimmunoassay. Because these methods are well-established and described (Brooker *et al.*, 1979; Brooker, 1988), they will not be further discussed here, except to note that total cAMP is measured, any compartmentation (Hayes & Brunton, 1982; Murray *et al.*, 1989) will be overriden, a significant number of cells must be destroyed for each time point, and the amount of tissue needs to be measured (typically as milligrams of protein) if the amount of cAMP (typically measured as picomoles) is to be normalized into concentration-type units. Thus such assays are analogous to measuring cellular Ca^{2+} by atomic absorption spectrophotometry or $^{45}Ca^{2+}$ equilibration.

11.2.2 Immunocytochemistry

Attempts have been made to use immunocytochemistry to localize cAMP in fixed or frozen tissues (Cumming, 1981). A basic problem is that cAMP should be reasonably diffusable, so that it is vulnerable to smearing or leaching from the tissue during the histological processing. These steps must involve permeabilization of the membranes and introduction of antibodies, so that relatively few positive results have been obtainable. Recently Barsony and Marx (1990) have reported that microwave fixation under very specific conditions traps cAMP immunoreactivity in interesting spatial patterns. Since it is unknown how microwaves could cross-link cAMP to tissue macromolecules and what sorts of cross-links would still allow recognition by the antibody, it is difficult to judge whether the cAMP distributions reflect the true pattern before fixation or whether they are somehow created or accentuated by fixation.

Considerably more success has been reported for immunocytochemistry of the macromolecules related to cAMP signalling, particularly of A-kinase subunits, since they are conventional proteins with respect to fixation. Detection can be accomplished either with antibodies or with exogenous fluorescein-labelled catalytic subunits (to detect free regulatory subunits; Fletcher *et al.*, 1988) or fluorescein-labelled protein kinase inhibitor (to detect free catalytic subunits; Byus & Fletcher, 1988). Several reviews may be consulted for further information (Cumming, 1981; Lohmann & Walter, 1984; Nigg, 1990).

11.2.3 Fluorescent analogues of cyclic nucleotides

Several fluorescent analogues of cyclic AMP have been synthesized, for example, 1,N^6-etheno-cAMP and 2′-*O*-(*N*-methylanthraniloyl)-cAMP (Hiratsuka, 1982). The latter contains an environmentally sensitive fluorophore and is about 20% as effective as unmodified cAMP as a substrate for beef heart phosphodiesterase. Despite such utility as substrate analogues for *in vitro* biochemical studies, it is not clear how such derivatives could be used to monitor endogenous unlabelled cAMP.

11.2.4 Fluorescently labelled cAMP receptor protein from *E. coli*

The cAMP receptor protein (CRP) from *E. coli* is a transcription factor that directly binds cAMP. Wu *et al.* (1974) found that CRP could be labelled with either of two environmentally sensitive fluorophores, acetamidoethylnapthalenesulphonate (AENS) or dansyl. The fluorescences of the resulting conjugates were somewhat sensitive to the binding of cAMP, which increased the emission intensity of the AENS label by 30%, whereas the dansyl derivative showed a 10%

decrease. Meanwhile, the emission wavelengths were hardly changed by cAMP. The apparent dissociation constant of the AENS derivative for cAMP was 10 μM, which is considerably higher than the concentrations for half-maximal activation of A-kinase, the mammalian cAMP-sensor. The AENS-labelled protein was used for temperature-jump studies of the rate of reaction with cAMP (Wu & Wu, 1974); no attempt was reported to try it in intact cells.

With modern imaging equipment and perhaps better labels, CRP might justify re-examination as the basis for a fluorescent probe for cAMP, especially for the upper range of cAMP concentrations. Unfortunately, there is not yet enough information to predict rationally what fluorescence change would be produced by cAMP, especially how one might generate wavelength shifts or dual-wavelength ratio changes, which are much more reliably quantified than mere intensity changes (Tsien & Poenie, 1986; Bright *et al.*, 1989; also see other chapters in this book).

11.3 ALTERNATIVE cAMP BINDING SITES

Before discussing A-kinase holoenzyme, the basis for our current cAMP probe, it is worthwhile discussing yet other sources of cAMP binding sites that might also be made into optical indicators.

11.3.1 Abiotic, totally synthesized receptors

Recently chemists have made remarkable progress in designing totally artificial binding sites for nucleotides, including cAMP (Deslongchamps *et al.*, 1992). A carbazole ring system provides π-stacking interaction with the adenine, bicyclic imides provide Watson–Crick and Hoogsteen basepairing, and a guanidinium group is poised to ion-pair with the phosphate. At present, selectivity for cAMP over cGMP is only modest, though this would normally not pose a biological problem since usually cAMP considerably exceeds cGMP in concentration. However, binding was assessed only as the ability to extract cAMP from an 0.11 M aqueous solution into dichloromethane. Therefore, it is unclear what affinity might be observed in aqueous media at more physiological concentrations of cAMP and whether any optical changes can be coupled to binding. Nevertheless, this approach offers much promise, since it offers about the only approach to probes that would not need microinjection. Also, even without optical sensitivity, artificial cAMP binding molecules could be very valuable as buffers or antagonists of cAMP signalling.

11.3.2 Regulatory subunit of A-kinase

Because the cAMP binding sites for A-kinase are on the regulatory (R) subunit, the latter would be an obvious target of labelling. However, there are at least three theoretical objections to using labelled R alone as a cAMP probe: (1) In the absence of the catalytic (C) subunit, the affinity of R for cAMP is extremely high because the dissociation rate becomes kinetically very slow (Døskeland, 1978; Rannels & Corbin, 1980). Therefore, exogenous R subunits in excess of C could well remain saturated with cAMP even at basal intracellular levels at which the endogenous holoenzyme would be largely inactive. (2) If under basal conditions the R subunits were *not* saturated with cAMP, they would tend to compete for any increase in cAMP and thereby inhibit the very transduction cascade being monitored. (3) Just as with bacterial CRP, there is no rational way to design a usable fluorescence change upon cAMP binding; one would simply have to rely on random trials of different labels.

11.3.3 *Dictyostelium* receptors for cAMP

The slime mould *Dictyostelium discoideum* uses extracellular cAMP signals to trigger aggregation and a new developmental programme during nutritional starvation (recently reviewed by Van Haastert, 1991). The cAMP concentrations needed for activation probably overlap those used intracellularly in animal cells. Unfortunately, the *Dictyostelium* cAMP receptors are integral membrane proteins belonging to the family with seven transmembrane helical domains and which act through G-proteins (Klein *et al.*, 1988). Therefore, it would be very difficult to get soluble injectable protein. Genetic incorporation would be useless unless the extracellular-facing binding site could be turned around to face the cytoplasm of the target cell.

11.4 PROPERTIES OF A-KINASE

This enzyme is the major intracellular receptor of cAMP in eukaryotic cells. The inactive holoenzyme is a tetramer consisting of two regulatory (R) subunits (present as a dimer) and two catalytic (C) subunits (Beebe & Corbin, 1986). Binding of two cAMP molecules to each R subunit results in the reversible dissociation to an R_2 dimer and free C subunits, according to the following equation:

$$\underset{\text{(inactive)}}{R_2C_2} + 4\ \text{cAMP} \rightleftharpoons R_2(\text{cAMP})_4 + \underset{\text{(active)}}{2\ C}$$

We have exploited this unusual activation mechanism for A-kinase to make a fluorescent sensor for cAMP, simply by labelling each of the R and C subunits with two different fluorophores which are capable of fluorescence resonance energy transfer (FRET) in the holoenzyme (Adams *et al.*, 1991). Upon binding cAMP, the subunits dissociate to effectively infinite distance, thus preventing FRET (Fig. 11.1).

11.4.1 Isozymes

In mammalian cells, there are two major types of A-kinase (types I and II), which are differentiated by their R subunits, R^I or R^{II}; the C subunits are identical (Beebe & Corbin, 1986; Taylor *et al.*, 1990a,b). R^I and R^{II} differ in amino acid sequence, the ability of R^{II} to undergo autophosphorylation, which may decrease its rate of recombination with C (Rangel-Aldao & Rosen, 1977; Rymond & Hofmann, 1982), and by the high-affinity binding of MgATP to type I holoenzyme. Several isoforms of each subunit have been described, which differ in their tissue distribution. α-Forms are generally more widespread, whereas the β-form is restricted to neural and reproductive tissues. Activation of the holoenzyme composed of $R^I{}_\beta$ is believed to occur at 3–7 times lower concentrations of cAMP than the $R^I{}_\alpha$-containing isoform (Cadd *et al.*, 1990). The crystal structure of the C subunit complexed with a peptide inhibitor was recently solved to 2.7 Å resolution (Knighton *et al.*, 1991a,b), revealing a bilobal ellipsoid (65 × 45 × 45 Å) with a deep dividing cleft containing the catalytic site. Both MgATP and peptide substrate bind to the cleft. No detailed structure of the R subunits or holoenzyme are yet available, although each R subunit contains a substrate-like autoinhibitor site that occupies the consensus peptide binding site in the cleft.

The apparent dissociation constant (of the subunits) for the holoenzyme (type I or II) is 0.2–0.7 nM (Hofmann, 1980); this increases by four orders of magnitude upon binding cAMP (Granot *et al.*, 1980). The intracellular concentration of A-kinase has been reported to be in the range of 0.2–2 μM in a variety of tissues, almost in the same concentration range as its activator, cAMP (Beavo *et al.*, 1974; Lohmann & Walter, 1984).

11.4.2 Sources

The traditional sources of the holoenzyme or the subunits have been bovine heart for the type II and rabbit or porcine skeletal muscle for type I (Beebe & Corbin, 1986). The two forms may be separated (and in fact were named) by their elution from DEAE-cellulose at different salt concentrations. The subunits may be isolated by cAMP-affinity chromotography, which retains the R subunits which can then be eluted with cAMP.

More recently, high-yielding *E. coli* expression systems for the recombinant subunits have been used. We use the recombinant murine C_α (Slice & Taylor, 1989) and bovine $R^I{}_\alpha$ (Saraswat *et al.*, 1986); purification of the latter involves an ammonium sulphate precipitation and elution from DEAE-cellulose. C subunit requires phosphocellulose chromatography and ammonium sulphate precipitation followed by gel-filtration. Yields range from 5 to 10 mg pure protein per litre of culture broth; each run produces tens of milligrams for approximately 1 weeks' labour. The recombinant subunits appear to be similar to

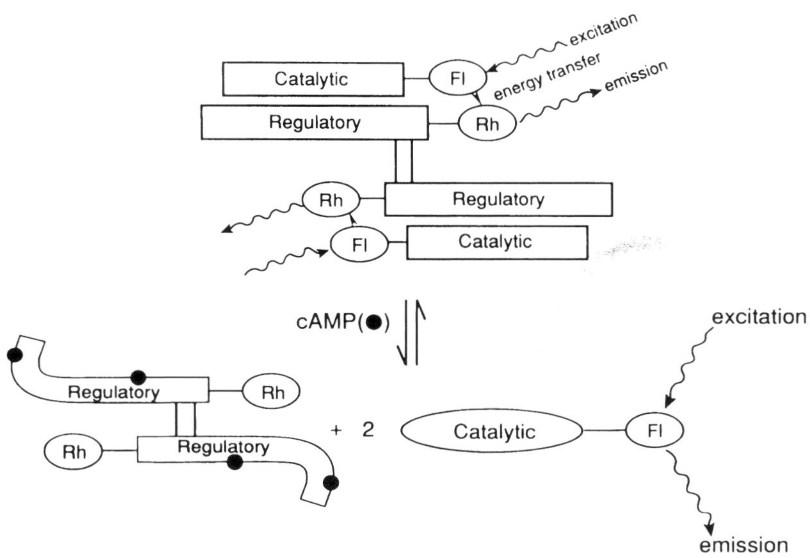

Figure 11.1 Diagram depicting the method for detecting cAMP using fluorescence resonance energy transfer (FRET) between the subunits of cAMP-dependent protein kinase (Adams *et al.*, 1991). When the cAMP concentration is low, most of the A-kinase is in the holoenzyme state, which permits FRET to occur between the fluorescein-labelled catalytic (C) subunits and the rhodamine-labelled regulatory (R) subunits. In high cAMP, the kinase is mostly dissociated and prevents energy transfer, so excitation of fluorescein results in fluorescein emission instead of rhodamine re-emission. (From Adams *et al.*, 1991, with permission.)

native enzyme in their capacity to form holoenzyme and be activated by cAMP. The recombinant C subunit lacks an *N*-myristoyl group and is slightly more sensitive to heat denaturation than the mammalian enzyme. However, an expression system that includes *N*-myristoylation has been devised (Duronio *et al.*, 1990). The purified R^{I} subunit contains two tightly bound cAMP molecules, whose removal requires strongly denaturing conditions (8 M urea). A recombinant system for expressing the $R^{II}{}_{\alpha}$ subunit has also been described (Scott *et al.*, 1990).

11.4.3 Subcellular distribution and function

Specific tissues usually contain predominantly one type of A-kinase or a mixture of both, although this may differ between species. For example, bovine heart contains predominantly type II, rat and mouse heart contain type I, and human and rabbit heart contain a mixture of both types (Beebe & Corbin, 1986). The rationale for such a variety is still obscure, although differences in sensitivity to cAMP (Cadd *et al.*, 1990), selective activation of the two types by hormones (see Beebe *et al.*, 1988 for review; Cho-Chung *et al.*, 1991; Jones *et al.*, 1991; Boshart *et al.*, 1991), and homone inducibility (Gross *et al.*, 1990) have been described. R^{II} has been reported to bind specifically to several intracellular proteins (e.g. Carr *et al.*, 1991; Lohmann *et al.*, 1988; Scott, 1991) and such binding probably localizes the kinase to specific sites. This binding, for example, may account for membrane- or cytoskeletal-associated A-kinase and influence substrate availability. The possibility that the R^{II} subunit could play a separate role from the C subunit in regulating gene transcription following an increase in intracellular cAMP levels, has been dismissed (Büchler *et al.*, 1990; reviewed by Nigg, 1990).

More recently, direct microinjection of fluorescently labelled C subunit (or holoenzyme) has directly confirmed the numerous (but often conflicting) reports (reviewed by Nigg, 1990) of translocation of the C subunit from the cytoplasm to the nucleus upon cAMP elevation. The injected holoenzyme and free R^{I} subunit appear to stay in the cytoplasm (Meinkoth *et al.*, 1990).

11.5 FLUORESCENT LABELLING OF A-KINASE

11.5.1 Fluorescence resonance energy transfer

The transfer of excitation energy of a donor fluorophore through space (non-radiatively) to a nearby acceptor chromophore or fluorophore is called fluorescence resonance energy transfer or FRET (Stryer, 1978; Lakowicz, 1983; Jovin and Arndt-Jovin, 1989; Herman, 1989). This process depends upon spectral overlap between the donor emission spectrum and the acceptor excitation (or absorbance) spectrum, the distance between the donor–acceptor pair (generally 2–7 nm), and the orientation of their respective dipoles. The efficiency of energy transfer is inversely proportional to the sixth power of the distance between donor and acceptor. FRET can be experimentally observed by a decrease in donor fluorescence in the presence of the acceptor, by sensitized acceptor re-emission when excited at the donor wavelength, and by a decrease in the fluorescence lifetime of the donor. There have been many attempts to use this phenomenon as a 'spectroscopic ruler' in biology (Stryer, 1978; Lakowicz, 1983).

11.5.2 Choice of fluorophores

Our method (Adams *et al.*, 1991) for converting A-kinase to a fluorescent cAMP sensor relies upon the reversible dissociation of the tetrameric holoenzyme to subunits upon binding cAMP. In the holoenzyme, the fluorophores are initially in close proximity. However, upon dissociation of the subunits by cAMP, the fluorophores become separated by effectively an infinite distance. Because the extent of energy transfer in the holoenzyme can be empirically determined, and the variable of interest is the percentage of holoenzyme left, it is not necessary to label unique sites on each subunit. This contrasts with the many applications of FRET in which a precise distance between two specified sites is being sought.

In general, the choice of donor–acceptor pair must at least partially fulfil certain criteria:

(1) The excitation spectrum of the acceptor should maximally overlap the emission spectrum of the donor.
(2) The fluorescent quantum yields of the donor and acceptor should be as high as possible as this results in the optimal acceptor re-emission.

More specifically for the cAMP sensor:

(3) The fluorophores should be environmentally insensitive so changes in quantum yield and emission wavelength do not occur upon minor conformational changes in the protein, or upon binding to other macromolecules within a cell. Thus any changes in fluorescence seen will result only from changes in the association and dissociation of the C and R subunits.

Our first choice of donor fluorophore was fluorescein because of its high quantum yield (0.90 in water, although this is generally reduced on protein surfaces; Tsien & Waggoner, 1990), excitation and emission at convenient visible wavelengths, availability with a variety of coupling chemistries, and relatively environmentally insensitive fluorescence. Disadvantages include a small Stokes's shift (the difference between absorption and emission wavelengths), relatively poor photostability (compared to rhodamine), and pH-sensitivity. Possible acceptor fluorophores include substituted fluoresceins, rhodamines, cyanines and phycobiliproteins. Rhodamines are the classical acceptor for fluorescein, with good photostability and spectral overlap (particularly tetramethylrhodamine or rhodamine B, but less ideal with rhodamine X or Texas Red chromophores). However, the quantum yields tend to be quenched when coupled to proteins (Tsien & Waggoner, 1990). Cyanines (Southwick *et al.*, 1990), borate-dipyrromethenes or Bodipy (Haugland, 1989) and substituted fluoresceins are more recently described alternatives which may offer some advantages. Phycobiliproteins (Glazer & Stryer, 1990), the fluorescent proteins from cyanobacteria, are highly fluorescent, show large Stokes's shifts and little photobleaching (White & Stryer, 1987) but their high molecular weight dwarfs the subunits of A-kinase and might very likely hinder holoenzyme formation and subunit diffusion in cells.

11.5.3 Coupling chemistry

The most chemically reactive groups in proteins are generally amines (from lysines and the N-terminal) and sulphydryls (cysteines). The latter are generally much less frequent in occurrence but more reactive and offer the ability to label specific sites. Of the two exposed cysteines in the C subunit, the most reactive is Cys^{199} near the active site which is essential for kinase activity. Cysteine modifications have been reviewed by Bramson *et al.* (1984). First and Taylor (1989) devised a procedure for selective labelling of the least reactive thiol, Cys^{343}, which requires temporary protection of the most reactive. The R^I subunit has no readily reactive cysteines (although R^{II} has several; Nelson & Taylor, 1983) so we therefore initially concentrated on labelling the more ubiquitous amines. Amines are far more common in proteins and hence it is harder to selectively modify one residue, but such selectivity is not a necessity for the cAMP sensor because the subunits undergo complete dissociation upon binding cAMP. Random labelling at a number of sites may yield more efficient energy transfer because of an average closer distance between the donor and acceptor fluorophores. The most-common 'amine-selective' groups are isothiocyanates, *N*-hydroxysuccinimidyl esters of carboxylic acids, sulphonyl halides and dichlorotriazinylamino (DTA) derivatives. From a chemical point of view, *N*-hydroxysuccinimidyl esters are probably the best, since they are the most selective reagents, are reactive at pH values close to 7, and form stable amides as products (Anjaneyulu & Staros, 1987). Isothiocyanates react more rapidly but reversibly with sulphydryl groups than with amines (Drobnica *et al.*, 1977), and sulphonyl halides and DTA derivatives are less reactive at lower pH (Haugland, 1989). Because of these complexities, we empirically tried a wide range of combinations of fluorophore and coupling chemistry.

11.5.4 Labelling strategy

Table 11.1 summarizes the results of our labelling attempts of A-kinase holoenzyme and subunits. All experiments exploit protection by MgATP of the catalytic subunit activity. The cAMP binding domains of the isolated R subunits (but not in the holoenzyme) are probably protected by bound cAMP. Initial attempts to label isolated C and R^I subunits with *N*-hydroxysuccinimidyl esters (OSu) were unsuccessful as holoenzyme could not be reformed. Unlabelled C could form holoenzyme with labelled R^I, suggesting we were labelling some site on C important in binding R^I. To protect these sites, we tried labelling the holoenzyme, separating the subunits by cAMP-affinity chromatography, and recombining with similarly prepared subunits labelled with a different fluorophore. Although this strategy appeared to work, it involved a moderately difficult separation and more steps than simply labelling the subunits. Interestingly, Leathers *et al.* (1990) have recently reported the successful labelling of C with carboxyfluorescein succinimidyl ester (FOSu) but only by labelling holoenzyme followed by separation. We also tried the isothiocyanates of fluorescein and tetramethylrhodamine (FITC and TRITC) and immediately found moderate energy transfer. Fortunately, these labels seemed to work as well on the isolated subunits, so that the complications of labelling the holoenzyme could be avoided. This combination of C labelled with FITC and R^I with TRITC has proven to be the most successful for energy transfer within type I holoenzyme, despite intensive testing of other fluorophores and coupling chemistries. Previous preparations of C-FITC as a cytochemical probe for R subunits (Fletcher *et al.*, 1988) stressed the need to label the intact holoenzyme and then separate subunits by DEAE-cellulose chromatography, but in our hands MgATP was sufficient to protect the C subunit.

The currently favoured labelling conditions for forming type I holoenzyme are as follows: recombinant C_α and $R^I{}_\alpha$ subunits (approximately 0.5–1.0 mg,

Table 11.1 Fluorescent labelling of cAMP-dependent protein kinase.

Fluorescent label				
Catalytic	*R^I*	*Holoenzyme formation*	*Energy transfer*	*Comments*
F.Osu[a]	TR.OSu[b]	—	—	
F.Osu[a]	TCF.OSu[b]	—	—	R^I cross linked
F.Osu[a]	Coum.OSu[d]	—	—	R^I unlabelled
F.Osu[a]	Unlabelled	—	—	
TR.OSu[b]	F.OSu[a]	—	—	
Unlabelled	F.OSu[a]	+	—	
Coum.OSu[d]	F.OSu[a]	+	—	C unlabelled
TCF.OSu[c]	F.OSu[a]	—	—	C cross-linked
Cy3.18.OSu[c]		nd	nd	C cross-linked
Resos.OSu[f]		nd	nd	C reacts with -SH
	Rhod6G.OSu[g]	nd	nd	R^I cross-linked
FITC[h]	XRITC[i]	+	+	40% ratio change
FITC[h]	TRITC[j]	+	+	34–60% ratio change
FITC[h]	EosinITC[k]	+	+	35% FITC quench
FITC[h]	ErythrosinITC[l]	—	—	
FITC[h]	RhodBITC[m]	+	+	
FITC[h]	TR.OSu[b]	—	—	
FITC[h]	TCFITC[n]	+	+	TCFITC quenched
FITC[h]	Unlabelled	+	—	
FITC[h]	RhodB.SO_2Cl[o]	—	—	RhodB.SO_2Cl cross-links
TRITC[j]	FITC[h]	—	—	
TRITC[j]	F-C_6.OSu[p]	+	+	<10% ratio change
XRITC[i]	FITC[h]	—	—	C precipitated
EosinITC[k]	FITC[h]	—	—	
ErythrosinITC[l]	FITC[h]	—	—	
TRITC[j]	Bodipy.OSu[q]	+	—	Bodipy quenched
Bodipy.OSu[q]	TRITC[j]	+	—	Bodipy quenched
DTAF[r]	TRITC[j]	+	+	40% ratio change
Type I holoenzyme				
TCF.OSu[c]		nd	nd	Cross-linked
F.OSu[a]		nd	nd	
FITC[h]		+	+	With TRITC
XRITC[i]		—	—	Precipitated
TRITC[j]		+	+	With FITC
Catalytic	*R^II*			
FITC[h]	TRITC[j]	—	—	
FITC[h]	TR.OSu[b]	+	+	75–100% ratio change
FITC[h]	RhodB.SO_2Cl[o]	+	+	35% ratio change

nd, not determined.
Labels (supplier, catalogue number):
[a] 5(6)-Carboxyfluorescein succinimidyl ester (MP, C–1311).
[b] 5(6)-Carboxytetramethylrhodamine succinimidyl ester (MP, C–1171).
[c] 5(6)-Carboxy-2′,4′,5′,7′-tetrachlorofluorescein succinimidyl ester (Adams & Tsien, unpublished results).
[d] Coumarin 343, succinimidyl ester (MP, C–1419).
[e] Cy3.18, succinimidyl ester (gift of A. Waggoner).
[f] *N*-(Resorufin-4-carbonyl)-piperidine-4-carboxylic acid, *N′*-hydroxy succinimide ester (BM, 1042 653).
[g] 5(6)-Carboxyrhodamine 6G succinimidyl ester (gift of R. Haugland).
[h] Fluorescein 5-isothiocyanate (MP, F–143).
[i] Rhodamine X isothiocyanate (MP, R–491).
[j] Tetramethylrhodamine-5-isothiocyanate (MP, T–1480).
[k] Eosin 5-isothiocyanate (S, E–2005).
[l] Erythrosin B isothiocyanate isomer II (A, 32,380–2).
[m] Rhodamine B isothiocyanate (S, R–1755).
[n] 2′,4′,5′,7′-Tetrachlorofluorescein 5-isothiocyanate (Adams & Tsien, unpublished results).
[o] Lissamine rhodamine B sulphonyl chloride (MP, L–20).
[p] Fluorescein 5(6)-carboxamido-6-hexanoic acid, succinimidyl ester (MP, F–2181).
[q] Bodipy propionic acid, succinimidyl ester (MP)
[r] 5-(4,6-Dichlorotrianzinyl)aminofluorescein (MP, D–16).

MP, Molecular Probes, Eugene OR; BM, Boehringer-Mannheim, Indianapolis, IN; S, Sigma Chemical Company, St Louis, MO; A, Aldrich Chemical Company, Inc., Milwaukee, WI.

as 0.5–2 mg ml^{-1}) were dialysed (Spectra/Por, MWCO 12 000–14 000; Spectrum Medical Industries, Inc., Los Angeles, CA) for 4 h at 4°C against 25 mM bicine, 0.1 mM EDTA, titrated to pH 8 with KOH, with several buffer changes. Following measurement of the absorbance of each solution at 280 nm (to determine approximate protein concentration using extinction coefficients of 45 000 and 48 000 M^{-1} cm^{-1} for C and R^{I} respectively), the solution of C was made 8 mM in $MgCl_2$ and 5 mM in ATP (from a stock solution titrated to about pH 8). R^{I} was labelled with 0.5 mM tetramethylrhodamine 5-isothiocyanate (TRITC) and C was labelled with 0.3 mM fluorescein 5-isothiocyanate (FITC) for 30 min at room temperature. Both dye reagents were from Molecular Probes (Eugene, OR) and were freshly prepared as 100 mM stocks in dry dimethyl formamide. The labellings were quenched by the addition of 5 mM glycine (pH 8) for 10–15 min. Excess dye was removed by passing each protein solution through a Sephadex G-25 column (3 ml), eluting with 25 mM potassium phosphate pH 6.8, 2 mM EDTA, 5 mM 2-mercaptoethanol and 5% glycerol at 4°C. The first coloured band was collected and dialysed (Spectra/Por 1, MWCO 6000–8000) overnight against the same buffer (2 litres) at 4°C. The dye:protein stoichiometries were determined by absorbance spectrometry (at pH 7.2) assuming the following extinction coefficients (in M^{-1} cm^{-1}) of 65 000 and 11 000 for protein-bound fluorescein (Haugland, 1989) at 495 and 280 nm, and 72 000 and 18 000 for protein-bound tetramethylrhodamine (Johnson *et al.*, 1984) at 550 and 280 nm respectively. Covalent attachment of most of the dye was verified by gel electrophoresis under denaturing conditions (SDS-PAGE). The subunits were then mixed at equal concentrations by weight (assessed by absorbance at 280 nm and Coomassie Blue staining of SDS-PAGE), typically 0.5 mg ml^{-1}, and dialysed against several changes of 25 mM potassium phosphate pH 6.7, 0.5 mM $MgCl_2$, 0.1 mM ATP, 5 mM 2–mercaptoethanol, 5% glycerol for 3–5 days at 4°C. Formation of holoenzyme was monitored by kinase assay (Cook *et al.*, 1982). For storage, the labelled holoenzyme was concentrated (using Centricon 30 microconcentrators; Amicon, Beverly, MA) to about 10 mg ml^{-1} in injection buffer (25 mM potassium phosphate pH 7.2, 1 mM EDTA, 0.5 mM 2-mercaptoethanol, and 2.5% glycerol). It has also been successfully stored frozen in buffer containing 50% glycerol.

More recently we have also labelled the R^{II} subunit (unpublished results). In contrast to R^{I}, labelling R^{II} with tetramethylrhodamine OSu (R^{II}-TROSu) gave the best energy transfer with C-FITC, whereas R^{II}-TRITC gave the least (Table 11.1). Recombinant R^{II} behaves similarly to R^{II} isolated from pig heart. Independently, Leathers *et al.* (1990) have prepared C-FOSu and R^{II}-Texas Red, reconstituted holoenzyme and detected energy transfer.

Our current protocol for preparing labelled type II holoenzyme is similar to that for type I described above except that 5(6)-carboxytetramethylrhodamine succinimidyl ester (TROSu; obtained from Molecular Probes) replaces TRITC as the label for R^{II} subunit. A lower concentration (0.1 mM) of TROSu is required in the labelling reaction of R^{II} (about 1 mg ml^{-1}) with a similar 30 min reaction time. The C labelled with FITC used for the type I holoenzyme is also used for the type II. The remaining procedures are the same for type II as type I except that MgATP is not required in the buffer for forming holoenzyme (2 mM EDTA replaces it) and only 1–2 days of dialysis are required. Holoenzyme formation can be monitored by a modified kinase assay that uses only 10 mM potassium MOPS and no KCl, since preliminary results indicate the labelled type II holoenzyme partially dissociates at the low protein concentrations used (1–10 nM) if normal salt levels are present.

11.6 PROPERTIES OF FlCRhR

As described above, the best combination for FRET in A-kinase has been C-FITC with R^{I}-TRITC or R^{II}-TROSu, typically with about one dye molecule per subunit. These are the only cAMP sensors we have used in living cells to measure intracellular free cAMP levels (Adams *et al.*, 1991; Gurantz *et al.*, 1991; Sammak *et al.*, 1992). Therefore, only the properties of these two preparations will be described in this section. For brevity we propose calling the labelled type I and II holoenzymes FlCRhR^{I} and FlCRhR^{II}, respectively, as acronyms for *Fl*uorescein-labelled *C*atalytic and *Rh*odamine-labelled *R*egulatory subunit type I or II (pronounced 'flicker').

11.6.1 Energy transfer

The change in emission spectrum of FlCRhR^{I} from low to high cAMP in a cuvette is shown in Fig. 11.2. In the holoenzyme (zero cAMP), the fluorescein emission at 520 nm is quenched by about 31% and the emission at 580 nm, mostly but not entirely due to rhodamine, is enhanced by about 13% compared to the respective amplitudes after full dissociation. Control experiments indicate that fluoresceinated catalytic subunit exhibits negligible change in fluorescence upon combination with unlabelled R^{I}, so that the partial quenching seen with rhodamine-labelled R^{I} represents energy transfer and not a conformational or environmental effect. The energy transfer efficiency

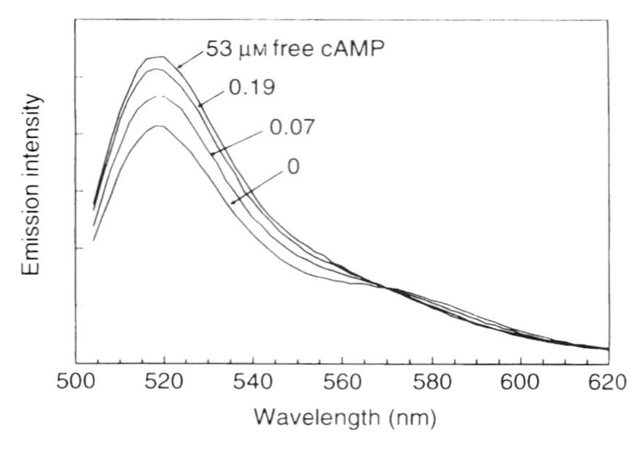

Figure 11.2 The response of the emission spectra of FlCRhR[I] upon titration with cAMP when the fluorescein label is excited at 495 nm. With increasing cAMP, the emission at 520 nm (from fluorescein) increases while the emission at 580 nm (mostly from tetramethylrhodamine) decreases. For clarity, only four concentrations of cAMP are shown. The cuvette contained 3 nM FlCRhR[I] in 130 mM KCl, 5 mM $MgCl_2$, 3 mM ATP, 10 mM MOPS, pH 7.2 at 22°C. Free cAMP was calculated by subtracting the calculated bound cAMP from the total cAMP assuming that two molecules of cAMP must bind to R^I to release C. Excitation and emission bandwidths were 1.8 and 4.6 nm respectively. (Reproduced from Adams *et al.*, 1991, with permission.)

in the holoenzyme is therefore about 0.31, which corresponds to an average distance of 5.8 nm between the donor and acceptor fluorophores (assuming random orientations for their transition moments). Although large, this distance is consistent with the known dimensions of the holoenzyme, whose Stokes's radius is 4–5 nm (Zoller *et al.*, 1979) and whose distance between adjacent C active sites is >5 nm (E. First, D. Johnson & S. Taylor, unpublished results). Possibly the ability of isothiocyanates to label sites distant to the C and R binding domains explains why these labels were the most compatible with reformation of holoenzyme. We have not yet determined if the labelling is truly random or (at least partly) site-specific. One fluorescein molecule appears to be incorporated faster than further ones into the C subunit, suggesting some selectivity (unpublished results).

Upon binding cAMP, the emission of FlCRhR[I] (when the fluorescein is excited at 495 nm) increases at 520 nm and decreases at 580 nm (Fig. 11.2). The ratio of emissions at 520 to 580 nm increases by a factor of 1.4 to 1.65, depending on the particular batch, upon changing the cAMP concentration from 0 to 50 μM. FlCRhR is therefore a ratiometric indicator, which is of particular advantage in intracellular measurements by fluorescence microscopy because ratioing allows correction for variations in probe concentration, optical pathlength, lamp intensity and emission collection efficiency (Tsien & Poenie, 1986; Bright *et al.*, 1989). The reason why the apparent change in emission intensity at 580 nm is smaller than that at 520 nm is due to at least two reasons: the rhodamine does not re-emit a photon every time it quenches a fluorescein excited state, and the emission at 580 nm contains a component due to the tail of the fluorescein emission spectrum, which is affected oppositely from the rhodamine.

FlCRhR[II] shows a larger change than FlCRhR[I] in the emission ratio of 520–580 nm, with changes of 1.6–2.2-fold being obtained in calibrations and in living cells. One price to be paid for this enhanced magnitude of response to cAMP is the slow dissociation of the holoenzyme when diluted to submicromolar concentrations even in the absence of cAMP. Perhaps in FlCRhR[II] the rhodamine label is closer to the R^{II}–C interaction site and weakens their interaction.

11.6.2 cAMP sensitivity and kinase activity

The optical sensitivity of FlCRhR to cAMP can be measured *in vitro* simply by recording the changes in fluorescence emission upon titration with cAMP as in Fig. 11.2. The free cAMP concentration can then be calculated for each curve by measuring the percentage activation and subtracting the calculated amount of cAMP bound to the holoenzyme. This requires knowing the holoenzyme concentration in the cuvette and assumes that two cAMP molecules bind per R subunit.

The kinase activity of FlCRhR and unlabelled holoenzyme can be measured by ^{32}P incorporation into a peptide substrate, Kemptide, Leu-Arg-Arg-Ala-Ser-Leu-Gly, or more conveniently by coupling the decrease in ATP (again with Kemptide) to a change in NADH using pyruvate kinase and lactate dehydrogenase (Cook *et al.*, 1982).

When the ratio of emissions at 520 nm to that at 580 nm of FlCRhR[I] is measured as a function of free cAMP concentration (Fig. 11.3, taken from Adams *et al.*, 1991), the resulting calibration curve is similar to the plot of kinase activity vs. cAMP for both the labelled and unlabelled enzyme. The activation constants, i.e. free cAMP concentrations that give half-maximal emission ratio or kinase activity, are all about 0.1 μM with slight positive cooperativity (Hill coefficient 1.2–1.8). These values, measured at vertebrate ionic strength in the presence of MgATP, vary slightly between batches but show no systematic difference between labelled and unlabelled holoenzyme. The kinase activity of C measured with Kemptide appears to be almost unchanged by labelling with FITC in the presence of MgATP. The advantage of using enzymically active FlCRhR in intracellular experiments is that cAMP binding to the exogenous

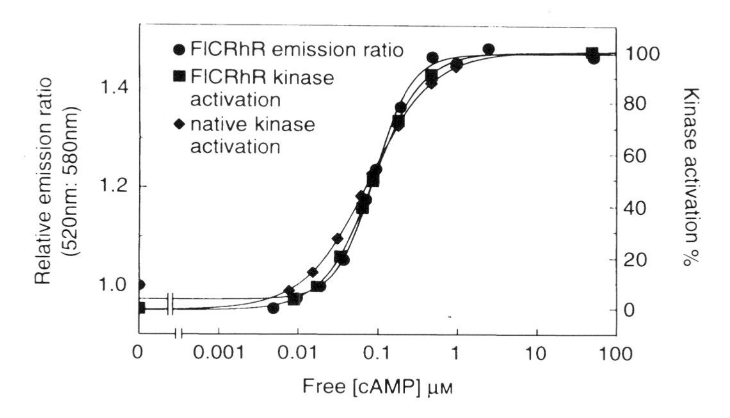

Figure 11.3 Typical calibration curve of FlCRhR^{I} relating the ratio of emissions at 520 nm to 580 nm to the free cAMP concentration (circles, left-hand scale). Also depicted is the activation of kinase activity of FlCRhR^{I} and unlabelled A-kinase holoenzyme by cAMP (squares and diamonds respectively, right-hand scale). The kinase activity was measured using the assay of Cook *et al.*, (1982) with the synthetic peptide substrate Kemptide. The *y*-axis are normalized to emission ratio or percentage kinase activation at zero cAMP for clarity. The data were fitted by least-squares to the Hill equation (using Graphpad Inplot 3.0, San Diego, CA); the cAMP concentrations giving half-maximal response were 88 nM, 86 nM and 76 nM, and the Hill coefficients were 1.8, 1.4 and 1.1 for emission ratio, labelled kinase, and native kinase respectively. The minor differences within these values are within the range shown by different batches of protein. (From Adams *et al.*, 1991, with permission.)

enzyme should still elicit normal patterns of phosphorylation, whereas if the sensor were catalytically inactive, its binding of cAMP would detract from the downstream physiological response. If enzymically inactive FlCRhR were desired, MgATP could be omitted during the labelling of C, or C could be subsequently reacted with sulphydryl reagents.

The emission ratio of FlCRhR^{II} can be calibrated in a similar way to FlCRhR^{I} except that at the enzyme concentrations used (typically <10 nM) preliminary experiments suggest that the labelled holoenzyme slowly dissociates in the absence of cAMP (unpublished results). By contrast, unlabelled type 2 holoenzyme has been reported to be 50% dissociated when diluted to 0.2–0.3 nM (Hofmann, 1980). However, when microinjected into cells, FlCRhR^{II} shows a stable emission ratio in unstimulated cells (see later) suggesting dissociation is not a problem at these higher protein concentrations, micromolar or greater. Calibration in a cuvette at these higher holoenzyme concentrations is inaccurate because of errors in converting from total cAMP to the more biologically relevant free cAMP as this requires subtracting two large, imprecise numbers. A better method is to calibrate in a microdialysis capillary on the fluorescence microscope (see Section 11.8.4). An alternative is to calibrate quickly in a cuvette before significant dissociation can occur. Both methods give similar values for activation constant, Hill coefficient, and percentage ratio change for the same batch of protein (0.4 μM, 1.0 and 70–80% respectively, measured at high ionic strength mimicking marine invertebrate cytoplasm).

11.6.3 Stoichiometry of labelling and subunits

The dye-to-protein ratio can be estimated from the absorbance spectrum of the labelled C or R subunit and the extinction coefficients for subunit and dye at 280 nm and dye absorbance maximum (Adams *et al.*, 1991). With FlCRhR^{I}, there is an average of one fluorophore per subunit, though some subunits may be unlabelled and some multilabelled. Optimal energy transfer would probably occur with as many labels per subunit as possible but empirically 1:1 appears to be the best. For example, overlabelling the R^{I} subunit with TRITC increases its susceptibility to precipitate and diminishes its ability to reform holoenzyme. Fluorophores that are in too close proximity (<2 nm) can quench each other by mechanisms other than FRET (Tsien & Waggoner, 1990).

When recombining the labelled subunits to form FlCRhR, we usually balance the subunits by determining concentrations by absorbance or by protein staining following SDS-PAGE, and by loss of kinase activity upon dialysis. Usually a slight excess of R over C is preferred to minimize free C-FITC, which can interfere in experiments studying nuclear translocation of C. Neither method is foolproof as some R may be unable to form holoenzyme because of labelling. A slight excess of R subunit appears to be less detrimental than free C, perhaps because R already contains bound cAMP and seems not to perturb cAMP signalling (Sammak *et al.*, 1992).

11.6.4 Concentration and storage

The holoenzyme is usually re-formed at protein concentrations of about 1 mg ml^{-1} or 6 μM in holoenzyme. For microinjection, 10-fold higher concentrations are usually required in the pipette so that injections of 5–10% of cell volume give final intracellular levels of 3–6 μM FlCRhR. FlCRhR solutions are concentrated by using centrifugal ultrafiltration (e.g. Centricon filters, Amicon Corp.), which causes minimal protein loss through adsorption to the membrane, although some protein precipitation may occur. We also change the buffer to one more suitable for microinjection and storage, e.g. 25 mM potassium phosphate, pH 7.3, 1 mM EDTA, 0.5 mM 2-mercaptoethanol, and 2.5% glycerol. The last two components are to help stabilize the protein; the EDTA is to discourage bacterial contamination, which nevertheless occurs eventually after dipping many non-sterile pipette tips and removal of many aliquots from one vial. The protein seems to be quite stable just kept at 4°C in a sealed vial protected from light; samples over 6 months old have still been responsive to cAMP in cells. Deterioration can be recognized by increasing difficulty in microinjection, even after extensive centrifugation at 16 000 × **g** at 4°C to remove any precipitated protein and particulates, and decreased ratio change in cells. Proteolysis is visualized by polyacrylamide gel electrophoresis under denaturing conditions.

11.7 INTRODUCTION OF FlCRhR INTO CELLS

11.7.1 Microinjection

A necessary step for biological applications of FlCRhR is the introduction of the protein into the cytosol of living cells. Because FlCRhR is a protein complex with an aggregate molecular weight of 167 or 172 kDa (type I or II respectively), it cannot be loaded via acetoxymethyl esters or iontophoresis as can low-molecular-weight cation indicators. For all applications to date we have relied on pressure microinjection into individual cells (Wang *et al.*, 1982). Microinjection is quick, direct and economical of protein. In recent years it has also become much easier and more widely applicable, thanks to automated micromanipulators, pressure pulse regulators, and piezoelectric stabbers.

Pressure injection of single cells requires micropipettes for penetrating cells and introducing the protein, a closed system for delivering pressure, and a means for monitoring injection volume. Perhaps the most important component for this technique is the micropipette. We use 1.0 mm thin-walled borosilicate glass which contains a microfilament along the inner surface. The microfilament provides a convenient and effective means of capillary back-filling the pipette with the minimum volume necessary. With these electrodes, volumes of <0.5 μl of the concentrated protein solution are more than sufficient. A reproducible and reliable pipette puller is also important. We use a Flaming-Brown type puller (Sutter Instruments, Inc., P80–PC). The actual shape and size of the pipette may depend on the type of cell being injected. For general-purpose injection of cells >20 μm in diameter, pipettes with tips of about 0.5 μm are used. Finer tipped pipettes would be preferable except that the concentrated high-molecular-weight protein tends to clog very small pipettes much more easily. For very delicate cells, the pipette tips may require bevelling as well.

Once pulled and filled, the pipettes are placed in a pipette holder (E.W. Wright, Guilford, CT) which contains a side-port to attach tubing. The holder is attached to a three–axis hydraulic micromanipulator (Narishige Corp. or Newport Corp.) in order to position the pipette stably and precisely on the cells. Most often, some means for isolating vibrations is necessary, e.g. an air suspension table. Occasionally, movement along the oblique axis of the pipette is useful, and can be achieved either with an additional hydraulic positioner or better yet with a piezoelectric manipulator (Inchworm no. IW-700, Burleigh Instruments). The latter allows rapid and reproducible axial movements of the pipette, valuable for fast penetrations of tough cell membranes. To deliver pressure, we have used various systems ranging from a 50 ml disposable syringe to regulated, high-precision automated pressure injection systems (Eppendorf no. 5242, or Medical Systems Corp., PLI-100). The automated systems generally provide three or more available pressures; a constant balance pressure, a timed injection pressure, and a clearing pressure for clogged tips. These systems provide reliable and reproducible injections with a touch of a button.

Monitoring the injection volume can be difficult. With a precision pressure injector, one can calibrate the time of injection pulse and the pressure required for a reproducible pipette geometry to obtain a known volume per pulse. Others have monitored the movement of the meniscus within the shaft of the pipette to estimate injection volume (Castellucci *et al.*, 1980). It is also possible to use the fluorescence of the probe itself to estimate the injection volume under constant imaging conditions. That is, one can compare the brightness of the fluorescence within the cell with *in vitro* samples of the protein at known concentrations. In this way, one can measure the dilution of the fluorescence after injection into the cell, and with an estimate of the volume of the cell, can calculate the volume introduced. This is perhaps the most practical approach, since the minimum volume that must be injected is that which provides a bright enough fluorescent signal for optical measurements. Therefore, we inject the minimum volume of protein for a good signal-to-noise ratio, which is readily monitored on the imaging microscope.

Two general approaches are used for loading cells. For flat, well adherent, easily penetrable cultured cells, a small constant pressure is applied to the pipette so that the pipette solution is continuously flowing while the pipette tip is lowered vertically into contact with the cell. The injection volume then depends on the amount of time the pipette is kept intracellularly. In the other protocol, a pressure pulse is applied to the pipette only after cell penetration. This approach requires more elaborate equipment but is best for cells that can only tolerate very limited injection volumes or are difficult to penetrate because they are not rigidly anchored everywhere and would tend to back away from a continuous outflow.

11.7.2 Alternatives to microinjection

There will be applications in which microinjection of individual cells is not applicable, for example in the smallest or most fragile cell types or when biochemical responses of a bulk population must be studied. Electroporation would seem to be the most promising approach (see, for example, Liang *et al.*, 1988). We have not tried this yet, mainly because electroporation in commercial equipment wastes relatively large amounts of protein merely to fill the chamber. However, protocols may well be developed once the protein or microscale electroporation apparatus become more easily available. Other techniques (see review by McNeil, 1989) such as scrape loading, hypotonic shock/resealing, or vesicle fusion (Straubinger *et al.*, 1985) have not been tried but may prove helpful.

11.8 IMAGING OF FlCRhR AND FREE cAMP

11.8.1 Imaging modes

Observation of FlCRhR in single cells requires a fluorescence microscope equipped for single-excitation (490 nm) and dual-emission (520 and 580 nm) wavelength recording. Dual-wavelength recording is essential because the intensity changes at any one wavelength are modest in amplitude, relatively slow (seconds to minutes), and because bleaching of the fluorescein is a significant concern. In excitation ratioing, one would illuminate with fluorescein and rhodamine excitation wavelengths in alternation, while continually monitoring the emission at rhodamine wavelengths. Unfortunately, the former signal decreases only 10–20% upon cAMP binding (as explained above in Section 11.6.1) and the latter signal changes negligibly. Therefore excitation ratioing would be very insensitive. By contrast, emission ratioing (with constant excitation of the fluorescein) makes use of both the 30–50% enhancement of fluorescein emission and 10–20% decrease in rhodamine emission, giving 35–65% increases in FlCRhR^{I}ratio upon binding of cAMP, and is therefore far superior.

11.8.2 Apparatus for emission ratioing

The simplest and perhaps most sensitive approach for dual emission detection uses an image-plane mask to isolate the region of interest. The emitted light from this region is split into two wavelength bands by a dichroic mirror and sent to two detectors whose outputs undergo background subtraction and then ratioing. Such simple, relatively inexpensive, and commercially available apparatus (e.g. Photoscan, Nikon Corp.) provides a time-course of spatially averaged fluorescence. However, to resolve the spatial segregation of the probe or its ratio changes, or to track moving cells, an imaging system must be used. Most ratio imaging systems have been set up for excitation ratioing because that mode best suits many popular probes for cations such as fura-2, BCECF, and SBFI (Tsien, 1989). Dual-emission ratio imaging is optically more troublesome than dual-excitation ratio imaging because emission-selecting optics must transmit as many photons as possible, preserve the optical quality of the image, and maintain accurate registration between the images at two wavelengths. By contrast, excitation wavelength-selecting optics can be relatively wasteful of photons and need handle only diffuse illumination. Dual-emission ratio imaging on a conventional microscope is generally accomplished either by mechanically switching barrier filters in front of a single imaging camera, or by splitting the fluorescence emission with a dichroic mirror and directing the separate emission wavelengths to two identical cameras. Both of these techniques suffer from the difficulty of accurately registering pixel information at each of the two wavelengths. The single-camera approach wastes half the photons, takes twice as long per wavelength pair, and is vulnerable to motion artifacts if the cell moves significantly between the two exposures. Further registration artifacts result from small spatial irregularities in barrier filter construction and inaccuracies of filter positioning from cycle to cycle. The use of two imaging cameras is costly and requires perfect matching of magnifications, freedom from geometrical distortions, and careful translational and rotational alignment to achieve registration. Adaptive warping algorithms have been described to correct digitally for misregistered images (e.g. Jerićevic *et al.*, 1989; Takamatsu & Wier, 1990), but they are still computationally complex and far from routine.

11.8.3 Confocal microscopy

Our preferred strategy to achieve dual-emission image registration is the use of a confocal laser scanning microscope (CLSM) equipped with two emission detectors (Tsien, 1990). Registration is automatically achieved by using an achromatic scanner both to deflect the exciting laser beam and to de-scan the returning emission. After the emission has passed back through the scanner and the confocal pinhole, the spatial properties of the beam are no longer of interest, so the light can simply be split with a dichroic mirror and fed to two non-imaging detectors, whose outputs inherently correspond to the same point of the raster scan (Tsien & Waggoner, 1990). Notice that such registration only holds for scanning with non-imaging detectors. Scanning-disk confocal microscopes that use cameras as detectors will have the same problems of registration as conventional wide-field microscopes. An independent advantage of confocal microscopy is that its optical sectioning capability helps in showing where the R and C subunits have migrated within the cell, as mentioned below (Section 11.9).

11.8.4 Calibrations *in vitro*

Ratio imaging of FlCRhR in single cells allows measurements of relative changes in A-kinase activity independently of the concentration of the probe, optical path length of the imaging system, and illumination intensity. One can readily derive that the calibration of an observed ratio R in terms of an absolute concentration of free cAMP [cAMP] is given by the following equation:

$$[\text{cAMP}] = K'_{\text{d}} \left(\frac{R - R_{\text{min}}}{R_{\text{max}} - R} \right)^{1/n}$$

Where R_{min} and R_{max} are the ratios observed at zero and saturating [cAMP] respectively; n is the Hill coefficient describing the cooperativity of cAMP binding to FlCRhR; and K'_{d} is an apparent dissociation constant corresponding to the cAMP concentration at which R is midway between R_{min} and R_{max}. K'_{d} is also given by the following expression:

$$K'_{\text{d}} = K_{\text{d}} \left(\frac{S_{\text{f2}}}{S_{\text{b2}}} \right)^{1/n}$$

Here K_{d} is the true dissociation constant for cAMP binding to the probe, and $(S_{\text{f2}}/S_{\text{b2}})$ is the ratio of emission intensities of the probe without and with cAMP measured at the denominator wavelength, i.e. at rhodamine emission wavelengths. These equations are closely analogous to the standard ones used for calibrating cation indicators with 1:1 binding stoichiometry (Grynkiewicz *et al.*, 1985), except that they have been modified here to allow for cooperativity, i.e. a Hill coefficient n that can take values other than 1.

In vitro, values for R_{min}, R_{max}, n, and K'_{d} are readily obtainable by direct titration of the probe with cAMP, providing that one has a way of controlling or at least determining free cAMP. Unfortunately, no reliable buffer systems for cAMP are available at present. The simplest method is to put FlCRhR at a known low concentration in a cuvette and add successive aliquots of cAMP between measurements of the emission spectrum, as in Fig. 11.3. Free cAMP is retrospectively calculated for each known concentration of total cAMP by measuring the percentage saturation of the probe and assuming that 4 moles of cAMP are bound per mole of holoenzyme that has been dissociated. This approach has a number of limitations, so we now prefer to suck a tiny droplet of FlCRhR into a microdialysis capillary that is permeable to solutes of molecular weight below a few kilodaltons (e.g. Spectra/Por hollow fibre microdialysis tubing, 150 μm diameter, 9 kDa cut-off). This capillary is then immersed in a chamber containing cAMP solutions mounted on the microscope. This approach has many advantages. Free cAMP inside the capillary is directly controlled by the external cAMP, so that the concentration of protein can be relatively high (as required for $\text{FlCRhR}^{\text{II}}$) yet no assumptions need to be made about the concentration or stoichiometry of binding sites. [cAMP] can be lowered as well as raised, so that a single sample of protein can be checked for hysteresis, stability, self-modification perhaps by autophosphorylation, or alterations due to ionic strength, ATP concentration, sulphydryl reductants, etc. Finally, the response is directly measured in the same microscope used for the imaging in cells, so that no corrections are required for differences between spectrofluorometer and microscope in spectral sensitivities and bandwidths.

11.8.5 Calibration in intact cells

In intact cells, methods have not yet been worked out to control the free cAMP concentration at precise intermediate concentrations as required to perform an *in situ* calibration. Perhaps selective permeabilization of the cells to low-molecular-weight solutes, together with inactivation of adenylate cyclase and phosphodiesterases, might be sufficient to clamp intracellular cAMP equal to extracellular. Until such methods are validated, the values for K'_d and Hill coefficient n can only be assumed to equal *in vitro* values. R_{min} and R_{max} are somewhat more accessible. R_{min} should be the ratio observed when no stimuli to raise cAMP are present and a pharmacological antagonist of cAMP is added. The only such antagonist currently available is R_pcAMPS (BioLog Life Science Institute, La Jolla, CA). These phosphorothioate analogues of cAMP (Botelho *et al.*, 1988) are supposed to be somewhat membrane-permeant, but to conserve expensive material and ensure delivery, we have mostly injected them directly into cells. In most cells that we have examined, before stimulation the emission ratio of FlCRhR is unaffected by addition of the antagonist, suggesting that the basal level of free cAMP is essentially zero. Likewise, R_{max} can be assessed in intact cells by saturating them with cAMP or an effective analogue, for example using forskolin to stimulate adenylate cyclase directly, IBMX to inhibit phosphodiesterase activity, permeant non-hydrolysable cAMP analogues like dibutyryl cAMP, direct injection of cAMP, or combinations thereof. Our impression is that R_{min} and R_{max} are up to 30% lower inside cells than in the microdialysis capillary, but that they are both changed by about the same factor.

11.8.6 Potential artifacts

11.8.6.1 Bleaching

Excessive illumination of FlCRhR at fluorescein excitation wavelengths causes a systematic fall in the ratio of fluorescein to rhodamine emissions, simulating a decrease in [cAMP]. This effect seems due to photobleaching of the fluorescein, which suffers more than the rhodamine both because the input light is primarily received by the fluorescein and because fluorescein is inherently more easily bleached. If the emission at rhodamine wavelengths were due only to FRET from fluorescein or the long-wavelength tail of fluorescein itself, fluorescein bleaching would affect the emissions at both wavelengths proportionately and cancel out in the ratio. However, the emission at rhodamine wavelengths includes a significant component due to direct excitation of rhodamine via the short-wavelength tail of its absorbance spectrum. This component does not get bleached, so the ratio decreases somewhat as fluorescein bleaches. The most obvious remedy is to decrease the frequency, duration or intensity of illumination periods. We generally have been able to collect several tens to hundreds of ratio images without too much bleaching. Obviously a less bleachable fluorophore would be desirable in place of fluorescein, but trials with alternatives with supposedly greater photostability, such as Bodipy or cyanines, have thus far been unsuccessful (see Table 11.1). Another approach might be to increase the spectral separation between the fluorophores to reduce direct excitation of the acceptor. Some small improvement might be obtained this way, but too great a spectral separation would decrease FRET efficiency.

11.8.6.2 Scrambling with endogenous kinase

The presence of endogenous unlabelled kinase subunits within the cell raises the concern that a transient elevation of cAMP would cause scrambling and reassociation of exogenous labelled subunits with unlabelled endogenous subunits. FRET would not be reconstituted in these hybrid holoenzyme complexes, so $[cAMP]_i$ would seem to remain elevated even when it had really fallen. Such an effect can be demonstrated in the microdialysis capillary by deliberately mixing FlCRhR with unlabelled holoenzyme. Nevertheless, in real cells FlCRhR has usually proven able to reconstitute energy transfer and to indicate a decline in apparent $[cAMP]_i$, often back to levels indistinguishable from the prestimulus state. The probable explanation is that the exogenous kinase is in considerable excess over the endogenous, so statistics favour the desired reassociation of FlC with RhR. However the possibility of scrambling must be kept in mind whenever apparent $[cAMP]_i$ fails to decrease as expected. It would be interesting to try cross-linking FlC and RhR with a flexible tether; this would discourage scrambling, but might also prevent cAMP from dissociating the subunits at all.

11.8.6.3 Perturbation of the cAMP response pathway by exogenous holoenzyme

The realization that microinjected FlCRhR is probably in considerable excess over endogenous kinase leads to concern that FlCRhR might buffer $[cAMP]_i$. Some such effect is probably unavoidable; its seriousness should be testable by deliberately introducing different amounts of FlCRhR, and minimizable by using the lowest concentrations that give adequate fluorescence and permit reconstitution of FRET upon reduction of $[cAMP]_i$. There is some evidence in hepatocytes (Corbin *et al.*, 1985) that kinase activation exerts negative feedback in cAMP levels; such a homeostatic mechanism might render $[cAMP]_i$ more resistant to perturbation by buffering. We have usually observed that strong stimulation of cAMP-linked receptors generates enough endogenous cAMP to saturate at least a few micromolar FlCRhR. In one cell type, angelfish melanophores, deliberate injections of tens of micromolar FlCRhR have been tried; at this level of enzyme, apparent $[cAMP]_i$ levels were depressed (Sammak *et al.*, 1992) compared to those observed with standard injections.

A distinct question is whether exogenous FlCRhR promotes or inhibits downstream events mediated by kinase activation. The strategy of using the native kinase was meant to minimize such perturbations, but may not be perfectly successful. If the cell can only make a limited amount of cAMP, then provision of extra kinase might be expected to maximize the efficacy of each cAMP molecule, resulting in accentuation of distal effects. But if the added FlCRhR is not quite as active as endogenous kinase or is in the wrong location with respect to substrates, it could divert cAMP into less productive binding and reduce distal phosphorylations. The fact that R subunits need two cAMP molecules to release C might also favour inhibition, since excess kinase would statistically favour non-productive single occupancy of R subunits at the expense of productive double occupancy. Obviously, it will be important to monitor downstream cell responses whenever possible. So far, responses such as rounding up in fibroblasts, aggregation and dispersal of pigment granules in melanophores (Sammak *et al.*, 1993), and increased excitability of *Aplysia* sensory neurons appear to be qualitatively unaffected by the usual levels of a few micromolar FlCRhR. However, tens of micromolar FlCRhR caused permanent aggregation of melanophore granules consistent with depression of $[cAMP]_i$ and overall kinase activity.

11.9 APPLICATIONS

To date (September 1991) we have successfully obtained cAMP recordings from REF52 fibroblasts, BC3H-1 smooth muscle-derived cells (Adams *et al.*, 1991), PC-12 pheochromocytoma cells, neonatal cardiac myocytes, primary and transformed osteoblasts, chick ciliary ganglion neurons (Gurantz *et al.*, 1991), rat Schwann cells, MA-10 Leydig tumour cells, fish melanocytes (Sammak *et al.*, 1992), and cultured *Aplysia* sensory neurons. Space does not permit detailed description of the results obtained, which have been or will be reported in appropriate primary research papers. However, one sample application can be described here, the visualization of the spread of cAMP within MA-10 Leydig tumour cells from sites of relatively localized hormone stimulation (Plate 11.1). This experiment, an extension of previous morphological observations (Podesta *et al.*, 1991), was performed by Brian Bacskai and Clotilde Randriamampita in collaboration with José Lemos, Ernesto Podesta and Mario Ascoli. Luteinizing hormone (LH) was locally applied from a puffer pipette placed sequentially at two locations near a cluster of four MA-10 cells injected with FlCRhR. Elevations in intracellular cAMP can be seen to spread from each site of local application. Further details can be found in the plate caption.

Besides the ability of FlCRhR to monitor dynamic changes in free cAMP concentrations in single cells, it is also possible to use the dual labelling of the protein subunits to independently track their positions within a cell. Spatial segregation and dynamic translocation of proteins within cells can crucially influence their biological effectiveness. In general, FlCRhR injected in the cytoplasm remains there and seems not to enter the nucleus as long as cAMP is low. After cAMP-induced dissociation of the holoenzyme, the C subunit translocates into or equilibrates with the nucleus, whereas the R subunits seem to remain in the cytoplasm (Meinkoth *et al.*, 1990). Upon lowering cAMP, the fluorescein-labelled C subunit leaves the nucleus and re-forms holoenzyme with the R subunit in the cytoplasm (Adams *et al.*, 1991). Labelled holoenzyme or subunits should also be a powerful tool for assessing and visualizing binding to cytoskeletal elements or other organelles.

Results such as these suggest that imaging of FlCRhR in single cells can provide unprecedented spatial and temporal resolution to study the distribution of cAMP and A-kinase and thereby to analyse the biochemical control and physiological function of this most important second messenger pathway. Furthermore, the ability of fluorescence resonance energy transfer to monitor protein–protein and protein–DNA interactions non-destructively in living cells might well be extended to many other analogous supramolecular combinations. Just within the field of signal transduction, a few of the obvious examples would include complexes of receptor tyrosine kinases with effector proteins, G-protein α subunits with $\beta\gamma$ subunits or with 7-transmembrane-segment receptors or effector proteins (e.g. Erickson & Cerione, 1991), calmodulin with its targets, transcription factor complexes (e.g. Patel *et al.*, 1990), and so on.

Enquiries regarding the commercial availability of FlCRhR should be directed to Atto Instruments, 1500 Research Boulevard, Rockville, MD 20850 USA, which has been licensed by the University of California to produce the material.

REFERENCES

Aass H., Skomedal T. & Osnes J.-B. (1988) *J. Molec. Cell Cardiol.* **20**, 847–860.

Adams S.R., Harootunian A.T., Buechler Y.J., Taylor S.S., & Tsien R.Y. (1991) *Nature* **349**, 694–697.

Akil M. & Fisher S.K. (1989) *J. Neurochem* **53**, 1479–1486.

Anjaneyulu P.S.R. & Staros J.V. (1987) *Int. J. Peptide Protein Res.* **30**, 117–124.

Ascoli M. (1981) *Endocrinology* **108**, 88–95.

Barsony J. & Marx S.J. (1990) *Proc. Natl. Acad. Sci. USA* **87**, 1188–1192.

Beavo J.W. & Reifsnyder D.H. (1990) *Trends Pharmacol. Sci.* **11**, 150–155.

Beavo J.W., Bechtel P.J. & Krebs E.G. (1974) *Proc. Natl. Acad. Sci. USA* **71**, 3580–3583.

Beebe S.J. & Corbin J.D. (1986) In *The Enzymes*, Vol. 17, *Control by Phosphorylation*, Part A, P.D. Boyer & E.G. Krebs (eds). Academic Press, New York, pp. 43–111.

Beebe S.J., Blackmore P.F., Chrisman T.D. & Corbin J.D. (1988) *Methods Enzymol.* **159**, 118–189.

Berridge M.J., Cobbold P.H. & Cuthbertson K.S.R. (1988) *Phil. Trans. R. Soc. Lond. B* **320**, 325–343.

Bode D.C. & Brunton L.L. (1988) *Molec Cell. Biochem.* **82**, 13–18.

Boshart M., Weih F., Nichols M. & Schütz G. (1991) *Cell* **66**, 849–859.

Botelho L.H.P., Rothermel J.D., Coombs R.V. & Jastorff B. (1988) *Methods Enzymol.* **159**, 159–172.

Bourne H.R., Sanders D.A. & McCormick F. (1991) *Nature* **349**, 117–127.

Bramson, H.N., Kaiser E.T. & Mildvan A.S. (1984) *CRC Crit. Rev. Biochem.* **15**, 92–124.

Bright G.R., Fisher G.W., Rogowska J. & Taylor D.L. (1989) *Methods Cell Biol.* **30**, 157–192.

Broach J.R. & Deschenes R.J. (1990) *Adv. Cancer Res.* **54**, 79–139.

Brooker G. (1988) *Methods Enzymol.* **159**, 45–50.

Brooker G., Harper J.F., Terasaki W.L. & Moylan R.D. (1979) *Adv. Cyc. Nucl. Res.* **10**, 1–33.

Büchler W., Meinecke M., Chakraboty T., Jahnsen T., Walter U. & Lohmann S.M. (1990) *Eur. J. Biochem.* **188**, 253–259.

Byus C.V. & Fletcher W.H. (1988) *Methods Enzymol.* **159**, 236–254.

Cadd G.G., Uhler M.D. & McKnight G.S. (1990) *J. Biol. Chem.* **265**, 19502–19506.

Carr D.W., Stofko-Hahn R.E., Fraser I.D.C., Bishop S. M., Acott T.S., Brennan R.G. & Scott J.D. (1991). *J. Biol. Chem.* **266,** 14188–14192.

Castellucci V.F., Kandel E.R., Schwartz J.H., Wilson F.D., Nairn A.C. & Greengard P. (1980) *Proc. Natl. Acad. Sci. USA* **77**, 7492–7496.

Cho-Chung Y.S., Clair T., Tortora G. & Yokozaki H. (1991) *Pharmac. Ther.* **50**, 1–33.

Cook P.F., Neville M.E. jr., Vrana K.E., Hartl F.T. and Roskoshi R. jr. (1982) *Biochemistry* **21**, 5794–5799.

Corbin J.D., Beebe S.J., Blackmore P.F. (1985) *J. Biol. Chem.* **260**, 8731–8735.

Cumming R. (1981) *Trends Neurosci.* **4**, 202–204.

Deslongchamps G., GalEn A., de Mendoza G. & Rebek J. Jr (1992) *Angew. Chem. Intl Ed.* **31**, 61–63.

Døskeland S.O. (1978) *Biochem. Biophys. Res. Commun* **83**, 542–549.

Drobnica Ľ., Kristián P., & AugusťJn J. (1977) In *The Chemistry of Cyanates and Their Thio Derivatives*, Part 2, S. Patai (ed.). J. Wiley, New York, pp. 1003–1221.

Duronio R.J., Jackson-Machelski E., Heuckeroth R.O., Olins P.O., Devine C.S., Yonemoto W., Slice L.W., Taylor S.S. & Gordon J.I. (1990) *Proc. Natl. Acad. Sci. USA* **87**, 1506–1510.

Erickson J.W. & Cerione R.A. (1991) *Biochemistry* **30**, 7112–7118.

First E.A. & Taylor S.S. (1989) *Biochemistry* **28**, 3598–3605.

First E.A., Johnson D.A. & Taylor S.S. (1989) *Biochemistry* **28**, 3606–3613.

Fletcher W.H., Ishida T.A., Van Patten S.M. & Walsh D.A. (1988) *Methods Enzymol.* **159**, 255–267.

Gilman A.G. (1990) *Adv. Sec. Mes. Phos. Prot. Res.* **24**, 51–57.

Glazer A.N. & Stryer L. (1990) *Methods Enzymol.* **184**, 188–194.

Granot J., Mildvan A.S., Hiyama K., Kondo H. & Kaiser E.T. (1980) *J. Biol. Chem.* **255**, 4569–4573.

Greenberg S.M., Bernier L. & Schwartz J.H. (1987) *J. Neurosci.* **7**, 291–301.

Gross R.E., Lu X. & Rubin C.S. (1990) *J. Biol. Chem.* **265**, 8152–8158.

Grynkiewicz G., Poenie M. & Tsien R.Y. (1985) *J. Biol. Chem.* **260**, 3440–3450.

Gurantz D., Harootunian A.T., Tsien R.Y., Dionne V.E. & Margiotta J.F. (1991) *Soc. Neurosci. Abstr.* **17**, 959 (Abstr. 384.5).

Haugland R.P. (1989) *Handbook of Fluorescent Probes and Research Chemicals.* Molecular Probes, Eugene, Oregon.

Hayes J.S. & Brunton L.L. (1982) *J. Cyc. Nucl. Res.* **8**, 1–16.

Herman B. (1989) *Methods Cell Biol.* **30**, 219–243 .

Hiratsuka T. (1982) *J. Biol. Chem.* **257**, 13354–13358.

Hofmann F. (1980) *J. Biol. Chem.* **255**, 1559–1564

Jerićevic Ž., Wiese B., Bryan J. & Smith L.C. (1989) *Methods Cell Biol.* **30**, 47–83.

Johnson D.A., Voet J.G. & Taylor P. (1984) *J. Biol. Chem.* **259**, 5717–5725.

Jones K.W., Shapero M.H., Chevrette M. & Fournier R.E.K. (1991) *Cell* **66**, 861–872.

Jovin T.M. & Arndt-Jovin D.J. (1989) In *Cell Structure and Function by Microspectrofluorometry*, E. Kohen & J.G. Hirschberg (eds). Academic Press, San Diego, pp. 99– 117.

Karin M. (1989) *Trends Genetics* **5**, 65–67.

Klein P.S., Sun T.J., Saxe C.L. III, Kimmel A.R., Johnson R.L. & Devreotes P.N. (1988) *Science* **241**, 1467–1472.

Knighton D.R., Zheng J., Ten Eyck L.F., Ashford V.A., Xuong N-h., Taylor S.S. & Sowadski J.M. (1991a) *Science* **253**, 407–414.

Knighton D.R., Zheng J., Ten Eyck L.F., Xuong N-h., Taylor S.S. & Sowadski J.M. (1991b) *Science* **253**, 414–420.

Lakowicz J.R. (1983) *Principles of Fluorescence Spectroscopy.* Plenum Press, New York, pp. 303–339.

Leathers V.L., Fletcher W.H. & Johnson D.A. (1990) *J. Cell. Biol.* **111**, 90a (Abstr. 393).

Liang H., Purucker W.J., Stenger D.A., Kubinec R.T. & Hui S.W. (1988) *BioTechniques* **6**, 550–558.

Lohmann S.M. & Walter U. (1984) *Adv. Cycl. Nucl. Prot. Phos. Res.* **18**, 63–117.

Lohmann S.M., De Camilli P. & Walter U. (1988) *Methods Enzymol.* **159**, 183–193.

McNeil P.L. (1989) *Methods Cell Biol.* **29**, 153–173.
Meinkoth J.L., Ji Y., Taylor S.S. & Feramisco J.R. (1990) *Proc. Natl. Acad. Sci. USA* **87**, 9595–9599.
Meinkoth J.L., Montminy M.R., Fink J.S. & Feramisco J.R. (1991) *Molec. Cell Biol.* **11**, 1759–1764.
Meyer T. & Stryer L. (1991) *Ann. Rev. Biophys. Biophys. Chem.* **20**, 153–174.
Montminy M.R. & Bilezikjian L.M. (1987) *Nature* **328**, 175–178.
Murray K.J., Reeves M.L. & England, P.J. (1989) *Mol. Cell. Biochem.* **89**, 175–179.
Nakamura T. & Gold G.H. (1987) *Nature* **325**, 442–444.
Nelson N.C. & Taylor S.S. (1983) *J. Biol. Chem.* **258**, 10981–10987.
Nigg E.A. (1990) *Adv. Cancer Res.* **55**, 271–310.
O'Dowd B.F., Lefkowitz R.J. & Caron M.G. (1989) *Ann. Rev. Neurosci.* **12**, 67–83.
Patel L., Abate C. & Curran, T. (1990) *Nature* **347**, 572–575.
Podesta E.J., Solano A.R. & Lemos J.R. (1991) *J. Mol. Endocrinol.* **6**, 269–279
Rall T.N. & Sutherland E.W. (1961) *Cold Spring Harbor Symp. Quant. Biol.* **26**, 347–354.
Rangel-Aldao R. & Rosen O.M. (1977) *J. Biol. Chem.* **252**, 7140–7145.
Rannels S.R. & Corbin, J.D. (1980) *J. Biol. Chem.* **255**, 7085–7088.
Robison G.A., Butcher R.W. & Sutherland E.W. (1971) *Cyclic AMP.* Academic Press, New York.
Rymond M. & Hofmann F. (1982) *Eur. J. Biochem.* **125**, 395–400.
Sammak P.J., Adams S.R., Harootunian A.T., Schliwa M., Tsien R.Y. (1992) *J. Cell Biol.* **117**, 57–72.
Saraswat L.D., Filutowicz M. & Taylor S.S. (1986) *J. Biol. Chem.* **261**, 11091–11096.
Schacher S., Castellucci V.F. & Kandel E.R. (1988) *Science* **240**, 1667–1669.
Scott J.D. (1991) *Pharmac. Ther.* **50**, 123–145.
Scott J.D., Stofko R.E., McDonald J.R., Comer J.D., Vitalis E.A. & Mangili J.A. (1990) *J. Biol. Chem.* **265**, 21561–21566.
Slice L.W. & Taylor, S.S. (1989) *J. Biol. Chem.* **264**, 20940–20946.
Southwick P.L., Ernst L.A., Tauriello E.W., Parker S.R., Mujumdar R.B., Mujumdar S.R., Clever H.A. & Waggoner A.S. (1990) *Cytometry* **11**, 418–430.
Straubinger R.M., Düzgünes N. & Papahadjopolous D. (1985) *FEBS Lett.* **179**, 148–154.
Stryer L. (1978) *Ann. Rev. Biochem.* **47**, 819–846.
Takamatsu T. & Wier W.G. (1990) *Cell Calcium* **11**, 111–120.
Taylor S.S., Buechler J.A. & Knighton D.R. (1990a) In *Peptides & Protein Phosphorylation* B.E. Kemp (ed.). CRC Press, Boca Raton pp. 1–41.
Taylor S.S., Buechler J.A. and Yonemoto W. (1990b) *Ann. Rev. Biochem.* **59**, 971–1005.
Terasaki W.L. & Brooker G. (1977) *J. Biol. Chem.* **252**, 1041–1050.
Tsien R.Y. (1989) *Methods Cell Biol.* **30**, 127–156.
Tsien R.Y. (1990) *Proc. Roy. Microsc. Soc.* **25**, S53 (Micro '90 Supplement).
Tsien R.Y. and Poenie M. (1986) *Trends Biochem. Sci.* **11**, 450–455.
Tsien R.Y. & Tsien R.W. (1990) *Annu. Rev. Cell Biol.* **6**, 715–760.
Tsien R.Y. & Waggoner A. (1990) In *Handbook of Biological Confocal Microscopy*, J. Pawley (ed.). Plenum Press, New York, pp. 169–178.
Van Haastert P.J.M. (1991) *Adv. Sec. Mess. Phos. Prot. Res.* **23**, 185–226.
Wang K., Feramisco J.R. & Ash J.F. (1982) *Methods Enzymol.* **85**, 514–562.
White J.C. & Stryer L. (1987) *Anal. Biochem.* **161**, 442–452.
Wu C.-W. & Wu F.Y.-H. (1974) *Biochemistry* **13**, 2573–2578.
Wu F.Y.-H., Nath K. & Wu C.-W. (1974) *Biochemistry* **13**, 2567–2572.
Zoller M.J., Kerlavage A.R. & Taylor S.S. (1979) *J. Biol. Chem.* **254**, 2408–2412.

PART V

Technology for Qualitative and Quantitative Detection of Optical Probes in Living Cells

CHAPTER TWELVE

Quantitative Digital Imaging of Biological Activity in Living Cells with Ion-sensitive Fluorescent Probes

W.T. MASON,[1] J. DEMPSTER,[2] J. HOYLAND,[1] T.J. MCCANN,[1] B. SOMASUNDARAM[1] & W. O'BRIEN[1]

[1] Life Science Resources Ltd, Cambridge, UK

[2] Department of Physiology and Pharmacology, University of Strathclyde, Glasgow, Scotland

12.1 INTRODUCTION

The development of chemical probes to image ions has made it possible to study biological activity in single living cells. Computer-controlled instrumentation for acquiring data from living tissue at fast rates has in turn made it possible to acquire ultra low light level data from these probes and to analyse either spatial or temporal changes in light emission, or both.

The ability to interface light detectors including photomultiplier tubes, fast CCD cameras and computer technology to the conventional microscope has made it possible not only to make qualitative observations, but also to derive quantitative image data from single cells, at speeds of up to *c.* 1000 frames per second if we use digital CCD cameras or confocal laser scanning technology, or many tens of thousands of samples per second if we use photon counting technology with photomultipliers.

Recent advances in imaging technology allow us to use digital cameras to acquire high precision data at rates of over 100 Hz. In this chapter, we shall discuss the means for applying these developments to previously intractable problems. Digital acquisition of information and subsequent processing is at the heart of virtually all of these new approaches. Here we focus mainly on a variety of technologies which can be used to study ionic gradients in living cells with optical probes. These include:

- precision digital CCD imaging of optical probes used as reporters of biological activity measured in live cells
- confocal imaging as an optical technique
- digital deconvolution or digital confocal to improve image resolution
- photometric measurements of optical probes in single cells
- integrated photometry and imaging.

Some specific applications from our laboratory using this technology will be used to illustrate the potential for such techniques.

FLUORESCENT AND LUMINESCENT PROBES, 2ND EDN
ISBN 0-12-447836-0

12.2 FLUORESCENT PROBES FOR LIVING CELL FUNCTION

12.2.1 Intracellular ions

The first measurements of ionic activity used photoproteins such as the calcium-sensitive molecule aequorin, which emitted light when combined with calcium ions. Calcium-sensitive dyes including arsenazo and murexide also provided advances. These dyes had some disadvantages in that they were accessible to only a few scientists, since introducing them into cells required microinjection. In 1982 the development of new fluorescent dyes by Tsien and colleagues provided the ability to investigate ionic activity in single cells (Grynkiewicz *et al.*, 1985). The dyes are sensitive to minute concentrations of intracellular ions such as occur in single living cells. The acetyoxymethyl ester form of the dye can be loaded into single cells, so virtually all scientists can use the dyes. Most cells contain endogenous esterases which rapidly (*c.* 5–30 minutes) hydrolyse the dye to form the free acid, which is trapped in the cell and is ion-sensitive. The wide range of dyes for imaging ions in cells is covered elsewhere in this volume in more detail, in Chapter 3.

A number of such dyes are shown in Table 12.1. These dyes have differing wavelengths of light output, but they all emit photons of light in the visible spectrum. Their light output is well-matched to available detectors, including both photomultipliers and intensified video cameras.

The two most commonly used dyes for measuring intracellular calcium are fura-2 (Fig. 12.1) and indo-1. The dyes have high quantum efficiency and are sensitive to calcium at concentration ranges from 30 nM to 5 μM. Fura-2 displays a single emission peak at 510 nm, but two calcium-dependent absorption maxima, one at 340 nm which increases with increasing ionized calcium, and a second at 380 nm which similarly decreases with a rise in ionized calcium. Indo-1 is generally excited by only a single wavelength of light (340–360 nm), but emits light at two different calcium-sensitive wavelengths (405 and 490 nm). With all such probes, choice of excitation wavelength will influence the wavelengths of light emitted and the specific dynamic range of the dye response.

If fluorescent signals are obtained as a pair at 340 and 380 nm (with fura-2 for instance) or at 405 and 490 nm (with indo-1 for instance) and the number of emitted photons or light intensity ratioed on a point-by-point basis with respect to time, the resulting 'ratio measurement' is proportional to Ca^{2+} concentration and reduces the chance of possible artifacts due to uneven loading or partitioning of dye within the cell, or varying cell thickness and dye concentration. In general, dye loading properties of cells vary widely. As an example, pituitary cells loaded with fura-2 load uniformly (Plate 12.1) whereas some other preparations may concentrate dye in the nucleus or other intracellular compartments, and

Table 12.1 Dual-emission and dual-excitation optical probes for imaging Ca^{2+} in single cells.

Dye	*Ionic specificity*	*Excitation wavelength (nm)*	*Emission wavelength (nm)*
Dual-excitation dyes			
Fura-2	Calcium	340 and 380	500–520
BCECF	Hydrogen (pH)	440 and 490	510–550
SNAFL-1	Hydrogen (pH)	470–530 and 550	600
SNARF-6	Hydrogen (pH)	490–530 and 560	600–610
SBFI	Sodium	340 and 380	500–520
PBFI	Potassium	340 and 380	500–520
Mag-fura-2	Magnesium	340 and 380	500–520
Dual-emission dyes			
Indo-1	Calcium	405 and 490	340–360
DCH	Hydrogen (pH)	435 and 520	405
SNAFL-1	Hydrogen (pH)	540–550 and 630	510–540
SNARF-2	Hydrogen (pH)	550 and 640	490–530
FCRYP-2	Sodium	405 and 480	340–350
Mag-indo-1	Magnesium	405 and 480	340–350

Note: These wavelengths are approximate. It is usual to employ interference filters for separation of discrete wavelengths with 7–15 nm half-bandwidth (i.e. the wavelength spread at half-maximum transmission for the filter centre frequency).

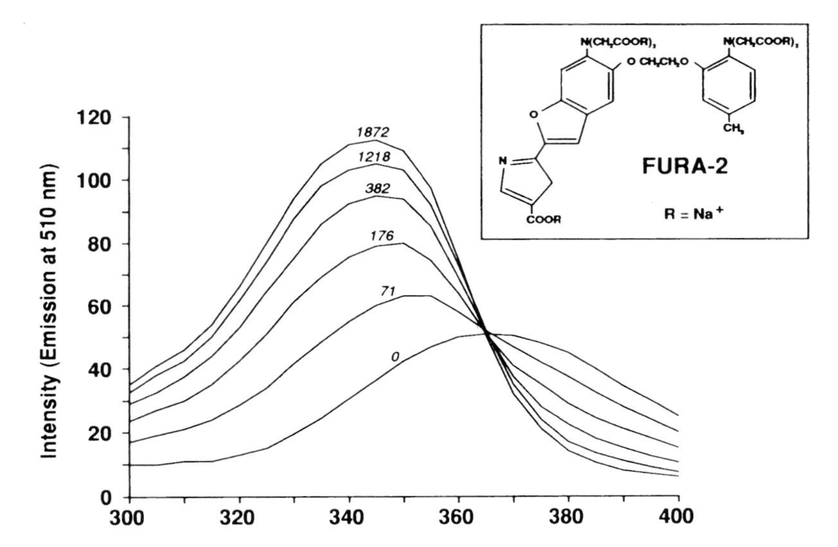

Figure 12.1 Fluorescence excitation spectra of fura-2 acid.

under these circumstances other loading methodologies may be more appropriate, such as loading at low temperatures or even microinjection.

Indo-1 (Fig. 12.2) is generally used for photometric measurements – it has a slightly faster time response than fura-2 in terms of dissociation time constant (typically estimated 5–20 ms as opposed to 30–40 ms for fura-2), and can be used with static optical beam-splitters to separate the emitted light and focus it on to two photomultiplier tubes as a continuous signal. Ratio measurements of the emitted photon channels can also be employed. This approach has the advantage that no movement need take place in order to change filter position, and so measurements can be fast and vibration free.

Considerable work is currently taking place to develop new probes for ionic activity in living cells. The field of calcium is attracting most attention and new probes for calcium ions include Calcium Green, Calcium Crimson and Calcium Orange, and Fura Red. Another potential combination of optical probes which enables measurements in the visible region is the combination of fluo-3 and Fura Red. Both are excited near the fluorescein excitation of about 490 nm. The former increases fluorescence output at 520 nm with increasing Ca^{2+} and the latter decreases fluorescence output above 560 nm with increasing Ca^{2+}. Because these dyes are excited near convenient visible laser lines, they have the potential to be simultaneously loaded into a single cell for ratiometric confocal imaging, or for dynamic video imaging with CCD cameras in systems not configured with UV-transmitting optics.

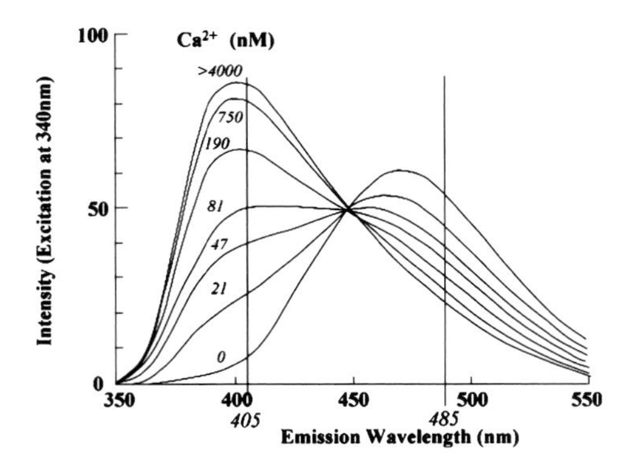

Figure 12.2 Fluorescence excitation spectra of indo-1 free acid.

Another dye, BCECF, measures intracellular pH as an optical signal, using similar dual-wavelength imaging technology. This dye is excited at 440 and 490 nm and measured at about 510–520 nm. pH rises increase the fluorescence of dye excited at 490 nm, but have little effect on 440 nm fluorescence. This dye is being successfully combined with fura-2 to provide simultaneous imaging of Ca^{2+} and pH. For this type of work, a four-position filter wheel is used to excite in turn the probes at 340, 380, 440 and 490 nm, having loaded both dyes into the single cell simultaneously. Fluorescence is measured with a 515 long-pass dichroic mirror and a 535 bandpass filter; this provides strong signals for both dyes with minimal overlap or interference of the two probes. The dye DCH and other probes of the SNARF and SNAFL family provide the potential to do either single-wavelength, non-ratiometric, or dual-wavelength emission ratiometric experiments.

Optical probes are also available for sodium (SBFI) and chloride (SPQ), but frustratingly little progress has been made in the development of a good, selective probe for potassium in living cells. Optical probes are also available for a wide range of heavy metals such as Zn^{2+}, Mg^{2+} (Mag-fura-2) and Cd^{2+}.

Other work with new calcium probes is focusing on indicators with higher and lower affinity for Ca^{2+}. Fura-5, for example, has an affinity for Ca^{2+} of 40 nm compared with fura-2 with an affinity of 135 nm at room temperature. Other probes are also being developed for magnesium ions, with excellent sensitivity and specificity.

12.3 OBSERVING BIOLOGICAL ACTIVITY IN 'REAL TIME'

The phrase 'real time' is often applied to specialized systems used for the study of biological systems. Most important events at the cellular level are dynamically changing on the order of seconds or less, and to capture these events in a meaningful way requires hardware and software which can work quickly – in real time – while the events themselves are occurring.

Response times of some of the ion-sensitive probes discussed here occur on the order of about 10 ms or less. The video cameras and photomultiplier tubes used as sensors can respond on a similar time scale. So the hardware and software used for capturing, processing and storing signals must be about as fast, or ideally even faster.

12.4 RATIOMETRIC IMAGING OF ION-SENSITIVE FLUORESCENT PROBES

Ratiometric imaging is at the heart of fluorescence microscopy for optical probes capable of detecting ions. If fluorescent images are obtained as a pair at 340 and 380 nm (with fura-2 for instance), and the images are ratioed on a point-by-point, or pixel-by-pixel basis (a pixel is the single resolving unit of a video camera, many thousand of which are combined together to give an overall image), the resulting 'ratio image' is proportional to ionized calcium concentration and reduces the chance of possible artifacts due to uneven loading or partitioning of dye within the cell, or varying cell thickness and dye concentration. Ratio imaging thus eliminates artifacts due to probe localization and cell geometry. Many of the best ion-sensitive and the new nucleotide-sensitive probes change spectral properties at two wavelengths. Ratio analysis of the two images produces accurate quantitation and reduces many artifacts associated with dye localization and cell thickness.

12.5 IMAGING STRATEGIES

12.5.1 Dynamic CCD ratio imaging of ions in cells

Digital imaging with charge-coupled device (CCD) cameras permit a second dimension of observation. Quantitative digital CCD imaging permits not only *temporal* measurements, but also *spatial* measurements of biological activity. The comparative characteristics of imaging and photometry are shown in Table 12.2. For most work with ion-sensitive fluorescent probes, either intensified analogue cameras or digital CCD cameras are used since the working levels of light emitted by these dyes are not detectable by normal video cameras alone. With this approach, a photosensitive array is used to image the cell or cells under study. Typically these arrays in analogue cameras might provide up to 768 × 512 pixel resolution, and can be used to capture up to 30 images per second. However, digital CCD cameras can function considerably above these rates with higher pixel resolutions. Frame transfer cameras, for example, can operate at 100 frames per second, and arrays up to 1600 × 1000 pixels or higher can be used.

Dynamic video imaging can resolve optical probes within cells in terms of both time and space. The MiraCal and MERLIN systems used in our laboratory are dedicated real-time imaging systems which enable this work. A typical configuration is shown in Plate 12.2. This system takes advantage of very recent developments in imaging technology by utilizing digital CCD cameras providing 12 or 14 bits of data, i.e. 4000–16 000 grey levels, at very fast speeds. Compared with analogue video cameras which have been used historically, these digital CCD cameras enable frame

Table 12.2 Comparison of characteristics of imaging and photometric measurements.

Imaging	*Photometry*
Spatial information	Limited spatial information
Slow (1–100 samples s^{-1})	Fast (up to 80 000 samples s^{-1})
Detectors are less sensitive	Detectors are highly sensitive
High data content (5–100 Mb)	Low data content (1–2 Mb)
Multiple parameter (current, voltage) difficult	Multiple parameter acquisition straightforward
Results typically off-line	Results can easily be on-line
Higher entry cost	Lower entry cost

rates of 100 images per second or faster, and can resolve very low contrast images because of the proportionately higher bit depth.

Living cells are loaded with fluorescent probe and mounted on a microscope, illuminated by a stable wide-spectrum light source such as a xenon lamp. An image of the specimen is projected from the microscope onto the face-plate of the digital CCD camera which produces a digital signal. Fluorescence is detected with programmable pixel resolution in both horizontal and vertical axes. Whereas previously intensified CCD cameras were used because the fluorescent image is very faint, modern CCD sensors from EEV, Kodak and others provide high quantum efficiencies and high dynamic range which may not require intensification and make them suitable for digital cameras. The faintness is due to compromises in the experimental arrangement: the dye concentration has to be low enough to avoid toxic effects to the cells under study, and the light source must not be so bright that it bleaches the dye.

12.5.1.1 Low light level cameras for fluorescence ratio imaging

Cameras employed for real-time fluorescence ratio imaging are similar to those used for astronomy. Several different types of detectors are discussed in this volume, consisting of either video frame rate detectors or digital CCD detector technology.

The signals emitted from optical probes are very faint and it is desirable to maintain excitation light at low levels to avoid bleaching of the probe. It is thus not possible to use standard video cameras for real-time fluorescence ratio imaging. Either intensified video cameras or cooled fast scan read-out cameras are employed. Both employ CCD detectors, but with different electronic outputs and different noise levels.

Intensified analogue video cameras are used for fast applications where video signals are required. They are generally two-stage, with an optically coated front end intensifier which governs the spectral sensitivity of the camera which in turn is coupled optically with a lens or with a fibreoptic taper to the video camera stage. Coupling with a fibreoptic taper is preferable to an optical relay lens as light loss is minimized. Typically, a relay lens coupled detector will be 5–10 times less sensitive than a fibreoptic coupled system. Most cameras for this work are custom designed. The first stage of the camera provides intensified input via a microchannel plate. Typically 10^5 or 10^6 lux is the light level required to be detected. A fibreoptic taper then reduces the image area onto a CCD image sensor. These devices generally put out a standard video signal which can be displayed on a television monitor and captured using video frame grabbers. The video signal is composed of odd and even lines (interlaced), generally producing 25 frames per second in Europe or 30 frames per second in North America. Many systems use only the odd or even lines, permitting a filter to be changed in between for very fast applications. For slower applications, both odd and even lines may be acquired. These detectors may be somewhat noisy due to the intensification process, but new generation CCD technology is impressive and single frame images obtained without signal averaging to reduce noise can contain very usable data. Because the data flow is high, high-quality frame grabbers with high-speed averaging are required if information flow is not to be lost. Averaging or integration is performed after an analogue image has been acquired, and following averaging the image is read out into computer memory through a high-speed analogue-to-digital converter, typically at 8-bit accuracy. Intensified cameras have limited dynamic range (about 103), but this is quite well suited to most available optical probes such as fura-2 which have dynamic ranges of about 30.

A more satisfactory type of detection technology is the cooled digital CCD camera. These have many advantages including a wide choice of chips with respect to resolution, output speed and general performance (see Plate 12.3 for examples). These cameras output a true digital signal into the computer, avoiding the need for further digitization as in the case of analogue cameras. 'Slow-scan' or 'fast-scan' digital CCD cameras typically consist of a surface mounted chip which is subjected to cooling to -20 to $-45°C$. This reduces dark current on the chip 10-fold for every 20 degree drop in temperature from ambient, and provides the capability to accumulate photon levels on the chip face for long periods of time without elevating the background signal. Unlike analogue cameras, which are typically 8-bit only, digital cameras provide much higher bit depth, from 10 to 20 bits digititation per pixel. 16-bit cameras are valuable for studying optical signals which do not vary greatly with time, but the 12- and 14-bit versions using new EEV and Kodak frame transfer, interline transfer and sequential transfer technology available from LSR AstroCam and other suppliers have numerous advantages. The cameras produce lower noise images and possess higher dynamic range and can be read out at frame rates faster than analogue TV cameras. Dynamic ranges of 10^5 to 10^6 are achievable by using 12- or 16-bit conversion, providing up to 65 536 grey levels.

12.5.1.2 Computer hardware for fluorescence ratio imaging

Given the complexity of current technology, it is seldom feasible for research scientists to develop their own software and hardware for imaging or photometry. Several manufacturers offer ready-made solutions which provide ready access to this technology. Our laboratory uses two approaches for fluorescence ratio imaging. Called MiraCal and MERLIN, their respective features are compared in Table 12.3. Whereas MERLIN is a fast, truly real-time image acquisition system capable of grabbing up to 30 frames per second from a video-based intensified CCD camera or in excess of 100 frames per second from a frame transfer digital CCD all at 12 bits resolution, MiraCal is based on a cooled slow scan read-out camera and can produce up to 8 images per second at 8-bit dynamic resolution, given availability of light emission. There is a cost differential of more than 2:1, with the complexity of faster real-time imaging costing substantially more.

MiraCal – digital ratiometric imaging for live cell experiments

MiraCal is configured on a high end PC running 32-bit Windows operating systems, making use of advanced graphics display technology. It uses a custom-designed, compact, cooled read-out camera with 8-bit analogue-to-digital conversion, and having over 32 000 pixels. Image integration and filter wheel control can be controlled from the icon-like interface, such that variable exposure times for each wavelength image can be requested to make full use of the camera dynamic range. Images are written either to hard disk or more rapidly to system memory, which can be up to 256 Mb. The system can capture up to 8 images per second or alternatively log data from up to 100 irregular regions of interest to a file available for graphing. The resulting data provides ratiometric images of more than adequate resolution for most biological applications.

MERLIN – fast fluorescence imaging with high-precision digital imaging cameras

There are many benefits to be gained from using a digital camera over analogue systems for the study of the ubiquitous secondary messenger intracellular calcium, involved in controlling processes as diverse as contraction secretion and nervous activity. The MERLIN system (Life Science Resources, Cambridge, UK) has the capability to digitize data at up to 14 bits at high speed, giving more than 16 000 grey levels per image. MERLIN enables both strong and weak fluorescence emissions to be studied in the same image. This system employs either a fast filter changer (RAINBOW) or a high-speed grating monochromator (SpectraMASTER) to effect rapid wavelength changes for multiple wavelength imaging. A general system schematic for MERLIN is shown in Plate 12.4.

The implications of these advances for ion imaging are several-fold. Firstly, neurons, for example, emit a strong signal from the cell body and a much weaker

Table 12.3 Technical comparison of MiraCal PRO and MERLIN real-time digital ion imaging systems.

	MiraCal PRO	*MERLIN*
Digital cameras		
Speed	8 images s^{-1}	100+ images s^{-1}
Pixel resolution	196 × 165 MIRA1000 TE	512 × 512, 1024 × 1024 frame transfer 768 × 512, 1317 × 1035, 1536 × 1024 shuttered
Pixel depth	8 bits (256 grey levels)	12 and 14 bit (4096/16 384 grey levels)
Pixel read-out rate	10^6 pixels s^{-1}	5.5×10^6 or 8×10^6 pixels s^{-1} maximum
Cooling	Peltier cooled 35°C below ambient	Peltier, −40°C
Analogue cameras		
Speed	8 images s^{-1}	25 images s^{-1}
Pixel resolution	768 × 576	768 × 576
Pixel depth	8 bits (256 grey levels)	8 bits (256 grey levels)
Excitation source	Rainbow	SpectraMASTER
Number of channels	4 wavelengths	8 wavelengths
System processor	Intel Pentium	Intel Pentium
RAM	32–64 Mb	96–256 Mb
Hard drive	2 Gb SCSI	4 Gb Fast SCSI (Barracuda)
Operating environment	Windows 95	Windows 95

signal from the axon. With an 8-bit camera one may have to choose from which part of the cell to collect data and to set the camera gain accordingly. Data from the other region of the cell will be lost! However, the MERLIN system allows all the data to be collected in the same image without loss or distortion. Secondly, the ability of high bit-depth cameras to discriminate low-contrast features in samples with an inherently low signal-to-noise ratio means that subtle changes in fluorescence, and therefore intracellular [Ca^{2+}], can now be detected. This feature permits the discrimination of more fundamental events in the generation of intracellular Ca^{2+} signals. Thirdly, an extremely wide range of light intensities can be detected. This offers the research scientist a very flexible camera which can image bright, transmitted light samples, dim fluorescent samples and even achieve photon counting sensitivity for luminescent applications. Fourthly, an all digital system means that data from other sources (such as membrane potential and current, from a patch-clamp amplifier) can also be read in to software at high speed and with high accuracy. MERLIN has the ability to acquire such analogue data in parallel with fluorescence imaging. The MERLIN system also has a very easy-to-use control interface which handles all aspects of acquisition, peripheral integration and analysis (Plate 12.5).

The sensitivity of the MERLIN cameras can be further enhanced by 'binning' data from several pixels into one 'super' pixel (Plate 12.6(a)). For example, 2 × 2 binning would combine the signals generated by light striking a 2 × 2 group of pixels (i.e. 4) on the CCD into a single data point. In this case sensitivity increases up to four-fold, as light striking four elements is combined to one data point. A further advantage of binning is a concomitant increase in frame acquisition rate (8 × 8 binning allows an image acquisition speed of 16 ms, or less). The resulting time resolution brings good temporal recording characteristics to imaging (Plate 12.6(b)).

So what does all this mean in practical terms for Ca^{2+} imaging? One advantage is that we now have much more choice in imaging technology, since MERLIN supports a wide range of digital cameras. Additionally, for those applications requiring the highest rates of data acquisition digital cameras allow MERLIN to easily double or triple frame acquisition rates, compared with analogue video cameras. The fastest camera (Neurocam) can actually acquire low-resolution images at 500 images s^{-1} (at full frame, 80 × 80 pixels) or faster (beyond 1000 images s^{-1}). This means that camera technology is now beginning to surpass the physical response times of the dyes under investigation. Finally, where high-quality, high-resolution images are required, the low noise levels of 12- and 14-bit digital cameras and their capacity for on-chip signal integration give far superior results to an analogue camera, whether or not it is intensified.

The MERLIN system has been used to acquire fluorescence images from spontaneously beating rat cardiac myocytes, loaded with the Ca^{2+}-sensitive dye, fluo-3, to quantify minute fluctuations in intracellular Ca^{2+}. The detection system used was a 12-bit, 512 × 512 pixel frame transfer digital camera from LSR AstroCam. The camera is based on a precision CCD sensor, but (unlike analogue video cameras) the image is digitized immediately it is read from the CCD chip. This generates much less noise than using a frame-grabber and the digitization can be up to 14-bits, giving 16 384 grey levels. Noise levels are further reduced by cooling the CCD to −40°C and by optimizing electronically the read-out signal. Immediately there are benefits to the user, as regions of a sample with widely different fluorescence intensities can be imaged easily and much smaller differences in signal intensity can be resolved. Plate 12.7 shows the resolution of Ca^{2+} 'sparks' with the MERLIN system, coupled with a CCD-based confocal laser scanning system shown later in this chapter. Other system components are a SpectraMASTER high-speed monochromator and a Pentium PRO workstation, to control acquisition and analysis. The MERLIN software is a comprehensive and easy to use package (Plate 12.5) which has all the features necessary for processing the data obtained.

Since digital cameras are not limited to a fixed frame rate or pixel resolution, they can provide images with more flexible read-out control than analogue video cameras and at much higher resolution (up to 1536 × 1024 pixels with MERLIN). The sensitivity of these cameras is also greater than their analogue counterparts, with detective quantum efficiencies (DQE) of 40% or more (compared with up to 10% DQE for analogue cameras).

12.6 DIGITAL IMAGE PROCESSING

Once images have been captured, they are processed prior to analysis. Background fluorescence is removed by capturing images at each desired wavelength, without cells, but in conditions identical to the experiment. These background images are subtracted pixel-by-pixel from each of the cell images before further processing. Another factor to allow for is uneven illumination of the field of view, i.e. 'shade correction'.

Ratioing to give a calibrated ion concentration involves applying a formula at every pixel, which takes into account an experimentally determined constant appropriate to the dye–ion interactions, dye quantum efficiency and system optics, intensity ratios for the pixel in each of the two images, and calibratable extremes of ratio intensity measured in the experimental arrangement.

Image analysers generally store a whole number at each pixel, having 256 possible values. Tables are used within the software to map the possible range of ion concentrations either linearly or logarithmically onto the range from 0 to 255. Ratioed images contain whole numbers representing ratios or ion concentrations and measurements made on these images use the tables to look up the true values. The tables also speed the computation of ratioed images: rather than performing a time-consuming division at every pixel the result is looked up in a table addressed by the two intensity values being ratioed. A pseudocolour look-up table is then mapped onto the range of grey values or ion concentrations.

12.7 DATA PRESENTATION

Software in the MERLIN and MiraCal imaging systems can present results in a wide variety of ways, which contribute to gaining insight into what really happened during an experiment. Examples include

- superimposing graphs of different regions, either of the same or of different cells, to compare their behaviour (Plate 12.8);
- histograms of the frequency of occurrence of ion concentrations at all of the pixels in a region;
- profiling pixel intensity along lines defined through cells and comparing these with profiles of the same lines from other images in the sequence, or plotting pixel profiles as a function of time;
- three-dimensional views of ion concentration profiles across a region (Plate 12.9);
- animating sequences to compare the changing ion profiles as a function of time;
- allowing tables of results to be analysed statistically, studied in spreadsheets, graphed in different ways, etc.

12.8 CONFOCAL LASER SCANNING MICROSCOPY – OPTICAL APPROACHES TO ENHANCED IMAGE RESOLUTION

Confocal imaging provides the means to gain improved resolution from microscope samples. Confocal laser scanning microscopy (CLSM) provides spatial and temporal resolution, although the method of imaging may vary widely from manufacturer to manufacturer. Images of 512 × 512 pixels may be obtained also, but they are constructed by sequential scanning a small point of laser light across the sample, and typically detecting emitted photons with photomultiplier tubes. CLSM technology has the advantage of providing a very small depth of field, eliminating out-of-focus photons and thus producing very fine detail. It may also permit optical sectioning through the cell, building up a 3-D image from serial optical sections. A disadvantage, however, is that images obtained with the technique tend to be rather noisy, so frame averaging is required to reduce noise. Although some systems can acquire confocal images at video frame rate and under excitation conditions where minimal bleaching occurs, it is usually necessary to image 20–100 images to obtain high-quality images. The latest CLSM technologies available from Zeiss, Leica, Noran and BioRad also permit work with UV-excited probes like indo-1. In general, the CLSM microscopes currently available divide between those best suited for obtaining 3-D information from fixed cells and tissues and those developed for high-speed temporal studies. In general, confocal technologies required to achieve high speeds are not wholly compatible with achieving high resolution, mainly owing to the nature of the optical characteristics of the respective systems.

CLSM technology offers several specific advantages over conventional bright-field or fluorescence microscopy:

(1) *Enhanced resolution.* With conventional light microscopy and fluorescent ratio imaging, the information sought is collected from a deep volume and is thus defocused, limiting resolution. With confocal technology, photons are collected only from a very small plane of interest, and successive fields can be compiled by 3-D reconstruction software to give precise localization of calcium waves, for instance.
(2) *Improvements in inaccuracy due to z-axis localization of ions or indicator dye.* Because the confocal microscope rejects out-of-focus light, a whole class of geometric artifacts is eliminated.
(3) *Elimination of background due to stray light.* Point scanning confocal microscopy is intrinsically insensitive to ambient light. Experiments can be performed in a lighted room. In conventional microscopy, care must be taken to prevent ambient light reaching the objectives. This point is of particular importance given the low light level emissions of the probes to be used, where stray light threatens the experiments.
(4) *Reduction in background due to medium.* Confocal microscopy is immune to medium fluorescence arising from, for example:

- indicator dye remaining from the loading process
- indicator dye leaking out of cells
- phenol red
- serum.

In conventional microscopy, efforts must be made to minimize each of these background signals. Spatial or temporal variation in any of these background components limit the precision of the experiment.

(5) *Reduction in haloing*. Conventional microscopic images show haloing: the cell boundary is difficult to distinguish clearly because of out-of-focus light from above and below the focal plane. Confocal microscopy provides clear definition of the cell edges, or subcellular organelle edges.

Both CLSM and conventional digital CCD imaging microscopy potentially provide a quantitative approach to imaging optical probes. Both approaches have the advantage that digital image analysis techniques permit a wide range of image information to be obtained and can yield both quantitative and qualitative data.

12.9 A NOVEL HIGH-SPEED DIGITAL CONFOCAL MICROSCOPE

We have co-developed, with an international optical manufacturer (Yokogawa Electric Corporation) together with a leading academic research institute (Ichihara *et al.*, 1996), a completely new approach to hardware-based optical confocal microscopy which potentially offers major advantages in terms of both high-resolution and high-speed confocal imaging. The optical and mechanical basis of this instrument is described elsewhere in this book (Chapter 30). In brief, the Ultra*VIEW* system is based on novel, patented, high optical throughput Nipkow disk technology packaged in a compact housing, accepting either laser or monochromatic or near-monochromatic light from a special optically modified and optimized monochromator or filter wheel. Because this system effectively produces a multipoint scanning pattern, the resulting image can be formed in real time and focused onto a CCD camera for image acquisition.

The system consists of a hardware-based confocal instrument combined with light source, high-precision digital CCD camera and software, and provides very flexible fast or slow image capture in a single package. A major advantage is the availability of either 12- or 16-bit imaging capability, and this system is the first 16-bit confocal microscope to become available. One of the objectives of the Ultra*VIEW* system is to address both research requirements (speed, resolution) in a single instrument, with potentially very high-speed, high-performance real-time imaging for true physiological applications in living cell research requiring high frame rates, as well as a host of high-resolution, high-dynamic-range applications including multiprobe applications requiring precision imaging on fixed or live cells. As such, the Ultra*VIEW* product family is positioned at the intermediate range of the market in price terms, but at the high end of the market in performance terms.

12.9.1 Advantages

The LSR Ultra*VIEW* Confocal Imaging System is a true optical confocal system which offers a number of advantages over current technology.

- It is highly compact (about 8 × 8 × 6 in; 20 × 20 × 15 cm). It requires no complex user alignment as all optical components are pre-set during manufacturing and it can operate with analogue or digital cameras.
- It is the first true flexible 12- or 16-bit confocal microscope.
- It uses a new generation of Nipkow disk technology with microlenses, offering significantly higher quantum efficiency than other instruments on the market, and thus lower bleaching rates and reduced phototoxicity.
- It has a scan speed of 360 frames per second, making it one of the fastest confocal products available today.
- It is compatible with digital imaging systems and with all modern optical microscopes, which are required to capture the confocal image from the attachment.
- It can be used with either a monochromatic light source or an Ar or Ar/Kr laser, and has fibreoptic delivery of light into the attachment. For multi-wavelength work, it can also function with a filter wheel on the output port. Use of a monochromator effectively enables the use of the confocal optics across a very wide range of wavelengths, providing superior performance improvements over a conventional laser.
- It provides a new optical acquisition technique to improve image quality from optical microscopes.
- The optical output of the confocal scanner is very fast and, when utilized together with acquisition and control technologies such as the MERLIN 12-bit imaging system, offers potentially more than 100 images per second. It is equally effective as a slow, high-resolution imaging system.
- It terminates in a standard C-mount.

- It has a measured point spread function of about 0.5 μm.
- Ultra*VIEW* can be transferred easily between microscopes without any need for realignment.
- Microlens system for improved optical throughput requires lower light levels for illumination and causes less photobleaching and phototoxicity
- Real-time viewing for fast focusing and set-up.

The LSR Ultra*VIEW* Confocal Imaging system provides 12- or 16-bit acquisition modes at high speeds and alternative higher resolution, slower data acquisition with LSR AstroCam precision cooled digital cameras employing Kodak 1400 or 1600 CCDs (12, 14 or 16 bits). A wide range of image processing and analysis features are also included in the control and acquisition software including live or offline temporal plotting versus intensity in grey level terms, image overlays for multiple label work, optional Z-control support for image stacks and a wide range of image processing.

Fluorescence images acquired with the Ultra*VIEW* confocal system are exceedingly sharp, even at submicrometre levels (Fig. 12.3). With high-speed imaging, this high resolution is maintained, allowing observation of submicrometre real-time events such as calcium 'sparks' in cardiac myocytes (Plate 12.7).

(a)

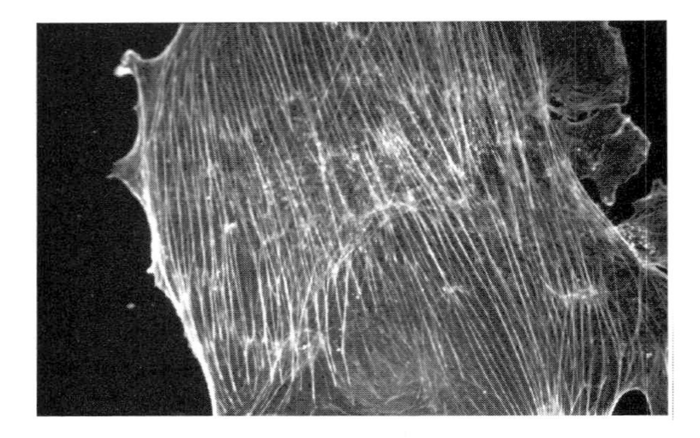

(b)

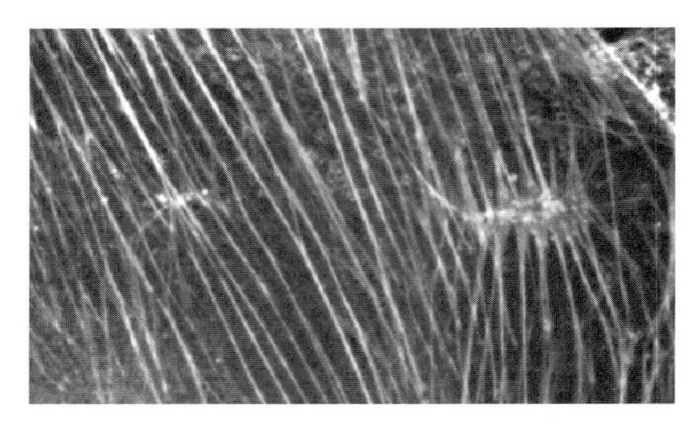

(c)

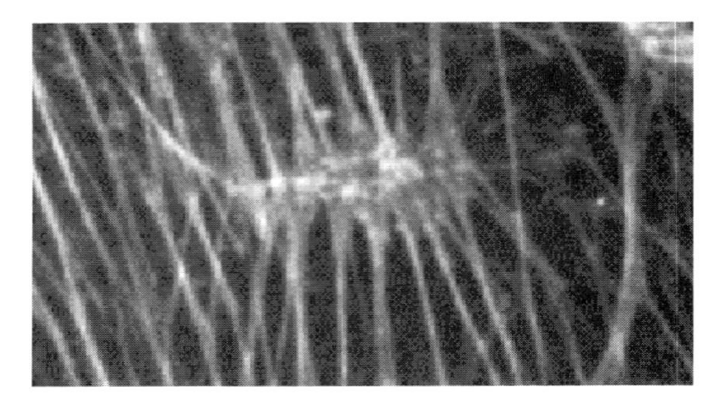

Figure 12.3 FITC-stained actin filaments imaged with the Ultra*VIEW* confocal imaging system at 16-bit resolution, and employing a progressive digital zoom in magnification. This shows that even at the highest resolution, optical resolution is still excellent even with the fibre diameter being significantly less than 0.5 μm.

12.9.2 Technical considerations with the Ultra*VIEW*

The method by which Ultra*VIEW* achieves collection of confocal optical images can be described briefly. About 20 000 microlenses on an upper disk focus collimated light from a laser or monochromatic wavelength-selective system on corresponding pinholes on a lower disk arranged in the same pattern as the microlenses on the upper disk (Fig. 12.4). Light passing through the pinholes is focused by the objective lens onto a spot in the specimen. Fluorescent light returns along the same path and is reflected by a dichroic mirror through a relay lens and onto a CCD camera. Upper and lower disks are tightly coupled, and thus the light beam effectively raster-scans the specimen. The instrument therefore performs as a true point scanner. Owing to the microlenses, the pinholes effectively collect 40% of the light incident on the surface of the upper disk. By means of this method, unlike a galvanometer, a constant, stable, and high-speed scanning can be achieved.

The essence of confocal technology is to attain maximum signal-to-noise images from the system. In the current instrument, we have manufactured all fundamental optical components including the microlens array, the pinhole array and dichroic mirrors with synthetic quartz and thus succeeded in minimizing fluorescent light in the system, yielding high signal-to-noise performance. Constant-pitch helical alignment of the confocal pinhole patterns provides uniform illumination and even scanning, unlike previous designs. Manufacturing variations are also minimal.

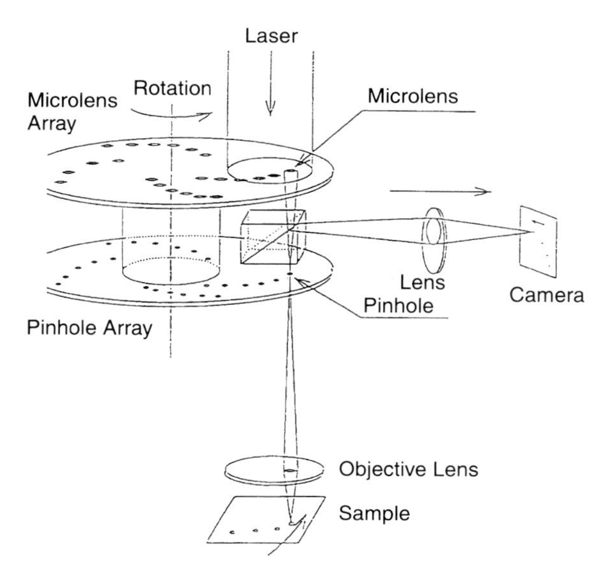

Figure 12.4 Optical arrangement of the LSR Ultra*VIEW* confocal system. Light from a laser or monochromator is focused onto the sample through an array of microlenses on the Nipkow disk and thence through the microscope objective. Fluorescence emission is passed back to the CCD camera or eyepiece through a pinhole, dichroic mirror and lens. This is effectvely a multibeam scanning instrument since all points in the sample are illuminated simultaneously.

Because in this system the primary image is focused on a pinhole array by the objective lens, a worst case semiconductor error in manufacturing of 0.5 μm leads to a corresponding error of only 50 nm specimen observation error with a 10× objective lens, and 5 nm with a 100× objective lens – small enough to be negligible in both cases

12.10 DIGITAL DECONVOLUTION AND DIGITAL CONFOCAL MICROSCOPY – 'SOFT' APPROACHES TO ENHANCED IMAGE RESOLUTION

All biological specimens viewed on microscopes are three-dimensional objects. This effect also impacts on specimens viewed with any optical device, such as a telescope used in astronomy. In the focal plane of a fluorescence microscope, the quality of image is affected by light from both above and below the actual true focal plane. This typically results in haziness, or blurring, of the image. Such effects occur in both 2-D and 3-D images.

A confocal microscope goes some way to eliminating such out-of-focus information, but even confocal images also contain significant out-of-focus information. Considerable mathematical modelling of this phenomenon has taken place over the last 10 years. It is clear from work in a number of laboratories that a variety of mathematical strategies exist for removing out-of-focus information from 2-D and 3-D images acquired from biological microscopes. The result is a marked improvement in image quality.

For many applications, particularly biological microscopy using precision cooled digital CCD cameras, this approach may be superior to confocal microscopy because in the absence of a confocal pinhole, the CCD camera collects all the image data, and this results in much shorter exposure times, less photobleaching and less phototoxicity. However, unlike optical confocal microscopy described in the section above, this is not a real-time technique due to the complexity of mathematical calculations. On the other hand, whereas early systems demanded high-powered and expensive workstations to implement the complex operations for deconvolution, newer high-performance PCs provide very satisfactory performance in general.

A system applying this functionality is called Ultra-VIS (Life Science Resources, Cambridge, UK). Improvements in image resolution with optical deconvolution may be gained by a variety of mathematical image processing modalities including nearest neighbour deconvolution, maximum entropy and other techniques, allowing finer features in the image to be acquired with greater accuracy and reliability (for a comparison of raw versus deconvolved images using FITC-labelled actin, see Fig. 12.5(a) and 12.5(b)). Because the inherent sensitivity of the CCD camera in such a system is greater than the PMT/laser combination on a confocal microscope, less optical probe can be used and conventional light sources as on a standard fluorescence microscope can be utilized.

It has also been shown that application of such techniques can result in marked improvements to images acquired on conventional confocal microscopes.

The technique of mathematical photon reassignment is variously called:

- digital confocal microscopy
- digital image deconvolution
- exhaustive photon reassignment
- image deblurring
- poor man's confocal.

The critical issue for performing such experiments is that the image quality obtained by application of image deconvolution is strongly dependent on the image quality input to the system; noise-in yields noise-out. It is thus preferable to acquire the best possible images using a cooled, high precision 12- or 16-bit digital CCD camera with the best resolution

(a)

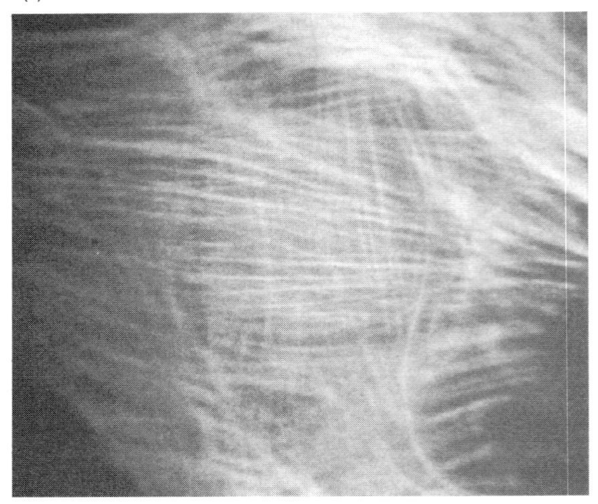

(b)

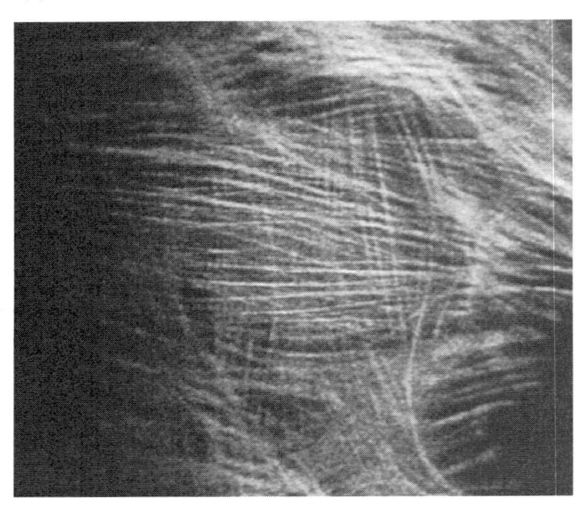

Figure 12.5 (a) Raw FITC-labelled actin filament captured with a 12-bit digital CCD camera. (b) Same image, but here deconvolved using the LSR UltraVIEW package with a blind deconvolution method.

possible. This combined with automated focusing control devices allows collection of 3-D digital information from biological specimens.

Various manufacturers provide systems which enable digital deconvolution. These include Applied Precision, Scanalytics, Vaytek (whose algorithms are incorporated in several packages) and Life Science Resources. Each package uses a variety of different approaches to identify out-of-focus information and remove it using mathematical algorithms.

12.11 FLUORESCENT MEASUREMENTS OF CYTOSOLIC IONS – COMBINED PHOTOMETRY WITH ELECTROPHYSIOLOGY

The ability to interface photomultiplier tubes (PMTs), fast video cameras and computer technology to the conventional microscope has made it possible not only to make qualitative observations, but to derive quantitative data from single cells, at speeds of up to 30 video frames per second if we use video cameras or confocal laser scanning technology, or many hundreds of samples per second if we use fast digital CCD cameras or photon counting technology with photomultipliers.

Every cell has ionic charge gradients generated by ions like Ca^{2+}, Na^{+} and K^{+}. Gradients in intracellular hydrogen ion concentration also are important to many biological processes. Ionic concentrations inside of cells change quickly and dramatically and underlie a wide range of cellular processes including development, growth, secretion and reproduction, so it is important to observe and understand them. As we have been able to apply image processing and low light level image capture from signals emitted by optical probes, it has also become clear that large standing ionic gradients occur within cells, and may persist for many seconds during stimulus or suppression of cellular activity. The source of ionic changes may also occur in widely different parts of the same cell, and even with small cells of only 10 μm or so, these ionic pools may be detected.

Photon acquisition and analysis is making it possible to study ionic gradients in living cells by using optical probes. Some specific applications and technology from our laboratory will be used to illustrate the potential for such techniques.

12.12 PHOTOMETRIC MEASUREMENTS IN SINGLE CELLS

A principal method of performing fast fluorescent measurements of ion-sensitive probes is by use of photomultiplier-based technology. This technology is covered in greater detail in Chapter 16 elsewhere in this book, which includes information about several different suppliers. Here we shall discuss one specific system used in our laboratory called PhoPRO, which embodies two functional Windows 95-based software systems called PhoCal and PhoClamp (Life Science Resources, UK). The comparative performance features of these systems are shown in Table 12.4. This approach is not restricted by the video frame rate of

Table 12.4 Technical comparison of PhoCal and PhoClamp high-speed photometric detection systems.

PhoCal	*PhoClamp*
• Photon counting	• Photon counting
• General purpose fluorescence and luminescence	• Advanced and general fluorescence/ luminescence
• Input/output trigger functions	• Input/output trigger functions
• Photometry with slow electrophysiology	• Photometry with fast electrophysiology
• On-line results display	• On-line results display
• Limited analysis/fitting with extensive measurement	• Statistical analysis/fitting and measurement package
• Windows 95 printer support	• Windows 95 printer support
• Supports all electrophysiology amplifiers	• Supports all electrophysiology amplifiers
• Single time base	• Dual time base
• Slow analogue (1000 s^{-1})	• Fast analogue (20 000 s^{-1})
• No waveform generator	• Digital waveform generator
• No automatic leak subtraction	• Automatic leak subtraction
• No superposition of records	• Superposition of records
• Triggered sweeps or continuous sweeps	• Triggered sweeps
• Chart recorder mode	• Repeated sweeps recording mode
• Single, dual and multiple emission and excitation	• Single or dual emission
• 1, 2, 3 or 4 wavelengths	• 1 or 2 wavelengths only
• Shuttering for bleach reduction	• Shuttering for bleach reduction
• Multiple probes	• Single probe work generally

normal cameras, but provides only temporal information with no spatial information. To obtain valuable data about the location as well as temporal changes which calcium undergoes, it is necessary to employ the more sophisticated technology of real time fluorescence ratio imaging.

Photometric detection can also be combined with electrophysiology, making it possible to accumulate fast electrophysiological signals (up to 80 kHz) while at the same time recording the somewhat slower responses of calcium ions, for example (Fig. 12.6).

In practical terms, an experimental system for measuring cytosolic Ca^{2+} using fluorescent probes has to cope with the following major tasks:

(1) measurement of the fluorescence emissions from the probe
(2) application of excitation light wavelength(s) to the probe
(3) measurement of cellular electrophysiological signals
(4) stimulation of cellular responses
(5) recording of experimental signals
(6) control and timing of the experiment
(7) display and analysis of results.

Although the detailed implementation of systems may vary, certain hardware items are required. Owing to the typically low levels of fluorescent light-emission, a light-measuring device of high sensitivity is required, such as a photomultiplier tube. If dual-excitation probes are to be used, a switching system capable of alternating the wavelength of the excitation light is required. Recording cellular membrane currents and voltage require the use of a voltage clamp device, either a switching voltage clamp or a patch voltage clamp. At least five different signal channels of experimental data are produced during the experiment, two fluorescent signals and their ratio per probe molecule, membrane current and voltage. These signals must be stored, typically in digital form on computer disk. The computer system and associated software plays a crucial role both in the management of these disparate measurement and stimulation devices during the experiment and the analysis of results afterwards. A generalized diagram of a typical combined fluorescence/electrophysiogical measurement system is shown in Fig. 12.7.

12.13 MEASUREMENT OF FLUORESCENT LIGHT – PHOTOMETRY VERSUS IMAGING

When designing (or choosing) a system for fluorescence measurements from single cells under a microscope, a basic choice has to be made between using an imaging device, such as a high sensitivity television camera, or a non-imaging photometric transducer such as a photomultiplier tube (PMT). The pros and cons of imaging versus photometric measurements

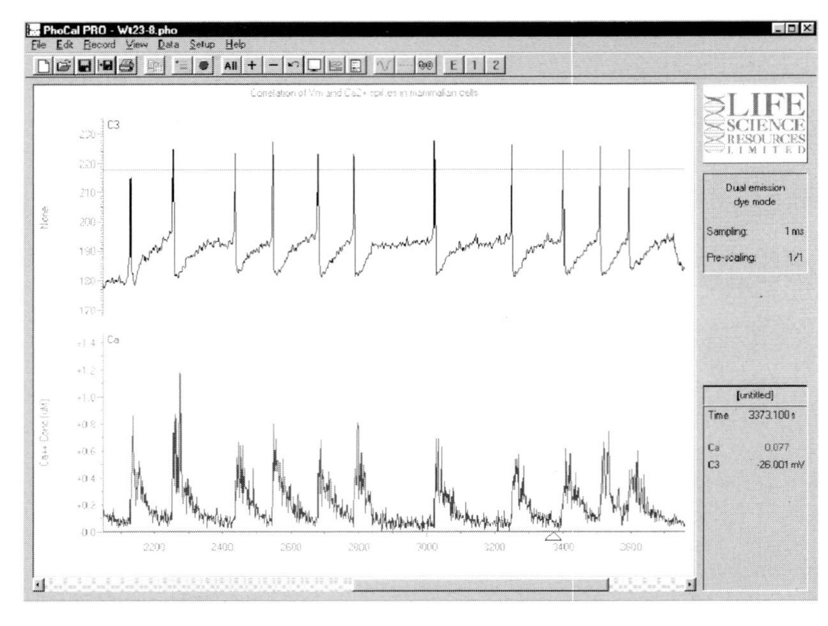

Figure 12.6 The user interface of the PhoPRO photometric system showing calcium transients in a pituitary cell loaded with indo-1 (bottom trace) and recorded with a patch-clamp electrode to detect voltage transients in the cell (top trace), which give rise to the calcium transients.

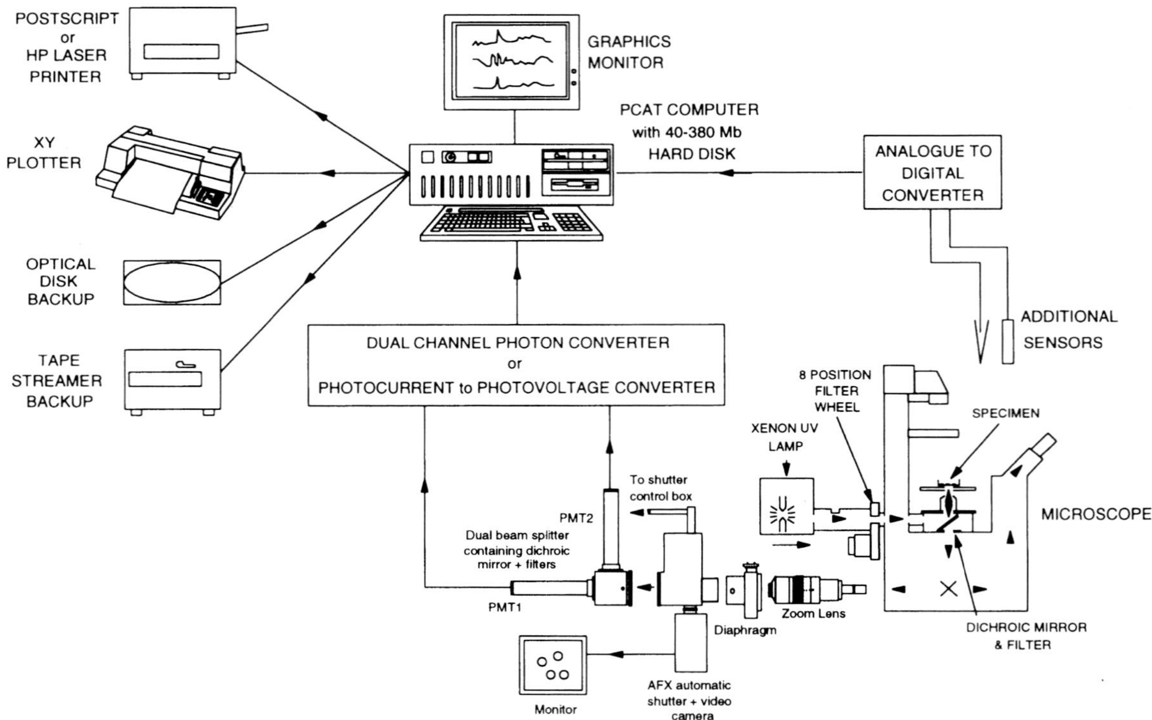

Figure 12.7 Schematic diagram of the PhoPRO system for photometric detection of single cell fluorescence, integrated with electrophysiology.

have already been presented in Table 12.2. PMTs produce a signal proportional to the amount of light falling upon their light-sensitive surface. They therefore provide a measure of the total fluorescence being emitted from the cell, rather than an image. The choice of camera or PMT depends upon the requirements of the experiment. If the spatial distribution of Ca^{2+} within the cell is important, or the differential responses of a number of cells within a wider visual field, then clearly an imaging device is required.

On the other hand, the PMT can be used at lower light levels and can acquire measurements at higher rates than cameras. For example, photometric devices can easily accumulate 1 data point each ms or faster, depending on light availability. Thus, the main factor which dominates the quality of low level light measurements is the quantal nature of light itself. Light emissions consist of a random stream of photons. In any given time interval, the number of photons captured by a sensor will fluctuate with a standard error equal to the square root of the mean photon count. A light sensor (PMT or camera) must accumulate enough photons to reduce these statistical fluctuations to acceptable levels. This, more than anything else, determines the rate at which light measurements can be made. With intracellular probe concentration and UV illumination levels found under typical experimental conditions, an acceptable photon count can be accumulated by a PMT within 5–10 ms. It takes much longer, however, to form an acceptable camera image since each sensing element of the camera is receiving photons from a much smaller area. For instance, a cell might occupy a camera image area consisting of 100 × 100 pixels. If this cell were emitting 1000 photons per ms (an entirely acceptable level for a PMT), each pixel would only accumulate an average of 1 photon each, in a 10 ms exposure.

Camera images also require a great deal more computer processing than the single measurement from a PMT. With a digital CCD camera, the contents of each pixel must be read out in sequence, digitized and transferred to the computer. Although there are means with some cameras to only digitize certain parts of image, and to accumulate the contents of blocks of pixels within the camera (a process called on-chip binning), nevertheless overheads of 30 ms to 1 s (depending on the type of camera) cannot readily be avoided. The PMT signal, on the other hand, is directly transferred to the data acquisition system with little overhead.

12.14 PHOTOMULTIPLIER TUBE TECHNOLOGY

The PMT is one of the few examples of thermionic tube technology which have not been replaced by semiconductors. It is an evacuated glass tube with a photosensitive surface painted onto one end. Photons striking this surface eject electrons which are accelerated along the tube by a high voltage (~1000 V) applied to a chain of metal electrodes (the dynode chain). As the fast-moving electrons strike the electrode surfaces, a cascade of further electrons is produced, getting larger with each step. This process of electron amplification results in a measurable pulse of current being produced at the PMT anode whenever a photon strikes the tube surface. The signal from a PMT is thus a series of these randomly occurring short duration (10–100 ns) current pulses and it is the rate of occurrence of the pulses which provides an indication of the amount of light falling upon the tube. The PMT is a very sensitive light-measuring device capable of detecting single photons of light when combined with an appropriate pulse counting device. Thus a complete PMT-based light measurement system requires:

- photomultiplier tube(s)
- stable high voltage (0–2000 V) source
- photon counting device
- data acquisition system
- PMTs, their associated power supplies and counting devices

PhoPRO and the new PhoTURBO systems use a custom-designed, high-performance photon counter board which is resident in the PC. Recent electronic processing advances have enabled photon counting to be operated simultaneously on two channels at rates up to 64 kHz.

12.15 PHOTON COUNTING VERSUS PHOTOCURRENT INTEGRATION

Two approaches can be taken to convert the raw stream of photon pulses produced by the PMT into a stable signal indicating light level. These are photon counting versus photocurrent measurement, and the pros and cons are presented in Table 12.5.

In the photon counting method, the current pulses are applied to a high-speed digital counter. The counter is allowed to accumulate pulses for a fixed period of time, at the end of which the number of counts is read out, stored, and the counter reset. The number of counts for successive count periods provides a measure of the light level. Photon counting requires a high-speed digital counter capable of responding to short duration pulses. The current pulses produced by a PMT are too small to be detected directly by the interface board and therefore an amplifier/discriminator built into the PMT is used to convert them into a digital pulse to match the counter input requirements.

The alternative to photon counting is to integrate the PMT current directly to produce an analogue voltage proportional to the average pulse frequency. Each photon current pulse adds a small amount of charge to the capacitor C, thus increasing the integrator output voltage. The resistor R, in parallel with the capacitor, causes the stored charge to ‘leak’ away with

Table 12.5 Pros and cons of photocurrent versus photon counting light detection.

Photocurrent	*Photon counting*
Low temporal resolution	High temporal resolution
Lower dynamic range	High dynamic range
Integrated current measurement	Bin counting
Time resolution depends on RC filter setting	Time resolution dependent only on bin width setting
Integrates many photons hence less sensitive	Single photon detection hence more sensitive
Limited at low light levels	Better for low light levels
Suited to large cells (>20 μm)	Ideal for small cells (5–20 μm)
Lower cost	Higher cost

a time constant $\tau=RC$. The integrator thus produces an output voltage which is proportional to the balance between the charge added from the photon pulses and that leaking away through R. In broad terms, the time constant τ is equivalent to the counting interval of the photon counter. This method can be faster than photon counting, and may be useful for some faster applications and faster probes. One example of this is recent work using fluorescent ion channel markers genetically expressed in cells; fluorescence changes occurring in this model are extremely fast, requiring measurements at intervals of 10–100 μs.

Overall, the photon counting method is preferable to photocurrent integration. PMTs produce a random background noise signal known as the 'dark current' upon which the photon current pulses are superimposed. The threshold discriminator of a photon counter can be set so that this signal is largely ignored, thus providing higher sensitivity and a better signal-to-noise ratio, compared with the integrator where its contribution cannot be avoided. Photon counting typically has a high dynamic range up to the number of photons able to be accumulated, typically in the millions to tens of millions before saturation occurs. The integrator, on the other hand, produces an analogue signal which when digitized with the typical 12-bit A/D converter has only a 12-bit range (1 to 4095).

12.16 EXCITATION FILTER SWITCHING

As with imaging, photometric systems designed to handle dual-excitation probes must have a means of switching the fluorescence excitation light between the absorbance peaks of the Ca-bound and free forms of the probe. In general, the excitation light wavelength is controlled by means of a rotating wheel containing two or more bandpass filters (e.g. for fura-2 340 nm (free) and 380 nm (Ca-bound)) or monochromator are mounted in the UV excitation light path. A 340/380 ratio measurement is obtained by rotating each filter, under computer (or specialized hardware) control, into the light path, taking a light measurement for each, storing the result, then computing the ratio. This must be done as quickly as possible to maximize the rate at which fluorescence ratio measurements can be made. The required rate depends on how fast the intracellular Ca^{2+} concentration is varying. In some cases this may be a matter of minutes and filter exchange times of 1–2 s can be used. However, some Ca^{2+} concentration changes, such as those associated with action potentials, are very rapid and in these situations ratio measurement rates of 500 Hz may be required. For such applications, dual-emission measurements are significantly faster since they do not require the filter wheel to be moved.

The complexity of the task of co-ordinating the rotation of the filter wheel, light measurements, and ratio calculations, requires that it be done under the control of a computer, or with specialized hardware designed for the purpose. For this reason most filter wheels are sold as part of complete fluorescence ratio measurement packages with appropriate hardware/software to control the wheel.

Filter wheels can vary quite radically in design and performance. The Life Science Resources RAINBOW filter wheel supported by the PhoPRO systems holds up to 8 filters and blanks controlled by a powerful DC servo motor. It is highly versatile, operating to 1 part in 2048 accuracy and capable of either rapid stepping (*c.* 100 ms) or smooth high-speed rotation up to about 50 revolutions per second, thus yielding as many as 1000 filter changes per second. This wheel is available with a UV-transmitting liquid light guide for remote operation. DC servo motors probably provide the best compromise of speed and smoothness overall.

Another strategy is use of a high-speed grating monochromator, like the SpectraMASTER, and this can offer computer-controlled wavelength changes at speeds up to 200 Hz, or less than 5 ms per filter change. The OSP-EXA unit from Olympus uses a galvanometer-driven mirror to alternate a light beam through dual interference filters, and can provide 400 filter changes per second.

12.17 DUAL-EMISSION PROBES

Many of the problems associated with the use of dual-excitation probes (complexity, limited ratio measurement rates, vibration) do not exist for dual-emission probes where it is the emission peak which is shifted by the binding of Ca^{2+}. No filter switching is required since these probes can be excited with a single wavelength. Instead, the emitted light is split into two components using a 45° wavelength-sensitive dichroic mirror and passed via bandpass filters to two PMTs. Light at the two characteristic peak emission wavelengths for the probe can thus be measured simultaneously, rather than alternately. This readily allows ratio measurement rates as high as 1 kHz, a limit set by the Ca^{2+}-dye dissociation time constant of the probe rather than any mechanical consideration. Sampling theory suggests that a probe with a dissociation time constant of 5 ms should be sampled about 4–5 times faster as a minimum to permit faithful reconstruction of the signal.

12.18 ELECTROPHYSIOLOGY COMBINED WITH PHOTOMETRY OR IMAGING

Fluorescence measurements using photometry or imaging can readily be combined with a wide range of electrophysiological measurements, ranging from cellular action potentials to single ion channels. The electrophysiological apparatus used in combination with fluorescence studies is quite standard and is discussed in Chapter 13 by Dempster.

The aims are generally to observe the effects of variations in an intracellular ion concentration (usually, Ca^{2+}, Na^{+}, H^{+}) on the behaviour of an ionic conductance system in the cell membrane, or vice versa. The whole cell patch-clamp technique is particularly well suited for use in fluorescence studies in that the patch pipette can be used to introduce the fluorescent probe into the cell cytoplasm, as well as to control the cell membrane potential. This has the distinct advantage that it allows well-controlled concentrations of the membrane impermeant acid form of the probe to be directly applied to the cell interior, rather than (as is normal) applying the permeant ester form to the bathing solution and relying on intracellular esterases to convert it inside the cell to the active form.

12.19 DATA ACQUISITION

Experiments combining two different measurement techniques can produce quite a large number of separate signal channels. At least five are produced with PhoCal or PhoClamp (providing two fluorescence channels, ratio, current, voltage and a marker channel). The only feasible contemporary approach to store this data is to digitize it 'on-line' by the computer system controlling the experiment and store it on computer disc. On-line computer-based fluorescence measurement systems are often simpler to use than non-digital acquisition systems such as tape recorders or even digital acquisition based on multiple computer workstations, as all major functions are controlled from one master program. They also can take better advantage of the high-precision results from digital photon counters.

With some limitations, electrophysiological analysis software can usually handle fluorescence signals. However, not being designed with fluorescence ratio measurements in mind, they often lack more sophisticated features such as calculating the actual ionic concentration from the ratio. Similarly, they cannot make use of digital photon counters.

The only package which addresses these problems is the PhoClamp system (part of the PhoPRO suite from Life Science Resources) which combines the simultaneous recording of fluorescence signals from dual-emission probes, using a photon counter, with the voltage pulse generation and recording features necessary for voltage clamp studies (Plate 12.10). Unlike most other programs, PhoClamp allows fluorescence and voltage clamp recording sweeps to be of different durations, allowing them to be better matched to the different time courses of Ca^{2+} currents (>1 ms) and concentration transients (5–20 s). The program also supports on-line leak current subtraction, a technique commonly used to separate voltage-activated Ca^{2+} currents from background membrane currents carried by other ions. PhoClamp also has a waveform generator which allows control of the patch-clamp amplifier for complex voltage clamp protocols, in conjunction with photometric experiments.

12.20 CONCORD – INTEGRATING IMAGING, PHOTOMETRY AND ELECTROPHYSIOLOGY ON A SINGLE WORKSTATION FOR ION IMAGING EXPERIMENTS

Unravelling the complex mechanisms which underlie cellular function warrants techniques that are capable of measuring, simultaneously, multiple biological events. CONCORD from Life Science Resources is an integrated system which provides imaging, photometric and electrophysiology data on a single platform, a growing trend of considerable interest for physiology, pharmacology and cell biology. The system is capable of obtaining ratiometric imaging data using a range of digital or analogue cameras and sequentially or simultaneously acquiring high-resolution temporal photometric data. CONCORD also allows simultaneous measurement of ion channel currents and membrane potentials using patch-clamp.

The versatility of the CONCORD system is demonstrated here with two specific examples. The first shows how fast and complex Ca^{2+} homeostasis mechanisms in cardiac myocytes can be analysed spatially, using images captured on camera and how the same work station can also capture sequential data of fast (1000 samples s^{-1}) temporal Ca^{2+} changes using photometry. The second example demonstrates simultaneous study of Ca^{2+} oscillations and related changes in membrane potential and voltage-gated calcium channel current in GH3 cells (a pituitary cell line). First we describe the CONCORD system in more detail.

12.20.1 Configuration of a combined imaging, photometric and electrophysiology system

CONCORD comprises three main components. The first component consists of the light detectors – the digital camera and two photomultiplier tubes. The second component is a fast monochromator light source (SpectraMASTER). The third component is the control unit which comprises a Pentium workstation, the CONCORD software and an analogue-to-digital acquisition board. With regard to software, the system comprises an integrated suite including functionality similar to that found in PhoPRO and MERLIN systems discussed earlier in this chapter.

The camera system used here was a 12-bit, 512 × 512 pixel frame transfer LSR AstroCam digital camera. This is described in an earlier section of this chapter in more detail and the identical approach is employed within the CONCORD system.

As an example, experiments on cardiac myocytes were undertaken on cells cultured for 4 days. For fura-2 experiments (imaging) cells were excited alternately at 340 and 380 nm and fluorescence images captured at 510 nm using the imaging module of CONCORD software. Indo-1-loaded cells (photometry) were excited at 340 nm and the emitted fluorescence was split by a dichroic mirror (450 nm), and detected simultaneously by two photomultiplier tubes, at 405 and 490 nm, using the photometric module of CONCORD software. Changes in Ca^{2+} levels were calculated as a change in ratio or Ca^{2+} concentration changes were determined by the method of Grynkiewicz *et al.* (1985). Ca^{2+} and electrophysiological measurements were made on GH3 cells 2–8 days after plating. The cells were loaded with indo-1/AM before nystatin-perforated patch recordings of the membrane potential were obtained by the method of Lledo *et al.* (1992). To activate voltage-gated calcium channels and record the influx of Ca^{2+} together with the underlying calcium current, the whole-cell patch-clamp configuration was used.

12.20.2 Calcium homeostasis in cardiac myocytes

The spatial changes in Ca^{2+} from a spontaneously contracting cardiac myocyte loaded with fura-2 are shown in Plate 12.5. The images were captured at a rate of 18 ratios per second, giving a ratio image every 55 ms. Ca^{2+} oscillations occurred at a 1.1 Hz frequency (peak-to-peak), but the duration of each transient is about 280 ms (Plate 12.6(b)). Such events could not be clearly resolved by an analogue video system, where the best ratio frame rate would generate only 1 or 2 data points for each transient.

The photometric module in CONCORD offers even higher temporal resolution of the changes in bulk Ca^{2+} levels. Plate 12.8 shows Ca^{2+} changes in spontaneously contracting cardiac ventricular myocytes loaded with indo-1 and detected using two photomultipliers. The Ca^{2+} changes have been calculated as a ratio of photons emitted at 405 nm and 490 nm when the sample was excited at 340 nm. Note that the CONCORD photometric system can acquire data at a rate of 1000 samples per second giving a data point every 1 ms. However, there is a trade-off between high temporal resolution and the loss of spatial information.

12.20.3 Calcium oscillation and ion channel activity in a pituitary cell model

Using CONCORD, changes in the membrane potential and intracellular Ca^{2+} in a GH3 cell have been measured simultaneously (Fig. 12.6 and Plate 12.10). The membrane potential showed repetitive depolarizations which were immediately followed by increases in intracellular Ca^{2+}. The increase in Ca^{2+} begins after the membrane potential has dropped to a more positive voltage, which suggests that the change in Ca^{2+} is a consequence of depolarization and activation of voltage-gated Ca^{2+} channels in the GH3 cells. To test this hypothesis, the whole cell Ca^{2+} current and the intracellular Ca^{2+} was measured simultaneously using the photometry/patch-clamp module in CONCORD. Because the kinetics of the voltage-gated Ca^{2+} channels are much faster than the kinetics of the Ca^{2+} homeostasis, the two parameters were measured at different rates and displayed using the unique dual time base of CONCORD (as data in Plate 12.10). The cell was subjected to 10 mV depolarizing voltage steps each lasting 100 ms, applied via the CONCORD pulse generator. The resulting Ca^{2+} current over the 100 ms period and the concurrent changes in Ca^{2+} over a 10 s period were measured. These data clearly demonstrate the causative role of Ca^{2+} channel activity in modulating intracellular Ca^{2+} levels.

Key advantages of such a combined approach are versatility and flexibility. CONCORD can meet the diverse needs of modern cell biologists and electrophysiologists with state-of-the-art performance in a single system. Additionally, more choice in imaging technology is possible since CONCORD supports a wide range of digital cameras. For those applications requiring the highest rates of data acquisition, digital cameras allow CONCORD easily to double or triple frame acquisition rates, compared with analogue video cameras. CONCORD also offers a unique solution correlating ion channel activity and ion concentration, with high-speed, dual time base patch-clamp and fluorescence recording and which increases the fluorescence sampling rate to 64 000 s^{-1} on each of two channels. This will exceed the performance of even the new generation of ultra-fast voltage-sensitive dyes, but will be particularly valuable for some newer probe technologies where protein–ion channel configurations probed with fluorescent tags may show relaxation kinetics well into the microsecond range.

12.21 CELL CULTURE AND LOADING OF FLUORESCENT PROBES

Routine cell preparation and culturing techniques are employed. The cells intended for study are plated onto thin glass coverslips, usually No. 1.5. This permits focusing from below using an inverted microscope and allows passage of the lower wavelengths of UV light required to excite many of the available fluorescent probes (<360 nm). Plastic media cannot be used as they absorb UV excitation light.

Cells may be loaded with ion-sensitive fluorescent probe in one of two ways. The free acid form of the probe may be directly loaded into the cell by micropipette or patch pipette. For fura-2, a concentration of around 50–100 μM has been found to be satisfactory. This method, however, requires techniques which are often unavailable in many laboratories. The great attraction of many of these probes is that they are available in the acetoxymethyl ester form. These non-polar ester derivatives may be added to the extracellular medium where they will diffuse across the membrane to be hydrolysed by non-specific cytoplasmic esterases, resulting in the membrane-impermeable free acid form. Generally, the ester form of the dyes is insensitive to changes in ion concentration. Unfortunately most plant cells studied so far appear to have low esterase activity, and this has made experiments difficult unless microinjection is used.

For experiments measuring intracellular free calcium using Fura-2 or measuring pH using BCECF or other probes for other ions, the probe can be initially made up as a stock solution of 2 mM in DMSO. Cells are normally incubated in a 4 μM solution of the acetoxymethylester form (fura-2/AM, BCECF-AM or other) made up in a standard extracellular medium for 30 minutes at room temperature or 37°C.

Data collection from cells should start within approximately 2 h of loading as they have been seen to lose responsiveness when loaded for extended periods.

12.22 CALIBRATION OF ION-SENSITIVE DYES IN LIVING CELLS AND IN SOLUTION

The following method uses as an example intracellular free calcium measurements using the dual-excitation fluorescent probe fura-2, which is available in both the free acid and acetoxymethyl ester forms. Many other probes for different ions are now available but the general methods are valid for any dual excitation probe. For pH-sensitive dyes such as BCECF, permeabilization with nigericin could be used in place of ionomycin, as in the example given below.

The MERLIN and CONCORD systems enable a very straightforward approach to dye calibration *in situ*. In these packages, a system called the Calibration WIZARD has been developed to streamline calibration procedures. Data required for calibration may be entered into the Calibration WIZARD software by one of two methods:

(1) entering the maximum (R_{max}) and minimum (R_{min}) ratios obtained into the equation below which is pre-programmed into the software; or
(2) by defining regions of interest around one or more cells in a field and automatically extracting data at the required wavelengths, which is in turn entered into the equations and applied to further data acquisition.

The equation relating the different quantities is:

$$\text{Ion concentration} = K_d \times b \times [(R - R_{min}) / (R_{max} - R)]$$

where K_d is the dissociation constant of the probe, b is (intensity at the upper wavelength at R_{min}) / (intensity of the upper wavelength at R_{max}), R is the measured ratio, R_{max} is the ratio when the probe is saturated with calcium ions, and R_{min} is the ratio with no free calcium ions present.

12.22.1 Free acid solution method

The simplest method but arguably least accurate method for initial calibration is to prepare two solutions of medium, one containing 5 mM $CaCl_2$ which will saturate the probe with free calcium ions and result in the maximum ratio obtainable. The other should have the $CaCl_2$ substituted by 1–10 mM EGTA which will bind all free calcium ions to result in the minimum ratio obtainable. Both must contain the free acid form of the probe at a concentration of about 50–100 μM, which is the approximate concentration found in cells loaded with the ester derivative.

12.22.2 Intracellular method

Cells are loaded with the acetoxymethyl ester derivative of the probe and the cell membrane permeabilized. Ionomycin at a concentration of about 2 mM is suitable for calcium. Some laboratories have also used low concentrations of digitonin or saponin to permeabilize cells for calibration.

The R_{max} measurement should be made in elevated extracellular $CaCl_2$. A concentration of 10 mM has been found to be sufficient to saturate the probe within the cell. Addition of the ionomycin causes a rapid, sustained rise in intracellular calcium. Measuring the R_{min} value on the same field by washing the cells two or three times with calcium-free medium containing 1–10 mM EGTA is the next step. Transport across the cell membrane – resulting in binding of free calcium ions – may require at least 15 minutes.

12.23 DERIVING SPECTRA DATA FROM OPTICAL PROBES *IN SITU* – THE SPECTRALWIZARD

A further development from Life Science Resources is the SpectralWIZARD software which works interchangeably with the SpectraMASTER monochromator. The design of the SpectraMASTER monochromator means that a new experimental protocol may be utilized, namely that of scanning through a range of wavelengths and producing graphs of fluorescence against wavelength. This data can then be used to characterize the behaviour of a fluorescent probe *in situ* and use this information to interpret data from the probe more accurately. In particular, the wavelengths used to excite the probe could be selected accurately. In addition, recent work suggests that certain artifacts associated with ratio dyes such as fura-2 (viscosity, path length differences) can be minimized by modifying the excitation wavelengths by subtle amounts based on the spectral properties of the optical probe *in situ*.

The Wizard takes the form of a Windows 95 tabbed dialogue box which is launched either stand alone (for off-line data analysis) or from within the host application. A spectral scanning experiment will typically consist of either a single spectral scan, or a number of scans taken at differing but known concentrations of the optical probe target (for example Ca^{2+} or pH).

Before running the SpectralWIZARD, the acquisition device, which would be the camera of an imaging system, or the photomultiplier tube of a photometric system, is set up to provide optimal results. In the case of an imaging system, regions of interest are placed over the parts of the image from which measurements are taken, up to 100 regions. The SpectralWIZARD remotely controls the parent application and reads from it the resulting grey level or intensity data directly.

Once acquired, scans can be saved to disk for future reference, and a number of important analyses may be performed. The most commonly used of these will be the 'Find peak' function, which will show the excitation wavelength of light that produces the brightest level of output fluorescence. The results of this calculation may then be used to select accurately the wavelengths used for subsequent ratio imaging experiments. Other analysis types include the ability to calculate the dye dissociation constant of a probe (K_d), and the isosbestic point can be estimated for ratio probes.

12.23.1 Potential Artifacts in Optical Imaging with Fluorescent Probes

In discussing the potential artifacts and pitfalls which may be encountered in optical imaging of fluorescent probes, it is important to look at the general applications for which fluorescent probes may be used. For example, the calcium-sensitive probe fura-2 may be used either as a single-wavelength qualitative probe to image the spatial distribution of calcium. Alternatively it may be used in dual-wavelength ratio mode in a calibrated system to quantify additionally calcium concentrations and fluxes. The potential artifacts will be very different in the two systems; indeed, simply utilizing fura-2 in its ratio mode corrects most of the problems associated with single-wavelength measurements.

Potential artifacts which may affect the measurement of ion concentrations are covered more fully in Chapters 7 and 8 earlier in this book.

12.24 SUMMARY

Considerable progress has been made in a range of important enabling technologies which are required to support quantitative digital imaging of optical probes in living cells. These include the development of improved probe and reporter technology, faster and more sensitive digital CCD cameras and photometric acquisition technology, and optical and mathematical approaches for optical and digital confocal imaging. All of these are contributing to our ability to study biological events underlying most cellular function on ever-faster time scales and with greater resolution and precision. New advances in probe technology will be of particular importance for the future as better, more specific probes are developed, not only for ions, but also for many second messenger pathways including enzymes, proteins and a variety of related small molecules.

REFERENCES

Grynkiewicz G., Poenie M. & Tsien R.Y. (1985) *J. Biol. Chem.* **260**, 3440–3450.

Ichihara A., Tanaami T., Isozaki K., Sugiyama Y., Kosugi Y., Mikuriya K., Abe M. & Uemura I. (1996) *Bioimages* **4**, 57–62.

Lledo P.-M., Somasundaram B., Morton A.J., Emson P. & Mason W.T. (1992) *Neuron* **9**, 943–954.

CHAPTER THIRTEEN

Fast Photometric Measurements of Cell Function Combined with Electrophysiology

J. DEMPSTER
Department of Physiology and Pharmacology, University of Strathclyde, Glasgow, UK

The development of highly fluorescent ion-sensitive dyes has revolutionized techniques for the measurement of intracellular ion concentrations. With these dyes, it has become possible to make quantitative measurements of intracellular Ca^{2+}, Na^{+} and H^{+} ion concentrations from a wide variety of cell types, in a much less invasive fashion than possible with earlier techniques such as ion-sensitive microelectrodes, or photoproteins such as aequorin.

A particular feature of these dyes is their potential for extracting absolute measurements of intracellular ion concentrations, independent of dye concentration, cell thickness, etc. The binding of inorganic ions to dyes, forming ion–dye complexes, changes the light emission or excitation properties of the dye. For instance, the binding of Ca^{2+} to the dye quin-2 increases its fluorescent emission by up to seven-fold. However, it is difficult to relate such changes directly to Ca^{2+} concentration since the overall magnitude of the signal is also dependent on dye concentration which may itself be changing with time (often due to photobleaching). These problems can be avoided by using dual-wavelength dyes such as the Ca^{2+}-sensitive fura-2 or indo-1 dyes, Na^{+}-sensitive SBFI and H^{+}-sensitive BCECF. In these dyes, the binding of the inorganic ion has a differential effect at separate parts of the excitation or emission spectrum.

For instance, the excitation spectrum of fura-2 is modulated by Ca^{2+} binding. In the absence of Ca^{2+}, the dye absorbs ultraviolet light strongest at wavelengths around 380 nm and hence produces the strongest emission when illuminated with light of that wavelength. When Ca^{2+} ions are bound to the molecule, the excitation peak shifts to 340 nm, with the position of the emission spectrum peak (at 510 nm) little affected (Grynkiewicz *et al.*, 1985). Although the absolute amplitude of the fluorescent emission from the dye is dependent on a multitude of factors, including dye concentration and excitation light intensity, the 340/380 ratio is dependent only on the Ca^{2+}-fura/fura ratio and can thus be used as a quantitative measure of Ca^{2+} concentration. Fura-2, and other dyes with ion-sensitive excitation spectra, are described as *dual-excitation* dyes.

Ca^{2+} has a similar dual-wavelength effect on indo-1, except that it is the emission spectrum which is sensitive to Ca^{2+} binding. The peak emission wavelength of indo-1 is 482 nm, while with a Ca^{2+} molecule bound it

FLUORESCENT AND LUMINESCENT PROBES, 2ND EDN
ISBN 0–12–447836–0

is 398 nm. In this case, the 398/482 emission ratio is proportional to Ca^{2+} concentration. Indo-1 is described as a *dual-emission* dye. Further details of the properties of fura-2 and indo-1 can be found in Grynkiewizc *et al.* (1985).

13.1 FLUORESCENT LIGHT MEASUREMENT

At least three different methodological approaches, each with their own advantages, can be taken to the measurement of fluorescent emissions from dye-loaded cells – spectrofluorometers, imaging systems, and microscope-based photometric systems.

Spectrofluorometric techniques are used in a variety of fields and general-purpose spectrofluorometers (e.g. Perkin-Elmer LS-5) for measuring the light emission and excitation properties of samples in liquid suspension are widely available. Such devices are equipped with light sources capable of producing controllable monochromatic light over a range of wavelengths. This capability makes them particularly suitable for use in studies with dual-excitation dyes such as fura-2. Their main limitation, as research tools in cell physiology, is that measurements are made from large populations of cells in suspension. Consequently, the properties of single cells cannot be studied. Also, the rate at which ratio measurements can be made is often limited to less than one per second.

In general, it is preferable to study the properties of single cells, in order to avoid problems with inhomogeneities in the response of cells within populations. This requires the use of a microscope capable of visualizing individual cells and a system for analysing the image. Within microscope-based systems, either a photometric or an imaging approach can be taken. Imaging is discussed in detail elsewhere in this volume, but essentially involves the use of a high sensitivity CCD camera or confocal microscope system to capture the fluorescence images of the cell and an image-analysis system to process the resulting pictures. The technique permits the spatial distribution of ions such as Ca^{2+} with the cell to be observed and quantified (e.g. Williams *et al.*, 1985). Inhomogeneities in the response of cells within small populations to a given stimulus can also be observed. It is hard to achieve more than about 10 ratio measurements per second and the large amount of data accumulated per image makes the storage and manipulation of series of images difficult. There may also be problems integrating such systems with other apparatus such as voltage clamps.

In the microscope-based photometric system, fluorescent light emission from the cell is channelled, via the microscope optics, to a photodetector instead of a camera. Thus, in terms of properties, it lies between the spectrofluorometer and the imaging system. Single cells can be studied but detailed spatial information is lost, with only the integrated light emission from the whole cell being collected. The distinct advantages of the system, however, are that significantly faster measurement rates (in the order of 100–1000 ratios s^{-1}) are possible, and the system can be more easily combined with other types of single cell measurement systems.

High-speed measurements of intracellular ion concentration changes are of interest in themselves, but the technique is further enhanced when combined with the simultaneous measurement of additional cell parameters such as muscle length and tension, and electrophysiological measurements such as membrane potential and current. This method has been used extensively in the study of muscle contraction (e.g. Baylor & Hollingsworth, 1988; Eisner *et al.*, 1989) and neurosecretion (Mollard *et al.*, 1989).

The desire to capture and analyse such a varied range of signals from a single cell places demands on the instrumentation for recording experimental signals and the production of stimuli – an area in which the computer has come to play a central role. This chapter discusses some of the issues involved in the development of such computer-based combined fluorescence/electrophysiological measurement systems.

13.2 A FLUORESCENCE/ELECTROPHYSIOLOGICAL RECORDING SYSTEM

A diagram of a typical measurement system is shown in Fig. 13.1. Cells or tissues are mounted in an experimental chamber on a microscope adapted for UV fluorescence work (most commonly used dyes have excitation and emission spectra in the UV range), such as the widely used Nikon Diaphot system (Nikon Corp., Tokyo, Japan). The cells are epi-illuminated with narrow-band UV light via the objective lens, as shown. A device (monochromator or rotating filter wheel) may exist within the excitation light path to allow excitation at different UV wavelengths, such as required when dual-excitation dyes are in use. The resulting fluorescent light emissions from the cells are projected via the microscope optics, through the camera port, onto photomultiplier tube (PMT) light measurement devices. For experiments with dual-emission dyes, two PMTs are used, combined with a wavelength-sensitive 45° dichroic mirror and band-pass filters to allow fluorescent emissions to be simultaneously measured at two different light wavelengths.

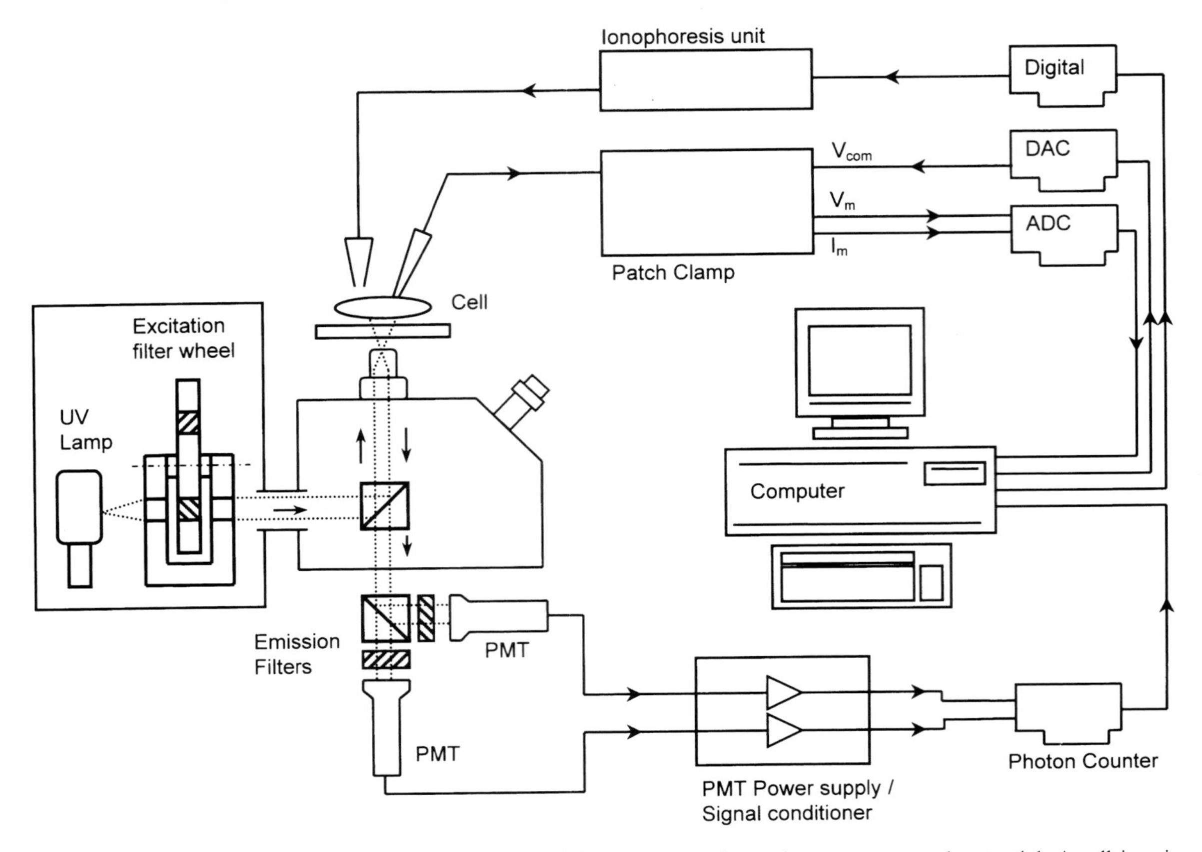

Figure 13.1 A system for the combined recording of cell fluorescence and membrane current and potential. A cell is epi-illuminated with UV light through the microscope objective lens. Fluorescent light emissions are captured via the microscope camera port and measure with PMTs. For dual-emission dyes, the emitted light at the peak emission wavelengths for the dye is measured by a pair of PMTs, after being split into two streams using a 45° dichroic mirror and bandpass filters. For dual-excitation dyes, a filter wheel placed in the excitation light path is used to provide alternating UV excitation wavelengths. Cell membrane potential and current are measured via a micropipette attached to the cell and drugs applied via an ionophoresis unit. Overall timing, application of stimuli and digital recording of signals are performed by a laboratory computer equipped with A/D, D/A converters and photon counting interface cards.

Electrophysiological measurements are made via glass micropipette electrodes attached to the cells. A variety of recording techniques are possible (e.g. see Standen *et al.*, 1987) but the most common one, in the context of fluorescence analysis, is the whole-cell patch-clamp method. A clean fire-polished glass micropipette with a tip diameter of around 1 μm can be made to form a tight seal when pressed against the cell membrane. If a suction pulse is applied to the pipette, the interior membrane under the pipette can be sucked away, providing low resistance access to the inside of the cell. When a patch-clamp amplifier (e.g. EPC-7 or EPC-9, HEKA Elektronic, Lambrecht/Pfalz, Germany; Axopatch, Axon Instruments, Foster City, CA, USA) is attached to the electrode, the current flowing across the cell membrane can be measured and the cell membrane potential set or 'clamped', to precisely controlled values. A detailed introduction to the patch-clamp method can be found in Sakmann & Neher (1983).

In general, some means is required to apply a variety of controlled stimuli to the cell under study. Many ion channels (e.g. Na^+, K^+, Ca^{2+}) are voltage-sensitive and can be activated by abrupt depolarizing changes in cell membrane potential. The patch-clamp has a command voltage input for this purpose, allowing externally generated voltage patterns to be applied to the cell. Other ion channels may be linked to receptors on the cell surface and activated by the binding of a specific agonist. Agonist may be applied to the cell either ionophoretically or by pressure ejection from a second micropipette, or by changing the solution flowing into the experimental chamber.

In summary, a combined fluorescence–electrophysiological recording system produces at least four output signals: two fluorescent light emission channels from the PMTs, cell membrane potential and current from the patch-clamp. Stimulus patterns must also be provided to the patch-clamp command voltage input and/or an agonist application system. Overall control of the experiment is usually handled by a computer system which generates the stimulus voltage patterns and records PMT and patch-clamp signals.

13.3 THE PHOTOMULTIPLIER TUBE

The fluorescent light emissions from ion-sensitive dyes, although having improved tremendously over the past decade, are nevertheless of a low level by normal standards, and require a light measurement device of high sensitivity. This requires the use of the photomultiplier tube rather than simpler solid-state devices such as the photodiode. A PMT is an evacuated glass tube containing an array of electrodes called the dynode chain, with a light-sensitive phosphor-coated surface at one end. A high voltage is applied to the dynode chain. When a photon strikes the sensitive surface, some electrons are knocked out of the phosphor. These electrons are accelerated by the electric field and when they strike the first electrode in the chain, they knock out more electrons which are swept down to the next electrode. A cascade of electrons occurs from one dynode to the next, each contributing more electrons until they finally appear at the anode as a measurable current. A PMT produces a current pulse for each photon striking the tube surface. PMTs with appropriate housings and high-voltage power supplies can be obtained from a number of suppliers, e.g. Electron Tubes Ltd. (Ruislip, Middlesex, UK) and Oriel Instruments (Stratford, CT, USA). (It is worth noting that PMT tubes can vary in terms of sensitivity and other operating factors. If work with dual-emission dyes is considered, the PMTs should be purchased as pairs with matching characteristics.)

Unlike most transducers used in the physiological sciences, which produce analogue voltages proportional to the quantity being measured (e.g. pressure, temperature), the PMT produces a series of random short-duration current pulses (50–100 ns) whose average *frequency* rather than amplitude, is a measure of the incident light level. One consequence of this is that the PMT cannot give an instantaneous measurement of light level. Rather, photon pulses must be accumulated over a finite period of time. Two approaches can be taken to converting the PMT pulse signals into a stable light level signal.

In the *photon counting* method, the current pulses are applied to a high-speed digital counter with each pulse incrementing the counter by one unit. The counter is allowed to accumulate counts for a fixed time interval and the average light level expressed as the number of counts acquired in that interval. Photon counting requires a high-speed digital counter capable of responding to short-duration pulses. A pulse discriminator is also required to detect the presence of the PMT current pulse.

Figure 13.2(A) shows a typical photon counting system using an Electron Tubes CT1 counter, a high-speed digital counter expansion board for IBM PC compatible computers. It provides two separate counters capable of accumulating over 16 million counts each and responding to pulses occurring randomly at rates up to 10 MHz. Under the control of a program running on the PC, the counters can be made to accumulate counts for preset periods, ranging from 0.5 ms to 20 s. The current pulses produced by the PMT are too small to be measurable by the CT1 board directly. An amplifier/discriminator is therefore

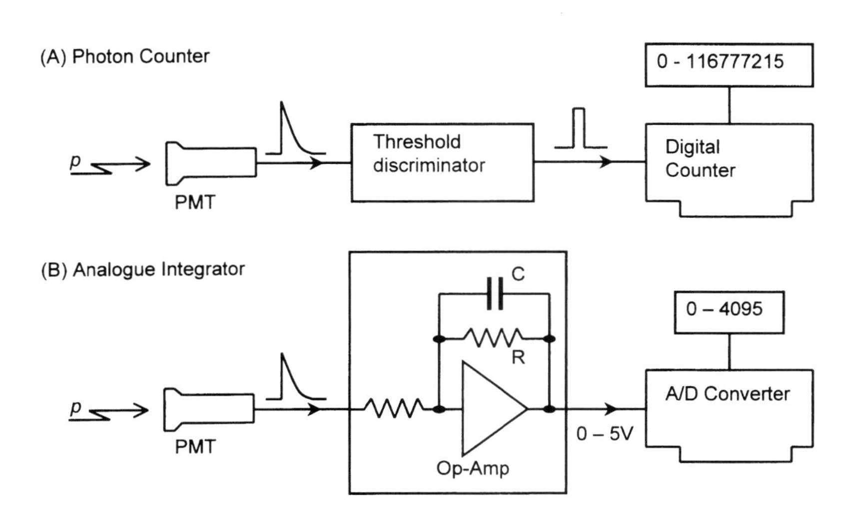

Figure 13.2 Photomultiplier tube light measurements methods. (A) Photon counting. Direct counting of PMT output produced by photons striking the tube surface. Pulses with amplitude greater than 1 mV are detected using a threshold discriminator, producing a TTL digital logic pulse which increments a high-speed digital counter in the computer. (B) Analogue integration. PMT output is fed into an integrator circuit which produces an analogue output voltage proportional to the photon pulse rate. The voltage level is then digitized by an A/D converter.

used to convert the current pulse into a standard ECL (emitter coupled logic) digital pulse. The discriminator also provides a detection threshold which separates the photon pulses from background noise.

The alternative to photon counting is to integrate directly the current output of the PMT to effect a pulse frequency-to-voltage conversion, as shown in Fig. 13.2(B). The photon current pulses from the PMT charge the capacitor C of the integrator. As the charge from each current pulse accumulates, the integrator output voltage increases. A resistor R, in parallel with the capacitor, causes the stored charge to leak away with a characteristic time constant of $\tau = RC$. The integrator, therefore, produces an average output voltage which is proportional to the balance between the rate at which the PMT supplies current and the integrator time constant. In broad terms, this time constant is equivalent to the counting interval of the photon counter. (A similar effect can be achieved by feeding the PMT output signal through a low-pass filter.)

Photon counting has a number of technical advantages in terms of sensitivity, dynamic range and immunity to noise. In addition to the intermittent current pulses produced by photons, PMTs produce a background noise current independent of any light input, and hence called the *dark current*. This dark current contributes to the signal measured using analogue integration. At very low light levels, where only a few photons are being collected, the integrated dark current overwhelms the light signal. The photon counting system, however, is largely insensitive to dark current noise since it can discriminate between the transient photon-produced current pulses and the background current. Using photon counting the light sensitivity can be extended down to single photons per counting period.

The photon counter can also handle a much wider range of light intensity than the analogue method. The 24-bit counter in the Electron Tubes CT1, for instance, can handle a range of 0–16 777 216 photons per counting period. On the other hand, the integrator output is an analogue voltage in the range 0–10 V. When this voltage is digitized, for measurement on a computer, using a 12-bit analogue-to-digital converter (as is commonly used for electrophysiological work), a resolution of only 0–4095 levels would be obtained.

Nevertheless, the performance of the analogue integrator is quite adequate for general applications. The range of fluorescence light levels usually produced by typical dyes do not require the full sensitivity of the photon counter. At present, both photon counting and analogue integration are widely used in fluorescence measurement systems, with some suppliers offering systems configured with either option.

13.4 DUAL-EMISSION DYE MEASUREMENT SYSTEMS

The detailed configuration of a fluorescence recording system depends on whether a dual-emission or excitation dye is being used. Experiments using dual-emission dyes are perhaps the simplest to perform, at least in terms of the recording instrumentation. A configuration such as in Fig. 13.3(A) is used. The fluorescent light, emitted from the dye-loaded cell, is split into two components using a 45° wavelength-sensitive dichroic mirror and passed, via narrow bandpass filters, to separate PMTs. The filter pass-bands are chosen to make each PMT sensitive to light of wavelengths close to the characteristic emission peaks for the free and Ca^{2+}-complexed forms of the dye.

Taking indo-1 as an example, 405 nm and 490 nm filters might be used. An estimate of intracellular Ca^{2+} concentration can be obtained from the ratio

$$R = \frac{F_{405} - F_{\text{back}}}{F_{490} - F_{\text{back}}} \quad [13.1]$$

where F_{back} is the background light emission from sources other than the dye within the cell, such as the inherent fluorescence of the cell itself and/or light from external sources. A measurement of F_{back} may be obtained by recording the fluorescent emission before the application of the dye to the cell. R is not linearly related to Ca^{2+} concentration but varies sigmoidally between a minimum value in the absence of Ca^{2+} and a maximum at high Ca^{2+} concentrations. The range of Ca^{2+} concentration measurement is thus somewhat limited.

A quantitative estimate of Ca^{2+} concentration can be obtained using the equation (Grynkiewicz *et al.*, 1985)

$$[Ca^{2+}] = K_d \frac{S_{f,490}}{S_{b,490}} \times \frac{R - R_{\min}}{R_{\max} - R} \quad [13.2]$$

where $R_{\min}$ and $R_{\max}$ are the minimum and maximum observed ratios, K_d is the equilibrium binding coefficient for Ca^{2+} and the dye. $S_{f,490}$ and $S_{b,490}$ are proportionality coefficients which relate emitted fluorescence intensity to dye concentration at the 490 nm wavelength for the Ca-bound form of the dye ($S_{b,490}$) and the free form $S_{f,490}$. A detailed discussion and derivation of this equation can be obtained in Grynkiewicz *et al.* (1985). $R_{\min}$, $R_{\max}$, $S_{f,490}$ and $S_{b,490}$ depend on a multitude of factors relating to the dye, UV filter bandwidths, transmission properties of the microscope optics, and PMT properties, but are essentially constant for a particular

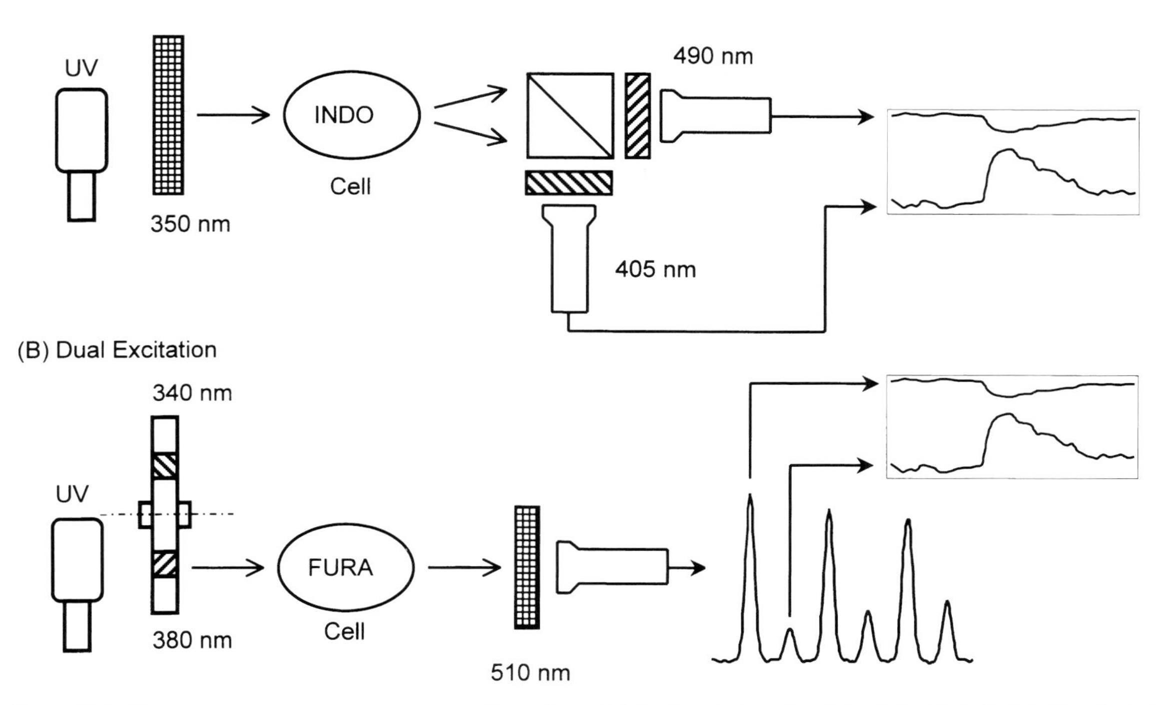

Figure 13.3 Fluorescence measurement system configurations. (A) Dual-emission. A cell loaded with indo-1 is illuminated with UV light at the peak absorption wavelength of the dye (350 nm). The spectral components of the fluorescent emissions, corresponding to the peak emission wavelengths for the Ca-bound (405 nm) and free (490 nm) dye forms, are separated using 45° partial transmission mirror and bandpass filters, and measured using a pair of PMTs. $[Ca^{2+}]$ is determined from the ratio of the PMT output signals. (B) Dual-excitation. A cell loaded with fura-2 is alternately illuminated with UV light at the peak absorption wavelengths for the Ca-bound (340 nm) and free (380 nm) dye forms, using a rotating filter wheel. Fluorescent emissions at the peak emission wavelength (510 nm) are measured using a PMT. A computer (or appropriate analogue hardware), synchronized to the rotation of the filter wheel, is used to store measurements from each filter and perform ratio and $[Ca^{2+}]$ calculations.

combination of dye, cell type, microscope and measurement system. For calibration purposes, the equation can be simplified to

$$[Ca^{2+}] = K_{eff} \frac{R - R_{min}}{R_{max} - R} \qquad [13.3]$$

where $K_{eff} = K_d\,(S_{f,490}/S_{b,490})$. The three parameters, K_{eff}, R_{min}, R_{max}, must be determined experimentally for the recording system and cell type being studied before equation [13.3] can be used. A variety of calibration methods are possible, one of the simplest being to measure values of R for a series of known Ca^{2+} concentrations using mixtures of Ca^{2+}, dye, and a buffer such as EGTA (Fabiato, 1991). R_{min}, R_{max} and K_{eff} may be obtained by fitting equation [13.3] to the R versus $[Ca^{2+}]$ curve using iterative non-linear curve fitting (as implemented in commonly available scientific graph plotting programs such as Jandel Sigmaplot or Graphpad Prism). Such techniques, however, do not necessarily account for the effects of dye binding to intracellular proteins (Highsmith *et al.*, 1986), and more sophisticated approaches such as the selective permeabilization of cells bathed in buffered solutions of known Ca^{2+} concentration may be preferable (Grynkiewicz *et al.*, 1985; Williams & Fay, 1990).

13.5 DUAL-EXCITATION DYE MEASUREMENT SYSTEMS

An essentially similar ratiometric approach can be applied to the dual-excitation dyes, by using a UV excitation light source which alternates between the peak excitation wavelengths for the bound and unbound forms of the dye. This may be achieved by placing a monochromator or a rotating filter changer in the excitation light path. Since the peak emission wavelength does not shift, only a single emission filter and PMT is required (Fig. 13.3(B)).

Filter wheels are currently in more common use than monochromators. Most wheels have at least four filter positions. This is particularly useful since it allows multiple ion measurements to be made when more than one type of dye is applied to the cell. The rate at which the filter wheel rotates determines the number of ratio measurements per second that can be achieved. If measurement rates greater than 1–2 per second are to be achieved, it is important to employ a filter wheel which can be made to rotate at a constant angular velocity, since it is more efficient to accelerate a wheel up to a high constant speed than to repeatedly start a wheel moving and bring it to a halt again at a new position. Rotating wheels, such as the Cairn Instruments (Faversham, Kent, UK) wheel shown in Fig. 13.4, can be made to rotate at frequencies as high as 300 revolutions per second. When electrophysiological measurements are to take place it is important to ensure that the vibration associated with filter wheel motion is isolated from cell and patch electrode. The excitation light source with the filter wheel is thus normally housed separately and coupled to the microscope using a fibreoptic or liquid light guide.

The computation of R is more complicated than for dual-emission dyes since the PMT signal is split into a series of discrete pulses as the filters rotate into and out of the light path (see Fig. 13.3(B)). To compute R, with equation [13.1], it is necessary to determine where each filter is and to integrate the light emission during the filter's period in the light path. After a filter change cycle has been completed the signal for each filter episode is retrieved and used to compute the ratio. This process requires that the filter wheel position can be determined by an external device – either specialized hardware modules with sample-and-hold

Figure 13.4 Cairn Instruments excitation filter wheel. This is a 6-position wheel capable of rotating at rates up to 300 revolutions per second.

facilities for integrating and storing the light signal, or by software within the computer controlling the system.

The more recently developed computer-controllable monochromator can also be used as a means of varying the wavelength of the excitation light. This kind of monochromator consists of a diffraction grating mounted on a galvanometer movement which allows rapid computer-controlled variation of the grating width. Suppliers producing suitable monochromators include: Cairn Instruments, Life Science Resources (Cambridge, UK), Photon Technology International (New Jersey, USA) and Till Photonics (Planegg, Germany). The output of such monochromators can be adjusted from one wavelength to another within 1–2 ms. The monochromator has the advantage of being able to produce a wider range of wavelengths on demand than the filter wheel with its fixed set of filters and can switch wavelengths at a comparable rate. However, it should be borne in mind that while only the illumination intensity changes as the filter wheel effects a wavelength switch, the monochromator illuminates the cell with intermediate wavelengths during the change. Thus if a monochromator is to be used for alternation of excitation wavelengths there must also be a means of shuttering or otherwise of disabling light measurement during transitions and of relaying the monochromator's state to the controlling hardware or computer.

The use of dual-emission or excitation dyes has an effect on the rate at which ratio measurements can be made. Dual-emission dyes allow simultaneous measurement at the characteristic emission peak wavelengths. Consequently, the rate at which ratio measurements can be made is limited only by the response time of the dye itself to changes in ion concentration, which are typically in the order of a few milliseconds. On the other hand, since dual-excitation dyes require a mechanically alternated light source, measurements are discontinuous, with the maximum rate determined by time taken to execute a complete filter change cycle. Currently, only the 300 rev s^{-1} Cairn Instruments wheel allows dual-excitation dye measurements at rates approaching those of dual-emission dyes.

Dual-emission dyes are thus often used in combination with electrophysiological measurements. The fast response time of the method proves valuable in the study of rapid Ca^{2+} transients evoked by action potientials in neurons. Figure 13.5 shows some results from such an experiment. The 405 nm and 490 nm fluorescence emissions were obtained using photon counting over a 2 ms period, the 405/490 ratio was computed using equation [13.1] and $[Ca^{2+}]$ using equation [13.4]. The cell is electrically active, producing a series of spontaneous action potentials, as can be seen

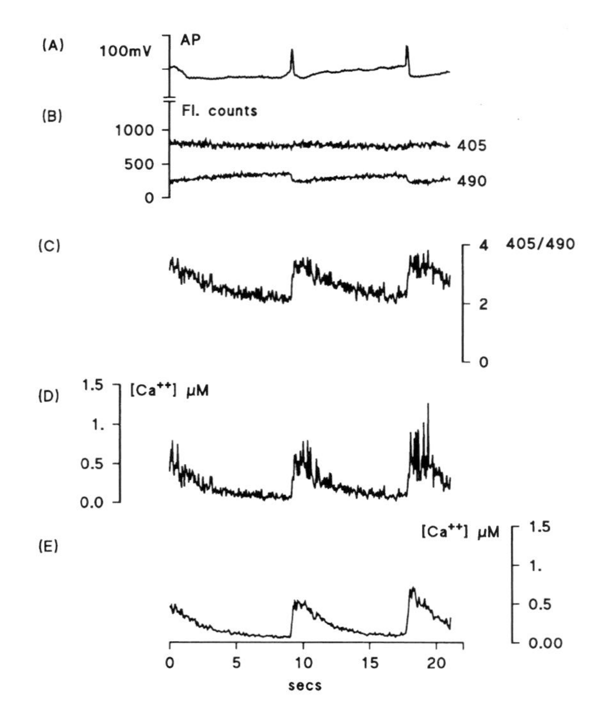

Figure 13.5 Membrane potential and ratiometric [Ca^{2+}] recordings made using a 'chart recorder'-type fluorescence measurement program. (A) Cell membrane potential from an indo-1 dye loaded, whole-cell patch-clamped pituitary cell. (B) Fluorescent emissions at 490 nm and 405 nm, using photon counting with a 2 ms counting period. (C) Ratio of 405/490 using equation [13.1]. (D) Computed [Ca^{2+}] using equation [13.3] with R_{min} = 0.46, R_{max} = 4.5, K_{eff} = 0.426 nM. (E) [Ca^{2+}] after smoothing of 405 nm and 490 nm signals using a 10-point running average. (Data courtesy of Dr W.T. Mason, Babraham Institute.)

from the membrane potential recording. Each action potential produces a transient influx of Ca^{2+} into the cell, increasing intracellular Ca^{2+} concentration.

Figure 13.5 also illustrates some of the difficulties associated with the ratiometric computation of [Ca^{2+}]. The fluorescence signals are relatively small and quite noisy, as a consequence of the low light levels and small number of counts accumulated during the short counting period. In general, the standard deviation of the measured light signal is proportional to the square root of the average number of photon counts. An average count of 1000 will exhibit fluctuations with a standard deviation of 31. As can be seen from Fig. 13.5 (C), taking the ratio of two such signals increases the noise yet again. However, compared with the ratio, the computed [Ca^{2+}] is enormously noisy. The variance of the [Ca^{2+}] signal is not constant, but increases dramatically as [Ca^{2+}] increases. This is a consequence of the mathematical form of equation [13.4]. As R approaches R_{max} with increasing [Ca^{2+}], the denominator in equation [13.4] becomes very small and any noise on the R signal becomes increasingly magnified.

The variance of the fluorescence signals can be reduced by using a longer photon counting period to accumulate larger numbers of counts which are subject to less statistical variation. An alternative approach, which can be applied after recording, is to average adjacent count periods. Figure 13.5(D) shows the [Ca^{2+}] computed after a series of 10 adjacent samples in the 405 and 490 signal have been averaged using a 10 point running mean algorithm.

13.6 ANALOGUE SIGNAL DIGITIZATION

If electrophysiological or other signals are to be recorded simultaneously with the fluorescence signals, an additional computer interface card is required for analogue signal digitization. An analogue-to-digital (A/D) converter is a computer controlled voltmeter which can be made to measure a voltage signal and return a number which can be stored in computer memory. Analogue signals are digitized by making a series of such measurements (samples) at fixed intervals and storing the stream of numbers in memory as a digital signal record.

A complete system providing the hardware for A/D conversion of 8–16 input channels, sample timing, trigger synchronization, and digital-to-analogue (D/A) conversion, is usually described as a *laboratory interface*. Laboratory interface cards are widely available for the IBM PC family, the most commonly used for electrophysiological work being the Axon Instruments Digidata 1200 and CED 1401 (Cambridge Electronic Design, Cambridge, UK). A more detailed discussion of laboratory interfaces can be found in Dempster (1993).

13.7 FLUORESCENCE MEASUREMENT SYSTEMS

The major task in developing a combined cellular fluorescence and electrophysiological measurement system is the integration of the measurement of different types of light and electrical signals with the control of a variety of experimental stimuli applied to the cell, usually under the supervision of some form of recording and analysis software. A number of suppliers (e.g. Cairn Instruments, Life Science Resources, Photon Technology International) now provide complete measurement system packages and it is also possible to develop custom systems within the laboratory from components derived from a variety of suppliers. It is important to appreciate that different approaches can be taken to the design of such systems and this can have a bearing on their suitability for particular types of experiment and on overall flexibility.

A complete system capable of recording and analysing combined cellular electrophysiology and fluorescence signals using either dual-emission or excitation dyes must be capable of performing the following functions:

- measurement of fluorescent light signals from one or two PMTs
- computation of fluorescence signal ratios
- control of excitation light wavelength
- measurement of cell membrane current and voltage signals
- voltage-clamp control of cell membrane potential
- application of drugs/solutions to the cell
- digitization of signals
- standard electrophysiological analysis functions
- computation of ion concentrations from fluorescence ratios.

All systems can be considered as consisting of four basic subsystems:

(1) Transducers and excitation devices (PMTs, voltage-clamps, filter wheels).
(2) Analogue signal conditioning to support the transducers.
(3) A/D and D/A digitization and computer hardware.
(4) Recording, control and analysis software.

System designs fall into two basic camps depending on the emphasis placed on particular subsystems. One approach focuses design effort on the analogue signal conditioning subsystem with the aim of providing analogue voltage outputs for all important measurement quantities, i.e. fluorescence signals and ratios, current and voltage. Relatively sophisticated signal conditioning electronics, involving a measure of analogue computing, is required for this approach, particularly when dual wavelength excitation must be supported. The Cairn Spectrophotometer system outlined in Fig. 13.6 is a good example of this approach. This is a modular system consisting of signal conditioning and analogue computation modules for handling the PMT signals, ratioing, and filter wheel control. All significant signals are expressed as analogue voltage outputs. Analogue integration and sample and hold circuitry is used to extract stable fluorescence signals associated with each filter of the rotating filter wheel.

The alternative approach is to use relatively simple analogue signal conditioning and perform the control and computation using computer software, as was illustrated in Fig. 13.1. The Life Science Resources PhoCal system is an example of this approach. The 'intelligence' of this system is based largely within the software running on the personal computer, with signal conditioning largely restricted to amplification and conversion of PMT output pulses into digital pulses. Fluorescence signals are measured using the photon counting method via a high-speed counter board located in the computer's expansion slot, analogue signals are acquired via an A/D converter board and the excitation wavelength controlled via a microprocessor-based filter wheel controller. Ratioing, decoding of fluorescent signals and other functions are all performed by the software after digitization.

The 'analogue/hardware-based' approach is a more open system, in that the analogue voltage signals that it provides can be easily recorded on standard paper chart recorders, instrumentation tape recorders (e.g. Biologic (Claix, France) DTR-1802, Instrutech (Great Neck, NY, USA) VR100B-8) as well as data acquisition systems. On the other hand, the 'digital/

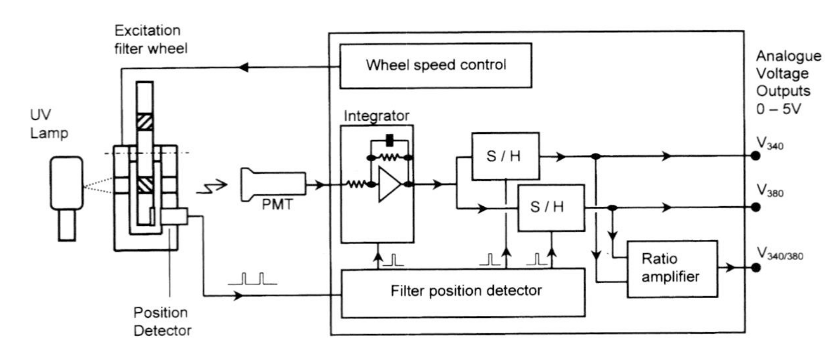

Figure 13.6 An analogue hardware-based approach to dual excitation measurements. Pulses from a position detector on the rotating filter wheel provide timing information on filter position. This is used to gate the operation of an analogue integrator measuring the PMT output. Sample and hold (S/H) devices (one for each filter) acquire and store the integrator voltage at the end of each filter pass. S/H outputs from a pair of filters is fed into a ratio amplifier which outputs the ratio of its two outputs.

software-based' approach provides a closer integration of hardware and software. The recording and analysis programs provided with the LSR package, for instance, are more powerful than that provided with the Cairn Instruments system, both in terms of analysis and being capable of controlling the filter wheel, excitation light shuttering and voltage-clamp command voltage. However, the Cairn system can be easily interfaced to electrophysiological data acquisition systems such as Axon Instruments' pCLAMP package which can be used to provide many of these functions. The LSR system can only be controlled by its own software. The choice of an appropriate system thus depends on a detailed appraisal of requirements.

13.8 SOFTWARE FOR RECORDING FLUORESCENCE SIGNALS

The computer software, which supervises the recording of signals from, and application of stimuli to, an experiment, plays a crucial role in modern experimentation. In many ways the type of work that can be performed is dependent upon the software being capable of supporting the required experimental protocols. Such software can be developed within the laboratory but it is a difficult and time-consuming process. It is also difficult, in practice, to produce a single program which is ideal for all types of cell fluorescence experiments. This is not due simply to computer memory limitations or other hardware factors, although these are issues which must be considered. Rather, it is that a variety of experimental paradigms are in use and the program features appropriate for one approach may conflict with those required for another.

13.9 THE 'CHART RECORDER' PARADIGM

Many experimental protocols require a continuous unbroken record to be made of the fluorescence, electrical and/or other signals, in much the same way as might be obtained with a chart recorder. Figure 13.5 is an extract from such a recording, in which six data channels were acquired over a period of 95 s. A key feature in such programs is the ability to handle data files of large size. For instance, the data in Fig. 13.5 was collected using a 300 Hz sampling rate which resulted in a file containing 1 Mbyte of data.

Continuous *recording-to-disk* methods are often required to collect such large records without gaps since many laboratory computers do not have sufficient RAM (random access memory) space to accommodate the whole data file. The incoming signals must therefore be copied to disk, while digitization is in progress, using techniques such as the double buffer method. Samples, as they are acquired from the A/D converter and photon counter, are stored in a small (e.g. 16 kbyte) temporary buffer in RAM which is continuously filled in a cyclic fashion. The buffer is split into two halves, and as each half becomes full its contents are written on to disk while samples continue to be stored in the other half.

Similarly, the program's analysis features must be able to handle such large records efficiently, to display magnified sections of the record, and provide cursor measurements. The chart recorder paradigm proves to be satisfactory for a wide range of experimental protocols where a continuous, relatively slow, record is required, and a small number of stimuli are applied at infrequent intervals. Much of the work to which intracellular ion measurements have so far been applied falls into this category and most commercially available measurement systems include programs which adhere to this paradigm.

It is also worth noting that a number of general-purpose chart recorder programs also exist; such as AxoTape from Axon Instruments, the MacLab Chart program running on the Apple Macintosh (A.D. Instruments, Sydney, Australia), and Cambridge Electronic Design's Chart or Spike 2 program for the CED 1401 interface. Such software can be coupled to systems like the Cairn Instruments Spectrometer which output the fluorescence signals in analogue form.

13.10 THE 'OSCILLOSCOPE' PARADIGM

While chart recorder programs are, probably, the most generally useful, there are categories of experiment where they have distinct limitations. This is particularly so where the prime focus of interest is the properties of voltage activated Ca^{2+} currents, and their effect on intracellular Ca^{2+} concentration. Such currents are normally studied using the whole-cell patch-clamp technique, and Ca^{2+} currents are evoked by depolarizing the cell membrane under the control of a voltage pulse applied to the voltage-clamp command input. Such current signals are transient and of brief duration (100–500 ms). The flow of Ca^{2+} ions into the cell through these channels results in changes in intracellular Ca^{2+} concentrations which are similarly transient but with a slower time course (1–10 s). In order to study the voltage sensitivity of the Ca^{2+} current and/or concentration transients, it is usually necessary to apply a range of voltage steps of varying amplitude and duration.

These voltage-clamp studies are distinctive in that the experimental protocol is split into a series of relatively short, discrete, recording sweeps, rather than a single long continuous record. Also, instead of a relatively small number of stimuli, separated by intervals of minutes, hundreds of precise voltage pulse stimuli are applied, at 10–20 s intervals. This style of experiment suggests the use of an 'oscilloscope' paradigm in the design of the program. Returning to the analogy with conventional recording devices, the oscilloscope is an instrument designed for recording repeatable high-speed transients, synchronized with some external trigger event. In general, electrophysiology software for recording voltage-activated currents, such as Axon Instruments' pCLAMP, follow this oscilloscope paradigm. The generation of repeated pulse stimuli, with a variety of heights and/or widths is itself a distinct task. It is now usual to generate pulse waveforms within the computer and apply them using a D/A converter.

A program designed for recording and analysing voltage-activated currents and fluorescence, therefore, has to perform the following functions.

- Record fluorescence sweep.
- Record current, voltage sweep.
- Generate voltage-clamp command pulses.
- Apply leak current subtraction.
- Display and analyse recordings.

A display screen from Life Science Resources' PhoClamp Pro, a combined voltage-clamp/fluorescence program, can be seen in Fig. 13.7. The screen is split into two panels, the upper panel showing the recording sweep of indo-1 dye (405 nm, 490 nm) fluorescence and computed [Ca^{2+}]. The lower panel shows the electrophysiological sweep with membrane current and potential. Due to the marked differences in the time course of the Ca^{2+} current and concentration signals, it is convenient to use different sweep durations: 500 ms for the current and 5 s for the fluoresence.

13.11 LEAK CURRENT SUBTRACTION

As an added complication, the current signals often require the process of digital leak current subtraction. One of the difficulties inherent in the voltage-clamp technique is ensuring that the current being measured is being mediated only by the ion channels under study. Cell membranes contain a wide variety of channel types, selective for most of the ions present in normal physiological salt solutions (Na^+, K^+, Ca^{2+}, Cl^-). Channel types may also have different voltage sensitivities and kinetics. During an experiment, it is normal practice to attempt to isolate a single channel type by blocking other channels using a variety of pharmacological agents. Such blocking strategies are rarely completely effective and a residual leak often remains in addition to the Ca^{2+} current. In many cells the leak current can be significant in comparison with the Ca^{2+} current under study.

In many circumstances, however, a digital subtraction process can be used to remove the leak current component from the current signal. Leak currents are often due to current flowing through imperfections in the pipette-membrane seal, or to passive K^+ or Cl^- channels in the cell. In either case, the conductance of

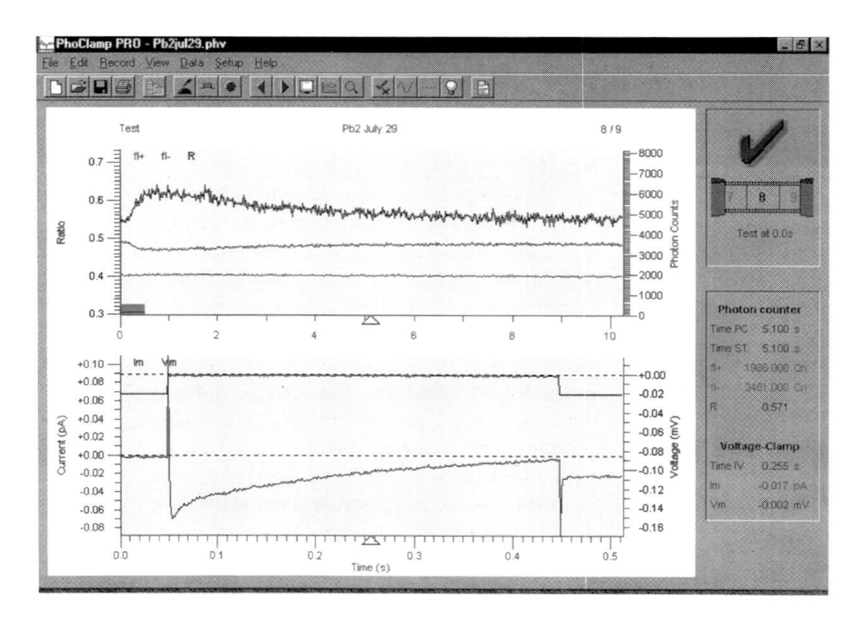

Figure 13.7 Display screen from the PhoClamp Pro fluorescence/voltage-clamp program, illustrating an 'oscilloscope' paradigm. The upper display panel shows a 10 s fluorescence sweep, 490 nm and 405 nm indo-1 fluorescence signals and 405/490 ratio. The lower panel shows the associated 500 ms voltage-clamp sweep with the Ca^{2+} current which was evoked by a 400 ms depolarizing voltage pulse.

the leak channels is constant and the current scales linearly with membrane potential. In contrast, the Ca^{2+} current is highly voltage- and time-dependent, absent in response to hyperpolarizing voltage steps, and requiring steps to potentials more positive than −40 mV before any current appears. Consequently, the current record, in response to a hyperpolarizing (or small depolarizing) step contains only the leak current component. The leak component can be removed from the Ca^{2+} current, evoked by large depolarizations, by scaling the leak record in proportion to the ratio of voltage step sizes and subtracting it from the Ca^{2+} current record. Further details of digital leak subtraction can be found in Dempster (1993).

The protocol for acquiring a leak subtracted Ca^{2+} current and the resulting intracellular Ca^{2+} concentration changes is as follows:

(1) Acquire a series of leak current records.
(2) Create average leak current and store on file.
(3) Start fluorescence recording sweep.
(4) After a delay, start membrane current, potential recording sweep.
(5) Generate voltage-clamp command step.
(6) On completion of fluorescence sweep, store on file.
(7) Scale and subtract leak current from Ca^{2+} current and store leak subtracted current on file.

13.12 COMPUTER SYSTEM DESIGNS

The combination of fluorescence and electrophysiological signal recording is quite a demanding task for a computer system. The computer must perform a complex task of generating the voltage-clamp command waveforms and co-ordinating these with the recording of current and voltage from the patch-clamp and the fluorescence signals. Ideally, incoming signals should also be displayed on screen as they are digitized and also the fluorescence ratio calculated on-line.

The timing requirements of each channel may differ, as seen in the PhoClamp Pro example in Fig. 13.7. A typical Ca^{2+} current may last 100–500 ms, and a 1000 sample record would require a sampling rate of 2 kHz per channel. Fluorescence signals, being of longer duration, generally require sampling at rates no more than 0.5 kHz per channel. Fluorescence measurements were made using an Electron Devices CT1 photon counter board and A/D and D/A conversion with a National Instruments (Austin, TX, USA) LAB-PC laboratory interface. Both of these devices have on-board clocks and are capable of a significant degree of autonomy in their operation. Overall timing and control of recording sweeps is co-ordinated by the PhoClamp Pro master program, running on the host computer, which displays results on screen as they are acquired and performs ratiometric calculations and other analysis functions.

The key to an effective design is to manage the transfer of data into and out of the computer with the minimum load on the central processing unit (CPU) of the computer. Fortunately, most personal computers are equipped with a variety of data transfer channels. The IBM PC-compatible family, for instance, has eight direct memory access (DMA) channels and 16 interrupt lines as standard features, which make it well suited to handling these multiple tasks.

DMA is a technique for automatically transferring data directly between a peripheral device, such as a laboratory interface, and the host computer's RAM memory without involving the CPU. It is very efficient, taking only 1–2 μs to transfer a byte of data, and leaves the CPU free for other functions. Most of the modern IBM PC-based laboratory interfaces used in the electrophysiological laboratory support the use of at least one DMA channel.

Interrupt lines provide a simpler, though less efficient, alternative to the DMA data transfer method. An interrupt line can be used to signal to the host computer that data are ready to be transferred (e.g. an A/D conversion has completed). It causes the CPU to interrupt the program which is currently executing and transfer control to an interrupt service routine – a subprogram which transfers the data from the interface into memory. Once the transfer is complete, control is returned to the original program. Compared with DMA, a considerable overhead is involved in the use of an interrupt, in that the state of the interrupted program (i.e. the contents of the CPU registers) must be saved before, and restored after, the execution of the interrupt service routine to allow the program to continue correctly. A discussion of the operation of DMA and interrupts can be found in Eggebrecht (1990).

A/D conversion of the current and voltage signals makes the highest demand on the system. The LAB-PC interface is programmed to collect a sweep of 256–2048 samples, alternating between two analogue inputs, digitizing the signals at strictly timed intervals under the control of its sampling clock. Samples are transferred into a storage buffer in computer memory using a DMA channel. Once programmed, the complete sequence of operations is totally independent of the CPU, and very efficient. Ideally, all of the three main input and output streams of data (A/D converter, photon counter, D/A converter) would be carried via their own DMA channel, leaving the CPU free for supervisory operations. However, although there are sufficient free DMA channels on the PC, the photon counter card in this system does not support DMA. However, it can be made to initiate an interrupt request whenever a count period has been completed. Fortunately, the sampling and update rates required for the photon counters are sufficiently low that the overheads incurred using interrupt lines are acceptable. Two interrupt-driven D/A channels are also used; one to prove the command voltage waveform, the other a synchronization pulse used to trigger the A/D sweep.

13.13 CONCLUSION

The development of a high-speed fluorescence measurement system is a complex task, involving some expertise in a number of quite disparate fields – fluorescence microscopy, properties of dyes, computer hardware and software. Developing such a system in-house can be a time-consuming process. It is often more practical to purchase a system from a supplier specializing in this area, who will provide a complete integrated system. It is important to have a clear idea about what such a system is expected to do. Does it have the required performance in terms of ratio measurements per second? If a computer-based system is being considered, does the supplied software support the type of experimental protocols envisaged? If an analogue system is being considered, can it be combined with existing data acquisition and analysis packages, such as Axon Instruments' pCLAMP? If commercially available packages, or combination of packages, do not support the envisaged experiments, then there may be little option but to tackle the development of a custom system within the laboratory. In such cases the development of the software may prove to be the greatest challenge, particularly if it must run under a modern operating system such as Microsoft Windows. Data acquisition software development environments such as National Instruments' Lab-View package may greatly simplify this task, bringing it within the programming skills range of suitably motivated researchers rather than professional programmers.

Fluorescent dyes are useful tools in many areas of cell research and new dyes with different ion sensitivity and emission/excitation wavelength properties continue to be produced. The measurement systems, and particularly the computer software, have by no means reached the ultimate in terms of performance and versatility. For example, the potential exists to study excitation–secretion coupling by combining the measurement of intracellular Ca^{2+} with the electrophysiological measurement of cell capacity (a measure of secretory vesicle fusion), using a lock-in amplifier (Lindau 1991; Zorec *et al.*, 1991). Similarly, experiments using intracellular concentration jump stimuli are possible using UV-activated caged Ca^{2+} buffers such as Nitr-5 and DM-nitrophen (Zucker, 1992). It is likely that the simplicity, sensitivity and flexibility of the microphotometric measurement system will guarantee it a continued place in the electrophysiological laboratory.

REFERENCES

Baylor S.M. & Hollingsworth S. (1988) *J. Physiol.* **403**, 151–192.

Dempster J. (1993) *The Computer Analysis of Electrophysiological Signals.* Academic Press, London.

Eggebrecht L.C. (1990) *Interfacing to the IBM Personal Computer*, 2nd edn. Howard W. Sams & Co., Indianapolis.

Eisner D.A., Nichols C.G., O'Neill S.C., Smith G.L. & Valdeolmillos M. (1989) *J. Physiol.* **411**, 393–418.

Fabiato A. (1991) In *Cellular Calcium: A Practical Approach*, J.G. McCormack & P.H. Cobbold (eds). IRL Press, Oxford, pp. 159–175.

Grynkiewicz G., Poenie M. & Tsien R.Y. (1985) *J. Biol. Chem.* **260**, 3440–3450.

Highsmith S., Bloebaum P. & Snowdowne K.W. (1986) *Biochem. Biophys. Res. Commun.* **138**, 1153–1162.

Lindau M. (1991) *Q. Rev. Biophys.* **24**, 75–101.

Mollard P., Guerineau N., Audin J. & Dufy B. (1989) *Biochem. Biophys. Res. Commun.* **164**, 1045–1052.

Sakmann B. & Neher E. (1983) *Single-channel Recording.* Plenum Press, New York.

Standen N.B., Gray P.T.A. & Whitaker M.J. (1987) *Microelectrode Techniques – The Plymouth Workshop Handbook.* Company of Biologists, Cambridge.

Williams D.A. & Fay R.S. (1990) *Cell Calcium* **11**, 75–83.

Williams D.A., Fogarty K.E., Tsien R.Y. & Gray F.S. (1985) *Nature* **18**, 558–561.

Zorec R., Henigman F., Mason W.T. & Kordas M. (1991) In *Methods in Neurosciences*, Vol. 4, P.M. Conn (ed). Academic Press, London, pp. 194–210.

Zucker R.S. (1992) *Cell Calcium* **13**, 29–40.

EQUIPMENT SUPPLIERS

AD Instruments Pty, Sydney, Australia. (*www.adinstruments.com*)

Axon Instruments Inc., Foster City, CA, USA. (*www.axonet.com*)

Cairn Instruments Ltd, Faversham, Kent, UK. (*www.cairnweb.com*)

Cambridge Electronic Design Ltd, Cambridge, UK. (*www.ced.co.uk*)

Heka Elektronic GmbH, Lambrecht/Pfalz, Germany. (*www.heka.com*)

Life Science Resources Ltd, Cambridge, UK. (*www.lsr.com*)

National Instruments Corp., Austin, TX, USA. (*www.natinst.com*)

Oriel Instruments Corp., Stratford, CT, USA.
(*www.oriel.com*)
Photon Technology International Inc., NJ, USA.
(*www.pti-nj.com*)
Electron Tubes Ltd, Ruislip, UK.
(*www.electech.demon.co.uk*)
Till Photonics Gmbh, Planegg, Germany.
(*www.till-photonic.de*)

CHAPTER FOURTEEN

Potentiometric Membrane Dyes and Imaging Membrane Potential in Single Cells

LESLIE M. LOEW
Department of Physiology, University of Connecticut Health Center, Farmington, CT, USA

14.1 INTRODUCTION

Dye indicators of membrane potential have been employed in numerous studies of cell physiology ever since their introduction in the early 1970s. L.B. Cohen and his co-workers (Davila *et al.*, 1973; Cohen *et al.*, 1974) have pioneered this effort leading to the discovery of a large number of organic dyes whose spectral properties are sensitive to changes in membrane potential. This enterprise was motivated by studies requiring multisite mapping of electrical activity in complex neuronal systems; a variety of applications in both cell biology and neuroscience have emerged, however (Loew, 1988). The aim of this chapter is to review the methodology in sufficient detail to provide the reader with an appreciation of the factors involved in appropriate choice of potentiometric molecular probes and implementation of the technology in particular applications. The size of the voltage-dependent signal, while certainly important, is by no means the only factor to be considered in choosing a dye. The intent here is to identify the important parameters which have to be understood about the chemistry of a dye if it is to be a useful and practical indicator of potential. Also introduced below are some brief generalities about the structures and physical properties of the various dye chromophores which have formed the backbones of potentiometric dyes. This information should help to take some of the mystery out of the chemistry of these physiological indicators and provide guidance toward the appropriate dye selection.

My laboratory has been engaged in the design of membrane-staining styryl dyes based on the theoretical expectation of an electrochromic mechanism for the potentiometric response (Platt, 1956; Loew *et al.*, 1978). Fluorescent indicators which operate via a potential-dependent redistribution across the membrane have also emerged from our efforts at dye development (Ehrenberg *et al.*, 1988; Farkas *et al.*, 1989; Loew, 1993, 1997). Particular emphasis, therefore, will be placed on these sets of potentiometric probes. The very different properties of these classes of dyes will help to illustrate the variety of considerations required for appropriate and successful application of this complex methodology. I will also present applications to problems of interest in this laboratory which

FLUORESCENT AND LUMINESCENT PROBES, 2ND EDN
ISBN 0-12-447836-0

demonstrate a merger of the probe technology with quantitative fluorescence microscopy. A primary application of potentiometric dyes is, of course, mapping of electrical activity in complex preparations of excitable tissue; this is described in this volume (Chapter 15).

14.2 OPTIMIZATION OF DYE INDICATOR SENSITIVITY

The first attempts to develop potentiometric dyes relied on serendipity and trial and error. By screening large numbers of dyes on the squid giant axon, Cohen and his colleagues (Cohen *et al.*, 1974; Ross *et al.*, 1977; Gupta *et al.*, 1981) were able to discover a series of highly sensitive dyes as well as develop a large database of information. The resultant database revealed several broad rules which could be used to help design new generations of probes. For example, it became clear that the large class of azo dyes were not going to be particularly suitable because of their photoinstability and their propensity for toxic and photodynamic damage to biological preparations. The dyes had to have some hydrophobic appendages to promote interaction with the membrane, but alkyl groups longer than about eight carbons imparted too much insolubility for some applications. It became apparent that several different molecular mechanisms were employed by different dyes to produce potential-dependent spectral changes. It also soon became clear that the same dye could behave very differently in different model and biological preparations (Ross & Reichardt, 1979; Loew *et al.*, 1985b).

Cohen & Salzberg (1978) divided potentiometric dyes into two simple classes – slow and fast. 'Fast' dyes are able to follow changes in the millisecond range – fast enough to monitor individual electrical events in excitable cells and tissue. They are membrane stains which are subtly perturbed by the transmembrane voltage resulting in a small spectral response. 'Slow' dyes can measure voltage changes that may accompany hormonal responses in non-excitable cells or the level of activity in energy-transducing organelles. They are charged and are pulled from one environment to another by the potential difference between them; the environmental dependence of their spectra (e.g. aqueous versus membrane-bound) underlies the typically large potentiometric response. Therefore, in addition to identifying the range of applications accessible to potentiometric indicators, mechanism and sensitivity are, to some extent, classified by the 'fast' and 'slow' designations.

14.2.1 Fast dyes

Generally, the mechanisms underlying the fast dye responses involve potential-dependent intramolecular rearrangements or small movements of the dye from one chemical environment to another. These reactions have the requisite speed but usually do not produce very large changes in the spectra of the dye. Typically

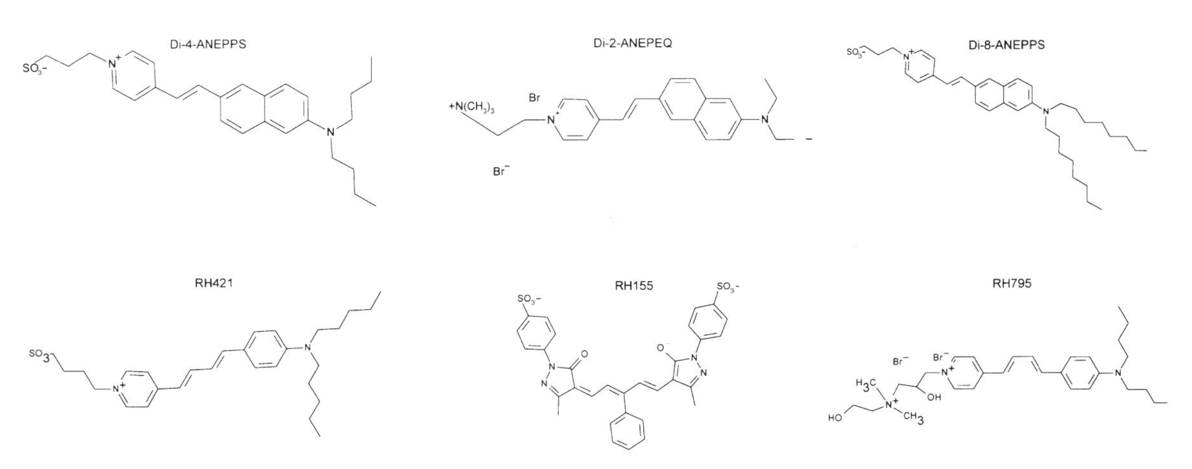

Figure 14.1 A representative sampling of fast dyes. Di-4-ANEPPS is a versatile fast styryl dye that can be used for dual-wavelength ratiometric measurements. Di-2-ANEPEQ, a more soluble analogue of di-4-ANEPPS, is the best ratiometric dye for microinjection. It also has good tissue penetration when applied externally, but may wash out quickly. Di-8-ANEPPS is more slowly internalized than di-4-ANEPPS and needs Pluronic F127 to stain. RH421 is a styryl dye that displays a 21% per 100 mV fluorescence change on a neuroblastoma cell preparation (Grinvald *et al.*, 1983). RH155 is an oxonol chromophore. It is a good choice for absorbance measurements. There are no fluorescence signals. RH795 is a good styryl dye for penetration into cortex (Grinvald *et al.*, 1994).

good fast dyes respond with fluorescence changes of only 2–10% per 100 mV. It should also be stressed again that a fast dye may give a large response in one biological preparation with a given set of experimental conditions but may be totally insensitive to potential changes when applied to a different preparation. The structures of a number of fast dyes are collected in Fig. 14.1. The properties of these and several others will be discussed below.

An effort to develop fast dyes that employ an electrochromic (Platt, 1956; Liptay, 1969; Loew *et al.*, 1978) mechanism has been a major focus of this laboratory. Briefly, electrochromism is possible if there is a large shift in electronic charge when a chromophore is excited from the ground to the first excited state; if the direction of charge movement lies parallel to an electric field, the energy of the transition will be sensitive to the field amplitude. Thus, if the chromophore is oriented so that the charge redistribution is perpendicular to the membrane surface, an electrochromic dye should be an indicator of membrane potential. The reason this mechanism was attractive to us was that it lent itself to a theoretical molecular orbital treatment (Loew *et al.*, 1978) which could aid in the design of appropriate targets for organic synthesis. The aminostyrylpyridinium chromophore ('styryl' hereafter) best fit the criterion of a large charge shift upon excitation and could easily be modified with side chains that would assure an appropriate orientation in the membrane. Variations on this chromophore have been the targets of synthetic efforts (Loew *et al.*, 1979; Grinvald *et al.*, 1982, 1983, 1994; Hassner *et al.*, 1984; Fluhler *et al.*, 1985; Tsau *et al.*, 1996). Figure 14.1 includes examples that have been particularly successful in a variety of applications. Among their special attributes are relatively good photostability, a high fluorescence quantum yield for the membrane-bound dyes and almost no fluorescence from aqueous dye. It should be noted that the potentiometric responses of these dyes in different biological preparations may be complex superpositions of several mechanisms, with only a minor contribution from electrochromism (Loew *et al.*, 1985b). Nevertheless, the styryl dyes have emerged as the most popular fast fluorescent potentiometric indicators.

Di-4-ANEPPS is a good, general-purpose fast probe which has several beneficial attributes and has been used on a large variety of preparations (Fluhler *et al.*, 1985; Gross *et al.*, 1986; Ehrenberg *et al.*, 1987; Lojewska *et al.*, 1989; Müller *et al.*, 1989; Chien & Pine, 1991; Rosenbaum *et al.*, 1991; Davidenko *et al.*, 1992; Herbele & Dencher, 1992; Loew *et al.*, 1992; Knisley *et al.*, 1993; Pertsov *et al.*, 1993; Wan *et al.*, 1993; Delaney *et al.*, 1994; Borst, 1995; Knisley, 1995). It has been reasonably consistent in giving a relative fluorescence change of ≈10% per 100 mV in a number of different cell types and experimental protocols. It can be used in a dual-wavelength ratiometric excitation mode to normalize out artifacts due to uneven staining or photobleaching (Montana *et al.*, 1989). As with other styryls, only the membrane-bound dye displays appreciable fluorescence. In several experimental protocols, it has been persistent for hours under continuous perfusion with dye-free medium.

Rational modifications can be made to optimize probe characteristics for particular experimental demands. Decreasing the size of the chromophore, as in di-4-ASPPS (Loew & Simpson, 1981; Fluhler *et al.*, 1985), results in a blue shift of the spectral characteristics of the response of about 30 nm; unfortunately a 40% reduction in sensitivity also results. Decreasing the chain length of the hydrocarbon tails, as in di-2-ANEPPS, increases the water solubility of the dye. This is necessary for thick tissue preparations where the dye must penetrate through many cell layers; the more water-soluble dyes, of course, will give less persistent staining of the preparation. The opposite situation will pertain to a dye like di-8-ANEPPS (Ehrenberg *et al.*, 1990; Bedlack *et al.*, 1992), where low solubility requires a high molecular weight surfactant like Pluronic™ F127 (BASF Corp.) to promote staining (Davila *et al.*, 1973; Lojewska & Loew, 1987). In addition to persistence, we have discovered that another attribute of di-8-ANEPPS is a very slow rate of internalization. In several cell types, di-4-ANEPPS was found to be incorporated into intracellular organelles over times as short as 10 minutes; di-8-ANEPPS is retained exclusively on the plasma membrane over periods of hours. This dye has now also been successfully applied to a variety of preparations (Bedlack *et al.*, 1992; Rohr & Salzberg, 1994a,b; Beach *et al.*, 1996; Hayashi *et al.*, 1996). Dual-wavelength ratio imaging of di-8-ANEPPS also permitted the mapping of membrane electric profiles along the surface of a cell (Bedlack *et al.*, 1992, 1994; Zhang *et al.*, 1998). Varying headgroup charge can subtly change the location of bound dye in the membrane with the effect of sometimes significant improvements in sensitivity (Grinvald, *et al.*, 1994). Also, it seems generally true that positively charged dyes are especially well suited for experiments requiring microinjection to localize dye to just one cell in a complex preparation (Grinvald, *et al.*, 1987). Recently, a hydrophilic positively charged styryl, di-2-ANEPEQ (also known as JPW1114), was developed specifically for microinjection and migration to remote processes in a large invertebrate neuron (Antic *et al.*, 1992; Zecevic, 1996). The opposite strategy was employed for the development of retrogradely transported dyes in a vertebrate preparation; here hydrophobic positively charged dyes could be applied externally and were transported over many hours to the distal end of the cell (Tsau *et al.*, 1996; Wenner *et al.*, 1996). Many analogues of styryl dyes have been synthesized to meet these various experimental demands in both this laboratory by Joseph Wuskell and in the laboratory of A. Grinvald by Rena Hildesheim.

Merocyanine and oxonol chromophores have formed the basis of potentiometric dyes which respond to voltage via changes in absorbance rather than fluorescence. Absorbance measurements are often desirable, especially for complex preparations with non-excitable satellite cells. This is because the large background fluorescence from the membranes of these uninvolved cells would attenuate the relative fluorescence response with a corresponding degradation of signal to noise; a transmitted light signal is used in absorbance and is not significantly affected by background staining.

The oxonol class of dyes has now emerged as the most sensitive for detecting fast potential-dependent absorbance changes, although some merocyanine dyes, the earliest class of successful fast indicators, are also still in use. The oxonol chromophore is defined by its delocalized negative charge. The mechanism for the potential dependent response has been determined for several of these probes; it involves a movement between a binding site on the membrane surface and an aqueous region adjacent to the membrane (Waggoner *et al.*, 1977; George *et al.*, 1988). An example of an oxonol dye, RH155, is given in Fig. 14.1. It was utilized in an elegant study of neuronal firing patterns during different behaviours in an invertebrate nervous system (Wu *et al.*, 1994).

A newly developed approach for fast dye signals employs fluorescence resonance energy transfer between a pair of dye molecules and has produced some of the largest relative fluorescence changes seen to date (Gonzalez & Tsien, 1995, 1997). In the most recent version of this scheme (Gonzalez & Tsien, 1997), a coumarin chromophore attached to a lipid on the outside of a cell is the donor to an oxonol chromophore that can migrate across the membrane in a voltage-dependent manner with submillisecond kinetics. The resultant modulation of energy transfer efficiency can cause a change in the ratio of donor and acceptor emission wavelengths of as much as 50% per 100 mV. This approach also carries the benefits of dual-wavelength ratio measurements. The necessity for high concentrations of two membrane-localized dyes may cause significant perturbations in some preparations, however.

14.2.2 Slow Dyes

Potential-dependent partitioning between the extracellular medium and either the membrane or the cytoplasm is the general process underlying the mechanisms of slow dyes. Unlike fast potentiometric indicators, slow dyes have to be charged, so that the voltage difference can pull the dyes from one compartment to another. The change in environment is what produces the voltage-dependent spectral change. Three chromophore types have yielded useful slow dyes: cyanines, oxonols and rhodamines. Each of these have special features that make them applicable to different kinds of experimental requirements. Figure 14.2 displays some of the more important slow dyes.

The cyanine class of symmetrical dyes with delocalized positive charges were originally introduced by Alan Waggoner and have proven to be extraordinarily

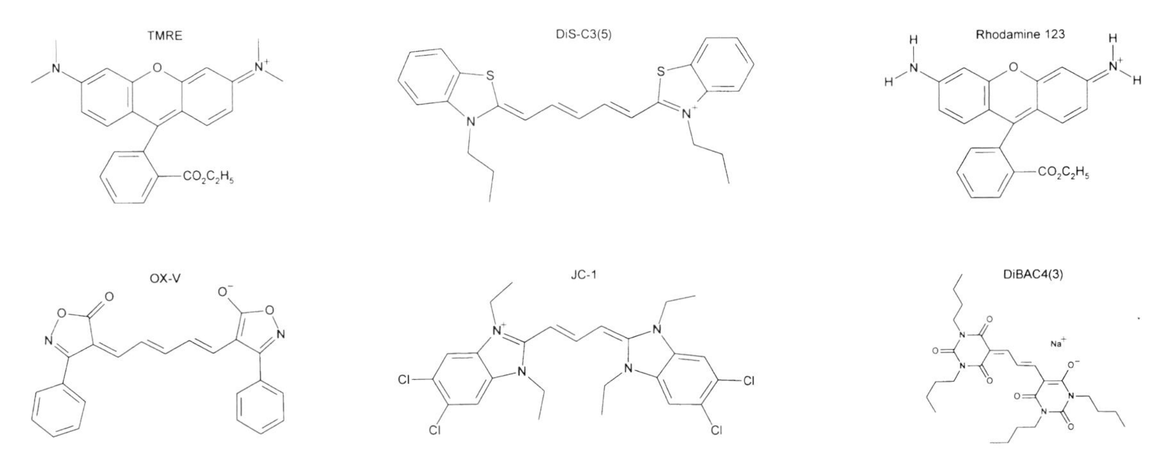

Figure 14.2 A representative sampling of slow dyes. TMRE is a rhodamine-redistribution dye for single cell and mitochondrial potential measurements. It is suitable for timescales of ~10 s. DiS-C3(5) is a very sensitive cyanine dye for cell suspensions (Sims *et al.*, 1974). Rhodamine 123 is a redistribution dye for mitochondrial staining. OX-V is an anionic oxonol redistribution dye for absorbance measurements. JC-1 is a cyanine redistribution dye with applications as a dual-emission mitochondrial membrane potential indicator. DiBAC4(3) is an anionic redistribution indicator also known as 'Bis-oxonol'.

sensitive probes for potential changes in populations of non-excitable cells (Sims *et al.*, 1974; Waggoner, 1979, 1985). A large number of these dyes with varying hydrocarbon chain lengths, numbers of methine groups in the bridging polyene, and heterocyclic nuclei are available. Depending on the nature of the dye and its concentration, potential-dependent uptake can effect either an increase or a decrease in fluorescence intensity. In general, the fluorescence of these dyes is enhanced upon membrane binding – thus accumulation of dye leads to fluorescence enhancement. However, at high dye:lipid ratios many of the dyes have a tendency to aggregate, resulting in self-quenching of fluorescence. The latter is the case for di-S-C_3(5) (Fig. 14.2), which can lose 98% of its fluorescence when a cell or vesicle preparation is polarized to 100 mV (Sims *et al.*, 1974; Loew *et al.*, 1983, 1985a). Di-O-C6(3) has less of a tendency to aggregate and displays an increased fluorescence quantum yield as it binds to the plasma and organelle membranes. Its lipophilicity is responsible for its use as a stain for mitochondria and endoplasmic reticulum (Terasaki *et al.*, 1984). However, dye association with intracellular organelles can lead to cytotoxicity and/or misinterpreted fluorescence changes (Korchak *et al.*, 1982) and appropriate controls must be performed to check for this possibility in experiments on bulk cell suspensions with cyanines. Chen's laboratory has developed a cyanine dye, JC-1, which forms J-aggregates when it is concentrated in highly polarized mitochondria; since the aggregate has a red emission and the monomer fluoresces green, the colour of the emitted light provides a direct indication of mitochondrial potential (Reers *et al.*, 1991, 1995; Smiley *et al.*, 1991); this technique is the subject of the chapter by Chen & Smiley in this volume.

Anionic oxonols also show enhanced fluorescence upon binding to membranes, but, because of their negative charge, binding is promoted by depolarization. More importantly, the negative charge lessens intracellular uptake of oxonol dyes, solving some of the difficulties encountered with the cyanines. Also, because delocalized anions are more permeable than cations, the oxonols can respond to voltage with kinetics approaching those of fast dyes. They are, however, less sensitive than the cyanines. Bis-oxonol – and its relatives with barbituric acid nuclei – have been used in fluorescence experiments on cell suspensions (Brauner *et al.*, 1984; Mohr & Fewtrell, 1987; Labrecque *et al.*, 1989; He & Curry, 1995). Oxonol V, depicted in Fig. 14.2, is among several similar oxonols developed by Chance, Smith & Bashford (Smith *et al.*, 1976; Bashford *et al.*, 1979; Smith & Chance, 1979) for dual-wavelength absorbance measurements on energy-transducing organelle suspensions.

The membrane potential of individual cells can be monitored with a fluorescence microscope and a cationic redistribution dye. Most of these have been qualitative studies; this is because of the difficulty in calibrating the fluorescence intensities determined from variably sized cells as well as the complexity of the fluorescence perturbations induced by the dynamic intracellular environment. A cyanine dye was used to distinguish between quiescent and cycling cells based on the difference in dye accumulation (Cohen *et al.*, 1981). Rhodamine-123 was originally introduced as a mitochondrial stain by Chen and co-workers (Johnson *et al.*, 1980, 1981; Chen, 1988). It has been used largely in qualitative studies of mitochondrial membrane potential and has been especially effective in flow cytometry applications.

In order to develop a quantitative assay of membrane potential in individual cells, it would be desirable to use a permeable redistribution dye with spectral characteristics having minimal environment sensitivity. Thus, the fluorescence intensity will reflect the degree of nernstian accumulation of dye only and can therefore be readily interpreted. This laboratory (Ehrenberg *et al.*, 1988) has synthesized two rhodamine dyes, TMRE and TMRM, which are very similar to rhodamine 123 but have the free amino groups substituted with methyl substituents. This makes these dyes more permeable than rhodamine 123 and also blocks any poorly reversible hydrogen bonding interactions with anionic sites in the mitochondrial inner membrane and matrix. This, combined with the general environmental insensitivity of rhodamine fluorescence and the low tendency of these dyes to aggregate, makes them good 'nernstian' indicators of membrane potential. That is to say, the ratio of fluorescence intensities measured in two compartments separated by a membrane, viz. F_{in}/F_{out}, when properly corrected for background dye binding, can be inserted into the Nernst equation to provide a direct measure of the potential difference between the compartments:

$$\Delta V_m = -\frac{RT}{F}\ln\frac{F_{in}}{F_{out}} \qquad [14.1]$$

where R is the ideal gas constant, T the absolute temperature and F is Faraday's constant. This approach was successfully applied to the plasma membrane potential of several different cell types (Ehrenberg *et al.*, 1988). It has also been used to follow changes in mitochondrial membrane potential via digital imaging microscopy (Farkas *et al.*, 1989; Loew *et al.*, 1993, 1994). These studies will be described in more detail in the following section.

14.3 MAPPING MEMBRANE POTENTIAL BY DIGITAL FLUORESCENCE MICROSCOPY

Appropriate choices of either the fast or slow dyes can make it possible to study phenomena involving membrane potential in individual cells under the microscope. As has been noted above, the primary application of the fast dyes has been in studies of the electrical activity in excitable systems. These are the subject of the chapter by Cohen & Wu (Chapter 15), elsewhere in this volume. In this section, I would like to describe the use of the fast dyes to map membrane potential variations along the surface of single cells. This depends on the ability of these membrane-staining dyes to sense local electric fields, rather than on their speed. Slow dyes have been mainly applied to studies of cell and organelle suspensions (Freedman & Laris, 1981, 1988; Waggoner, 1985). Later in this section, the ability of the 'nernstian' dyes to determine membrane potential in single cells will be described and their ability to quantitate membrane potential in individual mitochondria will be explored.

14.3.1 Membrane potential maps from dual-wavelength fluorescence ratio images

As detailed in Section 14.2, potentiometric indicators classified as 'fast' stain the plasma membrane. These dye molecules individually sense the electric field in the membrane with a resultant perturbation of their spectral properties. They are therefore capable of reporting rapid changes in membrane potential and can be used to follow the spread of an action potential down a single neuron (Grinvald, H. *et al.*, 1981; Krauthamer & Ross, 1984; Ross & Krauthamer, 1984; Shrager *et al.*, 1987; Shrager & Rubinstein, 1990; Fromherz & Müller, 1994; Zecevic, 1996). Similarly, they can be used to detect temporally constant spatial variations in membrane potential along the surface of cells. This latter application requires some special dye properties; among these is a potential-dependent spectral shift which can become the basis for dual-wavelength ratiometric measurements.

The success of dual-wavelength ratiometric methods for measurements with fluorescent cation indicators (Rink *et al.*, 1982; Tsien & Poenie, 1986) is largely due to the ability of this approach to eliminate artifactual variations in total dye fluorescence from the assay. Thus cells in suspension can be assayed without concern for small variations in the level of dye loading from sample to sample. More importantly, optically heterogeneous specimens, such as single cells under the fluorescence microscope, can have their cation distributions mapped via digital ratio imaging (Tsien & Poenie, 1986). The same advantages pertain for membrane potential measurements with membrane-staining dyes. This is because the fluorescence patterns can be heterogeneous owing to complex submicroscopic membrane morphologies or uneven staining; these would mask any variation in the electrical properties of the membrane. The dual-wavelength ratio normalizes away such non-potentiometric variations for a dye that undergoes a potential-dependent spectral shift.

The ANEP naphthylstyryl chromophore has an electric-field-induced spectral shift (Fluhler *et al.*, 1985) and can, therefore, be used to measure membrane potential by dual-wavelength ratio methods (Montana *et al.*, 1989). Using di-4-ANEPPS, we have demonstrated the viability of this idea with both a dual-wavelength spectrofluorometer on lipid vesicle suspensions and a fluorescence microscope equipped with dual-wavelength digital imaging equipment to study individual cells exposed to varying external electric fields (Montana *et al.*, 1989). Calibration of the ratio on a dye-stained cell can also be achieved with the valinomycin/K^+ method (Bedlack *et al.*, 1992; Loew, 1994). Most recently we have combined the imaging with a patch-clamp set-up to enable direct calibration of the ratio by voltage-clamp through a whole cell patch (Zhang *et al.*, 1998); Plate 14.1 and Fig. 14.3 display a calibration by this method of di-8-ANEPPS on a N1E-115 neuroblastoma cell. The dual-excitation ratio does provide

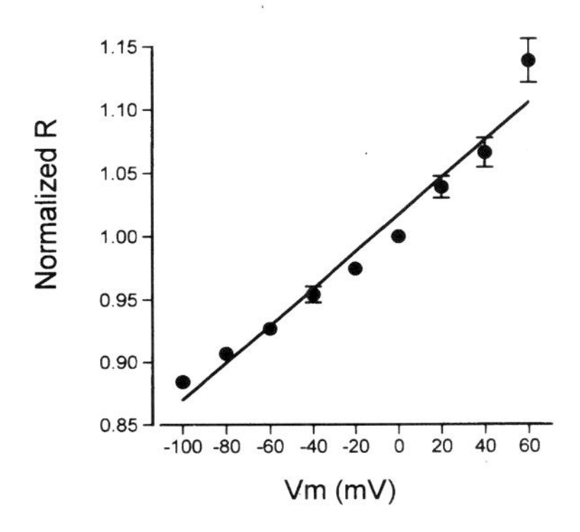

Figure 14.3 Dependence of R on voltage-clamped membrane potential. For each cell, R was normalized to the value obtained at the transmembrane potential of 0 mV. The normalized R was then plotted against the clamped transmembrane potential (Vm). The data were obtained from 40 cells and each data point corresponds to a mean ± SEM. The calibration line was derived from linear regression of the data, resulting in a change of 0.15 of normalized R per 100 mV. (Reproduced, with permission, from Zhang *et al.*, 1998).

a good indicator of membrane potential; more importantly, it is insensitive to sources of artifact such as photobleaching and non-potentiometric fluorescence changes due to variable levels of dye binding.

In one application of ratio imaging to map membrane potential, we studied electric field effects in cells with complex geometries where the induced membrane potential cannot be simply predicted from analytical solutions of Laplace's equation. The overall goal was to assess the ability of electric fields to direct neurite growth in differentiated N1E-115 neuroblastoma cells (Bedlack *et al.*, 1992). Since these cells have both complex geometries and voltage-gated channels, determination of the membrane potential distribution is only possible by direct experimental mapping using the dual-wavelength ratiometric approach. Di-4-ANEPPS was rapidly internalized into these cells; therefore, di-8-ANEPPS was used to stain the cells. The side of the cell facing the cathode is depolarized while the anodal side is relatively unaffected. Thus, the simple expectation of equal but opposite membrane potential changes on the opposite faces of the cell is not what is found experimentally. The cathodal depolarization is the first step in a chain of events that includes opening of voltage-dependent calcium channels, localized $[Ca^{2+}]_{cyt}$ increase, filopodial reorganization and, finally neurite outgrowth toward the cathode (Bedlack *et al.*, 1992). The technique of dual-wavelength ratiometric imaging of membrane potential may become an important tool in investigations of localized cellular responses to external stimuli.

Ratiometric images can also reveal intrinsic variations of membrane electrical properties along the surface of a cell. The non-homogeneous ratio in the same neuroblastoma cells has been investigated (Bedlack *et al.*, 1994) and attributed to variations in the membrane dipole potential. The dual-wavelength ratio of di-8-ANEPPS is quite sensitive to the dipole potential (Gross *et al.*, 1994), which sets up an intense electric field in the ester region of the lipid bilayer. In the differentiated neuroblastoma cells, the dye experiences an electric field ≈3 mV per Å more positive in the membrane of the neurites and growth cones than in the membranes of the somata. A map of the electrical potential along the surface of one of these cells is shown in Plate 14.2. This regional variation in the dipole potential may explain the difference in the activation kinetics of the voltage-gated sodium channel that was uncovered in single channel recordings from membrane patches derived from the soma and growth cone (Zhang *et al.*, 1996).

14.3.2 Nernstian dyes permit imaging of cell and mitochondrial membrane potential

As noted in Section 14.2, this laboratory has developed a pair of cationic rhodamine derivatives, TMRE (Fig. 14.2) and TMRM, with properties appropriate

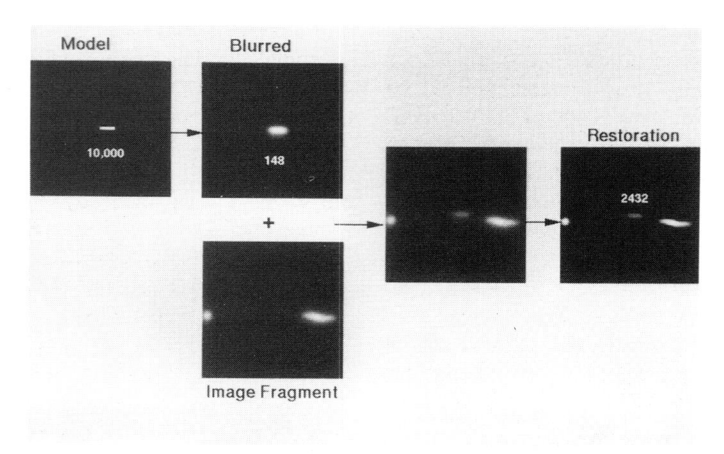

Figure 14.4 Model mitochondria are used to calibrate the restorations. A model cylindrical mitochondrion (upper left) is constructed with a length of 1000 nm and a diameter of 230 nm (from electron microscopy data) and an assigned intensity of 10 000. The model is blurred with a point spread function obtained from a 180 nm fluorescent microsphere on the coverslip which generated the data to be calibrated. The blurring operation is carried out at 3× the pixel resolution of the microscope to avoid pixelization errors. The resultant, blurred, model is reduced to the appropriate pixel dimensions and combined with a 3-D image fragment to introduce experimental noise. This image is then restored with the same point spread function used on the data. The maximum intensity of the restored model is compared with that of the original model to generate a calibration factor. The above images are all midplanes of 3-D data sets. (Reproduced, with permission, from Loew *et al.*, 1993.)

for quantitation of membrane potential in individual cells (Ehrenberg *et al.*, 1988). A microscope photometer was employed in this original study to obtain fluorescence intensities in the cytosol and in the extracellular medium as required by equation [14.1], above. To avoid contaminating signal from the very bright mitochondria, the measuring pinhole aperture was positioned over a mitochondria-free region (often the cell nucleus). The measured fluorescence intensities had to be corrected for a small amount of non-potential-dependent binding of the dye as determined from cells depolarized with K^+/valinomycin. Another important consideration in correctly determining F_{in}/F_{out}, is the depth of field of the fluorescence microscope. The problem is that the emission is collected from the entire thickness of medium in the bathing solution while only the thickness of the cell contributes to the cytosolic fluorescence. This problem was resolved by assessing the out of focus contribution to cell fluorescence from extracellular fluorescein–dextran added to the bathing solution.

These problems are not so easily solved if ordinary wide-field imaging microscopy is to be substituted for microphotometry. Specifically, the issue of out-of-focus contributions becomes intractable because of the necessarily wide field and measuring apertures in the imaging microscope optical path (Farkas *et al.*, 1989). However, confocal microscopy combines imaging with pinhole detection and can, therefore, directly and quantitatively measure the plasma membrane potential with nernstian redistribution indicators (Loew, 1993, 1997). This is because the ‘optical sectioning’ effect of the confocal microscope confines the measured fluorescence to a region smaller than the thickness of the cell. The fluorescence in the cytosol and the adjacent extracellular medium are sampled in the confocal image and equation [14.1] is applied to the corresponding fluorescence ratio to derive the membrane potential. The optical section is thin enough that the fluorescein–dextran correction procedure is usually unnecessary.

An extension of this approach toward measuring mitochondrial membrane potential has been a goal in our laboratory. A major problem is that the width of the optical slice is limited to about 0.6 μm even with the smallest confocal pinhole and the highest numerical aperture objective; mitochondria are typically ellipsoidal or cylindrical objects with their smallest dimensions in this same range. So even confocal microscopy may not be able to provide sufficiently narrow depth of field for objects as small as mitochondria. Most importantly, for a given mitochondrion in a given image there is currently no way of judging how serious this problem is. A second limitation of confocal microscopy involves the high excitation light intensities necessitated by the technique (at least in the current generation of confocal microscopes) because it is based on the principle of rejecting the fluorescence emission emanating from outside the plane of focus; thus, single measurements on living cells can be performed, but the accumulation of large image sets for either kinetics or 3-D reconstructions lead to fading of the dye and/or phototoxic effects.

An alternative to confocal microscopy for accurate 3-D imaging, is the restoration of out of focus contributions to widefield images by computer-based deconvolution techniques (Agard *et al.*, 1989; Fay *et al.*, 1989). Since this method restores out-of-focus light to its point of origin, it requires lower overall doses of excitation light than confocal microscopy. Using this method we investigated time-dependent changes in the membrane potential of individual mitochondria in the neurites of living differentiated N1E-115 neuroblastoma cells (Loew *et al.*, 1993). At each time point, a 3-D image consisting of a stack of seven planes was acquired, and up to eight time points could be collected before photo-degradation became apparent. TMRE fluorescence was analysed only in mitochondria that remained completely within the sampled volume; this ensured that the measured fluorescence reflected the concentration of TMRE and not the movement of a mitochondrion to the edge of the sampled volume.

The problem that remains is that typical mitochondria are below the resolution limit of the microscope, especially in the direction of the optical axis. Therefore, the mitochondrial intensity that is measured after restoration (or in a confocal microscope) is still diluted by a significant surrounding volume of low fluorescence. However, it is possible, based on a careful characterization of the microscope optics, to model the behaviour of objects such as mitochondria and thereby calibrate the measurements. The microscope optics are empirically characterized by the 3-D point spread function obtained by imaging a subresolution fluorescent microsphere. The calibration process is illustrated in Fig. 14.5 where an image of a mitochondrion is modelled as a cylindrical object with a length of 1 μm and a diameter of 230 nm (derived from analysis of mitochondrial dimensions in electron micrographs of the neurites of the neuroblastoma cells). This model is assigned a uniform intensity of 10 000 units and convolved with the point spread function of the optics to create an accurate representation of how such a mitochondrion would appear as imaged through the 3-D digital microscope. This image is then merged with a real image of a neurite to provide the appropriate noise background; finally, the merged 3-D image is deconvolved with the restoration procedure. The resultant image of the model mitochondrion has a maximum intensity of 2432 units; this provides a calibration factor for the mitochondrial intensities of

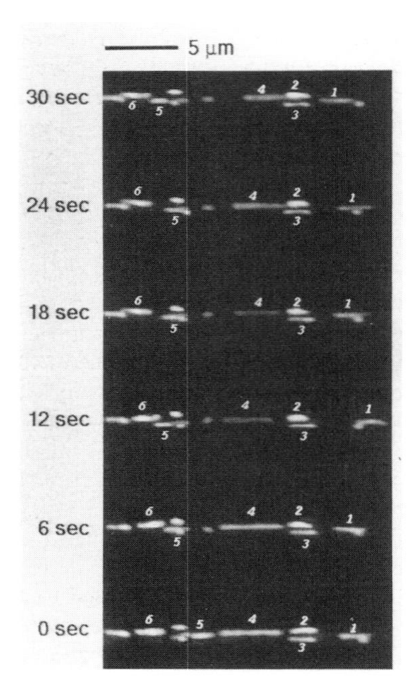

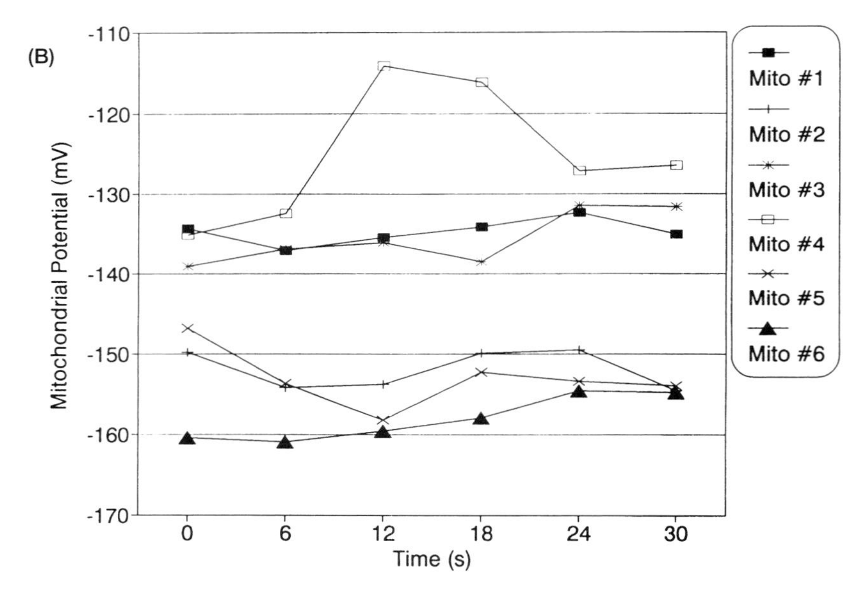

Figure 14.5 Mitochondria occasionally display large membrane potential fluctuations. (A) Time series of restored images taken at 6 s intevals. Notice the loss of intensity in mitochondrion #4 in the 3rd and 4th time points. (B) Mitochondrial potential is plotted for each of the labelled mitochondria, indicating the magnitude of the fluctuation in mitochondrion #4. (Reproduced, with permission, from Loew *et al.*, 1993.)

10 000/2432 = 4.1. This calibration factor, *cal*, may be determined for a given cell type if it has a relatively narrow mitochondrial size distribution, as was the case for the mitochondria in the neurite of the N1E-115 neuroblastoma cells. It can be used in a modified form of equation [14.1] to determine the membrane potential across individual mitochondria:

$$\Delta V_{\text{mit}} = -\frac{RT}{F} \ln \frac{cal \cdot F_{\text{mit}}}{F_{\text{cyt}}} \qquad [14.2]$$

Using this approach, it was possible to determine fluctuations in membrane potential in single mitochondria (Loew *et al.*, 1993). Figure 14.5 shows how one mitochondrion in the image of a neurite undergoes a transient depolarization of approximately 20 mV that lasts for about 20 s. During an observation period lasting 60 s, about 7% of several hundred mitochondria that were monitored undergo significant transient changes. Longer observation times could yield a better idea of the frequency and significance of these fluctuations. However, because of the limited tolerance to light exposure and the motion of mitochondria up and down the neurite, longer time series are currently technically precluded. It is clear, nonetheless, that such fluctuations could only have been discovered because of our ability to monitor the behaviour of single mitochondria. In a subsequent study (Loew *et al.*, 1994), mitochondrial membrane potential measurements with TMRE were combined with intracellular calcium measurements with fura-2. The ability to measure mitochondrial potential *in situ* within a living cell made it possible to establish that mitochondria are depolarized by about 20 mV during transient physiological increases in cytosolic free calcium.

14.4 CONCLUSION

The determination of membrane potential with potentiometric dyes requires an appropriate match of the biological preparation, indicator probe, optical detection scheme and the scientific question to be answered. This review has focused on the variety of potentiometric indicators that have been developed over the past 20 years and has attempted to provide sufficient information so that new experiments can be designed. This process was illustrated with applications of slow and fast dyes to digital imaging microscopy of membrane potential distributions in single cells.

This was not intended as an extensive review of the subject. Indeed, several exciting recent applications of fast dyes in multicellular excitable systems have employed digital video microscopy. These include a number of studies on functional neuronal connections within the olfactory system (Kauer, 1988; Delaney *et al.*, 1994; Cinelli & Kauer, 1995; Cinelli *et al.*, 1995a,b) and on patterns of electrical activity in intact heart tissue (Davidenko *et al.*, 1992; Pertsov *et al.*, 1993; Knisley *et al.*, 1994). In addition, several technical advances to the camera technologies have permitted extensions of the potentiometric dye technology to more complex and specialized questions (Borst, 1995; Gogan *et al.*, 1995; Baxter *et al.*, 1996; Bullen *et al.*, 1997). These studies have been generally confined to temporal resolutions of greater than 30 ms. For studies which require millisecond time resolution, arrays of 100–500 discrete detectors can be used to develop low spatial resolution maps of potential changes. Such methods are reviewed in the chapter by Cohen & Wu (Chapter 15) in this volume. However, at least one study has employed gated intensifier video camera technology to achieve microsecond time resolution (Hibino *et al.*, 1993), while another utilized an array of 128 × 128 photodiodes with a temporal resolution of 0.6 ms (Tanifuji *et al.*, 1994). Still, the current state of technology generally requires some compromise between optimal temporal and spatial resolution. As the imaging and dye technologies improve, such compromises may ultimately become unnecessary.

ACKNOWLEDGEMENTS

I am pleased to acknowledge my many collaborators who have made this work possible and so enjoyable. Special thanks to Britton Chance for introducing me to the potentimetric dyes and to Larry Cohen for his long-term collaboration and friendship. This work was supported by a grant from the USPHS, GM35063.

REFERENCES

Agard D.A., Hiraoka Y., Shaw P. & Sedat J.W. (1989) In *Methods in Cell Biology*, Vol. 30, D.L. Taylor & Y. Wang (eds). Academic Press, San Diego, pp. 353–377.
Antic S., Loew L., Wuskell J. & Zecevic D. (1992) *Biol. Bull.* **183**, 350–351.
Bashford C.L., Chance B., Smith J.C. & Yoshida T. (1979) *Biophys. J.* **25**, 63–85.
Baxter W.T., Davidenko J.M., Loew L.M., Wuskell J.P. & Jalife J. (1996) *Ann. Biomed. Eng.* **25**, 713–725.
Beach J.M., McGahren E.D., Xia J. & Duling B.R. (1996) *Am. J. Physiol.* **270**, H2216–H2227.
Bedlack R.S., Wei M.-d. & Loew L.M. (1992) *Neuron* **9**, 393–403.
Bedlack R.S., Wei M.-d, Fox S.H., Gross E. & Loew L.M. (1994) *Neuron* **13**, 1187–1193.
Borst A. (1995) *Z. Naturforsch.* **50**, 435–438.
Brauner T., Hulser D.F. & Strasser R.J. (1984) *Biochim. Biophys. Acta* **771**, 208.
Bullen A., Patel S.S. & Saggau P. (1997) *Biophys. J.* **73**, 477–491.
Chen L.B. (1988) *Ann. Rev. Cell Biol.* **4**, 155–181.
Chien C.-B. & Pine J. (1991) *Biophys. J.* **60**, 697–711.
Cinelli A.R. & Kauer J.S. (1995) *J. Neurophysiol.* **73**, 2033–2052.
Cinelli A.R., Hamilton K.A. & Kauer J.S. (1995a) *J. Neurophysiol.* **73**, 2053–2071.
Cinelli A.R., Neff S.R. & Kauer J.S. (1995b) *J. Neurophysiol.* **73**, 2017–2032.
Cohen L.B. & Salzberg B.M. (1978) *Rev. Physiol. Biochem. Pharmacol.* **83**, 35–88.
Cohen L.B., Salzberg B.M., Davila H.V., Ross W.N., Landowne

D., Waggoner A.S. & Wang C.H. (1974) *J. Membr. Biol.* **19**, 1–36.
Cohen R.L., Muirhead K.A., Gill J.E., Waggoner A.S. & Horan P.K. (1981) *Nature* **290**, 593–595.
Davidenko J.M., Pertsov A.V., Salomonsz R., Baxter W. & Jalife J. (1992) *Nature* **355**, 349–351.
Davila H.V., Salzberg B.M., Cohen L.B. & Waggoner A.S. (1973) *Nature New Biol.* **241**, 159–160.
Delaney K.R., Gelperin A., Fee M.S., Flores J.A., Gervais R., Tank D.W. & Kleinfeld D. (1994) *Proc. Natl. Acad. Sci. USA* **91**, 669–673.
Ehrenberg B., Farkas D.L., Fluhler E.N., Lojewska Z. & Loew L.M. (1987) *Biophys. J.* **51**, 833–837.
Ehrenberg B., Montana V., Wei M.-d, Wuskell J.P. & Loew L.M. (1988) *Biophys. J.* **53**, 785–794.
Ehrenberg B., Wei M. & Loew L.M. (1990) *Biophys. J.* **57**, 484a.
Farkas D.L., Wei M., Febbroriello P., Carson J.H. & Loew L.M. (1989) *Biophys. J.* **56**, 1053–1069.
Fay F.S., Carrington W. & Fogarty K.E. (1989) *J. Microsc.* **153**, 133–149.
Fluhler E., Burnham V.G. & Loew L.M. (1985) *Biochemistry* **24**, 5749–5755.
Freedman J.C. & Laris P.C. (1981) In *International Review of Cytology*, Suppl. 12. Academic Press, New York, pp. 177–246.
Freedman J.C. & Laris P.C. (1988) In *Spectroscopic Membrane Probes*, Vol. 3, L.M. Loew (ed). CRC Press, Boca Raton, pp. 1–50.
Fromherz P. & Müller C.O. (1994) *Proc. Natl Acad. Sci. USA* **91**, 4604–4608.
George E.B., Nyirjesy P., Pratap P.R., Freedman J.C. & Waggoner A.S. (1988) *J. Membr. Biol.* **105**, 55–64.
Gogan P., Schmiedel-Jakob I., Chitti Y. & Tyc-Dumont S. (1995) *Biophys. J.* **69**, 299–310.
Gonzalez J.E. & Tsien R.Y. (1995) *Biophys. J.* **69**, 1272–1280.
Gonzalez J.E. & Tsien R.Y. (1997) *Chem. Biol.* **4**, 269–277.
Grinvald A.S., Hildesheim R., Farber I.C. & Anglister J. (1982) *Biophys. J.* **39**, 301–308.
Grinvald A., Fine A., Farber I.C. & Hildesheim R. (1983) *Biophys. J.* **42**, 195–198.
Grinvald A., Salzberg B.M., Lev-Ram V. & Hildesheim R. (1987) *Biophys. J.* **51**, 643–651.
Grinvald A., Lieke E.E., Frostig R.D. & Hildesheim R. (1994) *J. Neurosci.* **14**, 2545–2568.
Grinvald H., Ross W. N. & Farber I. (1981) *Proc. Natl. Acad. Sci. USA* **78**, 3245–3249.
Gross D., Loew L.M. & Webb W.W. (1986) *Biophys. J.* **50**, 339–348.
Gross E., Bedlack R.S. & Loew L.M. (1994) *Biophys. J.* **67**, 208–216.
Gupta R.K., Salzberg B.M., Grinvald A., Cohen L.B., Kamino K., Lesher S., Boyle M.B., Waggoner A.S. & Wang C. (1981) *J. Membr. Biol.* **58**, 123–137.
Hassner A., Birnbaum D. & Loew L.M. (1984) *J. Org. Chem.* **49**, 2546–2551.
Hayashi Y., Zviman M.M., Brand J.G., Teeter J.H. & Restrepo D. (1996) *Biophys. J.* **71**, 1057–1070.
He P. & Curry F.E. (1995) *Microvasc. Res.* **50**, 183–198.
Herbele J. & Dencher N.A. (1992) *Proc. Natl Acad. Sci. USA* **89**, 5996–6000.
Hibino M., Itoh H. & Kinosita K. (1993) *Biophys. J.* **64**, 1789–1800.
Johnson L.V., Walsh M.L. & Chen L.B. (1980) *Proc. Natl Acad. Sci. USA* **77**, 990–994.
Johnson L.V., Walsh M.L., Bockus B.J. & Chen L.B. (1981) *J. Cell Biol.* **88**, 526–535.
Kauer J.S. (1988) *Nature* **331**, 166–168.
Knisley S.B. (1995) *Circ. Res.* **77**, 1229–1239.
Knisley S.B., Blitchington T.F., Hill B.C., Grant A.O., Smith W.M., Pilkington T.C. & Ideker R.E. (1993) *Circ. Res.* **72**, 255–270.
Knisley S.B., Hill B.C. & Ideker R.E. (1994) *Biophys. J.* **66**, 719–728.
Korchak H.M., Rich A.M., Wilkenfeld C., Rutherford L.E. & Weissmann G. (1982) *Biochem. Biophys. Res. Commun.* **108**, 1495–1501.
Krauthamer V. & Ross W.N. (1984) *J. Neurosci.* **4**, 673.
Labrecque G.F., Holowka D. & Baird B. (1989) *J. Immunol.* **142**, 236–243.
Liptay W. (1969) *Angew. Chem. Internat. Ed.* **8**, 177–188.
Loew L.M. (1988) *Spectroscopic Membrane Probes*. CRC Press, Boca Raton.
Loew L.M. (1993) In *Cell Biological Applications of Confocal Microscopy. Methods in Cell Biology*, B. Matsumoto (ed). Vol. 38, pp. 194–209. Academic Press, Orlando.
Loew L.M. (1994) *Neuroprotocols* **5**, 72–79.
Loew L.M. (1997) In *Cell Biology: A Laboratory Handbook*, J.E. Celis (ed). Vol. 3, pp. 375–379. Academic Press, San Diego.
Loew L.M. & Simpson L. (1981) *Biophys. J.* **34**, 353–365.
Loew L.M., Bonneville G.W. & Surow J. (1978) *Biochemistry* **17**, 4065–4071.
Loew L.M., Simpson L., Hassner A. & Alexanian V. (1979) *J. Am. Chem. Soc.* **101**, 5439–5440.
Loew L.M., Rosenberg I., Bridge M. & Gitler C. (1983) *Biochemistry* **22**, 837–844.
Loew L.M., Benson L., Lazarovici P. & Rosenberg I. (1985a) *Biochemistry* **24**, 2101–2104.
Loew L.M., Cohen L.B., Salzberg B.M., Obaid A.L. & Bezanilla F. (1985b) *Biophys. J.* **47**, 71–77.
Loew L.M., Cohen L.B., Dix J., Fluhler E.N., Montana V., Salama G. & Wu J.-Y. (1992) *J. Membr. Biol.* **130**, 1–10.
Loew L.M., Tuft R.A., Carrington W. & Fay F.S. (1993) *Biophys. J.* **65**, 2396–2407.
Loew L.M., Carrington W., Tuft R. & Fay F.S. (1994) *Proc. Natl Acad. Sci. USA* **91**, 12579–12583.
Lojewska Z. & Loew L.M. (1987) *Biochim. Biophys. Acta* **899**, 104–112.
Lojewska Z., Farkas D.L., Ehrenberg B. & Loew L.M. (1989) *Biophys. J.* **56**, 121–128.
Mohr C.F. & Fewtrell C. (1987) *J. Immunol.* **138**, 1564–1570.
Montana V., Farkas D.L. & Loew L.M. (1989) *Biochemistry* **28**, 4536–4539.
Müller W., Windisch H. & Tritthart H.A. (1989) *Biophys. J.* **56**, 623–629.
Pertsov A.M., Davidenko J.M., Salomonsz R., Baxter W.T. & Jalife, J. (1993) *Circ. Res.* **72**, 631–650.
Platt J.R. (1956) *J. Chem. Phys.* **25**, 80–105.
Reers M., Smith T.W. & Chen L.B. (1991) *Biochemistry* **30**, 4480–4486.
Reers M., Smiley S.T., Mottola-Hartshorn C., Chen A., Lin M. & Chen L.B. (1995) *Methods Enzymol.* **260**, 406–417.
Rink T.J., Tsien R.Y. & Pozzan T. (1982) *J. Cell Biol.* **95**, 189–196.
Rohr S. & Salzberg B.M. (1994a) *J. Gen. Physiol.* **104**, 287–309.
Rohr S. & Salzberg B.M. (1994b) *Biophys. J.* **67**, 1301–1315.
Rosenbaum D.S., Kaplan D.T., Kanai A., Jackson L., Garan H., Cohen R.J. & Salama G. (1991) *Circulation* **84**, 1333–1345.
Ross W.N. & Krauthamer V. (1984) *J. Neurosci.* **4**, 659–672.
Ross W.N. & Reichardt L.F. (1979) *J. Membr. Biol.* **48**, 343–356.
Ross W.N., Salzberg B.M., Cohen L.B., Grinvald A., Davila

H.V., Waggoner A.S. & Wang C.H. (1977) *J. Membr. Biol.* **33**, 141–183.

Shrager P. & Rubinstein C.T. (1990) *J. Gen. Physiol.* **95**, 867–890.

Shrager P., Chiu S.Y., Ritchie J.M., Zecevic D. & Cohen L.B. (1987) *Biophys. J.* **51**, 351–355.

Sims P.J., Waggoner A.S., Wang C.-H. & Hoffman J.F. (1974) *Biochemistry* **13**, 3315–3330.

Smiley S.T., Reers M., Mottola-Hartshorn C., Lin M., Chen A., Smith T.W., Steele G.D. & Chen L.B. (1991) *Proc. Natl Acad. Sci. USA* **88**, 3671–3675.

Smith J.C. & Chance B. (1979) *J. Membr. Biol.* **46**, 255.

Smith J.C., Russ P., Cooperman B.S. & Chance B. (1976) *Biochemistry* **15**, 5094–5105.

Tanifuji M., Sugiyama T. & Murase K. (1994) *Science* **266**, 1057–1059.

Terasaki M., Song J., Wong J.R., Weiss M.J. & Chen L.B. (1984) *Cell* **38**, 101–108.

Tsau Y., Wenner P., O'Donovan M.J., Cohen L.B., Loew L.M. & Wuskell J.P. (1996) *J. Neurosci. Meth.* **170**, 121–129.

Tsien R.Y. & Poenie M. (1986) *Trends Biochem. Sci.* **11**, 450–455.

Waggoner A.S. (1979) *Ann. Rev. Biophys. Bioeng.* **8**, 847–868.

Waggoner A.S. (1985) In *The Enzymes of Biological Membranes*, A.N. Martonosi (ed). Plenum, New York, pp. 313–331.

Waggoner A.S., Wang C.H. & Tolles R.L. (1977) *J. Membr. Biol.* **33**, 109–140.

Wan B., Doumen C., Duszynski J., Salama G., Vary T.C. & LaNoue K. (1993) *Am. J. Physiol.* **265**, H453–H460.

Wenner P., Tsau Y., Cohen L.B., O'Donovan M.J. & Dan Y. (1996) *J. Neurosci. Methods* **170**, 111–120.

Wu J.-y, Cohen L.B. & Falk C.X. (1994) *Science* **263**, 820–822.

Zecevic D. (1996) *Nature* **381**, 322–325.

Zhang J., Loew L.M. & Davidson R.M. (1996) *Biophys. J.* **71**, 2501–2508.

Zhang J., Davidson R.M., Wei M. & Loew L.M. (1998) *Biophys. J.* **74**, 48–53.

CHAPTER FIFTEEN

Fast Multisite Optical Measurement of Membrane Potential, with Two Examples

JIAN-YOUNG WU[1,2], LAWRENCE B. COHEN[3,4] & CHUN X. FALK[3,4]

[1] Georgetown Institute of Cognitive and Computational Sciences, Georgetown University, Washington, DC, USA
[2] WuTech, Inc., Potomac, MD, USA
[3] Department of Cellular and Molecular Physiology, Yale University School of Medicine, New Haven, CT, USA
[4] OptImaging, LLC, Fairfield, CT, USA

15.1 INTRODUCTION

An optical measurement of membrane potential using a molecular probe can be beneficial in a variety of circumstances. In particular, this kind of measurement offers the possibility of simultaneous measurements from many locations. This is important in the studies of the nervous system where many parts of cells, many cells, or many regions are simultaneously active.

All of the optical signals described in this paper are fast signals (Cohen & Salzberg, 1978). These signals are presumed to arise from membrane-bound dye; they follow changes in membrane potential with time courses that are rapid compared with the rise time of an action potential. Several voltage-sensitive dyes (see, e.g., Fig. 15.1) have been used to monitor changes in potential in a variety of preparations.

Two topics have been discussed in detail in earlier reviews (Cohen & Lesher, 1986; Grinvald *et al.*, 1988; Wu & Cohen, 1993; Loew, 1994) and will not be covered here. First, the evidence showing that these optical signals are potential-dependent. Second, the evidence that pharmacological effects and photodynamic damage resulting from the voltage-sensitive dyes are manageable (see e.g. Cohen & Salzberg, 1978; Waggoner, 1979; Salzberg, 1983; Grinvald *et al.*, 1988). We begin with a discussion of dyes, light sources, photodetectors, optics, and computer hardware and software. This concern about apparatus arises because the signals in optical measurements are often small and attention to detail is required to achieve optimal signal-to-noise ratios. While some of the discussion is most relevant to our own apparatus (which is commercially available via WuTech, Inc., and OptImaging, LLC), most aspects would apply to any optical measurement.

The second half of the chapter will describe the experimental details and examples of results obtained from two kinds of measurements. First, measurements of the spike signals from many individual neurons in an invertebrate ganglion. Second, measurements of population signals representing the average change in membrane potential from many cells and processes. Both kinds of optical recordings have provided information about the function of the nervous system that was previously unobtainable.

FLUORESCENT AND LUMINESCENT PROBES, 2ND EDN
ISBN 0–12–447836–0

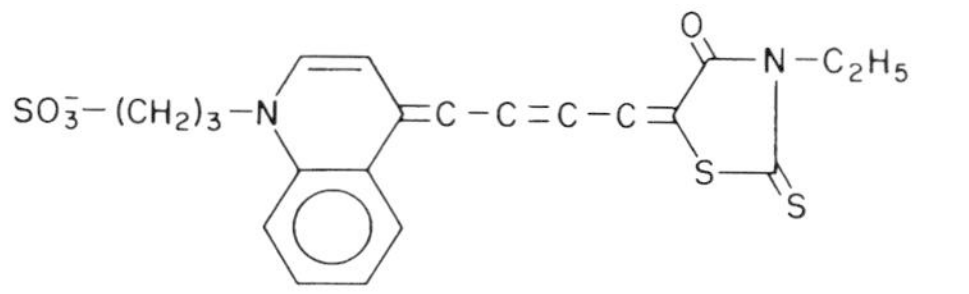

XVII, Merocyanine, Absorption, Birefringence

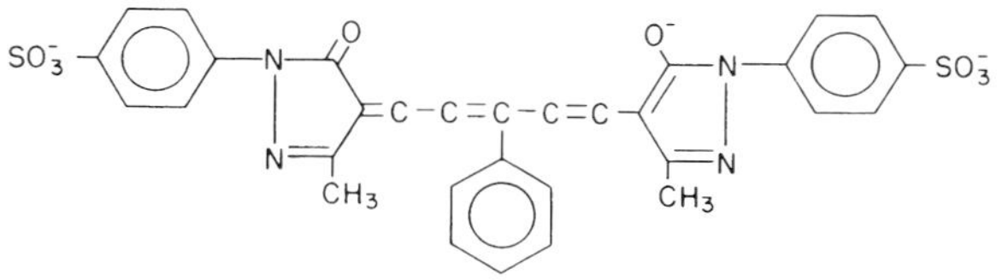

RH155, Oxonol, Absorption

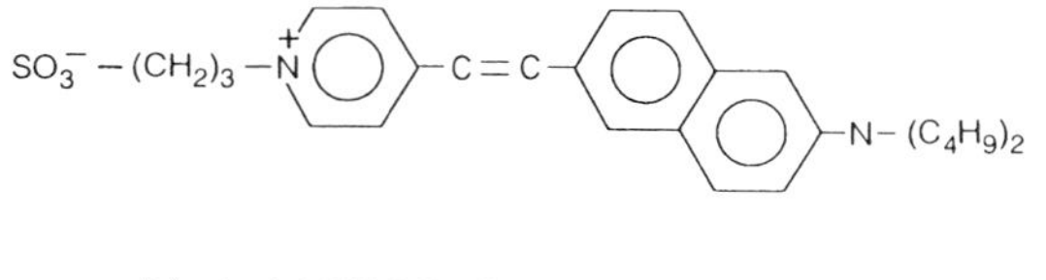

Di-4-ANEPPS, Styryl, Fluorescence

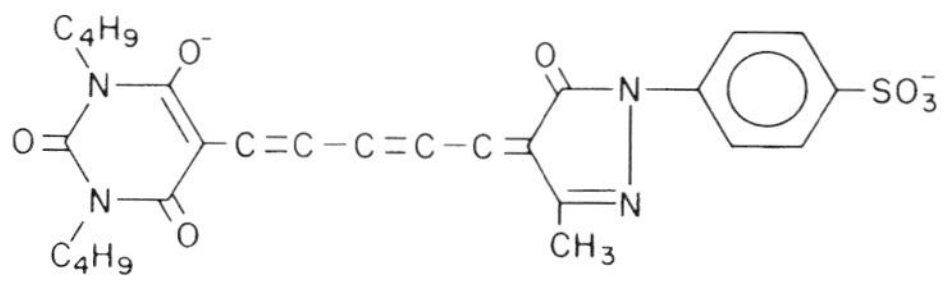

XXV, Oxonol, Fluorescence, Absorption

Figure 15.1 Examples of four different chromophores that have been used to monitor membrane potential. The merocyanine dye, XVII (WW375), and the oxonol dye, RH155, are commercially available as NK2495 and NK3041 from Nippon Kankoh-Shikiso Kenkyusho Co. Ltd, Okayama, Japan. The oxonol, XXV (WW781) and styryl, di-4-ANEPPS, are available commercially as dye R-1114 and D-1199 from Molecular Probes, Junction City, OR.

15.2 SIGNAL TYPE

Sometimes it is possible to decide in advance which kind of optical signal will give the best signal-to-noise ratio, but in other situations an experimental comparison is necessary. The choice of signal type often depends on the optical characteristics of the preparation. Birefringence signals are relatively large in preparations that, like giant axons, have a cylindrical shape and radial optic axis. However, in preparations with spherical symmetry (e.g. molluscan cell soma), the birefringence signals in adjacent quadrants will cancel (Boyle & Cohen, 1980). In one experiment where birefringence should have been tested, it was not (Shrager *et al.*, 1987).

Thick preparations (e.g. mammalian cortex) also dictate the choice of signal. In this circumstance transmitted light measurements are not easy (a subcortical implantation of a light guide would be necessary), and the small size of the absorption signals that are detected in reflected light (Ross *et al.*, 1974; Orbach & Cohen, 1983) meant that fluorescence would be optimal (Orbach *et al.*, 1985). Blasdel & Salama (1986) suggested that dye-related reflectance changes can be measured from cortex; however, Grinvald *et al.* (1986) suggest that these slow signals result from a change in intrinsic reflectance.

An important factor affecting the choice of absorption or fluorescence is that the signal-to-noise ratio in fluorescence is more strongly degraded by dye bound to extraneous material. Figure 15.2 illustrates a spherical cell surrounded by extraneous material. In Fig. 15.2(A), dye binds only to the cell; in Fig. 15.2(B), there is 10 times as much dye bound to extraneous material. To calculate the transmitted intensity we assume that there is one dye molecule for every 2.5 phospholipid molecules and a large extinction coefficient (10^5). The amount of light absorbed by the cell is still only 0.01 of the incident light and thus the transmitted light is 0.99 I_0. Thus, even if this dye completely disappeared because of a change in membrane potential, the fractional change in transmission, $\Delta I/I_0$, would be only 1% (10^{-2}). The amount of light reaching the photodetector in fluorescence will be much lower, say 0.0001 I_0. But, even though the light reaching the fluorescence detector is small, disappearance of dye would result in a 100% decrease in fluorescence, a fractional change of 10^{0}. Thus, in situations where dye is bound only to the cell membrane and there is only one cell in the light path, the fractional change in fluorescence is much larger than the fractional change in transmission.

However, the relative advantage of fluorescence is reduced if dye binds to extraneous material. When 10 times as much dye is bound to the extraneous material as was bound to the cell membrane (Fig. 15.2(B)), the transmitted intensity is reduced to approximately 0.9 I_0 but the fractional change in transmission is nearly unaffected. In contrast, the resting fluorescence

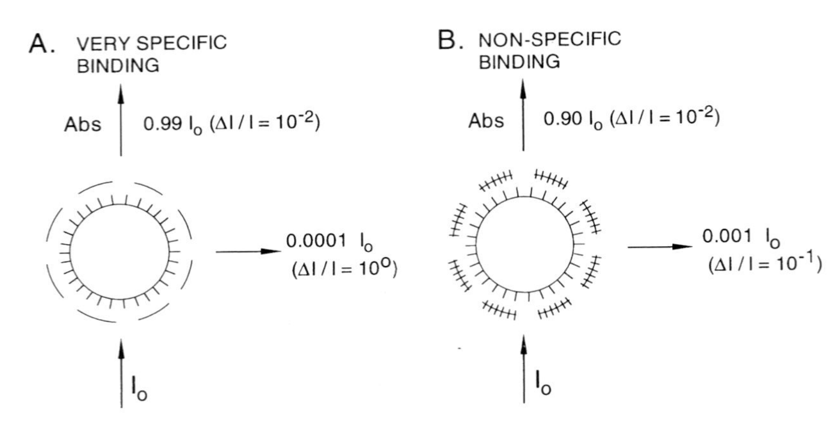

Figure 15.2 (A) The light transmission and fluorescence intensity when only a neuron binds dye and (B) when both the neuron and extraneous material binds dye. In (A), assuming that one dye molecule is bound per 2.5 phospholipid molecules, 0.99 of the incident light is transmitted. If a change in membrane potential causes the dye to disappear, the fractional change in transmission is 1%, but in fluorescence it is 100%. In (B), nine times as much dye is bound to extraneous material. Now the transmitted intensity is reduced to 0.9, but the fractional change is still 1%. The fluorescence intensity is increased 10-fold, and therefore the fractional change is reduced by the same factor. Thus, extraneously bound dye degrades fluorescence fractional changes and signal-to-noise ratios more rapidly. Redrawn from Cohen & Lesher (1986).

is now higher by a factor of 10, and the fractional fluorescence change is reduced by the same factor. It does not matter whether the extraneous material happens to be connective tissue, glial membrane, or neighbouring neuronal membranes.

In Fig. 15.2(B), the fractional change in fluorescence was still larger than in transmission. However, in fluorescence, the light intensity was about 10^3 smaller, which reduces the signal-to-noise ratio. Partly because of the signal degradation due to extraneous dye, fluorescence signals have most often been used in monitoring activity from tissue-cultured neurons, whereas absorption has been preferred in measurements from ganglia. Both kinds of signals have been used in brain slices. The discovery of methods for neuron type-specific staining (see below) would make the use of fluorescence more attractive.

15.3 DYES

The choice of dye is important in maximizing the signal-to-noise ratio. More than 1500 dyes have been tested for signal size in absorption and fluorescence using giant axons from the squid. This screening was made possible by synthetic efforts of three laboratories: Jeff Wang, Ravender Gupta and Alan Waggoner then at Amherst College; Rina Hildesheim and Amiram Grinvald at the Weizmann Institute; and Joe Wuskell and Leslie Loew at the University of Connecticut Health Center. Included in these syntheses were approximately 100 analogues of each of the four dyes illustrated in Fig. 15.1. In each of these four groups there were 10 or 20 dyes that gave approximately the same signal size on squid axons (Gupta *et al.*, 1981).

However, dyes that gave nearly identical signals on squid axons gave very different responses on other preparations, and, for example, tens of dyes had to be tested to maximize the signal in mammalian cortex (Orbach *et al.*, 1985; Grinvald *et al.*, 1994). Some dyes that worked well in squid did poorly in other preparations because they do not penetrate through connective tissue or along intercellular spaces to the membrane of interest.

15.3.1 Amplitude of the voltage change

All of the voltage-sensitive dye signals discussed in this article are simple changes in intensity (ΔI) or the fractional intensity change ($\Delta I/I$). These signals give information about the time course of the potential change but no direct information about its magnitude. In some instances, indirect information about the magnitude of the voltage change underlying the optical signal can be obtained (e.g. Orbach *et al.* 1985; Delaney *et al.*, 1994; Antic & Zecevic, 1995). Another approach is the use of ratiometric measurements at two independent wavelengths (Gross *et al.*, 1994). However, to determine the amplitude of the voltage change from a ratio measurement one must know the fraction of the fluorescence that results from dye not bound to active membrane, a requirement that can only be met in special circumstances (e.g. tissue culture).

15.4 MEASURING TECHNOLOGY

15.4.1 Noise

15.4.1.1 Shot noise

The limit of accuracy with which light can be measured is set by the shot noise arising from the statistical nature of photon emission from presently available light sources. Fluctuations in the number of photons emitted per unit time will occur, and if an ideal light source (tungsten filament at 3300°F) emits an average of 10^{14} photons ms^{-1}, the root-mean-square (RMS) deviation in the number emitted is the square root of this number, or 10^7 photons ms^{-1}. If the noise is dominated by shot noise, the signal-to-noise ratio is proportional to the square root of the number of measured photons and the bandwidth of the photodetection system (Braddick, 1960; Malmstadt *et al.*, 1974). The basis for the square root dependence on intensity is illustrated in Fig. 15.3. In Fig. 15.3(A) the result of using a random number table to distribute 20 photons into 20 time windows is shown. In Fig. 15.3(B) the same procedure was used to distribute 200 photons into the same 20 bins. Relative to the average light level there is more noise in the top trace (20 photons) than in the bottom trace (200 photons). On the right side of Fig. 15.3 the signal-to-noise ratios are measured and the improvement is similar to that expected from the square root relationship. This relationship is indicated by the dotted line in Fig. 15.4. In a shot-noise limited measurement, improvements in the signal-to-noise ratio can only be obtained by (1) increasing the illumination intensity, (2) improving the light-gathering efficiency of the apparatus, or (3) reducing the system bandwidth.

Because only a small fraction of the 10^{14} photons ms^{-1} emitted by a tungsten filament will be measured, a signal-to-noise ratio of 10^7 (see above) cannot be achieved. A partial listing of the light losses follows. A 0.7 NA lamp collector lens would collect 0.06 of the light emitted by the source. Only 0.2 of that light is in the visible wavelength range; the remainder is primarily infrared (heat). An interference filter of 30 nm width at half-height might transmit only 0.05 of the visible light. Finally, additional losses will occur at all air–glass interfaces. Thus, the light reaching the preparation might typically be reduced to 10^{10} photons ms^{-1} $pixel^{-1}$. If the light-collecting system has high efficiency, e.g. in a transmitted light measurement, about 10^{10} photons ms^{-1} will reach the photodetector, and if the photodetector has a quantum efficiency of 1.0, then 10^{10} photoelectrons ms^{-1} will be measured. The RMS shot noise will be 10^5 photoelectrons ms^{-1}; thus the relative noise is 10^{-5} (a signal-to-noise ratio of 100 dB).

In a measurement of fluorescence there will be additional losses. A 0.7 NA objective lens would collect only 0.06 of the light emitted by the preparation; there will be additional losses at the dichroic mirror. In addition, the collected light will be limited by the quantum efficiency of fluorescence emission. Thus, at best, 10^7 photoelectrons ms^{-1} will be measured. The RMS shot noise will be 3×10^3 photoelectrons ms^{-1}; and the relative noise is 3×10^{-3}.

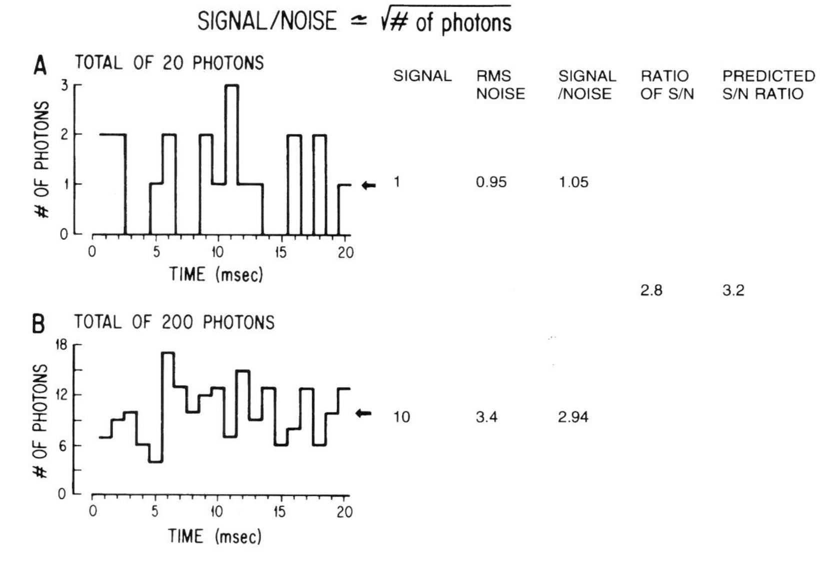

Figure 15.3 Plots of the results of using a table of random numbers to distribute 20 photons (top, A) or 200 photons (bottom, B) into 20 time bins. The result illustrates the fact that when more photons are measured the signal-to-noise ratio is improved. On the right, the signal-to-noise ratio is measured for the two results. The ratio of the two signal-to-noise ratios was 2.8. This is close to the ratio predicted by the relationship that the signal-to-noise ratio is proportional to the square root of the measured intensity. Redrawn from Wu & Cohen (1993).

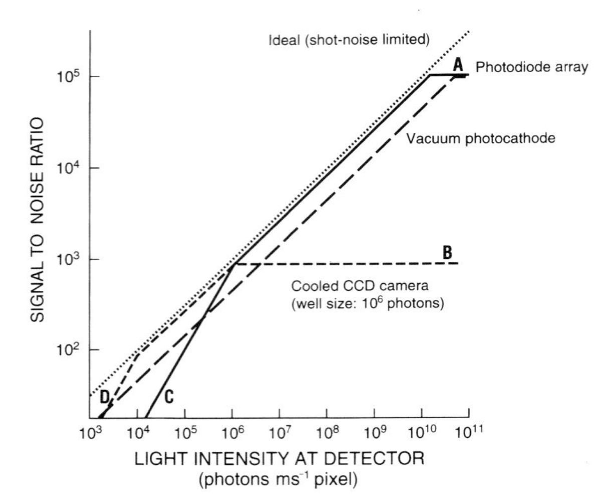

Figure 15.4 Signal-to-noise ratio as a function of light intensity in photons per millisecond per detector. The ideal case (shot-noise limited) and the approximate performance of three different kinds of imaging devices are indicated. The approximate light intensity in fluorescence measurements from a branch of an intracellularly stained neuron is 10^5 photons ms^{-1}. The approximate intensity in fluorescence measurements from *in vivo* or *in vitro* vertebrate brains is 10^8 photons ms^{-1}. The approximate intensity in absorption measurements in ganglia or brain slices is 10^9 photons ms^{-1}. The theoretical optimum (shot-noise limited) signal-to-noise ratio is shown in the dotted line. The signal-to-noise ratio obtained from a commercially available photodiode array system (NeuroPlex, OptImaging, LLC) is indicated by the solid line. The photodiode signal-to-noise ratio approaches the theoretical maximum at intermediate light intensities (10^6 to 10^{10} photons ms^{-1}) but falls off at low intensities (segment C) because of dark noise, and falls off at high intensities (segment A) because of extraneous noise. The signal-to-noise ratio for a cooled CCD camera with a well size of 10^6 electrons is shown in the shorter dashes. The maximum signal-to-noise ratio that can be achieved is 10^3 because the camera saturates at higher intensities. On the other hand, the cooled CCD has better performance than photodiode array systems at low light levels (10^3 to 10^6 photons ms^{-1}) mainly because cooling reduces the dark noise. The signal-to-noise ratio for a vacuum photocathode detector is indicated by the line with long dashes. At very low intensities ($<10^3$ photons ms^{-1}) the vacuum photocathode is better than both types of silicon-diode device because of its very low dark noise. At intermediate intensities it is not as good because of its lower quantum efficiency.

15.4.1.2 Extraneous noise

A second type of noise, termed extraneous or technical noise, is more apparent at higher light intensities where the sensitivity of the measurement is high because the fractional shot noise and dark noise are low. There are several sources of extraneous noise. One type is caused by fluctuations in the output of the light source (see below). Two other sources are vibrations and movement of the preparation. A number of precautions for reducing vibrational noise are described in Salzberg *et al.* (1977) and London *et al.* (1987). The pneumatic isolation mounts on vibration isolation tables are more efficient in reducing vertical vibrations than in reducing horizontal movements. We now use air-filled soft rubber tubes (Newport Corp, Irvine, CA) or a combination of pneumatic isolation mounts and tubes.

Noise due to movement of the preparation is a problem in *in vivo* measurements. Methods used in mammalian preparations are described below.

15.4.1.3 Dark noise

Dark noise will degrade the signal-to-noise ratio at low light levels.

15.4.2 Light sources

Three kinds of sources have been used. Tungsten filament lamps are a stable source, but their intensity is relatively low, particularly at wavelengths less than 480 nm. Arc lamps are somewhat less stable but can provide more intense illumination. Until recently measurements made with laser illumination have been substantially noisier.

15.4.2.1 Tungsten filament lamps

It is not difficult to provide a power supply stable enough so that the output of the bulb fluctuates by less than 1 part in 10^5. In absorption measurements, where the fractional changes in intensity are relatively small, only tungsten filament sources have been used. On the other hand, fluorescence measurements often have larger fractional changes that will better tolerate light sources with systematic noise, and the measured intensities are low, making improvements in signal-to-noise ratio from brighter sources attractive.

15.4.2.2 Arc lamps

Opti-Quip, Inc., Highland Mills, NY, provides 150 and 250 W xenon power supplies, lamp housings, and arc lamps with noise that is in the range of 1 part in 10^4. In our apparatus the 150 W bulb yielded 2–3 times more light at 520 ± 45 nm than a tungsten filament bulb and in turn the 250 W bulb was 2–3 times brighter than the 150 W bulb. The extra intensity is especially useful for fluorescence measurements from single neurons. If the dark noise is dominant, then the signal-to-noise ratio will improve linearly with intensity, and if the shot noise is dominant it will improve as the square root of intensity (Fig. 15.4).

15.4.3 Optics

15.4.3.1 Numerical aperture

The need to maximize the number of measured photons has also been a dominant factor in the choice of optical components. The number of photons collected by an objective lens in forming the image is proportional to the square of the numerical aperture (NA). In epi-illumination measurements, both the excitation light and the emitted light pass through the objective, and the intensity reaching the photodetector is proportional to the fourth power of NA. Accordingly, objectives with high numerical apertures have been employed. However, conventional microscope optics have very low numerical apertures at low magnifications. Salama (1988), Ratzlaff & Grinvald (1991), and Kleinfeld & Delaney (1996) took advantage of the high NA that can be achieved by using a camera or video lens in place of a microscope objective. In a microscope based on a 25 mm focal length, 0.95 f camera lens, the intensity reaching the photodetector was 100 times larger than with our conventional microscope at a magnification of 4×.

15.4.3.2 Objective efficiency

Direct comparison of the intensity reaching the image plane has shown that the light-collecting efficiency of an objective is not completely determined by the stated magnification and NA. We recommend empirical tests of several lenses for efficiency.

15.4.3.3 Depth of focus

Salzberg *et al.* (1977) determined the effective depth of focus for a 0.4 NA objective lens by recording an optical signal from a neuron when it was in focus and then moving the neuron out of focus by various distances. They found that the neuron had to be moved 300 μm out of focus to cause a 50% reduction in signal size. (This result will be obtained only when the diameter of the neuron image and the diameter of the detector are similar.) Using 0.5 NA optics, 100 μm out of focus in tissue led to a reduction of 50% (Kleinfeld & Delaney, 1996).

15.4.3.4 Light scattering and out-of-focus light

Light scattering can limit the spatial resolution of an optical measurement. Orbach & Cohen (1983) measured the scattering of 750 nm light in the salamander olfactory bulb. They found that inserting a 500 μm-thick slice of olfactory bulb in the light path caused light from a 30 μm spot (Fig. 15.5, part A) to spread so that the diameter of the circle

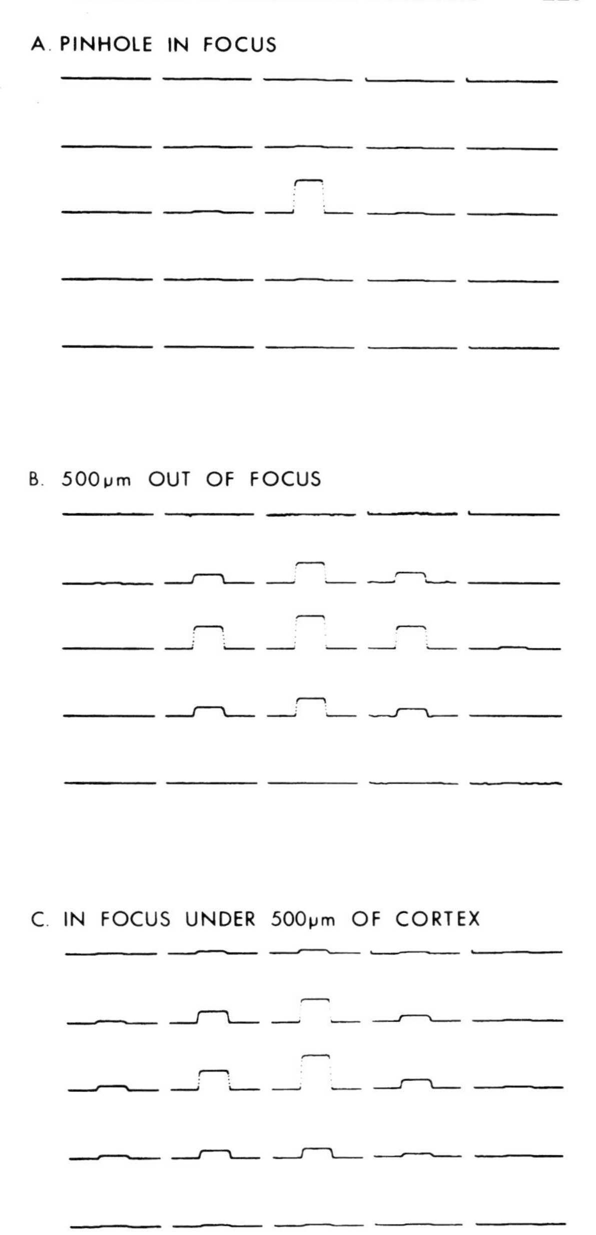

Figure 15.5 Effects of focus and scattering on the distribution of light from a point source onto the array. (A) A 40 μm pinhole in aluminium foil covered with saline was illuminated with light at 750 nm. The pinhole was in focus. More than 90% of the light fell on one detector. (B) The stage was moved downward by 500 μm. Light from the out-of-focus pinhole was now seen on several detectors. (C) The pinhole was in focus but covered by a 500 μm slice of salamander cortex. Again the light from the pinhole was spread over several detectors. A 10 × 0.4 NA objective was used. Kohler illumination was used before the pinhole was placed in the object plane. The recording gains were adjusted so the largest signal in each of the three trials would be approximately the same size in the figure. Redrawn from Orbach & Cohen (1983).

of light that included intensities greater than 50% of the total was roughly 200 μm, (Fig. 15.5, part C). Mammalian cortex appears to scatter more than the olfactory bulb. Thus, light scattering will cause considerable blurring of signals in adult vertebrate preparations.

A second source of blurring is signal from regions that are out of focus. For example, if the active region is a cylinder (e.g. a column) perpendicular to the plane of focus, and the objective is focused at the middle of the cylinder, then the light from the middle will have the correct diameter at the image plane. However, the light from the regions above and below are out of focus and will have a diameter that is too large. The middle section (part B) of Fig. 15.5 illustrates the effect of moving a small spot of light 500 μm out of focus. The light from the small spot is spread to about 200 μm. Thus, in preparations with considerable scattering or with out-of-focus signals, the actual spatial resolution may be limited by the preparation and not by the number of pixels in the imaging device.

15.4.3.5 Confocal and two-photon microscopes

The confocal microscope (Petran & Hadravsky, 1966) substantially reduces both the scattered and out-of-focus light that contributes to the image. More recently an optical sectioning method using two-photon excitation of the fluorophore has been developed. Two-photon excitation reduces out-of-focus fluorescence and photobleaching (Denk *et al.*, 1995). With both types of microscope one can obtain images from intact vertebrate preparations with much better spatial resolution than was achieved with ordinary microscopy. These microscopes have been used to monitor changes in calcium concentration inside small processes of neurons (Eilers *et al.*, 1995; Yuste & Denk, 1995; Svoboda *et al.*, 1997). While slower voltage-sensitive dye signals have been measured confocally (Loew, 1993), at present the temporal resolution and the sensitivity of these microscopes are relatively poor; there are no reports of their use to measure the fast and small signals obtained with voltage-sensitive dyes of the type discussed in this chapter.

15.4.4 Photodetectors

Because the signal-to-noise ratio in a shot-noise-limited measurement is proportional to the square root of the number of photons converted into photo-electrons (see above), quantum efficiency is important. As indicated in Table 15.1, silicon photodiodes have quantum efficiencies approaching the ideal. In contrast, only specially chosen vacuum photocathode devices (phototubes, photomultipliers or image intensifiers) have a quantum efficiency as high as 0.15 at wavelengths above 600 nm. Thus, in shot-noise-limited situations, a silicon diode will have a signal-to-noise ratio that is at least 2.5 times larger. This advantage is indicated in Fig. 15.4 by the fact that the photodiode and CCD curves are higher than the vacuum photocathode curve (dashed line) over much of the intensity range.

On the other hand, dark noise can degrade the signal-to-noise ratio from this theoretical limit. Also, dark noise is generally far larger in a silicon diode than in a vacuum photocathode device (Table 15.1). Thus, at low light levels, a vacuum photocathode will provide a larger signal-to-noise ratio. When the light level is reduced so that the shot noise is less than the dark noise (about 10^6 photons ms^{-1} for a silicon photodiode at room temperature), the signal-to-noise ratio decreases linearly with light intensity (segment C, Fig. 15.4). Cooling the detector will reduce the dark noise (CCD camera, Fig. 15.4).

15.4.5 Imaging devices

The emphasis in this section will be on systems that allow frame rates near 1 kHz. Many factors must be considered in choosing an imaging system. The most important considerations are the requirements for spatial and temporal resolution and dynamic range. Because the signal-to-noise ratio in a shot-noise-limited measurement is proportional to the number of measured photons, increases in either temporal or spatial resolution reduce the number of photons and thus reduce the signal-to-noise ratio. An important figure of merit for an optical recording system is dynamic range; this number can be specified in dB,

Table 15.1 Detector comparison

	Silicon diode (R.T.)	*Vacuum photocathode*
Quantum efficiency	0.9	0.15
Dark noise equivalent power	~10^6 photons s^{-1}	<10^2 photons s^{-1}
1/f noise	Some diodes	None

in bits, or as an exponent (e.g. 100 dB = 17 bits = 10^5). The dynamic range determines the size of smallest fractional change that can be measured. For example, a dynamic range of 100 dB would allow one to measure a signal with a fractional change ($\Delta I/I$) of 10^{-5}. The smallest signal that can be measured with a dynamic range of 60 dB is 10^{-3}. The dynamic range cannot be larger than the effective resolution of the analogue-to-digital converter but it can be considerably smaller if, for example, saturation in the photodetector limits the number of measured photons.

15.4.5.1 Film

One type of imager that has outstanding spatial and temporal resolution is movie film. But, because it is difficult to obtain quantum efficiencies of even 1% with film (Shaw, 1979), there would be a substantial degradation in signal-to-noise ratio. This and other difficulties, including frame-to-frame and within-frame non-uniformity of the emulsion, has discouraged attempts to use film.

15.4.5.2 Silicon diode arrays

Arrays of silicon diodes are attractive because they have nearly ideal quantum efficiencies.

Parallel read-out arrays

Diode arrays with 124–1020 elements are now in use in several laboratories (Grinvald *et al.*, 1981; Iijima *et al.*, 1989; Zecevic *et al.*, 1989; Nakashima *et al.*, 1992; Hirota *et al.*, 1995). In addition, Hammamatsu has constructed a system with 2500 elements (Hosoi *et al.*, 1997). These arrays are designed for parallel read-out; each detector is followed by its own amplifier whose output can be digitized at frame rates of 1 kHz. While the need to provide a separate amplifier for each diode element limits the number of pixels in parallel read-out systems, it contributes to the very large (10^5) dynamic range that can be achieved with these systems (Fig. 15.4). A discussion of amplifiers has been presented earlier (Wu & Cohen, 1993). Two parallel read-out array systems are commercially available (from OptImaging, LLC, Fairfield, CT, http://www.mindspring.com/~neuroplex/NeuroPlex.html; and from Hammamatsu Corp., Japan, http://www.hpk.co.jp/products/SYS/Arg5PdaE.htm).

Serial read-out arrays

By using serial read-out, the number of amplifiers can be reduced. Gen Matsumoto and his collaborators at the Electrotechnical Laboratory in Tsukuba City, Japan, together with Fuji Film, have implemented a 128 × 128 (16 384 pixel) CCD array that can be read out at a frame rate of 2 kHz (Ichikawa *et al.*, 1992). At 1 kHz, the data rate from this system is 32 Mbytes s^{-1} requiring relatively specialized hardware for recording. The commercial Fuji system (http://www.fujifilm.co.jp/eng/bio/hrd1700.html) is much less sensitive than parallel read-out arrays at the light levels obtained in fluorescence measurements from the nervous system. Furthermore, the Fuji system saturates at the light levels that can be obtained in absorption measurements and therefore must be used with reduced illumination (which reduces the signal-to-noise ratio). Thus, the dynamic range of this system is much smaller than the parallel read-out systems described above. The BrainVision Company (YPC, Tokyo, Japan) and PixelVision, Inc. (Beaverton, OR) have recently marketed fast CCD systems.

Several somewhat slower systems based on CCD cameras have been used to measure activity-dependent optical signals in neurobiological preparations (Lasser-Ross *et al.*, 1991; Poe *et al.*, 1994; Delaney *et al.*, 1994). Lasser-Ross *et al.* (1991) have modified the software for a Photometrics CCD camera to allow several choices of spatial and temporal resolution. Table 15.2, taken from their paper, illustrates these choices. The dynamic range of CCD cameras is limited by saturation and by the accuracy of the analogue-to-digital conversion. A dynamic range of 10^3 is not easily achieved. These cameras saturate at the fluorescence intensities that can be obtained from vertebrate brain preparations (Fig. 15.4); on the other hand, because they can be cooled their dark noise is reduced and they have a better signal-to-noise ratio for measurements from individual neurons stained with internally injected dyes.

15.5 TWO EXAMPLES

15.5.1 Action potentials from individual neurons in an *Aplysia* ganglion

Nervous systems are made up of large numbers of neurons; many of these are simultaneously active during the generation of behaviours. The original

Table 15.2 Spatial and temporal resolution from a photometrics camera (from Lasser-Ross *et al.*, 1991).

Frame rate (s^{-2})	*Pixels*
100	324
40	2500
20	10 000

motivation for developing optical methods was the hope that they could be used to record all of the action potential activity of all the neurons in simpler invertebrate ganglia during behaviours (Davila *et al.*, 1973). Techniques which use microelectrodes to monitor activity are limited in that they can observe single cell activity in only as many cells as one can simultaneously place electrodes (typically two or three neurons). Obtaining information about the activity of many cells is important for understanding the roles of the individual neurons in generating behaviour and for understanding how nervous systems are organized.

In the first attempt to use voltage-sensitive dyes in ganglia (Salzberg *et al.*, 1973), we were fortunate to be able to monitor activity in a single neuron because the photodynamic damage was severe and the signal-to-noise ratio small. Now, however, with better dyes and methods, the spike activity of hundreds of individual neurons can be recorded simultaneously (Zecevic *et al.*, 1989; Nakashima *et al.*, 1992). In the experiments described below we measured the spike activity of up to 50% of the approximately 1000 cells (Cash & Carew, 1989; Coggeshall, 1967) in the *Aplysia* abdominal ganglion. Opisthobranch molluscs have been a preparation of choice for this kind of measurement because their central nervous systems have relatively few, relatively large neurons, and almost all of the cell bodies are fully invaded by the action potential.

Optical recordings were made using a 464-element silicon photodiode array system with parallel read-out (Fig. 15.6) (Falk *et al.*, 1993; Wu & Cohen, 1993). The diode array was placed in the image plane formed by a microscope objective of 25 × 0.4 NA. A single-pole high-pass and a four-pole low-pass Bessel filter in the

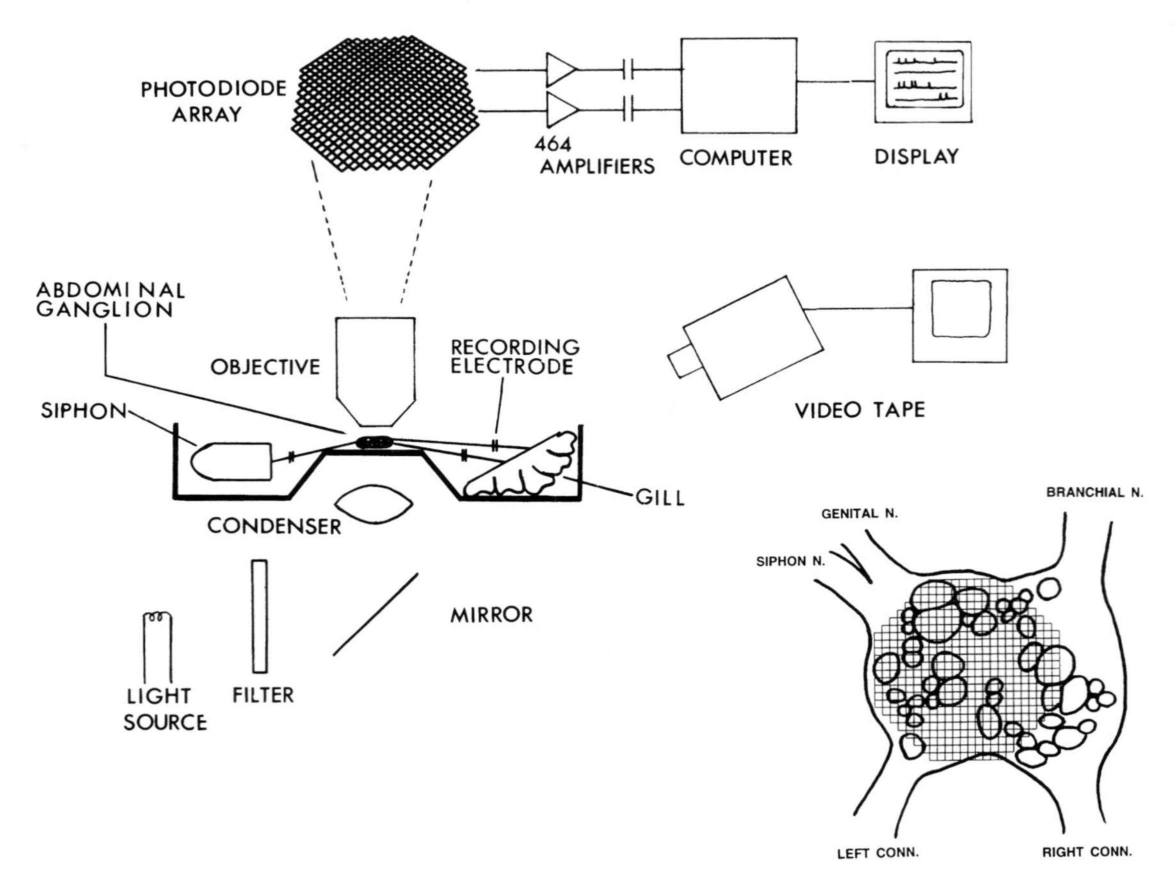

Figure 15.6 Schematic diagram of the apparatus used in the *Aplysia* measurements. Light from a tungsten halogen lamp passed through a 720 ± 25 nm interference filter and was focused on the preparation. We used a modification of Kohler illumination (Leitz Ortholux II microscope) where the condenser iris was opened so that the condenser numerical aperture would equal the objective numerical aperture. A 464-element photodiode array was placed at the plane where the objective forms the real, inverted image. The outputs from each diode were individually amplified. The amplifier outputs were digitized by a 512-channel, 12-bit, analogue-to-digital converter and stored in a PC computer. The figure illustrates the isolated siphon preparation (Kupfermann *et al.*, 1971). The relative size and position of the photodiode array and the image of an *Aplysia* abdominal ganglion is shown on the lower right. Redrawn from Falk *et al.* (1993).

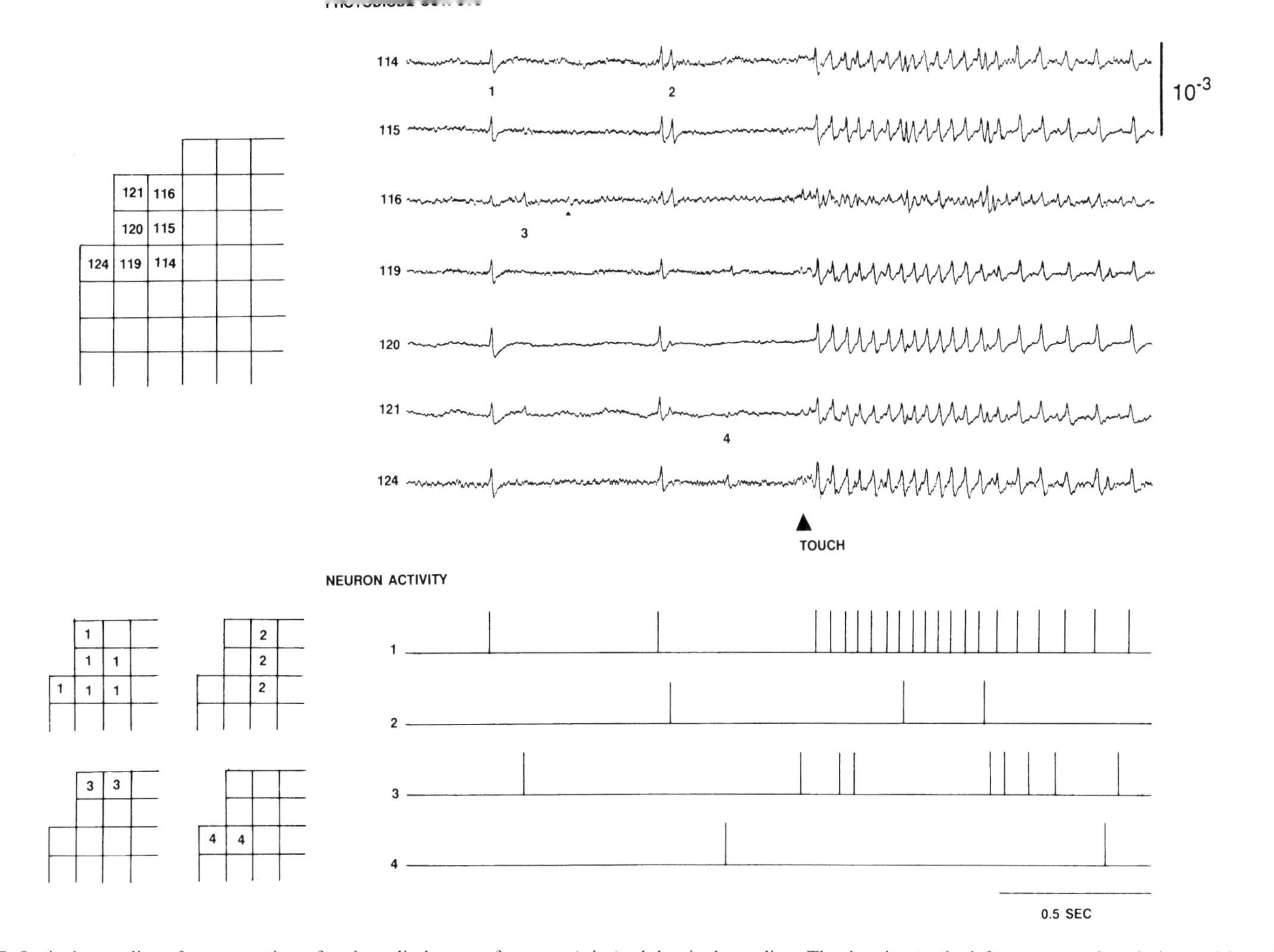

Figure 15.7 Optical recordings from a portion of a photodiode array from an *Aplysia* abdominal ganglion. The drawing to the left represents the relative position of the detectors whose activity is displayed. In the top right, the original data from seven detectors is illustrated. The line to the right of these traces indicates the size of a fractional intensity change ($\Delta I/I$) of 10^{-3}. Clearly, a dynamic range of more than 80 dB is required to measure these signals. The numbers to the left of each trace identify the detector from which the trace was taken. The bottom section shows the raster diagram illustrating the results of our sorting of this data into the spike activity of four neurons. At the number 1 in the top section there are synchronously occurring spike signals on all seven detectors. A synchronous event of this kind occurs more than 20 times; we presume that each event represents an action potential in one relatively large neuron. The activity of this cell is represented by the vertical lines on trace 1 of the bottom section. The activity of a second cell is indicated by small signals on detectors 119 and 124 at the time indicated by the number 4. The activity of this cell is represented by the vertical lines on trace 4 of the bottom section. The activity of neurons 2 and 3 was similarly identified. Modified from Zecevic *et al.* (1989).

amplifiers limited the frequency response to 1.5–200 Hz. We used the isolated siphon preparation developed by Kupfermann *et al.* (1971). Considerable effort was made to find conditions that maximized the dye staining while causing minimal pharmacologic effects on the gill-withdrawal behaviour. Intact ganglia were incubated in a 0.15 mg ml^{-1} solution of the oxonol dye, RH155 (Fig. 15.1) (or its diethyl analogue) using a protocol developed by Nakashima *et al.* (1992). A light mechanical stimulation (1–2 g) was delivered to the siphon.

Because the image of a ganglion is formed on a rectilinear diode array, there is no simple correspondence between images of cells and photodetectors (Fig. 15.6, lower right). The light from larger cell bodies will fall on several detectors and its activity will be recorded as simultaneous events on neighbouring detectors. In addition, because these preparations are multilayered, most detectors will receive light from several cells. Thus, a sorting step is required to determine the activity in neurons from the spike signals on individual photodiodes. In the top right of Fig. 15.7, the raw data from seven photodiodes from an array are shown. The line to the right of these traces indicates the size of a fractional intensity change ($\Delta I/I$) of 10^{-3}. Clearly a dynamic range of more than 80 dB is required to measure these signals. The activity of four neurons (shown in the raster diagram at the bottom) can account for the spike signals in the top section. Two problems are illustrated in this figure. Both arise from the signal-to-noise ratio. First, there may be an additional spike on detector 116 just before the stimulus (at the arrow), but the signal-to-noise ratio was not large enough to be certain. Second, following the stimulus, there is a great deal of activity which will obscure small signals.

The result of a complete analysis of one data set is shown in the raster diagram of Fig. 15.8. The mechanical stimulus occurred at the time indicated by the bar at the bottom. There are 135 neurons whose activity was detected optically. Similar results have been obtained by Nakashima *et al.* (1992). We estimated the recording illustrated in Fig. 15.8 was about 50% complete (Wu *et al.*, 1994). Thus, the actual number of activated neurons during the gill-withdrawal reflex was about 300.

We were surprised at the large number of neurons that were activated by the light touch. Furthermore, a substantial number of the remaining neurons are

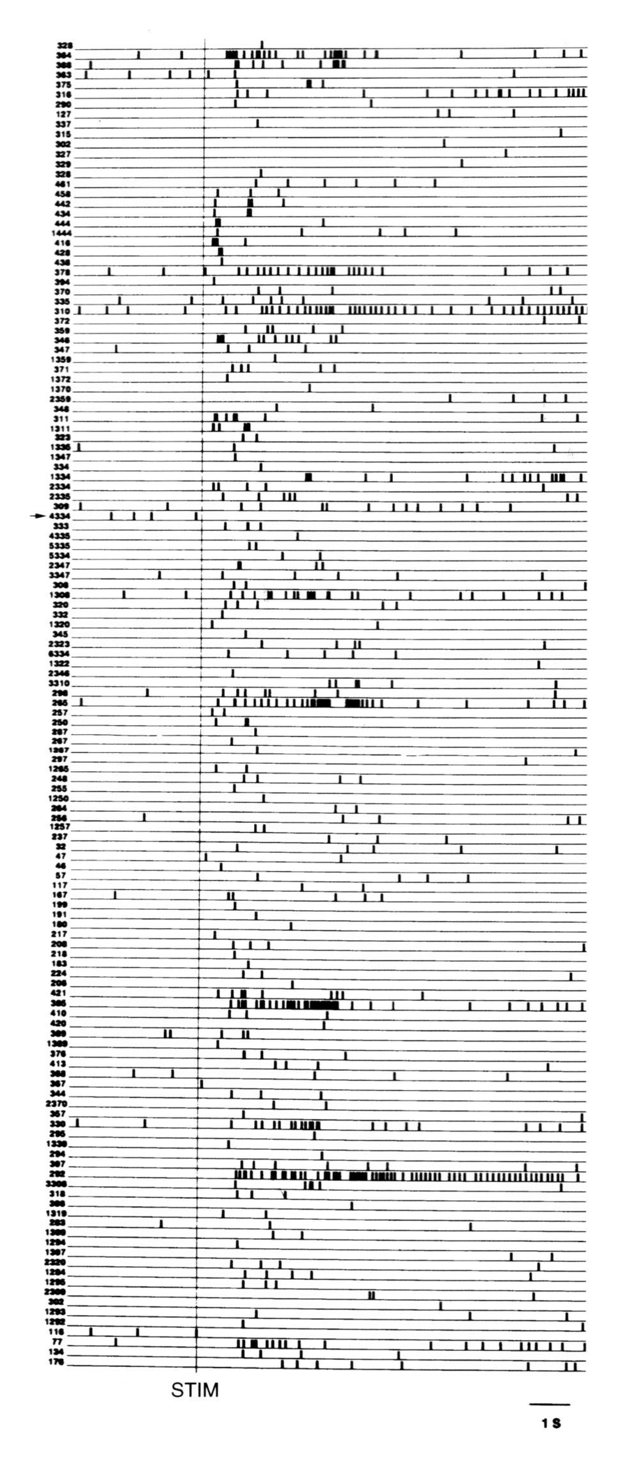

Figure 15.8 Raster diagram of the action potential activity recorded optically from an *Aplysia* abdominal ganglion during a gill-withdrawal reflex. The touch to the siphon occurred at the time of the bar labelled STIM. In this recording, activity in 135 neurons was measured. We think this recording is incomplete and that the actual number of active neurons was between 250 and 300. Most neurons are activated by the touch but one, #4334 (arrowed) This inhibition was seen in repeated trials in this preparation. Modified from Wu *et al.* (1994).

likely to be either inhibited by the stimulus or receive a large subthreshold excitatory input. It is almost as if the *Aplysia* nervous system is designed such that every cell in the abdominal ganglion cares about this (and perhaps every) sensory stimulus. In addition, more than 1000 neurons in other ganglia are activated by this touch (Tsau *et al.*, 1994). Clearly information about this very mild and localized stimulus is propagated widely in the *Aplysia* nervous system. These results force a more pessimistic view of the present understanding of the neuronal basis of apparently simple behaviours in relatively simple nervous systems. Elsewhere we present arguments suggesting that the abdominal ganglion may function as a distributed system (Tsau *et al.*, 1994; Wu *et al.*, 1994).

15.6 POPULATION SIGNALS FROM VERTEBRATE BRAIN

15.6.1 Turtle visual cortex

In a measurement from a vertebrate brain stained by superfusing a solution of the dye over the surface, each photodetector will receive light from a substantial number of neurons and neuronal processes. Because of scattering and out-of-focus light, this will be true even if the pixel size corresponded to a very small area of the object. As a result the voltage-sensitive dye signal will be a population average of the change in membrane potential of all of these neurons and processes. In the measurements from the turtle visual cortex described below, each detector received light from a 200 μm square area (an approximate volume of 0.1 mm^3) of cortex which includes thousands of neurons. Populations signals have been recorded from many preparations (e.g. Grinvald *et al.*, 1982a; Orbach & Cohen, 1983; Sakai *et al.*, 1985; Kauer, 1988; Cinelli & Salzberg, 1992; Albowitz & Kuhnt, 1993; Delaney *et al.*, 1994); the results from turtle are described because of our familiarity with them.

Prechtl (1994) and Prechtl & Bullock (1994) discovered that visual stimuli induced oscillations in the local field potential measured in the turtle visual cortex. They found that a moving stimulus would induce oscillations that were sometimes synchronous when comparing regions of dorsal cortex and the dorsal ventricular ridge. This synchrony at two locations would imply that there was a standing wave of depolarization. We measured the voltage-sensitive dye signal that accompanies these oscillations in the local field potential recordings.

The dorsal cortex was exposed by removing the overlying bone; the pial surface was incubated with a 0.25–0.8 mg ml^{-2} solution of the styryl dyes, RH-795 (Grinvald *et al.*, 1994) or JPW1114 (Antic & Zecevic, 1995), for 15 minutes. No other dyes were tested. The brain was partially isolated by sectioning the spinal cord and cutting cranial nerves IV–XII. To eliminate artifacts resulting from the heartbeat and the movement of blood cells through the brain, the turtle was perfused with a steady flow of an oxygenated artificial cerebrospinal fluid.

Two kinds of stimuli were used: a step increase in diffuse light or a looming stimulus; both had a duration of 3 s. Figure 15.9(A) (Prechtl *et al.*, 1997) shows the responses from a single photodetector to

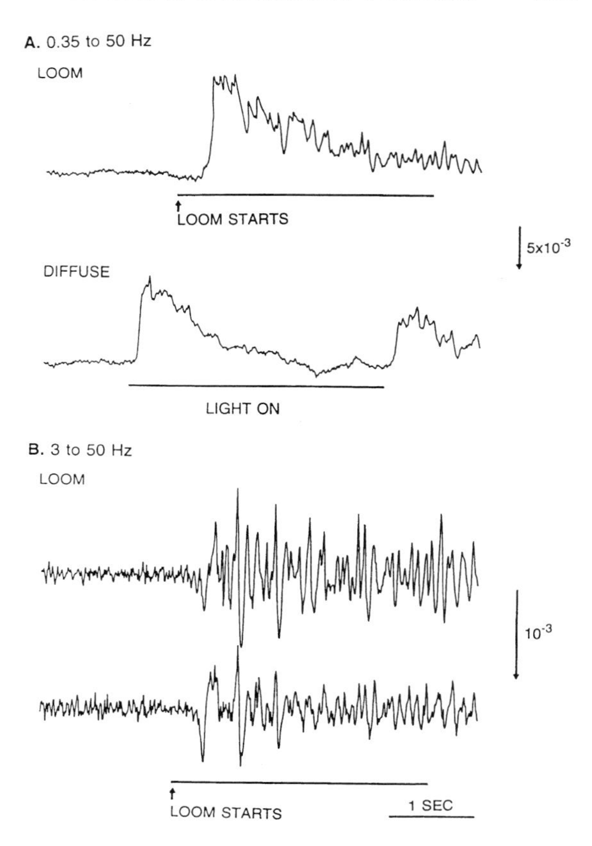

Figure 15.9 The outputs of individual detectors in a population signal measurement from turtle visual cortex. The responses to both looming and diffuse light stimuli include both very low frequency (<0.1 Hz) and higher frequency (1–30 Hz) components. (A) The response of the same detector to the two kinds of visual stimuli. (B) The response to a looming stimulus measured simultaneously in two different detectors. To emphasize the high-frequency components these signals have been digitally filtered to remove components below 3 Hz. The length of the vertical calibration on the right represents the stated value of the change in fluorescence divided by the resting fluorescence ($\Delta F/F$). A dynamic range of more than 80 dB is required to measure these signals. Modified from Prechtl *et al.* (1997).

both a looming (top) and a diffuse light (bottom) stimulus. The diode output was filtered with a bandpass of 0.35–50 Hz. For both stimuli, part of the response is a relatively large and long-lasting depolarization. The response to the looming stimulus has a longer latency and returned to the baseline more slowly. From the signals measured with the same dyes on other preparations where there was a direct comparison with an intracellular electrode recording; we infer that an upward deflection in this figure represents a depolarization.

Riding on top of the long-lasting components in Fig. 15.9(A) are higher-frequency signals; these were larger in the looming response than in the diffuse light response. These high-frequency components are emphasized by increasing the high-pass cutoff to 3.0 Hz and increasing the gain (Fig. 15.9(B)). The top trace in Fig. 15.9(B) is from the same data as that shown in the top trace in Fig. 15.9(A). The arrow to the right of these traces indicates the size of a fractional intensity change ($\Delta I/I$) of 10^{-3}. Again, it is clear that a dynamic range of more than 80 dB is required to measure these signals.

The bottom trace in Fig. 15.9(B) is from a second detector that received light from an area of cortex 1.2 mm away. Although a few events in these two detectors appear to be in phase, there are many instances where the two traces are quite different. This implies that the high-frequency signal does not result from either a simple standing wave or a repetitive travelling wave at a single frequency and suggests the possibility that the voltage-sensitive dye signals have a much better spatial resolution than the local field potential measurements.

We compared the spatial resolution provided by the two kinds of recordings. Figure 15.10 illustrates simultaneous local field potential and optical recordings from two positions on turtle visual cortex that were separated by 2.3 mm. The top pair of superimposed traces in Fig. 15.10 are the local field potential recordings from the two sites. There is considerable overlap in these two recordings; the correlation coefficient calculated for the two traces was 0.9. The bottom pair of traces are the voltage-sensitive dye recordings from the two sites. There is less overlap in the optical recordings; the correlation coefficient calculated for the two traces was 0.6. Thus, in comparison with the optical recordings, the local field potential recordings appear to blur spatial differences.

We estimated the relative spatial resolution of the

SPATIAL RESOLUTION OF LFP AND OPTICAL RECORDINGS; TWO LOCATIONS ON CORTEX SEPARATED BY 2.3 MM.

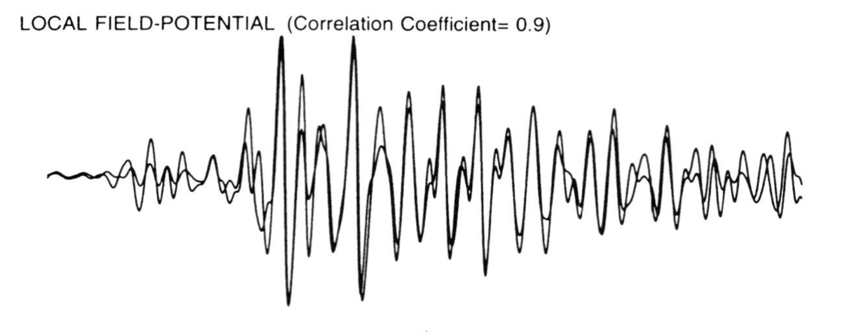

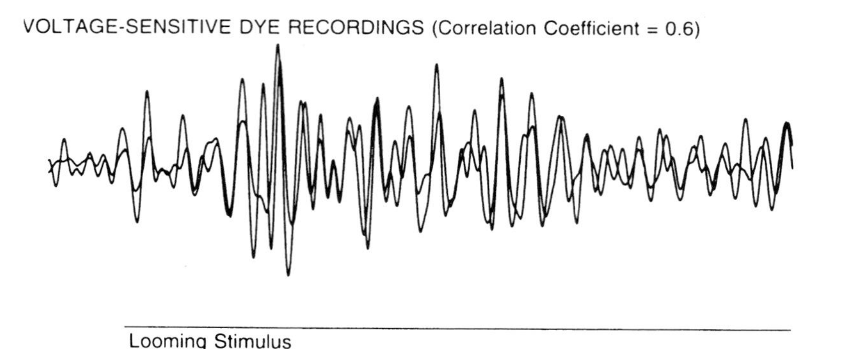

Figure 15.10 Comparison of voltage-sensitive dye and local-field potential recordings. Simultaneous optical and local field potential recordings from two positions on turtle visual cortex that were separated by 2.3 mm. The top pair of superimposed traces are the local field potential recordings from the two sites. The bottom pair of traces are the voltage-sensitive dye recordings from the two sites. There is much less overlap in the optical recordings. Thus, the voltage-sensitive dye recordings have better spatial resolution. Both sets of recordings were bandpass filtered (10–30 Hz). (L.B. Cohen, J. Prechtl & D. Kleinfeld, unpublished results).

two methods by determining the distance on cortex at which a pair of voltage-sensitive dye recordings would have a correlation coefficient equal to the local field-potential recordings. In six trials from two preparations the correlation coefficient for the optical measurement was equal to that for the local field potential measurement when the two optical measurements were a distance apart that was 0.21 ± 0.05 (SEM) of the distance of the two local-field potential electrodes. Thus, the optical measurement has a linear spatial resolution about five times better than the electrode measurement and a two-dimensional resolution about 25 times better.

Nonetheless, it is difficult to infer anything about the spatial spread of the events giving rise to the voltage-sensitive dye signals from an examination of the bottom two traces in Figs 15.9 or 15.10. Therefore, we also present a portion of the same measurement as a series of pseudocolour frames made using the data from all photodetectors in the array (Plate 15.1). Each frame is separated by 6 ms; red represents the largest depolarization and purple the largest hyperpolarization (using a single detector in the middle of the array to set the colour scale). The series starts at the top left and goes left-to-right and then top-to-bottom. The movement of the looming stimulus begins at the first frame (top left corner). After a delay of several hundred ms, the response consists of a very complex, non-repeating pattern of travelling and/or expanding and contracting waves of depolarization that continues after the end of the stimulus. The waves differ in their origins, direction and speed of travel, end points, and the area of cortex involved (Prechtl *et al.*, 1997).

The voltage-sensitive dye recording has provided a uniquely detailed picture of the spatial aspects of this oscillatory response. This response is quite complex and, indeed, more complex than had been anticipated from local field potential measurements.

15.6.2 *In vivo* mammalian brain

Measurements from *in vivo* mammalian preparations are more difficult than the perfused turtle because of additional sources of noise from the heartbeat and respiration and probably also because mammalian preparations are not as easily stained as those of lower vertebrates. Two methods for reducing the movement artifacts from the heartbeat and respiration are, together, quite effective. First, a subtraction procedure is used where two recordings are made but only one of the trials has a stimulus (Orbach *et al.*, 1985). Both recordings are triggered from the upstroke of the electrocardiogram so both should have similar heartbeat noise. When the trial without the stimulus is subtracted from the trial with the stimulus, the heartbeat artifact is substantially reduced. If respiration is triggered by the heartbeat, the subtraction procedure will also remove respiratory noise. Amiram Grinvald and collaborators have modified a Basile respirator so that it could be stopped and restarted under computer control. The following protocol is used for both recordings of the pair which makes up one trial:

- stop the respirator at a fixed position in the respiratory cycle
- identify the next electrocardiogram (ECG) upstroke
- restart the respirator
- begin the optical recording.

A second method for reducing the movements of the brain is to fix a chamber onto the skull surrounding the craniotomy (e.g. Blasdel & Salama, 1986). When this chamber is filled with silicone oil and closed, the movements due to heartbeat and respiration are substantially reduced. Using both methods reduces the noise from these movement artifacts enough so that it is no longer the main source of noise in the measurements.

15.7 FUTURE DIRECTIONS

Because the apparatus is already reasonably optimized, any improvement in the sensitivity of the voltage-sensitive dye measurements will need to come from the development of better dyes and/or investigating signals from additional optical properties of the dyes. The dyes in Fig. 15.1 and the vast majority of those already synthesized are of the general class named polyenes (Hamer, 1964), a group that is used to extend the wavelength response of photographic film. It is possible that improvements in signal size can be obtained with new polyene dyes (see Waggoner & Grinvald, 1977, and Fromherz *et al.*, 1991, for a discussion of maximum possible fractional changes in absorption and fluorescence). On the other hand, the fractional change on squid axons has not increased in recent years (Gupta *et al.*, 1981; L. B. Cohen, A. Grinvald, K. Kamino & B. M. Salzberg, unpublished results), and most improvements (e.g. Grinvald *et al.*, 1982b; Momose-Sato *et al.*, 1995; Tsau *et al.*, 1996; Antic & Zecevic, 1995) have involved synthesizing analogues that work well on new preparations.

The best of the styryl and oxonol dyes have fluorescence changes of 10% per 100 mV in a bilayer. Recently, Gonzalez & Tsien (1995) introduced a new scheme for generating voltage-sensitive signals using two chromophores and energy transfer. While these fractional changes were also in the range of 10% per 100 mV, more recent results are about 30%

(T. Gonzalez & R. Tsien, personal communication). However, because one of the chromophores must be very hydrophobic and does not penetrate into brain tissue, this pair of dyes has not generated signals in intact tissues (T. Gonzalez & R. Tsien; A. Obaid & B.M. Salzberg; personal communications).

Bouevitch *et al.* (1993) and Ben-Oren *et al.* (1996) found that membrane potential changed the non-linear second harmonic generation from styryl dyes in cholesterol bilayers and in fly eyes. Large (50%) fractional changes were measured. It is hoped that improvements in the illumination intensity will result in larger signal-to-noise ratios.

Ehrenberg & Berezin (1984) have used resonance Raman to study surface potential; these methods might also be applicable for measuring transmembrane potential.

15.7.1 Neuron-type-specific staining

Another novel direction is the development of methods for neuron-type-specific staining of vertebrate preparations. Three quite different approaches have been tried. First, the use of retrograde staining procedures has recently been investigated in the embryonic chick and lamprey spinal cords (Wenner *et al.*, 1996; Tsau *et al.*, 1996). An identified cells class (motoneurons) was selectively stained. In lamprey experiments, spike signals from individual neurons were measured (Hickie *et al.*, 1996). Further efforts at optimizing this staining procedure are needed. Second, is the use of cell-type specific staining developed for fluorescein (Nirenberg & Cepko, 1993). It might be possible to use similar techniques to stain cells selectively with voltage-sensitive or ion-sensitive dyes. Third, Siegel & Isacoff (1997) constructed a genetically encoded combination of a potassium channel and green fluorescent protein. When introduced into a frog oocyte, this molecule had a (relatively slow) voltage-dependent signal with a fractional fluorescence change of 5%.

Optical recording already provides unique insights into brain activity and organization. Clearly, improvements in sensitivity and/or selectivity would make these methods even more powerful.

ACKNOWLEDGEMENTS

The authors are indebted to their collaborators Vicencio Davila, Kohtaro Kamino, David Kleinfeld, Les Loew, Jim Prechtl, Bill Ross, Brian Salzberg, Alan Waggoner and Joe Wuskell for numerous discussions about optical methods. The experiments carried out in our laboratories were supported by NIH grants NS08437 and NSF grant IBN 9604356.

REFERENCES

Albowitz B. & Kuhnt U. (1993) *Brain Res.* **631**, 329–333.
Antic S. & Zecevic D. (1995) *J. Neurosci.* **15**, 1392–1405.
Ben-Oren I., Peleg G., Lewis A., Minke B. & Loew L. (1996) *Biophys. J.* **71**, 1616–1620.
Blasdel G.G. & Salama G. (1986) *Nature* **321**, 579–585.
Bouevitch O., Lewis A., Pinevsky I., Wuskell J. & Loew L. (1993) *Biophys. J.* **65**, 672–679.
Boyle M.B. & Cohen L.B. (1980) *Fed. Proc.* **39**, 2130.
Braddick H.J.J. (1960) *Rep. Prog. Phys.* **23**, 154–175.
Cash D. & Carew T.J. (1989) *J. Neurobiology* **20**, 25–47.
Cinelli A.R. & Salzberg B.M. (1992) *J. Neurophysiol.* **68**, 786–806.
Coggeshall R.E. (1967) *J. Neurophysiol.* **30**, 1263–1287.
Cohen L.B. & Lesher S. (1986). *Soc. Gen. Physiol. Ser.* **40**, 71–99.
Cohen L.B. & Salzberg B.M. (1978) *Rev. Physiol. Biochem. Pharmacol.* **83**, 35–88.
Davila H.V., Salzberg B.M., Cohen L.B. & Waggoner A.S. (1973) *Nature, New Biol.*, **241**, 159–160.
Delaney K.R., Gelperin A., Fee M.S., Flores J.A., Gervais R., Tank D.W., Kleinfeld D. (1994) *Proc. Natl Acad. Sci. USA* **91**, 669–673.
Denk W., Piston D.W. & Webb W. (1995). In *Handbook of Biological Confocal Microscopy*, J.W. Pawley (ed). Plenum Press, New York, pp. 445–458.
Ehrenberg B. & Berezin Y. (1984). *Biophys. J.*, **45**, 663–670.
Eilers J., Callawaert G., Armstrong C. & Konnerth A. (1995) *Proc. Natl Acad. Sci. USA*, **92**, 10272–10276.
Falk C.X., Jy Wu, Cohen L.B. & Tang C. (1993) *J. Neuroscience* **13**, 4072–4081.
Fromherz P., Dambacher K.H., Ephardt H., Lambacher A., Muller C.O., Neigl R., Schaden H., Schenk O. & Vetter T. (1991) *Ber. Bunsenges. Phys. Chem.* **95**, 1333–1345.
Gonzalez J.E. & Tsien R.Y. (1995) *Biophys. J.* **69**, 1272–1280.
Grinvald A., Cohen L.B., Lesher S. & Boyle M.B (1981) *J. Neurophysiol.* **45**, 829–840.
Grinvald A., Manker A. & Segal M. (1982a) *J. Physiol. (Lond.)* **333**, 269–291.
Grinvald A., Hildesheim R., Farber I.C. & Anglister L. (1982b) *Biophys. J.* **39**, 301–308.
Grinvald A., Lieke E., Frostig R.D., Gilbert C.D. & Wiesel T.N. (1986) *Nature* **324**, 361–364.
Grinvald A., Frostig R.D., Lieke E. & Hildesheim R. (1988) *Physiol. Rev.* **68**, 1285–1366.
Grinvald A., Lieke E.E., Frostig R.D. & Hildesheim R. (1994). *J. Neurosci.* **14**, 2545–2568.
Gross E., Bedlack R.S. & Loew L.M. (1994) *Biophysical Journal* **67**, 208–216.
Gupta R.K., Salzberg B.M., Grinvald A., Cohen L.B., Kamino K., Lesher S., Boyle M.B., Waggoner A.S. & Wang C.H. (1981) *J. Membr. Biol.* **58**, 123–137.
Hamer F.M. (1964) *The Cyanine Dyes and Related Compounds.* Wiley, New York.
Hickie C., Wenner P., O'Donovan M., Tsau Y., Fang J. & Cohen L.B. (1996) *Abstr. Soc. Neuroscience* **22**, 321.
Hirota A., Sato K., Momose-Sato Y., Sakai T. & Kamino K. (1995) *J. Neurosci. Methods* **56**, 187–194.
Hosoi S., Tsuchiya H., Takahashi M., Kashiwasake-Jibu M., Sakatani K. & Hayakawa T. (1997) In *Modern Optics,*

Electronics, and High Precision Techniques in Cell Biology, G. Isenberg (ed). Springer, New York, pp. 76–87.
Ichikawa M., Iijima T. & Matsumoto G. (1992) *Simultaneous 16,384-site Optical Recording of Neural Activities in the Brain*. Oxford University Press, New York.
Iijima T., Ichikawa M. & Matsumoto G. (1989) *Abstr. Soc. Neurosci.* **15**, 398.
Kauer J.S. (1988) *Nature* **331**, 166–168.
Kleinfeld D. & Delaney K. (1996) *J. Comp. Neurol.* **375**, 89–108.
Kupfermann I., Pinsker H., Castellucci V. & Kandel E.R. (1971) Science **174**, 1252–1256.
Lasser-Ross N., Miyakawa H., Lev-Ram V., Young S.R. & Ross W.N. (1991) *J. Neuroscience Methods* **36**, 253–261.
Loew L.M. (1993) *Methods in Cell Biology* **38**, 195–209.
Loew L.M. (1994) *Neuroprotocols* **5**, 72–79.
London J.A., Zecevic D. & Cohen L.B. (1987) *J. Neurosci.* **7**, 649–661.
Malmstadt H.V., Enke C.G., Crouch S.R. & Harlick G. (1974). *Electronic Measurements for Scientists*. Benjamin, Menlo Park, CA.
Momose-Sato Y., Sato K., Sakai T., Hirota A., Matsutani K. & Kamino K. (1995) *J. Membr. Biol.* **144**, 167–176.
Nakashima M., Yamada S., Shiono S., Maeda M. & Sato F. (1992) *IEEE Trans. Biomed. Engng.* **39**, 26–36.
Nirenberg S. & Cepko C. (1993) *J. Neurosci.* **13**, 3238–3251.
Orbach H.S. & Cohen L.B. (1983) *J. Neurosci.* **3**, 2251–2262.
Orbach H.S., Cohen L.B. & Grinvald A. (1985) *J. Neurosci.* **5**, 1886–1895.
Petran M. & Hadravsky M. (1966) Czechoslovakian patent 7720.
Poe G.R., Rector D.M. & Harper R.M. (1994) *J. Neurosci.* **14**, 2933–2942.
Prechtl J.C. (1994) *Proc. Natl. Acad. Sci. USA* **91**, 12467–12471.
Prechtl J.C. & Bullock T.H. (1994) *Electroencephalogr. Clin. Neurophysiol.* **91**, 54–66.
Prechtl J.C., Cohen L.B., Peseran B., Mitra P. & Kleinfeld D. (1997) *Proc. Natl. Acad. Sci. USA* **94**, 7621–7626.
Ratzlaff E.H. & Grinvald A. (1991) *J. Neurosci. Methods* **36**, 127–137.
Ross W.N., Salzberg B.M., Cohen L.B. & Davila H.V. (1974) *Biophys. J.* **14**, 983–986.
Sakai T., Hirota A., Komuro H., Fujii S. & Kamino K. (1985) *Brain Res.* **349**, 39–51.
Salama G. (1988) *SPIE Proc.* **94**, 75–86.
Salzberg B.M. (1983) In *Current Methods in Cellular Neurobiology*, J.L. Barker & J.F. McKelvy (eds). Wiley, New York, pp. 139–187.
Salzberg B.M., Davila H.V. & Cohen L.B. (1973) *Nature* **246**, 508–509.
Salzberg B.M., Grinvald A., Cohen L.B., Davila H.V. & Ross W.N. (1977) *J. Neurophysiol.* **40**, 1281–1291.
Shaw R. (1979) *Appl. Optics. Optical. Eng.* **7**, 121–154.
Shrager P., Chiu S.Y., Ritchie J.M., Zecevic D. & Cohen L.B. (1987) *Biophys. J.* **51**, 351–355.
Siegel M.S. & Isacoff E.Y. (1997) *Neuron* **19**, 735–741.
Svoboda K., Denk W., Kleinfeld D. & Tank D. (1997) *Nature* **385**, 161–165.
Tsau Y., Wu J.Y., Hopp H.P., Cohen L.B., Schiminovich D. & Falk C.X. (1994). *J. Neurosci.* **14**, 4167–4184.
Tsau Y., Wenner P., O'Donovan M.J., Cohen L.B., Loew L.M. & Wuskell J.P. (1996) *J. Neurosci. Methods* **70**, 121–129.
Waggoner A.S. (1979) *Ann. Rev. Biophys. Bioeng.* **8**, 47–68.
Waggoner A.S. & Grinvald A. (1977) *Ann. NY Acad. Sci.* **303**, 217–241.
Wenner P., Tsau Y., Cohen L.B., O'Donovan M.J. & Dan Y. (1996) *J. Neurosci. Methods* **70**, 111–120.
Wu J.Y. & Cohen L.B. (1993). In *Fluorescent and Luminescent Probes for Biological Activity* (Mason W.T., ed.). Academic Press, London, pp. 389–404.
Wu J.Y., Tsau Y., Hopp H.P., Cohen L.B., Tang A.C. & Falk C.X. (1994). *J. Neurosci.* 14, 1366–1384.
Yuste R. & Denk W. (1995). *Nature* **375**, 682–684.
Zecevic D., Wu J.Y., Cohen L.B., London J.A., Hopp H.P. & Falk C.X. (1989) *J. Neurosci.* **9**, 3681–3689.

CHAPTER SIXTEEN

Imaging Membrane Potential Changes in Individual Neurons

SRDJAN ANTIĆ & DEJAN ZEČEVIĆ
Department of Cellular and Molecular Physiology, Yale University School of Medicine, New Haven, CT, USA

16.1 INTRODUCTION

Understanding the biophysical properties of single neurons and how they process information is fundamental to understanding how the brain works. The main task of functional neuronal networks is to process information and that task depends critically on how exactly signals are integrated by individual nerve cells that are often anatomically and functionally complex. The principles of information processing in single neurons can only be determined by studying specific neuronal types in different experimental preparations.

With the development of new measuring techniques that allow more direct investigation of individual nerve cells, it has become widely recognized, especially during the last five years, that dendritic membranes of many vertebrate CNS neurons contain active conductances such as voltage-activated Na^+, Ca^{2+} and K^+ channels (e.g. Stuart & Sakmann, 1994; Spruston *et al.*, 1995; Magee & Johnston, 1995; Magee *et al.*, 1995). An important consequence of active dendrites is that regional electrical properties of branching neuronal processes will be extraordinarily complex, dynamic, and, in the general case, impossible to predict in the absence of detailed measurements.

To obtain such a measurement one would, ideally, like to be able to monitor, at multiple sites, subthreshold events as they travel from the sites of origin on neuronal processes and summate at particular locations to influence action potential initiation. It is important to be able to perform these measurements in at least partially intact neuronal structures (isolated invertebrate ganglia or tissue slices of vertebrate CNS) to insure that highly specific regional electrical properties of individual neurons (e.g. Tauc & Hughes, 1963; Llinas & Sugimori, 1980; Stuart & Häusser, 1994; Stuart & Sakmann, 1994) and characteristic synaptic connections (largely lost in cultures) are preserved. This goal has not been achieved in any neuron, vertebrate or invertebrate, owing to technical limitations of experimental measurements that employ electrodes.

To achieve better spatial resolution it is necessary to turn from direct electrical recording to indirect, optical measurements using voltage-sensitive dyes. Optical signals are useful since it is easy to monitor light

FLUORESCENT AND LUMINESCENT PROBES, 2ND EDN
ISBN 0–12–447836–0

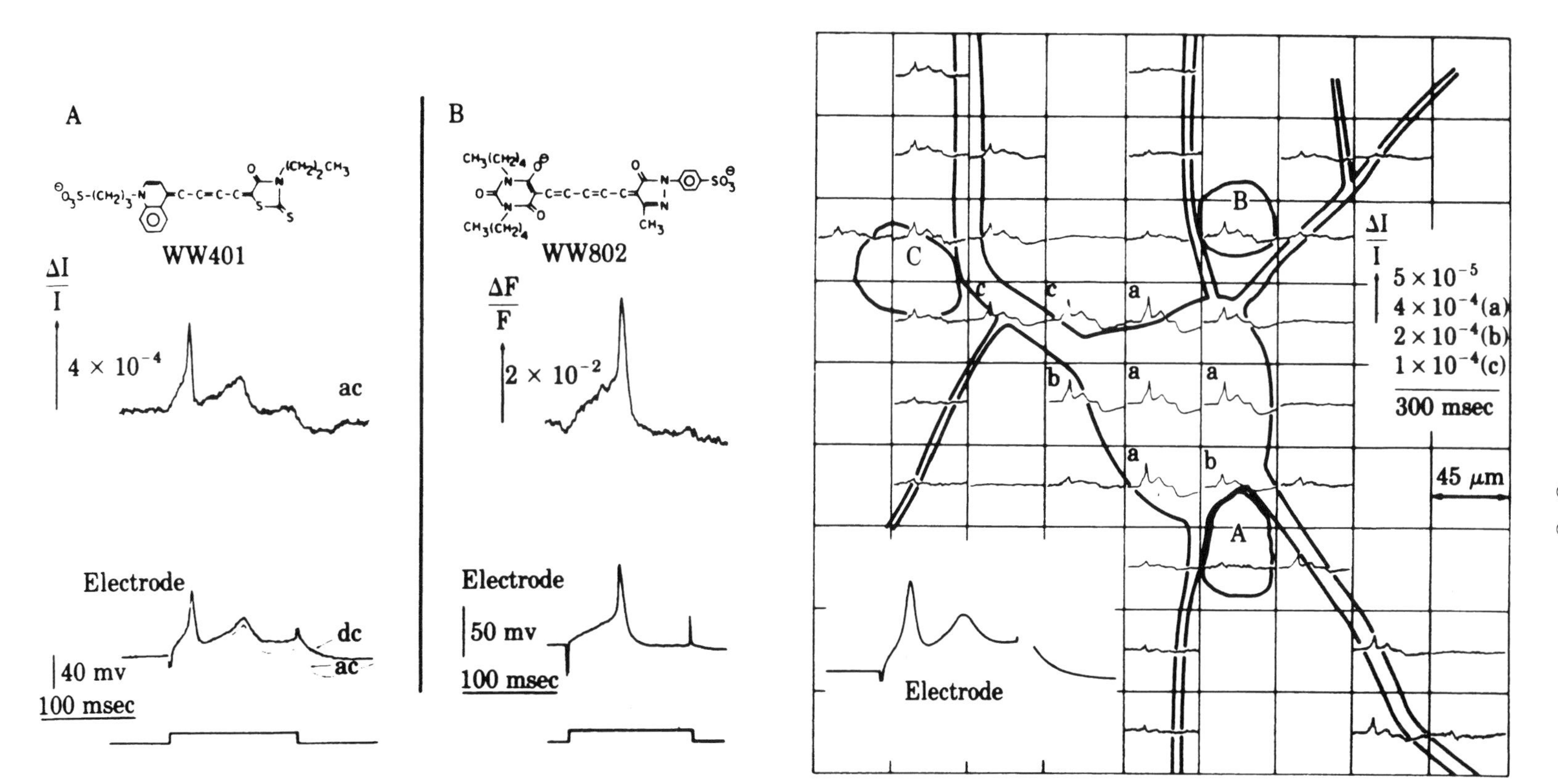

Figure 16.1 Transmission (A) and fluorescence (B) signals recorded with a single photodiode, without signal averaging, from an extracellularly stained neuroblastoma cell in culture. A multisite recording using a photodiode array is shown on the right. For details see Grinvald *et al.* (1981). Modified from Grinvald *et al.* (1981).

simultaneously at multiple sites. However, many of the limitations of this technique arise because the signals are small.

16.2 EXTRACELLULAR APPLICATION OF VOLTAGE-SENSITIVE DYES

The importance of multisite measurements from individual neurons was recognized earlier, and that problem has been approached by optical recording methods (for reviews see Cohen & Lesher, 1986). Grinvald, Ross & Farber (1981) demonstrated, on dissociated neurons in culture, that multisite optical measurements can be successfully applied to determine conduction velocity, space constants and regional variations in electrical properties of neuronal processes. Both absorption and fluorescence measurements were used in these experiments (Fig. 16.1) and the dissociated neurons in a dish (culture) were stained from the *extracellular* side by bath application of the dye. It has been possible, using the same approach, to study synaptic interactions between several interconnected neurons in culture (Parsons *et al.*, 1991).

However, monolayer neuronal culture is a low opacity system especially convenient for both absorption and fluorescence measurements. Unfortunately, primary culture of dissociated neurons is functionally very different from intact neuronal structures. Extending these methods to *in situ* conditions has been difficult. Non-selective binding of the bath applied dye to connective tissue and other extracellular membranes dramatically reduces the signal-to-noise ratio in absorption and especially in fluorescence measurements from both invertebrate ganglia and vertebrate brain slices (Ross & Krauthamer, 1984; Borst *et al.*, 1997). The relatively small signal size and the requirement for extensive averaging would make the use of this approach difficult in studying synaptic interactions and plasticity. Also, when many neurons are active in a densely packed neuropile it is difficult to determine the source of the signal if all the cells and processes are stained by extracellular dye application. Thus, the perspective of using extracellular staining in recording from individual neurons *in situ* is limited.

16.3 INTRACELLULAR APPLICATION OF VOLTAGE-SENSITIVE DYES

A different approach is to selectively stain particular neurons *in situ* by intracellular application of an impermeant voltage-sensitive dye. This approach is based on measurements carried out on the giant axon of the squid, demonstrating that optical signals may be obtained when the dye is applied from the inside (Davila *et al.*, 1974; Cohen *et al.*, 1974; Salzberg, 1978; Gupta *et al.*, 1981). Following the idea of selective staining by intracellular dye application Obaid, Shimizu & Salzberg (1982) and Grinvald, Salzberg, Lev-Ram & Hildesheim (1987) carried out experiments on leech neurons. The results demonstrated advantages (and difficulties) in using intracellular fluorescent probes. In these experiments the fluorescence measurements were used to record action potential and synaptic potential signals from single cells selectively stained by intracellular, iontophoretic injection with voltage-sensitive dyes. Due to the relatively low sensitivity of the available dyes (aminophenyl styryl dyes, e.g. RH437 and RH461; fractional change in fluorescence intensity between 0.01% and 0.1%), extensive averaging and correction procedures (Fig. 16.2) were necessary and spatial resolution was sacrificed to improve the signal-to-noise ratio. Furthermore, the recording was limited to one location. Because the signal size was small, it was not possible to use multisite optical recording to evaluate details of synaptic interactions at the level of neuronal processes of neurons in intact preparations; additional improvements in sensitivity were required.

However, these experiments articulated and emphasized the advantages of selective staining of particular

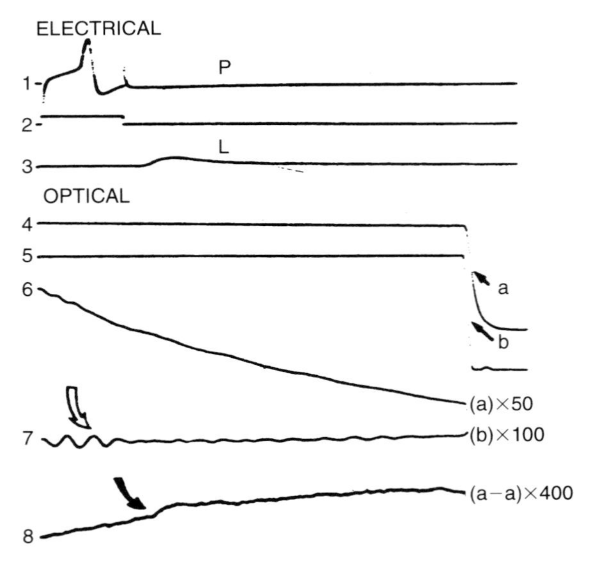

Figure 16.2 Fluorescence recording of the synaptic potential from an *in situ* leach neuron stained by an intracellular styryl dye. Recovery of the optical signal from raw data. Traces 1 and 3 are electrical recordings from the pre- and post-synaptic cells. Trace 8 shows the optically recorded synaptic potential obtained as the average of 14 pairs of trials in which sweeps without stimulus are subtracted from sweeps with stimulus. Traces 4, 5, 6 and 7 are used in the correction procedure. Modified from Grinvald *et al.* (1987).

neurons by intracellular application of the dye. The reported signal-to-noise ratio was the result of a relatively modest screening effort suggesting that better signals might be obtained by synthesizing and screening new molecules, by modifying the injection procedure, and by improving the recording apparatus.

Recently, the sensitivity of intracellular voltage-sensitive dye techniques for monitoring neuronal processes *in situ* has been dramatically improved (by a factor of roughly 150), allowing direct recording of subthreshold and action potential signals from the neurites of invertebrate neurons (Antić & Zečević, 1995; Zečević, 1996). Very encouraging results have also been obtained in preliminary studies on vertebrate CNS neurons in brain slices (Antić *et al.*, 1997). The improvement in signal-to-noise ratio is based on: (1) the advantages of fluorescence measurements over absorption in recording from small objects; (2) finding an intracellular dye that gives a relatively large fractional change in fluorescence, and (3) improvements in the apparatus to increase the incident light intensity, lower the noise, and to filter more efficiently.

16.4 DYE INJECTION

16.4.1 Invertebrate neurons

Sharp electrodes are used for the injection of the styryl dye (JPW1114 or its dimethyl analogue JPW3028; Molecular Probes Inc., Eugene, OR) in invertebrate neurons. The tip of the microelectrode is backfilled with the solution of a voltage-sensitive dye dissolved in distilled water. The dye solution is filtered before filling the electrodes to eliminate microscopic particles. We used Millex-GV_4 filters with 0.22 μm pore size (Millipore, Bedford, MA). Dyes are injected by applying repetitive, short pressure pulses to the microelectrode using a Picospritzer (General Valve Corp., Fairfield, NJ). The microelectrode tip size, pressure settings and pulse duration are adjusted for different cells (pressure was varied between 5 and 60 psi, pulse duration between 1 and 50 ms, and microelectrodes ranged from 2 to 10 MΩ when filled with 3 M KCl). After the injection was completed, the ganglia are routinely incubated for 12 hours before making optical recordings to allow for the spread of dye into the distal processes.

16.4.2 Vertebrate neurons

The dye is applied into vertebrate neurons using dark-field video-microscopy and patch-electrodes. The tip of the patch-pipette is backfilled with the solution of voltage-sensitive dye. Simple diffusion is sufficient to load the soma with an adequate amount of the dye. The major problem in injecting vertebrate neurons from patch-pipettes is the leakage of the dye from the electrode. Patching technique requires positive pressure to be applied to the patch-pipette during electrode manipulation through the tissue while approaching healthy neurons. This pressure ejects the solution from the electrode. To avoid extracellular deposition of the dye and the resulting large background fluorescence, the tip of the electrode has to be filled with dye-free solution and the electrode backfilled with the dye. It is possible, with practice, to load neurons without any leakage of the dye to the surrounding tissue. An example of the selective staining of only one neuron is shown in Fig. 16.5.

16.5 OPTICAL RECORDING

The recording system for fast, multi-site optical monitoring of membrane potential changes was originally developed in Larry Cohen's laboratory at Yale University. The system is based on an array of 464 silicon photodiodes; a detailed description has appeared earlier (Cohen & Lesher, 1986; Zečević *et al.*, 1989; Wu & Cohen, 1993; Antić and Zečević, 1995). The preparation is placed on the stage of a microscope and the image of the stained cell projected onto the photodiode array positioned at the primary image plane. A 250 W xenon, short-gap arc lamp is used as a light source. The best signals with styryl dyes are obtained using an excitation interference filter of 520±45 nm, a dichroic mirror with central wavelength of 570 nm and 610 nm barrier filter (a Schott RG610). The optical signals are recorded with a 464-element photodiode array (Centronix Inc., Newbury Park, CA). The output current of each diode is converted to voltage and individually amplified. High-frequency noise in the recording is limited by the 700 Hz cutoff frequency of the low-pass filter (4-pole Bessel). An RC filter with a cutoff frequency of 1.7 Hz is used to limit low-frequency noise. Amplifier outputs are digitized using a data acquisition system for an IBM-PC computer (Model DAP 3200e/214, Microstar Laboratories Inc., Bellevue, WA). The system provides for 512 analogue inputs and a 769 kHz throughput rate with 12-bit A/D resolution. The fastest acquisition rate, limited by a single conversion time of 1.3 μs, is 0.6 ms per full frame with 464 pixels. Faster sampling rates can be achieved by monitoring a subset of individual detectors selected under software control. This apparatus is available from OptImaging, LLC, Fairfield, CT.

16.6 MULTISITE RECORDING

A typical result of multisite optical recording from an individual invertebrate nerve cell *in situ* using the 464-element photodiode array is shown in Fig. 16.3(A). The image of the cell was projected by a 10×/0.4 NA objective onto the array as indicated by the drawing of the cell over the detectors in Fig. 16.3(A). Optical signals corresponding to evoked action potentials were clearly recorded from both soma and processes. These spikes were evoked by transmembrane current pulses (panel D) delivered through the recording microelectrode in the cell body. Each optical trace in panel (A) represents 100 ms of recording centred around the peak of the action potential as indicated by the time bar in panel D. Since this was the only cell stained in the ganglion, the source of the signal is certain and signals can be safely attributed to potential transients in the specific regions of the neuron. Optical signals associated with 85 mV action potentials, expressed as fractional changes in fluorescent light intensity (dF/F), were between 1 and 2% in recordings from neuronal processes. The fractional

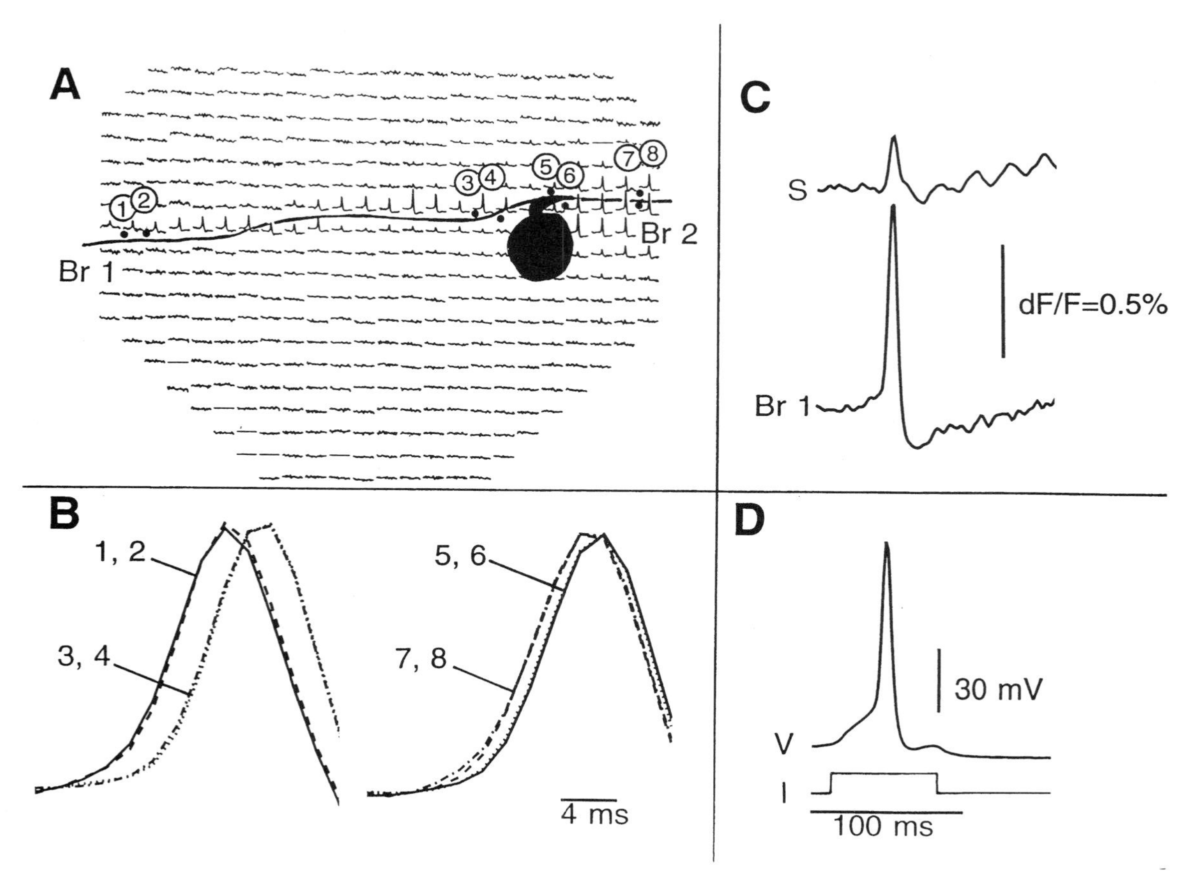

Figure 16.3 Spike initiation and propagation in main axonal branches. (A) Multiple site optical recording with the 464-element photodiode array (top five rows omitted from the figure). Each trace represents the output of one diode. The traces are arranged according to the disposition of the detectors in the array. Spikes were evoked by transmembrane current pulses (I, panel D), delivered through the recording microelectrode in the soma. The fluorescence recordings are divided by the resting light intensity to compensate partially for differences in signal size owing to uneven staining and to differences in membrane area imaged onto one detector. Each diode received light from a 50 × 50 μm area in the object plane. Twenty-four trials were averaged to improve the signal-to-noise ratio. (B) Recordings from individual detectors, marked (•) and scaled to the same height, at locations 1–4 show antidromic propagation of the spike toward the cell body. Recordings from individual detectors at locations 5–8 show that the action potential in Br2 also propagated antidromically toward the soma from a more distal site of origin. (C) Fractional changes in fluorescence light intensity, associated with the action potential of 85 mV, recorded from branch Br1 and the soma (S) is shown on the same dF/F scale. (D) Action potential recorded with microelectrode in the soma (V) evoked by transmembrane current pulse (I).

changes recorded from the cell body were about five times smaller but the signal-to-noise ratio was not reduced by this factor (see below).

With measurements like that shown in Fig. 16.3(A) it was straightforward to determine the direction and velocity of action potential propagation in the neuronal processes. In panel B of Fig. 16.3 recordings from individual detectors from different locations, scaled to the same height, are compared using an expanded time scale. In axonal branch Br1 the first spike was initiated in a distal axonal segment, more than 800 μm from the soma in a region distal to our recording, and it propagated antidromically toward the cell body. A propagation velocity of 0.25 m s^{-1} was determined by comparing recordings from locations 1 and 3. Recordings made from individual detectors at locations 5–8 show that the action potential in axonal branch Br2 also propagated antidromically toward the soma from a more distal site of origin to the right of our recording region, with an apparent propagation velocity of 0.20 m s^{-1}. These results, based on timing information derived from action-potential-related optical signals, show that there must be at least two separate spike trigger zones in the metacerebral interneuron.

16.7 COMPARISON OF OPTICAL AND ELECTRICAL SIGNALS

It is important to know how accurately the voltage-sensitive dye signals represent electrical events. For many voltage-sensitive dyes absorption and fluorescence changes are both fast and linear with membrane potential changes in a squid giant axon (Gupta *et al.*, 1981; Cohen & Lesher, 1986; Salzberg *et al.*, 1993). We confirmed this conclusion on metacerebral giant neurons injected with voltage-sensitive dyes by comparing electrical signals with optically recorded action potentials. Figure 16.4 shows optical recordings from selected detectors on an expanded time scale; the optical signals closely follow the waveform of electrically recorded action potentials.

Although the optical signals are similar to the microelectrode measurements, there are some discrepancies due to filtering the optical signals. Figure 16.4 shows, on an expanded time scale, a comparison of the unfiltered, microelectrode recordings of the action potential and a subthreshold hyperpolarizing response with the optical signals recorded with the bandwidth determined by a high-frequency cutoff of 1.0, 0.5 and 0.3 kHz. With the cutoff of recording set

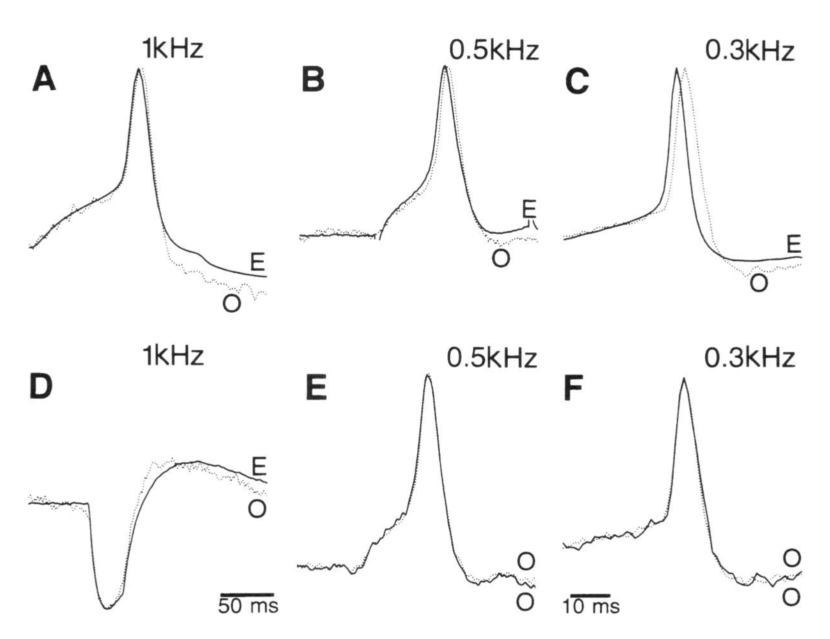

Figure 16.4 The comparison of electrical and optical signals. Traces from microelectrode recordings and optical detectors were scaled to the same height by adjusting the baseline level and *y*-axis gain for each trace separately. Filtering optical signals with a low-pass cutoff frequency of 1, 0.5 and 0.3 kHz progressively reduced the noise while the distortion of the action potential shape was negligible (A, B, C). Low-pass Bessel filters exhibit a delay between the input signal and the output signal that is inversely proportional to the cutoff frequency (most obvious at 0.3 kHz, (C)). The shape of the hyperpolarizing signal was also unaffected by low-pass filtering at a 1 kHz frequency (D). Panels (E) and (F) show superimposed optical recordings from two adjacent regions that appear identical if filtered at the same corner frequency (0.5 or 0.3 kHz), showing that the delay characteristics of the individual channel filters are similar. The faster time scale applies to the action potential recordings and the slower time scale to the hyperpolarizing response. E, electrical recording; O, optical recording.

at 1 kHz, the optically recorded action potentials lagged slightly behind microelectrode recordings (this effect could be removed by increasing the cutoff frequency to 3 kHz). While the distortion of the waveform at 1 kHz cutoff frequency was relatively small, this filtering only partially reduced the high-frequency noise. The effect of filtering on signal dynamics was more evident at 0.3 kHz; as expected, this effect was less pronounced with slower, subthreshold voltage changes. Since our conclusions are based on comparing optical signals with the same filtering, the small effects of these high-pass filters on the signal dynamics were neglected and all subsequent records were filtered with a high frequency cutoff between 0.3 and 1 kHz. In addition, digital smoothing was used (1–2–1 smoothing routine, London *et al.*, 1987), but only to an extent that did not change the timing information in the signal. Following the spikes in Fig. 16.4(A), (B) and (C) the optical signal appears to decrease more rapidly than the electrical signal. This was an effect of the high-pass filter (in the optical amplifiers) used to limit low-frequency noise. This difference between electrical and optical signals could be largely removed by digitally applying a similar high-pass filter to the electrical signals (not shown).

16.8 DYE SENSITIVITY (d*F*/*F*) AND SIGNAL-TO-NOISE RATIO

Optical signals associated with 85 mV action potentials, expressed as fractional changes in fluorescent light intensity (d*F*/*F*), were between 1 and 2% in recordings from neuronal processes. The fractional changes recorded from the cell body were about five times smaller but the signal-to-noise ratio was not reduced by this factor, because the light intensity from the soma was much higher. Presumably, the background fluorescence from dye not bound to surface membrane was higher in the soma. The soma might have higher proportion of internal membranes per unit volume in comparison to the axon, and/or a less favourable surface-to-volume ratio. The signal-to-noise ratio is a complex function of the light intensity in the optical recording (Wu & Cohen, 1993; Antić & Zečević, 1995). The fluorescence intensity was roughly 100 times higher in the cell body than in distal processes. At these high intensities the dark noise is insignificant and the fractional shot noise is low, but extraneous noise due to vibrations in the light path and movement of the preparation becomes dominant. On the other hand, the limiting factor in recording from distal neuronal processes was dark noise, since resting light levels are low. At intermediate intensities the signal-to-noise ratio is limited by shot noise.

With the sensitivity of the dye described and the level of noise in our recording system, relatively good signal-to-noise ratios could be obtained in single trial recordings from most of the processes. However, more distal regions required averaging to improve signal-to-noise. Usually 4–25 trials were averaged for action potential signals and 20–100 trials for 10 mV subthreshold responses.

16.9 CALIBRATION OF THE VOLTAGE-SENSITIVE DYE MEASUREMENTS IN TERMS OF MEMBRANE POTENTIAL

Voltage-sensitive dye recording of membrane potential transients belongs to a class of indirect measurements. The quantity that is being measured directly, by photodiodes, is light intensity, and the quantity that needs to be monitored (membrane potential) is derived from a known relationship between the light intensity and the membrane potential.

It is convenient that the relationship between light intensity and membrane potential is linear for many voltage-sensitive dyes (e.g. Gupta *et al.*, 1981; Wu & Cohen, 1993). For that reason, if the measurements are being made from one place, and if electrical measurement can be done simultaneously with optical measurements (as in squid giant axon experiments), then an absolute calibration is automatically obtained. The electrodes could then be removed and all further optical recordings would be precisely calibrated in terms of voltage.

A more complicated situation arises when optical measurements of membrane voltage are done from multiple (e.g. several hundred) sites on the object. Usually, most of these sites are not accessible to direct electrical measurements, and the simple calibration procedure described above is therefore not possible. In a multisite recording the change in light intensity is still proportional to voltage, but also, to a different extent at different sites, to two additional factors. The first of these factors is the amount of dye that contributes to the fluorescent light intensity recorded by each individual detector. The amount of dye will depend on two parameters: (a) the volume of the objects projected by a microscope objective onto each detector (surface area if depth of field is shallow) and, (b) the amount of dye bound to a unit volume (or surface area). The second factor is the ratio of the amount of dye that is bound to connective tissue or any other membranes that do not change potential (inactive dye; I-dye) to the amount of dye that is bound to the excitable membrane being monitored (active dye; A-dye). The I-dye contributes to the resting fluorescence only, and the light from A-dye contributes to the resting fluorescence and also carries the signal. Obviously, if all of the dye is in the excitable membrane, the optical signal expressed as the fractional fluorescence change ($\Delta F/F$) will be 10 times bigger, for the same change in membrane potential, than if there is 10 times more I-dye than A-dye. This is an essential consideration that explains why extracellular dye application is dramatically inferior, in terms of signal-to-noise ratio, to intracellular staining of individual nerve cells. Extracellular staining generates a very large excess of I-dye and large background fluorescence.

An important special case is the situation where all of the voltage-sensitive dye is bound to excitable membrane. In that case the differences in the sensitivity between detectors are caused solely by the differences in the resting intensity of the light projected to different detectors (due to differences in the amount of dye in different regions). It is intuitively clear that this factor can be eliminated by normalizing the signals to the resting light intensity for each particular detector. Dividing the signal (ΔF) by resting light intensity (F) for each detector will equalize the sensitivity of all elements in the photodetector array. In that case, calibration of the optical signal in terms of membrane potential from any one location (which is usually easy) by simultaneous optical and electrical recording will be valid for the whole array. One deviation from this rule would arise in situations where autofluorescence (not related to voltage-sensitive dye) from the object is not negligible and contributes significantly and to a different degree to the resting fluorescence recorded by each detector. If autofluorescence is not a problem and if all of the voltage-sensitive dye is bound to excitable membrane, it is possible to use dual-wavelength ratiometric methods, analogous to measurements with cation indicators (Grynkiewicz *et al.*, 1985), as a reliable measure of membrane potential (Montana *et al.*, 1989). Again, the ratiometric approach will provide reliable calibration only if all of the dye contributes to the voltage-related signal.

The special case where all of the dye is A-dye is rarely found. Generally, there will be both A-dye and I-dye contributing to fluorescence light. For example, if the preparation is stained by the extracellular application of the dye, I-dye would be the dye bound to the connective tissue or other components of the preparation. In experiments utilizing intracellular application of the dye, there will be I-dye bound to intracellular membranes and organelles.

Furthermore, it is a general rule that the ratio of A-dye to I-dye is unknown. This ratio will be different for different regions of the object. It follows that the sensitivity of the elements of a multidetector array that are recording from different regions of the object will be different and the calibration of all detectors cannot be done by calibrating the optical signal from any single site. The amplitude calibration of optical signals in terms of voltage in this situation will require separate calibration of each detector to determine the 'sensitivity profile' of the array.

Such a calibration is absolute and straightforward if a calibrating electrical signal that has a known amplitude at all locations is available. An all-or-none action potential signal that propagates throughout the region of interest is ideal for this purpose and can be used to create a 'sensitivity profile' of the measuring system. This type of calibration was used to scale the amplitudes of hyperpolarizing subthreshold signals in an interneurone from land snail (Antić & Zečević, 1995).

If action potentials are not available, another type of calibrating electrical signal that has a known amplitude at all locations is necessary. One idea is to take advantage of the fact that slow electrical signals spread over relatively long distances in neuronal processes because nerve cells are 'electrically compact' for slow voltage changes. This has been documented, by direct electrical measurement simultaneously from two locations on the neuron, for both invertebrate nerve cells (Tauc, 1962) and vertebrate neurons (Stuart & Häusser, 1994).

Finally, in many measurements the absolute calibration of optical signals in terms of voltage is not critical. Many conclusions depend on the comparison of relative amplitudes and timing information which is directly obtained from voltage-sensitive dye recordings.

16.10 PHARMACOLOGICAL EFFECTS AND PHOTODYNAMIC DAMAGE

Experiments on marine molluscs, *Navanax* and *Aplysia* (London *et al.*, 1987; Zečević *et al.*, 1989; Parsons *et al.*, 1991) showed that while some oxonol absorption dyes caused little or no photodynamic damage even after prolonged illumination, bleaching and photodynamic damage were substantially more rapid with fluorescent dyes. Also, in our initial experiments on *Helix* neurons using fluorescent styryl dyes, we were limited by photodynamic damage to averaging approximately 20 trials because of a pronounced reduction in the membrane input resistance and a progressive decline in action potential amplitude. In those experiments neurons were exposed to high-intensity light from the start of the averaging series to the end of the last trial. With improved amplifiers (Wu & Cohen, 1993; Antić & Zečević, 1995) that allow intermittent illumination we saw very little photodynamic damage. It was possible to average

100 trials with no noticeable change in action potential amplitude or waveform. The introduction of dark intervals in the series of measurements might have reduced photodynamic damage not only by reducing the absolute time the cell was exposed to high-intensity light, but also by introducing intermittent 'recovery' periods (Bonhoeffer & Staiger, 1988).

It was not possible to detect any pharmacological effects or photodynamic damage caused by the dye at the concentration we used. The absence of large pharmacological effects is evident from the fact that electrically recorded action potentials were essentially unchanged after staining and an incubation period of more than 12 hours. Photodynamic damage was likely to be either absent or insignificant since the time course of the electrically recorded action potential from the cell body and optically recorded signals from the soma and neuronal processes did not change over several averaging trials. However, it is not possible, at present, to be absolutely certain that staining or photodynamic damage does not influence regional properties of the neuron in some subtle way, because in the absence of voltage-sensitive dye, no independent measurements with adequate spatial resolution is available. Nonetheless, no evidence for such effects was found.

16.11 IMAGING SPIKE TRIGGER ZONE

Multisite optical recording of membrane potential changes, as shown in Fig. 16.3, revealed that at least two independent action potentials are initiated at separate locations in the axonal branching structure of the giant metacerebral neuron in response to direct soma stimulation. To understand the functional organization of this neuron, experiments were carried out to determine the position of the trigger zones in different processes. Plate 16.1 illustrates the results of the measurement carried out to investigate spike initiation in the branching structure of Br2. The fluorescent image of a metacerebral *Helix* neuron following 12 hours incubation at 15°C after injection with the fluorescent voltage-sensitive dye, JPW1114, is shown in Plate 16.1(A).

The image of the cell was projected by an objective onto the array of photodiodes as indicated in Plate 16.1(B), part (a). This panel represents multisite recording of action potential signals, evoked by a transmembrane current step (Plate 16.1 (B), part (d)), from axonal branches Br2, Br3 and Br4. Optical signals associated with action potentials, expressed as fractional changes in fluorescent light intensity ($\Delta F/F$), were about 1% in recordings from the processes. The signal size ($\Delta F/F$) dropped abruptly in the most distal axonal regions (Br3 in Plate 16.1(B), part (a)) because the concentration of the dye declined with distance from the soma and the resting light for the detectors that receive light from distal processes becomes dominated by autofluorescence. With these measurements it is straightforward to determine the direction and velocity of action potential propagation in neuronal processes. In Plate 16.1(B), part (b), recordings from different locations, scaled to the same height, are compared to determine the site of origin of the action potential. The earliest action potential, in response to soma stimulation, was generated near location 2, in the axonal branch Br4, situated in the cerebral-buccal connective outside the ganglion. The spike propagated orthodromically from the site of initiation toward the periphery in branch Br4, and antidromically toward the soma and into the branch Br3 in the external lip nerve. Although the difference in the timing of the spikes at locations 2 and 3 (Plate 16.1(B), part (b)) is small, the direction of propagation is clear from the colour-coded representation of the data (Plate 16.1(B), part (c)). This figure shows the potential changes in the branching structure at nine different times separated by 1.6 ms. The red colour corresponds to the peak of the action potential. The panels show the position of the action potential trigger zone at location 2 and ortho- and antidromic spread of the nerve impulse from the site of initiation. The earliest spike was evoked approximately 1 mm away from the soma. The spike initiation segment in the axon is roughly 300 μm in length. Under normal conditions slow depolarizing voltage pulses applied to the soma are electronically spread into the processes with little attenuation, owing to the very long axonal space constant for slow voltage changes. These depolarizing pulses initiate action potentials at remote sites in the processes that are characterized by higher excitability than that of neighbouring segments.

16.12 VERTEBRATE NEURONS

It is of considerable interest to apply the same technique, as developed for invertebrate neurons, to dendrites of vertebrate CNS neurons in brain slices. Our preliminary experiments (Kogan *et al.*, 1995; Antić *et al.*, 1997), are the only report of this kind of measurement.

Experiments were carried out on slices from the neocortex of P14-P18 rats. The fluorescent image of the cell was projected onto the photodiode array (Fig. 16.5). Optical signals associated with action potentials, expressed as fractional changes in fluorescent light intensity ($\Delta F/F$), wee between 1 and 2% in

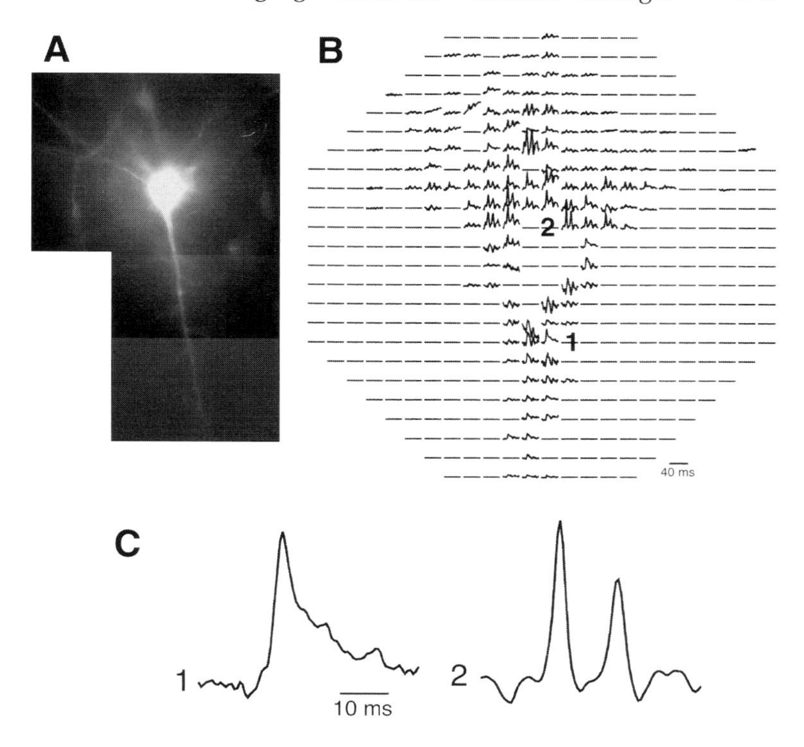

Figure 16.5 (A) Fluorescent image of a pyramidal neuron. The cell was stained with the voltage-sensitive dye JPW3028 (dimethyl analogue of JPW1114) via a patch pipette. Note that the surrounding tissue does not show any background fluorescence.
(B) Optical recordings of action potential signals from elements of a photodiode array. Each diode received light from a 14 × 14 μm area in the object plane. Each trace represents the output of one diode, when the neuron was stimulated synaptically, by a bipolar electrode positioned in the white matter at the edge of layer 6, to produce an action potential. Six trials were averaged. Optical signals corresponding to evoked action potentials were clearly recorded from both soma (the largest signals from the soma are eliminated from the figure for clarity) and processes.
(C) Optically recorded action potential signal from two locations on the same neuron modified from Antić *et al.*, 1997).

recordings from neuronal processes. The signal-to-noise ratio was sufficiently large that action potential signals could be monitored from multiple sites on dendritic processes in single trial measurements. An unexpected finding from these initial recordings, shown in Fig. 16.5(C), is that an electrical signal can have a very different shape in two spatially close regions of the same cell.

Also, these initial experiments provide several important methodological results. First, the results show that it is possible to deposit the dye into the cell without staining the surrounding tissue (Fig. 16.5). Also, we established that pharmacological effects of the dye were negligible since the action-potential characteristics did not change during staining. Furthermore, photodynamic damage, at the incident light intensities used, was not significant since the time course of the electrically recorded action potential from the cell body and optically recorded signals from the soma and neuronal processes did not change over many measurements. Finally, the sensitivity of the dye was comparable to that achieved in the experiments on invertebrate neurons (Zečević, 1996). In these preliminary experiments, the dye would spread about 200 μm into dendritic processes within one hour. A possible way to attain better staining of neuronal processes is to detach the patch electrode from the soma after the dye is applied and allow a longer period of time for the spread of the dye. Alternatively, it is possible to attach a dye-electrode to a distal region on a dendrite, as was done previously with calcium-sensitive dyes (e.g. Markram & Sakmann, 1994; Schiller *et al.*, 1995). This approach will shorten the time needed for the spread of the dye into terminal dendritic branches and might eliminate the need to re-impale neurons. However, patching the dendrites has a lower rate of success.

16.13 SUMMARY

Voltage-sensitive dye recording with the present sensitivity and temporal and spatial resolution is a powerful tool in the investigation of the principles of signal integration in single neurons. For example, multisite recording would significantly facilitate experiments on local modulations of excitability in restricted regions of the neuronal dendritic tree (Hoffman *et al.*, 1997) or experiments investigating synaptic effects on the precise localization of the spike trigger zone (Chen *et al.*, 1997).

REFERENCES

Antić S. & Zečević D. (1995) *J. Neurosci.* **15**, 1392–1405.
Antić S., Major G., Chen W., Wuskel J., Loew L. & Zečević D. (1997) *Biol. Bull.* **193**, 261.

Bonhoeffer T. & Staiger M. (1988) *Neurosci. Lett.* **92**, 259–264.
Borst A., Heck D. & Thormann M. (1997) *Neurosci. Lett.* **238**, 29–32.
Chen W., Midtgaard J. & Shepherd G. (1997) *Science* **278**, 463–467.
Cohen L.B. & Lesher S. (1986) *Soc. Gen. Physiol. Ser.* **40**, 71–99.
Cohen L.B., Salzberg B.M., Davila H.V., Ross W.N.D., Landown S., Waggoneer A. & Wang H. (1974) *J. Membr. Biol.* **19**, 1–36.
Davila H.V., Cohen L.B., Salzberg B.M. & Shrivastav B.B. (1974) *J. Membr. Biol.* **15**, 29–46.
Grinvald A., Ross W.N. & Farber I. (1981) *Proc. Natl. Acad. Sci. USA* **78**, 3245–3249.
Grinvald A., Salzberg B.M., Lev-Ram V. & Hildesheim R. (1987) *Biophys J.* **51**, 643–651.
Grynkiewicz G., Poenie M. & Tsien R.Y. (1985) *J. Biol. Chem.* **260**, 3440–3450.
Gupta R.K., Salzberg B.M., Grinvald A., Cohen L.B., Kamino K., Lesher S., Boyle M.B., Waggoner A.S. & Wang C.H. (1981) *J. Membr. Biol.* **58**, 123–137.
Hoffman D.A., Magee J.C., Colbert C.M. & Johnston D. (1997) *Nature* **387**, 869–875.
Kogan A., Ross W.N., Zečević D. & Lasser-Ross N. (1995) *Brain Res.* **700**, 235–239.
Llinas R. & Sugimori M. (1980) *J. Physiol.* **305**, 197–213.
London J.A., Zečević D. & Cohen L.B. (1987) *J. Neurosci.* **7**, 649–661.
Magee J.C. & Johnston D. (1995) *Science* **268**, 301–304.
Magee J.C., Christofi G., Miyakawa H., Christie B., Lasser-Ross, N. & Johnston D. (1995) *J. Neurophysiol.* **74**, 1335–1342.
Markram H. & Sakmann B. (1994) *Proc. Natl Acad. Sci. USA* **91**, 5207–5211.
Montana V., Farkas D.L. & Loew L.M. (1989) *Biochemistry* **28**, 4536–4539.
Obaid A.L., Shimizu H. & Salzberg, B.M. (1982) *Biol. Bull.* **163**, 388.
Parsons T.D., Salzberg B.M., Obaid A.L., Raccuia-Behling F. & Kleinfeld D. (1991) *J. Neurophysiol.* **66**, 316–333.
Ross W.N. & Krauthamer V. (1984) *J. Neurosci.* **4**, 659–672.
Salzberg B. (1978) *Biol. Bull.* 463–464.
Salzberg B.M., Obaid A.L. & Bezanilla F. (1993) *Jap. J. Physiol.* **43**, S37–S41.
Schiller J., Helmchen F. & Sakmann B. (1995) *J. Physiol.* **487**, 583–600.
Spruston N., Schiller Y., Stuart G. & Sakmann B. (1995) *Science* **268**, 297–300.
Stuart G. & Häusser M. (1994) *Neuron.* **3**, 703–712.
Stuart G.J. & Sakmann B. (1994) *Nature* **367**, 69–72.
Tauc L. (1962) *J. Gen. Physiol.* **45**, 1077–1097.
Tauc L. & Hughes G.H. (1963) *J. Gen. Physiol.* **46**, 533–549.
Wu J.Y. & Cohen L.B. (1993) In *Fluorescent and Luminescent Probes for Biological Activity*, W.T. Mason (ed.). Academic Press, London, pp. 389–404.
Zečević D. (1996) *Nature* **381**, 322–325.
Zečević D., Wu J.-Y., Cohen L.B., London J.A., Hopp H.-P. & Falk X.C. (1989) *J. Neurosc.* **9**, 3681–3689.

PART VI

Using Novel Indicators for Genetic, Molecular and Cellular Function

CHAPTER SEVENTEEN

Bioluminescent and Chemiluminescent Indicators for Molecular Signalling and Function in Living Cells

GRACIELA B. SALA-NEWBY[1], JONATHAN M. KENDALL, HELEN E. JONES, KATHRYN M. TAYLOR, MICHAEL N. BADMINTON, DAVID H. LLEWELLYN & ANTHONY K. CAMPBELL

Department of Medical Biochemistry, University of Wales College of Medicine, Cardiff, UK

[1] Also at Department of Surgery, Bristol Heart Institute, University of Bristol, Bristol, UK

17.1 THE NATURAL HISTORY OF BIO- AND CHEMILUMINESCENCE

17.1.1 What are bioluminescence and chemiluminescence?

Chemiluminescence is the emission of light as a result of a chemical reaction (Campbell, 1988, Chapter 1). The enthalpy of the reaction gives rise to an atom or molecule in a vibronically excited state; when the electron decays back to ground state a photon is emitted. Bioluminescence is visible light emission from luminous organisms. The term bioluminescence is also used to describe reactions extracted, or DNA cloned and engineered, from luminous organisms. Most of the other chapters in this book are concerned with fluorescence. It is important, therefore, to understand the difference between fluorescence, together with true phosphorescence, and chemiluminescence. The confusion that exists is compounded by the lay term 'phosphorescence' which in the Oxford English Dictionary is still defined as the property of shining in the dark.

The term 'luminescenz', associated with the prefixes photo-, electro-, thermo-, crystallo- and chemi-, was first coined by a German physicist in 1888, Eilhardt Weidemann, to distinguish phenomena which resulted in light emission without requiring the high temperatures necessary for incandescence. Luminescent phenomena do not obey the laws of black body radiation. The discovery of the electron in 1897, together with the development of quantum theory and electron spin, led to a clear understanding of the difference between chemiluminescence and the related phenomenon of photoluminescence. In photoluminescence, usually known either as fluorescence or phosphorescence, the energy for exciting the electron arises from absorption of electromagnetic radiation in the near IR, visible or UV region. In chemiluminescence, however, the energy for exciting the electron comes from the enthalpy of the chemical reaction. Both result in electronically excited states, and can involve intersystem crossing, i.e. to the triplet spin state, when

FLUORESCENT AND LUMINESCENT PROBES, 2ND EDN
ISBN 0-12-447836-0

the decay becomes phosphorescence. To produce orange-red light at 600 nm at least 47.6 kcal mol^{-1} (1 orange-red photon = 4.41×10^{-19} J) are required, or 63.5 kcal mol^{-1} for blue light at 450 nm. Thus chemiluminescence requires a reaction with an enthalpy of 50–100 kcal per mol. The two most common chemical mechanisms with sufficient energy to generate an excited state by chemiluminescence are electron transfer with radical annihilation and the cleavage of linear or cyclic peroxides.

There are two important differences between a chemiluminescent and a fluorescent compound. First, the actual emitter in chemiluminescence is different chemically from the initial substance. Secondly, and as a consequence of this, a chemiluminescent compound (C) can only produce a photon once. In contrast a fluor (F), in the absence of a photobleaching reaction, is identical chemically to the initial compound and can be excited again and again. Thus

Fluorescence

$$\begin{array}{c}\downarrow \text{light}\\ F \rightarrow F^* \rightarrow F + h\nu\end{array} \quad [17.1]$$

Chemiluminescence

$$\begin{array}{c}\downarrow \text{catalyst}\\ C + \text{reactants} \rightarrow Pr^* \rightarrow Pr + h\nu\end{array} \quad [17.2]$$

Yet fluorescence is associated with chemiluminescence. The initial product (Pr) of the reaction is, by definition, capable of fluorescence, but may only be transiently stable. The initial compound C is also capable of fluorescence, but its fluorescence spectrum will be quite different from Pr, since it is structurally different. Fluors can also act as energy transfer acceptors from Pr*, as they do in several luminous organisms and in the light sticks commercially available for decoration or use in golf balls at night. Blue, green, yellow and red fluors are found associated with bioluminescence in at least eight phyla (Bacteria, Dinophyta, Cnidaria, Arthropoda, Mollusca, Polychaeta, Echinodermata, Chordata). These occur for any one of these reasons. First, the luciferin may be fluorescent and stored in high concentration (e.g. dinoflagellates, copepods). Second, the oxyluciferin may accumulate in sufficient quantities to be dissected by fluorescence. Third, the fluor may be an energy transfer acceptor changing the colour or quantum yield of the emission (e.g. some bacteria, cnidaria and certain fish). The most important example of the latter is the green fluorescent protein (GFP) first isolated from the jellyfish *Aequorea forskalea* and found in several other luminous hydrozoans such as the hydroid *Obelia* and the sea pansy *Renilla*. GFP forms its fluor by cyclization of three amino acids (65SYG67) in the protein sequence. In the organism GFP shifts the emission from blue (*c.* 470–480 nm peak) to blue-green (508 nm peak), and sharpens the spectrum. In *Renilla* it also increases the apparent quantum yield by three-fold (Ward & Cormier, 1979). This type of energy transfer follows the Förster equation and is radiationless, not requiring the direct transfer of a photon.

Five parameters characterize a chemiluminescent reaction: intensity, speed of onset, decay of light intensity, colour, and polarization, if any. Since chemical reactions are much slower than electronic excitation (10^{-15} s) or electronic decay (10^{-9}–10^{-8} s for singlet), the first two of these are dependent on three factors: the rate of the chemical reaction, the efficiency of the chemical reaction in generating electronically excited molecules, and the efficiency of excited molecules in producing photons:

Thus the overall quantum yield of a chemiluminescent reaction, ϕ_{CL}, is given by:

$$\begin{aligned}\phi_{CL} &= \text{(total number of photons emitted) /}\\ &\quad \text{(numbers of molecules reacting)} \qquad [17.3]\\ &= \phi_C\, \phi_{EX}\, \phi_F\end{aligned}$$

where ϕ_C is the chemical yield, i.e. the fraction of molecules going through the chemiluminescence pathway; ϕ_{EX} is the yield of excited state molecules; and ϕ_F is the excited state quantum yield. Bioluminescent molecules usually have ϕ_{CL} in the range 0.01–1. For methods of standardization see Hastings & Weber (1963), Hastings & Reynolds (1966), Nakamura (1972), and Campbell (1988, chapter 2).

The analytical application of chemi- and bioluminescence arises from the ability to couple analytes of interest to one of the components of the reaction. The exquisite sensitivity of chemiluminescence analysis, at least down to 10^{-21} mol in a solution of 0.1–1 ml, derives from the low noise from normal solutions. Clean solvents such as H_2O, and cuvette housings painted black, produce no chemiluminescence, but they do fluoresce. Small though this fluorescence may be, it usually means that in the test tube chemiluminescence has superior sensitivity to fluorescence. As an example aequorin and GFP can be formed from cDNA *in vitro* using TNT^{TM}. The aequorin can generate millions of detectable chemiluminescent counts, but GFP fluorescence is virtually impossible to detect, except through energy transfer from aequorin (J.P. Waud & A.K. Campbell unpublished).

17.1.2 The chemical reactions

Bioluminescent reactions require a minimum of three components: the chemiluminescent substrate = the luciferin; the catalyst = the luciferase; and oxygen or one of its metabolites. Up to three other components may also be required: a metabolite such as NADH or ATP, a cation such as Cu^{2+}, Mg^{2+} or Ca^{2+}, and an energy transfer acceptor. However, even when ATP is required, as in the beetle system, the energy for light comes from oxidation of the luciferin, usually via a cyclic oxygen dioxetan or dioxetanone intermediate.

There is insufficient energy in ATP hydrolysis to generate a visible photon. Thus ATP is not required directly for most bioluminescent reactions. It is important to remember that the terms luciferin and luciferase, first introduced by Dubois in 1887, are generic terms (Harvey, 1952). Each luciferase has a unique amino acid sequence, though close homology obviously exists between related species. Five distinct chemical 'families' of luciferin have been identified so far (Fig. 17.1(A)), each resulting in a quite different chemical reaction (Fig. 17.1(B)). Benzothiazole bioluminescence has been found only in luminous beetles, yet imidazolopyrazine bioluminescence is the most common one in the sea, being found in eight distinct phyla (Campbell & Herring, 1990; Thomson *et al.*, 1997). In some of these, namely the radiolarians, cnidarians and ctenophores, the luciferin and oxygen can be tightly or covalently bound to the luciferase so that the complex can be isolated as a whole (Shimomura *et al.*, 1962; Shimomura & Shimomura, 1985). Shimomura, who has carried out such unique and distinguished work on the chemistry of bioluminescence for over 30 years, named this complex photoprotein. Addition of Ca^{2+} to the photoprotein triggers the chemiluminescent reaction (Fig. 17.1), hence it can be used as an indicator for intracellular Ca^{2+}. Once the reaction has taken place the product, the oxyluciferin, is no longer covalently linked to the protein. The apoprotein can then be reactivated to form photoprotein by addition of coelenterazine (Campbell *et al.*, 1988; Campbell & Herring, 1990), in the presence of oxygen and absence of Ca^{2+}. Some 30 coelenterazine derivatives have been synthesized (Shimomura *et al.*, 1988, 1989, 1990, 1993, 1995). The novel photoproteins generated from these have properties which differ from native aequorin. One has a bimodal spectrum where the ratio of 409/465 is linearly related to free Ca^{2+}, whereas others have changed kinetics, ϕ_{CL} and increased or decreased affinity for Ca^{2+}.

Fish, appedicularians, an annelid, squid, shrimp, ostracods, copepods, a chaetognate and anthozoans which also use imidazolopyrazine bioluminescence, use a simple O_2-luciferin–luciferase system, sometimes squirted out into the seawater. They all emit blue light, though the emission maximum can vary from 440 to 490 nm without energy transfer. As with similar changes in the colour of beetle emission from green, green-yellow, orange to red, just a few amino acid differences (Wood *et al.*, 1989a,b,c; Wood, 1991a,b), usually charged ones, are responsible for these colour shifts. Such spectral differences are not found so dramatically in other luciferin–luciferase systems, e.g. bacteria, unless intermolecular energy transfer occurs. It was this lesson from nature which led us to select imidazolopyrazine apophotoproteins and benzothiazole luciferases to engineer as indicators of molecular signalling in live cells (see Sections 17.4 and 17.5).

Most synthetic organic chemiluminescent reactions are also oxidations, using O_2, O_2^- or H_2O_2 as oxidant to produce a dioxetan intermediate which then spontaneously generates the excited carbonyl emitter. It is, however, possible to synthesize stable dioxetans (Adam & Cilento, 1982), which decompose to excited carbonyls on mild heating. One such compound of important application as an immunoassay and DNA probe is adamantylidene adamantane 1,2-dioxetan. A phosphate derivative of this is not chemiluminescent. However, catalytic removal of the phosphate by alkaline phosphatase results in generation of the chemiluminescent compound, and light emission (Schaap *et al.*, 1989).

17.2 THE ANALYTICAL POTENTIAL OF CHEMILUMINESCENT COMPOUNDS

The study of cell activation and cell injury, together with the study and detection of microbes, in the research and clinical laboratory requires the identification and quantification of chemical changes, both within the cell and in the surrounding fluid. Substances to measure include substrates and nutrients, metabolites, enzymes and other proteins, hormones and vitamins, ions, drugs and pathogens. The concentrations and amounts of these substances available for analysis can vary over some 11 orders of magnitude, from 140 mM for extracellular Na^+ to a few picomolar for some free hormones, from millimoles of K^+ extracted from a whole liver to a few attomoles of ATP extracted from a single bacterium. It is the combination of three features of chemiluminescent compounds which enables us to identify where they should be used; and where they have unique potential over coloured, fluorescent or radioactive probes:

(1) their sensitivity;
(2) the fact that they are non-radioactive; and
(3) their ability to produce a signal from within living cells.

Sensitivity of detection is determined by the signal-to-noise ratio. Synthetic chemiluminescent compounds such as isoluminol, acridinium esters and dioxetans

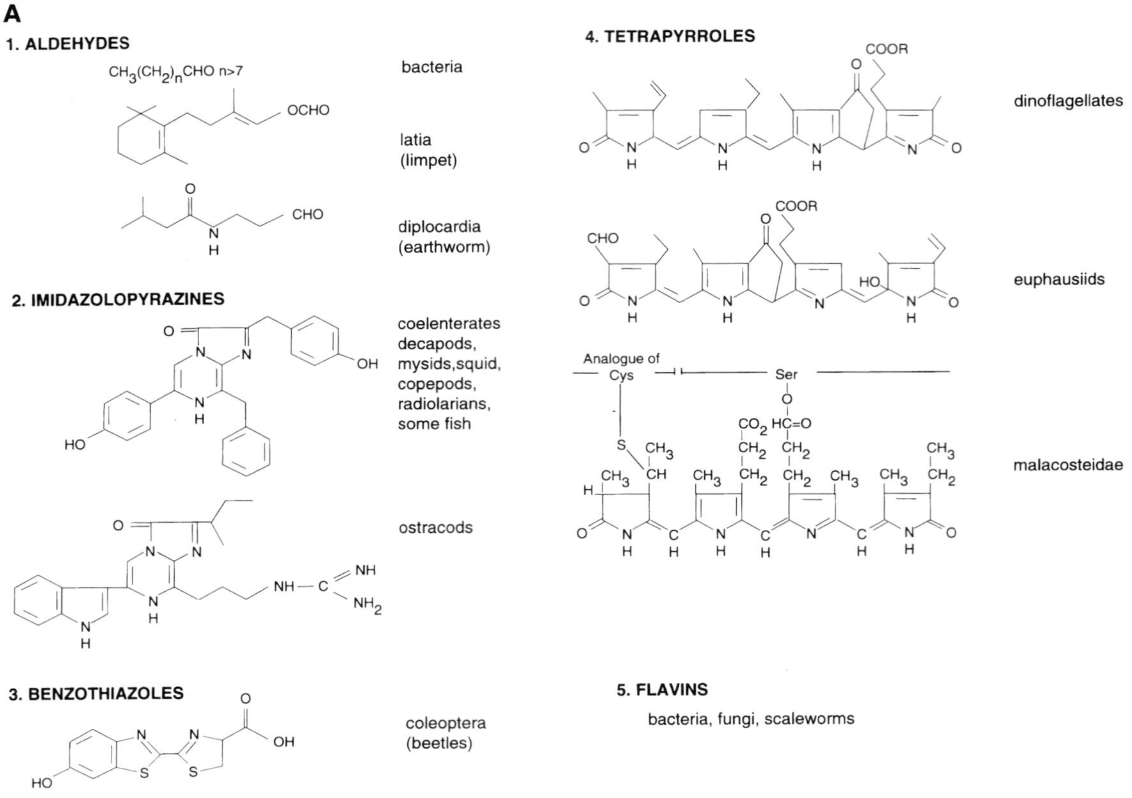

A

1. ALDEHYDES

$CH_3(CH_2)_nCHO$ n>7 — bacteria

OCHO — latia (limpet)

O, N, H, CHO — diplocardia (earthworm)

2. IMIDAZOLOPYRAZINES

O, N, N, OH, N, H, HO — coelenterates, decapods, mysids, squid, copepods, radiolarians, some fish

O, N, N, N, H, NH — C, NH, NH_2, N, H — ostracods

3. BENZOTHIAZOLES

O, N, N, OH, S, S, HO — coleoptera (beetles)

4. TETRAPYRROLES

COOR, O, O, N, H, N, H, N, H, N, O — dinoflagellates

COOR, O, CHO, O, N, H, N, H, N, HO, N, O, H — euphausiids

Analogue of Cys, Ser, O, CO_2 HC=O, S, CH_3, CH_2 CH_2, CH_3, CH_3 CH, CH_3, CH_2 CH_2, CH_3, CH_3 CH_2, H, O, N, C, N, C, N, C, N, O, H H H H H H — malacosteidae

5. FLAVINS

bacteria, fungi, scaleworms

B

1. Bacteria (no role for Ca_2+) - blue light

$$NADH + FMN \xrightarrow{\text{reductase}} NAD^+ + FMNH_2$$

$$FMNH_2 + RCHO + O_2 \xrightarrow[\text{luciferase}]{} FMN + RCO_2H + H_2O + \text{light}$$

2. Latia (a fresh water limpet): pale green light

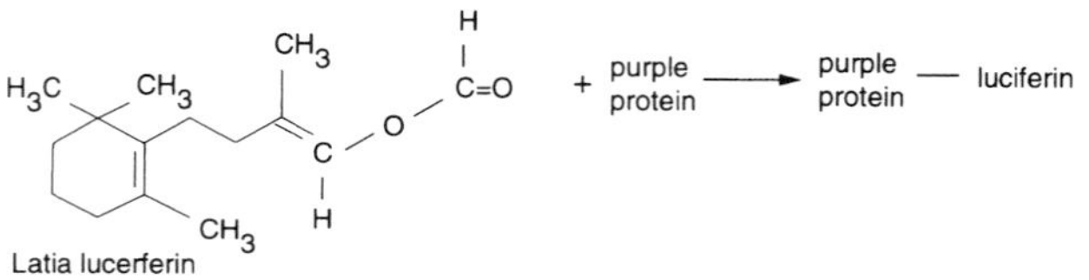

H_3C, CH_3, CH_3, H, C=O, O, C, H, CH_3 + purple protein → purple protein — luciferin

Latia lucerferin

$$\text{purple protein — luciferin} + O_2 \xrightarrow{\text{luciferase}} \text{light} + \text{purple protein} + \text{products}$$

3. Diplocardia (an earthworm) - blue - green light

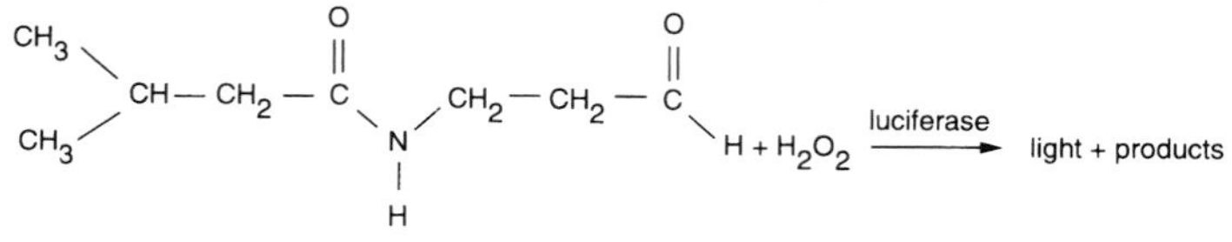

$$(CH_3)_2CH—CH_2—C(=O)—NH—CH_2—CH_2—C(=O)H + H_2O_2 \xrightarrow{\text{luciferase}} \text{light + products}$$

Diplocardia luciferin

4. Aequorea (a hydrozoan jellyfish): blue light (animal: blue-green light)

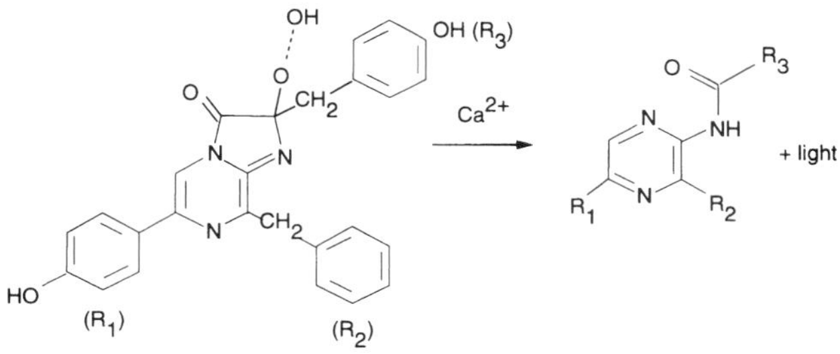

5. Photinus (the firefly) - yellow light

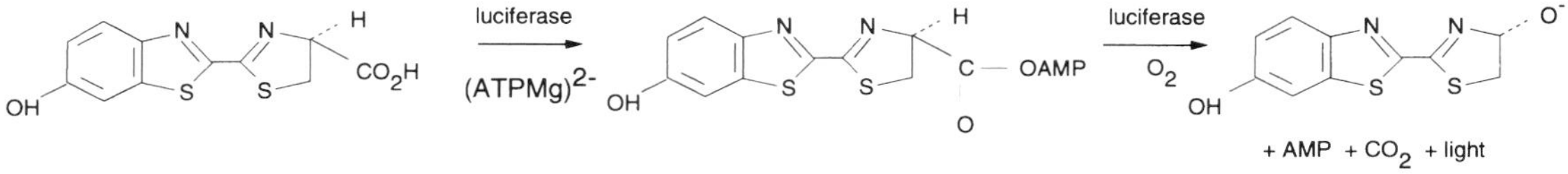

Figure 17.1 Chemistry of bioluminescence. (A) Chemical families in bioluminescence. (B) Some bioluminescent reactions.

can be detected easily in the range 10^{-15}–10^{-19} mol, and with enzyme amplification down to 10^{-21} mol. Similarly, firefly luciferase can be detected down to approximately 10^{-18} mol, and obelin or aequorin down to 10^{-21} mol. Isotopes such as ^{14}C and ^{3}H can be detected only down to 10^{-12}–10^{-15} mol, and ^{125}I or ^{32}P to around 10^{-18} mol. Furthermore, radioactive isotopes are hazardous, their shelf life is short for the more sensitively detectable, and they often require long counting times and a separation step in the procedure to quantify the analyte, i.e. they cannot be used in 'homogeneous' assays. In contrast, chemi- and bioluminescent labels are stable until triggered, apparently safe, can be analysed in seconds, and can be coupled to homogeneous immunoassays or DNA analysis (Weeks *et al.*, 1983a,b; Campbell & Patel, 1983; Nelson *et al.*, 1990).

Thus synthetic chemi- and bioluminescent assays are widely used for analysis of a range of substances of biological interest (Table 17.1). Firefly luciferase is the method of choice for assaying biomass via ATP. As little as 0.1 fmol can be detected in a solution of 0.1–1 ml. Several chemiluminescent labels including chemiluminogenic substrates for non-luminous enzymes

Table 17.1 Some metabolites measured by luminescent reactions

Bioluminescent system	*Chemiluminescent system*	*Analyte*
Firefly	N/A	ATP, ADP, AMP ATP utilizing enzymes CoA
Bacteria	N/A	O_2 NAD(P), NAD^+, FMN aldehydes, pheromones enzymes coupled to FMN, NAD(P)H proteases
Photoproteins (Aequorin, Obelin, Thalasicola)	N/A	Ca^{2+}
Renilla	N/A	PAP/PAP(S) phospho-adenosine phosphate (SO_4)
Pholasin	Lucigenin	O_2^-
	Luminol	OCl^-
Earthworm	Luminol	H_2O_2

systems are commercially available for a wide range of immunoassays and bioassays. Some can also be visualized from Southern, Northern and Western and DNA sequencing blots. They are:

(1) peroxidase detected by luminol-enhanced chemiluminescence (Matthews *et al.*, 1985);
(2) acridinium esters (Weeks *et al.*, 1983a, 1987; Nelson *et al.*, 1990);
(3) adamantylidene adamantane dioxetan containing substrates for alkaline phosphatase, β-galactosidase (βGal) and β-glucuronidase (Hummelen *et al.*, 1987; Geiger *et al.*, 1989; Schaap *et al.*, 1989; Bronstein *et al.*, 1996);
(4) bacterial luciferase (Balaguer *et al.*, 1989; Lindbladh *et al.*, 1991).
(5) aequorin (Stults *et al.*,1991; Lizano *et al.*, 1997);
(6) firefly luciferase (Kobatake *et al.*, 1993).

This chapter is focused principally on the application of chemi- and bioluminescence to measurements in live cells rather than to fluids and tissue extracts.

17.3 APPLICATION OF CHEMI- AND BIOLUMINESCENCE TO LIVING CELLS

17.3.1 Principles

What do we need to measure and are there appropriate chemi- and bioluminescent probes available? A full understanding of the mechanisms responsible for activating cells or for cell injury, induced, for example, by components of the immune system in rheumatoid arthritis or by viruses, requires elucidation of the complete molecular sequence, initiated at the plasma membrane or within the nucleus and ending in a cell response. This response begins with the generation of a signal such as Ca^{2+}, IP_3 or cyclic AMP, which then induces a structural or covalent modification in target proteins. If these modifications happen at the right time, in the necessary part of the cell and to the right extent, and if energy phosphate in the form of ATP and GTP is available, the cell undergoes an end response. These end responses include cell movement, secretion, division, transformation, defence by removal of the attacker, apoptosis and lysis.

The timing and magnitude of each step in the sequence varies considerably from cell to cell. Thus ultimately we can only follow the complete sequence using single cell analysis. For example, a calcium cloud, detected using aequorin, has to reach its target before exocytosis in a fertilized fish egg (Gilkey *et al.*, 1978) or gap junction closing in a *Chironomus* salivary gland (Rose & Lowenstein, 1976) occur. Using fura-2 imaging for Ca^{2+} or 2,7-dichloro-fluorescein for O_2^- and H_2O_2 we have shown that in a population of neutrophils activated by f-Met-Leu-Phe, the time course of O_2^- release is a reflection of individual neutrophils starting to release O_2^-, at different times up to 30 min after addition of the stimulus. Furthermore, four types of Ca^{2+} signal have been visualized in different cells (Hallett *et al.*, 1990), a rapid Ca^{2+} signal within 6 s, a delayed Ca^{2+} signal up to 60 s after f-Met-Leu-Phe, an oscillatory Ca^{2+} signal, and no signal at all in 30% of the cells. In contrast the Ca^{2+} signal induced by leukotriene B4 is synchronous, though, once again, not all of the cells respond (Davies *et al.*, 1991a,b). More recently it has become clear that a decrease in Ca^{2+} inside the endoplasmic reticulum can activate Ca^{2+} channels in the plasma membrane, nuclear gene expression, proteolysis and inhibition of secretion. Thus it is vital to measure free Ca^{2+} not only in the cytosol, but also in the ER and nucleus (Berridge, 1995; Clapham, 1995; Llewellyn *et al.*, 1996).

The behaviour, development and survival of the whole tissue and organism in health and disease is dependent on the timing and number of thresholds crossed by particular cells. In order to define these thresholds at the molecular level we need methods for quantifying and locating these chemical sequences, not only in live cells but in intact organs. We need to locate not only which cells have undergone a chemical change, and when, but also where in the cell it occurs, i.e. in the centre of the cytosol or close to the mitochondria, plasma membrane, in the endoplasmic reticulum or nucleus, or in the chloroplast or tonoplast in a plant cell. Chemiluminescent probes, in particular engineered bioluminescent proteins, have the ability to do this. A new generation of engineered fluorescent probes tagged with the green fluorescent protein (GFP) from *Aequorea victoria* can now also be used to monitor subcellular activity (Gerdes & Kaether, 1996; this volume). The aim is to measure in live cells within whole tissues eight stages in the sequence:

(1) The concentration and movement of intracellular signals, e.g. Ca^{2+}, IP_3, diacylglycerol and cyclic nucleotides.
(2) The concentration and locality of changes in intracellular energy supply and regulatory metabolites, e.g. ATP, GTP and AMP.
(3) Covalent modification catalysed by e.g. serine, threonine, tyrosine kinases and the activity of the regulatory enzymes they phosphorylate.
(4) The movement of regulatory proteins and subcellular structures to a target, e.g. calmodulin and secretory vesicles.
(5) The expression of new proteins at their correct site

(normally only possible by immunofluorescence on fixed tissue but now monitored by the use of GFP).
(6) The activation or inhibition within the nucleus of proteins, enhancers and promoters, controlling gene expression, cell priming and cell transformation.
(7) Cell and organelle viability and permeability changes in the plasma or organelle membranes.
(8) The end response, e.g. secretion, transformation or death.

17.3.2 The criteria

In order to achieve these objectives we need to find or design chemiluminescent indicators which, ideally, satisfy the following criteria:

(1) It must be possible to couple the substance or process of interest to a chemiluminescent reaction so that it is 'homogeneous', i.e. usable inside a cell without the need for a separation step.
(2) It must be sufficiently sensitive, ultimately for single cell analysis, and specific.
(3) It must be possible to incorporate it into the cell and it must distribute uniformly in the required compartment of the cell, and nowhere else.
(4) It must be possible to quantify and locate the indicator in the cell by a change in light intensity or colour, or possibly polarization of the light emitted, and within subcellular structures and cells from intact organs. It has been argued that for ease of quantification it is better to have a linear relationship between the analyte concentration and the signal. However, a power law relationship, as with aequorin and obelin light emission to free Ca^{2+}, is more easily detectable. A ratio of intensity at two wavelengths is also advantageous since it is independent of the amount of indicator injected or expressed inside the cell.
(5) It must be non-lethal, and not disturb significantly the phenomenon under investigation.
(6) It must be stable enough, and respond with the appropriate kinetics, to follow the phenomenon over the desired time course, sometimes milliseconds to seconds, sometimes hours to days.
(7) It should be readily available and not too expensive.

Considerable success has already been achieved using proteins extracted from luminous organisms and synthetic chemiluminescent reactions, but, as we shall see in Sections 17.4 and 17.5 the full potential of chemiluminescent indicators can now be realized using controlled expression and engineering of bioluminescent genes.

17.3.3 Detection and visualization of chemi- and bioluminescence

In practice chemi- and bioluminescent indicators produce a lower intensity of photon emission than fluors, particularly when the latter are excited with a high-intensity lamp or laser. Thus to realize the potential of chemiluminescence the most sensitive photomultiplier tubes (PMT) available are used (Campbell, 1988, Chapter 2). Battery-operated photodiodes are available in chemiluminometers for use in the field. However, their photon sensitivity is usually at least two orders of magnitude lower than that of a photomultiplier tube. A typical PMT is bialkali with 11 dynodes with a dark current of 0.05 nA at 1000 V, operating in true digital rather than analogue mode, with a recorded background of 10–20 cps. Photomultipliers with 13 dynodes produce a higher absolute signal but more noise. It is important to remember that the peak in spectral sensitivity for all photocathodes available is in the blue. For a bialkali tube some 5–10 more photons are required from firefly luciferase (λ_{max} 565 nm) than from aequorin (λ_{max} 465 nm) to generate the same recorded counts, though it may be possible to improve this slightly by careful selection of the photomultiplier. Taking into account sample geometry, PMT sensitivity (never >25%), discriminator setting and noise, a typical good chemiluminometer has an efficiency of 0.2–1% and a detection limit of approximately 100 emitted photons s^{-1}, giving a detection limit for a chemiluminescent indicator with $\phi_{CL} = 0.01$ of about 10^{-22} mol. Several good, true digital chemiluminometers are now available commercially, e.g. Berthold (Stanley, 1997), but we have constructed our own for greater flexibility, particularly in the sample housing.

Imaging of bioluminescence was pioneered by Reynolds (1972, 1978), using an EMI four-stage electromagnetically focused imaging intensifier, visualizing aequorin Ca^{2+} waves and clouds in single cells (Rose & Lowenstein, 1976; Eisen & Reynolds, 1985; Campbell, 1988, Chapter 2). These are no longer commercially available. Chemiluminescent labels for Southern or Western blots or immunoassay can be visualized with 20 000 ASA photographic film, with the appropriate colour sensitivity, by autoradiography (Matthews *et al.*, 1985; Kricka & Thorpe, 1987) and with phosphor imaging screens (Nguyen & Heffelfinger, 1995). However, to image individual luminous cells, particularly when observed in a microscope, a good imaging camera is required (Hooper *et al.*, 1990). The three-stage Photonics Science ISIS3M, with photocathode, microchannel plate and CCD, produces a video signal which can be processed by a computer, just as for fluorescent dyes. However, it is not sufficiently sensitive for low photon emissions, i.e.

down to a few cps for cells. But addition of a second or third microchannel plate, results in a photon-counting image intensifer capable of imaging transgenic plants, firefly luciferase in COS cells (Craig *et al.*, 1991; White *et al.*, 1995; Campbell *et al.*, 1996). Image Research and Photek have developed contact imaging cameras (Hooper *et al.*, 1990; Badminton *et al.*, 1996) where a fine fibreoptic bundle is in direct contact with a cell culture, enabling an image, albeit a fuzzy one, to be quantified from individual cells without the need of a microscope. This is important for observing cells with very low photon emission, e.g. the resting glow from aequorin at Ca^{2+} <0.1 μM, since at every optical surface in a microscope as many as 50% of the photons may be lost. Contact imaging obviously improves the solid angle considerably.

Liquid N_2-cooled CCD cameras appear to have the necessary sensitivity for imaging chemiluminescence from cells. They have very low noise in the detector, but still introduce noise in the signal processor and often require long accumulation times. This is all right for astronomy but too slow for real-time biological phenomena. These intensifiers require proper evaluation. The four-stage system used by Reynolds, though cumbersome, is still probably the most sensitive detector. For real-time bioluminescence imaging an intensified CCD camera provides the best signal-to-noise. With the appropriate software bioluminescent indicators can be imaged in single cells or whole organs for a range of a few seconds up to several hours. We have used imaging successfully to monitor free Ca^{2+} using aequorin in isolated cells, in smooth muscle and intact leaves or seedlings (Rembold *et al.*, 1996; Campbell *et al.*, 1996) and luminous bacteria and *Escherichia coli*. In order to maximize light capture a fibreoptic bundle is used. Each fibre (6 μm in diameter) is used in direct contact with a coverslip on which the cells are attached. Cells more than 50 μm apart show up in separate pixels. Using appropriate image processing digital events can be monitored in large numbers of individual cells (Plates 17.3 and 17.4). Plants produce sufficient light to monitor aequorin with a macro lens (Plate 17.1). High expressing cells can be visualized using a microscope. The camera must be set up for the true photon counting. Each pixel is thus 0 or 1 at 50 Hz giving a range of 0–20 photon per s per pixel. By gating the voltage the range of light intensities can be extended over five orders of magnitude. The Photek camera we have established has either 385 × 288 or 512 × 512 pixels. The software only saves the positive pixels as *x–y* co-ordinates, saving considerably on the memory required. We have also shown that it is possible to image changes in ATP in live cells (Plate 17.3; Sala-Newby *et al.*, 1998). Using two interference filters at 474 and 561 we have shown that two bioluminescent indicators emitting different colours can be imaged in the same cells, e.g. the luciferase–aequorin chimaera (Fig. 17.2) or two reporter genes emitting different colours. Bioluminescence spectra can also be imaged by spraying the spectrum from a diffraction grating on to the photocathode, as seen for firefly and *Renilla* luciferases (Plate 17.2). The photon counting imaging system can also image fluorescence and bioluminescence in the same sample. Photon counting imaging, coupled with bioluminescent indicators, provides a platform technology for drug discovery and clinical assays using imaging of microtitre plates.

17.3.4 Incorporation of indicators into cells and reactivation

Synthetic chemiluminescent indicators such as luminol and luciferin appear to diffuse into cells, though their location in the phagocyte has never been definitively proven. Bioluminescence requires a protein, which cannot diffuse across the plasma membrane. The photoprotein aequorin has been injected successfully into giant cells such as barnacle muscle (Ridgeway & Ashley, 1967), squid axon (Baker *et al.*, 1971) and many other cells (Blinks *et al.*, 1978, 1982) including photoreceptors, several invertebrate and vertebrate muscle and nerve cells, a variety of eggs, *Chironomus* salivary gland, slime mould, protozoa and a few plant cells. The elegant studies of Cobbold and his collaborators (Cobbold & Bourne, 1984; Woods *et al.*, 1986; Cobbold & Rink, 1987) have shown that it is also possible to microinject cells just a few micrometres in diameter, including single myocytes, hepatocytes, and adrenal chromaffin cells. Measurements of ATP in single cells injected with firefly luciferase have also been performed (Koop & Cobbold, 1993; Allue *et al.*, 1996). The availability of computerized microinjectors makes it possible to inject protein, mRNA or cDNA into cells as small as human neutrophils (*c.* 7 μm diameter). However since the genes/cDNA/mRNA coding for bioluminescent proteins have become available, the transfer of genetic material is the method of choice to deliver indicators to cells. Examples of the most popular methods of delivering indicators are listed below.

(1) Microinjection (Blinks *et al.*, 1978; Cobbold & Bourne, 1984);
(2) Transient permeabilization of the plasma membrane: electroporation (Maxwell & Maxwell, 1988);
(3) Liposomes (Campbell *et al.*, 1988; Lee & Huang, 1997);
(4) Use of agents that facilitate DNA uptake such as calcium phosphate and (DEAE)-dextran (Chang, 1994);

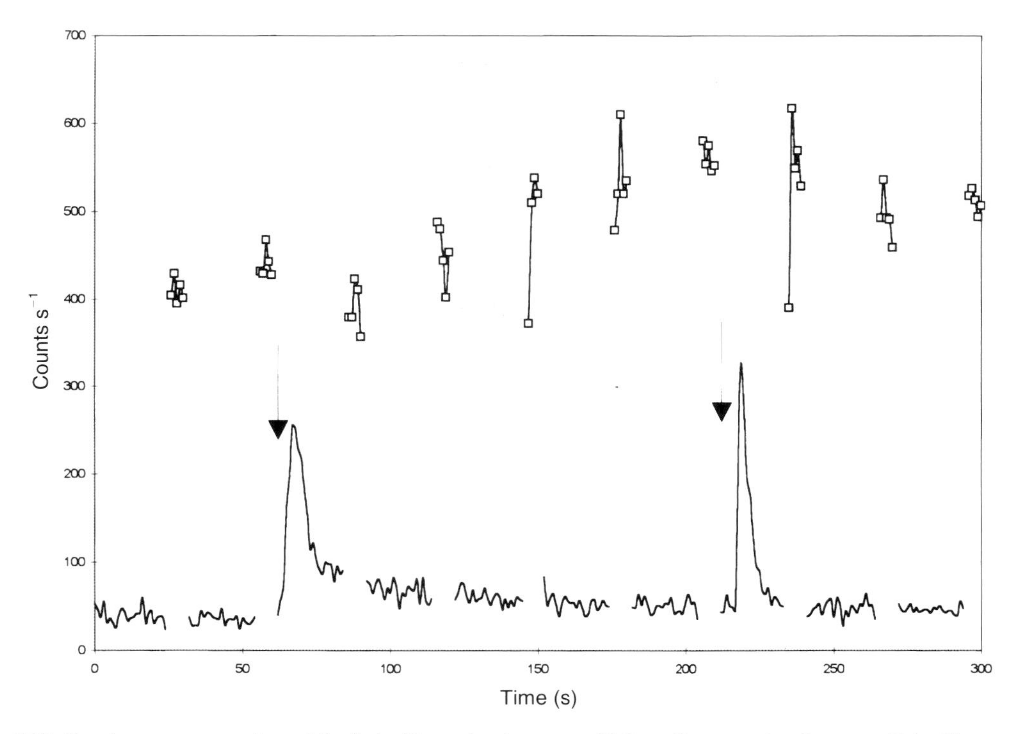

Figure 17.2 Simultaneous aequorin and firefly luciferase luminescence. HeLa cells expressing the cytosolic luciferase–aequorin chimaera were stimulated by 100 μM histamine (first arrow), perfused with Ca^{2+}-containing lysis buffer (indicated by second arrow) at the end of the experiment to calibrate the aequorin signal. Luminescence was measured using the Photek camera after passing through narrow bandpass interference filters of 474 and 561 nm positioned in front of a lens (switch-over readings are not included). The aequorin luminescence (—) monitored intracellular Ca^{2+} changes and the luciferase signal (□) the ATP levels.jklm

(5) Viral vectors: phages, replication-deficient adenovirus and adeno-associated virus and retrovirus (Leber *et al.*, 1996; Kendall *et al.*,1996);
(6) Generation of transgenic organisms (Knight *et al.*, 1993; Rosay *et al.*, 1997; Wei, 1997).

As already discussed, cofactors are also required for bioluminescence. Oxygen is always present, though its K_m for luciferases has been poorly characterized. The luciferin, or long-chain aldehyde in the case of bacterial luciferase, has to be added. Aldehydes are lipid-soluble but toxic. Evaporation from a hanging droplet has been used. Firefly luciferin, though charged with a $-CO_2^-$ group at physiological pH, penetrates into cells very rapidly. External pH may affect this, but light emission from firefly luciferase in COS cells is immediate after luciferin addition (Craig *et al.*, 1991; Sala-Newby & Campbell, 1992). A number of luciferin esters have been synthesized which enable maximum light emission from COS cells to be achieved at concentrations some 10–100 times lower than native luciferin (Craig *et al.*, 1991); however they require the presence of esterases in the cell to become substrates of the luciferase.

Coelenterazine is not needed for aequorin or obelin, but is required for the reactivation of the apoprotein, it penetrates biological membranes readily. However up to 24 h might be required for maximal activity (Shimomura *et al.*, 1988; Campbell & Herring, 1990). Coelenterazine is unstable, being susceptible to photo-oxidation, except when it is bound to photoprotein.

17.3.5 Application of extracted bioluminescent and synthetic chemiluminescent indicators in living systems

There have been three major applications in living systems: measurement of intracellular free Ca^{2+}, oxygen metabolites and ATP. But the most successful of the bioluminescent indicators so far have been the Ca^{2+}-activated photoproteins, aequorin and obelin (Ashley & Campbell, 1979; Blinks *et al.*, 1982; Cobbold & Rink, 1987). Until the successful use of the fluorescent indicators invented by Tsien (Tsien *et al.*, 1982; Grynkiewicz *et al.*, 1985), all the really new information about intracellular free Ca^{2+} in living cells came from studies using the photoproteins.

During the 1970s three ranges of free Ca^{2+} were defined, resting cells = 30–300 nM, stimulated cells ≤1–5 μM and dying cells ≥100 μM. Primary stimuli, such as action potentials, transmitters and hormones, were identified which worked via a rise in intracellular Ca^{2+}. Calcium-independent stimuli, e.g. particles triggering O_2^- in neutrophils (Hallett & Campbell, 1982), were also found even though manipulation of external Ca^{2+} sometimes affected the cell response. Secondary regulators, such as adrenaline in the heart and CO_2 in barnacle muscle, which work by modifying the primary Ca^{2+} transient, were defined. A key role for irreversible cell injury by complement was first identified using obelin (Campbell & Luzio, 1981; Campbell, 1987). The source of intracellular Ca^{2+} for cell activation, external or from an internal store, was first clarified using photoproteins. Ca^{2+} oscillations, Ca^{2+} waves and localized Ca^{2+} transients were first discovered using photoproteins in slime mould, eggs, nerve cells and hepatocytes (Rose & Lowenstein, 1976; Gilkey *et al.*, 1978; Woods *et al.*, 1986). The availability, ease of use and imaging potential of the fluors has apparently put them now as first choice.

Certainly, important new principles as well as much needed details in large numbers of cell types, have come from using the chemically synthesized fluors (Poenie & Tsien, 1986; Davies *et al.*, 1991a,b; this volume). But artifacts can occur. Not all cells load easily with these fluors. It is difficult to use them in bacteria and to target them specifically to organelles or sites within the cell. Also the fluors can overload into subcellular compartments, and binding to internal proteins can alter the affinity constants and other fluorescent characteristics. This cannot be overcome even with a ratiometric fluorescent indicator. The controlled expression and engineering of photoproteins has certainly widened their use as indicators of free Ca^{2+} in living systems (see Sections 17.4 and 17.5). However, a new challenge from the fluorescent field is mounting with the recent development of ratiometric indicators based on GFP chimaeras in which fluorescence resonance energy transfer (FRET) can occur depending on substrate binding (Cubitt *et al.*, 1995a; Heim & Tsien, 1996; Rosomer *et al.*, 1997; Persechini *et al.*, 1997; Miyawaki *et al.*, 1997).

Another area where chemi- and bioluminescent indicators have played an important role has been in identifying cells and organs which produce oxygen metabolites (O_2^-, H_2O_2, OCl^- and $^1O_2$) after exposure to various stimuli or toxins, and in defining the molecular mechanisms responsible (Adam & Cilento, 1982; Campbell, 1988, Chapter 6). Luminous bacteria were first used to monitor O_2 production during photosynthesis nearly a century ago (Beijerinck, 1902). True endogenous ultraweak chemiluminescence is an indicator of oxidative reactions which may damage DNA, protein or membranes. It has been observed in plant and animal cells, including root tips, phagocytes, platelets, liver, lung, brain and fertilized eggs. It has even been imaged. The infrared chemiluminescence of $^1O_2$ at 1268 nm is a useful criterion of singlet oxygen in biological systems. Addition of a chemiluminescent compound which reacts with one or more of the oxygen metabolites responsible for endogenous ultraweak chemiluminescence results in up to 10^4 times the light intensity. This indicator-dependent chemiluminescence can also be imaged. Lucigenin and pholasin from the boring mollusc *Pholas dactylus* are more selective for O_2^- than luminol, which requires access to peroxidase for maximum light emission. Pholasin is sensitive enough to detect O_2^- release from a single human neutrophil (Roberts *et al.*, 1987). It has also been injected into myocytes (Cobbold & Bourne, 1984). Indicator-dependent chemiluminescence is a very convenient method for detection and analysis of abnormalities in oxygen metabolite production in clinical samples.

Firefly luciferase has been extensively used to assess biomass in several industries, including general microbial detection in water, brewing, textiles, milk, fruit juice and food, and in urine. Dead cells have no ATP and also coenzymes are lost. It has even been used to search for life on Mars. These assays are non-specific. *E. coli* cannot be distinguished from *Salmonella* for example. Although apparently very sensitive, ATP from 1 neutrophil (*c.* 10^{-15} mol) and from 50 bacteria (5×10^{-16} mol) being detectable, clinical analysis requires detection of 10^2 bacteria per ml, which still requires filtration. In food, one *Salmonella* in 25 g must be detected if we are to be certain of avoiding food poisoning. Microinjection of firefly luciferase into single ventricular myocytes and hepatocytes has allowed the monitoring of intracellular ATP during metabolic poisoning. Attention to the effect of pH, ionic strength, protein concentration and pyrophosphate allows the correlation of the luminescent signal to the ATP concentration (Koop & Cobbold, 1993: Allue *et al.*, 1996).

Surprisingly, there have been no reports of using bacterial luciferase to monitor NAD(P)H, FMN or aldehydes in live cells. However, an ingenious application of this bioluminescent system has been reported to detect live bacteria as described in Section 17.4.3.2.

Some years ago we developed a homogeneous chemiluminescence immunoassay based on resonance energy transfer (Campbell & Patel, 1983; Campbell *et al.*, 1988; Campbell, 1989). Since the efficiency of energy transfer decreases with the sixth power of the distance we were able to establish the principle of chemiluminescence resonance energy transfer (CRET) as a homogeneous assay for ligand–ligand interaction, which was used to establish assays for cyclic AMP, cyclic GMP, steroids and protein antigens). ABEI cyclic AMP and cyclic GMP derivatives were synthesized that when bound to fluorescein-labelled antibody caused a shift in the colour of light emission from blue to green to occur. This was reversed by displacement with unlabelled cyclic nucleotide. The 'homogeneous' assay quantified cyclic nucleotide by the ratio of light at 460/525 nm over the range 0.1–100 pmol, or 10 nM–10 μM, just right for the cytosol. Unfortunately, ABEI works best at alkaline pH and requires 1 mM H_2O_2 outside the cell! Furthermore, excited states are very susceptible to interference from solvents and other solutes, particularly protein. What was needed, therefore, was a bioluminescent protein label, creating its own 'solvent' within the active centre, linked in an analyte-sensitive way to a fluor mimicking the natural energy transfer which occurs in some coelenterates and deep-sea fish. The availability of the DNA coding for bioluminescent proteins and the Aequorea victoria GFP should allow the recreation of this assay under physiological conditions. The analyte of interest is quantified by measuring the ratio of light emission at two wavelengths simultaneously. We have established the principle of CRET by constructing chimaeras between GFP and aequorin linked by a peptide containing a reactive site (J.P. Waud & A.K. Campbell, unpublished).

17.4 BIOLUMINESCENT REPORTER GENES

17.4.1 Isolation of the genes

Luminous bacteria are very abundant and widely distributed. They can be found free in seawater or growing on dead fish and meat, in the guts and in the light organs of fish and squid, and as parasites. They belong in three genera: *Vibrio* (sp. *harveyi* and *fischeri*), *Photobacterium* (sp. *phosphoreum*, *leiognathi* and *logei*), and the freshwater or soil species *Xenorhabdus* now *Photorhabdus* (*luminescens*) (Campbell, 1988).

Bacterial luciferases catalyse the reaction shown in Fig. 17.1(B) where R is a long aliphatic chain, usually C12–14, and the FMNH2 is generated by an NAD(P)H oxidoreductase present in most pro- and eukaryotic cells.

The luciferase is a 77 kDa chimaeric protein of two non-identical subunits (α and β) coded by adjacent genes luxA and luxB that form part of the regulated lux operon that also comprises the genes coding for fatty acid reductase (lux CDE) and others. The luxA and B genes and other members of the operons of *V. fischeri* (Cohn *et al.*, 1983; Baldwin *et al.*, 1989; Foran & Brown, 1988), *V. harveyi* (Cohn *et al.*, 1985; Johnson *et al.*, 1986), *X. luminescens* (Meighen 1991; Frackman *et al.*,1990; Xi *et al.*, 1991), *P. leiognathi* (Baldwin *et al.*, 1989) and *P. phosphoreum* genes have been cloned (Mancini *et al.*, 1988). There is a high degree of homology between the amino acid sequence of the α (54–88%) and β (45–77%) subunits of luciferase among the different species (Meighen, 1991). A fusion protein for luxAB has also been constructed with a spacer to ensure retention of bioluminescence (Boylan *et al.*, 1991). The lux operon is regulated by an auto-inducer (Swartzman *et al.*, 1990).

Several genes encoding for the coelenterazine containing bioluminescent proteins that require the presence of Ca^{2+} for luminescence have been cloned. Aequorin, the photoprotein from *Aequorea victoria* (189/196 amino acids) was cloned from plasmid cDNA libraries (Inouye *et al.*, 1985; Prasher *et al.*, 1985, 1987). Clones selected contained open reading frames that started at methionine and coded for seven amino acids more than the mature protein. The sequence of several clones and protein sequence data indicate that 24 residues are heterogeneous with two and sometimes more amino acid variants at each position. Three EF hands (Ca^{2+} binding domains) and a hydrophobic domain which contribute to the coelenterazine binding site were identified (Charbonneau *et al.*, 1985; Cormier *et al.*, 1989). The cDNAs for three other members of the Ca^{2+}-dependent photoprotein family have been isolated: clytin (Inouye & Tsuji, 1993), mitrocomin (Fagan *et al.*, 1993) and *Obelia longissima* (Illarionov *et al.*, 1995).

The luminescence in the marine ostracod *Vargula hilgendorfii,* is caused by the oxidation of another imidazolopyrazine luciferin (Fig. 17.1) by molecular oxygen, and catalysed by a luciferase composed of 555 amino acids. It is a secreted protein that has been cloned and expressed in COS cells (Thompson *et al.*, 1989).

Renilla reniformis luciferase (311 amino acids) emits light (λ_{max} = 480 nm) in the presence of coelenterazine and O_2. It has been cloned and the recombinant protein expressed in *E. coli* is indistinguishable from the native luciferase (Lorenz *et al.*, 1991).

Beetle luciferases catalyse the oxidative decarboxylation of a benzothiazole luciferin in the presence of Mg-ATP (Fig. 17.1). The best characterized of the beetle luciferases and first cloned is that from the firefly *Photinus pyralis* (550 amino acids) (De Wet *et al.*, 1985). The luciferase from *Luciola cruciata* (548 amino acids) was next to be cloned (Tatsumi *et al.*, 1989). A cDNA library from the abdominal light organ of the beetle *Pyrophorus phagiophthalamus* yielded four luciferases (553 amino acids) that emitted light with colours ranging from green to orange. (Wood *et al.*, 1989a). At present 12 beetle luciferases have been cloned. The colour of the light emitted by beetle luciferases ranges from green (λ_{max} = 543 nm) to red (λ_{max} = 620 nm)

17.4.2 Uses of bioluminescent genes as reporter genes

The regulation of transcription can be studied after analysis of the structure and concentration of the relevant mRNA, either *in situ* or after its isolation. Direct methods of analysis are time-consuming, destroy the tissue and quantification is laborious, especially for very low abundance or very unstable mRNAs. Only the most sensitive methods of detection work, e.g. ribonuclease protection assay and reverse transcriptase coupled to the polymerase chain reaction (Killen, 1997; Jung *et al.*, 1997). They have been steadily replaced by the use of reporter genes that offer the advantage of allowing the dissection of regulatory regions more closely. The regulatory DNA element under study is ligated 5′ or 3′ to the reporter gene's genomic or cDNA. The construct, usually a plasmid that grows in *E. coli*, includes antibiotic resistance genes, a bacterial origin of replication and a polyadenylation signal. Whether one includes a strong promoter with either wide or restricted tissue specificity, sequences for homologous recombination, a selectable marker, a strong transcription termination signal upstream of the promoter to avoid the generation of transcripts initiated at other sites in the vector will depend on the purpose of the investigation. There are many commercial sources of cassettes tailored to suit different strategies.

17.4.2.1 The ideal reporter gene

The ideal reporter gene should satisfy the following criteria:

(a) *Specificity*. The protein coded by the reporter gene should be absent from the tissue of interest. Several reporter genes have been used (Table 17.2). As shown genes coding for bacterial proteins such as chloramphenicol acetyl transferase (CAT), β-galactosidase (β-Gal), β-glucuronidase, luciferases; as well as eukaryotic genes for proteins such as *Drosophila* alcohol dehydrogenase and several luciferases. Proteins easily distinguishable from tissue proteins can also be used but only if they are present as an isoform or have a very restricted tissue distribution, e.g. globin (Treisman *et al.*, 1983) and placental alkaline phosphatase, or if they are normally secreted into the medium, e.g. human growth hormone and secreted alkaline phosphatase.

Although luminescent mammals do not exist, some cells produce weak chemiluminescence that could interfere with the assay of luminescent proteins, e.g. phagocytes. However, regarding specificity, luminescent proteins have distinct advantages. Heat treatment needs to be given to CAT extracts to remove interfering activities. β-Gal activity is expressed by gut epithelial cells (Sambrook *et al.*, 1989) and placental type alkaline phosphatase is expressed in tissues such as lung, testes, cervix and some malignant cells (Alam & Cook, 1990).

(b) *Sensitivity*. Just a few molecules of the reporter should be necessary to detect its presence. To define precisely the sensitivity of the different reporter genes is difficult, but usually their activity is arbitrarily compared with that of CAT driven by the same promoter in the same cells, with the activity expressed either as the minimum number of cells required in the assay or as the length of time after transfection needed for the reporter protein to be significantly different from background. Estimation of the minimum number of molecules detected by any assay is only valid when the stability of the reporter gene mRNA or its protein have been evaluated, as well as the effect of the extraction procedure, the presence of inhibitors, and the effect of the incubation temperature. The photoprotein aequorin exemplifies this point. A chemiluminometer can detect as little as 10^3 molecules of the protein, but when used as a reporter gene it was not better than CAT (Tanahashi *et al.*, 1990), having a limit of detection of at least 10^7 molecules (Alam & Cook, 1990). The lability of apoaequorin in the cell cytosol (half-life of 20 min) can explain this discrepancy (Badminton *et al.*, 1995). Optimal conditions for extraction and assay have reduced the limit of detection for firefly luciferase to 2000 molecules (Wood, 1991a,b), but in intact cells at least ten times more molecules would be required because anions and pH affect its activity and colour (DeLuca & McElroy, 1978).

(c) *Detectability*. It should be preferably detectable without the destruction of the transfected tissue. One of the most exciting aspects that reporter genes offer is the possibility of detecting low levels of gene expression and how transcriptional regulatory elements are controlled in individual live cells. This allows the examination of the effect of the integration site on expression, as well as quantification of cell heterogeneity and isolation of clonal populations (White *et al.*, 1995). Fluorescent and bioluminescent reporters have been successfully used for single cell analysis. Fluorogenic assays for β-Gal (Nolan *et al.*, 1988) and β-glucuronidase (Jefferson *et al.*, 1987) utilize fluoresceinated substrates which, when cleaved by the respective enzymes, generate fluorescein inside the cells; this enables FACS analysis to detect and

Table 17.2 Reporter genes.

Reporter gene	*Type of assay*	*Tissue preparation*	*Relative sensitivity*[a]	*References*
Chloramphenicol acetyl transferase	TLC/differential extraction/ scintillation counting	Cell extracts	1	Gorman *et al.* (1982), Sambrook *et al.* (1989), Alam & Cook (1990)
Firefly luciferase	Luminometric/scintillation counting	Cell extracts	30–100	De Wet *et al.* (1985), Brasier *et al.* (1989), Nguyen *et al.* (1988)
	Luminometric/ autoradiography	*In vivo*		Gould & Subramani (1988), Wood *et al.* (1989c)
	Imaging camera	*In vivo*		Hooper *et al.* (1990)
Human growth hormone	Radioimmunoassay	Supernatants	10	Selden *et al.* (1986)
Bacterial luciferase	Luminometric/scintillation counter/autoradiography	Cell extracts/ *in vivo*	>1	Legocki *et al.* (1986), Olsson *et al.* (1988), Rogowsky *et al.* (1987)
β-Galactosidase (*E. coli* lac Z gene product)	Spectrophotometric	Cell extracts	<1	Sambrook *et al.* (1989)
	Histochemical	Tissue		
	FACS	*In vivo*		Nolan *et al.* (1988)
	Luminometric	Cell extracts	»1	Bronstein *et al.* (1996)
β-Glucuronidase	Spectrophotometric	Cell extracts	<1	Sambrook *et al.* (1989)
	Histochemical	Tissue		Liu *et al.* (1990)
	Fluorometric	Cell extracts	>1	Jefferson *et al.* (1987)
	FACS	*In vivo*		
	Luminometric	Cell extracts		Bronstein *et al.* (1996)
Alkaline phosphatase (or its secretory variant)	Spectrophotometric/ histochemical	Cell extracts/ tissue	1	Henthorn *et al.* (1988)
	Luminometric	Cell extracts/ supernatants	>1	Schaap *et al.* (1989), Bronstein *et al.* (1996)
Aequorin	Luminometric/imaging camera	Cell extracts/ *in vivo*	1	Tanahashi *et al.* (1990), Rutter *et al.* (1996), Campbell *et al.* (1996)
Vargula luciferase	Luminometric/scintillation counter	Supernatants	>1	Thompson *et al.* (1990)
Renilla luciferase	Luminometric/imaging camera	Cell extracts/ *in vivo*	»1	Lorenz *et al.* (1996), Mayerhofer *et al.* (1995)
Green fluorescent protein		*In vivo*	n.d.	Chalfie *et al.* (1994)

[a] Relative sensitivity: (% sample needed to measure CAT)/(% sample needed to measure reporter); only measurements for cell extracts or supernatants can be compared.
Relative sensitivity is >1 means that the assay is more sensitive than that for CAT but the exact figure is not known or it is based on the number of molecules of pure reporter protein that can be detected being less than the 10^7–10^8 molecules that can be detected by the CAT assay. Relative sensitivity is <1 means the reverse.

isolate individual expressing cells. Single cells expressing firefly luciferase under the control of a *Vaccinia* promoter have been detected using CCD imaging (Hooper *et al.*, 1990). Firefly luciferase activity has also been detected *in vivo* in bacterial and yeast populations but not in single cells; in plants (Ow *et al.*, 1987) and in transfected cells (Rodriguez *et al.*, 1988; Pons *et al.*, 1990; White *et al.*, 1995). Some cells required permeabilization to allow in the luciferin (Gould & Subramani, 1988). When permeability to luciferin is a problem, it may be possible to solve this by using luciferin esters (Craig *et al.*, 1991), however we and others have not experienced any problems (Sala-Newby & Campbell, 1992; Koop & Cobbold, 1993; Allue *et al.*, 1996; Rutter *et al.*,1995). Bacterial luciferase has also been detected *in vivo* (Table 17.2).

(d) *Amenable to tissue localization*. The use of histochemical stains has allowed researchers to look at gene expression in individual cells, but only after fixation. β-Gal has been used extensively to study mammalian cells (see Nielsen & Pedersen, 1991 for references) and β-glucuronidase has proved useful in plants (Jefferson *et al.*, 1987). Alkaline phosphatase has also been used but its application is limited to tissues with low levels of endogenous expression or very strong promoters (Henthorn *et al.*, 1988), e.g. *Drosophila* alcohol dehydrogenase (Neilsen & Pedersen, 1991). Luminescent proteins can be detected *in vivo*. Transgenic tobacco plants carrying the firefly luciferase gene showed luciferase activity in various organs and whole plants after soaking with luciferin and exposure to X-ray film (Ow *et al.*, 1987). Bacterial luciferase can be expressed in plants (Table 17.2) but *in vivo* measurements are hampered by the low levels of $FMNH_2$ (Koncz *et al.*, 1987).

(e) *Ease of use*. The assay should be simple, inexpensive, use commercially available reagents, fast to perform, and the concentration of the reporter gene product should closely reflect transcriptional activity. Thus the reporter should monitor both the switching on and off of the response element being investigated. This requires a knowledge of the half-life of the indicator. Long-lived indicators such as CAT, β-Gal and GFP are not as flexible as bioluminescent reporters (Badminton *et al.*, 1995). CAT is still a widely used reporter gene, perhaps because of the widespread availability of constructs. But the assay is cumbersome, requires radioactive substrate and it is expensive. Alam & Cook (1990) have estimated the cost of the most common reporter genes used. Optimal conditions to assay firefly luciferase in extracts (Brasier *et al.*, 1989; Wood, 1991a,b) have been developed. The possession of a luminometer is not essential for the use of bioluminescent reporter proteins except when the photoprotein aequorin is used. All the luciferases and the most sensitive assays for alkaline phosphatase, β-Gal, β-glucuronidase (detect 10^3–10^4 molecules) by dioxetan-enhanced chemiluminescence (Schaap *et al.*, 1989; Bronstein *et al.*, 1996) can be measured using a scintillation counter (Nguyen *et al.*, 1988). However, the real potential of bioluminescent genes as reporters can only be realized by detection using a photon-counting image intensifier with a contact imaging system or attached to a microscope (Hooper *et al.*, 1990; Campbell *et al.*, 1996; White *et al.*, 1995; Rutter *et al.*, 1995).

17.4.1.2 *Use of dual reporter systems*

In order to standardize results from different experiments it is common practice to co-transfect with a second reporter. The reporters that code for secreted proteins, even if not completely specific, are useful to correct for transfection efficiency because they do not compete for precious transfected cells. To this group belong (Table 17.2) human growth hormone, secreted alkaline phosphatase and *Vargula* luciferase (the use of this luciferase is limited because the luciferin is not commercially available).

Combinations of firefly luciferase with CAT, β-Gal and β-glucuronidase are commonly used too. The development of very sensitive chemiluminescent assays for the last two should prove popular (Bronstein *et al.*, 1996) and they are commercially available from Clontech and Tropix. Promega has recently introduced a Dual luciferase™ system pairing firefly and *Renilla* luciferases, both detectable sequencially in a single sample using a luminometer. Since the colour of light emitted is different (565 nm and 480 nm, respectively) the method can also be applied to live cells using an imaging camera fitted with a spectrometer (Plate 17.2) or with interference filters to select two wavelengths (Fig. 17.2) making it a useful addition to the field of real time transcriptional studies.

Further applications of bioluminescent reporter gene technology are included in Section 17.4.3.2.

17.4.3 Examples of the many uses of bioluminescent genes

17.4.3.1 *As markers of recombinant events and gene delivery protocols*

A cloning vector has been developed based on the loss of bioluminescent phenotype. A multiple cloning site was inserted into the luxB gene in a vector containing the *luxA* and *B* genes from *V. harveyi* (pLUM), the *E. coli* ori and phage f ori that specifies single-stranded DNA synthesis. Transfection into *E. coli* generated luminous bacteria and the insertion of a foreign gene destroyed the luminous phenotype. Light generated by live bacteria was detected on X-ray film. This system seems superior to the IPTG/X-gal/β-Gal system (Sevigny & Gossard, 1990). The future of human gene therapy relies on the development of efficient and safe methods to deliver genes to a variety of cell types and bioluminescent indicators are useful to monitor the efficacy of a particular protocol (Yu *et al.*, 1996).

17.4.3.2 In vivo *indicators of cellular functions*

The application of bioluminescence in microbiology, endocrinology and toxicology, for research and industry, is a rapidly developing field hence some examples are appropriate to encourage further use.

The introduction of lux genes into bacteriophages followed by infection of a bacterial host leads to the expression of the genes and the generation of the luminescent phenotype. *Salmonella*, *Listeria* and enterics as a group have been targeted (references in Jassim *et al.*, 1990). Engineered bioluminescent bacteria have been used to monitor for biocides and sublethal injury and recovery because only live bacteria carry the metabolites required for bacterial luciferase luminescence and bacteriophages carrying the luxAB genes have been used to screen for virucidal activity (Jassim *et al.*, 1990). The same group engineered promoter elements for *Lactobacillus casei* upstream from the *luxA* and *B* genes from *V. fischeri*. Transfected *Lactobacillus* species became luminescent and have been used to monitor antimicrobial activity in milk and gene expression in lactic acid bacteria in fermentation systems *in situ* (Ahmed & Steward, 1991).

The use of a firefly luciferase-expressing strain of *Mycobacterium tuberculosis* to evaluate antimycobacterial activity in infected mice promises to speed up the *in vivo* monitoring of antibiotics for slow-growing pathogens (Hickey *et al.*, 1996). Intercellular communication between Gram(−) pathogens is mediated by small molecules which control virulence gene expression. Since autoinduction of the bacterial luciferase genes is also mediated by an *N*-acylhomoserine lactone, a recombinant *E. coli* biosensor based on lux genes was used to measure the presence of autoinducers in cultures of non-luminous *Vibrio* pathogens (Milton *et al.*, 1997).

Pathogenic and symbiotic plant bacteria have been given a luminous phenotype by transfection with a plasmid carrying the lux operon from *V. fisheri* without decreasing their virulence. Autoradiography was then used to monitor the infection of the plant hours or days before the symptoms became visible. It has also been proposed that genetically engineered microorganisms could be released into the environment tagged with *lux* genes to allow aerial detection (Shaw & Kado, 1986). Bacterial and firefly luciferases have also been used to detect genetically engineered microorganisms in soil, 10^2–10^3 cells ml^{-1} being detectable (Rattray *et al.*, 1990; Lindow, 1995). A non-invasive, non-destructive, rapid, *in situ* and population-specific method for naphthalene exposure and biodegradation has also been developed using bacterial lux genes. The lux genes have been inserted in the *nahG* gene encoding an enzyme induced by naphthalene. Bacteria carrying the plasmid developed luminescence when exposed to naphthalene and soil slurries containing polycyclic aromatic hydrocarbons (King *et al.*, 1991). The promoters of a series of genes responsible for the degradation of aromatic hydrocarbons were coupled to luciferase cDNA and the recombinant *E. coli* strains generated were used to detect the presence of benzene-related pollutants when incorporated into a fibreoptic device (Ikariyama *et al.*, 1997).

A recombinant baculovirus containing the firefly luciferase cDNA under the control of the polyhedrin promoter has been used to follow the dissemination of this virus in live caterpillars (Jha *et al.*, 1990). *Staphylococcus aureus* engineered to express firefly luciferase glow on addition of luciferin, and since they express protein A on their surface they can be used to detect immobilized antigen–antibody complexes (Steidler *et al.*, 1996).

Oestrogen receptor positive cell lines, stably transfected with a plasmid that allows expression of firefly luciferase under the control of the oestrogen regulatory element of the vitellogenin A2 gene, have been used to develop an *in vivo* test for oestrogen and anti-oestrogen molecules. Luciferase was detected in extracts and *in vivo* using a CCD camera (Pons *et al.*, 1990). For further developments in this field, see Joyeux *et al.* (1997).

An assay for the human immunodeficiency virus (HIV) and antiviral drugs has also been developed. A plasmid carrying firefly luciferase cDNA under the control of HIV was transfected into several cell lines. Challenging with HIV resulted in luciferase expression (Schwartz *et al.*, 1990).

In the process of targeting firefly luciferase to the cytosol it was found that the last seven amino acids at the *C*-terminus of firefly luciferase were not required for activity but the stepwise removal of the last 8–12 caused progressive loss of activity. Since replacement of the *C*-terminus by other peptides restored some activity, the potential of this region for engineering was realized (Sala-Newby & Campbell, 1994). The proposal was shown to be correct since a luciferase variant containing a thrombin recognition sequence LVPRES at the *C*-terminus was shown to lose activity when treated with the protease and a kinase indicator was also developed (Waud *et al.*, 1996; Section 17.5.2.1).

A firefly luciferase reporter system developed to investigate the expression of chlorophyll binding proteins in *Arabidopsis* seedlings allowed imaging their circadian rhythms of expression and helped in the understanding of their regulation (Millar *et al.*, 1995).

17.4.3.3 As labels for immunoassays and DNA

Several groups have coupled aequorin, firefly and bacterial luciferase to antibodies or protein A either chemically (Erikaku *et al.*, 1991) or by genetic manipulation (Casadei *et al.*, 1990; Zenno & Inouye, 1990; Lindbladh *et al.*, 1991; Kobatake *et al.*, 1993), or have biotinylated aequorin (Zatta *et al.*, 1991) for use in immunoassays. A simple ELISA method that can detect *Salmonella* contamination in food using biotinylated recombinant aequorin has been reported and compared favourably with other alternatives (Smith *et al.*, 1991).

Biotinylated recombinant firefly and *Renilla* luciferases and aequorin have been used as probes to detect proteins and nucleic acids in blots. Firefly luciferase gave the most sensitive results, comparable to using alkaline phosphatase (Stults *et al.*, 1991). A bioluminescent solid phase system using bacterial luciferase has been used to detect DNA and single base mutations obtained by polymerase chain reaction (Balaguer *et al.*, 1991).

Bioluminescent labels for immunoassay or DNA analysis can be at least as sensitive as the best synthetic labels, acridinium esters (Weeks *et al.*, 1983a,b; Woodhead & Weeks, 1991) or adamantane dioxetans substrates for alkaline phosphatase and other enzymes (Schaap *et al.*, 1989; Bronstein *et al.*, 1996). These are just as good or better than luminol-enhanced peroxidase, producing a glow which can be recorded on sensitive (20 000 ASA) photographic film.

The ability to produce 100 mg of recombinant bioluminescent protein from just a few litres of bacterial culture opens the way for exploitation of the exquisite sensitivity of these immunoassay and DNA probes in clinical assays (Zatta *et al.*, 1991; Lizano *et al.*, 1997).

17.4.3.4 As models in protein folding assays

Firefly and bacterial luciferases have been used in *in vitro* assays designed to disect the chaperoning activities present in the eukaryotic cytosol (Nimmesgern & Hartl, 1993; Kruse *et al.*, 1995).

17.5 BIOLUMINESCENT INDICATORS FOR MOLECULAR SIGNALLING IN LIVE CELLS

17.5.1 Calcium measurements using recombinant aequorin

17.5.1.1 Calcium in subcellular compartments measured using targeted aequorin

The mobilization and turnover of intracellular Ca^{2+} is an integral aspect of cell regulation during responses to a number of different classes of extracellular stimuli. Extensive use of Ca^{2+} chelator dyes (e.g. Fura-2 and BAPTA) has resulted in an enormous body of work both in populations and at the single cell level, part of which reveals that agonists produce many different patterns of change in cytosolic $[Ca^{2+}]$ (Berridge, 1993; Clapham, 1995). The spatial and temporal complexities of intracellular Ca^{2+} changes presumably reflect a need for specification of signals conveyed by a wide array of agonists which must operate in many different phenotypes (Berridge, 1997a,b). Control of cytosolic Ca^{2+} signals involves storage of Ca^{2+} inside intracellular organelles, stimulus-activated release and efficient uptake of Ca^{2+} from outside the cell (Berridge, 1995; Clapham, 1995).

The cloning of the cDNA for aequorin has given the opportunity to develop techniques based on the genetic transformation of bacteria, yeast, plants and animal cells with the gene encoding apoaequorin. The first measurements of cytosolic Ca^{2+} transients were made with recombinant photoprotein, to stimuli as diverse as chemotactic peptide, wind (plants), complement attack (bacteria) and mating pheromones (yeast) (Campbell *et al.*, 1988; Knight *et al.*, 1991a,b; Nakajima-Shimada *et al.*, 1991a,b). Light signals have been imaged and equated to cytosolic Ca^{2+} signals in transgenic multicellular organisms expressing recombinant aequorin (Cubitt *et al.*, 1995b; Campbell *et al.*, 1996; Plate 17.1), and in an organotypic context by restricting expression of the transgene to specific cell types (Rosay *et al.*, 1997). The rise in cytosolic free Ca^{2+} triggered by the formation of the membrane attack complex of complement on HeLa cells expressing the recombinant cytosolic fusion protein of firefly luciferase and aequorin was imaged (Badminton *et al.*, 1995; Plate 17.3). These experiments confirmed that the Ca^{2+} rise preceded any other changes by at least 60 s and showed the response of cells to be heterogeneous (also see Section 17.5.2.2).

However, the most exciting development to have arisen as a result of the cloning of the cDNA for apoaequorin, has been the development of methods to measure $[Ca^{2+}]$ in discrete subcellular domains such as the lumen of the mitochondria, nucleus and endoplasmic reticulum (ER). The advancement of biotechnology has resulted in an understanding of how proteins are efficiently targeted and imported both into organelles and to discrete subcellular localizations (Goldfarb, 1992). Targeting of aequorin to defined subcellular domains has been achieved either with minimal targeting sequences or with resident organellar proteins which already include the information necessary for its proper localization (Table 17.3). Chimaeras of organellar proteins or of targeting sequences and apoaequorin can be generated easily by standard cloning techniques and introduced into cells by transient or stable transfection (Rizzuto *et al.*, 1994). This offers clear advantages over alternative methods (e.g. fluorescent dyes) which diffuse throughout the cytoplasm, and display no selective localization. Table 17.3 lists details of the strategies employed to target aequorin to various organelles.

The first successful investigation of Ca^{2+} homeostasis in living cells at the intracellular organelle level was made in the mitochondria and the endoplasmic reticulum (Table 17.3). Rizzuto *et al.* (1992) showed that stimulation of receptors coupled to the generation of IP_3 produced increases in intramitochondrial

Table 17.3 Aequorin targeting strategies.

Subcellular localization	*Cell type*	*Target strategy*	*Reference*
Cytosol	COS7	None	
	COS7	Luciferase chimaera	Badminton *et al.* (1995)
	293	None	
Mitochondria	HeLa, CHO	Leader sequence of cytochrome *c* oxidase subunit VIII	Rizzuto *et al.* (1992), Rutter *et al.* (1996)
Nucleus	HeLa	Part of glucocorticoid receptor (117 residues)	Brini *et al.* (1993)
	HeLa	Nucleoplasmin chimaera	Badminton *et al.* (1996)
	Trypanosoma brucei	L25 signal motif (6 residues)	Xiong & Ruben (1996)
ER	COS7	Calreticulin signal motif (17 residues)/KDEL	Kendall *et al.* (1992b)
	293	Invariant chain of MHC-II	Button & Eidsath (1996)
	HeLa	Immunoglobulin heavy chain chimaera	Montero *et al.* (1995, 1997)
Sarcoplasmic reticulum	Cardiac myotubes	Calsequestrin chimaera	Brini *et al.* (1997)
Golgi	293	Galactosyltransferase chimaera	Button & Eidsath (1996)
Plasma membrane	*Xenopus* oocytes	5-HT_{1A} receptor chimaera	Daguzan *et al.* (1995)
	Ar57	SNAP-25 chimaera	Marsault *et al.* (1997)
	Cos7/HeLa	Connexin fusion	Martin *et al.* (1998)

[Ca^{2+}], required for the activation of intramitochondrial enzymes such as pyruvate dehydrogenase (Rutter *et al.*, 1996), which significantly exceeded the accompanying rise in cytosolic [Ca^{2+}]. Changes in intranuclear [Ca^{2+}] were also shown to be different from those occurring in the cytosol (Badminton *et al.*, 1996), and may be involved in controlling key nuclear events (Santella, 1996).

Another discrete subcellular domain of Ca^{2+} which was shown by targeted aequorin to be at a significantly higher concentration than that in the bulk cytosol was identified directly beneath the plasma membrane (Marsault *et al.*, 1997); this may play a role in intercellular communication via gap junctions (Martin *et al.*, 1998).

Despite these pioneering experiments, measuring [Ca^{2+}] either in the cytosol or in discrete subcellular locals with appropriately targeted recombinant aequorin has not been without problems. This is exemplified by the problems encountered when measuring [Ca^{2+}] in the ER. From the point of view of cellular homeostasis, the ER provides a major store of rapidly mobilizable $Ca^{2+}$and controls an array of important cellular functions including protein biosynthesis and degradation (Llewellyn *et al.*, 1996) and permeability of the plasma membrane to Ca^{2+} (Berridge, 1995). A number of different strategies have been adopted to find a suitable way to employ recombinant aequorin as an ER Ca^{2+} probe. Addition of the -KDEL sequence to the *C*-terminus of the protein aids ER targeting but causes an increase in the Ca^{2+}-independent light emission, rendering calibration of the light signal difficult (Kendall *et al.*, 1992b, 1994). However despite the adoption of alternative targeting strategies which do not involve modification of the *C*-terminus (Montero *et al.*, 1995; Button & Eidsath, 1996), the affinity of the protein for Ca^{2+} was too high to measure steady state ER [Ca^{2+}] values, even using an aequorin engineered to lower its affinity for Ca^{2+} (Kendall *et al.*, 1992a,b, 1994). In addition, a number of the characteristics of aequorin *per se* make it less amenable to single cell imaging (Rutter *et al.*, 1996), for example the Ca^{2+}-dependent light emission per molecule is low (approximately 0.2 photons per molecule compared with 0.8 for GFP (Cubitt *et al.*, 1995a); other factors include the Ca^{2+} dependent consumption of the photoprotein and the requirement of the exogenous cofactor, coelenterazine.

A variety of coelenterazine analogues have been synthesized and used to activate apoaequorin, generating semisynthetic aequorins which have different light-emitting properties and Ca^{2+} sensitivities from normal aequorin (Shimomura *et al.*, 1988, 1993). Knight *et al.* used *h*-coelenterazine, which generates an aequorin which is more sensitive to [Ca^{2+}], to monitor changes in [Ca^{2+}] in different tissues of transgenic *Nicotiana* plants which were too small to measure with conventional recombinant aequorin (Knight *et al.*, 1993). Conversely, *n*-coelenterazine, an analogue which lowers the affinity of aequorin for Ca^{2+}, has been used in conjunction with an engineered low Ca^{2+} affinity recombinant apoaequorin (Kendall *et al.*, 1992b), in an attempt to improve the stability of aequorin targeted to the ER (Montero *et al.*, 1997). Furthermore, another coelenterazine analogue (*e*-coelenterazine) confers a bimodal spectrum of luminescence emission on recombinant aequorin (Shimomura, 1995). The use of dual-wavelength luminometry then provides a simple method for calibration of Ca^{2+} concentrations without requiring information of how much aequorin has been expressed, reconstituted or consumed, unlike the calibration at a single wavelength for standard aequorin (Cobbold & Lee, 1991; Knight *et al.*, 1993). Despite the attraction of using semisynthetic aequorins to measure Ca^{2+} over a wider range than that encountered, for example in the cytosol, there has, however, been only limited application of these analogues as they are less stable, have shorter half-lives and significantly reduced quantum yields compared with natural aequorin (Shimomura *et al.*, 1988, 1989).

17.5.1.2 Calcium homeostasis in bacteria

Compared with eukaryotes, the role of Ca^{2+} in prokaryotes has not been well defined, although there is extensive but indirect evidence that Ca^{2+} plays a part in their cell biology (Norris *et al.*, 1996). Specific roles for Ca^{2+} in bacterial physiology have proved difficult to determine due to the problems arising from measuring and manipulating free Ca^{2+} in live bacteria. For example, fura-2 has been used to measure free Ca^{2+} in *E. coli* (Gangola & Rosen, 1987; Tisa & Adler, 1995) but the fluors are difficult to load, leak out of cells and may affect Ca^{2+} buffering. Recombinant aequorin has been used to monitor free Ca^{2+} in live bacteria (Knight *et al.*, 1991b; Watkins *et al.*, 1995) at aequorin concentrations of less than 10 μM, an order of magnitude less than normally used for fluors. The recombinant aequorin is reconstituted in the living bacteria with the addition of coelenterazine, providing a simple method for measuring bacterial cytosolic free Ca^{2+} in response to a variety of external stimuli. Furthermore, the multi-host-range vector pMMB66EH (Furste *et al.*, 1986; Watkins *et al.*, 1995) yielded good expression of aequorin in a variety of strains of *E. coli* including mutants defective in components of the Ca^{2+} handling system (H.A. Jones & A.K. Campbell, unpublished). Additionally, the β-lactamase signal peptide engineered onto the *N*-terminus of aequorin (Kendall *et al.*, 1992b) will be used to establish the role of the periplasmic space in Ca^{2+} regulation.

17.5.2 Firefly luciferase-based indicators

17.5.2.1 Indicators for cyclic AMP-dependent protein kinase (protein kinase A)

cAMP has been shown to increase in response to varied hormonal stimulation to control cell metabolism. The majority of its actions are mediated by a family of cAMP-dependent protein kinases (Walsh & VanPatten, 1994) that recognize determinants present in the primary structure of substrate proteins (Kennelly and Krebs, 1991). A consensus recognition sequence for protein kinase A: RRXS was engineered at the VRFS (217–220) site of firefly luciferase by the mutation V217 to R. The RRFS luciferase mutant had a specific activity some 80% less than the normal protein and emitted greener light. Phosphorylation by protein kinase A reduced the light intensity by >80%, which was reversed by addition of a phosphatase. This indicator expressed in COS7 cells provided for the first time a means to assess protein kinase activity in living cells (Sala-Newby & Campbell, 1991, 1992). However, it was not possible to insert other peptides into this site without destroying activity. Another indicator was obtained by incorporation of the protein kinase A recognition sequence, the heptapeptide kemptide (LRRASLG), at the *C*-terminus of firefly luciferase. Luciferase activity decreased when incubated with the catalytic subunit of the kinase (Section 17.4.3.2; Waud *et al.*, 1996).

17.5.2.2 Targeting luciferase to subcellular compartments to measure ATP

Firefly luciferase resides in the peroxisomes of the insect light organ and contains a peroxisomal targeting signal at the *C*-terminus, originally thought to reside in the last 12 amino acids (Keller *et al.*, 1988). Removal of just the last three amino acids, known to be the key signal (Gould & Subramani, 1987), did not affect activity (Sala-Newby *et al.*, 1990). Luciferase minus three amino acids expressed in COS cells was uniformly distributed throughout the cytosol (Sala-Newby & Campbell, 1992).

The fusion protein luciferase–aequorin was used to image and measure changes in ATP and cytosolic free Ca^{2+} in HeLa cells attacked by complement using a Photek camera. Nuclear fluorescence of propidium was used as a measure of permeability to small molecules, and luciferase activity imaged at the end of the experiment to assess lysis. The changes in luciferase light emission (Plate 17.4) were dependent on the concentration of the terminal component of complement (C9) and the cell responses were heterogeneous. The calculated decrease in ATP concentration was 93%, 58% or 56% at C9 doses of 4, 2, or 1 $\mu g\ ml^{-1}$, respectively. The ATP concentration of the control cells was 4.3 mM and assuming that did not change during the time of the experiment, these reductions corresponded to calculated final concentrations of ATP of 0.3, 1.8, or 1.9 mM, respectively (Sala-Newby *et al.*, 1998). This confirmed the original report of Allue *et al.* (1996) that the apparent K_m for ATP increased some 10-fold inside cells compared with the previously published values from luciferase *in vitro*.

Using the approach described in Section 17.5.1.1, firefly luciferase has been targeted to the cytosolic side of mitochondria to measure ATP in live yeast (Aflalo, 1990) and to the lumen of the ER to monitor the effect of the ATP levels on protein folding (Dorner & Kaufman, 1994); however, in both cases no calibration of the signal was attempted.

17.6 CONCLUSIONS AND FUTURE PROSPECTS

17.6.1 Chemiluminescence versus fluorescence

The debate as to whether a chemiluminescent or fluorescent probe is best for a particular application continues. Certainly, because of the poor turnover numbers of bioluminescent proteins (usually around 1 s^{-1}), the number of absolute photons emitted from bioluminescent indicators is less than that from fluors. However under certain circumstances bioluminescence can win out in terms of signal-to-noise. The luciferase reported genes are still the best for studying response element control, particularly because of their short half-lives compared with CAT, β-Gal and GFP. The strength of bioluminescence is also obvious in free solution assays in microtitre plates, where CRET is likely to be better than FRET. Only when fluors are in a microcuvette, such as a cell in the cell sorter, do fluors win out. CRET and FRET open the door to a platform technology for measuring signals such as cyclic AMP, IP_3 and IP_4, covalent modification such as phosphorylation or proteolysis, and protein–protein interactions by a change in energy transfer using a 'rainbow protein' (Campbell, 1989). Targeting of bioluminescent proteins has established how unique information can be gathered about the chemistry of living cells without the need to generate spatially resolved images. In view of the considerable activity with the targeting of GFP, its mutants and FRET, the question now arises whether it will be possible to develop an equivalent platform technology to that established using bioluminescence. This debate can only be resolved by experiment, but CRET combined with targeting certainly has considerable potential. The two indicator types work in symbiosis.

A further advantage of bioluminescence is the ability to measure and image global signals in large numbers of individual cells, whole organs and organisms. This has already been shown with *Drosophila*, fish and plants transgenic with aequorin or firefly luciferase. Imaging of Ca^{2+} signals from whole seedlings showed differential sensitivity to cold shock and a communication system between roots and leaves. This was only found because of the ability to image signals from the whole plant or leaf. Ca^{2+} signals in whole hearts and brains should be possible. However, two-photon confocal microscopy will allow imaging in tissue slices.

17.6.2 Simultaneous monitoring of more than one indicator in a cell

The development of multicoloured bioluminescent probes (rainbow proteins) such as luciferase–aequorin or GFP–aequorin open up the potential for ratiometric bioluminescent indicators and for monitoring more than one substance in the same cell at the same time. In order to exploit this, it must be possible to measure bioluminescence spectra in real-time, even when the light intensity is rapidly decaying. The use of a monochromator with an imaging camera enables this to be achieved (Plate 17.2).

17.6.3 The future

The engineering of bioluminescent proteins has established a new era for these indicators inside cells. New bioluminescent systems are likely to become available by cloning of proteins from deep sea organisms. The X-ray crystal structures of two bioluminescent systems, bacterial luciferase (Fisher *et al.*, 1996) and firefly luciferase (Conti *et al.*, 1996) are now available. This should ultimately provide a sound basis for developing a strategy for engineering sites into bioluminescent proteins. 3-D structural information will also establish whether the evolution of bioluminescence, and its puzzling phylogenetic distribution are simply dependent on the generation of the appropriate solvent cage. Bioluminescent proteins may thus provide a model for the first threshold in the evolution of an enzyme. Bioluminescence, and the probes developed from its components, offer great potential for investigating one of the key problems in biology – whether a process is digital or analogue.

REFERENCES

Adam W. & Cilento G. (1982) *Clinical and Biological Generation of Excited States.* Academic Press, New York.

Aflalo C. (1990) *Biochemistry* **29**, 4758–4766.

Ahmed K.A. & Steward G.S.A.B. (1991) *J. Appl. Bacteriol.* **70**, 113–120.

Alam J. & Cook J.L. (1990) *Anal. Biochem* **188**, 245–254.

Allue I., Gandelman O., Dementieva E., Ugarova N. & Cobbold P. (1996) *Biochem. J.* **319**, 463.

Ashley, C.C. & Campbell, A.K. (eds) (1979) The Detection and Measurement of Free CA^{2+} in Cells. Elsevier/North Holland, Amsterdam.

Badminton M.N., Sala-Newby G.B., Kendall J.M. & Campbell A.K. (1995) *Biochem. Biophys. Res. Commun.* **217**, 950–957.

Badminton M.N., Campbell A.K. & Rembold C.M. (1996) *J. Biol. Chem.* **271**, 31 210–31 214.

Baker P.F., Hodgkin A.L. & Ridgeway E.B. (1971) *J. Physiol.* **218**, 708–755.

Balaguer P., Terouanne B., Boussioux A.M. & Nicolas J.C. (1989) *J. Biolumin. Chemilumin.* **4**, 302–309.

Balaguer P., Terouanne B., Boussioux A.M. & Nicolas J.C. (1991) *J. Biolumin. Chemilumin.* **195**, 105–110.

Baldwin T.O., Devine J.H., Heckel R.C., Lin J.W. & Shadel G.S. (1989) *J. Biolumin. Chemilumin.* **4**, 326–341.

Beijerinck M.W. (1902) *Proc. Acad. Sci. Amsterdam* **4**, 45–49.

Berridge M.J. (1993) *Nature* **361**, 315–325.

Berridge M.J. (1995) *Biochem. J.* **312**, 1–11.

Berridge M.J. (1997a) *J. Physiol.* **499**, 291–306.

Berridge M.J. (1997b) *Nature* **386**, 759–760.

Blinks J.R., Mattingly, P.H., Jewel B.R., Van Leauvan M., Harren G.C. & Allen D.G. (1978) *Methods Enzymol.* **57**, 292–328.

Blinks J.R., Wier W.G., Hess P. & Prendergast F.G. (1982) *Prog. Biophys. Mol. Biol.* **40**, 1–114.

Boylan, M.O., Pelletier, S., Dhepagnon, S., *et al.* (1991) *Chemilum.* **4**, 310–316.

Brasier A.R., Tate J.E. & Habener J.F. (1989) *Biotechniques* **7**, 1116–1122.

Brini M., Murgi M., Pasti L., Picard D., Pozzan T. & Rizzuto R. (1993) *EMBO J.*, **12**, 4813–4819.

Brini M., De Giorgi F., Murgia M., Marsault R., Massimino M.L., Cantini M., Rizzuto R. & Pozzan T. (1997) *Mol. Biol. Cell* **8**, 129–143.

Bronstein I., Martin C.S., Fortin J.J., Olesen C.E.M. & Voyta J.C. (1996) *Clin. Chem.* **42**, 1542–1546.

Button D. & Eidsath A. (1996) *Mol. Biol. Cell* **7** 419–434.

Campbell A.K. (1987) *Clin. Sci.* **72**, 1–10.

Campbell A.K. (1988) *Chemiluminescence: Principles and Applications in Biology and Medicine.* Horwood/VCH, Chichester & Weinheim.

Campbell A.K. British Patent Application 8916806.6. International Patent Application PCT/GB90/01131 (1989) and (USA patent 5683888 in 1997).

Campbell A.K. & Luzio J.P. (1981) *Experientia* **37**, 1110–1112.

Campbell A.K. & Patel A. (1983) *Biochem J.* **216**, 185–194.

Campbell A.K., Patel A.K., Razavi, Z.S. & McCapra F. (1988) *Biochem. J.* **252**, 143–149.

Campbell A.K. & Herring P.J. (1990) *Mar. Biol.* **104**, 219–225.

Campbell A.K., Trewavas A.J. & Knight M.R. (1996) *Cell Calcium* **19**, 211–218.

Casadei J., Powell M.J. & Kenton J.H. (1990) *Proc. Natl Acad. Sci. USA* **87**, 2047–2051.

Chalfie M., Tu Y., Euskirchen G., Ward W.W. & Prasher D.C. (1994) *Science* **263**, 802–805.

Chang P.L. (1994) In *Gene Therapeutics*, J.A. Wolf (ed). Birkhauser, Boston, pp. 157–179.

Charbonneau H., Walsh K.A., McCann R.O., Prendergast F.G., Cormier M.J. & Vanaman T.C. (1985). *Biochemistry* **24**, 6762–6771.

Clapham D.E. (1995) *Cell*, **80**, 259–268.

Cobbold P.H. & Bourne P.K. (1984) *Nature (Lond.)* **312**, 446–448.

Cobbold P.H. & Lee J.A.C. (1991) In *Cellular Calcium: a Practical Approach*, J.G. McCormack & P.H. Cobbold (eds). Oxford University Press, New York, pp. 55–80.

Cobbold P.H. & Rink T.J. (1987) *Biochem J.* **248**, 313–318.

Cohn D.H., Mileham A.J., Simon M.I., Nealson K.H., Rausch S.K., Bonam D. & Baldwin T.O. (1985) *J. Biol. Chem.* **260**, 6139–6146.

Cohn D.H., Ogden R.C., Abelson J.N., Baldwin T.O., Nealson K.H., Simon M.I. & Mileham A.J. (1983) *Proc. Natl Acad. Sci. USA* **80**, 120–123.

Conti E., Franks N.P. & Brick P. (1996) *Structure* **4**, 287–298.

Cormier M.J., Prasher D.C., Longiaru M. & McCann R.O. (1989) *Photochem. Photobiol.* **49**, 509–512.

Craig F.F., Simmonds A.C., Watmore D. & McCapra F. (1991) *Biochem. J.* **276**, 637–641.

Cubitt A.B., Heim R., Adams S.R., Boyd A.E., Gross L.A. & Tsien R.Y. (1995a) *Trends Biochem.* **20**, 448–455.

Cubitt A.B., Firtel R.A., Fischer G., Jaffe L.F. & Miller A.L. (1995b) *Development* **121**, 2291–2301.

Daguzan C., Nicolas M-T., Mazars C., Leclerc C. & Moreau M. (1995) *Int. J. Dev. Biol.* **39**, 653–657.

Davies E.V., Campbell A.K. & Hallett M.B. (1991a) *FEBS Lett.* **291**, 135–138.

Davies E.V., Campbell A.K., Williams B.D. & Hallett M.B. (1991b) *Br. J. Rheum.* **30**, 443–448.

De Wet J.R., Wood K.V., Helinski D.P. & DeLuca M. (1985) *Proc. Natl Acad. Sci. USA* **82**, 7870–7873.

DeLuca M. & McElroy W.D. (1978) *Methods Enzymol.* **57**, 3–15.

Dorner A.J. & Kaufman R.J. (1994) *Biologicals* **22**, 103–112.

Dubois R. (1887) Comptes Rend. Seanc. Soc. Biol. **4**, 564–565.

Erikaku T., Zenno S. & Inouye S. (1991) *Biochem. Biophys. Res. Commun.* **174**, 1331–1336.

Eisen, A. & Reynolds, G.T. (1985) *J. Cell Biol.* **101**, 1522–1527.

Fagan T.F., Ohmiya Y., Blinks J.R., Inouye S. & Tsuji F.I. (1993) *FEBS Lett.* **333**, 301–305.

Fisher A.J., Thompson T.B., Thoden J.B. & Baldwin T.O. (1996) *J. Biol. Chem* **271**, 21 956–21 968.

Foran D.R. & Brown W.M. (1988) *Nucleic Acids Res* **16**, 177.

Frackman S., Anhalt M. & Nealson K.H. (1990) *Bacteriology* **172**, 5767–5773.

Furste J.P., Pansegrau W., Frank R., Blocker H., Schol Z.P., Bagdasarian M. & Lanka E. (1986) *Gene* **48**, 119–131.

Gangola O. & Rosen B.P. (1987) *J. Biol. Chem.* **262**, 12 570–12 574.

Geiger R., Hauber R. & Miske W. (1989) *Mol. Cel Probes* **3**, 309–328.

Gerdes H-H. & Kaether C. (1996) *FEBS Lett.* **389**, 44–47.

Gilkey J.C., Jaffe L.F., Ridgway E.B. & Reynolds G.T. (1978) *J Cell Biol* **76**, 448–466.

Goldfarb D.S. (1992) *Cell* **70**, 185–188.

Gorman C.M., Moffat L.F. & Howard B.H. (1982) *Mol. Cell. Biol.* **2**, 1044–1051.

Gould S.J. & Subramani S. (1988) *Anal. Biochem.* **175**, 5–13.

Grynkiewicz G., Poenie M. & Tsien R.Y. (1985) *J. Biol. Chem.* **260**, 3440–3450.

Hallett M.B. & Campbell A.K. (1982) *Nature* **295**, 155–158.

Hallett M.B., Davies E.V. & Campbell A.K. (1990) *Cell Calcium* **11**, 655–663.

Harvey E.N. (1952) *Bioluminescence.* Academic Press, New York.

Hastings J.W. & Reynolds G.T. (1966) In *Bioluminescence in Progress*, F.H. Johnson & Y. Haneda (eds). Princeton University Press, Princeton, pp. 45–50.

Hastings J.W. & Weber G. (1963) *J. Opt. Soc. Am.* **53**, 1410–1415.

Heim R. & Tsien R.Y. (1996) *Curr. Biol.* **6**, 178–182.

Henthorn P., Zervoz P., Raducha M., Harris H. & Kadesch T. (1988) *Proc. Natl Acad. Sci. USA* **85**, 6342–6346.

Hickey M.J., Arain T.M., Shawar R.M., Humble D.J.,

Langhorne M.H., Morgenroth J.N. & Stover C.K. (1996) *Antimicrob. Agents Chemother.* **40**, 400–407.
Hooper C.E., Ansorge R.E., Browne H.M. & Tomkins P. (1990) *J. Biolumin. Chemilumin.* **5**, 123–130.
Hummelen J.C., Luider T.M. & Wynberg H. (1987) Methods Enzymol **133**, 531–557.
Ikariyama Y., Nishigushi S., Koyama E., Kobatake E., Aizawa M., Tsuda M. & Nakazawa T. (1997) *Anal. Chem.* **69**, 2600–2605.
Illarionov, B.A., Illarionov, V., Bondar, V.S. & Vysotski, E.S. (1995) *Gene* **153**, 273–274.
Inouye S. & Tsuji F.I. (1993) *FEBS Lett.* **315**, 343–346.
Inouye S., Noguchi M., Sakaki Y., Takagi I., Miyata T., Iwanaga S. & Tsuji F.I. (1985) *Proc. Natl Acad. Sci. USA* **82**, 3154–3158.
Jassim S.A.A., Ellison A., Denyer S.P. & Stewart G.S.A.B. (1990) *J. Biolumin. Chemilumin.* **5**, 115–122.
Jefferson R.A., Kavanagh T.A. & Bevan M.W. (1987) *EMBO J.* **6**, 3901–3907.
Jha P.K., Nakhai B., Sridhar P., Talwar G.P. & Hasnain S.E. (1990) *FEBS Lett.* **331**, 25–30.
Johnson T.C., Thompson R.B. & Baldwin T.O. (1986) *J. Biol. Chem.* **261**, 4805–4811.
Joyeux A., Balaguer P., Germain M., Boussioux A.M., Pons M. & Nicholas A.C. (1997) *Anal. Biochem.* **249**, 119–130.
Jung R., Ahmad-nejad P., Wimmer M., Gerhard M., Wagener C. & Neumaier M. (1997) *Eur. J. Clinical Chem. Lab. Med.* **35**, 3–10.
Keller G.A., Gould S., DeLuca M. & Subramani S. (1987) *Proc. Natl Acad. Sci. USA* **84**, 3264–3268.
Kendall J.M., Sala-Newby G., Ghalaut V., Dormer R.L. & Campbell A.K. (1992a) *Biochem. Biophys. Res. Commun.* **187**, 1091–1097.
Kendall J.M., Dormer R.L. & Campbell A.K. (1992b) *Biochem. Biophys. Res. Commun.* **189**, 1008–1016.
Kendall J.M., Badminton M.N., Dormer R.L. & Campbell A.K. (1994) *Anal. Biochem.* **221**, 173–181.
Kendall J.M., Badminton M.N., Sala-Newby G.B., Wilkinson G.W.G. & Campbell A.K. (1996) *Cell Calcium* **19**, 133–142.
Kennelly P.J. & Krebs E.G. (1991) *J. Biol. Chem.* **266**, 15 555–15 558.
Killen A.A. (1997) *Clin. Lab. Med.* **17**, 1–21.
King J.M.H., DiGrazia P.M., Applegate B., Buriage R., Sanseverino J., Dunbar P., Larimer F. & Sayler G.S. (1991) *Science* **249**, 778–781.
Knight M.R., Campbell A.K., Smith S.M. & Trewavas A.J. (1991a) *Nature* **352**, 524–526.
Knight M.R., Campbell A.K., Smith S.M. & Trewavas A.J. (1991b) *FEBS Lett.* **282**, 405–408.
Knight M.R., Read N.D., Campbell A.K. & Trewavas A.J. (1993) *J. Cell. Biol.* **121**, 83–90.
Kobatake E., Iwait T., Ikariyama Y. & Aizawa M. (1993) *Anal. Biochem.* **208**, 300–305.
Koncz C., Olsson O., Langridge W.H.R., Schell J. & Szalay A.A. (1987) *Proc. Natl Acad. Sci. USA* **84**, 131–135.
Koop A. & Cobbold P.H. (1993) *Biochem. J.* **295**, 165–170.
Krika, L.J. & Thorpe, G.H.G. (1987) *Methods Enzymol.* **133**, 404–420.
Kruse M., Brunke M., Esccher A., Szalay A.A., Tropschug M. & Zimmerman R. (1995) *Biol. Chem.* **270**, 2588–2594.
Leber S.M., Yamagata M. & Sanes J.R. (1996) *Methods Cell Biol.* **51**, 161–183.
Lee R.J. & Huang L. (1997) Critical Reviews in Therapeutic Drug Carrier Systems **14**, 173–206.
Legocki R.P., Legocki M., Baldwin T.O. & Szalay A.A. (1986) *Proc. Natl Acad. Sci. USA* **83**, 9080–9084.
Lindbladh C., Mosbach K. & Bulow L. (1991) *J. Immunol. Methods* **137**, 199–207.
Lindow S.E. (1995) *Mol. Ecol.* **4**, 555–566.
Liu J., Prat S., Willmitzer L. & Frommer W.B. (1990) *Gen. Genet.* **1990**, 401–406.
Lizano S., Ramanathan S., Feltus A., Witkowski A. & Daunert S. (1997) *Methods Enzymol.* **279**, 296–303.
Llewellyn D.H., Kendall J.M., Sheikh F.N. & Campbell A.K. (1996) *Biochem. J.* **318**, 555–560.
Lorenz W.W., Cormier M.J., O'Kane D.I., Hua D., Escher A.A. & Szalay A.A. (1996) *J. Biolumin. Chemilumin.* **11**, 31–37.
Lorenz W.W., McCann R.O., Longiaru M. & Cormier M.J. (1991) *Proc. Natl Acad. Sci. USA* **88**, 4438–4442.
Mancini J.A., Boylan M., Soly R.R., Graham A.F. & Meighen E.A. (1988) *J. Biol. Chem.* **263**, 14 308–14 312.
Marsault R., Murgia M., Pozzan T. & Rizzuto R. (1997) *EMBO J.* **16**, 1575–1581.
Martin P.E.M., George C.H., Castro C., Kendall J.M., Capel J., Campbell A.K., Revilla A., Barrio L.C. & Evans W.E.J. (1998) *J. Biol. Chem.* **273**, 1719–1726.
Matthews J.A., Batki A., Hynds C.C. & Kricka L.J. (1985) *Anal. Biochem.* **151**, 205–209.
Maxwell O.J. & Maxwell F. (1988) *DNA* **7**, 557–562.
Mayerhofer R., Langridge W.H.R., Cormier M.J. & Szalay A.A. (1995) *Plant J.* **7**, 1031–1038.
Meighen E.A. (1991) *Microbiol. Rev.* **55**, 123–142.
Millar A.J., Carre I.A., Strayer C.A., Chua N.H. & Kay S.A. (1995) *Science* **267**, 1161–1163.
Milton D.L., Hardman A., Camara M., Chabra S.R., Bycroft B.W., Stewart G.S.A.B. & Williams P. (1997) *J. Bacteriol.* **179**, 3004–3012.
Miyawaki A., Lopis J., Heim R., McCaffery J.M., Adams J.A., Ikura M. & Tsien R.J. (1997) *Nature (London)* **388**, 882–887.
Montero M., Brini M., Marsault R., Alvarez J., Pozzan T. & Rizzuto R. (1995) *EMBO J.* **14**, 5467–5475.
Montero M., Barrero M.J. & Alvarez J. (1997) *FASEB J.* **11**, 881–885.
Nakajima-Shimada J.N., Iida H., Tsuji F.I. & Anraku Y. (1991a) *Proc. Natl Acad. Sci. USA* **88**, 6878–6882.
Nakajima-Shimada J.N., Iida H., Tsuji F.I. & Anraku Y. (1991b) *Biochem. Biophys. Res. Commun.* **174**, 115–122.
Nakamura T. (1972) *Biochem.* **72**, 173–177.
Nelson N.C., Hammond P.W., Wiese W.A. & Arnold L.J. (1990) In *Luminescence Immunoassay and Molecular Applications*, K. van Dyke & R. van Dyke (eds). CRC Press. Boston, pp. 293–312.
Nguyen G. & Heffelfinger D.M. (1995) *Anal. Biochem.* **226**, 59–67.
Nguyen V.T., Morange M. & Bensaude O. (1988) *Anal. Biochem.* **171**, 404–408.
Nielsen L.L. & Pedersen R.A. (1991) *Exp Zool* **257**, 128–133.
Nimmesgern E. & Hartl F.U. (1993) *FEBS Lett.* **331**, 25–30.
Nolan G.P., Fiering S., Nicholas J.F. & Herzenberg L.A. (1988) *Proc. Natl Acad. Sci. USA* **85**, 2603–2607.
Norris V., Grant S., Freestone P., Canvin J., Sheikh Toth I., Trinei M., Modha K. & Norman R.I. (1996) *J. Bacteriol.* **178**, 3677–3682.
Olsson O., Koncz C. & Szalay A.A. (1988) *Mol. Gen. Genet.* **215**, 1–9.
Ow D.W., Jacobs J.D. & Howell S.H. (1987) *Proc. Natl Acad. Sci. USA* **84**, 4870–4874.
Persechini A., Lynch J.A. & Romoser V.A. (1997) *Cell Calcium* **22**, 209–216.
Poenie M. & Tsien R.Y. (1986) *Trends Biochem Sci.* **11**, 340–355.

Pons M., Gagne D., Nicolas J.C. & Methali M. (1990) *Biotechniques* **9**, 450–459.
Prasher D., McCann R.O. & Cormier M.J. (1985) *Biochem. Biophy. Res. Commun.* **126**, 1259–1268.
Prasher D., McCann R.O., Longiaru M. & Cormier M.J. (1987) *Biochem.* **26**, 1326–1332.
Rattray E.A.S., Prosser J.I., Killham K. & Glover L.A. (1990) *Appl. Environ. Microbiol.* **56**, 3368–3374.
Rembold C.M., Kendall J.M. & Campbell A.K. (1996) *Cell Calcium* **21**, 69–79.
Reynolds G.T. (1972) *Q. Rev. Biophys.* **5**, 295–347.
Reynolds G.T. (1978) *Photochem. Photobiol.* **27**, 405–421.
Ridgeway E.B. & Ashley C.C. (1967) *Biochem. Biopys. Res. Commun.* **29**, 229–234.
Rizzuto R., Brini M. & Pozzan T. (1994) *Methods Cell Biol.* **40**, 339–358.
Rizzuto R., Simpson A.W.M., Brini M. & Pozzan T. (1992) *Nature* **358**, 325–327.
Roberts P.A., Knight J. & Campbell A.K. (1987) *Anal. Biochem.* **160**, 139–148.
Rodriguez D., Rodriguez J.R., Rodriguez J.F., Mcgowan E.B. & Esteban M. (1988) *Proc. Natl Acad. Sci. USA* **85**, 1667–1671.
Rogowsky P.M., Close T.J., Chimera J.A., Shaw J.J. & Kado C.I. (1987) *J. Bacteriol.* **169**, 5101–5112.
Rosay P., Davies S.A., Yu Y., Ali Sozen M., Kaiser K. & Dow J.A.T. (1997) *J. Cell Sci.* **110**, 1683–1692.
Rose B. & Lowenstein W.R. (1976) *J. Memb. Biol.* **28**, 87–119.
Rosomer V.A., Hinkle P.M. & Persechini A. (1997) *J. Biol. Chem.* **272**, 13 270–13 274.
Rutter G.A., White M.R.H. & Tavare J.M. (1995) *Current Biology* **5**, 890–899.
Rutter G.A., Burnett P., Rizzuto R., Brini M., Murgia M., Pozzan T., Tavare J.M. & Denton R.M. (1996) *Proc. Natl Acad. Sci. USA.*, **93**, 5489–5494.
Sala-Newby G.B., Kalsheker N. & Campbell A.K. (1990) *Biochem. Biophys. Res. Commun.* **172**, 477–482.
Sala-Newby G.B., Taylor K.M., Badminton M.N., Rembold C.M. & Campbell A.K. (1998) *Immunology* **93**, 601–609.
Sala-Newby G.B. & Campbell A.K. (1991) *Biochem. J.* **279**, 727–732.
Sala-Newby G.B. & Campbell A.K. (1994) *Biochem. Biophys. Acta.* **1206**, 155–160.
Sala-Newby G.B. & Campbell A.K. (1992) *FEBS Lett.* **307**, 241–244.
Sambrook J., Fritsch E.F. & Maniatis T. (1989) In *Molecular Cloning*, Vols 1–3. Cold Spring Harbor Laboratory Press, New York.
Santella L.J. (1996) *Membrane Biol.* **153**, 83–92.
Schaap A.P., Akhavan H. & Romano L.J. (1989) *Clin. Chem.* **35**, 1863–1864.
Schwartz O., Virelizier J.L., Montagnier L. & Hazan U. (1990) *Gene* **88**, 197–205.
Selden R.F., Burke Howie K., Rowe M.E., Goodman H.M. & Moore D.D. (1986) *Mol. Cell Biol.* **6**, 3173–3179.
Sevigny P. & Gossard F. (1990) *Gene* **93**, 143–146.
Shaw J.J. & Kado C.I. (1986) *Biotechnology* **4**, 560–564.
Shimomura O. & Shimomura A. (1985) *Biochem. J.* **228**, 745–749.
Shimomura O. (1995) *Biochem. J.* **306**, 537–543.
Shimomura O., Johnson F.H. & Saiga Y. (1962) *J. Cell Comp. Physiol.* **59**, 223–239.
Shimomura O., Musicki B. & Kishi Y. (1988) *Biochem. J.* **251**, 405–410.
Shimomura O., Musicki B. & Kishi Y. (1989) *Biochem. J.* **261**, 913–920.
Shimomura O., Inouye S., Musicki B. & Kishi Y. (1990) *Biochem. J.* **270**, 309–312.
Shimomura O., Musicki B., Kishi Y. & Inouye S. (1993) *Cell Calcium* **14**, 373–378.
Smith D.F., Stults N.L., Rivera H., Gehle W.D., Cummings R.D. & Cormier M.J. (1991) In *Bioluminescence and Chemiluminescence Current Status*, P.E. Stanley & L.S. Kricka (eds). John Wiley & Sons, Chichester, pp. 529–532.
Stanley P.J. (1997) *Biolumin. Chemilumin.* **12**, 61–78.
Steidler L., Yu W., Fiers W. & Remault E. (1996) *Appl. Environ. Microbiol.* **62**, 2356–2359.
Stults N.L., Stocks N.A., Cummings R.D., Cormier M.J. & Smith D.F. (1991) In *Bioluminescence and Chemiluminescence Current Status*, P.E. Stanley & L.S. Kricka (eds). John Wiley & Sons, Chichester, pp. 533–536.
Swartzmann, E., Kapoor, S., Graham, A.F. & Meighan, E.A. (1990) *Bacteriology* **172**, 6797–6802.
Tanahashi H., Ito T., Inouye S., Tsuju F.I. & Sakaki Y. (1990) *Gene* **96**, 249–255.
Tatsumi H., Masuda T., Kajiyama N. & Nakano E. (1989) *J. Biolumin. Chemilumin.* **3**, 75–78.
Thompson E.M., Nagata S. & Tsuju F.I. (1990) *Gene* **96**, 257–262.
Thompson E.M., Nagata S. & Tsuju F.I. (1989) *Proc. Natl Acad. Sci. USA* **86**, 6567–6571.
Thomson C.M., Herring P.J. & Campbell A.K. (1997) *J. Biolumin. Chemilumin.* **12**, 87–91.
Tisa L.S. & Adler J. (1995) *Proc. Natl Acad. Sci. USA* **92**, 10 777–10 781.
Treisman R., Green M.R. & Maniatis T. (1983) *Proc. Natl Acad. Sci. USA* **80**, 7428–7432.
Tsien R., Pozzan T. & Rink T.J. (1982) *J. Cell. Biol.* **94**, 325–334.
Walsh D.A. & VanPatten S.M. (1994) *FASEB J.* **8**, 1227–1236.
Ward, W.W. & Cormier, M.J. (1979) *J. Bio. Chem.* **254**, 781–788.
Watkins N.J., Knight M.R., Trewavas A.J. & Campbell A.K. (1995) *Biochem. J.* **306**, 865–869.
Waud J.P., Sala-Newby G.B., Matthews S.B. & Campbell A.K. (1996) *Biochim. Biophys. Acta.* **1292**, 89–98.
Weeks I., Beheshti I., McCapra F., Campbell A.K. & Woodhead J.S. (1983a) *Clin. Chem.* **29**, 1474–1479.
Weeks I., Campbell A.K. & Woodhead J.S. (1983b) *Clin. Chem.* **29**, 1480–1483.
Weeks, I., Sturgess, M., Brown, R.C. & Woodhead, J.S. (1987) *Methods Enzymol.* **133**, 366–387.
Wei L.N. (1997) *Ann. Rev. Pharmacol. Toxicol.* **37**, 119–141.
White M.R.H., Masuko M., Amet L., Elliot G., Braddock M., Kingsman A.J. & Kinsman S.M. (1995) *J. Cell Science* **108**, 441–455.
Wiedemann E. (1888) *Ann d Phsik u Chemie* **34**, 446–463.
Wood K.V. (1991a) In *Bioluminescence and Chemiluminescence: Current Status*, A.K. Kricka & L.J. Stanley (eds). John Wiley and Sons, Chichester, pp. 11–14.
Wood K.V. (1991b) *J. Biolumin. Chemilumin.* **5**, 107–114.
Wood K.V., Lam Y.A. & McElroy W.D. (1989a) *J. Biolumin. Chemilumin.* **5**, 31–39.
Wood K.V., Lam Y.A. & McElroy W.D. (1989b) *J. Biolumin. Chemilumin.* **5**, 289–301.
Wood K.A., Lam Y.A., Seliger H.H. & McElroy W.D. (1989c) *Science* **244**, 700–702.
Woodhead J.S. & Weeks I. (1991) *Clin. Chem.* **37**, 472.
Woods N.M., Cuthbertson K.S.R. & Cobbold P.H. (1986) *Nature* **319**, 600–602.
Xi L., Cho K.W. & Tu S.C. (1991) *Bacteriology* **173**, 1399–1405.
Xiong Z. & Ruben l. (1996) *Mol. Biochem. Parasitol.* **83**, 57–67.
Yu Z., Redfern C.S. & Fishman G.I. (1996) *Circ. Res.* **79**, 691–697.
Zatta P.F., Nyame K., Cormier M.J., Mattox S.A., Prieto P.A., Smith D.F. & Cummings R.D. (1991) *Anal. Biochem.* **194**, 185–191.
Zenno S. & Inouye S. (1990) *Biochem. Biophys. Res. Commun.* **171**, 169–174.

CHAPTER EIGHTEEN

Luminescence Imaging of Gene Expression in Single Living Cells

GUY A. RUTTER, AINSLEY A. CULBERT & JEREMY M. TAVARÉ
Department of Biochemistry, School of Medical Sciences, University Walk, University of Bristol, Bristol UK

18.1 INTRODUCTION

Regulated gene expression plays a central role in the development and viability of all organisms. Expression of genes can be controlled at the level of gene transcription itself, or by altering the rates of mRNA export, translation or stability (Derman *et al.*, 1991; Welsh *et al.*, 1985; Hedeskov, 1980). In higher eukaryotes, the activation and inactivation of gene transcription is central to intercellular communication and in processes such as cytokine and growth factor signalling, cell survival, metabolic regulation, defence against pathogens as well as in memory and learning (Docherty & Clark, 1994; Else *et al.*, 1994; Frank & Greenberg, 1994; Vaulont & Kahn, 1994).

Transcriptional regulation involves specific nucleotide sequences flanking the regulated gene at its 5′ end, termed promoter and enhancer regions. The interaction between these regions and transcription factors allows for very precise modulation in the rate of transcriptional initiation. Much attention, therefore, has been paid to the identity of the transcription factors that bind to these *cis*-acting sequences, and the signalling systems that impinge on them to allow regulation by extracellular stimuli (Treisman, 1990; Hunter & Karin, 1992; Denton & Tavaré, 1995).

In this chapter, we describe a system which combines single cell microinjection and cell imaging to non-invasively measure alterations in the activity of a promoter sequence in a single living cell. To do this we employ bioluminescent reporter proteins (luciferases), and ultra low light level photon-counting imaging with intensified cameras. Luminescence imaging using this type of camera has the advantage of sensitivity, low background interference and a high signal-to-noise ratio; qualities which are not provided by fluorescent methods.

Using this technique, it is possible to dissect which transcription factors and what signalling pathways (e.g. Ca^{2+} or protein kinase cascades) are used to control the activity of a particular promoter. This is because microinjection allows the introduction of a wide variety of reagents (e.g. antisense DNA, dominant negative/constitutively active proteins, peptides or antibodies), allowing the manipulation of specific

FLUORESCENT AND LUMINESCENT PROBES, 2ND EDN
ISBN 0-12-447836-0

signalling proteins or transcription factors. Thus the cell becomes the ultimate 'test tube'.

Central to this method is an ability to monitor independently two distinct luciferases within the same cell; one under the control of the regulated promoter, and the other controlled by a constitutive, usually viral, promoter. This allows us to measure the specific activity of the regulated promoter corrected for changes in microinjection volume and cell viability. Finally, digital imaging allows intercellular heterogeneity in a population of cells to be determined, or changes in the activity of promoters to be followed during the cell cycle in a non-synchronized population of cells (Millar *et al.*, 1995; Zwicker *et al.*, 1995, Hwang *et al.*, 1995).

Descriptions of the imaging of firefly luciferase in mammalian cells can be found in the literature (Hooper *et al.*, 1990; White *et al.*, 1990, 1995). Imaging of multiple photoproteins, and of the combined use of microinjection to analyse the intracellular signalling pathways leading to the activation of gene transcription, was first described in Rutter *et al.* (1995), Rutter & Kahn, (1996) and Kennedy *et al.* (1997), and is reviewed by Rutter *et al.* (1996), Tavaré *et al.* (1996a,b), White (1996a,b) and White *et al.* (1996). Luciferase imaging has also been performed in bacteria, plants and anaesthetized mice (Masuko *et al.*, 1991; White *et al.*, 1996; Millar *et al.*, 1995).

18.2 GENERAL CONSIDERATIONS

The development of reporter gene technology has revolutionized our understanding of transcriptional regulation (Alam & Cook, 1990). This involves fusing *cis*-acting promoter regions of a gene, upstream of a cDNA encoding a protein ('reporter'), with an activity that is conveniently measured in cell lysates. For single cell imaging, however, the most commonly used reporters (chloramphenicol acetyl transferase or CAT, and β-galactosidase) usually require cell lysis before activity can be determined. However, the recent development of firefly luciferase reporter technology has the advantage that its cofactor, luciferin, is membrane permeant, and that novel luciferases (e.g. from *Renilla reniformis*) with distinct membrane-permeant cofactors (i.e. coelenterazine) have been developed.

18.2.1 Cell types

Both firefly and *Renilla* luciferases can be expressed in all of the mammalian and insect cells we have investigated, as well as in bacterial (Masuko *et al.*, 1991), plant (White, 1996a; Millar *et al.*, 1995) and fish (Mayerhofer *et al.*, 1995) cells. Since microinjection is the preferred method to introduce plasmid copies, the feasibility of the single cell studies may depend on the ease with which individual cells may be microinjected. This itself depends upon how adherent the cell type of interest is, and the size and accessiblity of the nucleus. We have successfully injected a large variety of adherent cell types including immortalized cell lines (3T3, CHO, PC12, INS-1, MIN6, H4IIE and GH3) as well as primary cells (human dermal fibroblasts, pancreatic β-cells, and neurons from the dorsal root ganglion and superior cervical ganglia).

Indeed, a particularly attractive feature of this approach is that it permits studies on primary cells, e.g. human parathyroid or islet β-cells, which may be difficult to isolate and culture in large numbers, and on cells (e.g. 3T3 L1 adipocytes) that cannot be transfected using commonly available methodologies.

18.2.2 Reporter genes

The main criteria that a reporter gene must satisfy for single cell imaging are:

(1) No endogenous activity in the cells under investigation.
(2) The activity must be measurable without cell disruption and any cofactor used must be non-toxic.
(3) The activity of the protein should be easily quantifiable over a wide dynamic range.
(4) The protein should turnover rapidly so that oscillatory changes in transcriptional activity are reported without an excessive lag period.
(5) The reporter gene must be active immediately upon synthesis (i.e. cofactor binding, and not post-translational maturation, is rate-limiting).
(6) The reporter should generate a stable signal, dependent only upon the number of protein molecules.

In our laboratories we have developed methods for imaging in real-time the chemiluminescence of two distinct luciferases, each of which satisfies the above criteria. These are from the firefly *Photinus pyralis* (De Wet *et al.*, 1985) and from the sea pansy *Renilla reniformis* (Matthews *et al.*, 1977; Lorenz *et al.*, 1991). Firefly luciferase has a subunit molecular weight of 62 kDa (Wood *et al.*, 1985) and catalyses the oxidation of the cofactor, luciferin (M_r 478), in the presence of molecular oxygen and ATP, with the generation of oxyluciferin plus a photon of yellow green light (λ_{max} = 560 nm) (Fig. 18.1(A)).

Although not fully permeable to mammalian cells, luciferin will cross the plasma membrane at more than adequate rates at millimolar concentrations (White *et al.*, 1990; DeWet *et al.*, 1985; Craig *et al.*, 1991). The

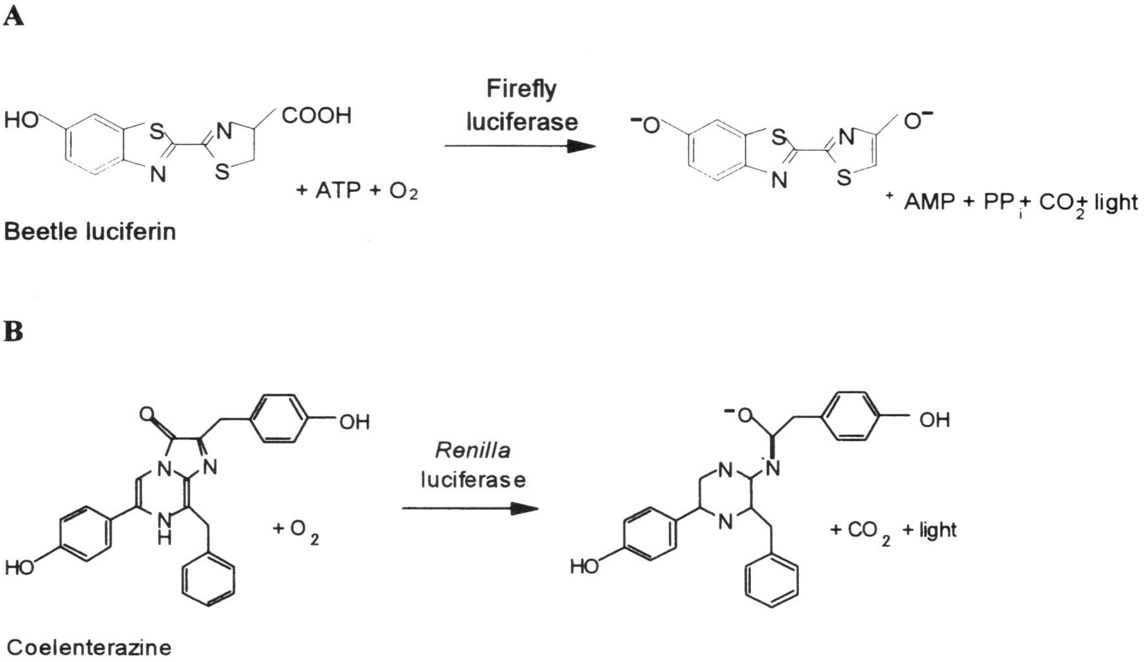

Figure 18.1 Structures and reactions of luciferin and coelenterazine.

levels of expression of luciferase usually achieved are well below those that would influence intracellular ATP or oxygen concentrations (K. Brindle, University of Cambridge, personal communication). Removal of the *C*-terminal peroxisomal targeting sequence (Ser–Lys–Leu), and optimization of codon usage for expression in mammalian cells, increases the expression of active firefly luciferase in mammalian cells some 10- to 50-fold (Sherf & Wood, 1994) (G.A. Rutter and J.M. Tavaré, unpublished). Firefly luciferase is particularly well suited as a reporter of gene expression, as a result of its short half-life ($t_{1/2}$) in mammalian cells (2–3 h) (White *et al.*, 1996) compared with reporters such as CAT ($t_{1/2}$ = 16 h) (White *et al.*, 1993).

Renilla luciferase is a 31 kDa protein that oxidizes the cofactor coelenterazine, which also readily permeates across the mammalian cell membranes, to generate blue light (λ_{max} = 460 nm)(Fig. 18.1(B)).

Plasmids possessing modified firefly luciferase (pGL3) and *Renilla* luciferase (pRL) are commercially available from Promega (Madison, WI). Recent reports describe the expression of *Renilla* luciferase in mammalian cells, either alone (Lorenz *et al.*,1996) or together with firefly luciferase (Rutter & Kahn, 1996). In living cells, the specificity of the firefly and *Renilla* luciferase for their respective substrates is essentially complete. Thus, addition to the cells of either luciferin or coelenterazine at a concentration sufficient to allow free permeation across the plasma membrane reveals only the activity of the 'cognate' enzyme. In each case, a steady glow from individual cells, which remains constant in the absence of a change in enzyme amount over days (firefly) or hours (*Renilla*), can be measured.

Thus the activity of two separate promoters, each driving firefly or *Renilla* production, can be assayed entirely independently *in vivo* by the sequential addition of the two cofactors, as will be described below. This method is preferred to the use of filters to discriminate the two activities due to the substantial light losses which are incurred with presently available filters.

We usually arrange it so that the weaker 'regulated promoter' drives the expression of the firefly enzyme. This is measured first, and then the *Renilla* activity is measured which is driven by the stronger constitutive promoter (i.e. the plasmid pRL-CMV according to Promega's nomenclature).

Two other potential reporters of transcriptional activity are worthy of mention. First, secreted alkaline phosphatase provides a sensitive and convenient assay of transcription, which does not require cell lysis. However, it is, as such, of little value for use with single cells.

We have also used the Ca^{2+}-sensitive photoprotein, aequorin, from the jellyfish *Aequoria victoria* (Inouye *et al.*, 1985). This enzyme resembles *Renilla* luciferase, in oxidizing coelenterazine to release blue light

(λ_{max} 460 nm) but is Ca^{2+}-dependent, allowing it to be detected separately from that of other reporters (see below). Indeed, aequorin was the forerunner of the *Renilla* enzyme in our early experiments (Rutter *et al.*, 1995).

18.2.3 Introduction of reporter genes into cells

Plasmid constructs, that possess the firefly luciferase gene under the control of a regulated promoter, can be introduced into cells using transfection with Ca^{2+} phosphate, DEAE-dextran or cationic lipids, or by electroporation (Sambrook *et al.*, 1989). We prefer to use direct intranuclear microinjection for three reasons: (1) it allows fine control over the number of plasmid copies introduced into a cell; (2) it allows us to cluster cells expressing luciferase into a single field of view of the camera for imaging purposes; and (3) it allows the introduction of two or more different plasmids at a defined and fixed ratio. This latter facility is important since it permits the activity of a regulated promoter to be normalized to that of a strong constitutive reporter within the same cell. In this way, small variations in the total amount of plasmid mixture injected, as well as inherent heterogeneities in the basal transcriptional and translational machinery of individuals cells, can be accounted for.

The ability to introduce a high number of plasmid copies (>1000) into an individual nucleus frequently permits the detection in single cells of a signal from the reporter protein when classical methods of transfection have failed (Rutter *et al.*, 1995). However, it should be noted that this can result in a phenomenon termed 'squelching' when endogenous transcription factors are titrated away from chromosomal DNA, and onto the promoter introduced in the plasmid, thus altering transcription rates of endogenous genes. For this reason, consideration must be paid to the resulting effect that squelching would have on cell physiology.

18.3 EQUIPMENT REQUIRED

18.3.1 Microinjection

18.3.1.1 Microelectrode pullers

Sutter Instruments provide a thermal puller (Flaming/Brown micropipette puller P-97), consisting of a thermal coil and a unit able to hold and pull a capillary at either end. This allows control of numerous parameters including temperature, duration and force of pull, each of which dictates the shape and fineness of the injection microelectrode. Generally, we aim to pull a 'needle' using capillaries with an inner filament and 1.2 mm outer diameter (GC120F-10; Clark Electromedical Instruments, Reading, UK) to provide a tip of 0.1–0.5 μm external diameter. However, perfectly adequate less expensive needle pullers are available (Campden Instruments Ltd, Loughborough, UK) but are slower, and allow less optimization of the needle's properties. Finally, pre-pulled sterile tips (FemptoTips™) are available from Eppendorf but these are expensive and, having a rather wide mouth, are not appropriate for all cell types (e.g. primary hepatocytes). Furthermore, because they are quite malleable, when they become blocked they are difficult to 'break' to free the flow.

18.3.1.2 Microscope

An inverted microscope is required. While the optical requirements for microinjection and digital imaging differ, we still use a single microscope performing both roles. A good quality antivibration table (Miles Griot, Cambridge, UK or TMC, Peabody, MA, also provided by Spindler and Hoyer, Milton Keynes, UK) is necessary to prevent microelectrode breakage during injection.

Suitable microscopes are available from Olympus (IX70), Zeiss (Axiovert 100TV or 135 TV), Nikon (Diaphot) and Leica (DMIRBE). The main requirements are a 32× phase contrast lens (Zeiss Achrostigmatic, 0.4 NA) or a suitable alternative (20× plus additional 1.5× element; Olympus LCAch 20×/0.40 and Leica 20×/0.40 N plan, Nikon 20 DL). In our experience the Zeiss slightly outperforms its competitors for image quality during microinjection, and for ease of attachment of the micromanipulator. For the Leica and Nikon microscopes, direct attachment to the stage is not possible, and a separate support must be constructed adjacent to the instrument. For microinjection purposes, the numerical aperture (NA) of the lens is relatively unimportant.

18.3.1.3 Microinjector

Pressure microinjection (Ansorge, 1982) involves locating a microelectrode containing the sample for microinjection adjacent to the cell to be microinjected. The material is then, preferably, introduced by automatic axial injection directly into the cells.

This is achieved with a micromanipulator capable of moving the tip in three planes (x,y,z), which is clamped to the microscope stage. We prefer the 5171 micromanipulator and 5246 microinjector from Eppendorf (Hamburg, Germany) which performs these functions semiautomatically. These devices control, respectively, the positioning of the microelectrode and the increase in air pressure during microinjection. More sophisticated systems, programmed using a video monitor, are also available (Eppendorf), though considering the cost (some six times that of the manual system) it is debatable whether these provide more advantages than drawbacks. Other microinjection systems are also available (IM-188, Narashige, Japan; Fukuda *et al.*, 1996).

18.3.2 Equipment for ultra low light level photon counting

18.3.2.1 Cameras

The camera used in our studies (Hayakawa *et al.*, 1986; Hayakawa, 1992) contains four principal elements (Fig. 18.2):

(1) The photocathode, which may be of the bialkali (peak sensitivity at 350 nm) or low-noise multi-alkali (S20) type (peak sensitivity at 450 nm), converts individual photons into an electron event;
(2) Two or three microchannel plates which amplify a single incident electron through one microchannel within a two-dimensional array (usually 512 × 512 pixels). Each microchannel consists of a hollow glass cylinder coated so as to release secondary electrons, thus acting like a tiny photomultiplier tube. As a result, a gain of between 10^6 and 10^9 compared with the original input electrons can be achieved, whilst maintaining the spatial coherence of the original image.
(3) A phosphor screen which reconverts electrons into visible light.
(4) A charge-coupled device (CCD) or conventional TV camera.

Photon counting relies on the fact that at very low intensities, light begins to behave as discrete packets (i.e. photons; Hayakawa, 1992). With an intensified camera, these photons can be detected individually and quantified precisely, since they can be discriminated from noise which occurs at much lower energies. This is not the case with cooled charge-coupled device cameras, where there is a less marked separation between noise and signal events (Hayakawa, 1992). The physics of the intensified system means that spatial information may be lost during the amplification of the electrons through the microchannel plate array. This phenomenon becomes more pronounced as the number of microchannel plates increases. However, it can be corrected mathematically in a process called 'centroiding', which calculates the likely spatial location of the original excitation (i.e. which pixel of the photocathode was struck). This allows the generation of a 'centre of gravity' image. Some cameras generate both a crude non-centroided as well as a centre of gravity image. Others produce only the centre of gravity image.

Importantly, the centre of gravity image is used for precise quantitation of the numbers of photons detected, and varies in linear proportion to the level of luciferase expression.

Both bialkali and S20 photocathodes are appropriate for imaging luciferases, with the S20 providing slightly greater sensitivity at the cost of a higher dark count rate. Typical dark counts are 60 and 600 photon min^{-1} $image^{-1}$ (512 × 512 pixels), for bialkali and low-noise S20 photocathodes, respectively. Comparisons of the sensitivity of different cameras should be based on: (a) signal-to-noise ratio (SNR), and (b) time required for detection of a signal from a weak source (i.e. a cell transfected with firefly luciferase, under the control of a regulatable promoter).

Typical values of SNR for a cell injected with a regulatable promoter (L-pyruvate kinase or collagenase) imaged with a bialkali photocathode, dual-microchannel plate intensifier configuration at 10× magnification (producing an area of illumination of about 25 × 25 pixels) are 50–500 photon per 15 min versus a background reading of about 10 photon/15 min. Equivalent values for *Renilla* luciferase (under CMV promoter control) are 1000–10 000 versus a background of 50–100 counts.

In some systems, rapid data acquisition (at video rates, i.e. one image per 40 ms) is possible. This function, called time resolved imaging, allows each photon event to be stored within a matrix, with a record of its incident time. Images can therefore be built up *post facto* over any interval, from these data. By contrast with other systems, the number and length of each image is preset by the operator before data acquisition.

Intensified photon counting cameras with the above specification are available from Hamamatsu Photonics (C2400–40, comprising a bialkali photocathode, two microchannel plates; Hamamatsu City, Japan, Bridgewater, NJ and Welwyn, Herts, UK) and Photek Ltd (ICCD 318, bialkali or low-noise S20 photocathode, three microchannel plates; St Leonards-on-Sea, Sussex, UK). The latter can be supplied as a cooled version (down to −20°C) which allows suppression of the dark noise from the more sensitive S20 photocathode.

Intensifiers with only one microchannel plate tend not to provide adequate sensitivity for luciferase imaging. Other suppliers include Photonic Science (Robertsbridge, Sussex, UK), and Videoscope International Ltd (Sterling, VA), who supply a higher resolution camera with a greater density of

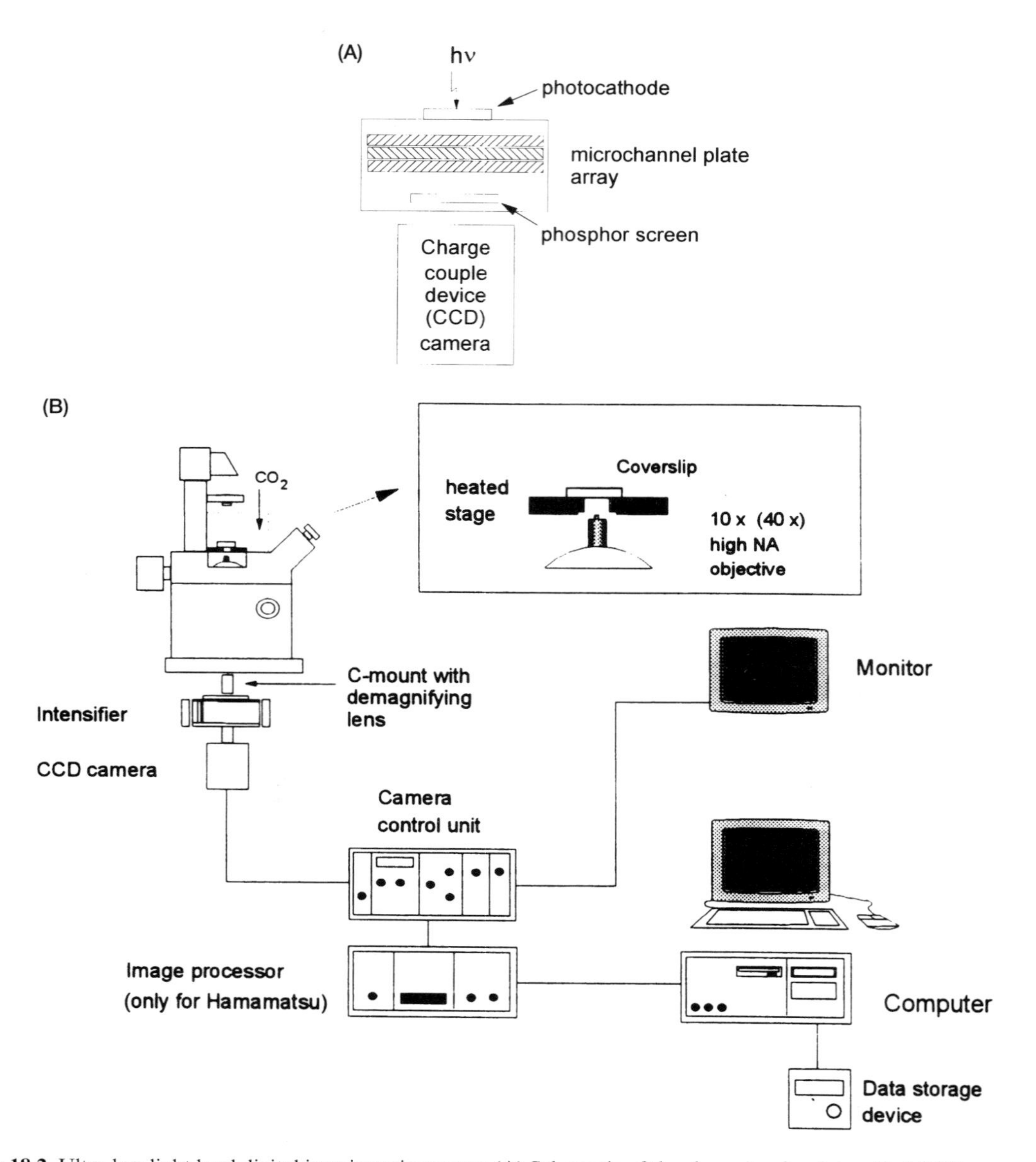

Figure 18.2 Ultra low light level digital imaging microscopy. (A) Schematic of the elements of an intensified CCD camera (see the text for further details). (B) The digital imaging microscope. Samples are located on the heated microscope stage, in a chamber which allows perfusion if required (inset). Alternatively, the stage can be enclosed allowing humidification and CO_2 gassing (see Section 18.3). The intensified CCD camera is attached to the lower port of the microscope, with a C-mount adapter. A camera control unit provides the high-tension voltage for the intensifier, and reads out the signal from the CCD camera. The image is output in real-time to a video monitor, and to the PC directly (Photek) or via an image processor (Hamamatsu). Data is transmitted to a frame-grabber card within the PC.

microchannels per plate (each having a diameter of *c.* 3 μm, compared with the usual 10 μm).

Alternative systems, involving cooled CCD cameras (Princeton Instruments, Harlow, Bucks, UK, and Trenton, NJ; Photometrics, Tucson, AZ), though less expensive, have in our hands been unable to provide comparable sensitivity at very low light levels and reasonable imaging times. However, these on-chip devices undoubtedly provide better spatial resolution, which may be useful in certain circumstances when using extremely strong viral promoters, for instance.

When comparing different camera systems, ensure that all other variables, i.e. microscope, objective, etc., are identical. These optical elements have a substantial impact on sensitivity and image quality.

18.3.2.2 Microscope

The collection of emitted light must be maximized. To image live cells, the microscope should have inverted optics, with the camera attached to a lower port allowing 100% of emitted light to be diverted straight into the camera without interference from mirrors or prisms.

Enhancements in sensitivity can be achieved by introducing a demagnifying lens (e.g. 0.6×) in the C-mount adapter to which the camera is attached. This allows fewer individual pixels within the photocathode to be excited by a larger number of photons per unit time. This provides increased sensitivity, although this is of course only achieved at the expense of some loss of resolution.

For maximum sensitivity, a high transmittance (high numerical aperture; NA) lens is required (note that light transmittance through the objective lens is proportional to NA^2/magnification^2). For most purposes, a 10× air objective, such as the Zeiss Fluar 10× 0.5 NA, provides the right balance between sensitivity and magnification, allowing a large number (hundreds) of individual cells to be imaged simultaneously. Long working distance lenses generally have low NA values and are not suitable.

For some applications, higher magnification, using a 40× 1.3 NA oil immersion lens may be preferred. Higher magnification still (63 or 100×) is generally undesirable, since light losses incurred by the additional objective elements become very significant.

Great care is always required to ensure that all the microscope optics are clean and regularly maintained. Suitable microscopes (see above) are available from Zeiss, Nikon and Olympus (but not Leica).

The microscope should be placed on an antivibration table modified so that a hole (*c*. 18 cm diameter) is located beneath the microscope to allow attachment of the intensified camera. Ideally, the microscope and table should be placed in complete darkness, for example within a dark room. The camera control unit and data acquisition computer can then be located outside, and operated whilst imaging takes place in total darkness. However, video monitors, as well as instrument LEDs, invariably thwart the best attempts to ensure that ambient light is zero, and the presence of this background light can interfere significantly with the detection of very weak luminescence signals.

18.3.2.3 Other equipment

Most imaging software for intensified cameras is PC-compatible only. A reasonably fast processor (e.g. Pentium 166 MHz), with ≥16 Mbyte RAM, and 2 GByte hard disk, running Windows 3.11 or Windows 95, is preferable. Access to a network (requiring a network card, e.g. Etherlink III) for data analysis on a remote computer is also important if maximum use of the equipment is to be achieved. Off-line copies of software are usually provided by image intensifier manufacturers, along with updates.

Data analysis packages such as Microsoft Excel, and presentation software (Freelance-for-Windows, Powerpoint or Adobe Photoshop) are also required for off-line image manipulation and presentation.

Even a small experiment may generate up to 1 Mbyte of data, thus a mass data storage device, such as a magneto-optical disk drive (Panasonic PD LF-1004 or LF-1000, Japan, or Fujitsu M2512A, Fujitsu, Kawasaki, Japan) or writable compact disk drive (Hewlett Packard 4020I, Loveland, CO) is essential.

Where imaging is required over long periods, the choice of the heating system and culture environment is important. Some cells will survive for considerable periods of time in Hepes-buffered medium (see below), but others will require gassing with CO_2 (e.g. a chamber from Carl Zeiss Ltd, Herts, UK). To maintain the cells at constant temperature a suitable heated stage is required (e.g. Zeiss TRZ 3700). Other heating systems (e.g. the Δ *T* system from Bioptechs Inc., Butler, PA) use dishes which can be directly heated, allowing the precise temperature at the surface of the coverslip to be maintained.

18.4 EXPERIMENTAL PROCEDURES

18.4.1 Microinjection

18.4.1.1 Preparation of samples to be injected

Plasmid DNAs, prepared by centrifugation on two successive CsCl gradients, are resuspended at 0.1–0.4 mg ml^{-1} in 2 mM Tris, pH 8.0, 0.2 mM EDTA ('injection buffer') and centrifuged for 30 min at 10 000 *g* to clear any remaining particulate material which would otherwise block the microinjection needle.

DNA prepared using commercially available ion exchange kits (e.g. Qiagen or Anachem) can also be used, although more problems are encountered with viscosity and hence microelectrode blockage.

Antibodies can be co-microinjected at a concentration of ≥1 mg ml^{-1} but should be dialysed extensively against injection buffer before use.

18.4.1.2 Culture of cells

Prior to cell culture, the coverslip or Petri dish is scratched to produce a square, usually about 1 mm across, using either a sterile diamond pen or needle. Coverslips with premarked grids are also available (Eppendorf CELLocate®). While experiments with cell suspensions are possible in theory (using a holding pipette), most success with microinjection is obtained when the adherent cells are firmly attached such that the nucleus is prominent.

Cells are seeded either on a glass coverslip (22 mm) or plastic Petri dish (35 or 60 mm diameter) to provide an adequate density (about 1000 cells per mm^2) on the day of injection. For fibroblasts this is usually obtained with a seeding density of about 1×10^6 cells per ml.

18.4.1.3 Microelectrode preparation

In our laboratory, microelectrodes are pulled by the Sutter electrode puller (settings; heat = 317/999; velocity = 10/255; time = 250/255; pull = 0), and should be used within 3 h of pulling. The DNA solution (~3 μl) is then transferred into the microelectrode with an automatic pipette (Finn/Gilson) using a capillary pipette tip (Eppendorf Cat. #5442 956.003).

18.4.1.4 Microinjection

Microinjection is performed with cells in the presence of normal culture medium (e.g. Dulbecco's modified Eagle's medium; DMEM), supplemented with 25 mM Hepes (pH 7.4) to prevent dramatic changes in pH due to loss of CO_2 during the injection procedure. Alternatively, the cells can be switched to DMEM with low bicarbonate (2 mM) for the duration of the injection procedure (typically 1 h). Otherwise, many cell types will survive short periods of time in Hepes-buffered salt solutions (e.g. Krebs/Ringer).

The following instructions apply to the Eppendorf semiautomatic microinjection system.

Locate the needle tip using the microscope optics, and the manipulator joystick (coarse control) so that it is gradually lowered to about a cell's depth above the first cell to be injected. Use a holding pressure of ~50 hPa to maintain a trickle of DNA from the tip (often this cannot be seen visually, unless a fluorescent dye is included with the DNA).

Under fine control, lower the tip until it touches the top of the cell, directly above the nucleus, and then move gently down so that the cell is very slightly squashed (a ring on the surface of the cell will be seen where the tip produces a small indentation). The tip of the needle is now at the correct final position for injection. This height is set on the controls of the micromanipulator unit as the minimum *z*-level for injection.

The injection and holding pressure are now optimized. The injection pressure depends on the size of the tip, and the viscosity of the liquid within it. Typically, it is around 900 hPa with a fresh electrode, but must be lowered to <100 as the tip atrophies.

The microelectrode is then retracted vertically to about a cell's depth above the cell for injection (until it goes slightly out of focus), and injection initiated (for convenience, by depressing a button at the top of the joystick, although a foot pedal can be used as an alternative). The needle then punctures the cell in an axial fashion, at which point there is an increase in pressure, allowing the introduction of DNA. A whitening of the nucleus is observed as the DNA enters, which can be readily followed by the operator. Injection is continued at the operator's discretion until a wave of liquid has passed across the nucleus. This can take anything from 0.2 to 2 s, and considerable practice is required to perfect this art.

Little if any permanent damage appears to be done to the cell which quickly recovers. During this process, a volume of liquid equivalent to 5–10% of the cell volume (i.e. 50–100 fl out of 1 pl, typically) enters the cell. The electrode is then retracted, by depressing the joystick button again, following the same path as during injection.

Typically 200–500 cells may be injected in this way per hour. But this number may be considerably less with cells which are more difficult to inject (e.g. primary islet β-cells or neurons). Productive injection is usually achieved at a level of about 60% by an experienced worker, although 10% is more normal for a beginner.

Cells are then transferred back to the incubator after the medium has been changed. Alternatively, cells can be maintained for imaging on the microscope stage at 37°C for real-time imaging. For studies of the effect of a hormone or other transcriptional regulator, the agent can be added to the medium at this point. Alternatively, the stimulus may be applied after a steady state level of basal gene expression has been achieved.

18.4.2 Imaging

Two protocols may be employed, depending upon whether constant monitoring of gene expression is to be performed.

18.4.2.1 Measurement of gene expression at a fixed time point

Mounting the sample on the microscope

Medium is removed and replaced with phosphate buffered saline (0.2 ml for a mounted coverslip or 0.75 ml for a 35 mm diameter Petri dish) supplemented with 1 mM beetle luciferin (K^+ salt, Promega).

The coverslip or Petri dish is placed on the microscope stage, and the area of injected cells identified using the grid, with the 10× objective.

Obtaining a bright-field image

This serves to locate the correct focal plane on the intensifier and subsequently to identify luminescent cells. The cells are illuminated with weak light (preferably with a neutral density or colour filter in place), and light diverted to the bottom port of the microscope, and hence the intensifier. With the camera on low gain (about 15% of maximum) an integrated image is taken, with the cells slightly out of focus to obtain enhanced contrast.

Imaging firefly luciferase

After refocusing, illumination is switched off and photon counting of firefly luciferase luminescence begun. The camera gain is set to maximum, and photon counting is performed typically for between 5 and 20 min in complete darkness (see Section 18.3 above). At the end of this integration, the intensifier is switched off, and the image saved to disk. This will give the activity of the regulated promoter.

Imaging Renilla *luciferase*

At this point, the plate can be rinsed to remove luciferin, which causes the cessation of light production by firefly luciferase, and 5 μM coelenterazine (Molecular Probes, Eugene, OR) added to reveal *Renilla* luciferase activity. Coelenterazine, which is relatively unstable, should be stored under nitrogen or argon at −80°C in small aliquots at a concentration of 2.5 mM in 60% methanol, 5 mM HCl, 7.5% glycerol. In practice, it is frequently unnecessary to remove luciferin, if light production by the *Renilla* enzyme (produced under strong CMV promoter control) is more intense than photon release from the firefly enzyme. The contribution of light resulting from the activity of the firefly enzyme is therefore small and can be removed mathematically *post facto* without introducing any large error.

After ensuring the position of cells in the field, and their correct focus, photon counting of the *Renilla* activity is then performed as for firefly luciferase, and the images stored on the hard disk or an optical device.

18.4.2.2 Repeated measurements

Here, the cells remain on the microscope in a humidified environment (see Section 18.3) and images are taken at preset intervals (e.g. 15 min integration every 2 h) with the Hamamatsu, or continuously in time resolved imaging mode (Photek). Cells are maintained at 37°C in normal serum-containing medium (this can be Hepes-buffered low bicarbonate medium if a CO_2 chamber is unavailable) plus other stimuli (hormone, nutrient or growth factor) and 1 mM beetle luciferin as necessary.

If necessary, *Renilla* activity can be measured immediately after firefly luciferase, by switching temporarily to coelenterazine-containing medium, using a perfusion system (see Fig. 19.2(B)). In this way, the activity of two genes can be monitored over several days, with medium changes allowing addition and withdrawal of different growth factors or other stimuli.

18.4.3 Data analysis

The activity of the regulated promoter is proportional to the activity of firefly luciferase, divided by *Renilla* luciferase activity (Rutter & Kahn, 1996; Kennedy *et al.*, 1997). This is calculated for each cell by determining the number of photon events produced by that cell during the period of integration in the presence firstly of luciferin, then of coelenterazine (or a luciferin/coelenterazine mixture).

A region of interest is marked around each luminescent cell on the centroided luminescence image (centre of gravity) obtained in the presence of luciferin. The computer calculates the number of photons detected within that region. A background region, with no cell apparent, should also be quantified. This list is then saved as an ASCII (text) file. This procedure is repeated with the image obtained in the presence of coelenterazine, and a second ASCII file generated. These files can then be loaded into a spreadsheet (e.g. Microsoft Excel). For each cell, the 'specific' firefly luciferase activity, and hence the activity of the regulated promoter, is given by the ratio (R) of firefly luciferase to *Renilla* luciferase. If luciferin is washed from the cells before the addition of coelenterazine, then this is calculated according to:

$$R = (FF - bg_1)/(Ren - bg_2) \quad [18.1]$$

where FF and Ren correspond to the number of photon events in the presence of luciferin and coelenterazine, respectively, and bg_1 and bg_2 represent the background luminescence in the presence of the two respective cofactors.

If *Renilla* luciferase is assayed without removal of luciferin, then,

$$R = (FF - bg_1)/(T - FF - bg_2) \quad [18.2]$$

where T is the number of photons detected in the simultaneous presence of luciferin and coelenterazine. These data can either be tabulated or displayed in bar graph form.

Individual colour images (512 × 512 pixels) can be enhanced and exported (e.g. as a TIFF or BMP file) for presentation in a suitable package. An example of the use of this method to analyse gene expression in 3T3-L1 adipocytes is shown in Plate 18.1.

18.5 CONCLUDING REMARKS

The high sensitivity of the assay for firefly and *Renilla* luciferase is the result of the high quantum efficiency of the luminescent reactions, the high sensitivity to individual photons of photon counting imaging cameras, and the absence of any luminescent background from non-expressing cells.

Unlike fluorescence detection, the photon counting system provides an absolute measure of photoprotein activity, without the need for complex calibration procedures.

A particular advantage over conventional investigations of gene expression in cell populations is that luciferase activity can be detected very early after microinjection (as little as 30 min). Thus, even very low levels of mRNA, which are difficult to detect by Northern or RNAase protection, are 'reported' by the expression of luciferase. Furthermore, heterogeneities in gene expression between different cells is readily identifiable (White *et al.*, 1995; Rutter *et al.*, 1995).

We have used this approach to investigate the regulation by extracellular stimuli of the transcriptional activity of several promoters. However, we have also used this approach to investigate the regulation of translation of *in vitro* translated and microinjected mRNA. The technique also provides a means to study several fundamental cell biological processes, including the cell cycle and mitosis, apoptosis, and differentiation in the context of the living cell.

Luciferase imaging outperforms that of the use of other *in situ* reporter genes such as green fluorescent protein and the recently described β-lactamase system (Zlokarnik *et al.*, 1998). In cells co-expressing firefly luciferase and GFP, we have been able to detect clearly a luciferase signal within 60 min, under conditions where detectable GFP fluorescence was not apparent for at least 4 h post injection. The lag before GFP fluorescence was detected, represents the time required for folding and formation of the GFP fluorophore by oxidation (Chalfie *et al.*, 1994) and the inherently greater sensitivity of photon counting luminescence imaging compared with fluorescence imaging, even with cooled CCD cameras (Wood, 1994). Thus, whilst signal-to-noise ratios for GFP imaging of the activity of a strong promoter are at best 20–25, those which can be achieved with firefly luciferase can approach 1000.

In a recent paper Zlokarnik *et al.* (1998) describe an elegant method for monitoring gene expression in single cells using the β-lactamase reporter gene and a fluorigenic substrate. The ability to monitor gene expression using a fluorescence-based approach is a very important advance, especially when combined with the use of fluorescence-activated cell sorting (FACS). However, for imaging rapid changes in gene expression at the single cell level it has some disadvantages compared with previously established techniques. In particular, the fluorescence approach is limited at present (1) to measurements with a single promoter, (2) by likely toxicity of the UV excitation wavelengths required (especially for long-term incubations), (3) by a rather limited dynamic range, and (4) low sensitivity, for example, with a 5-min period of integration we estimate that we can detect as few as 10^4 molecules of luciferase per cell. This compares with an estimated value of ~ 10^6 molecules of β-lactamase per cell over a 5-min incubation period with the fluorigenic substrate.

ACKNOWLEDGEMENTS

Work in the authors' laboratories was supported by the Wellcome Trust, The Medical Research Council (UK) the British Diabetic Association, the Biotechnology and Biological Sciences Research Council, and The Royal Society. J.M.T. is a British Diabetic Association Senior Research Fellow.

REFERENCES

Alam J. & Cook J.L. (1990) *Anal. Biochem.* **188**, 246–254.
Ansorge W. (1982) *Exp. Cell Res.* **140**, 31.
Chalfie M, Tu Y., Euskirchen G., Ward W.W. & Prasher D.C. (1994) *Science* **263**, 802–805.
Craig F.F., Simmonds A.C., Watmore D., Watmore D., McCapra F. & White M.R.H. (1991) *Biochem. J.* **276**, 637–641.
Denton R.M. & Tavaré J.M. (1995) *Eur. J. Biochem.* **227**, 597–611.
Derman E., Krauter K., Walling L., Weinberger C., Ray M. & Darnell J.E., Jr (1991) *Cell* **23**, 731–739.
De Wet J.R., Wood K.V., Helinski D.R. & DeLuca M. (1985) *Proc. Natl Acad. Sci. USA* **82**, 7870–7873.
Docherty K. & Clark A.R. (1994) *FASEB J.* **8**, 20–27.
Else K.J., Finkleham F.D., Maliszewski C.R. & Grencis R.K. (1994) *J. Exp. Med.* **179**, 347–351.
Frank D.A. & Greenberg M.E. (1994) *Cell* **79**, 5–8.
Fukuda M., Gotoh I., Gotoh Y. & Nishida E. (1996) *J. Biol. Chem.* **271**, 20024–20028.
Hayakawa T. (1992) *Image Analysis in Biology*, D. Hader (ed). CRC Press, Boca Raton, pp. 75–86.
Hayakawa T., Kinoshita S., Miyaki S., Fujiwake H. & Ohsuka S. (1986) *Photochem. Photobiol.* **43**, 95–97.
Hedeskov C.J. (1980) *Physiol. Rev.* **60**, 442–509.

Hooper C.E., Ansorge R.E., Browne H.E. & Tomkins P. (1990) *J. Biolumin. Chemilumin.* **5**, 123.
Hunter T. & Karin M. (1992) *Cell* **70**, 375–387.
Hwang A., Maity A., McKenna W.G. & Muschel R.J. (1995) *J. Biol. Chem.* **270**, 28419–28424.
Inouye S., Aoyama S., Miyata T., Tsuji F.I. & Sakaki Y. (1985) *J Biochem (Tokyo)* **105**, 473–477.
Kennedy B., Viollet I., Rafiq A., Kahn, A. & Rutter G.A. (1997) *J. Biol. Chem.* **272**, 20636.
Lorenz W.W., McCann R.O., Longiaru M. & Cormier M.J. (1991) *Proc. Natl Acad. Sci. USA* **88**, 4438–4442.
Lorenz W.W., Cormier M.J., O'Kane D.J., Hua D., Escher A.A. & Szalay A.A. (1996) *J. Biolumin. Chemilumin.* **11**, 31.
Masuko M., Hosoi S. & Hayakawa T. (1991) *Acta Histochem. Cytochem.* **29**, 156–157.
Matthews J.C., Hori K., Cormier M.J. (1977) *Biochemistry* **16**, 85–91.
Mayerhofer R., Araki K. & Szalay A.A. (1995) *J. Biolum. Chemilum.* **10**, 271–275.
Millar A.J., Carre I.A., Strayer C.A., Chua N.H. & Kay S.A. (1995) *Science* **267**, 1161–1163.
Rutter G.A. & Kahn A. (1996) *Diabetologia* **39**, A40.
Rutter G.A., White M.R.H. & Tavaré J.M. (1995) *Curr. Biol.* **5**, 890–899.
Rutter G.A., Burnett P., Rizzuto R., Brini M., Murgia M., Pozzan T., Tavare J.M. & Denton R.M. (1996) *Proc. Natl. Acad. Sci. USA* **93**, 5489–5494.
Sambrook J., Fritsch E.F. & Maniatis T. (1989) *Molecular Cloning: a Laboratory Manual*, 2nd ed. New York: Cold Spring Harbor Laboratory Press.
Sherf B.A. & Wood K.V. (1994) *Promega Notes* **49**, 14–21.
Tavaré J.M., Rutter G.A., Griffiths M.R., Dobson S.P. & Gray H. (1996a) *Biochem. Soc. Trans.* **24** 378–384.
Tavaré J.M., Dickens M., Dobson S.P., Rutter G.A. & Williams A.G. (1996b) *Acta Histochem. Cytochem.* **29**, 156–157.
Treisman R. (1990) *Cancer Biol.* **1**, 47–58.
Vaulont S., Kahn A. (1994) *FASEB J.* **8**, 28–35.
Welsh M., Nielsen D.A., MacKrell A.J. & Steiner D.F. (1985) *J. Biol. Chem.* **260**, 13590–13594.
White M.R.H. (1996a) *J. Biolum. Chemilum.* **11**, 53–54.
White M.R.H. (1996b) *Acta Histochem. Cytochem.* **29**, 156–157.
White M.R.H., Morse J., Boniszewski Z.A.M., Mundy C.R., Brady M.A.W. & Chiswell D.J. (1990) *Technique* **2**, 194–201.
White M.R.H., Braddock M., Byles E.D., Amet L., Kingsman A.J. & Kingsman S.M. (1993) In *Biotechnology Applications of Microinjection, Microscopic Imaging and Fluorescence*, P.H. Bach C.H. Reynolds J.M. Clark J. Mottley & P.L. Poole (eds). Plenum Press, New York, pp. 19–28.
White M.R.H., Masuko M., Amet L., Elliott G., Braddock M., Kingsman A.J. & Kingsman S.M. (1995) *J. Cell Sci.* **108**, 441–455.
White M.R.H., Wood C.D. & Millar A.J. (1996) *Biochem. Soc. Trans.* **24**, S411.
Wood K.V. (1994) *Curr. Opin. Biotechnol.* **6**, 50–68
Wood K.V., deWet J., Dewji N. & DeLuca M. (1985) *Biochem. Biophys. Res. Commun.* **124**, 592–596.
Zlokarnik G., Negulescu P.A., Knapp T.E., Mere L., Burres N., Feng L., Whitney M., Roemer K. & Tsien R.Y. (1998) *Science* **279**, 84.
Zwicker J., Gross C., Lucibello F.C., Truss M., Ehlert F. Engeland K. & Muller R. (1995) *Nucl. Acid Res.* **23**, 3822–3830.

CHAPTER NINETEEN

Enhanced Variants of the Green Fluorescent Protein for Greater Sensitivity, Different Colours and Detection of Apoptosis

STEVEN R. KAIN
Cell Biology Group, CLONTECH Laboratories, Inc., Palo Alto, CA, USA

The green fluorescent protein (GFP) from the jellyfish *Aequorea victoria* is a versatile reporter protein for monitoring gene expression and protein localization in a variety of systems. Applications using GFP reporters have expanded greatly due to the availability of mutants with altered spectral properties, including emission variants which produce different colours of light. Wild-type GFP, and most previously described GFP variants have limited utility, primarily due to relatively dim fluorescence and low expression levels attained in higher eukaryotes. To improve upon these qualities, we have combined several point mutations to alter the spectral properties of GFP with a synthetic gene sequence containing codons preferentially found in highly expressed human proteins. The combination of improved fluorescence intensity and higher expression levels yield enhanced GFP variants which provide greater sensitivity, and are suitable for multicolour detection by fluorescence microscopy and flow cytometry. The following chapter describes the properties, advantages, and applications of these 'enhanced' GFPs – EGFP, EBFP and EYFP.

19.1 INTRODUCTION

In the jellyfish *Aequorea victoria* (Fig. 19.1), light is produced when energy is transferred from the Ca^{2+}-dependent photoprotein aequorin to green fluorescent protein or GFP (Morin & Hastings, 1971; Shimomura *et al.*, 1962; Ward *et al.*, 1980). As illustrated in Fig. 19.2, these energy transfer reactions result in a shift in the emission maxima from 470 nm to 509 nm by the participation of GFP. This process occurs in specialized photogenic cells called photocytes located at the base of jellyfish umbrella, which contain GFP at an estimated concentration of 25 mg ml^{-1} (Carey *et al.*, 1996). The cloning of the wild-type GFP gene (wt GFP) (Inouye & Tsuji, 1994; Prasher *et al.*, 1992) and its subsequent expression in heterologous systems (Chalfie *et al.*, 1994; Wang & Hazelrigg, 1994) has established GFP a powerful tool for the analysis of gene expression and protein localization in a range of experimental designs. When expressed in either eukaryotic or prokaryotic cells and illuminated by blue or UV light, GFP emits a bright green fluorescence which is easily detected by fluorescence

FLUORESCENT AND LUMINESCENT PROBES, 2ND EDN
ISBN 0-12-447836-0

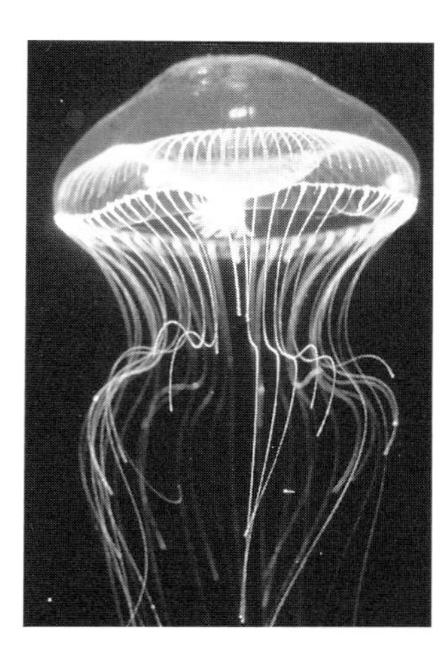

Figure 19.1 Photograph of the jellyfish *Aequorea victoria* which produces the green fluorescent protein (GFP). Photograph courtesy of Claudia Mills at the Friday Harbor Research Laboratories, Washington.

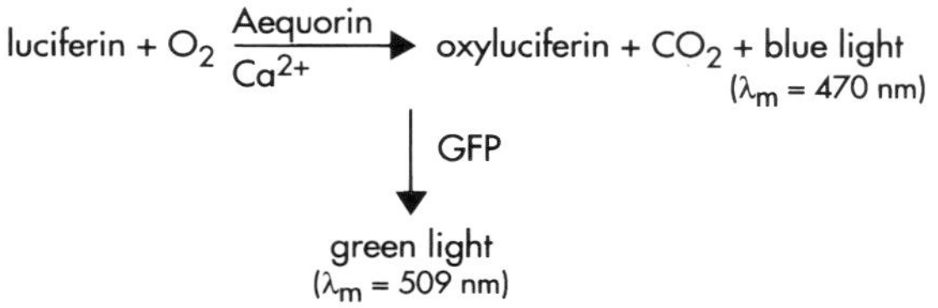

$$\text{luciferin} + O_2 \xrightarrow[Ca^{2+}]{\text{Aequorin}} \text{oxyluciferin} + CO_2 + \text{blue light } (\lambda_m = 470 \text{ nm})$$

$$\downarrow \text{GFP}$$

$$\text{green light } (\lambda_m = 509 \text{ nm})$$

Figure 19.2 Energy transfer reactions which occur inside the photocytes of *Aequorea victoria* leading to the emission of green light by GFP.

microscopy, flow cytometry, or other fluorescence imaging techniques. Light-stimulated GFP fluorescence is species-independent and does not require cofactors, substrates, or gene products from *A. victoria*. Additionally, detection of GFP and its variants can be performed in living samples, and is amenable to real time analysis of molecular events.

GFP, like the fluorophore fluorescein, has a fluorescence quantum yield (QY) of about 70–80% (Ward, 1981; Table 19.1) although the extinction coefficient (E_m) for wt GFP is much lower. Nevertheless, in fluorescence microscopy, wt GFP has been found to give greater sensitivity and resolution than staining with fluorescently labelled antibody (Wang & Hazelrigg, 1994). GFP has the advantage of being more resistant to photobleaching and of avoiding background caused by non-specific binding of primary and secondary antibodies to targets other than the antigen (Wang & Hazelrigg, 1994). Although binding of multiple antibody molecules to a single target offers a potential amplification not available for GFP, this is offset because neither labelling of the antibody nor binding of the antibody to the target is 100% efficient.

In addition to expression of GFP alone, GFP and its variants have also been used extensively to express GFP fusions with a variety of other proteins. In many cases, chimaeric genes encoding either *N*- or *C*-terminal fusions to GFP retain the normal biological activity of the heterologous partner, as well as maintaining fluorescent properties similar to native GFP (Flach *et al.*, 1994; Wang & Hazelrigg, 1994; Marshall *et al.*, 1995; Stearns, 1995). The use of GFP and its variants in this capacity provides a 'fluorescent tag' on the protein, which allows for *in vivo* localization of the fusion protein. GFP fusions can provide enhanced sensitivity and resolution in comparison to standard antibody staining techniques (Wang & Hazelrigg, 1994), and the GFP tag eliminates the need for fixation, cell permeabilization and antibody incubation steps normally required when using antibodies tagged with chemical fluorophores. Lastly, use of the GFP tag permits real-time kinetic studies of protein localization and trafficking (Flach *et al.*, 1994; Náray-Fejes-Tóth & Fejes-Tóth, 1996; Wang & Hazelrigg, 1994; Kaether & Gerdes, 1995; Carey *et al.*, 1996; Elowitz *et al.*, 1997).

19.1.1 The GFP chromophore

One of the features of GFP is that it is a naturally fluorescent protein, encoding the chromophore responsible for light emission within the primary amino acid sequence. The GFP chromophore consists of a cyclic tripeptide derived from Ser–Tyr–Gly at positions 65–67 in the protein (Fig. 19.3) (Cody *et al.*, 1993), and is only fluorescent when embedded within the fully folded, complete GFP protein. GFP is 238 amino acids in length (27 kDa), and there have been no reports of significant truncations or internal deletions in GFP which retain fluorescence. Limited truncation of the protein yields a fluorescent derivative, with the minimal domain required for fluorescence consisting of residues 7–229 (Li *et al.*, 1997). The crystal structures of GFP and a variant termed S65T (see discussion below) have revealed a tightly packed 'β-can' structure enclosing an α-helix containing the chromophore (Ormö *et al.*, 1996; Yang F. *et al.*, 1996). Truncations of GFP which impact the β-can are not tolerated (Li *et al.*, 1997). This structure provides the proper environment for the chromophore to fluoresce by excluding solvent and oxygen. Nascent GFP is not fluorescent, since chromophore formation occurs post-translationally (Heim & Tsien, 1996). The chromophore is formed by a cyclization reaction and an oxidation step that requires molecular oxygen (Heim *et al.*, 1994; Davis *et al.*, 1995). These steps are either autocatalytic or use factors that are ubiquitous, since fluorescent GFP forms in a broad range of cells and organisms. Chromophore formation may be

Table 19.1 Fluorescence properties of enhanced GFP variants

Variant	*Mutations*	*Excitation max. (nm)*	*Emission max. (nm)*	E_m	*QY*	*Comments*
Wild-type	None	395 (470)	509	21 000 (7150)	0.77	
EGFP	F64L S65T	488	507	~55 000	~0.7	Single 488 excitation peak; brighter fluorescence; human codon optimized
EBFP	F64L S65T Y66H Y145F	380	440	31 000	~0.2	Blue emission; resistance to photobleaching; human codon optimized
EYFP	T203Y S65G V68L S72A	513 (488)	527	36 500	0.63	Yellow-green emission; brighter fluorescence; human codon optimized

E_m = molar extinction coefficient in units of cm^{-1} M^{-1}.
QY = fluorescence quantum yield.

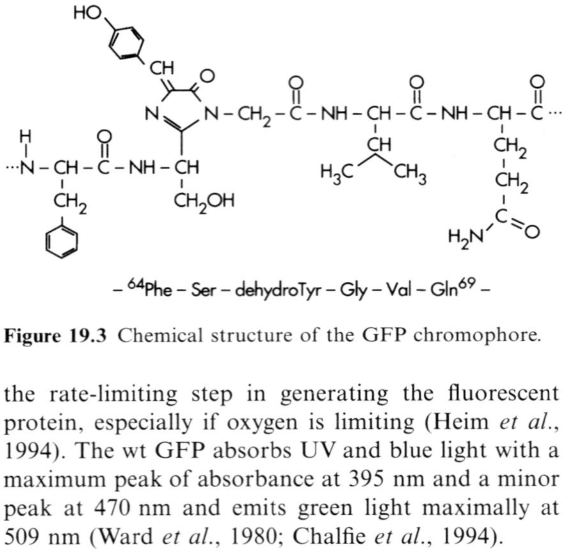

– ^{64}Phe – Ser – dehydroTyr – Gly – Val – Gln^{69} –

Figure 19.3 Chemical structure of the GFP chromophore.

the rate-limiting step in generating the fluorescent protein, especially if oxygen is limiting (Heim *et al.*, 1994). The wt GFP absorbs UV and blue light with a maximum peak of absorbance at 395 nm and a minor peak at 470 nm and emits green light maximally at 509 nm (Ward *et al.*, 1980; Chalfie *et al.*, 1994).

19.2 GFP VARIANTS

19.2.1 GFP variants with altered excitation spectra

A number of different 'red-shifted' variants of wt GFP have been described by several investigators, most of which contain one or more amino acid substitutions in the chromophore region of the protein. The red-shifted terminology refers to the position of the major fluorescence excitation peak, which is shifted for each of these variants towards the red, from 395 nm in wt GFP, to 488–505 nm. The emission spectra for such variants is largely unaffected, and these mutants still produce green light with a wavelength maxima of approximately 507–511 nm. The major excitation peak of the red-shifted variants encompasses the excitation wavelength of commonly used fluorescence filter sets (such as those used for fluorescein), so the resulting signal is much brighter relative to wt GFP. Similarly, the argon-ion laser used in most flow cytometry instruments and confocal scanning laser microscopes emits at 488 nm, so excitation of the red-shifted GFP variants is much more efficient than excitation of wt GFP. In practical terms, this means the detection limits are considerably lower with the red-shifted variants. The red-shifted excitation spectra also allows such variants to be used in combination with wt GFP in double-labelling experiments. Although the peak positions in the emission spectra of the red-shifted GFP variants are virtually identical, double-labelling can be performed by selective excitation of wt GFP and red-shifted GFP (Kain *et al.*, 1995; Yang T.T. *et al.*, 1996b).

The two most commonly used red-shifted GFP variants are S65T (Heim *et al.*, 1995; Heim & Tsien 1996; Brejc *et al.*, 1997) which contains a Ser-65 to Thr substitution in the chromophore, and GFPmut1 or EGFP (Cormack *et al.*, 1996; Yang T.T. *et al.*, 1996b), having the same S65T change plus a Phe-64 to Leu mutation (Table 19.1). GFPmut1 and EGFP have identical amino acid sequences, but the EGFP coding sequence has been further modified with 190 silent base changes to contain codons preferentially found in highly expressed human proteins (Haas *et al.*, 1996). The 'humanized' backbone used in EGFP contributes to efficient expression of this variant in mammalian and plant cells and subsequently very bright fluorescent signals. Based on spectral analysis of equal amounts of soluble protein, EGFP fluoresces approximately 35-fold more intensely than wt GFP when excited at 488 nm (Cormack *et al.*, 1996), owing to an increase in its extinction coefficient (E_m). The E_m for EGFP has been measured at ~55 000 cm^{-1} M^{-1} for 488 nm excitation (Patterson *et al.*, 1997), compared with 7000 cm^{-1} M^{-1} for wt GFP under similar conditions. Studies with wt GFP expressed in HeLa cells (Niswender *et al.*, 1995) have shown that the cytoplasmic concentration must be greater than ~1.0 μM to obtain a signal that is twice the autofluorescence. This threshold for detection is lower with the red-shifted variants: ~200 nM for S65T and ~100 nM for EGFP (Patterson *et al.*, 1997). For EGFP, the detection limit is equivalent to ~10 000 molecules per cell in the cytoplasm, or ~700 molecules per cell on the surface. In addition to improved sensitivity, other advantages of the EGFP and S65T variants over wt GFP include: (1) improved solubility; (2) more efficient protein folding in bacteria; (3) faster chromophore oxidation to form the fluorescent form of protein; and (4) reduced rates of photobleaching.

Perhaps the most significant advantage of the EGFP protein for mammalian cell expression is the thermotolerance provided by the Phe-64 to Leu mutation. EGFP exhibits equal fluorescence intensity when expressed at 28°C or 37°C as monitored by the percentage of chromophore folding (Patterson *et al.*, 1997). In contrast, wt GFP and S65T have poor folding properties at 37°C, and thereby show lower fluorescence when expressed in cells at this temperature. A similar degree of thermotolerance was observed for the variants GFPuv and EBFP (see discussion below and Patterson *et al.*, 1997), indicating these variants are also useful for studies conducted with mammalian cells propagated at 37°C.

19.2.2 GFP variant for UV excitation

The 'cycle 3' GFP variant (commercially available as GFPuv from CLONTECH) is optimized for maximal fluorescence when excited by UV light (360–400 nm) and for higher bacterial expression. The cycle 3 gene is a synthetic GFP gene in which five rarely used Arg codons from the wt gene were replaced by codons preferred in *Escherichia coli.* Consequently, this variant is expressed very efficiently in *E. coli*. The cycle 3 variant is ideal for experiments in which GFP expression is detected using UV light for chromophore excitation (e.g. for visualizing bacteria, yeast colonies, plants, or other macroscopic applications). The cycle 3 variant was developed using an *in vitro* DNA shuffling technique to introduce point mutations throughout the GFP coding sequence (Crameri *et al.*, 1996). Colonies were visually screened using a long-wave UV lamp (365 nm), and the brightest clones were pooled and subjected to another round of DNA shuffling. The cycle 3 variant emerged as the brightest GFP mutant after three rounds of shuffling and selection, hence the name 'cycle 3'. This variant contains three amino acid substitutions (Phe-99 to Ser, Met-153 to Thr, and Val-163 to Ala), none of which alter the chromophore sequence. These amino acid changes make *E. coli* expressing the cycle 3 variant fluoresce 18 times brighter than wt GFP (Náray-Fejes-Tóth & Fejes-Tóth, 1996). While these mutations dramatically increase the fluorescence of the cycle 3 variant, the emission and excitation maxima remain at the same wavelengths as those of wt GFP (Crameri *et al.*, 1996).

The development of EGFP and cycle 3 (GFPuv) may allow these variants to be used in combination for double-labelling experiments by selective excitation at 488 nm and 395 nm, respectively. Some promising applications for use of this dual reporter combination are: (1) microscopy of multiple cell populations in a mixed culture; (2) monitoring gene expression from two different promoters in the same cell, tissue, or organism; (3) monitoring the localization of two different protein fusions in the same cell, tissue, or organism; and (4) fluorescence-activated cell sorting (FACS) of mixed cell populations (e.g. cells expressing GFPuv, EGFP and non-fluorescent cells).

19.2.3 Blue and yellow fluorescent proteins

The properties of the red-shifted and UV-optimized GFP variants largely overcome the limitations of wt GFP for single reporter applications. However, one important feature shared by wt GFP and each of these variants remains the same – they all yield green fluorescence. As discussed above, several combinations of GFP variants can be used for dual reporter studies by virtue of different excitation maxima, but this process is complicated as the image collected from each reporter is green. Dual-colour images must be generated by pseudocolouring techniques (Heim & Tsien, 1996; Yang T.T. *et al.*, 1996b), or by depicting separate images for each variant. Moreover, use of a reporter protein that yields green light is limited in cases where cellular green autofluorescence is a concern, or when the reporter is used in conjunction with fluorophores such as fluorescein. For each of these reasons, it is desirable to have emission variants of GFP capable of producing distinct colours.

The first such examples of useful variants are the blue emission mutants of GFP, each of which contain an invariant Tyr-66 to His mutation in the GFP chromophore (Heim *et al.*, 1994). The initial variant of this type, referred to as P4, contains the single point mutation Tyr-66 to His, and yields a beautiful cobalt blue signal, but only dim fluorescence. An improvement in this variant was termed P4–3, which contains an additional Tyr-145 to Phe substitution. The P4–3 double mutant has a major shift in both the excitation and emission maxima, with values of 381 nm and 445 nm, respectively. The P4–3 mutant is approximately two-fold brighter than P4, primarily due to a higher E_m value which is similar to that of wt GFP. However, the fluorescence intensity of this variant is considerably less than that of S65T or EGFP. The P4–3 mutant has recently been used in conjunction with wt GFP in co-transfection studies to visualize simultaneously mitochondria and the nucleus in the same living mammalian cell (Rizzuto *et al.*, 1996). Such studies are facilitated by the fact that wt GFP and P4–3 excite with UV light, and allow the real-time analysis of mitochondrial and nuclear dynamics in proliferating cells.

In addition to weak fluorescence, another important disadvantage of the P4–3 variant is rapid photobleaching due to the perturbation of the GFP chromophore and the need to excite with UV light. In experiments with conventional epifluorescence microscopy we observe complete loss of the P4–3 signal in transfected mammalian cells within a few seconds. Moreover, all previous reports with blue emission variants such as P4–3 have used GFP genes containing the wild-type jellyfish codons. As stated above for EGFP, the use of a humanized GFP gene is certain to improve the mammalian cell expression of blue emission variants as well. To address these needs, we have recently developed a novel humanized blue emission variant termed EBFP, which contains four mutations: Phe-64 to Leu, Ser-65 to Thr, Tyr-66 to His, and Tyr-145 to Phe (Table 19.1). This variant contains the same 190 silent mutations found in EGFP, and consequently yields more efficient expression in mammalian cells. The E_m of EBFP is 31 000 cm^{-1} M^{-1} for 380 nm excitation, leading to a fluorescent signal that is brighter than that of P4–3. The fluorescence excitation and emission maxima of EBFP are similar to the other blue emission variants at 380 nm and 440 nm, respectively. Lastly, while detailed studies are yet to be completed, our results with EBFP transfected CHO-K1 cells indicate this variant is resistant to photobleaching for at least several minutes. Because of the brighter signal, photostability and elevated expression of EBFP, this mutant is likely to be the blue variant of choice for most applications in mammalian cells.

The analysis of the crystal structure for S65T has revealed a critical association between Thr-203 and the phenolic group on Tyr-66 in the GFP chromophore (Ormö *et al.*, 1996). Therefore, a targeted mutagenic strategy was used to replace Thr-203 with several other residues in hope of yielding GFP mutants with altered and desirable spectral properties. One such mutant, termed '10C', yielded dramatic red shifts in both the excitation and emission maxima at 513 nm and 527 nm, respectively (Ormö *et al.*, 1996). The 10C variant contains a total of four mutations: Thr-203 to Tyr, Ser-65 to Gly, Val-68 to Leu, and Ser-72 to Ala. The Val-68 to Leu and Ser-72 to Ala mutations improve the folding of this variant at 37°C in much the same fashion as Phe-64 to Leu, but are not significantly responsible for the shift in excitation or emission maxima (Yang T.T. *et al.*, 1996a). This variant has E_m and QY values similar to those for S65T, and therefore yields a bright fluorescent signal. We have incorporated the 10C mutations into the same humanized backbone utilized for EGFP and EBFP to produce a variant optimized for mammalian and plant cell expression termed EYFP (Table 19.1).

As indicated by the emission maxima for EYFP, this variant yields a yellow-green fluorescence which is discernible from wt GFP, S65T, EGFP, and EBFP. Moreover, because of the position of the EYFP excitation and emission peaks, it is possible to design microscopy and flow cytometry filter sets (Lybarger *et al.*, 1998) compatible with dual reporter applications using EYFP and several other GFP variants (refer to Fig. 19.4 for the overlap of the excitation and emission spectra for each enhanced GFP variant). In fact, it now appears possible to perform three-colour detection using the appropriate filter combination and reporter constructs expressing EYFP, EBFP and either EGFP or S65T. The combination of these distinct spectral variants opens up a wide range of applications such as the simultaneous analysis of multiple promoter elements, intracellular localization of several different proteins, real-time analysis of protein–protein interactions with fluorescence resonance energy transfer (FRET; Heim & Tsien, 1996), and monitoring the lineage of different cell populations (Table 19.2). As additional variants are likely to follow by virtue of the information derived from the GFP and S65T crystal structures, it seems likely that the palette of GFP colours will continue to expand in the near future.

19.2.4 Red fluorescent protein?

For investigators involved in the pursuit of GFP mutants, the holy grail of this endeavour has been a red-emission variant. Such a variant would represent the ideal partner to EGFP, EYFP, or other yellow/green emission variants as the red signal is well suited to existing hardware and filter sets (rhodamine filters)

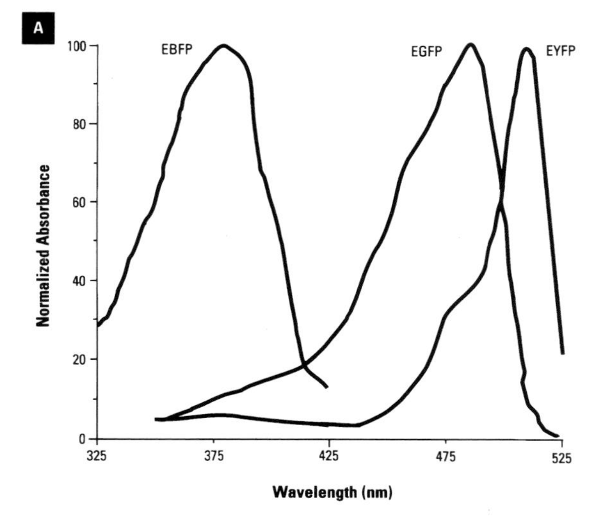

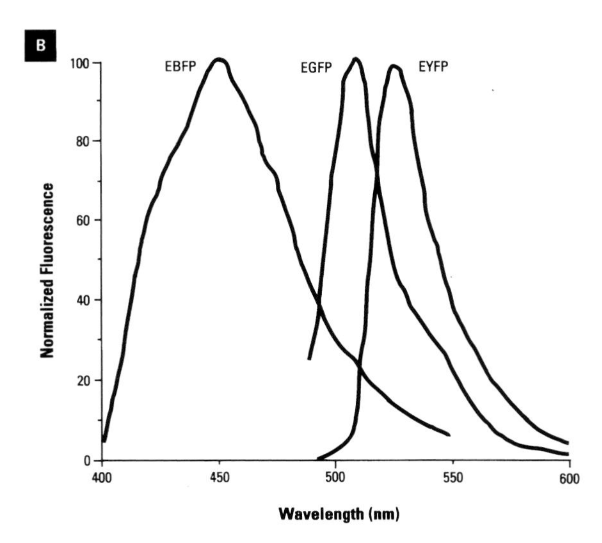

Figure 19.4 Fluorescence excitation (panel A) and emission (panel B) spectra of EGFP, EYFP, and EBFP.

used in microscopy and flow cytometry. Efforts to mutagenize the chromophore region of GFP and other domains within the protein have not yielded much in the way of long-wavelength fluorescence. At present, EYFP with an emission maxima of 527 nm is the closest thing available to 'RFP'. However, recent studies (Elowitz *et al.*, 1997) have revealed that a red-emitting form of GFP has been with us all along, it just required the appropriate conditions to blush.

Elowitz and co-workers have found that a green-absorbing, red-emission form of GFP can be generated by the photoactivation of pre-existing GFP with blue light under conditions of low oxygen. Photoactivation requires a fully formed chromophore. The results indicate that purified GFP protein, or GFP expressed in cells, can be converted to a red-emitting form by a simple microscope assay using a 488 nm laser. The resulting photoactivated form of the protein has an excitation maxima around 525 nm, and emission maxima of 560, 590, and 600 nm. The fluorescence from this form of the protein is easily distinquished from green fluorescence. The requirement for low oxygen is provided either with a nitrogen environment, by prolonged culture under sealed coverslips to deplete O_2, or by using an O_2-scavenging system consisting of glucose oxidase, catalase, and β-D-glucose. Remarkably, a variety of different GFP variants have been photoactivated when expressed in *E. coli*, including EGFP, GFPuv, S65T and GFPmut2 (Cormack *et al.*, 1996). In additon to bacterial expression, the GFPmut2 variant has been photoactivated to a red-emitting form in *Saccharomyces cerevisiae* and *S. pombe* (Sawin & Nurse, 1997), indicating the process may occur in a variety of cell types. The red-emitting form of GFP is stable for at least 24 hours once formed. While the requirement for a low O_2 environment may limit the practical utility of the photoactivation procedure, it is possible a more complete understanding of the photoactivation process can lead to the design of constitutively red mutants of GFP.

Table 19.2 Applications for emission variants of EGFP

- Intracellular localization of different proteins
- Separation of mixed cell populations by flow cytometry
- Simultaneous analysis of different promoters
- Monitoring different cell lineages in a single organism or mixed culture
- Dual-labelling with FITC-labelled reagents
- Fluorescence resonance energy transfer (FRET)

19.3 DETECTION OF APOPTOSIS WITH GFP

Apoptosis is an important and well-regulated form of cell death observed under a variety of physiological and pathological conditions. Inappropriate apoptosis has been implicated in many disease processes such as Alzheimer's disease, immune and autoimmune disorders, leukaemia, lymphomas, and several other malignancies (Duvall & Wyllie, 1986; Nicholson, 1996). Mammalian cell apoptosis can be triggered by a variety of stimuli, and proceeds by a regulated cascade of events beginning with activation of ICE-family proteases (caspases), and leading eventually to DNA fragmentation and rupture of the plasma membrane (Cohen *et al.*, 1992; Martin & Green, 1995). A relatively early event in the apoptotic cascade is the acquisition of cell surface changes by apoptotic cells that result in the recognition and uptake of these cells by phagocytes (Duvall *et al.*, 1985; Koopman *et al.*, 1994; Martin *et al.*, 1995). This recognition is due to the externalization of a plasma membrane phospholipid called phosphatidylserine (PS), an event which occurs regardless of the stimulus used to trigger apoptosis (Martin *et al.*, 1995). Most of the PS in healthy cells is localized to the interior (cytoplasmic) leaflet of the plasma membrane. Within one hour after the induction of apoptosis in most cell types, the PS redistributes to the outer leaflet of the plasma membrane. Based on the kinetics of this process, and the precedent for active transport of other membrane phospholipids, it is thought the PS is translocated by an energy dependent 'flipase' (Martin *et al.*, 1995). By assaying for PS externalization researchers can detect apoptosis as it begins, providing the means to analyse the early events which regulate programmed cell death.

19.3.1 Annexin V-EGFP

Annexin V is a calcium-dependent phospholipid-binding protein which preferentially binds to PS (Andree *et al.*, 1990; Thiagarajan & Tait, 1990; Martin *et al.*, 1995). Therefore, by conjugation of this protein to chemical fluorophores such as FITC, one can generate a useful probe for apoptotic cells which is compatible with both fluorescence microscopy and flow cytometry (Koopman *et al.*, 1994; Martin *et al.*, 1995; Zhang *et al.*, 1997). The principle of the annexin V assay is shown in Fig. 19.5. The assay is a simple one-step procedure that takes less than 10 minutes to perform. It is important to emphasize that the assay uses living cells, thereby allowing for recovery of the cells for analysis of later stages in apoptosis.

In order to improve sensitivity, and reduce complications for quantitative studies due to photobleaching and loss of signal, we constructed a genetic fusion between annexin V and EGFP and compared the performance of this reagent with that of annexin V-FITC. The results shown in Plate 19.1 illustrate that both annexin V-FITC (panel A) as well as annexin V-EGFP (panel B) yield the characteristic 'green halo' due to binding of the probe to cell surface PS. It is apparent from these results that annexin V-EGFP gives a brighter fluroescence signal, which was also more photostable following continued excitation with blue light (data not shown). This probe is also suitable to quantify the number of apoptotic cells by flow cytometry. As shown in Fig. 19.6, we observed that 65% of Jurkat cells treated with anti-Fas monoclonal antibody underwent PS externalization as measured by binding of annexin V-EGFP (compare panels A and B). The sensitivity provided by annexin V-EGFP in flow cytometric analysis of apoptosis was approximately 20-fold greater than the FITC conjugate (data not shown). These results suggest that annexin V-EGFP, as well as other conjugates such as EYFP, may provide adequate sensitivity for analysis

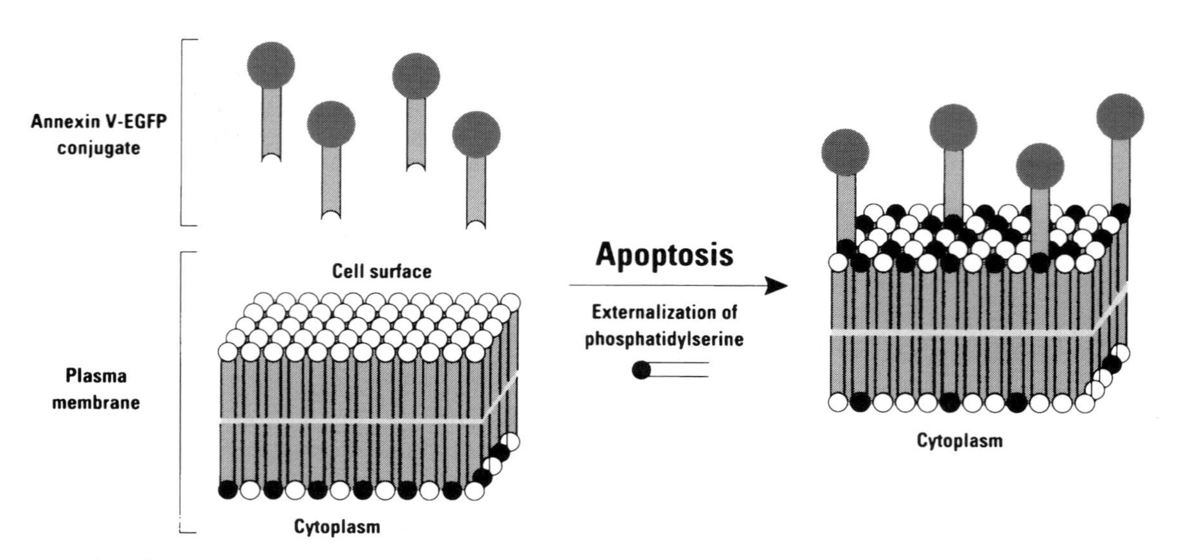

Figure 19.5 The principle of annexin V-EGFP detection of apoptosis. In healthy cells, PS is predominantly localized to the inner leaflet of the plasma membrane. Following induction of the apoptotic cascade, PS is translocated to the outer leaflet. Annexin V-EGFP binds to PS exposed on the cell surface in a Ca^{2+}-dependent fashion, thereby making the cell fluorescent.

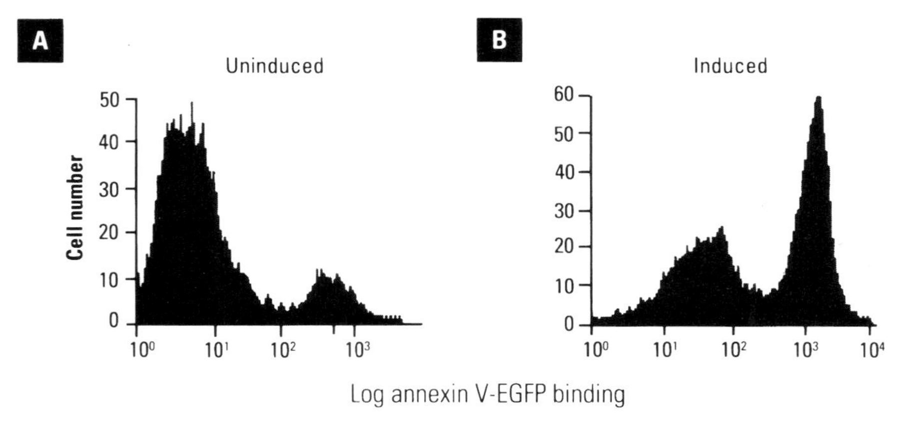

Figure 19.6 Detection of annexin V-EGFP binding by flow cytometry. Apoptosis was induced in Jurkat cells by incubation with 200 ng ml^{-1} anti-Fas monoclonal antibody (clone CH-11) for 6 hours (panel B). Cells were collected by centrifugation, and incubated with 0.4 μg ml^{-1} annexin V-EGFP for 10 minutes in the dark. (A) Uninduced control cells; (B) cells induced with anti-Fas antibody.

of apoptosis in multiwell plates designed for high-throughput screening and apoptosis-related drug discovery.

REFERENCES

Andree H.A.M., Reutelingsperger C.P.M., Hauptmann R., Hemker W.T. & Willens G.M. (1990) *J. Biol. Chem.* **265**, 4923–2928.

Brejc K., Sixma T.K., Kitts P.A., Kain S.R., Tsien R.Y., Ormö M. & Remington S.J. (1997) *Proc. Natl Acad. Sci. USA* **94**, 2306–2311.

Carey K.L., Richards S.A., Lounsbury K.M. & Macara I.G. (1996) *J. Cell Biol.* **133**, 985–996.

Chalfie M., Tu Y., Euskirchen G., Ward W.W. & Prasher D.C. (1994) *Science* **263**, 802–805.

Cody C.W., Prasher D.C., Westler W.M., Prendergast F.G. & Ward W.W. (1993) *Biochemistry* **32**, 1212–1218.

Cohen J.J., Duke R.C., Fadok V.A. & Sellins K.S. (1992) *Ann. Rev. Immunol.* **10**, 267–293.

Cormack B.P., Valdivia R. & Falkow S. (1996) *Gene* **173**, 33–38.24.

Crameri A., Whitehorn E.A., Tate E. & Stemmer W.P.C. (1996) *Nature Biotechnol.* **14**, 315–319.

Davis D.F., Ward W.W. & Cutler M.W. (1995) Proceedings of the 8th International Symposium on Bioluminescence and Chemiluminescence.

Duvall E. & Wyllie A.H. (1986) *Immunol. Today* **7**, 115.

Duvall E., Wyllie A.H. & Morris R.G. (1985) *Immunology* **56**, 351–358.

Elowitz M.B., Surett M.G., Wolf P-E. Stock J. & Leibler S. (1997) *Curr. Biol.* **7**, 809–812.

Flach J., Bossie M., Vogel J., Corbett A., Jinks T., Willins D.A. & Silver P.A. (1994) *Mol. Cell. Biol.* **14**, 8399–8407.

Haas J., Park E.-C. & Seed B. (1996) *Curr. Biol.* **6**, 315–324.

Heim R. & Tsien R.Y. (1996) *Curr. Biol.* **6**, 178–182.

Heim R., Prasher D.C. & Tsien R.Y. (1994) *Proc. Natl Acad. Sci. USA* **91**, 12501–12504.

Heim R., Cubitt A.B. & Tsien R.Y. (1995) *Nature* **373**, 663–664.

Inouye S. & Tsuji F.I. (1994) *FEBS Lett.* **341**, 277–280.

Kaether C. & Gerdes H.-H. (1995) *FEBS Lett.* **369**, 267–271.

Kain S.R., Adams M., Kondepudi A., Yang T.T., Ward W.W. & Kitts P. (1995) *Biotechniques* **19**, 650–655.

Koopman G., Reutelingsperger C.P.M., Kuijten G.A.M. Kechnen R.M.J., Pals S.T. & van Oers M.H.J. (1994) *Blood* **84**, 1415–1420.

Li X., Zhang G., Ngo N., Zhao X., Kain S.R. & Huang C.C. (1997) *J. Biol. Chem.* **272**, 28545–28549.

Lybarger L., Dempsey D., Patterson G.H., Piston D.W., Kain S.R. & Chervenak R. (1998) *Cytometry* **31**, 147–152.

Marshall J., Molloy R., Moss G.W.J., Howe J.R. & Hughes T.E. (1995) *Neuron* **14**, 211–215.

Martin S.J. & Green D.R. (1995) *Crit. Rev. Oncol. Hematol.* **18**, 137–153.

Martin S.J., Reutelingsperger C.P.M., McGahon A.J., Rader J.A., van Schie R.C.A.A., LaFace D.M. & Green D.R. (1995) *J. Exp. Med.* **192**, 1545–1556.

Morin J.G. & Hastings J.W. (1971) *J. Cell. Physiol.* **77**, 313–318.

Náray-Fejes-Tóth A. & Fejes-Tóth G. (1996) *J. Biol. Chem.* **271**, 15436–15442.

Nicholson D.W. (1996) *Nature Biotechnol.* **14**, 297–301.

Niswender K.D., Blackman S.M., Rohde L. Magnuson M.A. & Piston D.W. (1995) *J. Microbiol.* **180**, 109–116.

Ormö M., Cubitt A.B., Kallio K., Gross L.A., Tsien R.Y. & Remington S.J. (1996) *Science* **273**, 1392–1395.

Patterson G.H., Knobel S.M., Sharif W.D. Kain S.R. & Piston D.W. (1997) *Biophys. J.* **73**, 2782–2790

Prasher D.C., Eckenrode V.K., Ward W.W., Prendergast F.G. & Cormier M.J. (1992) *Gene* **111**, 229–233.

Rizzuto R., Brini M., De Giorgi F., Rossi R., Heim R., Tsien R.Y. & Pozzan T. (1996) *Curr. Biol.* **6**, 183–188.

Sawin K.E. & Nurse P. (1997) *Curr. Biol.* **7**, R606–R607.

Shimomura O., Johnson F.H. & Saiga Y. (1962) *J. Cell. Comp. Physiol.* **59**, 223–227.

Stearns T. (1995) *Curr. Biol.* **5**, 262–264.

Thiagarajan P. & Tait J.F. (1990) *J. Biol. Chem.* **265**, 17420–17423.

Wang S. & Hazelrigg T. (1994) *Nature* **369**, 400–403.

Ward W.W. (1981) In: *Bioluminescence and Chemiluminescence: Basic Chemistry and Analytical Applications*, M.

DeLuca & W.D. McElroy (eds). Academic Press, New York, pp. 235–242.
Ward W.W., Cody C.W., Hart R.C. & Cormier M.J. (1980) *Photochem. Photobiol.* **31**, 611–615.
Yang F., Moss L.G. & Phillips G.N. (1996) *Nature Biotechnol.* **14**, 1246–1251.
Yang T.T., Cheng L. & Kain S.R. (1996a) *Nucleic Acids Res.* **24**, 4592–4593.
Yang T.T., Kain S.R., Kitts P., Kondepudi A., Yang M.M. & Youran D.C. (1996b) *Gene* **173**, 19–23.
Zhang G., Gurtu V., Kain S.R. & Yan G. (1997) *Biotechniques* **23**, 525–529.

CHAPTER TWENTY

Rapid Detection of Microorganisms

R. HAGGART, N. BODSWORTH & N. FOOTE
Celsis Ltd, Cambridge Science Park, Cambridge, UK

Luminescence is a generic term covering a range of processes that produce light, but does not include hot bodies that are self-luminous solely because of their temperature, i.e. incandescence.

Luminescence is the emission of light by an atom or molecule as an electron is transferred to the ground state from a higher state. A luminescing system is losing energy and for light emission to continue indefinitely some form of energy must be supplied from elsewhere. The classification of various types of luminescence is partly based upon the source of the excitation energy (Table 20.1).

Luminescence detection of analytes offers an advantage over the widely used absorption method. Absorption methods involve a comparison of the intensity of light transmitted by the sample and by a suitable reference. Luminescence measurements in contrast, involve a direct measurement of emitted light.

In absorption spectrometry the concentration of analyte is proportional to the logarithm of the ratio of transmitted light (I) to incident light (I_0) and the instrument factor governing the minimum detectable concentration is the minimum detectable difference between I and I_0. With commercial instruments the overall precision with which $\log(I_0/I)$ can be measured is usually no greater than 0.001 units using a 1 cm cuvette optical depth (Williams, 1983). A molecule of medium size has an extinction coefficient ($E_{max.}$) usually not greater than 10^5. Using the Beer–Lambert law,

Table 20.1 The classification of various types of luminescence based upon the source of excitation energy

Classification	*Energy source*
Electroluminescence	Electrical current
Radioluminescence	Radioactive material
Chemiluminescence	Chemical reaction
Bioluminescence	Chemical reaction (within living organisms)
Thermoluminescence	Heat
Triboluminescence	Mechanical
Sonoluminescence	Sound
Photoluminescence (fluorescence and phosphorescence)	Radiation (UV, visible, IR)

FLUORESCENT AND LUMINESCENT PROBES, 2ND EDN
ISBN 0–12–447836–0

$$\log (I_0/I) = A = Ecd$$

where A is the absorbance (a dimensionless number), d is the sample depth (cm), c is the concentration (mol l^{-1} or M) and E is the extinction coefficient, the maximum concentration detectable is around 10^{-8} M (Clayton 1970).

The intensity of fluorescence emission is defined by the formula

$$F = I_0(1 - 10^{-Ecd})Q$$

where F is the intensity of fluorescence and Q is the quantum efficiency of fluorescence. This formula reduces to $F = I_0(2.3\ Ecd)Q$ when Ecd (absorbance) is small (Parker, 1968).

The sensitivity of a spectrofluorometer is limited in principle only by the maximum intensity of light available and not by the precision with which the light intensity can be measured. Under ideal conditions concentrations as low as 10^{-12} M can be detected (Parker, 1968). Mathies & Stryer (1986) have detected three molecules of the fluorescence molecule phycoerythrin obtained from photosynthetic organelles of cyanobacteria and red algae (E_{max} 2.4×10^6 at 562 nm due to its 34 bilin chromophores, $Q = 0.98$) using a modified flow cytometer. Nguyen *et al.* (1987) detected one molecule of phycoerythrin using similar techniques.

Although it can be demonstrated that the ultimate detection of single molecules can be achieved, the conditions used in these measurements are unique in that special care has been taken to set up such an instrument, assay conditions are optimal, and the solution of analyte is free from contaminating substances. Such assays are not possible when measuring analytes in body fluids, such as whole blood, serum or tissue samples, where native molecules can absorb and/or fluoresce at the wavelength of the analyte measurement. It is the effect of contaminating substances in the assay mixture that often limits the detection of analytes in biological samples. Most sensitive assay systems involve the physical separation of the analyte from the sample and in some way amplifying the signal obtained from the analyte.

The development of enzyme immunoassay provides a means of separating the analyte from the sample using specific trapping antibodies attached to a solid phase, then detecting the analyte using a second antibody conjugated to a label. If the label is an enzyme, it provides a means of amplifying the signal and increasing the sensitivity of the assay. The detection of enzymes using chemiluminescence substrates has achieved detection limits of 10^{-21} moles of alkaline phosphatase (A.P. Schaap, unpublished results).

Detection of the light emission from such assays is normally achieved using a photomultiplier tube (PMT) based luminometer. PMTs are sensitive detectors capable of counting single photons emitted from an assay sample. Although PMTs are not capable of spatial resolution in the sample, they can be incorporated into detection systems in which the samples are contained in tubes or microtitre plate wells. These are sequentially presented to the detector and the light emission from each sample measured.

Sensitive detection of chemiluminescence assays with spatial resolution can be achieved using imaging cameras. Imaging detectors have the advantage of being able to detect and quantitate individual emitted photons and represent this emission in the form of a two-dimensional image. Astronomers have used such detection systems to detect and quantitate the light emitted from stars and galaxies (Kristian & Blouke, 1982).

There are large numbers of chemiluminescence substrates that can be used in chemiluminescent-based assays (McCapra *et al.*, 1989). Detection of such chemiluminescence reactions using imaging detectors tends to be limited to those reactions which emit light using glow-type kinetics as opposed to flash kinetics. Chemiluminescence having glow kinetics allows the reaction to be initiated prior to imaging. Reactions exhibiting flash kinetics require the reaction to be initiated in the field of view of the detector. This can be achieved using injectors at the measuring head, as seen in commercial luminometers based on PMT detectors. However, luminometers tend to confine the reaction to tubes or wells of microtitre plates where the small area of the tube or well allows good mixing of reactants and the injectors are well away from the measuring head or field of view of the PMT.

It is difficult to get even mixing of reagents fast enough to initiate the reaction simultaneously over a large enough area to provide accurate, reproducible, quantitative results. Enhanced chemiluminescence reaction uses sodium luminol as the substrate emitting light in the blue area of the electromagnetic spectrum (425 nm). This reaction shows glow kinetics after an initial increase in light production over a 30 second period. The reaction depends on the peroxidase-catalysed oxidation of the sodium luminol producing the light-emitting 3-aminophthalate ion (Thorpe & Kricka, 1986). The light-emitting species is formed at the HRP reactive site, making it ideal for imaging. It can be shown that removal of the enzyme HRP immobilized on to a solid-phase polystyrene bead from the glowing reaction mixture extinguishes the light from the reaction mixture. The light can be replenished by placing the bead back into the same reaction mixture (Haggart & Leaback, 1989).

The sensitivity of detection of the chemiluminescence assay is dependent on the wavelength of the light emitted and the spectral response of the detector.

Image-intensified charge-coupled device (CCD) cameras are sensitive cameras allowing real-time imaging of chemiluminescence reactions. The spectral response of the camera is dependent on the photocathode material of the image intensifier. As the majority of the photocathodes are bialkali, the peak quantum efficiency (22%) is in the blue region of the spectrum (~400 nm). Photocathodes with peak quantum efficiencies in the green (525 nm) are available but their higher noise characteristics limits their performance. Chemiluminescent reactions emitting in the blue end of the spectrum are optimal for this type of imaging camera.

Cooled slow-scan CCD cameras have no image intensifier and the light collected by the lens focuses the light directly on the silicon-based CCD detector. The peak quantum efficiency of these detectors is around 50% in the red end of the spectrum, ~700 nm, owing to the light absorbance characteristics of the silicon. CCDs coated with a UV/blue fluorescent film or special thinned-back illuminated CCDs allow quantum efficiencies as high as 40% in the blue region of the spectrum and as much as 90% efficiency in the red end of the spectrum (Mackay, 1991). Cooling these cameras (−45°C) is essential for low light level performance. Their high sensitivity and good spatial resolution allowed the detection of cellular analytes such as oestrogen receptors in sections of breast tissue (Haggart & Leaback, 1989; Fig. 20.1(A)), and the detection of chemiluminescent-based immunoassays (Leaback & Haggart, 1989; Fig. 20.1(B)).

The most common method of detecting microbes is to incubate the sample with a nutrient medium such as a broth or agar plate, and detect growth by change in turbidity of the broth or the presence of colonies on the surface of the agar plate. These methods typically involve incubation periods between 2 and 5 days.

All living cells, microbial or eukaryotic, contain adenosine triphosphate. The presence of ATP can be detected using firefly (*Photinus pyralis*) bioluminescence. The luciferase enzyme from the firefly reacts with ATP, together with the co-substrate D-luciferin, in a reaction which gives rise to the emission of light.

$$\text{ATP} + \text{D-luciferin} + O_2 \xrightarrow{\text{firefly luciferase}} \text{AMP} + \text{PPi} + CO_2 + \text{oxyluciferin} + \text{light}$$

The colour of the emitted light is green (λ_{max} 562 nm). The remarkable property of the reaction is that the quantum yield is virtually 100%; i.e. almost every molecule of ATP that is acted upon by luciferase gives rise to a photon.

However, even the most sensitive bioluminescence reagents are not capable of detecting a single bacterium. The drawback of relying on detection of a metabolite such as ATP is that an organism contains only a limited pool, and the analyte is destroyed by the luciferase reaction. No more than one photon can be obtained from each ATP molecule. On the other hand, if the target is an enzyme, each molecule can be made to produce many molecules of product in a short period of time. This gives a time-dependent signal amplification. When the product of the enzyme reaction is ATP it can be measured using bioluminescence but with the increased sensitivity given by enzymatic cycling. In effect, the microbial ATP is amplified.

The most suitable enzyme has been identified as adenylate kinase (AK). Its reaction is a special case of the kinase mechanism in which a phosphate group is reversibly transferred to ADP from a phosphorylated compound, forming ATP. The general reaction scheme of 'X kinase' is:

(A) 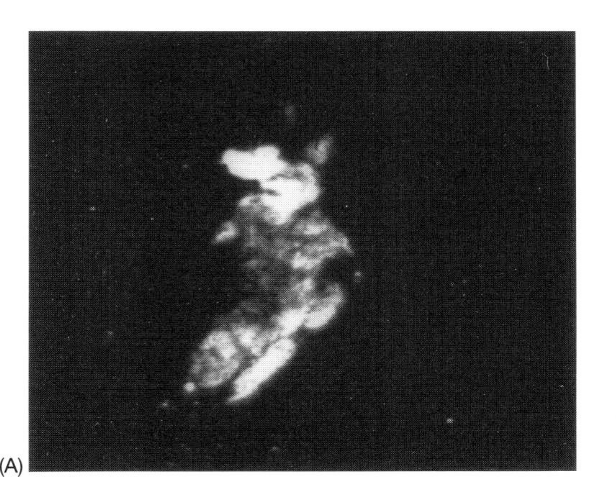

(B) 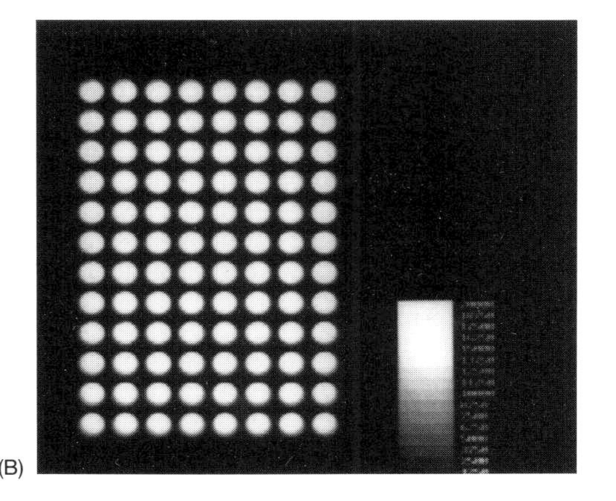

Figure 20.1 Cooled slow-scan CCD images of the light emission from the enhanced chemiluminescence reaction. (A) The detection of oestrogen receptor present in a section of breast tumour. (B) Light emitted from 96 wells of a microtitre plate.

$$ADP + X\text{-}P \xleftrightarrow{\text{X kinase}} ATP + X$$

In the case of adenylate kinase, the substrate 'X-P' is actually a second molecule of ADP and the reaction can be written:

$$2ADP \xleftrightarrow{\text{adenylate kinase}} ATP + AMP$$

Published values vary, but as a guideline a healthy bacterium may contain approximately 1 attomole (10^{-18} moles) of ATP. In addition, it contains enough AK to produce 40 attomoles of ATP per minute under optimum conditions. Therefore, intrinsic microbial ATP can be amplified 40-fold in one minute. Since the reaction is linear with time, a 25 minute reaction – still within the time frame of 'rapid microbiology' – can be expected to give a 1000-fold amplification.

The amounts of both ATP and AK within a microorganism will depend on the type of organism and its physiological state and may vary greatly. However, in all of the assays performed so far where ATP bioluminescence has been compared directly with an AK assay, the latter has consistently proved to be more sensitive.

The most common method of enumeration of bacteria present in water is to filter the sample to trap the organisms on the surface of the membrane and place the membrane on a nutrient source such as an agar plate. Using a combination of amplification of the ATP content by using the cells, native AK and a low decay rate bioluminescence reagent (glow kinetics), together with a sensitive imaging camera, it is possible to reduce the time required for detection from days to hours.

The imaging camera is based on an image intensified CCD camera (Millipore RMDS Microstar) constructed specifically to detect organisms filtered onto membranes using bioluminescence or chemiluminescence. A schematic diagram of the detection system is shown in Figure 20.2.

We describe a method to detect rapidly and accurately the presence of microrganisms filtered on to membranes. This reduces the detection time from 5 days to 2.5 hours.

20.1 MATERIALS AND METHODS

Cell suspensions of *Escherichia coli* and *Candida albicans* were prepared from broth cultures incubated overnight at 37°C, by diluting them in Maximum Recovery Diluent (Oxoid, CM733).

To induce a stressed state similar to conditions found in processed water samples, a suspension of *Pseudomonas aeruginosa* was diluted in water and left for 14 days. Aliquots, containing 30–80 cells, were removed and added to a large volume of sterile water which was then filtered through pre-wetted 0.45 μm partitioned membrane (Millipore, MSHV 04720).

Membranes were assayed immediately or placed onto an agar plate (TSA or R2A) for a predetermined

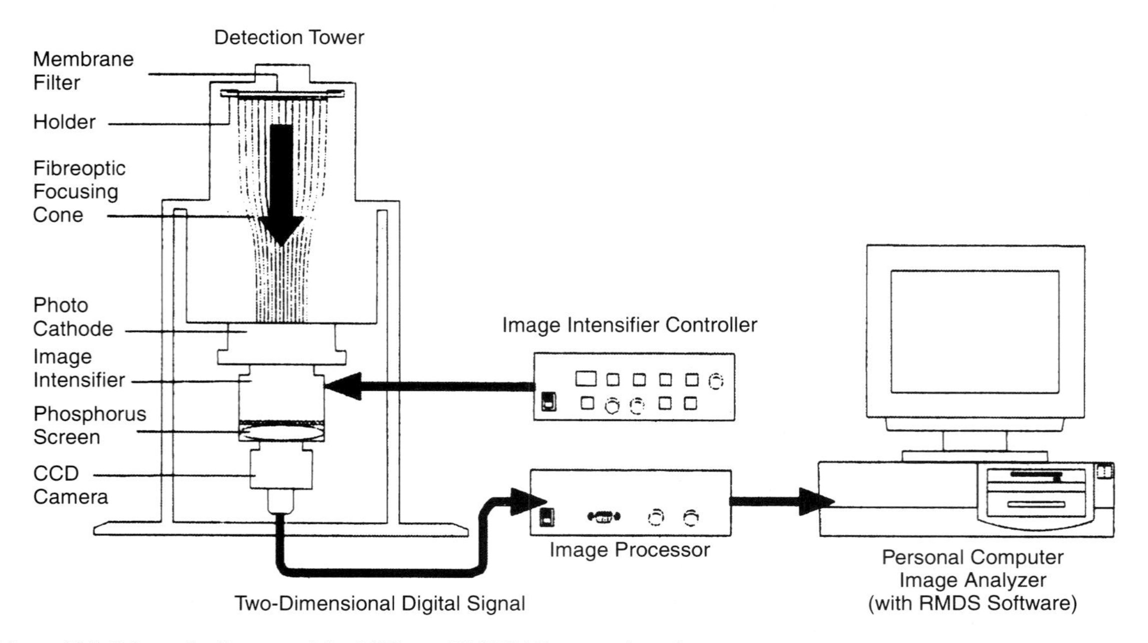

Figure 20.2 Schematic diagram of the Millipore RMDS Microstar detection system.

time to resuscitate the stressed organisms before assaying.

A solution to induce cell permeability and containing adenosine diphosphate (ADP) was applied, followed by a cocktail of low decay rate bioluminescence reagent (Celsis.Lumac). Images of the light emitted from the organisms over a 2-minute integration period were recorded by the Millipore RMDS Microstar. Duplicate membranes were grown for a further 2–5 days and the colonies formed were counted by eye.

To understand the variation of sensitivity across the imaging surface, a solution of the bioluminescence cocktail (3 ml) was premixed with 10 μl of 2×10^{-6} M ATP solution and poured into a 50 mm diameter Petri dish. The imaging surface was levelled prior to imaging the solution. The intensity of the light emitted from this homogeneous solution was recorded after a 10-second integration period by the Millipore RMDS Microstar. The level of the imaging surface was checked again after imaging.

A green B-light, in a microtitre plate format (calibrated at the National Physical Laboratory, Teddington, λ_{max} = 560 nm, 1.6×10^9 photons s^{-1} sr^{-1}), was used to calibrate the Millipore RMDS Microstar by a process of substitution. The NPL B-light was placed into a microtitre plate luminometer (Millipore, MicroCount) and the light intensity from a 1-second integration was recorded. A second green B-light (λ_{max} = 560 nm) was placed in the Microstar and the light intensity from a 1-second integration was recorded. This second B-light was then placed into a microtitre plate luminometer (Millipore Microcount) and the light intensity from a 1-second integration was recorded. Using this substitution method it was possible to calculate and convert the intensity of light recorded by the Microstar as relative light units (RLU) to photons s^{-1} sr^{-1}.

20.2 RESULTS

A pseudo-colour flat-field image of the glowing homogeneous solution is shown in Fig. 20.3. The software of the Microstar did not allow the measurement of the variation of the light intensity across the image. It is evident from the image that the light intensity across the image was not even, a lower intensity of light was recorded on the left side of the image. This detected variation did not seem to affect the performance for microbial detection

Table 20.2 shows the results from the calibration experiment. The light intensity from the NPL B-light is 1.6×10^9 photons s^{-1} sr^{-1}. The Millipore Microcount gave an average reading of 558 587 RLU s^{-1}. Each RLU s^{-1} is equivalent to 2864 photons s^{-1} sr^{-1}. The second B-light gave a reading of 3018 RLU s^{-1} when measured in the Millipore MicroCount, which is equivalent to 9×10^6 photons s^{-1} sr^{-1}. The light intensity from the second B-light measured from the image recorded in the Millipore Microstar was 36 040 units s^{-1}. Therefore each unit s^{-1} intensity of light (λ_{max} = 560 nm) measured from the images recorded by the Millipore RMDS Microstar is equivalent to 239 photons s^{-1} sr^{-1}.

Figure 20.3 Flat-field image recorded by the Millipore RMDS Microstar.

Table 20.2 Results from the calibration experiment

Second B-light recorded in the Millipore RMDS Microstar (units s^{-1})	*Second B-light recorded in the Millipore Microcount (RLU s^{-1})*	*NPL B-light recorded in the Millipore Microcount (RLU s^{-1})*
36 040	3018	558 587

Table 20.3 shows a comparison of numbers of colonies of *E. coli* counted as visible colonies on membranes incubated on agar plates (TSA, 2 days at 37°C) with the counts detected from the image taken by the Millipore RMDS Microstar (Fig. 20.4) after the membrane was incubated on the agar plate for 2.5 hours. There was a very good correlation between the colonies visualized on the images recorded by the Microstar (Fig. 20.5) after 2.5 hours resuscitation with the colonies visible to the eye after 2 days growth (r = 0.97).

A cell suspension of *C. albicans* was filtered onto partitioned membranes treated with cell permeabilization solution containing ADP, followed by bioluminescence cocktail, before being imaged for 2 min in the Millipore RMDS Microstar. Figure 20.6 shows the image recorded. A count of 46 from the image compared with 41 colonies visible on a duplicate membrane grown for 3 days on TSA plate at 37°C.

P. aeruginosa were stressed by transferring to water and left for 14 days. The stressed organisms were filtered then imaged immediately or placed on an

Table 20.3 Comparisons of the number of colonies counted on membranes grown on agar plates for 2 days with counts from images taken by the Millipore RMDS Microstar after 2.5 hours resuscitation on agar plates

Membrane filtration	*Microstar counts*
79	45
75	53
40	37
43	36
23	18
26	26
7	7
16	14
10	8
7	6
6	3
4	6
2	2
1	2
0	2
0	0

agar pate (R2A) for a few hours prior to imaging (Fig. 20.7). A graph was plotted of the total light intensity from the membranes imaged after no growth, 3 h and 5 h growth. This is shown in Fig. 20.8.

20.3 DISCUSSION

The Millipore RMDS Microstar is an image-intensified CCD camera for the detection of bacterial colonies on membranes using a chemiluminescence or bioluminescence-based assay system.

The analysis of the light detected across the imaging surface using a homogeneous light-emitting solution indicated an unevenness, as shown by the pseudo-colour image. This may be due to a number of factors, variation in the light-collecting ability of the optical fibres, different sensitivity of the microchannels, or variations in the sensitivity of the photocathode or phosphor screen. It would be preferable if the software were to allow flat-field correction of images to take into account any variations of the sensitivity across the image.

The calibration of the Millipore RMDS Microstar using the green B-lights shows that one of the Microstar light units is equivalent to 239 photons s^{-1} sr^{-1}, giving a light collection efficiency of 0.4% at a λ_{max} of 560 nm. A PMT-based luminometer has an estimated light collection efficiency of around 0.25% at λ_{max} = 560 nm (Stanley & McCarthy, 1989) and approximately 0.7% at λ_{max} = 460 nm (Haggart & Leaback, 1990).

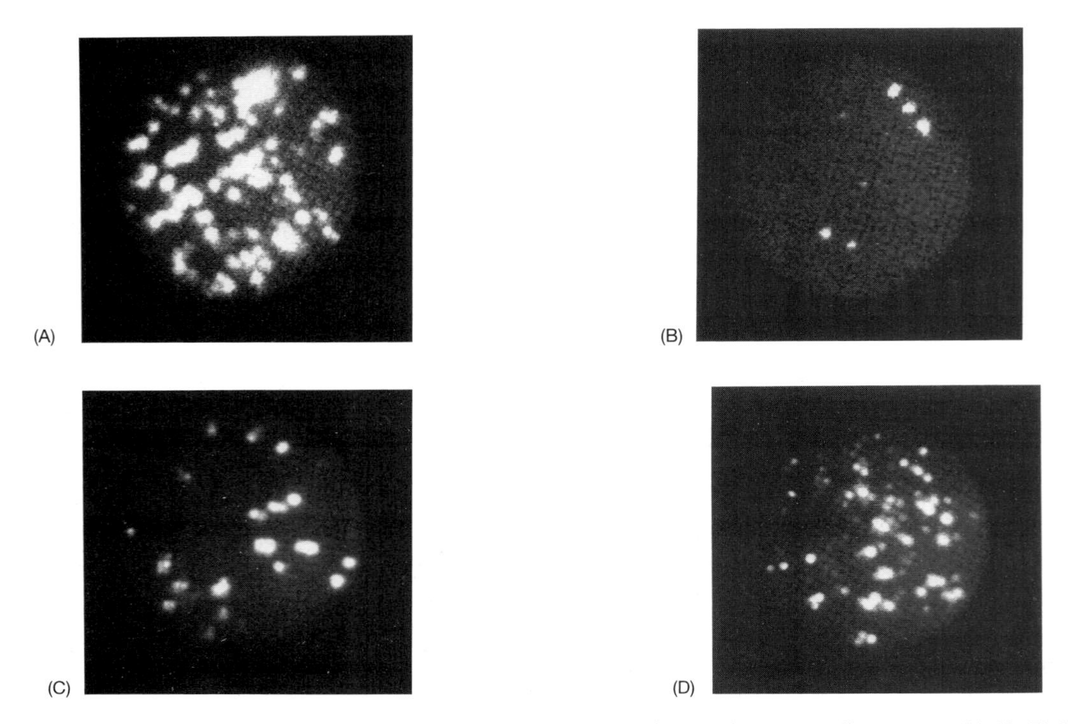

Figure 20.4 Images of *E. coli* after 2.5 hours resuscitation on agar plates and corresponding counts. (A) 53 (79,75), (B) 37 (40,43), (C) 18 (23,26), (D) 6 (6,4). The number of colonies counted by eye, after 2 days growth on agar plate, present on duplicate membranes are shown in brackets.

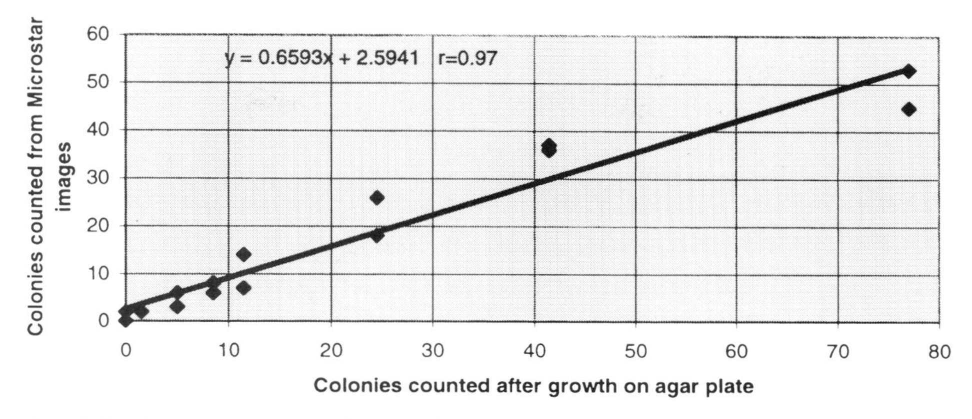

Figure 20.5 Correlation between the counts of *E. coli* visualized on the images recorded by the Microstar after 2.5 hours resuscitation with the colonies visible to the eye after 2 days growth.

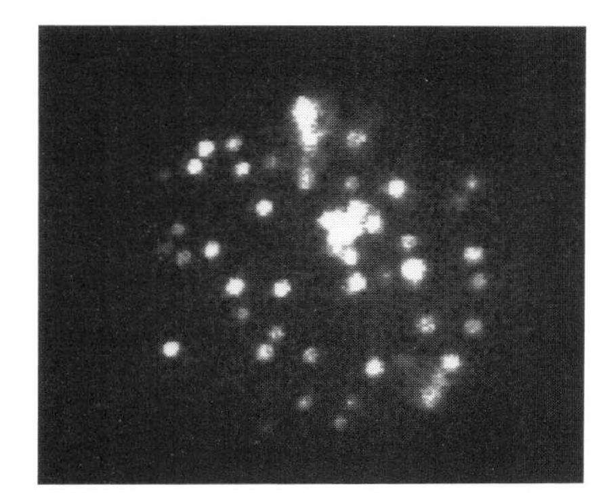

Figure 20.6 Millipore RMDS Microstar image of *C. albicans* cells (46) imaged immediately after filtration of the sample. A duplicate membrane grown for 3 days before the colonies could be visualized gave a count of 41.

The most common method of enumeration of bacteria present in water is to filter the sample to trap the organism on the surface of the membrane and place the membrane on a nutrient source such as an agar plate. Depending on how stressed the organism is, it can take between 2 and 5 days before the colonies are of a sufficient size to be visible to the human eye and be counted.

Using a combination of amplification of the ATP content by using the cells' native adenylate kinase and a low decay rate bioluminescence reagent, together with a sensitive imaging camera, it was possible to reduce the time required for resuscitation from days to hours.

The yeast *C. albicans* could be detected immediately after filtration onto the membrane. These cells are 100 times larger than bacteria and contain more ATP and approximately 100 times more adenylate kinase. This increases the amounts of ATP present and the amount of light that can be produced by each cell, making it more likely to be detected. Duplicate sample filtered

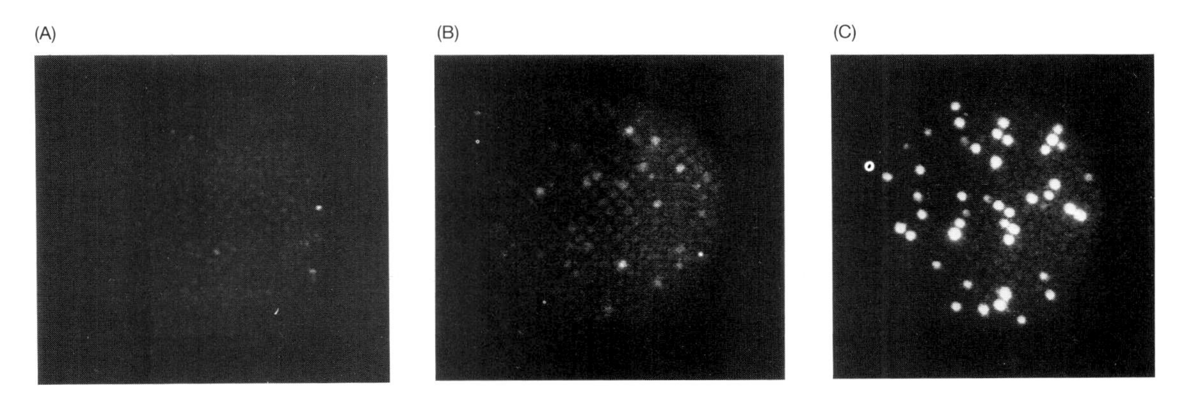

Figure 20.7 Images of membranes containing *P. aeruginosa* taken after (A) no resuscitation period (6 colonies) and (B) 3 hours (12) and (C) 5 hours resuscitation on an agar plate (44). A similar membrane grown on an agar plate for 5 days gave a count of 62 colonies.

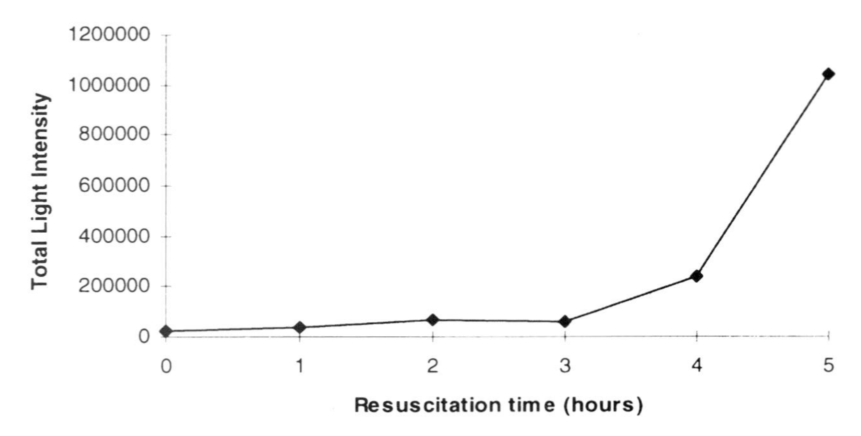

Figure 20.8 The relationship of the length of resuscitation time for stressed *P. aeruginosa* with the intensity of light measured from the images recorded by the Millipore RMSD Microstar.

onto membranes and grown for 2 days gave a similar count to that achieved by the Millipore RMDS Microstar.

Analysis of dilutions of a suspension of *E. coli* again confirmed that the Millipore RMDS Microstar could detect bacteria on a membrane more quickly (2.5 hours) and accurately ($r = 0.97$) as compared with counting colonies by eye, after a 2-day growth period on an agar plate.

P. aeruginosa are common contaminants of processed waters. Because the processed water contains limited nutrients, these bacteria tend to be in a highly stressed state. In this condition it takes a long time for these organisms to recover their normal growth cycle. Special agars such as R2A are used to obtain full recovery. Organisms that had been left in purified water for 14 days to induce a stressed state took 5 days to grow enough to produce colonies that were visible to the eye. The Millipore RMDS Microstar could detect comparable numbers of cells on a duplicate membrane after only 5 hours growth. A graph of total light detected from membranes grown on the agar for different times shows the dramatic increase in light emission after 5 hours growth.

The use of a sensitive imaging camera together with a novel bioluminescence detection system allows a fast and accurate method of detecting contamination in samples such as processed water. The method of filtering a large sample volume is simple to do and follows a conventional well-proven method of isolating organisms in processed water samples. The vast improvement in detection times, a factor of 24 times better, means that products that rely on processed waters can be released much more quickly, thus cutting costs associated with waiting for the agar plate test results before a product can be released.

Work is in progress to decrease or eliminate the recovery time required to detect organisms.

REFERENCES

Clayton K.R. (1970) *Light and Living Matter: A Guide to Photobiology*, Vol. 1: The Physical Part. McGraw-Hill, New York.

Haggart R. & Leaback D.H. (1989) *Anal. Chim. Acta* **227**, 257–265.

Haggart R. & Leaback D.H. (1990) *Bioluminescence and Chemiluminescence: Current Status.* John Wiley & Sons Ltd, New York, p. 365.

Kristian J. & Blouke M. (1982) *Sci. Am.* **247**, 67–74.

Leaback D.H. & Haggart R. (1989) *J. Biolumin. Chemilumin.* **4**, 512.

Mathies R.A. & Stryer L. (1986) In *Applications of Fluorescence in the Biomedical Sciences*, Liss, New York, p. 129.

Mackay C.D. (1991) *A Technical Guide from Astromed Ltd.*, Vol. 1. Astromed Ltd.

McCapra F., Watmore D., Sumun F., Patel A., Beheshti I., Ramakrishnam K. & Branson J. (1989) *J. Biolumin. Chemilumin.* **4**, 51–58.

Nguyen C.D., Keller R.A., Jett J.H. & Martin J.C. (1987) *Anal. Chem.* **59**, 2158.

Parker C.A. (1968) *Photoluminescence of Solutions.* Elsevier, Amsterdam, p. 21.

Stanley P.E. & McCarthy B.J. (1989) *Society of Applied Bacteriology, Technical Series No.* **26**, 73.

Thorpe G.H.G. & Kricka L.J. (1986) *Methods Enzymol.* **133**, 331.

Williams P.W. (1983) In *Biochemical Research Techniques*, J.M. Wrigglesworth (ed). John Wiley & Sons Ltd, New York, p. 49.

PART VII

Applications of Confocal Microscopy for Optical Probe Imaging

CHAPTER TWENTY-ONE

Confocal Microscopy – Principles, Practice and Options

C.J.R. SHEPPARD[1,2]

[1] Physical Optics Department, School of Physics, University of Sydney, NSW, Australia

[2] Australian Key Centre for Microscopy and Microanalysis, University of Sydney, NSW, Australia

21.1 INTRODUCTION

The main advantage of confocal microscopy (Sheppard and Shotton, 1997) is its ability to produce three-dimensional (3-D) images of thick objects. This is possible because of its so-called optical sectioning property, which allows sections to be imaged with minimal blur from other parts of the sample. At present its major application areas are in biology and medicine, where it is widely used in a fluorescence mode; and in industry, for inspection of surfaces and coatings, and measurement of small structures, for which it is used in a bright-field reflection mode.

It is generally accepted that the confocal principle was invented by Marvin Minsky, who filed a patent in 1957. Then, after being rediscovered several times, the first commercial instrument was launched in 1982. This instrument was primarily intended for the semiconductor device industry. The confocal microscope became widely known to biological researchers after strong publicity in 1987, and since that date several thousand microscopes have been sold worldwide. At present there are more than a dozen different manufacturers of instruments based on different technologies, so market competition is very strong.

The confocal microscope is based on the principle of the scanning optical microscope. In the most widely used and simple form of the instrument, laser light is focused by an objective lens (Fig. 21.1) on to the object, and the reflected, or fluorescent, light focused on to a photodetector via a beamsplitter. An image is built up by scanning the focused spot relative to the object, and is usually stored in a computer imaging system for subsequent display. To make the instrument operate as a confocal microscope, a confocal aperture, or pinhole, must be placed in front of the photomultiplier tube (or other) detector.

21.2 ADVANTAGES OF SCANNING

Broadly the advantages of scanning optical microscopy stem from two main properties (Sheppard, 1987). First is the fact that the image is measured in the form of an electronic signal, which allows a whole range of electronic image processing techniques, both

FLUORESCENT AND LUMINESCENT PROBES, 2ND EDN
ISBN 0-12-447836-0

analogue and digital, to be employed. These include image enhancement techniques such as frame averaging, contrast enhancement, edge enhancement, and image subtraction to show changes or movement; image restoration techniques for resolution enhancement and noise reduction; and image analysis techniques such as feature recognition and cell sizing and counting.

Second is the property that imaging in a scanning microscope is achieved by illuminating the object with a finely focused light spot. This allows a number of novel optical imaging modes to be employed such as confocal imaging or differential phase contrast, but also introduces the possibility of imaging modes in which the incident light spot produces some related effect in the specimen which can be monitored to produce an image.

A particularly important class of such methods occurs when the wavelength of the detected radiation differs from that of the illumination. An advantage of using scanning techniques for such spectroscopic imaging is that imaging methods, performed with the incident radiation, is separated from wavelength selection and analysis of the emitted radiation, thus simplifying system design and resulting in superior performance. The detection system, because it does not have to image, may also have greater sensitivity. This is of great advantage in fluorescence microscopy, which also results in the further advantage that the resolution is determined by the shorter incident wavelength, rather than the longer fluorescence wavelength.

Fluorescence microscopy can give information concerning spatial variations in excitation states, binding energies, band structure, molecular configuration, structural defects, and the concentration of different atomic and molecular species.

Use of a pulsed laser allows investigation of transient effects such as the lifetime of excited states, and capture and emission cross-sections. Other examples of spectroscopy which may be performed using scanning techniques include absorption spectroscopy, Raman spectroscopy, resonance Raman spectroscopy, coherent anti-Stokes Raman spectroscopy (CARS), two-photon fluorescence, photoelectron spectroscopy, and photoacoustic spectroscopy.

21.3 CONFOCAL MICROSCOPY

In a conventional microscope the object is illuminated using a large-area incoherent source by way of a condenser lens, and each individual point of the object is imaged by the objective lens. It is thus the objective which is responsible for determining the resolution of the system. If the image is measured point-by-point by a small detector, the image is unchanged providing the detector is small enough to be considered point-like. In a scanning microscope, on the other hand, we have a point source and a large area detector, rather than a large area source and a point detector. From reciprocity arguments, image formation is identical in a scanning microscope and a conventional microscope, but in the scanning microscope it is the first lens (also termed the objective, but sometimes the projector) which determines the resolution.

In the confocal microscope (Fig. 21.1) we use both a point source and a point detector, achieved by placing a pinhole in front of the detector, so that in this case imaging properties are completely different: both lenses take part equally in the image formation process with the result that the resolution is improved by a factor of about 1.4 times. This worthwhile improvement in resolution is an important advantage of confocal microscopy. Figure 21.2 shows the theoretical image of a single point object, illustrating the sharper image produced by the confocal microscope. Here the co-ordinate v is a dimensionless optical co-ordinate, related to the true radius r in the image by

$$v = (2\pi r\ n \sin \alpha)/\lambda \qquad [21.1]$$

where $n \sin \alpha$ is the numerical aperture of the objective and λ the wavelength of the illumination. The curves apply both for bright-field and also for fluorescence if the wavelength of the fluorescent light is approximately equal to that of the incident light. Also shown is the confocal image for a system where an annular mask is placed in the pupil of the objective

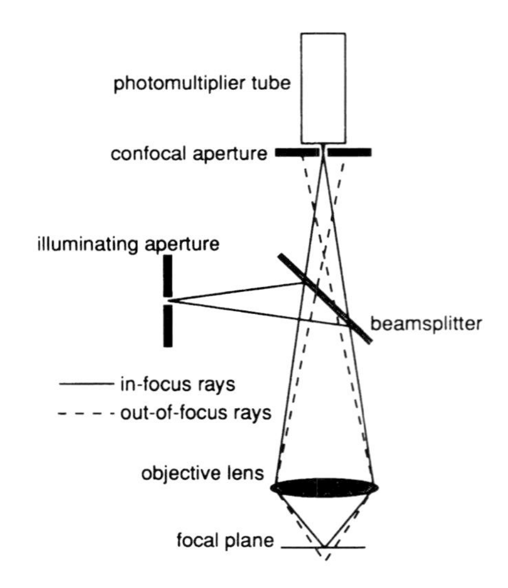

Figure 21.1 Schematic diagram of a confocal microscope. Light originating from points away from the focal plane is defocused at the confocal aperture and thus detected weakly.

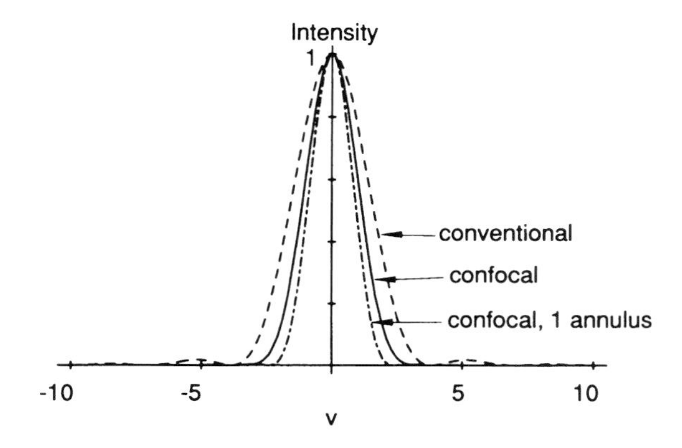

Figure 21.2 The image of a single point in conventional and confocal microscopes with circular pupils, and also in a confocal microscope with one circular and one thin annular pupil. The confocal image is a factor of 1.4 sharper than the conventional.

for any one of the two passes through the objective, thus resulting in an even sharper image. Details of confocal image formation can be found in Wilson & Sheppard (1984).

The principal advantage of confocal microscopy is, however, that out-of-focus information is rejected by the pinhole, so that an optical section is imaged. In addition, unwanted scattered light is also rejected by the pinhole, thus improving contrast. Light emanating from regions of the specimen distant from the focal plane are defocused at the pinhole plane and hence rejected, thus resulting in an optical sectioning effect. By scanning in the depth direction, sequential sections can be studied and a complete 3-D image built up. The 3-D data can, for example, be stored in a computer for subsequent processing and display.

Of course, 3-D images are difficult to display directly, so an alternative is to extract sections oriented in some arbitrary direction. For example xz images are sections parallel to the system axis. We can also produce projections in an arbitrary direction, in which the depth information is suppressed. Projections can be produced either by summation of sections (called an average projection), or alternatively by detecting the peak signal in depth (called a peak projection). Either approach can be achieved using digital or analogue methods: an average projection image can also be generated directly by photographic integration on a cathode ray tube. Two projections produced at slightly differing angles result in a stereoscopic image pair. Projections in different directions can be computed and stored for animated display, thus resulting in improved depth cues.

In addition to recording the peak signal in depth, as in the peak projection, we can record its position in depth. This depth information can be displayed as a surface profile image in grey-levels or colour, or can be combined with the peak information to produce colour-coded images or reconstructed views. The sensitivity of the depth measurement is about a nanometre. This surface profiling method is widely used for non-contact inspection in a range of industrial applications.

By using a mirror as a test specimen, the imaging performance in the depth direction can be investigated by axial scanning and observation of the resulting defocus signal. This can also be performed either by tilting the mirror slightly, and examining the line-scan signal when scanning in a transverse plane, or by producing an xz image and again extracting a line-scan. The defocus signal gives much information about the performance of the optical system. Ideally it should be a smooth, narrow response with weak side lobes. In practice it is made broader as a result of the finite size of the pinhole, and often degraded by poor alignment or the presence of aberrations. In particular the presence of strong side lobes or auxiliary peaks is very detrimental to 3-D imaging performance. In fact confocal microscopy is very sensitive to the presence of small amounts of spherical aberration which can be introduced by the use of incorrect coverglass thickness or when focusing deep into the specimen. In these cases observation of the defocus signal allows optimization of the imaging performance by adjustment of a correction collar if fitted to the objective.

The transverse resolution of the microscope is proportional to the numerical aperture of the objective. However, the axial imaging performance depends more strongly on objective aperture. The axial imaging performance can be quantified by observing the defocus signal. The width of the peak of the axial response varies considerably with aperture, so that in order to get good axial imaging it is necessary to use the largest possible aperture. Figure 21.3 shows the width of the axial response in wavelengths at which the intensity has dropped to one half (the full-width at half-maximum, or FWHM). The importance of using as high an aperture as is possible is apparent. The variation in width for both dry and oil immersion objectives are shown. It should be noted that the response is sharper for a dry, rather than an oil immersion, objective of the same numerical aperture, although of course higher numerical apertures are available for immersion objectives. It should also be remembered that immersion objectives also have better aberration correction than dry objectives.

At present most commercial confocal microscopes are designed for fluorescence operation, so that to many people confocal implies fluorescence. However, it should be stressed that there are many useful non-fluorescent applications using confocal bright-field

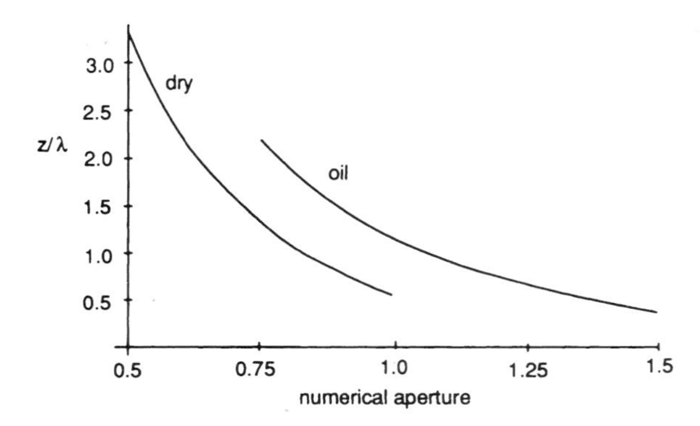

Figure 21.3 The width of the axial response in a confocal microscope from a plane reflector at which the intensity has dropped to one half (FWHM).

techniques. The main advantages of the fluorescence-mode of confocal microscopy are that specific stains can be introduced. Most commercial confocal fluorescence systems use an argon-ion laser to provide a main line at 488 nm which is close to the absorption peak of the fluorescent dye FITC. The other strong line of the argon-ion laser at 514 nm can be used to excite rhodamine or Texas red. The krypton-ion laser can give a large number of lines throughout the visible and ultraviolet. Alternative lasers for fluorescence work are the helium–cadmium laser (442 and 325 nm), helium–neon (633 nm) and the frequency-doubled neodymium YAG laser (532 nm).

Confocal microscopy in the bright-field (non-fluorescence) reflection mode has many applications in the materials science and industrial areas, and can also be exploited in biological studies, such as investigation of skin, hard tissue or the structure of the eye. Reflected bright-field has the avantage of giving extremely sharp depth imaging, sharper than in confocal fluorescence, but in some cases there can be difficulties in interpretation as a result of coherent noise (speckle). This is not usually a problem when the specimen consists of reflecting surfaces, but can be important when investigating a semitransparent object. The coherent noise is reduced by producing a projection (particularly by the average projection algorithm), and in some cases the visibility of the image features can be improved by digital filtering.

A further important technique of confocal reflection microscopy is the use of immunogold probes. If the gold particles are spatially well separated coherent interactions are not important, and the gold particles can be located in 3-D space. Scanning microscopes can be very sensitive for detecting scattering from these gold particles, and preparations with particles as small as 5 nm in diameter, without silver enhancement, can be imaged. In many cases, however, the lower limit to the size of the particles is set by the requirement that the scattering by the particles is strong compared with scattering by the refractive index variations in the specimen itself.

Although much of the early work on confocal microscopy was performed in the transmission mode, this is now very rarely used. The reason for this is that there is extreme difficulty in aligning, and maintaining alignment, during scanning, as a result of refractive effects in the specimen. This can be to some degree alleviated by using a double-pass method, where the transmitted light is relfected back through the specimen and detected using the usual reflected light detector. Confocal transmission retains the resolution improvement of confocal imaging, and also results in some improvement in depth imaging, but not to such a marked degree as in confocal reflection.

As described earlier, resolution of a scanning fluorescence microscope is superior to that in a conventional fluorescence microscope. There is a further improvement for confocal fluorescence microscopy, so that for example if the fluorescence wavelength is 1.5 times the primary wavelength the cutoff in the spatial frequency response is 1.5 times as great in a scanning compared with a conventional fluorescence microscope, but 2.5 times as great in a confocal fluorescence system. The price one pays for this dramatic improvement in resolution is the decreased signal strength of the confocal arrangement, although we argue later that the reduction in signal is caused by the rejection of unwanted, out-of-focus light.

21.4 PRACTICAL ASPECTS

In a scanning microscope the object is illuminated with a focused light spot which is scanned relative to the object. This can be achieved either by scanning the light spot, or by scanning the object itself. Most commercial instruments scan the light spot. The most widely used method employs galvo-mirror scanners for both the x- and y-scan. Galvo-mirrors can be either of the feedback-stabilized type, which can scan at a line frequency of about 1 kHz, or of the resonant variety which can oscillate at high speed but are fixed frequency devices. Alternative beam scanning systems include polygon mirror scanners, which achieve high scanning speeds but are difficult to synchronize, and acousto-optic scanners, which allow TV scanning rates but need special imaging geometries for fluorescence applications because of chromatic effects.

The main advantage of scanning the object rather than the beam is that the optical system is then completely invariant during scanning, so that the imaging properties do not vary across the field. Beam scanning systems, on the other hand, can exhibit brightness variations across the field, a fall-off in resolution at the edges of the field, and also noticeable curvature of field. Object scanning is thus preferable for quantitative work. The disadvantages of object scanning are that imaging is rather slow, taking perhaps a few seconds to record a single frame, and is not suitable with large or non-rigid samples, or when using electrical probes. The speed consideration is perhaps not so important when it is realized that often long scan times are necessary in order to collect sufficient light to produce a low-noise image.

A number of alternative arrangements are used for beam-scanning microscopes. Usually a scan lens, of similar design to a microscope eyepiece, is used to produce a focus which is subsequently imaged by the objective lens. In order that the beam fills the objective lens aperture, the axis of rotation of the scanning mirrors must be situated close to the plane of the entrance pupil of the scan lens. A single mirror placed here and scanned in both the x and y directions is the simplest design. Alternatively two separated galvo-mirrors can be used, coupled by a telecentric system. Finally, two close-coupled galvo-mirrors can be used. Sometimes the axis of rotation of one is offset so that it both translates and rotates.

In an object scanning system, the optical system can be simplified as the system is operated in an on-axis condition. It is found that it is possible to optimize performance by small corrections to the effective tube length of the objective. Mechanical object scanning is achieved using electromechanical devices, stepper motors or piezoelectric actuators. One feature which must receive some attention is to ensure that the plane of scanning is accurately located. This can be achieved by mounting the specimen stage on leaf springs or stretched wires.

An optical system only gives its best performance if it is accurately aligned. In particular a confocal microscope must be aligned correctly or it will exhibit artifacts, especially when a small pinhole is used. When the system is first set up, first of all the beam expanding system must be adjusted and the beam collimated so that the objective is used at its correct tube-length. The easiest way of aligning the system is to examine a planar object with a pinhole of large diameter. Without refocusing the object the pinhole size is reduced and its axial and transverse position adjusted for maximum signal. Alternatively, the 3-D image of a point-like object can be observed.

Various types of detector can be used for scanning optical microscopy. If signal level is high, as for example in transmission or in reflection from a surface such as bone or tooth, a simple photodiode can be employed. For confocal fluorescence applications, signal level is low, especially when imaging very thin optical sections. In this case highly sensitive detectors such as photomultiplier tubes, perhaps cooled, and sometimes with photon counting are advantageous.

Conventional photomultiplier photocathodes have a quantum efficiency (the fraction of light photons detected) of only about 20%. Silicon detectors of either the photodiode, avalanche photodiode or CCD variety can have quantum efficiencies greater than 80%. Avalanche photodiodes in principle combine high quantum efficiency with the ability to photon count, and with pulse height analysis for rejection of dark current. However, a problem is that they often cannot cope with the high count rates needed for images with many grey levels. CCDs combine high quantum efficiency with the opportunity for signal integration within the device, together with the possibility of cooling to improve noise performance. CCDs are usually of the area or linear array variety, and in some cases additional information can be extracted from this spatial information.

We are now beginning to see special objectives developed for confocal microscopy. These are water immersion objectives, designed for a limited field of view, but with long working distance, and sometimes good performance into the UV. The water immersion is important in order to avoid spherical aberration when focusing through a dielectric interface, for example from glass to a watery biological sample. There are also possibilities for other special designs. For example, for laser microscopy the objective only need be corrected for one wavelength (or two, including a fluorescent wavelength), which should allow lenses of increased numerical aperture or working distance to be developed. There is a need for low magnification lenses of high aperture, to give good collection efficiency. These are not necessary for use with on-axis microscopes as then one objective can be used to cover the whole range of magnifications, simply by altering the amplitude of scan. In microscopes with on-axis scanning, off-axis aberrations are unimportant, giving further flexibility in the design. For example, an immersion lens, designed for one wavelength and with a limited field of view for on-axis scanning, having a numerical aperture of 1.4 combined with a working distance of 1 mm seems feasible.

Achromat lenses have some advantages over apochromat lenses for some applications as they have lower loss, as well as being considerably cheaper. We have found fluorite lenses to be a good compromise for on-axis scanning. However for beam scanning the off-axis aberrations and field curvature are usually too large. Confocal microscopes are more sensitive to aberrations than conventional instruments, particularly for imaging in the depth direction. So incorporation of a correction collar is useful.

It is worth pointing out that many of the early scanning laser microscopes did not use a computer for image manipulation, but rather used analogue electronics together with a long persistence display. The main impetus for employing digital techniques was to provide image storage to improve the real-time observation of images, but in fact very high-quality images with several thousands of lines resolution can be produced by photographing directly the cathode-ray tube display. Similarly many forms of image processing such as contrast enhancement, filtering, image addition and subtraction, and so on can be achieved using analogue methods. The use of digital methods of course greatly extends the flexibility of the system, but nevertheless analogue processing facilities are worth retaining in order to improve the dynamic range of the recorded data.

A single 512 × 512 image contains a quarter of a megabyte of information. This means that a 3-D image consisting of many sections requires large memory and also processing time for 3-D manipulation. In fact, good stereoscopic effects can be obtained from only a few sections, and it is also possible to store projections directly, rather than sections, thus also greatly reducing the quantity of data. Personal computers nowadays have large enough memories to permit direct storage of the images in their RAM. PCs can provide most of the functions necessary for scanning microscopes, but are not fast enough for real-time manipulation of 3-D images. A range of standard image enhancement methods can be used with advantage in scanning optical microscopy. These include contrast enhancement by linear stretching or histogram equalization, edge enhancement by filtering in either the spatial or Fourier domain, low-pass or median filtering to reduce noise in images, image averaging and so on. Most of these methods can be employed with 3-D as well as 2-D data sets. Once a 3-D image has been stored, projections in arbitrary directions can be produced by image rotation. This is a computationally intensive process, and for small rotations a simpler alternative is to stack sections with an appropriate pixel offset between adjacent sections.

21.5 SYSTEM PERFORMANCE

With a small pinhole the system behaves as a true confocal microscope, but for large pinholes imaging performance is identical to that in a conventional microscope. Thus, between these two limits, the size of the pinhole affects the imaging performance greatly. In practice, microscope users often open up the size of the pinhole in order to get more signal from a weakly fluorescent or scattering object, so it is important to consider what effect this will have on the various imaging properties.

Figure 21.4 shows the effect of pinhole normalized radius v_d, defined by equation [21.1] for the *image* space, on the strength of the signal from a planar in-focus object, and the axial resolution of a planar object, assuming the system is shot-noise-limited (Sheppard *et al.*, 1995). The curves apply for a confocal fluorescence system. As the pinhole size is increased the signal increases, but the axial resolution decreases. It is seen that the signal rises quickly with increasing pinhole size, reaching about 70% for a value of v_d of 4. Thereafter there is little increase in signal level from a planar object. One might then ask why microscope users sometimes open up the pinhole much larger than this? The answer is that by using a larger pinhole the axial resolution is decreased, and hence for a thick object we collect a larger signal, simply because the optical sectioning effect is weaker. This means that the increased signal does not carry useful information.

The behaviour of the confocal fluorescence system is shown for both pinhole and slit forms of confocal aperture. It has been claimed that slit apertures can give improved signal strength, as indeed is apparent from Fig. 21.4, but in fact this occurs only because a thicker section is imaged, i.e. it is a direct result of the decreased axial resolution. Overall, therefore, there seems little to be gained from using a slit aperture from the point of view of imaging performance.

These effects can also be considered in terms of signal-to-noise ratio. If the system were assumed to be shot-noise-limited, the signal-to-noise ratio would increase monotonically with pinhole size. But in practice there is always stray light present, which also contributes to the noise observed. This greatly influences the noise behaviour of the system.

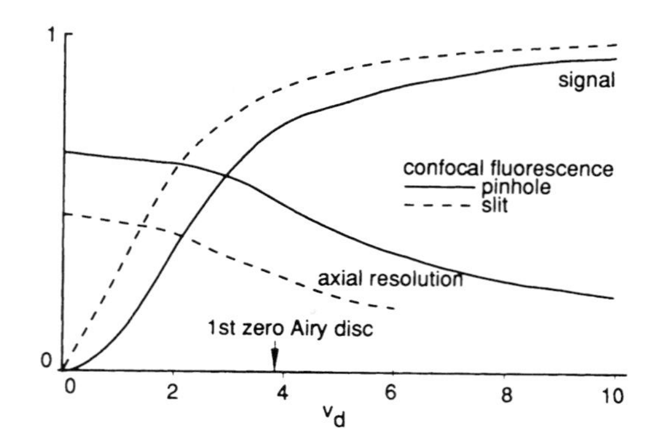

Figure 21.4 Comparison of the axial resolution and signal strength for a confocal fluorescence system with pinholes and slits of varying sizes. The first zero of the Airy disk in the pinhole plane is shown for comparison.

We have made measurements of signal and strength of stray light using an object scanning microscope. Very good agreement was found between the variation in the strength of the signal with pinhole size measured and predicted by theory. These measurements were used to calibrate the optical coordinate v_d. If the stray light is assumed to be uniformly distributed over the pinhole plane, its strength is proportional to the pinhole area, which was found to give a good fit to the experimental data. The stray light varied between 10^{-9} and 10^{-6} of the total signal for the pinhole sizes used. For small pinhole sizes the ratio signal/stray light became constant at about 2.5×10^7. This means that, in this particular microscope, the signal can be as low as 4×10^{-8} relative to the incident power before it is equal in strength to the stray light.

As the stray light is constant in strength throughout the image, it can be subtracted from the signal electronically to improve contrast. However, its presence nevertheless reduces the signal-to-noise ratio, which thus rises from zero for small pinholes, and eventually drops for large values of pinhole diameter as more stray light is detected. Interestingly, as the signal strength decreases, there is less latitude in the choice of pinhole size. The optimum pinhole size for maximizing signal-to-noise ratio also decreases. For a relative signal strength of 4×10^{-8}, for example, the optimum pinhole size has reduced to $v_d = 2.3$. This argument suggests that opening up the pinhole to detect more light may not be desirable, as more stray light will be detected also.

The standard design of confocal microscope images single sections at a rate of perhaps several per second. In order to investigate quickly varying objects it is desirable to be able to image at TV rates or faster. This can be achieved using resonant galvo-mirrors, but with fast scanning signal level can be very low, resulting in unacceptable images. There are two fundamental limitations to the signal level that can be detected in fluorescence microscopy: photobleaching and saturation. Photobleaching is an irreversible decay in fluorescence which occurs after exposure to a particular total illumination energy. It is known to be sensitive to the environment of the fluorescent molecule, particularly so to the presence of oxygen. The dependence on the exposure time is not so strong, and, indeed is not well established. Photobleaching thus sets the fundamental limit to the number of photons that can be detected, irrespective of the time taken to produce an image. Saturation, on the other hand, results from the non-linear variation in fluorescence yield with illumination power. In order to achieve a linear dependence between signal level and dye concentration, the illumination power must be kept below about 100 μW. This sets the limit to the speed of image capture, for if the scan time is reduced the power must be increased beyond an acceptable value, thus reducing fluorescence yield and increasing photobleaching. The way to overcome this problem is to illuminate many points simultaneously. This can be done by scanning using a rotating aperture disk, or by illuminating with a line, rather than a point, of light. Some designs allow direct observation of the resulting confocal images by eye. Unfortunately the imaging performance is necessarily compromised to some degree, but such instruments are particularly suited for fast imaging. If, on the other hand, one wants to obtain the best possible image, single-point scanning systems are preferable. There are numerous different confocal microscopes on the market, based on different designs, the relative merits depending on the particular application.

REFERENCES

Sheppard C.J.R. (1987) In *Advances in Optical and Electron Microscopy*, Vol. 10, R. Barer & V.E. Cosslett (eds). Academic Press, London, pp. 1–98.

Sheppard C.J.R. & Shotton D.M. (1997) *Confocal Laser Scanning Microscopy*. BIOS Scientific Publishers, Oxford.

Sheppard C.J.R., Gan X.S., Gu M. & Roy M. (1995) In *The Handbook of Biological Confocal Microscopy*, J. Pawley (ed.). Plenum Press, New York, pp. 363–371.

Wilson T. & Sheppard C.J.R. (1984) *Theory and Practice of Scanning Optical Microscopy*. Academic Press, London.

CHAPTER TWENTY-TWO

Dual-excitation Confocal Fluorescence Microscopy

KAY-PONG YIP[1] & IRA KURTZ[2]

[1] Department of Molecular Pharmacology, Physiology and Biotechnology, Brown University, Providence, RI, USA

[2] Division of Nephrology, Department of Medicine, UCLA School of Medicine, Los Angeles, CA, USA

22.1 INTRODUCTION

The biological problems which are studied with confocal fluorescence microscopy can be categorized into two groups: (1) the study of living or fixed preparations whose fluorescence properties do not change with time; and (2) the study of living preparations whose fluorescence changes temporally and/or spatially in up to three dimensions. A number of investigations have focused on the former category, but there are few reports which fall into the latter category. An example of studies which involve the change of fluorescence temporally and spatially in a living cell is the measurement of ion activity such as pH and intracellular calcium using fluorescent probes. Although these probes have been used to measure ion activity in single cells using regular epifluorescence microscopy, only recently have measurements of ion activity been performed using confocal microscopy. Biological systems often are in the form of cylindrical structures, for example, gastric glands, nephrons and blood vessels. Extracting the dynamics of the fluorescence signal from a cylindrical structure based on epifluorescence imaging is complicated by out-of-focus fluorescence, which limits interpretation of the data. One major advantage of confocal fluorescence microscopy is that the out-of-focus fluorescence can be excluded to a great extent, allowing the dynamics of specific cellular phenomena to be acquired from specific anatomical locations in a heterogeneous epithelium.

22.2 MEASUREMENT OF pH_i WITH BCECF

Since the introduction of carboxyfluorescein (CF) by Thomas *et al.* in 1979, fluorescent techniques for measuring intracellular pH (pH_i) have virtually replaced microelectrode techniques (Thomas *et al.*, 1979). Optical measurements of pH have a rapid response time, high sensitivity and provide spatial information. Because of the excess leakage rate of CF, Tsien and colleagues developed BCECF in 1982 (Rink *et al.*, 1982). This probe is presently the most widely used pH probe. BCECF has a pK of approximately 6.9 and a very slow leakage rate. To load cells with the dye, the acetoxymethyl ester derivative is used which

FLUORESCENT AND LUMINESCENT PROBES, 2ND EDN
ISBN 0–12–447836–0

is lipid-permeable and therefore crosses cell membranes easily. Inside the cell, esterases cleave the probe, leaving the charged form of the dye trapped in the cytoplasm.

BCECF is excited at two wavelengths: 490 nm (pH-sensitive wavelength) and at 440 nm (isosbestic wavelength). The 490/440 nm excitation ratio varies with pH and is linear between pH 6.5 and 7.4 in most cells. The dye is calibrated *in vivo* using the high K^+ nigericin technique (Thomas *et al.*, 1979). Using a microfluorometer coupled to either a photomultiplier tube or a two-dimensional detector, pH can be measured in single cells (Tanasugarn *et al.*, 1984; Wang & Kurtz, 1990; Yip & Kurtz, 1995). However, if the epithelium is heterogeneous in the depth dimension, e.g. kidney tubule or gastric glands, measurements of pH in single cells becomes less accurate because of the potential acquisition of out-of-focus fluorescence information. This occurs because of the poor resolution in the *z* dimension of regular epifluorescence microscopes. The use of regular epifluorescence microscopy also makes it difficult to study spatial pH differences within different compartments in a single cell, i.e. cytoplasm versus nucleus, endocytotic vesicles. The cortical collecting duct, a portion of the kidney tubule studied in our laboratory, is cylindrical in shape and possesses at least two cell types: intercalated and principal cells (Fig. 22.1). The cylindrical shape and heterogeneous nature of this preparation necessitated the development of a new optical approach for monitoring pH in this preparation.

The confocal microscope has improved resolution in the *z* dimension (approximately 0.5 μm) (Zimmer *et al.*, 1988; Wells *et al.*, 1989). Confocal microscopes are excited with either a laser or a white light source (xenon, mercury arc lamp). These two types of confocal microscopes differ in their sensitivity, speed of data acquisition and wavelength selection. White-light-based systems offer a wide range of excitation wavelength choice, have rapid rate of image acquisition (30 per second), and are less sensitive than the laser-based systems. The latter have limited excitation wavelength capability and slower image acquisition rates. Most laser-based systems utilize a low-powered air-cooled argon laser which emits at two wavelengths: 488 nm and 514 nm.

In our initial efforts to measure pH confocally, a laser-based system was chosen because of its greater sensitivity. The argon 488 nm line excites the pH-sensitive wavelength of BCECF. However, argon lasers do not emit at the isosbestic wavelength of BCECF. Other pH dyes such as carboxy-SNARF (semi-naphthorhodafluor) and carboxy-SNAFL (semi-naphthofluorescein) are excited at a single laser line, 514 nm, and emit at two peak wavelengths whose intensity varies inversely with pH (*Bioprobes*, 1991). Given the ability to measure two emission wavelengths using two photomultiplier tubes, the initial logical approach in measuring pH with a laser scanning confocal microscope appeared to be to use carboxy-SNARF-l. However, studies with the dye revealed an excessive leakage rate. In addition, the p*K* of the dye is approximately 7.6, making it difficult to measure pH up to 6.5. Therefore, we chose to use BCECF. In order to excite BCECF at its isosbestic wavelength a 15 mW helium–cadmium laser was also coupled to the confocal microscope. By alternating the two excitation sources, pH could be measured in real-time.

Figure 22.1 Pseudo-Nomarski image of an isolated perfused rabbit cortical collecting duct (40 × objective, 8 × zoom). The small round cells are intercalated cells and the larger polygonal cells are principal cells. (Adapted from Wang & Kurtz, 1990.)

22.3. DESIGN OF A DUAL-EXCITATION LASER SCANNING CONFOCAL FLUORESCENCE MICROSCOPE FOR MEASURING pH_i WITH BCECF

In the first design of the instrument, an MRC-500 scanning unit (Bio-Rad) was interfaced with an inverted Nikon Diaphot microscope (Wang & Kurtz, 1990). More recently an MRC-600 scanning unit has replaced the original unit (Fig. 22.2). The emission port was used to excite and collect emitted fluorescence from the kidney tubule perfused on the microscope stage. A polarized 25 mW argon laser (model 5425A, Ion Laser Technology) and a polarized 15 mW helium–cadmium laser (He-Cd) (model 4214B, Liconix) were coupled to the MRC-600 scanning unit. The two laser beams were combined by reflecting the 442 nm He-Cd laser with a 100% mirror (Oriel) onto a dichroic mirror (Omega Optics) inserted at an angle of 45° in front of the argon laser. The dichroic mirror reflected the 442 nm light and transmitted 488 and 514 nm light to the scanning unit. An electronic shutter (Vincent Assoc.) was placed at an angle in front of each laser to prevent laser light from reflecting backward into the laser cavity. The shutters were opened and closed alternately under computer control. The duration of shutter opening and time between the opening of one shutter and closing of the second shutter were software-selectable. In all experiments a 40 × fluorite objective was used. This objective has a high optical throughput and minimal longitudinal chromatic aberration (Keller, 1989). The excitation filter cube in the MRC-600 scanner reflected 442 and 448 nm light to the microscope.

Photomultiplier tube no.1 was used for fluorescence imaging and photomultiplier no. 2 was used for bright-field imaging. A fibre bundle was used to transfer transmitted light to photomultiplier tube no. 2 in the MRC-600 scanner. Software was written (1) to control the timing parameters of the shutters, (2) to store digitally fluorescence images sequentially at 442 nm and 488 nm excitation, and (3) to extract the emission ratio from selected areas of the digitized images retrospectively. A zoom factor of 3 or 4 times was used. Up to 15 excitation ratios from spatially distinct locations in the preparation (*xy* plane) could be monitored in real-time with or without background subtraction.

The optical properties of the initial MRC-500 based system were characterized. To minimize curvature of field and radial chromatic aberration the data were acquired as close to the optical axis of the objective as possible (Wells *et al.*, 1989). A fluorite objective was used to minimize chromatic aberration (Keller,

(A)

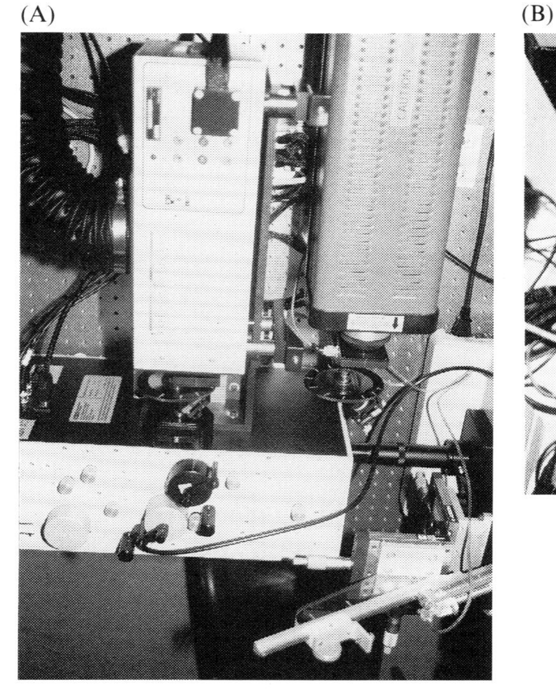

(B)

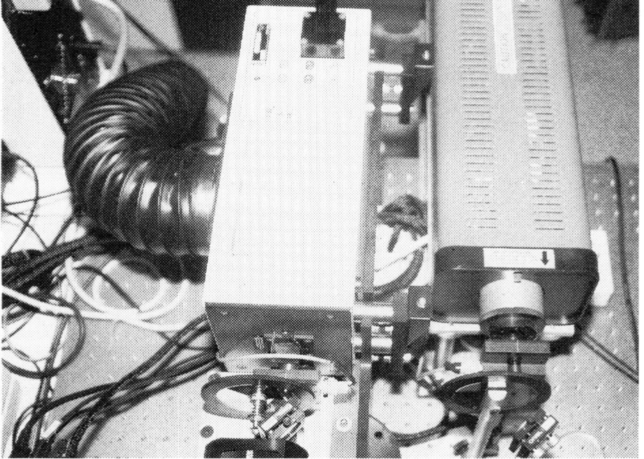

Figure 22.2 (A) View of the coupling of a 25 mW argon laser (right) and a 15 mW He-Cd laser (left) to the MRC-600 scanning unit. (Adapted from Kurtz & Emmons, 1993.) (B) Close-up view of the optics in front of each laser (He-Cd laser (right), argon laser (left)).

1989). The z-axis resolution in the reflected light mode at 488 nm with the pinhole at 0.96 nm was approximately 1 μm. In the fluorescence mode, the excitation and emitted wavelength are not the same. Chromatic aberration prevents the excitation and emission wavelengths from following the same optical path (Wells *et al.*, 1989). The emitted light will not be imaged at the detector pinhole which decreases the z-axis resolution. At 488 nm excitation with the detector pinhole closed maximally, the signal-to-noise ratio from cells loaded with BCECF was too low. Therefore, the pinhole was increased to 1.68 nm. The z-axis resolution at this pinhole size was 1.8 μm at 488 nm excitation and 1.1 μm at 442 nm excitation.

22.4 MEASUREMENT OF pH_i IN THE CORTICAL COLLECTING TUBULE

As depicted in Fig. 22.1 the cortical collecting tubule possesses more than one cell type, i.e. principal and intercalated cells. The cylindrical shape of the epithelium and its heterogeneous nature make this preparation difficult to study with conventional epifluorescence methodology.

In studies of this preparation, the tubule is cannulated with micropipettes which permits the luminal as well as the basolateral solutions to be changed rapidly. Movement of the preparation is particularly problematic when using a confocal microscope. However, a laminar flow chamber and floating table have ameliorated this problem. We measure pH in single principal and intercalated cells with BCECF. An example of an experiment measuring pH_i in a single principal cell is depicted in Fig. 22.3. Principal cells were found to have a basolateral Na^+/H^+ antiporter, a Na^+/base cotransporter and a Na^+-independent Cl^-/base exchanger (Wang & Kurtz, 1990). More recently, studies from our laboratory have demonstrated that the majority of intercalated cells have both apical and basolateral Cl^-/base exchangers (Emmons & Kurtz, 1994). This finding suggested that the current model of two types of intercalated cells, i.e. α-cells (basolateral Cl^-/base exchange) and β-cells (apical Cl^-/base exchange), needed to be modified. The cells with bilateral Cl^-/base exchangers were called γ-cells (Emmons & Kurtz, 1994).

The cortical collecting duct is a major site of NH_3 secretion in the nephron. However, the NH_3 permeability (pNH_3) properties of principal cells and intercalated cells and their contribution to passive NH_3 flux was unknown. Conventional approaches using fluid collection methodology could only estimate the whole tubule transcellular pNH_3 without differentiating the contribution by principal cells and intercalated cells. We determined the basolateral and apical pNH_3 of principal cells and intercalated cells by measuring the changes of pH_i with fluorescence confocal microscopy. Pairs of fluorescence images (440 nm and 488 nm excitation) were sampled and stored at 4 Hz. The time course of changes in pH_i of individual cells were extracted retrospectively from the stored images (Fig. 22.4). The rate of cellular NH_3 influx was calculated from the time course of increase in pH_i when the tubules were exposed to 20 mM NH_4Cl (Yip & Kurtz, 1995). After correction for membrane folding, the basolateral and apical pNH_3 of principal cells were 5.0±0.7 and 34±3 μm s^{-1}, respectively. The basolateral and apical pNH_3 of intercalated cells were 9.0±1.0 and 47±5 μm s^{-1} after membrane folding

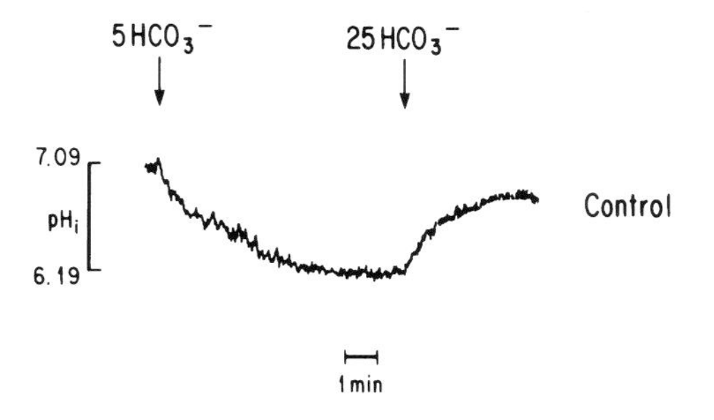

Figure 22.3 Measurement of pH_i in a single principal cell. Basolateral HCO_3^- was decreased from 25 to 5 mM and increased after several minutes to 25 mM. By measuring the rate of change of pH_i under different experimental conditions, the mechanisms of H^+/base transport across the apical and basolateral membrane were determined. (Adapted from Wang & Kurtz, 1990.)

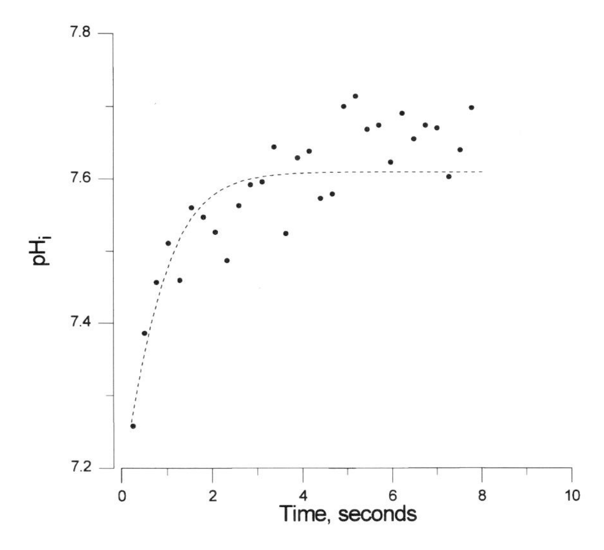

Figure 22.4 Time course of changes in pH_i of principal cells when 20 mM NH_4Cl was added to the bath perfusate at time = 0. (Adapted from Yip & Kurtz, 1995.)

correction. The results demonstrated that the apical surface was more permeable than the basolateral surface in both cell types. In addition, intercalated cells were more permeable to NH_3 than principal cells across both membranes. The data suggested that the contribution of principal cells and intercalated cells to transtubular NH_3 secretion would not be dependent on the difference in transcellular pNH_3 values but on the net proton secretory rate of each cell type.

22.5 MEASUREMENT OF INTRACELLULAR Ca^{2+} IN THE PERFUSED AFFERENT ARTERIOLE

Microperfused afferent arterioles of the glomerulus have cellular heterogeneity in the *z*-axis dimension, i.e. a single layer of smooth muscle cells surrounding an endothelial cell layer, rather than heterogeneity in the *xy* plane as in the cortical collecting duct. In a separate study we adapted the confocal imaging techniques developed for the perfused renal tubule experiments, to study afferent arterioles. This approach permitted the differentiation of fluorescence signals arising from vascular smooth muscle cells and endothelial cells. We determined the temporal changes of intracellular Ca^{2+} concentration ($[Ca^{2+}]_i$) in smooth muscle cells and endothelial cells during myogenic constriction, triggered by a step increase in the transmural pressure. $[Ca^{2+}]_i$ was measured by the emission ratio generated with a mixture of calcium indicator dyes fluo-3 and Fura Red, a new method recently introduced for use in a confocal microscopy using an argon laser (Lipp & Niggli, 1993). When excited at 488 nm, fluo-3 exhibits an increase in green fluorescence (525 nm) upon Ca^{2+} binding, while Fura Red shows a decrease in red fluorescence (640 nm). The emission ratio of fluo-3/Fura Red can then be used to monitor the changes in $[Ca^{2+}]_i$. The emission ratio is independent of the changes of local dye concentration that occur during contraction and relaxation of the vessel. Both indicators were simultaneously loaded into the endothelium via a luminal perfusate, or into the vascular smooth muscle layer via the bathing solution. Confocal fluorescence images were acquired at the mid-plane of the perfused arterioles, so that the cross-section of the endothelial cells and vascular smooth mucle cells could be differentiated. Because of the symmetrical geometry of the arteriole, passive dilation and subsequent constriction during the myogenic response did not change the position of the middle plane, which permitted scanning of the same area throughout the experiment. Local lateral dislocation of the sampling area due to the response of the vessel could be circumvented by redefining the sampling area in each image during the retrospective analysis of the acquired images. The time course of changes in the fluo-3/Fura Red ratio in smooth muscle and endothelium in response to a step increase in perfusion pressure is shown in Fig. 22.5. It had been suggested that the myogenic constriction of arterioles was due to an increase of $[Ca^{2+}]_i$ in smooth muscle, plus a decrease in the production of

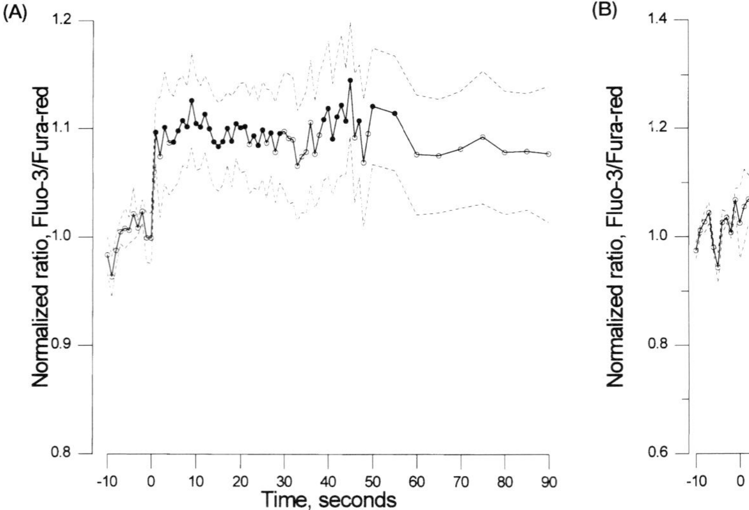

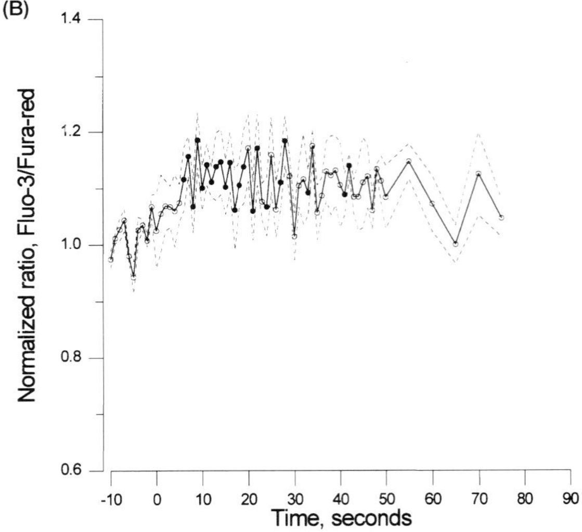

Figure 22.5 Normalized time course of changes in fluo-3/Fura Red emission ratio of smooth muscle cells (A) and endothelial cells (B). The transmural pressure was stepped up from 80 to 120 mmHg at time = 0. Dotted lines are mean ± SEM. Key: ●, significant difference from prepressurized baseline; ○ no difference from prepressurized baseline (n = 9, P<0.05). (Adapted from Yip & Marsh, 1996.)

nitric oxide from endothelium due to a decrease of $[Ca^{2+}]_i$ in the endothelial cells (Yip & Marsh, 1996). The results of these studies indicated that myogenic constriction in afferent arterioles was associated with an increase of $[Ca^{2+}]_i$ in vascular smooth muscle cells but did not require a simultaneous decrease of $[Ca^{2+}]_i$ in endothelial cells to decrease production of nitric oxide from endothelium.

22.6 FUTURE DEVELOPMENT

The development of confocal laser scanning microscopes designed for indicators with ultraviolet excitation wavelengths has made it possible to measure $[Ca^{2+}]_i$ with indo-1 (Niggli *et al.*, 1994). Recently the resolution of laser scanning microscopes has been improved with the development of non-linear laser scanning microscopes (Williams *et al.*, 1994; Xu *et al.*, 1996; Maiti *et al.*, 1997). These microscopes are capable of two-photon and three-photon excitation, allowing fluorescence dyes which normally require UV excitation to be excited with visible wavelengths. The elimination of UV excitation can reduce cellular damage in living tissue. Non-linear excitation provides superior three-dimensional resolution for imaging and avoids out-of-focus information. This technique has been used to image $[Ca^{2+}]_i$, NADH, and serotonin in living cells (Williams *et al.*, 1994; Xu *et al.*, 1996; Maiti *et al.*, 1997).

A recent important advance has been the development of near-field confocal microscopy with photon density feedback which permits diffraction limited studies of living cells (Haydon *et al.*, 1996). By integrating wide-field fluorescence microscopy with confocal microscopy and near field microscopy, it is possible to resolve fluorescent pH and Ca^{2+} changes beneath the cell membrane. It should also be possible to detect, with a finer level of resolution than was previously possible, the behaviour of fluorescently tagged molecules in the cell membrane.

ACKNOWLEDGEMENTS

This work was supported in part by NIH grants DK41212 and 851 IG-4, the Iris and B. Gerald Cantor Foundation, the Max Factor Family Foundation, the Verna Harrah Foundation, the Richard and Hinda Rosenthal Foundation, and the Fredricka Taubitz Foundation. Dr Kurtz is an Established Investigator of the American Heart Association.

REFERENCES

Bioprobes (1991) Molecular Probes Inc., Eugene, OR.
Emmons C. & Kurtz I. (1994) *J. Clin. Invest.* **93**, 417–423.
Haydon P.G., Marchese Ragona S., Basarsky T., Szulczewski M. & McCloskey M. (1996) *J. Microsc.* **182**, 208–216.
Keller H.E. (1989) In *The Handbook of Biological Confocal Microscope*, J. Pawley (ed). IMR, Madison, WI, pp. 69–77.
Kurtz I. & Emmons C. (1993) *Methods Cell. Biol.* **38**, 183–193.
Lipp P. & Niggli E. (1993) *Cell Calcium* **14**, 359–372.
Maiti S., Shear J.B., Williams R.M., Zipfel W.R. & Webb WW. (1997) *Science* **275**(5299): 530– 532.
Niggli E, Piston D.W., Kirby M.S., Cheng H., Sandison D.R., Webb W.W. & Lederer W.J. (1994) *Am. J. Physiol.* **266**, C303–C310.
Rink T.J., Ysien R.Y. & Pozzan T. (1982) *J. Cell Biol.* **95**, 189–196.
Tanasugarn L., McNeil P., Reynolds G.T. & Taylor D.L. (1984) *J. Cell Biol.* **98**, 717–724.
Thomas J.A., Buchsbaum R.N., Zimniak A. & Racker E. (1979) *Biochemistry* **18**, 2210–2218.
Wang X. & Kurtz I. (1990) *Am. J. Physiol.* **259**, C365–C373.
Wells K.S., Sandison D.R., Strickler J. & Webb W.W. (1989) In *The Handbook of Biological Confocal Microscope*, J. Pawley (ed). IMR, Madison, WI, pp. 23–35.
Williams R.M., Piston D.W. & Webb W.W. (1994) *FASEB J.* **11**, 804–813.
Xu C., Zipfel W., Shear J.B., Williams R.M, & Webb W.W. (1996) *Proc. Natl Acad. Sci. USA* **93**, 10763–10768.
Yip K.-P. & Kurtz I. (1995) *Am. J. Physiol.* **269**: F545–F550.
Yip K.-P. & Marsh D.J. (1996) *Am. J. Physiol.* **271**: F1004–F1011.
Zimmer F.J., Dryer C. & Hausen P. (1988) *J. Cell Biol.* **106**, 1435–1444.

CHAPTER TWENTY-THREE

High-speed Confocal Imaging in Four Dimensions

Y.S. PRAKASH, MATHUR S. KANNAN & GARY C. SIECK
Departments of Anesthesiology and Physiology & Biophysics, Mayo Clinic and Foundation, Rochester, MN, USA and Department of Veterinary PathoBiology, University of Minnesota, St Paul, MN, USA

23.1 INTRODUCTION

Advances in the speed and sensitivity of image acquisition have transformed the confocal microscope from an instrument for examining cell structure into a state-of-the-art tool for qualitative and quantitative assessment of cellular dynamics. In this context, the expression 'real-time' has become the buzzword for vendors when emphasizing the acquisition rate capabilities of their confocal imaging systems. Furthermore, since a cell is necessarily three-dimensional (3-D) in nature, several, if not all, vendors emphasize 3-D acquisition and analysis. However, until recently, limitations in detector sensitivity and the need for frame averaging, as well as computer hardware and software limitations had allowed enhanced 3-D acquisition only at the expense of considerable reductions in frame acquisition rates. These limitations were only compounded in four dimensions (4-D), i.e. repetitive 3-D imaging of cellular events over time. Accordingly, realistic and reliable 3-D and 4-D real-time confocal microscopy was possible only under conditions where cellular dynamics were slow enough to be repetitively captured through the volume of a cell. However, recent developments in scanning technology have made it possible for some confocal imaging systems to attain high-sensitivity, faster-than-video rates of acquisition in one optical plane, and combine this high-speed imaging with rapid 3-D optical sectioning, thus allowing rapid 4-D acquisitions. It is reasonable to expect that other confocal systems will also soon develop such capabilities for high-speed high-sensitivity imaging in 2-D, 3-D and 4-D. These exciting developments hold the promise for new investigations on a variety of cellular dynamics across a wide range of spatial and temporal response patterns.

The goal of this chapter is to emphasize the importance of spatial and temporal resolution in 2-D, 3-D and 4-D data acquisition using rapid confocal imaging techniques. Specific cases will be used to illustrate the effects of resolution on the quality of images, and the limitations of interpretations based on these images. The technical limitations of confocal imaging *vis-à-vis* issues of resolution will be discussed.

FLUORESCENT AND LUMINESCENT PROBES, 2ND EDN
ISBN 0–12–447836–0

23.2 KEY QUESTIONS RELATING TO DATA SAMPLING

As with other imaging technologies, high-speed confocal imaging has come at a price: data content. Despite the considerable expansion in the capabilities of image acquisition and computer hardware and software, and the sharp fall in their cost, researchers are still faced with the somewhat gargantuan task of probing through megabytes, if not gigabytes, of images in order to sometimes obtain a single datum number. Furthermore, given the 2-D nature of computer displays, manipulation and presentation of stacks of image data can be a daunting task. As the reader can imagine, these problems are only compounded when dealing with 4-D data. Under these circumstances, an informed decision on an optimal set of image acquisition parameters that will not compromise data reliability sounds warranted. In this regard, the key questions are:

(1) How sharply should individual cellular features be identified within an image? (2-D spatial resolution)
(2) How quickly should each optical section be acquired? (2-D temporal resolution)
(3) How thick should each optical section be? (3-D spatial resolution)
(4) How quickly should the volume of the cell be scanned? (3-D temporal resolution)
(5) How frequently should the volume scans be repeated? (4-D temporal resolution)

All of the above questions address a critical issue that spans all forms of data acquisition: *sampling frequency*. Sampling frequency translates to resolution in that decreasing sampling frequency decreases resolution, and vice versa. The lower limit for sampling frequency is given by the Nyquist criterion (Oppenheim *et al.*, 1983; Webb & Dorey, 1995) which can be roughly stated as:

> In order to detect an event occurring at a frequency X, samples must be obtained at least at a frequency of $2X$.

This concept is nothing new to confocal imaging, since most researchers and manufacturers recognize the need for adequate sampling in both the spatial and temporal domains in order to detect intracellular events. However, detection of events alone is rarely adequate. Accordingly, it is essential to choose a sampling frequency well beyond the Nyquist limit, thus allowing not only detection, but also detailed analysis of intracellular events. As indicated by the above questions, the issue of sampling frequency gets further complicated with high-speed confocal imaging where choices have to be made in two, three and four dimensions.

There is obviously no single set of answers to the above questions on spatial and temporal sampling frequencies that will satisfy every intracellular event. Furthermore, even for a particular intracellular event, the set of answers is likely to be a trade-off based on several other factors above and beyond the capabilities of the confocal microscope alone, such as signal-to-noise ratio, acquisition hardware and analysis software. In the following sections, we mostly use case studies from ongoing projects in our laboratory to demonstrate the effects of confocal imaging at different spatial and temporal resolutions on the observed changes in intracellular Ca^{2+} concentration ($[Ca^{2+}]_i$) in smooth and cardiac muscles. Other data has been obtained from test samples provided by confocal manufacturers.

23.3 NORAN ODYSSEY XL REAL-TIME CONFOCAL SYSTEM

Our studies were performed on an Odyssey XL real-time laser confocal system (Fig. 23.1) manufactured by Noran Instruments (Middleton, Wisconsin USA). The system is equipped with a three-line Ar-Kr laser, and mounted on a Nikon Diaphot (inverted) microscope. The confocal system is controlled through a Silicon Graphics Indy workstation and manufacturer-supplied software (InterVision).

The Odyssey system uses a slit aperture design instead of a pinhole aperture for optical sectioning, thus producing a confocal image (Draiijer & Houpt, 1988a,b). The slit is actually a series of slit apertures in a chrome-on-glass plate, and a selected slit can be positioned in the light path via software, thus varying the optical section thickness based on the microscope objective lens being used, and the morphology of the cell being studied (Table 23.1).

The confocal slit is an important design choice for high-speed imaging of intracellular dynamics. In comparison to pinhole apertures, slit apertures provide better signal-to-noise (Amos & White, 1995). Furthermore, compared with the point pinhole used in a number of confocal systems, the slit aperture is ideally suited for high-speed confocal imaging. On the other hand, the slit design also suffers from a lower axial resolution (along the z-axis or focus axis of the microscope) (Amos & White, 1995). However, axial resolution can be controlled by a judicious combination of a proper microscope objective lens and a suitable slit width. Obviously, the best axial resolution is achieved with the narrowest slit. However, this occurs at the

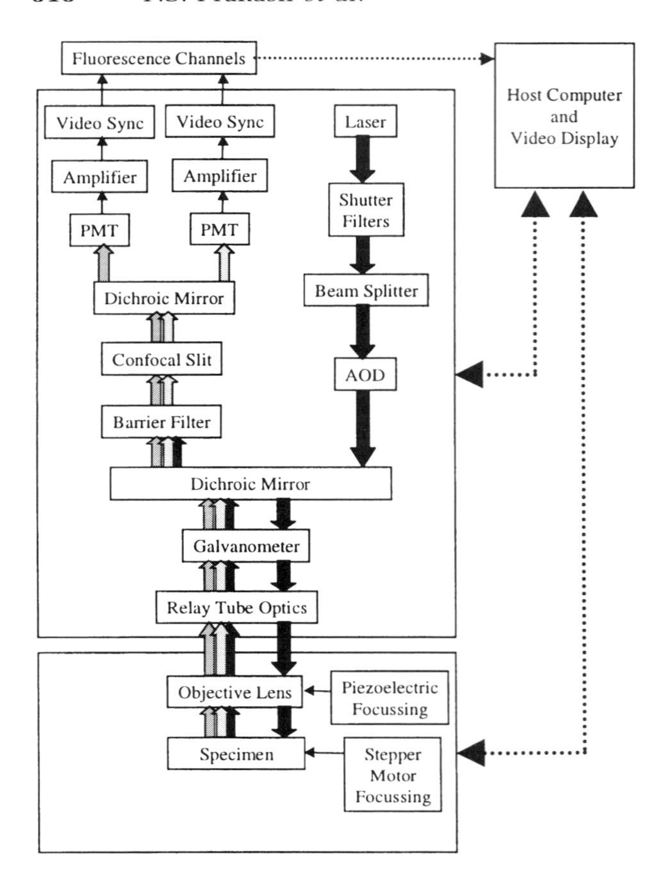

Figure 23.1 Schematic of the Noran Odyssey XL real-time confocal imaging system. The Odyssey uses a slit design for the confocal pinhole and optical sectioning, and an acousto-optical device to generate a raster scan. As with other confocal systems, appropriate dichroic mirrors and barrier filters are used to channel fluorescence to and from the specimen. (Adapted from the Odyssey User's Manual from Noran Instruments Inc., Middleton, WI.)

Table 23.1 Slit widths for optimal confocality.

Lens magnification	*Numerical aperture*	*Slit width (μm)*
Nikon 10× (dry)	0.30	10
Olympus 20× (dry)	0.70	10
Olympus 40× (dry)	0.85	15
Olympus 40× LWD (dry)	0.55	25
Nikon 40× (oil)	1.30	10
Olympus 100× (oil)	1.25	25

cost of lower signal-to-noise. Therefore, the photomultiplier tubes of the Odyssey system are designed to be high-sensitivity, low-noise.

23.4 SPATIAL AND TEMPORAL RESOLUTIONS IN 2-D IMAGING

23.4.1 2-D spatial resolution

With a light microscope objective lens, the 2-D spatial resolution under confocal conditions is generally dictated by the numerical aperture of the objective lens and the wavelength of incident light (Inoue & Oldenburg, 1994; Inoue, 1995). Although the theoretical limit using commercially available objective lenses and visible light sources is 0.2 μm, aberrations in the optical path are likely to limit the resolution to ~0.3 μm.

With any confocal imaging system, the optical image within an optical section is digitized into pixels. The pixel dimension along any axis is purely the ratio of the total length scanned by the imaging system to the total number of pixels along that axis. Accordingly, the smallest useful pixel dimension or pixel resolution is ~0.3 μm. In this regard, it must always be kept in mind that pixel dimension need not necessarily match optical resolution (see Webb & Dorey, 1995, for a detailed discussion). With low magnification lenses such as a 10× or 20×, the scanned areas are large, but the total number of pixels is fixed. Accordingly, the pixel resolution is likely to be worse than the optical resolution, and there is considerable data loss. With increasing magnification, pixel resolution may equal the optical resolution, thus optimizing the digitization of the optical image. Magnifications beyond this point result in oversampling (empty magnification) that decrease pixel dimensions but do not alter the resolution itself. Such magnifications of the image do not alter the minimum area that can be resolved, but only the number of pixels that represent this area. Under certain conditions, this oversampling may actually be used advantageously by limiting the scanned area but maintaining the total number of pixels (hardware zoom).

The maximum image size on the Odyssey system is 640×480 pixels. Therefore, assuming the smallest pixel dimensions, an ~190×140 μm area can be scanned. Further magnification can be achieved by scanning a smaller area of specimen and displaying the image over the 640×480 pixel area.

23.4.2 2-D temporal resolution

Most commercially available confocal microscopes that generate an *xy* raster to form a 2-D image require a digital frame buffer where the image can be collected and stored in computer memory. This frame buffer image is statically displayed until a subsequent image is collected and replaces the buffer contents. Given the inherent limitations in raster scanning and memory updates, the acquisition rates are typically slow (<5 frames per s). The Odyssey XL system differs from these slow-scan systems in that the specimen is scanned at video rates according to the RS-170 or PAL video standards, and does not require the video scanning segment of the hardware to be connected to a digital frame buffer. The Odyssey system uses acousto-optical deflectors to provide an ~15 kHz horizontal scan of the laser beam. The acousto-optical deflector is a glass slab with bonded piezoelectric transducers on either end. The transducers are driven by RF frequencies causing acoustic waves to be propagated through the glass. The waves act as diffraction gratings for the incident laser beam. The diffracted beam is then made to strike a galvanometer mirror that produces the vertical part of the scan by oscillating at a slower rate (60 Hz). With interleaved scanning, a 30 frames per s video output is produced that can be connected directly to a TV monitor to form a live image. Accordingly, frame-grabbing image processors that are intended for TV cameras can also be connected to the video scan module, thus allowing for display of both live as well as static images. In the case of the Odyssey system, a frame grabber in the host computer is used to store images rapidly.

The wide range of acquisition rates in the Odyssey system is achieved in a rather innovative fashion. The total area of the video display (field of view) is fixed at 640 × 480 pixels which is updated at a fixed rate of 30 frames per s (video rate) unless any on-line image processing such as frame-averaging or pixel integration is being performed. At relatively slow acquisition rates ranging from 0.5 to 30 frames per s, the entire field of view is used as a single image and the frames are updated at the appropriate rate (Table 23.2). At faster acquisition rates, ranging from 120 to 480 frames per s, the total field of view is still updated at a fixed rate of 30 frames per s, but the area is split into 4, 8 or 16 parts. Thus, by updating each part at 30 frames per s, effective rates of 120, 240 or 480 frames per s are achieved (Table 23.2).

23.4.3 2-D resolutions with high-speed imaging

Optical spatial resolution is dependent only on the inherent properties of the optics involved. However, pixelation of the image results in an interdependency between 2-D spatial and 2-D temporal resolutions. This interdependency arises from technical limitations in the process of digitization, pixelation, display and storage. These issues become particularly important with high-speed imaging, where higher acquisition rates may be achieved at the cost of decreased spatial resolution.

At the present time, there are a number of commercially available confocal systems that are also capable of rapid scanning over a wide range of acquisition rates. However, an important difference between these systems and a system such as the Odyssey is that in other systems, acquisition rates at ~500 frames per s are limited to single line scans in the *x* direction, with a resolution of just one pixel in the *y* direction. In contrast, even at ~500 frames per s, 60 pixels in the *y* direction are available on the Odyssey system. The obvious advantage of this feature is that by appropriate hardware zoom, it should be possible to follow at least the 2-D dynamics of an extremely rapid intracellular event (e.g. a Ca^{2+} transient in skeletal muscle) perhaps

Table 23.2 Examples of scanning and display options on the Odyssey

Acquistion rate (fps)	*Acquisition time (ms)*	*Pixel dwell time (ns)*	*Display size (pixels)*
0.5	2140	6400	640×480
7.5	134	400	640×480
30	33	100	640×480
30	33	400	320×240
120	8.35	100	640×120
120	8.35	100	320×420
240	4.17	100	640×60
240	4.17	100	320×120
480	2.08	100	640×30
480	2.08	100	320×60

with sufficient temporal and spatial resolution for data analysis above and beyond just the detection of the event.

Most commercially available confocal systems including the Odyssey achieve slow acquisition rates using a 'full-frame' image (maximum number of pixels) of the scanned area. Faster acquisition rates are achieved in one of several ways: (1) decreased number of pixels per frame with no change in scanned area; (2) decreased scanned area with no change in the number of pixels per frame; or (3) decreased scanned area as well as decreased number of pixels per frame. Each of these options results in different effects on spatial resolution (Fig. 23.2). With no change in scanned area, a decrease in the number of pixels per frame obviously results in decreased spatial resolution. A decreased area with no change in the number of pixels per frame is essentially not different from a zoomed area and the higher speed is achieved by faster scanning of the area. However, the effects on spatial resolution are difficult to predict. For example, if pixel resolution was previously worse than that possible with the optics, the faster acquisition is achieved with repeated sampling of the same area and an improvement in pixel resolution. On the other hand, if pixel resolution was already optimal, the faster acquisition results in no change in pixel resolution and empty magnification. With both decreased scan area as well as the number of pixels per frame, the effects on pixel resolution depend on the relative changes in space and time. Thus, the tradeoff between

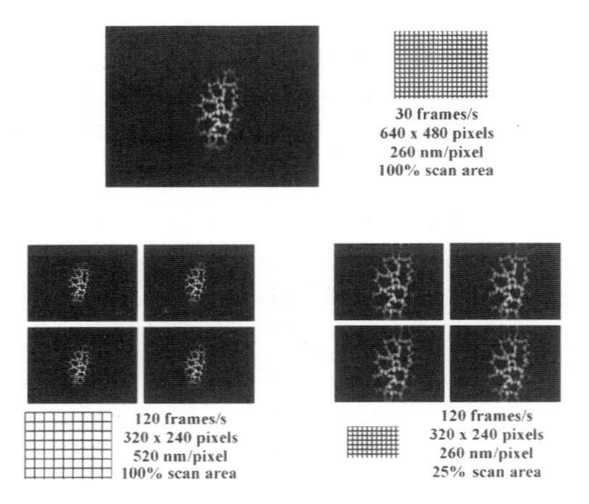

Figure 23.2 Relationships between image acquisition rate, spatial resolution and scan area of the Odyssey XL confocal system. 2-D images of a fluorescently-labelled pollen grain are shown. The Odyssey displays and stores full-frame images at a fixed rate of 30 frames per s (video rate). Faster acquisition is achieved by splitting the full frame into 4, 18, or 16 equal parts, but updating display and storage at a fixed rate of 30 frames per s. However, depending on the mode selected, faster acquisition is achieved at a loss of either spatial resolution or scan area.

spatial and temporal resolutions is likely to differ between confocal systems, and may or may not be beneficial.

With the Odyssey system, the total video display area is fixed at 640 × 480 pixels. Faster acquisitions are achieved with a trade-off in image area, which is reduced to 1/4, 1/8 or 1/16 of the total area (Table 23.2; Fig. 23.2). However, this loss in pixel resolution can be potentially offset by a hardware zoom of the specimen area, albeit at the risk of a higher rate of photobleaching. The fixed video display area also facilitates real-time display even at maximum acquisition rates since several acquisition frames are simultaneously displayed and updated at a fixed video rate. Real-time data storage is achieved with a Cosmo compression board that is commercially available and is compatible with the Silicon Graphics computers that are necessary to control the Odyssey system.

Maximum utilization of the rapid acquisition capabilities of a system such as the Odyssey is only possible if every single image has enough contrast and detail and does not require any averaging to improve visualization and/or analysis. There are at least two limiting factors: fluorescence input from the specimen and photomultiplier sensitivity. In most experiments involving living cells, it is advantageous to limit the amount of fluorescence dye within the cell and, more importantly, the intensity of laser light incident on the cell. Both these situations lead to a low level of fluorescence light reaching the photomultiplier, and necessitates acquisition of several consecutive image frames to 'build up' image contrast. This can be achieved in several ways including pixel-by-pixel summation of intensities, exponential filtering, Kalman filtering, etc. (see Russ, 1995, to learn more about image filtering and processing). An obvious consequence of frame averaging is a direct reduction of the net acquisition rate. One approach to minimizing the need for frame averaging is to use a high-sensitivity (low-noise) photomultiplier (available with the Odyssey system), thus making maximum use of the rapid acquisition capabilities of the confocal system.

In addition to limitations in scanning techniques, display capabilities also play an important role at higher acquisition rates. In fact, in some cases these limitations have dictated the overall capabilities of the confocal system for rapid acquisition. However, with the rapid developments in video hardware and software and faster computer display terminals, problems relating to rapid display are not likely to remain. Furthermore, access to cheap and extremely rapid mass data storage devices have also alleviated data storage problems. For example, in the Odyssey system, the fast video display controllers of the

Silicon Graphics workstation allow for real-time visualization of the fluorescence images. The use of a data compression board then allows for on-line recording of data. However, even with these capabilities, it is not uncommon to encounter non-uniform video display updates and/or recording of data. These irregularities can wreak havoc on the interpretation of the biological data, and should be meticulously weeded out.

23.5 CASE STUDIES ON 2-D RESOLUTIONS

A major focus of our laboratory is the regulation of intracellular Ca^{2+} concentration ($[Ca^{2+}]_i$) in smooth, cardiac and skeletal muscles. In these cell types, $[Ca^{2+}]_i$ regulation involves multiple mechanisms including second messenger production, sarcoplasmic reticulum (SR) Ca^{2+} release and reuptake (which may themselves be feedback regulated by $[Ca^{2+}]_i$ levels), and Ca^{2+} influx and efflux across the cell membrane. Differences in these $[Ca^{2+}]_i$ regulatory processes exist across cells (intercellular heterogeneity). Furthermore, within a single cell, spatial heterogeneity in $[Ca^{2+}]_i$ regulation may arise from variations in the distribution of membrane receptors, production and/or diffusion of second messengers, and SR and membrane Ca^{2+} channels. These spatial heterogeneities may also give rise to temporal heterogeneities in $[Ca^{2+}]_i$ regulation, leading to non-synchronized elevations in $[Ca^{2+}]_i$ in different parts of the cell.

Real-time confocal microscopy of fluorescent Ca^{2+} indicators is an ideal technique for studying $[Ca^{2+}]_i$ regulation in smooth, skeletal and cardiac muscle. In this regard, 2-D spatial resolution is critical, given the relatively small cell sizes (e.g. a typical porcine tracheal smooth muscle (TSM) cell is 5–7 μm wide and 20–50 μm long). Furthermore, the size of the region of interest (ROI) over which measurements are averaged relative to cell volume will influence the interpretation of spatial heterogeneities in $[Ca^{2+}]_i$ regulation. $[Ca^{2+}]_i$ kinetics and dynamics are also known to be different between different muscle types. Therefore, 2-D temporal resolution also becomes critical. To illustrate the influence of 2-D spatial and temporal resolutions on the apparent interpretation of confocal $[Ca^{2+}]_i$ data, we present three case studies: (1) agonist-induced $[Ca^{2+}]_i$ oscillations in porcine TSM cells; (2) spontaneous $[Ca^{2+}]_i$ waves in rat cardiac myocytes; and (3) spontaneous 'Ca^{2+} sparks' in cardiac myocytes. The sufficiently different $[Ca^{2+}]_i$ dynamics between smooth and cardiac muscle cells will illustrate the limitations of current technology for high-speed imaging.

23.5.1 Case Study 1: $[Ca^{2+}]_i$ oscillations in TSM cells

A number of previous studies have reported agonist-induced $[Ca^{2+}]_i$ oscillations in different cell types (see Amundson & Clapham, 1993; Berridge & Dupont, 1994, for reviews). These oscillations may occur via a number of mechanisms including repetitive Ca^{2+} release and reuptake from intracellular Ca^{2+} stores. Given the variations in the distribution of intracellular stores and the lack of synchronization of release across different parts of a cell, intracellular spatial heterogeneity in $[Ca^{2+}]_i$ oscillations is likely to exist, and may underlie the apparent propagation of the oscillations. Since the Ca^{2+} kinetics and dynamics are also likely to differ between cell types, both spatial and temporal heterogeneities in $[Ca^{2+}]_i$ regulation can only be addressed using real-time imaging. Rapid image acquisition using systems such as the Odyssey offer a distinct advantage in this regard.

We used freshly isolated porcine TSM cells loaded with 5 μM fluo-3/AM (Molecular Probes, Eugene, OR) and plated on collagen-coated glass coverslips to study $[Ca^{2+}]_i$ oscillations (Prakash *et al.*, 1997). The cells were visualized using a Nikon 40×/1.3 oil-immersion objective lens. The optical section thickness was set to 1 μm by controlling the slit size on the Odyssey system. ROIs within the boundaries of individual cells had a fixed dimension of 5 × 5 pixels (1.5 μm²). Therefore, $[Ca^{2+}]_i$ measurements were obtained from a volume of 1.5 μm³. Each ROI represented 0.05–0.10% of the volume of a TSM cell. To determine intracellular heterogeneity in the $[Ca^{2+}]_i$ response, up to seven ROIs were defined. The distances between these ROIs were measured using the length calibration for the 40× lens. One larger ROI was defined to include the entire cell so that the global $[Ca^{2+}]_i$ response could be evaluated.

A fixed combination of laser intensity and photomultiplier gain were set to ensure that pixel intensities within ROIs ranged between 25 and 255 grey levels. To calibrate $[Ca^{2+}]_i$, cells were exposed to 10 μM of a Ca^{2+} ionophore (4-Br-A23187) at varying levels of extracellular Ca^{2+} ranging from 0 (HBSS with EGTA) to 10 μM, and fluorescence intensities were measured. The relationship between fluorescence intensity and $[Ca^{2+}]_i$ was found to be linear from 10 nM to 10 μM. The resolution of the Ca^{2+} measurement was ~10 nM.

Images of TSM cells were acquired at sampling frequencies ranging between 1 and 480 frames per s. Image size and pixel resolution varied depending on acquisition rate (see Table 23.2). On-line $[Ca^{2+}]_i$ measurements from ROIs were made for acquisition rates of 30 frames per s or less, while measurements at higher acquisition rates were made *post hoc* using the same ROIs from acquired images using an image processing software package (ANALYZE, Mayo Biomedical Imaging Resource; Robb *et al.*, 1989; Robb, 1994). Propagation velocity of $[Ca^{2+}]_i$ oscillations was estimated from the time difference of $[Ca^{2+}]_i$ peaks between adjacent ROIs and the distance between these ROIs.

By analysing images acquired at different acquisition rates, we found that, within an intracellular ROI, ACh induces $[Ca^{2+}]_i$ oscillations in TSM cells ranging in frequency from 8 to 20 oscillations per min. Therefore, in order to only detect the occurrence of $[Ca^{2+}]_i$ oscillations, an acquisition rate of just 1 frame per s appeared sufficient. However, at acquisition rates of 1 to 7.5 frames per s, the propagation of $[Ca^{2+}]_i$ oscillations could not be distinguished (Plate 23.1). When the cells were imaged at 30 frames per s or greater, the propagation of $[Ca^{2+}]_i$ oscillations could be clearly observed (Plate 23.1).

Propagation was detectable at a resolution of 4 pixels (corresponding to 1 μm at a pixel resolution of ~0.25 μm). The resolution was limited by noise in each image and the need to measure distances accurately using software tools. Accordingly, at acquisition rates of 30, 60, and 120 frames per s, the maximum detectable propagation velocities were ~30, 60 and 120 μm s^{-1}. From images taken at 30, 60 and 120 frames per s, the propagation velocity was calculated to range from 8 to 25 μm s^{-1}. Therefore, the propagation of $[Ca^{2+}]_i$ oscillations in TSM cells could be detected and measured reliably using 30 frames per s.

From images acquired at different rates, the rise time of an individual oscillation was calculated to range from 500 ms to 2 s. When normalized for the amplitude of an individual oscillation, the rise time was ~0.5 to 6.5 ms nM^{-1}. Therefore, to detect changes in $[Ca^{2+}]_i$ at a resolution of 50 nM (~50% for the minimum amplitude of oscillations observed in our studies) required 25 to 325 ms. According to the Nyquist criterion, this corresponded to acquisition rates of ~80 to 2 frames per s. These data indicate that with a maximum acquisition rate of 480 frames per s, the current 2-D temporal capabilities of the Odyssey system were more than sufficient for both the detection and detailed analysis of the dynamics of propagating $[Ca^{2+}]_i$ oscillations in TSM cells. Detailed analysis of several TSM cells revealed that in >80% of the cases, an acquisition rate of 30 frames per s was more than sufficient to analyse propagating $[Ca^{2+}]_i$ oscillations.

The propagating nature of the $[Ca^{2+}]_i$ oscillations had considerable impact on the apparent local versus global $[Ca^{2+}]_i$ response of a TSM cell. At 30 frames per s acquisition, the propagating wave appeared as distinct $[Ca^{2+}]_i$ oscillations within an ROI, and displayed a biphasic pattern (Fig. 23.3). Upon initial exposure to ACh, the $[Ca^{2+}]_i$ oscillations had higher frequency and lower amplitude, while the steady-state response had lower frequency but higher amplitude. In contrast, when the global $[Ca^{2+}]_i$ response was measured, no oscillations were present (Fig. 23.3). However, the $[Ca^{2+}]_i$ response continued to appear biphasic with an initial higher peak followed by a lower, steady-state value. Thus, it appeared that the $[Ca^{2+}]_i$ oscillations within localized regions had been integrated across the cell, yielding a global $[Ca^{2+}]_i$ response that masked the underlying spatial heterogeneity.

The differences in local versus global $[Ca^{2+}]_i$ responses were further accentuated when acquisition rate was also altered. $[Ca^{2+}]_i$ oscillations were undersampled at less than 5 frames per s, and were absent at 1 frame per s. However, the global $[Ca^{2+}]_i$ response was not considerably altered with changes in acquisition rate.

In a separate analysis, we simulated the effects of frame averaging on the apparent $[Ca^{2+}]_i$ response

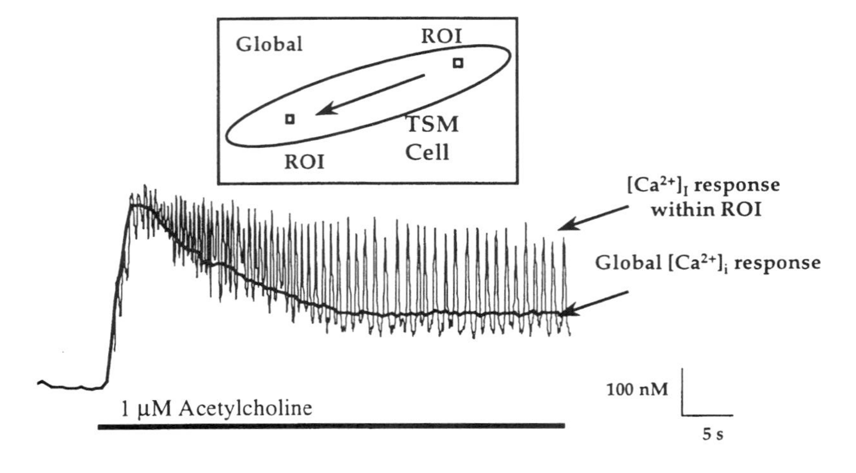

Figure 23.3 Effect of spatial integration on the apparent $[Ca^{2+}]_i$ response of TSM cells to acetylcholine. Measurement of $[Ca^{2+}]_i$ within a small region of interest (ROI) of the TSM cell revealed the oscillatory nature of the $[Ca^{2+}]_i$ response. However, when the global $[Ca^{2+}]_i$ response of the cell was measured, only a 'biphasic' response with no oscillations was observed.

(representing situations of low photomultiplier sensitivity and/or fluorescence levels). This analysis differed from a simple change in acquisition rate (where the time taken to acquire a single image is fixed, but the gap between successive acquisitions is long) in that the time taken per image was extended by averaging sets of 4, 8 and 16 frames prior to display and/or analysis. Under these conditions, the oscillatory $[Ca^{2+}]_i$ response (within an ROI), as well as the propagating nature of $[Ca^{2+}]_i$ oscillations, were no longer observed, and only a 'biphasic' elevation of $[Ca^{2+}]_i$ with agonist exposure was observed.

Based on the above analysis, one could conclude that in order to examine propagating $[Ca^{2+}]_i$ oscillations in cell systems that are not faster than TSM cells in their $[Ca^{2+}]_i$ responses, an acquisition rate of 30 frames per s is sufficient. However, when the $[Ca^{2+}]_i$ dynamics are fast enough to require acquisitions greater than 30 frames per s, a system such as the Odyssey decreases frame size and/or pixel resolution (see Table 23.2). Accordingly, the accuracy with which spatial measurements can be made also decreases. Fortunately, in case of TSM cells, which are only a few micrometres wide and 20–30 μm long, we found that even at 240 frames per s the spatial resolution was sufficient to measure propagation velocity accurately. However, the percentage error in these measurements was substantially higher. Other parameters such as rise and fall times could be measured with greater accuracy, as expected.

23.5.2 Case Study 2: spontaneous $[Ca^{2+}]_i$ waves in cardiac myocytes

Spontaneous $[Ca^{2+}]_i$ waves in cardiac myocytes have been reported previously (Lopez *et al.*, 1995; Cheng *et al.*, 1996a,b), although their physiological significance is not clear. However, the high speed of propagation of $[Ca^{2+}]_i$ waves in cardiac myocytes makes these waves a suitable 'sample' to test rapid imaging systems.

We used fluo-3-loaded isolated rat cardiac myocytes to study spontaneous $[Ca^{2+}]_i$ waves. The settings for cell loading and confocal imaging were similar to those used in TSM cells. ROIs within the boundaries of individual cells had a fixed dimension of 1.5 μm^3 for an optical section thickness of 1 μm. Each myocyte was ~25 μm wide and 200 μm long. Therefore, $[Ca^{2+}]_i$ measurements were obtained from 0.002–0.005% of the cell volume. As with the TSM cells, seven small ROIs and one large ROI were defined. The fluo-3 response was calibrated using similar techniques. Images were acquired at different acquisition rates and analysed.

Cardiac myocytes displayed considerable variations in the pattern of $[Ca^{2+}]_i$ waves, ranging from 'simple' propagated waves along the long the axis of the cell (Plate 23.2), through colliding and annihilating waves with multiple origins, to local swirls and eddies that were limited to less than 25% of the cell volume. $[Ca^{2+}]_i$ waves recurred with frequencies ranging from 11 to 240 per min. Therefore, in order to just detect the initiation of a wave, an acquisition rate of 8 frames per s was necessary (compared with 1 frame per s for TSM cells). However, as with TSM cells, at low acquisition rates (4–15 frames per s), wave propagation could not be clearly distinguished. In cells with simple propagated waves, the velocity of wave propagation ranged from 20 to 200 $\mu m\ s^{-1}$. Indeed, even at 30 frames per s, propagation was observed only in some cases (unlike in TSM cells) and in these cells, propagation velocity was found to be less than 30 $\mu m\ s^{-1}$ (e.g. Plate 23.2). In cells with faster propagation, we attempted to follow the $[Ca^{2+}]_i$ dynamics at 240 frames per s. Interestingly, even at a 4 ms resolution, we found that in myocytes with complex Ca^{2+} movements, the location of the $[Ca^{2+}]_i$ wavefront between consecutive image frames was disjointed, with the wavefront having advanced by ~5 μm within 4 ms (Plate 23.3). These data suggest that Ca^{2+} can move within a local region of a cardiac myocyte at velocities of ~1250 $\mu m\ s^{-1}$. From these data, one can conclude that in order to detect propagating $[Ca^{2+}]_i$ waves reliably in cardiac myocytes, even an acquisition rate of 480 frames per s may be insufficient.

23.5.3 Case Study 3: spontaneous $[Ca^{2+}]_i$ sparks in cardiac myocytes

Spontaneous, localized Ca^{2+} transients of small amplitude (<100 nM), termed 'Ca^{2+} sparks', have been observed in skeletal fibres, cardiac myocytes, and vascular and tracheal smooth muscle cells (e.g. Cheng *et al.*, 1996b; Schneider & Klein, 1996; Sieck *et al.*, 1997). Ca^{2+} sparks are thought to represent unitary and synchronous Ca^{2+} release from small groups of RyR channels (Cheng *et al.*, 1996b; Schneider & Klein, 1996; Sieck *et al.*, 1997). Since $[Ca^{2+}]_i$ waves in cardiac myocytes involved RyR channels, it is likely that there is a spatiotemporal relationship between the small-amplitude Ca^{2+} sparks and large-amplitude $[Ca^{2+}]_i$ oscillations and waves.

The temporal aspects of Ca^{2+} sparks have been characterized in previous studies using line scans with confocal microscopy (Cheng *et al.*, 1993, 1996a; Klein *et al.*, 1996; Schneider & Klein, 1996), which provide important information on the incidence of sparks. However, there are serious limitations to using line scans for evaluating the amplitude of Ca^{2+} sparks. An inherent assumption in that technique is that the spatial or 2-D location of the peak of successive sparks does not change such that a line scan through the peak of the first spark will continue to provide information on the peak amplitude. However, as shown below, this may not necessarily be the case. Furthermore, in order to evaluate the relationship between spontaneous Ca^{2+} sparks and $[Ca^{2+}]_i$ oscillations and waves, measurements must be made from different parts of the cell at high spatial resolution, which is clearly not possible using line scans.

We used fluo-3-loaded rat cardiac myocytes to examine Ca^{2+} sparks. The settings for cell loading and confocal imaging were similar to those used in other case studies. However, in this case, $[Ca^{2+}]_i$ measurements were obtained from 0.0005–0.0001% of the cell volume by defining smaller ROIs. Furthermore, the ROIs were systematically placed in a grid fashion in close proximity to each other so that the spatial pattern of the sparks could be followed. Images were again acquired at different acquisition rates and analysed.

Individual Ca^{2+} sparks were localized to <0.005% of the cell volume. In >80% of the cells examined, multiple foci for Ca^{2+} sparks (3–5 per myocyte) were present. The incidence of individual Ca^{2+} sparks was coupled in 65% of adjacent ROIs that were separated by <2 μm. Within a single focus of a Ca^{2+} spark, the amplitude was relatively constant over time. However, across ROIs within a cell, the amplitudes of Ca^{2+} sparks displayed considerable variance, ranging from 30 to 90 nM. The rise time of Ca^{2+} sparks ranged from 10 to 35 ms (Plate 23.4). This corresponded to a rise time of ~0.1 to 1.2 ms nM^{-1}. Therefore, detection of a 50 nM rise in $[Ca^{2+}]_i$ required ~5 to 60 ms. According to the Nyquist criterion, this corresponded to acquisition rates of 400 to 33 frames per s. Therefore, even with its 480 frames per s acquisition capability, the Odyssey system appears barely to have the temporal resolution to obtain images for the analysis of Ca^{2+} sparks in cardiac myocytes, although the detection of these events does not appear to be limited. The incidence of Ca^{2+} sparks ranged from 5 to 25 per min within an ROI. In several cases, individual Ca^{2+} sparks summated into larger $[Ca^{2+}]_i$ elevations that initiated a single propagating wave throughout the myocyte.

We also examined variations in the spatial distribution of Ca^{2+} sparks within the same locus. We found that although the general area over which consecutive sparks were distributed did not change appreciably, the 2-D peak of each spark was located at least 2–3 pixels away from a previous peak, i.e. the peak did not occur in the same location every time (Plate 23.5). This finding has significant implications for the interpretation of previous reports on Ca^{2+} sparks where line scans (along the same line) have been used to measure the amplitude of Ca^{2+} sparks. Our data indicate that the apparent amplitude of a Ca^{2+} spark may be considerably underestimated using only a line scan, and that a 2-D image of Ca^{2+} distribution at the peak of a spark is essential in determining the true amplitude.

23.6 SPATIAL AND TEMPORAL RESOLUTIONS IN 3-D IMAGING

23.6.1 3-D spatial resolution

It is generally recognized that axial resolution (along the *z*-axis) is dependent on the numerical aperture of the objective lens and the refractive index of the surrounding medium (Rayleigh criterion), and is typically twice that of the lateral 2-D resolution (Inoue, 1995). Therefore, at a 0.2 μm 2-D lateral resolution, the 3-D spatial resolution for optical sectioning is likely to be 0.4 μm. However, several factors such as lens magnification and numerical aperture, and mismatches in refractive indices, decrease the practical axial resolution to ~0.8 to 1 μm.

As with a number of commercially available confocal systems, the Odyssey system is equipped with a stepper motor that controls the fine focus knob of the objective lens turret (inverted microscope configuration). Most stepper motors have a step size resolution of 0.05 to 0.2 μm, which is more than sufficient to attain the practical optical section thickness or 3-D spatial resolution of light microscope objective lenses. Using appropriate software controls, images of each optical section can be easily obtained.

There are two key issues relating to 3-D spatial resolution *vis-à-vis* optical sectioning with stepper motors. First, it is essential to match the step size for each image (confocal slice thickness) with the section thickness determined by the optics. As can be imagined, a step size smaller than the optical section thickness results in repeated sampling of the same optical plane and an overestimation of the thickness of the specimen. A step size larger than the optical section thickness results in an undersampling of the specimen thickness. The second issue relates to the accuracy of stepper motor movements. Several technical parameters such as step resolution, reproducibility of step movements and backlash are involved.

Several empirical approaches to determining the 3-D spatial resolution of a specific confocal system have been previously published (see Cogswell *et al.*, 1990; Prakash *et al.*, 1993; Cogswell & Larkin, 1995). A number of these techniques essentially determine the error in estimating the shape and size of objects, such as fluorescently coated latex microspheres where the dimensions have been independently verified.

Other techniques determine the point-spread function for every objective lens (the apparent image of an infinitesimally small point source of brightness distorted by limited optics). Using such approaches, we have empirically determined the 3-D spatial resolution of the Odyssey system in our laboratory and found it to range from 0.45 (100× lens) to 1.2 μm (20× lens), respectively. However, it must be emphasized that the 3-D spatial resolution needs to be determined for each individual confocal system in order to optimize data sampling in this dimension.

23.6.2 3-D temporal resolution

Several currently available confocal systems allow for automated, sequential acquisition of optical sections between preset depths through a specimen. The 3-D temporal resolution of such protocols is dependent on how quickly the imaging system can achieve the desired step changes, scan the desired area, and acquire, display and store a single image before moving on to the next optical section. It is difficult to determine the 3-D temporal resolution for an imaging system directly. However, an average time per image can be obtained by measuring the time taken for collecting several consecutive images.

As the reader can imagine, several factors influence 3-D temporal resolution. First, the resolution of the focus motor and the time taken to achieve a single step of minimum size both determine the speed at which the plane of focus is changed following data acquisition from the previous plane. By virtue of the mechanical linkages involved, this aspect of 3-D data acquisition is the limiting factor for 3-D temporal resolution. Under unloaded conditions (with motor not attached to the focusing knob of the microscope, or attached to the microscope but with no lenses or specimen stage), most focus motors can achieve a single step in ~20 ms. However, most focus motors also have extremely fine step resolutions (50 nm in the case of the Odyssey system). Since the practical optical section thickness is on the order of 0.5–1 μm, several single steps are necessary. Accordingly, it is not uncommon for the total time for change in focal plane to reach hundreds of ms. Software controls and communication between the computer and the focus motor may prolong this time.

Following the positioning of the focus motor at the desired plane of focus, the time taken for image acquisition, display and storage contribute to 3-D temporal resolution. With the advent of new video display technology and computer hardware, single images of reasonable size (e.g. 1024 × 1024 pixels) can be easily displayed and saved within 30–50 ms. The time for image acquisition obviously varies between confocal systems and is dependent on factors such as the frame acquisition rate and the number of frames to average. For example, with the Odyssey system, at an acquisition rate of 480 frames per s with no frame averaging, a single frame is acquired in <3 ms. This time is insignificant compared with the delays introduced by focus motor positioning and image display and storage. However, with an acquisition rate of 30 frames per s and averaging of four frames (comparable to an option available with several video-rate confocal systems), the time taken for acquiring a single image increases to 133 ms.

Based on the above reasoning, it appears that even with extremely high 2-D image acquisition rates, limitations in mechanical devices will severely limit the 3-D resolution of confocal systems. Large improvements in 3-D resolution in the future are likely to depend on the availability of faster stepper motors. However, faster motors are also likely to suffer from oscillation artifacts due to their rapid starts and stops ('ringing') and, as a result, considerable delay may be introduced while the oscillations are damped out. A potential option for improving 3-D resolution is a continuous motion of the focus motor over the range of acquisition with image acquisition synchronized to the starting and stopping of the focus motor, with a recording of position at regular intervals. The advantage of 3-D acquisition during continuous focusing is that the interaction between the acquisition computer and the focusing system is minimal in the duration of acquisition, except at the starting and stopping positions. Accordingly, considerably less time is lost since the focusing system is not starting and stopping to capture an image at each position. Obviously, the raw image set obtained by this technique is a motion-blurred representation of the 3-D volume. However, it may be possible to deblur the images using mathematical transforms that are currently being used to correct motion artifacts in MRI and CT images.

23.7 4-D TEMPORAL RESOLUTION

4-D resolution can be defined as the time resolution at which the occurrence of an intracellular event in a given optical plane can be repeatedly sampled during the collection of a set of 3-D images. If there is no time delay between the collection of sequential sets of 3-D optical sections, then the time elapsed before a given optical plane is sampled again is simply the total time taken to collect one 3-D set of images. Under these conditions, the 4-D resolution obviously depends on the number of 3-D sections within a set and the 3-D temporal resolution of the confocal system. When the different factors contributing to time delays in sequential 3-D image acquisition are considered (see above), it may be quite difficult to attain a 3-D temporal resolution better than 150–200 ms. Accordingly, even with just five optical sections per 3-D set (typical of many cultured cells for example), the 4-D resolution is already at 1 s. Moreover, there is currently no commercially available confocal system that is capable of acquiring sequential 3-D data sets with no time lag between sets. These time lags are dependent on both hardware as well as software limitations, and are likely to be in the hundreds of ms. These issues severely limit the possibility of using focus motors to obtain adequate 4-D resolution of repetitive, fast intracellular events occurring on the order of even several seconds.

In addition to the focus stepper motor, the Odyssey system is equipped with a single-axis piezoelectric positioning system that is attached directly to the objective lens. Essentially, the expansion of the piezoelectric system, which depends on the voltage applied to the piezoelectric crystal, results in a physical elevation of the objective lens. There are several advantages of this innovative technique. First, the objective lens is substantially lighter than the focusing system and the objective turret of an inverted microscope (in case of an upright microscope, the lens is also lighter than the specimen stage). This allows for considerably faster movement along the z-axis without loading the positioning system. The rigidity of the system also minimizes hysteresis during repeated or bidirectional movements. Third, the resolution of the positioning system is in the submicrometre level, which is well-suited for a variety of objective lenses.

In the Odyssey system, we have used the piezoelectric positioning system in conjunction with a Nikon 40×/1.3 oil-immersion objective lens. The position of the lens is controlled via software from the manufacturer. Under software control, the focus is stepped at the selected interval and an image is acquired (at any acquisition rate) before the focus is changed. The system is capable of both unidirectional and bidirectional data acquisition. The advantage of bidirectional data acquisition is that the time taken to reset the lens position during repetitive unidirectional acquisition is utilized for data acquisition. However, the bidirectional mode suffers from non-uniform 4-D resolution (see below).

Presently, with no frame averaging, the positioning system is capable of acquiring 15 frames per s in one direction when stepped at any resolution (ranging from 0.5 μm to 2 μm). Accordingly, five optical sections per 3-D set will require ~330 ms. Therefore, in a unidirectional acquisition mode (the motor flies back to the starting point before beginning a second acquisition), a 4-D resolution is also ~330 ms, which is a three-fold improvement over the resolution with a focus stepper motor. However, in a bidirectional mode, the 4-D resolution is non-uniform since the last position in a 3-D set becomes the first to be sampled during the acquisition in the reverse direction. Accordingly, even with no lag between acquisition of consecutive 3-D sets, the 4-D resolution is ~330 ms at one position of the 3-D set and twice this time at the other end of the set.

At the present time, the 4-D resolution of the piezoelectric focusing system is at least three times better than that expected from a stepper-motor-based system. Accordingly, the piezoelectric focusing system operating in the unidirectional mode holds the promise for faster real-time acquisition in 4-D. However, it also appears that at the present time, the improved resolution is only barely sufficient to capture events that are recurring on the order of 1 s. The major hardware limitation appears to be the load (albeit smaller than that with the focus motor) on the piezoelectric system which is a combined effect of the weight of the objective lens and the specimen holder (in the case of an oil-immersion lens) and the surface tension of the immersion oil being used. These limitations are somewhat difficult to overcome within the framework of available optical systems. A potential option is to use low-viscosity immersion oil, or utilize a dry objective lens of high magnification. However, the latter option also suffers from the disadvantage of lower numerical aperture and reduced efficiency for collecting fluorescence signals. It may also be possible to improve 3-D acquisition rates and thus indirectly improve 4-D resolution. As mentioned above, a potential option is a continuous motion of the objective lens (via the piezoelectric system) over the range of focal planes with synchronized recording of the objective position at regular intervals. The motion-blurred raw image set can then be processed using mathematical transforms.

23.8 3-D AND 4-D IMAGE PROCESSING, DISPLAY AND STORAGE

A computer or paper display is necessarily 2-D in nature. Accordingly, an issue that is of considerable interest and research focus is the display and manipulation of 3-D data sets. Most image processing software packages with capabilities for 3-D reconstruction make use of algorithms such as voxel gradient shading, depth gradient shading, surface shading, and maximum intensity projections (see Russ, 1995, for a review). Using these algorithms, the 3-D reconstructions can be viewed at any desired angle and resectioned if necessary to view the interiors of the specimen.

Plate 23.1 2-D temporal resolution and propagated $[Ca^{2+}]_i$ oscillations in porcine tracheal smooth muscle. When TSM cells were imaged at 7.5 frames per s, the propagating nature of the $[Ca^{2+}]_i$ response (observed at 30 frames per s or greater) was absent.

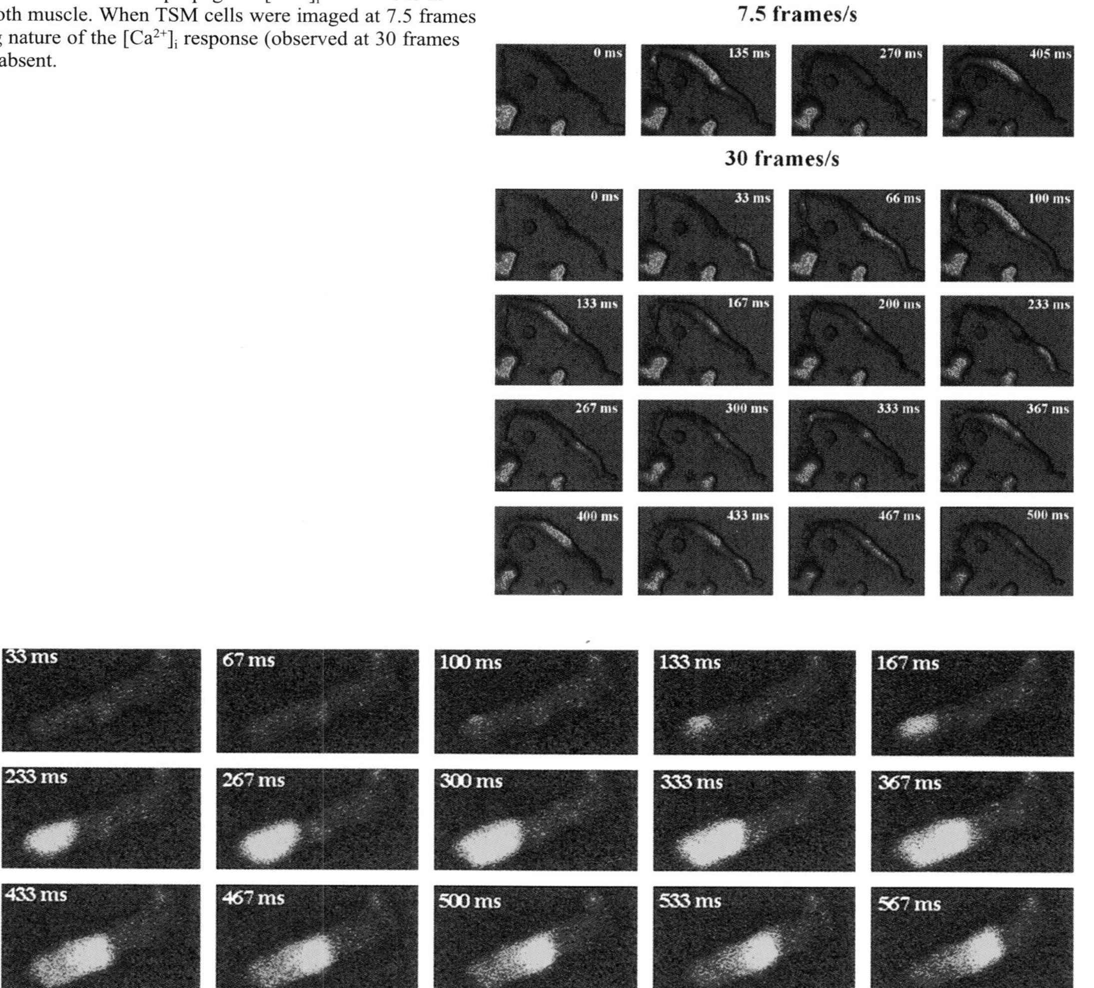

ate 23.2 Time series 2-D images of a relatively slow, spontaneous $[Ca^{2+}]_i$ wave in rat cardiac myocyte. Images were acquired at 30 frames per s d represent an optical section of ~1 μm. The spatiotemporal heterogeneity in the $[Ca^{2+}]_i$ response underlines the need for adequate 2-D spatial as ell as temporal resolution.

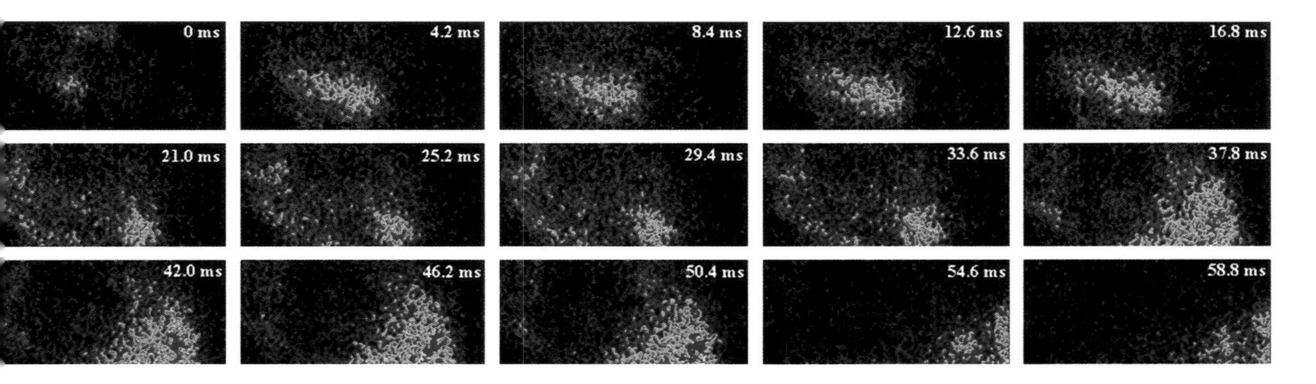

ate 23.3 Time series 2-D images of a rapidly propagated $[Ca^{2+}]_i$ wave in rat cardiac myocyte. Although the images were acquired as fast as 240 ames per s, the smooth spatial movements of $[Ca^{2+}]_i$ could not be captured, and were indicated by sudden changes in the spatial distribution of $a^{2+}]_i$ (e.g. at 16.8 and 37.8 ms).

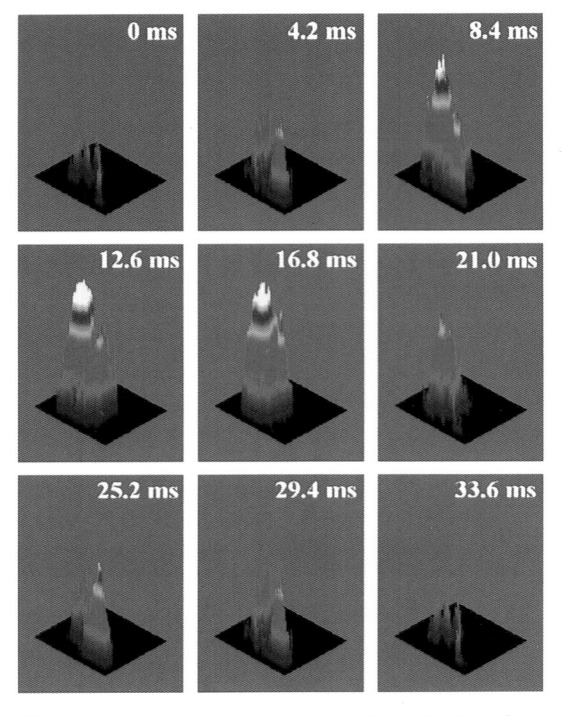

Plate 23.4 Time series images of the 2-D distribution of a Ca^{2+} spark in rat cardiac myocyte. Colour coding represents varying levels of $[Ca^{2+}]_i$. Even at an acquisition rate of 240 frames per s, the temporal changes in the distribution of Ca^{2+} within the spark was barely captured.

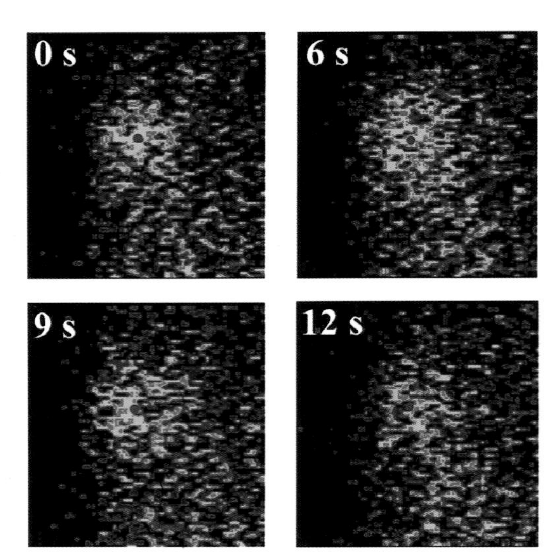

Plate 23.5 Variations in the 2-D distribution of Ca^{2+} at the peak of a spark in a cardiac myocyte. Images were acquired at 240 frames per s. Each image represents the distribution of Ca^{2+} at the approximate time points for the peak of each spark. The dot in each image is the location of the peak of the spark at time $t = 0$ s. Note the variation in the location of the peak of the spark relative to the dot in successive images.

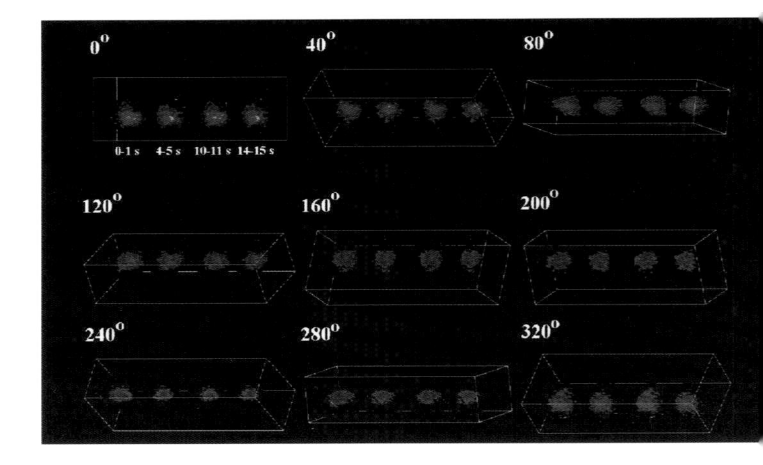

Plate 23.6 Rotational views of temporal changes in the 3-D volume distribution of Ca^{2+} within a spark in a TSM cell. Such rotational views of 3-D reconstructions allows for representation of 4-D data sets such that temporal changes in all three spatial dimensions can be evaluated. In this example, the volume distribution of the Ca^{2+} spark does not appear to change considerably over time. The time gaps in the 3-D data sets are due to software and hardware limitations of the confocal systems.

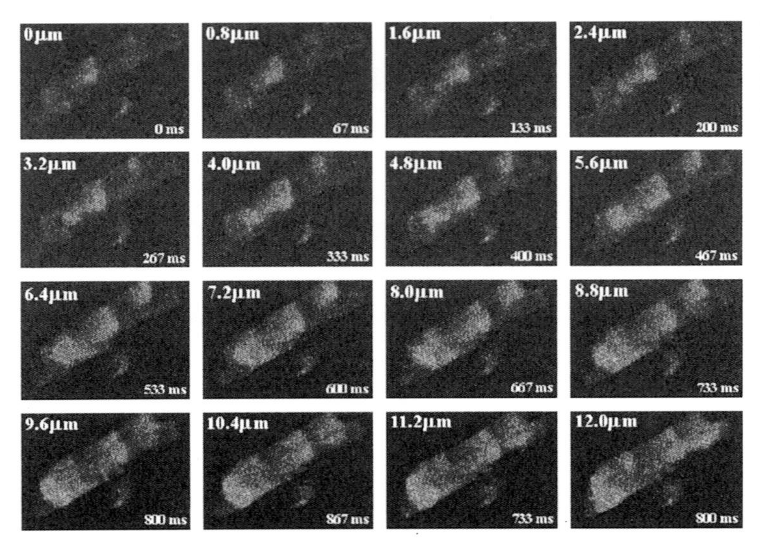

Plate 23.7 3-D propagation of an $[Ca^{2+}]_i$ wave through a cardiac myocyte. 1 µm thick optical sections taken at a frequency of 15 frames per s allow for smooth tracking of $[Ca^{2+}]_i$ as it moves through the depth of the cell.

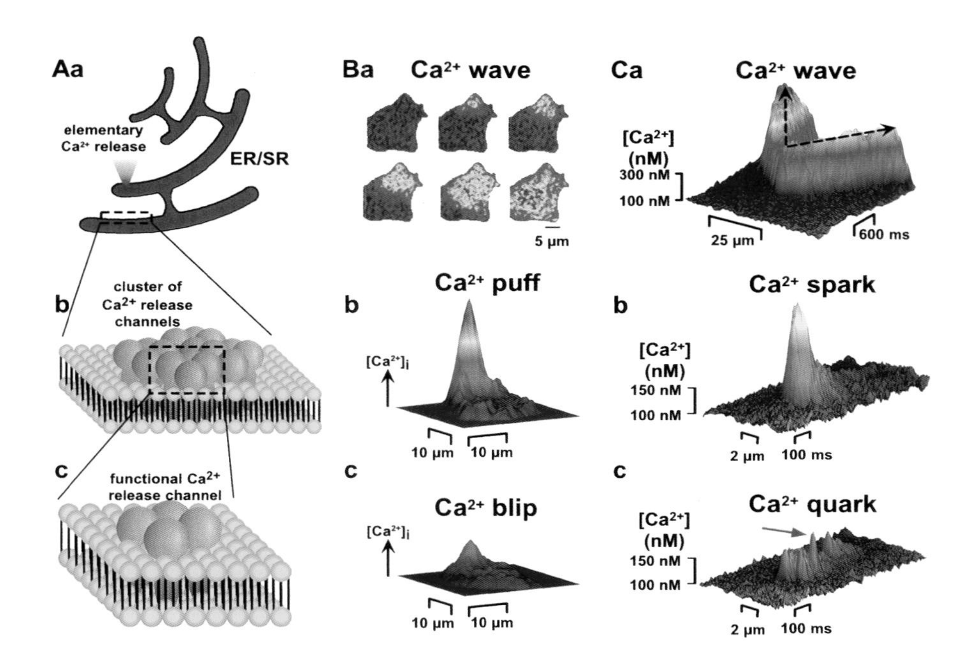

Plate 25.1 Examples of elementary Ca^{2+} signals. This figure illustrates different levels of elementary Ca^{2+} signalling in electrically non-excitable (B; HeLa cell) and excitable (C; cardiac myocyte) cells. Panel (Aa) shows a cartoon of intracellular Ca^{2+} stores. The region bounded by the dashed lines is expanded to depict a cluster of $InsP_3Rs$ or RyRs (Ab), and an isolated Ca^{2+} channel (Ac). Ca^{2+} signals corresponding to the different levels of magnification in (A) are shown in (B) and (C). (Ba) and (Ca) illustrate global Ca^{2+} signals, i.e. Ca^{2+} waves. (Bb) and (Cb) illustrate a Ca^{2+} puff (Ca^{2+} release from a cluster of $InsP_3Rs$) and a Ca^{2+} spark (Ca^{2+} release from a cluster of RyRs), respectively. (Bc) and (Cc) show a Ca^{2+} blip (arising from the gating of a single $InsP_3R$) and a series of Ca^{2+} quarks (single RyRs), respectively. Note that the figures in (B) were captured using a Noran Odyssey running in image mode (for (Ba), the frame rate was 7.5 Hz), whereas those figures in (C) represent linescan images acquired with a BioRad MRC600. All images were obtained using fluo-3-loaded cells. The Ca^{2+} signals were triggered using either histamine (B) or two-photon photolysis of caged Ca^{2+} (C).

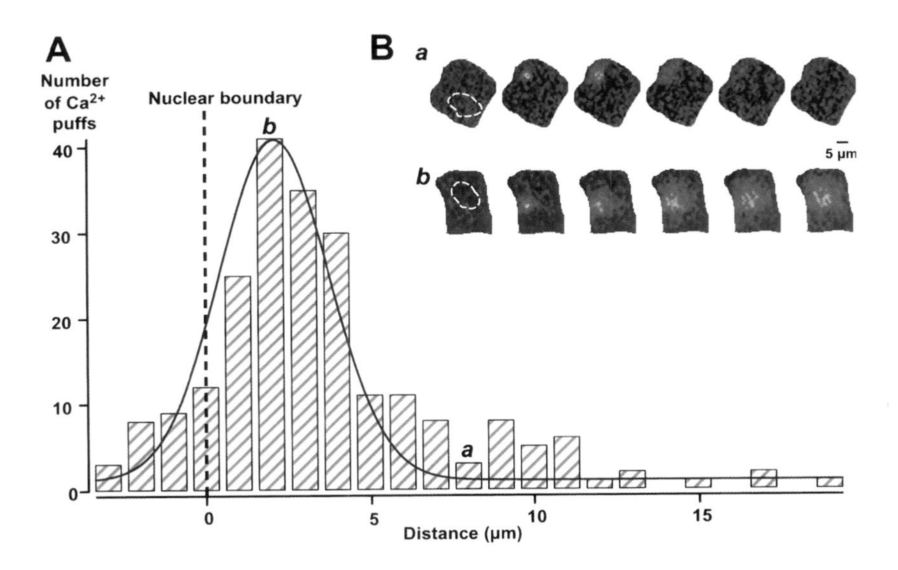

Plate 25.2 Example of the spatial utility of elementary Ca^{2+} signals; Ca^{2+} puffs distributed around the nuclear envelope. Fluo-3-loaded HeLa cells were stimulated with threshold histamine concentrations (0.5–1 μM) to evoke elementary Ca^{2+} release events. The histogram in (A) describes the occurrence of such release signals ($n = 233$) in 1 μm-wide sections either side of the nuclear envelope. The dashed vertical line indicates the relative position of the nuclear envelope. The curve was fitted assuming a gaussian distribution (FWHM ~3.8 μm; peak at 2 μm from the nuclear envelope). (B) Time-series of confocal images (7.5 Hz; acquired using a Noran Odyssey) showing a remote Ca^{2+} puff (Ba, ~6 μm from nucleus) and a perinuclear Ca^{2+} puff (Bb, ~2 μm from nucleus). The Ca^{2+} signal resulting from the perinuclear Ca^{2+} puff displayed a longer time course and occupied a larger volume in comparison to the remote release event. For further description, see Lipp *et al.* (1997) *EMBO J.* **16**, 7166–7173.

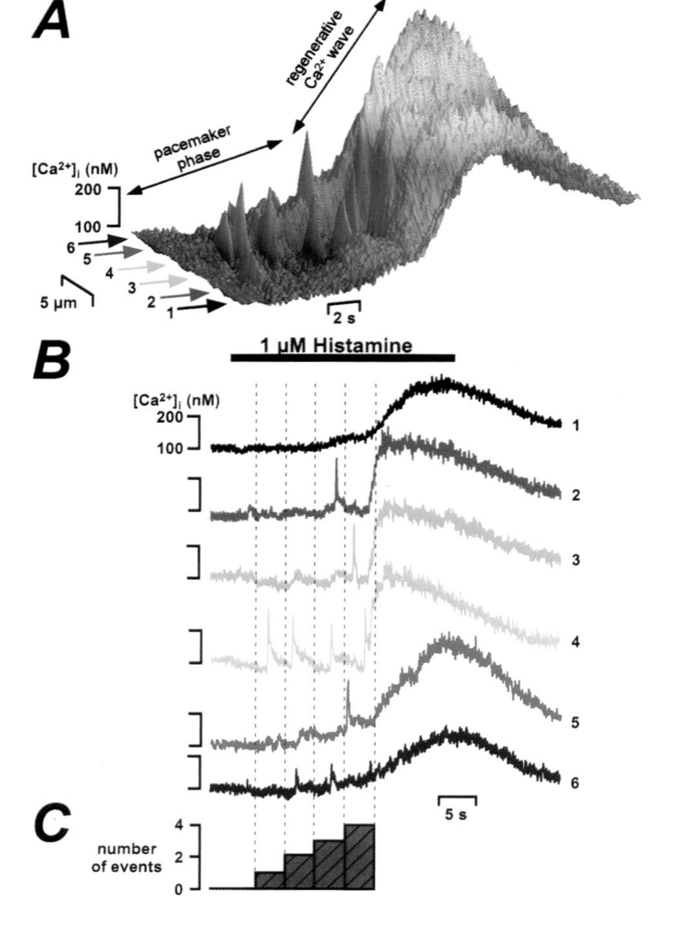

Plate 25.3 Recruitment of elementary Ca^{2+} release events during the onset of a regenerative Ca^{2+} wave. (A) Surface representation of a linescan image of a fluo-3-loaded HeLa cell (4 ms per line; acquired using Noran Odyssey) obtained during stimulation with histamine (histamine application shown by filled bar). Five independent elementary Ca^{2+} release sites were evident in this plot. $[Ca^{2+}]_i$ is coded in both the height and the colour of the surface. The $[Ca^{2+}]_i$ at these release sites (traces 2–6 in (B)), and at one remote site (trace 1), was calculated by averaging the signal in 2 µm wide bands along the lines centred around the arrowheads in (A). The colours and the labelling of the arrowheads in (A) correspond to the line plots in (B). Panel (C) shows the increasing cumulative frequency of elementary signals, derived by counting the events occurring in 4 s long segments. For further description, see Bootman *et al.* (1997) *Cell* **91**, 367–373.

Plate 25.4 (below) Ca^{2+} sparks during cardiac excitation–contraction coupling. (A) Linescan image of the Ca^{2+} response in a guinea pig ventricular myocyte to a step depolarization from –50 mV to +5 mV (2 ms per line; acquired using a BioRad MRC600). The homogeneous Ca^{2+} transient illustrated in panels (Aa)–(Ac) was reduced to spatially isolated Ca^{2+} sparks following partial inhibition of the L-type Ca^{2+} current with 5 µM verapamil (Ba–Bc). Reproduced by permission from Lipp & Niggli (1996) *Prog. Biophys. Mol. Biol.* **65**, 265–296.

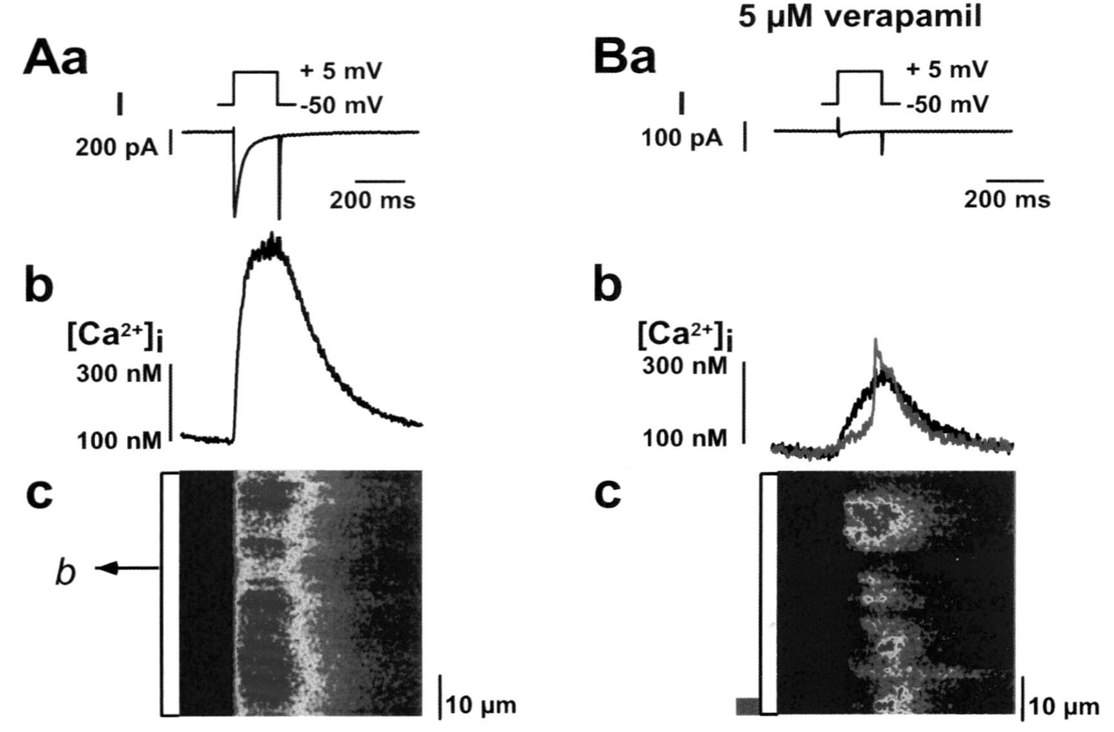

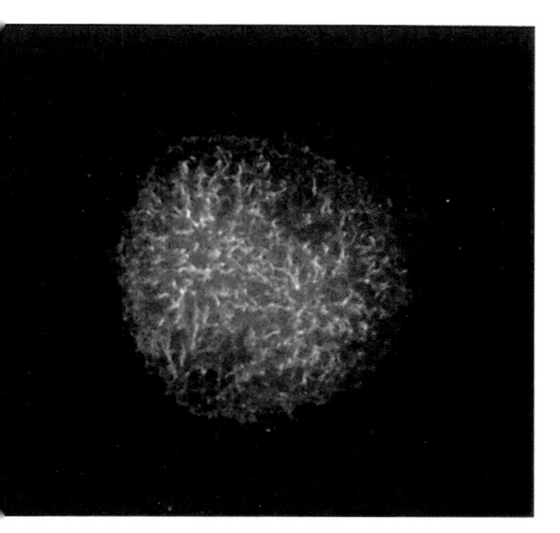

Plate 26.1 Colour photographs of a series of optical sections of whole-mount sea urchin embryonic microtubules at the first mitotic stage stained with BODIPY. The images were taken at the same spot but with gradually shifting z-axis focusing.

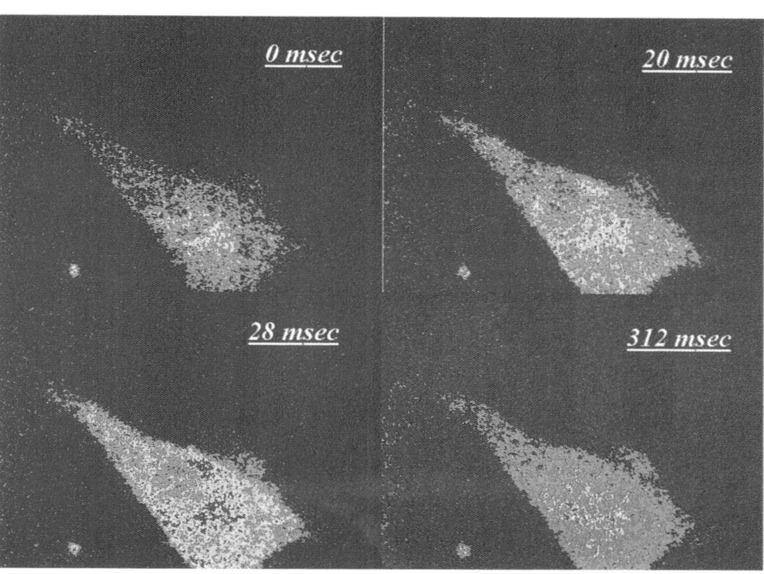

Plate 26.2 High-speed observation of Ca^{2+} changes in cardiac myocytes (Ishida, 1996). Rapid changes in the Ca^{2+} level at spontaneous beating of cultured cardiac myocytes were recorded using high-speed ICCD camera (1000 HR, Kodak) at 4 ms per frame. Cardiac myocytes isolated from mouse fetus were loaded with 10 μM fluo-3 AM (Molecular Probes) for 20 min at 37°C.

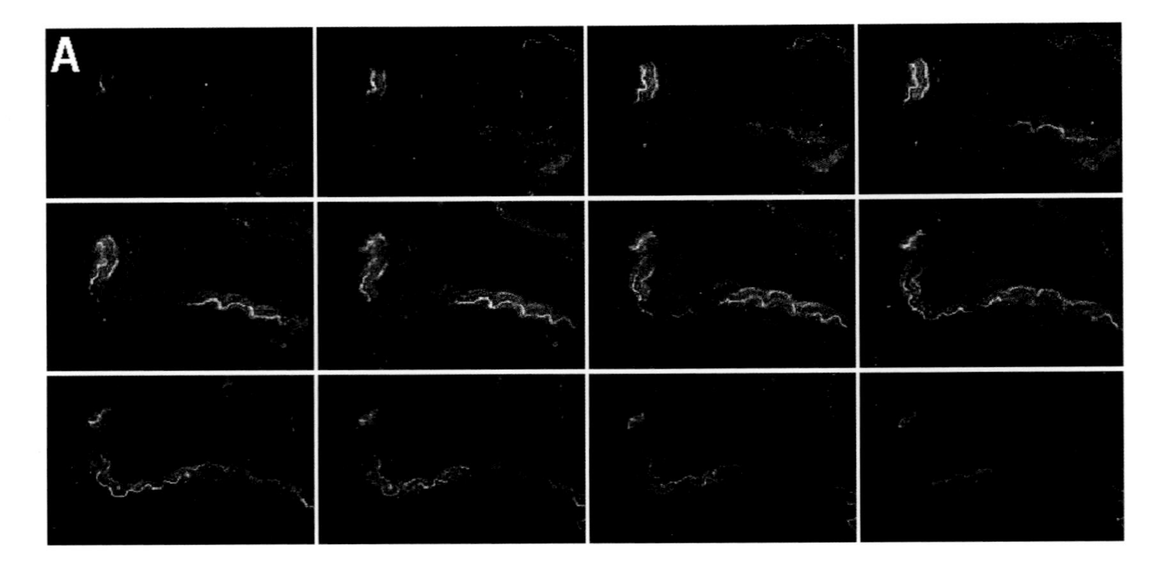

Plate 27.1 (A) Serial optical sections (*z*-series) through a whole mount of a rat trachea that has been labelled with an NOS antibody. In any of the individual optical sections it is difficult to gain an impression of the overall 3-D structure of the nerves. (B) A maximum intensity projection of the data set in (A) depicts the 3-D structure of the nerves. Serial optical sections and 3-D rendering such as the maximum intensity projections are quick and simple procedures with confocal microscopy. (Spalding, Rigby & Goldie, unpublished).

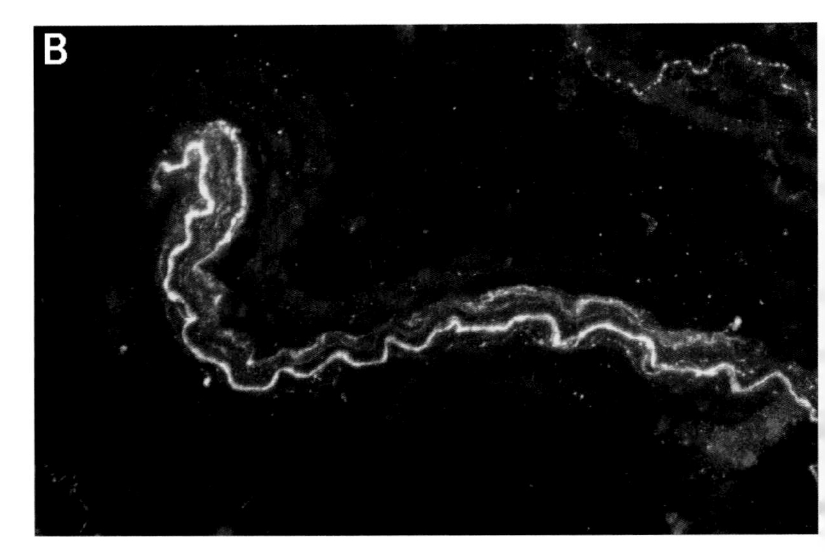

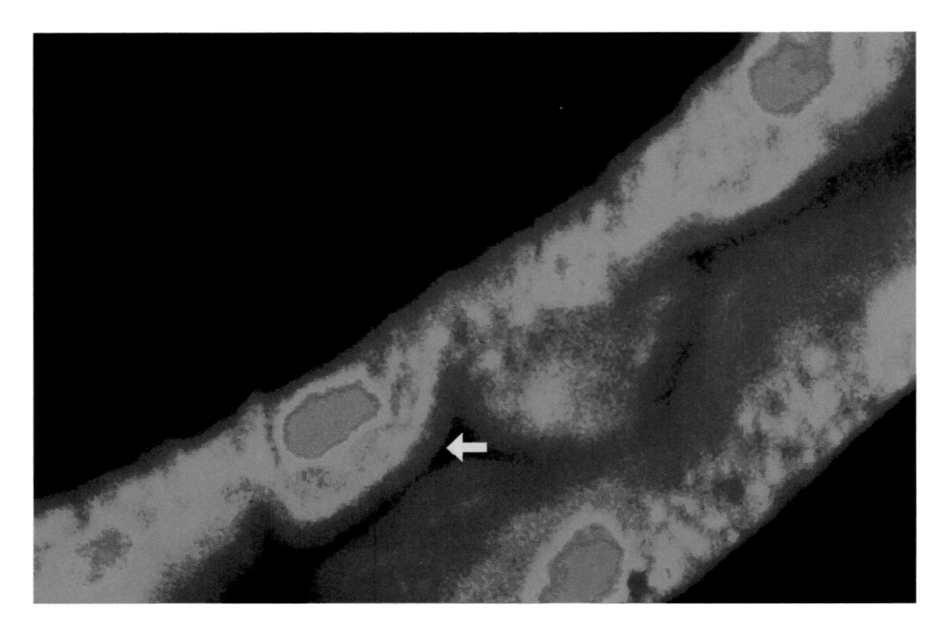

Plate 27.2 Isolated rabbit proximal kidney tubule loaded with the pH-sensitive dye BCECF (Harris, 1991). As the intracellular fluorescent signal is heterogeneous, it is tempting to conclude that these kidney cell have a highly varied intracellular pH patter within them. However, such conclusions from single-wavelength dyes are fraught w danger. The fluorescent signal from the bru border (arrow) is less bright than that from the cytosol due to pathlength errors (see Fi 27.4). The apparent higher pH of the nuclei probably represents a combination of highe BCECF accumulation and altered behaviou of fluorophore in the nuclear environment (compare with Fig. 27.5). Scale bar = 5 μm

Plate 29.1 Example of the application of line illumination in a Raman microspectroscopic experiment. (a) Line illumination of an erythrocyte (top view and cross-section). (b) Image captured by the CCD-chip: signal collection time, 30 s; laser power 25 mW; microscope objective, see Fig. 29.14. (c) Spectral information (Raman spectrum of haemoglobin in the erythrocyte) is obtained along the vertical axis. (d) Spatial information, in this case the distribution of haemoglobin in the cell, is found along the horizontal axis. This graph is based on the integrated intensity of the Raman signal in the spectral interval 1450–1650 cm^{-1}.

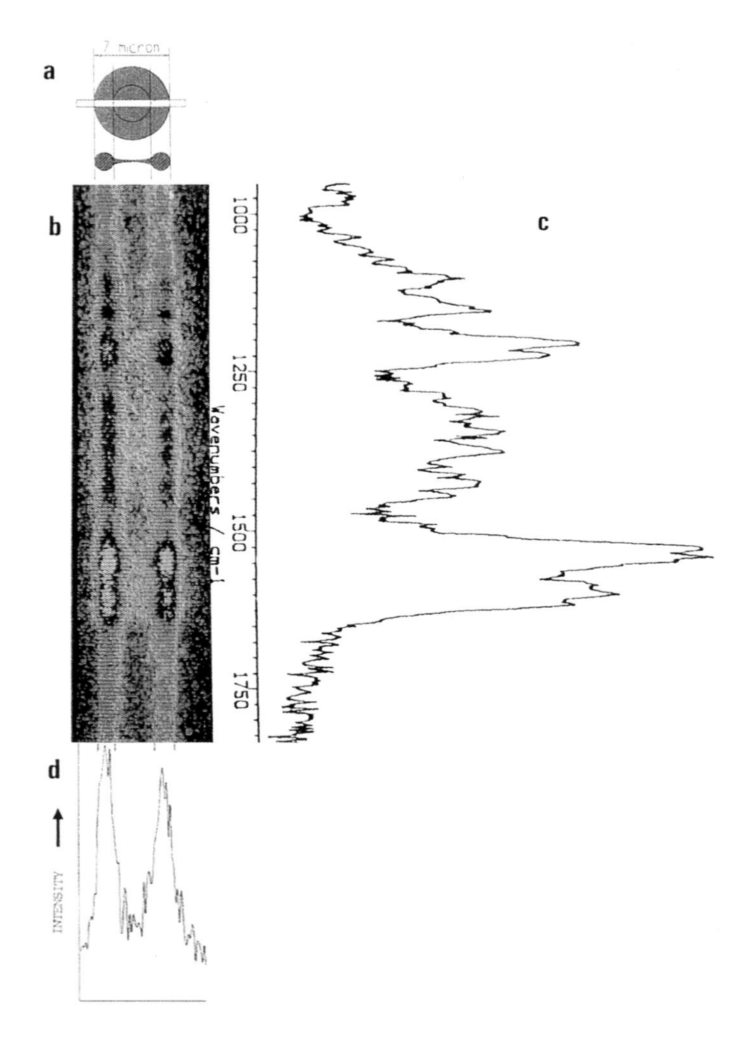

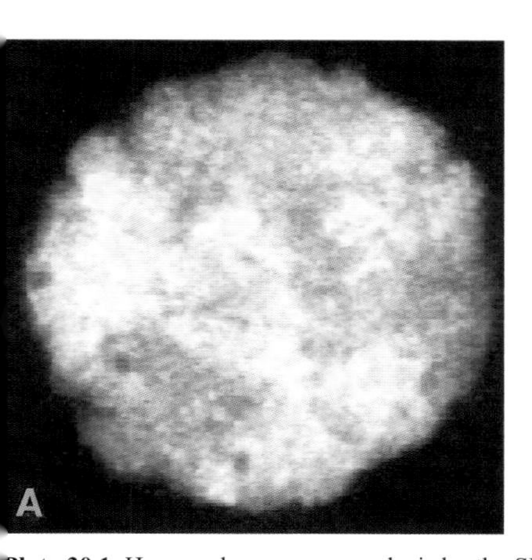

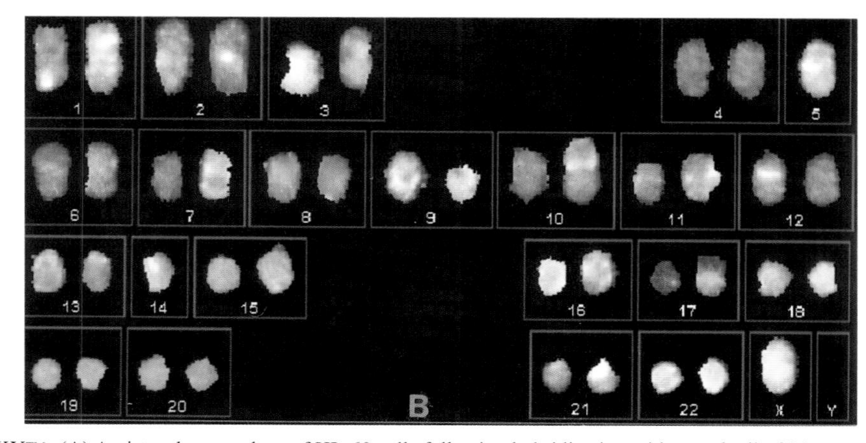

Plate 30.1 Human chromosome analysis by the SKY™. (A) An interphase nucleus of HL-60 cells following hybridization with a cocktail of 24 painting probes for all human chromosomes. Pseudo-colours were assigned to each chromosome according to the specific spectral signature of the chromosome. (B) A karyogram of a HL-60 metaphase; the chromosomes were automatically arranged by the system in the karyotypic table. The karyogram reveals a chromosomal translocation in chromosome 10. Other chromosomal aberrations are seen such as the absence of chromosome 5, 14 and X.

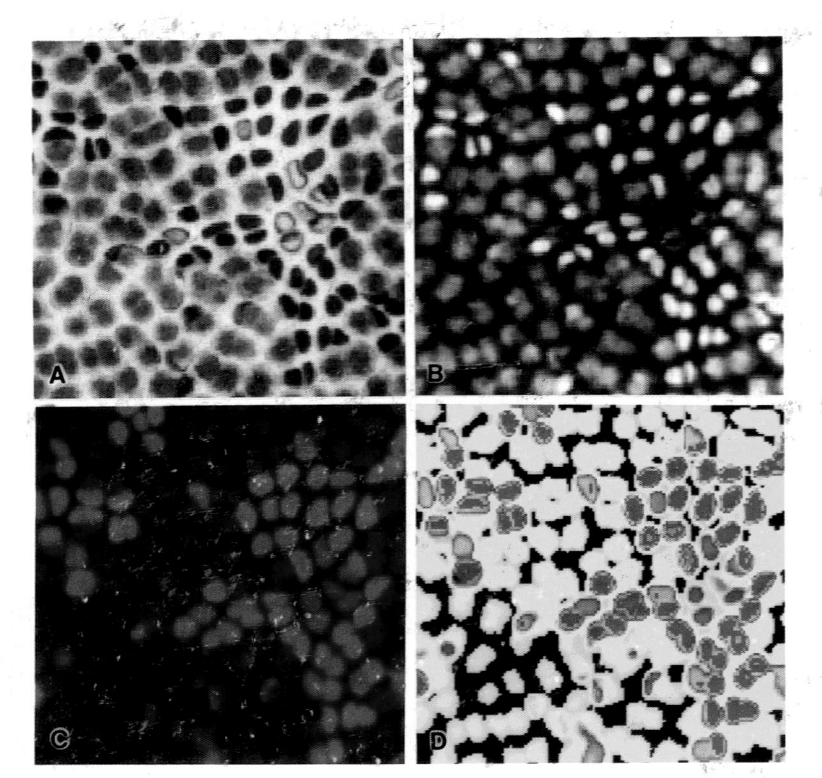

Plate 30.2 Large and small vegetative cells of live *Porphyra linearis.* (A) A transmitted light image of the vegetative cells. (B) An absorbance image of the same area formed by the application of the Beer–Lambert equation. (C) Fluorescence of the pigments excited at 406 nm. (D) The classified image obtained by spectral similarity mapping based on the fluorescence spectra as shown in (A).

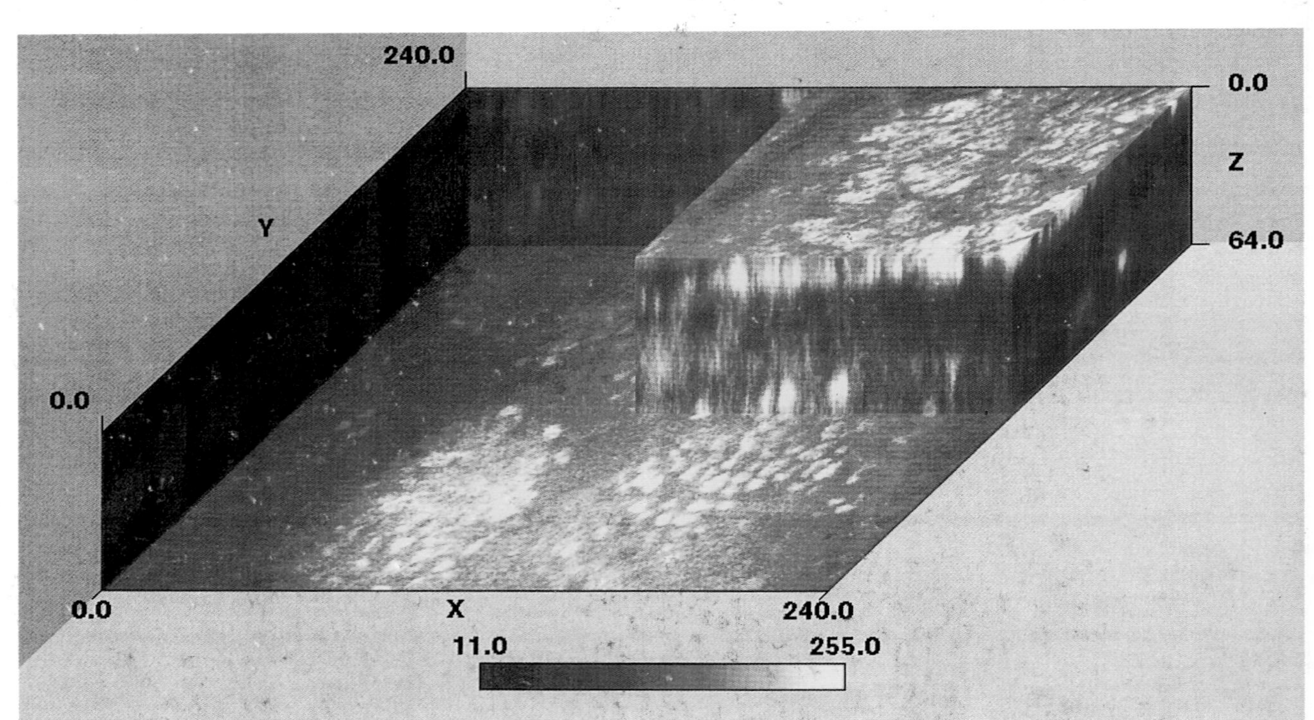

Plate 31.1 Three-dimensional reconstructions of *in vivo* human skin. The colour bar under the figure indicates the relative intensity of reflected light. Dark red indicates low intensity level and bright yellow indicates a high intensity level. Reconstructed volumes are 240 μm × 240 μm in area and 64 μm thick. Reconstruction is pseudo-coloured to indicate the intensity of scattered and reflected light from the optical sections. Intensity colour map is below the reconstruction. Image shows the surface of the skin; the thickness of the stratum corneum; the dark, round nuclei of the stratum spinosum; the bright, oval nuclei of the basal cells; and the papillary dermis.

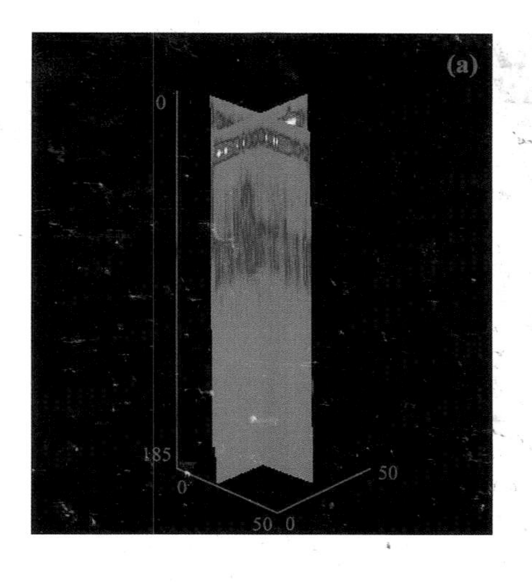

0 255

Plate 31.2 Three-dimensional reconstruction of multiphoton excitation microscopy of *in vivo* human skin. (a) 730 nm excitation; (b) 960 nm excitation. *x* and *y* orthogonal slices are shown. The axis dimensions are in μm. In (a), the top bright layer corresponds to the stratum corneum surface. The second bright band at a depth of 80–100 μm is the basal cell layer at the top of the papillary dermis. In (b), the top bright band extends throughout the stratum corneum.

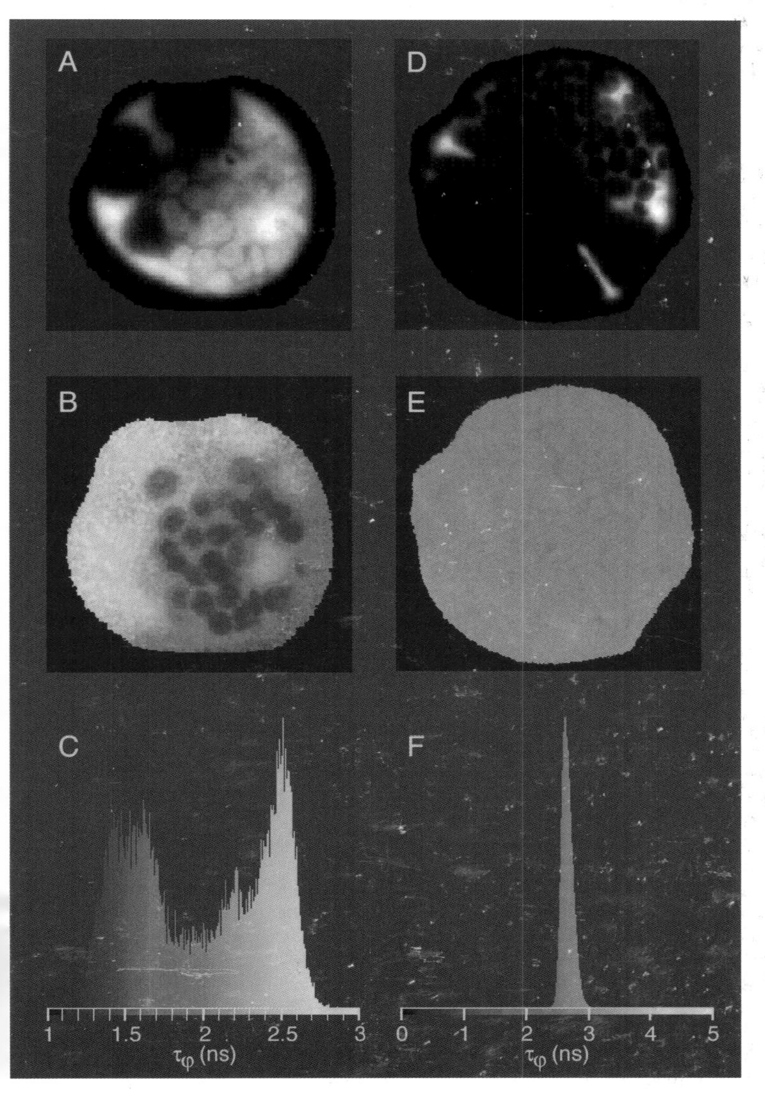

Plate 34.1 FLIM analysis of living Cowpea protoplasts expressing S65T mutated GFP. Cowpea (*Vigna unguiculata* L.) mesophyll protoplasts were prepared and transfected as described (van Bokhoven *et al.*, 1993). The protoplasts were transfected with the pMON30060 dicot expression vector encoding S65T*pgfp* with optimized codon usage for expression in plants (Pang *et al.*, 1996). Protoplasts expressing GFP were used for microscopy 24 h after transfection and excited with the 488 nm Ar laser line modulated at 40.0 MHz. A ×100 PL Fluotar oil immersion objective (NA 1.3), a 505 nm DRLP dichroic mirror (Omega Optical Inc., Brattleboro, VT, USA), and a long-pass GG 515 (Schott, Mainz, Germany) (A–C) or a bandpass Omega 525DF30 (D–F) emission filter were used to collect the fluorescence. One blank and 20 consecutive phase images separated by 18° were recorded (0.5 s exposure time each, no binning on the CCD). Twenty additional phase images were recorded with reversed phase increments -18° to correct for possible photobleaching as described (Gadella Jr *et al.*, 1997). Data analysis and processing was performed as described (Gadella Jr *et al.*, 1994, 1997). (A) and (D) represent the calculated steady-state fluorescence intensities (F_{DC} or D_0 of equation [34.14]). The green colour shows pixels not analysed due to intensity thresholding. (B) and (E) are the pseudo-coloured calculated lifetime ($\tau\varphi$) images (equation [34.16]), and (C) and (F) are the corresponding temporal histograms of the $\tau\varphi$ pixel values. For both analyses, erythrosin B in water with a lifetime of 80 ps was used as a reference compound. The reference modulation M_R as defined in equation [34.12] was 89.3%, and total bleaching was <1%. The actual imaged area is 54 × 54 μm for images (A) and (B), and 92 × 92 μm for images (D) and (E).

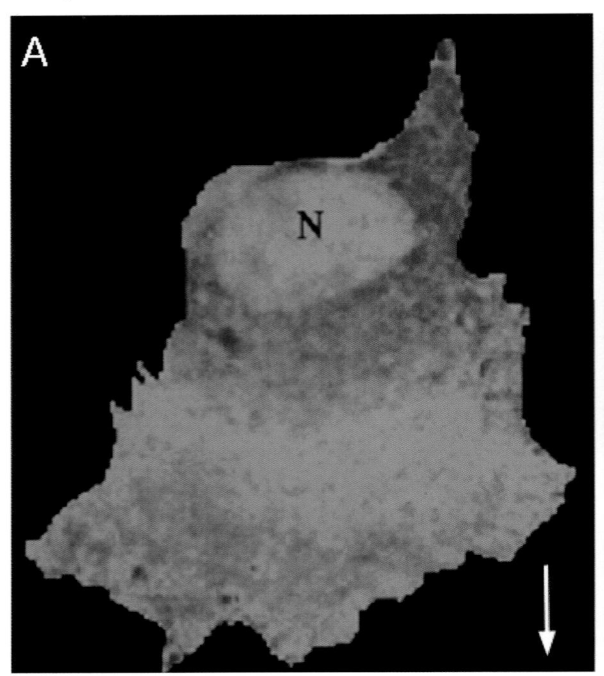

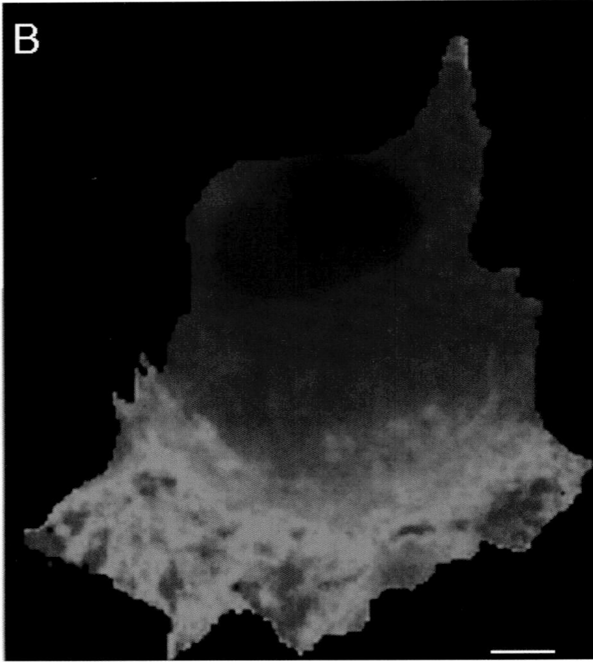

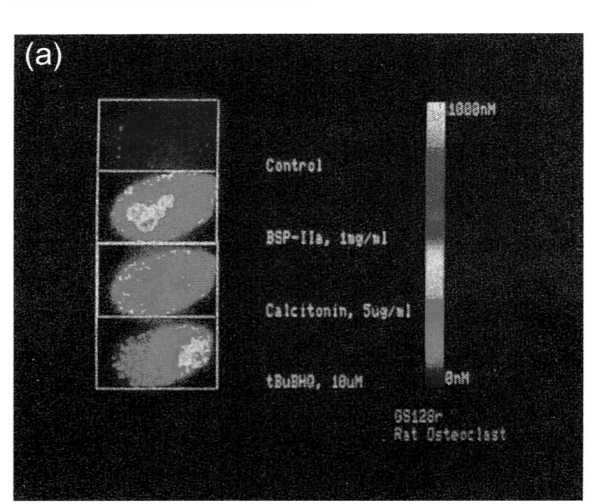

Plate 41.1 (above) Myosin II phosphorylation biosensor in a polarized, migrating cell. The energy transfer phosphorylation biosensor and a cascade blue 10 kDa dextran (a cell volume marker) were microinjected into wound-healing fibroblasts. Image (A) is the rhodamine/fluorescein emission ratio of the biosensor, showing the distribution of phosphorylated, activated myosin II in this cell; image (B) is the rhodamine/cascade blue emission ratio, showing the relative concentration of myosin II in this cell. These images were pseudocoloured so that warmer, red colours represent higher levels of serine 19 phosphorylation of the regulatory light chain (A) or myosin II concentration (B), while cooler blue colours represent lower levels of phosphorylation (A) or myosin II concentration (B). This cell is moving in the direction of the white arrow; N denotes the nucleus of the cell. (Reprinted with permission from *Molecular Biology of the Cell.*)

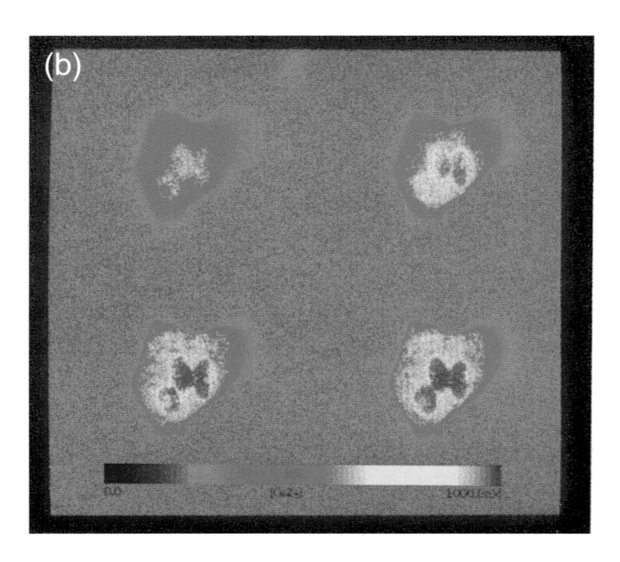

Plate 43.1 (a) Spatial characteristics of the intracellular calcium signal in rat osteoclasts, imaged using conventional fura-2 non-confocal ratiometric imaging. Shown here is a single multinucleated osteoclast imaged with an intensified CCD camera. This image shows a comparison of the spatial characteristics of the calcium signal in response to bone sialoprotein (BSP-IIA), salmon calcitonin and *t*BuBHQ, all additions being preceded by a buffer wash. BSP-IIA, a candidate for an endogenous ligand for the vitronectin receptor used as the stimulus, appears to elicit a response in the nuclear or perinuclear region of the cell. Calcitonin elicits a generalized rise in intracellular calcium which depends on extracellular calcium. *t*BuBHQ inhibits ATP-dependent sequestration of calcium into the endoplasmic reticulum and releases ER calcium without affecting nuclear calcium uptake (b) Preliminary data obtained with a Noran confocal laser scanning microscope of osteoclasts loaded with the non-ratiometric calcium probe fluo-3. Images are taken in a single wavelength and run from top left to bottom right in time, being separated in time by about 2 s. BSP-IIA was added after frame one, and this gave rise to an immediate sustained nuclear calcium signal.

Plate 44.1 Intercellular communication in carboxyfluorescein-labelled HT cells. The centre cell (no. 3) was bleached, and fluorescence in both neighbouring cells (nos 1 and 2) and the entire triplet (no. 4) was monitored. Three pseudo-colour plots show fluorescence before bleaching, immediately after the bleach and after recovery, with the summary of the time-dependent recovery also shown for all four areas monitored.

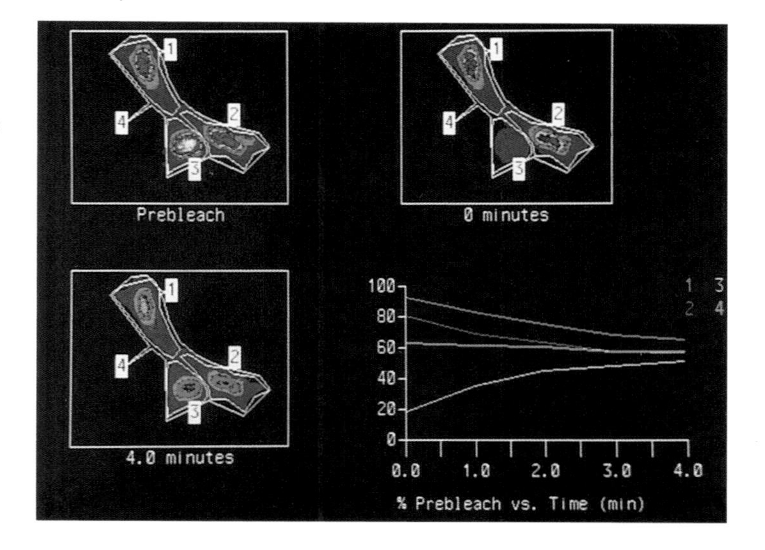

Plate 44.2 Digitized pseudo-colour images of SB-3 cells under control conditions (A), and after treatment with 16 nm TPA for 1h (B). Images (225 × 225 μm) of cells prior to, immediately after, and 16 min after photobleaching selected cells are shown. A plot with the percentage recovery of fluorescence at 4 min intervals is displayed for each of the cells selected. The plotted data were corrected for background decline of fluorescence using the non-photobleached cell in area no. 1. The single cell in area no. 2 served as a negative control for fluorescence redistribution.

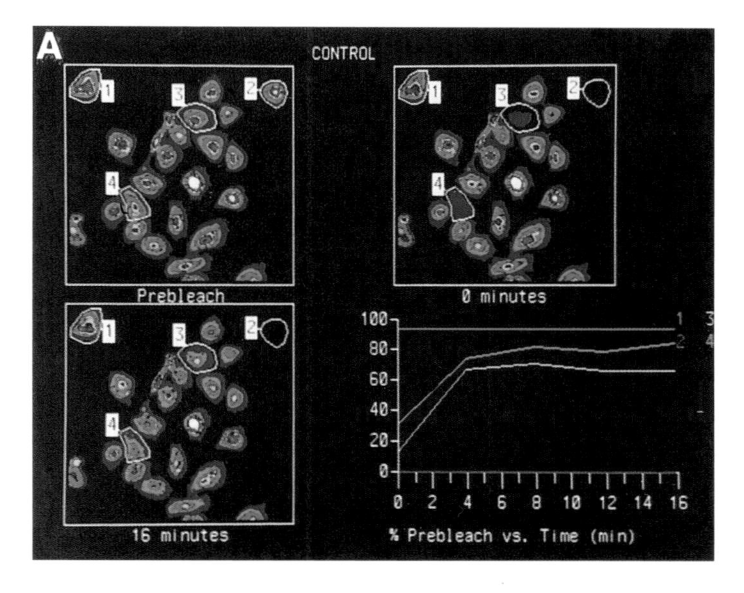

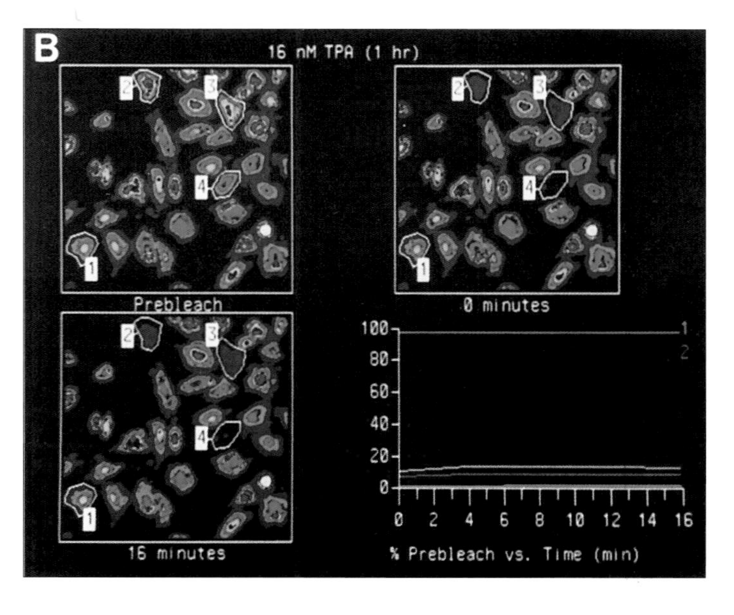

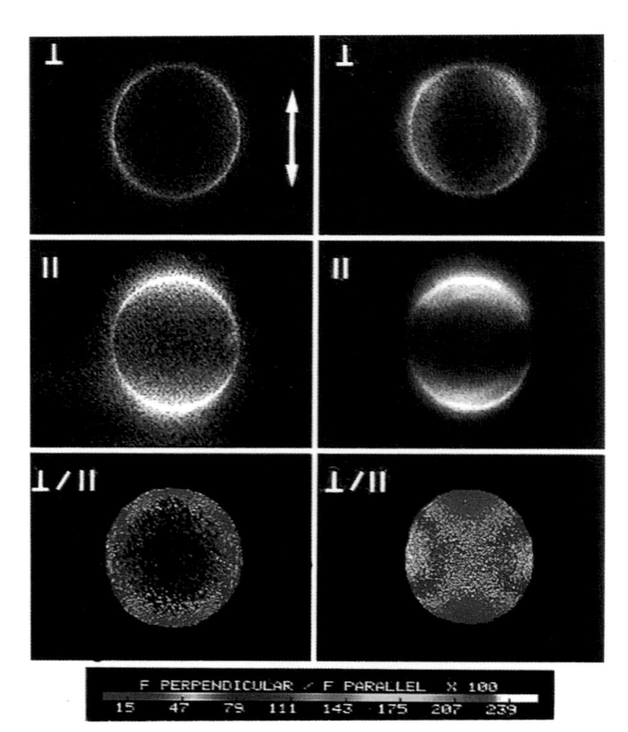

Plate 46.1 Polarized fluorescence of TMA-DPH-labelled liposomes composed of fluid-phase POPC (left) or gel-phase DPPC (right). Top and middle rows: Fluorescence images, after background subtraction, obtained with the emission polarizer oriented perpendicular and parallel, respectively, to the excitation light polarization direction (indicated by the arrow). The top, right side, bottom, and left side of each image corresponds to $\rho = 0°, 90°, 180°$, and $270°$, respectively, and $\gamma = 0°$ (see Fig. 46.1). Bottom row: Pseudo-colour polarization ratio images, after mapping. Liposome diameters are 31 μm (left) and 38 μm (right).

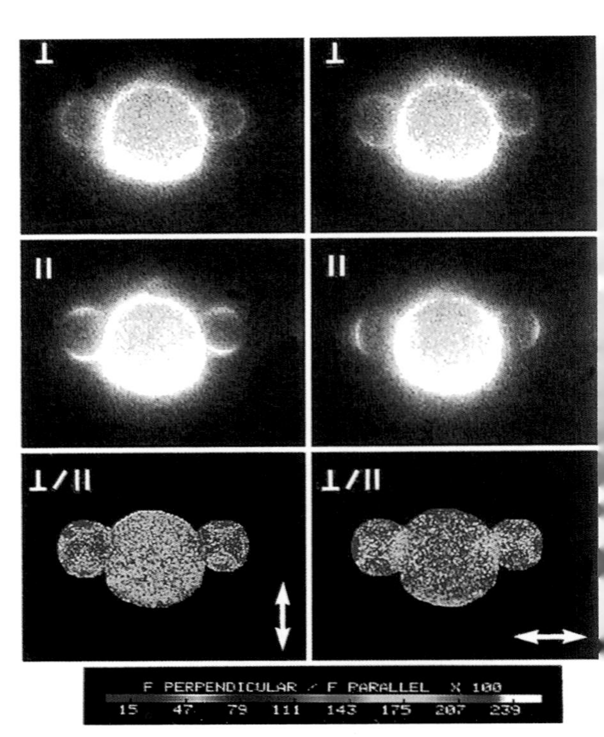

Plate 46.2 Fluorescence polarization measurement of a blebbing rat hepatocyte labelled with TMA-DPH. Top and middle rows: Fluorescence images, after background subtraction, obtained with the emission polarizer oriented perpendicular and parallel, respectively, to the excitation light polarization direction for each of two excitation polarizer orientations (indicated by the arrows). The focal plane passes through the centres of the two plasma membrane blebs. Around the bleb perimeters, $\rho = 0°$ and $180°$ corresponds to the top and bottom, respectively, of the bleb images in the left-hand column and the right and left sides, respectively, of the bleb images in the right-hand column. Bottom row: Pseudo-colour polarization ratio images, after mapping. Cell diameter is 25 μm. (Reprinted with permission from *The FASEB Journal*.)

While the visualization and manipulation problem appears to be well in control for 3-D data sets, adequate representation of 4-D data is still an issue without resolution, although some impressive options exist. One option is to reconstruct each 3-D data set and provide a 2-D view of the reconstruction at an arbitrarily defined angle. The result is a 'time' sequence of 3-D reconstructions at a certain angle. These reconstructions can then be visualized in a movie sequence where successive images are played back akin to video frames. However, it must be recognized that with only one angle of view, this form of 4-D representation is actually only 3-D with two spatial and one temporal dimension. On the other hand, this technique can be extended by forming 3-D reconstructions of each 3-D data set at different angles and placing them next to each other to form a spatial sequence of 2-D views of the reconstructions. These rotational sets can then be constructed for successive 3-D data sets. Finally, the sequence of (multiple) 3-D reconstructions can be played back as a movie. Alternatively, the 4-D data sets can be resampled into 3-D sets such that two of the dimensions are spatial and the third is temporal. Thus, a '3-D set' is actually a time sequence of 2-D images. The 3-D set can then be reconstructed for successive sets (now representing the third spatial dimension) and played back as a movie. The advantage of this visualization technique is that the spatial variation in the intracellular dynamics along one dimension can be followed as if in time. The choice of an appropriate visualization procedure is dependent on the point of emphasis. In order to emphasize the temporal changes in the volume occupied by the structure (or even ion such as Ca^{2+}) of interest, temporal sequences of 3-D reconstructions either at one or several angles would be highly appropriate. However, for structures that may be temporally variant only in one spatial dimension, a resampling of the 4-D data sets may be more appropriate.

A potential limiting factor in rapid confocal imaging is the tremendous rate at which data needs to be visualized as well as stored. Obviously, the rate of image acquisition needs to be chosen based on the intracellular dynamics to be examined. However, as with the above case, even rates as high as 480 frames per s may sometimes be inadequate. Therefore, future developments in confocal technology need to focus on allowing for even higher acquisition rates. Innovative techniques for resizing the image during rapid acquisition while maintaining the same overall rate of data storage, such as that in the Odyssey, are extremely useful under conditions when data storage capabilities are a limitation. For example, even at an acquisition rate of 480 frames per s, the net rate at which data is written is still 30 full frames per s. However, this is clearly a compromise, since the resizing of the image can result in considerable loss of spatial resolution. Furthermore, once the data has been stored, it still needs to be imported into image processing packages such as ANALYZE, VoxelView and NIH Image. While image processing software is making giant steps in 3-D data handling capabilities, 4-D images are likely to introduce a new dimension of complexity, especially in terms of computer and video memory and hard disk access. However, since visualization is also likely to be limited to 2-D views from reconstructions of single 3-D data sets, the problem of 4-D visualization is essentially the same as that of 3-D visualization, which is already well advanced in many software packages. Thus, the real limitation to storage of high-speed confocal data may be computer-related, and is likely to be resolved with ongoing technological developments.

23.9 CASE STUDIES ON 3-D AND 4-D RESOLUTIONS

23.9.1 Case Study 3: optical sectioning of pollen

Removal of out-of-focus information allows for simple 3-D reconstruction of confocal optical sections and estimation of 3-D parameters such as volume and surface area. However, accurate 3-D measurements are possible only if the step size for each image (dependent on the stepper motor or other device being used to change the focal plane) is matched to the optical section thickness (dependent on the optics, especially the objective lens). The example below illustrates the effect of mismatching step size to optical section thickness, and the potential for recovery of lost optical sections by interpolation.

A slide of fluorescently labelled pollen (commonly provided by confocal manufacturers as a test sample) was used. A single pollen grain was optical sectioned using a Nikon 40×/1.3 NA oil-immersion objective lens. The optical section thickness at a fixed confocal slit setting was determined empirically using previously published procedures and found to be ~0.8 µm. Accordingly, the step size was set to 0.5, 0.8 and 1.2 µm, and the pollen grain was repeatedly sectioned to obtain 3-D image stacks. Using a volume rendering tool in the ANALYZE software package, the confocal images were reconstructed in both the *xy* plane (optical section plane; 'face on' view or 0° by convention) and the *xz* plane ('side' view at a 90° *y*-rotation of the *xy* plane).

As shown in Fig. 23.4, when the step size was matched to optical section thickness at 0.8 µm, the *xz* view of the pollen grain displayed a more or less

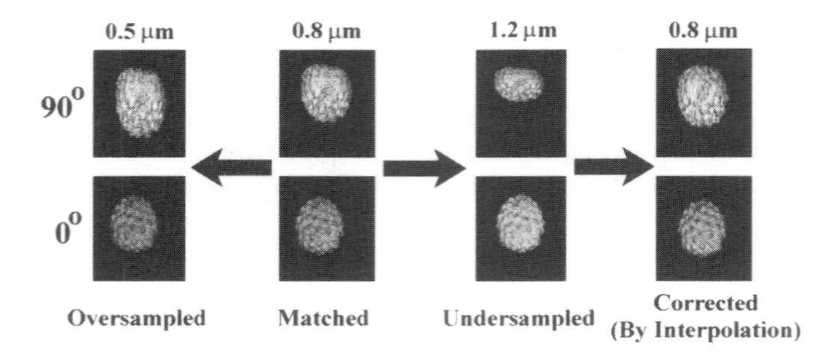

Figure 23.4 Effect of mismatched step size vs. optical section thickness on 3-D reconstructions of pollen grains. A step size smaller (0.5 μm) than the optical section thickness (0.8 μm) results in oversampling and overestimation of the apparent *z*-axis specimen thickness. In contrast, a step size larger than the optical section thickness results in an underestimation of the specimen thickness. Image resampling of the under- or oversampled specimen can potentially correct these errors.

spherical profile. When the step size was smaller than the optical section thickness (0.5 μm), the specimen thickness was grossly overestimated, leading to a considerably oblong profile. However, the 'face on' or *xy* view did not appear different from that for the 0.8 μm thick sections. The volume of the oversampled specimen, estimated using a voxel-connecting algorithm in ANALYZE, was found to be ~30% greater than that obtained when step size and optical section thickness were matched. In contrast, when the step size was greater than the optical section thickness (1.2 μm), the specimen thickness was grossly underestimated, and the *xz* profile appears squished. Again, the *xy* profile appeared unchanged. The estimated volume was found to be ~45% greater than that at 0.8 μm. However, if the 1.2 μm spaced optical sections were resampled at 0.8 μm by linear interpolation, and the new set of images reconstructed, the side profile was restored to the more or less spherical shape, and the estimated volume was only ~10% greater than that of the true data set obtained at 0.8 μm.

23.9.2 Case Study 4: optical sectioning of Ca^{2+} sparks in TSM cells

Ca^{2+} sparks represent spatial and temporal heterogeneity in Ca^{2+} regulation (Sieck *et al.*, 1997) (also see Plate 23.5). Since cells have a 3-D structure, it is of interest to follow $[Ca^{2+}]_i$ regulation in the 3-D domain. Given the limitations in 3-D resolution of current confocal systems, there is currently no information on 3-D $[Ca^{2+}]_i$ regulation in cells.

In this case study, we were particularly limited by the relatively slow 3-D acquisition rate of the piezoelectric focusing system on the Odyssey. As discussed in Case Study 3, Ca^{2+} sparks in cardiac myocytes can barely be analysed at 2-D acquisition rates of 240 frames per s. It is likely that a comparable 3-D acquisition rate will be required to examine sparks in 3-D. Accordingly, a 3-D acquisition rate of 15 frames per s was too slow to examine sparks in cardiac myocytes. Compared with sparks in cardiac and skeletal muscles, Ca^{2+} sparks in TSM cells display relatively slower kinetics, making them compatible with the slow 3-D acquisition rate of the piezoelectric focusing system on the Odyssey system. Therefore, we examined sparks in fluo-3-loaded porcine TSM cells.

Most settings for cell loading and confocal imaging were similar to those used in other case studies, except that no ROIs were outlined, and images were processed *post hoc* using ANALYZE. Although the Odyssey system is capable of acquiring images at 480 frames per s, the piezoelectric focusing system is only capable of 15 frames per s. Accordingly, there is no advantage in acquiring images at >15 frames per s in the 3-D mode since there is unnecessary resampling of the same optical plane, albeit of different time points. Therefore, images were acquired at a fixed rate of 15 frames per s. Step size was set at 0.75 μm, the closest multiple of 0.25 μm (minimum step size of piezoelectric focusing system) to the actual optical section thickness of 0.8 μm. Since the Ca^{2+} sparks were spontaneous as well as irregular, acquisition was not synchronized with their occurrence. Instead, the acquisition was repeated using the unidirectional mode of the focusing system, with no time delays beyond those introduced by software and hardware. These image sets were then processed with ANALYZE and the relevant images were extracted for 3-D and 4-D reconstruction.

In TSM cells, Ca^{2+} sparks occurred at a frequency of 5–15 per min (which necessitated the acquisition of several 3-D sets containing no sparks). Each spark had a detectable 3-D distribution, occupying three to four optical sections (2.2–3 μm) along the *z*-axis and 4–5 pixel-wide areas in the *xy*-plane, which corresponded to ~0.1% of the volume of a TSM cell. The 3-D volume occupied by the Ca^{2+} spark did not vary considerably over time. Views of the 3-D reconstructed data set at different *x*, *y* and *z* angles also did not display appreciable variation (Plate 23.6). Therefore, it appears that Ca^{2+} sparks, at least in TSM cells, are more or less time-invariant in their 3-D distribution, supporting the idea that they represent unitary Ca^{2+} release from a fixed set of relatively synchronous RyR channels.

23.9.3 Case Study 5: optical sectioning of spontaneous $[Ca^{2+}]_i$ waves in cardiac myocytes

As with TSM cells, variations in the distribution of intracellular Ca^{2+} stores and asynchronous Ca^{2+} release from different parts of a cardiac myocyte lead to $[Ca^{2+}]_i$ waves. Since the myocyte has a 3-D structure, these heterogeneities are also likely to be 3-D in nature. Therefore, it is of interest to follow the pattern of $[Ca^{2+}]_i$ waves in 3-D. Given the limitations in 3-D resolution of current confocal systems, and the relatively high speed of wave propagation, there is currently no information on the 3-D distribution pattern of $[Ca^{2+}]_i$ waves in cardiac myocytes.

We used fluo-3-loaded rat cardiac myocytes to examine the 3-D pattern of $[Ca^{2+}]_i$ waves, and their variation with time (4-D pattern). Images were acquired at a fixed rate of 15 frames per s. Step size was set at 0.75 μm. Since the $[Ca^{2+}]_i$ waves were spontaneous but regular, acquisition was timed so that baseline images prior to wave initiation and following wave completion were included. The acquisition was repeated using the unidirectional mode of the focusing system, with no time delays beyond those introduced by software and hardware limitations. These image sets were used for 4-D reconstructions.

At an acquisition rate of 15 frames per s, the 3-D propagation pattern of $[Ca^{2+}]_i$ waves could be followed through the depth of a cardiac myocyte (Plate 23.7) when the propagation velocity was less than ~20 μm s^{-1} (measured in the *xy* plane). This corresponded to the movement of ~1 μm of $[Ca^{2+}]_i$ in both the *xy* plane as well as the *z* plane, which was comparable to the thickness of a single optical section. In other words, every step change in focus corresponded to the actual movement of $[Ca^{2+}]_i$ as it propagated through the TSM cell. Accordingly, a 3-D volume reconstruction of a set of optical sections represented the volume of the cardiac myocyte that underwent a change in $[Ca^{2+}]_i$ during the propagation of the $[Ca^{2+}]_i$ wave. By performing these reconstructions on consecutive 3-D data sets, we found that during successive $[Ca^{2+}]_i$ waves, different intracellular

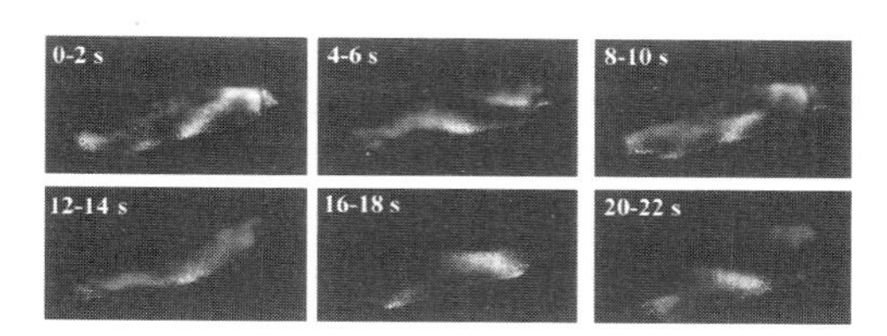

Figure 23.5 Temporal sequence of 3-D volume distribution of Ca^{2+} during a propagated $[Ca^{2+}]_i$ wave in rat cardiac myocyte. Representation of 4-D data as a sequence of 3-D reconstructions allows for evaluation of the temporal changes in the volume of an object or intracellular event. The 3-D rotations may optionally be rotated to yield a true 4-D representation of the data.

regions were involved (Fig. 23.5). This is in contrast to Ca^{2+} sparks that were more or less temporally invariant in their 3-D volume distribution.

As mentioned previously, the patterns of $[Ca^{2+}]_i$ waves displayed considerable heterogeneity. For several myocytes displaying $[Ca^{2+}]_i$ waves, even with 2-D acquisitions at high rates, 3-D and 4-D imaging did not yield reliable or consistent information on the 3-D distribution of $[Ca^{2+}]_i$ within the myocyte. In fact, in several cases, the times of image acquisition happened to coincide with the lack of elevated $[Ca^{2+}]_i$ within the focal plane and the 3-D reconstruction displayed no elevation of $[Ca^{2+}]_i$! Therefore, it appears that while the impressive kinetics of the piezoelectric focusing system allow real-time 4-D image acquisition of intracellular dynamics, these 4-D studies may be limited when dealing with $[Ca^{2+}]_i$ events such as rapidly propagated $[Ca^{2+}]_i$ waves in myocytes.

23.10 FUTURE DIRECTIONS

The above discussion and specific case studies point towards some key directions for future developments in confocal imaging technology. First, it appears that even 2-D acquisition rates as high as ~500 frames per s may be insufficient for real-time imaging of rapid intracellular events such as Ca^{2+} transients in excitable tissue. The need for even more rapid imaging is not only limited to Ca^{2+}, since a number of investigators use fluorescent dyes for imaging membrane potential, pH and even other ions, and could benefit from rapid acquisition capabilities. Developments in scanning technology may provide viable solutions in the very near future.

The second issue at hand is improvements in the technology for real-time 3-D imaging. Clearly, technologies such as the piezoelectric focusing system are impressive and innovative and could potentially be used in several applications. However, further advances in their somewhat limited temporal resolution is necessary before they can be used for real-time 3-D imaging of rapid intracellular events. These advances may come from the removal of limitations such as the need for stopping the focus motor for the acquisition of an image at each focal plane, and minimizing the focus motor flyback time during unidirectional acquisitions. These improvements will also result in enhanced 4-D temporal resolution.

ACKNOWLEDGEMENTS

This work is supported by grants HL34817. HL37680, HL057498 and GM57816 from the National Institutes of Health, the Mayo Foundation, and the University of Minnesota. The authors gratefully acknowledge the assistance of Mark Cody, Laurel Wanek and Thomas Keller in the case studies on Ca^{2+} dynamics. The technical input from Noran Instruments is also gratefully appreciated.

REFERENCES

Amos W.B. & White J.G. (1995) In *Handbook of Biological Confocal Microscopy*, J.B. Pawley (ed). Plenum Press, New York, pp. 403–415.

Amundson J. & Clapham D.E. (1993) *Curr. Opinion Neurobiol.* **3**, 375–382.

Berridge M.J. & Dupont G. (1994) *Curr. Opinion Cell Biol.* **6**, 267–274.

Cheng H., Lederer W.J. & Cannell M.B. (1993) *Science* **262**, 740–744.

Cheng H., Lederer M.R., Lederer W.J. & Cannell M.B. (1996a) *Am. J. Physiol.* **270**, C148–159.

Cheng H., Lederer M.R., Xiao R.P., Gomez A.M., Zhou Y.Y., Ziman B., Spurgeon H., Lakatta E.G. & Lederer W.J. (1996b) *Cell Calcium* **20**, 129–140.

Cogswell C.J. & Larkin K.G. (1995) In *Handbook of Biological Confocal Microscopy*, J.B. Pawley (ed). Plenum Press, New York, pp. 128–138.

Cogswell C.J., Sheppard C.J.R., Moss M.C. & Howard C.V. (1990) *J. Microsc.* **158**, 177–185.

Draiijer A. & Houpt P.M. (1988a) *Scanning Imaging Technol.* **809**, 85–88.

Draiijer A. & Houpt P.M. (1988b) *Scanning* **10**, 139.

Inoue S. (1995) In *Handbook of Biological Confocal Microscopy*, J.B. Pawley (ed). Plenum Press, New York, pp. 1–17.

Inoue S. & Oldenburg R. (1994) In *Handbook of Optics*, M. Bass (ed). McGraw-Hill, New York, Ch. 17.

Klein M.G., Cheng H., Santana L.F., Jiang Y.H., Lederer W.J. & Schneider M.F. (1996) *Nature* **379**, 455–458.

Lopez J.R., Jovanovic A. & Terzic A. (1995) *Biochem. Biophys. Res. Commun.* **214**, 781–787.

Oppenheim A.V., Willsky A.S. & Young I.T. (1983) *Signals and Systems*. Prentice-Hall, Englewood Cliffs, New Jersey.

Prakash Y.S., Smithson K.G. & Sieck G.C. (1993) *NeuroImage* **1**, 95–107.

Prakash Y.S., Kannan M.S. & Sieck G.C. (1997) *Am. J. Physiol.* **272**, C966–C975.

Robb R.A. (1994) *Three Dimensional Biomedical Imaging – Principles and Practice*. VCH Publishers, New York.

Robb R.A., Hanson D.P., Karwoski R.A., Larson A.G., Workman E.L. & Stacy M.C. (1989) *Comput. Med. Imaging Graph.* **13**, 433–454.

Russ J.C. (1995) *The Image Processing Handbook*. CRC Press, Boca Raton, FL.

Schneider M.F. & Klein M.G. (1996) *Cell Calcium* **20**, 123–128.

Sieck G.C., Kannan M.S. & Prakash Y.S. (1997) *Can. J. Physiol. Pharmacol.* **75**, 878–888.

Webb R.H. & Dorey C.K. (1995) In *Handbook of Biological Confocal Microscopy*, J.B. Pawley (ed). Plenum Press, New York, pp. 55–67.

CHAPTER TWENTY-FOUR

Multiphoton Fluorescence Microscopy

GUY COX[1] & COLIN SHEPPARD[2]

Australian Key Centre for Microscopy and Microanalysis, University of Sydney, NSW, Australia
[1] Also with Electron Microscope Unit, University of Sydney
[2] Also with Physical Optics Department, School of Physics, University of Sydney

24.1 INTRODUCTION

Fluorescence microscopy has been a practical tool for cell biologists for around 50 years. Early fluorescence microscopes illuminated the sample from below and were cumbersome, inefficient and dangerous to use. J.S. Ploem's invention of epifluorescence (Ploem, 1967) made the technique more practical and very much safer, and this advance in hardware, coupled with the introduction of fluorescent antibody labelling techniques, led to fluorescence microscopes becoming standard tools in every cell biology laboratory. Ten years ago came the confocal revolution, when the combination of a laser to provide the illumination and a pinhole to exclude out-of-focus light gave the biologist much clearer fluorescence images which could also be three-dimensional (Cox, 1993). Confocal microscopy had actually been invented many years earlier (Egger & Petràn, 1967; Davidovits & Egger, 1969; Minsky, 1988), but it was the introduction of commercial, user-friendly systems that gave good fluorescence performance which made them indispensable tools in many branches of biology.

24.2 PRINCIPLES OF TWO-PHOTON FLUORESCENCE

Many dyes which had become popular with cell biologists were not suitable for confocal microscopy, since confocal microscopes did not offer excitation in the violet and ultraviolet range. The introduction of ultraviolet lasers was a partial solution, but these are bulky and expensive and still do not offer the right wavelengths for popular calcium ratio-imaging dyes. They also present a major optical problem to the microscope designer. In confocal microscopy the incident light must be focused on precisely the spot which the objective lens is imaging. But microscope lenses are not corrected to have the same foci for ultraviolet and visible light. An ultraviolet confocal system therefore requires compensation optics which are tuned specifically to individual objectives (Bliton & Lechleiter, 1995). Only one or two lenses, supplied by the manufacturer, can be used with such a system. To a biologist accustomed to having a choice of oil, dry and water lenses, at a wide range of magnifications, this is a serious limitation. The cost and complexity of a

FLUORESCENT AND LUMINESCENT PROBES, 2ND EDN
ISBN 0-12-447836-0

system, which in the end is only a (not very good) compromise, has limited the popularity of ultraviolet lasers in confocal microscopy.

Two-photon excitation provides a radically different approach to the same problem. It overcomes the problems of ultraviolet excitation, and provides many new advantages of its own, but of course requires other trade-offs. To understand the principle we need to look a little at fluorescence itself. Fluorescence is the use of light at one wavelength to stimulate the emission of light at another. The physics of the process is well known. The incoming light gives its energy to an electron, knocking it into an excited state, a higher energy level which it can only occupy transiently. Soon, in nanoseconds, it loses this energy and drops back to its rightful place in the molecule. The energy is emitted as a quantum of light, a photon. Since some energy is lost in the process, the photon produced has less energy than the one which originally excited the molecule. Lower energy corresponds to a longer wavelength, so this gives us Stokes' law, which says that the emitted wavelength is longer than the exciting one. The difference in wavelength is known as the Stokes shift, and varies from one fluorescent dye to another. For conventional fluorescence microscopy it is convenient for a dye to have a large Stokes shift since it is then easy to separate the incoming photons from the outgoing ones. (The excess energy is eventually dissipated as heat, but this is not normally sufficient to affect the sample.)

Two-photon excitation is completely different: it excites fluorescence with photons of longer wavelength than the emitted light. However it does not violate any of the laws of physics since it knocks the electron into its excited state with two photons of half the required energy, arriving simultaneously. In most cases the electron ends up in exactly the same (S1) excited state before it drops down to the ground state, so that the fluorescence emitted is identical to that given off by normal single-photon excitation (Xu & Webb, 1996). However, the excitation pathway is not always identical since different selection rules can apply to two-photon excitation. Although, as a rule of thumb, a fluorochrome will be excited by two-photon events at approximately twice the wavelength required for single-photon excitation, the two-photon excitation spectrum is often found to be broadened and/or blue-shifted. In principle a totally symmetrical molecule can only be two-photon excited to a different energy state than that reached by single-photon excitation, so that in this case both emission and absorption spectra will shifted.

In practice, therefore, we expect molecules with considerable symmetry to behave rather differently between two-photon and single-photon excitation, whereas non-symmetrical molecules are likely to behave similarly in both cases. Experimental results bear this out. Xu & Webb (1996) showed that fluorochrome molecules with no centre of symmetry (Cascade Blue, Coumarin, Indo-1 and Lucifer Yellow) have two-photon excitation spectra which closely resemble the single-photon spectra, but at twice the wavelength. However, two-photon excitation spectra for more symmetrical dyes (Fluorescein, Rhodamine B and DiI) were substantially blue-shifted relative to twice the single-photon excitation spectrum. Two-photon excitation maxima for fluorescein (782 nm) and rhodamine (840 nm) were therefore at much shorter wavelengths than twice the single-photon maxima (490 nm and 528 nm, respectively) (Fig. 24.1).

The excitation cross-section (that is, the probability that excitation will occur) does not have any linear relationship with the single-photon excitation cross-section (Xu & Webb, 1996; Xu *et al.*, 1996) so a dye which is very effective for single-photon fluorescence may be much less so in two-photon mode (or vice versa). For example, the presence of a 'triplet state' with an energy level at around half the single-photon excitation level can facilitate two-photon excitation (even though the electron does not actually occupy this state in the process!). It is likely, therefore, that eventually dyes will be designed specifically for two-photon microscopy, though to date only conventional fluorochromes have been used in biological work.

The process of two-photon excitation has long been known to physical chemists (Friedrich & McClain, 1980), and the idea that it could be used in microscopy was first proposed by one of the present authors (Sheppard & Kompner, 1978) as one of a variety of non-linear optical processes on which microscopy could be based. At that time experiments were performed on the similar technique of second harmonic microscopy (Gannaway & Sheppard, 1978), which can be regarded as a coherent form of two-photon microscopy. However, it was a team at Cornell University, New York, who first made a working two-photon microscope, around 10 years later (Denk *et al.* 1990, 1991).

To have two photons arriving simultaneously at a single molecule obviously requires a very large flux of photons impinging on the specimen. In fact, the probability of two photons arriving at a point at the same moment depends on the probability of one photon being there multiplied by the probability of another photon being there. In other words, it depends on the square of the flux density of photons. This has an interesting consequence. If we focus light to a small diffraction-limited spot, the likelihood of two-photon excitation occurring depends on the square of the intensity distribution at the focus. This is exactly the same as the intensity distribution detected in a confocal microscope, so that, in other words, two-photon excitation produces the same imaging proper-

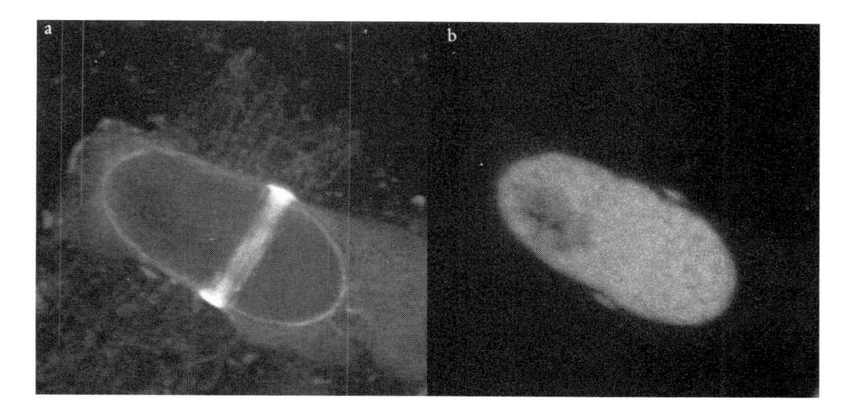

Figure 24.1 Two-photon, two-channel image of meristem cells in wheat root. Excitation at 800 nm from Ti-sapphire laser. Because the excitation maximum for FITC is substantially blue-shifted in two-photon mode both FITC and DAPI can be excited simultaneously at this wavelength. (a) Green fluorescence of FITC-anti-tubulin showing microtubules arranged in a pre-prophase band and cortical array. (b) Blue fluorescence of DAPI showing the nucleus – individual chromosomes are just becoming visible. Projection of 15 serial optical sections. (Micrograph by Guy Cox with grateful thanks to Dr Teresa Dibbayawan, Dr Min Gu, Mr Steven Schilders and Olympus Australia.)

ties as confocal microscopy without any need for a pinhole in front of the detector. Thus two-photon microscopy exhibits an optical sectioning property analogous to that in confocal microscopy. This property was demonstrated many years ago for the similar technique of second harmonic microscopy (Gannaway & Sheppard, 1978).

Looking at it in simpler terms, we can see that since two-photon excitation can only occcur when large numbers of photons are present in a small volume, it only occurs at the focus of the illuminating light. Above and below this point the photons are not sufficiently concentrated, and two-photon events are vanishingly rare. So, like the confocal microscope, a two-photon fluorescence microscope only forms images of the plane that is in focus.

24.3 THE COSTS AND THE BENEFITS

We consider how the performance of the two-photon microscope with wide-field detection compares with that of the single-photon confocal instrument. In formal terms, the imaging properties are similar, except that in practice the wavelength is longer. The resolution will therefore be worse – how much worse, for the biologist, depending in practice on the dyes used.

For the same label there is always a loss in resolution (Sheppard & Gu, 1990). Table 24.1 lists some comparisons. In the (hypothetical) worst case where exciting and emitted wavelengths were identical, two-photon microscopy would exhibit a resolution half that of confocal single-photon microscopy. However, the emitted wavelength will in fact always be longer than the single-photon excitation wavelength (Stokes' law) – how much so depending on both the Stokes shift of the dye and the laser wavelengths and filters available to us. In two-photon microscopy it is only the excitation wavelength which determines the resolution, whereas in single-photon confocal both excitation and emission wavelength are involved. In conventional wide-field fluorescence microscopy only the emission wavelength is relevant. In the real world, then, Table 24.1 will be unduly optimistic for both wide-field and (to a lesser extent) confocal. Since the optimum wavelength for two-photon excitation is typically shorter than twice the single-photon optimum, as we saw in Section 24.2, Table 24.1 will also be unduly pessimistic for the two-photon case. The actual resolution, with the same label, will therefore be a little better than half the single-photon figure.

Since one of the principal purposes for using two-photon excitation is to make use of fluorochromes which normally require ultraviolet excitation, the assumption of the same emission wavelength will frequently not be true. For example, in a confocal microscope using single-photon excitation, with an argon–krypton laser one might use TOTO-3 to label DNA, excited with the red line at 647 nm, with emission in the far red, around 720 nm. However in two-photon mode one could use DAPI, excited at 700 nm (equivalent to single-photon excitation at 350 nm), the resolution in both cases being approximately the same.

Table 24.1 Comparison of resolution in single- and two-photon fluorescence microscopes. Values for half-width (half-width at half maximum, or HWHM) of images of a point and a planar object are given in dimensionless optical co-ordinates for the case of equal emission wavelength for single- and two-photon excitation.

	Conventional single-photon	*Confocal single-photon*	*Conventional two-photon*	*Confocal two-photon*
Half-width of transverse point image, $v_{1/2}$	1.62	1.17	2.34	1.34
Half-width of axial point image, $u_{1/2}$	5.56	4.01	8.02	4.62
Half-width of axial plane image, $\Delta u_{1/2}$		4.3	8.6	5.1

There are some benefits to be gained from the potential loss in resolution Firstly, bleaching of the fluorochrome is restricted to the plane of focus, for bleaching is caused by excitation, which can only take place there. In the confocal microscope, on the other hand, bleaching takes place throughout the cone of illumination, even though the image comes only from the plane of focus. Sensitive fluorochromes can be bleached seriously while optical sections in other planes are taken. So, while two-photon excitation does not eliminate bleaching, it does ensure that only the areas being imaged are bleached. In practice the specimen can give usable images for very much longer in a two-photon microscope than in either confocal or full-field fluorescence microscopes.

Secondly, there is a gain in depth of penetration into tissue. Longer wavelengths are scattered much less than shorter ones. There is a limit to this effect since long infrared rays are absorbed by water, but in the practical range for most work, for example using wavelengths between 900 nm and 1100 nm for exciting FITC or rhodamine derivatives, the gain over the single-photon equivalents is very substantial. In addition, as a pinhole is not used in two-photon microscopy, scattering of the emitted light does not degrade the image: the light collection system only has to detect the emitted light, rather than image it. Combined with the reduction in bleaching the result is that approximately twice the depth is usable for three-dimensional imaging.

The third benefit is the effect on living cells. Ultraviolet light is very harmful to cells, and blue light is also harmful to some extent. Red, far-red and near-infrared has much less effect. DAPI is a vital stain: in low concentrations it can stain DNA with no obvious effect on cell viability. Illuminating the cell with 350 nm light, however, so that the DAPI can actually be seen, is fatal in a few minutes. Using, instead, two-photon excitation at around 700 nm, cells can be followed through several divisions with no loss of viability (Piston *et al.*, 1993). In fact, even three-photon excitation (Hell *et al.*, 1996) of DAPI, using wavelengths around 1050 nm, has been shown to be possible (Wokosin *et al.*, 1996). The resolution, however, for the same dye will be worse than that obtained with two-photon excitation (Sheppard, 1996), though it will be better than for two-photon excitation with the same laser wavelength. Analogous conclusions hold for even higher orders of multiphoton excitation. However, these do offer exciting possibilities for non-destructively imaging important physiological cell constituents which are naturally autofluorescent, but which have single-photon excitation wavelengths in the short ultraviolet. This has been used to assay NAD(P)H in pancreatic islet cells (Bennett *et al.*, 1996) and may also be useful for tracking neurotransmitters such as serotonin, dopamine and bradykinin (Shear *et al.*, 1996; Maiti *et al.*, 1997).

The final conclusion is that with two-photon excitation we can image deeper into cells, with less fading of our fluorochromes, and with less damage to living cells. The only disadvantage is a certain trade-off in resolution.

24.4 HARDWARE FOR TWO-PHOTON MICROSCOPY

The microscope itself is relatively simple. The beam is scanned across the specimen just as in a confocal microscope with single-photon excitation, but unlike the conventional system there is no need to descan the beam or focus it on to a small pinhole. In practice commercial systems retain conventional confocal optics so that they can image in both single- and two-photon modes, but the pinhole (and sometimes the scanning mirror) is by-passed for two-photon collection. In fact two-photon microscopy can also be performed in a confocal mode (Stelzer *et al.*, 1994), the resolution being improved (Gu & Sheppard, 1993, 1995a,b; Nakamura, 1993) compared with conventional two-photon microscopy, but at the expense of loss of light detection efficiency. Details are given in Table 24.1. Resolution is slightly worse than for single-photon confocal microscopy. Microscope lenses are corrected into the red region of the spectrum, so the illuminating beam has a wavelength in a range where the lenses perform at their best, but since there is no need to focus the emitted fluorescence to a precise spot, the colour correction of the lens is no longer of great importance. Only one wavelength, the excitation one, needs to be focused precisely.

The light source, on the other hand, is anything but simple. If we were to bombard our sample continuously with a strong enough beam of light to generate two-photon excitation, it would receive so much energy that it would probably suffer considerable damage. Two-photon excitation is a very weak process, but in fact some materials are not as inefficient as two-photon dyes as might be expected (Scharf & Band, 1988). But nevertheless the sensitivity is weak, so that to overcome this the light is delivered in very short pulses (Sheppard & Kompner, 1978), spaced so that the total energy is no greater than in single-photon excitation. For a given total overall power, the shorter the pulses the better, up to a limit set by the relaxation time of the molecule (Denk *et al.*, 1990).

Use of short pulses also has the added benefit of allowing time-resolved imaging to be performed (Piston *et al.*, 1992). This can be used to separate emission from different fluorophores, as well as to investigate variations in the environment of the dyes. Time-resolved studies can also be used to study fluorescence anisotropy, by observing depolarization with polarized illumination. It is found that orientation dependence is stronger for two-photon excitation, and even stronger for higher order multiphoton processes (McClain, 1971, 1972). Another interesting effect is that the bandwidth of the short pulse illumination also results in a small improvement in resolution (Gu *et al.*, 1995).

The lasers used to produce these short pulses are usually solid-state titanium–sapphire lasers, delivering pulses in the hundred femtosecond range (1 fs = 10^{-15} s). These are tunable, typically between ~700 and ~1100 nm. This obviates another problem of confocal microscopy (especially UV confocal microscopy), that the wavelengths available from the laser are never exactly as required. It also avoids other disadvantages of UV microscopy (Stelzer *et al.*, 1994) mentioned previously. Unfortunately, tuning the laser from one wavelength to another requires a skilled operator. It also takes some time, so that it is not practically possible to image a field with two different wavelengths. One might hope that in the future this problem will be solved and, for example, excitation ratio imaging of calcium concentration will be possible. At present calcium imaging has to be performed using a single excitation wavelength (Piston & Webb, 1991).

A Ti-sapphire laser has to be pumped with another laser of quite substantial power. In the past this has required an expensive and temperamental water-cooled argon laser, but more recently pumping with solid-state lasers has become a possibility, reducing the bulk and complexity of the installation substantially, and improving system reliability. Coupling the laser to the microscope is also not simple: the obvious choice, an optical fibre, tends to spread the pulses by dispersion so that they emerge with a longer duration. The choices are to compensate for this with special optics, or to couple the laser indirectly. Even then the microscope objective, which contains a large thickness of glass, causes dispersion which lengthens the pulses. This can be overcome by using pairs of prisms to compensate for the dispersion (Fork *et al.*, 1984).

The result of all these points is that the total system is both large and expensive. An alternative, simpler approach is not to aim for perfection but to accept a compromise, using an Nd YLF or similar solid-state laser. These are compact, stand-alone lasers giving a single wavelength of around 1060 nm and pulses in the picosecond range. The resulting system is smaller and cheaper but rather less versatile. Nevertheless, with brightly labelled samples, and if excitation of dyes with single-photon excitation wavelengths in the UV is not a requirement, the compromise is worth considering.

At present two-photon microscopy has much to offer the biologist. Its future development is largely dependent on advances in lasers: with a cheap, compact, femtosecond laser, and computer-controlled fast tuning, the possibilities would be almost limitless. It is interesting to note that semiconductor lasers with similar properties would be highly desirable for optical communications applications, so these advances may well take place.

ACKNOWLEDGEMENTS

We are very grateful to Eleanor and Scott Kable for valuable discussion and for sharing with us their knowledge of the literature.

REFERENCES

Bennett B.D., Jetton T.L., Ying G., Magnuson M.A. & Piston D.W. (1996) *J. Biol. Chem.* **271**, 3647–3651.

Bliton A.C. & Lechleitner J.D. (1995) In *Handbook of Biological Confocal Microscopy*, J. Pawley (ed). Plenum Press, New York & London, pp. 431–444.
Cox G.C. (1993) *Micron* **24**, 237–247.
Davidovits P. & Egger M.D. (1969) *Nature* **223**, 831.
Denk W., Strickler J.P. & Webb W.W. (1990) *Science* **248**, 73.
Denk W., Strickler J.P. & Webb W.W. (1991) US Patent 5 034 613 (filed 1989).
Egger M.D. & Petràn M. (1967) *Science* **157**, 305–307.
Fork R.L., Martinez O.E. & Gordon J. P. (1984) *Opt. Lett.* **9**, 150.
Friedrich D.M. & McClain W.M. (1980) *Ann. Rev. Phys. Chem.* **31**, 559.
Gannaway J.N. & Sheppard C.J.R. (1978) *Opt. Quant. Elect.* **10**, 435.
Gu M. & Sheppard C.J.R. (1993) *J. Mod. Opt.* **40**, 2009.
Gu M. & Sheppard C.J.R. (1995a) *J. Microsc.* **177**, 128.
Gu M. & Sheppard C.J.R. (1995b) *Opt. Commun.* **114**, 45.
Gu M., Tannous T. & Sheppard C.J.R. (1995) *Opt. Commun.* **117**, 406.
Hell S., Bahlmann K., Schrader M., Soini A., Malak H., Gryczynski I. & Lakokowicz J.R. (1996) *J. Biomed. Opt.* **1**, 71.
McClain W.M. (1971) *J. Chem. Phys.* **55**, 2789.
McClain W.M. (1972) *J. Chem. Phys.* **57**, 2264.
Maiti S., Shear J.B., Williams R.M., Zipfel W.R. & Webb W.W. (1997) *Science* **275**, 530–532.
Minsky M. (1988) *Scanning* **10**, 128–138.
Nakamura O. (1993) *Optik* **93**, 39.
Piston D.W. & Webb W.W. (1991) *Biophys J.* **59**, 156.
Piston D.W., Sandison D.R. & Webb W.W. (1992) *Proc. SPIE* **1640**, 379.
Piston D.W., Summers R.G. & Webb W.W. (1993) *Biophys. J.* **63**, A110.
Ploem G.S. (1967) *Z. Wiss. Microsk.* **68**, 129.
Scharf B.E. & Band Y.B. (1988) *Chem. Phys. Lett.* **144**, 165.
Shear J.B., Brown E.B. & Webb W.W. (1996) *Anal. Chem.* **68**, 1778–1783.
Sheppard C.J.R. (1996) *Bioimaging* **4**, 124.
Sheppard C.J.R. & Gu M. (1990) *Optik* **86**, 104–106.
Sheppard C.J.R. & Kompner R. (1978) *Appl. Opt.* **17**, 2879–2882.
Stelzer E.H.K., Hell S. & Lindek S. (1994) *Opt. Commun.* **104**, 223.
Wokosin D.L, Centonze V.E., Crittenden S. & White J. (1996) *Bioimaging* **4**, 208–214.
Xu C. & Webb W.W. (1996) *J. Opt. Soc Am. B*, **13**, 481–491
Xu C., Zipfel W., Shear J.B., Williams R.M. & Webb W.W. (1996) *Proc. Natl Acad. Sci.* USA **93**, 10 763–10 768.

CHAPTER TWENTY-FIVE

High-resolution Confocal Imaging of Elementary Ca^{2+} Signals in Living Cells

PETER LIPP[1] & MARTIN D. BOOTMAN[1,2]
[1] Laboratory of Molecular Signalling, The Babraham Institute, Babraham, Cambridge, UK
[2] Department of Zoology, University of Cambridge, Downing Street, Cambridge, UK

25.1 CALCIUM AS AN INTRACELLULAR MESSENGER

Calcium is a ubiquitous intracellular signal, controlling a wide array of cellular processes including secretion, contraction, cell proliferation and cellular metabolism (Berridge, 1993; Clapham, 1995). Upon stimulation of cells, the average cytosolic Ca^{2+} ion concentration ($[Ca^{2+}]_i$) can rise from basal levels of 10–100 nM up to several μM, depending on the cell type. Such Ca^{2+} signals often have a complex temporal and spatial arrangement. Owing to the toxicity of elevated $[Ca^{2+}]_i$, tonic increases do not occur in many cells; instead cytoplasmic Ca^{2+} is usually increased in a phasic manner. Well-known examples are the repetitive Ca^{2+} spikes which drive the beating heart, the rapid sub-plasmalemmal $[Ca^{2+}]_i$ increases that occur during depolarization of excitable cells, and the regular Ca^{2+} spikes (or oscillations) observed during hormonal stimulation of many non-excitable cell types. In some cell types, the spatial correlate of a Ca^{2+} spike is a Ca^{2+} wave, in which Ca^{2+} is initially elevated in a discrete region of the cell before spreading throughout the cell as a regenerative increase (Lechleiter *et al.*, 1991; Thomas *et al.*, 1991).

The action of Ca^{2+} as an intracellular messenger is mediated by the binding of Ca^{2+} to intermediary proteins, e.g. calmodulin and troponin C, or by direct interaction with effectors such as ion channels. Information can be encoded in the amplitude, frequency and spatial location of $[Ca^{2+}]_i$ increases (Berridge, 1993; Petersen *et al.*, 1994; Clapham 1995; Berridge, 1997). Fine control of intracellular Ca^{2+} signals thus provides a mechanism for regulating specific cellular activities.

25.1.1 Global versus elementary Ca^{2+} signals

The Ca^{2+} waves and spikes mentioned above are considered to be 'global' signals, since they usually invade the entire cell, including cytoplasmic and nucleoplasmic compartments, and some organelles e.g. mitochondria. With such global signals, the differential modulation of cellular processes has to occur via changes in frequency and/or amplitude of the $[Ca^{2+}]_i$ increases. A more subtle method of controlling Ca^{2+}-dependent activities is to elevate $[Ca^{2+}]_i$ in the locality

FLUORESCENT AND LUMINESCENT PROBES, 2ND EDN
ISBN 0-12-447836-0

of the target proteins. In recent years, there has been a growing awareness that Ca^{2+} functions as a local signal in a wide variety of cell types (Petersen *et al.*, 1994; Clapham, 1995; Bootman & Berridge, 1995; Lipp & Niggli, 1996a; Berridge, 1997).

Cells utilize two sources of Ca^{2+} for generating signals: Ca^{2+} release from intracellular stores and Ca^{2+} entry across the plasma membrane. Both Ca^{2+} release and entry mechanisms can give rise to local, or 'elementary', Ca^{2+} signals. Most of the elementary events described so far are associated with the release of Ca^{2+} from intracellular stores (the endoplasmic reticulum, ER; or sarcoplasmic reticulum, SR), which is controlled by two families of channel: ryanodine receptors (RyR) and inositol 1,4,5-trisphosphate receptors ($InsP_3R$) (Berridge, 1997). Examples of elementary Ca^{2+} signals arising via Ca^{2+} release are the Ca^{2+} puffs in *Xenopus* oocytes and HeLa cells (Yao *et al.*, 1995; Bootman *et al.*, 1997a), and the Ca^{2+} sparks in cardiac myocytes (Cheng *et al.*, 1993; Lipp & Niggli, 1994; López-López *et al.*, 1995) (Plate 25.1).

Although Ca^{2+} puffs and Ca^{2+} sparks arise via the opening of different intracellular Ca^{2+} channels, they are functionally analogous Ca^{2+} signals. Ca^{2+} puffs are somewhat larger (2–7 μm), but much more sluggish ($t_{1/2}$ for increase > 100; $t_{1/2}$ for decay > 250) than Ca^{2+} sparks (1–3 μm; $t_{1/2}$ for increase <30 ms; $t_{1/2}$ for decay ~100 ms) (Bootman, 1996; Lipp & Niggli, 1996a). Calculation of the ionic fluxes underlying Ca^{2+} puffs and Ca^{2+} sparks indicates that they arise via the opening of a cluster of $InsP_3Rs$ or RyRs on the ER or SR. Furthermore, Ca^{2+} puffs and Ca^{2+} sparks have been deconvolved into smaller Ca^{2+} signals, representing the opening of single $InsP_3R$ or RyR, denoted 'Ca^{2+} blips' and 'Ca^{2+} quarks' (Lipp & Niggli, 1996b; Bootman *et al.*, 1997a; Lipp & Niggli, 1998). The basic concept, therefore, is that elementary Ca^{2+} release signals arise via the co-ordinated activity of one or a few Ca^{2+} channels scattered around the surface of intracellular stores (Plate 25.1).

There are fewer examples of elementary events associated with Ca^{2+} entry. However, the spontaneous openings of voltage-operated Ca^{2+} channels within discrete regions of the squid giant synapse give rise to localized Ca^{2+} signals described as Quantum Emission Domains (Sugimori *et al.*, 1994). Quantal Ca^{2+} entry signals have also been seen following NMDA receptor activation in cortical neurons (Murphy *et al.*, 1994). In *Drosophila* photoreceptors, individual photons evoke discrete electrical events, called quantum bumps (Hardie, 1995), representing unitary Ca^{2+} entry signals.

The properties of elementary Ca^{2+} entry signals are often significantly different from those observed for Ca^{2+} release (Bootman & Berridge, 1995). Their kinetics can be up to 1000-fold faster, and the amplitude of the Ca^{2+} signal is several orders of magnitude greater. These differences relate to the sensitivities of the cellular processes that are controlled by Ca^{2+} entry and Ca^{2+} release. For example, neurotransmitter secretion requires a rapid subplasmalemmal Ca^{2+} elevation, in the order of tens of micromolar (Monck *et al.*, 1994). In contrast, calmodulin, which mediates many of the effects of Ca^{2+} released from intracellular stores, has a $K_d \approx 1$ μM.

25.1.2 Examples of the utility of elementary Ca^{2+} signals

Elementary Ca^{2+} signals, whether derived from Ca^{2+} release or Ca^{2+} entry, provide the cell with a hugely flexible signalling system. One advantage gained by employing elementary Ca^{2+} signals is that the local Ca^{2+} concentration is much higher than in the bulk cytoplasm. The gradient of Ca^{2+} between the locus of release and the cytosol provides a mechanism for differential control of cellular activities. For example, several recent studies have shown that mitochondria accumulate Ca^{2+} during hormone-evoked Ca^{2+} spikes (Rizzuto *et al.*, 1994; Hajnóczky *et al.*, 1995). The affinity of mitochondrial Ca^{2+} uptake is too low for them to detect global $[Ca^{2+}]_i$ elevations (usually ≤ 1 μM), but if they are sufficiently close to sites of Ca^{2+} release, they can respond to elementary Ca^{2+} signals with an increase in respiratory metabolism (Hajnóczky *et al.*, 1995).

In HeLa cells, the elementary Ca^{2+} signals were found to largely occur within 1–3 μM of the nucleus (Lipp *et al.*, 1997) (Plate 25.2). The Ca^{2+} from such perinuclear elementary events diffused anisotropically into the nucleus, but not into the cytoplasm, thus elevating nucleoplasmic Ca^{2+}. The repeated firing of perinuclear elementary Ca^{2+} signals gave rise to a sustained almost nuclear-specific Ca^{2+} signal, which barely affected cytoplasmic Ca^{2+} levels (Lipp *et al.*, 1997). Such signals could, for example, modulate gene expression.

Arterial smooth muscle cells also utilize elementary Ca^{2+} signals to control their excitability. In these cells, global elevation of $[Ca^{2+}]_i$ triggers contraction, whereas Ca^{2+} sparks produced near the sarcolemma caused the cells to relax (Nelson *et al.*, 1995). The relaxation is due to the activation of potassium channels, which serve to hyperpolarize the cell, thereby inhibiting Ca^{2+}-dependent contractile events. This striking example of Ca^{2+} exerting a bidirectional effect illustrates how elementary events localized to specific cellular regions can have signalling functions distinct from those mediated by global signals.

It should also be pointed out that elementary Ca^{2+} signals such as Ca^{2+} puffs and Ca^{2+} sparks are not only used for local signalling, but are also the building blocks from which global Ca^{2+} signals, i.e. waves and oscillations, are composed (Berridge, 1997). Both $InsP_3$ and RyRs, which are responsible for generation of Ca^{2+} puffs and Ca^{2+} sparks, respectively, display the autocatalytic feature of Ca^{2+}-induced Ca^{2+} release (CICR), whereby a small trigger signal can be amplified by further Ca^{2+} release. When elementary Ca^{2+} events occur at low frequency or in spatial isolation, rapid cellular Ca^{2+} sequestration/buffering mechanisms remove the signals before they have a chance to build up. However, increases in frequency or amplitude of the elementary Ca^{2+} signals, or greater spatial coupling between individual elementary Ca^{2+} release sites, can cause the signals to overwhelm the buffering capacity of the cytoplasm. In these situations, CICR is activated and usually causes a global Ca^{2+} signal (Bootman *et al.*, 1997b) (Plate 25.3). Such a response can be observed in many different cell types, particularly in non-excitable cells. Excitation–contraction coupling in skeletal and cardiac muscle tissue also involves the recruitment of elementary Ca^{2+} release sites (Ca^{2+} sparks), but by a different mechanism (López-López *et al.*, 1995; reviewed in Lipp & Niggli, 1996b) (Plate 25.4).

25.2 INVESTIGATING ELEMENTARY Ca^{2+} SIGNALS

The sections above illustrate the characteristics and physiological significance of elementary Ca^{2+} signals. In the following discussion, we will highlight the optical methodology and experimental approach to visualizing such signals. Although some elementary Ca^{2+} signals have been visualized using conventional fluorescence video imaging (see Parker *et al.*, 1996), the thorough investigation of such signals requires the resolution afforded by confocal microscopy, whereby a cell can be non-invasively optically sectioned. The remainder of this chapter will therefore consider the advantages and pitfalls of using confocal imaging to investigate elementary Ca^{2+} signals. The following discussion briefly pinpoints some of the major considerations for confocal microscopy of elementary Ca^{2+} signals, and largely avoids technical details. For a more thorough coverage and critique of the different approaches to confocal microscopy, we would direct the reader to the excellent text by Pawley and co-authors (1995).

25.2.1 Ratiometric versus non-ratiometric Ca^{2+} indicators

Ratiometric and non-ratiometric Ca^{2+} indicators are commercially available. Choosing the correct indicator is important when considering the visualization of subcellular Ca^{2+} signals. With confocal microscopy, regardless of the type of confocal microscope used, ratiometric indicators are useful for observing Ca^{2+} signals in subcellular volumes where the indicator distribution may change over the course of the experiment. The most prominent reason for such a change of indicator distribution is movement of the cell under the microscope. This can include rapid cell movements, such as muscle contraction, but also slower movements such as migration. The effect of movement is that the subcellular volume under observation does not stay the same. Changes in the fluorescence cannot therefore be unequivocally related to changes of Ca^{2+} concentration, but may in fact be due to a change in the indicator concentration. Ratiometric Ca^{2+} measurements take this into account, at least up to a certain degree, since the estimation of the Ca^{2+} concentration does not depend on the indicator concentration.

Since most commercially available confocal microscopes rely on the use of lasers for the indicator excitation, ratiometric Ca^{2+} measurements with the temporal resolution required for the visualization of subcellular Ca^{2+} signals are not possible with dual-excitation indicators such as fura-2. Although theoretically possible, such rapid changes of the excitation wavelength have not yet been realized for existing confocal microscopes. At present, there are no ratiometric indicators available that work in the visible range of wavelengths. Therefore, for most applications, the use of ratiometric indicators is limited to emission-ratiometric indicators, such as the commonly used indicator Indo-1.

25.2.2 Indo-1 and UV excitation

With respect to confocal microscopy, the use of UV light, which is required for the excitation of indo-1, presents major optical problems, such as chromatic aberration. The microscope objectives generally used are only chromatically corrected for visible wavelengths. Essentially, chromatic aberration can result in an excitation plane and a confocal detection plane that are separated. Obviously, such aberration prevents the microscope from achieving optimal detection.

There are several options that may solve this problem. The easiest is to use microscope objectives that are optically corrected for the use of UV light. Indeed, UV-corrected objectives are commercially available, although they are very expensive. It should be stressed that one should not confuse objectives that possess correction for such chromatic aberration with those objectives that are simply optimized for UV transmission (so-called UV objectives). Other solutions involve additional optical components within the microscope that compensate for the different chromatic properties of the objectives.

Having solved the problem of aberration, there still remains another important consideration – the rapid photobleaching of indicators, such as indo-1. Bleaching may not be a problem with ratiometric dyes, since both wavelengths diminish at the same rate. However, it has to be realized that the detection of rapid and small changes of Ca^{2+} in a spatially restricted subcellular volume requires a very good signal-to-noise ratio. If the absolute indo-1 fluorescent drops to low levels during the progression of an experiment, it will not allow the reliable detection of such minute Ca^{2+} changes.

Photobleaching (and indeed light-induced cell damage) arises via the production of highly reactive oxygen radicals. Such damage may therefore be avoided, or at least reduced, by the inclusion of oxygen-scavenging compounds. Candidates for such a role include ascorbate and Trolox (a vitamin E analogue). The latter has been shown to protect indo-1 from photodegradation in intact cells (Scheenen *et al.*, 1996).

25.2.3 Ratiometric measurements using two visible indicators

As discussed above, ratiometric measurements of elementary Ca^{2+} signals is not simple with UV-excitable indicators. One solution is to use a mixture of two visible-wavelength indicators with overlapping excitation, but different emission spectra (Lipp & Niggli, 1993). Standard confocal microscopes are equipped with an argon-ion laser, which usually limits the available excitation wavelengths to either 488 nm or 514 nm. Consequently, one has to search for indicators that share a usable excitation spectrum around 500 nm, such as fluo-3 or the Calcium Green family of fluorescent Ca^{2+} indicators as one part of the mixture. These indicators display maximal fluorescence emission in the green part of the spectrum, and are suitable for combination with indicators that emit at longer (red) wavelengths as the second partner in the indicator mixture.

Ideal choices for the second longer-wavelength indicator are either Ca^{2+}-sensitive indicators, such as Fura Red, or Ca^{2+}-insensitive dyes such as carboxy-SNARF (actually a pH indicator). The advantage of using a mixture of two Ca^{2+}-sensitive indicators, such as fluo-3 and Fura Red, is an improvement in the signal-to-noise ratio; when a Ca^{2+} change occurs fluo-3 will exhibit an increase in fluorescence emission, which will be mirrored by a decrease of Fura Red emission. On the other hand, by using a Ca^{2+}-insensitive indicator, the signal-to-noise ratio can be enhanced quite significantly, since the concentration of an indicator, such as carboxy-SNARF, can be increased substantially without increasing the overall Ca^{2+} buffer capacity of the indicator mixture.

If the two single-wavelength indicators are loaded into cells using their acetoxymethyl esters, the ratio of the indicators will vary considerably between cells. This variable ratio can make calibration rather problematic, since the apparent K_d will also change. One way of avoiding this dilemma is to introduce the two indicators at a known ratio using microinjection or a whole-cell patch pipette. A further problem is that the indicators within such mixtures will probably not photobleach at the same rate. Therefore the fluorescence ratio of the mixture will artifactually deviate from the original level. Another potential pitfall with mixtures of indicators is that their emission spectra can overlap. Cross-talk between different channels therefore has to be tested and corrected for.

25.2.4 Single-wavelength indicators

We are now left with the single-wavelength, non-ratiometric Ca^{2+} indicators, the choice of which has considerably increased in recent years. There are again two major decisive factors of such indicators that have to be taken into consideration when trying to visualize elementary Ca^{2+} signals in living cells: photo-bleaching and signal-to-noise ratio. In our hands, the best indicator is fluo-3, which exhibits a dynamic range (ratio of F_{max}/F_{min}; fluorescence at a saturating Ca^{2+} against fluorescence at zero Ca^{2+}) of > 10 in living cells. Furthermore, fluo-3 displays a low fluorescence at resting Ca^{2+} concentrations. The low basal fluorescence of indicators such as fluo-3 can be quite important, since most confocal microscopes are limited to 8-bit images which allows only 256 grey levels. If the fluorescence is already quite high at resting Ca^{2+} concentrations, as for example with the Calcium Green dyes, there are not many grey levels left for the actual signal. In other words, the dynamic range is already limited. An advantage of indicators with a higher fluorescence at low Ca^{2+} concentrations is that cell boundaries can be more easily defined. This could be useful, for example, in trying to visualize fine structures such as the dendrites of neurons. In terms of photobleaching, there are some single-wavelength indicators available, e.g. Oregon Green BAPTA 488, which have been designed to bleaching more slowly than fluo-3. Unfortunately, some of these indicators exhibit a limited signal-to-noise ratio since their dynamic range is poor.

25.2.5 Indicator affinity

Another factor that has to be considered is the Ca^{2+} affinity of the indicator. For subcellularly restricted, low-amplitude Ca^{2+} signals, the Ca^{2+} affinity should not be too low, since only part of the dynamic range of the indicator is used. On the other hand, an indicator with too high an affinity may help to resolve elementary Ca^{2+} signals, but will saturate if larger signals such as Ca^{2+} waves are triggered. Some estimation of the amplitude of the Ca^{2+} signals to be measured is required before choosing an indicator. In our hands, the most convenient indicator is fluo-3, which shows slow photobleaching, an appropriate K_d value (measured *in vivo*) of ~800 nM, and certainly has the best *in vivo* dynamic range of the presently available visible wavelength indicators.

25.2.6 Indicator distribution

One of the problems that has to be considered is inhomogeneous distribution of the Ca^{2+} indicators, especially when loaded via the AM ester form of the dyes. The so-called self-ratio method, originally suggested by Minta *et al.* (1989), is able to account for most of the problems associated with the distribution problem. With this method, the actual fluorescence is normalized by the fluorescence under resting conditions. It is assumed that the distribution of Ca^{2+} in a cell at rest is homogenous, an assumption that is usually applicable to most of the cells, but should be checked in case of doubt. The normalized fluorescence, or self-ratio, is then largely independent of the concentration and uneven distribution of the indicator, and can be calibrated to give absolute Ca^{2+} concentrations by using parameters obtained during an *in vivo* calibration.

25.3 THE TYPE OF CONFOCAL MICROSCOPE AND THE MODE OF SCANNING

In principle, there are three commercially available types of confocal microscopes which differ substantially in the methods by which the excitation light is scanned onto the specimen, and the emitted light is descanned to form an image: (1) systems using two galvanometric mirrors (e.g. BioRad, Zeiss or Leica); (2) systems using one galvanometric mirror and an acousto-optical deflector (AOD) (e.g. Noran); and (3) disk-scanning devices, such as those using a Nipkow disk (e.g. Life Science Resources).

The first of these types of confocal is the most common, and uses two mirrors to address all locations during the scanning process; the horizontal mirror swings back and forth multiple times, during one sweep of the vertical mirror. The consequence of this design is that the speed of image formation is limited by the mechanical properties of the horizontally scanning mirror. Such confocals usually have the best spatial resolution, but they are relatively slow, achieving scanning rates of ~10 Hz with small window sizes.

The second type of confocal listed above uses one mirror for (the slower) vertical scanning, and an AOD for horizontal scanning. An AOD is essentially a crystal that changes its angle of light deflection as different sound frequencies are applied. The AOD does not physically move and can instantaneously address random points, which means that the horizontal scanning process can be done very rapidly. With such a mechanism, images can be collected at up to 480 Hz, using even bigger window sizes than for 10 Hz image collection with galvonometer-driven systems. One of the problems with AODs is that they are dispersive (light of different wavelengths are differentially deflected); for this reason they are only suited for use with monochromatic light. However, as mentioned above, dual-excitation confocal microscopes are not yet widely available, so this is not a serious limitation.

The third type of microscope essentially uses multiple pinholes to construct a confocal image. The specimen is illuminated by light that passes through thousands of pinholes arranged in a geometrically precise pattern (e.g. an Archimedes spiral) on a rapidly spinning (~360 rpm) disk. The effect is that all parts of the specimen are illuminated quasi-simultaneously. The emitted light passes back through the same pinholes, thus excluding out-of-focus light originating from other points of the specimen. The emitted light is not descanned as with other confocals, but instead forms a real-time colour image that can be viewed via an eyepiece or camera. It is also practical to use this type of system with non-laser light sources, thus expanding the range of useable fluorophores. Disk-scanning devices are considerably faster than the two-galvonometric-mirror systems.

Various makes of galvonometer- and AOD-based systems have been used successfully for studies of elementary Ca^{2+} signals (Lipp & Niggli, 1996b; Bootman *et al.*, 1997a,b). A disk scanning confocal microscope has also been used to visualize Ca^{2+} sparks in cardiac myocytes (W.T. Mason, Life Science Resources, personal communication).

25.3.1 Line scanning

Sacrificing one spatial dimension (the vertical axis) can increase the temporal resolution of galvonometer- and AOD-driven systems. This so-called line-scan mode, where only a single line of the entire confocal image is chosen and repetitively scanned, allows a temporal resolution close to 1 kHz for galvonometer-driven confocal microscopes, and more than 10 kHz for those microscopes using an AOD. In order to record fast events, such as elementary Ca^{2+} signals, it might seem obvious to use the line-scan mode, but it has to be taken into consideration that because only a single line is scanned, the interpretation of the line-scan images has to be done very carefully.

In cardiac myocytes, where the elementary Ca^{2+} release sites are rather evenly distributed, the interpretation of line-scan images is relatively simple. The situation is certainly much more complex when looking into spatially restricted Ca^{2+} release events in neurons or non-excitable cells. For neurons the problem is quite obvious since in order to be able to use the line-scan mode, the cell has to have a sufficient part of its cell body aligned along a straight axis, which is normally not the case. In electrically non-excitable cells, the situation is slightly different, since their shapes are usually quite regular, at least in comparison to the dendritic trees of some neurons. However, it is still difficult to predict where elementary Ca^{2+} release events will occur, and by scanning only a single line through the cell it is possible to miss all such Ca^{2+} signals.

It is clear, therefore, that for many situations, scanning the entire field of view instead of choosing a single line is beneficial. However, as discussed above, scanning an entire image is much slower than line scanning. Therefore, the scanning speed of the confocal microscope and the time course of the Ca^{2+} signals have to be considered. Usually, global Ca^{2+} signals, such as Ca^{2+} waves, display a moderate propagation velocity of between 10 and 80 $\mu m\ s^{-1}$, and acquisition rates of 10 Hz are certainly sufficient to record such events. In general, elementary Ca^{2+} signals are much faster, and need greater temporal resolution. Ca^{2+} puffs, for example, have an overall lifetime of seconds, and events such as Ca^{2+} sparks are even shorter (see above). The situation becomes even more difficult with smaller signals such as Ca^{2+} quarks and Ca^{2+} blips, which have lifetimes of between 100 ms and 500 ms. At the present time, only real-time confocal microscopes using an AOD or disk-scanning can provide sufficient resolution in image mode.

25.4 ADDITIONAL CONSIDERATIONS

25.4.1 Data storage

In the discussion above, we have shown that recording elementary Ca^{2+} release events requires rapid scanning. To achieve a good spatial resolution, large window sizes are favourable. For our studies of elementary Ca^{2+} signals in cardiac myocytes, neurons and epithelial cells, we use window sizes of 256 × 256 pixels (horizontal × vertical) or greater. The acquisition of a large number of these images raises the problem of data storage, since one can easily produce several gigabytes of data in one experimental day. Usually the data are stored temporarily on the hard drive of the controlling computer, but need to be removed very rapidly. In principle, there are several options for permanent data storage: (1) rewritable optical media; (2) write-once read-many optical media; (3) magnetic mass storage such as big hard drives or removable hard drives; and (4) mass storage on tapes, such as DAT tapes. The cheapest solution is certainly the DAT tape, which can hold several gigabytes of data on a single cartridge, but access times are usually quite slow. In our hands, the best solution in terms of cost-efficiency, reliability and accessibility are CD-ROMs. The medium is reasonably cheap, and can hold up to 650 megabytes. Furthermore, CD-ROM burners are readily and cheaply available.

25.4.2 Data analysis and presentation

Usually all confocal microscopes are bundled with software that allows the recording of data, and some limited analysis. In our laboratory, we do all analysis off-line using Apple Macintosh workstations running various software packages. The initial data analysis is performed in a modified version of NIH-Image (National Institutes for Mental Health, USA), whereby the time course of fluorescence change in various regions of interest can be examined, and average pixel intensities exported in ASCII format. In addition some image processing, such as self-ratioing and surface plots, can be performed using NIH-Image. Secondary analysis and compilation of images to make figures for publication is performed with programs such as Canvas, CorelDraw, Igor, Photoshop and PageMaker.

25.4.3 Cost

The choice of a confocal microscope is best determined by the nature of the work to be undertaken. Those systems using two galvonometer-driven mirrors are usually best for image resolution, whilst the AOD-based systems are supremely fast. However, both of these types of device are expensive. The commercially available disk-scanning technologies offer similar spatial resolution and intermediate temporal resolution, when compared with galvonometer- or AOD-based systems, but are cheaper.

25.5 SUMMARY

There are many factors that need to be considered when using confocal microscopy to study elementary Ca^{2+} signals in living cells. The most serious are the choice of indicator and the type of image to be collected (image mode versus line-scanning) and the speed of data acquisition. In addition, there are several smaller details, such as data analysis and storage, that must be thought through prior to the experiment.

REFERENCES

Berridge M.J. (1993) *Nature* **361**, 315–325.
Berridge M.J. (1997) *J. Physiol.* **499**, 291–306.
Bootman M.D. & Berridge M.J. (1995) *Cell* **83**, 675–678.
Bootman M.D. (1996) *Cell Calcium* **20**, 97–104.
Bootman M.D., Niggli E., Berridge M.J. & Lipp P. (1997a) *J. Physiol.* **499**, 307–314.
Bootman M.D., Berridge, M.J. & Lipp P. (1997b) *Cell* **9**, 367–373.
Cheng H., Lederer W.J. & Cannell M.B. (1993). *Science* **262**, 740–744.
Clapham D.E. (1995). *Cell* **80**, 259–268.
Hajnóczky G., Robb-Gaspers L.D., Seitz M.B. & Thomas A.P. (1995) *Cell* **82**, 415–425.
Hardie R.C. & Minke B. (1995) *Cell Calcium* **18**, 256–274.
Lechleiter J., Girard S., Clapham D. & Peralta E. (1991). *Nature* **350**, 505–508.
Lipp P. & Niggli E. (1993) *Cell Calcium* **14**, 359–372.
Lipp P. & Niggli E. (1994). *Circ. Res.* **74**, 979–990.
Lipp P. & Niggli E. (1996a). *Mol. Biol.* **65**, 265–296.
Lipp P. & Niggli E. (1996b) *J. Physiol.* **492**, 31–38.
Lipp P. & Niggli E. (1998) *J. Physiol.* **508**, 801–809.
Lipp P., Thomas D., Berridge M.J. & Bootman M.D. (1997) *EMBO J.* **16**, 7166–7173.
López-López J.R., Shacklock P.S., Balke C.W. & Wier W.G. (1995). *Science* **268**, 1042–1045.
Minta A., Kao J.P.Y. & Tsien R.Y. (1989) *J. Biol. Chem.* **264**, 8171–8178.
Monck J.R., Robinson I.M., Escobar A.L., Vergara J.L. & Fernandez J.M. (1994) *Biophys. J.* **67**, 505–514.
Murphy T.H., Baraban. J.M., Wier, W.G. & Blatter L.A. (1994) *Science* **263**, 529–532.
Nelson M.T., Cheng H., Rubart M., Santana L.F., Bonev A., Knot H. & Lederer W.J. (1995) *Science* **270**, 633–637.
Parker I., Choi J. & Yao Y. (1996) *Cell Calcium* **20**, 105–121.
Pawley J.B. (1995) *Handbook of Biological Confocal Microscopy* Pawley J.B. (ed.). Plenum Press, New York.
Petersen O.H., Petersen C.C.H. & Kusai H. (1994). *Annu. Rev. Physiol* **56**, 297–319.
Rizzuto R., Bastianuto C., Brini M., Murgia M. & Pozzan T. (1994) *J Cell. Biol.* **126**, 1183–1194.
Scheenen W.J.J.M., Makings L.R., Gross L.R., Pozzan T. & Tsien R.Y. (1996) *Chem. Biol.* **3**, 765–774.
Sugimori M., Lang E.J., Silver R.B. & Llinàs R. (1994) *Biol. Bull.* **187**, 300–303.
Thomas A.P., Renard D.C. & Rooney T.A. (1991) *Cell Calcium* **12**, 111–126.
Yao Y., Choi J. & Parker I. (1995). *J. Physiol.* **482**, 533–553.

CHAPTER TWENTY-SIX

Confocal Fluorescent Microscopy Using a Nipkow Scanner

AKIRA ICHIHARA[1], TAKEO TANAAMI[2], HIDEYUKI ISHIDA[3] & MIZUHO SHIMIZU[2]

[1] Yokogawa Research Institute, Tokyo, Japan
[2] Yokogawa Electric Corporation, Tokyo, Japan
[3] Department of Physiology, School of Medicine, Tokai University, Kanagawa Prefecture, Japan

26.1 INTRODUCTION

In the last several decades, the focus of life science research has been shifting from studies at the basic molecular level to more complicated studies of the mechanism of life at the cellular and organ levels. While with conventional microscopy it is generally necessary to fix and slice biological samples for high-resolution observation, a number of new technologies related to microscopy have been developed, especially in the last decade, which have opened up ways to study living specimens without the need for fixing or slicing. For example:

- The confocal microscope (Wilson, 1990) has become one of the most powerful tools with which to study living biological specimens, enabling observation in 3-D of intact living cells and tissues.
- In only 3 years personal computers have become faster by 10-fold, from 20 MHz to 200 MHz and beyond. As a result of this and improvements in the design of CCD chips, the read-out speed of CCD cameras has also increased nearly 10-fold. High-resolution cameras with more than one million pixels are readily available now while cameras with four million pixel are coming onto market. Improvements in the scanning speed and resolution will continue to be made, as will improvements in the sensitivity of cameras.
- Two Japanese groups (Sase *et al.*, 1995, 1996; Funatsu *et al.*, 1995) have separately succeeded in observing a single fluorescent molecule with fluorescent microscopy and with a scanning camera at 30 frames per second, which suggests that dynamic observation of even a single molecule inside living cells is possible. One of the key technologies in achieving this was the minimization of the background light (flair) in the microscope so as to improve the signal-to-noise ratio (SNR) of the observation system.
- Developments in microscope techniques as represented by the video-enhanced contrast microscopy method (Inoué, 1986) have also given rise to greatly improved microscope images.
- Three-dimensional image reconstruction from sections by volume rendering has been developed in image processing technology used in X-ray CT

FLUORESCENT AND LUMINESCENT PROBES, 2ND EDN
ISBN 0-12-447836-0

(computer tomography) and MRI (magnetic resonance imaging) and other techniques, which can be directly applicable to the processing of optical microscope images.

- Development of optical probes for biological activity advanced greatly to make possible the targeting of specific components of living cells; in particular, probes at the near-infrared wavelength range will allow non-invasive observation of very thick biological specimens. Moreover, recently developed GFP (green fluorescent protein) technology has opened up new possibilities to label genetically certain proteins, and then trace the behaviour of the expressed proteins in living cells.

While each of these innovations has greatly contributed to improvements in microscope technology, the total integration of the most advanced of these technologies has the potential to provide a revolutionary approach to microscopy which might allow studies on dynamic events in living cells from a single molecule level to the whole-mount organ level, and in real-time. Taking such factors in mind, we aimed at developing a system which could track in real-time very rapid dynamic physiological processes in living cells and organs, such as that of the nervous system, at 1 ms speeds. To date, we have developed a Nipkow disk confocal scanner unit, named CSU10, which allows 3-D observation of living cells and tissues at a scanning rate of 360 frames per second with high SNR (Ichihara *et al.*, 1996).

26.2 CONSTRUCTION OF THE CSU10

Figure 26.1 shows an overview of the construction of the CSU10. This system was based on the Nipkow disk scanner, which was invented by Paul Nipkow in 1884 as an optical scanner using rotation of a disk with pinholes to produce an image. The Nipkow disks were used at both the camera and the monitor in the first stage of the development of television (Mellors & Silver, 1951; Ramberg & Zvorykin, 1961). While a Nipkow disk retains many advantages such as fast scanning speed and simple construction, its illumination efficiency is very low owing to the very low area ratio of the pinhole apertures to the disk surface. To attain a confocal effect, it is necessary to place pinholes distant enough from each other: if the ratio of pinhole diameter to pinhole pitch is 1:10, the aperture ratio of pinholes is only 1% of the disk area, which means only 1% of the light can pass the pinhole. At the same time, 99% of the impinging light cannot pass the pinhole, but rather is reflected at the disk surface, which results in high background noise. Moreover, in a conventional

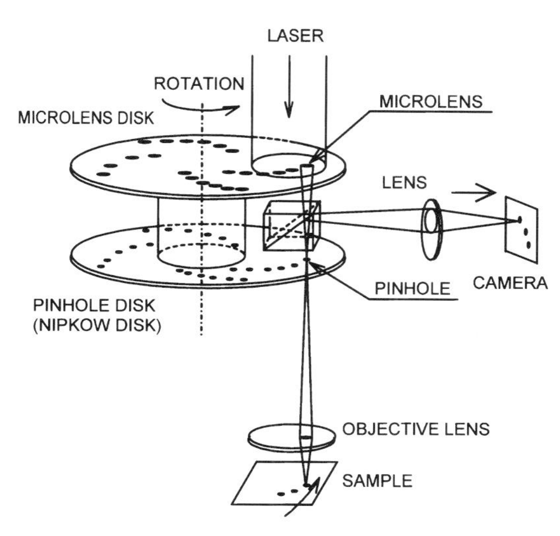

Figure 26.1 Overview of the CSU10.

Nipkow disk scanner, the detector is placed on the light-source side of the pinhole array, which results in high background light and thus low SNR.

The CSU10 has two disks; light incident on about 20 000 microlenses in the upper disk are focused by the microlenses onto corresponding pinholes on the lower disk, which are arranged in the same pattern as the microlenses on the upper disk. The presence of the microlenses means that the pinholes effectively collect about 40% of the light incident on the surface of the upper disk. The light passing through the pinholes is focused by an objective lens on a spot in the specimen. Fluorescent light from the specimen returns along the same path through the objective lens and pinhole, and is reflected by a dichroic mirror through a relay lens to the imaging point in a camera. The upper disk containing the microlenses is mechanically connected to the lower disk containing the pinholes, and both are rotated by an electrical motor. Thus the light beams raster-scan the specimen. The pinhole pattern on the Nipkow disk is designed so that excitation light beams uniformly scan a whole plane of the specimen.

To increase the SNR of the system, it is also important to minimize the background light inside the system. In the CSU10, fundamental glass parts such as the microlens array, pinhole array and dichroic mirror are made of synthetic quartz, and the antireflection films inside the system are designed so as not to emit non-specific fluorescence.

26.3 CHARACTERISTICS OF THE CSU10

Pinhole-type confocal microscopy can be divided into two distinctly different types: the single-beam scanning type using galvanometer mirrors, and the multiple-beam scanning type using a Nipkow disk, such as the CSU10. There are three main differences between the two systems. In galvanometer mirror scanners: (1) a single pinhole is fixed, (2) only the fluorescent light beam passes through the confocal pinhole (non-common path), and (3) the confocal images are collected as zero-dimensional signal (single pinpoint image) by a photomultiplier tube (PMT). In a Nipkow-type scanner, as opposed to the above: (1) multiple pinholes rotate to raster scan the specimen, (2) both the illumination light beam and fluorescent light beam pass through the same pinhole (common path), and (3) confocal (optical section) images are produced as two-dimensional images, just like optical microscope images. Because of the differences in the mechanism, the Nipkow-type scanner has the following major advantages over single-beam scanners:

- no aberration in the produced images,
- direct viewing of real-colour confocal images,
- fast scanning rate, and
- resistance to vibration.

26.3.1 Aberration in the image position

The principle of confocal fluorescence microscopy is to measure the intensity of fluorescence in 3-D space for each point in a specimen by scanning the 3-D space and visualizing it. Thus for high-quality image resolution it is most important to prevent aberrations between the real fluorescent image of the specimen and the actual image produced. In single-beam scanners, a single laser beam scans the specimen by using two mirrors, one for the x-scan and one for the y-scan, and the fluorescent light intensity of each point captured by the PMT is displayed with equivalent spatial position and relative brightness on a computer monitor; such complex positional conversion could give rise to aberration in images. In contrast, in the CSU10 a Nipkow disk is placed between the objective lens and the ocular lens, and the pinhole window containing about 1000 light beams within one 7 mm × 10 mm visual frame move to raster-scan the specimen to produce an image which is focused by the objective lens. Thus there is no factor to cause aberration between the actual and produced images.

26.3.2 Direct viewing of the confocal images

Scanning at 360 frames per second is fast enough to produce continuous images to the human visual system so that the CSU10's confocal images can be viewed with the eye. As opposed to the images produced only on computer monitors, images viewed directly are much more easily interpreted by the human brain, and it is easier for researchers to focus and locate the specimen with a direct view. More importantly, it is possible to watch fast changes occurring within the specimen in real-time.

26.3.3 Real colour observation

Emission colour provides important spectrum information in an observation system. A single-beam confocal can only collect images with a monochromatic PMT, so that it is necessary to manipulate the image information in various ways such as inserting filters before the PMT to obtain spectrum information. But with the CSU10, observation with either direct viewing or with a colour CCD camera can easily give spectrum information without the need for image processing by a computer.

26.3.4 Scanning speed

In the CSU10, the Nipkow disk rotates to raster-scan the specimen with about 1000 beams at 1800 rpm. The pinhole is designed to capture 12 frames per rotation, so that the image-capturing speed is 360 frames per second. As observation of living specimens becomes more and more important, high-speed scanning is a great advantage. The next generation models will rotate at up to 5000 rpm to enable scanning at 1000 frames per second.

26.3.5 Resistance to vibration

In the CSU10, the focal point of the excitation beam which impinges onto the pinhole is far enough away such that pinhole–plate vibration of up to 1 mm is permissible, and thus the effects of vibration in the axial direction are negligible. This is in contrast to the much larger vibration effects that occur with a single-beam confocal system, in which it is necessary to install the system onto an anti-vibration desk so as to avoid aberration of images during scanning. Moreover, in the common path system in the CSU10, the excitation light beam and the reflected fluorescent light beam pass through the same pinhole even when vibration occurs in the radial direction. As a consequence, the CSU10 can be installed in the same way as a conventional optical microscope, and is very insensitive to vibration.

26.4 SYSTEM INTEGRATION OF THE CSU10

The CSU10 can be installed onto most of the currently used upright and inverted microscopes at their standard camera port. One can view full colour confocal images in real-time at the eyepiece of the CSU10. The confocal images can be recorded as real colour photographs. Dynamic confocal images produced with CSU10 can be acquired by a charge-coupled device (CCD) camera at video rate (30 frames per second) and recorded either in a videotape or captured as a digital file in a personal computer. For 3-D image reconstruction, a series of optical sections can be acquired by gradually moving the microscope's stage at a constant pitch with a *z*-axis control unit such as a motor drive on the microscope's focus knob or by using a piezoelectric translator. From optically sectioning and acquiring an image series, 3-D images can be reconstructed by using image processing software.

26.4.1 CCD camera

While there are many kinds of CCD cameras, those using silicone diodes as a photodetector are most popular. To acquire dynamic images, standard video rate (30 frames per second) cameras are readily available, but they only provide acceptable image quality at around the 300 000 pixels level. To acquire very dark fluorescent images at high sensitivity, an image intensifier with a multiplication factor of 1000 to 10 000, coupled with a standard CCD camera (ICCD camera), is used. A cooled digital CCD camera of more than 1–4 million pixels is used to capture high-quality images of static specimens, often by extending exposure time to acquire high-resolution and high-sensitivity images. This latter type of operation is employed in the use of the UltraVIEW system described elsewhere in this volume (see Chapter 12 by Mason *et al.*).

26.4.2 Three-D image processing

From a series of optical sections taken by a confocal system, a 3-D image can be reconstructed using image processing software. While it used to be necessary to use a work station for 3-D image analysis, a variety of personal computer systems are readily available today. For example, 78 (as of September 1, 1998) 3-D software packages were mentioned on the Internet (http://biocomp.stanford.edu/3dreconstruction/), with a wide range of platforms such as SGI, Sun, HP, DEC, PC and Mac. The basic functions necessary for 3-D software are: (1) to have a driver for *z*-axis controller which can decide the point of start and stop at an optional position, the pitch of the axial movement, and the number of images to be acquired, and (2) can utilize the acquired images as a stack file. Moreover, recent software may have various advanced functions, such as manipulating display angles or transparency and rotating the stereo images by using a volume-rendering algorithm. Some can do 3-D analysis such as distance and volume of the stereo images. As a matter of course, before using 3-D software, it is necessary to understand and prepare such basic conditions as matching the refractive index of the immersion oil and specimen, correction of the *z*-axis stroke, determining the relationship between the pixel size of the CCD camera and the axial distance of each section, etc., to reconstruct a 'correct' stereo image

26.5 APPLICATIONS

The greatest advantage of the CSU10 over a single-pinhole confocal system is its high temporal resolution. The CSU10 is designed to scan 12 frames in one rotation, and rotates at 1800 rpm, thus its resolution velocity is 360 frames per second. Because of this faster-than-video-rate scanning speed, the CSU10 can produce real-time and real-colour images which one can observe at the eyepiece, or record not only via a CCD camera but also as full colour photographs. In general, direct viewing of confocal images is possible with specimens which are bright enough to be observed with conventional fluorescent microscopy, as long as the excitation and emission wavelength are suitable. Even if the specimens are not bright enough for direct viewing, it is possible to capture images using either an image-intensified camera (ICCD camera) at video rate (30 frames per second) or a CCD camera at slower speed.

26.5.1 Observation of whole-mount specimens

Optical sections of whole-mount sea urchin embryonic microtubules at the first mitotic stage are shown in Plate 30.1. Sea urchin embryos were treated with dispase, washed with sea water, extracted in a buffer solution containing surfactants, and fixed with formaldehyde. Monoclonal antitubulin antibody (Amersham) was used as the first probe for tubulin. Embryos were then stained with the secondary antibody, BODIPY-FL-labelled antimouse IgG antibody (Molecular Probes). An argon laser with a 488 nm line wavelength was used for excitation, and photographs were taken through an FITC filter. The magnification of the objective lens was ×63 and ASA 1600 film was used (Uemura, 1996).

26.5.2 Real-time observation

Some examples of video rate or faster application on living specimens are:

- calcium imaging in living cells or organs,
- observation of microcirculation in living animals or perfused whole-mount organs,
- observation of endocytosis and tracking of individual microinjected or GFP-expressing cells.

Plate 26.2 shows images of rapid changes in the Ca^{2+} level at spontaneous beating of cultured cardiac myocytes recorded using a high-speed ICCD camera (1000 HR, Kodak) at 4 ms per frame (Ishida *et al.*, 1996). Further examples with digital CCD cameras are shown in Chapter 12 in this volume.

26.5.3 3-D confocal imaging

The most typical application of a confocal microscope is 3-D image processing. Figure 26.2 shows a 3-D image of sea urchin embryo. Optical sections of sea urchin embryo were taken continuously using the CSU10, and the 3-D image was reconstructed using image analyser software contained in the Insight IQ System (Meridian); the alignment of the microtubules can be clearly seen inside the cell (Uemura, 1996).

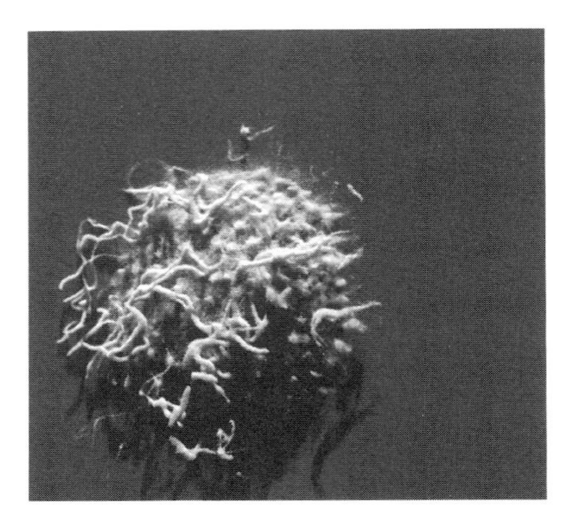

Figure 26.2 3-D image of sea urchin embryonic microtubules stained with BODIPY. The image was reconstructed from 0.1 μm optical sections (×63 objective lens).

While the reconstructed stereo images can only be observed as digital data, direct viewing of confocal sections can give an idea of the actual spatial structure of specimens in real-time.

26.5.4 Real-time 4-D confocal imaging

A real-time 4-D system can be constructed by using a piezoelectric microscope focusing drive and ICCD camera in combination with the CSU10, which can take z-axis section series with an optimal time interval.

26.6 FUTURE PROSPECTS

The basic features necessary to allow observation of a single molecule in real-time in a living cell have been developing gradually, but we still need to wait for the development of peripheral technologies as discussed below before this becomes a reality.

26.6.1 Light source

One of the limitations of the confocal microscope is its restricted choice of laser light source. Currently, gas lasers such as argon, krypton or helium–neon lasers are typically used. The problem with gas lasers includes the necessity for high electrical power, the high level of heat generation and noise, the short life and high cost, on top of the limited choice of excitation wavelength. Recently, various new lasers such as solid state lasers, SHG (Second Harmonic Generator) or LD (Laser Diode) have been developed, which will be applied as light sources for confocal microscopes in the near future.

26.6.2 Near-infrared application

One of the important applications of a confocal microscope is the observation of whole-mount specimens. Since light scattering is in inverse proportion to the fourth power of the wavelength, the longer the wavelength, the deeper light can penetrate into a specimen. Moreover, autofluorescence of a biological sample becomes smaller at the near-infrared range. Thus, it may be desirable under some conditions to use a near-infrared light source to observe thick specimens. Even though near-infrared fluorescent dyes such as Cy7 (excitation at 743 nm with 767 nm emission) have been developed, there is no optimal laser for such dyes. Recent developments in two- or multiphoton technology could be one solution, but there is still a big problem in the availability of a convenient laser source.

26.6.3 High-performance cameras

For the observation of living specimens, it is necessary to minimize not only the photobleaching of the fluorophore, but also the toxicity of the fluorophores or laser light which results in a low light level. Therefore, sensitivity of photodetectors is critical. In this regard, progress has been made rapidly in various types of CCD cameras, and we can expect further improvements in both temporal and spatial resolution. Highly sensitive image intensifiers, so-called Gen3 or Gen4 ICCDs, using GaAs phototubes, have become available recently – these can capture invisibly dark images at video rate. Back-illuminated CCD cameras have also been developed whose quantum efficiency has been doubled by introducing light from the back of the array. This option is offered with the UltraVIEW system from Life Science Resources (Cambridge, UK). Scientific grade CCD cameras with about 1 million pixels have also become popular, and those with even 4 million pixels have been commercialized. Regarding temporal resolution, a high-speed camera that can capture 1000 images per second is still very expensive, but it is possible to capture faster than video-rate images by limiting the read-out of CCD camera with techniques such as binning or line scanning. Again, Life Science Resources offer frame transfer camera options which can readily acquire low light level images at rates between 100 and 1500 frames per second.

26.6.4 Polarization of laser beam

In general, a laser beam is linearly polarized, and thus for regular fluorescent microscopy it is necessary to make it either circularly polarized light or random light. However, polarization confocal microscopy, by using the linearly polarized laser light source, may become a meaningful tool in situations such as observing polarized specimens conjugated with polarized fluorophores, or super-molecules which often are polarized.

26.6.5 Improvement in personal computers and software

Unceasing improvements in PCs have made complicated image processing possible with PCs. At the same time, the transfer speed of image data across electronic networks has become faster following the development of Internet technology, and this will continue with the development of digital and high-resolution television.

26.7 CONCLUSION

Confocal microscopy technology has contributed greatly to advances in life science research by enabling studies on intact living specimens. As discussed here, it is the result of integration of such technologies as fluorescence microscopy, computers, lasers, imaging devices, and image processing software together with advanced optical and mechanical engineering. CSU10, our first commercial model, has made real-time and real-colour 4-D observation possible. By taking advantage of rapid advances in all relevant technologies, we expect further development in confocal microscopy to the extent that we shall be able to facilitate the observation of a single molecule in living specimens in real-time.

REFERENCES

Funatsu T., Harada Y., Tokunaga M., Saito K. & Yanagida T. (1995) *Nature* **374**, 555–559.

Ichihara A., Tanaami T., Isozaki K., Sugiyama Y., Kosugi Y., Mikuriya K., Abe M. & Uemura I. (1996) *Bioimages* **4**, 57–62.

Inoué S. (1986) *Video Microscopy*. Plenum Press, New York.

Ishida H., Genka C., Nakazawa H. & Tanaami T. (1996) *Proceedings of 18th Meeting of Japan Society of Laser Microscopy*, pp. 48–51.

Mellors R.C. & Silver R. (1951) *Science* **114**, 356–360.

Ramberg E.G. & Zvorykin V.K. (1961) In *Photoelectricity and Its Applications*, 4th edn. John Wiley, New York, ch. 16, pp. 366–367.

Sase I., Miyata H., Corrier J.E.T., Craik J.S. & Kinoshita K. (1995) *Biophys. J.* **69**, 323–328.

Sase I., Corrier J.E.T., Craik J.S. & Kinoshita K. (1996) *Bioimages* **4**, 36.

Uemura I. (1996) http://www.comp.metro-u.ac.jp/~iuemura/.

Wilson T. (1990) *Confocal Microscopy*. Academic Press, London.

CHAPTER TWENTY-SEVEN

Optimizing Confocal Microscopy for Thick Biological Specimens

STEPHEN H. CODY & DAVID A. WILLIAMS

Confocal and Fluorescence Imaging Group, Department of Physiology, University of Melbourne, Parkville, Victoria, Australia

27.1 INTRODUCTION

The purposes of this chapter are two-fold: to describe in general terms the advantages of confocal microscopy for live cell work with illustrative biological examples; and to expound the virtues of some important optical principals, which are too often ignored by biologists, to the detriment of the quality of their confocal microscopy imaging. The present chapter is directed at those who have little or no experience with confocal microscopy. The principle of confocal microscopy has been described elsewhere in this volume (see Chapter 21). A good understanding of this principle is required to make the most of this chapter, or indeed to make the most of confocal microscopy itself. Other texts that may prove useful to review the principle of confocal microscopy are Inoué (1995) and Goldie *et al.* (1997).

It is often said that confocal microscopy fills a niche somewhere between conventional light microscopy and electron microscopy. This statement is very misleading, giving unreal expectations to the novice user. Confocal microscopy is a form of light microscopy that imparts some benefits compared with conventional light microscopy, in the same manner that phase contrast is a form of light microscopy with some advantages. Confocal microscopy bestows many benefits to biomedical research, especially when working with thick biological specimens. While it is true there is an increase in resolution in the *xy*-plane, this increase is minimal when compared to the resolution of electron microscopy. While increased *xy*-resolution is of some benefit, the great advantage of confocal microscopy to biomedical research is the increase in axial resolution (i.e. resolution along the *z*-axis or optical axis). The tangible advantage of increased optical axis resolution is the rejection of out-of-focus blur or, perhaps more simply, the ability of the confocal microscope to produce optical sections. The optical sectioning capability of the confocal microscope is the principal reason why this form of microscopy is so popular in biomedical research. The main purpose of this chapter is to describe some of the general principles that affect the optical sectioning efficiency of confocal microscopes and to give some practical hints on how to maximize the efficiency of optical sectioning.

FLUORESCENT AND LUMINESCENT PROBES, 2ND EDN
ISBN 0-12-447836-0

27.2 ADVANTAGES OF CONFOCAL OVER WIDE-FIELD MICROSCOPY, ESPECIALLY FOR THICK BIOLOGICAL SPECIMENS

When observing a thick specimen with conventional light microscopy, it is difficult to gain any useful impression of the structures within. The microscope is focused on a single plane within the specimen. While this plane is in focus, it may be difficult to discern from the out-of-focus blur from planes above and below the plane of focus. It is for this reason that during routine histology thin sections are cut, typically around 6 μm. The out-of-focus blur is physically removed by the process of thin sectioning. While histological techniques have a long history, are highly refined and successful, there are some disadvantages. It is not possible to observe living cells and tissues with routine histological techniques. The tissue must be fixed, dehydrated and embedded prior to sectioning, which alters the cellular appearance. Serial sectioning and 3-D reconstruction is extremely time-consuming, hence expensive. To avoid some of the problems associated with fixation and dehydration artifacts, frozen sections may be used. However, these also have problems with ice-crystal damage.

In confocal microscopy a small aperture in front of the detector prevents most of the light from the out-of-focus planes (above and below the plane of focus) being imaged. Only light from the in-focus plane reaches the detector, and as a result an optical section is created (see Chapter 21). The collection of serial optical sections (a *z*-series) is a simple matter of moving the specimen or the objective lens along the optical axis. As confocal microscopy allows the efficient collection of optical sections from thick specimens, many of the problems associated with routine histology may be avoided. Probably the most revolutionary change brought about by confocal microscopy is the ability to obtain clear images from living specimens. The benefits of optical sectioning are clearly illustrated in Fig. 27.1. The conventional epifluorescent image (Fig. 27.1(A)) of the single cardiac cell shows very little detail, as the out-of-focus blur masks the plane of the cell that is in focus. The confocal image of a similar cardiac cell (Fig. 27.1(B)) shows much more detail; the two nuclei and perinuclear regions are clearly imaged, and bands that correspond to alternating bands of contractile proteins and mitochondria are evident. The benefits of removing out-of-focus blur are even more evident when large tissue pieces or whole organs are examined. Clear cellular detail may be imaged when large tissue pieces (Chapter 5) and indeed whole organs (e.g. Plate 27.1 and Fig. 27.2) are mounted on a confocal microscope. With conventional fluorescence microscopy virtually no detail at all may be obtained from thick biological specimens.

Given the ability to collect fluorescent images at such fine detail (resolution), it is possible to use a multitude of fluorescent techniques in conjunction with confocal microscopy. The high spatial resolution of fluorescence imaging is a great advantage whether studying the localization of ions (e.g. Figs 27.1, 27.2, 27.5, 27.7, Plate 27.2), macromolecules (e.g. Plate 27.1), or the three-dimensional organization of organelles and cells. The ability to collect extensive serial optical sections within a rapid period of time can be extremely useful. Within a thin two-dimensional slice (e.g. Plate 27.1(A)) it is often difficult to interpret the three-dimensional structure. Too often, with conventional microscopy the overall 3-D structure is completely missed. Serial sectioning using histological methods, followed by 3-D reconstruction, is an elaborate, time-consuming

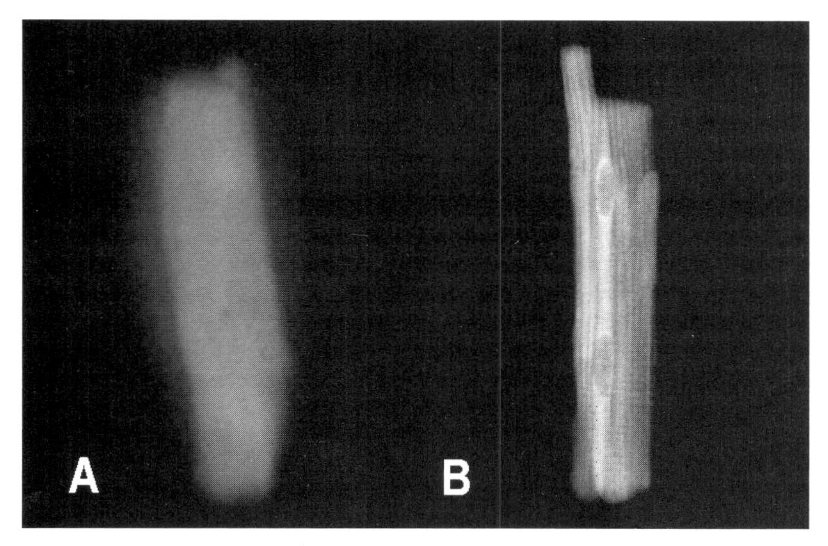

Figure 27.1 (A) Isolated cardiac myocyte loaded with the fluorescent Ca^{2+} dye fluo-3, imaged with conventional epifluorescence microscopy and a SIT camera. The blur is due to out-of-focus light being included in the image and is typical of a non-confocal fluorescent image. (B) Similar cell loaded with the same fluorescent dye, imaged with an MRC-500 confocal microscope (Bio-Rad). The confocal image is much sharper due to the rejection of the out-of-focus light; the nuclei and the striated pattern of the muscle are quite clear. The perinuclear regions represent the cytosol where there is little contractile proteins due to the presence of the nuclei.

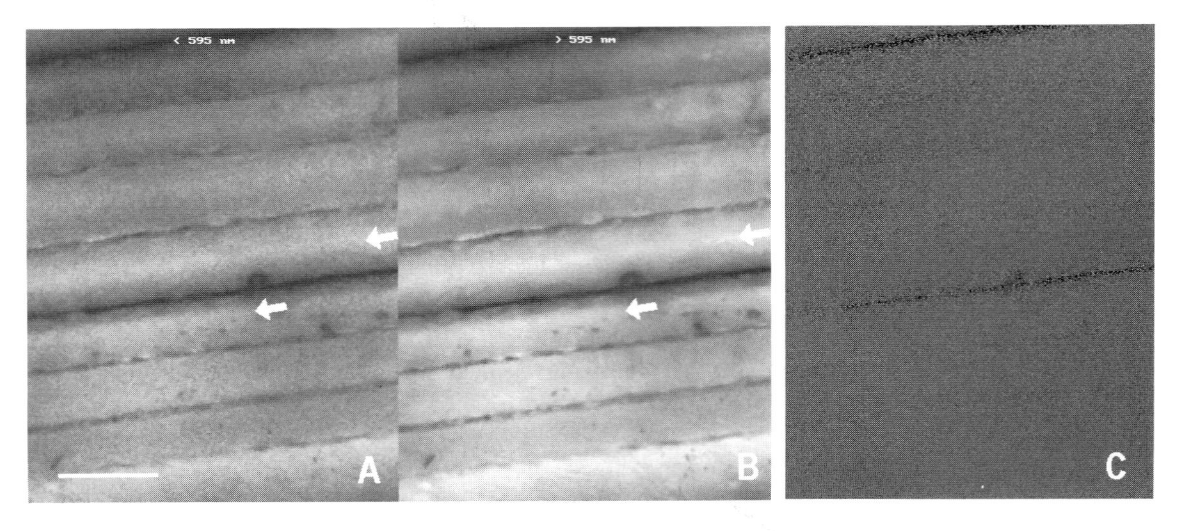

Figure 27.2 Dual-emission image of a rat skeletal muscle loaded with the pH dye carboxy-SNARF-1 fluorescence. This image was collected using a ×10, NA 0.3 dry objective lens on an MRC-600 (Bio-Rad) confocal microscope. The low NA of this lens causes the optical sections to be quite thick. Due to the thick optical sections and the curved surface of the muscle fibres, the fibres are brighter along their central axis (arrow). The central axis of the fibre is thicker, causing the bright band, which is a pathlength error. Scale bar = 100 μm. (Reproduced with permission from Cody *et al.*, 1993.)

alternative. A simple digital method (image projection) may be used with a confocal data set to represent the whole 3-D structure within a single 2-D image (e.g. Plate 27.1(B)). Other methods of 3-D rendering, such as animations and red/green anaglyphs, may also be employed to enhance our 3-D interpretation (see Goldie *et al.*, 1997).

27.3 PATHLENGTH ERRORS

Confocal microscopy is often used in conjunction with fluorescent probes to monitor the intracellular concentration of various physiologically important ions and organic molecules (see Haugland, 1996; see also Chapter 3). One of the major advantages of using confocal microscopy for quantitative determination of fluorescence intensities is that the depth of tissue that is imaged may be controlled. In conventional fluorescence microscopy, the intensity of a specimen is not just related to the intracellular concentration and brightness of the probe, but also to the thickness or 'pathlength' of the specimen. The optical sectioning properties of confocal microscopy allow light to be collected from a discrete volume of the cell, and thus pathlength errors are generally avoided. However, the microscopist should be aware there are some circumstances where pathlength errors may occur even with confocal microscopy. If the specimen or parts of the specimen are not as thick as the optical section, it is still possible to have pathlength errors. Pathlength errors are most frequently encountered when using a relatively low numerical aperture (NA) lens and/or a large pinhole, as the optical sections will be relatively thick. The outermost fibres of an isolated whole rat skeletal muscle preparation loaded with the pH-sensitive fluorescent dye are shown in Fig. 27.2 (A&B). This confocal image was obtained using a low NA dry lens (NA = 0.3). Some of the fibres in Fig. 27.2(A) appear brighter along their central axis. If the data collected consisted of only one emission wavelength, it may be concluded, incorrectly, that some of the fibres have a central cytosolic band that is at a different pH level. However, the most likely cause of this bright band is a pathlength artifact. As the fibres are circular in cross-section, if a relatively thick optical section glances the top of a fibre, then a pathlength error will result (see Fig. 27.3).

Although pathlength errors are more likely to occur

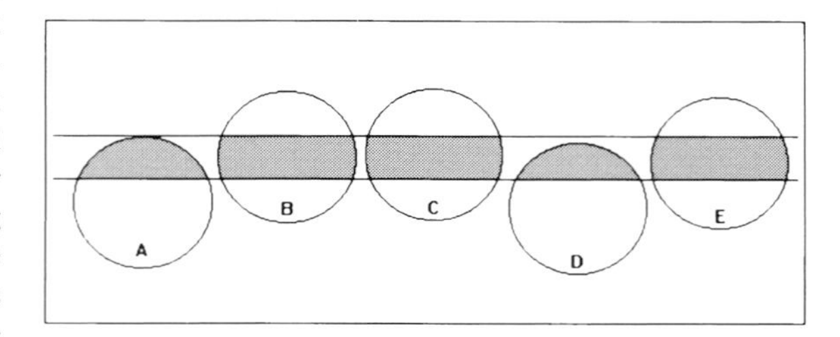

Figure 27.3 Diagrammatic longitudinal section through the fibres in Fig. 27.2. The optical section is shown passing through the middle of some fibres in which there will be no pathlength error. However, the optical section 'glances' other fibres. In these it is clear that the central portion will appear brighter due to pathlength error. (Reproduced with permission from Cody *et al.*, 1993.)

when using relatively thick optical sections, they may also occur when high NA lenses and small pinholes are employed to achieve relatively thin optical sections. The confocal image of a freshly dissected rabbit proximal kidney tubule loaded with the pH fluorescent indicator BCECF (Harris, 1991) (Plate 27.2), shows a weakly fluorescent brush border adjacent to the lumen. This could be erroneously interpreted as indicating that the pH of microvilli is lower than that of the cytosol. However, when the 3-D structure of the cell is considered, it is apparent that the difference may reflect a pathlength error caused by luminal fluid interspaced between the microvilli (Fig. 27.4).

27.4 DYE CONCENTRATION AND PHOTOBLEACHING ARTIFACTS

Another artifact that may be encountered when using fluorescent probes is that of dye concentration heterogeneity. When comparing the fluorescent intensities between cells or different areas within the same cell, it must be remembered that the intensity of the fluorescence is not only related to the concentration of the ion under investigation, but also to the concentration of the dye itself. There is no certainty when loading cells with the esterified form of dyes (see Chapter 5 by Bowser *et al.*) that different cells will take up the same amount of dye. Even within an individual cell there may be discrepancies in the dye concentration from organelle to organelle. The intense fluorescence in the kidney tubule cells in Plate 27.2 may be interpreted as either a higher nuclear pH or higher dye concentrations within the nuclei. It is difficult to make any quantitative conclusions from such data. A further complication when analysing single-wavelength data of this type is that the fluorescent dye will photobleach (fade) after exposure to light. This can render any analysis of time course data erroneous.

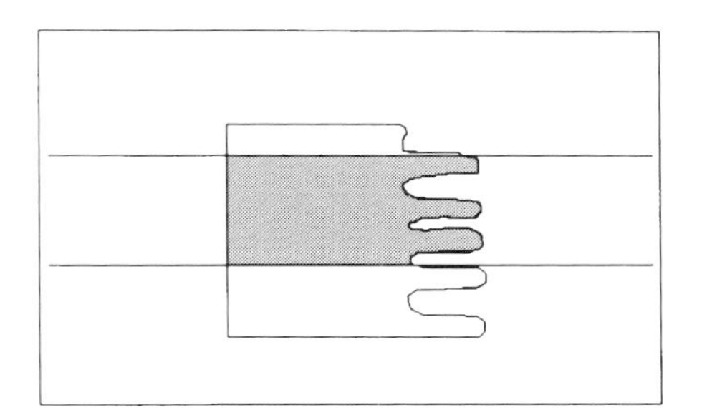

Figure 27.4 Diagrammatic representation of an optical section passing through a cell with microvilli. The shaded region indicates the fluorescent signal that will be detected by the confocal microscope. It is clear that the brightness of the signal will be less in the brush border region due to a pathlength artifact.

27.5 RATIOMETRIC ANALYSIS

To correct for dye concentration (dye loading), pathlength and photobleaching artifacts, a ratiometric method should be employed where possible. For a review of some of the more commonly used ratiometric fluorophores, see Chapter 3. Briefly, a dye that is able to be used in a ratiometric fashion must have a shift associated with its excitation or emission spectra. For simplicity we shall only consider fluorescent probes with a shift in emission spectra, as these are more easily employed with confocal microscopy. However, dyes that exhibit a shift in excitation spectra are also commonly used in fluorescence microscopy (e.g. fura-2) and some may also be used in conjunction with confocal microscopy (e.g. indo-1). A simple explanation of an emission shift is that the emitted fluorescent light changes colour when a target ion binds to the fluorophore. The pH-sensitive fluorescent dye SNARF-1 (Molecular Probes) is a perfect example of this type of indicator. As each molecule of SNARF-1 binds a hydrogen ion the fluorescence changes from a red (630 nm) to an orange (580 nm) colour. Within any given cell the proportion of SNARF-1 molecules in the bound and unbound state will of course be dependent on the $[H^+]$ (pH) of the cell. The emission spectra of SNARF-1 (see Chapter 3) are perhaps simpler to understand if we keep these colour shifts in mind. The dual-emission image shown in Fig. 27.5 is of another rabbit kidney proximal tubule. However, this tubule is loaded with the pH indicator SNARF-1. Like the cells shown in Plate 27.2, the brush border may be seen fluorescing less brightly than the cytosol in both Figs 27.5(A) and (B). The nuclei in Fig. 27.5(A) are less fluorescent than the cytosol, but in Fig. 27.5(B) the nuclei and the cytosol are of similar intensity. A ratio image (Fig. 27.5(C)) may be created by performing a pixel-by-pixel division of the individual images (e.g. Fig. 27(A) by 27.5(B)). The ratio image created is a quantifiable map of intracellular pH. This map is totally independent of dye concentration, photobleaching and pathlength errors. The pH map indicates that the pH of the nuclei is slightly higher than that of the surrounding cytosol. However, the microvilli have the same pH as the rest of the cytosol, thus illustrating the advantages of a ratiometric approach to intracellular ion measurements. To quantify intracellular pH, the ratio values must be compared with a standard curve created on the same microscope with exactly the same acquisition settings and optical components (for further details,

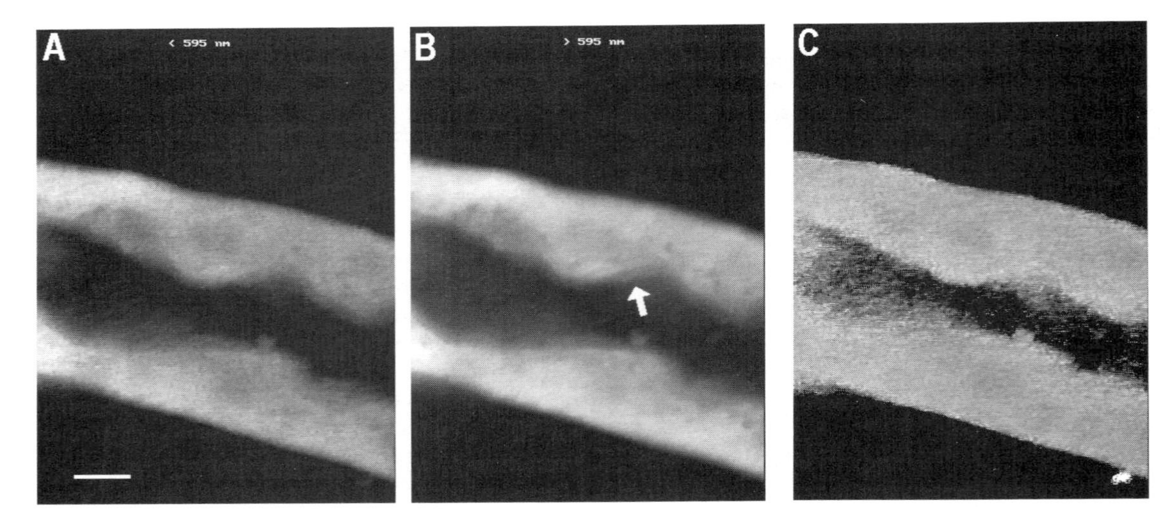

Figure 27.5 Another isolated rabbit proximal kidney tubule in this case loaded with the dual-emission pH-sensitive dye carboxy-SNARF-1. (A) Emission wavelengths greater than 595 nm. As the pH increases the brightness of this channel will increase. (B) Emission wavelengths less than 595 nm. As the pH increases the brightness of this channel will decrease. (C) Ratio image created by the pixel-by-pixel division of (A) by (B). The ratio image is a pH distribution map that is totally independent of dye loading, pathlength, and photobleaching artifacts. Scale bar = 5 μm.

see Cody *et al.*, 1993; Dubbin *et al.*, 1993; see also Chapter 5). Ideally the calibration should be performed intracellularly, preferably with the same cells that were used in the experiments.

27.6 ABERRATIONS AND CONFOCAL MICROSCOPY

The major difference between confocal microscopy and conventional microscopy is the pinhole, which ideally excludes only light that is out of focus. In conventional microscopy, it can be very difficult to discern structures of thick specimens due to the contribution of out-of-focus blur. Confocal microscopy offers a solution by removing the out-of-focus light. With both conventional and confocal microscopy, light is scattered by the specimen as the focal plane is moved further into a thick biological specimen. When examining such a specimen with confocal microscopy, the intensity of the signal may decrease very quickly with depth into the specimen. Part of the reduction in intensity is caused by light scatter by the specimen. However, the effect is often worse than with conventional microscopy. It appears that aberrations may cause the pinhole to exclude in-focus light, so that the instrument performs poorly in certain circumstances. For this reason it is essential that biomedical researchers using confocal microscopy have at least a rudimentary understanding of aberrations, their common causes and methods that can be used to avoid them. An understanding of the general principles of confocal microscopy is essential in understanding how an optical aberration may effect image formation (see Chapter 21; Inoué, 1995; Inoué & Spring, 1997).

The single most important component of any light microscope is the objective lens. Great care is taken by the microscope manufacturers in the design and manufacture of their objective lenses, particularly expensive high NA lenses. It follows, therefore, that to produce the best images these lenses must be used in the manner for which they were designed. To do otherwise may cause the most important component of the confocal microscope system to perform poorly, with the resulting images perhaps being less than satisfactory. While there are many different optical aberrations that will affect confocal imaging, the most deleterious are probably spherical and chromatic aberrations. In order to obtain the best possible result when using confocal microscopy, with the great variety of specimens that a microscopist is likely to encounter, spherical and chromatic aberrations must be understood so that their effects may be minimized.

27.7 SPHERICAL ABERRATION

For a lens to produce sharp images it must be designed and manufactured so that an infinite number of rays from the excitation source are all focused to a point (Fig. 27.6(A)). Severe spherical aberration will be induced if an incorrect medium is introduced into the optical path. Most microscopists have an oil immersion lens as their highest NA and most expensive lens and the expectation is that that this lens should produce outstanding images. This expectation is not unreasonable if the lens is used correctly. The reason why we use immersion oil at all has to do with the refractive index (RI) of glass. It is by no way a coincidence that the RI of glass, immersion oil, and traditional mountants such as DPX are the same (RI≈1.51). The rays of light may pass from the lens through the immersion oil, through the coverglass, and into the DPX-infiltrated specimen without refracting, thus minimizing aberrations (Brenner, 1994; Inoué, 1995; Keller, 1995). However, if the mountant is replaced with an aqueous mountant (RI≈1.33) for instance, the rays will bend (refract) upon entering the water layer. The rays of light from the excitation source will no longer form a focused point (as in Fig. 27.6(A)), but illuminate points at various depths along the optical axis (Fig. 27.6(B)). The more we attempt to focus into the aqueous environment, by bringing the lens closer to the coverglass, the worse the problem becomes (Fig. 27.6(C)). Once aberrations are induced by refractive index mismatch, the images appear blurred as each optical section contains a contribution from layers other than the plane of focus. A comparison of images of isolated rat cardiac cells mounted in saline, collected with two high numerical aperture lenses – a distilled water immersion lens NA 1.2 (Fig. 27.7(A)) and an oil immersion lens NA 1.4 (Fig. 27.7(B)) – illustrates the effect of spherical aberration. The image collected with the oil immersion lens (Fig. 27.7(B)) is not as sharp, most likely due to out-of-focus light. The aberrations also cause distortions along the optical axis, so that any 3-D measurements will be inaccurate. A 3-D data set of a sphere collected with an optical system with spherical aberration would produce 3-D representations of the sphere that would appear elongated along the optical axis. It is important to realize, however, that directly at the surface of the coverglass, these effects may be negligible (Hell *et al.*, 1993; Brenner, 1994; Inoué, 1995). The further the plane of focus is moved into a specimen, the more detrimental aberrations caused by RI mismatch become.

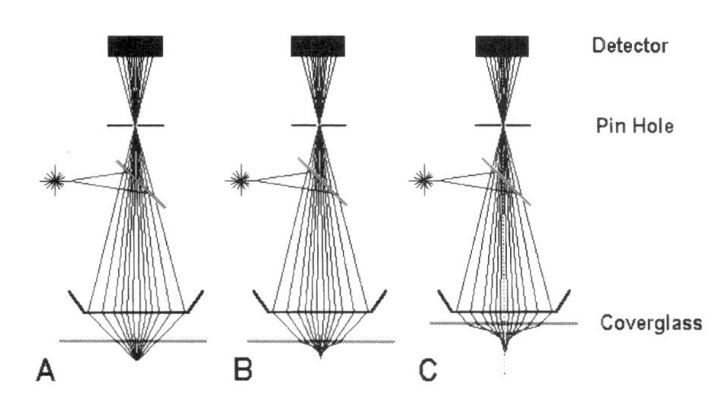

Figure 27.6 (A) Ray diagram showing an ideal oil immersion lens with no spherical aberration; all rays converge to a single point. (B) If the correct mounting media is replaced with water, the difference in the RI between the immersion oil and glass (RI≈1.51) and that of water (RI≈1.33) will cause the light rays to bend or refract, causing spherical aberration. The amount of refraction will depend on the angle that each ray meets the glass–water interface. It is clear that the rays no longer converge on a single point, but rather they focus at various depth along the optical axis. (C) When focusing even further into the water layer the rays refract to a greater extent, i.e. the spherical aberration becomes worse.

27.8 CHROMATIC ABERRATION

With fluorescence microscopy the wavelength of emitted light is different from the excitation wavelength, therefore chromatic aberration as well as spherical aberration may result from refractive index mismatches. This has very serious consequences for confocal microscopy, which is usually used as a fluorescence microscope (Hell *et al.*, 1993; Brenner, 1994; Inoué, 1995). The ray diagram in Fig. 27.8(A) is a simplified version of that in Fig. 27.6(A). The wavelength of emitted light from a fluorescent molecule will be longer than the excitation light. Therefore, as the fluorescent light meets the glass coverglass, the degree of refraction will be less than that of the excitation beam (Fig. 27.8(B)); this is termed chromatic aberration. In a conventional fluorescent microscope this would be relatively inconsequential, since good quality lenses are designed to correct most chromatic aberration. However, the pinhole of a confocal microscope, which is designed ideally to exclude only out-of-focus light, may also exclude in-focus, chromatically aberrant fluorescent signal. The effects of chromatic aberration are usually most noticeable when focusing deep within a thick specimen (Fig. 27.8(C)). The intensity of the fluorescent signal will drop very quickly with increasing depth into the specimen. The combined effects of refractive index mismatch from spherical and chromatic aberrations, resulting in blurred images and poor tissue penetration, may cause a confocal microscope to perform less efficiently than a conventional fluorescence microscope. To enable the confocal microscopist to obtain images of the highest quality, especially at depth within a thick specimen, the cause and effect of these aberrations must be fully appreciated.

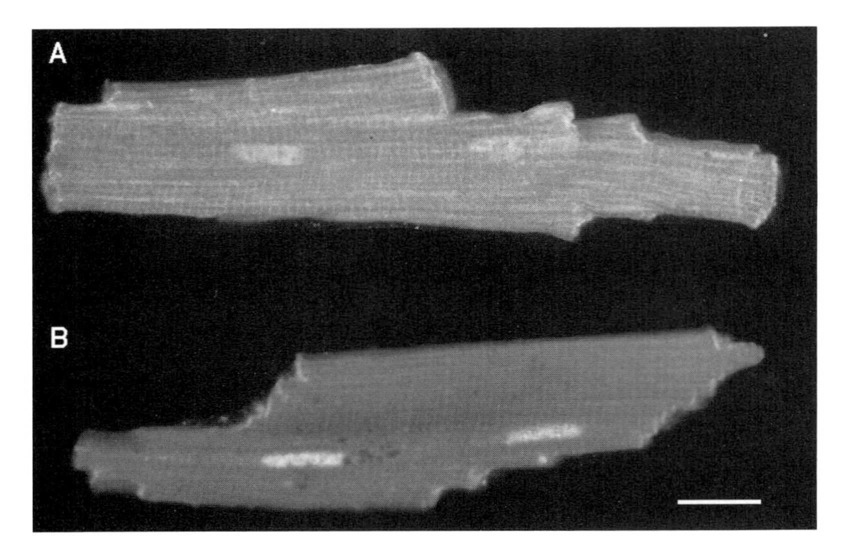

Figure 27.7(A) A living, isolated cardiac myocyte loaded with the Ca^{2+}-sensitive dye Fura Red, the cell is bathed in physiological saline and imaged with a ×60, 1.2 NA, distilled water immersion lens, with correct compensation for coverglass thickness, in conjunction with an MRC-1000 (Bio-Rad) confocal microscope. (B) A similar cardiac myocyte also loaded with Fura Red. This cell is imaged with a ×60, 1.4 NA, oil immersion lens. The image is not as sharp as the cell in (A), most likely because of spherical aberration, which allows light from out-of-focus planes through the pinhole to the detector. Scale bar = 10 μm.

27.9 AVOIDING REFRACTIVE INDEX MISMATCH

There are many ways in which the performance of a confocal microscope may be maximized. The most important consideration is to choose the appropriate objective lens for the specimen in order to minimize RI mismatch. Using high NA oil immersion lenses with specimens that have been infiltrated with traditional mountants such as DPX minimizes optical aberrations such as spherical and chromatic aberration. However, most specimens presented for confocal microscopy are probably not mounted in this manner. There are two main reasons why microscopists have avoided using DPX as a mountant for fluorescence experiments: (1) because DPX is a non-polar mountant, specimens must be fixed, dehydrated, and then infiltrated with DPX before mounting, so it is not a suitable mountant for live cell work; (2) specimens stained with fluorescein (probably the most commonly used fluorescent label for immunohistological methods) must be mounted in a polar, buffered solution at a high pH (pH 9.0), and therefore cannot be mounted in DPX. There is now a strong culture amongst immunohistologists that all fluorescently labelled specimens should be mounted in one of a myriad of glycerol- or water-based mountants. It should be remembered that these mountants only became important because of the extensive use of fluorescein. However, it is now becoming more common to use newly developed fluorescent labels for immunohistochemistry that are insensitive to the pH and polarity of the mounting medium (e.g. the BODIPY range of dyes by Molecular Probes and the Cy range by Amersham). Given the considerable advantages of using thick sections in conjunction with confocal microscopy for 3-D reconstructions, we would like to encourage microscopists to reconsider the use of DPX as a mountant by considering the use of dyes that are compatible with DPX.

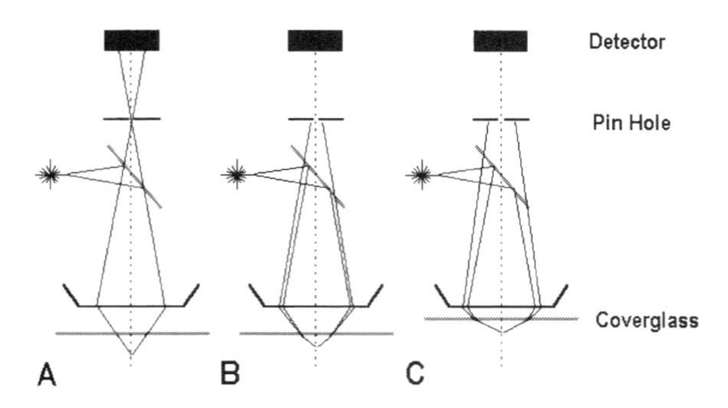

Figure 27.8 (A) Ray diagram showing rays focused to a point in ideal conditions where there is no RI mismatch. (B) The fluorescent signal, being a longer wavelength than the excitation light, will refract less than the excitation light. The confocal pinhole will then exclude much of the in-focus, though chromatically aberrant light. (C) As the plane of focus is moved further into the specimen the effect becomes worse, and more light is excluded by the pinhole.

The use of glycerine immersion lenses is another way of avoiding RI mismatch when specimens are mounted in a glycerine-based mounting media. It was found in preliminary investigations (S.H. Cody & M.H. Zheng, unpublished) that satisfactory optical sections could be obtained at far greater depths into specimens mounted in glycerine-based mountants when using glycerine immersion lenses (Nikon ×20 UV-F, NA 0.8 and Nikon ×10 UV-F, NA 0.5) compared with dry, oil or water immersion lenses (Fig. 27.9). This effect was even more marked when using

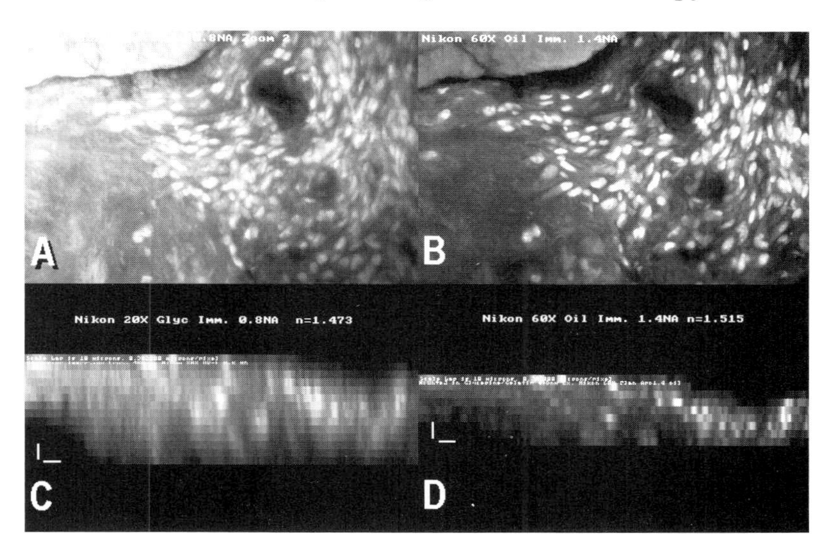

Figure 27.9 Projections of serial optical sections through a thick section of a sheep long bone, stained with the fluorescent dye acridine orange and mounted in a glycerine-based mountant. (A) Images collected with a glycerine immersion lens (Nikon ×20, 0.8 NA, UV-F). (B) Images collected using an oil immersion lens (Nikon ×60, 1.4 NA, PlanApo). (C) and (D) Isotropic *z*-sections computed from the data sets shown in (A) & (B) respectively. While individual cells are much clearer with the oil immersion lens (B), the depth of optical sectioning is greater using the glycerine lens, as seen in the *z*-sections (C). All scale bars represent 10 μm.

UV confocal microscopy. In a series of investigations into the role endothelin plays in inflammatory diseases of the airway (Fernandes *et al.*, 1998; Goldie *et al.*, 1997) glycerine immersion lenses were employed to reduce aberrations when examining whole mounts or thick sections of rat trachea mounted in glycerine-based mounting media (e.g. Plate 27.1). The depth at which satisfactorily sharp images could be collected when the specimens were mounted in glycerine-based media was greatly enhanced using glycerine immersion lenses compared with dry, oil, or water immersion lenses.

When working with thick living preparations, the water immersion lens is the lens of choice, as the refractive indices of the immersion fluid, saline and specimen are very similar. Water immersion lenses fall into two categories. To avoid confusion between the two types of lenses we have termed those designed to be used in conjunction with a coverglass, with a small drop of distilled water between the lens and the coverglass, as 'distilled water immersion lenses'. Those lenses designed for dipping directly into saline we designate 'saline-immersible lenses' (Cody & Williams, 1997). Distilled water immersion lenses are used in a similar manner as oil immersion lenses, but a small drop of water, instead of oil, is placed between the lens and the coverglass. Distilled water immersion lenses must have correction optics built into them to compensate for the aberrations induced by the RI mismatch caused by the introduction of the coverglass in the optical path. The exact compensation for aberrations is highly dependent on the thickness of the coverglass (Brenner, 1994; Inoué, 1995). However, even within the same packet the thickness of coverslips is highly variable. High numerical aperture distilled water immersion lenses are equipped with correction collars that must be accurately adjusted for the thickness of each individual coverglass (Inoué & Spring, 1997).

There are two practical ways of adjusting the collar for the coverglass thickness. The first is perhaps the simplest and most reproducible. Before experiments, the actual thickness of each 170 μm (nominal) coverglass can be measured with Vernier callipers, then the coverglasses may be sorted into empty boxes based on their thickness. It is then possible to do experiments by taking a coverglass and setting the correction collar on the distilled water immersion lens according to the premeasured thickness. Multiple experiments may also be conducted with other coverslips of the same thickness, without having to reset the collar. This method has the advantages that it may save considerable amounts of time, and results may be more reproducible as the collar does not need to be reset, between experiments. An alternative method to adjust the correction collar is to place a 170 μm (nominal) coverglass (it need not be measured) with the specimen on the stage of the confocal microscope. Whilst scanning the specimen, the collar should be adjusted until the sharpest image is obtained. This method has the advantage that it does not just correct for aberrations induced by the coverglass, but for aberrations caused by the mounting medium and specimen itself. The collar may even be readjusted at each depth increment when collecting a *z*-series, to obtain the sharpest possible images. However, readjusting the collar at different depths or after changing specimens may introduce an uncontrolled variable and may, in some circumstances, represent poor experimental practice. The advantages and disadvantages of the two techniques should be considered carefully as a part of the experimental design.

The use of a saline-immersible lens avoids the problems associated with coverglass-induced aberrations altogether. These lenses were designed for live cell work and are intended to be used on upright microscopes dipped directly into the biological buffer solution. They are typically designed with a fairly low NA (e.g. 0.4–0.7) and have long working distances, which is a great advantage when working with live specimens. However, most live cell microscopy is conducted on inverted microscopes and until recently it was not possible to use saline-immersible lenses on these microscopes. There has been a tendency for microscopists to use dry lenses when long working distances are required for live cell experiments. However, dry lenses have been designed presuming that the specimen is in contact with the coverglass, or is immersed in a mounting medium with a RI the same as glass (Inoué, 1995). The development of the 'bottomless bath' (Cody & Williams, 1997; Minamikawa *et al.*, 1997) allows the microscopist to take advantage of the benefits of the saline-immersible lenses whilst using inverted microscopes.

If the primary research aim is to section optically deep within a living specimen, then the intermediate-NA saline-immersible lens may be of great benefit. In addition to the extraordinary working distance of saline-immersible lenses (e.g. the Olympus ×40 WI NA 0.7 has a working distance of 3.3 mm) and the removal of aberrations associated with a coverglass, the intermediate-numerical aperture (e.g. NA 0.7) saline-immersible lenses may also bring some optical advantages with confocal microscopy. In order to determine the best approach to section optically deep within a whole rat skeletal muscle, we collected *z*-series images with a variety of lenses. Cross-sections (*z*-sections) of these data sets were computed and used to compare the various lens types with regard to the 'maximum section depth' (MSD, the maximum depth into the tissue that effective optical sections may be obtained). With the dry objective lens it was only possible to image the outermost fibres of the intact muscle (Cody *et al.*, 1993). The *z*-section created from serial optical sections, collected with the ×10 dry objective (Fig. 27.10(A)), shows poor optical sectioning and is typical of low NA (0.3), dry objectives. The poor depth discrimination (*z*-resolution) and the rapid attenuation of the fluorescent signal with depth into the tissue is the result of the low numerical aperture of the lens and refractive index mismatch.

The novel design of the 'bottomless bath', enabled the use of the ×40, 0.7 NA (Olympus, WPlanFL ×40 UV) saline-immersible lens on the inverted microscope, and produced a dramatic improvement in the MSD and image quality (Fig. 27.10(B)). Using the ×40 saline-immersible lens, fibres from the second and third layer from the periphery of the muscle were clearly detectable. The individual fibres at the periphery of the muscle (Fig. 27.10(B)) appear circular in cross-section and relatively aberration free. However, fibres deeper into the muscle in the central part of the image seem to be elongated in the axial direction, possibly caused by a slight mismatch between the refractive index of water and that of the muscle. Despite these aberrations, the images produced with the 'bottomless bath' and the ×40, 0.7 NA (Olympus WPlanFL ×40 UV) saline-immersible lens (Fig. 27.10(B)), are clearly superior in resolution and aberration correction when compared with images obtained with the 'dry' objective (Fig. 27.10(A)), and produced the greatest MSD of the four lens types used.

The ×60, 1.2 NA (Nikon, CF N Plan Apochromat) distilled water immersion lens produced images with the best *z*-resolution (Fig. 27.10(C)) when the microscope was focused close to the coverglass. As the objective was moved closer to the specimen, and the plane of focus was moved further into the tissue, the muscle fibres appeared to become elongated, producing 'tear drop' shaped cross-sections (Fig. 27.10(C)). While fibres from the second layer, and possibly the third layer, were imaged, the intensity of the fluorescent signal attenuated rapidly with depth into the tissue, suggesting that the aberrations may have prevented most of the fluorescent signal from passing through the confocal aperture. It was not possible to image fibres deeper into the intact muscle using the distilled water immersion lens owing to aberrations possibly caused by a slight mismatch between the refractive index of the specimen and that of distilled water (Inoué, 1995; Inoué & Spring, 1997). The relatively high NA of the Nikon, ×60, 1.2 NA, CF N Plan Apochromat distilled water immersion lens would produce very thin confocal optical sections, and it is perhaps for this reason that the high NA lens may be more sensitive to refractive index mismatch than the lower NA saline-immersible lens.

Distortions to the muscle fibre cross-section as a result of aberrations are evident in the *z*-section through the muscle imaged with an oil immersion lens (Fig. 27.10(D)), and are most likely due to refractive index mismatch between the immersion oil and the aqueous bathing medium and specimen. The relatively small working distance of oil immersion lenses is further reduced by optical aberrations at the coverglass–water interface, limiting their usefulness for physiological investigations.

Long working distance, high numerical aperture, saline-immersible lenses are of great benefit for optically sectioning live tissue using confocal microscopy, especially during physiological experiments. There is no requirement to make accurate optical corrections for the aberrations induced by the coverglass,

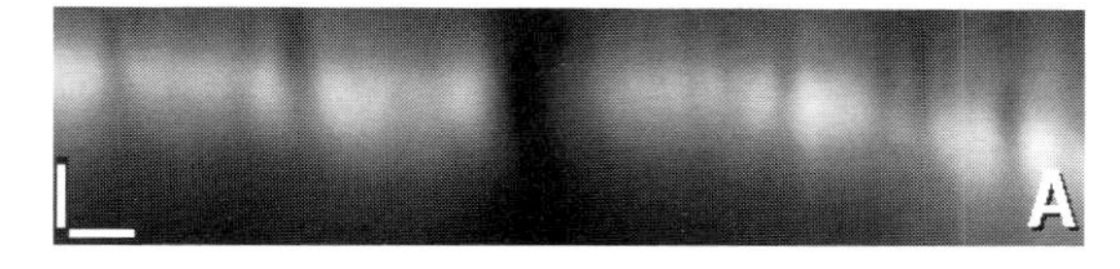

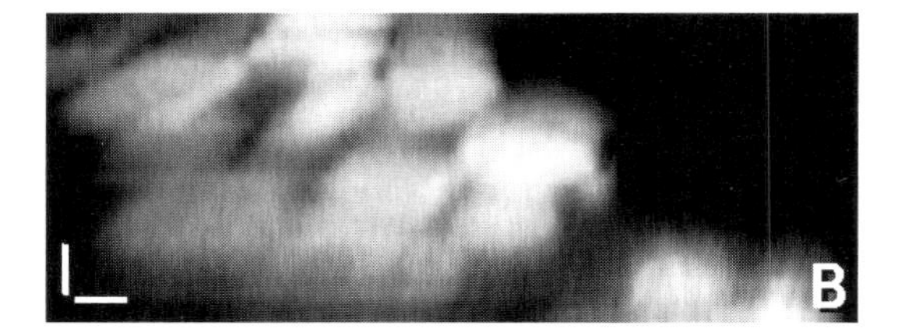

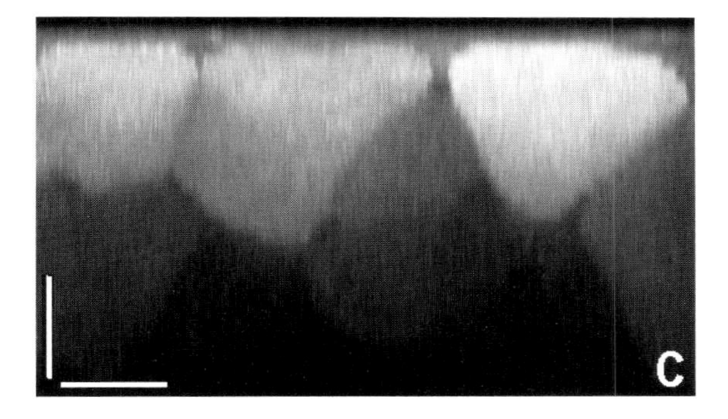

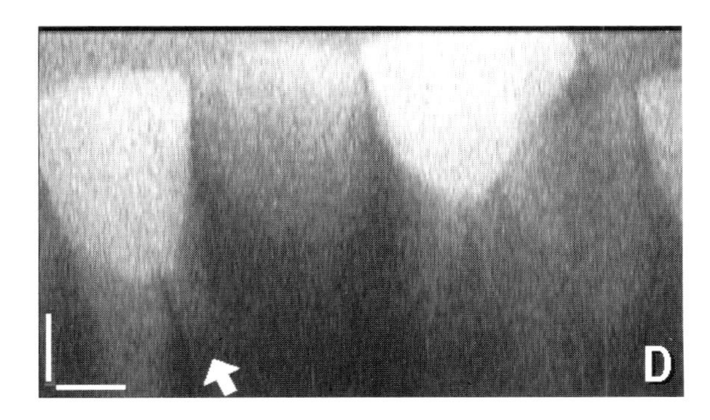

Figure 27.10 Computer-generated isotropic cross sections (z-sections) of z-series through whole, living rat skeletal muscles, mounted in physiological saline and loaded with SNARF-1.
(A) Imaged with an Olympus ×10, 0.3 NA, 'dry' objective lens, individual fibres on the outside of the muscle may be seen. However, resolution is poor and the intensity of the image drops quickly with depth, so that the entire depth of a single fibre cannot be imaged.
(B) Imaged with the aid of a saline-immersible lens (Olympus, WPlanFL ×40 UV). Although the image is not of the highest resolution, satisfactory images (for the experimental purpose) were obtained to a comparatively great depth. The individual fibres appear circular in cross-section, indicating that deleterious aberrations have been minimized.
(C) The distilled water immersion lens (Nikon ×60, 1.2 NA CF N Plan Apochromat) produced the sharpest images of the muscle preparation close to the coverglass. However, as the plane of focus was moved further into the muscle, the fluorescent signal diminished rapidly. The non-circular nature of the fibres indicates deleterious aberrations, most probably induced by the muscle itself.
(D) The oil immersion lens (Nikon ×40, Fluor, 1.3 NA) produced aberrations in the image of the muscle (arrow). The distortions are most likely due to RI mismatch. The refraction caused by the RI mismatch also causes a drastic reduction of the working distance of this lens. All scale bars represent 25 μm. (Parts A and B reproduced with permission from Cody & Williams, 1997.)

a necessity with distilled water immersion lenses. However, probably the greatest benefit, especially for physiological experiments, is that removal of the coverglass means that lens movement is no longer restricted, and the full working distance of the lens may be utilized. The novel design of the 'bottomless bath', used in conjunction with saline-immersible lenses on inverted microscopes, is highly suited to physiological experiments with live specimens, especially for optically sectioning thick specimens. It must be emphasized that although the saline-immersible lens produced a greater depth of optical sectioning in the muscle preparation, the image it produced was not as sharp as that produced by the distilled water immersion lens. In most experimental situations with living specimens it would be expected that a high numerical aperture distilled water immersion lens would produce the clearest images. Perhaps there is a market for a multi-immersion lens with a collar to adjust for coverslip thickness, and a second collar to adjust a diaphragm at the entrance pupil, so that the desired NA of the lens may be chosen as well. For particularly difficult specimens, such as whole skeletal muscle, where the preparation itself induces aberrations, a compromise between image resolution and MSD may be made by setting the lens to a lower NA, thereby increasing the 'tolerance' of the confocal pinhole.

ACKNOWLEDGEMENTS

We would like to thank Dr Guy Cox, Guy Cox Software, Australia, for writing and donating the '*z*-section' software utility; Dr Chris Pudney, Department of Pharmacology, University of Western Australia, for advice and assistance with the linear interpolation calculations; Trevor Parker, FSE Pty. Ltd, Australia, and Nikon, Japan, for the loan of the Nikon water immersion objective; and Roger Wallis, formerly of Faulding Imaging, Australia, for the loan of the Olympus water immersion objective. Funded by the National Health and Medical Research Council of Australia.

REFERENCES

Brenner M. (1994) *Am. Lab.* **26**, 14–19.

Cody S.H. & Williams D.A. (1997) *J. Microsc.* **185**, 94–97.

Cody S.H., Dubbin P.N., Beischer A., Duncan N.D., Hill J., Kaye A. & Williams D.A. (1993) *Micron* **24**, 573–580.

Dubbin P.N., Cody S.H. & Williams D.A. (1993) *Micron* **24**, 581–586.

Fernandes L.B., Henry P.J., Spalding L.J., Cody S.H., Pudney C.J. & Goldie R.G. (1998) *J. Cardiovasc. Pharmacol.* **31**, S222–S224.

Goldie R.G., Rigby P.J., Pudney C.J. & Cody S.H. (1997) *Pulm. Pharmacol. Ther.* **10**, 175–188.

Harris P.J. (1991) *Today's Life Sci.* **13**, 30–38.

Haugland R.P. (1996) *Handbook of Fluorescent Probes and Research Chemicals*, 6th edn. Molecular Probes catalogue.

Hell S., Reiner C., Cremer C. & Stelzer E.H.K. (1993) *J. Microsc.* **169**, 391–405.

Inoué S. (1995) In *Handbook of Biological Confocal Microscopy*, 2nd edn, J.B. Pawley (ed). Plenum Press, New York, pp. 1–17.

Inoué S. & Spring K.R. (1997) *Video Microscopy. The Fundamentals*, 2nd edn. Plenum Press, New York.

Keller H.E. (1995) In *Handbook of Biological Confocal Microscopy*, 2nd edn, J.B. Pawley (ed). Plenum Press, New York, pp. 111–126.

Minamikawa T., Cody S.H. & Williams D.A. (1997) *Am. J. Physiol.* **272**, H236–H243.

CHAPTER TWENTY-EIGHT

Redox Confocal Imaging: Intrinsic Fluorescent Probes of Cellular Metabolism

BARRY R. MASTERS[1] & BRITTON CHANCE[2]

[1] Department of Anatomy and Cell Biology, Uniformed Services University of the Health Sciences, Bethesda, MD, USA

[2] Department of Biochemistry and Biophysics, University of Pennsylvania, Philadelphia, PA, USA

28.1 INTRODUCTION

28.1.1 What is redox fluorometry?

This chapter describes redox imaging of cells, tissues and organs based on intrinsic fluorescent probes of cellular metabolism. Cellular metabolism may be non-invasively interrogated through the 'optical method' based on the fluorescence intensity of intrinsic probe molecules (Chance, 1991). The intrinsic fluorescent probes which report on cellular metabolism are the reduced pyridine nucleotides, NAD(P)H, and the oxidized flavoproteins.

The basis of redox fluorometry is that the quantum yield of the fluorescence, and hence the intensity, is higher for the reduced form of NAD(P) and lower for the oxidized form (Fig. 28.1). The reverse is true for the flavoproteins; the quantum yield, and hence the intensity, is higher for the oxidized form and lower for the reduced form. Measurement of the fluorescence intensity of either NAD(P)H or flavoproteins, or their ratio, is an optical indicator of cellular metabolism.

The main advantage of redox fluorometry is that it is a non-invasive optical method with high spatial and temporal resolution. The temporal resolution is of the order of milliseconds. The spatial resolution is given by the spot size of the excitation light beam, and may be a diffraction-limited spot of a few micrometres in diameter.

Redox fluorometry has numerous applications in cell biology, tissue physiology and organ physiology. Applications include such diverse samples as single cells, muscle tissue and organs such as the eye and the brain. The method is not limited to surface fluorometry of cells and organs; the widespread use of fibre optics in minimally invasive surgery and diagnostics, permits the use of redox fluorometry inside the body. An example of this approach is the use of redox fluorometry to monitor the metabolism of myocardium during open-heart surgery. In this experimental approach, the laser excitation light is transmitted to the *in vivo* myocardium through a fibre optic, and the NAD(P)H fluorescence is transmitted to the redox fluorometer through another fibre optic.

In addition to the use of redox fluorometry to monitor cellular metabolism, it can also be utilized as an optical probe of oxygen concentration in cells,

FLUORESCENT AND LUMINESCENT PROBES, 2ND EDN
ISBN 0–12–447836–0

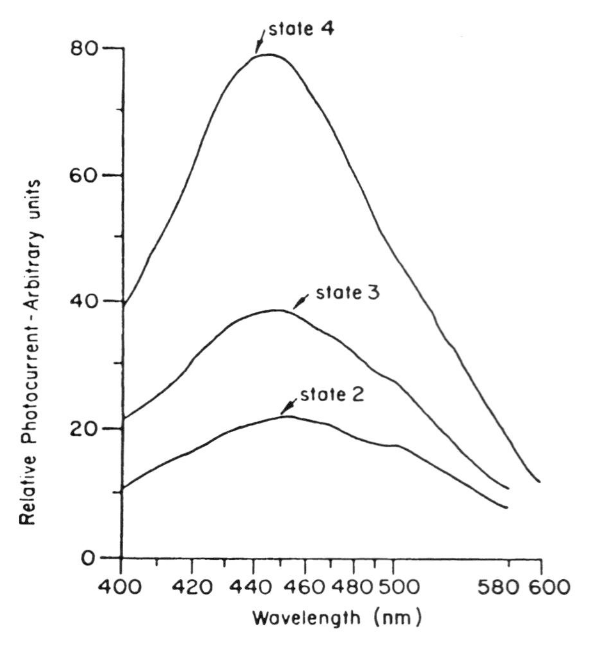

Figure 28.1 Fluorescence spectra of rat liver in various redox states. The excitation is at 366 nm. The upper curve is characteristic of the hypoxic state. The lower curve is characteristic of normoxic tissue.

tissues and organs. The intensity of NAD(P)H fluorescence varies with oxygen concentration; this is the basis of non-invasive monitoring of corneal mitochondrial oxygen utilization (Masters, 1988).

Redox fluorometry must be used with proper calibration and an understanding of its technical limitations. As a technique, it has the following limitations: (1) it lacks an absolute scale which relates the fluorescence intensity of the intrinsic probes to a unique index of cellular metabolism; (2) there are contributions of fluorescence from both the mitochondrial space and the cytoplasmic space (in the case of NAD(P)H fluorescence) to the total fluorescence intensity (Sies, 1982); (3) there are several factors which can affect the fluorescence intensity of the NAD(P)H signal; and (4) redox fluorometry based on intensity measurements is dependent on the geometry of the sample and the optical instrument (Rost, 1991).

Recent developments in fluorescent redox imaging based on fluorescent lifetimes can eliminate this geometrical dependence (Lakowicz, 1992). For specific tissues, e.g. the cornea, the NAD(P)H fluorescence intensity from the mitochondrial space dominates the total NAD(P)H fluorescence intensity. When used with proper controls, redox fluorometry provides a simple, non-invasive, optical probe of cellular metabolism.

28.1.2 Chapter scope

This chapter presents a critical, self-contained evaluation of redox confocal imaging. The coverage includes historical developments, biochemical and photophysical basis of the methodology, instrumentation, applications to various tissues and organs, and a comparison with other non-invasive techniques. The subjects of chemiluminescence and bioluminescence, whilst within the subject domain of this chapter, are discussed elsewhere in this book.

The authors have decided to present a critique of the methodology which puts the advantages and disadvantages of the technique in perspective. It is hoped that this approach will enable the investigator to use redox fluorometry in a manner that maximizes its utility and minimizes artifacts. The subjects not covered in depth are cited in the references. The coverage of applications will concentrate on ocular tissue and the emphasis is on two-dimensional confocal redox imaging. Previous reviews have provided extensive coverage of redox applications with other cells, tissues and organs (Masters, 1984a; Chance, 1991).

In summary, this chapter discusses new developments in the methodology. Confocal microscopy provides high-resolution optical sectioning yielding two-dimensional images and the concomitant increase in spatial resolution. The application of two-photon confocal fluorescence microscopy to the redox imaging of the eye is a dramatic new development in two-dimensional redox imaging. *In vivo*, real-time confocal microscopy combined with redox fluorometry provides the basis of functional imaging; *in vivo* confocal microscopy provides two-dimensional images of cell and tissue morphology, and redox fluorometry provides the metabolic imaging (Masters, 1993b). The reader of this chapter will find a self-contained review of the theory, methodology and instrumentation.

28.1.3 Previous reviews

An excellent, recent book on spectrofluorometry of cells and tissues is edited by Kohen and Hirschberg. (1989). This volume contains comprehensive coverage of the following topics: history of fluorescence microscopy, image spectroscopy of living cells, fluorescence microscopy in three dimensions, frequency-domain fluorescence spectroscopy and fluorescence probes. In addition, cytometry and cell sorting as well as bioluminescence are covered.

Another recent review is the chapter by Balaban and Mandel (1990) on optical methods for metabolic studies of living cells. This review covers the authors' work and details the problem of redox measurements of the surface of a beating heart. They provide a technical solution which consists of an internal fluorescent standard in addition to the NAD(P)H measurements (Balaban & Mandell, 1990).

Reviews which are more specific to studies of ocular tissue include the following: Masters *et al.* (1981, 1982a) and Masters (1984a). The latter review is a comprehensive study of instrumentation and biochemical studies on which redox fluorometry is based. Much of the instrumentation and techniques that are of use for studies of cells in tissue culture are discussed. The method of freeze-trapping tissues and organs, and mechanical sectioning prior to two-dimensional redox imaging is described. Applications to *in vitro* and *in vivo* organs and tissues are covered in a previous major review of redox fluorometry (Masters, 1984a).

28.2 HISTORY OF THE USE OF INTRINSIC PROBES TO MONITOR CELLULAR METABOLISM

While it is widely acknowledged that the development of the optical microscope is intimately linked with the development of the science of pathology, the relationship of the spectroscope to medical science is uncommon knowledge. In fact, Leeuwenhoek was able to demonstrate the existence of capillary circulation, first described red blood cells, observed bacteria in dental plaque, and described lenticular fibres in the ocular lens as well as the striations of skeletal muscle fibres. All of these observations were made 350 years ago with his single lens microscope.

The history of intracellular respiration is described in its historical context in a fascinating work by Keilin (1966). The key people in this theme include Harvey, Malpighi, Leeuwenhoek, Hooke, Lower, Mayov, Priestly, Lavoisier, Spallanzani, Pflüger, Ludwig, Berthelot, Liebig, Pasteur, Bernard, Hoppe-Seyler, Berzelius, Traube, Cagniard-Latour and Appert (Keilin, 1966).

In 1880 an important book with the title *The Spectroscope in Medicine* was published in London (MacMunn, 1880) and in 1914, MacMunn published another book, *Spectrum Analysis Applied to Biology and Medicine.* Work on the colour of cells and tissues, and their absorption and fluorescence properties begins with the studies of haemoproteins by MacMunn, continues with the classic work of Keilin on cytochromes, and continues in this century with the modern advances in the development of redox fluorometry.

By the mid-1800s the following three important empirical observations had been made: (1) the fluorescence emission is always at a longer wavelength than the exciting light (known today as Stokes's law); (2) a body must first absorb the light before it can emit fluorescence; and (3) the fluorescent spectrum, which is the intensity of light emitted as a function of wavelength, can be used to characterize specific substances. It is the last point which is the basis of spectral characterization of specific molecules. In 1934 Haitinger coined the term 'fluorochrome' to describe the use of fluorescence dyes which results in fluorescent staining of tissues.

David Keilin had great success with a low-dispersion prism spectroscope fitted to a microscope which was manufactured by Carl Zeiss. This instrument was used by Keilin to study slices of plant and animal tissues as well as suspensions of bacteria and yeast. Keilin was able to observe the oxidation and reduction of cytochrome within living tissues and cells based on changes in the absorption spectrum of visible light. This work then led to the next generation of tissue spectrometers developed in Sweden (Caspersson, 1950, 1954) and in the United States (Chance & Thorell, 1959; Chance, 1951).

The advantage of microspectrographic methods is that cellular and intracellular organelles can be studied in the living state, in real-time with a non-invasive optical technique. This means that cellular organelles do not have to be removed from their natural surroundings or destroyed in the process of studying their metabolism. Prior to 1945, the early work tended to investigate the nucleus and chromosome structure in normal and tumour cells. After the Second World War research shifted towards optical studies of oxidative metabolism in cells and organelles. The studies of intact cells (yeast and bacteria) accelerated, and optical studies of mitochondria became a more active research area.

A historical survey of the origins of modern fluorescence microscopy and fluorescence is useful reading for every fluorescence microscopist (Kasten, 1989). The following listing of historical events is taken from the work of Kasten. Although the term fluorescence was coined by George Stokes in 1852, Kasten credits David Brewster with the first description of the phenomenon of fluorescence in 1838. August Köhler invented the first ultraviolet absorption microscope in Jena in 1904. In 1910, H. Lehmann of Carl Zeiss, Jena used a modification of the liquid filters developed by Robert Wood, a professor of experimental physics, at Johns Hopkins University, and in the next two years developed the first fluorescence microscope. The reader may find it of interest that the light sources were typically Siemens arc lamps with a power of 2000–3000 W!

Han Stübel made the first microscopic observations of autofluorescence in 1911. He investigated the autofluorescence of animal organs, bacteria and protozoa. Further development rapidly followed. In 1929,

Ellinger, a pharmacologist at Heidelberg University, developed the intravital fluorescence microscope, which Carl Zeiss, Jena then manufactured. This new instrument utilized vertical or epi-illumination, a water-immersion objective, and a series of filters for excitation wavelength selection which were mounted in a rotating wheel. A steady gravity flow of physiological solutions superfused the living organs for the duration of their microscopic observation.

A microscope which could photographically record the fluorescence spectrum induced on tissues, cells or fluid samples was developed by Borst and Köningsdörffer in 1929. This important technical development was used in numerous studies of porphyrin autofluorescence in patients with porphyhria. This important milestone antedated the work of van Euler on riboflavin fluorescence, as well as Caspersson's work on the ultraviolet absorption of nucleic acids and proteins. In 1959, rapid kinetic studies in mitochondria of living cells were performed with a differential microspectrofluorometer (Chance & Thorell, 1959).

Historical and personal perspectives on the development of microspectroscopy and the optical method contain important landmark developments in the field of non-invasive studies of cell and tissue metabolism (Chance, 1989, 1991). The development of the dual-wavelength spectrophotometer and its influence on cell and tissue optical studies is presented here by Dr Chance who was present during this development.

The origins of intracellular microspectrofluorometry started with the work of Warburg who discovered NAD(P)H fluorescence. Dr Bo Thorell and Britton Chance, and Baltscheffsky then discovered that the cell autofluorescence was due to NAD(P)H fluorescence from the mitochondrial matrix space. Together they were able to measure the contribution of the NADH/ NADPH in the cytosol to the mitochondrial NAD(P)H signal. These measurements were of fundamental importance in the development of the use of mitochondrial fluorescence as a sensitive indicator of mitochondrial hypoxia in tissues. This study succeeded in measuring the fluorescence from a single mitochondrion within a single cell (Chance & Thorell, 1959).

28.3 THE BIOCHEMICAL BASIS OF INTRINSIC FLUORESCENT PROBES IN LIVING CELLS

There are two intrinsic fluorescent probes which can serve as non-invasive intrinsic probes of cell metabolism: the oxidized flavoproteins and the reduced pyridine nucleotides.

The fluorescence intensity from intrinsic oxidized flavoproteins which are present in mitochondria is a non-invasive measure of cellular metabolic function. The main advantage of measuring the fluorescence intensity from the oxidized flavoproteins is that the fluorescence is localized in the mitochondrial space. Redox imaging based on NAD(P)H fluorescence originates in both the mitochondrial and the cytoplasmic spaces (Chance & Baltscheffsky, 1958; Chance *et al.*, 1978, 1979). Every tissue is different with respect to the ratio between NAD(P)H fluorescence contributions from the mitochondrial space and from the cytoplasmic space (Chance & Williams, 1955).

The fluorescence intensity from oxidized flavoproteins in cells, e.g. the cornea epithelium, occurs in the region from 520 nm to 590 nm with a broad maximum at 540 nm (Chance *et al.*, 1968). The light absorption of oxidized flavoproteins has a broad maximum at 460 nm and extends from 430 to 500 nm (Fig. 28.2). The fluorescence intensity of the oxidized flavoproteins in cells is usually much lower than the fluorescence intensity from the reduced pyridine nucleotides.

Fluorescence from the intrinsic oxidized flavoproteins in the cornea has been spectroscopically characterized in rabbit corneas frozen to 77K (Chance & Lieberman, 1978). Other evidence that the corneal epithelial cell fluorescence in both *in vitro* and *in vivo* corneas is due to oxidized flavoproteins comes from studies by Masters (1984a,b). The distribution

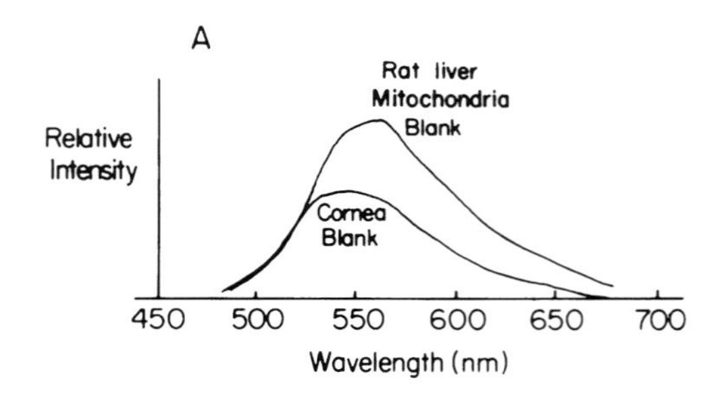

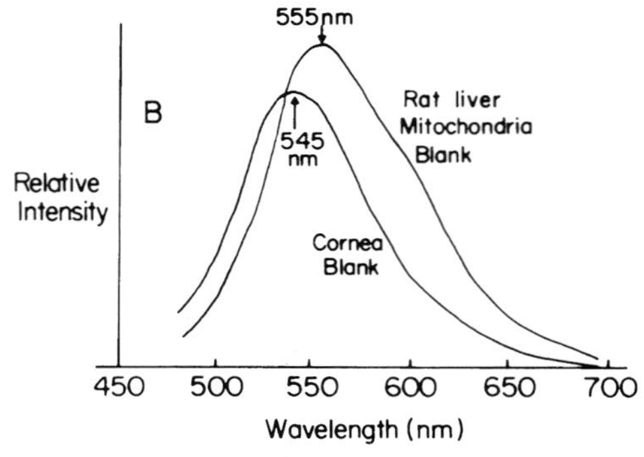

Figure 28.2 Fluorescence spectra of flavoproteins from cornea and rat liver mitochondria. The excitation is at 460 nm.

of mitochondria, stained with the cationic dye rhodamine 123, in the basal epithelial cells of the rabbit cornea has been studied with a confocal laser scanning fluorescence microscope (Masters, 1993b).

The alteration of the fluorescence intensity of the oxidized flavoproteins in the *in vivo* rabbit corneal epithelial cells as a function of cellular hypoxia has been demonstrated (Masters *et al.*, 1982a). A HeCd laser at 442 nm was used to excite the corneal epithelial cells in a living rabbit and the flavoprotein fluorescence intensity was measured in the wavelength region of 550 nm (Masters *et al.*, 1982b). The fluorescence intensity was reduced in the presence of a flow of hydrated nitrogen (tissue hypoxia) and the effect was reversed in the presence of hydrated air. These studies are in agreement with *in vitro* studies of flavoprotein fluorescence from corneal epithelial cells conducted in both *in vitro* rabbit perfused corneas at 37°C, and in freeze-trapped rabbit cornea at 77K.

The fluorescence from the naturally occurring reduced pyridine nucleotides in cells is an indicator of cellular respiration. The fluorescence of reduced pyridine nucleotides is excited with light of 364 nm and has a fluorescence emission in the range of 400–500 nm. Cellular hypoxia is associated with an increased ratio between reduced and oxidized pyridine nucleotides, and therefore increased fluorescence intensity in the region 400–500 nm. Since the quantum efficiency, and thus the fluorescence intensity, of the reduced pyridine nucleotides is significantly greater than that of the oxidized pyridine nucleotides, the fluorescence intensity monitors the degree of cellular hypoxia. This non-invasive technique is called redox fluorometry. We have demonstrated that the 400–500 nm fluorescence excited at 364 nm is due to the reduced pyridine nucleotides (Masters, 1984a). While the fluorescence intensity of the cornea has been investigated using optically sectioning microscopes to monitor the degree of cellular hypoxia, it was not previously possible to obtain single cell images of the reduced pyridine nucleotide fluorescence. However, two-dimensional redox imaging has been demonstrated by other investigators using isolated cardiac myocytes in tissue culture (Eng *et al.*, 1989).

The fluorescence from NAD(P)H is an intrinsic probe which can be used to study cellular metabolism (Chance & Thorell, 1959). The fluorescence intensity from these intrinsic probes provides a non-invasive optical method to monitor cellular respiration. The fluorescence from NAD(P)H has been used to study cellular metabolism in many tissues and organs due to the strong fluorescence intensity. The NAD(P)H fluorescence intensity occurs in two compartments, the mitochondrial and the cytosolic; this complicates the interpretation of the fluorescence studies. However, in some tissues, e.g. rat cardiac myocytes, the NAD(P)H fluorescence is predominantly from the mitochondrial space.

Two-dimensional images of the fluorescence intensity from NAD(P)H have been studied in brain slices and in isolated perfused hearts (Chance, 1991). Kohen *et al.* (1989) have developed instruments for rapid redox mapping of cells in culture. At the cellular level NAD(P)H imaging of isolated rat cardiac myocytes have been studied with a standard fluorescence microscope (Balaban & Mandel, 1990). These authors demonstrated that the fluorescence images are mainly due to mitochondrial NAD(P)H fluorescence (Eng *et al.*, 1989). Two-dimensional imaging of the NAD(P)H fluorescence intensity of *in vitro* corneal endothelial cells has been studied with an ultraviolet confocal laser scanning fluorescence microscope (Masters, 1993b).

28.4 INSTRUMENTATION FOR THE USE OF LOW-LIGHT-LEVEL FLUORESCENT IMAGING OF LIVING CELLS AND TISSUES

Masters (1984a) reviewed various instrumental setups for redox imaging of organs and tissues. Major advances in instrumentation include confocal microscopy and two-photon confocal microscopy for redox imaging.

The 'gold standard' for two-dimensional redox imaging of freeze trapped specimens is the automated, milling, redox ratio-scanning instrument (Quistorff *et al.*, 1985). This instrument is only used with frozen specimens; however, its spatial resolution and quantitative fluorescence as well as a histogram output of redox state homogeneity is of great value (Figs. 28.3 and 28.4).

The confocal microscope is an optical device which can be used to observe a single focal plane of thick objects with high resolution and contrast as compared to standard microscopes (Wilson, 1990; Masters & Kino, 1990). The confocal microscope used in the fluorescent mode has the excitation at one wavelength, and the fluorescent image is formed at a longer wavelength. This differs from the reflected light mode in which the confocal image is formed at the same wavelength as that of the laser illumination. The depth resolution of fluorescence mode confocal scanning optical microscopes is reduced as compared to that in the reflected imaging mode. The advantages of UV confocal microscopy include increased resolution and a reduced depth of focus as compared to visible light confocal microscopy. These advantages depend on a microscope objective which is corrected for the UV.

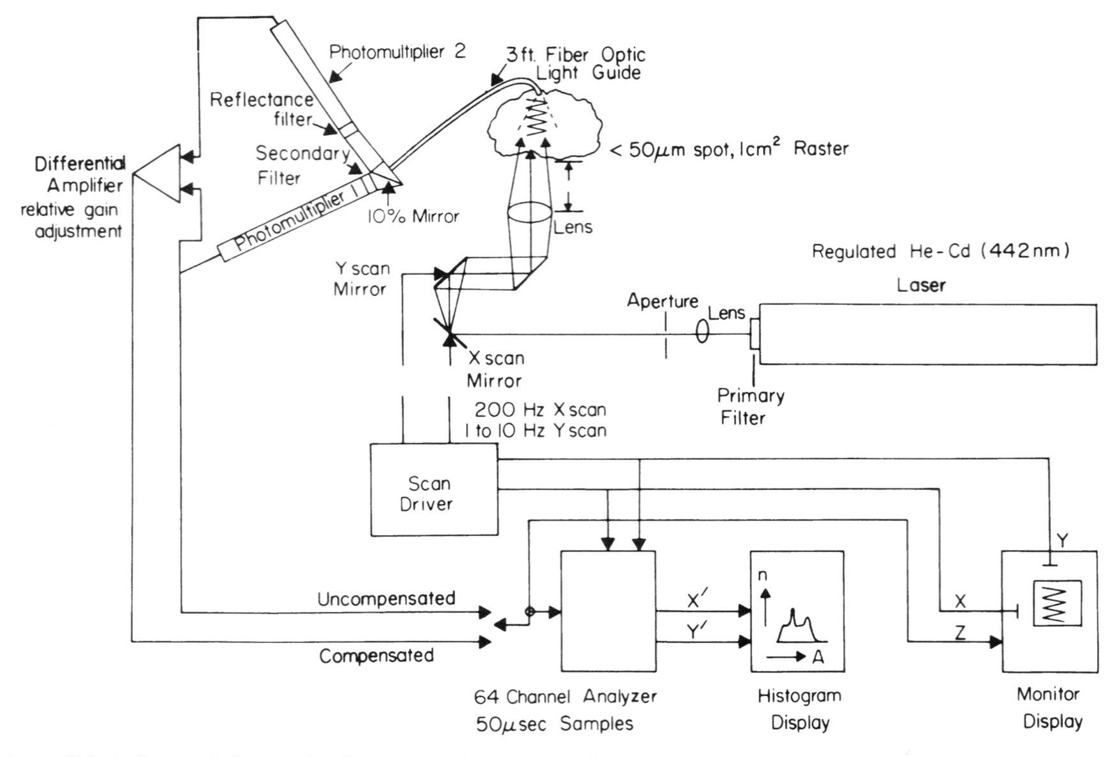

Figure 28.3 A fibre optic laser redox fluorometer for *in vivo* work.

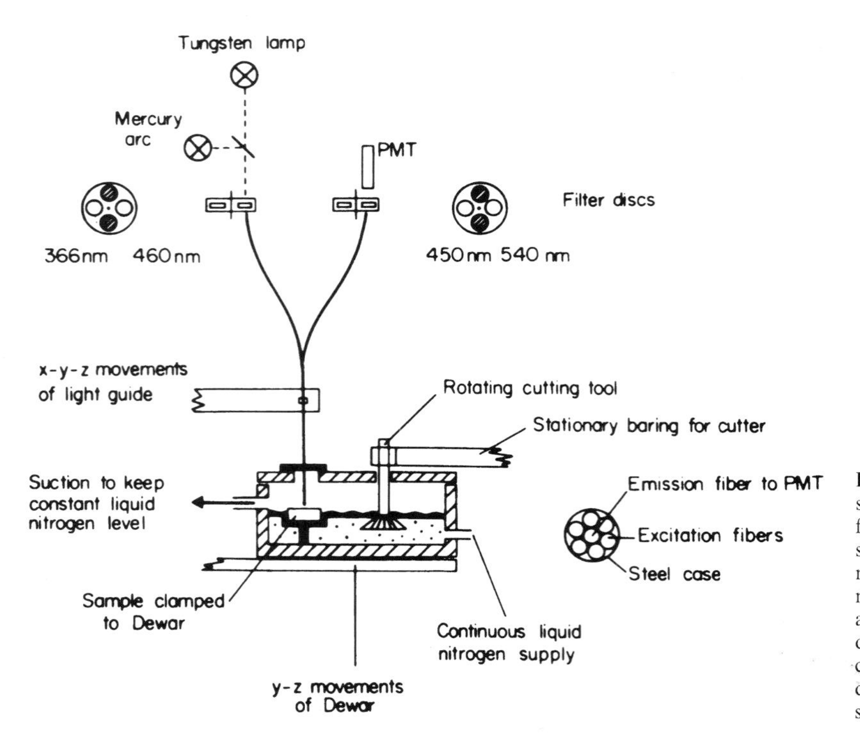

Figure 28.4 A low-temperature scanning, milling redox fluorometer. The freeze-trapped specimen is scanned at liquid nitrogen temperature, then the next layer is milled off the tissue and a new scan is made. The two-dimensional redox scans can be combined to form a three-dimensional image of tissue redox states.

The use of the confocal microscope to section the cornea optically has been demonstrated by Lemp *et al.* (1986). The fine structure of the *in vitro* cornea has been shown with both the one-sided Nipkow disk confocal microscope and with the laser scanning confocal microscope (Masters & Paddock, 1990a). Three-dimensional volume reconstruction from serial confocal optical sections of the *in vitro* cornea has been demonstrated (Masters & Paddock, 1990b).

The laser scanning confocal microscope (Zeiss, UV confocal LSM) permits two-dimensional confocal imaging of the redox fluorescence intensity of corneal endothelial cells (Kapitza & Wilke, 1988). Thus, a two-dimensional image or map of cellular hypoxia can be obtained. A combination of the reflected light images of cell morphology and the redox fluorescence images of cellular metabolism may be used to construct a multi-modality three dimensional image of cell structure and function.

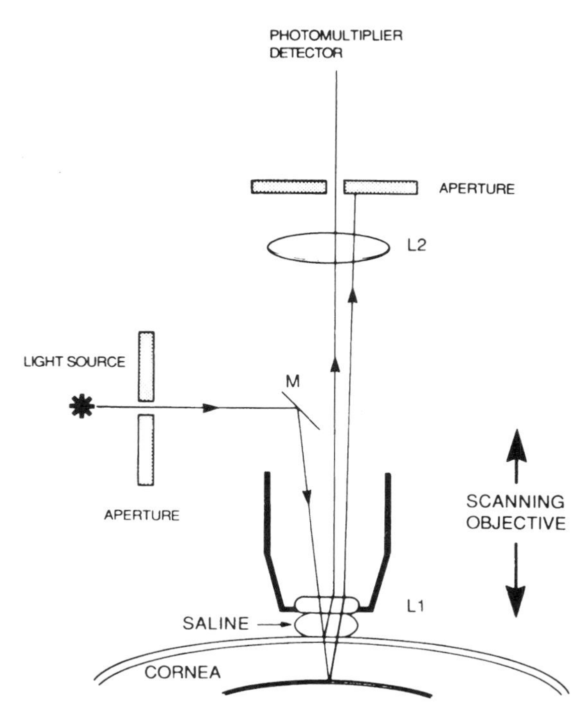

Figure 28.6 The principle of confocal microscopy. Schematic diagram of the z-scan confocal microscope developed for optically sectioning the living eye. The light source is connected to the instrument via a quartz fibre optic light guide. There are two conjugate slits which make the microscope a confocal microscope. One slit is imaged onto the object, and the second slit in front of the detector forms an image of itself on the object. The z-scan confocal microscope has a scanning microscope objective which moves on an axis under computer control via a piezoelectric driver.

28.4.1 One-dimensional confocal redox fluorometer

Figures 28.5 and 28.6 illustrate the mechanical and optical components of a confocal redox fluorometer. The device is a confocal microscope because it contains two slits located in conjugate planes, one for the illumination and one for the image plane. The confocal microscope is used in the vertical mode for work on tissue cultures and for studies of *in vitro* eyes, and in the horizontal mode for *in vivo* studies on animals or human subjects. This confocal redox fluorometer can be used to measure fluorescence from NAD(P)H, from oxidized flavoproteins, or from extrinsic probes, i.e. mitochondrial or nuclear stains, and back-scattered light. The unique feature of the confocal redox fluorometer is that the microscope objective is a scanning objective (Masters, 1988). A piezoelectric driver scans the microscope objective which gives rise to a depth profile across the cell or tissue. The spatial resolution of this system is demonstrated in Fig. 28.7. Figure 28.8 illustrates the *in vivo* use of the confocal redox fluorometer developed by Masters. This example shows the changes in the NAD(P)H fluorescence intensity of the *in vivo* rabbit eye due to contact lens-induced corneal hypoxia.

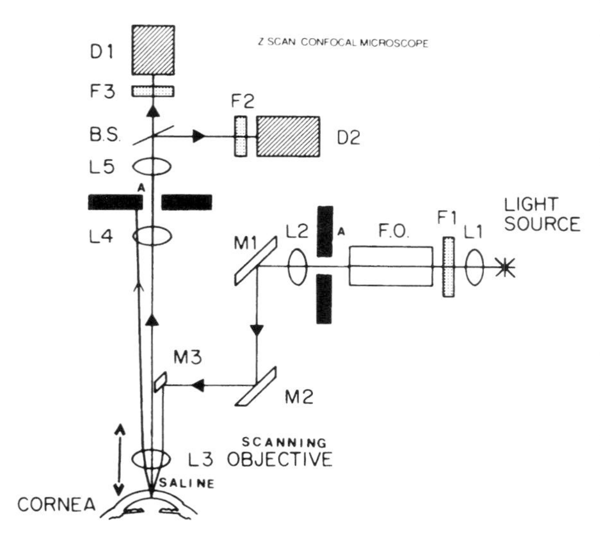

Figure 28.5 Schematic diagram of the scanning one-dimensional confocal microscope showing a light ray path. The light source is either a laser or a mercury arc lamp connected to the microscope by a fibre optic. F1 and F2 are narrow band interference filters to isolate the excitation wavelengths. F3 is a narrow band interference filter to isolate the emission light. M1, M2 and M3 are front surface mirrors, and B.S. is a quartz beam splitter. L3 is the scanning objective 50×, NA 1.00. The piezoelectric driver scans the microscope objective along the optic axis of the eye. This confocal microscope is suitable for use with tissue culture or in the horizontal mode for use with living animal or human subjects.

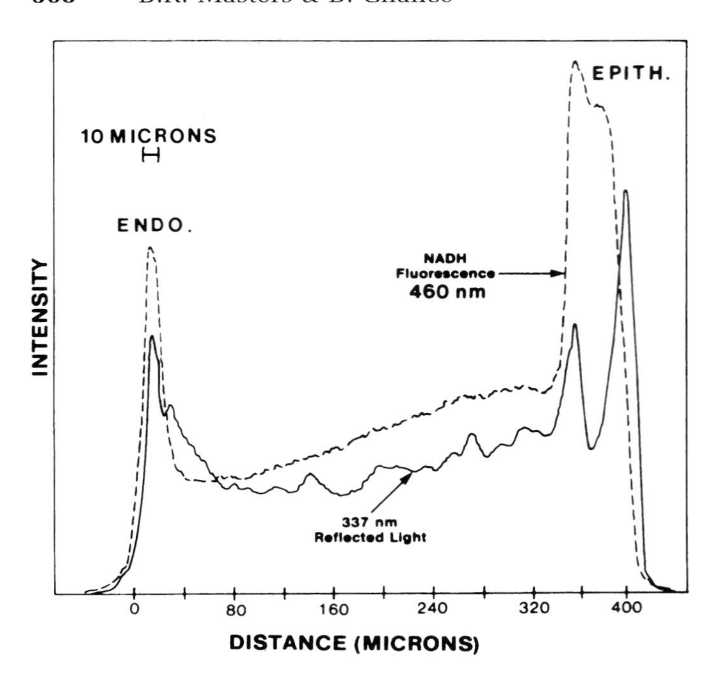

Figure 28.7 One-dimensional *in vivo* redox confocal imaging of the living eye. An optical section through a rabbit cornea illustrating the range resolution for the back-scattered light (solid line) and the NAD(P)H fluorescence emission (broken line). The intensity of the back-scattered light is 10 times that of the fluorescence. The tear film is on the right side of the scan and the aqueous humour is on the left side of the figure.

28.4.2 Confocal redox NAD(P)H imaging

A Zeiss confocal scanning laser microscope (LSM 10 UV, Carl Zeiss, Oberkochen, Germany) has been adapted for UV fluorescence confocal microscopy (Kapitza & Wilke, 1988; Masters, 1993b; Masters *et al.*, 1991). In addition to the argon ion laser (488 and 514 nm) and the HeNe laser (543 nm), another argon laser (364 nm) was added to the microscope. The exact wavelength was 333.6 nm. The UV argon ion laser was a Spectra Physics, Type 2016, with a variable output power of 20–100 mW. In addition to the third laser there were other changes to the microscope; the antireflection coatings on the *x–y* scanner were optimized for UV light.

A Zeiss water-immersion objective of 25×, NA 0.8, corrected for UV, was used to measure the fluorescence from optical sections of a freshly enucleated rabbit eye. The microscope objective was able to focus across the full 400 μm thickness of the *in vitro* rabbit cornea and digital images of 256 × 265 pixels and 256 grey levels were produced. The back-scattered light mode of the microscope at a wavelength of 364 nm was used to locate the endothelial cells which are situated about 400 μm below the anterior surface of the cornea. The microscope was focused about 2 μm into the corneal endothelial cells, and a light confocal image formed (Fig. 28.9). Then the microscope was switched to the fluorescence mode (Fig. 28.10). The excitation wavelength was 364 nm and the emission filters collected light in the region of 400–500 nm. Eight images were averaged to improve the signal-to-noise ratio of the final image. The wavelengths used for the laser excitation (364 nm) and the fluorescent emission (400–500 nm) correspond to the excitation and emission wavelengths of the reduced pyridine nucleotides. Several confocal images of the *in vitro* cornea were made using the reflected light mode with the laser source at 488 nm to demonstrate the optical resolution and contrast of the modified microscope.

28.4.3 Two-photon confocal NAD(P)H imaging

Two-photon laser scanning fluorescence microscopy has been used to image the NAD(P)H fluorescence of the basal epithelium of an *in vitro* rabbit cornea (Denk *et al.*, 1990). The technique has two important advantages over single-photon confocal fluorescence imaging. The point spread function varies as z to the fourth power, where z is the distance from the focal plane. In a standard confocal microscope z varies as the second power. Since the illumination is at 700 nm, photobleaching only occurs in the focal plane; with the standard confocal microscope photobleaching occurs in all the planes scanned by the laser beam. The application of two-photon fluorescence confocal microscopy to redox imaging is a major advance. Its disadvantages are complexity and cost; these may change in the future.

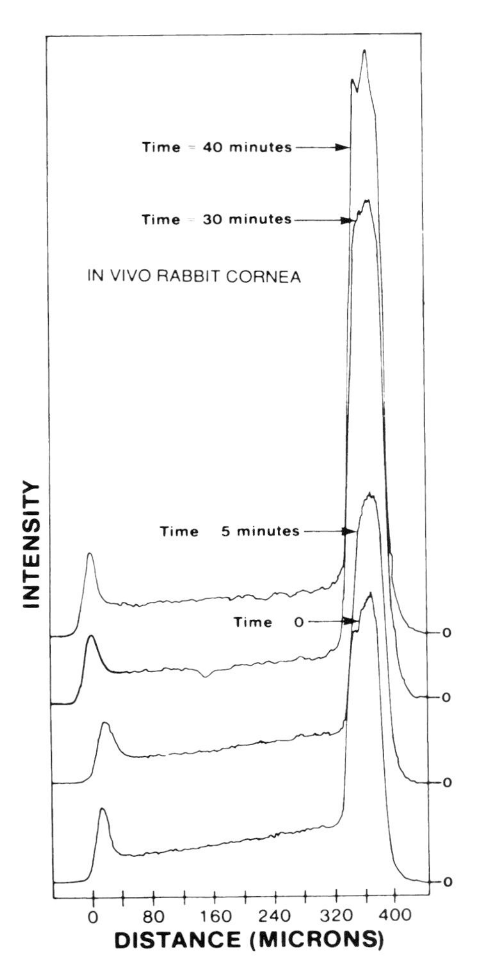

Figure 28.8 *In vivo* redox confocal imaging of the rabbit cornea. Time dependence of the effect of a PMMA contact lens on the NAD(P)H fluorescence intensity of the rabbit cornea. The peaks on the left side are from the corneal endothelium. The larger peaks on the right side are from the corneal epithelium. Time represents the duration that the contact lens was on the eye.

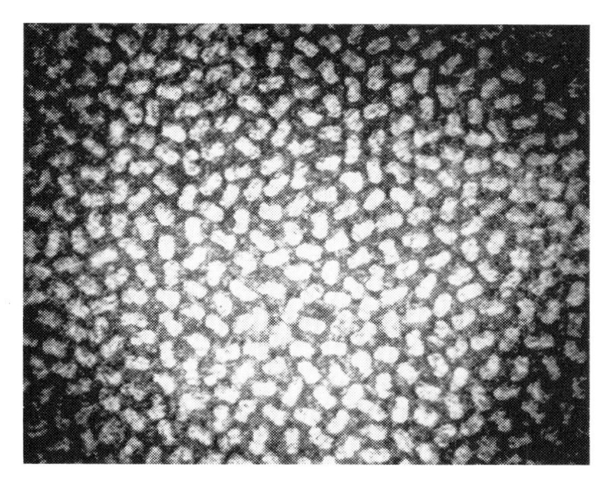

Figure 28.9 A confocal image formed in back-scattered light of the *in vitro* rabbit cornea. The image shows corneal endothelial cells. The bright regions are the cell nuclei. The image was formed with a Zeiss laser scanning confocal microscope.

28.4.4 Confocal redox flavoprotein imaging

The study by Masters and Kino (1990) used a laser scanning confocal microscope (BioRad MRC-600) to image freshly enucleated rabbit eyes. The microscope uses two scanning mirrors moved by scanning galvanometers to scan the laser beam across the microscope objective. The reflected light retraces the incident path and is collected by a set of curved mirrors and sets of flat mirrors. The scan mirrors are located at conjugate aperture planes. An adjustable pinhole is placed in front of the photodetector.

The freshly enucleated eye was transferred to a black plastic chamber containing Ringer's solution and placed on the stage of the confocal microscope. The bicarbonate Ringer's solution completely immersed the eye and the tip of the microscope objective. Every 10 min the Ringer's solution was exchanged for fresh aerated solution.

The BioRad laser scanning microscope used an air-cooled 25 mW argon ion laser to provide the 488 nm wavelength. This wavelength produced excitation in the reflected light mode. The low reflectivity of the cornea necessitated maximum amplification of the signal from the detector. At this high amplification, a bright spot of stray light appeared in the centre of each image. A black disk or square was used partially to mask this reflection and to designate its position. Kalman averaging was used to average ten frames to reduce the noise in the final image.

The fluorescence intensity in the wavelength region longer than 515 nm was used to image the flavoprotein fluorescence. The excitation filter centred at 488 nm with a 10 nm band-pass was used to isolate the 488 nm line from the argon ion laser. A dichroic reflector (BioRad DR 510 LP) reflected the 488 nm laser line onto the microscope objective. A BioRad barrier filter CG 515 (yellow) blocked all light with wavelengths shorter than 515 nm from reaching the detector. The emission spectrum of the corneal epithelial cells extends from 500 nm to 600 nm with a maximum at 545 nm. The absorption spectrum of the flavoproteins is in the range 400–500 nm, with a maximum at 550 nm. The argon ion excitation line at 488 nm, while not optimal, is a suitable laser line for fluorescence excitation of the oxidized flavoproteins.

28.5 APPLICATIONS OF INTRINSIC FLUORESCENT REDOX PROBES TO CELLULAR METABOLISM

To illustrate the numerous applications of the technique to *in vitro* and *in vivo* tissues the authors have selected references which illustrate significant studies of cell function. As stated in the Introduction, the applications are focused on redox fluorometry of the cornea.

28.5.1 Redox imaging based on NAD(P)H fluorescence

28.5.1.1 Confocal NAD(P)H redox imaging

In order to demonstrate the advantages of confocal microscopy in monitoring of corneal morphology we have produced images of the *in vitro* cornea both perpendicular to the corneal surface and in the plane of the cornea using 488 nm laser light in the reflected light confocal mode. Figure 28.11 illustrates the quality of *z*-scan of the cornea produced by back-scattered confocal microscopy. Figure 28.12 illustrates a three-dimensional reconstruction of a stack of serial optical sections of the rabbit cornea.

We have found that the image quality of two-dimensional redox images in dependent on the depth within the tissue. Images at the anterior of the cornea have superior resolution compared with those images made 400 μm into the cornea. The image quality at the posterior surface of the cornea, which is made through 400 μm, is shown in Fig. 28.9. The fluorescence of the reduced pyridine nucleotides is shown in Fig. 28.10. These images represent two-dimensional maps of cellular metabolic function.

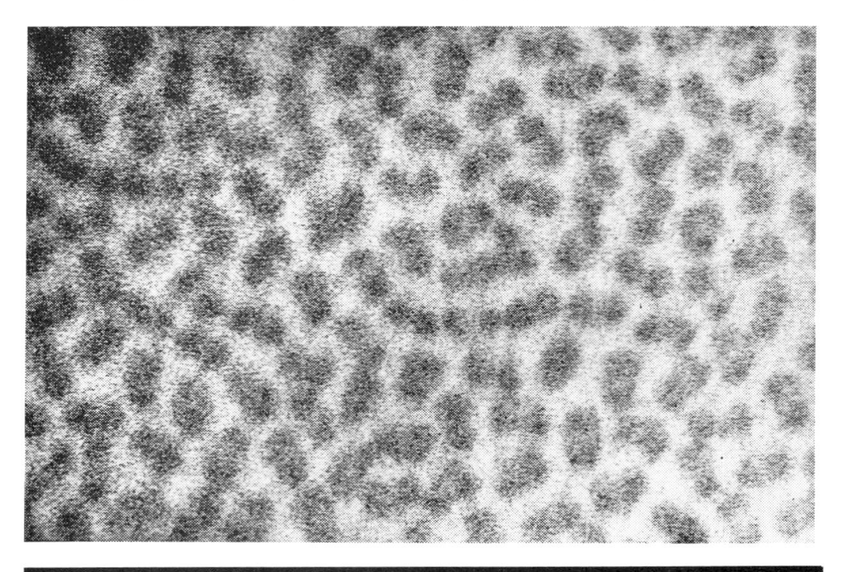

Figure 28.10 A confocal redox image of endothelial cells of the *in vitro* rabbit eye formed from NAD(P)H fluorescence. The bright regions correspond to the NAD(P)H fluorescence. The large dark regions are the endothelial cell nuclei.

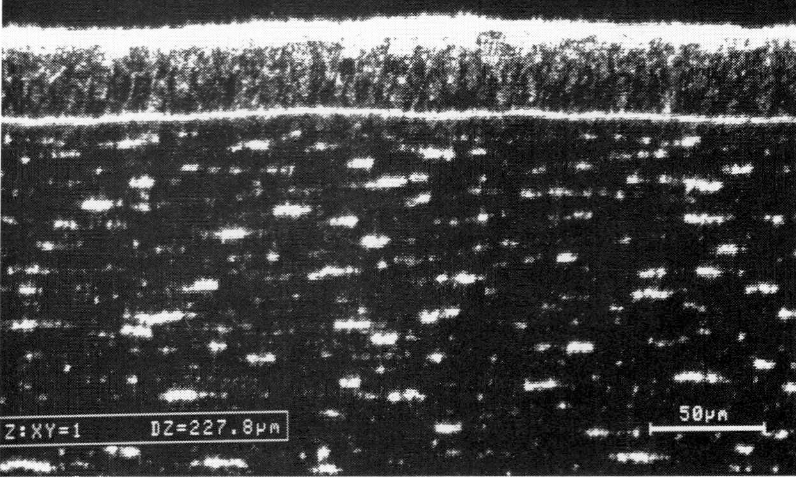

Figure 28.11 A *z*-scan of the cornea of a freshly enucleated rabbit eye obtained using a laser scanning confocal microscope in the back-scattered mode.

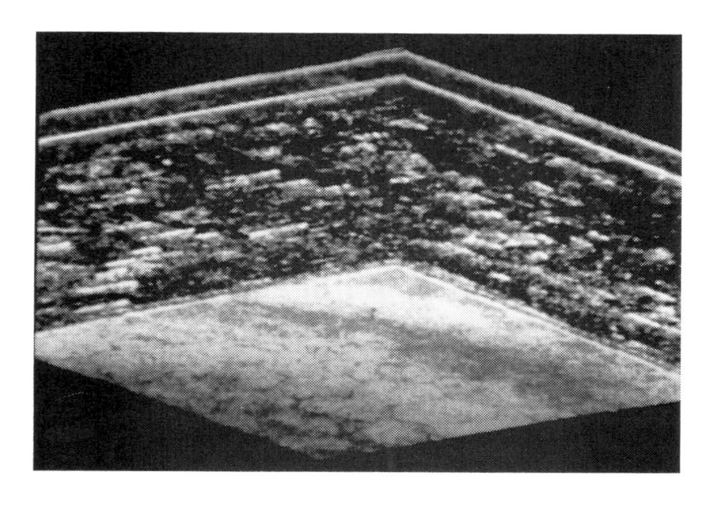

Figure 28.12 Three-dimensional volume reconstruction of a serial stack of two-dimensional back-scattered light confocal microscopic images of the full thickness of the cornea from an *in vitro* rabbit eye. The rectangular section is located in the central region of the cornea. The thickness of the cornea is 400 μm. The image was formed in a computer from the stack of two dimensional images using the volume rendering reconstruction technique. The bright line at the top of the figure is the superficial epithelium. The bright line 40 μm below is the reflection from the basal lamina. The horizontal lines are nuclei of stromal keratocytes.

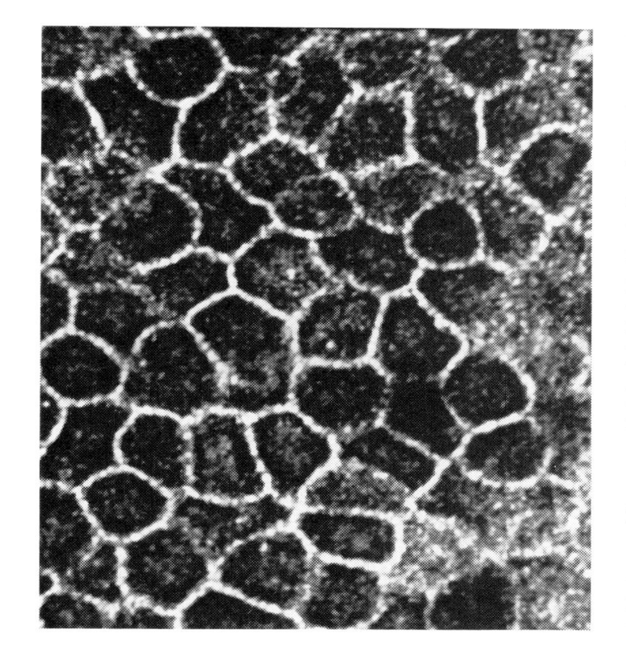

Figure 28.13 A confocal image of basal epithelial cells formed in back-scattered light of the *in vitro* rabbit cornea. The cell borders and the round cell nuclei can be seen.

28.5.1.2 Two-photon NAD(P)H redox imaging

To illustrate the superb signal-to-noise ratio obtained with two-photon confocal redox imaging the technique was applied to the *in vitro* cornea. Figure 28.13 is an image of the basal epithelium in the back-scattered light mode. Figure 28.14 shows the two-dimensional NAD(P)H imaging of the basal epithelium of the cornea.

28.5.2 Redox imaging based on flavoprotein fluorescence

The confocal microscopic optical section of the basal epithelial cells and their nuclei is shown in Fig. 28.13. The confocal microscope formed the image with light of 488 nm in the back-scattered mode. The contrast of the images is due to local differences in refractive index, e.g. between the cell cytoplasm and the cell nuclei, which generates image contrast. The focal plane was centred at the approximate centre of the height of the basal epithelial cells and in the centre of the cell nuclei. We have used the 488 nm line of the laser scanning confocal microscope to image oxidized flavoprotein fluorescence.

28.5.3 Discussion

In most cells and tissues, the signal from the NAD(P)H is more intense than that from the oxidized flavoproteins. Both signals are subject to photobleaching.

28.5.3.1 NAD(P)H redox imaging

Fluorescent images of pyridine nucleotide fluorescence in basal epithelial cells of the cornea, and cells 400 μm below the surface of the cornea (corneal endothelium) have been shown here. What do the images represent and why are the images of the surface cells different from those of the deeper cells? A comparison of the images obtained using 364 nm reflected light and 364 nm excitation/400–500 nm emission with 364 nm reflected light and 364 nm excitation/400–500 nm emission reveals the following differences. First, the images of the cells on the corneal surface have higher contrast and resolution that those 400 μm below the

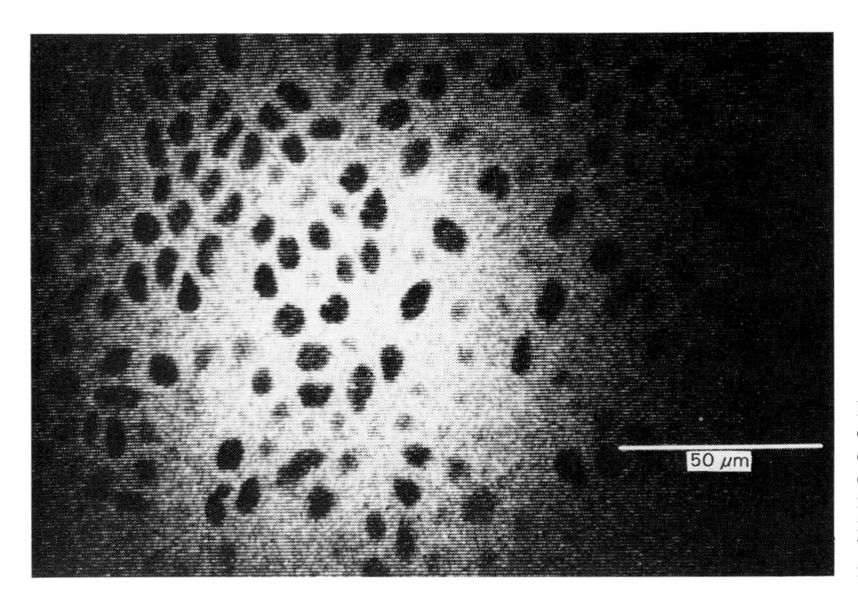

Figure 28.14 A two-photon confocal redox image of basal epithelial cells of the *in vitro* rabbit eye formed from NAD(P)H fluorescence. The bright regions indicate NAD(P)H fluorescence. The dark oval regions are cell nuclei.

surface. This was observed for both the reflected light modes and the fluorescent modes. Secondly, the resolution of the confocal images is lower than that of the images made in reflected light. This is consistent with calculations of the microscope point spread function for reflected light imaging and fluorescence light imaging; the latter shows a wider point spread function (Kimura & Munakata, 1989, 1990).

28.5.3.2 Flavoprotein redox imaging

Metabolic imaging of the flavoproteins has certain advantages over metabolic imaging of the reduced pyridine nucleotides in corneal cells. The oxidized flavoproteins are specifically located within the mitochondria and are not found in the cytoplasmic space of the cells. In contrast, the NAD(P)H is located in both the mitochondria and cytoplasmic spaces, which introduces an extra complexity for quantitative analysis and interpretation.

The previous work on freeze-trapped rabbit cornea showed an emission band with a peak at 545 nm (excitation at 442 nm) which is similar to that observed in mitochondrial preparations. The similarity of the emission spectra for the freeze-trapped rabbit cornea and the isolated mitochondria provides evidence for the identification of oxidized flavoproteins in the cornea (Chance & Lieberman, 1978).

The fluorescence intensity due to the flavoproteins in the living rabbit corneal epithelium has been investigated (Masters *et al.*, 1982a). The fluorescence emission in the region 550 nm (50 nm band-pass) due to excitation at 442 nm was investigated. The excitation light of the laser grazed the corneal epithelium of a living rabbit and the effect of a gentle flow of either hydrated air or hydrated nitrogen was studied. The nitrogen flow resulted in a decrease of fluorescence intensity which was reversible with the passage of air. These studies did not involve two-dimensional fluorescence imaging. Similar results were observed in mitochondrial suspension from rat liver. This study on the epithelium in a living rabbit and the studies of rabbit corneas conducted in freeze-trapped preparations supports the conclusion that the fluorescence is due to oxidized flavoproteins. In perfused rabbit corneas the dose–response curves of the effect of mitochondrial respiratory inhibitors at Sites I, II, and III on fluorescence intensity yielded results consistent with the characterization of the fluorescence signal (450 excitation, 550 nm emission) as due to fluorescence from the oxidized flavoproteins (Masters, 1984a).

In order to visualize the functional anatomy of the living cornea we have merged the three-dimensional volume reconstruction with a second volume reconstruction based on redox imaging of flavoproteins. The resulting three-dimensional reconstruction shows the morphology and the mitochondrial function of the cells in the living cornea (Fig. 28.12). The dynamic functional anatomy of the living cornea could be obtained through a time-series of merged three-dimensional functional and morphological volumes.

28.6 COMPARISON WITH OTHER NON-INVASIVE TECHNIQUES

Where are we to place the optical method involving intrinsic fluorescence probes in comparison to other non-invasive techniques (Chance, 1991)? This technique has been developed from the microscopic studies of organelles and cells, to studies on whole organs both *in vitro* and *in vivo*. Applications to the *in vivo* human eye with spatial resolution at the single cell level purport a bright future for the optical method in ophthalmology.

In addition, the use of optical fibres gives the observer a fibre link to the interior of the body, inside vessels of the heart, inside interior cavities, and with the rapid advances of endoscopic diagnostic and surgical techniques this tool promises numerous applications in both research and clinical diagnostics.

Other diagnostic techniques such as CT, PET, SPECT, MRI are either invasive or significantly more complex and costly. These methods involve ionizing radiation or high cost or both.

In times of major financial crisis in the provision of health care cost is a significant factor. The optical method is relatively simple, inexpensive, and can easily be coupled to a variety of endoscopic fibre optic probes. We predict that the optical method using the intrinsic fluorescence probe will continue to contribute to our understanding of cellular function and aid in clinical monitoring (during surgery and during intensive care) and diagnosis of cellular hypoxia.

28.7 SUMMARY AND CONCLUSIONS

Redox fluorometry based on intrinsic fluorescent probes of cellular metabolism is an evolving microscopic technique to investigate tissue hypoxia. Its advantages over other non-invasive methods are that it is a real-time, non-invasive optical technique, and that relatively inexpensive and simple instrumentation is required to perform the measurements.

Several disadvantages associated with the technique include problems associated with calibration and quantitation, and the multivariate nature of tissue fluorescence; metabolic substrate utilization versus oxygen concentration and utilization versus autofluorescence of tissue and cells not related to NAD(P)H or flavoproteins. The penetration of ultraviolet light is usually limited to a depth of 1 mm or less, except for transparent tissue such as the cornea or ocular lens of the eye. Therefore, the depth of tissue that can be physically located between the tip of the objective or optical fibre is about 1 mm.

There are several areas which warrant further development. The recent advances in two-dimensional redox imaging of tissues and cells based on laser confocal microscopy should be further developed. Two-photon fluorescence confocal microscopy of intrinsic redox probes, NAD(P)H or flavoproteins is a promising technical development with the advantages of minimal photobleaching of the intrinsic fluorescent probes, and the high spatial resolution and optical sectioning capability of confocal microscopy. More precise tissue localization is now possible with the continued development of confocal microscopes.

Another technological development is the coupling of fibre optic endoscopes to external redox fluorometers. This technology permits redox imaging of internal organs and tissues.

A very important development is the use of solid-state imaging devices (Masters, 1989). Large size (4000 × 4000 pixels) charge-coupled devices (CCDs) are being developed with very high detector quantum efficiencies These back-illuminated, thinned, slow-scanned CCD detectors are useful as area integrating detectors with a large dynamic range. These cooled, slow-scanned CCDs have two major advantages over other imaging detectors such as video cameras: (1) they show no geometric distortions, and (2) once calibrated they are linear photometric detectors, which is a critical requirement for quantitative measurements. For those experimental cases where motion is a problem and integration would result in blurring of the redox image, a rapid line scan which can measure a profile of the fluorescence intensity may be preferable. Newly developed linear diode array detectors permit the rapid acquisition of emission spectral information. All of these instruments discussed so far involve measurements of fluorescence intensity. Further developments of imaging devices based on other fluorescence parameters, such as fluorescence lifetimes, will continue to improve the specificity of redox microfluorescence techniques. The 'optical method' is under vigorous development and continues to provide a non-invasive method to monitor cellular metabolism.

ACKNOWLEDGEMENTS

This work was supported by a grant (BRM) from NIH EY-06958. The authors acknowledge the assistance of Dr W. Webb, and Dr D. Piston, Department of Applied Physics, Cornell University, in obtaining two-photon confocal images of the cornea.

REFERENCES

Balaban R.S. & Mandel L.J. (1990) In *Noninvasive Techniques in Cell Biology*, J.K. Foskett & S. Grinstein (eds). Wiley-Liss, New York, pp. 213–236.

Caspersson T. (1950) *Exp. Cell Res.* **1**, 595–598.

Caspersson T. (1954) *Exp. Cell Res.* **7**, 598–600.

Chance B. (1951) *Fedn. Proc.* **10**, 171.

Chance B. (1989) In *Cell Structure and Function by Microspectrofluorometry*, E. Kohen & J.G. Hirschberg (eds). Academic Press, New York, pp. 53–69.

Chance B. (1991) In *Annual Review of Biophysics and Biophysical Chemistry*, Vol. 20, D.M. Engelman (ed.). Annual Reviews, Palo Alto, CA, pp. 1–28.
Chance B. & Baltscheffsky H. (1958) *J. Biol. Chem.* **233**, 736–739.
Chance B. & Lieberman M. (1978) *Exp. Eye Res.* **26**, 111–117.
Chance B. & Thorell B. (1959) *J. Biol. Chem.* **234**, 3044–3050.
Chance B. & Williams G.R. (1955) *J. Biol. Chem.* **217**(1), 409–427.
Chance B., Mela L. & Wong D. (1968) In *Flavins and Flavoproteins, 2nd International Congress on Flavins and Flavoproteins*, K. Yagi (ed.). University Park Press, Baltimore, pp. 102–121.
Chance B., Barlow C., Haselgrove J., Nakase Y., Quistoroff B., Matschinsky F. & Mayevsky A. (1978). In *Microenvironments and Metabolic Compartmentation*, (P.A. Sere & R.W. Estabrook (eds). Academic Press, London, pp. 131–148.
Chance B., Schoener B., Oshino R., Itshak F. & Nakase Y. (1979) *J. Biol. Chem.* **254**, 4764–4771.
Denk W., Strickler J.H. & Webb W.W. (1990) *Science*, **248**, 73–76.
Eng J., Lynch R.M. & Balaban R.S. (1989) *Biophys. J.* **55**, 621–630.
Kapitza H.G. & Wilke V. (1988) *Proc. Soc. Photo-Optical Instrument Engineers (SPIE)* **1028**, 173–179.
Kasten F.H. (1989) In *Cell Structure and Function by Microspectrofluorometry*, E. Kohen, J.G. Hirschberg & J.S. Ploem (eds). Academic Press, New York, pp. 3–50.
Keilin D. (1966) *The History of Cell Respiration and Cytochrome.* Cambridge University Press, London.
Kimura S. & Munakata C. (1989) *Appl. Optics* **6**, 1015–1019.
Kimura S. & Munakata C. (1990) *Appl. Optics*, **29**, 489–494.
Kohen E. & Hirschberg J.G. (1989) *Cell Structure and Function by Microspectrofluorometry.* Academic Press, New York.
Kohen E., Kohen C., Hirschberg J.G., Fried M. & Prince J., (1989) *Optical Engng* **28**(3), 222–231.
Lakowicz J.R. (1992) *Principles of Fluorescence Spectroscopy.* Plenum Press, New York.
Lemp M.A., Dilly P.N. & Boyde A. (1986) *Cornea* **4**, 205–209.
MacMunn C.A. (1880) *The Spectroscope in Medicine.* Churchill, London.
MacMunn C.A. (1914) *Spectrum Analysis Applied to Biology and Medicine.* Longmans, Green & Co., London.
Masters B.R. (1984a) In *Current Topics in Eye Research*, Vol. 4, J. Zadunaisky & H. Davson (eds). Academic Press, London, pp. 139–200.
Masters B.R. (1984b) *Curr. Eye Res.* **3**, 23–26.
Masters B.R. (1986) In *The Precorneal Tear Film In Health, Disease and Contact Lens Wear*, F. Holly (ed.). Dry Eye Institute, Lubbock, TX, pp. 966–970.
Masters B.R. (1988) In *The Cornea: Transactions of the World Congress on the Cornea III*, H.D. Cavanagh (ed.). Raven Press, New York, pp. 2810–3860.
Masters B.R. (1989) In *New Methods in Microscopy and Low Light Imaging*, J.E. Wampler (ed.). *Proc. Soc. Photo-Optical Instrument Engineers (SPIE)* **1161**, 350–365.
Masters B.R. (1990a) In *Confocal Microscopy*, T. Wilson (ed.). Academic Press, London, pp. 305–324.
Masters B.R. (1990b) In *Noninvasive Diagnostic Techniques in Ophthalmology*, B.R. Masters (ed.). Springer-Verlag, New York, pp. 223–247.
Masters B.R. (1991) *Machine Vision and Applications* **4**, 227–232.
Masters B.R. (1992) *J. Microsc.* **165**, 159–167.
Masters B.R. (1993a) In *3-D Visualization in Microscopy*, A. Kriete (ed.). VCH, Germany, pp. 183–203.
Masters B.R. (1993b) *Appl. Optics* (in press).
Masters B.R. & Chance B. (1980) *Proc. Int. Soc. for Eye Res.* **1**, 30.
Masters B.R. & Kino G.S. (1989) *Proc. Institute of Physics, Electron Microscopy and Analysis Group*, **98**, 625–628.
Masters B.R. & Kino G.S. (1990) In *Noninvasive Diagnostic Techniques in Ophthalmology*, B.R. Masters (ed.). Springer-Verlag, New York, pp. 152–171.
Masters B.R. & Paddock S.W. (1990a) *J. Microsc.* **158**, 267–275.
Masters B.R. & Paddock S.W. (1990b) *Appl. Optics* **29**, 3816–3822.
Masters B.R., Fischbarg J., Chance B. & Lieberman M. (1980) *Invest. Ophthalmol. Vis. Sci.* (Suppl.) **19**, 63.
Masters B.R., Chance B. & Fischbarg J. (1981) *Trends Biochem. Sci.* **6**, 282–284.
Masters B.R., Chance B. & Fischbarg J. (1982a) In *Noninvasive Probes of Tissue Metabolism*, J.S. Cohen (ed.). John Wiley & Sons, New York, pp. 79–118.
Masters B.R., Falk S. & Chance B. (1982b) *Curr. Eye Res.* **1**, 623–627.
Masters B.R., Riley M.V., Fischbarg J. & Chance B. (1983) *Exp. Eye Res.* **36**, 1–9.
Montag M., Kükulies J., Jörgens R., Gundlach H., Trendelenburg M.F. & Spring H. (1991). *J. Microsc.* **163**, 201–210.
Quistorff B., Haselgrove J.C. & Chance B. (1985) *Analyt. Biochem.* **148**, 389–400.
Rost F.W.D. (1991) *Quantitative Fluorescence Microscopy.* Cambridge University Press, Cambridge.
Sies, H. (1982) *Metabolic Compartmentation.* Academic Press, New York.
Wilson T. (1989) *J. Microsc.* **154**, 143–156.
Wilson T. (1990) In *Confocal Microscopy*, T. Wilson (ed.). Academic Press, London, pp. 93–141.
Xiao G.Q., Kino G.S. & Masters, B.R. (1990) *Scanning* **12**, 161–166.

PART VIII

Advanced Imaging and Light Detection Approaches for Optical Probe Applications

CHAPTER TWENTY-NINE

Confocal Raman Microspectroscopy

G.J. PUPPELS

Laboratory for Intensive Care Research and Optical Spectroscopy, Erasmus University Rotterdam and University Hospital Rotterdam 'Dijkzigt', Rotterdam, The Netherlands

Fluorescence methods are employed with great success in cell and tissue biology. Their fundamental disadvantage is, however, that with the exception of those cases where intrinsic chromophores are investigated, probes have to be introduced into the system under study. The fact that the manner in which system and probe influence each other is often unknown can hamper the interpretation of experimental results. For this reason the development of direct analysis techniques is important.

Raman spectroscopy is a technique that provides molecular structural and compositional information without using probes. The sensitivity of Raman instrumentation has advanced to a level that, despite the generally low signal levels, studies of chromosomes and single living cells are possible.

29.1 INTRODUCTION

Many of the questions posed by modern cell biology are tackled by means of fluorescence techniques. As amply illustrated in other chapters of this volume, the key to solving these questions often lies in the development of suitable fluorescent probes. However, if the aim is to obtain information about the biological system it should always be kept in mind that fluorescence methods work indirectly. The introduction of a probe means that the information obtained is about the system with probe. Unless it is precisely known what influence the probe has on the system and, vice versa, how the system affects probe characteristics, the interpretation of experimental results has to be made with great caution.

This point is illustrated by the difficulties that were encountered in the interpretation of some well-known discoveries made by fluorescence microscopy.

In 1968 Caspersson and co-workers found that staining of fixed metaphase chromosomes with quinacrine-mustard gave rise to a fluorescent banding pattern on the chromosomes (Caspersson *et al.*, 1968). The question was, of course, what caused these banding patterns. The initial idea was that the mustard moiety would preferentially interact with DNA rich in guanine (G)–cytosine (C) base pairs. Experiments with quinacrine, however, led to the same results, so this explanation could not be upheld. Instead it was found that *in vitro* adenine

FLUORESCENT AND LUMINESCENT PROBES, 2ND EDN
ISBN 0–12–447836–0

(A)–cytosine (C)-rich DNA enhanced quinacrine fluorescence, while GC-rich DNA quenched quinacrine fluorescence (Weisblum & de Haseth, 1972; Pachman & Rigter, 1972). The Q-banding pattern, as it was called, therefore appeared to reflect local variations in chromosomal DNA base compositon, AT-rich regions fluorescing brightly. But that turned out to be only a partial explanation, for it was found that, for example, the centromeric heterochromatin of mouse chromosomes, containing AT-rich satellite DNA, showed dimmer fluorescence than the less AT-rich chromosome arms (Rowly & Bodmer, 1971). During interphase the centromeric heterochromatin fluoresced brightly (Natarajan & Gropp, 1972). It became clear that the presence of proteins exerted a strong influence on the quinacrine fluorescence of chromosomes, necessitating further studies of the binding mechanisms and fluorescence characteristics of the dye (Sumner, 1982). Even now, after more than 20 years of intensive research and the discovery of many other types of metaphase chromosome banding patterns, our understanding of this phenomenon is far from complete.

A more recent discovery was the binding of fluorescently labelled anti-Z-DNA antibodies to polytene chromosomes (Nordheim *et al.*, 1981; Arndt-Jovin *et al.*, 1983). This attracted a great deal of attention because it could indicate that this left-handed double helical DNA form is present in nature and possibly serves a specific function. A clear positive staining of the chromosomes was only observed, however, after fixation of the chromosomes in (3:1) alcohol–acetic acid or after incubation of unfixed chromosomes at low pH, prior to their incubation with the labelled anti-Z-DNA antibodies (Robert-Nicoud *et al.*, 1984). The question that remains is, therefore, whether the absence of, or only marginal, antibody binding to unfixed chromosomes means that there is no Z-DNA prior to fixation, and that fixation or low-pH treatment induce this DNA structure. An alternative explanation could be that the Z-DNA in the unfixed chromosomes is not accessible to the antibodies, e.g. because it is already complexed with a Z-DNA binding protein, or because the antibodies cannot penetrate the compact chromosomal chromatin. Another question that can be asked with respect to the use of anti-Z-DNA antibodies is to what extent they can be screened against other (unknown) DNA structures that may be induced by the chromosome fixation (Puppels *et al.*, 1994a).

Ideally, the introduction of a probe in a system should not lead to any perturbation of the system nor of the probe. In mathematical terms:

$$\text{(System)} + \text{(Probe)} = \text{(System + Probe)} \qquad [29.1]$$

The two examples above underline, however, that this is not usually the case and that despite the versatility of fluorescence methods, a definite need remains for direct analysis techniques, i.e. techniques that do not require the use of probes. This chapter deals with one such technique: confocal Raman microspectroscopy. Continuing progress in instrumentation has brought this technique to a level of sensitivity where it can be applied in studies of single living cells and chromosomes (Puppels *et al.*, 1990b).

29.2 RAMAN SPECTROSCOPY

Raman spectroscopy provides information about molecular composition and structure and about interactions between molecules. The technique is based on an inelastic light scattering process, predicted in 1923 by Smekal (Smekal, 1923) and first observed by the Indian scientists Raman and Krishnan in 1926 (Raman & Krishnan, 1928). A theoretical treatise of Raman scattering has been given in a number of textbooks (e.g. Koningstein, 1971; Long, 1977) and will be omitted here.

29.2.1 Raman scattering

The processes of Stokes, anti-Stokes and resonance Raman scattering, of Rayleigh and fluorescence scattering and of infrared absorption are schematically depicted in the energy level diagrams of Fig. 29.1. Rayleigh scattering and other elastic light scattering processes result from an interaction between a radiation field and a molecule, characterized by the absence of energy exchange. Stokes Raman scattering occurs when the interaction promotes the molecule to a higher vibrational energy level. In that case the scattered photon possesses less energy than the incident photon. This results in the appearance of 'Raman' bands in the spectrum of the scattered light at wavelengths longer than that of the incident light. In the anti-Stokes Raman process the opposite occurs, which results in 'Raman' bands at wavelengths shorter than that of the incident light. If the radiation field is monochromatic – nowadays lasers are used – Raman scattering gives rise to a spectrum of discrete lines, the positions of which correspond to the energies needed to excite vibrational modes. The shift of a Raman line relative to the wavelength of the exciting radiation is expressed in relative wavenumbers:

$$\Delta\ \text{cm}^{-1} = (1/\lambda_0 - 1/\lambda_s)\ 10^{-2} \qquad [29.2]$$

where $[\lambda] = \text{m}$, λ_0 = excitation wavelength, λ_s = wavelength of scattered light.

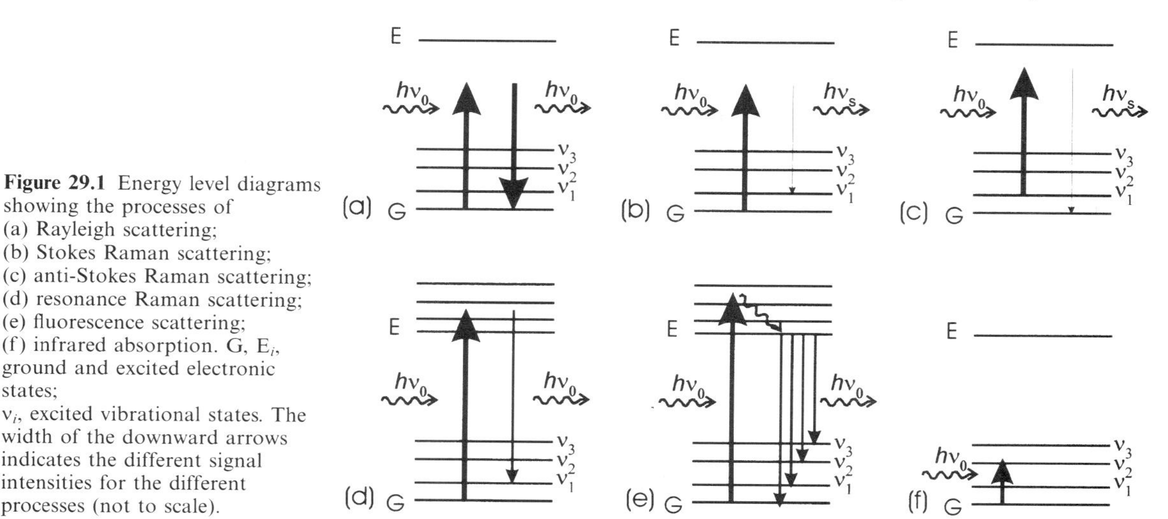

Figure 29.1 Energy level diagrams showing the processes of
(a) Rayleigh scattering;
(b) Stokes Raman scattering;
(c) anti-Stokes Raman scattering;
(d) resonance Raman scattering;
(e) fluorescence scattering;
(f) infrared absorption. G, E_i, ground and excited electronic states;
ν_i, excited vibrational states. The width of the downward arrows indicates the different signal intensities for the different processes (not to scale).

A molecule of N atoms possesses $3N$ degrees of freedom. After substraction of translational and rotational modes $3N - 6$ independent vibrational modes remain ($3N - 5$ for a linear molecule). Those modes that can be coupled to the electric field component of the incident radiation via a change in molecular polarizability are 'Raman-active', i.e. will give rise to a line in the Raman spectrum.

The intensity of Raman scattered light is many orders of magnitude lower than that of Rayleigh scattered light. A resonance enhancement of the Raman signal can be obtained when the wavelength of the exciting light coincides with an absorption band of a molecule (see the above-mentioned textbooks on Raman spectroscopy as well as books on biological applications of Raman spectroscopy by Carey (1982), Clark & Hester (1986) and Spiro (1987–1988)). An additional advantage of resonance Raman spectroscopy is that different molecular subgroups can be separately resonantly excited and investigated (see e.g. Salmaso *et al.*, 1994).

The molecular structural information in a Raman spectrum is contained in the position, intensity, polarization and width of the lines.

(1) The position of a Raman line corresponds to the frequency of a molecular vibration, which depends on molecular structure, on the masses of the atoms involved in the vibrational mode, and on their chemical bonds.
(2) The intensity of a Raman line depends linearly on the number of scatterers and on laser light intensity. In resonance Raman experiments it depends on both ground state and excited state properties.
(3) Raman scattering may be anisotropic due to the fact that the molecular polarizability is a tensor. The polarization properties of Raman lines depend on the symmetry of the molecules. The polarization of a line can be a valuable aid in assigning that line to a particular vibration (see e.g. Salmaso *et al.*, 1994).
(4) The width of a Raman line depends on molecular structural heterogeneity and is influenced by dynamical processes and molecular re-orientational motions taking place within a time-frame of about 0.1–10 ps (Bartoli & Litovitz, 1972).

The parameters mentioned often depend in a subtle way on the precise chemical microenvironment of the molecules under study.

In thermal equilibrium conditions the relative number of molecules in ground state and higher vibrational states is given by the Boltzmann distribution function.

$$N_1/N_2 = \exp(-h\nu_{vib}/kT) \qquad [29.3]$$

where h = Planck's constant, k = Boltzmann's constant, T = absolute temperature, N_1 = number of molecules in a higher vibrational state, and N_0 = number of molecules in the ground state.

The intensity ratio of the anti-Stokes Raman and Stokes Raman signals therefore decreases almost exponentially with frequency shift. At 1000 cm^{-1} the anti-Stokes Raman signal is about two orders of magnitude weaker than the corresponding Stokes Raman signal. For this reason usually only the Stokes signal is measured.

Fluorescently scattered light also undergoes a Stokes shift and often hinders the recording of a Raman spectrum. Because fluorescence scattering is a process with a much higher quantum yield than Raman scattering, it is usually the Raman spectroscopist whose efforts are frustrated. Even trace amounts of luminescing sample impurities can make it difficult to obtain a Raman spectrum of acceptable signal-to-noise. In flow cytometry, on the other hand, the lower level of fluorescence that can be detected is determined by water Raman scattering (Steen, 1990; Jett *et al.*, 1990). The simplest solution to these problems is for a Raman spectroscopist to choose an excitation wavelength away from the excitation spectrum of the fluorophore (see, for example, Yu *et al.*, 1987, p. 56; see also Section 29.4.2.2) and for the flow cytometrist to choose a fluorophore with an emission spectrum away from the strong water bands in the Raman spectrum. Other technologically more involved methods use phase-sensitive detection (to separate Raman from fluorescence signal), based on the fact that Raman scattering is an instantaneous process, whereas in fluorescence excited state lifetimes are often in the order of nanoseconds (Genack, 1984).

Infrared spectroscopy is a technique that is closely related to Raman spectroscopy. It is based on absorption of photons of low energy by molecules, which also results in the excitation of a molecular vibration. Therefore an infrared spectrum provides information about the vibrational energy levels of molecules, which can be translated into information on molecular composition and molecular structures. An absorption band due to a particular molecular vibrational mode may be visible in the infrared spectrum, while no band is visible in the Raman spectrum, and vice versa. This is due to the fact that different selection rules apply to determine whether or not a molecular vibration is infrared and/or Raman active (Parker, 1983). The techniques therefore provide complementary information. The advantage of infrared spectroscopy is that it is relatively easy to obtain spectra with a very high signal-to-noise ratio. A disadvantage is that water is a strong absorber in the infrared spectral region, which complicates measurements on hydrated samples.

29.2.2 Instrumentation

The instrumentation needed for a Raman experiment is conceptually simple. A schematic is shown in Fig. 29.2(A). The Raman effect is very weak and the technique became practicable and popular only after the invention of the laser, because of the possibility of tightly focusing the laser light, thus producing a high photon density in a sample. Scattered light is collected and its spectrum analysed by means of a grating spectrometer. A prerequisite for a successful Raman experiment is a strong suppression of the Rayleigh line at zero wavenumber shift. The detectors that are used are either photomultiplier tubes, operated in photon counting mode (single-channel detection), or array detectors such as intensified photodiode arrays or

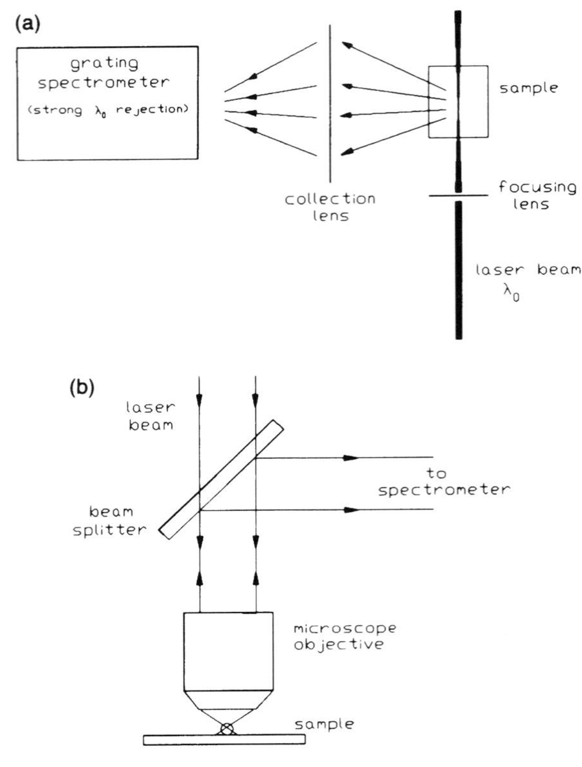

Figure 29.2 (a) Schematic of a Raman spectroscopy set-up; (b) Raman microspectroscopy set-up.

unintensified slow-scan charge-coupled device (CCD) cameras (multichannel detection). Especially when multichannel detection is employed, where a relatively low wavelength dispersion is desired in order to capture a large part of the Raman spectrum simultaneously, efficient Rayleigh line suppression demands some creativity (Mathies & Yu, 1978; Flaugh *et al.*, 1984; Baek *et al.*, 1988; Puppels *et al.*, 1990a; Rich & Cook, 1991).

In 1975 two groups independently reported the development of Raman microspectrometers (Rosasco *et al.*, 1975; Delhaye & Dhamelincourt, 1975). The design of Delhaye & Dhamelincourt, in which a microscope and a spectrometer are optically coupled, has found widespread acceptance (Fig. 29.2(B)). In a Raman microspectrometer laser light is focused on a sample by means of a microscope objective. The light that is scattered by the sample is collected by the same objective and focused onto the entrance of the spectrometer. In this way it is possible to study very small (micro-sized) samples, or to probe larger objects at specific sites and so combine spatial (morphological) information with spectroscopical information. Early applications included environmental particulate pollution studies and the identification of foreign body inclusions in tissue (Etz & Blaha, 1980).

29.2.3 Raman spectroscopy and molecular biology

Since the late 1960s and early 1970s Raman spectroscopy has been extensively used to study biological (macro)molecules such as nucleic acids, proteins and protein-bound chromophores, lipids and polyenes. For reviews and textbooks see Carey (1982), Parker (1983), Clark & Hester (1986) and Spiro (1987–1988).

The complexity of the spectra of these large molecules is high, owing to the large number of often overlapping Raman bands. It is nevertheless possible to extract a great deal of information from such spectra. The assignment of lines to particular vibrations and the interpretation of spectra in terms of molecular structure and composition of a sample is achieved by normal mode calculations (e.g. Abe *et al.*, 1978; Krimm, 1987), combined X-ray diffraction and Raman spectroscopic experiments (e.g. Erfurth *et al.*, 1975; Goodwin & Brahms, 1978; Benevides & Thomas, 1983), isotope substitution experiments, which lead to shifts in Raman line frequencies (e.g. Lord & Thomas, 1967; Benevides & Thomas, 1985; Oertling *et al.*, 1988) and, in the case of large molecules or complexes, by obtaining spectra of molecules or molecular subgroups separately. A brief overview of the type of information about biological molecules that can be obtained by means of Raman spectroscopy and some characteristic applications are given in Table 29.1.

29.2.4 Raman spectroscopy of single cells

Raman spectroscopy can provide a wealth of information and microspectrometers have, in principle, a diffraction-limited spatial resolution. Nevertheless, for a long time the technique was little used for studies at the level of a single cell. The reason for this was the small Raman scattering cross-section of many biological molecules, including nucleic acids, proteins and lipids, and the resulting low Raman

Table 29.1 Raman spectroscopic information about biological molecules and characteristic applications.

	Information	*Applications*
Nucleic acids	Sugar–phosphate backbone conformation Glycosidic bond orientation Nucleoside sugar pucker Base composition Base stacking Hydrogen bonding interactions	Study of DNA secondary structure and transitions (e.g. B to Z) Characterization of DNA melting and premelting phenomena Monitoring of DNA protonation and denaturation at low pH Characterization of natural and synthetic DNAs/RNAs Influence DNA–protein interactions on DNA structure
Protein	Protein main chain secondary structure Presence & conformation of disulfide, SH and CS groups Presence & microenvironment of aromatic amino acids: tyrosine – residue acting as H-bond donor or donor & acceptor tryptophan – residue buried or exposed	
Protein-bound chromophores, e.g. metallo-porphyrins	Oxidation state & spin state of the core metal Type of axial ligands and peripheral substituents Ground and excited state geometry of the molecule	Determination structure–function relations Monitoring of *in vitro* and *in vivo* enzymatic processes Identification of reaction intermediates
Lipids	Conformation of hydrocarbon chains Microenvironment of head and tail group molecules	Monitoring of phase transitions in membranes Characterization of membrane fluidity

See Carey (1982), Parker (1983), Clark & Hester (1986), Spiro (1987–1988).

signal intensity levels. The applicability of Raman spectroscopy in *in situ* studies of molecules in a single cell has long depended on favourable circumstances such as the possibility of resonance enhancement of the Raman signal. In this manner, for example, Barry & Mathies (1982) obtained spectra of visual pigments in single photoreceptor cells of toad and goldfish. The spectra of haemoglobin in a red blood cell (Jeannesson *et al.*, 1986) and spectra of single bacterial cells (Dalterio *et al.*, 1986) were also measured in this way. The very high concentration of DNA in sperm cells allowed a spectrum of DNA inside a single salmon sperm head to be obtained (Kubasek *et al.*, 1986). Single cell spectra were furthermore recorded of very large cells such as a muscle fibre (Pézolet *et al.*, 1980) and a giant lymph node cell (Abraham & Etz, 1979).

In order to obtain resonance enhancement for the study of nucleic acids and proteins, UV-laser excitation has to be employed. A UV-Raman microspectrometer has been developed that enables the recording of single cell spectra with submicrometre spatial resolution in this way (Sureau *et al.*, 1990). Using resonance excitation in single cell studies is not without risk, however, because of the danger of photodegradation. In studies of isolated purified compounds or of model systems in solution, this risk can be decreased by constantly refreshing the sample in the laser beam, e.g. by employing a flow chamber or a spinning cell (Carey, 1982). These protective measures cannot be taken in single cell studies, of course.

Surface-enhanced Raman spectroscopy has been used to detect very low concentrations of antitumour drugs in the nucleus and cytoplasm of single living cells (Nabiev *et al.*, 1991). This technique relies on the strong enhancement of the Raman signal of molecules close to a metal surface. In this study small silver hydrosols were introduced into the cell by means of endocytosis. Drug concentrations down to 10^{-10} M have been detected in this way.

In cases where these signal enhancement techniques cannot be applied, the only other way to go is optimization of sensitivity of the Raman instrumentation. This approach led to the development of the confocal Raman microspectrometer described in the next section.

29.3 THE CONFOCAL RAMAN MICROSPECTROMETER (CRM)

The CRM, shown in Fig. 29.3, enables the recording of high-quality Raman spectra of single cells and chromosomes. In brief, this section deals with three important instrumentation aspects of the CRM: sensivity, spatial resolution, and laser light-induced damage to cells and chromosomes.

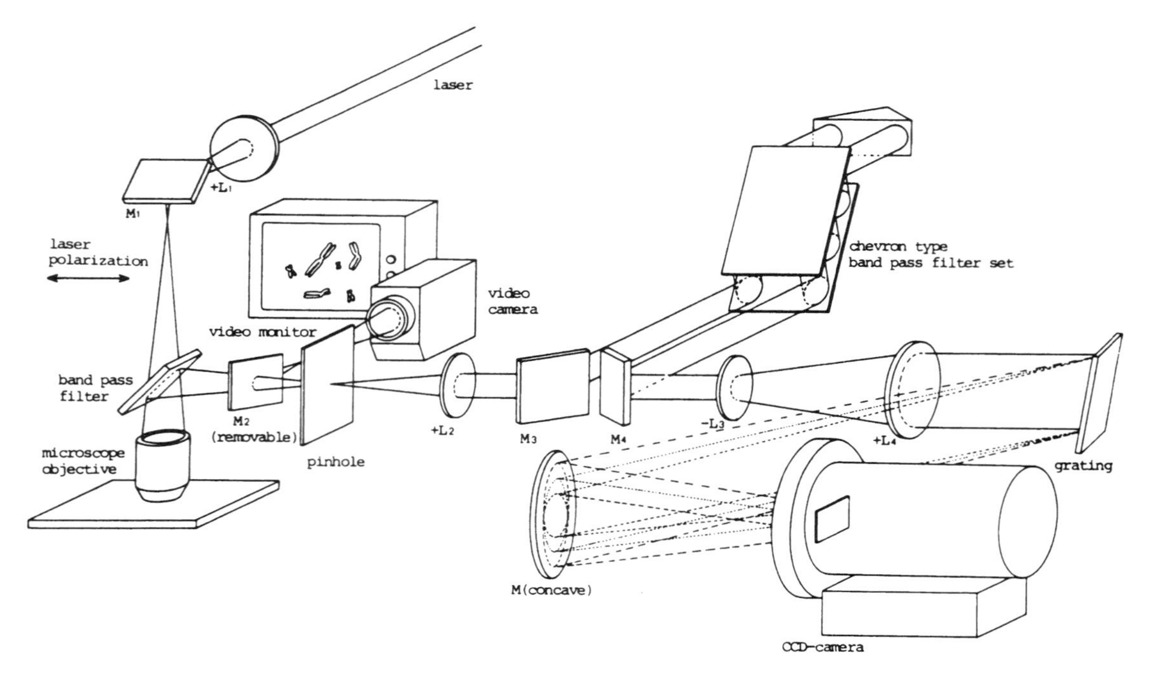

Figure 29.3 The confocal Raman microspectrometer. (From Puppels *et al.*, 1991a, reproduced by permission of John Wiley & Sons Ltd.)

29.3.1 Sensitivity

The signal detection efficiency of the original version of the CRM was optimized to the extent that up to 15% of the signal collected by the microscope objective was actually detected. This was achieved by designing the spectrometer for use with one fixed laser excitation wavelength only. All components of the set-up were optimized for Raman experiments with 660 nm laser excitation. A narrow-band transmission filter was used to couple microscope and spectrometer optically; 80% of the incoming laser light was transmitted, whereas Raman scattered light between 300 and 3000 cm^{-1} was reflected with an efficiency ≥99%. The chevron-type Raman notch filter set suppressed the intensity of the scattered laser light. The beam of light entering the spectrometer is reflected back and forth between two bandpass filters. At each reflection >80% of the laser light is transmitted, and in that way effectively separated from the Raman-scattered light, which is efficiently reflected (R>99% 600–2600 cm^{-1}). After 12 such reflections laser light intensity is suppressed by a factor 10^8, at the cost of only 10–20% of the Raman-scattered light (Puppels *et al.*, 1990a). This efficient blocking of laser light allowed the use of a ruled blazed grating for wavelength disperson, despite the fact that this type of grating has a much higher stray light level than holographic gratings. The advantages of ruled gratings over holographic gratings (in the UV and visible regions of the electromagnetic spectrum) are their much higher efficiency, and the fact that their performance is much less polarization-dependent. The Raman spectrum was imaged onto a liquid-nitrogen-cooled slow-scan front-illuminated CCD camera (Wright Instruments Ltd, EEV P8603 chip). These two-dimensional detectors are ideal for Raman spectroscopic purposes. They combine a high quantum efficiency (~40% at 700 nm for this type) with virtually noiseless operation. The cooling of this chip to below 170 K eliminates dark current and read-out noise is limited to about 10 electrons per channel. This means that at virtually any signal intensity level the signal-to-noise ratio is photon-noise-limited.

In later versions of the CRM, as used in Salmaso *et al.* (1994) and Bakker Schut *et al.* (1997), the use of back-illuminated CCD-chips for signal detection doubled signal detection efficiency (as defined above) to 30%. Holographic notch filters for laser light suppression became available, which could replace the chevron type notch filter and thereby considerably simplify instrument design. Confocal Raman microspectrometers are commercially available now from all major Raman instrument manufacturers (e.g. Dilor, Renishaw, and Kaiser Optical Systems). A niche for chevron-type notch filters still exists, in that they enable Raman measurements very close to the laser line, on Stokes and anti-Stokes sides simultaneously (Puppels *et al.*, 1994b).

29.3.2 Spatial resolution

High numerical aperture microscope objectives are used. This does not only ensure a submicrometre lateral spatial resolution but also signal collection under a wide solid angle. Raman signal is emitted not only by the object under investigation but also by its surroundings, e.g. the substrate on which a chromosome is deposited and the buffer in which it is immersed. This effect is comparable to the blurring of an image in a fluorescence microscope due to out-of-focus fluorescence. The background Raman signal can of course be measured separately, and subtracted, but adds photon shot-noise to the Raman signal of interest. These background signal contributions can be suppressed by means of confocal detection of the Raman signal (Fig. 29.4) (Brakenhoff *et al.*, 1979, see also Chapter 24, this volume). In the CRM in general a pinhole with a diameter of 100 μm is used, because it was found that in this way a significant reduction of background signal was obtained without loss of Raman signal of the object in focus. The effect of the pinhole is illustrated by the measurements in Fig. 29.5. Lateral spatial resolution is not improved by a pinhole of this size, so that it is determined by the size of the laser focus. The dimensions of the actual measuring volume of the CRM, equipped with a ×63 Zeiss Plan Neofluar water immersion objective (NA 1.2), were

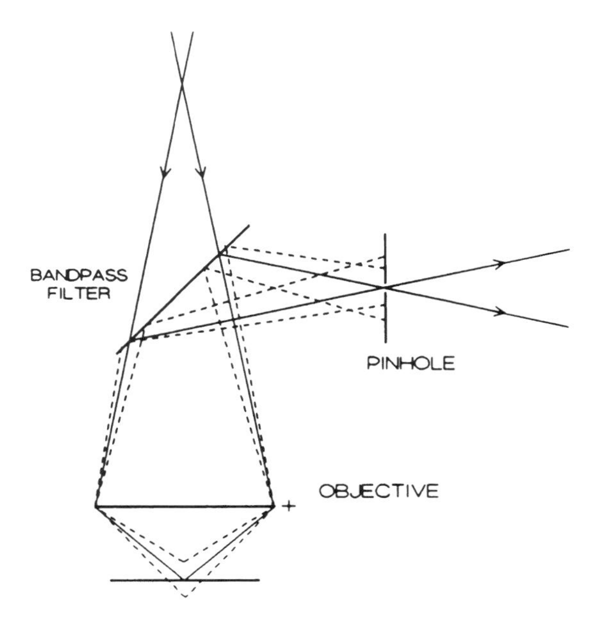

Figure 29.4 Suppression of out-of-focus signal contributions by means of confocal signal detection.

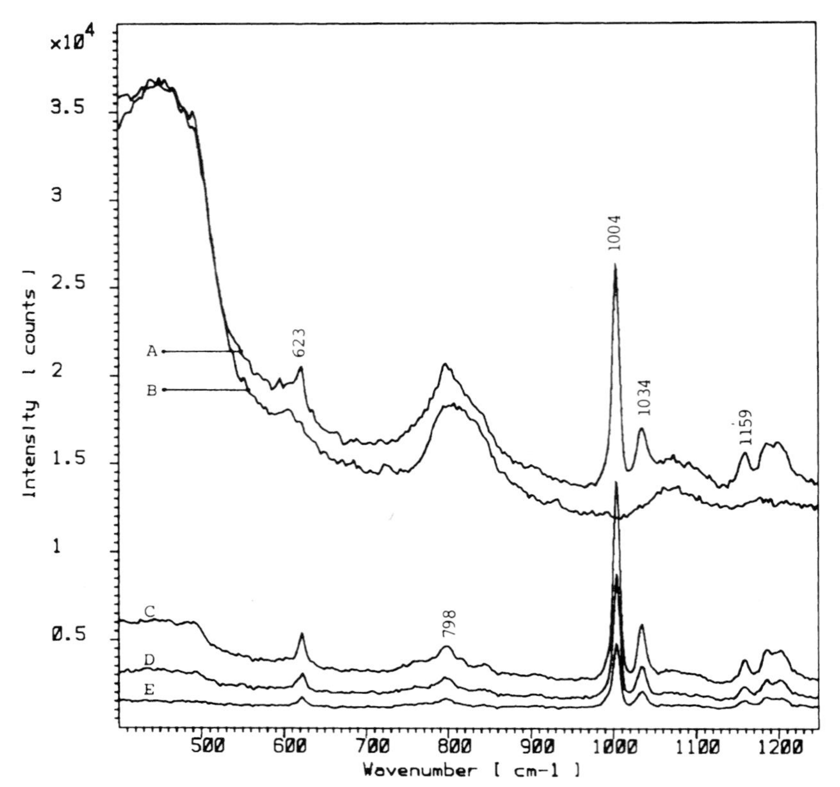

Figure 29.5 The influence of the size of the pinhole in the image plane of the objective on the Raman signal obtained from a 0.5 μm polystyrene bead on a fused silica substrate in water. (A) Measurement without pinhole; (B) background signal without pinhole; (C) 100 μm pinhole (diameter); (D) 50 μm pinhole; (E) 25 μm pinhole. It can be seen that the 100 μm pinhole strongly reduces background signal contributions without affecting the Raman signal from the object in focus. Smaller pinholes further reduce background signal intensity but also the signal contributions from the polystyrene bead. (From Puppels *et al.*, 1991a, reproduced by permission of John Wiley & Sons Ltd.)

determined by scanning a small polystyrene bead (0.22 μm) through the laser focus and measuring the intensity of the polystyrene Raman signal as a function of the position of the bead (Puppels *et al.*, 1991a). The FWHM values of the resulting curves showed a lateral spatial resolution of 0.45 μm (both directions) and an axial resolution of 1.3 μm. In theory FWHM values of 0.31 μm (lateral) and 0.8 μm (axial) should be possible for this system (Van der Voort & Brakenhoff, 1990).

29.3.3 Laser light-induced damage

The CRM was originally designed to work with laser light of 514.5 nm (Greve *et al.*, 1989; Puppels *et al.*, 1989). It turned out, however, that all samples investigated – cells, metaphase and polytene chromosomes – were invariably damaged by the laser light (see Fig. 29.6), even at laser powers below 1 mW. This damage became visible under a light microscope in the form of a 'paling' of the sample at the site of the laser focus. Simultaneously with this paling the intensity of the Raman signal decreased. Interestingly, a line due to a guanine ring vibration at 1487 cm^{-1} was often found to diminish faster than the other lines of the spectrum (Fig. 29.7). Experiments with concentrated DNA and histone solutions showed that the 514.5 nm laser light did not affect the main chromatin constituents in their purified uncomplexed form (Puppels *et al.*, 1991b). It was also shown that excessive heating of the samples due to laser light absorption, or indirectly due to substrate heating, did not play a role. Effects due to potential multiphoton absorption could also be excluded.

The choice of the laser wavelength used for excitation was found to be crucial. Whereas at wavelengths below 514.5 nm the same paling of samples was observed, this was not the case upon irradiation with laser light of 632.8 nm and above. Moreover, the Raman spectra of chromosomes and cells remained constant upon prolonged irradiation with 600 nm laser light of the same site (Puppels *et al.*, 1991b). Also, cells were found to be able to survive irradiation with much higher laser powers at 632.8 nm or 660 nm than at 514.5 nm (Fig. 29.8). Of course the 'safety' of 660 nm laser light is not a general rule, but it holds true for 'chromatin samples' such as chromosomes and cell nuclei. When (strongly) absorbing chromophores, such as haem proteins, are present in a cell, laser light intensity and signal acquisition time have to be carefully balanced in order to avoid artifacts due to photochemical reactions.

Photochemical effects appear to be the most plausible cause of the observed sample degradation at 514.5 nm. Since purified DNA and histones are not

(a)

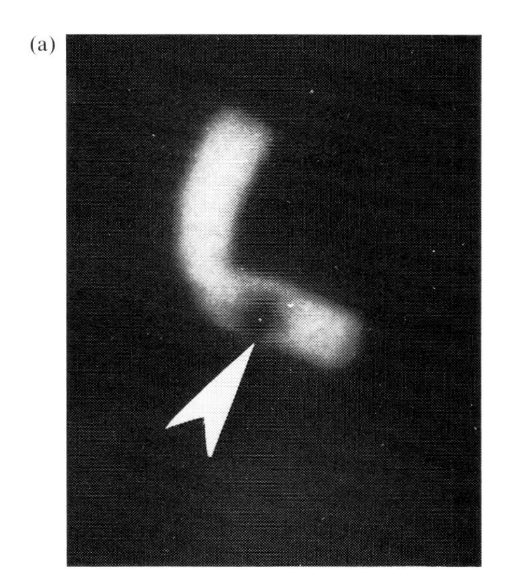

(b)

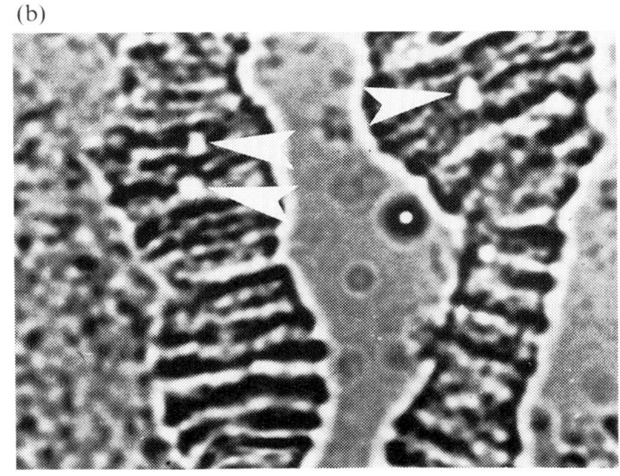

Figure 29.6 Paling of chromosomes by 514.5 nm laser light. (a) Metaphase chromosome (unfixed, magnification ×2000); (b) polytene chromosome (fixed, magnification ×1500). (From Puppels *et al.*, 1991b, reproduced by permission of Academic Press Inc.).

susceptible to this radiation damage, other compounds that act as photosensitizers must be present in chromosomes and cells. DNA bases (especially guanine) and a number of amino acids (methionine, tyrosine, tryptophan, histidine and cysteine) can be degraded in photodynamical processes, in which reactive oxygen species are formed (Foote, 1976). These could cause the observed radiation damage through oxidation of the DNA bases or amino acids, leading to lesions in these molecules. It appears that these, as yet unidentified, sensitizers are excited by blue or green laser light, but not by the 660 nm laser light used in the CRM. Recently, Peticolas *et al.* (1996) reported that no decomposition of chromosomal material occurred when measurements, using 514.5 or 442 nm laser light, were performed on polytene chromosomes inside the nucleus of a salivary gland cell.

29.4 APPLICATIONS

The CRM enables investigation of microscopically small biological objects. It can be used to determine which molecules are located where and in what form, conformation and quantity at a certain moment in time, inside a living cell or a chromosome.

A marked difference between working with intact biological structures and experiments with model systems is that, especially in the case of cells, the molecular composition of the sample is not precisely known. The first step in analysing the spectra is, therefore, to identify the molecules most prominently contributing to the Raman signal. After that an analysis of the structure of these molecules in their natural environment can be made and compared with that of the isolated compound or a model system under controlled conditions. Another approach is to regard a Raman spectrum as a spectroscopic fingerprint that is highly specific for a certain type of cell, and which can be used for cell identification or characterization, even without further interpretation in terms of molecular composition (Section 29.5.1).

Raman spectroscopy as a direct technique lacks the selectivity of, for example, immunofluorescence techniques. All molecules present in the measuring volume contribute to the Raman signal. This can lead to very complex spectra that are difficult to interpret. On the other hand it can also put immunofluorescence results in a different perspective. In a Raman spectroscopic study of the effects of low pH treatments of polytene chromosomes, needed to evoke the binding of anti-Z-DNA antibodies, it was found that these treatments not only led to Z-DNA immunoreactivity, but also to large-scale structural changes in chromosomal DNA structure (Puppels *et al.*, 1994a; see also Section 29.4.1.2). This example illustrates that rather than competitive techniques, Raman and (immuno) fluorescence methods should be considered as complementary techniques. In fact, it is possible to combine the two, by choosing fluorescent labels and laser excitation wavelength in Raman experiments in such a way that in the Raman experiments no fluorescence is excited. As shown below (Section 29.4.2.2), specific cell subsets can be made available for Raman spectroscopic investigation by labelling cells with a specific antibody conjugated to a fluorescent label and sorting them on a flow cytometer on the basis of their fluorescence signal. Such labelling could also be used to facilitate the localization of specific structures, e.g. sites of transcription of replication in a cell by

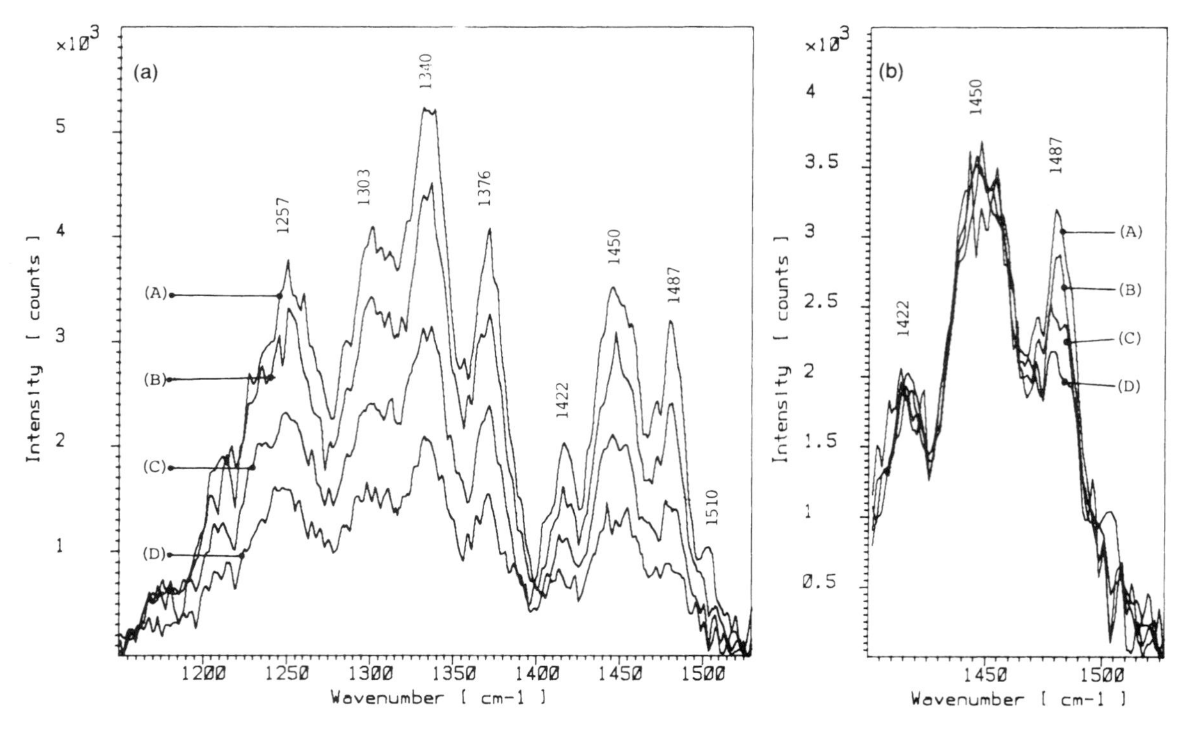

Figure 29.7 Effect of laser light (514.5 nm)-induced sample degradation on the Raman spectrum of metaphase chromosomes. Raman line assignments are given in Table 29.2. (a) Decrease of the intensity of the Raman signal, as recorded during measurements on isolated (unfixed) single (Chinese hamster lung cell) metaphase chromosomes. Shown are spectra obtained during (A) 1st minute of irradiation; (B) 3rd minute; (C) 5th minute; and (D) 9th minute of irradiation. (Averages of 10 measurement, for more details see Puppels *et al.*, 1991b.) (b) Illustration of the sensitivity of the 1487 cm^{-1} guanine line to 514.5 nm light irradiation. Detail of the spectra shown in Fig. 29.7(a). The spectra were scaled to have equal intensity in the spectral interval 1150–1460 cm^{-1}. (From Puppels et al., 1991b, reproduced by permission of Academic Press Inc.)

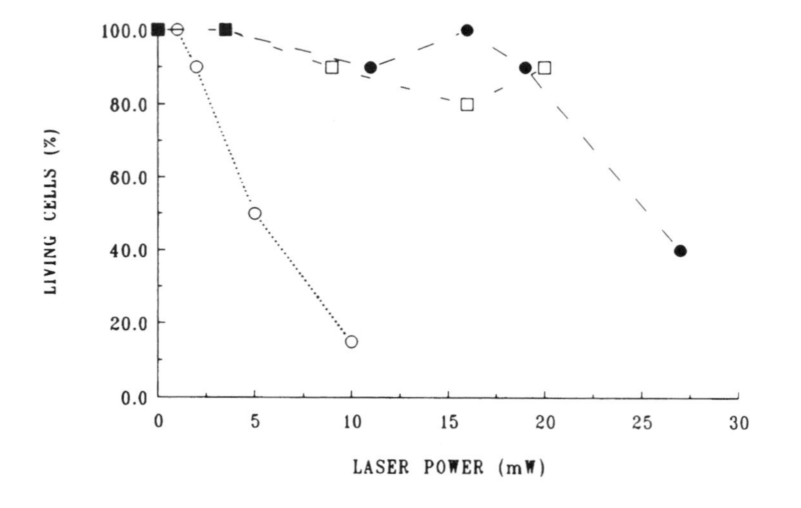

Figure 29.8 Percentage of cells (human lymphocytes) surviving a 5 min laser light irradiation as a function of laser light power for 514.5 nm, 632.8 nm and 660 nm. ○ · · · ○ 514.5 nm, ×63 Zeiss Plan Neofluar water immersion objective (NA 1.2); focus diameter ~ 0.5 μm. □ · · · □ 632.8 nm, ×40 Nikon E Plan objective; focus diameter ~ 1 μm. ● · · · ● 660 nm, same objective as for 514.4 nm. The laser beam was focused in the centre of the cells. Each point in the graph represents 10 irradiated cells (20 for 660 nm, 25 mW). Control samples contained > 90% living cells. (From Puppels *et al.*, 1991b, reproduced by permission of Academic Press Inc.)

means of fluorescence microscopy, prior to a Raman measurement.

29.4.1 Chromatin in chromosomes and cells

DNA in cells is complexed to proteins. Proteins fold and package the DNA (histones) and are essential in such fundamental processes as replication and transcription of the DNA (non-histone proteins). In non-dividing, transcriptionally inactive cells, the non-histone protein (NHP) content of chromatin is usually low (Bradbury *et al.*, 1981; Mathews & Van Holde, 1990). The way chromatin is organized strongly influences the transcriptional activity of the DNA. It is a well-known fact that the (inactive) heterochromatin in cells is much more condensed than (active) euchromatin. In R-banded metaphase chromosomes the stained bands are relatively rich in transcribed genes (Bickmore & Sumner, 1989). The fact that metaphase chromosome regions react differently to the R-banding treatment in a manner that correlates with transcriptional activity during interphase is another reflection of the fact that active and inactive chromatin are organized in a different way. In polytene chromosomes intense transcription of a gene leads to local puffing, i.e. swelling of the chromosome at the position where the gene is located.

The CRM was employed to characterize (local) chromatin composition in a number of systems. Shown

Table 29.2 Raman band assignments for (B-)DNA–protein samples.

Line position (cm^{-1})	*Assignment*	
622	Phenylalanine	
645	Tyrosine	
669	Thymine	
681	Guanine ring breathing	
729	Adenine ring breathing	
749	Thymine ring breathing	
782 } 787	Cytosine ring breathing	
790	DNA:PO_2 symmetric stretching	
833	DNA:PO_2 asymmetric stretching, tyrosine	
853	Tyrosine	
896	DNA:backbone	
925	DNA:backbone	
932	Protein:α-helix C–C skeletal mode	
1004	Phenylalanine	
1017	DNA:backbone C–O stretching	Protein: C–C and
1032	Phenylalanine	C–N stretching modes
1057	DNA:backbone C–O stretching	
1094	DNA:PO_2^- symmetric stretching	
1126	Protein: C–N stretching	
1176	Tyrosine, phenylalanine	
1211	Thymine, tyrosine, phenylalanine	Protein: Amide III
1240	Thymine	β-Sheet: ~1230–1240
1255	Cytosine	Random coil: ~1240–1250
		α-helix: ~1260–1300
1304	Adenine, cytosine	Protein: CH_2/CH_3
1340	Adenine	deformations
1376	Tymine, adenine, guanine	
1422	Adenine	
1449	Protein:CH_2/CH_3 deformations	
1487	Guanine, adenine	
1511	Adenine	
1578	Guanine, adenine	
1606	Phenylalanine, tyrosine	
1617	Tyrosine, phenylalanine	α-Helix: ~1645–1655
1640–1700	Protein: amide I	Random coil: ~1660–1670
1670	Thymine 30	β-Sheet: ~1665–1680

From Erfurth & Peticolas (1975), Thomas *et al.* (1977, 1986), Tu (1982), Prescott *et al.* (1984), Yu *et al.* (1987).

below are results of measurements of polytene chromosomes, metaphase chromosomes and nuclei in intact human white blood cells. Table 29.2 gives a listing of DNA and protein vibrations and their assignments.

29.4.1.1 Polytene chromosome banding pattern

Polytene chromosomes are a strongly amplified form of interphase chromosomes, found, for example, in salivary gland cells of *Drosophila* and *Chironomus*. They arise through many rounds of DNA replication without subsequent separation of daughter chromatids. The chromatids run through the entire length of a chromosome. Degrees of polyteny (the number of chromatids in a chromosome) of up to 8000 are encountered.

These giant chromosomes possess characteristic patterns of alternating dark bands and light interbands, which are readily observable under a light microscope (see Fig. 29.9). The fact that an order-of-magnitude correspondence exists between the number of genes and the number of bands and interbands has given rise to the concept that the banding pattern reflects functionally different chromatin domains. Beermann (1952) found that puffing of a chromosome region was the result of intense transcription. Based on that work others hypothesized that genes reside in bands. Evidence for transcriptional activity in up to 70% of all bands of *Drosophila melanogaster* was found (Ananiev & Barsky, 1978). However, evidence to the contrary was also reported, showing transcriptional activity primarily in interbands, diffuse bands and puffs (Fujita & Takamoto 1963; Semeshin *et al.*, 1979; Jamrich *et al.*, 1977). A more recently forwarded hypothesis states that the band–interband pattern is a reflection of local transcriptional activity rather than of the distribution of genic and non-genic material (Zhimulev *et al.*, 1981; Hill & Rudkin, 1987).

Only little is known about the organization of band and interband chromatin in terms of DNA and protein concentration, although it is known that in general the compaction of chromatin does influence transcriptional activity. Studies of the band–interband pattern of *Chironomus thummi thummi* chromosomes with the CRM indicate that their characterization by means of Raman spectroscopy is a very promising approach. Spectra were recorded of (3:1 ethanol:acetic acid) fixed polytene chromosomes from squash preparations and of unfixed chromosomes, isolated under physiological conditions (Robert-Nicoud *et al.*, 1984). In general, squash preparations are used in many experiments because of the very distinct and stable banding pattern of the chromosomes. The isolation of unfixed chromosomes is more time-consuming and the contrast between bands and interbands, as observed under a microscope, is lower. However, unfixed chromosomes provide much more information about the native state of the chromosomes.

Raman spectra were recorded of eight neighbouring band and interbands of the fixed chromosome shown in Fig. 29.9. Spectra of different bands were virtually indistinguishable. Spectra of different interbands showed large variations both in absolute signal intensity and in relative peak intensities. Figure 29.10 shows four of the spectra that were obtained. On the basis of the intensities of the 1093 cm^{-1} DNA backbone line and the 1448 cm^{-1} line due to protein CH_2

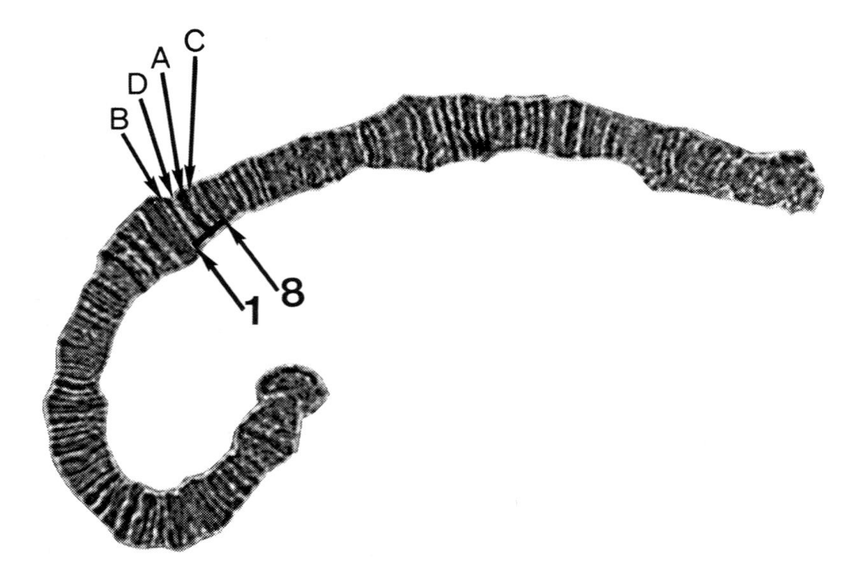

Figure 29.9 *Chironomus thummi thummi* chromosome II. Indicated are the positions of the measurements shown in Fig. 29.10 (arrows A, B, C and D) and the eight bands and interbands (between arrows 1 and 8) for which relative protein and DNA concentrations, protein–DNA ratio and relative chromatin concentration were determined (Fig. 29.11). Bar denotes 10 μm. (From Puppels *et al.*, 1991d, copyright 1991 by San Francisco Press, Inc., with permission.)

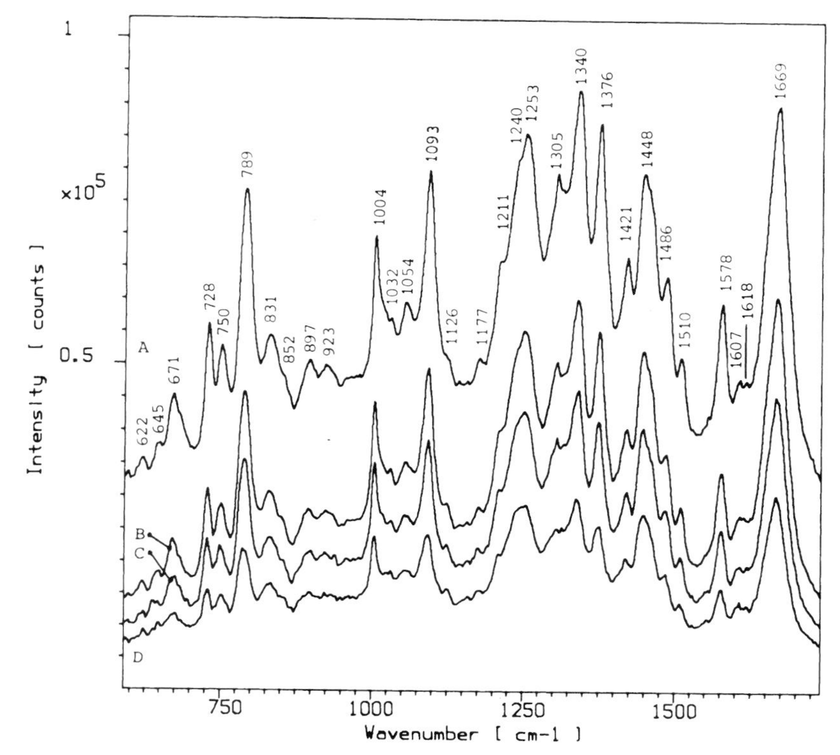

Figure 29.10 Raman spectra obtained from the fixed chromosome shown in Fig. 29.9; (A) band, (B, C, D) interbands. Laser power: 15 mW (660 nm); signed integration time: 10 min; ×63 Zeiss Plan Neofluar water immersion objective (NA 1.2, 0.12–0.22). The spectra have been shifted along the ordinate for clarity. Background (phosphate-buffered saline and fused silica) signal has been subtracted. (From Puppels *et al.*, 1991d, copyright 1991 by San Francisco Press, Inc., with permission.)

and CH_3 deformations, DNA and protein concentrations can be determined (method described in Puppels *et al.*, 1991c).

Shown in Fig. 29.11 are relative protein and DNA concentrations for the eight bands and interbands as well as the protein–DNA ratio (mass/mass) and total relative chromatin concentration. It shows clearly that DNA and protein concentrations are highest and nearly invariable in the bands. Both concentrations are very variable in the interbands. Parts (D) and (E) of Fig. 29.11 illustrate a very intriguing aspect: the local chromatin concentration in polytene chromosomes appears to be inversely related to the protein–DNA ratio.

In Fig. 29.12 the results of measurements on bands and interbands of unfixed chromosome are shown. These confirm the fixed chromosome results. Total chromatin concentration is lower in interbands (lower Raman signal intensity) and protein–DNA ratio is higher in interbands (Table 29.3). The position of the amide I and III lines in the spectra provide information about protein secondary structure. The maximum of the amide I line for bands is at 1662 cm^{-1} and for interbands at 1666 cm^{-1}. In the difference spectrum the amide I maximum is found at 1669 cm^{-1}. This shift of the amide I vibrations to higher wavenumbers indicates that on average the interband proteins contain a higher relative amount of β-sheet or random coil domains than the band proteins. The absence of a strong line below 1240 cm^{-1} in the amide III region in Fig. 29.12(C) which would be indicative of β-sheet structure, makes clear that it is especially the relative amount of random coil domains that is higher in interbands than in bands. This can be due either to actual local variations in the secondary structure of the chromosomal proteins, or to local variations in chromosomal protein composition. Electron microscopic (Ananiev & Barsky, 1985) and immunofluorescence (Bustin *et al.*, 1977; Kurth *et al.*, 1978) studies have shown that the DNA in both bands and interbands is organized in a nucleosomal form. The protein–DNA mass ratio in nucleosomes is 1:1. As indicated in Table 29.3, this means that bands contain much less non-histone proteins than interbands. Therefore it would seem likely that the differences in protein secondary structure are due to differences in protein composition. The high intensity of the phenylalanine line at 1004 cm^{-1} in Fig. 29.12(C) could be a further indication of this, as it is most likely explained by a higher relative amount of phenylalanine residues in the interband proteins.

A study in which atomic force microscopic images and Raman spectra were obtained of the same bands and interbands indicated that the difference in

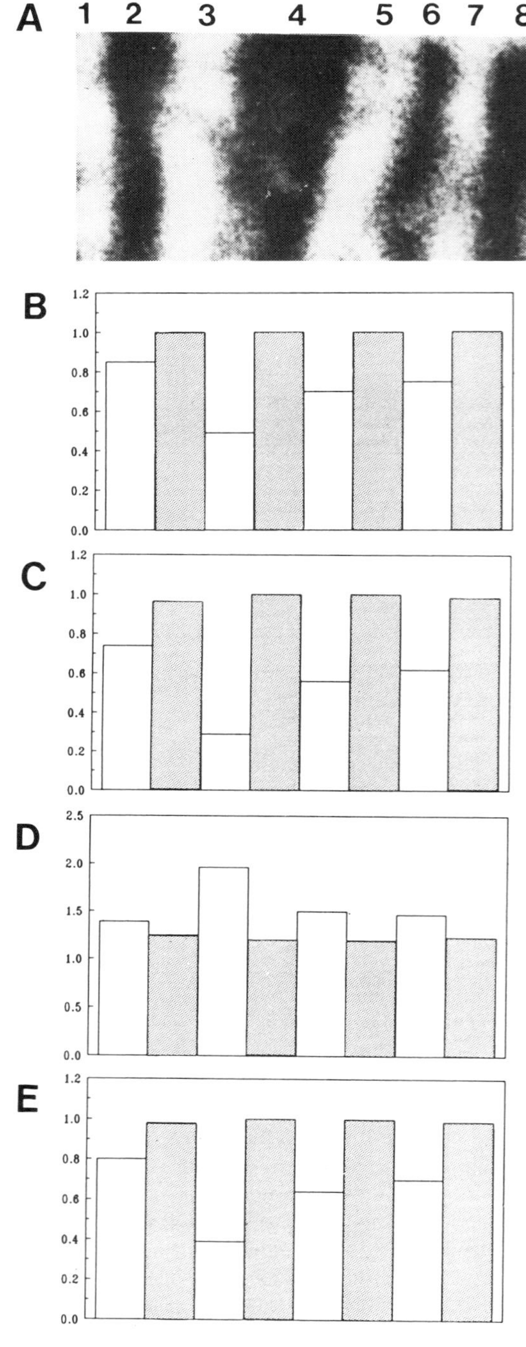

Figure 29.11 Variation in protein and DNA concentrations in eight adjoining bands and interbands of the chromosome of Fig. 29.9. (A) Magnification of Fig. 29.9; (B) protein concentration (normalized to band 4); (C) DNA concentration (normalized to band 4); (D) protein–DNA ratio (absolute values); (E) total chromatin concentration (protein + DNA; normalized to band 4). (B, C, E) error ± 5%, (D) error ± 19% (for absolute values), ± 10% (for normalized values). (From Puppels *et al.*, 1991d, copyright 1991 by San Francisco Press, Inc., with permission.)

molecular composition of bands and interbands coincides with differences in ultrastructural organization of the chromatin (Puppels *et al.*, 1993a).

29.4.1.2 Low-pH-induced changes in DNA structure of polytene chromosomes

Raman spectroscopy was employed to carry out a detailed study of the effects of low-pH treatments on molecular composition and DNA structure of polytene chromosomes. The direct incentive for this work was the fact that in many different immunofluorescence studies it was found that fluorescently labelled antibodies, raised against non-B-DNA structures (Z-DNA, triplex DNA, DNA-cruciforms), stained chromosomes and cell nuclei much more intensely after incubation at low pH for a period of time; staining in general became more intense and homogeneous with longer exposure to low pH, and lower pH (see e.g. Hill *et al.*, 1984; Robert-Nicoud *et al.*, 1984). The interpretation of these results, in terms of whether or not they are evidence of the existence of, and a possible function for these non-B-DNA-structures in nature, is the subject of much debate and speculation.

Raman spectra of polytene chromosomes provided a detailed picture of the changes in chromosomal DNA structure that occur at low pH. Detailed discussion of the spectral changes that occur are beyond the scope of this chapter (see Puppels *et al.*, 1994b). Table 29.4 lists the changes in DNA structure as a function of pH. It shows that already above pH 3.6 protonation of adenine and guanine–cytosine base pairs commences. This leads to unpairing of the adenine and thymine bases. The guanine–cytosine base pair, on the other hand, undergoes a change in conformation. Evidence was found that the new conformation is a protonated Hoogsteen base pair. This conformational change was found to be completely reversible upon return to neutral pH. A tempting hypothesis would be that these protonated Hoogsteen base pairs exist in equilibrium with Watson–Crick base pairs, and that they also exist in very low concentrations at neutral pH.

Spectra obtained from chromosomes upon return to neutral pH, after exposure to low pH, provided information about irreversible changes in molecular composition and DNA structure of the chromosomes as a function of pH. Table 29.5 shows how the amount of protein that is extracted from the chromosomes increases when the chromosomes are exposed to lower pH. Exposure of chromosomes to pH 2.2 or lower was found to lead to irreversible changes in DNA structure, in particular unstacking adenine and thymine residues (affecting at least 10% of the AT base pairs) and changes in DNA backbone structure. The fact that exposure to pH 2.6 or higher does not lead to

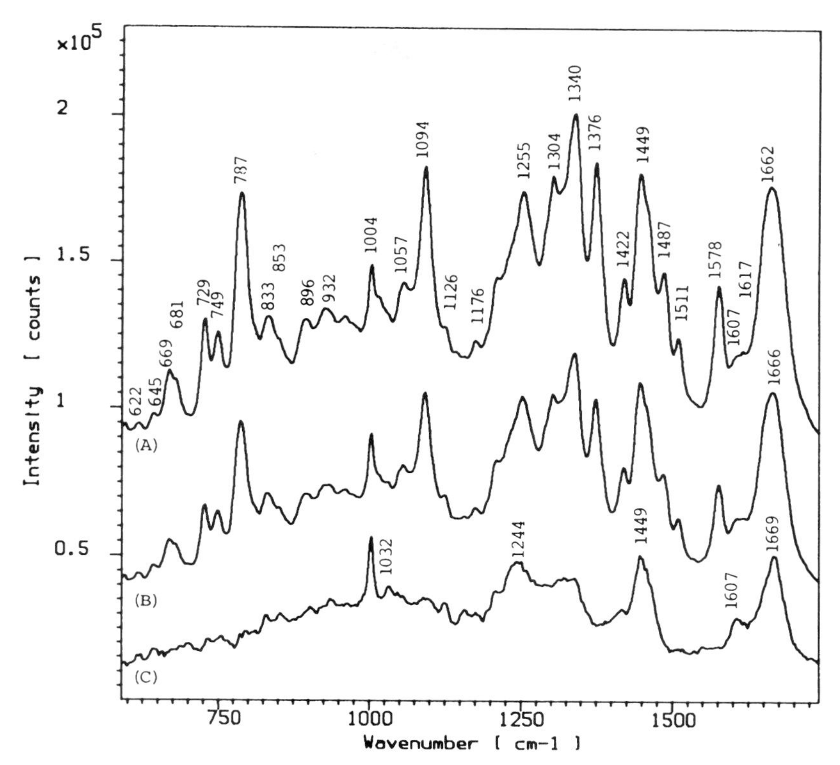

Figure 29.12 Raman spectra of unfixed *Chironomus thummi thummi* polytene chromosomes. (A) Average for four different bands; (B) average of four different interbands; (C) difference spectrum B − A after normalizing (A) and (B) with respect to the 1094 cm^{-1} DNA line (shown 3 times enlarged). Laser power: 15 mW (660 nm); measuring time: 10 min; microscope objective ×63 Zeiss Plan Neofluar water immersion (NA 1.2, 0.12–0.22). Spectra shifted along ordinate for clarity. (From Puppels *et al.*, 1991e, reproduced by permission of The Royal Society of Chemistry.)

Table 29.3 Protein and DNA concentrations of bands and interbands of unfixed *Chironomous thummi thummi* polytene chromosomes.

	Protein[a] (± 5%)	*DNA*[a] (± 5%)	*Protein:DNA*[b] (*mass/mass*) (± 19%)	*Total chromatin*[a] (± 5%)	*NHP:histone*[b,c] (*mass/mass*)
Bands	1.00	1.00	1.3	1.00	0.3 (± 0.3)
Interbands	0.79	0.65	1.6	0.76	0.6 (± 0.3)

[a] Normalized values.
[b] Absolute values.
[c] Assumption: DNA:histones = 1:1 (mass/mass).

irreversible changes in DNA structure implies that for such changes to occur unpairing at low pH of both AT base pairs and GC base pairs appears to be a prerequisite (see Table 29.4).

These results underline the unique possibilities of Raman microspectroscopy for studying biomolecular structure *in situ*.

29.4.1.3 Metaphase chromosomes

During mitosis the chromatin in a cell condenses into metaphase chromosomes, and transcription stops. The chromosomes consist of two identical chromatids connected in the centromere (see Fig. 29.6(A)). Metaphase chromosome banding was mentioned in the introduction of this chapter. Unlike the case of polytene chromosomes, here physicochemical treatments or endonuclease digestion of the chromosomes followed by staining are needed to make the banding pattern visible. Since the discovery of quinacrine-banding (Caspersson *et al.*, 1968), many other types of banding patterns have been discovered (reviewed in, for example, Babu & Verma, 1989). Although the banding pattern cannot be observed on untreated, unstained chromosomes, it is clear that they must reflect one or more properties of the chromosomes in their native state.

Raman microspectroscopy may play an important role in elucidating the origin of the banding patterns. The experiments with polytene chromosomes show

Table 29.4 Overview of the changes in polytene chromosomal DNA structure occurring at low pH

	pH				
	Neutral	*3.6*	*2.6*	*2.2*	*1.8*
A protonation	– – —	—	—	— – –	
T unstacking	– – —	—	—	— – –	
A unstacking		– – —	—	—	— – –
GC base pair protonation	– – —	—	—	— – –	
G conformational change	– – —	—	— – –		
C unstacking			– – —	—	— – –
AT unpairing	– – —	—	—	— – –	
GC unpairing			– – —	—	— – –
DNA backbone conformational changes					
~835 ↓		– – — AT sites	—	— – – AT and GC sites	
~877 ↑		– – —	—	— – –	
~803 ↑			– – —	—	— – –

[a] Legend: (– – —) starts in this interval, (—) continues in this interval, (— – –) completed in this interval.

Table 29.5 Changes in chromosomal DNA–protein ratio (weight/weight) induced by low-pH treatment

	DNA:protein	
Treatment	*Absolute (±19%)*	*Normalized (±9%)*
No	0.7	1
pH 3.6	0.7	1
pH 2.6	0.8	1.1
pH 2.2	0.9	1.3
pH 1.8	1.3	1.8
45% acetic acid	1.7	2.3

that it is a sensitive monitor of local variations in chromatin composition and structure. Figure 29.13 shows that good Raman spectra can also be obtained of the much smaller metaphase chromosomes. Spectrum (a) was obtained of a metaphase chromosome isolated under near physiological conditions, spectrum (b) of a 3:1 methanol:acetic acid-fixed chromosome from a squash preparation dried in air. Comparison of the two spectra shows that Raman spectroscopy is also a sensitive monitor of the molecular structural changes that take place as a result of the treatments a chromosome undergoes.

The shift of the amide I maximum to 1668 cm^{-1} in spectrum (b) indicates a change in protein secondary structure, which is probably due to protein denaturation by the fixative. The intensity increase of the phenylalanine line at 1004 cm^{-1} could have the same cause, indicating a change in the microenvironment of the phenyl rings. However, it cannot be excluded that during the fixation and squashing of the cells the chromosomes are contaminated with, for example, cytoplasmic proteins.

The fixation and drying of the chromosomes also alters DNA structure. Whereas in the unfixed chromosome the DNA is in a double helical right-handed B-form as evidenced by the characteristic vibrations at 1094 cm^{-1} and 835 cm^{-1} (backbone) and 784, 750, 731 and 682 cm^{-1} (ring breathing vibrations of C, T, A and G, respectively (see Table 29.2), characteristic of C_2-endo sugar pucker), in the spectrum of the fixed chromosome small lines at 708, 690, 660 and 645 cm^{-1} could indicate the presence of a small amount of A-DNA, a double helical right-handed DNA form found under low humidity conditions (Erfurth *et al.*, 1975). The high intensity of 1245 cm^{-1} in the fixed chromosome spectrum may be due to a partially denaturated state of the DNA leading to a hypochromic effect owing to unstacking of T-rings. Further evidence of this is the increased intensity of the A-line at 731 cm^{-1} which may have the same cause, i.e. unstacking of A-rings. We assume that the strong line at 928 cm^{-1} in spectrum (b) is due to acetic acid bound to the chromosome. Its intensity varies widely between different preparations.

We expect that a systematic Raman spectroscopic study of the local chromatin composition of metaphase chromosomes and the changes therein caused by physicochemical banding treatments will provide valuable information concerning the origin of the banding patterns.

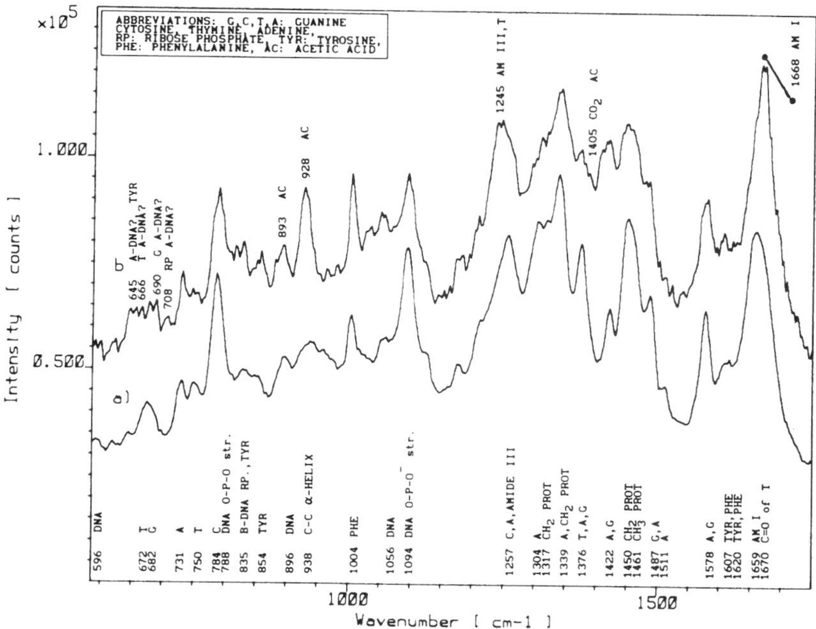

Figure 29.13 Raman spectra of single Chinese hamster lung cell chromosomes, obtained on a chromatid arm midway between centromere and telomere. (a) Unfixed chromosomes in hypotonic buffer; (b) fixed chromosomes, dry in air (enlarged by a factor of 2.5). Laser power: 10 mW (660 nm); measuring time: (a) 15 min, (b) 40 min; microscope objective ×63 Zeiss Plan Neofluar water immersion (NA 1.2). (From Puppels *et al.*, 1990c, reproduced by permission of John Wiley & Sons Ltd.)

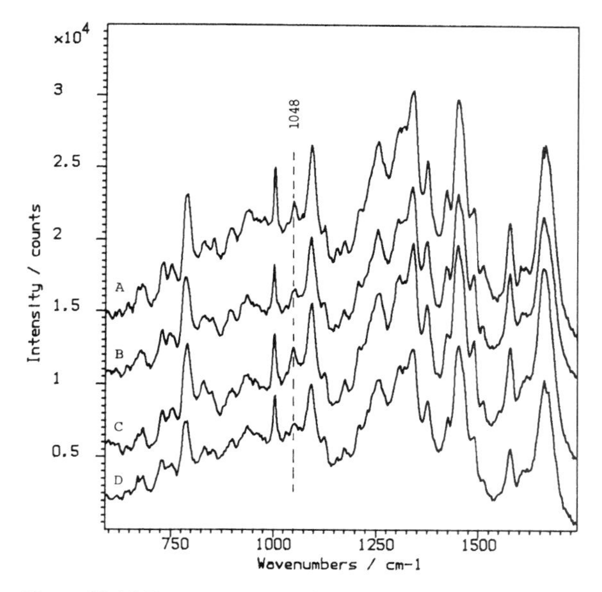

Figure 29.14 Raman spectra of the nucleus of single whole human leukocytes. (A) Lymphocyte, (B) neutrophil, (C) eosinophil, (D) basophil. Laser power: 6 mW (660 nm); measuring time: 150 s per measurement; microscope objective: ×63 Zeiss Plan Neofluar water immersion (NA 1.2). (From Greve & Puppels, 1993, reproduced by permission of John Wiley & Sons Ltd.)

29.4.1.4 Cell nuclei

The spatial resolution of the CRM makes it possible to obtain spectra of chromatin in the nucleus of an intact cell free from signal contributions of the cytoplasm (Puppels *et al.*, 1991c; Greve & Puppels, 1993; Takai *et al.*, 1997). Figure 29.14 shows spectra obtained from the nucleus of human white blood cells. As in the chromosome spectra, all lines except the 1048 cm^{-1} line can be assigned to DNA and protein vibrations (see Table 29.2). Possible RNA or phospholipid signal contributions are not discernible. The spectra are very repoducible. Only slight vibrations (~ 10%) in the intensity ratio of the 1449 cm^{-1} and 1094 cm^{-1} lines are found (indicating small variations in protein–DNA ratio). A conspicuous aspect is that the protein–DNA ratio is high (~ 2.3:1). This indicates the presence of a large amount of non-histone proteins. An unexpected result is the presence of a line at 1048 cm^{-1}. This line is not present in spectra of free DNA, nucleosomes, isolated chromatin, polytene or metaphase chromosomes (Erfurth & Peticolas, 1975; Thomas *et al.*, 1977; Goodwin & Brahms, 1978; Thomas *et al.*, 1980, 1986; Savoie *et al.*, 1985; Hayashi *et al.*, 1986) and is also not found in cytoplasmic spectra. Moreover, it is the only line that varies strongly in intensity in the nuclear spectra, sometimes being absent altogether. This may indicate that it originates from a compound or molecular complex unevenly distributed in or close to the nucleus. A possible candidate is taurine. Its spectrum contains an intense line at 1048 cm^{-1} (Caillé *et al.*, 1987). Taurine has been reported to be present in high concentrations up to 22–26 mM in human leukocytes (Huxtable, 1992). In neutrophilic granulocytes it is thought to play a role in the self-defence of the cells against the hypochlorous acid that is produced by the myeloperoxidase–hydrogen peroxide–chloride system (Learn *et al.*, 1990, see also Section 29.4.2.1 below).

29.4.2 The cytoplasm

The molecular composition of the cytoplasm of a cell is very complex and the distribution of molecules inhomogeneous. Because all molecules in the measuring volume contribute Raman signals this could easily give rise to unreproducible and uninterpretable spectra. Therefore, if the aim is to study a particular molecule in its natural environment, Raman microspectroscopy can be most successfully applied when that molecule is present in much higher quantities than all other molecules, either throughout the cytoplasm or in specific, identifiable cytoplasmic organelles, or when one or a few types of molecules give a much stronger Raman signal than all others. Below are given examples that illustrate how Raman spectroscopy can be used to obtain detailed information about the structure of the active-site haem ring of peroxidases in human granulocytes, and to locate carotenoids in lymphocytes.

29.4.2.1 *The study of haem molecules* in situ *and* in vivo

Haem molecules have been and continue to be a source of inspiration for Raman spectroscopists, both in a theoretical and in an experimental sense. The vast majority of studies are carried out on isolated, purified compounds. Detailed structure–function relations are being proposed on the basis of these studies. It is important, however, to be able to also study these molecules *in situ*, i.e. in their natural environment, and *in vivo*, in order to study structural changes that occur during cellular processes.

Myeloperoxidase in neutrophilic granulocytes
Neutrophils are very motile cells that respond to chemotactic stimuli. Their main functions are phagocytosis, killing and digestion of bacteria and other microorganisms. They possess some 200 cytoplasmic granules of about 0.2 mm in diameter which contain a wide variety of oxidative metabolites and digestive enzymes. Two types of granules are distinguished: azurophilic and specific granules. A prominent enzyme of the azurophilic granules is myeloperoxidase (MPO), a haem protein (Klebanoff & Clark, 1978). A neutrophil contains about 6 pg of MPO (Schultz & Kaminker, 1962). This enzyme catalyses the oxidation of chloride to hypochlorous acid, which is cytotoxic to bacteria

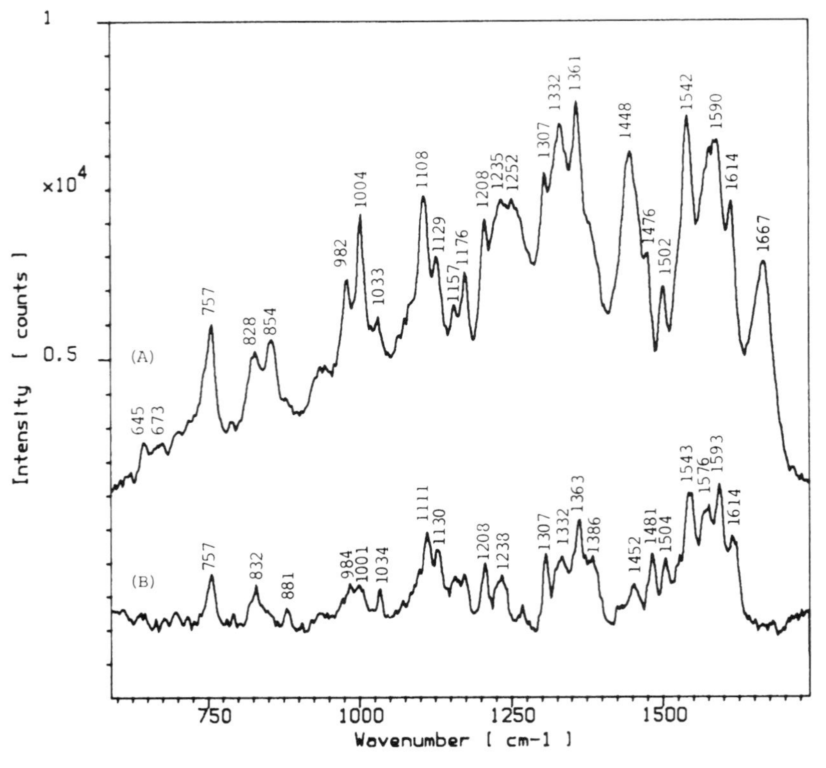

Figure 29.15 (A) Human neutrophil cytoplasmic Raman spectrum, averaged over five measurements on different cells. Laser power: 6 mW (660 nm); measuring time: 30 s per measurement; microscope objective: see Fig. 29.14. (B) Raman spectrum of isolated native (oxidized) human MPO. Sample concentration: 230 μM in 200 mM phosphate buffer; laser power: 10 mW (660 nm); measuring time 18 × 100 s; sample measured in a flow system. (From Puppels *et al.*, 1991c, reproduced from the *Biophysical Journal* by copyright permission of the Biophysical Society.)

(Parker, 1984). After phagocytosis, cytoplasmic granules fuse with the phagosome and degradation of the ingested object commences. The active-site haem groups of MPO are strong Raman scatterers, even when excited off-resonance. A comparison of spectra (A) and (B) of Fig. 29.15 makes clear that the cytoplasmic Raman spectrum of neutrophils is dominated by signal contributions of the MPO-haem groups (Puppels *et al.*, 1991c). This means that changes in haem structure as a result of activation and inactivation of this enzyme, during the course of phagocytosis and digestion of a foreign object, may be studied *in vivo* (Sijtsema, 1997; see also paragraph on eosiniphil peroxidase below).

Eosinophil peroxidase in human eosinophilic granulocytes

Eosinophilic granulocytes form a small fraction (~ 2%) of the circulating white blood cells. They have a bilobed nucleus and their cytoplasm contains many granules (~200) of roughly 1 μm in diameter. They are part of the humoral immune system, but much remains to be learned about their precise role and functions. One of their tasks is to protect the body against invading parasites. They do this by attaching to or by phagocytosis of a parasite, after which a range of bactericidal substances, contained in cytoplasmic granules, are brought into play to kill and digest the invading organism. One of these substances is the enzyme eosinophil peroxidase (EPO), which in the presence of hydrogen peroxide and halides produces bleach-like hypohalous acids. It is a protein which possesses an active-site protoporphyrin IX haem ring. EPO is found in the matrix of the granules, which surrounds a crystalloid core consisting entirely of the so-called Major Basic Protein. When activated in the presence of hydrogen peroxide and bromide or chloride, this enzyme produces hypohalous acid. Concentrations of 15 pg EPO per cell have been reported (Gleich & Adolphson, 1986). With a MW of about 72 kDa (Bolscher & Wever, 1984) this amounts to a concentration of the order of 1 mM in the cytoplasmic granules.

This high concentration makes it very easy to record Raman spectra of EPO *in situ*, i.e. in the cytoplasmic granules in intact eosinophils. This is illustrated in Fig. 29.16, which shows that the Raman spectrum obtained from cytoplasmic regions filled

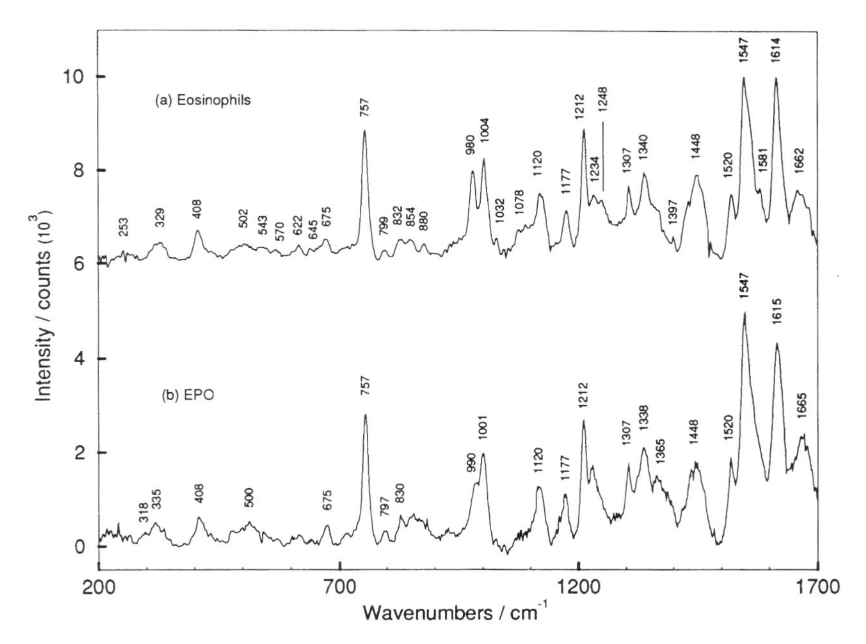

Figure 29.16 Raman spectra of eosinophil peroxidase, using a confocal Raman microspectrometer and 660 nm laser excitation in a charge transfer absorption band of the enzyme (see Fig. 29.17). (a) Spectrum obtained from granules in the cytoplasm of living human eosinophilic granulocytes. Average of 20 single cell measurements. Cells in Hank's buffered salt solution on fused silica substrates. Laser power: 6 mW, signal integration time: 30 s per measurement, ×63 Zeiss Plan Neofluar water immersion objective.
(b) Spectrum obtained from 100 μM sample of isolated native (oxidized) human EPO in a 200 mM potassium phosphate solution (pH 7.2) and 0.5% Tween. Average of 10 measurements. Laser power: 15 mW, signal integration time: 300 s per measurement. (Reproduced from Salmaso *et al.*, 1994, by permission of the Biophysical Society.)

(a)

PROTOPORPHYRIN IX

(b)

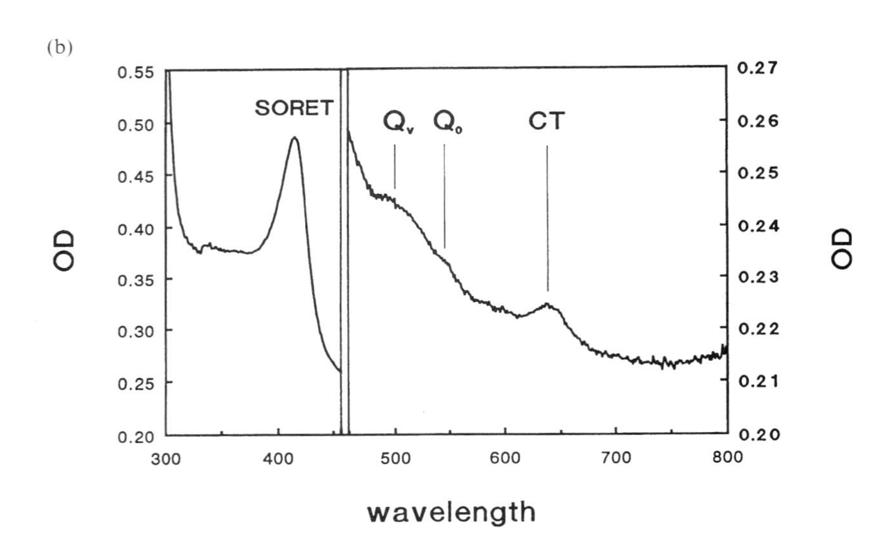

Figure 29.17 Active site haem group of EPO: protoporphyrin IX. (b) Absorption spectrum of a suspension of human eosinophilic granulocytes (2×10^7 cells ml^{-1}), obtained with an integrating sphere spectrophotometer.

with granules is virtually exclusively due to EPO (Salmaso *et al.*, 1994). Using laser excitation in the Soret band (413 nm), in the Q_V-band (514.5 nm) and in a charge transfer absorption band (660 nm) of EPO, the intensity of Raman bands can be selectively enhanced (see Figs 29.17–29.19). The detailed vibrational spectroscopic information that is obtained in this way, combined with the information contained in the polarization ratios of Raman bands, enables a fairly complete and accurate assignment of Raman bands to haem vibrational modes (Table 29.6). From this detailed information about haem-ring structure of the enzyme *in situ* can be extracted (Salmaso *et al.*, 1994).

It was found that the active site of EPO in the resting cell is a ferric, high-spin, 6-co-ordinated protoporphyrin IX, in agreement with literature on the isolated enzyme. The pyrrole-nitrogen Fe distance was determined to be 2.04 Å, with the central iron displaced from the porphyrin plane. The haem ring has a histidine residue with a strong imidazolate character as the proximal ligand and a weakly bound water molecule as the distal ligand. The low wavenumber region of the EPO spectrum has an uncommon appearance, when compared with that of other protoporphyrin IX haem proteins. The vinyl bending modes found with Soret excitation at 337 and 410 cm^{-1} are unusually strong and a propionate

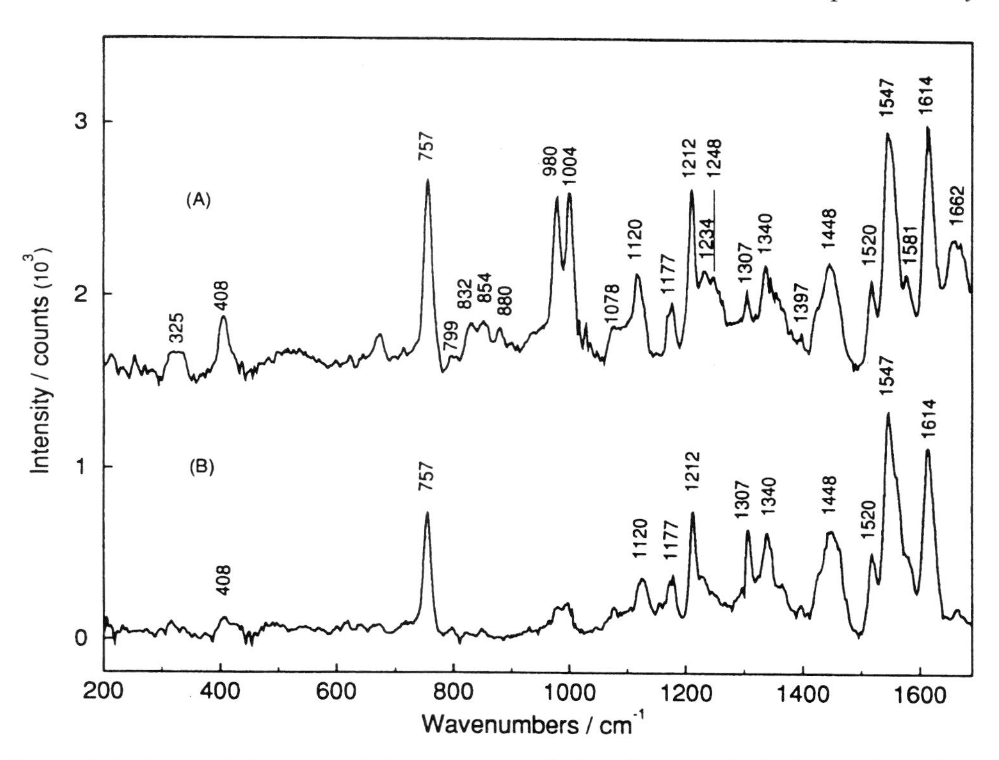

Figure 29.18 Polarized (A) and depolarized (B) Raman spectra obtained from granules in the cytoplasm of human eosinophilic granulocytes, using 660 nm laser excitation. Laser power: 7 mW. Spectra averaged over 30 (A) and 40 (B) single cell measurements on different cells. Other experimental conditions: see caption of Fig. 29.16. (Reproduced from Salmaso *et al.*, 1994, by permission of the Biophysical Society.)

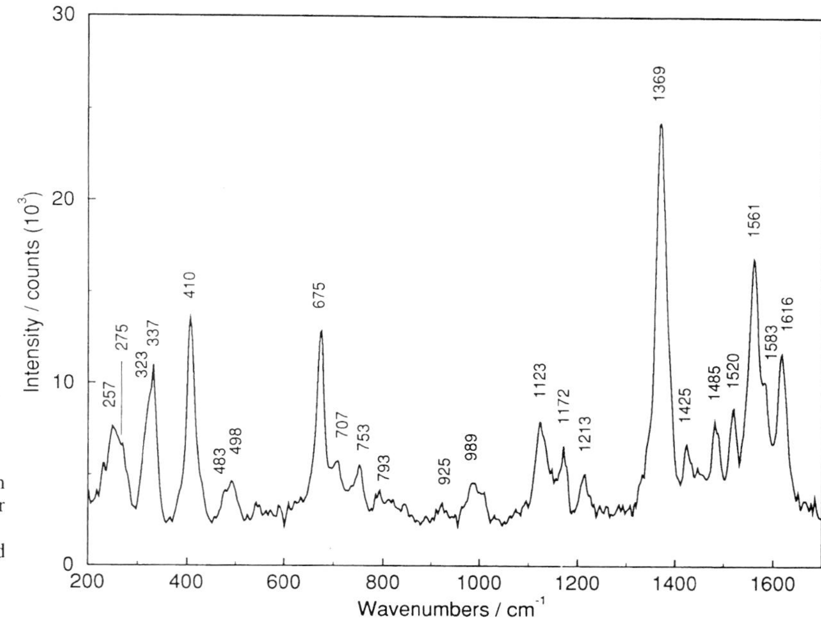

Figure 29.19 Raman spectrum obtained from granules in the cytoplasm of human eosinophilic granulocytes with 413.1 nm laser excitation in the Soret absorption band. Spectrum averaged over 10 single cell measurements on different cells. Laser power on sample: 400 μW, signal integration time: 10 s per measurement. Other experimental conditions: see caption of Fig. 29.16. (Reproduced from Salmaso *et al.*, 1994, by permission of the Biophysical Society.)

Table 29.6 Assignment of EPO Raman bands

λ_{exc} (nm)						Assignment				
660		*514.5*		*413.1*		*Porphyrin modes*				
Freq. (cm^{-1})	*Depol. ratio*	*Freq. (cm^{-1})*	*Depol. ratio*	*Freq. (cm^{-1})*	*Depol. ratio*	*Mode number*	*Symm. D_{4h}*	*Main contribution*	*Vinyl modes*	*Other*
1662	0.2	1659								Amide I
				1616	0.3				$\nu(C{=}C)$	
1614	0.8	1616	0.6			ν_{10}	B_{1g}	$\nu(C_aC_m)$		
1581		1581		1583	0.3	ν_{37}	E_u	$\nu(C_aC_m)$		
				1561	0.2	ν_2	A_{1g}	$\nu(C_bC_b)$		
		1558	0.8			ν_{19}	A_{2g}	$\nu^1(C_aC_m)$		
1547	0.8					ν_{11}	B_{1g}	$\nu(C_bC_b)$		
1520	0.8	1520	0.5	1520	0.2	ν_{38}	E_u	$\nu(C_bC_b)$		
				1485	0.3	ν_3	A_{1g}	$\nu(C_aC_m)$		
1448	0.8	1450	0.75							CH_2/CH_3
		1424	0.8	1425	0.7				$\delta_s(=CH_2)_{(1)}$	
		1369	0.4	1369	0.3	ν_4	A_{1g}	$\nu(C_aN)$		
1340	0.85	1343							$\delta_s(=CH_2)_{(2)}$	
1307	2	1307	1.2			ν_{21}	A_{2g}	$\delta^1(C_mH)$		
									$\delta(CH{=})$	
1234		1231				ν_{13}	B_{1g}	$\delta(C_mH)$		
1212	0.75	1212	0.4	1213	0.3	$\nu_5 + \nu_{18}$				
177?	0.6	1177?		1172	0.5	ν_{30}	B_{2g}	$\nu^1(C_bS)$		
									$\nu(C_b–C_a)_{(1)}$	
		1157								

1120	0.6	1120	0.4	1123	0.3	ν_{44}	E_u	$\nu(C_bS)$		
									$\nu(C_b{-}C_a)_{(2)}$	
1032										Phenylalanine
1004		1004	0.2							Phenylalanine
				989		ν_{45}	E_u	$\nu^1(C_3N)$		
980		980	0.2							
757	0.65	757	0.4			ν_{16}	B_{1g}	$\delta(C_aNC_a)$		
675	0			675		ν_7	A_{1g}	$\delta(C_bC_aN)$		
622										Phenylalanine
~500				498				Pyrrole fold		
				483						
408	0.25			410					$\delta(C_bC_\alpha C_\beta)_{(1)}$	$\nu_{34} + \nu_{35}$
										metal–ligand
~330				337					$\delta(C_bC_\alpha C_\beta)_{(2)}$	
				323		ν_8	A_{1g}	$\delta(C_bS)$		
				275		ν_{52}	E_u	$\delta(C_bS)$		
253				257						ν(Fe-His)

Symmetry species are for D_{4h} symmetry; C_a, C_b, C_m denote the a, b and meso carbon atoms of the porphyrin ring, respectively; S: peripheral substituent; ν^1 and δ^1 denote antisymmetric stretching and deformation. For the vinyl assignment: $\delta_s(=CH_2)$, in-plane CH_2 symmetric bending; $\delta(CH=)$, CH deformation; $\nu(C_b{=}C_a)$, ring-vinyl stretch; $(C_\beta C_\alpha C_\beta)$, vinyl bending mode; (1) and (2) denote in-phase and out-of-phase vinyl modes. (Reproduced from Salmaso *et al.* (1994), by permission of the Biophysical Society.)

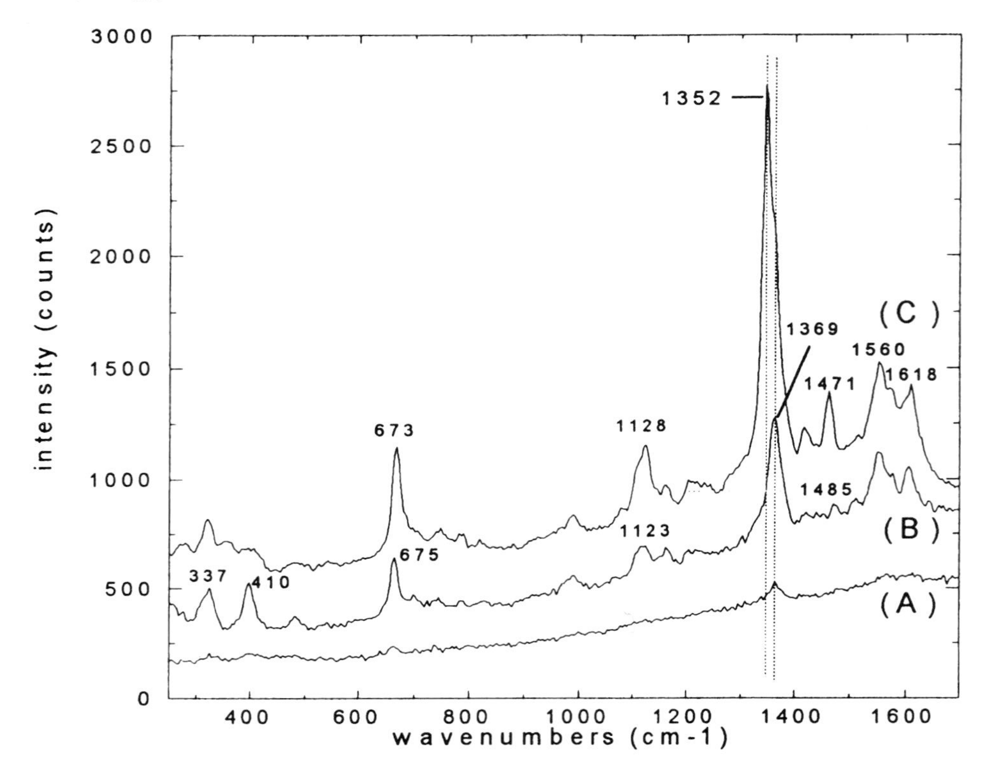

Figure 29.20 Raman spectra obtained in the cytoplasm of human eosinophilic granulocytes. (A) Spectrum of a resting eosinophil. (B) Spectrum of an eosinophil after a 15 minute stimulation with 5×10^{-10} M interleukin 5. (C) Spectrum of an eosinophil that has attached to a 2 μm opsonized polystyrene sphere (opsonization procedure described in Bloom *et al.*, 1992). Experimental conditions: laser power: 0.5 mW (413.1 nm), signal integration time: 10 s. Cells in Hank's buffered salt solution on fused silica slides. (Reproduced from Puppels *et al.*, 1995, by permission of Elsevier Science BV.)

bending mode usually found around 380 cm^{-1} is absent. These features appear to be common to mammalian peroxidases and are probably due to interactions between the haem and the protein, which are specific for this class of enzymes. The orientation of the vinyl groups with respect to the haem plane may be constrained by the protein enabling a stronger than usual conjugation to the porphyrin π-electron system. It has been suggested that the orientation of vinyl substituents influences their electron-withdrawing ability, thus providing a mechanism for controlling the reduction potential (Reid *et al.*, 1986), or that it may play a role in determining ligand binding affinity of haems (Rousseau *et al.*, 1983).

The possibility to determine haem structure *in situ*, and the sensitivity with which structure–function relations can be determined, as illustrated above, implies that changes in haem-structure, thought to occur during activation and inactivation of enzymes during cellular processes, can possibly be monitored by Raman microspectroscopy. First results of experiments aiming at this are promising (Puppels *et al.*, 1995; Sijtsema, 1997). Figure 29.20 shows the first results of such a study with eosinophils. It compares spectra of a resting (not-activated) eosinophil, an eosinophil that was stimulated with interleukin 5, and an eosinophil that had attached to an opsonized polystyrene bead (mimicking the events occurring during parasite killing). Clear differences are seen. From spectrum (B) it is clear that stimulation with interleukin 5 leads to exocytosis of the granule contents, so that little or no EPO remains in the cell. Spectrum (C) shows a spectacular change in the eosinophil spectrum when compared with spectrum (A). The oxidation state marker band (ν_4) has shifted from 1369 cm^{-1} to 1352 cm^{-1}. The ν_3 band has shifted from 1485 cm^{-1} to 1472 cm^{-1}. Also in the low wavenumber region marked changes have occurred. Apparently the haem molecule has changed from a Fe^{3+}, 6-co-ordinated, high-spin structure, characteristic of the resting cell, to an Fe^{2+}, 5-co-ordinated high-spin structure (Kitagawa, 1986; Sharma *et al.*, 1988).

29.4.2.2 Carotenoids in lymphocytes

The task of lymphocytes, which can be divided in many different subgroups, each with their own function, is to establish the immune response of the body against virus-infected cells and cancer cells. Carotenoids appear to play a beneficial role in this part of the immune system. The results of most, but not all, prospective and retrospective epidemiological studies, and animal experimental studies, point in that direction (Peto *et al.*, 1981; Ziegler, 1989; Bendich, 1990). Many cancer patients have a lower blood plasma level of carotenoids than healthy individuals (Wald *et al.*, 1980). Experimental studies have shown that, for example, tumour development is slower in mice fed a β-carotene-supplemented diet than in mice given the same diet without β-carotene (Tomita *et al.*, 1987). In humans it was found that increased intake of β-carotene increased the number of $OKT4^+$ lymphocytes (helper/inducer T-lymphocytes) in peripheral blood (Alexander *et al.*, 1985). Carotenoids could exert their influence on the immune system via their provitamin-A activity, via their strong antioxidant capacities, or via other, as yet unknown, mechanisms that modulate the immune system. Review papers on this subject are Bendich (1990) and De Vet (1990).

In order to investigate the mode of action of carotenoids it is necessary, first of all, to know their location. In view of their role in the immune system lymphocytes are obvious candidates. Raman spectra of pellets of lymphocytes indeed show strong carotenoid lines (Del Priore *et al.*, 1984). Because confocal Raman microspectroscopy enables single cell measurements with a high spatial resolution, it is possible to locate precisely the carotenoids inside the lymphocytes.

The location and concentration of carotenoids in lymphocytes was found to be highly dependent upon lymphocyte phenotype (Puppels *et al.*, 1993b). A very high (~1 mM) concentration is present in Gall bodies (Fig. 29.21). The Gall body is a little known cytoplasmic spherule, usually less than a micrometer in diameter, that was discovered in 1936 (Gall, 1936; Bessis, 1973). It can easily be identified in unstained cells (Fig. 29.22). It is found in helper/inducer-T-lymphocytes and in cytotoxic/suppressor-T-lymphocytes, and is known to be the focal point of acid α-naphthylacetate esterase activity (Boesen, 1984). It is often surrounded by clustered primary lysozomes (Marcus, 1982) and thought to be a lipid droplet, which would explain how such a high concentration of carotenoids can be present.

In the Golgi complex of Natural Killer cells, cytotoxic/suppressor-lymphocytes and Tγ/δ-lymphocytes, a carotenoid concentration was found that was about one order of magnitude lower than in the Gall body (Fig. 29.21). No carotenoids were found in B-lymphocytes, although they possess a well-developed Golgi complex. The highly specific subcellular and phenotypic distribution of carotenoids suggests a specific function in the immune system.

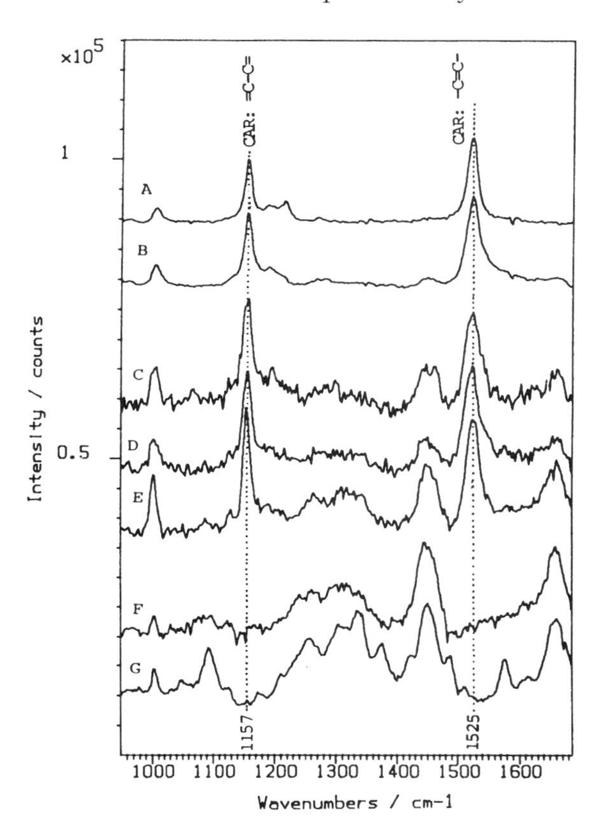

Figure 29.21 Raman spectra of β-carotene, human T-lymphocytes, B-lymphocytes and Natural Killer cells. (A) β-carotene solution in chloroform (0.5 mg ml^{-1}). (B) Helper/inducer T-lymphocyte; spectrum obtained from a Gall body (see Fig. 29.23). (C) Natural Killer cell; spectrum obtained in cytoplasm (most likely Golgi complex). (D) Cytotoxic/suppressor T-lymphocyte; spectrum obtained in cytoplasm (most likely Golgi complex). (E) Tγδ-lymphocyte; spectrum obtained in cytoplasm (most likely Golgi complex). (F) B-lymphocyte; characteristic cytoplasmic Raman spectrum (average of 9 measurements), lacking carotenoid signal contributions. (G) Average of 10 measurements in nuclei of different lymphocytes and Natural Killer cells. No carotenoid signal contributions are present. Experimental conditions: laser power, 5–6 mW; signal collection time, 30 s per measurement; cells on poly-L-lysine coated fused silica substrates and immersed in Hank's buffered salt solution. Intensity scale is for spectra (A) and (B); spectra (C–G) multiplied by factor 8. (Reproduced from Puppels *et al.*, 1993b, by permission of Wiley-Liss, Inc.)

The lymphocyte subtypes were isolated by means of a fluorescence activated flow cytometer, after staining of the cells with specific antibodies, labelled with a fluorescent dye. Phycoerythrine and fluorescein were used. Illumination of these compounds with 488 nm laser light during flow sorting resulted in intense fluorescence. No fluorescence interference was encountered in the Raman experiments because laser light of 660 nm was used, far away from the absorption bands of these molecules.

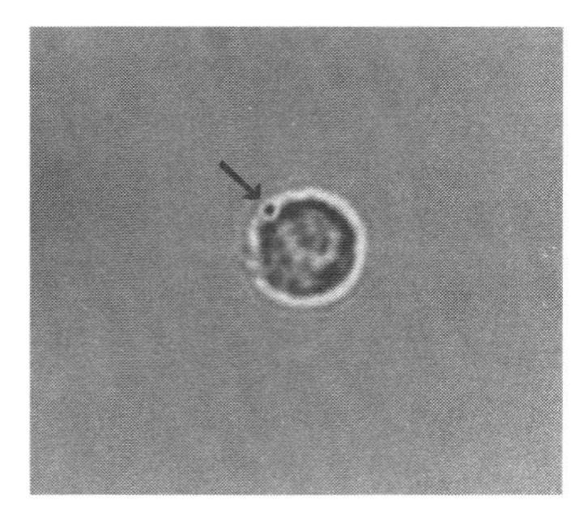

Figure 29.22 Photograph of a Gall body (arrowed) in a human lymphocyte.

A recent study has provided further support for an important role of carotenoids in the immune system. The carotenoid content of lymphocytes of healthy donors was compared with that of lung cancer patients, through Raman measurements of Gall bodies in helper/inducer-T-lymphocytes and in cytotoxic/suppressor-T-lymphocytes (Bakker Schut *et al.*, 1997). The experiments showed that in many cases the intensity of the carotenoid Raman signal obtained from Gall bodies in lymphocytes of lung cancer patients was dramatically low in comparison with that of healthy individuals. In a comparison of age-matched groups of healthy individuals and lung cancer patients a significantly lower carotenoid content was found in Gall bodies of lung cancer patients, especially in cases of adenocarcinoma. Another finding was that Gall body carotenoid levels in healthy individuals appeared to decrease with age. Preliminary results from a study, in which blood serum carotenoid levels were raised by carotenoid supplementation, indicate that only a weak correlation exists between blood plasma carotenoid levels and carotenoid Raman signal intensity obtained from Gall bodies in lymphocytes.

29.4.2.3 Cytotoxic granules in lymphocytes

Takai *et al.* (1997) employed confocal Raman microspectroscopy to study the lipid structure of cytotoxic granules in living human cytotoxic T-lymphocytes. Upon immunological recognition of a target cell, the cytotoxic T-lymphocyte attaches itself to the target cell. Lytic granules in the cytoplasm of the cytotoxic T-lymphocyte then move toward the target cell, fuse with the membrane of the lymphocyte, and secrete their contents in the cleft between lymphocyte and target cell. The granules contain many small internal vesicles and membrane cortices as well as membrane-binding proteins (perforin) and a variety of proteases (granzymes), which are lethal to the target cell. One of the open questions is why the granules are not toxic to the cytotoxic T-lymphocyte itself.

Takai and co-workers obtained Raman spectra of cytotoxic granules and plasma membranes of human cytotoxic lymphocytes. Using a range of triacylglycerols a Raman spectroscopic model was built that predicts avarage acyl chain unsaturation from the intensities and intensity ratios of a number of Raman bands. In that manner detailed information could be obtained about the lipids of cytotoxic granules, which were characterized by a significantly higher degree of unsaturation than plasma membrane lipids. About 1.5 C=C bonds per acyl chain were found for isolated (not clustered) granules, increasing to about 2.2 C=C bonds per acyl chain in clustering granules (Fig. 29.23). In plasma membrane lipids about 1 C=C bond per acyl chain was found. It was hypothesized that this large difference in lipid saturation may play a role in the self-defence of cytotoxic T-lymphocytes against their cytotoxic granules.

29.5 FUTURE DEVELOPMENTS

Venturing into the field of cell biology an exciting future lies ahead for Raman (imaging) microspectroscopy. As illustrated in this chapter, the enormous body of data, compiled in nearly 25 years of Raman studies of biological molecules, enables fruitful application of this technique in the investigation of complex biological structures such as chromosomes and single living cells.

29.5.1 Cell identification and characterization

In all the examples given above, spectra were interpreted in terms of molecular composition of an object, or in terms of molecular structural information. As mentioned in Section 29.4, another approach is to regard a Raman spectrum of a cell as a highly specific spectroscopic fingerprint, which can be used for cell identification or characterization, even without further interpretation in terms of molecular compositon. Multivariate statistical spectrum analysis techniques are employed to this purpose (see Chapter 32, this volume). This approach is used to develop a technique that will enable rapid identification and antibiotic/antifungal agent susceptibility testing of clinically relevant microorganisms. Routine analysis of patient material takes up to two or three days, mainly because of the time it takes to culture enough microorganisms for present-day identification and drug susceptibility testing techniques. First results in our laboratory show that there are good possibilities to differentiate microorganisms on the basis of their Raman spectrum, to the extent that antibiotic-sensi-

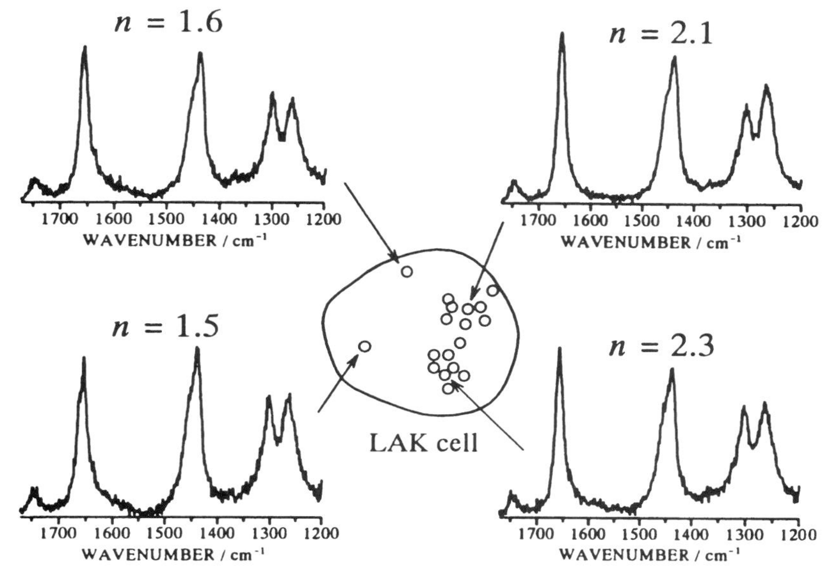

Figure 29.23 Dependence of the Raman spectrum of cytotoxic granules on their location in a living lymphokine-activated killer cell. The Raman spectra of granules provide evidence for lower lipid unsaturation in isolated granules (1.5–1.6 C=C bonds per acyl chain) than in clustered granules (2.1–2.3 C=C bonds per acyl chain). Drawing based on actual micrograph of living, unstained LAK-cell. (Reproduced from Takai *et al.*, 1997, by permission of Elsevier Science BV.)

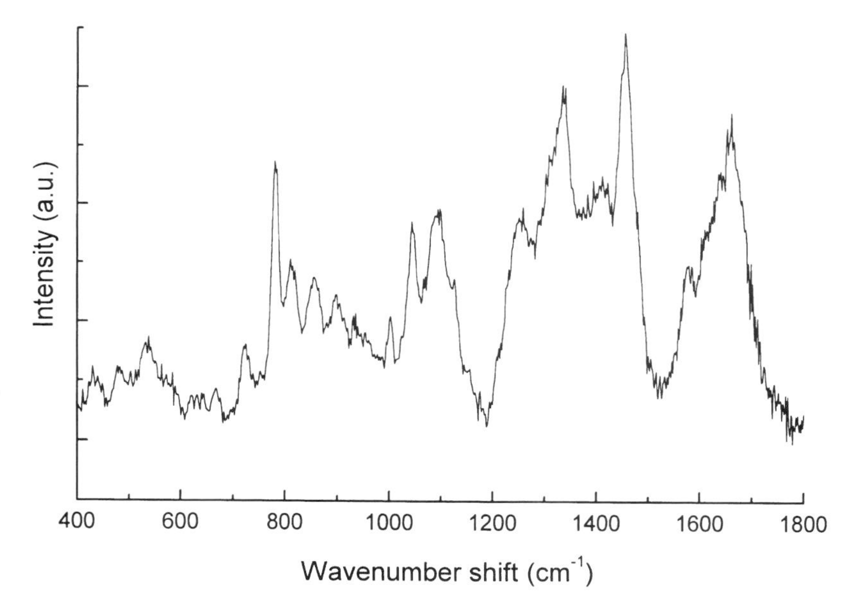

Figure 29.24 Raman spectrum of a microcolony of *Staphylococcus aureus* on Mueller–Hinton solid culture medium, obtained after ~6 h of culturing. Experimental conditions: laser power, 80 mW; laser wavelength, 856 nm; ×80 Olympus NIR objective; signal collection time, 5 min.

tive and resistants strains can be distinguished. Figure 29.24 shows that by means of confocal Raman spectroscopy, good spectra can be obtained from microcolonies of microorganisms on agar culture plates, on the basis of which microorganism classifications could be made after a culturing time of only 6–8 hours.

29.5.2 Raman imaging

Raman spectroscopy can be used as the basis for chemical imaging of samples. Many different approaches have been proposed. Since CCD cameras are two-dimensional array detectors, a simple extension of measurements at a single point in a sample is the use of line illumination (Bowden *et al.*, 1990). Figure 29.25 shows how, using one dimension of the CCD chip for spatial information and the other dimension for spectral information, spectra can be collected simultaneously along the illumination line on the sample. An example of Raman line imaging is given in Plate 29.1. An erythrocyte on a fused silica substrate in phosphate-buffered saline was illuminated as depicted in (a). The haemoglobin Raman signal was detected by the CCD camera and the image it captured is shown in (b). Different colours signify

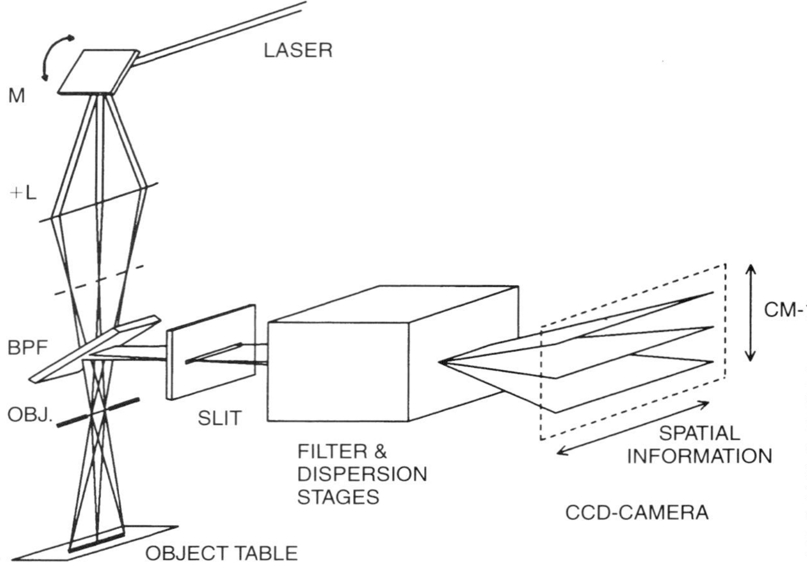

Figure 29.25 Raman microspectroscopy employing line illumination, using one dimension of a CCD camera for spectral information and one dimension for spatial information. (From Greve & Puppels, 1993, reproduced by permission of John Wiley & Sons, Ltd.)

different signal intensities. Spectral information is found along the vertical axis (c), spatial information along the horizontal axis. The spatial information is in this case the distribution of haemoglobin in the cell (d), which is determined by the doughnut-like shape of the erythrocyte. Other examples of the application of Raman line imaging of biological samples are given in Otto *et al.* (1997) and de Grauw (1997).

Two-dimensional Raman images can be obtained in a variety of ways. A chemical map of an object can be achieved by stepping it through the laser focus and collecting a Raman spectrum at each step. After the whole object has been sampled in this way, an image of the sample can be reconstructed based on the intensity of any band or combination of bands in the spectrum. The advantage of this approach is that all spectral information is available to form chemical compositional maps of an object. The disadvantage is that it usually results in very long signal collection times, especially for most biological objects. An alternative approach is provided by direct Raman imaging techniques. Here one Raman line, which is characteristic of a certain compound, is selected from the light that is scattered by the object, by for example a monochromator (Delhaye & Dhamelincourt, 1975), an ultra-narrow band dielectric filter (Batchelder *et al.*, 1991; Puppels *et al.*, 1993c; Sijtsema *et al.*, 1996), or a liquid crystal tunable filter (Kline & Treado, 1997). An image of the object is then formed directly on the CCD camera. An example of a Raman image of carotenoids in a human lymphocyte is given in Fig. 29.26.

A confocal direct imaging Raman set-up was recently developed by Sijtsema and co-workers (Sijtsema, 1997), enabling even 3-D Raman images of biological objects to be obtained.

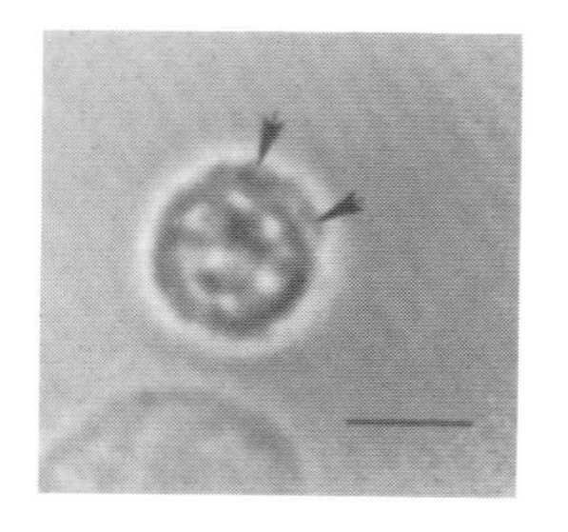

Figure 29.26 Raman image of carotenoid location in a human lymphocyte. (A) Bright-field image of a human lymphocyte. Bar denotes 5 μm. Arrows indicate the location of carotenoids (Fig. 29.27(b)). (B) Raman image obtained with the microscope tuned to the 1525 cm^{-1} carotenoid −C=C− stretching band (1500 cm^{-1} background image subtracted). Cell in phosphate-buffered saline on fused silica substrate. Total laser power on the sample, 25 mW (intensity ~2×10^4 W cm^{-2}); signal integration time, 150 s; ×63 Zeiss Plan Neofluar water immersion objective (NA 1.2). (Reproduced from Puppels *et al.*, 1993c, by permission of the Society for Applied Spectroscopy.)

ACKNOWLEDGEMENTS

The author wishes to thank A. Loenen, L.-P. Choo-Smith and K. Maquelin for help in the preparation of the manuscript. This chapter is a revised and updated version of the chapter on confocal Raman spectroscopy in the 1st edition of *Fluorescent and Luminescent Probes*, which was co-authored by Drs C. Otto and J. Greve of the University of Twente, and by M. van Rooijen.

REFERENCES

Abe M., Kitagawa T. & Kyogoku Y. (1978) *J. Chem. Phys.* **69**, 4526–4534.

Abraham J.L. & Etz E.S. (1979) *Science* **206**, 716–718.

Alexander M., Newmark H. & Miller R.G. (1985) *Immunol. Lett.* **9**, 211–214.

Ananiev E.V. & Barsky V.E. (1978) *Chromosoma* **65**, 359–371.

Ananiev E.V. & Barsky V.E. (1985) *Chromosoma* **95**, 104–112.

Arndt-Jovin D.J., Robert-Nicoud M., Zarling D.A., Greider C., Weimer E. & Jovin T.M. (1983) *Proc. Natl Acad. Sci. USA* **80**, 4344–4348.

Babu A. & Verma R.S. (1989) *Int. Rev. Cytol.* **108**, 1–59.

Baek M., Nelson W.H., Britt D. & Sperry J.F. (1988) *Appl. Spectrosc.* **42**, 1312–1314.

Bakker Schut T.C., Puppels G.J., Kraan Y.M., Greve J., Van der Maas L.L.J., Figdor C.G. (1997) *Int. J. Cancer (Pred. Oncol.)* **74**, 20–25.

Barry B. & Mathies R. (1982) *J. Cell Biol.* **94**, 479–482.

Bartoli F.J. & Litovitz T.A. (1972) *J. Chem. Phys.* **56**, 401–411.

Batchelder D.N., Cheng C., Müller W. & Smith B.J.E. (1991) *Makromol. Chem. Macromol. Symp.* **46**, 171–179.

Beerman W. (1952) *Chromosoma* **5**, 139–152.

Benevides J.M. & Thomas Jr. G.J. (1983) *Nucl. Acids Res.* **11**, 5746–5761.

Benevides J.M. & Thomas Jr. G.J. (1985) *Biopolymers* **24**, 667–682.

Bendich A. (1990) In *Carotenoids: Chemistry and Biology*, N.I. Krinsky, M.M. Mathews-Roth & R.F. Taylor (eds). Plenum Press, New York, pp. 323–336.

Bessis M. (1973) *Living Blood Cells and Their Ultrastructure*. Springer, Berlin.

Bickmore W.A. & Sumner A.T. (1989) *Trends Genet.* **5**(5), 144–148.

Bloom M., Tool, A. & Verhoeven, A. (1992) *J. Immunol.* **149**, 3672–3677.

Boesen A.M. (1984) *Scand. J. Haematol.* **32**, 367–375.

Bolscher B.G.J.M. & Wever R. (1984) *Biochim. Biophys. Acta* **791**, 75–81.

Bowden M., Gardiner D.J., Rice G. & Gerrard D.L. (1990) *J. Raman Spectrosc.* **21**, 37–41.

Bradbury E.M., Maclean N. & Matthews H.R. (1981) *DNA, Chromatin and Chromosomes*. Blackwell, Oxford.

Brakenhoff G.J., Blom P. & Barends P. (1979) *J. Microsc.* **117**(20), 219–232.

Bustin M., Kurth P.D., Moundrianakis E.N., Goldblatt D., Sperking R. & Rizzo W.B. (1977) *Cold Spring Harbor Symp. Quant. Biol.* **XLII**, 379–388.

Caillé J.P., Pigeon-Gosselin M., Pézolet M. (1987) *Can. J. Physiol. Pharmacol.* **65**, 1415–1420.

Carey P.R. (1982) *Biochemical Applications of Raman and Resonance Raman Spectroscopies*. Academic Press, New York.

Caspersson T., Farber S., Foley G.E., Kudynowski J., Modest E.J., Simonsson E., Wagh U. & Zech L. (1968) *Exp. Cell Res.* **49**, 219–222.

Clark R.J.H. & Hester R.E. (eds) (1986) *Spectroscopy of Biological Systems*. John Wiley & Sons, Chichester.

Dalterio R.A., Nelson W.H., Britt D., Sperry J.F. & Purcell F.J. (1986) *Appl. Spectrosc.* **40**, 271–272.

Delhaye M. & Dhamelincourt P. (1975) *J. Raman Spectrosc.* **3**, 33–43.

Del Priore L.V., Lewis A. & Schat K.A. (1984) *Membr. Biochem.* **5**, 97–108.

De Vet H.C.W. (1990) PhD thesis, University of Maastricht, Maastricht, The Netherlands.

Erfurth S. & Peticolas W.L. (1975) *Biopolymers* **14**, 247–264.

Erfurth S.C., Bond P.J. & Peticolas W.L. (1975) *Biopolymers* **14**, 1245–1257.

Etz E.S. & Blaha J.J. (1980) *NBS Special Publication* **533**, 153–197.

Flaugh P.L., O'Donnell S.E. & Asher S.A. (1984) *Appl. Spectrosc.* **38**, 847–850.

Foote V.S. (1976) In *Free Radicals in Biology*, Vol. II, W.A. Pryor (ed.). Academic Press, New York, pp. 85–133.

Fujita S. & Takamoto K. (1963) *Nature* **200**, 494–495.

Gall E.A. (1936) *Am. J. Med. Sci.* **191**, 380–388.

Genack A.Z. (1984) *Anal. Chem.* **56**, 2957–2960.

Gleich, G.J. & Adolphson, C.R. (1986) In *Advances in Immunology*, Vol. 39, F.J. Dixon (ed.). Academic Press, Orlando, pp. 177–253.

Goodwin D.C. & Brahms J. (1978) *Nucl. Acids Res.* **5**(3), 835–850.

Grauw C.J. de (1997) Ph.D. thesis, University of Twente, The Netherlands.

Greve J., Puppels G.J., Olminkhof J.H.F., Otto C. & De Mul F.F.M. (1989) In *Spectroscopy of Biological Molecules*, A. Bertoluzza, C. Fagnano & P. Monti (eds). Societa Editrice Esculapio s.r.l., Bologna, pp. 401–404.

Greve J. & Puppels G.J. (1993) In *Biomolecular Spectroscopy*, Part A, R.J.H. Clark & R.E. Hester (eds). Wiley, Chichester, pp. 231–265.

Hayashi H., Nishimura Y., Katahira M. & Tsuboi M. (1986) *Nucl. Acids Res.* **14**(6), 2583–2596.

Hill R.J. & Rudkin G.T. (1987) *BioEssays* **7**(1) 35–40.

Hill R.J., Watt F. & Stollar B.D. (1984) *Exp. Cell Res.* **153**, 469–482.

Huxtable R.J. (1992) *Physiol. Rev.* **72**, 101–163.

Jamrich M., Greenleaf A.L. & Bautz E.K.F. (1977) *Proc Natl Acad. Sci. USA* **74**, 2079–2083.

Jeannesson P., Angiboust J.F., Jardillier J.C. & Manfait M. (1986) In *Proceedings IEEE/8th Annual Conference of the Engineering in Medicine and Biology Society*, 1404–1406.

Jett J.H., Keller R.A., Martin J.C., Nguyen D.C. & Saunders G.C. (1990) In *Flowcytometry and Sorting*, 2nd edn, M.R. Melamed, T. Lindmo & M.L. Mendelsohn (eds). Wiley-Liss, New York, pp. 381–396.

Kitagawa T. (1986) In *Spectroscopy of Biological Systems*, R.J.H. Clark & R.E. Hester (eds.). John Wiley & Sons, Chichester, pp. 443–481.

Klebanoff S.J. & Clark R.A. (1978) *The Neutrophil: Function and Disorders*. North-Holland, Amsterdam.

Kline N.J. & Treado P.J. (1997) *J. Raman Spectrosc.* **28**, 119–124.

Koningstein J.A. (1971) *Introduction to the Theory of the Raman Effect*. D. Reidel Publishing Company, Dordrecht.

Krimm S. (1987) In *Biological Applications of Raman Spectroscopy*, Vol. I, T.G. Spiro (ed.). John Wiley & Sons, New York, pp. 1–45.

Kubasek W.L., Wang Y., Thomas G.A., Patapoff T.W., Schoenwaelder K.-H., Van der Sande J.H. & Peticolas W.L. (1986) *Biochemistry* **25**, 7440–7445.

Kurth P.D., Moundrianakis E.N. & Bustin M. (1978) *J. Cell Biol.* **78**, 910–918.

Learn D.B., Fried V.A., Thomas E.L. (1990) *J. Leukocyte Biol.* **48**, 174–182.

Long D.A. (1977) *Raman Spectroscopy*. McGraw-Hill, New York.

Lord R.C. & Thomas Jr. G.J. (1967) *Spectrochim. Acta* **23A**, 2551–2591.

Marcus J.N. (1982) *Lab. Invest.* **46**, 52A.

Mathews C.K. & Van Holde K.E. (1990) *Biochemistry*. Benjamin/Cummings, Redwood.
Mathies R. & Yu N.-T. (1978) *J. Raman Spectrosc.* **7**, 349–352.
Nabiev I.R., Morjani H. & Manfait, M. (1991) *Eur. Biophys. J.* **19**, 311–316.
Natarajan A.T. & Gropp A. (1972) *Exp. Cell Res.* **74**, 245–250.
Nordheim A., Pardue M.L., Lafer E.M., Möller A., Stollar B.D. & Rick A. (1981) *Nature* **294**, 417–422.
Oertling W.A., Hoogland H., Babcock G.T. & Wever R. (1988) *Biochemistry* **27**, 5395–5400.
Otto C., De Grauw C.J., Duindam J.J., Sijtsema N.M. & Greve J. (1997) *J. Raman Spectrosc.* **28**, 143–150.
Pachman U. & Rigter R. (1972) *Exp. Cell Res.* **72**, 602–608.
Parker C.W. (1984) In *Fundamental Immunology*, W.E. Paul (ed.). Raven Press, New York, pp. 697–747.
Parker F.S. (1983) *Application of Infrared, Raman and Resonance Raman Spectroscopy in Biochemistry*. Plenum Press, New York.
Peticolas, W.L., Patapoff, T.W., Thomas G.A., Postlewait, J. & Powell, J.W. (1996) *J. Raman Spectrosc.* **27**, 571–578.
Peto R., Doll R., Buckley J.D. & Sporn M.B. (1981) *Nature (Lond.)* **290**, 201–208.
Pézolet M., Pigeon-Gosselin M., Nadeau J. & Caillé J.-P. (1980) *Biophys. J.* **31**, 1–8.
Prescott B., Steinmetz W. & Thomas Jr. G.J. (1984) *Biopolymers* **23**, 235–256.
Puppels G.J., Olminkhof J.H.F., Otto C., De Mul F.F.M. & Greve J. (1989) In *Spectroscopy of Biological Molecules*, A. Bertoluzza, C. Fagnano & P. Monti (eds). Societa Editrice Esculapio s.r.l., Bologna, pp. 357–358.
Puppels G.J., Huizinga A., Krabbe H.W., De Boer H.A., Gijsbers G. & De Mul F.F.M. (1990a) *Rev. Sci. Instrum.* **61**(12), 3709–3712.
Puppels G.J., De Mul F.F.M., Otto C., Greve J., Robert-Nicoud M., Arndt-Jovin D.J. & Jovin T.M. (1990b) *Nature* **347**, 301–303.
Puppels G.J., Otto C., De Mul F.F.M. & Greve J. (1990c) In *Proceedings of the 12th International Conference on Raman Spectroscopy*, J.R. Durig & J.F. Sullivan (eds). John Wiley & Sons, Chichester.
Puppels G.J., Colier W., Olminkhof J.H.F., Otto C., De Mul F.F.M. & Greve J. (1991a) *J. Raman Spectrosc.* **22**, 217–225.
Puppels G.J., Olminkhof J.H.F., Segers-Nolten G.M.J., Otto C., De Mul F.F.M. & Greve J. (1991b) *Exp. Cell Res.* **195**, 361–367.
Puppels G.J., Garritsen H.S.P., Segers-Nolten G.M.J., De Mul F.F.M. & Greve J. (1991c) *Biophys. J.* **60**, 1046–1056.
Puppels G.J., Otto C. & Greve J. (1991d) In *Microbeam Analyses*, D.G. Howitt (ed.). San Francisco Press, San Francisco, pp. 85–87.
Puppels G.J., Otto C., Greve J., Roberts-Nicoud M., Arndt-Jovin D.J. & Jovin T.M. (1991e) In *Spectroscopy of Biological Molecules*, R.E. Hester & R.B. Girling (eds). The Royal Society of Chemistry, Cambridge, pp. 301–302.
Puppels G.J., Putman C.A.J., De Grooth B.G. & Greve J. (1993a) *SPIE* **1922**, 145–155.
Puppels G.J., Garritsen H.S.P., Kummer J.A. & Greve J. (1993b) *Cytometry* **14**, 251–256.
Puppels G.J., Grond M. & Greve J. (1993c) *Appl. Spectrosc.* **47**, 1256–1267.
Puppels G.J., Otto C., Greve J., Robert-Nicoud M., Arndt-Jovin D.J. & Jovin T.M. (1994a) *Biochem.* **33**, 3386–3395.
Puppels G.J., De Grauw C.G., Te Plate M.B.J. & Greve J. (1994b) *Appl. Spectrosc.* **48**, 1399–1402.
Puppels G.J., Bakker Schut T.C., Sijtsema N.M. Grond, M., Maraboeuf F., De Grauw C.J., Figdor C.G. & Greve J. (1995) *J. Mol. Struct.* **347**, 477–483.
Raman C.V. & Krishnan K.S. (1928) *Nature* **121**, 501.
Reid L.S., Lim A.R. & Mauk A.G. (1986) *J. Am. Chem. Soc.* **108**, 8197–8201.
Rich C.V. & Cook D. (1991) *SPIE Proceedings*, 1461.
Robert-Nicoud M., Arndt-Jovin D.J., Zarling D.A. & Jovin T.M. (1984) *EMBO J.* **3**(4), 721–731.
Rosasco G.J., Etz E.S. & Cassatt W.A. (1975) *Appl. Spectrosc.* **29**, 396–404.
Rousseau D.L., Ondrias M.R., LaMar G.N., Kong S.B. & Smith K.M. (1983) *J. Biol. Chem.* **258**, 1740–1746.
Rowley J.D. & Bodmer W.F. (1971) *Nature* **231**, 503–506.
Salmaso B.L.N., Puppels G.J., Caspers P.J., Floris R., Wever R. & Greve J. (1994) *Biophys. J.* **67**, 436–446.
Savoie R., Jutier J.-J., Alex S., Nadeau P. & Lewis P.N. (1985) *Biophys. J.* **47**, 451–459.
Schultz J. & Kaminker K. (1962) *Arch. Biochem. Biophys.* **96**, 465–467.
Semeshin V.F., Zhimulev I.V. & Belyaeva E.S. (1979) *Chromosoma* **73**, 163–177.
Sharma K.D., Andersson L.A., Loehr T.M., Terner J. & Goff H.M. (1988) *J. Biol. Chem.* **264**, 12772–12779.
Sijtsema N.M. (1997) Ph.D. thesis, University of Twente.
Sijtsema N.M., Duindam J.J., Puppels G.J., Otto C. & Greve J. (1996) *Appl. Spectrosc.* **50**, 545–551.
Smekal C.V. (1923) *Naturwissenschaften* **11**, 873.
Spiro T.G. (ed.) (1987–1988) *Biological Applications of Raman Spectroscopy*, Vols I–III. John Wiley & Sons, New York.
Steen H.B. (1990) In *Flowcytometry and Sorting*, 2nd edn, M.R. Melamed, T. Lindmo & M.L. Mendelsohn (eds). Wiley-Liss, New York, pp. 11–25.
Sumner A.T. (1982) *Cancer Genet. Cytogenet.* **6**, 59–87.
Sureau F., Chinsky L., Amirand C., Ballini J.P., Duquesne M., Laigle A., Turpin P.Y. & Vigny P. (1990) *Appl. Spectrosc.* **44**, 1047–1051.
Takai Y., Masuko T., Takeuchi H. (1997) *Biochim. Biophys. Acta* **1335**, 199–208.
Thomas Jr. G.J., Prescott B. & Olins D.E. (1977) *Science* **197**, 385–388.
Thomas Jr. G.J., Prescott B. & Hamilton M.G. (1980) *Biochemistry* **19**, 3604–3613.
Thomas Jr. G.J., Benevides J.M. & Prescott B. (1986) In *Biomolecular Stereodynamics*, R.H. Sarma & M.H. Sarma (eds). Adenine Press, Guilderland, New York, pp. 227–253.
Tomita Y., Himeno K., Nomoto K., Endo H. & Hirohata T. (1987) *J. Natl. Cancer Inst.* **78**, 679–680.
Tu A.T. (1982) *Raman Spectroscopy in Biology: Principles and Applications*. John Wiley & Sons, New York.
Van der Voort H.T.M. & Brakenhoff G.J. (1990) *J. Microsc.* **158**(1), 43–54.
Wald N., Idle M., Borcham J. & Baily A. (1980) *Lancet* **2**, 813–815.
Weisblum B. & de Haseth P.L. (1972) *Proc. Natl Acad. Sci. USA* **69**, 629–632.
Yu N.-T., DeNagel D.C., Jui-Yuan Ho D. & Kuck J.F.R. (1987) In *Biological Applications of Raman Spectroscopy*, Vol. 1, T.G. Spiro (ed.). John Wiley & Sons, New York, pp. 47–80.
Zhimulev I.F., Belyaeva E.S. & Semeshin V.F. (1981) *Crit. Rev. Biochem.* **11**, 303–340.
Ziegler R.G. (1989) *J. Nutr.* **119**, 116–122.

CHAPTER THIRTY

Spectral Imaging of Autofluorescence Molecules and DNA Probes

CHANA ROTHMANN[1], IRIT BAR-AM[2], ZVI DUBINSKY[1], SHLOMIT KATZ[1] & ZVI MALIK[1]

[1] Life Sciences Department, Bar Ilan University, Ramat-Gan 52900, Israel
[2] Applied Spectral Imaging, Migdal HaEmek, Israel

Spectral imaging opens up new possibilities in biology and medicine for the analysis of chromosomes, single cells, tissue sections and tumours. The SpectraCube™ system combines spectroscopy with imaging and is based on Fourier transform multipixel spectral analysis. Spectral imaging can be operated in combination with microscopy in a transmitted or fluorescence mode. The light intensity at any wavelength is measured from multiple points of an image and stored in a 'spectra cube' file whose appellate signifies the two spatial dimensions of a flat sample (*x* and *y*), while the third dimension, the spectrum, represents the light intensity for every wavelength. By mathematical analysis of the cube database, it is possible to perform spectral-similarity mapping (SSM) which demarcates areas occupied by the same type of material. Spectral similarity mapping constructs new images of the specimen, revealing areas with similar stain-macromolecule characteristics and enhancing subcellular features. The SKY technique utilizes the advantages of the SpectraCube™ for multiprobe FISH and chromosome karyotyping, identifying marker chromosomes, detecting subtle chromosome translocations and clarifying complex karyotypes. On the cellular level, spectral imaging enables a new insight into nuclear organization and chromosomal compartmentalization. Therefore, differentiation stages as well as apoptotic and necrotic conditions are easily quantified.

30.1 INTRODUCTION

Fluorescence microscopy reveals the distribution of fluorochromes in biological specimens, whether they are introduced externally as probes or occur naturally, as chlorophylls and porphyrins. Fluorescence imaging, in its conventional form, reveals biological structures and processes, but is limited by the number of fluorochromes that can be distinguished in one measurement. Serious attempts to increase the numbers of discernible fluorescence dyes were initiated by Nederlof and colleagues in 1989 using DNA probes labelled with three different fluorochromes (Nederlof *et al.*, 1989): AMCA (blue fluorescence), FITC (green fluorescence) and rhodamine (red fluorescence). The colour discrimination was based on the use of fluorochrome-specific optical filters. In 1990

FLUORESCENT AND LUMINESCENT PROBES, 2ND EDN
ISBN 0-12-447836-0

the same group succeeded in labelling more discernible targets than the numbers of fluorochromes (Nederlof *et al.*, 1990). This was achieved by a combinatorial labelling strategy, in which one probe was not only labelled with a pure fluorochrome but also with a combination of two. However, because of the limited number of spectrally separated fluorescence dyes, the discrimination between 24 labelled targets (the number of human chromosomes) became possible only recently by multiple exposures through single band-pass imaging (Speicher *et al.*, 1996).

Recently developed spectral imaging methods have introduced an important technical improvement to the field of fluorescence microscopy. Spectral imaging measures light wavelengths from multiple points of an image. This method is based on the ability of molecules to absorb, reflect, and emit photons of different wavelengths in characteristic ways. Spectroscopy records the interactions of light with matter, while imaging records and provides spatial information of the object. Spectral imaging has been used until recently only in space technology, to map earth surface resources and to monitor the environment, and was not suitable for microscopy in biology and medicine. The development of the SpectraCube™ (ASI, Applied Spectral Imaging, Migdal HaEmek, Israel) and the collaborative efforts of ASI, Bar-Ilan University (Israel) and the National Center for Human Genome Research of NIH (Washington DC) have resulted in the development of new applications of spectral imaging for biomedicine which highlight cellular features and processes (Malik *et al.*, 1996a).

30.2 THE SPECTRACUBE™ SYSTEM AND DESIGN

The SpectraCube™ system provides Fourier transform multipixel spectral imaging for light microscopy (Malik *et al.*, 1996a; Garini *et al.*, 1996; Simon-Blecher *et al.*, 1996; Katz *et al.*, 1997). In dramatic contrast to conventional epifluorescence microscopy, in which fluorochrome discrimination is based on the measurement of a single intensity through a fluorochrome-specific optical filter, spectral imaging enables the measurement and analysis of the full spectrum of light at all pixels of an image. The spectral image presents a three-dimensional array of data which combines precise spectral information with two-dimensional spatial correlation. The analysis of a spectral image creates a unique database, which enables the extraction of features and the evaluation of quantities from multipoint spectral information, impossible to obtain otherwise.

Theoretically, the most straightforward method for building a spectral image is to successively filter the light focused on a digital camera (e.g. a CCD) with a number of narrow-band filters for wavelength separation; for example, 40 filters are needed for a spectral resolution of 10 nm in the range of 400 to 800 nm. This is prohibitively expensive and cumbersome to build and use, and is in fact an inefficient utilization of the available photons. An improvement can be achieved with Liquid Crystal Tunable Filters (LCTF) and Acousto-Optic Tunable Filters (AOTF) (Morris *et al.*, 1994), because they eliminate the need for moving elements, but the low throughput of the filter method still remains. In cases where the noise is independent of signal, or when the spectrum has narrow regions of a signal higher than the average, and the noise is proportional to the square root of the signal, Fourier spectroscopy has a sensitivity advantage over any other filter, grating, AOTF and LCTF-based dispersion technique.

The SpectraCube™ is mounted on an upright or inverted microscope and the radiation emitted by the source is collected, collimated by an objective lens, spectrally dispersed, and then imaged onto a CCD by focusing optics. A triangular 'common path' type Sagnac interferometer is utilized as the dispersion element in the SpectraCube design. While the dispersion element is scanned, CCD frames are collected, and interferograms for all the pixels of the image are acquired simultaneously. The interferogram is measured individually at each pixel in a CCD array (Morris *et al.*, 1994). Fourier transformation thus yields a distinct spectrum at each pixel. Approximately 100 000 interferograms are acquired within a time frame of a few seconds to two minutes, depending on the intensity of the signal and the required spectral resolution. Interferometric Fourier spectroscopy enjoys several important advantages in comparison to other spectral imaging methods; it provides: (1) high optical accuracy; (2) high and variable spectral resolution; (3) wide spectral range, and (4) mechanical/thermal stability (Vane *et al.*, 1988; Chamberlin, 1979).

In interferometric spectral imaging, the spectral resolution is selected by the user without changing the hardware. All the interferograms are then Fourier transformed to obtain the spectra as a 'spectral cube' file. The data 'cube' of a spectral image is ordered as an intensity function which depends on three independent parameters: the two spatial dimensions of a sample (x and y) and a third dimension, the light intensity at any wavelength (λ). The spectral range of a transmission or fluorescence measurement is a function of the illuminating source, the spectral transmission of the optics, and the spectral sensitivity of the CCD. Typical ranges are 400 to 800 nm for fluorescence and

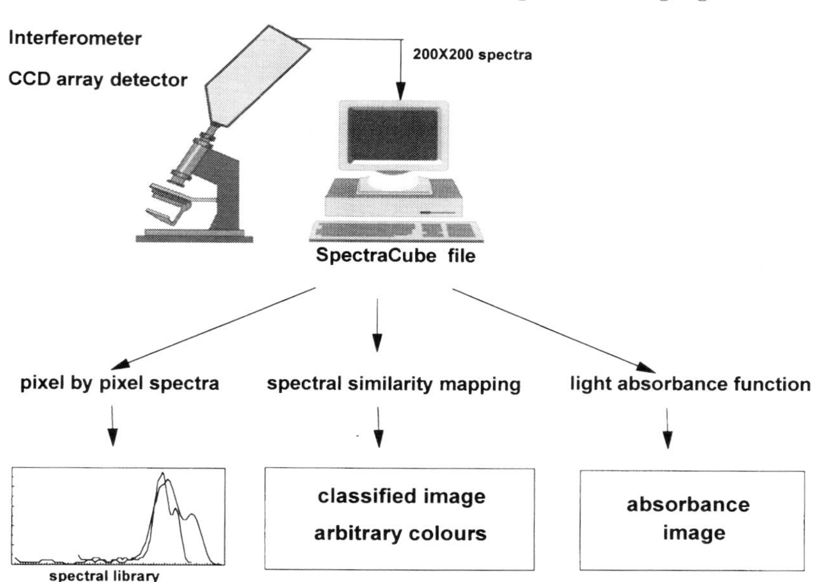

Figure 30.1 The SpectraCube™ system. The system is mounted on a conventional epifluorescence microscope. The light emitted from the object is collimated in the front lens and sent to the Sagnac interferometer, which generates a variable optical path difference (OPD). The light leaving the interferometer is focused on a cooled CCD camera. Repeated measurements with distinctly changed OPDs allow one to generate an interferogram for each pixel. Fourier transformation yields a distinct spectrum at each pixel of the image. The analysis of a spectral image can be performed by spectral-similarity mapping or by the construction of an absorbance image.

550 to 1000 nm for transmission or reflection of a standard halogen light source.

In principle, the number of pixels in which the spectrum can be measured equals the number of detector elements of the CCD, 1024 × 1024. However, it is not practical to measure more than 350 × 350 pixels at the maximum spectral resolution, since photobleaching of the fluorophores is a major concern during long acquisition times. At present, a multiple colour fluorescence measurement of 128 frames is collected with an acquisition time of 800 ms per CCD frame. The total acquisition time of an image at 12 nm spectral resolution, with a signal-to-noise ratio of 20, is 100 s, and calculation time of the interferogram cube is of the order of 10 s (Fig. 30.1).

30.3 THE ANALYSIS OF A SPECTRAL IMAGE

The correlation between spectrum and composition is easy in the study of known molecules; in this case the spectra are known and their interpretation is straightforward. In a composite biological material, the interpretation of the spectra is usually complicated. Any single position in space defined by a pixel is actually a volume, which is large compared to molecular size. Thus, the spectrum derived from that pixel contains contributions from a number of different types of molecules, each with its own complex spectrum and modified by the interactions between them. As a result, the most commonly adopted analytical methods are empirical and based on phenomenological approaches.

Spectral similarity mapping is useful in a situation in which the sample is composed of a number of spatially separated components, each characterized by a known and unique spectrum, and the task is to detect and map all components. The steps of this algorithm are as follows: first, the known set of spectra of the components are stored as a 'spectral library'; second, for every pixel of the cube, a comparison is made between its measured spectrum and all the spectra of the library; third, each pixel is identified with the component whose spectrum is most similar to the pixel's spectrum; fourth, each pixel is displayed in a previously established colour identifying the specific component, forming a so-called 'classified image'. The advantage of a classified spectral image is that it highlights and enhances small differences which are difficult to distinguish in a conventional image.

A comparison formula for the second step of the similarity mapping algorithm is as follows: n functions $f''x,y$ (n is the number of components present) are defined for every pixel of spatial co-ordinates x and y as follows:

$$f''_{x,y} = \left(\int_{\lambda_1}^{\lambda_2} [I_{x,y}(\lambda) - I_n(\lambda)]^2 \, d\lambda \right)^{1/2} \qquad [30.1]$$

where the integral over λ stands for an integral over a predetermined spectral range $\lambda_1 - \lambda_2$, $I_{x,y}(\lambda)$ is the spectrum of the pixel in question, and $I_n(\lambda)$ are the known spectra of the n components. n artificial colours are associated with each of these components. In this analysis scheme, each pixel is displayed in the colour which corresponds to the component for which equation (30.1) is minimum. The resulting 'classified image' reveals areas with similar chromophore compositions. In some cases, additional quantitative assessments of the sizes of the classified regions are needed, either in absolute or relative terms. Grey shades can be added according to the value of the selected $f''_{x,y}$. When there is only one component (n = 1) with which to compare each pixel, a black and white image is all that is needed, and the grey levels according to the values of $f^1_{x,y}$ are used for a good rendition of the image. The presence of the component in question is then easily seen in the bright pixels, whereas the dark ones represent its absence; the grey levels may correspond to different amounts of the component.

Another approach for the enhancement of the information obtained from a biological specimen in a transmitted light mode operation is by the calculation of its absorbance spectra. Application of the Beer–Lambert law for absorbance: log (I_0/I_t) = $\varepsilon cl/2.3$, serves to construct absorbance images. I_0 is the illuminating incident light in the microscope, I_t is the transmitted light passing through the specimen at a specific site, ε is the absorption coefficient of macromolecules, c their concentration and l the sample thickness. The εcl values cannot be determined for a cellular compartment and therefore they were replaced by an arbitrary constant k, standing for $\varepsilon cl/2.3$. The equation defines optical density relative values rather than absolute absorbance.

30.4 APPLICATIONS OF SPECTRAL IMAGING IN BIOLOGY

The SpectraCube™ has been applied to various biological studies. The subcellular localization of porphyrins in cancer cells during photodynamic therapy (PDT) was determined using the known fluorescence spectra of the porphyrins as a marker (Malik *et al.*, 1996b, 1997). The localization of the natural protoporphyrin was determined in skin cancer patients treated with ALA-PDT (Malik *et al.*, 1996a). Nuclear organization in normal erythropoiesis and apoptosis was studied by the transmitted light spectra of stained cytological specimens (Rothmann *et al.*, 1997). In addition, a novel karyotyping method was developed using the SKY™ technique in order to detect chromosomal abnormalities in solid tumours (Schrock *et al.*, 1996) and leukaemias (Veldman *et al.*, 1997).

30.4.1 DNA probes

Chromosome karyotyping is an important clinical diagnostic tool, especially when combined with specific fluorescence *in situ* hybridization (FISH) techniques. Since the discovery that chromosomal abnormalities are a leading cause of genetic diseases, their detection has been used extensively over the years for improved diagnosis and therapy. Chromosomal structure rearrangements have also gradually begun to be correlated with tumour aggressiveness, morphology and staging, and therefore can be of prognostic value. In addition, the identification of abnormal chromosome regions, as suggested by cytogenetic analysis, has led to the isolation of new genes that are involved in specific diseases (Solomon *et al.*, 1991; Heim & Mitelman, 1995).

The most common method used today for cytogenetic analysis is the G-banding technique. This method is used in routine screening of cells for the detection of numerical as well as structural aberrations associated with specific genetic syndromes or with certain types of cancers. However, G-banding sometimes yields insufficient information because of difficulties encountered in analysing complex karyotypes or subtle chromosomal rearrangements. FISH makes use of DNA probes covalently bonded to fluorescent dyes to permit rapid characterization of genetic aberrations at the cellular level (Lichter & Ward, 1990; Tkachuk *et al.*, 1991; Ledbetter, 1992).

Until now, the use of FISH has been restricted to cases in which only one or a few specific abnormalities are investigated in one sample, and there is prior knowledge of which chromosomes are involved. The analysis has therefore been directed at finding evidence of specific changes that are suspected to be present. At the same time, the cytogeneticist's task has been difficult and prone to mistakes due to the limited number of fluorescence probes available and the fact that their spectra overlap significantly. Simultaneous screening would require the hybridization of the same sample with a large number of probes for different genes, each labelled with different fluorophores.

Spectral Karyotyping™ (SKY™), developed by Applied Spectral Imaging (ASI), combines the resolution and power of FISH with the advantages of conventional cytogenetic methods (Schrock *et al.*, 1996). SKY™ uses the SpectraCube™ technology, integrated with a modified fluorescence microscope (Simon-Blecher *et al.*, 1996; Katz *et al.*, 1997). Once chromosomes extracted from a metaphase nucleus are hybridized with 24 whole chromosome probes, each one labelled and emitting a different fluorescence spectrum, the SKY™ system with its high spectral resolution can identify and display all the pixels of the image in enhanced pseudo-colours of each chromosome. In this manner, a colour karyotype can be created, enabling easy screening of cells for the detection of numerical and structural aberrations (Garini *et al.*, 1996; Liyanage *et al.*, 1996; Schrock *et al.*, 1996; Veldman, *et al.*, 1997). The advantage of SKY™ was also shown recently by the groups of Ried and Rowley (Veldman *et al.*, 1997). The two groups analysed haematological malignancies to detect hidden chromosomal abnormalities. Fifteen cases were analysed and in all instances SKY™ provided additional cytogenetic information that could not have been detected using the conventional banding technique, including the identification of marker chromosomes, the detection of subtle chromosome translocations and the clarification of complex karyotyping. The conclusions from this work are that 'SKY has the potential to become an important cytogenetic technique for the identification of subtle translocations, marker chromosomes and the delineation of complex chromosomal aberrations without requiring any preconceived notion of the abnormalities involved' (Veldman *et al.*, 1997).

Plate 30.1 presents an interphase nucleus and a spectral karyotype of HL-60 myeloid leukaemic cells prepared according to the SKY™ protocol. Plate 30.1(A) reveals the distinct chromosomal domains in the interphase nucleus; each domain is presented in pseudocolour identifying a specific fluorescence DNA probe. Plate 30.1(B) is the spectral karyotype of an HL-60 cell demarcating chromosomal translocation in chromosome 10. Other chromosomal aberrations are seen such as the absence of chromosome 5, 14 and X.

SKY™ can also be very useful in the analysis of solid tumour material. Very little is known about specific rearrangements in solid tumours (carcinomas, sarcomas and melanoma), despite the fact that these cancers contribute significantly more to morbidity and mortality than the haematological neoplasms. This is mainly due to technical difficulties in analysing solid tumour material with their complicated karyotypes and the often suboptimal morphology of the chromosomes. SKY™ can help overcome these limitations and can contribute to the correct identification of specific chromosomal alterations, which may be of diagnostic and prognostic value.

30.4.2 Cancer phototherapy: spectral imaging of protopophyrin in cancer cells

Photodynamic therapy (PDT) of malignant melanotic-melanoma poses practical and theoretical difficulties with regard to the photosensitizer localization in the cells and the effect of melanin on the photochemical reactions. A strong correlation between the degree of tumour pigmentation and the degree of regression following PDT has been found, with the lighter tumours responding much better than darker tumours. It was concluded that pigmented melanomas in humans do not respond satisfactorily to PDT, whereas amelanotic melanoma (such as of the iris) does respond positively (Favilla *et al.*, 1991). On the other hand, the remarkable effectiveness of ALA-induced PDT and its results with experimental melanoma have opened up new possibilities for the development of PDT (Malik & Lugaci, 1987; Kennedy & Pottier, 1992; Schoenfeld *et al.*, 1994; Malik *et al.*, 1995). Protoporphyrin biosynthesized in cancer cells (Malik & Lugaci, 1987; Schoenfeld *et al.*, 1994; Malik *et al.*, 1995) from the natural precursor ALA is a highly potent photosensitizer for the destruction of these cells, even by low light-doses. ALA-PDT has been applied successfully to human patients for the selective eradication of skin tumours (Kennedy & Pottier, 1992), especially basal cell carcinoma, as well as for internal solid tumours (Peng *et al.*, 1992). Topical ALA application, or its systemic injection, has been shown as highly selective both in demarcating the tumour and in its photodynamic destruction (Kennedy & Pottier, 1992; Peng *et al.*, 1992). These results are a direct consequence of the elevated protoporphyrin biosynthesis and accumulation in the rapidly dividing transformed cells in comparison with the surrounding normal tissue (Malik & Lugaci, 1987; Peng *et al.*, 1992).

Since practically all photosensitizers act via the $^1O_2$ pathway, only targets with significant sensitizer concentrations can be damaged. The localization of porphyrins in cytoplasmic organelles plays a major role in cell damage due to the limited distance of $^1O_2$ migration during its short lifetime. Spectral imaging has been used for the subcellular localization of endogenous protoporphyrin in single living melanoma cells during photosensitization (Malik *et al.*, 1996b). The fluorescence image of ALA-treated cells revealed protoporphyrin all over the cytosol with a vesicular distribution, which represent mitochondria and endoplasmic reticulum compartments. Two main spectral fluorescence peaks were demonstrated at 630 and 670 nm, of monomeric and aggregated protoporphyrin, with intensities that differed from one subcellular site to another. Light irradiation of the cells induced point-specific subcellular fluorescence spectral changes and demonstrated photoproduct formation. Spectral image reconstruction revealed the subcellular distribution of porphyrin species in single photosensitized cells. Multipixel spectroscopy of exogenous protoporphyrin revealed an endosomal-lysosomal compartment in aggregated states, whereas monomeric porphyrin species were localized mainly on the outer membrane. Photoproducts could be

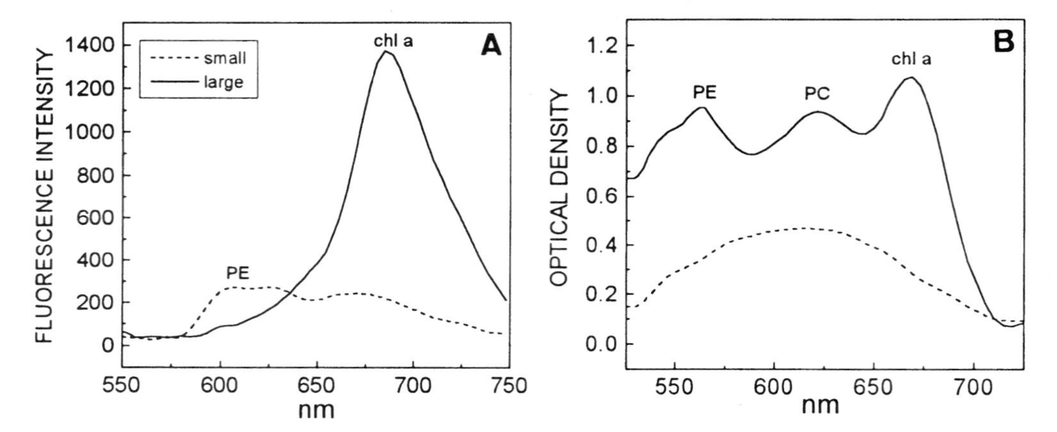

Figure 30.2 Fluorescence and absorption profiles of live *Porphyra linearis* in large and small vegetative cells of the thallus. (A) Excitation by blue light at 406 nm elicited fluorescence at 606 nm in the small cells representing phycoerythrin (PE) and the typical chlorophyl a (Chl a) at 685 nm in the large cells. (B) The absorption spectra of the vegetative cells reveal three peaks for the large cells: PE at 564 nm, phycocyanine (PC) at 625 nm and Chl a at 670 nm.

visualized at sites of formation in subcellular compartments.

30.4.3 *In vivo* spectral imaging

In vivo porphyrin localization in human skin lesions was performed by the SpectraCube™ combined with a photographic camera lens replacing the microscope. Fluorescence measurements were performed on patients with basal cell carcinomas (BCC) and acnes (Orenstein *et al.*, 1998). BCC lesions were examined after a 16-h topical application of ALA; for the acne lesions, fluorescence was studied without any additional treatment. The results of fluorescence microscopy of BCC lesions revealed red fluorescence in the stratum corneum, epidermis, pilosebaceous units and in the tumour sites. The *in vivo* macroscopic examination showed higher protoporphyrin fluorescence in BCC lesions and acnes as compared to surrounding normal tissue. An enhanced spatial resolution with demarcation of the lesions' borders was obtained using the spectral similarity-mapping function.

30.4.4 Spectral imaging of chlorophyll autofluorescence in plant cells

Spectral imaging can be used in the analysis of autofluorescence of plants. The thallus of the red algae *Porphyra linearis* consists of a monolayer of nonoverlapping cells. Therefore, it is possible to analyse the pigmentation of single cells by combining light absorbance with natural fluorescence data. Katz *et al.* (1997) used spectral imaging for the determination of the characteristic pigments for each region in the algae *Porphyra linearis*. From the image of each cell in the vegetative (Plate 30.2(A)), male and female reproductive, and holdfast regions, more than 4×10^4 fluorescence and absorbance spectra were obtained (Plate 30.2(B) & (C), respectively). The subcellular localization of the specific pigments in the different regions was resolved by spectral similarity mapping (Plate 30.2(D)). The results showed that the vegetative and female reproductive cell types had a significantly higher content of phycoerythrin than of phycocyanine, and quite similar chlorophyll a levels (Fig. 30.2). Most of the holdfast cells were poorly pigmented, but had more chlorophyll a than phycoerythrin or phycocyanin. The male reproductive cells contained only traces of pigments. Thus, by using Fourier transform multipixel spectroscopy, it was possible to characterize the pigmentation and to follow the distribution patterns of the different pigments on the subcellular level along the differentiation gradient of the alga.

REFERENCES

Chamberlain J. (1979) *The Principles of Interferometric Spectroscopy*, Wiley, New York.

Favilla I., Barry W.R., Gosbell A., Ellims P. & Burgess F. (1991) *Br. J. Ophthalmol.* **75**, 718–721.

Garini Y., Macville M., du Manoir S., Buckwald R.A., Lavi M., Katzir N., Wine D., Bar-Am I., Schrock E., Cabib D. & Ried T. (1996) *Bioimaging* **4**, 64–72.

Heim S. & Mitelman F.J. (1995) *Cancer Cytogenetics*, 2nd edn. John Wiley & Sons, New York.

Katz S., Dubinsky Z., Friedlander M., Rothmann C. & Malik Z. (1997) *J. Phycology* **33**, 222–229.

Kennedy J.C. & Pottier R.H (1992) *J. Photochem. Photobiol. B.* **14**, 275–292.

Ledbetter D.H. (1992) *Hum. Mol. Genet.* **1**, 297–304.
Lichter P. & Ward D.C. (1990) *Nature* **345**, 93–99.
Liyanage M., Coleman A., du Manoir S., Veldman T., McCormack S., Dickson R.B., Barlow C., Wynshaw-Boris A., Janz S., Wienberg J., Fergyson-Smith M.A., Schrock E. & Ried T. (1996) *Nat. Genet.* **14**, 312–315.
Malik Z. & Lugaci H. (1987) *Br. J. Cancer* **56**, 589–595.
Malik Z., Kostenich G., Roitman L., Ehrenberg B. & Orenstein A. (1995) *J. Photobiol. Photochem. B* **28**, 213–220.
Malik Z., Cabib D., Buckwald R.A., Talmi Y., Garini Y. & Lipson S.G. (1996a) *J. Microsc.* **182**, 133–140.
Malik Z., Dishi M. & Garini Y. (1996b) *Photochem. Photobiol.* **63**, 608–614.
Malik Z., Amit I. & Rothmann C. (1997) *J. Photochem. Photobiol.* **65**, 389–396.
Morris H.R., Hoyt C.C. & Treado P.J. (1994) *Appl. Spectrosc.* **48**, 857–864.
Nederlof P.M., Robinson D., Abuknesha R., Wiegant J., Hopman A.H., Tanke H.J. & Raap A.K. (1989) *Cytometry* **10**, 20–27.
Nederlof P.M. van der Flier S., Wiegant J., Raap A.K., Tanke H.J., Ploem J.S. & van der Ploeg M. (1990) *Cytometry* **11**, 126–131.
Orenstein A., Kostenich G., Rothmann C., Barshack I. & Malik Z. (1998) *Lasers Med. Sci.* **13**, 112–118.
Peng Q., Moan J., Warloe T., Nesland J.M. & Rimington C. (1992) *Int. J. Cancer* **52**, 433–443.
Rothmann C., Cohen A.M. & Malik Z. (1997) *J. Histochem. & Cytochem.* **45**, 1097–1107.
Schoenfeld N., Mamet R., Shafran M., Nordenberg Y., Babushkina T. & Malik Z. (1994) *Int. J. Cancer* **56**, 106–112.
Schrock E., du Manoir S., Veldman T., Schoell B., Wienberg J., Ferguson-Smith M.A., Ning Y., Ledbetter D.H., Bar-Am I., Soenksen D., Garini Y. & Ried T. (1996) *Science* **273**, 494–497.
Simon-Blecher N., Achituv Y. & Malik Z. (1996) *Marine Biol.* **126**, 757–763.
Solomon E., Borrow J. & Goddard A.D. (1991) *Science* **254**, 1153–1156.
Speicher M.R., Ballard S.G. & Ward D.C. (1996) *Nature Genet.* **12**, 368–375.
Tkachuk D.C., Pinkel D., Kuo W.L., Weier H.U. & Gray J.W. (1991) *Gata* **8**, 67–74.
Vane G., Chrien T.G., Reimer J.H. Green R.O. & Conel J.E. (1988) *Proc. SPIE-Recent Adv. Sensors, Radiometry Data Process. Remote Sens.* **924**, 168–178.
Veldman T., Vignon C., Schrock E., Rowley J.D. & Ried T. (1997) *Nat. Genet.* **15**, 406–410.

CHAPTER THIRTY-ONE

Multiphoton Excitation Microscopy and Spectroscopy of Cells, Tissues and Human Skin *In Vivo*

BARRY R. MASTERS[1], PETER T. C. SO[1,2] & ENRICO GRATTON[1]
[1]Laboratory of Florescence Dynamics, Department of Physics, University of Illinois at Urbana, Illinois, USA
[2]Department of Mechanical Engineering, Massachusetts Institute of Technology, Cambridge, Massachusetts, USA

31.1 INTRODUCTION

31.1.1 What is multiphoton excitation microscopy?

Non-linear optical techniques have been exploited to develop a new generation of optical microscopes with unprecedented capabilities. These new capabilities include the ability to use near-infrared light to induce absorption, and hence fluorescence, from fluorophores that absorb in the ultraviolet wavelength region. Other capabilities of non-linear microscopes include improved spatial and temporal resolution without the use of pinholes or slits for spatial filtering, improved signal strength, deeper penetration into thick, highly scattering tissues, and confinement of photobeaching to the focal volume.

Non-linear two-photon absorption processes were well known for over 30 years in molecular spectroscopy (Shen, 1984; Siegman, 1986; Yariv, 1991; Birge, 1983, 1986). If the frequency of the electric field is equal to the resonance frequency of an atom, then a single-photon absorption process can occur. However, if the frequency of the electric field is one half of the resonance frequency of the atom, and the intensity is sufficiently high, then two photons can simultaneously induce an absorption process. The critical point is that the two photons must interact simultaneously with the molecule. The rate of the two-photon absorption process is a function of the square of the instantaneous intensity of the excitation light. An example of this non-linear process is two-photon-induced fluorescence in which the simultaneous absorption of two photons of red light causes the subsequent emission of blue light.

In practical terms this new technology has been implemented in non-linear optical microscopes that use pulsed near-infrared light to induce absorption of fluorophores (ultraviolet and visible) in cells, tissues and organs. The microscope objective focuses the light pulses into a diffraction limited volume; only in this volume is the light intensity sufficient to induce absorption and fluorescence from the fluorophores at a rate suitable for microscopy. Photobleaching and phototoxicity are limited to the focal region of the microscope objective. Outside the focal volume there is insufficient intensity for the non-linear multiphoton process to occur with a high probability.

FLUORESCENT AND LUMINESCENT PROBES, 2ND EDN
ISBN 0-12-447836-0

Therefore, the optical sectioning of the non-linear, multiphoton excitation microscope is a consequence of the physics of the excitation process and does not require pinholes or slits for spatial filtering as in a confocal microscope.

31.1.2 Chapter scope

This chapter describes the history, theory, instrumentation and cell and tissue applications of non-linear multiphoton microscopy. We include a discussion of laser sources, detectors, scanners and microscope objectives. Experimental applications are presented in order to illustrate the unique capabilities of these innovative techniques. Applications are presented from studies on cells in culture, and studies of the *ex vivo* cornea which is an example of a thick, transparent tissue. The emphasis is on the application of multiphoton excitation microscopy to the functional imaging of *in vivo* human skin. These two examples, of transparent and turbid media, illustrate the utility of multiphoton excitation microscopy in thick specimens. The role of processes which cause photodamage to cells, tissues and organs is discussed and methods to mitigate these processes are described.

Figure 31.1 Dr Maria Göppert-Mayer and her daughter. (Reproduced, with permission, from AIP Emilio Segré Visual Archives, Stein Collection.)

31.1.3 Previous reviews

The reader is referred to several previous reviews of topics included in this chapter. Functional imaging of cells and tissues based on intrinsic autofluorescence (Chance & Thorell, 1959; Chance & Schoener, 1966; Chance, 1976; Chance *et al.*, 1979; Masters & Chance, 1984, 1999). Confocal microscopy is reviewed in a new book of reprinted selected historical papers and patents (Masters, 1996b). An introduction to confocal microscopy and its applications is developed in the following sources: Wilson (1990), Masters (1995a), Pawley, (1995), Corle & Kino (1996), Gu (1996b), Sheppard & Shotton (1997). Reviews of multiphoton excitation microscopy include: Williams *et al.* (1994), Denk *et al.* (1995), Denk (1996), Hell (1996), Piston (1996), So *et al.* (1996), Wokosin *et al.* (1996a,b). The patent from the Cornell group of Webb and coworkers which describes multiphoton excitation microscopy is critical reading (Denk *et al.*, 1991).

31.2 HISTORY OF TWO-PHOTON EXCITATION MICROSCOPY

31.2.1 Theory predicting two-photon absorption processes

Multiphoton excitation processes were predicted by a German physicist in 1931. In 1931, Maria Göppert-Mayer submitted her doctoral dissertation in Göttingen on the theory of two-photon quantum transitions in atoms (Göppert-Mayer, 1931). Figure 31.1 shows a photograph of Maria Göppert-Mayer. In 1963, she and J. Hans D. Jensen shared the Nobel Prize in physics for their independent work on nuclear shell theory.

31.2.2 Experimental verification of multiphoton processes

Non-linear optics may have begun with the work by Franken and his group in 1961 on second harmonic generation of light (Franken *et al.*, 1961). They showed that if a ruby laser pulse at frequency ω propagates through a quartz crystal, then light at the second harmonic frequency 2ω is generated. Light from the ruby laser at 694 nm which was incident on a quartz crystal generated light at 347 nm.

In 1963, a few weeks after the publication of the paper by Franken and his group, Kaiser and Garret published the first report on two-photon excitation of $CaF_2:Eu^{2+}$ (Kaiser & Garret, 1961). These authors later demonstrated that two-photon excitation could occur in organic molecules. Following this many examples of two-photon excitation processes in molecular spectroscopy were reported (McClain, 1971; McClain & Harris, 1977; Friedrich & McClain, 1980). Two-photon spectroscopy is an important tool in the investigation of the electronic structure of excited states of molecules (Friedrich, 1982; Birge, 1986). Thirty-two years after two-photon absorption processes were predicted by Maria Göppert-Mayer they were experimentally verified.

By analogy with the two-photon processes, three-photon excitation spectroscopy has been described. Three-photon absorption processes were first reported by Singh & Bradley (1964). Since then others have demonstrated three-photon excitation processes (Rentzepis *et al.*, 1970; Gryczynski *et al.*, 1995a,b); Szmacinski *et al.*, 1996; Lakowicz *et al.*, 1997). In the case of three-photon excitation, three near-infrared photons are simultaneously absorbed by a chromophore and a blue photon is emitted. For example, light of 900 nm could excite a chromophore with an absorption band centred at 300 nm with subsequent emission at 450 nm. Sometimes two- and three-photon excitation processes occur in the sample during the same measurement. Gu investigated the resolution in three-photon fluorescence scanning microscopy (Gu, 1996b). Both two- and three-photon detection were investigated as a means to increase the resolution of far-field microscopy (Hell *et al.*, 1995). The term multiphoton excitation includes two and higher numbers of photon excitation processes.

31.2.3 Microscopic implementation of nonlinear spectroscopy

The early implementations of non-linear optical microscopy used a laser scanning microscope to image second harmonic generation in crystals (Hellwarth & Christensen, 1974; Gannaway & Sheppard, 1978; Sheppard & Kompfner, 1978; Wilson & Sheppard, 1979).

However, it was the seminal work of Denk, Strickler and Webb, published in *Science* in 1990, that launched a new revolution in non-linear optical microscopy (Denk *et al.*, 1990). On 23 July 1991, a United States Patent on 'Two-photon laser microscopy' was assigned to the Cornell Research Foundation, Inc., Ithaca, New York (Denk *et al.*, 1991). The patent lists Winfried Denk, James Strickler and Watt W. Webb as the inventors. By integrating a laser scanning microscope (scanning mirrors, photomultiplier tube detection system) and a mode-locked laser, which generates pulses of near-infrared light, they succeeded in demonstrating a new type of microscope based on two-photon excitation of molecules. The pulses of red or near-infrared light (700 nm) were less than 100 fs in duration and the laser repetition rate was about 100 Hz. The patent states that 'focused subpicosecond pulses of laser light' are used (Denk *et al.*, 1991). These pulses have sufficiently high peak power to achieve two-photon excitation at reasonable rates at an average power less than 25 mW, which is not extremely damaging to biological samples. The benefits of two-photon excitation microscopy include: improved background discrimination, reduced photobleaching of the fluorophores, and minimal photodamage to living cell specimens. The inventors proposed the application of two-photon excitation microscopy for optical sectioning three-dimensional microscopy and for uncaging of molecules within cells and tissues.

New implementations of non-linear excitation processes are being developed. The possibility of two-photon single-molecular scanning microscopy has been realized (Mertz *et al.*, 1996; Plakhotnik *et al.*, 1996). The application of three-photon excitation to microscopy has also been demonstrated (Wokosin *et al.*, 1996a; Yuste & Denk, 1995; Xu *et al.*, 1996b,c; Masters *et al.*, 1997c).

31.3 PHYSICS OF MULTIPHOTON EXCITATION PROCESSES

31.3.1 Analytical description of two-photon excitation processes

We now compare the expressions for the rates of one photon and two-photon absorption processes for a single fluorophore (Fig. 31.2). For a one-photon absorption process, the rate of absorption is the product of the one photon absorption cross-section and the average of the photon flux density. For a two-photon absorption process, in which two photons are simultaneously absorbed by the fluorophore, the rate of absorption is given by the product of the two-photon absorption cross-section and the average squared photon flux density. In respect for the work of Maria Göppert-Mayer, who predicted the existence of two-photon absorption processes, the units of two-photon absorption cross-section are measured in GM (Göppert-Mayer) units. One GM unit is equal to 10^{-50} cm^4 s per photon.

Sufficient excitation intensity at the focal volume of the microscope can be provided by using a mode-locked dye laser. Typical laser pulses are in the red region of the spectrum (630 nm) with a duration of less than 100 fs and a repetition ration of 80 MHz. These pulses are of sufficient intensity to generate fluorescence in a fluorophore which usually absorbs at 315 nm. Alternatively, two photons in the infrared

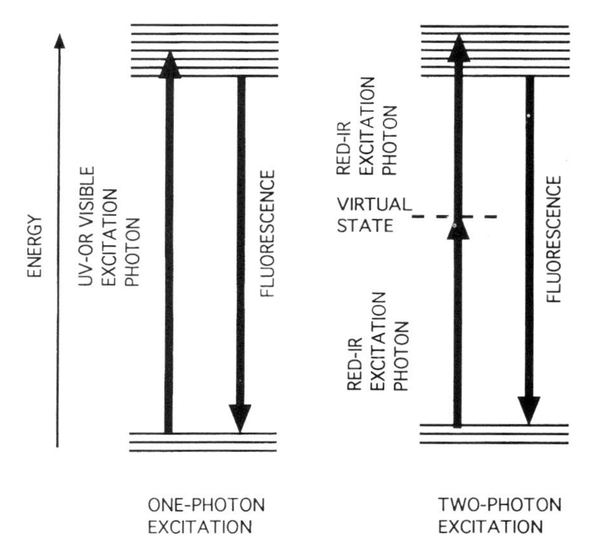

Figure 31.2 Jablonski diagram of two-photon process versus single-photon process. For the two-photon excitation process there is a short-lived virtual state shown by a dashed line.

region at 1070 nm could excite a fluorophore which usually absorbs at 535 nm in the visible region. The two photons must interact simultaneously with the molecule.

The temporal relation between the laser pulses, the single pulse width and the fluorescence decay time are shown in Fig. 31.3. In this example the pulse width is 10^{-13} s, the fluorescence decay time is 10^{-9} s, and pulse separation time is 10^{-8} s. At a laser pulse repetition rate of 100 MHz, there is a laser pulse every 10^{-8} s, or once every 10 ns.

Two-photon excitation processes do not require that the two photons that are simultaneously absorbed be of the same wavelength. Two different wavelengths can be combined by superimposing pulsed light beams of high peak powers. The two wavelengths can be chosen by use of the following equation:

$$1/\lambda_{ab} = 1/\lambda_1 + 1/\lambda_2 \qquad [31.1]$$

where λ_{ab} is the short wavelength of the absorber, and λ_1 and λ_2 are the incident beam wavelengths.

The rate of two-photon excitation can be described analytically as shown in equation [31.2]. This rate is expressed as the number of photons absorbed per fluorophore per pulse and is a function of the pulse duration, the pulse repetition rate, the photon absorption cross-section and the numerical aperture of the microscope objective which focuses the light (Denk *et al.*, 1990, 1991). The derivation of this equation assumes negligible saturation of the fluorophore and that the paraxial approximation is valid. Note that n_a,

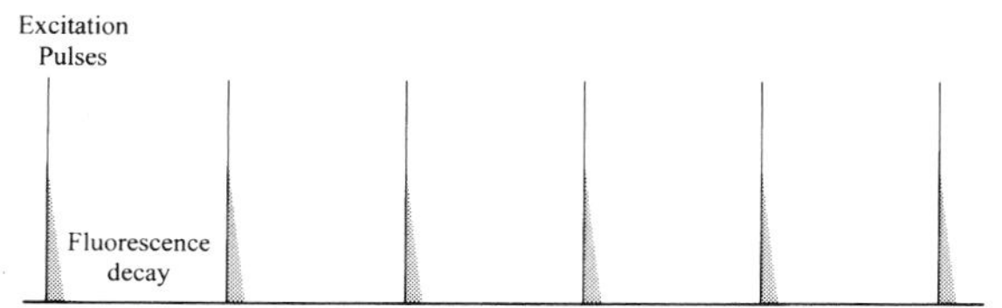

Figure 31.3 The temporal relation between the laser pulses, the single pulse width and the fluorescence decay time.

the number of photons absorbed per fluorophore per pulse, is inversely related to τ_p, the pulse duration.

$$n_a \approx \frac{p_0^2\delta}{\tau_p f_p^2}\left(\frac{\pi(\mathrm{NA})^2}{hc\lambda}\right)^2 \qquad [31.2]$$

where τ_p is the pulse duration, f_p is the repetition rate, p_0 is the average incident power, δ is the photon absorption cross-section, h is Planck's constant, c is the speed of light, NA is the numerical aperture of the focusing lens, and λ is the wavelength.

31.3.2 Optical sectioning

In a two-photon excitation process the rate of excitation is proportional to the square of the intensity of the incident light. This quadratic dependence follows from the requirement that the fluorophore must simultaneously absorb two photons per excitation process. The laser light in a two-photon excitation microscope is focused by the microscope objective to a focal volume. Only in this focused volume is there sufficient intensity to generate appreciable excitation. The low photon flux outside the focal volume results in a negligible amount of fluorescence signal. In summary, the origin of the optical sectioning capability of a two-photon excitation microscope is due to the nonlinear quadratic dependence of the excitation process and the strong focusing capability of the microscope

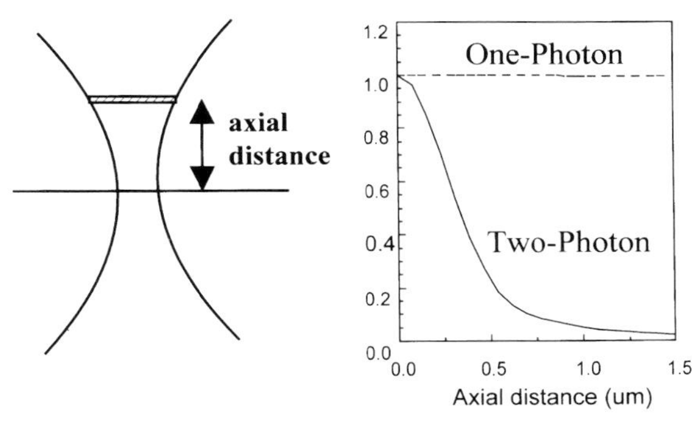

Figure 31.4 Diagram of the z-dependence of excitation processes in single-photon and two-photon excitation processes.

objective. Figure 31.4 illustrates the z-dependence for a single-photon and a two-photon excitation process.

How is three-dimensional microscopy implemented with a multiphoton excitation microscope? The quadratic dependence of the excitation probability on the intensity of the excitation light is the physical basis of the optical sectioning capability in a multiphoton excitation microscope. In a strongly focused excitation beam, the excitation probability outside of the focal volume rapidly decreases outside this volume. Only red or near-infrared light passes through the specimen. The focusing of the microscope objective results in a small focal volume in which the two-photon process occurs. Fluorescence can only occur if the fluorophore absorbs in the ultraviolet. It is possible to move the focused volume through the thickness of the sample and thus achieve optical sectioning in three dimensions. It is important to stress that the optical sectioning in a two-photon excitation microscope occurs during the excitation process. The emitted fluorescence can then be detected, without the requirement of descanning, by placing an external photon detection device as close as possible to the sample (Denk *et al.*, 1995; Piston, 1996; So *et al.*, 1996).

31.3.3 Laser pulse spreading due to dispersion

The laser pulses have a pulse width of 10^{-13} s as they emerge from the mode-locked laser. But what happens to the pulse within the optics of the microscope and in the thick, highly scattering sample? As the short laser pulses propagate through the glass and multilayer dielectric coatings they are spread out in time. This effect is due to a phenomenon called group velocity dispersion. Since each individual laser pulse consists of a distribution of optical frequencies, the wave packets will propagate at different velocities as determined by their group velocities.

Why is this dispersive laser pulse spreading important? From equation [31.2], we observe that n_a, the number of photons absorbed per fluorophore per pulse, is inversely related to τ_p, the pulse duration. Therefore, an increase in the laser pulse duration, due to group velocity dispersion from propagation through the microscope optics and the specimen, results in a decrease in the number of photons absorbed per fluorophore per pulse. The net effect is a decrease in the fluorescence due to multiphoton excitation. Fortunately, group velocity dispersion can be corrected experimentally. There are pulse compression techniques, also called 'prechirping', which can be used to compensate for group velocity dispersion (Fork *et al.*, 1984; Wolleschensky *et al.*, 1997).

31.3.4 Biophysical study of absorption cross-sections

It is important to chose or synthesize fluorophores with the following properties for use with multiphoton excitation microscopy. Ideal fluorophores would have a large absorption cross-section, a high quantum yield of fluorescence, low photobleaching and low phototoxicity to living cells and tissues. An important new development is the synthesis of new fluorophores with very large multiphoton absorption cross-sections (Bhawalkar *et al.*, 1996).

The difficulty is that a knowledge of the single-photon absorption cross-section of a fluorophore does not permit a quantitative prediction of the two-photon absorption cross-section. The quantum-mechanical selection rules for a two-photon absorption in an atom are different from those for a single-photon absorption process. A simple experimental 'rule-of-thumb' is that if the absorption band is centred at wavelength λ, then the appropriate wavelength for a two-photon excitation process is 2λ (Denk *et al.*, 1990, 1995).

Xu *et al.* have measured the two-photon excitation spectra of 25 ultraviolet and visible absorbing fluorophores from 690 to 1050 nm (Xu, 1966; Xu & Webb, 1996; Xu *et al.*, 1995, 1996a). Two-photon excitation cross-sections are given as a function of the two-photon excitation wavelength for many common fluorescent probes (e.g. fluorescein, rhodamine B and green fluorescent protein), for several calcium fluorescent indicator dyes, and for cellular endogenous fluorophores NADH and FMN (flavin mononucleotide).

31.4 COMPARISON WITH CONFOCAL MICROSCOPY

31.4.1 Optical sectioning and spatial resolution

The three-dimensional imaging properties of two-photon and single-photon fluorescence microscopes have been compared (Sheppard & Gu, 1990; Gu & Sheppard, 1995; Gu, 1996a,b). A confocal microscope achieves optical sectioning of a thick specimen with a set of conjugate pinholes. In one type of implementation a microscope objective focuses a laser beam into a diffraction-limited volume within the specimen. A pinhole, which acts as a spatial filter, is placed in a conjugate plane in front of a photodetector. The spatial filter restricts the light that reaches the photodetector to that which originates in the diffraction-limited volume of the focused laser beam. Three-dimensional confocal microscopy is achieved by passing the diffraction-limited volume across the thickness of the sample. If the sample is filled with fluorophores, then there is

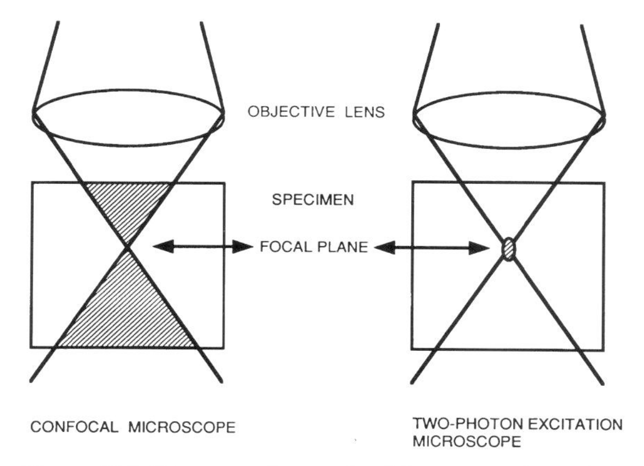

Figure 31.5 Diagram of a confocal laser scanning microscope and a two-photon excitation microscope. In a confocal laser scanning microscope the photobleaching and the photodamage occurs throughout the right circular cone of the excitation beam. In a two-photon excitation microscope the photobleaching and the photodamage is limited to the focal volume.

fluorescence in all regions of the right circular cone. At the focal volume, located at the vertex of the right circular cone, the fluorescence is maximum.

A two-photon excitation microscope does not require a spatial filter in front of the photon detector in order to achieve optical sectioning. The optical sectioning is strictly a consequence of the physics of the two-photon excitation process. Therefore, the emitted fluorescence can be detected in any position around the fluorophore. In contrast to a confocal microscope, in a two-photon excitation microscope the fluorescence is predominately limited to the focus of the red or near-infrared light (Fig. 31.5).

In the z-direction of a two-photon excitation microscope, the excitation probability falls off with the fourth power of the distance on the optic axis. This can be compared with single-photon excitation probability, which falls off with the square of the distance on the optical axis. The depth discrimination of a multiphoton excitation microscope is similar to that of an ideal confocal laser scanning microscope with conjugate pinholes for spatial filtering.

The resolution of a microscope depends on the wavelength of the illumination. A confocal microscope with ultraviolet illumination will have superior spatial resolution, increased by a factor of two, as compared with a two-photon excitation microscope with red or near infrared illumination. The resolution of a two-photon excitation microscope is limited by the size of the excitation volume of the focused light in the specimen. As previously explained, the emitted fluorescence need not be focused, and therefore it is less affected by scattering and chromatic aberration.

31.4.2 Depth of penetration

One advantage of the use of two-photon excitation microscopy is that the longer wavelengths of the red and near-infrared laser illumination afford deeper penetration into thick, highly scattering tissues. In order to excite ultraviolet absorption fluorophores, a confocal laser scanning microscope usually uses an ultraviolet laser. The tissue scattering coefficient for ultraviolet light is higher than for near-infrared light, and therefore the intensity diminishes rapidly as a function of depth.

Near-infrared light was able to penetrate the full thickness (400 μm thickness) of an *ex vivo* rabbit cornea (Piston *et al.*, 1995; Masters, 1996a,c; Masters *et al.*, 1998a,b,c). Potter *et al.* reported that for a variety of living and fixed specimens they were able to image two or three times deeper with two-photon excitation microscopy as compared with laser scanning confocal microscopy (Potter *et al.*, 1996).

31.4.3 Photobleaching

As seen in Fig. 31.5, in a confocal microscope there is fluorescence in the focal volume and also in both lobes of the right circular cone of illumination. Consequently the process of photobleaching occurs in the entire illuminated volume within the specimen. In two-photon excitation microscopy the fluorescence, and therefore the photobleaching, is predominately restricted to the focal volume (Gu & Sheppard, 1995).

31.4.4 Photodamage

For the single-photon process, fluorescence excitation typically uses continuous wave lasers and requires an average power of about 100 μW. For the two-photon microscope, lasers with 100 MHz and 100 fs pulses are often used and requires an average power of about 10 mW. This corresponds to a peak power on the order of 1 kW. The use of a laser pulse picker can significantly decrease the average laser power used by decreasing the laser pulse repetition rate and increasing the laser peak power. The use of laser pulse picker to mitigate thermal damage will be discussed in Section 31.5.

The processes that result in photodamage to the specimen are complex (Gu & Sheppard, 1995; König *et al.*, 1996, 1997). For multiphoton excitation microscopy, oxidative photodamage is reduced by delivering the laser pulse energy into a small focal volume. With a confocal laser scanning microscope, more significant photodamage occurs, resulting from single photon absorption throughout the total volume. Since typical two-photon excitation microscopy requires milliwatt average power laser pulse trains consisting of nanojoule pulses, severe thermal damage can occur as a result of one-photon excitation. This situation is particularly severe for highly pigmented cells such as melanophores (Masters *et al.*, 1999). Furthermore, the high peak power required for two-photon excitation may produce cell and tissue damage through dielectric breakdown mechanisms.

During a two-photon excitation process, it is important to realize that three-photon excitation can also be generated. Although the emission filters may be selected for fluorescence detection from the two-photon process, there may still be photodamage from the unintentional three-photon excitation process. When intentionally using three-photon excitation for tissue imaging, it is important to realize that three-photon excitation microscopy often requires an order of magnitude higher incident laser power than two-photon excitation in order to achieve similar rates of excitation. It is important to study the associated three-photon cellular damage mechanisms and identify damage threshold because of the high laser power required. Nevertheless, three-photon excitation microscopy provides a useful technique to excite those fluorophores with single-photon absorption bands in the ultraviolet and short-ultraviolet wavelengths.

31.4.5 Summary

How does multiphoton excitation microscopy compare with confocal laser scanning microscopy? The main drawbacks of two-photon excitation microscopy include: (1) multiphoton excitation microscopy is only suitable for fluorescent imaging; reflected light imaging is not currently available, and (2) the technique is not suitable for imaging highly pigmented cells and tissues which absorb near-infrared light. Multiphoton excitation microscopy has the following important advantages: (1) reduced phototoxicity, (2) reduced photobleaching, (3) increased penetration depth, (4) ability to perform uncaging or photobleaching in a diffraction-limited volume, (5) ability to excite fluorophores in the ultraviolet without an ultraviolet laser, (6) the excitation and the fluorescence wavelengths are well separated, and (7) no spatial filter is required.

31.5 INSTRUMENTATION FOR MULTIPHOTON EXCITATION MICROSCOPY, SPECTROSCOPY AND LIFETIME MEASUREMENT

31.5.1 Components of a multiphoton excitation microscope

The instrumentation and design of a basic multiphoton microscope has been described in a number of previous publications (Denk *et al.*, 1995; Wokosin *et al.*, 1996a,b) (Fig. 31.6). The most critical component of a multiphoton microscope is the laser light source. As discussed, the two-photon excitation cross-section is very small and negligible excitation is generated unless in the presence of a very high light flux. Simple calculation demonstrates that the fluorescence excitation rate is insufficient for scanning microscopy applications unless femtosecond (fs) or picosecond (ps) pulsed lasers are used. This does not mean that

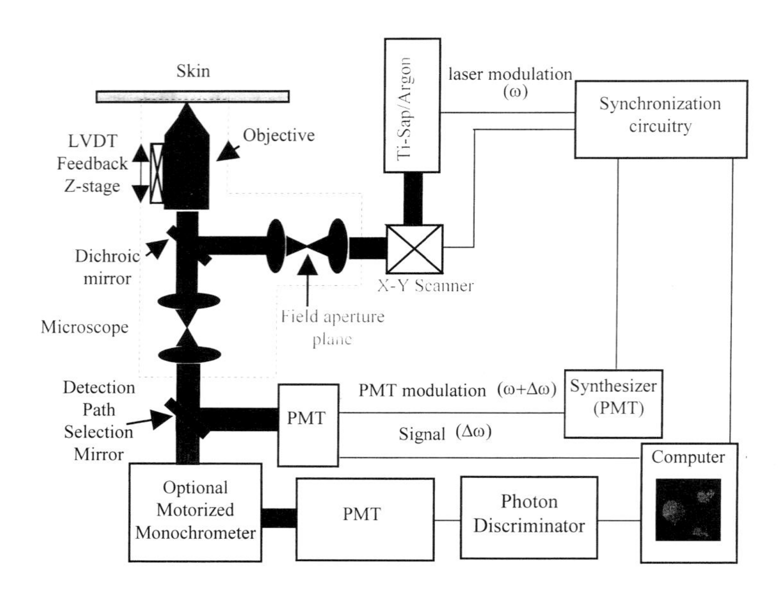

Figure 31.6 Diagram of multiphoton instrumentation, source, microscope and detector.

one cannot generate two-photon excitation using continuous wave lasers. It has been demonstrated that two-photon processes can be initiated, but the high average power needed is detrimental to vital cell or tissue imaging (König *et al.*, 1996). Although two-photon microscopy imaging was first demonstrated with a dye laser in a colliding pulse geometry, this system is rarely used today because of the difficulties involved in maintaining the system. Today, the most common laser systems used are titanium–sapphire laser systems. These systems provide high average power (1–2 W), high repetition rate (80–100 MHz), and short pulse width (80–150 fs). They are also robust and easy to maintain. Titanium–sapphire lasers require pump lasers. The older systems use argon-ion lasers, but the new ones use solid-state diode pumped Nd:YAG lasers. The newer lasers are basically turnkey systems. Ultra compact (the size of a shoe box), single-wavelength femtosecond lasers combining diode, Nd:YAG, and titanium–sapphire lasers are becoming commercially available. Furthermore, other single-wavelength solid-state systems such as diode pumped Nd:YLF lasers and diode pumped erbium-doped fibre laser systems have also become widely available. The choice between femtosecond and picosecond light sources for multiphoton excitation remains controversial. As discussed, multiphoton excitation can be achieved with both types of lasers. To generate the same level of fluorescence signal, the picosecond laser systems will need a significantly higher average input power, while the femtosecond laser systems has a much higher peak power. The choice between these laser systems depends on what is the predominant photodamage mechanism. If thermal damage is the major factor, as in the melanophores, it is desirable to minimize average power. However, if the predominant damage mechanism is related to high peak power, it is preferable to reduce peak power. For typical cells and tissues, the two-photon photodamage pathways still need to be better determined.

The other critical element is a high-throughput microscope system with beam scanning electronics. High numerical aperture objectives are critical for efficient two-photon excitation and the detection of low-level signals. As compared to a confocal system, chromatic aberration is not a crucial factor as the different colour emission light does not have to be descanned. However, new objective designs that maximize infrared transmission would be extremely helpful in this area. Maximizing infrared transmission decreases the scattering of the high-power excitation light which interferes in detecting the low-level fluorescence signal. Most implementations of multiphoton scanning microscopes incorporate infrared femtosecond light sources with an existing scanning confocal microscope (Denk *et al.*, 1990; Piston *et al.*, 1995). This implementation is fairly straightforward where only mirrors in the scanning head have to be modified to efficiently reflect infrared laser light. It is also critical to modify the emission beam path such that the high-loss descanning optics are not used. Since the mechano-optics involved in multiphoton microscopes is much simpler than in the confocal system, the modifications needed to convert a high-throughput fluorescence microscope for multiphoton scanning are actually quite simple. The details needed of these modifications have been described (So *et al.*, 1995). Changes mainly involve modification of the excitation light path to incorporate a scan lens and the provision of an electronic interface to synchronize the scanner mirror system and the data acquisition electronics.

Properly designed detection electronics is essential for a high-performance multiphoton scanning microscope. Typical scanning systems use photomultiplier tubes with analogue detection circuitry. Photomultiplier tubes have good quantum efficiency (10–25%) in the blue–green region, but poor (less than 1%) quantum efficiency in the red. Although analogue circuity does not provide very sensitive detection at very low signal levels, it can handle fairly high intensity signals without saturation. It is also critical that analogue conversion circuitry has sufficient dynamic range in cellular imaging applications where signal strength can vary greatly. Typically, a dynamic range of at least 12 bits is desirable. Another common detection method uses a single-photon counting scheme which is very sensitive at extremely low light situations and has excellent dynamic range, but suffers from the difficulty of easy saturation at high intensity. Other detectors that are promising for multiphoton microscopes are high-sensitivity single-photon counting avalanche photodiode detectors which have excellent quantum efficiency (over 70%) in the red spectral range and very good efficiency (30–50%) in the green range.

31.5.2 Measurement of the fluorescence emission spectrum in the microscope

Wavelength-resolved studies are an important extension of multiphoton microscopic imaging, with numerous important biomedical applications. The most common application involves resolving distinct cellular and tissue structures; this has found important applications in fluorophores with different emission spectra. Other important applications involve the use of ratiometric imaging to measure important cellular biochemical functions such as calcium distribution (Piston *et al.*, 1995; So *et al.*, 1995) and membrane fluidity. The instrumentation for wavelength-resolved spectroscopy measurements in multiphoton microscopy is essentially similar to the technology used in one-photon systems with two exceptions. First, since the two-photon excitation wavelength is red shifted to twice the normal spectra region, there is always good separation between two-photon excitation and emission spectral regions. This wide separation allows the emission intensity to be easily isolated from the high intensity excitation. More importantly, the overlap between the excitation and emission spectra in the one-photon case prevents the shorter wavelength region of the emission spectra to be studied, but this is not a problem in the two-photon-case. Second, two-photon excitation at the properly chosen wavelength region can simultaneously excite fluorophores in the near-ultraviolet, blue, green and red ranges. In a multiple labelling experiment, this capability allows all the fluorescent probes to be simultaneously excited and their individual distributions mapped by using wavelength-resolved spectroscopy.

The instrumentation involved in performing wavelength-resolved spectroscopy is basically similar to that used in one-photon microscopy. For a rough wavelength separation, as in applications where a number of chromophores need to be resolved, different wavelength channels can be constructed using a number of dichroic filters (Xu *et al.*, 1996a). When finer wavelength resolution is needed to study spectral features, single point measurement using monochromometers or spectrographs can be performed (Masters *et al.*, 1997c).

31.5.3 Measurement of the fluorescence lifetime in the microscope

Lifetime resolution is another powerful spectroscopy technique that has been integrated with multiphoton microscopy. Lifetime imaging can resolve multiple structures in cells and tissues similar to wavelength-resolved imaging (So *et al.*, 1995). It can also be used to quantitatively measure cellular metabolite concentration such as calcium using non-ratiometric probes such as calcium green (So *et al.*, 1995). However, the most important use of this technology is in the study of cellular biochemical processes that have no wavelength-resolved probes. One important example is the use of the changes in the lifetime of fluorescein–bovine serum albumin complex during proteolysis inside the vacuoles of macrophages to study the kinetics of this important immunological process (French *et al.*, 1997). Another important example is to use NAD(P)H lifetime as a measure of the cellular oxidative state (König *et al.*, 1997; Masters *et al.*, 1997c). The instrumentation involved in lifetime imaging inside a multiphoton microscope is very similar to that used in a confocal system. In the confocal system, lifetime measurements have been performed using both time domain and frequency domain techniques. Since multiphoton microscopy is still a relatively new approach, only the incorporation of frequency domain techniques has been attempted (So *et al.*, 1995; Masters *et al.*, 1997d).

31.5.4 Measurement of the fluorescence intensity dependence on excitation power

We have measured the fluorescence intensity as a function of excitation power to determine whether the excitation at 960 nm is due to two-photon excitation or involves higher photon processes. A 0.5 mm thick slice of human skin from the second digit of the right hand was excised. The sample was sandwiched between a piece of coverglass and a microscope slide. The fresh sample was imaged immediately. The incident power was controlled by a polarizer. Fluorescence images of the same area were collected at power levels ranging from 1 to 10 mW with 960 nm excitation. Equivalent portions (10×10 pixel areas) of these images were averaged and compared.

31.5.5 Pulse width measurement and pulse compensation techniques

In Section 31.3.3 we discussed the problem of pulse dispersion both in the microscope and the specimen. Pulse dispersion is a serious problem in multiphoton excitation microscopy since it results in a large reduction in the probability of multiphoton excitation. This is observed as a strong reduction in the intensity of the fluorescence signal from the fluorophores in the specimen. There is an experimental technique called 'prechirping' the laser pulses, which can be used to compensate the laser pulse dispersion (Fork *et al.*, 1984; Valdmansis & Fork, 1986). A 'prechirp' unit can be constructed with two prisms and a mirror. This system compensates for the group dispersion and causes all of the different wavelengths in each pulse to arrive simultaneously at the specimen after propagating through the microscope optics.

An important paper describes how to measure the group velocity dispersion for a microscope with a variety of microscope objectives (Wolleschensky *et al.*, 1997). A Michelson-type interferometric autocorrelator was attached to the microscope which was used to measure the pulse length of the laser pulses at the specimen. For example, a COHERENT MIRA 900-F titanium–sapphire femtosecond laser system

(780 nm, 72 MHz, 92 fs) was adapted to a Zeiss confocal laser scanning microscope LSM 410 with a C-APOCHROMAT ×40, NA 1.2 water immersion microscope objective. The measured pulse length at the specimen was 355 fs. Recompression of the pulse width to 135 fs at the specimen was achieved with the use of a 'prechirp' unit consisting of a pair of prisms.

31.6 APPLICATION TO CELLS AND TISSUES

31.6.1 Use of fluorescent probes with multiphoton excitation microscopy

Many of the commonly used molecular fluorescent probes can be imaged with multiphoton excitation microscopy. The following fluorescent probes represent a small sample of the probes that have been imaged with multiphoton excitation microscopy: NAD(P)H, flavoproteins, and serotonin (autofluorescence); rhodamine, Bodipy, Di-I, fluo-3, indo-1, GFP, Lucifer Yellow, Calcium Green, DAPI, Hoechst 33342 and DAPI.

31.6.2 Uncaging with multiphoton excitation microscopy

The uncaging of photolabile caged compounds is another important application of multiphoton excitation microscopy. Photolabile caged compounds are inert precursors of biologically active molecules that can be stimulated with near-UV light to release, or 'uncage', the active compounds. Caged compounds are applied extracellularly or they are incorporated into cells. With light-induced uncaging the physiology or the pharmacology of the cells could be manipulated on the order of milliseconds (Gurney, 1993). Many of the available caged compounds have an absorption band near 350 nm.

The major advantage of using a two-photon excitation process to uncage chemical effector compounds in cells and tissues are two-fold (Denk *et al.*, 1990, 1991; Silberzan *et al.*, 1993). First, most common caged compounds are activated by ultraviolet light. The need for high ultraviolet power for these studies makes ultraviolet-light-induced cellular damage a major concern. The ability to effect chemical reaction using infrared light dramatically reduces this problem. Second, the ability to release chemical compounds in a subfemtolitre volume has enabled a number of novel microscopy applications. An important example is the use of caged neurotransmitters to map acetylcholine receptor distributions (Denk, 1994). The ability to release neurotransmitter at a selected position provides the investigators a means to activate receptors within the local region of the cell membrane. The number of receptors within the interrogated volume is proportional to the whole cell current, which can be monitored by traditional patch clamp technique. Another important example is the use of two-photon technique to uncage calcium in a local region of the cell. Local calcium signalling events such as calcium release, which may be associated with intracellular structures, cannot be studied unless the local calcium concentration within a femtolitre volume can be manipulated.

One of the major difficulties associated with two-photon uncaging is that typically higher power is required than needed for two-photon induced fluorescence. This higher laser power requirement has two problems. First, greater cellular damage will result from the higher power needed and may negate the advantage of using infrared light. Second, it has been observed that high-intensity light can trigger a cellular response. For example, cellular calcium transients can be triggered with sufficiently high-power infrared light even in the absence of caged compound. These false positive responses are a serious concern. These difficulties originate from two technical problems. On the one hand, typical caged compounds have a particularly low two-photon quantum efficiency. On the other hand, high power is needed to uncage sufficient amounts of chemicals to trigger localized cellular responses before diffusion broadens the reaction volume.

31.6.3 Functional redox imaging of cellular metabolism

To demonstrate the efficacy of multiphoton excitation microscopy for the imaging of thick tissues with minimum photodamage, we studied the NAD(P)H fluorescence from an *ex vivo* rabbit cornea. The rabbit cornea is a thick (central corneal thickness is 400 μm), semitransparent, scattering tissue.

Functional imaging of cellular metabolism and oxygen utilization using the intrinsic fluorescence has been extensively studied in cells (Masters & Chance, 1984). A comprehensive review of the instrumentation, techniques and experimental results based on fluorescence measurements of the autofluorescence of cells is presented in Chapter 28 of this volume and in the first edition of this book (Masters & Chance, 1994, 1999). Specific studies based on redox fluorometry include the following: redox measurements of *in vivo* rabbit cornea based on flavoprotein fluorescence (Chance & Lieberman, 1978; Masters *et al.*, 1981; Masters, 1993a); monitoring of oxygen tension under the contact lens in live rabbits (Masters, 1988, 1990);

chemical analysis of nucleotides and high-energy phosphorus compounds in the various layers of the rabbit cornea (Masters *et al.*, 1989); and redox fluorescence imaging of the *in vitro* cornea with ultraviolet confocal fluorescence microscopy (Masters *et al.*, 1993).

Redox fluorometry is a non-invasive optical method to monitor the metabolic oxidation–reduction (redox) states of cells, tissues and organs. It is based on measuring the intrinsic fluorescence of the reduced pyridine nucleotides, NAD(P)H and the oxidized flavoproteins of cells and tissues (Chance & Thorell, 1959; Chance & Schoener, 1966). Both the reduced nicotinamide adenine dinucleotide, NADH, and the reduced nicotinamide adenine dinucleotide phosphate, NADPH, are denoted as NAD(P)H. Redox fluorometry is based on the fact that the quantum yield of the fluorescence, and hence the intensity, is greater for the reduced form of NAD(P)H, and lower for the oxidized form (Masters, 1990). For the flavoproteins, the quantum yield, and hence the intensity, is higher for the oxidized form and lower for the reduced form (Chance & Schoener, 1966). The reduced pyridine nucleotides are located in both the mitochondria and in the cytoplasm (Chance & Schoener, 1966). The flavoproteins are uniquely localized in the mitochondria (Chance & Schoener, 1966). The fluorescence from the reduced pyridine nucleotides is usually measured in tissue investigations since the measured fluorescence is higher than the flavoprotein fluorescence (Chance & Lieberman, 1978). Redox fluorometry has been applied to many physiological studies of cells, tissues and organs (Chance, 1976; Chance *et al.*, 1979; Masters, 1990). The review of redox fluorometry in the study of ocular tissue is particularly comprehensive (Masters, 1988, 1990, 1993a, 1995b, 1996a,c; Masters *et al.*, 1981, 1989, 1998a).

31.6.4 Ultraviolet confocal fluorescence microscopy

It is instructive to compare redox imaging of the NAD(P)H fluorescence from the cornea with both confocal and two-photon excitation microscopy.

The cornea is the transparent tissue on the front surface of the eye. The epithelium which constitutes the 50 μm surface is a renewable optical surface. The various cell layers of the *in vivo* human cornea have been studied with a new real-time confocal microscope (Masters & Thaer, 1994a,b). Differentiation of the basal epithelial cells into wing cells and their subsequent migration to the ocular surface results in the continuous renewal of the corneal surface (Masters & Thaer, 1995). This process is analogous to the differentiation of basal cells in skin and the renewal of the surface of the skin.

In the confocal studies, ultraviolet confocal laser scanning fluorescence microscopy was used to investigate the *ex vivo* cornea (Masters *et al.*, 1993). The laser scanning confocal microscope used was equipped with an argon-ion laser with an output of 364 nm. The excitation wavelength of 364 nm and the emission at 400–500 nm were used to image the fluorescence of the reduced pyridine nucleotides, NAD(P)H. A Zeiss water immersion microscope objective, ×25, with a numerical aperture of 0.8 and corrected for the ultraviolet, was used for imaging the fluorescence.

Rabbit eyes were obtained from New Zealand white rabbits. The rabbits were euthanized with an injection into the marginal vein of the ear with ketamine hydrochloride (40 mg kg^{-1}) and xylazine (5 m kg^{-1}). The eyes were freed immediately of adhering tissue and swiftly enucleated. The intact eyes were placed in a container filled with bicarbonate Ringer's solution, which contained 5 mM glucose and 2 mM calcium, and were kept at room temperature. A specimen chamber held the eye immobilized during microscopic imaging (Masters, 1993b). The eyes were imaged within 1–4 h after their removal from the rabbit.

Analysis of the resulting images showed photobleaching of the NAD(P)H during the course of the measurements. There was also image degradation due to the sample thickness. The microscope objective was designed for a refractive index of water; across the full 400 μm of corneal thickness the refractive index is changing. While the average refractive index of the cornea is known, the profile of refractive index is unknown. The microscope objective had chromatic aberrations for excitation and emission wavelengths. These effects resulted in reduced image quality in the confocal microscope.

The NAD(P)H fluorescence was imaged from the superficial cells of the corneal epithelium and from the corneal endothelium. Due to a poor signal-to-noise ratio we were not able to image the NAD(P)H fluorescence from the stromal keratocytes. The two-dimensional imaging of the corneal endothelial cells based on NAD(P)H functional redox imaging presents a new technique for investigating the heterogeneity of the cellular respiration in the corneal cell layers.

31.6.5 Two-photon excitation microscopy

Two-photon excitation microscopy at 730 nm and fluorescence emission spectroscopy were used to image the cells of the *ex vivo* rabbit cornea and to characterize spectrally the fluorescence (Piston *et al.*, 1995; Masters *et al.*, 1998d).

The *ex vivo* rabbit cornea was maintained in a

physiological state in a specially designed specimen chamber (Masters, 1993b). Two-dimensional images of cellular metabolic oxidation/reduction of cells in the cornea (wing cells, basal epithelial cells and stromal keratocytes) were obtained by imaging the fluorescence of reduced pyridine nucleotides, NAD(P)H. The emission spectrum had a peak centred near 450 nm and a bandwidth at half-height of 100 nm. The spectroscopic data suggest that the reduced pyridine nucleotides, NAD(P)H, are the primary source of the cornea autofluorescence. We found that the use of a two-photon excitation microscope without descanning in the detection light path can result in enhanced sensitivity; this is critical for tissue imaging of weak fluorophores such as NAD(P)H.

In the first study of this system, Masters and Piston, working in the laboratory of Professor Webb at Cornell succeeded in obtaining three-dimensionally resolved NAD(P)H cellular metabolic redox imaging of the *in situ* cornea with two-photon excitation laser scanning microscopy (Piston *et al.*,1995). The microscope used at Cornell University descanned the fluorescence emission using the traditional confocal detection light path with the pinhole kept open. The weak fluorescence from the stromal keratocytes in their normal physiological state could not be imaged with this instrument. However, the application of cyanide which resulted in a doubling of the concentration of NAD(P)H in the stromal keratocytes, and hence a doubling of the fluorescence intensity, permitted the microscope to image the keratocytes. On the other hand, the more recent experiment which used external detection of the fluorescence emission by Masters and So in the laboratory of Professor Gratton at Urbana-Champaign has shown that in the absence of descanning optics, stromal keratinocyte imaging was successful under physiological conditions. Figure 31.7 shows a multiphoton excitation microscopic image of basal epithelial cells from *ex vivo* cornea. These investigations illustrate the use of multiphoton excitation microscopy for the investigation of non-invasive functional redox imaging of a thick, semi-transparent tissue.

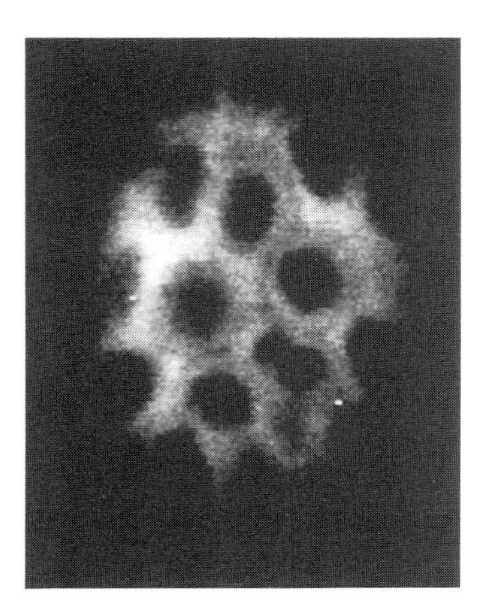

Figure 31.7 Multiphoton excitation microscopy of *ex vivo* cornea. The image shows basal epithelial cells. Horizontal field width is 100 μm.

31.7 APPLICATION TO *IN VIVO* FUNCTIONAL IMAGING OF HUMAN SKIN: AN EXAMPLE OF A THICK, HIGHLY SCATTERING *IN VIVO* TISSUE

In vivo human skin is a thick, highly scattering tissue. A multiphoton excitation microscopy and spectroscopy study of human skin is an excellent example of the technique's utility and capability. We have demonstrated the excellent depth penetration, three-dimensional functional imaging of cellular metabolism based on NAD(P)H and flavoprotein fluorescence (Masters *et al.*, 1997c). The autofluorescence was characterized by both emission spectroscopy and lifetime measurements (So *et al.*, 1995, 1996). These techniques may also have clinical diagnostic potential as a non-invasive optical biopsy (Masters *et al.*, 1997c,d).

31.7.1 Three-dimensional structure of human skin

The skin is divided into three layers: the epidermis, the dermis, and the subcutaneous tissue. The epidermis is the outermost portion of the skin and is composed of stratified squamous epithelium. The anatomy of the skin is illustrated in Fig. 31.8, which shows a vertical section of human skin. A three-dimensional visualization of *in vivo* human skin, based on reflected light confocal microscopy (Masters, 1996d; Masters *et al.*, 1997a,b), is shown in Plate 31.1.

The epidermis is a self-renewing multilayered tissue which continuously differentiates to produce stratified layers of resultant dead cells, the corneocytes, whose function is to protect the body against external insults (barrier function). Except for the palms and soles, the thin epidermis, which covers the body, comprises the following layers from the surface to the dermis:

- a stratum corneum, which contains 15–20 layers of flat, anucleate, pentagonal-shaped dead cells (corneocytes),
- the stratum lucidum, which contains 1–2 layers of corneocytes and marks the transition to the living cellular domain,
- the stratum granulosum, which contains about 2 layers of flattened cells with flat nuclei,

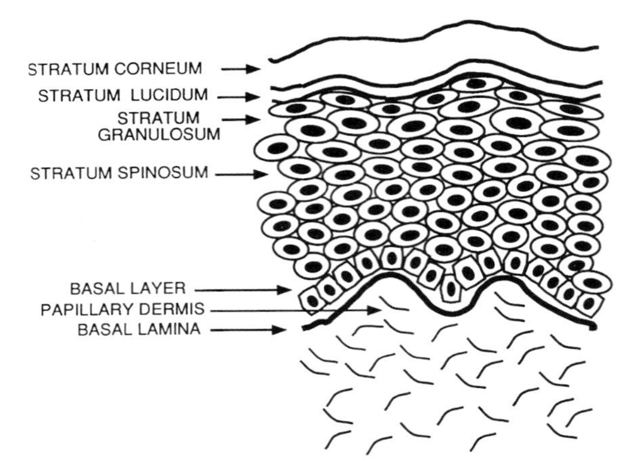

Figure 31.8 Schematic drawing of a vertical section of skin showing the following cell layers from the skin surface to the dermis: stratum corneum, stratum lucidum, stratum granulosum, stratum spinosum, basal layer, and the papillary dermis. Individual cells forming the stratum corneum and the thin stratum lucidum are not shown in the drawing.

- the stratum spinosum, which contains several layers of polyhedral keratinocytes with spherical nuclei,
- the stratum basale (germinative layer), which consists of a single layer of cuboidal cells, with ellipsoidal nuclei adhering to the basement membrane of the dermis.

Four types of cells are located within the living epidermis: (1) keratinocytes; and a few per cent of dendritic cells, (2) Langerhans cells, (3) melanocytes and (4) rare Merkel cells. Keratinocytes are located in all strata, melanocytes are located within the stratum basale, Langerhans cells are mostly located in the stratum spinosum, and Merkel cells are in or adjacent to stratum basale.

The dermoepidermal junction is comprised of structures at the interface between the epidermis and the dermis. As viewed with the light microscope, we observe this boundary as an undulating pattern of rete ridges (downward projections of the epidermis) and dermal papillae (upward projections of the dermis into the epidermis).

The single layer of basal cells located at the dermoepidermal junction are the source of new keratinocytes (by their differentiation and migration to the surface) in the renewal of the epidermis. The process of differentiation and migration towards the surface is analogous in both skin and cornea (Beebe & Masters, 1996). Understanding the regulation of proliferation and differentiation in stratified epithelia (skin and cornea) may yield new insights for future diagnostic techniques and therapies.

31.7.2 Multiphoton excitation microscopy of *in vivo* human skin: functional redox imaging

In order to interpret correctly the results of applying multiphoton excitation microscopy to three-dimensional functional imaging of thick, highly scattering *in vivo* tissue such as human skin, it is necessary to demonstrate the following. First, images of the cells based on the cellular autofluorescence are obtained and these images are correlated with similar images of the same tissue obtained with reflected light confocal microscopy. This use of correlative microscopy is important to demonstrate the histology of the tissue based on images acquired with multiphoton excitation microscopy. Second, the correct interpretation of skin physiology using multiphoton microscopy and spectroscopy requires that the tissue fluorophores responsible for the fluorescence contrast be correctly identified. Since only a few fluorophores in tissues have been studied, the correct identification of the fluorescent species required a knowledge of its excitation and emission spectra, which in turn requires knowing whether the fluorophores are excited by either two or higher photon processes. The characterization of the multiphoton process can be performed in the following manner. For a two-photon excitation process, it is necessary to show a plot of the quadratic dependence of the fluorescence intensity on the excitation power. A plot of the logarithm of the fluorescence intensity versus the logarithm of the excitation power should have a slope of 2 for a two-photon excitation process. Similarly, a slope of 3 indicates a three-photon process. Finally, for each type of cell or tissue studied, the emission spectra and the lifetime of the fluorescence should be determined to confirm the nature and biochemical state of the fluorophore, e.g. NAD(P)H or flavoprotein.

Multiphoton excitation microscopy at 730 nm and 960 nm was used to image *in vivo* human skin autofluorescence. The lower surface of the right forearm (of one of the authors) was placed on the microscope stage where an aluminium plate with a 1 cm square hole is mounted. The square hole is covered by a standard coverglass. The skin was in contact with the coverglass to maintain a mechanically stable surface. The upper portion of the arm rested on a stable platform, which prevented motion of the arm during the measurements. The measurement time was always less than 10 minutes. The estimated power incident on the skin was 10–15 mW. The photon flux incident upon a diffraction-limited spot on the skin is on the order of 10 MW cm^{-2}.

We observed individual cells within the thickness of the skin at depths from 25 to 75 μm below the skin surface. No cells were observed in the stratum corneum. These results are consistent with studies

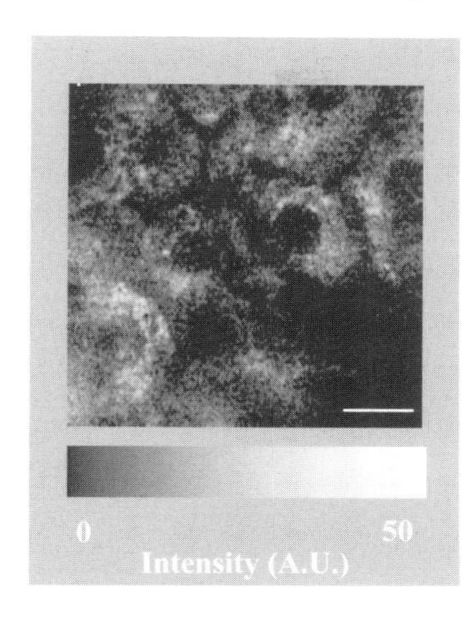

Figure 31.9 Multiphoton excitation microscopy of human skin. The scale bar corresponds to 10 μm.

using reflected light microscopy (Masters *et al.*, 1997b).

Figure 31.9 shows multiphoton excitation microscopy of human skin with an excitation wavelength of 730 nm. Cells of 15–20 μm diameter were imaged at an approximate depth of 40–50 μm. Cells of about 10 μm were observed at a depth of 30 μm below the skin surface. In the cytoplasm of the larger cells we observed punctated fluorescence. Similar findings were reported previously (König *et al.*, 1996, 1997). These fluorescent organelles are likely to be mitochondria with a high concentration of NAD(P)H.

In order to show the three-dimensional distribution of the autofluorescence we acquired optical sections with the two-photon excitation microscope and formed a three-dimensional visualization across the thickness of the *in vivo* human skin. Two representative three-dimensional reconstructions of *in vivo* human skin obtained with two-photon excitation microscopy are shown in Plate 31.2.

Two- and three-photon excitation processes have been shown in human skin (Fig. 31.10). At 730 nm excitation, only the two-photon process was observed. At 960 nm excitation, fluorescence intensity shows a quadratic dependence at low excitation power, but becomes a cubic dependence at high power.

31.7.3 Multiphoton excitation microscopy of *in vivo* human skin: emission spectroscopy and fluorescence lifetime measurements

It is important to characterize the source of the fluorescence that is imaged with multiphoton excitation microscopy. Two types of measurements are useful in the characterization of the fluorophore: emission spectroscopy and lifetime measurements. We measured these characteristics at selected points on the skin. Fluorescent spectra were obtained close to the stratum corneum (0–50 μm) and deep inside the dermis (100–150 μm). Measurements were made for both 730 nm and 960 nm excitation wavelengths corresponding to one-photon excitation wavelengths of about 365 nm and 480 nm, respectively. Figure 31.11 shows the emission spectra of *in vivo* human skin with multiphoton excitation microscopy. Table 31.1 shows the spectral characteristics from the measurements.

The fluorescence lifetimes were measured at selected points on the skin to complement the fluorescent spectral data obtained. Table 31.2 shows the fluorescence lifetimes measured in *in vivo* human skin. The lifetime results support NAD(P)H as the primary source of the autofluorescence at 730 nm excitation. The chromophore composition responsible for the 960 nm excitation is more complicated.

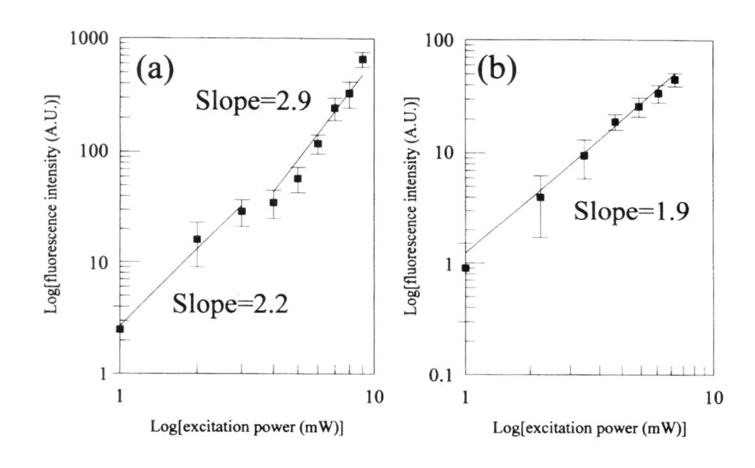

Figure 31.10 Fluorescence emission intensity as a function of laser excitation power at 960 nm (Masters *et al.*, 1997c).

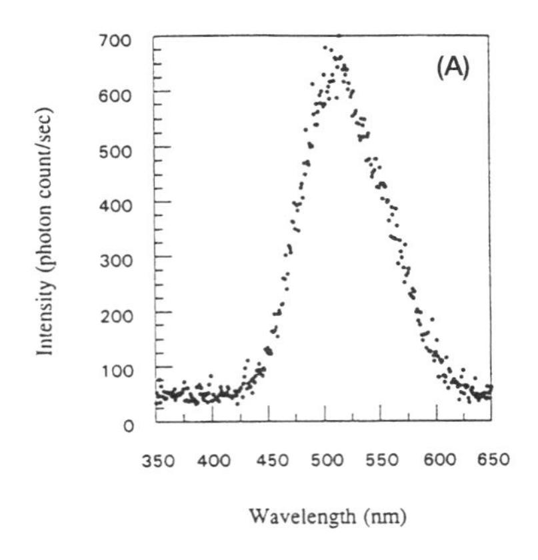

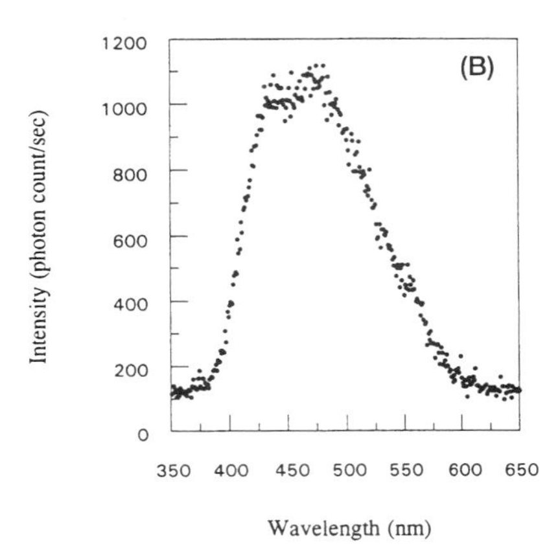

Figure 31.11 Emission spectra of *in vivo* human skin with multiphoton excitation microscopy. (A) 960 nm excitation; (B) 730 nm excitation, NADP(H) emission spectrum.

31.7.4 Optical microscopy for the investigation of *in vivo* human skin

Skin lesions which result in alterations of the surface of the tissue are readily observed with low-power microscopes (Goldman, 1951). The ability to obtain images from deeper layers of skin, together with three-dimensional visualization computer techniques, is the rationale for the use of confocal microscopic imaging systems.

Several types of confocal microscopes have been developed which use different designs for forming a two-dimensional image of a thin optical section within a thick specimen. The details of these scanning systems have been reviewed (Wilson, 1990; Masters, 1995a,b, 1996b). A tandem scanning confocal microscope was developed by Petran and co-workers to section optically thick, highly scattering tissues in real-time (Petran *et al.*, 1968). This microscope has been adapted for the *in vivo* examination of skin by several researchers. In particular, Corcuff and co-workers have advanced the development of a real-time confocal microscope for skin imaging (Corcuff *et al.*, 1993, 1996; Corcuff & Lévêque, 1993; Bertrand & Corcuff, 1994).

An *in vivo* confocal scanning laser microscope was developed for video-rate imaging of human skin. This instrument, which operates at video rates, has the capacity of reflected light imaging with wavelengths of 488 nm, 514 nm, 647 nm and 800 nm (Rajadhyaksha *et al.*, 1995). The use of 800 nm light in this instrument has the advantage of deeper penetration into the skin and it can image below the dermoepidermal junction. A new advance in the development of optical microscopy for the investigation of human skin is the simultaneous imaging of skin with both multiphoton excitation microscopy (fluorescence) and single-photon (reflected light) confocal microscopy of the same field of view (Kim *et al.*, 1998; So *et al.*, 1998a).

31.8 MITIGATION OF PHOTODAMAGE WITH MULTIPHOTON EXCITATION MICROSCOPY

As previously discussed, typical two-photon microscopy uses infrared lasers as a train of pulses at a repetition rate on the order of 100 MHz and pulse width on the order of 200–400 fs. Typically, an average power between 1 and 30 mW is needed for

Table 31.1 Fluorescence emission spectra measured *in vivo* in human skin (Masters *et al.*, 1997c).

Excitation (nm)	*Depth (μm)*	*Component 1*		*Component 2*	
		Centre (nm)	*FWHM (nm)*	*Centre (nm)*	*FWHM (nm)*
730	0–50	440	75	480	75
730	100–150	440	75	480	75
960	0–50	420	30	520	100
960	100–150	420	30	520	100

Table 31.2 Fluorescence lifetimes measured *in vivo* in human skin at 80 MHz (Masters *et al.*, 1997c).

Excitation (nm)	*Depth (μm)*	*τ phase (ns)*	*τ mod (ns)*
730	0–50	0.5±0.4	1.7±0.7
730	100–150	1.0±0.6	2.1±0.7
960	0–50	0.23±0.2	3.1±2.0
960	100–150	0.55±0.3	2.7±2.0

Note: The fluorescence lifetime is determined using frequency domain spectroscopy techniques (So *et al.*, 1995). The fluorescence lifetime is determined at a modulation frequency of 80 MHz and is calculated from both the phase (τ phase) and the modulation (τ modulation) values.

two-photon imaging. Significant thermal damage can occur in cells or tissues with a high concentration of pigment molecules, such as melanin, which absorb in the near-infrared. We have demonstrated a technique to reduce the infrared absorption while maximizing two-photon excitation efficiency (Masters *et al.*, 1999). The experimental verification of this method has been demonstrated in living specimens – frog melanophores in culture. The melanocytes are highly absorbing cells with a high concentration of melanin. These melanophores express green fluorescent protein (GFP) in their melanin-containing vacuoles. We have obtained GFP images of melanophores using a minimum average excitation power of 200 nW at a pulse repetition rate of 2 kHz.

For the successful study of these melanophores, it is critical to reduce thermal damage by reducing power deposition from individual laser pulses as well as reducing heat accumulation from the pulse train. The power management of the laser pulses is accomplished by using an attenuator to decrease laser pulse energy. With energy sufficiently attenuated such that thermal damage from individual pulses is insignificant, heat accumulation from the pulse train remains a major problem. It is possible to reduce further the laser pulse energy to minimize the effect of heat accumulation. However, given a constant pulse width, reducing the pulse energy rapidly decreases fluorescence excitation efficiency. It is preferable to reduce the laser pulse repetition rate by using a laser pulse picker which maximizes the ratio between the peak power and the average power (Fig. 31.12).

Theoretically, reducing the average laser power by a given factor, N, using the pulse picker will result in a two-photon fluorescence intensity which is a factor of N higher than that obtained using an attenuating filter by the same factor.

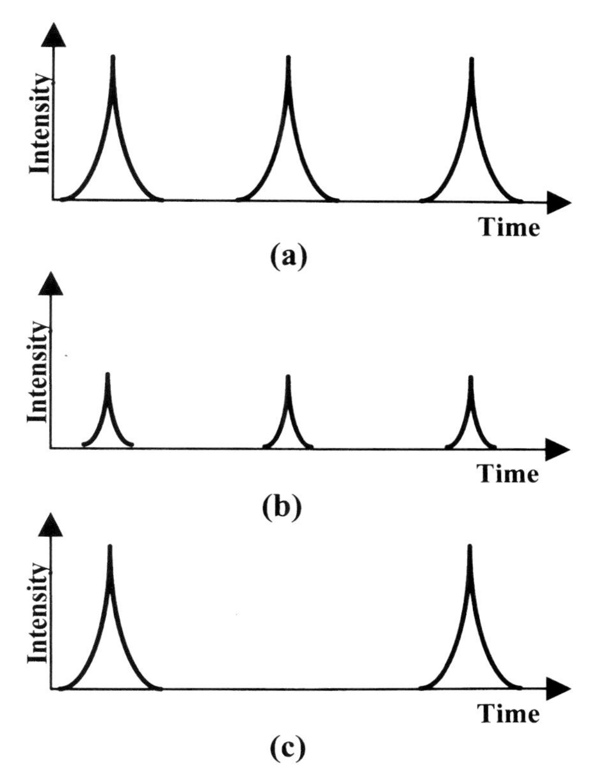

Figure 31.12 (a) Diagram of laser pulse sequences, (b) the attenuation of laser power, (c) the reduction of the number of laser pulses in the sequence.

31.9 DISCUSSION

One limitation of multiphoton excitation microscopy is that it is only implemented for fluorescence. We have developed a new technique to provide simultaneous laser scanning confocal microscopy and multiphoton excitation microscopy (Kim *et al.*, 1998). This is an important advance since it is sometimes difficult to determine the exact location of the focal plane during multiphoton excitation microscopic studies. This problem is more acute if the fluorophore is restricted to a particular local region of cells in a thick tissue, e.g. *in vivo* human skin. By moving the focal plane or volume through the thick specimen and simultaneously recording two stacks of optical sections, one in reflected light, and one in fluorescence light, two three-dimensional reconstructions of the cells and tissue could be produced. A careful comparison of the two stacks of optical sections could uniquely locate the position within the specimen.

The use of multiphoton excitation microscopy results in laser pulses with peak power of gigawatts incident on the cells and tissues. This peak power is usually in the red or infrared wavelengths. If the biological specimen is highly pigmented then this energy could be absorbed and result in thermal damage to the specimen.

31.10 SUMMARY AND CONCLUSIONS

Multiphoton excitation microscopy is an important new optical technique with many applications in biology. It has advantages over confocal laser scanning microscopy and wide-field microscopy. We have chosen two thick and highly scattering tissues, the *ex vivo* cornea and *in vivo* human skin, to illustrate the efficacy of the technique in non-invasive functional imaging based on NAD(P)H and flavoproteins. We strongly recommend the use of emission spectroscopy, lifetime measurements, and log–log plots of fluorescence intensity versus excitation power to characterize the fluorophore(s) and the nature of the multiphoton excitation process. Our experience with multiphoton excitation microscopy indicates that some of the most important applications involve studies of the intrinsic fluorescence, e.g. NAD(P)H of cells and tissues.

The high peak power of the laser pulses can result in cell damage in highly pigmented cells, and this damage can occur even with a single pulse of light. We have described techniques to mitigate photodamage and photobleaching to cells and tissues. Multiphoton excitation microscopy can be exploited to capitalize on its improved penetration, restriction of photodamage and photobleaching to the focal volume, as compared with laser scanning confocal microscopy. However, each biological specimen, from cells to tissues to *in vivo* studies, must be carefully studied with the techniques described in this chapter.

ACKNOWLEDGEMENTS

This work was supported by a grant from NIH EY-06958 (BRM). The Laboratory for Fluorescence Dynamics, Department of Physics, is supported by the National Institutes of Health (RR03155). We thank Steven D. Miklasz, Director, Immunological Resource Center, University of Illinois at Urbana-Champaign, for expert surgical assistance and supplying the biological samples.

REFERENCES

Beebe D.C. & Masters B.R. (1996) *Invest. Ophthalmol. Vis. Sci.* **37**, 1815–1825.

Bertrand C. & Corcuff, P. (1994) *Scanning* **16**, 150–154.

Bhawalkar J., Swiatkiewicz J., Pan S., Samarabandu J., Liou W., Berezney R., Cheng P. & Prasad P. (1996) *Scanning* **18**, 562–566.

Birge R.R. (1983) In *Ultrasensitive Laser Spectroscopy*, D.S. Kliger (ed). Academic Press, New York, pp. 109–174.

Birge R.R. (1986) *Acc. Chem. Res.* **19**, 138–146.

Chance B. (1976) *Circ. Res. (Suppl. 1)* **38**, I-31–I-38.

Chance B. & Lieberman M. (1978) *Exp. Eye Res.* **26**, 111–117.

Chance B. & Schoener B. (1966) In *Flavins and Flavoproteins*, E.C. Slater (ed). Elsevier, Amsterdam, pp. 510–528.

Chance B. & Thorell B. (1959) *J. Biol. Chem.* **234**, 3044–3050.

Chance B., Schoener B., Oshino R., Itshak F. & Nakase Y. (1979) *J. Biol. Chem.* **254**, 4764–4711.

Corcuff P. & Lévêque J.-L. (1993) *Dermatology* **186**, 50–54.

Corcuff P., Bertrand C. & Lévêque J.-L. (1993) *Arch. Dermatol. Res.* **285**, 475–481.

Corcuff P., Gonnord G., Pierard G.E. & Lévêque J.-L. (1996) *Scanning* **18**(3), 351–355.

Corle T.R. & Kino G.S. (1996) *Confocal Scanning Optical Microscopy and Related Imaging Systems*, Academic Press, San Diego.

Denk W. (1994) *Proc. Natl Acad. Sci. USA* **91**, 6629–6633.

Denk W. (1996) *J. Biomed. Opt.* **1**, 296–304.

Denk W.J., Strickler J.P. & Webb W.W. (1990) *Science* **248**, 73–76.

Denk W.J., Strickler J.P. & Webb W.W. (1991). US Patent, 5 034 613, 23 July 1991.

Denk W., Delaney K.R., Gelperin A., Kleinfield D., Stowbridge B.W., Tank D.W. & Yuste R. (1994) *J. Neurosci. Methods* **54**, 151–162.

Denk W.J., Piston D.W. & Webb W.W. (1995) In *Handbook of Biological Confocal Microscopy*, 2nd edn, J.B. Pawley (ed). Plenum Press, New York, pp. 445–458.

Fork R.L., Martinez O.E. & Gordon J.P. (1984) *Optics Lett.* **9**(5), 150.

Franken P.A., Hill A.E., Peters C.W. & Weinreich G. (1961) *Phys. Rev. Lett.* **7**, 118.

French T., So P.T.C., Weaver D.J. Jr, Coelho-Sampaio T., Gratton E., Voss E.W. Jr & Carrero J. (1997) *J. Micros.* **185**, 339–353.

Friedrich D.M. (1982) *J. Chem. Educ.* **59**, 472.

Friedrich D.M. & McClain W.M. (1980) *Annu. Rev. Phys. Hem.* **31**, 559–577.

Gannaway J.N. & Sheppard C.J.R. (1978) *Opt. Quant. Electron.* **10**, 435–439.

Goldman L. (1951) *J. Invest, Dermatol.* **16**, 407–427.

Göppert-Mayer M. (1931) *Ann. Phys. Leipzig* **9**, 273–295.

Gryczynski I., Malak H. & Lakowicz J.R. (1995a) *Chem. Phys. Lett.* **245**, 30–35.

Gryczynski I., Malak H. & Lakowicz J.R. (1995b) *Biospectroscopy* **2**, 9–15.

Gu M. (1996a) *Optics Lett.* **21**, 988–990.

Gu M. (1996b) *Principles of Three-Dimensional Imaging in Confocal Microscopy*, World Scientific, Singapore.

Gu M. & Sheppard C.J.R. (1995) *J. Microsc.* **177**, 128–137.

Gurney A.M. (1993) In *Fluorescent and Luminescent Probes for Biological Activity*, W.T. Mason (ed). Academic Press, London, pp. 335–348.

Hell S.W. (1996) *Bioimaging* **4**, 121–123.

Hell S.W., Soukka J. & Hänninen P.E. (1995) *Bioimaging* **3**, 64–69.

Hellwarth R. & Christensen P. (1974) *Opt. Commun.* **12**, 318–322.

Kaiser W. & Garrett C.G.B. (1961) *Phys. Rev. Lett.* **7**, 229–231.

Kim K.H., So P.T.C., Kochevar I.E., Masters B.R. & Gratton E. (1998) In *Optical Investigations of Cells In Vitro and In Vivo*, D.L. Farkas, R.C. Leif & B.J. Tromberg (eds). *Proc. SPIE* **3260**, 46–57.

König K., So P.T.C., Mantulin W.W., Tromberg B.J. & Gratton E. (1996) *J. Microsc.* **183**, 197–204.
König K., So P.T.C., Mantulin W.W. & Gratton E. (1997) *Optics Lett.* **22**, 135–136.
Lakowicz J.R., Gryczynski I., Malak H., Schrader M., Engelhardt P., Kano H. & Hell S.W. (1997) *Biophys. J.* **72**, 567–578.
Masters B.R. (1988) In *The Cornea: Transactions of the World Congress of the Cornea III*, H.D. Cavanagh (ed). Raven Press, New York, pp. 281–286.
Masters B.R. (1990) In *Noninvasive Diagnostic Techniques in Ophthalmology*, B.R. Masters (ed). Springer-Verlag, New York, pp. 223–247.
Masters B.R. (1993a) In *Medical Optical Tomography: Functional Imaging and Monitoring*, G. Müller, B. Chance, R. Alfano, S. Arridge, J. Beuthan, E. Gratton, M. Kaschke, B.R. Masters, S. Svanberg, P. van der Zee (eds). The International Society for Optical Engineering, Bellingham, Washington, pp. 555–575.
Masters B.R. (1993b) *Scanning Microsc.* **7**(2), 645–651.
Masters B.R. (1995a) *Comments Mol. Cell. Biophys.* **8**(5), 243–271.
Masters B.R. (1995b) *Opt. Engng.* **34**(3), 684–692.
Masters B.R. (1996a) In *Analytical Uses of Fluorescent Probes in Oncology*, E. Kohen & J.G. Hirschberg (eds). Plenum Press, New York, pp. 205–211.
Masters B.R. (1996b) *Selected papers on confocal microscopy*, B.R. Masters (ed). SPIE – The International Society for Optical Engineering, Bellingham, Washington, USA.
Masters B.R. (1996c) In *Biomedical Optical Spectroscopy and Diagnostics*, Vol. 3. E. Sevick-Muraca & D. Benaron (ed). Optical Society of America. Washington, DC, pp. 157–161.
Masters B.R. (1996d) *Bioimages* **4**(1), 13–19.
Masters B.R. & Chance B. (1984) In *Current Topics in Eye Research*, J.A. Zadunaisky & H. Davson (eds). Academic Press, London, pp. 140–200.
Masters B.R. & Chance B. (1999) In *Fluorescent and Luminescent Probes for Biological Activity*, W.T. Mason (ed). Academic Press, London, pp. 361–674.
Masters B.R. & Thaer A.A. (1994a) *Appl. Opt.* **33**(4), 695–701.
Masters B.R. & Thaer A.A. (1994b) *Microsc. Res. Tech.* **29**, 350–356.
Masters B.R. & Thaer A.A. (1995) *Bioimages* **3**, 7–11.
Masters B.R., Falk S. & Chance B. (1981) *Curr. Eye Res.* **1**, 623–627.
Masters B.R., Ghosh A.K., Wilson J. & Matschinsky F.M. (1989) *Invest. Ophthalmol. Vis. Sci.* **30**, 861–868.
Masters B.R., Kriete A. & Kukulies J. (1993) *Appl. Optics* **32**(4), 592–596.
Masters B.R. Aziz D.J., Gmitro A.F., Kerr J.H., O'Grady T.C. & Goldman L. (1997a) *J. Biomed. Optics* **2**(4), 437–445.
Masters B.R., Gonnord G. & Corcuff P. (1997b) *J. Microsc.* **185**, 329–338.
Masters B.R., So P.T.C. & Gratton E. (1997c) *Biophys. J.* **72**, 2405–2412.
Masters B.R., So P.T.C. & Gratton E. (1997d) *Cell Vision* **4**, 130–131.
Masters B.R., So P.T.C. & Gratton E. (1998a) *Ann. NY Acad. Sci.* **838**, 58–67.
Masters B.R., So P.T.C. & Gratton E. (1998c) *Lasers Med. Sci.* **13**, 196–203.
Masters B.R., So P.T.C., Mautulin W. & Gratton E. (1998d) In *Biomedical Optical Spectroscopy and Diagnostics*, E. Sevick-Muraca & J.A. Izatt (eds). Opt. Soc. Am., Washington, DC, 43–45.
Masters B.R., Mantulin M.M., Gratton E. & So P.T.C. (1998e) *Simultaneous Two-Photon Fluorescence and Reflected Light Confocal Imaging of In Vivo Human Skin*, 22–26 February 1998, 42nd Annual Meeting of the Biophysical Society, Kansas City, Kansas.
Masters B.R., So P.T.C., Dong C.Y., Bulhur C., Mantulin W.M., Gratton E., Rogers S., Tuma C. & Gelfand V. (1999) in preparation.
McClain W.M. (1971) *J. Chem. Phys.* **55**, 2789.
McClain W.M. & Harris R.A. (1977) In *Excited States*, E.C. Lim (ed). Academic Press, New York, pp. 1–56.
Mertz J., Xu C. & Webb W.W. (1996) *Optics Lett.* **20**, 2532–2534.
Pawley J.B. (1995) *Handbook of Biological Confocal Microscopy*, 2nd edn, J.B. Pawley (ed). Plenum Press, New York.
Petran M., Hadravsky M., Egger M.D. & Galambos R. (1968) *J. Opt. Soc. Am.* **58**, 661–664.
Piston D.W. (1996) In *Fluorescence Imaging and Microscopy*, Vol. 137, X.F. Wang & B. Herman (eds). John Wiley & Sons, New York, pp. 253–272.
Piston D.W., Masters B.R. & Webb W.W. (1995) *J. Microsc.* **178**, 20–27.
Plakhotnik T., Walser D., Pirotta M., Renn A. & Wild U.P. (1996) *Science* **271**, 1703–1705.
Potter S.M., Wang C.M., Garrity P.A. & Fraser S.E. (1996) *Gene* **173**, 25–31.
Rajadhyaksha M., Grossman M., Esterowitz D., Webb R.H. & Anderson R. (1995) *J. Invest. Dermatol.* **104**, 946–952.
Rentzepis P.M., Mitschele C.J. & Saxman A.C. (1970) *Appl. Phys. Lett.* **7**, 229–231.
Shen Y.R. (1984) *The Principles of Nonlinear Optics*. John Wiley & Sons, New York.
Sheppard C.J.R. & Gu M. (1990) *Optik* **86**(3), 104–106.
Sheppard C.J.R. & Kompfner R. (1978) *Appl. Opt.* **17**, 2879–2882.
Sheppard C.J.R. & Shotton D.M. (1997) In *Confocal Laser Scanning Microscopy*, C.J.R. Sheppard & D.M. Shotton (eds). Springer-Verlag, New York.
Siegman A.E. (1986) *Lasers*. University Science Books, Mill Valley, California.
Silberzan I., Williams R.M. & Webb W.W. (1993) *Biophys. J.* **63**, A109.
Singh S. & Bradley L.T. (1964) *Phys. Rev. Lett.* **12**, 612–614.
So P.T.C., French T., Yu W.M., Berland K.M., Dong C.Y. & Gratton E. (1995) *Bioimaging* **3**, 49–63.
So P.T.C., French T., Yu W.M., Berland K.M., Dong C.Y. & Gratton E. (1996) In *Fluorescence Imaging and Microscopy*, X.F. Wang & B. Herman (eds). Chemical Analysis Series, Vol. 137: John Wiley & Sons, New York, pp. 351–374.
So P.T.C., Masters B.R., Gratton E. & Kochevar I.E. (1998a) In *Biomedical Optical Spectroscopy and Diagnostics*, E. Sevick-Muraca & J.A. Izatt (eds). Opt. Soc. Am., Washington, DC, 40–42.
Szmacinski H., Gryczynski I. & Lakowicz J.R. (1996) *Biophys. J.* **70**, 547–555.
Valdmansis J.A. & Fork R.L. (1986) *IEEE J. Quant. Electron.* **QE-22**, 112–118.
Williams R.M., Piston D.W. & Webb W.W. (1994) *FASEB J.* **8**, 804–813.
Wilson T. (ed). (1990) *Confocal Microscopy*. Academic Press, London.
Wilson T. & Sheppard C.J.R. (1979) *Opt. Acta* **26**, 761–770.
Wokosin D.L., Centonze V.E., Crittenden S. & White J. (1996a) *Bioimaging* **4**, 208–214.
Wokosin D.L., Centonze V.E., White J., Armstrong D., Robertson G. & Ferguson A.I. (1996b) *IEEE J. Select. Top. Quant. Electr.* **2**(4), 1051–1065.

Wolleschensky R., Feurer T. & Sauerbrey R. (1997) *Proc. SCANNING 97* **19**, 150–151.

Xu C. (1966) PhD thesis, Cornell University, Ithaca, New York.

Xu C. & Webb W.W. (1996) *J. Opt. Soc. Am. B* **13**, 481–491.

Xu C., Guild J., Webb W.W. & Denk W. (1995) *Optics Lett.* **20**, 2372–2374.

Xu C., Williams R.M., Zipfel W. & Webb W.W. (1996a) *Bioimaging* **4**, 198–207.

Xu C., Zipfel W. & Webb W.W. (1996b) *Biophys. J.* **70**, A429.

Xu C., Zipfel W., Shear J.B., Williams R.M. & Webb W.W. (1996c) *Proc. Natl Acad. Sci. USA* **93**, 10763–10768.

Yariv A. (1991) *Optical Electronics*, 4th edn. Saunders College Publishing, Philadelphia, PA.

Yuste R. & Denk W. (1995) *Nature* **375**, 682–684.

CHAPTER THIRTY-TWO

Raman Spectroscopic Methods for *In Vitro* and *In Vivo* Tissue Characterization

R. WOLTHUIS[1], T.C. BAKKER SCHUT[1], P.J. CASPERS[1], H.P.J. BUSCHMAN[1,2], T.J. RÖMER[1,2], H.A. BRUINING[1] & G.J. PUPPELS[1]

[1] Laboratory for Intensive Care Research and Optical Spectroscopy, Erasmus University Rotterdam & University Hospital Rotterdam 'Dijkzigt', Rotterdam, The Netherlands
[2] Department of Cardiology, Leiden University Medical Center, Leiden, The Netherlands

32.1 INTRODUCTION

Raman spectroscopy is an optical, vibrational spectroscopic technique that provides detailed information about molecular composition and molecular structure (see Chapter 29). In recent years Raman spectroscopic tissue characterization and its potential application to *in vivo* diagnosis of diseases is attracting increasing attention.

Pathological conditions involve changes in molecular composition of tissue, both as a cause and as a consequence of disease. Raman spectroscopy enables detection of these changes in a non/minimal-invasive, non-destructive manner. Infrared spectroscopy provides information of the same nature as Raman spectroscopy (see Chapter 29). However, in contrast to infrared spectroscopy, which is hampered by the strong infrared absorption of water in tissues, Raman spectroscopy is particularly suited for in vivo tissue analysis.

The development of Raman spectroscopy in the biomedical field has been accelerated by the rapid development of optical instrumentation, such as CCD detectors, which are sensitive in the near-infrared spectral region, high-quality notch-filters and band-pass filters, stable high-power and narrow-band diode lasers, and fibreoptic probes. Suitable fibreoptic probes form an important part of instrumentation for *in vivo* Raman spectroscopy. Infrared spectroscopy is again at a disadvantage, in this respect; whereas flexible, high-throughput optical fibres are readily available for the visible and NIR regions, in which the Raman signal is detected, this is not the case for the 3–12 μm IR-region.

This chapter will discuss aspects of instrumentation, calibration and signal analysis, which are relevant to the development of clinical applications of Raman spectroscopy. Such applications require rapid, *in vivo* signal collection – preferably in the order of seconds – and on-line data analysis, using databases of reference spectra. The data that can be obtained from the Raman spectra could then guide further clinical action.

Examples are given of applications in the fields of skin research, of atherosclerosis, of detection of (pre-)malignant tissue, and of transplant liver characterization.

FLUORESCENT AND LUMINESCENT PROBES, 2ND EDN
ISBN 0–12–447836–0

32.2 INSTRUMENTATION

In this section a number of instrumental and methodological aspects will be discussed, which are critically important to the development of medical applications of Raman spectroscopy.

Raman spectroscopy holds great potential for *in vivo* applications. However, for many applications this is true only if spectra of sufficient signal-to-noise ratio can be obtained within signal collection times of a few seconds (and preferably faster). This means that the development of Raman instrumentation for (*in vivo*) medical applications, is dictated by optimization of signal detection efficiency and minimization of noise contributions. The choice of laser wavelength used to illuminate the sample, and the factors governing this choice will be discussed in Section 32.2.1. Design criteria for Raman instrumentation for *in vivo* tissue characterization and practical implementations of these are the subject of Section 32.2.2.

Databases of reference spectra will be at the heart of most medical applications. These reference spectra will be obtained over extended periods of time (months to years) and will most likely be regularly updated thereafter. Moreover, clinical applicability requires that they are transferable from one instrument to another. This means that both reference Raman spectra and spectra that are measured during an actual clinical application of the technique should be free of any signatures of the instrument by which they are obtained.

32.2.1 Choice of laser excitation wavelength

Detection of Raman scattering can be severely hindered by sample fluorescence (see Chapter 29; see also Frank *et al.*, 1994; Yu *et al.*, 1996). Excitation of fluorescence can be avoided by choosing an excitation wavelength in the NIR. As is illustrated by the results of Raman experiments in which a 1064 nm Nd:Yag laser was used, essentially no fluorescence background is present when excitation wavelengths above 1000 nm are used (Schrader *et al.*, 1995, 1997; Gniadecka *et al.*, 1997a,b). However, there are severe drawbacks to using such high excitation wavelengths. First, Raman signal intensity N_{Raman} is strongly wavelength dependent. With signal intensity measured in detected photons (Stevenson & Vo-Dinh, 1996):

$$N_{\mathrm{Raman}} \sim \lambda^{-3} \qquad [32.1]$$

Second, when a laser excitation wavelength of 1064 nm is used, the most informative region of the Raman spectrum (~200–3500 cm^{-1}) will be located between ~1.1 and 1.7 μm. In the 200–1100 nm region, the signal detection performance of CCD detectors comes close to that of an ideal detector. It couples multichannel detection and high quantum efficiency (up to 80%) to a read-out noise in the order of a few electrons (equivalent to a few detected photons, see Chapter 37). The noise-performance of gallium arsenide and germanium photodetectors for the 1.0–2.0 μm region, on the other hand, is much worse. A wavelength multiplexing technique, such as used in Fourier transform Raman instruments (Hendra, 1996), is needed to minimize this problem. At the heart of these instruments is a Michelson interferometer. Photons, scattered in the whole wavelength range in which the Raman signal is located, are detected by a single detector. The length of the reference arm of the interferometer determines how the different wavelengths are (intensity-)encoded. Signal intensities are recorded as a function of the length of this reference arm. The Raman spectrum of the sample is obtained by an inverse Fourier transformation of these signal intensities. This signal *multiplexing* is often confused with *multichannel* detection. However, it can be shown that, apart from secondary effects, multiplexing will at best lead to an instrument performance that is similar to that of a single-channel instrument with an ideal detector, in which signal intensities at different wavelengths are measured consecutively (Bell, 1972; Puppels *et al.*, 1993).

This implies that Raman spectra can be obtained much faster with dispersive (grating-based) multichannel spectrometers equipped with CCD detectors (see Fig. 32.1; for a more detailed description see Chapter 29) than is possible with FT-Raman spectrometers (Shim & Wilson, 1997). Since the speed with which spectra can be obtained will be the all-determining factor for many potential medical applications, dispersive, CCD-based spectrometers are necessarily the instruments of choice for (*in vivo*) Raman spectroscopic tissue characterization. Real advantages of Fourier transform Raman spectroscopy – such as lower fluorescence intensity background, better spectral resolution, the fact that signal can be obtained from larger sample areas than in dispersive instruments without sacrificing spectral resolution (the so-called Jacquinot advantage), and the inherent wavenumber-calibration of FT-Raman spectra – do not outweigh this disadvantage (Bell, 1972; Hendra, 1996).

Therefore, a trade-off has to be made between sample fluorescence minimization by employing a laser excitation wavelength far into the NIR region of the spectrum, and the wavelength dependence of the quantum efficiency of CCD cameras (limited to about 1050 nm for NIR-optimized cameras, e.g. those

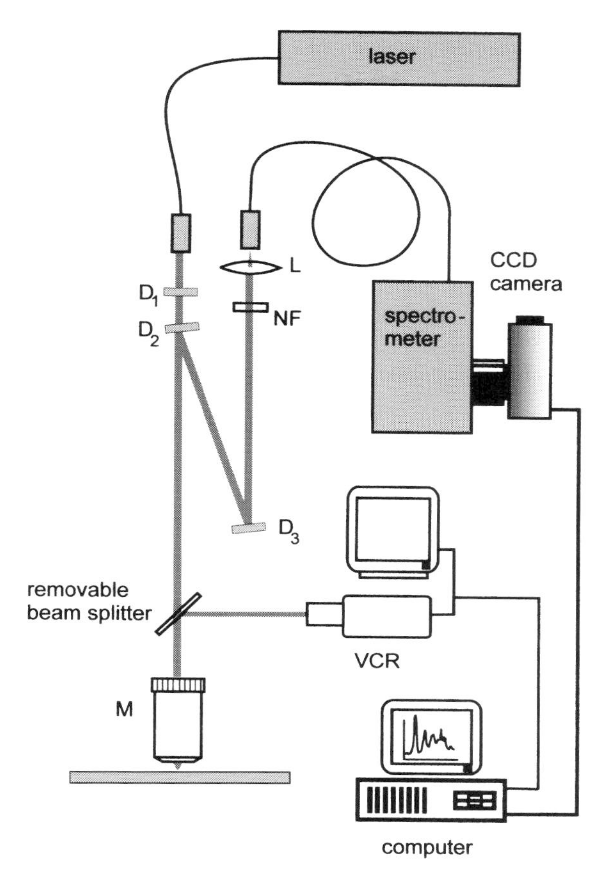

Figure 32.1 Schematic of the Raman microspectrometer currently in use at the Erasmus University Rotterdam. The instrument is built around a Leitz microscope stand. The output of a titanium–sapphire laser (Spectra Physics, Mountain View, CA, Model 3900S) is coupled into a mono-mode fibre (Point Source, UK) which delivers the laser light to the microscope stand. The output of the fibre is filtered twice by means of dielectric short-pass filters D_1 and D_2 (Omitec, UK, $T_{850\ nm} = 0.9$, $R_{> 200\ cm^{-1}} = 0.99$). The laser beam is focused onto the sample by means of a NIR-optimized microscope objective M.

Light that is scattered by the sample is collected and collimated by the same objective and directed towards an optical fibre by dielectric filters D_2 and D_3. A holographic notch filter, NF, (OD > 6, Kaiser Optical Systems, Ann Arbor, MI, USA) is used for rejection of scattered laser light. Lens L (f = 40 mm for non-confocal operation or f = 149 mm for confocal operation; NIR-achromats, Opto Sigma, Santa Ana, CA) focuses the Raman scattered light onto the core of a multimode tapered fibre of 10 m in length (100 μm core diameter at the sample side, 50 μm core diameter at the spectrometer side, Fiberguide Industries, NJ, USA). The spectrometer is shown in more detail in Fig. 32.2.

A pellicle beamsplitter (Melles Griot) can be inserted in the optical pathway to view the sample, by means of a video camera.

equipped with (thinned back-illuminated) deep-depletion CCD chips). This means that an excitation wavelength of 830 to 850 nm is more or less an optimum, if Raman signal is to be detected throughout the important biological 'fingerprint' region (~0–1800 cm^{-1}). Of course, lower excitation wavelengths are favoured in those cases where sample fluorescence is not an issue. Future improvement in the noise characteristics of gallium arsenide and germanium detectors, and the development and availability of other types of detectors that are sensitive in the NIR, will enable the use of higher excitation wavelengths.

32.2.2 Raman instrumentation for tissue characterization

32.2.2.1 Instrumentation for in vitro experiments

Many different sampling geometries can be used in Raman spectroscopic *in vitro* tissue studies. Raman microspectroscopy, in which high NA microscope objectives are employed (see Fig. 29.2(b)), enables measurements at high spatial resolution and with high signal collection efficiency. The *in vitro* tissue Raman spectra shown in this chapter were obtained with such a set-up (Fig. 32.1). Other optical configurations have been designed to sample larger tissue volumes (Liu *et al.*, 1992; Brennan *et al.*, 1997).

32.2.2.2 Fibreoptic probes for in vivo *studies*

A good design of a fibreoptic probe for *in vivo* Raman applications should meet the following criteria (Lewis & Griffiths, 1996; Puppels & Greve, 1996):

- The output of the laser delivery fibre has to be filtered to prevent the intense Raman signal induced in the fibre material to reach the sample under investigation, and in that way give rise to an intense background signal.
- The light that is scattered by a sample consists mainly of (diffusely) reflected and Rayleigh-scattered laser light. It should therefore be filtered by means of a laser-blocking filter before it enters the signal collection fibres; again in order to avoid Raman signal contributions of the fibre material.
- The signal collection efficiency should be high in order to minimize measurement times.
- The specific application will set limitations on size, rigidity and shape of fibreoptic Raman probes: intravascular use demands thin probes (< 2 mm in diameter) with a high flexibility, while laparoscopic application allows a diameter of up to 10 mm and a rigid housing.
- The probe should be either sterilizable or cheap (disposable).

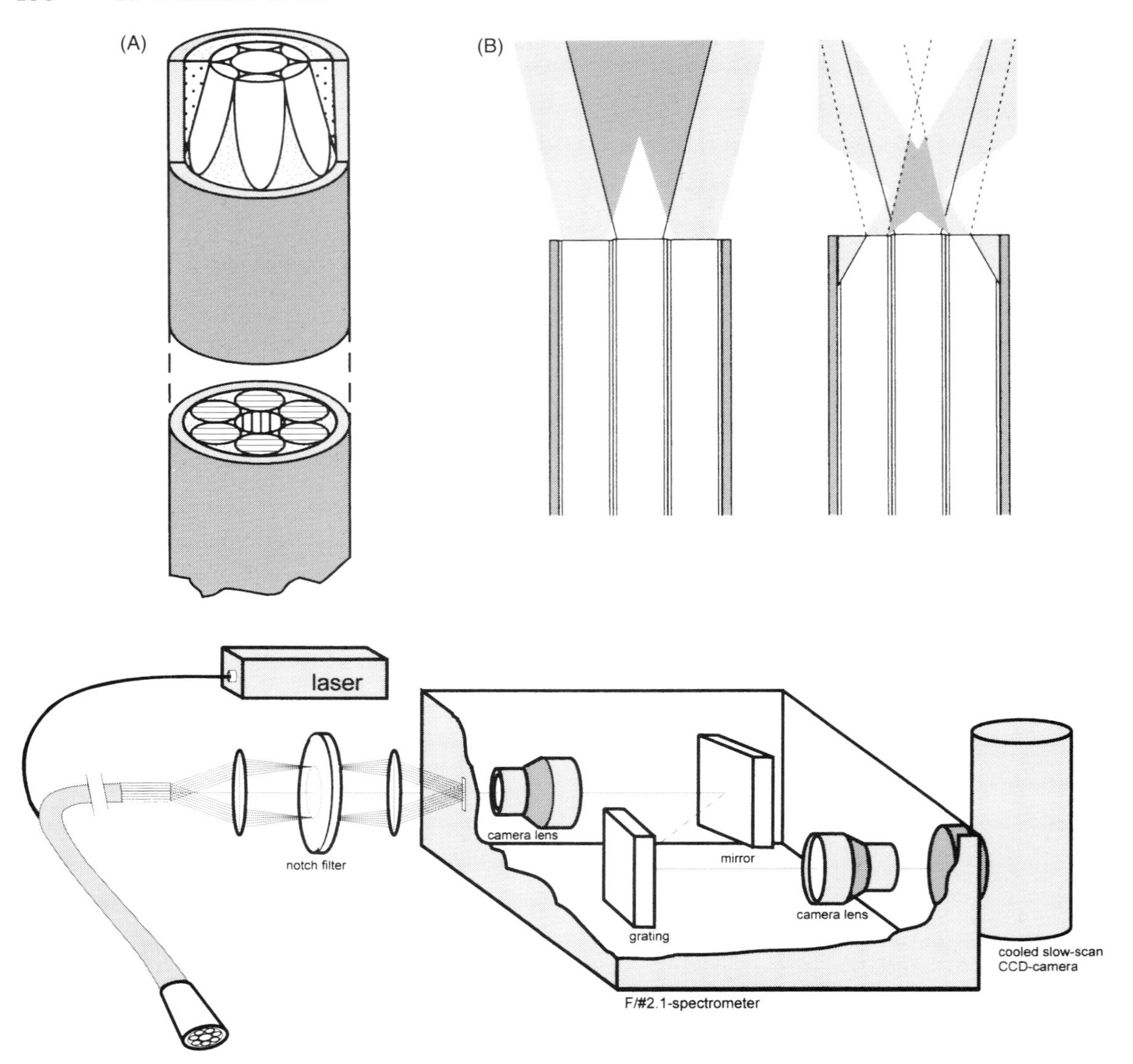

Figure 32.2 Layout of the Raman set-up for *in vivo* experiments. (A) Probe tip of the Enviva Biomedical Raman Probe (Visionex, Inc., Warner Robins, GA): The probe has a 7-around-1 fibre configuration design, with one 400 μm central fibre delivering laser light to the tissue, and seven 300 μm signal collection fibres, each of 0.22 numerical aperture. The rigid probe tip measures 1.4 mm in diameter and 40 mm in length. Fibre material Raman signal contributions are strongly suppressed. The centre fibre incorporates a dielectric narrow-bandpass filter (approximately 25 mm from the fibre end face), which transmits the laser light but blocks Raman signal from the fibre material (vertically hatched). In a similar architecture, the seven signal collection fibres incorporate a dielectric long-pass filter, which achieves an OD ≅ 3 at the laser wavelength and a transmittance of Raman scattered light up to 90% above 300 cm^{-1} (horizontally hatched). The central laser fibre is optically insulated from the seven signal collection fibres, thus preventing cross-talk. Moreover, the probe features an efficient overlap of the tissue volume that is most intensely illuminated and the volume from which Raman scattered light is collected through the application of beam-steering techniques.

(B) Illustration of the improvement in overlap between illuminated sample volume and signal collection volume by means of a beam-steering technique. Right: classical probe design (Schwab & McCreery) with flat fibre end faces. The dark grey area denotes the overlap between the illuminated sample volume and the sample volumes from which the signal collection fibres receive Raman signal. Left: illustration of Visionex's Gaser Light Management system (a beam-steering technique): the collection fibres are partly wedged. The wedged sides are made reflective. The dark grey area denotes the (most intensily illuminated) sample volume from which Raman signal is collected in this probe configuration, but from which no signal is received in the classical probe design.

(C) The Raman fibre probe, with the seven signal collection fibres arranged in a linear array, was coupled to a laboratory built F/ #2.1 imaging spectrometer. The filter stage, which suppresses the intensity of scattered laser light, images the fibre array onto the 100 μm entrance slit of the single grating dispersion stage. The dispersion stage images the Raman spectra from the seven signal collection fibres spatially separated, onto a back-illuminated deep-depletion CCD chip (Princeton Instruments, Trenton, NJ). The Raman signals from the seven signal collection fibres are read out separately. These seven spectra can then be individually corrected for the wavelength dependence of the signal detection efficiency of the set-up (see Section 32.3), and for fixed pattern noise and etaloning effects within the back-illuminated CCD chip, before being added.

Recently a fibreoptic probe that comes close to fulfilling most of these requirements became commercially available. A schematic of this probe, the Enviva biomedical Raman probe of Visionex (Warner Robins, GA, USA), is shown in Figs 32.2(A) and (B).

The numerical aperture of the Raman spectrometer, which traditionally was low, needs to be matched to the NA of the optical fibres that are used (usually NA=0.22 or higher). Figure 32.2(C) is a schematic of the instrument for fibreoptic *in vivo* Raman experiments realized in our laboratory.

The *ex vivo* tissue spectra shown in Fig. 32.3 and the *in vivo* spectra shown in Fig. 32.4 were obtained with this fibreoptic Raman probe. They illustrate that Raman spectra can already be obtained within reasonable signal collection times. Figure 32.5(A) shows a spectrum of rat muscle obtained *in vivo* with a fibreoptic probe and a signal collection time of 10 s. A comparison with spectrum (B), which has a similar signal-to-noise ratio, but was obtained *in vitro* with a Raman microspectrometer in 1 s, shows that there is still room for improvement of the signal collection efficiency of fibreoptic probes. Such improvement will bring *in vivo* signal collection times into the seconds range needed for medical applications.

32.3 CALIBRATION OF RAMAN SPECTRA

A Raman spectrometer records the intensity of the light that is scattered by the sample under investigation as a function of wavelength (wavenumber). This implies that both intensity scale and wavenumber scale need to be precisely calibrated. Only in this way is it possible to compare spectra that are obtained in different experiments and/or on different instruments.

As already mentioned above, especially for the development of medical applications of Raman spectroscopy, in which reference databases play an important role, correct instrument calibration is crucial.

This section treats the basic principles of calibration and gives a short overview of the methods in use. Wavenumber calibration and instrument response correction will be treated separately.

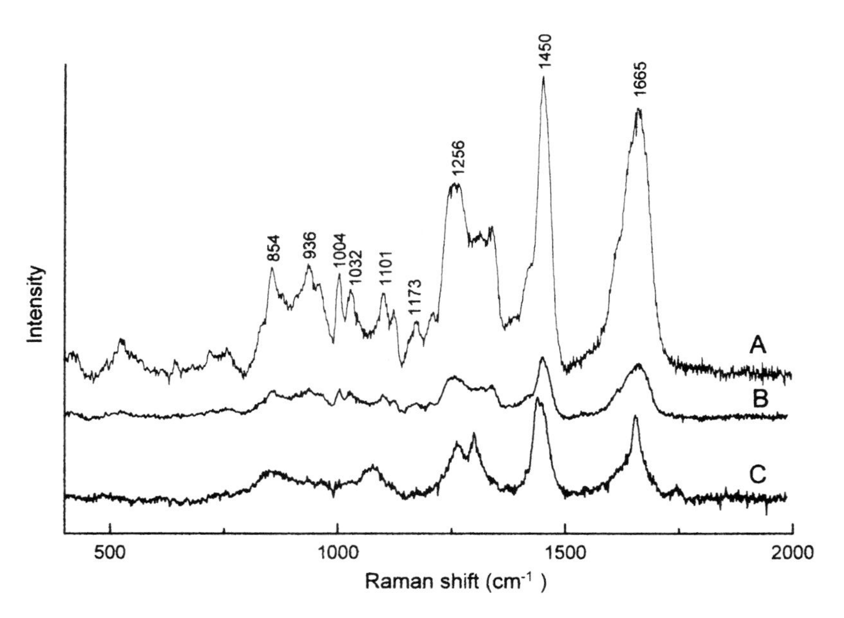

Figure 32.3 *Ex vivo* Raman spectra of rat aorta. A short piece of the aorta, about 2 cm in length, was resected *post mortem*. Blood was rinsed away with a physiological salt solution. The probe tip was inserted into the resected piece of aorta and kept in contact with the luminal side of the aorta during the measurements. (A) Spectrum taken at the luminal side of the aorta; signal collection time 50 s. (B) & (C) Spectra taken at the luminal side of rat aorta; signal collection time 10 s. The main signal contributions in spectra (A) and (B) resemble those of elastin and collagen. Spectrum (C), on the other hand, was apparently obtained from a lipid-rich region of the aorta wall (Manoharan *et al.*, 1992). Experimental conditions: laser power, 100 mW; laser wavelength, 785 nm. Residual fibre probe Raman signal and a fluorescence background were subtracted. Spectra were corrected for wavenumber dependence of instrument response. (Reproduced by permission of SPIE, from Puppels *et al.*, 1998.)

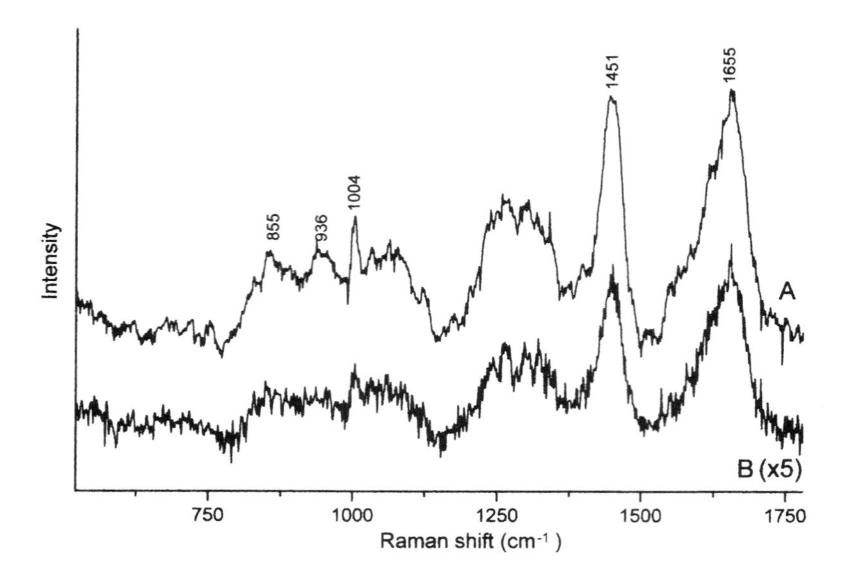

Figure 32.4 *In vivo* Raman spectra of rat oesophagus. (A) Spectrum taken at the luminal side of the oesophagus; signal collection time 50 s. (B) Spectrum taken at the luminal side; signal collection time 10 s. Other experimental conditions: see Fig. 32.3. (Reproduced by permission of SPIE, from Puppels *et al.*, 1998.)

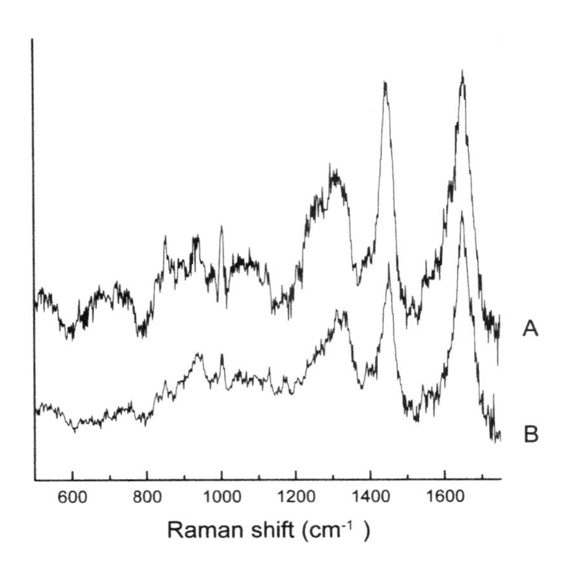

Figure 32.5 Raman spectra of rat muscle. (A) *In vivo* Raman spectrum, taken with Enviva Biomedical Raman probe (Visionex, Inc.). Laser wavelength, 785 nm; laser power on sample, 100 mW; signal collection time, 10 s. (B) *In vitro* Raman spectrum, taken with Raman microspectrometer of Fig. 32.1, equipped with ×20 NIR-microscope (NA 0.35). Laser wavelength, 850 nm; laser power on sample, 100 mW; signal collection time, 1 s.

32.3.1 Wavenumber calibration

The aim of a wavenumber calibration in multichannel dispersive instruments is to assign a wavenumber shift to each individual camera channel. An external calibration source is used for this purpose. This can be a calibration lamp or a set of reference compounds whose Raman peak positions have been precisely determined (Hamaguchi, 1988). Spectral calibration lamps (e.g. low-pressure gas discharge argon–mercury lamps) have very narrow emission lines at well-determined wavelengths. Raman shifts are not expressed in wavelengths but in relative wavenumbers (see expression [29.2] in Chapter 29). To transform the wavelength calibration into a wavenumber calibration the wavelength of the incident laser light has to be known accurately. For lasers operating at a fixed wavelength this is a good method; it has been incorporated in some commercially available instruments.

For tunable lasers, the use of Raman standards can be more appropriate. There the calibration is done directly in relative wavenumbers, independent of the exact excitation wavelength. Recently a set of ASTM (American Society for Testing and Materials) approved Raman standards became available (McCreery, 1996; ASTM E 1848). The Raman band positions of eight different substances (sulfur, 4-acetamidophenol, 1,4-bis(2-methylstyryl)benzene, cyclohexane, polystyrene, naphthalene, benzonitrile, 50/50 toluene/acetonitrile) were measured on FT-Raman spectrometers in six different laboratories. The averages were taken as reference values. A weak point is the low number of bands that are available for cm^{-1} calibration in the 1800–2700 cm^{-1} region. However, this region is usually of little interest in studies of biological samples, since very few Raman bands are found in this region.

The set of calibration points (camera pixels with assigned wavenumber shifts) obtained in this way, which should preferably be equally distributed over the whole spectral region of interest, is then fitted by a polynomial. This polynomial is used as the cm^{-1}

calibration function relating wavenumber shifts to camera pixels; a second- or third-order polynomial is usually sufficient. A calibration accuracy better than 0.2 channel can be obtained (Hamaguchi, 1988).

32.3.2 Correction of Raman spectra for the wavenumber-dependent instrument response

The quantum efficiency of the detector and the transmission and/or reflectivity of all optical elements in a Raman instrument are wavelength dependent. Therefore, the Raman signal detection efficiency is wavenumber dependent. This implies that the shape of the measured spectra depends on both molecular composition of the sample and the instrument response profile. If left uncorrected, this prohibits comparison of spectra measured on different instruments, since these will have different instrument response profiles, or even comparison of measurements obtained with one instrument (due to day-to-day variations caused by instrument drift, small changes in optical alignment or replacement of optical components).

Several correction methods are proposed in the literature (Fryling *et al.*, 1993). They are all based on measurement of the spectrum of a source that emits a known spectrum. In many cases a tungsten strip lamp (Preston, 1963; Tyler *et al.*, 1965) is used, emitting at a known temperature (determined with a pyrometer). Its emitted spectrum can be calculated from Planck's radiation law and the emissivity (de Vos, 1953, 1954) of tungsten:

$$dI = \varepsilon(\lambda, T)\,\frac{8\pi hc}{\lambda^5}\left(\frac{e^{-hc/\lambda kT}}{1-e^{-hc/kT}}\right)d\lambda \qquad [32.2]$$

where ε is the emissivity of tungsten, λ is the wavelength, h is Planck's constant, k is Boltzman's constant, and T is temperature.

Calibrated tungsten halogen lamps are a commercially available alternative (re-calibration is necessary after a few hundred hours of operation).

Calculating the ratio of the measured spectrum and the known spectrum emitted by the calibration source results in an instrument response profile that can be used to correct the spectra obtained of actual samples.

In the implementation of these correction methods it should be realized that Raman spectrometers usually count the number of detected photons instead of measuring intensities. Also, the detector pixels have a finite spectral width $d\lambda$, which is non-constant over the spectral interval in which measurements take place. The calculation of the theoretical reference spectrum needs to take this into account. The expression for the number of photons per channel, $N(n)$, is written as:

$$N(n) = \int_{\lambda}^{\lambda+d\lambda} (\lambda I(\lambda)/hc)\,d\lambda \qquad [32.3]$$

The correction factor C, for each detector channel n, then becomes:

$$C(n) = N_{std}^{obs}(n)/N_{std}(n) \qquad [32.4]$$

It is important to realize that Raman bands are (partially) polarized, and that polarization ratios in principle may vary between 0 and ∞ (see Chapter 29). Moreover, the detection efficiency of a set-up will differ, in a wavelength-dependent manner, for the different polarization directions. This is due primarily to the grating in the spectrometer, which can have very different diffraction efficiencies for different polarization directions. On the other hand, the light emitted, e.g. by a tungsten band lamp, is virtually unpolarized. Therefore, in order for the correction for wavelength-dependent signal detection efficiency to work properly, the polarization information in the Raman signal must be scrambled. In the set-up shown in Fig. 32.1, this polarization scrambling is brought about by the tapered optical fibre that connects the microscope to the spectrometer. Alternative approaches would be to use circularly polarized laser light, or to detect separately Raman scattered light in the two polarization directions.

It should be noted that the light distribution from a calibration lamp, and therefore the way its light enters the spectrometer, can be different from the situations in a Raman experiment. Depending on the design of the Raman instrument, the wavelength dependence of the instrument's signal detection efficiency may be (slightly) different for the two situations (i.e. light from a calibrated lamp, or Raman scattered light entering the spectrometer), so that the calculated correction factor C may not be precisely correct.

An alternative may then be to use standards that emit a fluorescence spectrum of a precisely known (calibrated) shape (Iwata *et al.*, 1988; Ray & McCreery, 1997). To obtain a fluorescence standard, its fluorescence spectrum is measured and then corrected using a lamp or black body radiation source. This fully corrected fluorescence spectrum can serve as a secondary intensity calibration spectrum.

The advantage of such intensity calibration standards is that a calibration is obtained using the exact same optical configuration as in the Raman experiments. Disadvantages are that they are derived standards, that the spectrum depends on the excitation wavelength, and that the fluorescent material may be sensitive to ageing.

Hamaguchi (1988) proposed the use of the rotation-vibration spectrum of deuterium. The relative band intensities of this spectrum are precisely known. The disadvantage of this method is that the spectral range for which it can be applied is limited (0 to 800 cm^{-1}) and that it consists of a set of discrete, narrow spectral lines. Fixed pattern noise, due to pixel-to-pixel variations in detector efficiency, cannot be corrected for. It is therefore primarily suited for the monotonic response function envelopes of single-channel systems.

32.4 DATA ANALYSIS

The success of *in vivo* clinical diagnosis by Raman spectroscopy will strongly depend on the performance of online data analysis techniques that rapidly extract the desired clinical information from the spectra. This implies the use of fast algorithms that project the measured spectrum in some way on a set of reference values (or reference spectra), resulting in a diagnosis. This set of references will be determined by a (multivariate) statistical model, developed on the basis of a database of supervised input spectra. For a robust model, it is necessary that the supervised input includes all normal biological and pathological spectral variance that can be encountered during *in vivo* application of the technique. The applicability of reference databases in the spectral analysis depends on good calibration transfer between measurements and instruments, and on maximal elimination of noise and artifacts from offset and scaling.

In most clinical applications the aim of the spectrum analysis will either be classification or quantification. The goal of a classification is to discriminate between healthy and diseased tissue, e.g. between different premalignant stages in the development of cancer. Quantification aims at determination of the concentration of one or more molecular compounds present in a tissue or body fluid.

In many cases it is hard to find specific Raman bands in spectra that can be used for discrimination between different types of tissue, or for the calibration of specific analyte concentrations. Multivariate statistical techniques can then be used to use simultaneously the information present in the whole spectrum, or in specific parts of the spectrum. Good textbooks and review articles about multivariate data analysis can be found in the chemometrics literature (Martens & Naes, 1989; Mobley *et al.*, 1996; Workman Jr *et al.*, 1996; Bro *et al.*, 1997). Another extensive source of information is the Internet (e.g. http://newton.foodsci.kvl.dk/chematoz, http://gepasi.dbs.aber.ac.uk/roy/sites/chemsite). Several specialized software packages for multivariate spectral analysis (mostly for concentration-calibration) are commercially available: The Unscrambler (Camo AS, Trondheim, Norway), a PLS-toolbox for use with MatLab (Eigenvector Technologies, Manson, Washington, USA), Pirouette (Infometrix, Seattle, Washington, USA), Grams 386 PLSplus (Galactic Industries, Salen, New Hampshire, USA), SIMCA (Umetri, Umea Sweden & Umetrics, Winchester, USA). Most of these have been developed for absorption or reflectance spectroscopy but are applicable to the analysis of Raman spectra as well.

Thanks to their ever-increasing computational power, multivariate statistical analysis of large amounts of (collinear) variables, such as encountered in Raman spectra, is feasible even on standard personal computers. Multivariate analysis can be used to extract the clinical relevant information from Raman spectra and therefore constitutes an important building block in the development for Raman spectroscopic methods of *in vivo* clinical diagnosis.

Here we will give a short overview of the most commonly used multivariate statistical methods for spectrum analysis, aimed at quantification and classification of spectra. Prior to building a statistical model, the input of the model should be pretreated in such a way that non-informative signal contributions and noise contributions to the spectra are minimized. Normalization of the spectra can be applied to ensure that all spectra contribute equally to the model. These data pretreatment methods will be discussed first. In Section 32.5, several examples will be given of applications of the data analysis techniques discussed below.

32.4.1 Data pretreatment

32.4.1.1 Noise and background

Filtering and other data pretreatment methods can be applied, prior to the statistical analysis, to eliminate the influences of noise and of uninformative, or even disturbing, (fluorescence) backgrounds.

The influence of shot noise, which increases with shorter (*in vivo*) measurement times (resulting in lower signal intensity levels), can be diminished by filtering techniques, either in the wavenumber domain or in the Fourier domain. Noise leads to pixel-to-pixel variations in signal intensity, whereas even the sharpest features in a Raman spectrum will cover several pixels. Low-pass filtering can therefore be applied. Several types of filters are used, such as average, moving average, spline, Savitzky–Golay and Fourier filters (Savitzky & Golay, 1964; Press *et al.*, 1993; Iwata & Koshoubu, 1994). A very recent development in filtering is the use of wavelet de-noising techniques (Donoho, 1993).

Fluorescence backgrounds have very broad spectral features compared with the Raman features. These can be removed by detrending, i.e. by subtraction of a suitable baseline (higher-order polynomials), by using a high-pass filter, or by using derivatives of the spectrum instead of the spectrum itself. A literature overview of different data pretreatment techniques is given by Mobley *et al.* (1996). Comparisons of different data pretreatment methods for multivariate problems in NIR spectroscopy have been published both for quantification and classification problems (Blanco *et al.*, 1997; Carlsson & Janne, 1989).

32.4.1.2 Normalization

The use of statistical models that project experimental data on reference values or reference spectra also implies the need for a suitable normalization procedure that eliminates the influence of wavenumber-dependent instrument response, and of variations in absolute signal intensity.

One possibility is the use of an internal standard that is present in most biological samples, e.g. the band at 1450 cm^{-1} due to CH-deformation vibrations of the methylene group in proteins, which is insensitive to protein conformation. This form of normalization has the obvious advantage of scaling directly on a Raman feature, but it can give false results if peak shape or peak position are altered by overlapping Raman bands of other compounds.

A more general normalization method is the Standard Normal Variate (SNV) (Barnes *et al.*, 1989) or autoscaling procedure. In this transformation every spectrum is mean-centred (so that the average of the spectral intensities in all wavenumber channels is zero) and scaled to have a standard deviation of one (the standard deviation of the spectral intensities calculated over all wavenumber channels in the spectrum becomes one). This transformation can be applied to all spectra and it requires no prior knowledge. Care should be taken in subtracting the fluorescence background before the SNV procedure in order to scale on the Raman features only.

For *in vivo* applications the data pretreatment procedure should be automated (unsupervised) and optimized to yield maximum enhancement of informative Raman features in the model spectra within acceptable computation times.

32.4.2 Multivariate spectrum analysis

32.4.2.1 Principal component analysis

After appropriate pretreatment and normalization procedures, a statistical model has to be applied to extract the desired clinically relevant data from the Raman spectra. The choice for a particular model strongly depends on the particular clinical question at hand. The information on which such a model is built is found in the spectral variance that is present in the database of reference spectra. Many spectral features have the same source of variation, so that their variation is collinear. This implies that the same information is present at different locations in the spectrum. A suitable data transformation can then be applied to remove this redundancy by finding the independent sources of variation in all spectra. The most frequently used method to achieve this is Principal Components Analysis (PCA) (Jolliffe, 1986).

PCA finds combinations of variables, called factors, which describe the major trends (sources of independent variation) in the data. Put in mathematical terms, PCA is an eigenvector decomposition of the variable correlation matrix. The principal components (PCs) are the eigenvectors and the corresponding eigenvalues are a measure of the amount of spectral variance captured in that PC. The first eigenvector is in the direction of the largest spectral variance in the database (the average of all spectra in the database), the second is orthogonal to the first one, and in the direction of the largest residual variance, and so on. The last PCs, with the smallest eigenvalues, often only represent noise and can be omitted in further analysis.

PCA is an unsupervised data transformation procedure that creates new variables in the directions of maximal variation, but not necessarily in the directions that are most useful for diagnosis. In many cases further modelling is required to achieve the latter. The main advantages of using PCA prior to further analysis, apart from data compression, are the removal of all collinearity and the inherent signal-averaging aspects. Because of the collinear nature of spectra, the amount of data reduction can be large; often a reduction of more than 95% can be achieved without the loss of useful information. The data compression and the orthogonality of the PCs can facilitate and speed up the subsequent steps in data analysis.

32.4.2.2 Multivariate calibration

In multivariate calibration, the concentration of one or more analytes present in a sample is predicted, using a model based on spectra from reference samples with known concentrations of the analytes of interest. The concentrations in these reference samples are determined by another method, which serves as 'gold standard'. The reference spectra must be obtained of samples in which is present all variation in molecular composition that may be encountered in practice. In this way, all possible spectral interference that is caused by such variation in molecular composition can be taken into account during the development of the model. The simplest multivariate model is Multiple Least Squares (MLS), in which the spectrum of the sample under investigation is fitted with a number of reference spectra. This method assumes that there is no collinearity between the pure component spectra.

A number of calibration models have been developed to deal with the problem of collinearity, ranging from Classical Least Squares (CLS) and Inverse Least Squares (ILS) to the methods that are mostly used nowadays, Principal Component Regression (PCR) and Partial Least Squares (PLS)-calibration.

In CLS, the pure component spectra are calculated from the calibration samples. Spectra of prediction samples are projected on these pure component spectra. CLS is a factor analysis method that directly transforms spectral space into analyte-concentration space. This new co-ordinate system need not be orthogonal and the scores of the samples in this new co-ordinate system represent the concentrations of the individual analytes. CLS uses the full information present in the spectrum but it has the disadvantage that all other compounds that could be encountered in prediction samples need to be identified and included in the model.

In ILS, also referred to as Multiple Linear Regression (MLR), or indirect or partial calibration, the error in the model is presumed to be in the component concentrations of the calibration samples and a transformation is calculated that minimizes the squared errors of these component concentrations. The main advantage of ILS is that a partial calibration can be performed. Only the concentrations of the analytes of interest in the calibration samples have to be known. Other compounds that could be present in prediction samples should be present and modelled during calibration, but do not have to be identified. The main disadvantage of ILS is that it must generally be restricted to a small number of spectral frequencies, since the number of frequencies considered may not be larger than the number of calibration samples in this type of calibration model.

PCR and PLS are both hybrid models that combine the advantages of CLS and ILS. The full spectrum is used as input and the calibration can be partial without complete knowledge of all interfering compounds. PCR is PCA followed by a regression step in which the errors in spectral space are minimized. The resulting new co-ordinate system is not specifically related to the pure analytes. In PLS the reduction of errors in spectral space is not optimal, but the errors in the concentrations are also minimized. In that respect, PLS is believed to be the more optimal method for concentration prediction.

Two excellent articles on the mathematics and use of these multivariate calibration methods have been written by Haaland *et al.* (Haaland & Thomas, 1988a,b). Some applications of multivariate calibration in Raman spectroscopy have been reported. Brennan *et al* used least squares methods to quantify the amounts of cholesterol, cholesterol esters, triglycerides, phospholipids and calcium salts present in human coronary artery samples (Brennan *et al.*, 1997; see also Section 32.5.1). Le Cacheux *et al.* (1996) used PLS to calibrate partially mixtures of cholesterol and cholesterol esters, in the presence of interfering compounds.

32.4.2.3 *Multivariate classification*

In multivariate classification, group membership for a certain spectrum is predicted by comparing the spectrum to a number of reference spectra by using some spectral distance measure, and classifying it to the most similar ones. The most straightforward way is to perform a cluster analysis of some form. Because of the high degree of collinearity in Raman spectra it is often useful to perform a PCA prior to cluster analysis. The clusters that are obtained can be inspected to establish the relationship between the cluster analysis results and the actual group membership of the samples. This procedure has the advantage that no information is put into the analysis (unsupervised); the identification is possible after modelling.

In most cases, however, the principal components space is not optimal for the separation of the desired groupings. Multivariate methods should then be applied that incorporate information about the origin of the samples in the model (supervised classification). This information is used to find the directions in spectral space that provide maximal distances between groups. The most often used method to achieve this is Linear Discriminant Analysis (LDA), or some derived form of this method such as Multiple Discriminant Analysis (MDA) or Factorial Discriminant Analysis (FDA).

LDA finds the direction in spectral space that is optimal for separating two groups. It selects the linear combination of vectors in spectral space that give the maximum value for the ratio between intergroup variance and intragroup variance. MDA is the multigroup analogue of LDA – it finds the $(n - 1)$ orthogonal directions that best separate n groups. LDA and MDA have the disadvantage that the number of input variables (e.g. wavenumbers) should not exceed the number of samples, in order to obtain a reliable model. FDA solves this problem by performing the discriminant analysis on the factors (PCs) extracted by PCA. FDA has the advantage of the data compression and signal-averaging aspects of PCA, while still using all relevant information in the spectrum for the discriminant analysis. The use of LDA and FDA for NIR-spectroscopic applications has been illustrated in a number of publications (e.g. Downey *et al.*, 1990; Choo *et al.*, 1995).

32.4.2.4 New developments

All methods discussed here are linear methods, although PLS and PCR can model some non--linearity. Progress in multivariate data analysis is expected from the incorporation of non-linearity into the multivariate models (Naes, 1992). If the calibration relation is non-linear but reasonably smooth, techniques like Locally Weighted Regression or Locally Weighted Least Squares can be used for multivariate calibration (Naes *et al.*, 1990).

More recently, newer methods like Artificial Neural Networks and Genetic Algorithms have also been shown to be useful in multivariate spectral analysis. Neural Networks can be trained to perform multivariate classification and calibration tasks: their prediction performance seems to be better than the standard linear multivariate techniques, especially when the models become more non-linear (Gemperline *et al.*, 1991; Borggaard & Thodberg, 1992; Lewis *et al.*, 1994; Goodacre *et al.*, 1996). Genetic Algorithms can also be used to explicitly incorporate non-linearity into the model by searching for the (non-linear) combinations of parameters that give the best results in regression models for data with fluctuating baselines and spectral overlap (Parakdar & Williams, 1997). When using Neural Networks or Genetic Algorithms, care should be taken of proper validation of the models to avoid overfitted solutions.

Improvement in the model can be made by selection of the part(s) of the spectrum to be included in the analysis. In many cases the information that is useful for the analysis is limited to certain regions of the spectrum; the rest of the spectrum adds noise and non-informative or even interfering signal contributions, which may worsen classification or quantification. Several search strategies techniques, such as Forward Selection, Backward Elimination, 'Branch and Bound', Simulated Annealing and Genetic Algorithms, can be used to select the optimum wavenumber ranges (McShane *et al.*, 1997). Genetic Algorithms in particular seem to constitute suitable and powerful search mechanisms for this application (Lucacius *et al.*, 1994).

32.5 EXAMPLES OF TISSUE CHARACTERIZATION BY RAMAN SPECTROSCOPY

32.5.1 Raman spectroscopy for detection of (pre-)malignant tissue

The application of Raman spectroscopy to study the molecular changes in cancer tissue has been described in a number of publications. These indicated that differences exist between the Raman spectra of healthy and cancerous tissues. However, most of this work was carried out on very small numbers of samples. Because of the absorption and the autofluorescence of tissue in the visible and near-ultraviolet (UV) wavelength range, the excitation wavelength is limited to the deep UV or the near-infrared (NIR) (Puppels & Greve, 1996). The use of UV excitation light offers higher Raman scattering cross-sections (and resonance effects) when compared with NIR excitation, but the possible mutagenic effects of UV radiation hinder its application in *in vivo* diagnostic studies.

An extensive overview of all work in the field of NIR-Raman spectroscopy for the detection of cancer and precancers has been published by Mahadevan-Jansen & Richards-Kortum (1996). Here we will only briefly review the results of these Raman spectroscopic studies, with an emphasis on the spectral differences between the various healthy and cancerous tissues that may prove to be useful for *in vivo* diagnosis. An example is then presented of a multivariate classification model for the detection of adenocarcinoma in Barrett's oesophagus using Raman spectroscopy.

Many Raman spectroscopic studies that have addressed the differences between cancerous and normal tissue have been focused on breast cancer. Spectra of normal breast tissue, benign breast tumour tissue and malignant breast tumour tissue were measured by FT-Raman spectroscopy by Alfano *et al.* (1991). The results suggested that the ratio between the spectral intensities at 1445 (C–H deformation) and 1651 cm^{-1} (amide I) can be used for classification between normal, benign tumour and malignant tumour. In a study by Frank *et al.* (1995) spectra with a much higher signal-to-noise ratio were presented. The difference between normal and diseased breast tissue samples was clearly visible in the ratio between the bands at 1654 cm^{-1} and at 1439 cm^{-1}, probably due to a lower total lipid content and a higher protein content (probably collagen) in cancerous tissue. Differences between benign and malignant tumour tissue were less prominent, but appeared to be present in the regions 850–950 cm^{-1} and 1200–1400 cm^{-1}.

Liu *et al.* (1992) were the first to present spectra of normal and cancerous tissues of the gynaecologic tract, obtained by NIR-FT-Raman spectroscopy. Spectral differences between normal tissue, and benign and malignant tumour tissue of the cervix, uterus, endometrium, and ovary were found in the CH deformation band at 1445 cm^{-1}, the amide III band at 1262 cm^{-1} and the amide I band at 1659 cm^{-1}.

Colorectal carcinomas (mostly adenocarcinomas) have been studied using UV-resonance Raman spectroscopy by Feld *et al.* (1994) and Manoharan *et al.* (1995). Nucleic acid bands at 1335 cm^{-1} and 1485 cm^{-1} were found to be more pronounced in adenocarcinoma. Feld *et al.* (1995) also used NIR-Raman spectroscopy to identify the differences between normal colon tissue and colon adenocarcinoma. Nucleic acid vibrational modes at 1340, 1458, 1576 and 1662 cm^{-1} were more intense in carcinomas, indicating a higher nuclear content; the normal samples showed more intense lipid signal contributions. Bladder cancers have also been studied by Feld *et al.* (1995) using NIR-Raman spectroscopy. Raman spectra of the bladder are dominated by protein bands. The spectral differences between normal bladder tissue and adenocarcinoma were similar to those observed in the studies of colon tissue.

Different types of brain tumours were investigated using FT-Raman spectroscopy by Mizuno *et al.* (1994). Glioma grade II resembled normal grey matter; glioma grade III tissue showed an intense band at 856 cm^{-1} (thought to be due to polysaccharides). One case of acoustic neurinoma displayed carotenoid bands that were not found in normal brain tissue. Central neurocytoma displayed an intense band at 960 cm^{-1}, characteristic for hydroxyapatite. In another study by Schrader *et al.* (1995), in which neurogenic sarcoma tissue, cultured in nude mice, was compared with normal human brain tissue, the spectra of the tumour tissue were characterized by a larger relative signal contribution of proteins and the absence of cholesterol Raman bands.

Abnormalities in different benign and malignant skin lesions were studied with NIR-FT-Raman spectroscopy by Gniadecka *et al.* (1997a). Skin lesions could be easily discriminated from normal skin, and histologically different lesions showed specific combinations of changes in the lipid bands, the phenyl ring breathing mode, and the amide bands. Using an artificial neural network with the spectral intensities at only five different wavenumbers as input, normal skin could be distinguished from basal cell carcinoma for a set of 32 samples (Gniadecka *et al.*, 1997b).

32.5.1.1 *Detection of (pre-)malignant tissue in the oesophagus*

Raman spectroscopy can be used for tissue diagnosis if the spectral differences between healthy and (pre)cancerous tissues are reproducible for most patients, and can be discerned from other, non-informative, spectral variance. In the Raman spectroscopic studies described above, the information to separate normal tissue from diseased tissue appeared to be present in a few bands. However, validation of these bands as markers for tumour tissue will require much larger numbers of samples than used in these studies. Multivariate statistical analysis of the resulting spectral databases will then be an indispensable tool to extract the spectral features with the highest possible discriminative power.

Here we present a multivariate model, based on the full Raman spectra that were obtained from resection material from three patients with an adenocarcinoma in a Barrett's oesophagus. Barrett's oesophagus is an acquired condition caused by chronic reflux of gastroduodenal juices into the oesophagus. This irritates and damages the normal stratified squamous epithelium that covers the lumen, which is then replaced by columnar lined epithelium that resembles the epithelium of the stomach. Because of the associated 30 times higher incidence of adenocarcinoma in a Barrett's oesophagus, it is considered to be a premalignant condition. The incidence rate of adenocarcinoma in the oesophagus is rapidly increasing in the Western world. The mortality rate is high, primarily due to the fact that the carcinoma is often discovered in a late stage. Intervention at an earlier stage (i.e. before an invasive tumour has developed) would significantly increase survival rates. Reliable screening methods are therefore necessary. Nowadays, an endoscopic examination during which tissue biopsies are taken is used for the screening of patients with a Barrett's oesophagus. However, early stages of malignant degeneration of the Barrett's epithelium cannot be recognized in the video-endoscopic images, so that biopsies are taken at more or less random locations. Methods that could guide the clinician taking the biopsies to suspicious tissue areas would much improve the efficacy of the screening procedure. In our laboratory, we are developing a Raman spectroscopic method for this purpose. *In vitro* studies are carried out to determine whether or not different stages in the development of adenocarcinoma can be identified on the basis of tissue Raman spectra.

Results are shown of a model that was developed to discriminate between normal epithelium, Barrett's epithelium and adenocarcinoma. To this end, tissue samples from three patients were classified by a pathologist, after which spectra were obtained of

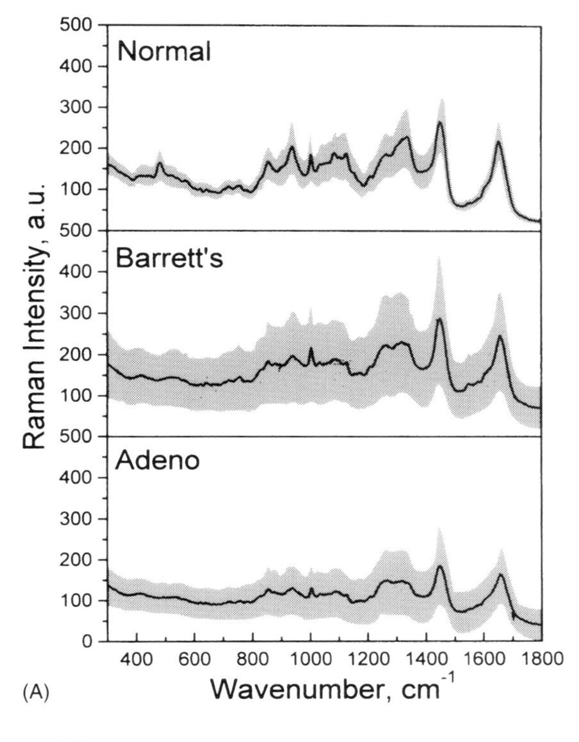

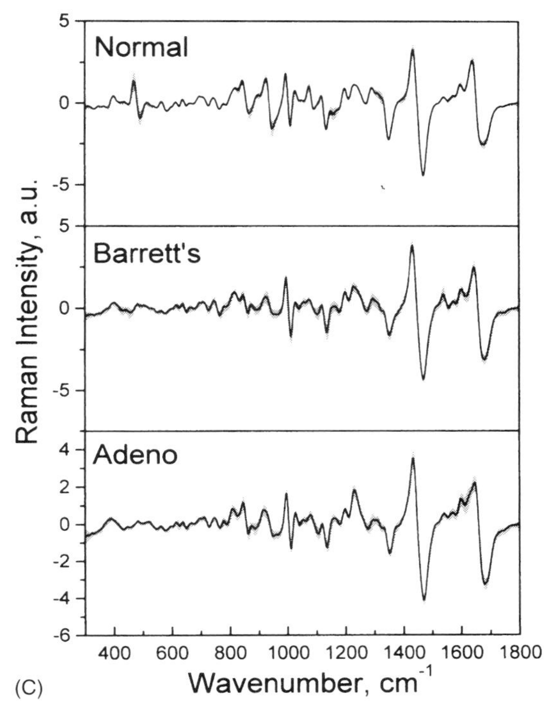

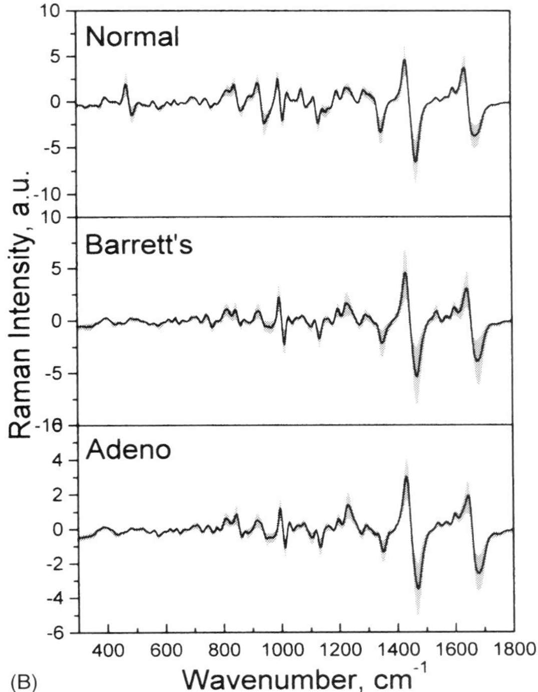

Figure 32.6 Raman spectra of normal oesophagus, Barrett's oesophagus and adenocarcinoma. Black lines indicate averages, grey area denotes one standard deviation of the averages. (A) Measured Raman spectra, corrected for wavenumber-dependent instrument response. (B) Savitzky–Golay differentiated spectra (first derivatives of spectra shown in A). (C) Result of SNV-normalization of spectra shown in B.

these samples by means of the Raman microspectrometer described in Fig. 32.1. A ×20 microscope objective was used to focus 100 mW of laser light (850 nm) onto the sample. In total 320 spectra were measured (80 of normal tissue, 120 of Barrett's tissue and 120 of adenocarcinoma). Of these 320 spectra, six spectra were excluded upon visual inspection (clear outliers). The remaining 314 spectra were used as input for a multivariate FDA classification model, implemented in a Matlab environment (MatLab 5, The MathWorks, Inc., Natick, MA, USA), using the PLS_toolbox from Eigenvector Research (PLS_toolbox 1.5, Eigenvector Research, Inc. Manson, WA, USA).

Prior to the FDA, the first-order derivatives of the spectra were calculated using a Savitzky–Golay procedure (Savitzky & Golay, 1964). After differentiation, the spectra were normalized using the SNV-procedure (Barnes *et al.*, 1989).

Figure 32.6 shows the raw spectra (A), the Savitzky–Golay differentiated spectra (B) and the SNV normalized first-derivative spectra (C) for the three types of tissue. As can be seen, the variation within the three groups diminishes with differentiation and normalization (see Section 32.4.1). The first step in developing the FDA model was a PCA on all differentiated and normalized data. Figure 32.7 shows the

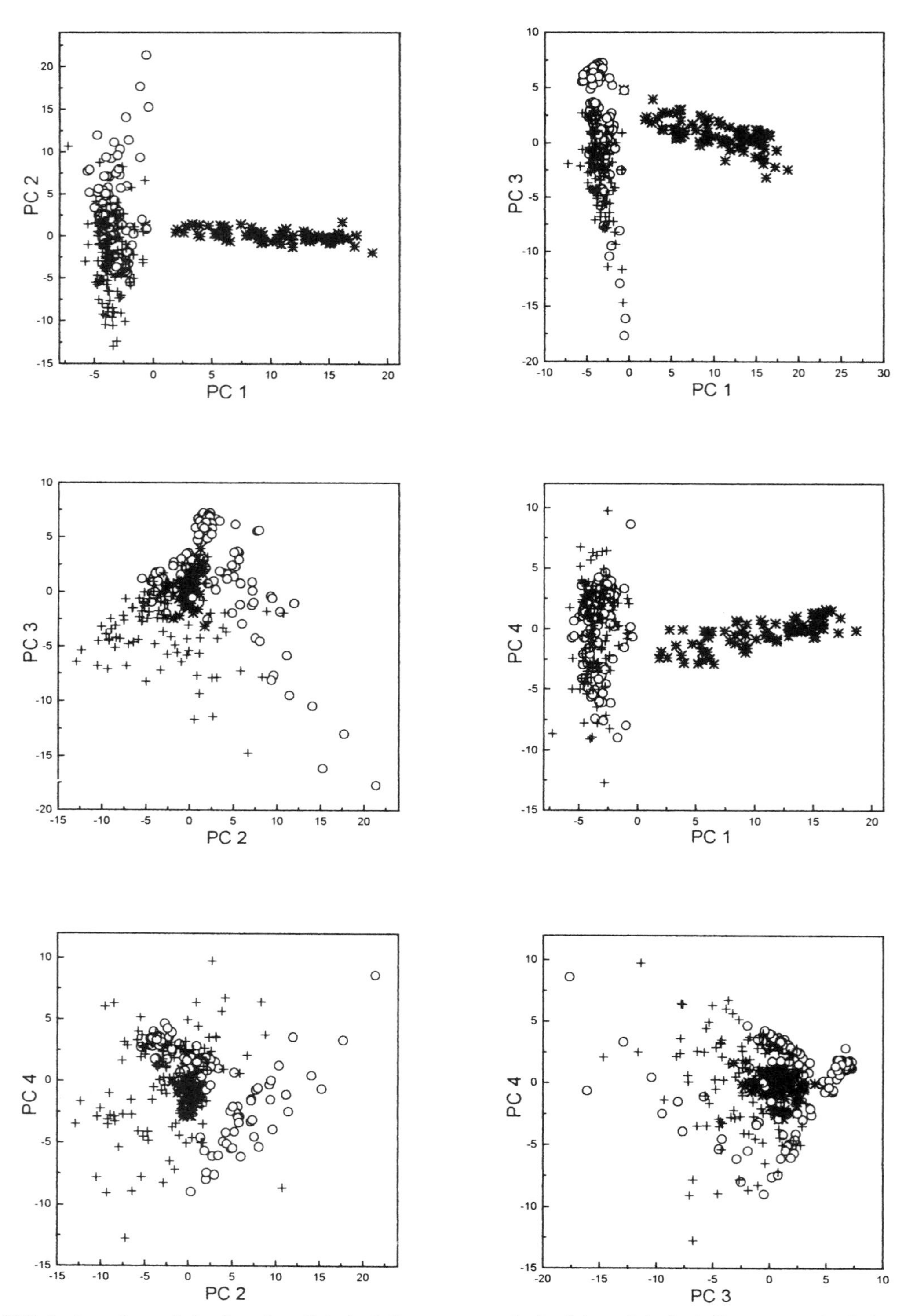

Figure 32.7 Scatter plots of the first four Principal Components obtained in a Principal Component Analysis on the differentiated and normalized data shown in Fig. 32.6. Symbols denote different tissue types: *, normal; ○, Barrett's; +, adenocarcinoma.

scores on the first four Principal Components (PCs), plotted against each other, for all 314 spectra. The first 20 PCs contained almost 95% of the total spectral variance. As can be seen from the scatter plots, the normal samples are almost completely separated from the diseased samples in the first PC. However, the separation between Barrett's and adenocarcinoma is less clear. To obtain a better separation, the 20 first PCs were used as input for an LDA model.

Figure 32.8 shows the results for the LDA analysis using all 314 spectra to build the model. To validate the model, a leave-one-out test was performed. In this test procedure, the classification of a single sample from the database is predicted by a model based on all other spectra of the database. This means that for N samples, N different models are built and tested. Table 32.1 gives the results for the leave-one-out validation of the overall model shown in Fig. 32.8.

From the clearly separated populations in Figure 32.8 and the validation given in Table 32.1, it can be

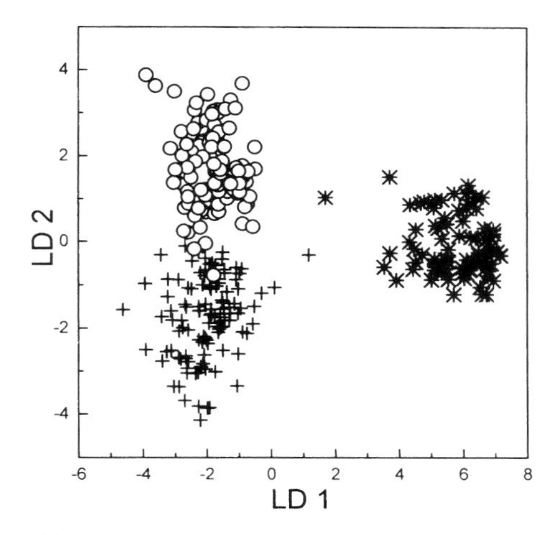

Figure 32.8 Scatter plot of the two disciminants obtained in Linear Discriminant Analysis on the first 20 Principal Components that were obtained from the data shown in Fig. 32.6. Symbols denote different tissue types: *, normal; ○, Barrett's; +, adenocarcinoma.

Table 32.1 Histopathological classification versus Raman spectroscopic classification of oesophageal tissues.

	Histology		
Raman	*Normal*	*Barrett's*	*Adenocarcinoma*
Normal	78	0	0
Barrett's	1	117	3
Adenocarcinoma	0	3	112

concluded that the FDA model gives good results for the limited database used here. In order to use it for *in vivo* diagnosis, the model has to be extended, using data of different patients until all possible variation in molecular composition of oesophageal tissue is represented in the model.

32.5.2 Atherosclerosis

Although several therapies are available to treat atherosclerosis, diagnostic methods that reliably predict lesion progression do not exist. The current opinion is that the chemical composition of plaque, rather than its volume or the severity of stenosis, is the most important determinant for the development of acute coronary artery syndromes (Falk *et al.*, 1995). Therefore, knowledge of plaque composition is of major importance for diagnosis, and for selecting and evaluating the effects of various interventional therapies. Current clinical methods, however, provide too little information about lesion morphology or biochemistry. Raman spectroscopy can deliver the desired information. In the last few years instrumentation has advanced to a level where spectra from tissue can be obtained in seconds, opening the way for obtaining information about the chemical composition of artery tissue *in situ*. Examples of Raman spectra from three different types of human coronary artery are shown in Fig. 32.9. The top spectrum was obtained from a sample of non-atherosclerotic (normal) coronary artery, the middle from a non-calcified atheromatous plaque, and the bottom from a calcified plaque. The spectra from these different artery types are distinct and provide clear features for determination of the chemical composition and for a histological classification of the arterial wall. For example, the normal coronary artery spectrum is dominated by protein features such as the amide I and III modes at ~1650 and 1250 cm^{-1}, respectively, and the CH_2 bending modes at ~1450 cm^{-1}. In non-calcified atheromatous plaques, spectral features of cholesterol and cholesterol esters constitute the major part of the spectrum. The symmetric stretch at 960 cm^{-1} of the calcium hydroxyapatite phosphate group dominates the spectrum of calcified plaques.

For the extraction of clinically useful information from artery Raman spectra, the group of Feld at MIT has developed a number of analytical models (Brennan *et al.*, 1997; Römer *et al.*, 1998a; Buschman *et al.*, 1998). In these models an artery spectrum is viewed as a linear superposition of spectra of individual components or of basis spectra. The models differ in the set of basis spectra. Two models will be discussed: a model based on the Raman spectra

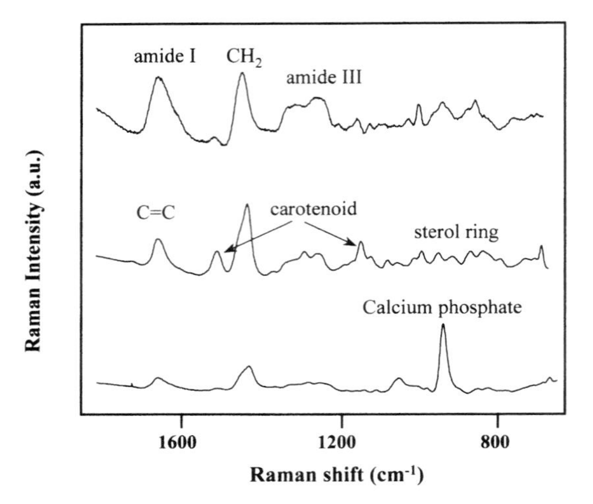

Figure 32.9 Raman spectra of three different types of coronary artery: (top) normal tissue; (middle) non-calcified atheromatous tissue; (bottom) calcified tissue.

collected from chemical components isolated from artery wall, and a model that is presently being developed which is based on the Raman spectral differences between morphological structures in artery wall.

32.5.2.1 Coronary artery chemical composition and histopathological classification based on Raman spectroscopy

The chemical model uses Raman spectra of individual chemical components present in normal and atherosclerotic artery wall as basis spectra. It was found that spectra of seven arterial components are needed to model adequately all of the measured coronary artery spectra. These components are free cholesterol (FC), cholesterol esters (CE), calcium salts (CS), triglycerides and phospholipids (TG&PL), two delipidized artery segments (DA) and β-carotene. The contribution of some components in a coronary artery spectrum, such as triglycerides or proteins, is difficult to model with spectra obtained from commercially available chemicals because these components are present in the form of mixtures in the artery. Therefore, these components were extracted from the artery wall itself. The DA spectra were obtained from delipidized artery samples, one from non-atherosclerotic tissue (DA I) and another from non-calcified atherosclerotic tissue (DA II). Linear superpositions of the seven components modelled the measured coronary artery spectra well, judged by the residuals of the fits that were

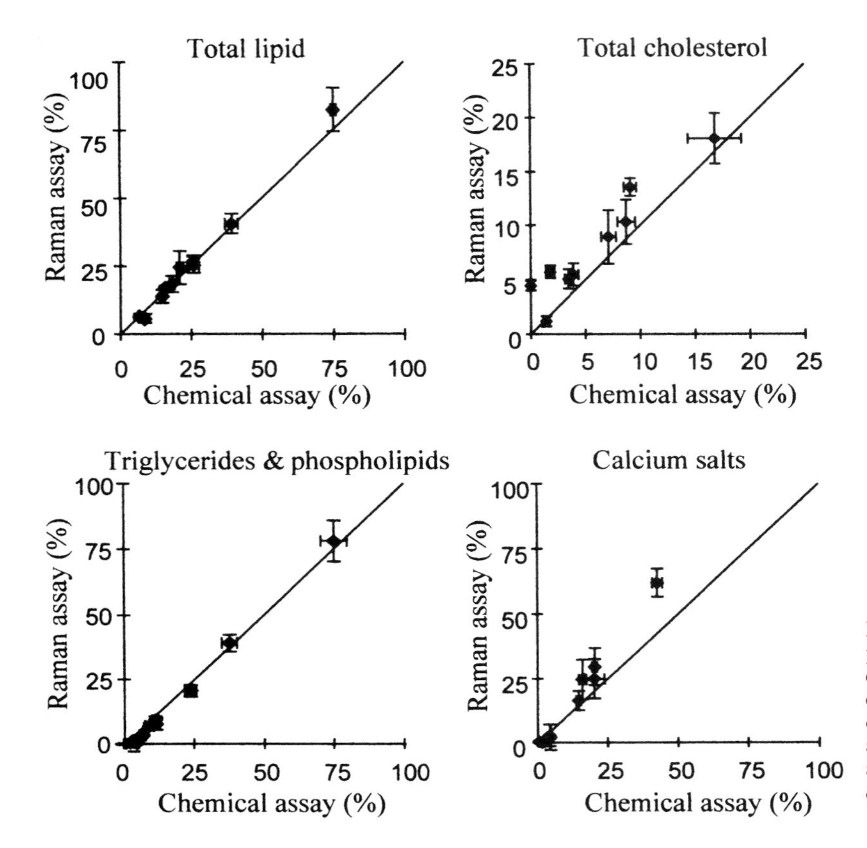

Figure 32.10 Comparison of the percentage weights of lipids and calcium salts measured in coronary artery minces, as determined by Raman spectral analysis and standard chemical assays. (Adapted from Brennan *et al.*, 1997, with permission from *Circulation*.)

obtained by subtracting the model fit from the artery spectrum.

The spectra of the seven model compounds were intensity scaled, in accordance with their relative Raman scattering cross-sections. The overall spectral model was validated by comparing chemical concentrations in coronary artery minces calculated from the Raman spectra to the actual concentrations measured with standard assay techniques. Excellent agreement was reached between the relative weights calculated with Raman spectroscopic techniques and those determined with standard assays conducted on the minces (Fig. 32.10). Details of the model and its validation with standard chemical assays have been described previously (Brennan, 1995; Brennan *et al.*, 1995, 1997).

The seven-component model enabled the extraction of quantitative chemical information from Raman spectra of coronary artery expressing different stages of atherosclerosis. This information was correlated with standard histopathological tissue classification. Figure 32.11 shows examples of Raman spectra (dots) of intimal fibroplasia, non-calcified atheromatous plaque and calcified atherosclerotic plaque that were chemically modelled (line). Figure 32.11(A) shows a spectrum of intimal fibroplasia. It is dominated by protein and TG features visible at 1650, 1250 and 1450 cm^{-1}. The TGs located in the adventitial layer are stronger Raman scatterers than the proteins in the intima and media. Therefore, their spectral features dominate the artery spectrum, although the relative weight percentage of TG is lower than that of proteins. In the spectrum of an atheromatous plaque shown in Fig. 32.11(B), spectral features from the sterol rings of FC and CE are visible below 1000 cm^{-1}. The Raman spectral model calculated a 12% relative weight of FC and a 6% relative weight of CE. Raman spectra obtained from calcified plaque are distinguishable by the symmetric stretch vibration of the phosphate group (960 cm^{-1}), present in calcium salts (mainly calcium hydroxyapatite). A large relative weight of CS was calculated from the spectrum of a highly calcified atheromatous plaque shown in Fig. 32.11(C).

It was found that cholesterol and calcium salts contents, as determined from Raman spectra, were particularly useful for classifying an artery as non-atherosclerotic tissue, non-calcified plaque, or calcified lesion. Using these two chemical parameters, diagnostic algorithms were constructed that could calculate the probability that an area of interest in a coronary artery falls into one of these three categories. These algorithms were successful in separating ~170 coronary artery samples into their proper diagnostic categories, as determined by the pathologist. This study suggests that the pathological state of a coronary artery site can be assessed successfully from its chemical composition determined with Raman spectra (Römer *et al.*, 1998a).

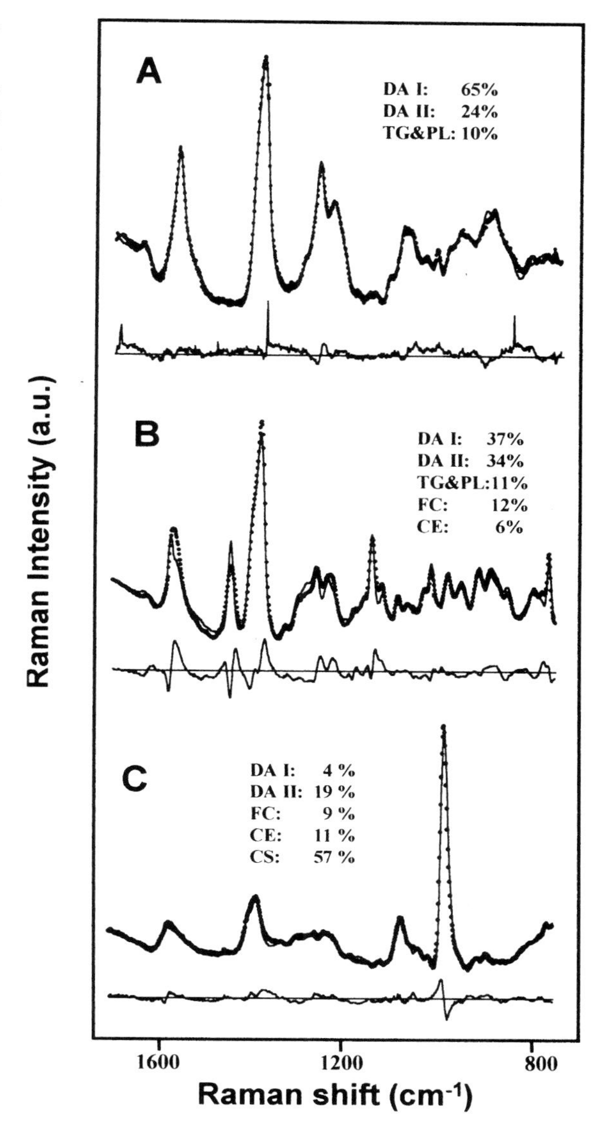

Figure 32.11 Raman spectra (dotted lines) of intimal fibroplasia (A), atheromatous plaque (B) and calcified plaque (C) were fitted with a set of spectra from individual components (solid lines) to quantify the chemical composition of the artery wall. The curve under each spectrum shows the difference of the spectrum and the model fit, i.e. the residual spectral features that could not be fitted. (Adapted from Brennan *et al.*, 1997, with permission from *Circulation*.)

32.5.2.2 *Quantitation of chemical composition in intact tissue*

For validation of the chemical model described above, spectra were obtained from homogeneous minces of coronary artery. Raman spectra collected from intact, inhomogeneous plaques, however, are more difficult to interpret. The intensity and shape of a spectrum measured at the surface of an artery wall is determined by the concentrations of the various artery wall constituents, and signal attenuation due to light scattering and absorption in the tissue. This signal attenuation will depend in a complex manner on the location below the tissue surface of a Raman scatterer, on the nature of the overlaying tissue, and on the signal excitation/collection geometry. The effect of the depth at which a cholesterol deposit is located in the artery wall on the relative weight percentage of cholesterol, as calculated from Raman spectra obtained from the luminal side of intact artery wall, was determined in a separate study (Römer *et al.*, 1998b).

This calculated relative weight percentage was found to decrease roughly exponentially with the depth at which the cholesterol deposit was located below the artery surface. A 300 μm layer of non-atherosclerotic tissue attenuated the Raman signal of plaque cholesterol by ~50% at 850 nm excitation, which is in agreement with earlier results (Baraga *et al.*, 1992). NIR-Raman spectroscopy can detect subsurface structures that are ~1–1.5 mm beneath the artery surface and therefore should be capable of detecting atherosclerotic deposits under thick fibrous caps. Atherosclerotic plaques in coronary arteries vary in thickness and may reach a fibrous cap thickness of 200–300 μm and an underlying core thickness of ~400 μm (Tracy & Kissling, 1987).

32.5.2.3 Morphological information from Raman spectra

The second method of spectrum analysis exploits the capabilities of Raman spectroscopy to identify morphological structures instead of chemical components (Buschman *et al.*, 1997). In this method a Raman spectrum is modelled as a linear superposition of the Raman spectra of the different morphological structures that can be distinguished in artery tissue (Buschman *et al.*, 1998). The use of this morphological model may provide direct quantitative information about the presence of morphological structures in the tissue, which can be used to give insight into the pathological condition of the artery. Raman spectra were obtained with a confocal microspectroscopic system similar to that described earlier. Thin (6 μm) unstained tissue sections were cut on a microtome, placed on a sample holder, and covered with saline. Using bright-field illumination, specific morphological structures were selected for spectroscopic examination. The laser spot was focused to a spot of about 1 μm on these structures by means of a ×63 objective,

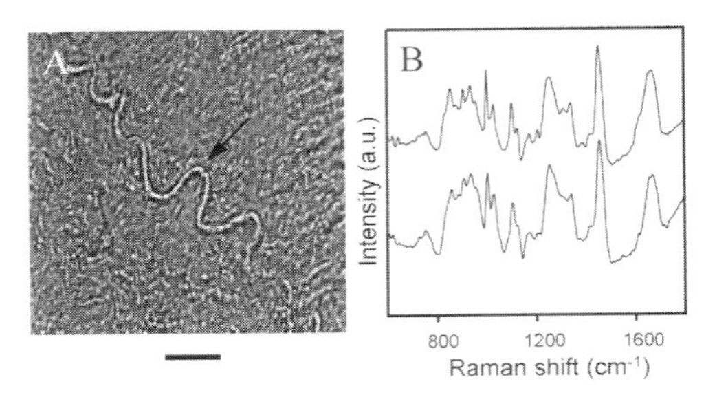

Figure 32.12 (A) A microphotograph of the internal elastic lamina (arrow) in an unstained coronary artery section of 6 μm thickness. Scale bar = 20 μm. (B) Raman spectra of this morphological structure (top) and of elastin (bottom).

after which Raman spectra were obtained. Raman spectra were recorded from the internal elastic lamina (IEL), collagen fibres, foam cells, adventitial fat, necrotic core, cholesterol crystals, calcium mineralizations, and β-carotene mineralizations. An example of a Raman spectrum obtained from an IEL is shown in Fig. 32.12. Comparison with a spectrum obtained of pure elastin confirms that this compound is the predominant constituent of IELs.

The Raman spectra from macroscopic artery samples expressing different stages of atherosclerosis ($n = 97$) were modelled with the morphological model, and showed excellent fits. With the fit-contribution of the morphological basis spectra to each macroscopic Raman spectrum we constructed an algorithm to calculate the probability that these samples belonged to one of the three diagnostic categories described above (Section 32.5.2.1). The accuracy of this algorithm was similar to that of the algorithm based on chemical components, which suggests that Raman spectroscopy can provide morphological information from intact tissue, which in turn can be used for pathological classification of artery tissue.

32.5.3 Skin

The human skin serves to prevent water loss and forms a barrier to physical, microbial and chemical assault. Impairment of these functions can have cosmetic and clinical consequences (Wertz & Downing, 1991; Lavrijsen *et al.*, 1995). Although crucial to a healthy skin, its impermeability is a major problem for dermal and transdermal drug administration.

The skin is a highly heterogeneous tissue composed of layers that differ in morphology and molecular composition. Conventionally it is divided in two major layers, the dermis and epidermis, and several sublayers. The barrier function of the skin mainly depends on the most superficial layer of the epidermis:

the *stratum corneum*. This layer consists of flat, closely packed cells with a high content of the insoluble protein *keratin*. Layers of lamellar lipid structures fill the intercellular spaces. The composition and conformation of these stratum corneum lipids are recognized to play an important role in maintaining the barrier function of the skin. Another factor involved in the water-preservation properties of the skin is the cellular water-binding capacity of the stratum corneum. It depends on the presence of a highly hygroscopic mixture of free amino acids, amino acid derivatives and salts, called natural moisturizing factor (NMF). The precise mechanisms involved in formation and maintenance of the water barrier of the skin are complex and not yet fully understood. The fact that Raman spectroscopy is a non-destructive technique that can be used non-invasively to obtain detailed information about molecular composition, structure and interactions makes it particularly interesting for *in vivo* skin research.

Fourier transform (FT) Raman spectroscopy has been used in several studies to investigate isolated stratum corneum *in vitro*. Williams *et al.* (1994) considered spectral differences between human, pig and reptilian stratum corneum in relation to differences in drug permeability. FT-Raman spectra of isolated stratum corneum showed dissimilarities between snake and mammalian skin in keratin conformation, ordering of stratum corneum lipids and sulfur content. Anigbogu *et al.* (1995) studied interactions between the penetration-enhancer dimethyl sulfoxide and stratum corneum. FT-Raman spectra of human stratum corneum samples revealed structural changes in stratum corneum proteins and lipids after treatments with dimethyl sulfoxide. *In vitro* FT-Raman spectra of healthy skin and of different pathological states were recorded by Gniadecka *et al.* (1997a,b) and Schrader *et al.* (1997). Only a few reports have touched upon the possibility of *in vivo* Raman spectroscopic measurements of human skin (Williams *et al.*, 1993; Schrader *et al.*, 1997; Shim & Wilson, 1997). Shim *et al.* and Schrader *et al.* used fibreoptic probes to record the spectra. The use of fibreoptics will be essential for clinical applications, but signal collection efficiency and spatial resolution will have to be improved.

A confocal Raman microspectrometer allows recording of Raman spectra with high spatial resolution (see Chapter 29). This technique can be used *in vivo* to record a Raman spectrum from a selected skin layer while signal contributions from the surrounding layers are suppressed. Using a confocal Raman microspectrometer and an inverted microscope, built in-house, we recorded *in vivo* Raman spectra of stratum corneum at various anatomical regions. Raman spectra of stratum corneum of the dorsal surface of the finger and the volar aspect (i.e. inner side) of the forearm are shown in Fig. 32.13. Prominent features in the spectra are the two bands around 1300 and 1655 cm^{-1}. These are assigned to protein vibrational modes involving the amide bonds (amide III and amide I respectively) and mainly originate from keratin, which is the most abundant protein in stratum corneum. The band positions provide information about protein secondary structure (Tu, 1982) and show that α-helical structure predominates, in accordance with existing literature (Barry *et al.*, 1992; Anigbogu *et al.*, 1995). Another prominent feature is the band at 1450 cm^{-1}, assigned to the collective vibration of the CH_2 groups. This is a common spectral feature of proteins and lipids. Clear spectral differences between the stratum corneum of the finger and the arm are observed in the 800–950 cm^{-1} and 1300–1500 cm^{-1} spectral regions.

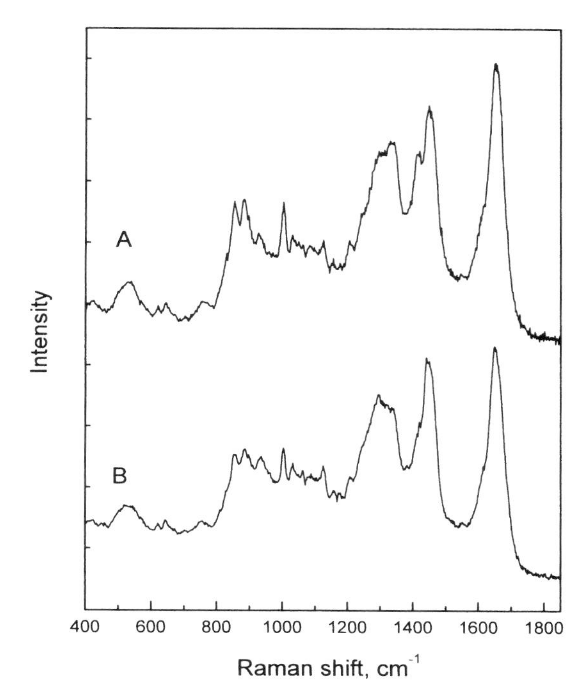

Figure 32.13 *In vivo* Raman spectra of stratum corneum at different anatomical regions. (A) Dorsal surface of the middle finger, (B) volar aspect of the forearm. Experimental conditions: laser wavelength, 850 nm; laser power, 100 mW; signal collection time, 120 s. Microscope objective, Leica Plan Fluotar ×63/0.70.

Figure 32.14 displays difference spectra, calculated from the stratum corneum spectra of the finger and forearm of two individuals. They are the result of a multiple regression fit to the finger spectra. The Raman spectra obtained from the stratum corneum of the arm, from the major amino acid constituents of NMF (serine, pyroglutamate, glycine, citrulline and urocanic acid), from urea, which is a sweat constituent, and from ceramide, representing the most abundant class of lipids

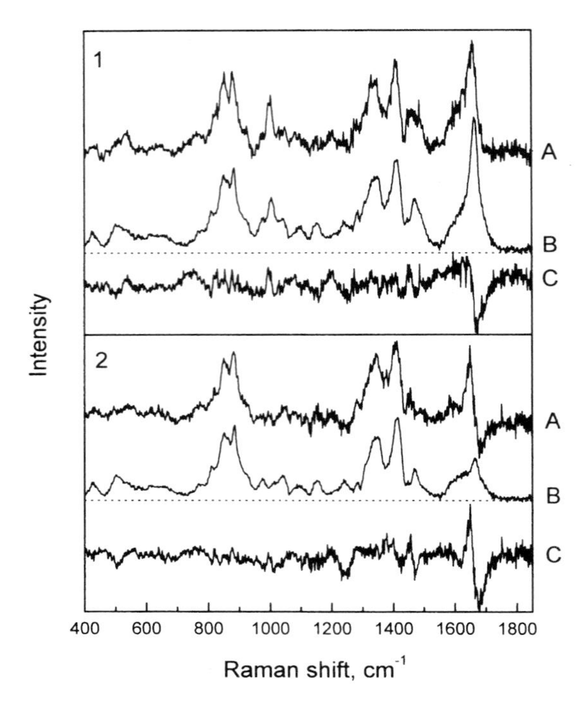

Figure 32.14 Difference spectra calculated from stratum corneum spectra of the finger and forearm of two individuals (1 and 2). (A) Difference spectra: finger minus forearm; (B) multiple regression fits of (A) (see text); (C) residual spectra (A − B).

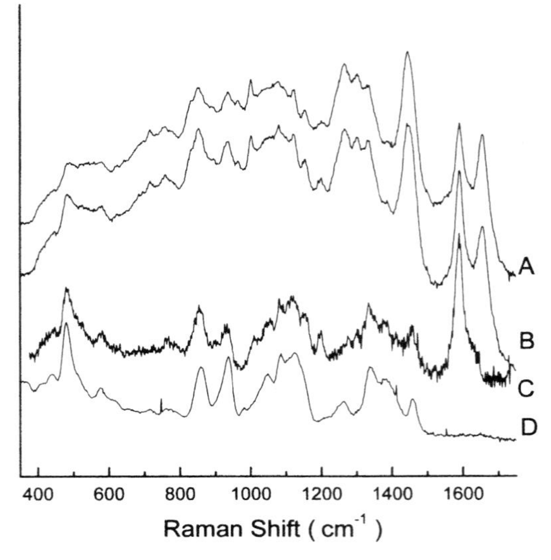

Figure 32.15 Raman spectra of rat livers. (A) Liver of a normally fed rat. (B) Liver of a rat fasted for 36 h and subsequently given 20% glucose to drink for 24 h leading to an increased liver glycogen level. (C) Difference spectrum: B − A, 3× enlarged. (D) Spectrum of a glycogen solution in water. Experimental conditions: laser wavelength, 850 nm; laser power, 75 mW; Raman microspectrometer of Fig 32.1 equipped with ×20 NIR-microscope objective. Spectra shown are averages of 25 measurements with 60 s signal integration time each.

in stratum corneum, were used to fit the finger spectra. Also shown are the summed contributions of the spectra of amino acids, urea and ceramide that result from the multiple regression fit. The residual spectra show that, apart from a large feature in the amide I region around 1655 cm^{-1}, for which no clear explanation exists at present, most spectral differences between the thin stratum corneum of the volar aspect of the arm and the thicker stratum corneum of the dorsal side of the finger can be explained in terms of differences in NMF and lipid content (Caspers *et al.*, 1998).

Presently Raman spectroscopy is the only technique that is able to provide this information regarding the molecular composition of stratum corneum *in vivo* and in a non-invasive manner. It has every potential, therefore, to develop into a powerful research technique for pharmaceutical, cosmetic and clinical skin characterization.

32.5.4 Determination of glycogen concentration in liver

The success of a liver transplantation depends on several parameters. Apart from immunological aspects, the concentrations of glycogen and lipids in the donor liver appear to play an important role. Low glycogen levels lead to an increase of preservation injury (Adam *et al.*, 1993) and high lipid content correlates with a high risk of primary non-functioning, or poor functioning of the graft (Ploeg *et al.*, 1993). A low glycogen content correlates with a reduced initial functioning of the liver after transplantation. However, due to the short allowable liver preservation times, these parameters are not routinely determined.

Below the results of a pilot experiment are shown. The goal of the experiment was to determine if it would be possible to detect differences in glycogen content in liver by means of Raman spectroscopy and suitable spectrum analysis methods. The use of data-pretreatment and Partial Least Squares regression to obtain quantitative information is illustrated. Raman spectroscopy could be a useful tool because it does not demand any sample preparation, and suitable fibreoptic probes should enable *in situ* measurements.

Expected difficulties in analysing the data are a large variation in fluorescent background and a relative low contribution of glycogen to the total Raman spectrum.

Figure 32.15 shows spectra obtained of two rat livers: one from a liver of a normally fed rat (A) and one of a rat kept on a diet that led to a high liver glycogen content (B). The spectra were scaled on the

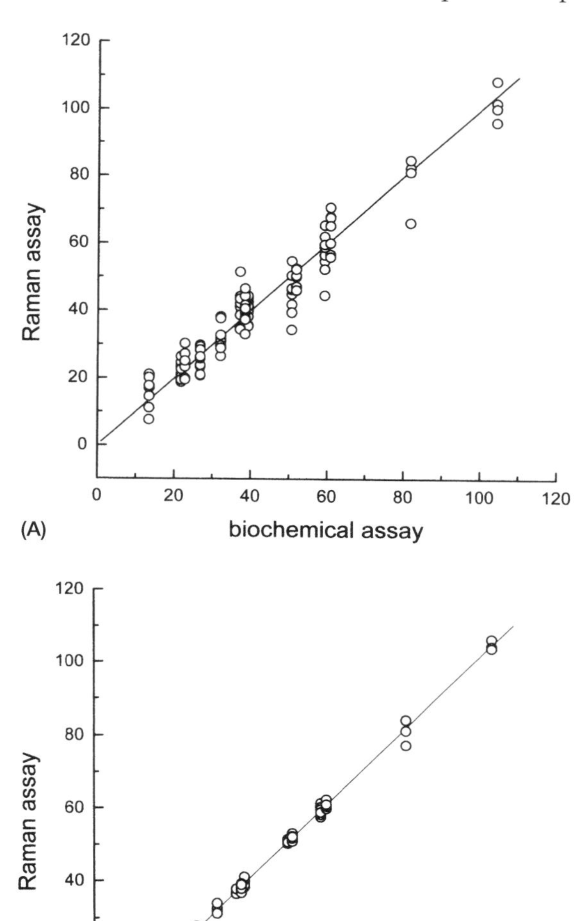

Figure 32.16 (A) Result of a PLS regression on a total of 109 spectra obtained of 14 samples spiked with different amounts of glycogen. (B) Result of a PLS regression on the Savitsky–Golay smoothed second derivatives of the spectra used in (A).

CH_2/CH_3 deformations band at 1450 cm^{-1} and subtracted. The difference spectrum (C) resembles the spectrum of glycogen (D); additional bands are attributed to differences in bilirubin content. This shows that information about differences in liver glycogen content can be obtained from Raman spectra. The next step is quantification of glycogen content.

To mimic a set of liver samples with different glycogen levels, the livers of two starved rats were homogenized and divided into 14 portions. To each sample, a different amount of glycogen was added. After measuring Raman spectra of each sample, the glycogen content was measured by a biochemical assay.

To find a relation between the Raman spectra and the glycogen concentration a Partial Least Square regression (PLS) was applied. As can be seen in Fig. 32.16(A), a reasonable result was obtained. Part of the variation in the spectra is due to variation in fluorescent background. This variation is not expected to contribute to the information present in the spectra because glycogen does not fluoresce. Taking a second derivative and simultaneously smoothing the spectra effectively removes the offset and slowly varying fluorescence signal contributions from the spectra. With this data set of second-derivative spectra as input for the PLS regression the fit improved considerably, as is clear from Fig. 32.16(B).

The results of this experiment, in which glycogen concentrations in liver that span the physiological range were determined, indicate that quick Raman spectroscopic determination of glycogen levels in transplant livers should be feasible.

32.6 CONCLUSION

The development of clinical applications of Raman spectroscopy has gained momentum in recent years. It is clear that the chemical information contained in Raman spectra can provide the clinician with new, or more complete, information. The facts that the technique is non-destructive, that no contrast-enhancing agents are involved, and that it can be applied through the use of fibreoptics, mean that it is highly suitable for *in vivo* tissue characterization, e.g. as a technique to guide clinical interventions. For this reason *in vivo* instrumentation is being rapidly developed.

Many clinical applications require the use of reference databases, and therefore a much more stringent control is needed of all instrumental parameters that can affect the measured Raman spectra than is required for most other applications. The clinical applications are also a strong stimulus for the use of multivariate statistical analysis methods in Raman spectroscopy.

As the examples in Section 32.5 illustrate, the prospects for successful implementation of Raman spectroscopic techniques in a wide range of clinical fields are excellent.

ACKNOWLEDGEMENTS

This work was supported by grants from The Netherlands Heart Foundation (R93.310 and 95.134), from the Interuniversity Cardiology Institute of The Netherlands (ICIN: D96.2158/MH), from The Netherlands Organisation for Scientific Research (NWO: 901–38–055), and from Philips Research.

REFERENCES

Adam R., Reynes M., Bao Y.M., Astarcioglu I., Azoulay D., Chiche L. & Bismuth H. (1993) *Transplant. Proc.* **25**, 1536–1537.

Alfano R.R., Liu C.H., Sha W.L., Akins D.L., Cleary J., Prudente R. & Cellmer C. (1991) *Lasers Life Sci.* **4**, 23–28.

Anigbogu A.N.C., Williams A.C., Barry B.W. & Edwards H.G.M. (1995) *nt. J. Pharm.* **125**, 265–282.

ASTM E 1848, American Society for Testing Materials, 100 Barr Harbor Drive, West Conshohocken, Pennsylvania, USA, 19 428–2959.

Baraga J.J., Feld M.S., Rava R.P. (1992) *Proc. Natl Acad. Sci. USA* **89**, 3473–3477.

Barnes R.J., Dhanoa M.S. & Lister S.J. (1989) *Appl. Spectrosc.* **43**, 772–777.

Barry B.W., Edwards H.G.M. & Williams A.C. (1992) *J. Raman Spectrosc.* **23**, 641–645.

Bell R.J. (1972) *Introductory Fourier Transform Spectroscopy.* Academic Press, New York.

Blanco M., Coello J., Iturriaga H., Maspoch S. & de la Pezuela C. (1997) *Appl. Spectrosc.* **51**(2), 240–246.

Borggaard C. & Thodberg H.H. (1992) *Anal. Chem.* **64**(5), 545–551.

Brennan III J.F. (1995) Ph.D. thesis, Massachusetts Institute of Technology, Cambridge, USA.

Brennan III J.F., Römer T.J., Tercyak A.M., Wang Y., Fitzmaurice M., Lees R.S., Kramer J.R., Dasari R.R., Feld M.S. (1995) *Proc. SPIE* **2388**, 105–109.

Brennan III J.F., Römer T.J., Lees R.S., Tercyak A.M., Kramer Jr J.R. & Feld M.S. (1997) *Circulation* **96**, 99–105.

Bro R., Workman Jr J., Mobley P.R. & Kowalski B.R. (1997) *Appl. Spectrosc. Rev.* **32**, 237–262.

Buschman H.P.J., Deinum G., Van der Laarse A., Bruschke A.V.G., Dasari R.R. & Feld M.S. (1997) *Cardiologie* **10**, 482.

Buschman H.P.J., Deinum G., Van der Laarse A., Bruschke A.V.G., Manoharan R. & Feld M.S. (1998) Histopathology of human coronary atherosclerosis by Raman spectroscopic morphological modeling. Manuscript in preparation.

Carlsson A.E. & Janne K.L.R. (1989) *Appl. Spectrosc.* **49**, 1037–1040.

Caspers P.J., Lucassen G.W., Wolthuis R., Bruining H.A. & Puppels G.J. (1998) *Biospectroscopy*, **4**, 531–539.

Choo L.-P., Mansfield J.R., Pizzi N., Somorjai L., Jackson M., Halliday W.C. & Mantsch H.H. (1995) *Biospectroscopy* **1**, 141–148.

Donoho D. (1993) In *Different Perspectives on Wavelets, Proceeding of Symposia in Applied Mathematics*, Vol. 47, I. Daubechies (ed). *Am. Math. Soc.*, Providence, RI, pp. 173–205.

Downey G., Robert P., Bertrand D. & Kelly P.M (1990) *Appl. Spectrosc.* **44**, 150–155.

Falk E., Prediman K.S. & Fuster V. (1995) *Circulation* **92**, 657–671.

Feld M.S., Manoharan R., Wang Y. & Dasari R.R. (1994) In *Proc. XIVth Int. Conf. on Raman Spectroscopy, Hong Kong*, N.-T. Yu & X.-Y. Li (eds). Wiley, Chichester, pp. 194–195

Feld M.S., Manoharan R., Salenius J., Orenstein-Carndona J., Römer T.J., Brennan III J.F., Dasari R.R. & Wang Y. (1995) *Proc. SPIE* **2388**, 99–105.

Frank C.J., Redd D.C.B., Gansler T.S., McCreery R.L. (1994) *Anal. Chem.* **66**, 319–326.

Frank C.J., McCreery R.L. & Redd D.C.B (1995) *Anal. Chem.* **67**, 777–783.

Fryling M., Frank C.J., McCreery R.L. (1993) *Appl. Spectrosc.* **47**, 1965–1974.

Gemperline P.J., Long J.R. & Gregoriou V.G. (1991) *Anal. Chem.* **63**, 2313–2323.

Goodacre R., Timmins E.M., Rooney P.J., Rowland J.J. & Kell D.B. (1996) *FEMS Microbiol. Lett.* **140**, 233–239.

Gniadecka M., Wulf H.C., Nielsen O.F., Christensen D.H. & Hercogova J. (1997a) *Photochem. Photobiol.* **66**, 418–423.

Gniadecka M., Wulf H.C., Nymark Mortensen N., Feurskov Nielsen O. & Christensen D.H. (1997b) *J. Raman Spectrosc.* **28**, 125–130.

Haaland D.M. & Thomas E.V. (1988a) *Anal. Chem.* **60**, 1193–1202.

Haaland D.M. & Thomas E.V. (1988b) *Anal. Chem.* **60**, 1202–1208.

Hamaguchi H. (1988) *Appl. Spectrosc. Rev.* **24**, 137–174.

Hendra P.J. (1996) In *Modern Techniques in Raman Spectroscopy*, J.J. Laserna (ed). John Wiley & Sons, Chichester, pp. 73–108.

Iwata T. & Koshoubu J. (1994) *Appl. Spectrosc.* **48**, 1453–1456.

Iwata K., Hamaguchi H. & Tasumi M. (1988) *Appl. Spectrosc.* **42**, 12–14.

Jolliffe I.T. (1986) *Principal Component Analysis* (Springer Series in Statistics). Springer-Verlag, Heidelberg, Germany.

Lavrijsen A.P., Bouwstra J.A., Gooris G.S., Weerheim A., Bodde H.E. & Ponec M. (1995) *J. Invest, Dermatol.* **105**, 619–624.

Le Cacheux P., Ménard G., Nguyen Quang H., Weinmann P., Jouan M. & Nguyen Quy Dao (1996) *Appl. Spectrosc.* **50**, 1253–1257.

Lewis I.R. & Griffiths P.R. (1996) *Appl. Spectrosc.* **50,** 12A–30A.

Lewis I.R., Daniel Jr N.W., Chaffin N.C. & Griffiths P.R. (1994) *Spectrochim. Acta* **50A**(11), 1943–1958.

Liu C.-H., Das B.B., Sha Glassmann W.L., Tang G.C., Yoo K.M., Zhu H.R., Akind L., Lubicz S.S., Cleary J., Prudente R., Celmer E., Caron A. & Alfano R.R. (1992) *J. Photochem. Photobiol. B.* **16**, 187–209.

Lucacius C.B., Beckers M.L.M. & Kateman G. (1994) *Anal. Chim. Acta* **286**, 135–154.

Mahadevan-Jansen A. & Richards-Kortum R. (1996) *J. Biomed. Optics* **1**, 31–70.

Manoharan R., Baraga J.J., Rava R.P., Dasari R.R., Fitzmaurice M. & Feld M.S. (1992) *J. Photochem. Photobiol. B* **16**, 211–233.

Manoharan R., Wang Y., Dasari R.R., Singer S., Rava R.P. & Feld M.S. (1995) *Lasers Life Sci.* **6**, 217–227.

Martens H. & Naes T. (1989) *Multivariate Calibration.* John Wiley & Sons, Chichester.

McCreery R.L. (1996) *Modern Techniques in Raman Spectroscopy*, J.J. Laserna (ed). John Wiley & Sons, Chichester, pp. 41–72.

McShane M.J., Coté G.L. & Spiegelman C. (1997) *Appl. Spectrosc.* **51**, 1559–1564.

Mizuno A., Kitajima H., Kawauchi K., Maraishi S. & Ozaki Y. (1994) *J. Raman Spectrosc.* **25**, 25–29.

Mobley P.R., Kowalski B.R., Workman Jr J. & Bro R. (1996) *Appl. Spectrosc. Rev.* **31**, 347–368.

Naes T. (1992) In *Near Infrared Spectroscopy*, K.I. Hildrim, T. Isaksson, T. Naes, & A. Tandberg (eds). Ellis Horwood Ltd, New York, pp. 51–60.

Naes T., Isaksson T. & Kowalski B. (1990) *Anal. Chem.* **62**, 664–673.

Parakdar R.P. & Williams R.R. (1997) *Appl. Spectrosc.* **51**, 92–100.

Ploeg R.J., D'Allessandro A.M., Knechtle S.J., Stegall M.D., Pirsch J.D., Hoffmann R.M., Sasaki T., Sollinger H.W., Belzer F.O. & Kalayoglu M. (1993) *Transplantation* **55**, 807–813.

Press W.H., Flannery B.P., Teukolsky S.A. & Vetterling W.T. (1993) *Numerical Recipes in C.* Cambridge University Press, Cambridge.

Preston J.S. (1963) *Br. J. Appl. Phys.* **14**, 43–45.

Puppels G.J. & Greve J. (1996) In *Advances in Spectroscopy*, Vol. 25: Biomedical Applications of Spectroscopy (Clark R.J.H. & Hester R.E., eds). John Wiley & Sons, Chichester, pp. 1–47.

Puppels G.J., Grond M., Greve J. (1993) *Appl. Spectrosc.* **47**, 1256–1267.

Puppels G.J., Van Aken T., Wolthuis R., Caspers P.J., Bakker Schut T.C., Bruining H.A., Römer T.J., Buschman H.P.J., Wach M.L. & Robinson J.S. Jr (1998) *Proc. SPIE* **3257**, 78–83.

Ray K.G. & McCreery R.L. (1997) *Appl. Spectrosc.* **51**, 108–116.

Römer T.J., Brennan III J.F., Fitzmaurice M., Feldstein M.L., Deinum G., Myles J.L., Kramer J.R., Lees R.S. & Feld M.S. (1998a) *Circulation* **97**, 878–885.

Römer T.J., Brennan III J.F., Bakker Schut T.C., Wolthuis R., Van den Hoogen R.C.M., Emeis J.J., Van der Laarse A., Bruschke A.V.G. & Puppels G.J. (1998b) *Atherosclerosis*, **141**, 117–124.

Savitzky A. & Golay M.J.E. (1964) *Anal. Chem.* **36**, 1627–1639.

Schrader B., Keller S., Löchte T., Fendel S., Moore D.S., Simon A. & Sawtzki J. (1995) *J. Mol. Struct.* **348**, 293–296.

Schrader B., Dippel B., Fendel S., Keller S., Löchte T., Riedl M., Schulte R. & Tatsch E. (1997) *J. Mol. Struct.* **408/409**, 23–31.

Schwab S.D. & McCreery R.L. (1984) *Anal. Chem.* **56**, 2199–2204.

Shim M.G. & Wilson B.C. (1997) *J. Raman Spectrosc.* **28**, 131–142.

Stevenson C.L. & Vo-Dinh T. (1996) *Modern Techniques in Raman Spectroscopy*, J.J. Laserna (ed). John Wiley & Sons, Chichester, pp. 1–39.

Tracy R.E. & Kissling G.E. (1987) *Arch. Pathol. Lab. Med.* **111**, 957–963.

Tu A.T. (1982) *Raman Spectroscopy in Biology.* John Wiley & Sons, Ltd, New York.

Tyler R.W., de Palma J.J. & Saunders S.B. (1965) *Photogr. Sci. Eng.* **9**, 190–196.

Vos J.C. de (1953), Thesis Amsterdam (VU).

Vos J.C. de (1954) *Physica* **20**, 690–714.

Wertz P.W. & Downing D.T. (1991) In *Physiology, Biochemistry and Molecular Biology of the Skin*, L.A. Goldsmith (ed). Oxford University Press, New York, pp. 205–236.

Williams A.C., Barry B.W., Edwards H.G.M. & Farwell D.W. (1993) *Pharm. Res.* **10**, 1642–1647.

Williams A.C., Barry B.W. & Edwards H.G.M. (1994) *Analyst* **119**, 563–565.

Workman Jr J., Mobley P.R., Kowalski B.R. & Bro R. (1996) *Appl. Spectrosc. Rev.* **31**, 73–124.

Yu N.-T., Li, X.Y. & Kuck J.F.R. (1996) In *Biological Applications of Raman Spectroscopy* (Spiro T.G., ed.). John Wiley & Sons, New York, pp. 143–184.

CHAPTER THIRTY-THREE

In Vivo Semiquantitative NADH-fluorescence Imaging

G.J. PUPPELS, J.M.C.C. COREMANS & H.A. BRUINING
Laboratory for Intensive Care Research and Optical Spectroscopy, Erasmus University Rotterdam & University Hospital Rotterdam 'Dijkzigt', Rotterdam, The Netherlands.

33.1 INTRODUCTION

Sufficient oxygenation of cells, tissues and organs and unimpaired oxidative phosphorylation are prerequisites for life. Ischaemia and reoxygenation may cause damage, which impairs tissue viability or organ functioning. Therefore, methods that enable assessment of the adequacy of oxygen delivery to mitochondria in tissue, and methods that can provide information regarding mitochondrial redox states and respiratory chain (dys)function, are crucial to (patho)physiological studies of tissues and organs. Optical techniques based on intrinsic optical properties of tissue, as well as optical techniques in which extrinsic fluorescent or phosphorescent probe molecules are employed, play an important role in this field.

Absorption and reflectance spectroscopy have been used to determine oxygen saturation of haemoglobin and to monitor the redox state of chromophoric groups in the respiratory chain in mitochondria (Pittman, 1986; Frank & Kessler, 1992; Chance, 1991; Jöbsis, 1972; Balaban *et al.*, 1980; Hoffman & Lübbers, 1986; Heinrich *et al.*, 1987; Coremans *et al.*, 1993). Phosphorescence-quenching techniques have been developed to determine intravascular Po_2, i.e. Po_2 in blood serum (Wilson *et al.*, 1993; Torres del Filho *et al.*, 1996; Sinaasappel & Ince, 1996; Shonat & Johnson, 1997). From there O_2 diffuses into the tissue, a process driven by the O_2 gradient between the serum and the mitochondria in the cells.

In mitochondria, the major net energy conversion is based on electron transport from reduced nicotinamide adenine dinucleotide (NADH) to molecular O_2 via a series of catabolic reactions in the respiratory chain (Fig. 33.1). NADH is a co-enzyme that carries electrons, which are released in the tricarboxylic acid cycle, to the respiratory chain. Provided that sufficient O_2 is present, NADH donates these electrons to the respiratory chain and is oxidized to NAD+. The respiratory chain converts the potential energy of the electrons into a proton gradient across the mitochondrial membrane, which drives ATP production. Mitochondrial NADH content is determined by NADH generation and NADH turnover. NADH generation in the tricarboxylic acid cycle, and by β-oxidation of fatty acid carnitine esters, depends on the availability of substrates and various feedback mechanisms.

FLUORESCENT AND LUMINESCENT PROBES, 2ND EDN
ISBN 0–12–447836–0

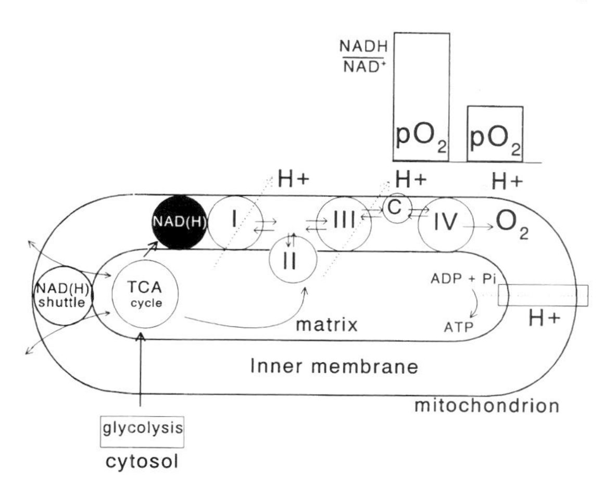

Figure 33.1 The respiratory chain.

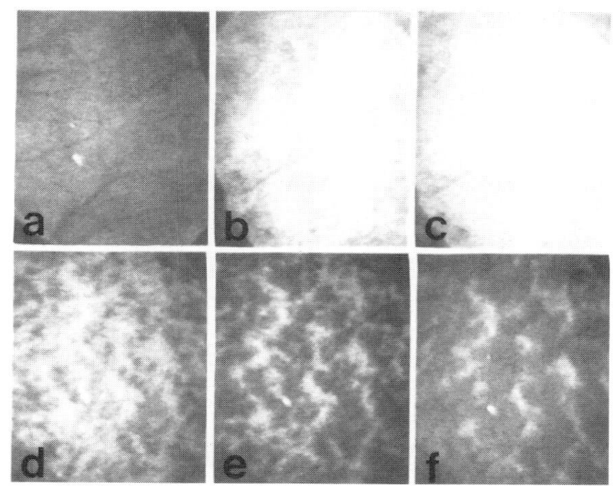

Figure 33.2 NADH-fluorescence images of an isolated beating rat heart in a Langendorff perfusion set-up obtained with the NADH videofluorometer set-up shown in Fig. 33.5(A). The images shown were obtained: (A) during perfusion with oxygen-saturated Tyrode; (B) after 40 s of anoxic perfusion with deoxygenated Tyrode; (C) after 120 s of anoxic perfusion with deoxygenated Tyrode; (D) 3 s after restoration of perfusion with oxygen-saturated Tyrode; (E) 5 s after restoration of perfusion with oxygen-saturated Tyrode; (F) 7 s after restoration of perfusion with oxygen-saturated Tyrode.

NADH turnover is determined by the rate of oxidative phosphorylation, which in turn is governed by ATP demand, NADH redox state, O_2 supply (Balaban, 1990), and the functional state of the respiratory chain complexes (which may be impaired, e.g. due to ischaemia-reperfusion damage). Unlike its conjugated electron acceptor NAD+, NADH fluoresces in the blue spectral region (emission centred around 460 nm) upon excitation with near-ultraviolet light (365 nm). It is for this reason that NADH-fluorometry may be used to monitor changes in the NADH/NAD+ ratio. This is illustrated in Figs 33.2(A)–(C), which show the increase in NADH-fluorescence intensity, as a result of hypoxia, in the myocardium of a beating rat heart in a Langendorff perfusion set-up.

This chapter will focus on NADH imaging of tissues and whole organs and will describe the methodology and instrumentation for *in vivo* semiquantitative NADH-fluorescence imaging. Examples will be given of measurements on beating rat heart in a Langendorff perfusion set-up, of measurements on rat transplant kidneys, and of *in vivo* measurements on rat.

The appearance of NADH images of tissue can be very inhomogeneous, as can be the dynamics of changes in the NADH/NAD+ ratio in response to metabolic challenges. This illustrates the important advantage of imaging, which can detect this heterogeneity, over single point measurements by means of fibreoptics.

An example is given in Figs 33.2(D)–(F). It shows that when anoxic beating rat heart in a Langendorff perfusion set up is returned to normoxic conditions, the myocardial NADH levels that built up homogeneously during ischaemia (Figs 33.2(A)–(C)) return to their original normoxic levels in a characteristic inhomogeneous fashion. This phenomenon was first described by Barlow & Chance (1976) and by Steenbergen *et al.* (1977), using NADH-fluorescence photography. More recently, Ince *et al.* (1993) showed that this hypoxic state heterogeneity is determined by heterogeneity in oxygen supply at capillary level. Apparently, steep oxygen gradients can occur in tissue.

33.2 ORIGIN OF UV-EXCITED (365 nm) TISSULAR FLUORESCENCE

The blue autofluorescence of tissues, observed under near-UV excitation, can only be used as an indicator of cellular hypoxia when mitochondrial NADH is the dominant intrinsic fluorescent probe. NADH is also present in the cytosol. However, its fluorescence quantum yield is strongly reduced due to quenching by cytosolic glyceraldehyde 3-phosphate dehydrogenase (O'Connor, 1977).

For heart tissue it has been convincingly proven that the NADH fluorescence signal is primarily of mitochondrial origin (Jöbsis & Duffield, 1967; Kanaide *et al.*, 1982; Nuutinen, 1984; Eng *et al.*, 1989). Moreover, fluorescence signal contributions of NADPH are negligible, due to the fact that it is present in much smaller amounts (Klingenberg *et al.*, 1959), and because its fluorescence quantum yield is three to four times lower than that of NADH (Avi-Dor *et al.*, 1962; Estabrook, 1962: Jöbsis & Duffield, 1967).

For other organs this has been less well established, although it has been reported that a 100-fold difference exists in the oxido-reduction state of the NAD+/NADH couple between the cytosolic and mitochondrial compartments in hepatocytes (Williamson *et al.*, 1967). Mitochondria occupy about 30% of the volume of the renal cortex, comparable to heart muscle (Pfaller & Rittner, 1980; Page *et al.*, 1971).

In studies of kidney and liver it is generally assumed that changes in fluorescence intensity in response to changes in metabolic conditions can be ascribed to mitochondrial NADH (Thorniley *et al.*, 1994; Okamura *et al.*, 1992; Balaban & Mandel, 1980, 1988; Franke *et al.*, 1976).

UV-induced blue fluorescence of collagen and fat is independent of tissular redox/oxygenation state and therefore does not disturb the evaluation of the metabolic state of tissue.

33.3 METHODOLOGY OF SEMIQUANTITATIVE *IN VIVO* NADH-FLUORESCENCE IMAGING

Figure 33.3 depicts the radiation transfer phenomena that play a role in tissue fluorescence measurements. At the tissue surface part of the incident UV excitation light (I_{ex}, 365 nm) is reflected (specular reflection: $R_{sp,365}$). Diffuse reflection of incident UV excitation light ($R_{diff,365}$) is caused by multiple scattering of photons in the tissue. In the tissue the UV excitation light can be absorbed by NADH molecules

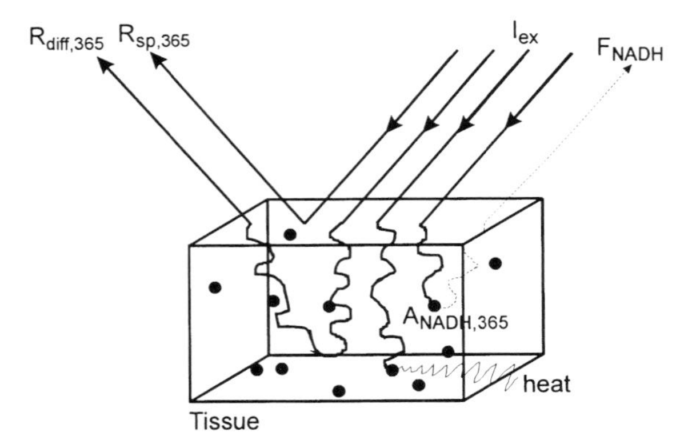

Figure 33.3 Radiation transfer phenomena involved in NADH-fluorescence measurements of tissue. Abbreviations: I_{ex}, intensity of UV (365 nm) excitation light; $R_{sp,365}$, intensity of specularly reflected excitation light; $R_{diff,365}$, intensity of diffusely reflected excitation light; $A_{NADH,365}$, absorption of UV excitation light by NADH molecules; F_{NADH}, NADH-fluorescence intensity.

($A_{NADH,365}$). The excited NADH molecules can return to the ground state through the fluorescence pathway, emitting light in the blue spectral region (F_{NADH}). When excitation light is absorbed by tissue pigments (e.g. cytochromes or myoglobin) or blood (haemoglobin), the absorbed energy is dissipated in the form of heat. Emitted fluorescence light is also absorbed by blood and tissue pigment. Absorption of excitation and emitted light causes a significant reduction in fluorescence intensity from the NADH molecules in the tissue.

A method was developed and validated by Coremans *et al.* (1997) to correct NADH-fluorescence intensity levels measured *in vivo* for a number of factors that influence the effective UV excitation intensity in the tissue. These factors are the absorption of UV excitation light by chromophores in the tissue, such as haemoglobin, myoglobin and cytochromes, and tissue motion. A change in oxygenation or redox state of the chromophores is accompanied by a change in their UV-absorption properties. A well-known example is haemoglobin; deoxy-haemoglobin has a higher absorption coefficient at 365 nm than oxy-haemoglobin. Therefore NADH-fluorescence intensity is influenced by changes in the spectral characteristics of these chromophores related to metabolic state (oximetric effect). Blood perfusion, and consequently haemoglobin concentration in organs and tissues with intact blood supply, is actively regulated. Changes in this concentration will influence the effective UV excitation intensity in the tissue and thereby NADH-fluorescence intensity (haemodynamic effect). This last effect is illustrated in Fig. 33.4. Lastly, tissue movement may alter the amount of specular reflection of UV excitation light at the air–tissue interface and thereby, again, the effective UV excitation intensity in the tissue.

Changes in tissular absorption characteristics at the fluorescence emission wavelength were found to be of minor importance in rat heart, exerting little influence on NADH-fluorescence intensity (Coremans *et al.*, 1993).

The theory of Kubelka and Munk for diffuse light scattering (Kubelka, 1948, 1954) and the theory of front-face fluorometry (Eisinger & Flores, 1979) were combined to derive a simple linear relation between the NADH concentration in tissue and the ratio of NADH-fluorescence intensity (F_{NADH}) and diffusely reflected UV excitation light ($R_{diff,365}$).

Front-face fluorometry refers to an experimental configuration in which the excitation light and emitted fluorescence light enter and leave the same sample face. For NADH fluorescence measurements this is the case when the UV excitation light (intensity, I_{ex}) is absorbed at a depth below the tissue surface that is much smaller then the thickness of the tissue under investigation (d). The optical penetration depth

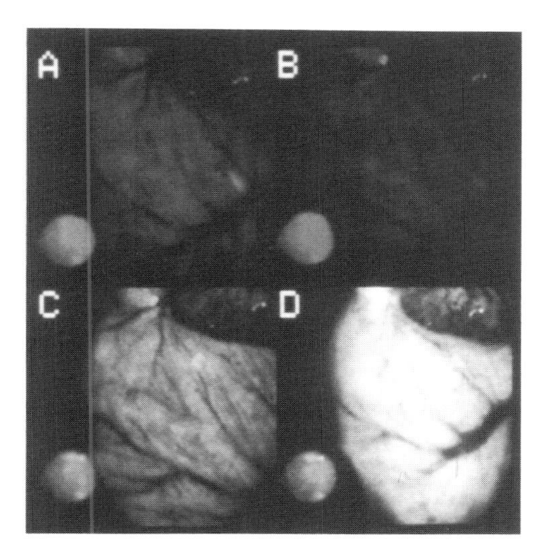

Figure 33.4 NADH-fluorescence images of isolated beating rat heart illustrating the effect of haemoglobin concentration on NADH-fluorescence intensity in myocardial tissue.
(A) Perfusion with oxygenated blood. (B) Perfusion with oxygenated blood supplemented with the vasodilator adenosine at a concentration of 10 μM. (C) Perfusion with deoxygenated blood in a dilated vascular system. (D) NADH image after a 2 minute interruption of blood supply. Images are displayed with relative gains A:B:C:D = 1:1:0.58:0.37 (compare the intensities of the uranyl calibration glass in the lower left corner).
Comparison of images (A) and (B) shows that the increase of the amount of blood, caused by vasodilation in the well-oxygenated myocardial tissue, leads to increased absorption of UV excitation light by blood and a decrease in absolute NADH-fluorescence intensity. Perfusion with deoxygenated blood leads to an upshift in the mitochondrial NADH/NAD+ equilibrium (image C). Occlusion of blood supply reduces the amount of blood in the myocardial tissue, resulting in decreased absorption of UV excitation light by blood and consequently an increase of NADH-fluorescence intensity (image D). (Reproduced from Coremans *et al.*, 1993, by permission of the Society of Photo-Optical Instrumentation Engineers.)

for the UV excitation light (δ) is defined as the distance over which the total fluence rate ($W\ mm^{-1}$) drops to e^{-1} (Svaasand & Ellingsen, 1983). The condition that must be fulfilled is therefore:

$$d/\delta \gg 1 \qquad [33.1a]$$

In terms of optical properties this becomes:

$$d/\delta = d\ \{3\mu_a\ (\mu_a + (1 - g)\ \mu_s)\}^{1/2} \gg 1 \qquad [33.1b]$$

where δ is given by $\{3\mu_a(\mu_a + (1 - g)\mu_s)\}^{-1/2}$, with μ_a the absorption coefficient (mm^{-1}), μ_s the scattering coefficient (mm^{-1}), and g the anisotropy factor (Ishimaru, 1978). For most blood-containing tissues δ will be of the order of one hundred to a few hundred micrometres, so that condition [33.1a] will usually be fullfiled in measurements on whole organs and tissues more than a millimetre in thickness.

NADH-fluorescence intensity of tissue (F_{NADH}) is then proportional to the ratio $A_{NADH,365}/A_{T,365}$, where $A_{T,365}$ is total UV absorbance and $A_{NADH,365}$ is UV absorbance by NADH (Eisinger & Flores, 1979):

$$F_{NADH} = G\phi\ (A_{NADH,365}/A_{T,365})\ I_{ex} \qquad [33.2]$$

with G a geometrical factor characteristic for the instrument, and ϕ the fluorescence quantum yield of NADH.

From the theory of Kubelka and Munk (Kubelka, 1948, 1954) a relationship can be derived between the total sample absorbance $A_{T,365}$, scattering S and diffuse UV reflectance $R_{diff,365}$ (Kessler & Frank, 1992):

$$A_{T,365}/S = 0.5\ R_{diff,365}/I_{ex} + I_{ex}/R_{diff,365} - 1 \qquad [33.3]$$

At sufficiently high absorbance, the right-hand side of equation [33.3] will be dominated by the second term and the ratio of absorption to scattering becomes inversely proportional to the reflectance intensity:

$$A_{T,365}/S \propto I_{ex}/R_{diff,365} \qquad [33.4]$$

Under the assumption that the basic scattering properties of the tissue are constant (Hoffmann *et al.*, 1983, 1984; Heinrich *et al.*, 1987; Kessler & Frank, 1992), equation [33.4] implies that the diffuse reflectance intensity can be used to monitor changes in tissue absorbance.

Combination of expressions [33.2] and [33.4] leads to:

$$F_{NADH}/R_{diff,365} \propto C\ A_{NADH,365} \qquad [33.5]$$

with S, G and ϕ incorporated in the constant C. Because $A_{NADH,365}$ is proportional to the NADH concentration, it follows that:

$$[NADH] \propto F_{NADH}/R_{diff,365} \qquad [33.6]$$

This linear relationship between [NADH] and $F_{NADH}/R_{diff,365}$ makes this ratio a highly suitable measure for semiquantitative *in vivo* NADH fluorometry.

Moreover, since $A_{T,365}$ is eliminated in the expression for $F_{NADH}/R_{diff,365}$, this ratio is independent of haemodynamic and oximetric changes. Because $F_{NADH}/R_{diff,365}$ is also independent of I_{ex}, fluctuations in effective excitation intensity caused by UV source instability or tissue movements do not affect the determination of relative NADH concentration changes.

33.4 INSTRUMENTATION

The basic set-up for *in vitro* NADH-fluorescence imaging is shown in Fig. 33.5(A). It is a videofluorometer consisting of a fluorescence unit B2-RFCA of an Olympus BH2 microscope, a 100 W Hg-arc lamp or a 200 W Xe(Hg) lamp providing 365 nm UV excitation light, and a second-generation intensified CCD video camera (MXRi, Adimec Advanced Image Systems, Eindhoven, The Netherlands), equipped with an ultraviolet blue-sensitive S20 photocathode (Philips, Eindhoven, The Netherlands) and a 105 mm macro lens. Images are stored on videotape and are analysed off-line. The images of Figs 33.2, 33.4 and 33.7 were obtained with this set-up.

The original set-up for simultaneous recording of NADH fluorescence images and UV diffuse reflectance images is shown in Fig. 33.5(B). It is an adaptation of the set-up shown in Fig. 33.5(A). A biprism (3° wedge) is used to enable simultaneous measurement of NADH fluorescence and diffuse UV reflectance images of rat heart. One half of the biprism is covered with an interference filter for transmission of NADH fluorescence (470 nm, bandwidth at half-maximum: 10 nm, Melles Griot, Zevenaar, The Netherlands). The other half of the biprism was covered with an UG-1 filter for selection of reflected UV excitation light. To prevent contributions of specular reflected UV light to the diffuse reflectance image, a crossed polarizer/analyser combination was used (Polaroid Type HNP'B Linear Polarizer; Polarizer Division, Hertfordshire, UK). The results shown in Figures 33.6 and 33.7 were obtained with this set-up.

A new instrument for semiquantitative NADH-fluorescence imaging was developed recently. It can record up to four images at different wavelengths simultaneously, and is adapted to the requirements for applications in an operating theatre (Fig. 33.5(C)). It has a field of view of ~100 cm^2. The light source is a 1000 W Xe(Hg) lamp (Oriel). The near-UV line (365 nm) is selected by means of filters and coupled into a liquid light guide. The output of the light guide is filtered by a UG1 colour glass filter and is linearly polarized. The illumination optics were chosen such that a fairly homogeneously illuminated area of about 100 cm^2 is obtained. The intensity of the UV excitation light on the sample is of the order of 1–2 mW cm^{-2}. Lens L1 produces an image of the sample in the plane of field stop FS. Dichroic mirrors DM are used to separate the light that is collected from the sample in different wavelength intervals. At

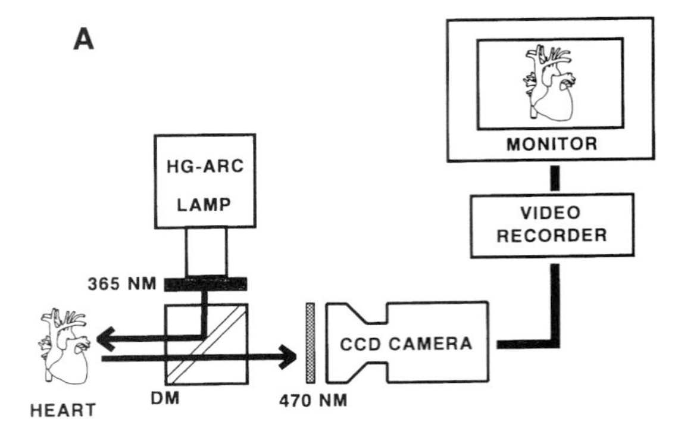

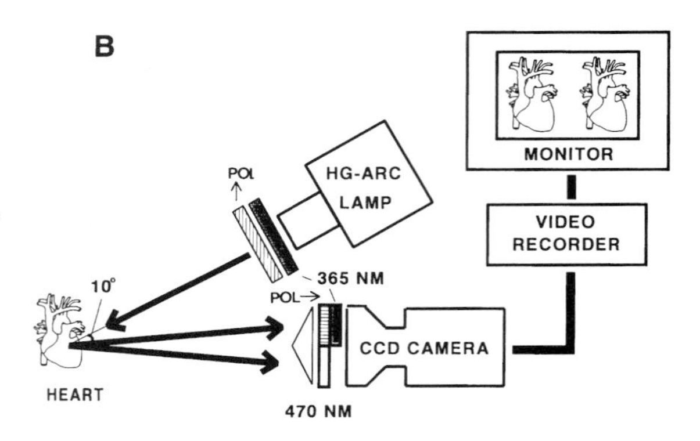

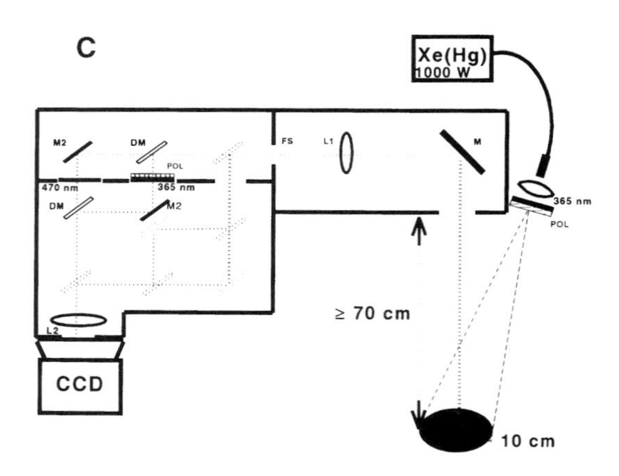

Figure 33.5 Schematics of NADH-fluorescence instrumentation (for details see Section 33.4): (A) NADH-fluorescence videofluorimeter (DM, dichroic mirror). (B) Set-up for simultaneous recording of NADH-fluorescence images and diffuse reflectance images (at 365 nm) (POL, polarizer). (C) Schematic of instrument for *in vivo* semiquantitative NADH-fluorescence imaging. (Fig. 33.5(B) reproduced from Coremans *et al.*, 1997, by permission of the Biophysical Society.)

present only two images are recorded simultaneously: a diffuse reflectance image of the sample at 365 nm and an NADH-fluorescence image in the 460–480 nm interval. The dichroic mirrors DM1 and DM2 reflect the UV excitation light and transmit fluorescent light. The 365 nm light path further contains a 365 nm narrow-bandpass filter and a polarizer P2, which forms a crossed polarizer/analyser pair with polarizer P1, ensuring that only diffusely reflected light is detected and that specular reflections do not contribute to the reflectance image. The fluorescence light path contains a 460–480 nm dielectric bandpass filter, by which primarily NADH-fluorescence is selected. The mirrors M2 (and dichroic mirror DM2) are positioned at slightly different angles. In this way the fluorescence and reflectance images can be focused at different positions on the chip of the unintensified Peltier-cooled CCD camera (Princeton Instruments, Trenton, NJ). The field stop FS ensures that no overlap between these images occurs on the CCD chip. It furthermore contains a pair of crossed wires, which are visible in both the fluorescence and reflectance images; these allow the images to be superimposed precisely before a $F_{NADH}/R_{diff,365}$ ratio image is calculated. The intensities in these ratio images are linearly related to local NADH levels in tissues and organs under investigation. Intensity changes due to changes in physiological/oxygenation conditions are proportional to changes in NADH concentration. The images and results shown in Fig. 37.8 were obtained with this instrument.

33.5 EXPERIMENTAL VERIFICATION OF THE LINEAR RELATIONSHIP BETWEEN [NADH] AND $F_{NADH}/R_{diff,365}$

In order to verify the linear relationship between [NADH] and the $F_{NADH}/R_{diff,365}$ ratio, experiments were carried out with tissue phantoms consisting of mixtures of haemoglobin and Intralipid® in cuvettes, in order to mimic the optical properties of heart tissue (Coremans *et al.*, 1997). Figure 33.6(A) demonstrates the linearity between NADH concentration and $F_{NADH}/R_{diff,365}$ ratio over the physiological NADH concentration range from 20 to 25 μM during normoxia to 300–450 μM during anoxia (Chance *et al.*, 1965; Katz *et al.*, 1987; Bessho *et al.*, 1989) in tissue phantoms composed of Intralipid®-3% and 50 μM haemoglobin. The positive intercept of the fluorescence axis is caused by Intralipid®, which also fluoresces around 470 nm upon excitation with UV light.

$F_{NADH}/R_{diff,365}$ as a function of total sample absorbance is shown in Fig. 33.6(B). As expected from equations [33.3] and [33.4], stability of the $F_{NADH}/R_{diff,365}$ ratio improves with increasing sample absorbance. Only for tissue phantoms containing less than 65 μM haemoglobin, i.e. with absorption coefficients that are significantly smaller than the μ_a of even

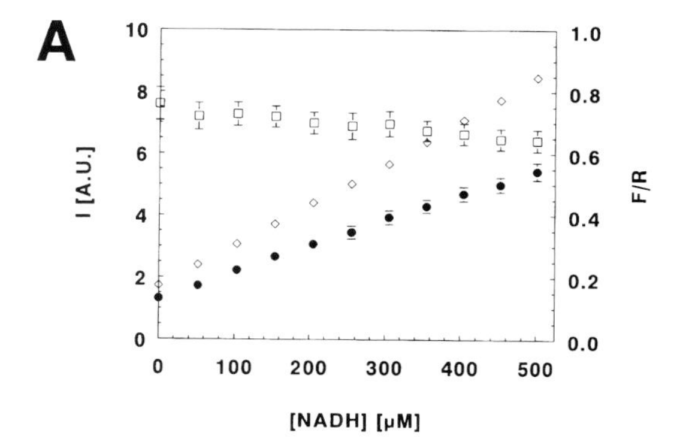

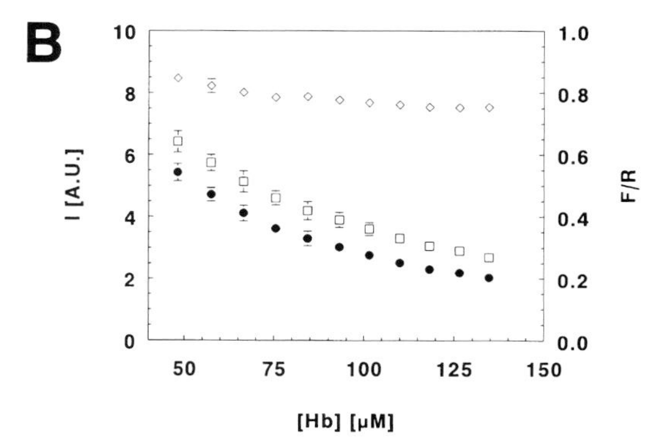

Figure 33.6 F_{NADH}-intensity (•), $R_{365,diff}$-intensity (□) and $F_{NADH}/R_{diff,365}$ ratio (◇) as: (A) a function of NADH concentration in a tissue phantom composed of 50 μM haemoglobin and Intralipid®-3%. The positive intercept of the fluorescence axis is caused by Intralipid®, which also produces signal contributions around 470 nm due to fluorescence upon excitation with UV light. (B) A function of increasing hemoglobin concentration with 400 μM NADH added to the tissue phantom. AU: arbitrary units. (Reproduced from Coremans *et al.*, 1997, by permission of the Biophysical Society.)

(virtually) bloodless rat heart, the $F_{NADH}/R_{diff,365}$ ratio deviates more than 5% from the stable value reached at higher μ_a values. Similar results were obtained with phantoms containing Intralipid®-6% and with whole blood instead of purified haemoglobin (for further experimental details, see Coremans *et al.*, 1997).

33.6 BIOMEDICAL AND CLINICAL APPLICATIONS OF NADH-FLUORESCENCE IMAGING

Adequate tissue oxygenation and respiratory chain functioning in the mitochondria is compromised in many clinical conditions. In the operating theatre a technique that can detect tissular ischaemia and/or its alleviation through surgical intervention would be of great value. Potential applications include monitoring of myocardial ischaemia during open heart surgery and catheterization (Duboc *et al.*, 1986; Horvath *et al.*, 1994; Tardiff & Juneau, 1996), monitoring of the alleviation of ischaemia and determination of tissue viability during surgical treatment of volvulus, mesenteric thrombosis and strangulated hernia (Dyess *et al.*, 1991), and determination of the viability of anastomoses in the gastrointestinal tract (during and after surgery), which depends critically on adequate oxygenation of the wound area (Pierie *et al.*, 1994).

Much research is being carried out to determine whether NADH fluorometry could play a role in the assessment of kidney and liver graft viability after hypothermic storage and transplantation and in determining the quality of transplant organs, i.e. predicting post-transplantation function (e.g. Thorniley *et al.*, 1994, 1995; Okamura *et al.*, 1992). It is expected that the pool of potential donor organs could be considerably enlarged, e.g. through inclusion of organs from non-heart-beating donors, were such an objective test to become available. Figure 33.7 shows results of *in vitro* experiments in which the NADH-fluorescence response of isolated perfused rat kidneys is measured upon switching from an N_2-saturated perfusate to an O_2-saturated perfusate. In isolated kidneys perfused with deoxygenated perfusate the mitochondrial NADH pool is reduced and a relatively high NADH fluorescence is observed. Upon reoxygenation with O_2-saturated perfusate, oxidative phosphorylation results in NADH oxidation, leading to a decrease in NADH-fluorescence intensity. It is clear from Fig. 33.7, that the decrease in NADH-fluorescence intensity in kidneys kept after excision in a preservation medium at 37°C for 1 h, prior to NADH-fluorescence measurements, occurs at a much slower rate than in fresh kidneys that did not undergo this period of warm ischaemia before the measurements. This was found consistently in experiments with over 30 kidneys (Coremans *et al.*, 1998). The explanation for this difference may be impaired oxidative phosphorylation due to ischaemic damage to mitochondria (Mittnacht & Farber, 1981; Saris & Eriksson, 1995; Ferrari, 1996) or to one or more complexes (in particular complex I) of the respiratory chain (Rouslin, 1983; Rouslin & Ranagathan, 1983; Hardy *et al.*, 1991; Thorniley *et al.*, 1994; Gonzalez-Flecha & Boveris, 1995; Cairns *et al.*, 1997; Steinlechner-Maran *et al.*, 1997). Another explanation may be impaired tissue perfusion at the microcirculatory level in kidneys that have suffered ischaemic damage (Lennon *et al.*, 1991). However, overall perfusate flow was equal in kidneys with and without ischaemic damage during the NADH-fluorescence experiments. It is being investigated if, on the basis of this difference in NADH-fluorescence intensity kinetics, a prediction can be made of post-transplantation kidney functioning.

Another field of research in which NADH fluorometry could play an important role, and which is of great clinical significance, is monitoring of the events that take place during septic shock. This complication, which is the most important cause of late death in intensive care units, and which is caused by the presence of bacterial products and inflammatory mediators in the blood, is characterized by a large drop in blood pressure and disturbance of the normal blood circulation. Compensatory mechanisms are activated to ensure adequate oxygen supply to the brain, heart and kidneys. This leads to insufficient oxygenation and consequent dysfunction of peripheral organs (Deitch, 1992). The most vulnerable of these peripheral organ systems is the splanchnic region, which is usually the first to become ischaemic (Noldge-Schomburg *et al.*, 1996). Therefore, an early warning technique for the development of sepsis in patients at risk could be envisioned based on *in vivo* NADH fluorometry. Sepsis can occur following trauma and infection, but also during procedures such as open-heart surgery, where extracorporeal circulation through heart–lung machines causes ischaemia of the gastrointestinal tract and activation of inflammatory processes.

The low blood pressure is a result of the excessive production of nitric oxide, a potent vasodilator, in endothelial and smooth muscle cells. However, as was shown by Avontuur *et al.* (1995) using NADH videofluorometry, the use of inhibitors of inducible NO-synthesis, such as analogues of L-arginine, to restore vascular tone may work counterproductively in terms of tissue oxygenation.

Currently experiments are being carried out aimed at monitoring the progress and treatment of sepsis in a rat model by means of semiquantitative *in vivo* NADH fluorescence imaging. Figure 37.8 shows the results of a pilot experiment, in which the effects of anoxic ventilation on NADH concentration in organs of an artificially ventilated rat were measured. Rats were anaesthetized with N_2O and enflurane. Anoxic ventilation was effected by replacing O_2 by N_2 in the ventilation mixture at time $t = 0$ s. Normal ventilation was restored at time $t = 200$ s. Anoxic ventilation causes progressive deoxygenation of the circulating blood, whereas restoration of normal (oxygen) ventilation leads to progressive reoxygenation of the blood.

Anoxic ventilation results in tissue hypoxia and an increase in [NADH] in all organs (Fig. 33.8(D)). However, after ~120 s of N_2 ventilation, NADH-fluorescence intensity of the liver decreased, whereas it reached a new equilibrium for heart, kidney and gut. The response of the liver to renewed oxygen ventilation (at $t = 200$ s) also differed from that of

(A)

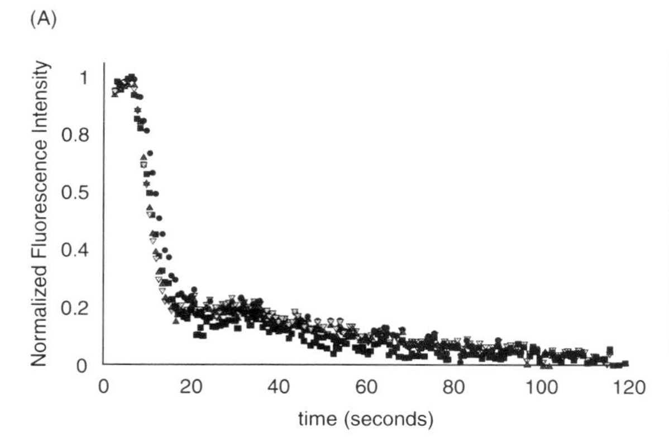

(B)

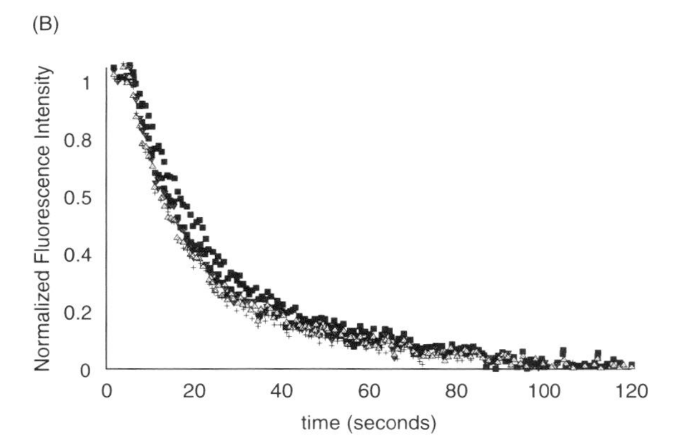

(C)

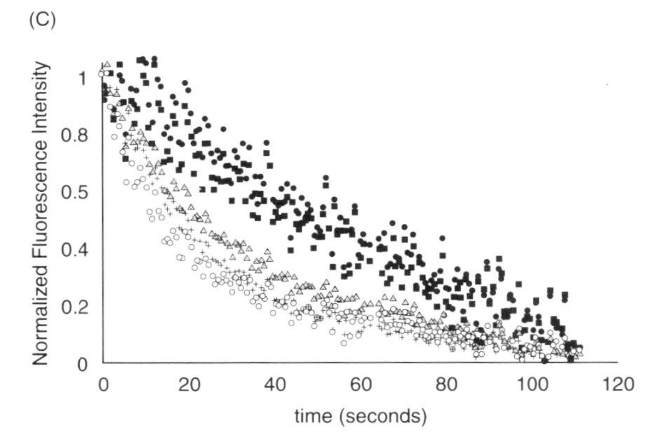

(D)

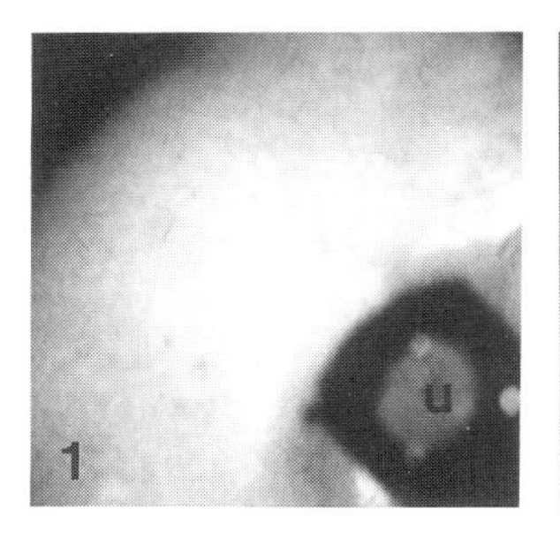

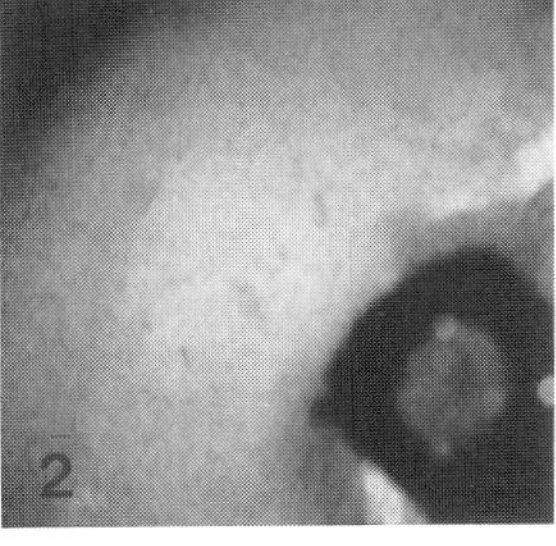

(E)

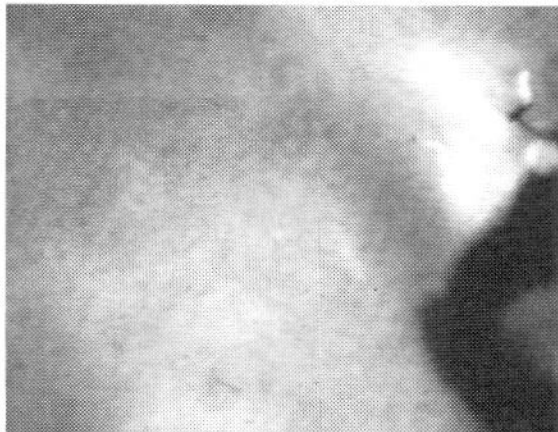

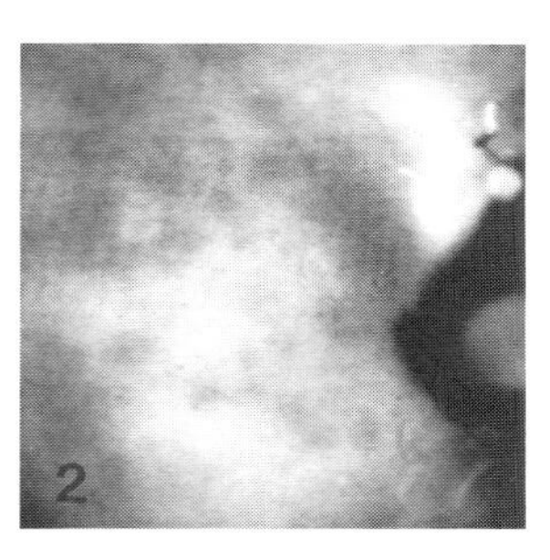

Figure 33.7 Change of NADH-fluorescence intensity of isolated perfused rat (male WagRij) kidneys in response to a change from anoxic perfusion with N_2-saturated perfusate to perfusion with O_2-saturated perfusate. The changes in NADH fluorescence intensity were normalized. Fluorescence intensity during anoxic perfusion was put at 100%. Fluorescence intensity measured 100 s after the start of perfusion with O_2-saturated perfusate was put at 0%.

(A) NADH oxidation kinetics of a freshly isolated kidney. Different symbols correspond to the response measured at different areas on the kidney surface. (B) NADH oxidation kinetics of a kidney kept in preservation medium at 37°C for 1 h, prior to the NADH-fluorescence experiment. Different symbols again correspond to the response measured at different areas on the kidney surface. (C) NADH oxidation kinetics of a kidney kept in preservation medium at 37°C for 1 h, prior to the NADH-fluorescence experiment. A marked inhomogeneous response for different areas on the kidney surface is observed in this case. (D) NADH-fluorescence images of the freshly isolated rat kidney, just before (1) and 20 s after (2) the change from perfusion with N_2-saturated perfusate to perfusion with O_2-saturated perfusate. (E) NADH-fluorescence images of a rat kidney that underwent a 1 h period of warm ischaemia just before (1) and 20 s after (2) the change from perfusion with N_2-saturated perfusate to perfusion with O_2-saturated perfusate. Male WagRij rats were used, weighing between 275 and 325 g. Preservation medium: Na_3-citrate, 78 mM; citrate, 84 mM; $MgSO_4$, 40 mM; mannitol, 100 mM; pH 7.1 (Howden, 1983). Perfusion medium: citrate solution supplemented with ureum, 4.0 mM; glucose, 5.0 mM; $CaCl_2$, 2.0 mM; cysteine, 0.5 mM; glycine, 2.3 mM; Na-pyruvate, 2.0 mM; Na-acetate, 1.22 mM; Na-glutaminate, 2.106 mM; Na-propionate, 0.208 mM; inosine, 1.0 mM, alanine, 5.0 mM; α-ketaglutaric acid, 1.15 mM; ascorbate, 2 mg l^{-1}; Na-lactate, 0.32 mg l^{-1}; cholinechloride, 1.0 g l^{-1}; pH 7.1. Perfusion pressure was 90 mmHg. Perfusion flow during the experiments was 7–10 ml min^{-1}. A uranyl fluorescence calibration glass (U) was used to correct for differences/fluctuations in 365 nm illumination intensity.

other organs. Fluorescence levels for heart, kidney and gut returned to their initial values (similar to the response observed *in vitro*; see e.g. Fig. 33.7), whereas in the liver first an increase in fluorescence intensity was observed, before return to its initial value.

The origin of this response of the liver to hypoxia and reoxygenation, which is consistently observed, is not yet understood. In *in vitro* measurements on

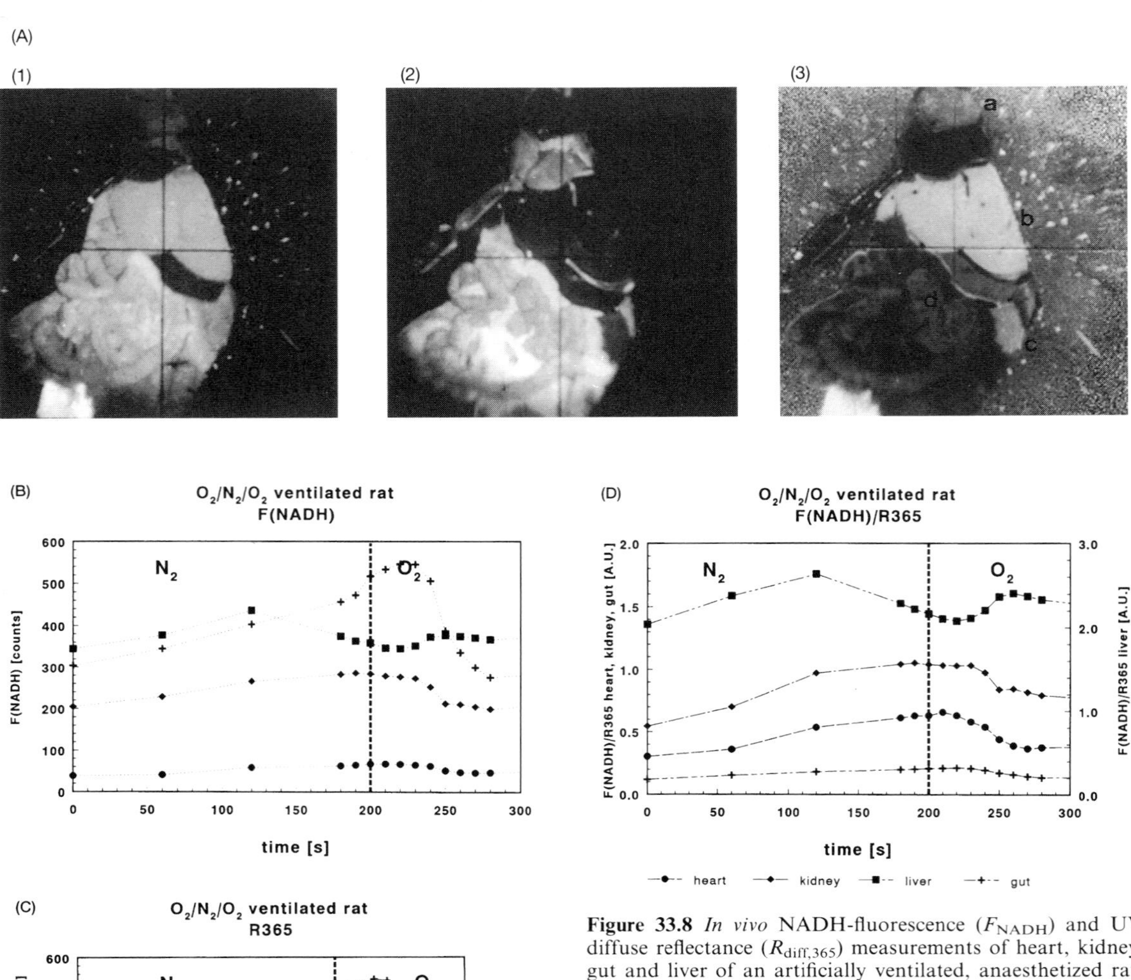

Figure 33.8 *In vivo* NADH-fluorescence (F_{NADH}) and UV diffuse reflectance ($R_{diff,365}$) measurements of heart, kidney, gut and liver of an artificially ventilated, anaesthetized rat. At time $t = 0$ s anoxic ventilation was commenced; O_2 in the ventilation mixture was replaced by N_2. At time $t = 200$ s normal ventilation was restored. (A) *In vivo* images obtained at time $t = 100$ s: (1) F_{NADH}-image (dark current and detector offset subtracted, a logarithmic intensity scale was used in view of the large differences in intensity); (2) $R_{diff,365}$ image (dark current and detector offset subtracted, logarithmic intensity scale); (3) ratio image $F_{NADH}/R_{diff,365}$ (logarithmic intensity scale). (B) Graph showing F_{NADH} intensity of heart, kidney, gut and liver as a function of time. (C) Graph showing $R_{diff,365}$ intensity of heart, kidney, gut and liver as a function of time. (D) Graph showing the $F_{NADH}/R_{diff,365}$ ratio of heart, kidney, gut and liver as a function of time.

isolated liver, an NADH-fluorescence response to anoxic perfusion and reoxygenation similar to that of other organs is found (Okamura *et al.*, 1992). Changes in systemic blood circulation that result from the anoxic ventilation could play a role. A possible explanation is given below for the *in vivo* experimental results found here. It is important to note that the UV diffuse reflectance of the gut (Fig. 33.8(C)) and the $F_{NADH}/R_{diff,365}$ ratio of the liver (Fig 33.8(D)) show opposite behaviour during the experiment. When hypoxia develops, this leads to vasoconstriction in the intestine (Ince *et al.*, 1995). The increase in UV diffuse reflection intensity is evidence of this (Fig. 33.8(C)); the decrease in intestinal blood volume leads to reduced absorption of UV light and therefore increased diffuse reflectance intensity. This means that the amount of blood that the liver receives through the portal vein will be strongly reduced. This is normally (partly) compensated by increased blood supply through the hepatic artery, in which

Po_2 is higher. The fact that UV diffuse reflection intensity of the liver shows very little change during hypoxia and reoxygenation (in contrast with that of gut) is in agreement with this. If the reduced blood supply through the portal vein were not compensated in this way, an increase in UV diffuse reflectance intensity of the liver would have been expected, similar to the decrease observed for intestines. The liver can extract O_2 very efficiently from blood supplied by the hepatic artery, which, moreover, is better oxygenated than blood supplied to the liver by the portal vein, and it is able to maintain a constant O_2 extraction over a wide Po_2 range (Arvidsson *et al.*, 1991; Samsel *et al.*, 1991; Tokuka *et al.*, 1996). In this way, an increase in hepatic arterial blood flow in response to decreased blood flow through the portal vein would improve liver oxygenation and could offer an explanation for the NADH-fluorescence response of liver to decreasing blood oxygenation.

When O_2 ventilation is recommenced, the development of liver NADH-fluorescence intensity is again different from that in other organs. Whereas after about 40 s (t = 240 s) NADH levels in heart, kidney and intestines decrease, the NADH level in liver first increases before returning to its initial value. Blood perfusion of the intestinal tissue is restored, as is clear from the drop in UV diffuse reflectance intensity, starting at t = 240 s. The concomitant decrease in $F_{NADH}/R_{diff,365}$ ratio to its initial value indicates that the intestinal tissue is sufficiently oxygenated. This means that the balance between portal vein and hepatic artery blood perfusion of the liver is again changed, the liver receiving more blood from the portal vein and less from the hepatic artery. Because blood from the hepatic artery is its primary source of oxygen, the liver could as a result temporarily become (slightly) hypoxic, until blood oxygen content has been sufficiently restored. This hypothesis is the subject of further research.

These *in vivo* results on rat liver during anoxic ventilation were not observed in earlier work by Mayevsky & Chance (1982). There an *in vivo* NADH-fluorescence response to anoxia of rat liver was recorded that was similar to that of other organs (brain, kidney, testis), using single point measurements by means of fibreoptic probes. This is most likely due to the fact that in their work anoxic ventilation was terminated after 120 s. Figures 33.8(B)–(D) show that the NADH-fluorescence response of liver differs from that of other organs for anoxic periods longer than 120 s.

33.7 OUTLOOK

It is clear that the (patho)physiological data, which can be obtained by optical methods, could be highly relevant in clinical decision-making. It is expected that the development of methodology and instrumentation for *in vivo* semiquantitative NADH-fluorescence imaging, as described in this chapter, will further the integration of this optical non-invasive technique in clinical practice. It eliminates sampling errors that are inherent to single point measurements made with fibreoptic probes, and it is able to monitor changes in [NADH] quantitatively, without interference from changes in blood content of tissue or changes in redox state of tissue pigments.

ACKNOWLEDGEMENTS

Part of this work was financially supported by The Netherlands Heart Foundation (projects 90.298 and D95.021) and by the Dutch Kidney Foundation (project C95.1433). The help of J. Ashruf MD and M. van Aken in the preparation of the manuscript is gratefully acknowledged.

REFERENCES

Arvidsson D., Rasmussen I., Almqvist P., Niklasson F. & Haglund U. (1991) *Surgery* **109**, 190–197.

Avi-Dor Y., Olson J.M., Doherty M.D. & Kaplan N.O. (1962) *J. Biol. Chem.* **237**, 2377–2383.

Avontuur J.A.M., Bruining H.A. & Ince C. (1995) *Circ. Res.* **76**, 418–425.

Balaban R.S. (1990) *Am. J. Physiol.* **258**, C377–C389.

Balaban R.S. & Mandel L.J. (1980) *J. Physiol. London* **304**, 331–348.

Balaban R.S. & Mandel L.J. (1988) *Am. J. Physiol.* **254**, F407–F416.

Balaban R.S., Soltoff S.P., Storey J.M. & Mandel L.M. (1980) *Am. J. Physiol.* **238**, F50–F59.

Barlow C.H. & Chance B. (1976) *Science* **193**, 909–910.

Bessho M., Tajima T., Hori S., Satoh T., Fukuda K., Kyotani S., Ohnishi Y. & Nakamura Y. (1989) *Anal. Biochem.* **182**, 304–308.

Cairns C.B., Ferroggiaro A.A., Walther J.M., Harken A.H. & Banerjee A. (1997) *Circ.* **96** (9 Suppl.): II–II2605.

Chance B. (1991) *Ann. Rev. Biophys. Chem.* **20**, 1–28.

Chance B., Williamson J.R., Jamieson D. & Schoener B. (1965) *Biochem. Zeitschr.* **341**, 357–377.

Coremans J.M.C.C., Ince C. & Bruining H.A. (1993) In *Medical Optical Tomography: Functional Imaging and Monitoring*, G. Müller, B. Chance, R. Alfano, S. Arridge, J. Beuthan, E. Gratton, M. Kaschke, B. Masters, S. Svanberg & P. van der Zee (eds). SPIE Optical Engineering Press, Washington, Vol. IS 11, pp. 589–617.

Coremans J.M.C.C., Ince C., Bruining H.A. & Puppels G.J. (1997) *Biophys. J.* **72**, 1849–1860.

Coremans J.M.C.C., Bruining H.A. & Puppels G.J. (1998) *Adv. Exp. Med. Biol.*, in press.

Deitch E.A. (1992) *Ann. Surg.* **216**, 117–134.

Duboc D., Toussaint M., Donsez D., Weber S., Guerin F.,

Degeorges M., Renault G., Polianski J. & Pocidalo J.J. (1986) *Lancet* **2**(8505), 522.
Dyess D.L., Bruner B.W., Donnell C.A., Ferrara J.J. & Powell R.W. (1991) *Southern Med. J.* **84**, 966–974.
Eisinger J. & Flores J. (1979) *Anal. Biochem.* **94**, 15–21.
Eng J., Lynch R.M. & Balaban R.S. (1989) *Biophys. J.* **55**, 621–630.
Estabrook R.W. (1962) *Anal. Biochem.* **4**, 231–245.
Ferrari R. (1996) *J. Cardiovasc. Pharmacol.* **28**(Suppl. 1), S1–S10.
Frank K. & Kessler M. (eds) (1992) *Quantitative Spectroscopy in Tissue*. Pmi-Verlagsgruppe, Frankfurt am Main.
Franke H., Barlow C.H. & Chance B. (1976) *Am. J. Physiol.* **231**, 1082–1089.
Gonzalez-Flecha B. & Boveris A. (1995) *Biochim. Biophys. Acta* **1243**, 361–366.
Hardy L., Clark J.B., Darley-Usmar V.M., Smith D.R. & Stone D. (1991) *Biochem. J.* **274**, 133–137.
Heinrich U., Hoffmann J. & Lübbers D.W. (1987) *Pflügers Arch.* **409**, 152–157.
Hoffmann J. & Lübbers D.W. (1986) *Adv. Exp. Med. Biol.* **200**, 125–130.
Hoffmann J., Wodick R., Hannebauer F. & Lübbers D.W. (1983) *Adv. Exp. Med. Biol.* **169**, 831–839.
Hoffmann J., Heinrich U., Ahmad H.R. & Lübbers D.W. (1984) *Adv. Exp. Med. Biol.* **180**, 555–563.
Horvath K.A., Schomacker K.T., Lee C.C. & Cohn L.H. (1994) *J. Thorac. Cardiovasc. Surg.* **107**, 220–225.
Ince C., Ashruf J.F., Avontuur J.A.M., Wieringa P.A., Spaan J.A.E. & Bruining H.A. (1993) *Am. J. Physiol.* **264**, H294–H301.
Ince C., Van der Sluijs J.P., Sinaasappel M., Avontuur J.A.M., Coremans J.M.C.C. & Bruining H.A. (1995) *Adv. Exp. Med. Biol.* **361**, 105–110.
Ishimaru A. (1978) *Wave Propagation and Scattering in Random Media*, Vol. 1. Academic Press, New York.
Jöbsis F.F. (1972) *Fed. Proc.* **31**, 1404–1413.
Jöbsis F.F. & Duffield J.C. (1967) *J. Gen. Physiol.* **50**, 1009–1047.
Kanaide H., Yoshimura R., Makino N. & Nakamura M. (1982) *Am. J. Physiol.* **242**, H980–H989.
Katz A., Edlund A. & Sahlin K. (1987) *Acta Physiol. Scand.* **130**, 193–200.
Kessler M. & Frank K. (1992) In *Quantitative Spectroscopy in Tissue*, K. Frank & M. Kessler (eds). PMI Verlag, Frankfurt am Main, Germany, pp. 61–74.
Klingenberg M., Slenczka W. & Ritt E. (1959) *Biochem. Zeitschr.* **332**, 47–66.
Kubelka P. (1948) *J. Opt. Soc. Am.* **38**, 448–457.
Kubelka P. (1954) *J. Opt. Soc. Am.* **44**, 330–335.
Lennon G.M., Ryan P.C., Gaffney E.F. & Fitzpatrick J.M. (1991) *Urol. Res.* **19**, 259–264.
Mayevsky A. & Chance B. (1982) *Science* **217**, 537–540.
Mittnacht Jr S. & Farber J.L. (1981) *J. Biol. Chem.* **256**, 3199–3206.
Noldge-Schomburg G.F., Priebe H.J., Armbruster K., Pannen B., Haberstroh J. & Geiger K. (1996) *Intensive Care Med.* **22**, 795–804.
Nuutinen E.M. (1984) *Basic Res. Cardiol.* **79**, 49–58.
O'Conner, M.J. (1977) In *Oxygen and Physiological Function*, F.F. Jöbsis (ed). Professional Information Library, Dallas, pp. 90–99.
Okamura R., Tanaka A., Uyama S. & Ozawa K. (1992) *Transplant Int.* **5**, 165–169.
Page E., McCallister L.P. & Fower B. (1971) *Proc. Natl Acad. Sci. USA* **68**, 1465–1466.
Pfaller W. & Rittner M. (1980) *Int. J. Biochem.* **12**, 17–22.
Pierie J.P.E.N., De Graaf P.W., Poen H., Van der Tweel I. & Obertop H. (1994) *Eur. J. Surg.* **160**, 599–603.
Pittman R.N. (1986) *Ann. Biomed. Engineer.* **14**, 119–137.
Rouslin W. (1983) *Am. J. Physiol.* **244**, H743–H748.
Rouslin W. & Ranagathan S. (1983) *J. Mol. Cell. Cardiol.* **15**, 537–542.
Samsel R.W., Cherqui D., Pietrabissa A., Sanders W.M., Roncella M., Emond J.C. & Schumacker P.T. (1991) *J. Appl. Physiol.* **70**, 186–193.
Saris N.E.L. & Eriksson O. (1995) *Acta Anaesthesiol. Scand.* **39**(Suppl 107), 171–176.
Shonat R.D. & Johnson P.C. (1997) *Am. J. Physiol.* **272**, H2233–H2244.
Sinaasappel M. & Ince C. (1996) *J. Appl. Physiol.* **81**, 2297–2303.
Steenbergen C., de Leeuw G., Barlow C., Chance B. & Williamson J.R. (1977) *Circ. Res.* **41**, 606–615.
Steinlechner-Maran R., Eberl T., Kunc M. Schöcksnadel H., Margreiter R. & Gnaiger E. (1997) *Transplantation* **63**, 136–142.
Svaasand L.O. & Ellingsen R. (1983) *Photochem. Photobiol.* **38**, 293–299.
Tardiff J.C. & Juneau M. (1996) *Can. J. Anaesth.* **43**, 989–994.
Thorniley M.S., Lane N.J., Manek S. & Green C.J. (1994) *Kidney Int.* **45**, 1489–1496.
Thorniley M.S., Simpkin S., Fuller B., Jenabzadeh M.Z. & Green C.J. (1995) *Hepatol.* **21**, 1602–1609.
Tokuka A, Kitai T., Tanaka A., Yanabu N., Sato B., Mori S., Inomoto T., Shinohara H., Yamaoka Y., Uemoto S., Tanaka K., Someda H., Fulimoto M. & Moriyasu F. (1996) *Hepato-Gastroenterol.* **43**, 1203–1211.
Torres del Filho I.P., Kerger H. & Intaglietta M. (1996) *Microvasc. Res.* **51**, 202–212.
Williamson D.H., Lund P. & Krebs H.A. (1967) *Biochem. J.* **103**, 514–527.
Wilson D.F., Gomi S., Pastuszko A. & Greenberg J.H. (1993) *J. Appl. Physiol.* **74**, 580–589.

CHAPTER THIRTY-FOUR

Fluorescence Lifetime Imaging Microscopy (FLIM): Instrumentation and Applications

THEODORUS W.J. GADELLA Jr
MicroSpectroscopy Center, Department of Biomolecular Sciences, Wageningen Agricultural University, Wageningen, The Netherlands

34.1 INTRODUCTION

Fluorescence lifetime imaging microscopy (FLIM) is a technique for imaging excited state lifetimes of luminophores by means of a fluorescence microscope. Lifetime images produced by FLIM are two- or three-dimensional digital images in which each pixel or voxel value represents the fluorescence lifetime (τ). As opposed to conventional steady-state fluorescence microscopy, FLIM images report on a kinetic parameter (τ) which, unlike fluorescence intensity, is independent of probe concentration or excitation light intensity.

Under non-saturating (i.e. normal) excitation light intensity conditions, steady-state fluorescence intensities are determined by the product of (a) excitation light intensity, (b) the absorption cross-section of the fluorophore, (c) the local concentration of the fluorophore, (d) the fluorescence quantum yield Q ($0 < Q < 1$) of the fluorophore (defined as the number of photons emitted by the fluorophore per absorbed photon), and (e) the detection efficiency of the microscope and (imaging) detector (Jovin & Arndt-Jovin, 1989b; Jovin *et al.*, 1990). Only (c) and (d) cannot be independently determined in a steady-state fluorescence microscope. For quantitative fluorescence microscopy, aiming at imaging probe concentration rather than intensity, this is a major problem. Since the fluorescence quantum yield (Q) is linearly proportional to the fluorescence lifetime (τ), FLIM enables independent estimation of Q, providing true quantitative microscopic capability. More importantly, τ (or Q) is greatly dependent on the direct molecular environment of the fluorophore. Hence, by measuring fluorescence lifetimes, detailed information on this microenvironment can be obtained. This property has been exploited extensively in time-resolved fluorescence spectroscopy (Bastiaens & Visser, 1993; Jameson & Reinhart, 1989; Lakowicz, 1991). By extracting such information from intact living cells with the spatial resolution achievable with optical microscopes, FLIM enables high-resolution imaging of molecular properties *in situ*, and thereby opens a new and very exciting area in fluorescence microscopy.

FLUORESCENT AND LUMINESCENT PROBES, 2ND EDN
ISBN 0-12-447836-0

34.2 EXPERIMENTAL REALIZATIONS OF LIFETIME-RESOLVED IMAGING MICROSCOPY

As for time-resolved fluorescence spectroscopy, FLIM has been realized following different strategies. The two most widely used modes of implementation are the time-domain and frequency-domain approach (Anderson, 1991). Single-point time-resolved fluorescence measurements in a microscope have been performed since 1958 (Venetta, 1959). Roughly 30 years later, the first imaging applications for lifetime-resolved microscopy have been described. The microscopy systems were (partly independently) developed in different laboratories following the time-domain (Cubeddu *et al.*, 1991; Kohl *et al.*, 1993; Minami & Hirayama, 1990; Ni & Melton, 1991; Oida *et al.*, 1993; Schneckenburger *et al.*, 1993; Wang *et al.*, 1991) or the frequency-domain approach (Clegg *et al.*, 1990, 1992; Gratton *et al.*, 1990; Lakowicz & Berndt, 1991; Morgan *et al.*, 1990; Piston *et al.*, 1992; Wang *et al.*, 1989). Confocal systems for both approaches to lifetime imaging have also been developed (Buurman *et al.*, 1992; Morgan *et al.*, 1992). More recently, instruments have been described utilizing two-photon (Piston *et al.*, 1992; So *et al.*, 1995; Sytsma *et al.*, 1998) or pump-probe (Buist *et al.*, 1997; Dong *et al.*, 1995) laser systems with increased time and spatial resolution. In the next sections the principles of these different types of implementation are described.

34.2.1 Time-domain FLIM

34.2.1.1 Principle of operation

In the time-domain, a short pulse (generally of ps–ns duration) of excitation light is used to excite the sample, and the exponentially decaying fluorescence emission is recorded. For detection, usually two time windows with different delays with respect to the excitation pulse are used (see Fig. 34.1(A)).

If a pulse with a time-dependent profile $E(t)$ excites fluorophores with multiexponential decay (with J components, with lifetimes τ_j and pre-exponential factors of α_j), the time-dependent fluorescence $F(t)$ is described as a convolution of the multiexponential decay and the (experimental, instrument-dependent) pulse profile (equation [34.13]):

$$F(t) = \int_{T=0}^{p} E(T) \sum_{j=1}^{J} \alpha_j \exp\left(-(t - T)/\tau_j\right) dT \qquad [34.1]$$

For a single exponential decay with lifetime τ_s this equation simplifies to:

$$F^s\,(t > p) = \alpha'_p \exp(-t/\tau_s) \qquad [34.2]$$

The pre-exponential factor α'_p is a function of the excitation profile. If the fluorescence is detected by applying a time gate Δt_j, then the detected signal D_j is an integration over time of the fluorescence intensity multiplied by the detector gain $(D(t))$. Since each detector has a certain rise and fall time, this integral can be considered as the sum of three integrals corresponding to the 'rise' time (δ_R), the 'on' time (Δt) and the 'fall' time (δ_F) of the detector according to equation [34.3].

$$D_j = \int_{t_j}^{t_j+\Delta t_j} f(t)D(t)\mathrm{d}t \approx \int_{t_j-\delta_\mathrm{R}}^{t_j} f(t)D_\mathrm{R}(t)\mathrm{d}t + D_\mathrm{on}\int_{t_j}^{t_j+\Delta t_j} f(t)\mathrm{d}t + \int_{t_j+\Delta t_j}^{t_j+\Delta t_j+\delta_\mathrm{F}} f(t)D_\mathrm{F}(t)\mathrm{d}t \qquad [34.3]$$

This technique has been implemented in non-confocal (Wang *et al.*, 1992b; Oida *et al.*, 1993; Ni & Melton, 1996) and confocal (Buurman *et al.*, 1992) microscopes. In the non-confocal implementation, the detector is a fast gateable image intensifier coupled to a CCD. The gate on the image intensifier is applied on the photocathode (Wang *et al.*, 1992b). In the confocal implementation the laser beam is focused in a single point in the specimen and spatial information is obtained by scanning the beam. The detector is a gateable (microchannelplate-)photomultiplier (Buurman *et al.*, 1992). For image-intensifier-based lifetime imaging set-ups, the rise and fall times are about 2–9 ns and hence not negligible since they are comparable to prompt fluorescence lifetimes of most fluorophores (Wang *et al.*, 1992a; Oida *et al.*, 1993; Periasamy *et al.*, 1996). For single-point detection, as for confocal FLIM applications, (microchannelplate) photomultipliers can be used with typical fall and rise times 0.1–0.8 ns (Buurman *et al.*, 1992; Wang *et al.*, 1992a).

34.2.1.2 Data handling and calculation of fluorescence lifetimes

For a single exponentially decaying compound, one can extract the fluorescence lifetimes by acquiring two images at two delay times with respect to the excitation pulse, provided that the integration times (Δt) are identical, and the fall and rise characteristics of the detector are identical at both delay times. The detected intensities for a single exponentially decaying compound under these conditions at delay times $t = t_1$ (D_1^s) and $t = t_2$ (D_2^s) are given in equation [38.4]:

$$\left.\begin{aligned} D_1^s &= \alpha' \int_{t_1-\delta_R}^{t_1} [\exp(-t/\tau_s) D_R(t)]\mathrm{d}t \\ &+ \alpha' \int_{t_1}^{t_1+\delta_F} [\exp(-t/\tau_s) D_F(t)]\mathrm{d}t + \alpha' D_{on} \\ &\int_{t_1}^{t_1+\Delta t_1} \exp(-t/\tau_s)\mathrm{d}t \\ D_2^s(t_2 &= t_1 + \Delta T;\ D_{R,1} = D_{R,2};\ D_{F,1} \\ &= D_{F,2};\ \Delta t_1 = \Delta t_2) = D_1^s \exp(-\Delta T/\tau_s) \end{aligned}\right\} \quad [34.4]$$

Under these conditions the lifetime of a single exponential decaying compound (τ_s) can be found by simply ratioing the detected integrated intensities D_1^s and D_2^s according to equation [34.5]:

$$\tau_s = \frac{t_2 - t_1}{\ln(D_1{}^s/D_2{}^s)} \quad [34.5]$$

For multiexponential decays (as generally present in biological samples), deconvolution over time of excitation pulses and detector characteristics is required to obtain accurate lifetimes. Although according to equation [34.5] an average lifetime can be simply determined, this average lifetime is very much dependent on instrumental characteristics such as the excitation pulse shape ($E(t)$), the time window settings (t_1, t_2), and the detector response characteristics ($D_R(t)$, $D_F(t)$). Generally, these parameters are different for each instrument, which may complicate quantitative interpretation.

34.2.2 Frequency-domain FLIM

34.2.2.1 Principle of operation

For frequency-domain FLIM, the excitation light intensity is modulated at radiofrequencies (RF), typically in the 10–100 MHz region. Due to the delay between excitation and emission, directly related to the fluorescence lifetime of a fluorophore, the fluorescence emission is demodulated and phase shifted with respect to the excitation light (see Fig. 34.1(B)). For excitation, any repetitive signal (sinusoidal, block shaped or pulse shaped) can be used. For reasons of simplicity, let us assume a pure sinusoidal signal with modulation depth M_E (= AC/DC) at an angular frequency ω (= $2\pi f$), and average (DC) intensity E_0:

$$E(t) = E_0[1 + M_E \sin(\omega t + \varphi_E)] \quad [34.6]$$

Then the fluorescence emission ($F(t)$) will also be sinusoidally modulated at the same frequency, demodulated by a factor M ($0 < M < 1$), and phase shifted by $\Delta\varphi$ radians ($0 < \Delta\varphi < \pi/2$) with respect to the excitation light, depending on the fluorescence lifetime:

$$\begin{aligned} F(t) &= F_0[1 + M_F \sin(\omega t + \varphi_F)] \\ &= F_0[1 + M M_E \sin(\omega t + \varphi_E + \Delta\varphi)] \end{aligned} \quad [34.7]$$

The phase shift $\Delta\varphi$ and the demodulation M are directly related to the lifetime according to:

$$\Delta\varphi = \tan^{-1} \omega\tau_\varphi = \tan^{-1} \frac{\sum_j^J \frac{a_j \omega \tau_j^2}{1 + (\omega\tau_j)^2}}{\sum_j^J \frac{\alpha_j \tau_j}{1 + (\omega\tau_j)^2}} \quad [34.8]$$

$$\begin{aligned} M &= \sqrt{\frac{1}{\omega^2 \tau_M^2 + 1}} \\ &= \sqrt{\frac{\left(\sum_j^J \frac{\omega a_j \tau_j^2}{1 + (\omega\tau_j)^2}\right)^2 + \left(\sum_j^J \frac{a_j \tau_j}{1 + (\omega\tau_j)^2}\right)^2}{\left(\sum_j^J a_j \tau_t\right)^2}} \end{aligned} \quad [34.9]$$

In equation [34.8], a_j is the pre-exponential factor belonging to the jth decaying species with lifetime τ_j. τ_φ and τ_M are defined as the lifetime determined from the phase and from the modulation, which in the case of a single exponential decay ($J = 1$) are identical. For a general theoretical description of phase fluorometry from which the equations are derived, the reader is referred to Spencer & Weber (1969), Weber (1981), and Jameson *et al.* (1984).

The intensity-modulated fluorescence, $F(t)$, is detected by a detector that is gain-modulated at frequency ω', which can be identical to ω (homodyne technique) or slightly different ($\omega' = \omega + \Delta\omega$, heterodyne technique) according to equation [34.10].

$$A(t) = F_0[1 + M_A \sin(\omega' t + \varphi_A)] \quad [34.10]$$

As a consequence, the detected signal ($D(t) = A(t) \times F(t)$) contains a cross-correlation signal at this low

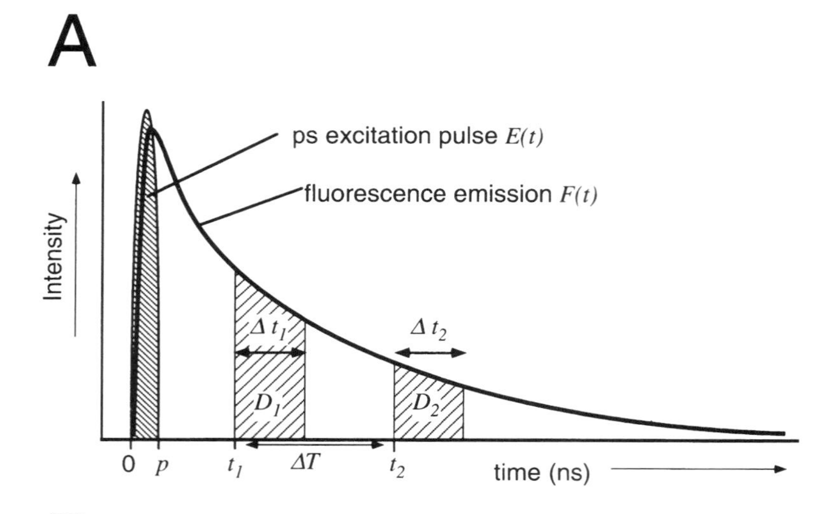

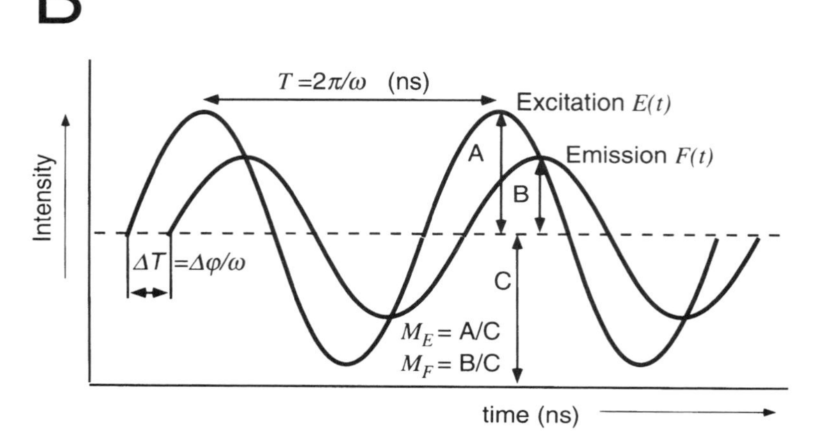

Figure 34.1 Principle of time-domain (A) and frequency-domain (B) fluorescence lifetime estimation. For further explanation see Sections 34.2.1.1 and 34.2.2.1.

frequency ($\Delta\omega$). At typical RF modulation frequencies (e.g. f = 40 MHz), the higher-frequency components in the detected signal (at ω and 2ω) are completely averaged out after integrating for a few µs. In this situation, only the cross-correlation signal as given in equation [34.11] is detected. It is of note that ($\Delta\omega = 0$) in the case of homodyne detection.

$$D(t) = D_0[1 + M_D \cos(\Delta\omega t + \varphi_D)] \qquad [34.11]$$

with

$$D_0 = F_0 A_0; \quad M_D = {}^1\!/_2\, M_A M_E M; \quad \varphi_D = \varphi_A - \varphi_E - \Delta\varphi$$

It is apparent that the high-frequency phase ($\Delta\varphi$) and modulation (M) characteristics, carrying all the lifetime information, are retained in the cross-correlation signal.

As for time-domain FLIM, the technique has been implemented on confocal (Morgan *et al.*, 1992) and wide-field microscopy systems (Lakowicz & Berndt, 1991; Clegg *et al.*, 1992; Gadella Jr. *et al.*, 1993, 1997; So *et al.*, 1994). For wide field systems, the detector is a gain-modulated image intensifier coupled to CCD camera (Gadella Jr *et al.*, 1997), and for confocal systems modulated PMTs are used.

For heterodyne detection, images are acquired by integrating $D(t)$ at several time points, phase-locked to the low-frequency cross-correlation signal. The phase locking can be achieved by coupling video cameras to an RF-modulated image intensifier, operating at video frame rates of $K \times \Delta f$ Hz ($\Delta f = \Delta\omega/2\pi$, $K \geq 4$). Hence K time images are acquired, which can be stored in K frame buffers and integrated over several $\Delta\omega$ periods (So *et al.*, 1994). Alternatively, simultaneous RF modulation of the microchannelplate (MCP) of an image intensifier (at ω') combined with boxcar gating on the photocathode at a frequency of $\Delta\omega$ also provides a heterodyne detection scheme (Gadella Jr *et al.*, 1993; Clegg *et al.*, 1994).

In the case of homodyne detection, a steady state signal is integrated for the desired time. Several phase images are acquired by varying the phase of the detector gain-modulation (φ_A), so that one high-frequency (ω) period can be reconstructed (Gadella Jr *et al.*, 1994, 1997; Szmacinski *et al.*, 1994). Although the theory for homodyne or heterodyne detection is equivalent, there may be some practical advantages for choosing either one of the two detection schemes, depending on the available types of (imaging) detectors and RF signal generators (Gadella Jr *et al.*, 1994).

For frequency-domain FLIM, a reference measurement has to be made to correct for the absolute detector and excitation modulation and phase settings. If one measures $D(t)$ for scattered light (i.e. $\tau_R = 0$; $\Delta\varphi = 0$; $M = 1$), then the detected modulation M_D is identical to the reference modulation M_R and the detected phase φ_D is identical to the reference phase φ_R according to equation [34.12]:

$$M_R = \tfrac{1}{2} M_A M_E; \quad \varphi_R = \varphi_A - \varphi_E \qquad [34.12]$$

For a reference measurement, a homogeneous fluorophore solution with a single exponential decay with known lifetime can also be used. In this case, the detected reference phase and modulation are corrected for this lifetime (Gadella Jr *et al.*, 1994).

For a true (biological) sample, the parameters M and $\Delta\varphi$, are obtained by combining equations [34.12] and [34.11]:

$$M = M_D / M_R; \quad \Delta\varphi = \varphi_D - \varphi_R \qquad [34.13]$$

34.2.2.2 Data handling and calculation of fluorescence lifetimes

The easiest way of data handling is by Fourier analysis of the acquired time or phase images. If one considers K time images spaced by $1/(K \times \Delta f)$ s (in case of heterodyne detection), or K phase images spaced by $2\pi/K$ radians (in case of homodyne detection), then three images F_{sin}, F_{cos} and F_{DC} can be reconstructed from these time/phase images:

$$\left.\begin{aligned} F_{sin} &= \left(\frac{2}{K}\right) \sum_{k=0}^{K-1} \sin\left(\frac{2\pi k}{K} - \varphi_R\right) . D_k \\ &= D_0 M_D \sin\Delta\varphi \\ F_{cos} &= \left(\frac{2}{K}\right) \sum_{k=0}^{K-1} \cos\left(\frac{2\pi k}{K} - \varphi_R\right) . D_k \\ &= D_0 M_D \cos\Delta\varphi \\ F_{DC} &= \left(\frac{1}{K}\right) \sum_{k=0}^{K-1} D_k = D_0 \end{aligned}\right\} \qquad [34.14]$$

D_k stands for the image corresponding to the time point for which $t_k = k/(K \times \Delta f)$ (heterodyne) or the phase image obtained at the detector phase setting of $\varphi_{A,k} = \varphi_A + 2\pi k/K$ (homodyne) in equation [34.11]. From the defined images the phase image ($\Delta\varphi$) and modulation image (M) (equations [34.8] and [34.9]) can be easily calculated according to Clegg *et al.* (1994):

$$\left.\begin{aligned} \Delta\varphi &= \tan^{-1}(F_{sin}/F_{cos}) \\ M &= \sqrt{(F_{sin}^2 + F_{cos}^2)}/M_R F_{DC} \end{aligned}\right\} \qquad [34.15]$$

In the case of a reference measurement, the reference phase (φ_R) and modulation (M_R) (equation [38.12]) can be found by substituting φ_R for 0, $\Delta\varphi$ for φ_R, M_D for M_R, M for M_R, and M_R for 1 in equations [34.14] and [34.15]. One reference measurement can be used for the analysis of multiple FLIM experiments.

Combining equations [34.8], [34.9] and [34.15] yields:

$$\left.\begin{aligned} \tau_\varphi &= (F_{sin} / F_{cos})/\omega \\ \tau_M &= \frac{1}{\omega}\sqrt{\frac{M_R^2 F^2{}_{DC}}{F_{sin}^2 + F_{cos}^2} - 1} \end{aligned}\right\} \qquad [34.16]$$

Even in the case of non-sinusoidal excitation or detector gain modulation, equations [34.6]–[34.16] are applicable. The equations are then referring to the fundamental component of the Fourier series expansion of the actual signals. In this case, however, it is important to sample the fundamental sine wave with enough phase (in the case of homodyne) or time (in the case of heterodyne) points D_k in order to prevent aliasing effects due to unresolved higher-frequency components (Hamming, 1973). For more details and the theoretical basis of the above statement, the reader is referred to Clegg & Schneider (1996).

Apart from its simplicity, the Fourier analysis has several specific advantages. Firstly, irrespective of how many phase or time images are acquired, the actual lifetime image calculation can be done on three images (equation [34.16]). Moreover, after the Fourier summations, the τ_φ image calculation is a simple ratioing procedure requiring only two images.

Secondly, phase and modulation images at the n times higher harmonic frequencies ($n \times \omega$) (if present in the detected signal $D(t)$), can be obtained by the same method, simply by substituting k for $n \times k$ in equation [34.14] and ω for $n \times \omega$ in equation [34.16]. In this way, at each harmonic frequency, lifetime images are available, which can be used for fitting to equations [34.8] and [34.9] to determine multiple lifetime components (Weber, 1981; Gratton *et al.*, 1984).

Thirdly, the $F_{\sin}$, $F_{\cos}$ and F_{DC} images defined in equation [34.14] are linear combinations of digitized fluorescence intensities with positive values (since $0 < \Delta\varphi < \pi/2$). This enables image deconvolution for removing out-of-focus fluorescence from *z*-sections of phase or time point images. Deconvolution methodology has been used extensively to deblur *z*-stacks of steady state confocal or wide-field microscopic images (Shaw, 1995). Such methodology is also applicable to *z*-stacks of $F_{\sin}$, $F_{\cos}$ and F_{DC} images, produced by either confocal or non-confocal FLIM systems.

34.2.3 Comparison between time- and frequency-domain FLIM

For FLIM, there are several clear advantages of the frequency-domain approach. It is important to note that the only instrumental parameters in equation [34.16] are the frequency and depth of modulation. These parameters can be exactly determined and compared for different FLIM systems. Irrespective of how complex the actual fluorescence decay in the time-domain will be, one always can define a lifetime from the phase and modulation and theoretically relate them to all the lifetime components (equations [34.8] and [34.9]). Furthermore, a multiexponential case can be very easily recognized since in this case $\tau_{\varphi} < \tau_{M}$. By phase suppression techniques, fluorophores with a certain lifetime can be selectively made invisible in an image whereas fluorophores with a longer or shorter lifetime remain visible (Lakowicz & Berndt, 1991; Clegg *et al.*, 1994). The technique can be very sensitive for small lifetime differences, and in principle it is easier to study rapidly decaying compounds. On the other hand, the time-domain approach has advantages in the case of existence of a long lived component. For this reason, the time-domain approach is the method of choice for recording delayed fluorescence or phosphorescence (see Section 34.5.5). Furthermore, the time-domain approach is conceptually easier to understand and requires only two images, whereas the frequency-domain method requires at least three or more images (D_k) in the case of absence of pure sinusoidal excitation or detector gain modulation.

34.2.4 Two-photon FLIM

The two-photon excitation technique can be conveniently combined with FLIM. Two-photon excitation occurs by simultaneous absorption of two photons, each contributing to half of the energy difference between the ground and excited states of a fluorophore (Denk *et al.*, 1990). Hence one can excite UV dyes (e.g. absorption band at 350 nm) by using a two-photon laser source emitting at 700 nm. There are several advantages of two-photon excitation. The excitation volume is very small since the two-photon cross-section depends on the squared spatial distribution of the excitation light. As a result, fluorophores are only excited within this very small excitation volume, and hence no out-of-focus fluorescence is induced. For this reason, two-photon microscopy does not require spatial filtering of the emitted light by a confocal pinhole as for conventional confocal microscopy. As a consequence, the detection efficiency is increased, since scattered fluorescence emission can also be detected, and no out-of-focus bleaching occurs. Because of the lower optical density of biological specimen at the near-IR, two-photon microscopy enables imaging of thick samples (Nakamura, 1993). For FLIM an attractive feature of two-photon laser sources is that they typically are pulsed at RF frequencies. This is an optimal profile for performing frequency-domain FLIM. It is for this reason that the first two-photon FLIM systems were implemented using the frequency method (Piston *et al.*, 1992; So *et al.*, 1995). Recently, a time-gating two-photon FLIM system has been constructed (Sytsma *et al.*, 1998). To enable optimal lifetime performance in the latter system, the high pulse rate of the two-photon source was reduced to < 8 MHz by a pulse picker (Sytsma *et al.*, 1998).

34.2.5 Pump-probe approach

For studying ultrafast kinetics (fs–ps scale), pump-probe techniques have been very useful in biological spectroscopy, e.g. for studying rhodopsin or photosynthetic reaction centres (Hochstrasser & Johnson, 1988). Generally, for pump-probe spectroscopy, a high-power pulsed laser source is beam-split and recombined in the sample by using optical paths of slightly different lengths. In this way an optical delay of the two pulses can be achieved of about 3 ps for each mm of path length difference (given the light speed of 3×10^{8} m s^{-1}). The first pulse (pump) should be sufficiently strong to deplete the ground state population of the fluorophores. The absorption of, or the additional fluorescence induced by the second pulse (probe) is then related to the number of fluorophores that returned to the ground state in the time period between the pulses. By varying the path lengths, a complete decay curve with very high time resolution can be reconstructed, ultimately limited by the pulse width of the laser source.

Following this approach, a FLIM system has been

constructed recently (Buist *et al.*, 1997). Advantages are the high achievable time resolution and the lack of need for a high-speed detector. In its current state of implementation, there are still some specific disadvantages such as high bleaching rates inherent to ground state depletion induced by high-power laser intensities, decreased resolution caused by the concomitant increased off-focal excitation intensity, and relatively low lifetime contrast (<1% at pulse delays >2 ns).

Another, very interesting, and alternative implementation of pump-probe FLIM comes from the group of Gratton. They avoid ground state depletion and bleaching by using a novel stimulated emission or light-quenching approach (Dong *et al.*, 1995). Light-quenching is induced by probing with a laser tuned at the emission wavelength of the fluorophore (Kusba *et al.*, 1994). By using slightly different RF repetition rates for the pump and the probe laser (i.e. asynchronous pump-probe), the decrease in fluorescence emission (caused by light quenching) can be detected as a low-frequency cross-correlation signal, avoiding the need for high-speed detectors. The phase and modulation of the cross-correlation signal are directly related to the fluorescence lifetime. In fact, the theory for lifetime estimation proves to be similar to that described in Section 34.2.2. Another advantage of this approach is the increased spatial resolution as the point spread function (PSF) of the cross-correlation signal is the product of the PSFs of the two pump-probe laser beams (Dong *et al.*, 1995). Practical disadvantages of the approach are the (costly) need for two pulsed laser sources and the strong demands on the emission light filtering in order to avoid detecting the pumping or probing laser sources.

34.2.6 Future developments

As is clear from the above, FLIM can be implemented in many different ways, each with specific advantages and disadvantages. FLIM greatly benefits from a variety of developments in fast detectors, non-linear spectroscopy, new pulsed light sources and improved scanning optics. The use of very bright modulated LEDs, standing wave acousto-optical tunable filters (AOTFs) and modulated CCDs are very promising for FLIM implementations on wide-field microscopes (Morgan *et al.*, 1997). In the future they may provide superior sensitivity at relatively low cost. Also the coupling to image restoration and deconvolution algorithms as described in Section 34.2.2.2 will be of great importance for fast FLIM in three spatial dimensions (Shaw, 1995). Such algorithms are also applicable for time-domain FLIM images as D_1^s and D_2^s in equation [34.5] prior to ratioing.

Unfortunately, the construction of a FLIM instrument requires considerable specific expertise and since no commercial instruments are yet available, application of FLIM is currently restricted to a selective number of research groups. Given (1) the many published applications of FLIM for solving specific scientific problems (see Section 34.5), (2) the expected future increase of such applications, and (3) the increasing interest of large companies producing optical microscopes, it is likely that FLIM will be available commercially within the next couple of years. Commercial instruments with compact and easy-to-use design are technically possible and certainly will greatly accelerate research in many fields.

34.3 IMPLEMENTATION OF A WIDE-FIELD FREQUENCY-DOMAIN FLIM SYSTEM

To give the reader an impression on how a FLIM system can be constructed, the frequency-domain FLIM apparatus in Wageningen is briefly described below. The equipment is described in detail elsewhere (Gadella Jr *et al.*, 1997). Following the light path (see Fig. 34.2), the first component is a laser source (Coherent, Innova spectrum 70 Ar/Kr mixed gas laser) producing a single (user-selectable) line which is passed through an acousto-optic modulator (AOM). The AOM modulates the laser light intensity in the MHz frequency range (10–150 MHz). One of the diffraction spots produced by the AOM is selected by a series of diaphragms. A rotating light-scrambling device ensures that laser speckles (interference patterns) are completely averaged out and produces an expanded laser beam which is incorporated into a Leica DMR/BE epifluorescence microscope. Inside the microscope, the intensity and field width can be adjusted. A dichroic mirror reflects the excitation light onto the sample. The filtered fluorescence emission is focused onto the photocathode of an RF frequency gain-modulated dual-stage MCP image intensifier (Hamamatsu, model C5825). The light emitted from the phosphor screen of the image intensifier is focused onto the chip of the CCD camera (Photometrics, model CH250 with an SI 502/AB grade 1 thinned and back-illuminated CCD sensor, Tucson, AZ, USA) by means of a relay lens system. Two computer-controlled frequency synthesizers (Programmed Test Sources model 310, Littleton, MA, USA) drive the modulation on the AOM and on the image intensifier gain, respectively. Phase image acquisition by the CCD, excitation shutter control (TTL) and phase settings on the frequency synthesizers (GPIB) are computer controlled and integrated within the IPLab spectrum image acquisition/processing software

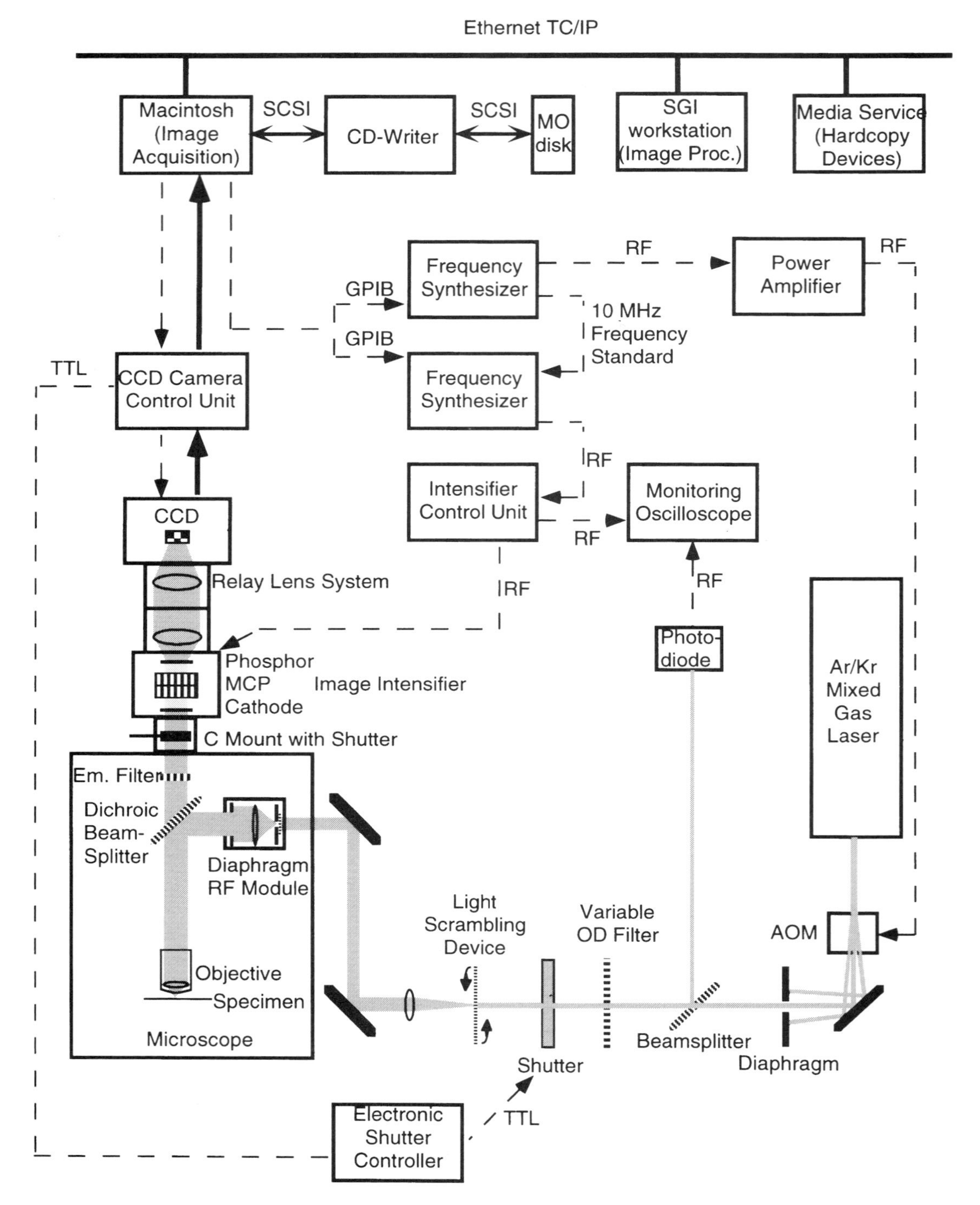

Figure 34.2 Schematic representation of a frequency-domain FLIM system. The dashed arrows indicate control signal flow (e.g. TTL shutter control or RF sinusoidal signals) and the solid arrows indicate image data flow. For more details, see Section 34.3.

package (Signal Analytics, Vienna, VA, USA). Image analysis for constructing lifetime images from the phase images is done on an SGI workstation using home-written software (Gadella Jr. *et al.*, 1994). Raw data are conveniently stored using rewritable magneto-optical (MO) disks as data buffer and recordable compact discs (CD) as final storage medium. Hard copies on paper (Pictography 3000, Fuji, Tokyo) or on 35 mm slides are generated at the MediaSevice Department of our University.

34.4 EXAMPLE OF FLIM MEASUREMENT: GFP IN LIVING CELLS

In Plate 34.1, an example is given of the data that can be generated by a FLIM system. Cowpea leaf protoplasts were transfected with a plasmid encoding for enhanced green fluorescent protein (E-GFP) (S65T) (Pang *et al.*, 1996). As a result, GFP is transiently expressed and accumulates in the cytosol after 24 h. Plate 34.1(D)–(F) represent a protoplast imaged using a bandpass emission filter transmitting only GFP fluorescence. Plate 34.1(D) several structures within the protoplast are visible. As clearly is visible from image (E), the E-GFP fluorescence lifetime is very homogeneous throughout the protoplast. The homogeneity is reflected by the very narrow lifetime distribution as depicted in the temporal image histogram (F), which is centred at a value of 2.67 ns for E-GFP with a very small coefficient of variation (Gadella Jr *et al.*, 1994) of only 3.5%. The small empty structures in Plate 34.1(D) represent the chloroplasts from which the GFP is excluded, and the large empty structure occupying half of the protoplast is the vacuole also excluding GFP. If a long-pass filter is used (images (A)–(C)) the chlorophyll fluorescence that is observed has a much shorter fluorescence lifetime (close to 1 ns). The chloroplasts can be clearly distinguished from the other protoplast compartments by the purple coloured short fluorescence lifetimes (image (B)). It is also clear that E-GFP appears in the protoplast nucleus; this is the bright structure surrounded by chloroplasts in Plate 34.1(A), which has a longer fluorescence lifetime (yellow-orange colour in (B)). The nuclear localization is expected given the size of E-GFP in relation to that of the nuclear pores. For both experiments, the average standard error in the fluorescence lifetime (τ_φ) for each individual pixel was only 0.07 ns, demonstrating that very accurate quantitative parameters can be extracted from living cells expressing GFP by FLIM. This will have major implications for the future use of GFP in various FLIM applications, e.g. for studying molecular proximity (see Sections 34.5.1 and 34.6) or probing of molecular microenvironment (see Section 34.5.2) in single cells.

As can be inferred from this example, lifetime images are always compared with the steady state fluorescence intensity images. Combined they provide detailed information on the localization and photophysical origin of the observed fluorescence.

34.5 APPLICATIONS OF FLIM

The fluorescence lifetime is a fundamental spectroscopic parameter, like wavelength or intensity, and hence lifetime-resolved images can be used for many applications. Below, published applications of FLIM are reviewed.

34.5.1 Studying molecular proximity in single cells by FRET-FLIM

The most powerful application of FLIM is arguably the determination of spatially resolved efficiencies of fluorescence resonance energy transfer (FRET). FRET occurs between a donor fluorescent molecule and an acceptor chromophore (which can also be fluorescent) when: (1) the donor fluorescence spectrum overlaps with the absorption spectrum of the acceptor, (2) the donor emission and acceptor absorption dipole moments are not perpendicularly oriented, and (3) the distance between donor and acceptor is smaller than roughly 1.5 times R_0. R_0 is the Förster distance for FRET, which for common donor-acceptor pairs has values of 3–6 nm (Wu & Brand, 1994). For a detailed theoretical description of FRET the reader is referred to Förster (1949) and Clegg (1996). For reviews of biological applications of FRET the reader is referred to Stryer (1978) and Clegg (1995).

Demonstration of FRET implies that two labelled molecules are not separated by more than say, 9 nm, which generally means that they interact with one another. FRET is manifested by a decrease in the donor fluorescence lifetime, which can be imaged independently of donor concentration (or intensity) by FLIM. In this way FLIM makes it possible to study whether and where molecules do interact, with high specificity and resolution, even in the complex environment of the living cell. FRET-microscopy can also be done by a triple ratio imaging procedure or by studying photobleaching rates (Jovin & Arndt-Jovin, 1989a; Gadella Jr & Jovin, 1995), but the ratioing procedure is very sensitive to errors, and the photobleaching method (despite its sensitivity and accuracy) is inherently destructive and slow. Both these drawbacks limit applicability to living cells.

FRET-FLIM has been applied to the study of membrane fusion in cells by monitoring FRET between NBD- and *N*-lissamine-rhodamine-B-sulfonyl-labelled phosphatidylethanolamines (Oida *et al.*, 1993). Furthermore, proteolytic processing of the signalling enzyme protein kinase C (PKC) into a catalytic and regulatory domain ending up at different locations in the cell has been visualized in single cells with FLIM (Bastiaens & Jovin, 1996). FRET between Cy-3-labelled PKC (regulatory domain) and a Cy-5-labelled specific antibody against the catalytic domain of PKC were used. FRET-FLIM has also been successfully applied for studying oligomerization of the membrane surface receptors for epidermal growth factor (EGF). By simultaneous administration of fluorescein- and rhodamine-labelled EGF to cells, FRET between EGF-receptor dimers could be imaged on single cells depending on the incubation temperature (Gadella Jr & Jovin, 1995). Based on the results an alternative model for tyrosin kinase receptor activation was proposed.

34.5.2 Probing the molecular microenvironment

Fluorescence lifetimes can be very sensitive to the direct chemical environment of the fluorophore. An example is ethidium blue (EB), which drastically increases its fluorescence lifetime upon binding to DNA. This property has been employed for discriminating between DNA-bound and free or membrane-bound EB in a single cell (Dong *et al.*, 1995). Similarly, the binding of fluorescent substrates to enzymes can be studied if binding of the substrate is accompanied by a change in fluorescence lifetime. For instance, the lifetime of the electron carrier NADH increases from 0.5 to 1 ns upon binding to proteins, enabling imaging of the free and protein-bound NADH in cells by FLIM (Lakowicz *et al.*, 1992b). The same strategy can be followed for the visualization of protein conformation or substrate binding by autofluorescent enzymes. Certain redox enzymes contain fluorescent cofactors (e.g. flavin). After binding of the substrate, the protein conformation of these enzymes changes, which can alter the autofluorescence flavin lifetime of the enzyme. This property has been successfully used for imaging the percentage of enzymes carrying a substrate (Gadella Jr, 1997).

34.5.3 Ion and oxygen imaging

Several reports have appeared showing novel ways of detecting ion concentrations by employing fluorescent indicators that change their lifetime rather than their fluorescence spectrum. For conventional ratiometric determination of Ca^{2+} concentrations, using Fura or Indo (Haugland, 1996), UV excitation has to be employed. UV is generally hazardous to living cells, which can limit applicability. UV excitation can be avoided by using non-ratiometric lifetime-based fluorescent Ca^{2+} indicators absorbing at longer wavelengths such as Calcium-Green (blue excitation) (Lakowicz *et al.*, 1992a; Sanders *et al.*, 1994a; So *et al.*, 1995) or Calcium-Crimson (orange excitation) (Periasamy *et al.*, 1996). Lifetime imaging of pH using c-SNAFL-1 has also been reported (Sanders *et al.*, 1994b).

Many other lifetime-based fluorescence indicators that can be ester-loaded into living cells are available, e.g. for detecting Mg^{2+} (Magnesium Green), Na^{+} (Sodium Green), Cl^{-} (*N*-ethoxycarbonylmethyl-6-methoxyquinolinium bromide (MQAE)) and heavy metals (bis-BTC), but still await application in lifetime-resolved imaging (Haugland, 1996; Szmacinski *et al.*, 1994).

By employing the ability of oxygen to decrease fluorescence lifetimes by collisional quenching, FLIM has been successfully applied in imaging oxygen concentrations even in living cells (Gerritsen *et al.*, 1997). The sensitivity for oxygen quenching increases with increased excited state lifetimes. Therefore, long-lived ruthenium dyes such as ruthenium-tris(2,2′-dipyridyl)dichloride-hydrate (τ = 800 ns) (Gerritsen *et al.*, 1997), or Tris(1,10′-phenantroline) ruthenium(II)chloride ($\tau > 1$ μs) (Hartmann *et al.*, 1997) are preferred for oxygen imaging by FLIM.

34.5.4 Quantitative microscopy

Once the fluorescence lifetime (image) is determined, one can correct the fluorescence intensities in a micrograph for local differences in quantum yields. Hence, for studies aiming at imaging of probe distribution rather than luminescence intensity, FLIM is an important tool that provides quantitative imaging. For studying migration of fluorescent glycolipids in plasma membranes of individual sperm cells, FLIM analysis could exclude alternative explanations for observed changes in fluorescence intensity (e.g. caused by quantum-yield changes induced by microaggregation or excimer formation) (Gadella Jr *et al.*, 1993; Gadella *et al.*, 1995).

34.5.5 High contrast generation by delayed fluorescence and phosphorescence imaging

For slowly decaying luminescent probes, the time-domain approach is the method of choice. In biological samples, extremely high contrast can be achieved by monitoring luminescence at a delay time of >1 μs after the excitation pulse, because most autofluorescing molecules have lifetimes <5 ns. Hence, short-lived autofluorescence is completely suppressed in a late time window. Because the timescale is generally several orders of magnitude slower than for prompt fluorescence, much cheaper mechanical time-gating procedures (e.g. by means of light choppers) can be employed (Marriott *et al.*, 1991, 1994a; Beverloo *et al.*, 1992; Seveus *et al.*, 1992). Essential for the method is the availability of specific long lived luminophores. Several probes have been recently developed such as metal–ligand complexes containing ruthenium, osmium or rhenium (Lakowicz *et al.*, 1997), europium ions complexed to fluorescent chelators (Seveus *et al.*, 1992; Marriott *et al.*, 1994b), as well as inorganic phosphorcrystals (Beverloo *et al.*, 1990). Coupling of the probes to antibodies or biotin for routine cytology has been described (Beverloo *et al.*, 1992; Marriott *et al.*, 1994b; Lakowicz *et al.*, 1997). Apart from high contrast generation by off-gating the prompt autofluorescence, the slow decaying probes can also be used to study microsecond rotational dynamics or domain motions in proteins (Lakowicz *et al.*, 1997).

34.5.6 Medical diagnosis

Several groups have found that tumours can be detected by FLIM in a very early stage of development after pretreatment with porphyrin photosensitizers. The tumour cells display a distinctly increased fluorescence lifetime as compared with normal slow dividing cells (Cubeddu *et al.*, 1993; Kohl *et al.*, 1993; Schneckenburger *et al.*, 1994). Recently endoscopes have been designed with fluorescence lifetime imaging capabilities (fluorescence lifetime imaging endoscopy, FLIE). Using natural autofluorescence, distinct differences between luminescence lifetimes in tumour cells as compared with normal cells can be detected. The endoscopes are all implemented using the frequency-domain method (Mizeret *et al.*, 1997; Schneider & Clegg, 1997; Wagnières *et al.*, 1997). The use of phase suppression techniques (Clegg *et al.*, 1994; Szmacinski *et al.*, 1994) can be combined with these endoscopes, enabling specific suppression of the luminescence emitted from normal cells. Recently such a FLIE instrument was built in which data acquisition, processing and display were implemented in real-time (Schneider & Clegg, 1997). Particularly for the specific detection of early stages of cancer, FLIE may be very useful. Since high-power IR lasers can be simultaneously incorporated into the endoscope, detected tumours can be instantly burned away, providing a promising fast and relatively non-invasive method for cancer diagnosis and therapy.

34.5.7 Industrial applications

Early stages of caries infection on teeth can be specifically detected by using FLIM (Birmingham, 1997; Schneckenburger *et al.*, 1994). This can be applied for designing new toothpaste (Birmingham, 1997). Ni and co-workers have demonstrated the use of lifetime imaging in the analysis of fuel combustion and the study of spatially resolved gas temperatures (Ni & Melton, 1996). Lifetime imaging can also be used in parallel analysis of many samples such as lifetime-based assays on microtitre plates or microcuvettes (Birmingham, 1997; Gadella Jr, 1997; Schneider & Clegg, 1997).

34.6 CONCLUDING REMARKS

FLIM is a sensitive, robust, fast (a typical experiment can be completed in a few seconds) and very useful technique that is available in an increasing number of laboratories, implemented in different ways. By exploiting the continuous development of new fluorescent probes, FLIM will provide unprecedented contrast, very low detection thresholds, and novel fluorescent indicator-based assays for a variety of ions and molecular processes. The revolution that has taken place since Chalfie and co-workers published details of the use of the green fluorescent protein (GFP) from the jellyfish *Aequorea victorea* for gene expression in 1994 (Chalfie *et al.*, 1994) will also have an increasing impact on applications for FLIM. Point mutations in or near the chromophore region of GFP have led to the development of fluorescent proteins with various and improved spectral properties fluorescing with blue, cyan, green or yellow colours, designated as BFP, CFP, GFP and YFP, respectively (Heim & Tsien, 1996; Ormö *et al.*, 1996). Using these different coloured mutants of GFP, FRET-based assays for studying molecular proximity have become available to any laboratory having access to molecular biological techniques. Successful demonstrations of FRET between BFP and GFP have been published (Heim & Tsien, 1996; Mitra *et al.*, 1996).

More recently, a novel Ca^{2+} indicator fusion protein was constructed consisting of calmodulin, the calmodulin binding peptide M13, and at each end two mutant GFPs emitting at different wavelengths. After binding of Ca^{2+} the conformation of the fusion protein is changed, and as a result the FRET efficiency between BFP and GFP or between CFP and YFP at the two ends of the chameleon is a direct indicator of the local [Ca^{2+}] (Miyakawa *et al.*, 1997). This paper also reports on FRET detected in cells co-transfected with the (separated) M13-YFP and CFP-calmodulin fusion proteins in a Ca^{2+}-dependent manner. Especially in case of unknown or variable donor-to-acceptor ratios within cells (as is generally the case with co-transfected or otherwise double-labelled cells), FLIM is an indispensable tool for correctly determining FRET efficiencies. As shown in Plate 34.1, GFP fluorescence lifetimes can be very accurately determined in living cells, allowing quantitative estimation even of very low FRET efficiencies (e.g. as small as 5%). This confirms that GFP is an excellent probe for FRET-FLIM. Hence, in the near future, it is likely that there will be an important role for FLIM in the detection of molecular interactions in single living cells using GFP fusion proteins, which undoubtedly will shed new light on subcellular signalling, communication and trafficking mechanisms.

ACKNOWLEDGEMENTS

TWJG is supported by the Royal Netherlands Academy of Sciences. The FLIM system as described in Fig. 34.2 was financed by an investment grant of The Netherlands Organization for Scientific Research (NWO). Ir. G. van der Krogt is gratefully acknowledged for carrying out the transfection of the Cowpea protoplasts. I am grateful to my colleagues A. van Hoek, A.J.W.G. Visser and T. Bisseling for proofreading the manuscript and for their unceasing interest and support for the FLIM project.

REFERENCES

Anderson S.R. (1991) *J. Biol. Chem.* **266**, 11 405–11 408.
Bastiaens P.I.H. & Jovin T.M. (1996) *Proc. Natl Acad. Sci. USA* **93**, 8407–8412.
Bastiaens P.I.H. & Visser A.J.W.G. (1993) In *Fluorescence Spectroscopy. New Methods and Applications*, O.S. Wolfbeis (ed). Springer-Verlag, Heidelberg, pp. 49–63.
Beverloo H.B., van Schadewijk A., van Gelderen-Boele S. & Tanke H.J. (1990) *Cytometry* **11**, 561–570.
Beverloo H.B., van Schadewijk A., Bonnet J., van der Geest R., Runia R., Verwoerd N.P., Vrolijk J., Ploem J.S. & Tanke H.J. (1992) *Cytometry* **13**, 561–570.
Birmingham J.J. (1997) *J. Fluorescence* **7**, 45–54.
Buist A.H., Müller M., Gijsbers E.J., Brakenhoff G.J., Sosnowski T.S., Norris T.B. & Squier J. (1997) *J. Microsc.* **186**, 212–220.
Buurman E.P., Sanders R., Draaijer A., Gerritsen H.C., van Deen J.J.F., Houpt P.M. & Levine Y.K. (1992) *Scanning* **14**, 155–159.
Chalfie M., Tu Y., Euskirchen G., Ward W.W. & Prasher D.C. (1994) *Science* **263**, 802–805.
Clegg R.M. (1995) *Curr. Opin. Biotechnol.* **6**, 103–110.
Clegg R.M. (1996) In *Fluorescence Imaging Spectroscopy and Microscopy*, X.-F. Wang & B. Herman (eds). John Wiley & Sons, Inc., New York, pp. 179–252.
Clegg R.M. & Schneider P.C. (1996) In *Fluorescence Microscopy and Fluorescence Probes*, J. Slavík (ed). Plenum Press, New York, pp. 15–33.
Clegg R.M., Marriott G., Feddersen B.A., Gratton E. & Jovin T.M. (1990) *Biophys. J., 34th Annual Meeting of the Biophysical Society* **57**, 375a.
Clegg R.M., Feddersen B., Gratton E. & Jovin T.M. (1992) *Proc. SPIE* **1640**, 448–460.
Clegg R.M., Gadella Jr T.W.J. & Jovin T.M. (1994) *Proc. SPIE* **2137**, 105–118.
Cubeddu R., Canti G., Taroni P. & Valentini G. (1991) *Proc. SPIE* **1525**, 17–25.
Cubeddu R., Canti G., Taroni P. & Valentini G. (1993) *Photochem. Photobiol.* **57**, 480–485.
Denk W., Strickler J. & Webb W. (1990) *Science* **248**, 73–76.
Dong C.Y., So P.T.C., French T. & Gratton E. (1995) *Biophys. J.* **69**, 2234–2242.
Förster T. (1949) *Z. Naturforsch.* **5**, 321–327.
Gadella B.M., Lopes-Gardozo M., Colenbrander B., van Golde L.M.G. & Gadella Jr T.W.J. (1995) *J. Cell Sci.* **108**, 935–945.
Gadella Jr T.W.J. (1997) *Eur. Microsc. and Anal.* **47**, 9–11.
Gadella Jr T.W.J. & Jovin T.M. (1995) *J. Cell Biol.* **129**, 1543–1558.
Gadella Jr T.W.J., Jovin T.M. & Clegg R.M. (1993) *Biophys. Chem.* **48**, 221–239.
Gadella Jr T.W.J., Clegg R.M. & Jovin T.M. (1994) *Bioimaging* **2**, 139–159.
Gadella Jr T.W.J., van Hoek A. & Visser A.J.W.G. (1997) *J. Fluorescence* **7**, 35–43.
Gerritsen H.C., Sanders R., Draaijer A., Ince C. & Levine Y.K. (1997) *J. Fluorescence* **7**, 11–15.
Gratton E., Jameson D.M., Rosato N. & Weber G. (1984) *Rev. Sci. Instrum.* **55**, 486–494.
Gratton E., Feddersen B. & vande Ven M. (1990) *Proc. SPIE* **1204**, 21–25.
Hamming R.W. (1973) *Numerical Methods for Scientists and Engineers.* Dover Publications, Inc., New York.
Hartmann P., Ziegler W., Holst G. & Lübbers D.W. (1997) *Sensors and Actuators B* **38–39**, 110–115.
Haugland R.P. (1996) *Handbook of Fluorescent Probes and Research Chemicals*, 6th edn. Molecular Probes, Eugene, OR.
Heim R. & Tsien R.Y. (1996) *Curr. Biol.* **6**, 178–192.
Hochstrasser R.M. & Johnson C.K. (1988) In *Ultrashort Laser Pulses*, W. Kaiser (ed). Springer Verlag, New York, pp. 357–417.
Jameson D.M. & Reinhart G.D. (1989) In *Fluorescent Biomolecules: Methodologies and Applications*, Plenum Press, New York, pp. 461.
Jameson D.M., Gratton E. & Hall R.D. (1984) *Appl. Spectrosc. Rev.* **20**, 55–106.
Jovin T.M. & Arndt-Jovin D.J. (1989a) In *Cell Structure and Function by Microspectrofluorometry*, E. Kohen & J.G. Hirschberg (ed). Academic Press, New York, pp. 99–117.
Jovin T.M. & Arndt-Jovin D.J. (1989b). *Annu. Rev. Biophys. Chem.* **18**, 271–308.
Jovin T.M., Arndt-Jovin D.J., Marriott G., Clegg R.M., Robert-Nicoud M. & Schormann T. (1990) In *Optical Microscopy for Biology*, B. Herman & K. Jacobson (eds). Wiley-Liss, New York, pp. 575–602.
Kohl M., Neukammer J., Sukowski U., Rinneberg H., Wörle

D., Sinn H.-J. & Friedrich E.A. (1993) *Appl. Phys. B* **56**, 131–138.
Kusba J., Bogdanov V., Gryczynski I. & Lakowicz J.R. (1994) *B. J. Phys. Chem.* **98**, 334–342.
Lakowicz J.R., (1991) In *Topics in Fluorescence Spectroscopy*, Vols 2 (*Principles*) and 3 (*Biochemical Applications*). Plenum Press, New York.
Lakowicz J.R. & Berndt K.W. (1991) *Rev. Sci. Instrum.* **62**, 1727–1734.
Lakowicz J.R., Szmacinski H., Nowaczyk K. & Johnson M.L. (1992a) *Proc. SPIE* **1640**, 390–404.
Lakowicz J.R., Szmacinski H.S. & Nowaczyk K. (1992b) *Proc. Natl Acad. Sci. USA* **89**, 1271–1275.
Lakowicz J.R., Terpetschnig E., Murtaza Z. & Szmacinski H. (1997) *J. Fluorescence* **7**, 17–25.
Marriott G., Clegg R.M., Arndt-Jovin D.J. & Jovin T.M. (1991) *Biophys. J.* **60**, 1374–1387.
Marriott G., Heidecker M., Diamandis E. & Yan-Marriott Y. (1994a) *Biophys. J.* **67**, 957–965.
Marriott G., Jovin T.M. & Yan-Marriott Y. (1994b) *Anal. Chem.* **66**, 1490–1494.
Minami T. & Hirayama S. (1990) *J. Photochem. Photobiol. A: Chemistry* **53**, 11–21.
Mitra R.D., Silva C.M. & Youvan D.C. (1996) *Gene* **173**, 13–17.
Miyawaki A., Liopis J., Heim R., McCaffery J.M., Adams J.A., Ikura M. & Tsien R. (1997) *Nature* **388**, 882–887.
Mizeret J., Wagnières G., Stepinac T. & van den Bergh H. (1997) *Lasers Med. Sci.* **12**, 209–217.
Morgan C.G., Mitchell A.C. & Murray J.G. (1990) *Trans. R. Microsc. Soc.* **1**, 463–466.
Morgan C.G., Mitchell A.C. & Murray J.G. (1992) *J. Microsc.* **165**, 49–60.
Morgan C.G., Mitchell A.C., Murray J.G. & Wall E.J. (1997) *J. Fluorescence* **7**, 65–73.
Nakamura O. (1993) *Optik* **20**, 39–42.
Ni T. & Melton L. (1991) *Appl. Spectrosc.* **45**, 938–943.
Ni T. & Melton L.A. (1996) *Appl. Spectrosc.* **50**, 1112–1116.
Oida T., Sako Y. & Kusumi A. (1993) *Biophys. J.* **64**, 676–685.
Ormö M., Cubbit A.B., Kallio K., Gross L.A., Tsien R.Y. & Remington J. (1996) *Science* **273**, 1392–1395.
Pang S.-Z., DeBoer D., Wan G., Layton J.G., Neher M.K., Amstrong C.L., Fry J.E., Hinchee M.A.W. & Fromm M.E. (1996) *Plant Physiol.* **112**, 893–900.
Periasamy A., Wodnicki P., Wang X.F., Kwon S., Gordon G.W. & Herman B. (1996) *Rev. Sci. Instrum.* **67**, 3722–3731.
Piston D.W., Sandison D.R. & Webb W.W. (1992) *Proc. SPIE* **1604**, 379–389.
Sanders R., Gerritsen H.C., Draaier A., Houpt P.M. & Levine Y.K. (1994a) *Bioimaging* **2**, 131–138.
Sanders R., Gerritsen H.C., Draaijer A., Houpt P.M., van Veen S.J.P. & Levine Y.K. (1994b) *Proc. SPIE* **2197**, 56–62.
Schneckenburger H., König K., Kunzi-Rapp K., Westphal-Frösch C. & Rück A. (1993) *J. Photochem. Photobiol. B* **21**, 143–147.
Schneckenburger H., König K., Dienersberger T. & Hahn R. (1994) *Opt. Eng.* **33**, 2600–2606.
Schneider P.C. & Clegg R.M. (1997) *Rev. Sci. Instrum.* **68**, 4107–4119.
Seveüs L., Väisälä M., Syrjänen S., Sandberg M., Kuusisto A., Harju R., Salo J., Hemmilä I., Kojola H. & Soini E. (1992) *Cytometry* **13**, 329–338.
Shaw P.J. (1995) In *Handbook of Biological Confocal Microscopy*, 2nd edn, J.B. Pawley (ed). Plenum Press, New York, pp. 373–387.
So P.T.C., Grench T. & Gratton E. (1994) *Proc. SPIE* **2137**, 83–92.
So P.T.C., French T., Yu W.M., Berland K.M., Dong C.Y. & Gratton E. (1995) *Bioimaging* **3**, 49–63.
Spencer R.D. & Weber G. (1969) *Ann. Acad. Sci.* **158**, 361–376.
Stryer L. (1978) *Annu. Rev. Biochem.* **47**, 819–846.
Sytsma J., Vroom J.M., de Grauw C.J. & Gerritsen H.C. (1998) *J. Microsc.* **191**, 39–57.
Szmacinski H., Lakowicz J.R. & Johnson M.L. (1994) *Methods Enzymol.* **240**, 723–748.
van Bokhoven H., Verver J., Wellink J. & van Kammen A. (1993) *J. Genr. Virol.* **74**, 2233–2241.
Venetta B.J. (1959) *Rev. Sci. Instrum.* **30**, 450–457.
Wagnières G., Mizeret J., Studzinski A. & van den Bergh H. (1997) *J. Fluorescence* **7**, 75–83.
Wang X.F., Uchida T. & Minami S. (1989) *Appl. Spectrosc.* **43**, 840–845.
Wang X.F., Uchida T., Coleman D.M. & Minami S. (1991) *Appl. Spectrosc.* **45**, 360–366.
Wang X.F., Periasamy A. & Herman B. (1992a) *Crit. Rev. Anal. Chem.* **23**, 369–395.
Wang X.F., Uchida T. & Minami S. (1992b) *Proc. SPIE* **1604**, 433–439.
Weber G. (1981) *J. Phys. Chem.* **85**, 949–953.
Wu P. & Brand L. (1994) *Anal. Biochem.* **218**, 1–13.

CHAPTER THIRTY-FIVE

Detection of Fluorescently Labelled Proteins following Gel Electrophoresis: A Key Enabling Technology for Proteomics

GERRY SKEWS & TERRY McCANN
Life Science Resources, Great Shelford, Cambridge, UK

35.1 INTRODUCTION

Electrophoresis is a key analytical technique for the separation and analysis of proteins, DNA carbohydrates and other biomolecules. It is an essential prerequisite for many techniques used in molecular biology and, alongside chromatography and spectroscopy, is a fundamental procedure in the laboratory. However, unlike these other two techniques, which are already highly sophisticated and automated, electrophoresis is manual and labour intensive. Now, with the growth in the availability and range of fluorophores, there is a real opportunity to enhance and develop the technique. Here we discuss and review the technique, summarize traditional and novel labelling procedures, and consider new digital imaging systems capable of highly sensitive detection and analysis. In isolation each of these developments makes a positive contribution, but when considered together, they contribute a multiplier effect which should establish fluorescence-based electrophoresis as the most exciting and powerful separation, preparative and highly resolving bioanalytical technique.

The technique itself has been in routine use since the 1930s and was revolutionized in the 1960s by the introduction of polyacrylamide as the separating media. There has been little change since and the technique is generally considered antiquated but effective.

As the need for fast and accurate data becomes more pressing with the explosion in quantitative molecular biology techniques, there is now considerable interest in evaluating new technologies which overcome the historical limitations of the process. It is clear that fluorescence will play an increasingly important role in electrophoresis and the market will develop chemistry, running systems and advanced imaging technology in response to the demand.

35.2 GENERAL APPLICATION AREAS FOR FLUORESCENT IMAGING IN ELECTROPHORESIS

The ability to address DNA, protein and carbohydrate separation applications is particularly important. The majority of applications that use histochemical,

FLUORESCENT AND LUMINESCENT PROBES, 2ND EDN
ISBN 0-12-447836-0

radioactive or chemiluminescent staining are likely to be replaced with single, or multiple, fluorescent labelling in the short term. Investment in fluorescent technology to date has been focused on DNA sequencing in response to the human and plant genome projects. However, the next phase of investigation, the so-called 'Proteome' stage, or study of protein expression arising from the genome, requires high-performance protein electrophoresis systems. Sequence data indicates gene position and structure, but gene function translates to protein expression and regulation. It is the presence, abundance and activity of protein that controls cell function and disease. The activity of proteins will therefore be critical and fundamental to understanding biological function. High-resolution electrophoresis and databasing will advance the study of protein activity. Proteins will continue to be sequenced by mass spectromety and HPLC, but advanced electrophoresis will be the method of isolation and identification. The instrument described here provides interesting potential for high-throughput multiwavelength imaging applications. Today it will be applied to routine electrophoresis analysis, but tomorrow it may well find itself being used in large-scale proteomics which itself may overshadow even the current genome initiative.

Protein, DNA and carbohydrate samples are all amenable to electrophoresis. Microbiology, epidemiology, toxicology, clinical research, diagnostics and drug discovery applications use electrophoresis as a routine technique to extend and enhanced understanding of the structure and function of these biomolecules. In forensic science, new electrophoresis techniques could improve the sensitivity, specificity and certainly speed of many tests. For environmental science applications, electrophoresis provides information about organisms down to the species and subspecies level. Clinical research requires protein profiling techniques which reveal characteristic data patterns that can be compared between normal and diseased states to improve diagnosis and prognosis.

35.3 WHAT IS ELECTROPHORESIS?

Electrophoresis is a technique for separating biomolecules by charge and mass using an electric field (Figure 1). Electrophoresis has a particularly wide dynamic range in molecular terms, being able to separate comparatively small molecules such as polypeptides, to extremely large molecules, such as carbohydrates and double-stranded DNA. It is in routine use in most biochemical and molecular laboratories as well as finding many clinical applications in hospitals and public health laboratories. It is used for relatively simple quantitative PCR analysis and to show the highest number of discrete proteins on a 2-D gel (Fig. 35.1). All of these techniques may well move towards fluorescence and enhanced fluorescence chemistry in the near future.

Electrophoresis is quite simple. Gel material, made from either a synthetic (polyacrylamide) or natural (agarose) source, is placed in a gel-running system that allows liquid buffer reservoirs to contact either end of the gel. After the sample is loaded at one end, an electric field is applied across the gel. Because the samples have a charge (often conferred by sodium dodecyl sulfate), they will move through the gel at a rate determined by the size and charge characteristics of the molecule and the strength of the electric field. In general, molecules of a larger size or smaller charge move more slowly than smaller or more highly charged ones. The separation is stopped by removing the voltage or when molecules reach their isoelectric point (*p*I), an equilibrium at which frictional and electrical forces are balanced.

This basic technique has been around for almost 70 years and many would claim that it is simple, uninteresting and has no real need of close examination or improvement. Indeed, electrophoresis has probably reached a limit that requires advances in technology in order to move on. The incorporation of fluorescence probes and automated imaging strategies are

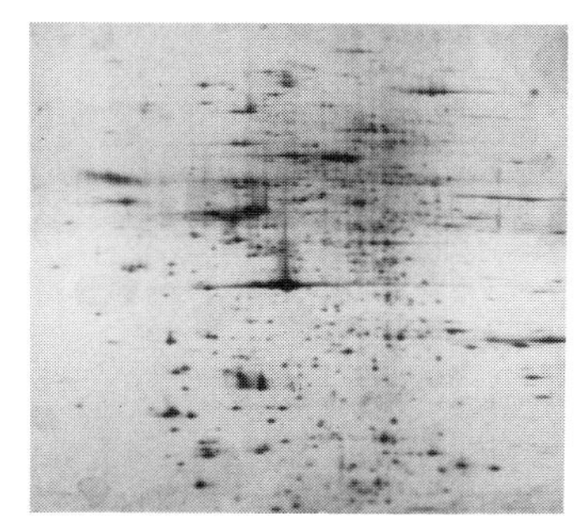
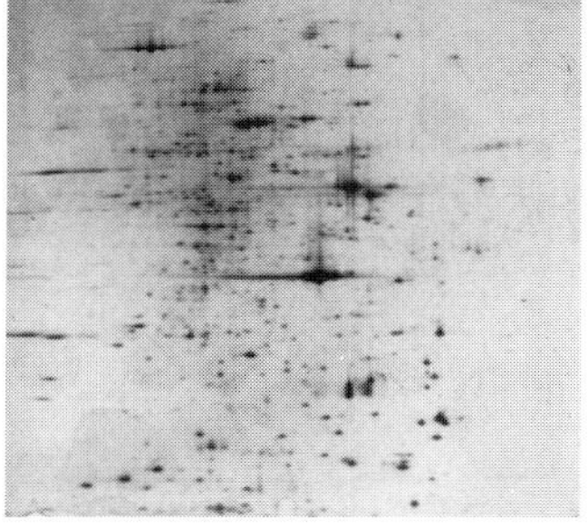

Figure 35.1 Typical images of silver-stained 2-D gels.

enabling this technique to do precisely that and will now extend our understanding of protein function in humans, animals and plants.

35.4 THE USE OF SLAB GELS

Recent advances in chemical and process technologies allow for some automation of gel processing. The growth in the availability, quality and use of precast gels has contributed towards improving reproducibility of electrophoresis. It is interesting to note that much of this progress is based on slab gel electrophoresis, which clearly offers advantages over capillary and other techniques; multiple samples can be run in parallel, standards can be run without difficulty to calibrate gels, and samples can be recovered relatively easily for further analysis. The principle advantage of capillary electrophoresis is speed but with the introduction of thin and micro slab gels this advantage may well disappear. Slab gel electrophoresis is one of the original techniques but it seems to have stood the test of time and with the improvements mentioned perhaps it will continue as the separation medium of choice in the ever expanding range of applications. Developments in slab gel chemistry as described will only serve to increase the demand for high-sensitivity, high-throughput gel reading technology.

The highest possible resolution of proteins on a gel is achieved using two-dimensional gel electrophoresis (2-D-PAGE). Here a protein sample is initially run on a thin rod or strip gel, which contains a pH gradient. The pH gradient can be established by a pre-electrophoresis run using ampholytes or making use of a precast immobilized pH gradient gel. Typically this might be a broad pH 4–10 gradient, or a higher resolution acidic or basic gradient from 4–7 or 6–10 pH units. This initial gradient separates protein molecules on the basis of isoelectric point and is called isoelectric focusing (IEF). Following a denaturing step which breaks the disulfide bridges between proteins, the gel rod or strip is laid on to the slab gel so that the proteins can be run out in a second dimension. This separation is principally on the basis of molecular weight. This two-step process yields a more finely resolved sample in two dimensions. The net result can be a gel with several hundred to many thousand protein spots. There are many non-linearities in the process which affect resolution and reproducibility and great care is needed to develop a system which effectively compensates or takes into account such variations.

Visualization of the gel is the next important step and this is achieved by post-labelling the proteins (widely used at present), or pre-labelling (described below).

35.5 VISUALIZATION OF PROTEINS

After electrophoresis is complete the gel is 'processed' to reveal the precise positions of the separated fragments in the gel. This allows an evaluation of the positions of the stained bands or dots. If a standard sample is run on the gel at the same time, the bands from the samples can be evaluated by comparing their measured positions to those of the standard.

In many applications, a simple chemical stain such as Coomassie Blue is used to identify abundant proteins within a few hours. Coomassie is a common but generally insensitive stain. Silver staining, which is the most common colloidal stain, uses silver nitrate precipitation to localize proteins and is also popular and more sensitive than Coomassie Blue. Radiographic techniques are also common since ^{35}S or ^{32}P isotopes can be incorporated into proteins during synthesis or post-translationally and the gels exposed to film or visualized using a radioactive imaging system such as a phosphor imager.

Each of these techniques has its merits and drawbacks, and no single technique successfully addresses all applications. However, current work may prove fluorescence to be the technique of choice.

Common in cell biology, flow cytometry and some areas of molecular biology, fluorescent labelling has many advantages that include speed, sensitivity, specificity and safety. However, it has yet to find broad acceptance in electrophoresis, other than in DNA sequencing, for two principle reasons:

- the lack of fluorescent dyes which bind to protein prior to electrophoresis to enable detection without affecting mobility of the protein on gels; and
- the lack of a comprehensive imaging solution.

35.6 FLUORESCENT PRELABELLING OF PROTEINS

Fluorescent proteins and glycoproteins can be detected on gels using labelling protocols either before or after running the proteins on a gel. A number of suitable fluorophores are offered on the market. Post-labelling is widely used for single sample protocols, but it does not allow multiplexed protein samples to be run on a single gel and differentiated at the detection stage. Post-labelling of gels eliminates many of the health and safety issues associated with radioactive or standard chemical labels, and offers the advantages of sensitivity and specificity over other techniques. However, it still requires handling of gels after electrophoresis.

Our co-workers have developed protein pre-labelling techniques that allow a fluorophore to be added during a 15-minute pre-electrophoresis treatment. Pre-labelling eliminates the intermediate staining step required for other techniques, prior to spot detection, and decreases the time taken to extract data from the gel (Fig. 35.2).

The basic procedure used for pre-labelling proteins in 2-D electrophoresis (although there seems absolutely no reason why these procedures should not be applied across the whole gamut of application areas) is described in US Patent 5 320 727. In this document the procedure is described thus:

> Labelling is conveniently effected by reacting protein in lysis buffer with monobromobimane or monochlorobimane added in solution in acetonitrile, by mixing and incubating at room temperature for between 2 and 60 minutes e.g. 10 minutes which results in production of a highly fluorescent derivative. The derivatives show absorption maxima at 370 to 385 nm and emit light at 477 to 484 nm. The fluorescent quantum yields are in the range 0.1 to 0.3.
>
> Some proteins include free sulphydryl groups and can be reacted directly with the label material. Generally however, free sulphydryl groups are produced by reduction in a known manner of disulphide bridges (cystinyl residues). Reducing agents such as DTE and DTT or tributylphosphine are conveniently used for this purpose.

In 1991 Urwin & Jackson published the technique in *Analytical Biochemistry* **195**, 30–37. In this work they showed comparisons of different labelling approaches of 2-D gels.

The gel is then imaged using fluorescent macro-imaging technology as described below. The ability to place a gel directly on an imager saves many hours when compared with traditional gel-staining and detection techniques. While our work has used the bimane dye approachs, Amersham Pharmacia Biotech have focused on the cyanine dyes. All of these dyes can bind to proteins and appear to have minimal effects on mobility compared with post-labelling technologies. Moreover, sophisticated design of the dyes can produce molecules that will label cysteine or other groups, and potentially *N*- or *C*-terminus reactive species (Fig. 35.3).

Such pre-labelling techniques have another major advantage, which is the possibility to multiplex, or to run two or more samples on a gel at the same time, using labelling with multiple fluorophores. This approach can have major advantages for quantitative 2-D gel electrophoresis, and can enable studies of up- or down-regulation of proteins from closely related samples. Such approaches are likely to be a fundamental enabling technology for the drug discovery industry.

Figure 35.2 Pre-labelled gel with monobromobimane.

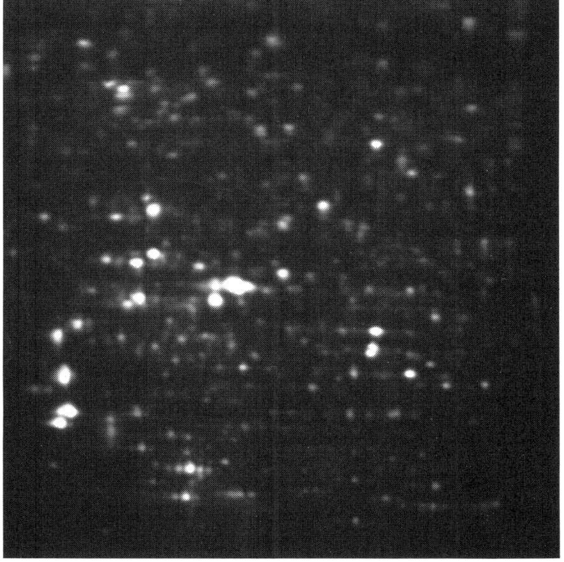

Figure 35.3 Fluorescent pre-labelling of 2-D gels gives high sensitivity for low-abundance proteins.

35.7 IMAGING FLUORESCENT PROTEINS IN GELS

As a major component of this work we have also developed a new instrument, using a high-performance cooled charge-coupled device (CCD) detector which can illuminate the gel from below or above (Fig. 35.4). Trans-illumination systems in broad or narrow UV and white light excite the sample from below, while from above, two fibreoptic probes can deliver monochromatic excitation light.

For high-performance detection of pre-labelled protein gels, a fish tail fibreoptic probe pipes light into the edge of the gel (called translucive illumination (Fig. 35.5)). This technique, borrowed from our group's earlier work on automated DNA systems, concentrates the photon path through the gel so that each fluorescent molecule is excited many times, maximizing fluorescence output. Approximately 10 times more sensitivity is achieved using the pre-labelling/translucive illumination technique, than a standard fluorescent post-labelling approach. Furthermore, these alternative illumination schemes allow pre- and post-labelled fluorescent gels as well as autoradiographs and silver-stained gels to be imaged with one instrument.

There are now nearly 3000 fluorophores commercially available for labeling biomolecules, so for an instrument to be versatile, it must have a broad spectral capacity. A xenon arc lamp provides a broad excitation range. Filter wheels optimize both the excitation and emission paths for the fluorophore in use. By way of contrast, other fluorescent imaging devices, such as those from Molecular Dynamics, use a laser for illumination and a photomultiplier tube for detection. Because a laser has a very limited number of wavelengths available, this limits the number of possible fluorophores available for detection.

Optical fibers deliver the illumination to the gel imaging chamber and carry back the emission through a filter arrangement, and the image is then focused onto a sensitive charge-coupled device (CCD) camera, cooled to reduce electrical noise and improve sensitivity. A high-precision x–y stage drives the detection system so it can cover the entire scanning area with 5 μm positional accuracy, yielding 50 μm spot resolution on the gel. The camera is a 16-bit digital output device, which provides a very high level of dynamic range, and this image is finally passed to a host computer for image analysis. The advantage of this high-bit-depth camera is that potential artifacts due to saturation of the labelling signal are avoided, and preliminary data reveal that a higher proportion of spots are revealed than with normal silver or dye staining. The technique provides images very rapidly and is suitable for high-throughput applications; scanning times of only a few minutes are required for typical gels. Being a digital camera based technology, software control of the imager provides the ability to adjust and optimize pixel resolution, speed and sensitivity. It is possible with this approach to perform multiple scans at different exposure times on the same sample. This effectively extends the dynamic range and

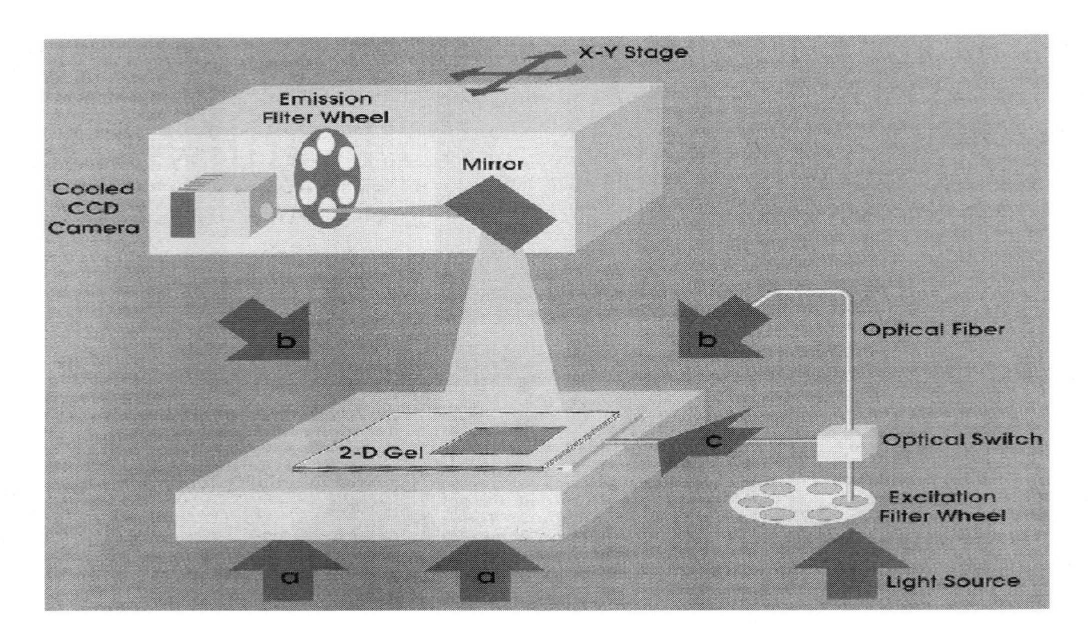

Figure 35.4 Schematic diagram of gel imager.

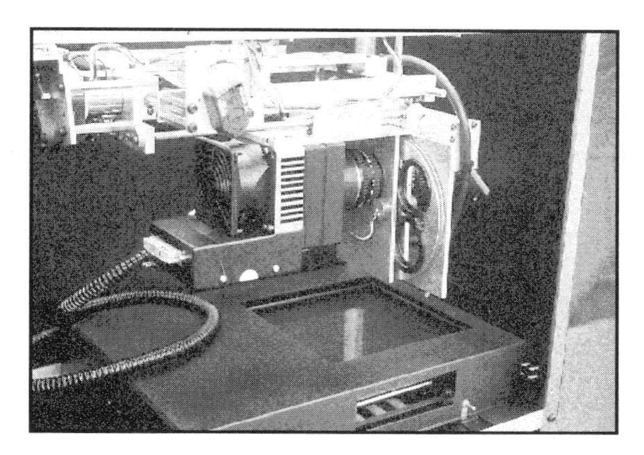

Figure 35.5 Interior of scanner showing illumination pathways.

sensitivity to several orders of magnitude beyond the 16-bit/65 500 grey level specification. The design of the instrument allows for the detector to be physically close to the sample. The resultant low demagnification provides a benefit of sensitivity proportional to the square of the distance. Early work with the instrument seems to support this design concept and extremely low concentrations of end-labelled DNA, for example, can be visualized with relative ease. Detailed applications work is currently underway to establish the actual limits of detection.

Obviously autoradiography of radiolabelled gels requires significantly longer exposure times, so the fluorescence approach seems to offer several advantages. As the availability of dyes increases, more applications will become attractive to this type of imaging.

35.8 ALTERNATIVE STAINING AND VISUALIZATION TECHNIQUES

Fluorescence is highly likely to become the technique of choice in many film-less applications – the advantages are far too strong to ignore. However, chemiluminescence is also widely used in immunoblotting applications, using film and long exposure times.

The design of the imager described here also lends itself well to these more traditional application areas and may well provide the means to move from these traditional approaches to the more adventurous and novel applications emerging out of the development of new dyes and techniques.

35.9 IMAGING AREA

The instrument can image an area of up to 280 × 230 mm, which covers the size of gels in most electrophoresis applications. To achieve a flat focal plane across the entire viewing area, the gel sits on a carrier that is kinematically mounted on a platen and electrically positioned in the imaging chamber. The instrument has been set to have scanning resolution of 50 μm. Although this resolution over the ~650 cm^2 area produces a 40 MB image file, most images will be

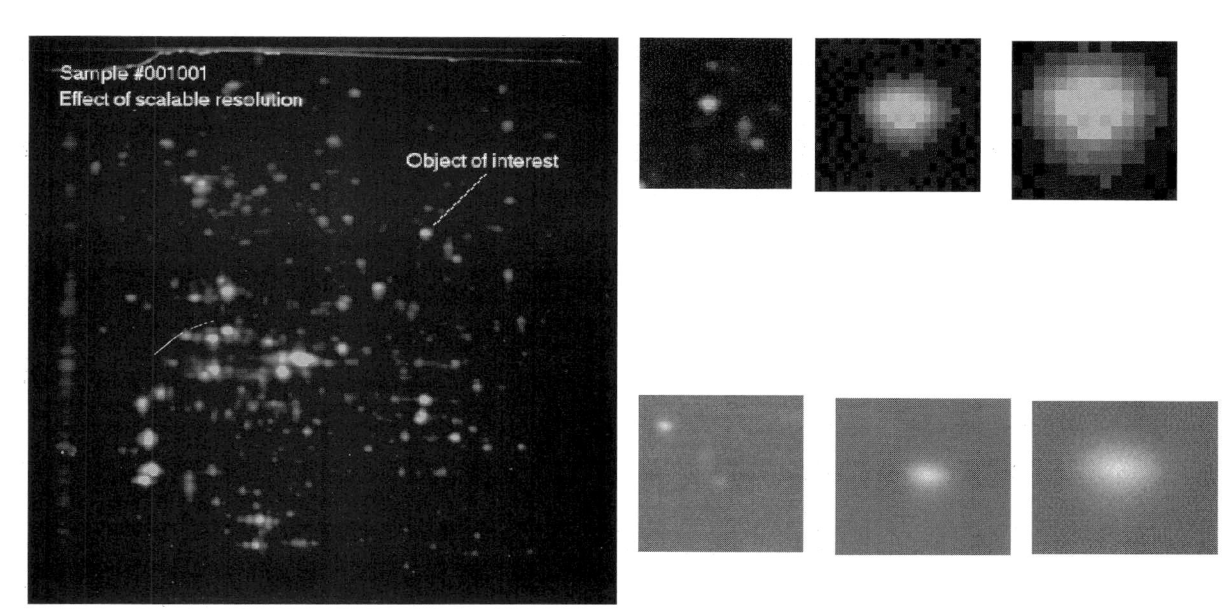

Figure 35.6 2-D gel scanned at 200 and 50 μm resolution showing pixelation effects.

smaller in area and/or resolution. The binning capabilities of the CCD allow for scalable resolution (Fig. 35.6).

The only rate-limiting step in the process is the image exposure time – typically a few seconds per image, but fully software-controllable. The longer the exposure, the greater the sensitivity of the scan. An initial, rapid scan of a whole gel can take only a few minutes, and the investigator may then return to particularly interesting areas for more detailed analysis.

35.10 HIGH-THROUGHPUT ACQUISITION

One of the most critical areas of use is to allow high-throughput data acquisition. The design of our gel imaging apparatus includes a broad range of features that allow laboratories to modify the instrument, under software control, for their own application. This makes the system adaptable to different fluorophores, gel formats and experimental conditions, as well as to completely different applications.

There are two main advantages of macro-imaging electrophoresis gels using fluorescent pre-labelled molecules. Firstly, the removal of any handling times for the gel following electrophoresis, thus avoiding problems in handling delicate gels, and the reduction in the overall time required to obtain images of the gel. The second main advantage is the availability of high-quality digital images, which allows superior quantification and facilitates archiving and inter-gel comparisons.

The gel reader is being used in a variety of applications, among them the new proteomics initiatives. Proteome research seeks to provide the final connection between genetic code and protein function. High-resolution protein maps can provide up to 2000–3000 discrete protein spots per experiment. By running hundreds of experiments it will be possible to link protein to DNA via post-translational modifications. Fluorescence is at the very heart of this activity as increased speed, sensitivity and specificity are essential for image analysis and data processing to be most effective. Recently published work by the Centre for Proteome Analysis in Denmark shows progress to date in this area (Fey *et al.*, 1997).

The advances of fluorescent labelling and a sensitive and easy-to-use detection system should increase the analytical power of electrophoresis in many fields such as forensics, toxicology, epidemiology, clinical diagnosis and disease development and prognosis. Examples of some of these are described below.

35.11 EXAMPLES OF SPECIFIC APPLICATION AREAS

35.11.1 Proteomics

As mentioned above, proteomics is the study of the proteome – the complete protein expression pattern resulting from activity of the genome. This is effectively the phenotypic expression of the genetic make-up of an organism. More than 40 000 genes have already been sequenced in the human, of a possible total of 80 000–100 000. Of major interest is the relationship between our gene composition and disease, and this is most accurately and meaningfully studied by examining quantitatively the protein expression of an organism. Whereas the gene sequence does not change acutely in diseased states, the protein expression levels do. The protein expression patterns of an organism are not coded by the gene composition, and therefore scientists are turning to direct analysis of protein expression levels, or proteomics, in order to understand health and disease, and more effectively understand drug action. The technologies involved include:

- protein separation, typically by 2-D gel electrophoresis
- imaging and spot detection
- bioinformatics to construct databases of proteins expressed in an organism from protein separation patterns
- robotics to enable individual protein spots to be recovered and further studied by mass spectrometric techniques in order to yield sequence information.

Pre-labelling chemistry protocols and the imager technology described above provides a central plank for such studies, by yielding a molecular profile of cells, tissues or organisms which reveal not only information about genetic coding but also about control of gene expression. This area will have major implications for drug discovery in the pharmaceutical industry.

35.11.2 Paternity testing

Initial indications show that analytical time could be reduced from days to hours. One typical laboratory conducts 50 000 tests per year and there are many such laboratories world-wide. Overall savings in time and money due to increased throughput could be significant. As most paternity laboratories are commercially motivated, there is a significant opportunity for cost savings.

35.11.3 Forensic science

Many well-publicized trials depending on forensic evidence have proved the inadequacies of contemporary electrophoretic techniques for DNA fingerprinting. The instrument gives accurate, quantifiable data on band positions. There is significant investment in this area but nothing addresses the whole procedure. There are 350 laboratories world-wide, each undertaking over 20 000 cases per year. When the genome is fully sequenced we will know where the 'eye colour' gene is, or the 'hair colour' or the 'age indicator' or the ethnic group or the height, or what diseases are present in the sample. So DNA profiling may take a whole new direction. A multiplexing instrument such as the gel imager described here could have a large impact in this area by building an interpretation from a panel of probes run on the same sample.

35.11.4 Microbiology

Today a positive species identification of a microorganism takes weeks. Cultures are isolated, plated out, grown in selective media, incubated and subjected to optical microscopic observation. The High Resolution Automated Microbial Initiative (HRAMI) is a European-wide project looking at better technologies for this process. The consensus is that DNA genotyping is definitive and fast, but no automated system exists. Most of the world's public health laboratories located at major population centres will have use for a system that could perform definitive analysis using high-speed electrophoresis.

35.11.5 Epidemiology

The World Health Organization is justifiably concerned at the re-emergence of tuberculosis as a major disease. There is a pressing need to understand where the strains are coming from, the demographic implications, the drug resistance characteristics and rapid identification to the subspecies level. The systems and approaches described here will be able to provide this information, not only for bacteriology but also in the fight against HIV, hepatitis and other infectious and highly mobile diseases.

35.11.6 Taxonomy

This is a research area requiring family trees (dendograms) showing the relationship between organisms. This application will find increasing acceptance in plant genetics and biotechnology. Both genotyping and phenotyping will be used to establish similarities between competing varieties. High-performance fluorescent electrophoresis could provide the 'definitive' test of similarity.

35.11.7 Clinical diagnostics

This area represents the future of the technology. Presently many tests are carried out *in situ* with scientists peering down microscopes searching for genetic and cellular abnormalities. This will ultimately move towards molecular detection as detection reagents improve and high-throughput devices come onto the market. The gene does not lie, and the protein expression of key indicators is strongly correlated to disease condition and development. Today only blood chemistry and gross biochemical indicators use electrophoresis, but in the future quantitative fluorescent electrophoresis will be used as a definitive signpost technique. There is a great deal of excitement in the clinical world regarding electrophoresis techniques in DNA, protein and carbohydrate chemistry, but there are also major issues still to address.

35.12 SUMMARY

Electrophoresis has been a popular technique capable of sorting molecules on the basis of charge and size. Recent chemical developments have yielded new fluorescent dyes capable of labelling proteins for high-resolution detection. These, combined with new

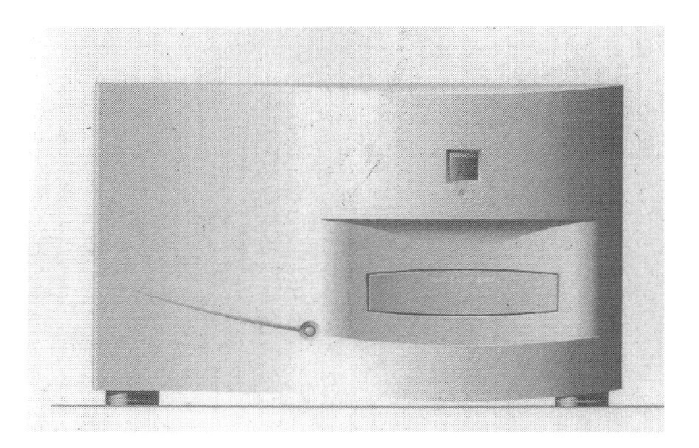

Figure 35.7 Commercial variant of the multi-wavelength imager.

high-throughput automated techniques for detection and quantification of such molecules in gels using high-resolution imaging (Fig. 35.7), are likely to have dramatic effects on the future of studies focused on protein expression.

REFERENCES

Fey *et al.*, (1997) *Electrophoresis* **18**, 1361–1372.

United States Patent number 5,320,727 Filing date Feb 12 1992.

Urwin V.E. & Jackson P. (1991) *Analytical Biochemistry* **195**, 30–37.

PART IX

CCD Cameras: Key Enabling Technologies for Optical Probe Imaging

CHAPTER THIRTY-SIX

Properties of Low-light-level Intensified Cameras

PATRICIA TOMKINS & ADRIAN LYONS
Photonic Science, Robertsbridge, East Sussex, UK

36.1 INTRODUCTION

The very significant advances that have been made with slow-scan technology over the last decade have resulted in the use of directly illuminated CCD cameras being used for an increasing range of applications. However, slow-scan is, by definition, relatively slow, and a great deal of work has been done, with some success, towards increasing read-out speed without losing all the benefits of this technology. We have seen 'fast slow-scan' technology emerge, and it is a real contributor to detector science.

Unfortunately, the faster you read out a CCD, the higher the noise. Also, given that one has a fundamental noise floor of say 10 electrons r.m.s. read-out noise (and that would not be at a high read-out rate), this means that with such a camera you simply *cannot* meaningfully take images with less than 10 detected photons per pixel. For temporally resolved work, such as fast ratio imaging, the read-out noise is often considerably higher, say 30 electrons r.m.s. per pixel or even more. A signal level of 30 electrons per pixel is actually quite high! Coupled with the fact that integration times must be kept short to preserve the temporal information, this means the actual flux is no longer so small – in other words, if you want temporal data, the sample must be rather bright.

In intensified CCD video cameras, the image intensifier converts the incoming photons to electrons, amplifies these (a spatially coherent parallel processing) and produces a secondary image up to thousands of times brighter than the incoming (primary) image. With good design, even a single detected photon can produce a signal in the CCD greater than the read-out noise. This means that such a system is no longer limited in its detectivity (or sensitivity) by the read-out noise of the CCD, or alternatively, you can run a great deal faster before a limit is reached.

By extensions of the technology that created intensified CCD video cameras, this new generation of 'scientific' CCD cameras can be fitted with image intensifiers to bring many advantages of the progress in electronics, along with the ultimate sensitivity that can be achieved by using image intensifiers. Such a 'hybrid' camera can offer extremely impressive performance, since the sensitivity advantages of the 'fast slow-scan' camera combine with the features of the

FLUORESCENT AND LUMINESCENT PROBES, 2ND EDN
ISBN 0-12-447836-0

intensifier to provide an imaging system with significant benefits over the more conventional TV-type CCD read-out. Of particular interest is matching the number of electrons produced in the CCD per detected primary photon to be approximately equal to the read-out noise. This way you combine the single-photon detection capability provided by the intensifier with the high dynamic range of the 'scientific' CCD – much greater than the 8-bit range traditionally available with video technology.

The latest directly illuminated CCD cameras will perform certain functions entirely adequately, such as bright fluorescence imaging, and video-rate intensified CCDs still offer excellent temporal performance with far greater sensitivity. However, where increased image acquisition speeds are needed or the signals are low, the use of non-video, 'scientific' CCDs as read-out stages can be very beneficial. For example, a 200 frames per second, high gain intensified CCD camera can have as low a noise level as 10 electrons *per image*, meaning that most pixels have zero signal in them, and are quite capable of recording as little as *one* photon event per pixel! Compare this with the plus-or-minus 30 events of the directly illuminated CCD. Our discussion here is taking us towards the realm of 'photon counting imaging' – where the measurement accuracy is defined by quantum statistics rather than read-out noise. A decade ago this technique was the reserve of the astronomer, but now such methods are being introduced to biology, with a growing number of researchers pushing the lower limits of detection by employing the very highest sensitivity cameras. Specially designed *intensified* CCD cameras achieve such performance.

The military equipment manufacturers, who only 10 years ago were so focused on military night vision products, have turned their undeniably creative development efforts towards civilian markets. In spite of a contraction of image intensifier manufacture world-wide, those manufacturers that have survived this change have made available devices now specifically aimed at the scientific user, whereas before this was the exception rather than the rule. A whole new range of high-performance electro-optical devices are now coming to the market, amongst which are image intensifiers whose performance exceeds their earlier military predecessors such that the specifications bear no resemblance. Coupled with targeted design, volume manufacture for the scientific market has created exciting opportunities to achieve technical excellence.

A fashion trend towards directly illuminated CCDs for every application has started to reverse, as the scientific market has begun to appreciate that the use of image intensifiers is not 'old technology'. Rather, by combining modern image intensifiers (particularly by using the latest fibreoptic techniques) with the latest 'digital' cameras, a more advanced family of imagers is coming into existence. In many ways, the new fast slow-scan cameras provide ideal 'back-ends' to sophisticated non-video ICCDs. It is interesting to observe that rather than having two competing technologies, the 'sharp end' of research is turning to an alloying of both approaches which exploits the benefits of both to bring state-of-the-art technical advantage. This exciting synergy of technology promises to open new possibilities in the field of biological research as well as to extend existing techniques to the quantum limits.

36.2 IMAGE INTENSIFIER: FIRST AND SECOND GENERATION

Image intensifiers depend on the photoelectric effect to convert a visible image to an electron image. An image intensifier reproduces at the output an image identical with one presented at the input. A net gain in intensity is produced without drastically affecting the contrast of the image.

An important characteristic is detective quantum efficiency (DQEt). This is the fraction of incident photons actually detected, which varies with design or suitability of the intensifier.

Primary photoelectrons produced by the photocathode may be used in two ways.

(1) *First-generation tubes.* Here primary photoelectrons, which are accelerated by a high electric field, impinge on a phosphorescent screen. The energy gained by the electrons in the field is released as a flash of light in the screen. With modern screens, several hundred photons may be released for a single energetic electron impact.
(2) *Second-generation tubes.* Here the primary photoelectrons are multiplied by a microchannel plate, which consists of an array of microscopic channel electron multipliers. The spatial coherence of the image is preserved during this amplification process. Many thousands of electrons may be released for each electron which strikes the microchannel plate. These electrons are then accelerated onto a phosphor screen, where light is produced as in first generation devices. Each photoelectron entering the microchannel plate releases up to hundreds of thousands of photons from the phosphor screen. The electron multiplication enables far higher intensification to be achieved than with first-generation devices.

36.3 MAXIMIZING SIGNAL AND MINIMIZING NOISE

To detect really low-light-level images, high sensitivity is absolutely necessary. It is important to make sure that your weak signal is not buried in noise, so low noise characteristics are also necessary. Ideally one wants to maximize the 'signal-to-noise' ratio. Regrettably, this is measured differently by just about everyone, and no real standard allowing intercomparison is conveniently available for the systems we are discussing. But the principle is valid; the best camera for any application is one that produces more signal and less noise, or a higher ratio of the two, for a given situation.

Fundamental to sensitivity is the efficient use of the precious few photons that are available. Image intensifiers use a light-sensitive input coating called a 'photocathode' to convert photons to electrons. These electrons are later amplified inside the intensifier. The percentage of photons that give rise to photoelectrons is called the quantum efficiency of the photocathode. The higher the quantum efficiency at the wavelength you are interested in, the higher the sensitivity of the overall detector is likely to be. Note that photocathode quantum efficiency is not the overall efficiency of the detector – just an important factor. There are other parameters (such as how efficiently the photoelectrons are themselves detected within the intensifier) that go together to make the overall efficiency always less than the photocathode quantum efficiency. Since the photocathode quantum efficiency always limits the theoretical maximum possible overall efficiency, this should always be maximized wherever possible.

Two important definitions are the quantum efficiency (QE) of the photocathode and the detective quantum efficiency (DQE) of the overall detector system. The DQE is the fraction (expressed as a percentage) of incident photons that are detected by the system. In an ideal electro-optical detector, the DQE equals the QE. In reality, internal losses cause the DQE to be less than the QE, sometimes by a surprisingly large factor in systems not designed for scientific applications. Quantum efficiency, when applied to photocathodes, is sometimes called responsive quantum efficiency (RQE). DQE can be related to RQE by a constant of proportionality, which could be called 'internal efficiency' of the detector. RQE is a strong function of the wavelength at which the measurement is made. The internal efficiency, which is a result of internal electron optical design, is wavelength-independent. This means that knowledge of the RQE of the photocathode of a device is sufficient to indicate how the DQE varies as a function of wavelength of input light. It is not easy to find out the absolute DQE without sophisticated measurement. It is often the difference in this value that distinguishes a 'good' camera from a 'bad' one.

36.4 THERMAL NOISE

Both photocathodes and semiconductor detectors are subject to random emission or production of electrons in the absence of light. This is a result of thermal effects within the sensitive material, and is given the name thermal noise. Without analysing the origins, it is sufficient for us to know that thermal noise increases rapidly with temperature. Thermal noise doubles for (approximately) every 7–8°C rise in temperature.

Similarly, this noise reduces by a factor of two for every 7–8°C. The cooling of the detector is the principle of noise reduction for slow-scan CCDs. There can be a tremendous range in the amount of thermal noise produced by different photoemissive materials (photocathodes), as well as marked variations in the rate of change of thermal noise with temperatures. As a general rule, photocathodes that respond well in the red and near-infrared will be noisier than photocathodes where the peak response is in the blue. This is because long-wavelength photons have less energy than short-wavelength ones ($e = h$). Therefore, electrons are released more easily in the red cathode than in the blue one, so thermal effects give rise to more emission. Typical photocathodes at room temperature can give rise to anywhere between 100 and 10 000 electrons cm^{-2} s^{-1}. The actual value depends very much on the specific manufacturing process, and is not only a function of spectral response, but also a combination of this with material formulation, manufacturing process and some, still not fully understood, 'seemingly random' variables. For this reason, it is important that devices are manufactured for demanding scientific applications. Low-noise photocathodes with typically 70 mA W^{-1} are now available.

36.5 NON-THERMAL NOISE

If a detector is cooled and its thermal noise plotted, it will be seen to decrease, as described above, but at a certain point the rate of decrease will virtually cease, and an almost constant noise level, almost indistinguishable from thermal noise or true photon events, will be found. The origin of these events is internal to the detector structure and in a photoemissive device, originates from two main sources. First, from internal micro-discharges, which emit very low light levels and are seen by the photocathode, thus producing electrons. Secondly, from direct electron emission from microscopic high points on the photoemissive surface, this being called field emission. The two can be distinguished since internal discharges produce a non-thermal noise component that covers an area, or possibly all of the image, whereas field emission appears as bright fixed spots. A third source of noise that can be eliminated by design is faint corona discharge around high-voltage structures external to the vacuum, whose light can fall on to the photocathode. Improved manufacturing of scientific tubes has reduced this noise resulting in quiet, low-noise, high-contrast intensifiers.

36.6 TEMPORAL CHARACTERISTICS OF PHOTOCATHODE NOISE

If a detector based on a photoemissive material sensitive to visible light, such as a photomultiplier or image intensifier, is exposed (when not operating) to a high level of illumination and then placed in complete darkness, the noise measured will be much higher than had the detector been kept in darkness prior to the measurement. This effect may produce increases in noise of several orders of magnitude. The noise, so induced, decays with a time constant of seconds to tens of seconds, depending on the particular photocathode. The magnitude of the effect is dependent upon the intensity of illumination, to some extent duration of illumination, and particularly upon the illuminating wavelength. The effect is far more pronounced when UV radiation is allowed to fall on the photocathode.

This has important practical consequences as regards experimental technique when low-light-level images are to be detected, particularly when low and stable background noise is required. The guidelines, therefore, are:

(1) Wherever possible, keep detectors in a dark environment prior to use.
(2) When changing samples, do not simply switch off the detector and expose it to room light, or other strong light sources. Protect the photocathode from such exposure.
(3) Never, if using fluorescence techniques, allow the exciting UV radiation to illuminate or to fall on the photocathode.
(4) If you need to expose the photocathode to high light levels (for installation or cleaning purposes), do this in incandescent lighting. Daylight contains much UV and fluorescent lighting emits a surprisingly large quantity of UV radiation.

It should be noted that this 'charging effect' is quite reversible and in no way damages the detector. It simply raises the noise. If this effect is noted, the best solution is to leave the detector running in total darkness with power applied, until the noise has fallen to a reasonable level.

36.7 GAIN USE AND ABUSE

If an ordinary video camera such as a CCD is operated at normal light levels, perfectly acceptable image results. At 1% of this light level, a perfectly acceptable image can be obtained from the detector by putting an image intensifier with a gain of 100 in front of it. Indeed, this arrangement is 100 times as sensitive as the original. It is tempting to believe that by increasing the gain more and more, ever-greater sensitivity will be achieved. This is not the case, as when more and more gain is employed at fainter and fainter light levels, the image will be seen to become progressively more grainy. These grains are, in fact, the amplified result of individual detected photons. At these light levels the quantum nature of light becomes apparent and adding further gain will simply make the speckle resulting from a photon appear brighter. In fact, providing that each detected photon speckle is clearly visible at the output of the camera, then there is no advantage in increasing the gain further. In these low-light regimes, extra sensitivity can only be achieved with detectors of higher QE, i.e. detectors that detect a larger fraction of the new photons in the input image. The quantum nature of light, therefore, imposes a useful limit to the maximum gain that can be employed in an intensified camera with good effect. One should therefore beware of claims for cameras with incredibly high gains – unless there is a design reason for this (e.g. operation in the photon counting mode) – since very high gains can actually limit rather than enhance performance.

For the following discussion of gain, we shall consider how the term may be applied to the image intensifier of an intensified CCD (ICCD) camera. There are several ways of defining gain. The most common is 'luminous gain'; this parameter, which is largely a hangover from military technology, is not particularly useful for scientific purposes. It is the ratio of output signal when measured with an eye-corrected photometer, using a tungsten lamp as the light source. Typical image intensifiers have luminous gains in the order of 10 000–30 000 (for a second-generation device incorporating a microchannel plate electron multiplier). This figure is misleadingly high since most intensifiers respond to the near-infrared of the tungsten lamp to which the eye-corrected photometer is blind. Thus the input signal is undersampled and a large figure obtained for the ratio.

A far more useful quantity is the radiant power gain or photon gain. The former, measured in units of watts per watt, is the ratio of optical power out of the intensifier to optical power in. Expressed as photon gain, it is a measure of the number of photons out per photon in. Clearly this is a function of wavelength and is best defined at the peak emission wavelength of the output phosphor, which is usually in the green.

Typical second-generation intensifiers have radiant power gains or photon gains in the order of 1000–2000. A third kind of gain is the photoelectron gain. This is the number of photons emitted at the screen per detected photon, i.e. per emitted photoelectron. Since only a small fraction (RQE) of the incident photons gives rise to photoelectrons (i.e. are detected), then the photoelectron gain is clearly the photon gain divided by the DQE.

The latter two types of gain interrelate in a way that is most important as regards image quality. The radiant power gain is the product of the photoelectron gain and DQE. Consider the following three image intensifiers:

DQE	*Photoelectron gain*	*Radiant power gain*
10%	1000	100
1%	10 000	100
0.1%	100 000	100

These three intensifiers would all have the same radiant power gain, and could have exactly the same luminous gain – in other words the published specifications for the three devices could be identical. Now consider how the three devices respond if given a signal of 1000 photons. The first device will give an output image of 100 speckles, each of 1000 photons. The second device will produce 10 output speckles each of 10 000 photons, and the third simply 1 bright spot of 100 000 photons.

The first intensifier would produce a basic image; the second limited spatial information and the third no image information at all. Therefore, it is vital to understand how photon gain is distributed between DQE and photoelectron gain in order to assess the potential imaging performance of a device – and none of this information is usually available in intensifier manufacturers' information. As we will discuss later, the only way you can tell if a camera designer has done a good job is to test the end product in your application.

This said, a given camera would have a constant DQE at a given wavelength, and if the gain is controllable then all the better. For higher light levels a low gain is required, and at the lowest light levels the maximum gain compatible with clear detection of individual photon events is needed. For this reason, a wide user-controllable gain range makes a camera far more versatile as regards the light level over which it can do meaningful science. In intensified CCD cameras this is not just an electronic gain but, as described above, is the factor by which electron showers are actively multiplied (truly, in number, within the tube) before reaching the image read-out element.

36.8 SPECTRAL RESPONSE

We have described previously how quantum efficiency of photocathodes varies with the wavelength of the incoming radiation. If the response is plotted as a function of wavelength, a 'spectral response curve', whose form is roughly an upturned 'U', is obtained. The peak will correspond to a particular part of the spectrum, and it is important to try to match this to the science, or at least ensure that there is reasonable response at the wavelength of interest.

The spectral response can also allow comparison of response at different wavelengths, important, for example, if ratioing images of orange and blue fluorescent dyes.

There is a very limited choice of response curves available in imaging devices, unlike photomultipliers, for which a wider choice has long been available.

The most common photocathodes in image intensifiers as used in ICCDs are 'S25' and 'S20'. Typical curves are shown in Fig. 36.1. These curves are not absolute and there is tremendous variation between devices and this again highlights the importance of careful selection for particular applications. The curve represents the average. The difference between S25 and S20 is basically one of photocathode formulation and process (details that do not need to concern us here, since we are more interested in the performance)

Terms you will encounter when reading manufacturers' literature are:

(1) *'Luminous sensitivity'* – expressed in microamps per lumen. In this mode the intensifier or other device is run as a photocathode and the current that can be drawn per unit of incident light is recorded. Unfortunately, this is done with a 2854 K tungsten lamp, which emits predominantly in the infrared, so photocathodes that have responses longer than 700 mm will have higher luminous sensitivity figures than ones which do not.
(2) *'Radiant sensitivity'* – This is measured as luminous sensitivity, but the light source is a monochromatic one of known optical power.

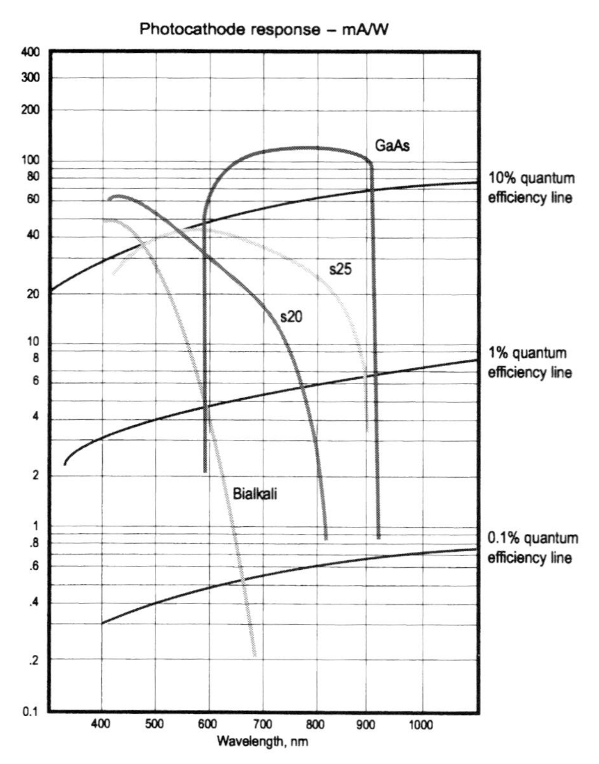

Figure 36.1 Typical photocathode spectral sensitivities.

The wavelength is varied and the photocathode current drawn is plotted as a function of wavelength. Radiant sensitivity is given in milliamps per watt and does at least provide real information about the response of the device.

(3) *'Quantum efficiency'* – A much more meaningful measurement for scientific purposes, the quantum efficiency as a percentage is plotted as a function of wavelength. This is related directly to the radiant sensitivity, but allows for the difference in energies between photons of different wavelength.

Often, spectral response curves will be shown in milliamps per watt with lines of equal quantum efficiency shown on them. Again, it should be noted, that a higher response (shown in milliamps per watt or %QE), at the wavelength you are working at, gives better detection efficiency and more signal.

36.9 COMPARISON BETWEEN SECOND- AND THIRD-GENERATION INTENSIFIER TUBES

Many scientists looking to maximize the sensitivity of their intensified cameras ask about third-generation image intensifiers, and want to consider what advantages the military image intensifier could offer. This chapter discusses the major difference between second-and third-generation intensifiers, their problems and advantages.

A second-generation intensifier, often called a wafer tube, consists of a photocathode in close proximity to the microchannel plate, with these two pieces held in close proximity to the phosphor of the output screen optic. One of the main differences between second- and third-generation tubes is that the photocathode of the third-generation tube is made of a layer of gallium arsenide (GaAs). This photocathode was specifically developed to provide a spectral response in the red and near infrared, which is ideal for gun sights and experiments within the 750–950 nm range. However, normal third-generation tubes generally have a poorer response over most of the visible region and do not respond well in the bluer wavelengths.

Recent developments in the GaAs photocathode has included an enhanced blue response which is considerably better than the earlier versions for response, in the visible part of the spectrum. Its main area of performance is still in the far red.

By comparison, second-generation intensifiers can be manufactured with the scientific user in mind and not simply as 'spin-offs' from military or security programmes. These tubes can be manufactured with a high green/blue response, ideally suited to the biologist working in calcium ratio imaging. These intensifiers can also be manufactured to have a very low background noise (the background speckle when an intensifier is on but not illuminated) and so provide better signal to noise for the overall image when in use.

Further developments in the technology of second-generation image intensifier tubes manufacture has been the improvement of QE in the blue/green part of the spectrum. Low noise blue S20 photocathodes have reached levels of efficiency which equal the bialkali photocathode.

One major problem with GaAs photocathodes is that ions event within the tube can actually poison the photocathode, and to prevent this problem all third-generation intensifiers have an 'ion barrier' film in front of the microchannel plate. This film is between the GaAs photocathode and the microchannel plate and has another side-effect in that it absorbs some of the electrons emitted by the photocathode. The barrier film therefore reduces the quantum efficiency of the photocathode such that the number of electrons actually available for subsequent amplification is often comparable to that in conventional second-generation devices. Some camera manufacturers quote the quantum efficiency of the third-generation photocathode in isolation, that is to say, measured as a simple diode, and avoid taking into account the subsequent usage of the emitted electrons in the functional intensifier environment. This argument produces impressive numbers for 'sensitivity' and 'quantum efficiency' (true values, but only part of the whole story), yet is often designed to mislead, by claiming advantages that are not truly realizable.

The photocathode sensitivity, photocathode quantum efficiency and related values are *not* absolute parameters that can be used to compare the sensitivity or detective quantum efficiency of different cameras unless the internal structures operate in identical ways. The basic internal structures of second- and third-generation devices are subtly different, with second-generation structures being more 'electron efficient'. Unfortunately, no 'overall' sensitivity measurement has been agreed, so again the only way to be sure is to test the product in the specific application. There is *no* guarantee that a camera with 30% photocathode quantum efficiency will outperform one with only 15% unless everything following that photocathode is identical!

Whilst third-generation intensifiers have some limited advantages (when the specimen is emitting primarily in the red), supply is at best difficult and can be limited due to these items being of a military nature, often leading to protracted delivery time scales. The only manufacturer of third-generation is in the USA, and export licensing (since they are components for gunsights) is a nightmare. The same problem applies to any service or repair should this become necessary. The authorities apply this control to cameras with third-generation intensifiers in them as well – the very fact that they have been 'incorporated' into a civilian application in no way exempts them from onerous control. A further side-effect of limited supplies is the likelihood of blemishes that would not be acceptable in second-generation tubes being passed on as ' well that's what's available' – resulting in a much reduced ability to select for real quality, and extreme difficulty and delay arising if such intensifiers are rejected on quality grounds.

36.10 HIGH-RESOLUTION INTENSIFIERS

This is an intensifier in which a high-resolution microchannel plate is used, often with other modification to the internal design (gap widths and voltages), to increase the spatial resolution of the intensifier. Early 'wafer' type intensifiers had resolutions of around 30 line pairs per mm (lp mm^{-1}), whereas the latest modern devices can have in excess of 60 lp mm^{-1}.

The techniques to increase resolution have nothing to do with the GaAs photocathode of third-generation products and can be equally well applied to standard second-generation to produce outstanding performance products with equally high resolution. Unfortunately, the military 'generation game' does not really adapt itself to classing these latter products, though some writers have coined the term 'gen 2+' or 4th gen.

Finer channel size and improved channel number density has seen improvements of resolution so that 40–57 lp mm^{-1} can be achieved. Changes in manufacturing techniques have reduced non-thermal noise, resulting in a second-generation intensifier with vastly improved characteristics.

The new higher-resolution intensifiers now make an appropriate high-resolution front input stage for high-resolution CCD camera, which opens up a whole range of modern digital, high-resolution ICCDs.

36.11 FIBREOPTICS OR LENSES FOR IMAGE TRANSFER?

For most people fibreoptics are usually 'something to do with telecommunications'. However, coherent optics are quite different from telecom fibres. To describe the material, fibreoptic glass is an array of many thousands of small parallel glass fibres set rigidly in another matrix of glass. Light entering the fibres at one end undergoes multiple total internal reflection as it travels down each individual fibre. Light from an image will enter thousands of individual fibres all running parallel to each other. As the fibres maintain the same spatial relationship to each other, an image placed at one end will appear on the surface of the other end of the fibreoptic component (see Fig. 36.2).

A property of glass is that when heated it can

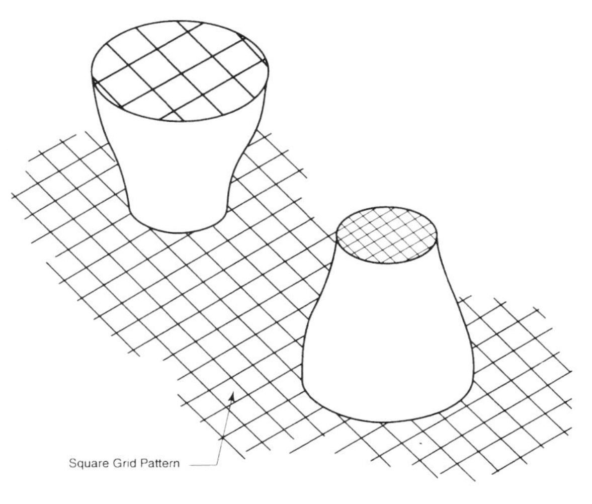

Figure 36.2 Magnification and demagnification with fibreoptic tapers.

undergo 'plastic deformation'. A parallel piece of fibreoptic glass can be heated and pulled in such a way that the fibres stretch and, at the same time, become thinner. This results in each individual fibre within the matrix now having a conical profile along its length – such a component is described as a fibreoptic taper. The fibres at the large end are the original size and taper down to a smaller diameter. Tapers are often characterized by the ratio of the large to small end size, i.e. a ratio of 3:1 (see Fig. 36.2). However, apart from being interesting technology, what advantage does this offer the biologist?

In most circumstances fibreoptic glass is a very efficient means of transferring light, especially when used in the demagnifying mode. The importance is that image intensifiers are usually 18 mm in diameter at the output and the CCD can be very small, typically 7 mm, and often smaller. A lens used to demagnify an image from the intensifier to the CCD may transfer as little as 1% of the light, whereas a fibreoptic taper can transfer as much as 10% of the incoming light to the CCD. The important factor is that the CCD will receive a much brighter image when fibreoptic coupling is used. However, a camera system that is lens coupled is inefficiently using the light output from the intensifier. This can be overcome by using a higher gain setting, and/or higher light levels to get the same output signal level and result in the intensifier being used harder than would otherwise be necessary, thus reducing its lifetime. More importantly, for a given intensifier, the use of fibreoptic coupling can increase the achievable sensitivity by an order of magnitude over a lens-coupled design.

In cameras that are designed for faster than video rate imaging, the CCD may be very small, since high clock speeds dictate smaller electrode capacitance and higher frame rates are more easily achieved with fewer pixels in the array. This results in the size of the CCD often being little more than a few mm^2, requiring considerable demagnification to transfer the image from intensifier to the CCD. In a design incorporating a relay lens, the losses associated with such inefficient coupling, along with the significantly smaller signal associated with such frame rates, makes such a design impractical.

With regard to fibreoptic components for transferring images from intensifier to CCD, light will be lost if more than one component is used. Some cameras are made using a straight fibreoptic block bonded to the CCD. A taper is then placed between intensifier and fibreoptic CCD. Since the fibres in each component are not directly matched fibre to fibre, this results in a 50% loss of signal at this junction, as well as a reduction in image contrast. The optimum is a single tapered fibreoptic component between the image intensifier and the silicon surface of the CCD. Good design ensures that image contrast is preserved by delivering light to each CCD pixel with a reasonable number of fibres.

36.12 THE CHARGE-COUPLED DEVICE

The CCD operating principle is called 'charge-coupling'. Finite amounts of electrical charge, often called 'packets', are created in specific locations in the silicon semiconductor material. Each specific location, a 'pixel', is created by the field of a pair of gate electrodes very close to the surface of the silicon at that location. By placing the storage elements in a line, for instance, voltages on the adjacent gate electrodes can be alternately raised and lowered and cause the individual charge packets beneath them to be passed from one storage element to the next.

A photoelectric effect, by which free electrons are created in a region of silicon, results when illuminated by photons in the approximate spectral range of 300 (ultraviolet) to 1100 (near-infrared) nm wavelength. Absorption of such incident radiation in the silicon generates a linearly proportional number of free electrons in the specific area illuminated. In a CCD, the number of free electrons generated in each site will be directly proportional to the incident radiation on that specific site. A focused image from an optical system creates charge packets in the finite photosite array that will be a faithful reproduction of the scene projected on its surface. After an appropriate exposure time, charge packets are simultaneously transferred by charge coupling under an adjacent single long gate electrode, to a parallel CCD analogue transport shift register.

Each charge packet corresponds to a picture element (pixel) and continues to represent faithfully the total sensed radiant energy, which was absorbed in the specific photosite. The transfer gate is immediately returned to the non-transfer clock level (LOW) so photosites can begin integrating the next line of incident image information. At the same time, now loaded with a line of picture information in the form of charge packets from a line of sensor sites, is rapidly clocked to deliver the picture information, in serial format, to the device output circuitry.

There are three types of CCD architecture, of which two types have an aluminium layer covering part of the CCD's surface, to provide a store area for rapid transfer and storage of CCD charge. The third type is a full-frame sensor, which does not have on-chip storage of charge for rapid read-out.

36.13 FRAME AND LINE TRANSFER CCDs

Interline and frame transfer sensors are both types of 2-D arrays of photosites, but functionally they are divided into image zones and storage zones. During 'exposure' the image falls in the image zone and builds up charge. At the end of the field time, this pattern is transferred to the storage zone, and is read out during the next field time, whilst the subsequent field is integrated in the image zone. A frame transfer sensor has (for example) an active area of 512 vertical by 1024 horizontal photosites or pixels. At the end of integration the whole image is moved 512 pixels vertically into the storage zone. The storage zone is protected from charge generated by light by covering it with an optically opaque mask such as an aluminium coating. Whilst the storage area is being read out, the next exposure is being made on the active imaging pixels. Hence, frame and interline sensors can work at 25 frames per sec with little dead-time between frames. The required resolution in the vertical direction is achieved by changing the electrode bias between odd and even fields to displace the potential wells effectively, such that twice the number of vertical elements are achieved in effect.

A line transfer sensor has the correct number of pixels, but each vertical column has an adjacent storage column. This means that only half the image area is actually light sensitive. This does not change the resolution, but it does reduce the sensor's quantum efficiency.

The only difficulty arises when time-resolved work is done. The frame transfer sensor actually integrates first one field and then the next, each with only 512 elements resolution. In photon-counting-type applications, where a photon 'speckle' may last less than a field time, it may end up being imaged only in one field. So a grabbed frame may show lots of speckles which are crossed with a blank line, since the events are imaged in one field only and when the subsequent interlacing field arrives at that spot, the image has gone. If this might cause a problem, use a line transfer sensor.

Line transfer sensors can either integrate each field for a field duration, or integrate each field for a frame duration, reading out 'out of phase'. Higher temporal resolution is achieved with field integration at the cost of a factor of 2 in effective sensitivity.

These CCDs can be cooled to remove thermal noise and the drive electronics designed for low read-out noise. Such cameras can be read out at 10- or 12-bit dynamic range. A cooled CCD can also be used to integrate charge on the CCD, since the incoming signal is still well above the thermal noise in the cooled CCD.

With larger CCDs, such as 1024^2 pixels, read out rates are about 10 MHz and slower than 25 frame/sec of video rate cameras. The read-out rate is often limited by the time required for accurate digitization of such signals.

36.14 FULL-FRAME CCDs

This third type of CCD architecture does not have a storage area on the CCD and a shutter is required to stop light from falling onto the CCD during read out. This type of CCD has traditionally been used in the scientific CCD camera, where speed of imaging is less important than quality of image and total detector sensitivity. When such CCDs are used as the output stages of intensified cameras, the image intensifier itself can be 'gated' to act as a highly efficient electronic shutter, which is, incidentally, capable of operating orders of magnitude faster than mechanical shutters, and is totally free of vibration. When a rapid sequence of images is to be taken, the mechanical vibration from a classical shutter can render the measurements useless.

The key point with full-frame CCDs is that the whole sensor area is 'active', increasing the sensitivity by about a factor 2 over line transfer devices. These sensors (particularly large-area versions) can make ideal read-out devices for very high performance ICCDs.

36.14.1 Large-area scientific sensors

True large-area CCDs have large pixels and can hold correspondingly more charge. Given that read-out noise is not pixel-size-dependent, this results in an increased dynamic range being available. For example, a small pixel CCD with 7 μm square pixels might have a full-well potential of 30 000 electrons, and (read slowly) a read-out noise of 10 electrons, so the dynamic range is 3000:1 and 12-bit digitization is entirely adequate. A large pixel CCD (the Thomson TH7896, for example) with 19 μm square pixels, has typically 450 000 electrons full-well potential, and with the same noise (under some operating conditions) the dynamic range becomes 45 000:1, i.e. 15 times better than the small pixel CCD, and able to justify 16-bit digitization.

36.15 CCD PIXEL SIZE AND DYNAMIC RANGE

Scientific CCDs usually have large pixels to handle the higher amount of charge from a bright point within the image in order to maintain a wide dynamic range within the image. CCDs with smaller pixels may not hold sufficient charge to enable meaningful digitization to 4096 grey scales (12-bit), let alone 65 536 grey scales for 16-bit! Whilst you can digitize the output of a CCD with small pixels to 12-bit or greater, the maximum signal that the pixels can hold may not be as high as 4096 times the CCD noise, hence higher levels of digitization are simply not justified. Over-digitization of small pixel CCDs is a 'specman-ship' game which can be very misleading. Cameras using small and inexpensive CCDs may be offered with 12-, 14- or 16-bit digitization, yet the silicon photosites of the CCD are simply not capable of yielding such performance. This is not to say that you cannot digitize such a CCD output to 16-bit, merely that you may have 5 bit of random noise and 11 bit of meaningful information, so there really is no point.

When a CCD is used as the output device in an intensified CCD camera, it is critical to ensure that the device is itself capable of the dynamic range required (irrespective of how many 'bits' the A/D converter may have). Unfortunately, fast read-out means high clock speeds, which requires lower device capacitance which in turn implies smaller physical structures. For this reason, providing similar resolution (pixel number) devices are considered, the faster the read-out (frame rate) the less the achievable dynamic range due to pixel size. Also, nature conspires to reduce achievable dynamic range from the A/D converter with increasing speed too. What this infers is that for an ICCD with high dynamic range, a large-area (large pixel) CCD has definite advantages. Some manufacturers argue that 'binning' (see Section 36.16) overcomes this problem, but most CCDs have output register pixel sizes similar to those of the main array, so the argument is simply untrue in most cases.

36.16 READ-OUT FEATURES

36.16.1 Binning

The read-out electronics may be programmed to combine two or more rows of charge into the serial register and/or to combine the contents of two or more pixels of the serial register at the output amplifier. During this process a 1024^2 pixel CCD will behave as if it has 512^2 pixels (binning 2×2). Binning offers a trade-off between reduced resolution and increased sensitivity, since each new super pixel in this example would have four times the charge of an individual CCD pixel.

36.16.2 Regions of interest

CCD cameras can be programmed to process only those pixels in a selected subarray or 'region of interest' fully, rapidly discarding the unwanted pixel information outside of this area. This technique is important to slow-scan camera operation, where read-out rates of around 200 MHz can be considerably speeded up. However, for faster read-out cameras that offer 12-bit at 5 or 10 MHz, the region of interest subarray may be less important as the limiting factor is the maximum CCD clocking speed and not the A/D conversion speed.

Both binning and regions of interest are used as techniques for increasing the frame rate. However, when the pixel clock speed is limited by the maximum clock speed allowable for the horizontal (serial) register of the CCD, the only gain in speed results from the vertical binning factor or the vertical faction of the whole array occupied by the region of interest.

36.17 HIGH-RESOLUTION DIGITAL ICCDs

Traditionally, an ICCD would be an intensifier coupled to a video camera providing a TV standard read-out, with 750 TV lines per picture width. However, development taking place within the electro-optics industry is such that many new products are now available and the range and diversity of both image intensifiers and CCDs is both vast and growing. Improvements in image intensifier technology, with the development of high-resolution image intensifiers and the improvement in low-noise performance that have been achieved, permits the construction of meaningful high-resolution ICCDs which can fully use the 1024×1024 pixels, whilst justifying digitization to better than 8-bit.

High-resolution intensifiers have become available due to improvement in the manufacture of the micro-channel plate, which is the gain stage in the intensifier. The important factor is the size and density of the channels, which have become smaller (6 μm) and more closely spaced. This component, when used in an intensifier manufactured for high resolution, improved the resolution of the intensifier from typically 34 lp mm^{-1} to about 57 lp mm^{-1}. The high-resolution intensifier can now be used as the front end to a high-resolution CCD camera, and would be wasted on a camera with less than 1000 pixels.

Normal TV standards lack the vertical resolution to exploit the benefits of these high-resolution image intensifiers fully. This is because standard CCIR video had only 625 lines vertical resolution, and the US EIA standard had even less, with only 525 lines. In scientific imaging systems where vertical and horizontal resolution should be the same, this makes TV standards inappropriate for high-resolution imaging systems.

The low-noise, high-resolution image intensifier is at a point where it can use the higher-resolution CCDs, previously used in slow-scan scientific cameras, as a suitable read-out stage, producing a fusion of intensified high-resolution cooled digital 10- or 12-bit camera, with a diverse range of features from both intensified and slow scan origin.

A recent innovation is the progressive scan camera which uses interline transfer CCDs with 1024 × 1024 pixels, where the CCD is clocked out, not as a TV standard, but as a stream of pixel data, producing a camera that produces typically 10 frames/sec. This type of camera is well suited to the high-resolution image intensifier. When the camera is cooled and, if designed to offer low-noise electronics such that the signal can be read out at 10- or 12-bit, the digitization to 10- or 12-bit is justified.

Another option as the read-out stage of a high-resolution intensifier is the full-frame scientific sensor which has been traditionally used as a slow-scan CCD, and cooled to reduce noise and blemishes. However, this type of CCD has the advantage of being physically a large sensor with large pixel size, which can be digitized to 12-bit. When used as the read-out for an intensified camera, the intensifier can be gated to act as a shutter to the CCD, hence offering a full-frame sensor with maximum detection area, to maximize the capture of signal from the intensifier. A further advantage of using a large CCD is that the fibreoptic component, used to conduit light from the screen of the intensifier to the sensor, could be a coupler of straight fibres and not a taper, which increases the amount of light transmitted. This type of ICCD is both new and highly sensitive, offering the best features of the slow-scan technology, with the speed and resolution now available with the modern intensifier.

36.18 FAST READ-OUT ICCDs

In recent years, developments in camera technology have allowed increasingly faster recording of transient events in calcium ion ratio imaging. Such cameras can provide images far faster than the traditional TV rate of 25 frames a second.

Cameras designed for such applications can provide up to 800 frames a second. However, the trade-off tends to be resolution, i.e. 200 frames per second with 256 × 256 pixels or 800 frames per second with 128 × 128 pixels, typically offering 8 bit of dynamic range. The limit, at this time, is mostly trying to stream the image data to some form of image capture system, usually through the bus of a computer. The camera technology is somewhat ahead, though computer architecture is rapidly catching up.

ICCDs with fast frame rates in the order of 200 plus frames a second have several characteristics, which need to be considered when assessing such a camera for microscopy applications. Fast decay time phosphors, sufficiently high intensifier gain, and efficient light transfer within the camera (fibreoptic coupling rather than lenses) are all aspects that need to be considered when designing a system for high-speed applications. It is not simply sufficient to replace the CCD of a conventional low-light camera with a faster one!

As discussed earlier in this chapter, at high frame rates it is absolutely necessary to use an appropriate fast decay time phosphor as the output of the image intensifier A slower phosphor would produce lag in the following frames, reducing the signal-to-noise of the camera and distorting subsequent results. Unfortunately, high-speed phosphors tend to be less efficient than traditional ‘laggy’ ones, and this results in reduced gain, which often needs to be compensated for by other mechanisms. Most military intensifiers feature deliberately slow phosphors to reduce the perceived ‘speckle’ when used in nightsights, and are therefore particularly inappropriate for high-speed camera use.

Usually, when imaging at a higher frame rate, the experiments have been undertaken at higher light level, especially in physics, since a camera running at 200 frames a second is going to capture only 1/10 of the flux per frame than the same experiment run at 20 frames a second. However, for the biologist, the amount of calcium ions that can be placed within the cell may be limited and furthermore, ‘turning the wick up’ on the microscope illumination may not be good for the specimen (or for the lifetime of the camera!).

To image weak signals meaningfully at high speed, the camera needs to have very high gain, and this has to be provided within the intensifier of the camera. This can be achieved by the use of special tubes, either using hybrid technology or in some cases two microchannel plates. Since there is little signal, it is an advantage to have a very-low-noise intensifier.

Initially this is not obvious, since the noise per

frame is clearly reduced in proportion to the frame time. However, when the gain is increased, the quantum nature of the signal becomes evident, and it is advantageous to have the sparse image data corrupted as little as possible by noise events. Simultaneously, it is an advantage to optimize the light transfer between the image intensifier and the CCD, otherwise unachievably high gains may be needed from the intensifier, resulting in unsatisfactory design compromises. The use of a tapered, coherent fibreoptic light guide is the method of choice. Ideally, this component should be bonded at its small end, directly to the active silicon surface of the CCD. Many manufacturers do not have this bonding technology, and instead purchase CCDs with a small fibreoptic window already attached by the CCD manufacturer, to which a tapered fibreoptic may be attached in turn. This approach overcomes the need for advanced technology in this part of the camera's manufacture, but brings with it a loss of 50% of the light and reduces the contrast in the image.

One could say that the use of anything other than direct bonding of the taper virtually halves the frame rate achievable for the same quality of image.

With the image captured, the problem becomes what to do with the vast quantity of data! An appropriate frame grabber and software will be necessary along with a large amount of computer memory into which the data can be streamed. Such systems already exist and have the image processing for fast calcium ratio imaging and control software for fast frame rate ICCDS.

As developments in fluorescence ratio imaging have occurred, the requirement to see a faster sequence of events has also grown. However, there is a trade-off between resolution and speed of getting an image to read out; the data rate is the limiting factor. 512^2 pixels at 25 frames/second can become 256^2 at 100 frames/second. The sensor architecture and dedicated read-out electronics often limit what can be achieved, the main limit being data quantity even at 8-bit.

Intensified CCDs with fast read-out need the ability to have high gain to collect the few photons available and a fast phosphor with a rapid decay time. Fibreoptic coupling also becomes very important in order not to throw away any photons.

For faster temporal dynamics, gating becomes a further option. This allows the selection of short temporal events, which can then be seen to progress in snapshots at 200 frames a second. The intensifier will need a special power supply and gating unit.

Special attention needs to be given to supporting software and hardware, which must take a very fast data stream and save to disk all frames. Frame grabbers and commercially available software and hardware for capturing these images is now easily available.

36.19 DIGITAL ICCDs WITH INCREASING BIT RANGE

Slow-scan large-area CCDs with digital read-out have also made considerable advances, to a point where a 12-bit digital image can be read out at speeds up to 10 MHz. This high-quality camera system with cooled CCD and a large number of pixels (typically 1024^2 or larger) produced a high-resolution low-noise image with sensor defects and thermal noise chilled out.

Recent developments of high-resolution low-noise intensifiers have resulted in a suitable front end appropriate to the high-resolution digital camera. Now the old obstacles of poor resolution and noisy intensifier have been removed, permitting the coupling of these two different techniques into a sophisticated new range of digital ICCDs.

New developments in quality fibreoptics allow the large area scientific sensor to be coupled to the intensifier. Since the larger scientific CCDs are often compatible in size to the image intensifiers output image size, a straight coupler can be used. This results in an efficient transfer of light, which is far better than a lens or fibreoptic taper.

The combination of these new technologies results in a 12-bit digital ICCD with fast read-out when compared with slow-scan, offering high-resolution, high-quality imaging at low light levels. Features such as binning and windowing (regions of interest) can improve sensitivity and increase read-out rate.

This type of camera does, however, need a digital frame grabber, and some control software. This code is available for those who wish to put their own systems together, since the software drives control camera functions.

More commonly, a range of frame grabbers and software is widely available to run such cameras. With the development of core modules and drivers at very low prices, these can sit alongside other third party software, providing a cost-effective platform for controlling the camera independent of the third party image analysis.

36.20 ANALOGUE VIDEO ICCDs

The most commonly available ICCD is usually a second-generation image intensifier coupled to a video camera, which has the advantage of producing a stream of images and can read out the CCD in either frame rate (25 frames/sec) or field rate of (50 fields/sec). TV rate CCDs typically have 750 plus pixels in the horizontal and read out by either frame transfer or interline transfer. Some modern cameras chill the CCD to remove CCD defects, i.e. bright pixels and thermal noise. In recent years, improvements in making low-noise image intensifiers, along with improved resolution, has changed the quality available to the scientist. This usually reflects in an improved modulation transfer function, which is an important factor for clarity of the image. For many biologists the science may not demand resolution better than 512 pixels in the horizontal. However, the need for a sharp unfoggy image is all-important.

What really matters is the modulation transfer function (MTF) rather than limiting resolution. This function is a plot of contrast (as percentage) as a function of spatial frequency when looking at a black and white (100% contrast) target. The contrast seen through the camera will fall as the spatial frequency increases, down to 4% at the 'limiting' resolution. The contrast at low spatial frequencies tells you how clearly you will see what is 'resolved', or whether 'black and white' has degenerated to shades of misty grey.

Factors that improve MTF are varied but the resolution of the intensifier, which should be higher than the CCD number of pixels, the quality of the lens or fibreoptic used to transfer the image from intensifier to CCD will also effect MTF.

In Fig. 36.3 we show the MTF curves of two cameras which have the same resolution: camera A will give clear, crisp images, whilst camera B will look as if it was always seeing through fog. For instance at 250 TV lines per picture width, camera A would give 80% contrast between black and white, while camera B would give only 30%. This is an important consideration for some (but only some) applications, yet it is information that is almost never published. Yet again, the only way to test the suitability of a camera for a particular application is to test it. The MTF curve may at least give some insight into why some cameras work better than others do, even though the 'resolution' is the same.

36.21 PHOTON COUNTING IMAGING

When the light levels become really low, and gain is as high as is useful, the image will simply consist of a succession of bright 'speckles' on the screen. This kind of image, rather like looking into a snowstorm, results from the quantum nature of light. The speckles each result from the detection of an individual 'photon' or light quantum. Increasing gain further will only make the individual speckles brighter. The only improvement that can be made in these situations is to ensure maximized DQE in order to obtain the maximum number of speckles for a given number of photons.

This kind of image, which can be summed, integrated or averaged in a conventional frame store to enhance signal-to-noise significantly, can be treated in quite another way. Rather than thinking of the quantized nature of the image as a difficulty, it can be turned into an advantage. Most simply, one takes a frame of video and looks at each pixel and decides whether there is a bright speckle in it or not. In the computer memory, each pixel has an address, and if, for the frame in question, there was a speckle present in several pixels, the number 'one' is added to the content of each corresponding memory location. This is repeated frame after frame, for each (memory location) pixel adding 'one' when a bright speckle is seen in the corresponding place, 'zero' otherwise. This

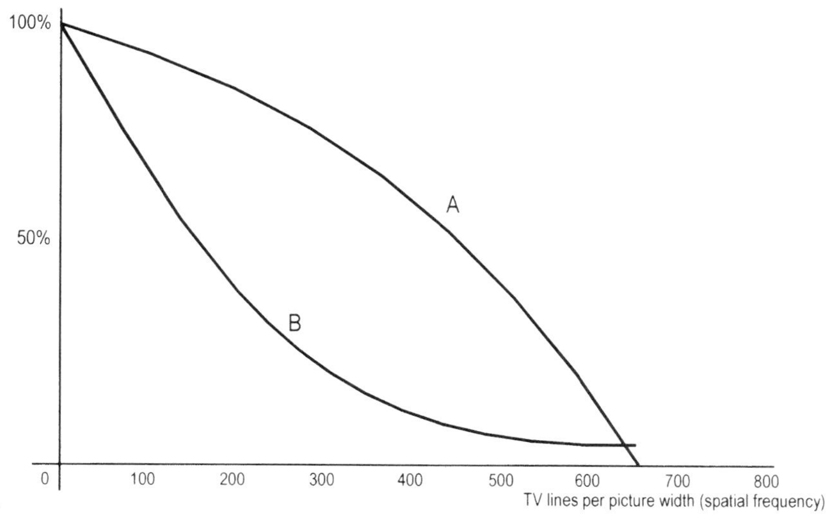

Figure 36.3 MTF curves of two cameras with the same resolution.

can be likened to a rather large array of photon-counting photomultipliers, and indeed, the technique has been used very successfully in astronomy and nuclear physics, where a huge 2-D array of PMT would be quite impractical.

The method is very simple in principle, but rather more difficult to implement in practice. The main problems occur because the 'speckly' sizes are not all the same in terms of intensity, even though they each represent precisely one photon. To gain advantage from this technique, it is necessary to reject the smaller speckles that represent noise arising 'later' in the camera's gain mechanism. By so doing, signal-to-noise ratios can be improved, even at these ultra-weak signal levels. This will only work if the 'spread' of speckle brightness somehow enables a distinction to be made between real photon-induced events and noise; otherwise rejecting the small events will throw out as large a fraction of the signal as it does of the noise, and no improvement will result. For this reason, a camera must be designed rather carefully to work in this mode, and normal ICCDs just will not perform.

This technique will not enable you to overcome photocathode dark (or 'thermal') noise, since a photoelectron emitted by the photocathode as a result of dark noise is indistinguishable from one released by a photon. Other problems arise since speckles rarely fall only in one pixel. Even if speckles were exactly the same brightness, then there would be quite a range in pixel value between the occasions where a speckle fell directly on a pixel and the cases where the event occurred at the junction of four pixels (simplistically, a 4:1 ratio). Sophisticated software is needed to prevent this latter case being recorded as four events.

There are, in most intensifier-based cameras, a few bright events that are significantly more intense than the average. These arise from ions, and can be discriminated against by rejecting very bright events.

In essence all this is accomplished by selecting a 'window' function for what qualifies as a photon event, a lower threshold to reject low-amplitude noise, and an upper threshold to reject ion events. This is virtually identical to the techniques long established for photomultipliers, except that here we are only working with a 2-D array of between 250 000 and one million channels! There are clearly trade-offs to be made between maximizing DQE and minimizing noise, as a result of window width and position.

To do this intelligently it is vital to understand the amplitude spread of the photon events, called the 'pulse-height distribution'. Ordinary intensified cameras have a negative-exponential form to their pulse-height distributions, and it is clear that setting of discriminator thresholds in this case is quite arbitrary. To work well, it is necessary to have a distinct 'photon peak' in the distribution, and it is to this end that specially designed cameras become essential.

Attempts have been made to ascribe significance to the speckle size other than 'one photon or zero', by say, 'one, two or three?' etc. with each frame. It is very doubtful whether present cameras can generate such narrow photon peaks. Here we see a compromise between photon counting and analogue 'total charge' integration, but such techniques can still contribute a little to noise reduction.

The technique has been called 'pseudo photon counting'. At its best, it is a very powerful technique, and at its worst, is truly 'pseudo' in as much as it will perform the same as analogue techniques at very low light levels, but will probably cost you far more for the equipment.

This kind of imaging is just making its way into biomedical applications and should be given a chance to prove itself in this area as it has already done in the physical sciences. As biologists, readers should take great care, however, to see that they are buying a genuine technical advantage for their science, and not just paying heavily for the latest gimmick. Yet again, the way to do this is by 'hands-on' evaluation in the specific application.

36.22 SYSTEMS INTEGRATION

36.22.1 Analogue cameras

36.22.1.1 Integration of a video ICCD into an image processing system

When connecting a video camera to a frame grabber there are five important questions that should be asked:

(1) Does the camera need to run in to a 75 Ω load and, if so, is the frame-grabber input configured this way?
(2) Will the frame grabber synchronize from the video signal, or does it need another separate sync input?
(3) Does the computer/frame-grabber need to input its sync for the camera to follow?
(4) What signal level does the frame-grabber call 'black' and is the camera set up so that its black level corresponds?
(5) What signal level does the camera call peak white and is the frame-grabber set up to accept this?

36.22.1.2 Synchronization

Virtually all video cameras provide 'composite video', which contains sync information. Most frame-grabbers are able to separate the sync from the video, and then use this to synchronize digitization and data storage. Some frame-grabbers cannot do this or cannot do it very well, and in this case it is necessary to feed it separate sync pulse from the camera to the frame-grabber. Most cameras (or their controllers) provide a 'compound' or 'mixed' sync output and this can be fed to the 'sync in' input on the frame-grabber. This signal will probably be at TTL levels and is best fed by a coaxial cable.

In this configuration, the camera is the 'master clock' and controls the timing of the frame-grabber. In some cases, particularly with fussy computer systems, the frame-grabber is given a disproportionately high level of importance and is configured to drive the camera. In this case the frame-grabbers' 'mixed sync out' should be fed to the 'external sync input' of the camera.

36.22.1.3 Setting black level

Many cameras produce the standard video signal with an arbitrary DC offset, so the bottom of the sync pulse is, for example, at +1.5 V DC. In most cases, sophisticated frame-grabbers are happy with this, since they will AC couple the signal and then change it to a DC level suitable to their internal circuitry. There are, however, a few otherwise very good products around that are not so accommodating. Such devices tend to take DC values as significant, so a black level at 0.5 V DC will result in the computer assigning a mid-grey image to its interpretation of black! Some cameras will let you adjust the DC offset. Almost all frame-grabbers that are not AC coupled will have an adjustment potentiometer on the board to set the black level. A few 'difficult products' will allow you to do this only via software.

Ideally, check if your frame-grabber produces a black image (i.e. near 0) for a dark input. Do not pursue a precise 'zero' too hard, as there may be a slight 'pedestal' when the active video line is slightly above the black level.

If not, and the manual says the input is DC coupled, find out the accepted range of the frame-grabber. This will probably be levels in the range +0.3 to 0.8 V DC or something similar. Now, with the camera connected, feed dark images to the frame-grabber and digitize, adjusting the potentiometer or software setting until you get reducing values and eventually zeros. If this cannot be achieved, the camera DC offset is probably too great. Check this on an oscilloscope with the input DC coupled. Look in the camera manual on how to change the DC offset, or call the camera manufacturer. Set the camera's DC output level to within the accepted range of the frame-grabber, then make fine adjustments to either or both, until black gives a digital value of near zero.

36.22.1.4 Setting peak white

Having set black = 0 (roughly), it is now necessary to determine that saturated white = 255 (with 8-bit digitization) or near enough. Again, check that the camera output is specified as '1 V peak-to-peak into 75 Ω' and that this is what the frame-grabber is looking for, and also that the 75 Ω load is connected. With an oscilloscope, adjust camera gain and/or image brightness so that a peak signal level of just over 0.7 V above the black level is obtained.

Now grab a frame and check the pixel values corresponding to the peak part of the signal. These values should be around 255. If not, then there is a problem. Check the manual for the frame-grabber, if there is no potentiometer to adjust, follow the procedures described or adjust the analogue-to-digital converter gain in the software.

These four steps can be summarized as:

(1) Impedance: Matching camera to frame-grabber
(2) Synchronization
(3) Black level: Setting up so black = 0
(4) Peak white: Setting up so peak white = 255

When you have achieved these steps, the camera should be integrated successfully into an image-acquisition system.

36.22.2 Digital cameras

36.22.2.1 Integration of a digital ICCD into an image processing system

A digital camera combines the CCD and its drive circuitry and an A/D converter into one unit, providing a digital signal out of the camera. This reduces analogue noise and signal deterioration in interconnecting cables. Such settings as range and offset will have been optimized by the camera manufacturer, so there is no set-up required to optimize the image.

Digital camera images can only be displayed via a computer, which is fitted with a digital frame grabber and suitable software. Typically you only get an image when the software and frame-grabber drivers are configured correctly. This can make digital systems harder to configure, as 'nearly right' or 'miles off' give the same result – no image!

Video contains a lot of data, so to provide 'live' video, fast data rates are required. This would typically require a 100 MHz + Pentium PC with PCI bus or a computer with similar data throughput.

A frame-grabber is still required to get the digital data into the computer. Digital frame-grabbers are often sold as add-on boards for analogue frame grabbers. Dedicated digital frame-grabbers are also available, although, strangely, at the time of writing, being mostly far more expensive than frame grabbers with their own A/D (i.e. analogue input boards).

Most digital cameras use differential parallel data since this is fast and resistant to noise. Maximum data speed is inversely proportional to cable length, so the faster the camera, the shorter the cables must be. The digital cable will contain many wires. A 12-bit camera would require typically 24 data lines 6 sync lines + control lines. There is no truly standard connector. It is necessary to get a special cable for each frame grabber/camera combination.

There are some serial digital cameras, but until recently they were unable to transfer data at a sufficient rate to give live images. This is likely to change in the near future due to the introduction of a new high-speed serial link called 'FireWire'. This is likely to become a new standard, making digital cameras much easier to connect and run on computers, and putting an end to the need for ridiculously expensive 'digital' frame grabbers.

36.23 FLAT-FIELD NORMALIZATION – THE MAGIC TOUCH

In a real camera, you will have some of the defects described above. There is a technique, developed by astronomers, for simultaneously removing the effects of shading, chicken wire and grey spots. The image is not sensitive to where in the camera these problems arise, and all the various contributions from different components can be treated at one time.

Precise details depend on the configuration of the computer system but the principle is the same: project onto the camera an image of a uniform featureless object, weakly illuminated, whose brightness is about the same as a typical integration, and look at the image. It might be awful, showing all the shading, chicken wire, dark defects, etc., but this is precisely what you need to see. Make sure that the integration is long enough to iron out quantum statistics, and thus giving a really good image of all the defects. Store this image. Now run some statistics on the image and find the average pixel value. It might vary considerably over the image, and, providing the featureless, uniform object truly was featureless and uniformly illuminated, this represents the variation of response of the camera, but results in obtaining one (and of course only one) average for the whole image. Then divide the whole image, on a pixel-by-pixel basis, by this scaler quantity (the average). Having obtained an 'image' where everything is close to unit (so in practice some care will be needed to avoid loss of accuracy), a pixel in the new image that was at the average value in the original will have a value of 1.00. A pixel in the new image corresponding to a grey pixel in the original of one-half average brightness will have a value of 0.5. This new image can be called the 'correction image'.

The next step is to divide any real image pixel-by-pixel by the correction image, and all the defects seen earlier will virtually disappear.

This technique, 'old hat' to astronomers, can do more than correct for the camera defects. Put your featureless, uniformly illuminated object under the microscope and perform the technique with the camera viewing in this way, and you take out shading in the microscope too!

Extensions of the technique can correct for uneven illumination, optical vignetting and a wide range of other shading problems. The technique will not handle true black and white spots, which it can often just reverse. Pixel fixing is still required. Neither, of course, can this method correct for distortion. It is, however, when properly applied, a virtually universal approach to shading problems.

CHAPTER THIRTY-SEVEN

Properties of Low-light-level Slow-scan Detectors

R. AIKENS
Photometrics, Ruscon, Arizona, USA

37.1 INTRODUCTION

Electronic image acquisition has undergone phenomenal advances over the last 20 years. The major imaging technology innovation of the last few years has been the introduction of a detector called the charge-coupled device (CCD) which replaces bulky, less sensitive scanning camera tubes such as vidicons. This discussion focuses on the theory of CCD imagers, slow-scan CCD camera implementations and the utility of CCDs in light microscopy.

37.2 CONTEMPORARY IMAGE-ACQUISITION TECHNOLOGY

37.2.1 Video cameras

The most familiar form of an image-acquisition system is the hand-held video camera. A modern video camera employs a CCD and generates a standard NTSC video signal which is compatible with a multitude of recording and display devices. A video camera operates at 30 frames per second and with a fixed format. Typically, a video image is composed of 525 lines, each with 250–600 elements of horizontal resolution, depending on the quality of associated electronics.

Video cameras have been used as a tool in light microscopy for several years (Inoue, 1986). An impressive array of video-based equipment is available which permits the user to digitize, process, display and archive images obtained with video cameras. However, video cameras have their limitations. Since they must operate within rigid established constraints, their performance is limited. Modern video cameras employ CCDs which are specifically designed to meet the requirements of video imaging by trading-off features which the CCD might otherwise offer. There is a new generation of video components emerging based on high-definition television (HDTV) which will offer higher resolution and movie-like image quality. The components which come out of HDTV will undoubtedly find utility in scientific imaging; however, price and performance will be tuned for the high-volume consumer which will limit the usefulness of these devices for quantitative imaging.

FLUORESCENT AND LUMINESCENT PROBES, 2ND EDN
ISBN 0–12–447836–0

37.2.2 Intensified video cameras

There is a class of low-light-level video cameras which employ image-intensifier stages in front of the basic sensor. The most common of these, the silicon intensified target vidicon (SIT), employs a traditional vidicon target camera tube with an integral image-intensification section. An image is focused on a photocathode and photoelectrons are accelerated toward a standard silicon target in the vidicon section of the tube. Conventional video electronics are used to render a standard NTSC video signal. The intensifier–vidicon combination allows individual photoelectrons to be detected but at the expense of dynamic range. The quantum efficiency of photocathodes is low, especially in the red region of the spectrum, which limits the useful range of these cameras. SIT cameras find utility where light levels are low and when it is required to observe a scene in real time. SITs and intensified SITs (ISIT) are expensive as well as fragile and are being replaced in many applications by high-performance cooled CCD cameras. Low-light-level video cameras will continue to find utility in many situations for years to come (Spring *et al.*, 1989).

37.2.3 Cooled slow-scan CCD cameras

The CCD is capable of remarkable performance when operated in conjunction with ideal electronics. In the 1980s, astronomers and scientists at several national laboratories and science centres made enormous progress on the development of essentially ideal CCDs for use in research and industrial environments. Most of that CCD development (Janesick, 1980–91) was not commercialized until the mid-1980s. Innovative CCD cameras were designed to capitalize on the performance of the CCD instead of conforming to video or television standards. A modern cooled CCD camera is capable of spatial resolution up to 16 million pixels, ultra high intrascene dynamic range in excess of 50 000 to 1, and exceptional sensitivity with the ability to integrate for hours in low-light-level applications. Several manufacturers now produce premium-quality slow-scan cooled CCD cameras which find utility in light microscopy and other disciplines.

37.3 HOW CCDs WORK

37.3.1 Characteristics of silicon

CCDs are processed in silicon much like other MOS integrated circuits. They range from a few square millimetres up to several square centimetres in area. It is useful to examine the properties of silicon to better understand CCD principles.

Silicon is a semiconducting element found between phosphorus and boron in the periodic table of elements. It can be purified and formed into large single crystals and sliced into very thin sheets for fabrication of integrated circuits. Photons of energy greater than about 1 eV, which penetrate silicon, will break the

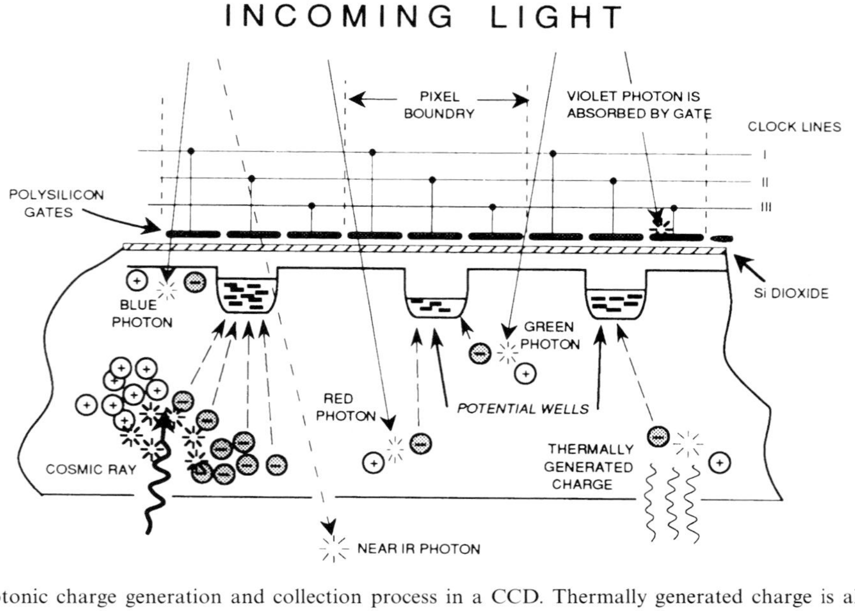

Figure 37.1 Photonic charge generation and collection process in a CCD. Thermally generated charge is also collected in potential wells.

covalent bonds that bind atoms together in the lattice, thus creating hole–electron pairs. Energy in the form of heat, even at ambient room temperature, can also break bonds, creating charge pairs. High-energy particles and cosmic events which are always present, produce multiple hole-electron pairs. Silicon is opaque to light of wavelength shorter than 400 nm and becomes transparent at 1200 nm.

37.3.2 The potential well concept

It is possible to capture the electronic charge generated by photon interaction in silicon before it recombines and to store it in potential wells. Potential wells can be produced near the surface of the silicon through the application of appropriate voltages to a series of insulated gate electrodes. A device capable of integrating photon-generated charge can be configured, as illustrated in Fig. 37.1. In this instance, the gates are positioned to create adjacent, but uncoupled potential wells. Note that light must pass through the electrically conductive polysilicon gates which limits the wavelength response of this device.

After photonic charge has accumulated in the discrete potential wells, it may be transferred through the silicon along channels as illustrated in Fig. 37.2. Photonic charge is swept along by drift fields as the gate potentials are changed in sequence to propagate the well to the right. This process is continued until charge from individual well sites is transferred to an output amplifier for measurement. This charge transfer mechanism is the fundamental principle underlying CCD operation.

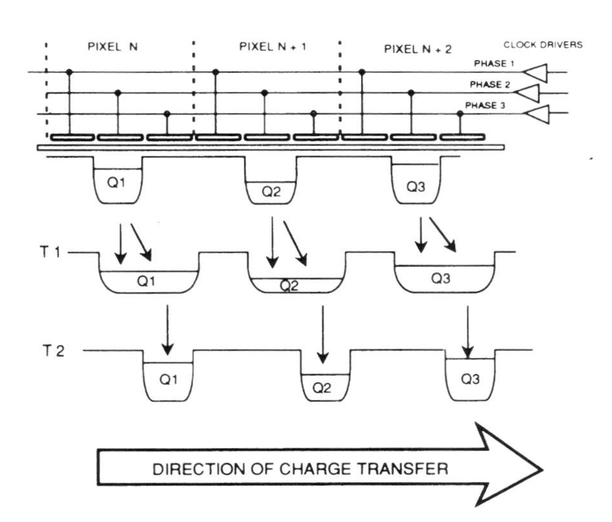

Figure 37.2 Integration of photonic charge and charge transfer concept underlying CCD operation.

37.3.3 The two-dimensional CCD

Figure 12.3 illustrates how the CCD concept can be extended to a two-dimensional structure. The imaging matrix is made up of light-sensitive CCD elements called picture elements or pixels. Note that the exposure of the CCD to light is a separate and distinct operation which precedes the application of the CCD principle to the read-out of integrated charge. After the termination of an exposure, entire rows of accumulated charge may be transported toward a serial CCD register. In the serial CCD register, individual charge packets are transported to a read-out amplifer where a measurable signal proportional to the amount of charge from each pixel is produced. The read-out process progresses row after row until the entire CCD has been cleared of charge.

Clearly, unless the charge transport process is extremely efficient, electronic charge will be lost or will corrupt adjacent pixels. In fact, modern CCDs exhibit outstanding transfer efficiencies as high as 0.999 999, where 1.0 is ideal. The CCD output amplifier has been developed to yield ultra low noise and high photometric linearity over a wide range of operating levels.

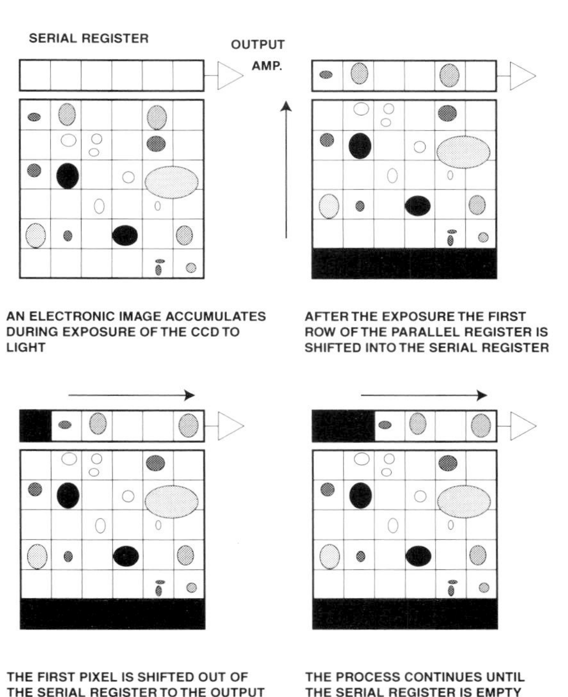

Figure 37.3 A two-dimensional CCD imaging device. The exposure and read-out of integrated charge are two distinctly separate processes.

37.3.4 CCD architecture

There are three basic CCD architectures, two of which are illustrated in Fig. 12.4. A full-frame CCD imager with four output registers allowing simultaneous read-out of four imaging quadrants is shown in Fig. 37.4(a). This configuration is used when a short read-out time is required. The frame transfer CCD in Fig. 37.4(b) is really two CCDs in the same package. One is used for imaging, while the other is used for temporary storage and read-out. Photonic charge is allowed to integrate in the active imaging area. After the integration is over, the resulting electronic image is rapidly shifted up into the masked storage register. This shift operation typically takes 2 μs per row, or about 1 ms for a 512 × 512 element CCD. While the storage area is being read out in conventional CCD fashion, the next exposure is made in the image area, thus there is little dead time between frames. A frame transfer CCD may be operated at 30 frames per second to produce a television image. The frame transfer CCD suffers from image smearing which occurs during the image transfer in the presence of light. A synchronous rotary shutter can be employed to eliminate this problem.

A third style of imager, the interline transfer CCD, employs masked transport registers interleaved with the imaging area to allow continuous imaging in a television mode. This CCD design suffers from a loss in sensitivity because light striking the covered transport registers does not contribute any signal. The advantage of this architecture is that it is inexpensive to produce since the total required area is that of only one CCD. Many variations of the interline transfer CCD have been used in television cameras; however, this device is rarely used for scientific imaging because of its poor sensitivity and venetian blind spatial sampling properties.

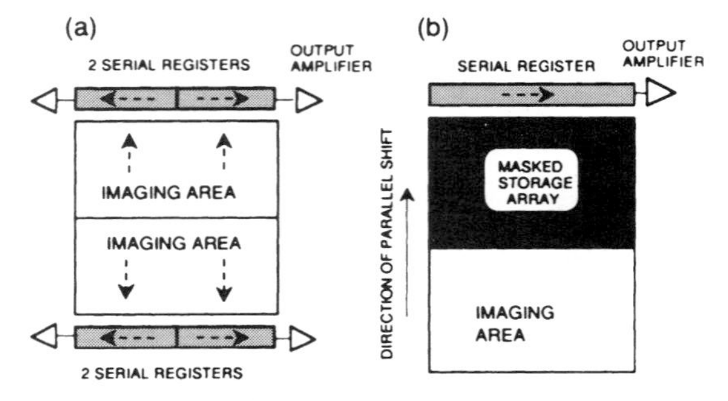

Figure 37.4 Two popular CCD architectures. (a) Full-frame CCD with four read-out registers. This permits high speed read-out, an especially useful feature in large CCD arrays. (b) Frame transfer CCD.

37.4 THE HIGH-PERFORMANCE SLOW-SCAN CCD CAMERA

When the CCD is removed from the video realm, it can be operated at any speed and in any read-out manner which fits the application. The CCD may be cooled to prevent the formation of thermally generated charge, which is indistinguishable from that created by light. Charge so generated, called dark current, can significantly degrade CCD performance by using up well capacity and creating undesirable noise. Since a slow-scan CCD camera need not conform to video standards, read-out speed can be reduced to realize ultra-low-noise performance. Novel masking and read-out modes permit very high frame rates through the use of a portion of the CCD for short-term electronic image storage. A well-designed cooled slow-scan CCD camera offers the user a variety of read-out speeds, cooling options, digitizing levels and CCD array sizes.

37.4.1 Subarray read-out

A slow-scan cooled CCD camera may be programmed during read-out so that only pixels in a selected region of interest (ROI) are processed. The CCD exposure is made as usual, but after it is terminated the parallel and serial registers are programmed to slew through regions outside the ROI and process only those pixels within it. Several ROIs may be read out after a single exposure. Since the time to process and digitize each pixel is fixed, the smaller ROIs allow higher frame rates. A 100 × 100 pixel ROI in a 512 × 512 pixel CCD may be read in 10 ms at a pixel read frequency of 1 MHz.

37.4.2 Charge grouping

During the CCD read-out, a slow-scan camera may be programmed to combine rows of charge into the serial register, or to combine two or more charge packets at the output amplifier. This process is commonly referred to as charge grouping or binning. During the binning process, the CCD is operated at reduced resolution in exchange for increased dynamic range and higher frame rate. A 512 × 512 pixel CCD binned 2 × 2 yields an image which has 256 × 256 super-pixels, each with four times the charge capacity of the individual CCD pixels. When programmed as in the example, the entire CCD may be read out in approximately one-quarter the time required for a full resolution image.

37.4.3 Camera implementations

Figure 37.5 is a block diagram of a contemporary slow-scan cooled CCD camera system. The term 'slow-scan' as used here means that the read-out speed of this camera is significantly lower than that of a video camera. The pixel-read frequencies for slow-scan cameras range from 50 kHz up to 2 MHz. The slow-scan read-out allows sufficient time for charge to be effectively transferred in the CCD registers and for the resulting output signals to be digitized to a high degree of accuracy. A host computer is shown in the illustration, because it is an important integral component of a slow-scan CCD camera system. The host computer affords the user the option of setting up a variety of operating modes under program control. Data processing and archiving are also under control of the host computer. The host is used to operate instrument accessories such as shutters, filter wheels and light sources. The slow-scan CCD camera components shown are the camera controller, camera electronics unit and camera head.

The camera controller contains the logic which causes camera action based on input from the host computer. It also passes digitized pixel data to the host computer, usually over a DMA channel. The camera controller generates all the sequences required for clocking the CCD phases and timing for the analogue processing circuits.

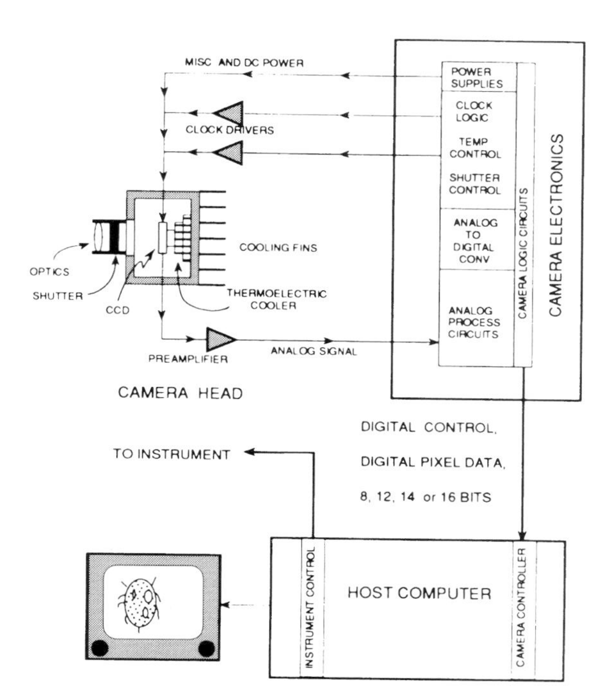

Figure 37.5 A typical slow-scan CCD camera and host computer configuration. An important system element is the software which must efficiently acquire, process and archive large amounts of data.

The camera electronics unit transforms digital commands and signals into active CCD clocking levels and sequences. It also performs analogue processing on the CCD output signal prior to digitization. The analogue-to-digital converter is usually contained in the camera electronics unit, although it may, in some cases, be located in the camera head in close proximity to the CCD. A 16 bit digitizer requires very high-performance analogue electronics and low-noise circuit design to effectively utilize the available CCD dynamic range. Digitizers from 8 to 16 bits are commonly used in slow-scan cameras depending on speed and dynamic range requirements.

The camera head contains the CCD and associated cooler. The cooler may be a Peltier device, or in extreme cases where exposure times are long, liquid nitrogen may be used. The CCD is usually enclosed in a hermetically sealed chamber to prevent the formation of frost or condensation. A window is provided to allow the entry of light.

A local preamplifier is employed to provide a strong signal which can drive a low-impedance line to the analogue electronics. The analogue electronics, including the analogue-to-digital converter (ADC), may be located in the camera head, especially if the read frequency is over 2×10^6 pixels per second. A shutter is employed with full-frame CCDs to prevent light from corrupting the CCD image during the read-out process.

CCDs are extremely sensitive to over voltage conditions or electrical transients, therefore, the camera system must be designed to protect the CCD from outside electrical disturbances.

37.5 SLOW-SCAN CCD CAMERA PERFORMANCE

37.5.1 Noise considerations

Noise is composed of undesirable signal components which arise from a variety of sources. There are four significant noise sources in CCD cameras which merit discussion. These are:

(1) KTC noise;
(2) noise from thermally generated charge (dark current);
(3) noise from the CCD amplifier; and
(4) photonic noise.

37.5.1.1 KTC noise

KTC noise is generated during the CCD reset process and can be eliminated with carefully designed analogue processing circuits.

37.5.1.2 Dark current noise

Thermally generated charge is indistinguishable from that generated by photons, hence it is a corrupting factor in a CCD signal. Dark current has several components (Janesick, 1989) and it is beyond the scope of this text to discuss them in detail. A dark current image has a spatial pattern which is often punctuated with 'hot' pixels that are many times brighter than the average background. Dark current should be well-behaved and reproducible in a high-quality scientific grade CCD. At temperatures between ambient and −60°C, dark current decreases by a factor of two for every 6°C decrease in temperature, so a modest amount of cooling can effect an appreciable dark current reduction.

Since the dark current image is reproducible, it can be 'subtracted off' the object image. This is done by acquiring a 'dark frame' with the same exposure time as the object image and then performing a pixel by pixel dark current subtraction. There is, however, a random noise component in the dark current which may degrade image quality. The noise component of the dark signal is equal to the square root of that signal in electrons. As an example, if a CCD has a dark current of 100 electrons per pixel per second, then in a 1 s exposure the dark signal will be 100 electrons, and the RMS dark noise will be 10 electrons. The noise will increase with integration time and may limit the usefulness of the CCD, especially at longer exposure times. The only recourse when long exposures are necessary is further cooling, which may require the use of a cryogen like liquid nitrogen. In general, for most light microscopy applications a Peltier cooler is satisfactory and it is not necessary to cool the CCD below −40°C.

A recent innovation has resulted in a 40-fold reduction in dark current in CCDs. Multi-pinned phase (MPP) operation (Janesick, 1989) has been implemented in CCD products by several CCD manufacturers. When using an MPP CCD, under moderate light level conditions where exposure times are short, a cooling system may not be necessary at all.

37.5.1.3 CCD amplifier noise

When a charge packet from a pixel reaches the output of the CCD serial shift register, it is placed on a capacitive node producing a voltage proportional to the quantity of delivered charge. Scientific CCDs are designed to produce between 1 and 5 μV per output electron. The charge-sensitive capacitive node is connected to the gate of an 'on chip' source follower which provides a relatively low output impedance. The noise from the source follower and associated circuitry combine to create a 'noise floor'. The noise floor is invariant in a given camera configuration. A high-quality CCD, coupled with sound camera design, should result in a noise floor of only a few electrons.

37.5.1.4 Photonic noise

This noise component arises because of the quantum nature of light. As with the dark current, photonic noise, often called shot noise, is the square root of the photonic signal in electrons. Photon statistics can limit CCD performance when light levels are low or exposure times are too short to capture a significant number of photons. When a CCD camera is photon noise limited in a given exposure time, there is no way to improve the situation except by increasing the light level or the CCD photon-to-electron conversion efficiency (quantum efficiency). It is important to make the photonic noise much larger than both the dark noise and the noise floor so that their effects are minimized. A useful figure of merit by which the cumulative effect of these noise sources can be quantified, is the signal-to-noise ratio (SNR).

Given all three noise sources, the total noise for a CCD may be written as their quadrature sum:

$$\text{Total noise (NT)} = \sqrt{\text{NA}^2 + \text{ND}^2 + \text{NP}^2}$$

where NA is the amplifer noise, ND is the dark noise, and NP is the photon noise.

The SNR may be written in terms of all the signal and noise components.

$$\text{SNR} = \frac{\text{QE} \times \text{IP}}{\sqrt{\text{NA}^2 + \text{ID} + \text{IP}}}$$

where IP is the total number of captured photons, QE is the quantum efficiency of the CCD, ID is the total dark in electrons, and NA is the amplifier noise.

It is always desirable to maximize the SNR. Since the SNR is a function of several variables, there are always trade-offs which will yield the best CCD performance in a specific application. For example, if light levels are high, then the noise floor and dark current noise may not be relevant and the CCD can be operated at elevated temperatures in a photon noise limited mode. Figure 37.6 is a graphical representation of the major noise sources in a CCD as a function of light level. Note that as soon as the signal exceeds the square of the noise floor, the total noise is dominated by photon statistics.

The sensitivity of a CCD is determined by the properties of silicon and the CCD configuration. The polysilicon gates covering the parallel register are opaque to all wavelengths from 120 to 400 nm. In order to achieve higher sensitivity in the blue and violet portion of the spectrum, CCD manufacturers developed methods for thinning CCDs so that illumination could be brought to the backside of the device. Figure 37.7 shows typical quantum efficiency curves for both front and backside illuminated CCDs. Phosphors which absorb ultraviolet wavelengths and emit in a sensitive region of the CCD response have been very effectively utilized to enhance short wavelength response.

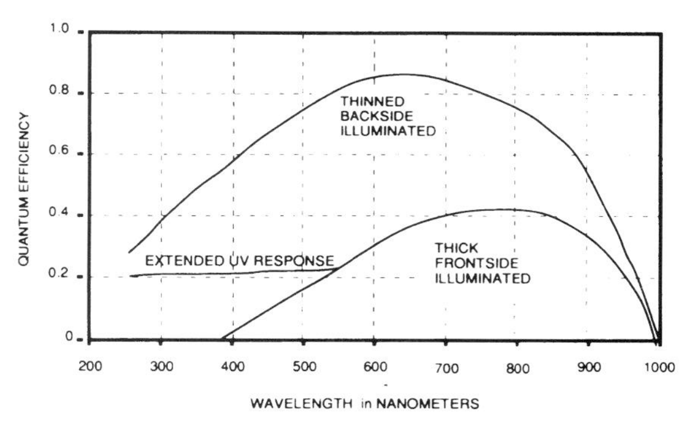

Figure 37.7 Quantum efficiencies of front and back illuminated CCDs. The back illuminated CCD offers nearly ideal response over a large wavelength range.

37.5.2 Photometric linearity

The charge generation mechanism in a CCD is intrinsically linear, so that the output signal should be precisely proportional to the integrated charge at each pixel. The output amplifier is the first place where a departure from linearity may occur. The correct choice of operating potentials can reduce the non-linear effects of the output amplifier to negligible proportions. A high-quality CCD will exhibit linear photometric behaviour to better than 0.1% of full scale from the noise floor up to several hundreds of thousands of electrons.

Photometric linearity is of concern when linear operators are applied to CCD data during image processing. Simple arithmetic operators such as an image ratio, or more complex image reconstruction algorithms which operate in linear transform domains, will produce spurious and erroneous results if the raw input data are produced by a non-linear detector system.

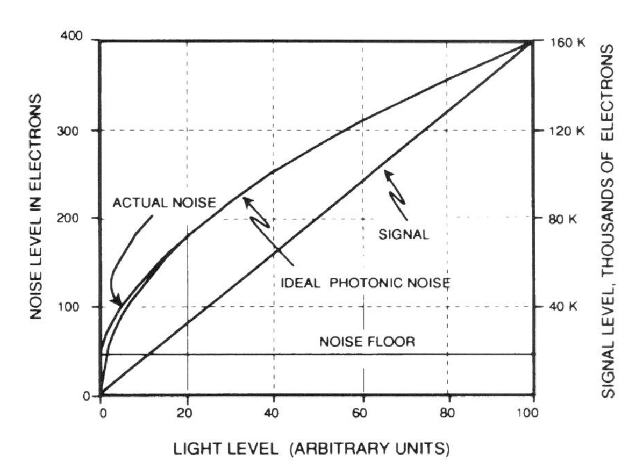

Figure 37.6 Graphical representation of signal and noise in a CCD. Noise floors as low as 5 electrons are possible.

37.5.3 Spatial resolution

The resolution of a CCD camera is determined by the geometry of the CCD pixels, which range in dimensions from 6 to 30 μm on a side. The pixels in a scientific-grade CCD are square and contiguous with no dead space between them. Since pixel size and format are fixed, a CCD image has no geometrical distortion. Square CCD pixels exhibit nearly ideal behaviour in terms of their spatial sampling properties, and it is possible to operate near the Nyquist sampling limit.

To avoid aliasing, an insidious phenomenon which occurs when an image is undersampled, the highest spatial frequencies in the image must be less than half the CCD pixel spatial frequency. Serious image degradation will occur if this rule is not observed. When an image with a periodic spatial pattern is undersampled, aliased components may easily be detected by the presence of a Moiré pattern. The same distortion, however, will occur when imaging an amorphous form, although it may go undetected. When imaging with a light microscope, the diffraction spot should cover several CCD pixels in order to completely avoid aliasing problems. Once an image is sampled with false aliased components, valuable information has been irrevocably lost.

The choice of CCD resolution will depend on the application. Large-format CCDs with 4×10^6 pixels are becoming commonplace, but there are systems implications when considering large CCDs. Each pixel may be digitized to between 8 and 16 bits so that a single image could require 8 Mbytes of storage. With the decreasing cost of memory, optical disk drives and other storage media, image storage is becoming less of a problem. The use of data compression algorithms can lead to further reductions in storage requirements. Another important consideration when choosing CCD resolution is image-processing time. A shading correction requires a pixel-by-pixel subtraction and a division which could require up to 16 million arithmetic operations. The time to process images must be appraised when doing an overall system design. Fortunately, very high-performance computer workstations are becoming available at reasonable prices which also eases this problem

37.5.4 Temporal resolution

A slow-scan camera is a still imaging device similar in nature to a snapshot film camera. Exposure times may range from nanoseconds to hours, depending on the application. The time to read the CCD and digitize its output signal determines maximum pixel read rate. Slow-scan pixel read-out times vary from about 20 μs to 500 ns. CCD performance drops off at pixel read frequencies above 2 MHz. Charge transfer efficiency in the serial register begins to degrade at the higher speeds, especially when the CCD is cooled.

Analogue-to-digital converters impose speed limitations. A 16 bit slow-scan CCD camera cannot operate faster than about 500 kHz per read-out register using current technology. The use of multiple serial register devices can yield a very short overall frame time at the expense of additional digitizers and interface electronics. The analogue processing circuits must settle out to 16 bit precision and a true 16 bit analogue-to-digital conversion takes a significant amount of time. Since CCD images may contain up to 8 Mbytes of digital data, frame rates may be also limited by the ability to move and archive large blocks of data.

37.5.4.1 Digital resolution versus speed

Slow-scan CCD cameras digitize pixels to between 8 and 16 bits depending on the application. Generally, precision must be traded-off against increasing frame rates. The user should decide on the minimum acceptable dynamic range and choose the digitizing level accordingly. Table 37.1 gives typical digital dynamic range as a function of pixel read time for single-channel contemporary slow-scan CCD cameras.

Table 37.1 Digital resolution as function of pixel read frequency.

Number of bits per pixel	*Pixel read frequency*
16	20 kHz–300 kHz
14	50 kHz–500 kHz
12	100 kHz–2 MHz
10	500 kHz–8MHz
8	1 MHz–10 MHz

Analogue-to-digital converters are available which operate considerably faster than the rates given in the table. The read frequencies given in Table 37.1 take into account the time for the CCD and associated analogue electronics to settle to the specified digital accuracy prior to digitizing. Multiple serial register devices will offer a speed increase proportional to the number of available output channels. A 1 k × 1 k CCD with four output registers operating at 2 MHz can operate at 8 frames per second.

In summary, temporal resolution is affected by a number of factors, and the user must decide how to trade-off frame rate and dynamic range against SNR to arrive at the 'best' combination of those factors to achieve the desired result.

37.6 APPLICATIONS OF SLOW-SCAN CCD CAMERAS

37.6.1 High-speed framing

Slow-scan cameras by their nature might appear to be poorly suited for high time resolution applications. There are trade-offs (Ross *et al.*, 1991) which permit frame rates greater than 100 frames per second. As an example, a large format CCD may be programmed to be read out over an ROI of 100 × 100 pixels. If the camera pixel rate is 1 MHz, then the image-acquisition time for the subarray is 10 ms. In addition to the read-out time, the exposure time plus the time required to discard pixels outside the ROI must be taken into account. Figure 37.8 illustrates the subarray read-out principle. the rapid scan overhead,

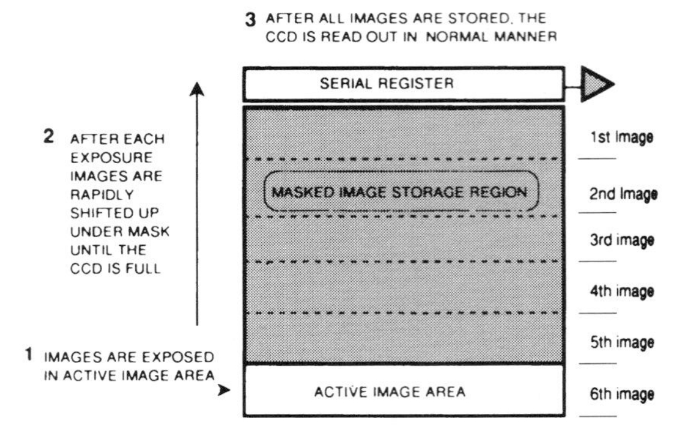

Figure 37.8 Subarray read-out. High frame rates over small regions of interest can be achieved by limiting the field of view. The regions outside the image area are rapidly scanned, and only the pixels in the image are digitized.

in addition to the image read-out time, is typically 20 μs per row and 0.5 μs per serial shift in a large CCD. If the CCD is a full-frame imager, some method for gating the input image must be employed so that light does not strike the CCD during read-out.

It is possible to obtain a series of images very rapidly by using a portion of the CCD as a storage device. Figure 37.9 illustrates this concept. A mask prevents light from striking all but the lower portion of a full-frame CCD. The mask need not be placed directly on the CCD but may be located at another location in the system by using appropriate transfer optics. It is possible to acquire several images in rapid succession by integrating in the open area of the CCD and then shifting the resulting electronic image under the mask. This may be done until the CCD is filled up with images. The CCD is then read out in normal slow-scan fashion. It is feasible to acquire a limited number of images at an effective frame rate of several thousand frames per second using this technique. When only two images in rapid succession are required, the frame transfer CCD is a good choice. This device is useful in ratio imaging for measuring pH levels and calcium concentrations in cells. The frame transfer CCD is particularly useful for ratio imaging since two separate images may be taken within a millisecond.

37.6.2 Shading corrections

When an object is imaged with a CCD and a microscope, CCD background and responsivity variations, system vignetting and scattered background light may mask important information which is present in the image. Because the CCD exhibits near perfect linearity over its entire dynamic range, it is practical, through the use of precise calibration images, to computationally remove shading and responsivity errors and render an image which is radiometrically precise and free of spurious intensity errors. Figure 37.10 shows the sequence of steps which must be taken to perform a shading correction. A dark frame containing CCD structure and dark current background is taken at the same exposure time as the image of interest.

A uniform source flat field image which contains shading errors is then acquired. The most critical step in the shading error correction process is the creation of the uniform source. Not only must it be uniform, it must also be imaged in precisely the same fashion as the object image to be corrected. This is difficult in the case of a fluorescent image, because the microscope stage must be moved away from the object of interest to a region of uniform fluorescence which may be above or below the focal plane. If the microscope is refocused, the illumination pattern will change and the flat field correction will introduce shading errors instead of correcting them. The shading correction process is wavelength-dependent which compounds the problem.

After a flat field shading correction is performed, each pixel becomes an individual radiometrically precise photometer.

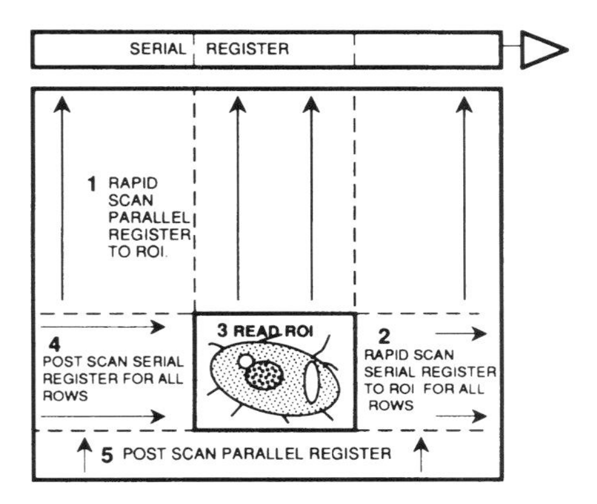

Figure 37.9 High-speed frame transfer imaging. The effective frame rate is determined by the number of unmasked CCD rows.

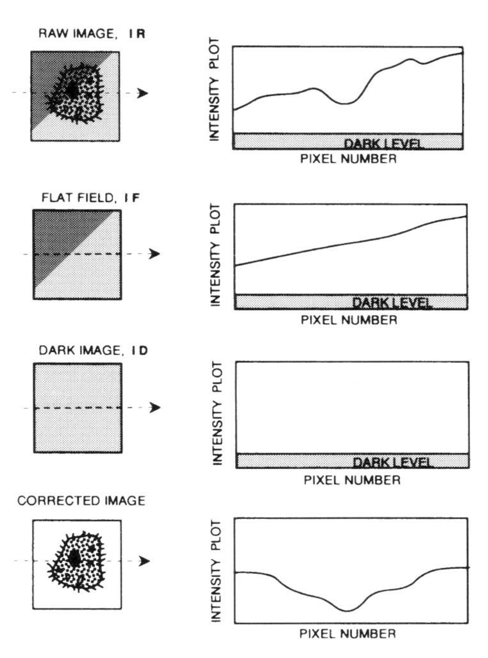

Figure 37.10 Shading correction. This process can be accomplished with a single flat field image. Since the CCD is linear over the entire dynamic range the correction works at all intensities.

37.6.3 Low-light-level imaging

The slow-scan CCD camera is a nearly perfect integrator, and in low-light-level circumstances exposures up to several hours may be taken. During exposures longer than a few minutes, high energy particles and cosmic rays will begin to degrade image quality. Long exposures should be broken up into a sequence of several shorter exposures so that uncorrelated events may be detected and removed from the images. Dark current, system noise and light level should all be considered, and the exposure time must be selected to yield an acceptable SNR when using the CCD at low light levels. In the final analysis, it comes down to the total number of captured photons, which ultimately determines performance. Clean and efficient optics coupled with high quantum efficiency are necessary in low-light-level situations.

37.6.4 Image reconstruction

Digital images produced by a CCD and a light microscope can be enhanced through the application of image-reconstruction techniques. These methods have been successfully employed in spectroscopy, astronomy and microscopy (Agard *et al.*, 1989). An object image is first spatially sampled by the CCD and digitized. The combined effects of the system, including the microscope, CCD and camera electronics are then characterized by imaging a point source which simulates a delta function. The point source image is captured in precisely the same manner as the object image. The recorded object image is the convolution of the system transfer function with the original input image. The corrupting effects introduced by the system may be removed through the application of powerful mathematical transforms. By individually transforming the object image and system transfer function into Fourier space, it is possible to deconvolve the system effects through a simple algebraic operation. Fast Fourier transform software and hardware can deconvolve a one million pixel image in a few seconds. Just as in the flat field problem, the most difficult task is obtaining a calibration image which is a faithful representation of the system transfer function. Through the careful application of Fourier-based image-processing techniques, out-of-focus images may be de-fuzzed and resolution approaching the diffraction limit of the microscope can be achieved.

The successful application of transform techniques places severe demands on camera performance. The camera used to acquire both the object image and the system transfer function must be linear over several orders of intensity level, and the image must be spatially oversampled to ensure that all the information is captured. If the entire process is not carried out very carefully, the results will contain false, spurious image components which can easily be misinterpreted. Attempts to perform image reconstruction with conventional video cameras have never been totally successful because of the poor quality of video image data.

37.7 SUMMARY

The slow-scan CCD camera is a powerful tool which permits light microscopy to be taken to new levels of excellence. Images required with a well-designed slow-scan CCD camera ‘stand up’ to the severe requirements of image-processing algorithms, which fail when applied to conventional video images. The large dynamic range, superb linearity and photometric integrity of the data produced by slow-scan CCD cameras give the imaging scientist the ability to make measurements which were heretofore impossible. While this technology opens up new opportunities for discovery, video cameras and image intensifiers are still needed to provide total imaging capability. No one type of camera can do everything and the user must decide where the attributes of a slow-scan CCD camera are best utilized. While this discussion has been slanted toward microscopy, there are other equally exciting life sciences applications for slow-scan CCD cameras. Electrophoresis, chromatography, X-ray diffraction, transmission electron microscopy (TEM) and radiology are a few of the other disciplines where slow-scan CCD cameras are being successfully applied.

REFERENCES

Agard D.A. *et al.* (1989) *Methods Cell Biol.* **30**, 353–377.
Inoue S. (1986) *Video Microscopy.* Plenum, New York.
Janesick J. (1980–91) Informal Notes, Jet Propulsion Laboratory, California Institute of Technology, Pasadena, California.
Janesick J. (1989) *Proc. SPIE*, 1071–1015.
Ross N.L. *et al.* (1991) *J. Neurosci. Methods* **36**, 253–261.
Spring Kenneth R. *et al.* (1989) *Methods Cell Biol.* **29**, 269–289.

CHAPTER THIRTY-EIGHT

High-speed Digital CCD Cameras – Principles and Applications

CRAIG D. MACKAY
Institute of Astronomy, University of Cambridge, UK

38.1 INTRODUCTION

Digital CCD cameras are now the detector of choice of the great majority of life science imaging applications. For many years the sort of performance needed for precision scientific applications forced users to make do with slow-scan detectors. Chapter 37 gives a detailed account of the properties of low-light-level slow-scan detectors. In recent years, however, there has been a greatly increased demand for high-speed imaging that has forced camera designers to make faster and faster systems that are still able to satisfy the requirements of low noise and wide dynamic range.

This chapter will look at the changes that have become possible, the reasons why they are required, and the technology that is now available to allow them to be achieved.

38.2 THE NEED FOR HIGH-SPEED DIGITAL CCD CAMERAS

There are many reasons why the emphasis in imaging of many life scientists has moved towards fast digital CCD cameras. In most respects they reflect the increasing sophistication of research carried out with microscopes and the demand for an ever higher quality of image data. Some of these reasons are listed below.

38.2.1 High-resolution imaging

It is now possible to buy CCDs with a very large number of pixels. Devices with 2000 × 2000 pixels are now being used in significant numbers, and CCDs with a resolution in excess of 4000 × 4000 pixels are beginning to become available. Although the user of a fast digital imaging system should always try to use the minimum number of pixels possible, the image quality of modern microscopes increasingly justifies much higher resolution across the field of

FLUORESCENT AND LUMINESCENT PROBES, 2ND EDN
ISBN 0-12-447836-0

view. In addition techniques such as confocal microscopy and the new digital deconvolution methods are able to provide images of very high resolution indeed. With CCDs of this size the read-out time can be a significant issue. A CCD of 4000 × 4000 pixels that is read out at a pixel rate of 50 kHz will take in excess of five minutes to read out a single image. This is clearly a serious problem in the laboratory, particularly when the CCD itself is capable of being read out at a very much higher rate.

38.2.2 Faster frame rates

Faster frame rates are firstly desirable as a matter of convenience in that they allow more rapid focus and alignment of the microscope. There are also an increasing number of applications where there is a need to measure temporal changes in the sample. A great deal of interesting things happen to living cells on timescales of a small fraction of a second and it is not surprising that life scientists wish to measure and study these accurately.

As computers become more powerful scientists are able to take larger data volumes. For example when using digital deconvolution methods the microscope stage is stepped in the z-direction and a sequence of images is taken, one at each step, to provide a three-dimensional data set. It is quite common for the scientist to produce a data volume consisting of many tens of images and it is always a concern that the sample might change its properties while the sequence of images is being taken, for example by photobleaching or by a gradual drift in the alignment of the optical system. The ability to take images quickly is clearly likely to minimize this problem.

38.2.3 Multicolour imaging

It is sometimes forgotten that even at the lowest light levels there is as much colour as there is at ambient lighting conditions. Increasingly researchers wish to study samples with a wide range of colour filters. The researcher may wish to use multiple fluorophores, each working at a different emission wavelength. These techniques are useful because they allow different features on a sample to be studied simultaneously and for the spatial relationship of one feature to be compared with another. It is now possible to buy filter-changing mechanisms or scanning monochrometers that allow very rapid wavelength changing in the excitation and the emission wavelength.

CCDs are available covered in a colour mask so that different pixels on the CCD are sensitive separately to the red, green and blue signals, but it is greatly preferred to use the whole of the CCD for each colour of the image so as to preserve full resolution in each of the three colours. Three-chip colour TV cameras may be purchased but the standard RGB filters that are used for television cameras do not usually match the colour requirements of the microscopist in fluorescence studies. For example, when combining three fluorophores one may find the need to detect green, yellow-green and yellow light, which are rather hard to separate using red, green and blue filters. With a fast change filter wheel or monochromater a precise selection of excitation and emission wavelengths may be used and multicolour images may be created by the combination of these three separate images taken sequentially. In order to make the image registration as good as possible it is often important to take the three colour images in rapid succession so that the sample does not move or the alignment of the optical system does not change. This again emphasizes the need for fast imaging with CCD cameras.

38.2.4 CCD performance

Many of the earlier scientific CCD devices were manufactured to give good performance at slow rates. The increasing demand for high-definition TV and similar applications has forced CCD manufacturers to design their products to give a low read-out noise and a high dynamic range at high read-out speeds. It is now possible to purchase CCD chips which themselves are capable of giving a full well capacity that is 10 000 times the read-out noise they give at 20 MHz pixel rate. It may be very hard to achieve that performance electronically, but at least the CCDs themselves now have that capability.

38.2.5 Electronic system advances

It had been believed for long time that the principal electronic method used to suppress noise in a scientific CCD imaging system, the method known as double correlated sampling, simply did not work at pixel rates in excess of perhaps 250 kHz. In the early 1990s, however, this barrier was broken when AstroCam Ltd, Cambridge, UK, produced the 4100 controller which could operate at a pixel rate in excess of 5 MHz with true double correlated sampling. The ‘secret’ in making these high-speed controllers was to use radio-frequency engineering techniques and electronic components such as those used in radar systems to give very clean electronic systems with the absolute minimum of noise. Other manufacturers now offer systems that also achieve this.

There were also problems in the past with the availability of wide dynamic range analogue-to-digital converters. Although there were many devices available to provide 8- to 10-bit digitization, parts to provide much higher precision were scarce and expensive. It is now possible to purchase 14-bit A/D converters at an acceptable price that will work at pixel rates of 10 MHz. This has had a great impact on the ability of manufacturers to build cameras at reasonable prices.

38.2.6 Data transfer rates and computer architecture

Even if the camera manufacturer had been able to produce a high-speed camera, there were considerable difficulties in getting large volumes of data into the computer system at high speed. Although this was possible with top-end computers which were available to the largest organizations, budget constraints in the typical life science laboratory made it critical to have systems running on desktop personal computers. The old ISA bus that was used in the IBM/PC machines could achieve a maximum pixel acceptance rate of about 400 kHz. The use of frame grabber cards in the computer did allow fast data transfer rates from the camera to the grabber card, but the ISA bus always ended up being the bottleneck, limiting the amount of data that could be captured to the amount of data that could be handled by the buffer memory contained on the card. This buffer memory was very expensive, often much more expensive than memory for the computer system itself.

An alternative was the SCSI interface. This provided slightly higher data throughput, but the SCSI bus was also shared by other peripherals such as disk drives and tapes. This made the data capture unreliable unless a substantial amount of image buffer was provided in the camera hardware, again increasing the cost.

The advent of the PCI bus has greatly simplified the design of CCD imaging systems. The PCI bus allows a data transfer rate in excess of 15×10^6 pixels per second in a sustained way direct to the memory of the computer without involving significant computer power itself. This means that the computer is still able to carry out a variety of tasks even while a large sequence of data is being taken by the system, always assuming that the operating system does in fact allow this. In addition, PCI interface cards are relatively inexpensive to manufacture. They require only a minimum of on-board buffer memory, relying almost entirely on the memory of the host computer. The PCI bus is now also standard on a number of different makes of computer, allowing the camera manufacturer to provide camera systems for several different computer systems without the considerable cost of additional interface card development.

38.3 OTHER TECHNOLOGICAL ADVANCES

There are number of additional advances in the technology of CCDs themselves as well as the electronic systems that read them out which are having a significant effect on the application of CCDs in the life sciences.

38.3.1 Enhanced quantum efficiency

In comparison with other scientific applications, fluorescence microscopy is not a particularly low-light application. Exposure times are often only a fraction of a second in order to give a good high-quality image with excellent dynamic range. Unfortunately if very much shorter exposure times are desired, then fluorescence microscopy very quickly turns into a real low-light application. In addition, as with all camera systems, running the system faster gives a greater read-out noise level, further reducing the signal-to-noise ratio of the final image. Anything that can be done, therefore, to improve the quantum efficiency (a measurement of the light-gathering efficiency of the CCD), is potentially very important. With most CCD architectures the light is passed through the covering electrodes of the chip before it is absorbed in the silicon underneath. This is described in detail in Chapter 37. The covering electrodes are made of a material called polysilicon, which acts as a longpass blue filter, only passing light that is yellow or of longer wavelengths. As a result most CCDs have a low quantum efficiency in the region of 400 nm and a peak quantum efficiency in the region of 600–700 nm. Techniques have been developed, principally for astronomers, to get around these problems. These are described below.

The CCD is manufactured on a thin wafer of silicon which may perhaps be a fraction of 1 mm thick (Fig 38.1). If the silicon is thinned from the rear surface so that it is only 10–15 μm thick, then the light can enter from the underside of the device which is thinned silicon and which absorbs light over a wide range of wavelengths extremely efficiently. With suitable surface treatment (such as an anti-reflection coating which can be important in achieving the very best efficiency) a quantum efficiency in excess of 90% may be achieved. Thinning these devices and subsequently mounting them in a package is clearly not easy, so these CCDs are inevitably much more expensive than their unthinned counterparts. Nevertheless,

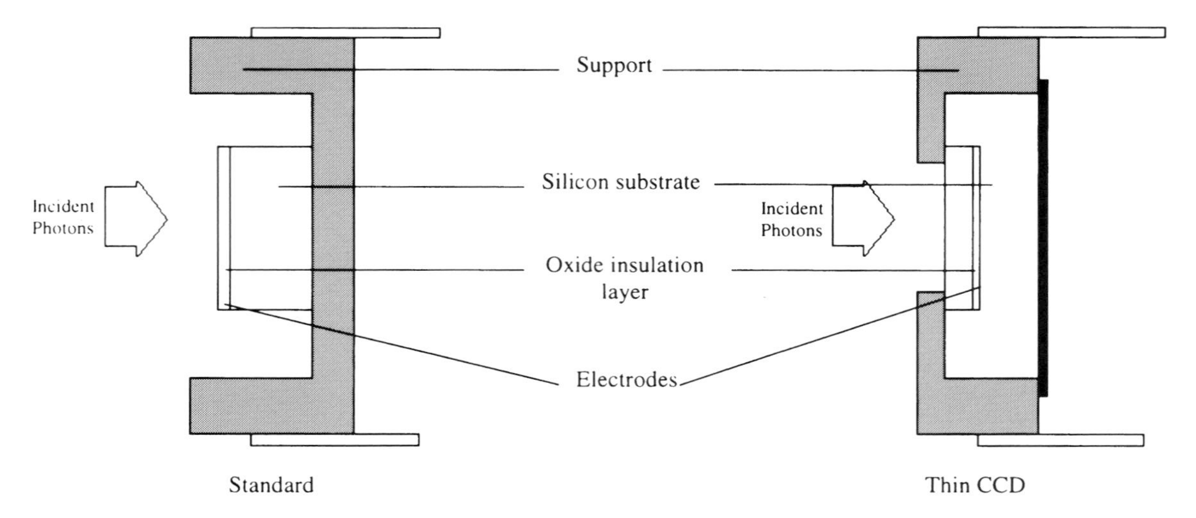

Figure 38.1 Line drawing of a thinned CCD showing structure and the way light enters from the chip underside.

for the user who does need that very high quantum efficiency, and particularly for those who need good sensitivity at the blue end of the spectrum where a normal CCD has a low or no sensitivity at all, this can be very important (Fig 38.2).

Another method of achieving blue sensitivity, but at much lower cost, was also developed by astronomers. The surface of a normal CCD can be coated with a very thin layer of inorganic phosphor that absorbs blue and ultraviolet light and re-emits it at a wavelength of about 560 nm. This material is called luimogen and the layer need only be about 1 μm thick. Because the re-emitted light travels in all directions only about half of it enters the CCD and therefore one can hope for a quantum efficiency approaching half that of the CCD at that wavelength. In practice the quantum efficiency that is achieved in the blue part of the spectrum is typically only 10–15% (Fig. 38.2). For many applications to have at least some blue sensitivity is preferred, and so this solution is very satisfactory as well as being very much cheaper than having to use thinned CCDs. The coating method also has the advantage that the coated device has sensitivity right out to very short wavelengths (at least down to 70 nm). For applications that require sensitivity in the far-UV this approach is probably the only one that is going to give fairly good, reliable sensitivity.

The situation is completely different when considering luminescence applications. In virtually all luminescence applications one is working at very low signal levels. Even with enhanced chemiluminescence, it is usually necessary to take exposures that are quite long (many seconds or many tens of seconds),so high quantum efficiency is extremely important. It is also the case, however, that with luminescence one is not usually looking at very fast phenomena and therefore slow-scan camera systems may be used. These systems are capable of the lowest possible noise performance.

38.3.2 Dark current suppression techniques

In order to achieve good scientific performance it is necessary to cool the CCD chip itself. The reason for this is that the dark current (the signal that builds up on a CCD in the absence of any light falling upon it) is normally too high to permit exposures of even a few seconds with standard CCDs. Developments in the design and manufacture of CCD chips has led to a new technology, known as a multiphase pinned (MPP) or inverted mode operation (IMO), which is able to deliver a dramatic reduction in the level of dark current achieved by a CCD chip at any temperature. As a technique it is least successful with thinned CCDs; they will show a reduction in dark current of only about a factor of 30. In the case of other CCDs the dark current suppression achieved may be as high as a factor of 1000. A factor of 10 in dark current corresponds to a temperature difference of 20°C, so dark current suppression by a factor of 1000 is equivalent to cooling the CCD by 60°C. It is clear that it is often possible to make a satisfactory precision scientific imaging CCD system that is in fact uncooled. Unfortunately it is not quite as simple as that! For all the CCDs that have this kind of dark current suppression, the dark current quoted by manufacturers usually refers to the mean dark current. In practice all real devices suffer from individual pixels which have significantly higher dark current. The number of pixels affected is usually quite small and it is always easy to correct for these 'hot' pixels so that they are effectively removed. Uncooled cameras do,

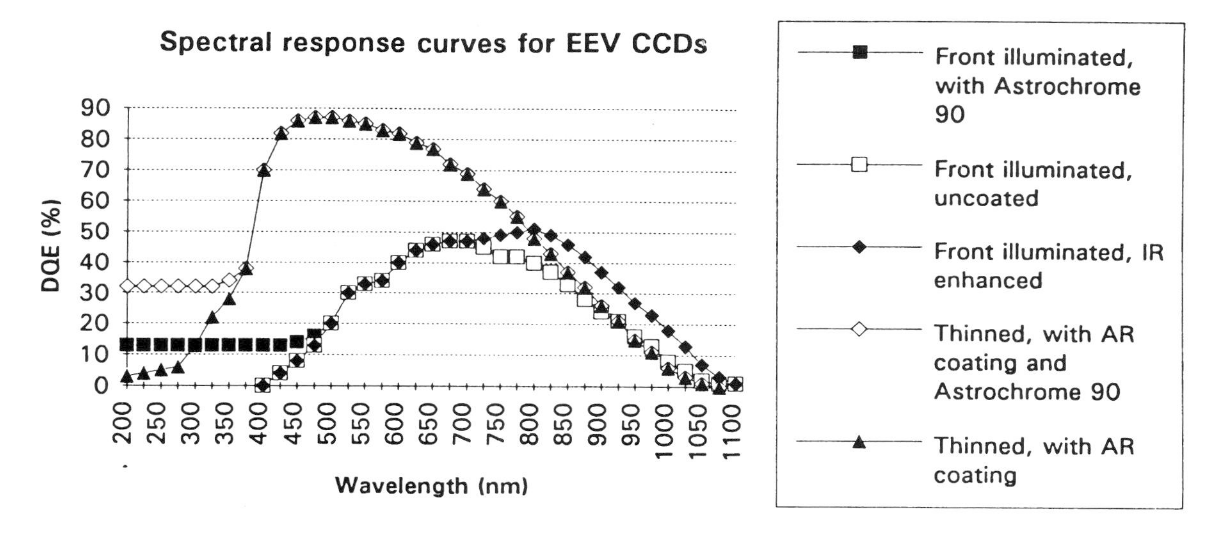

Figure 38.2 The detective quantum efficiency of standard, thinned and coated standard CCDs.

however, give an image which at first sight does appear to be a little bit more spotted than might otherwise be the case. For this reason, although it may be formally true that an uncooled system would be entirely satisfactory, users are often prepared to buy a cooled system because it does give them that extra margin and reduces the need to correct for hot pixels. One approach that is becoming more popular is to use a much reduced level of cooling. By cooling the device thermoelectrically to perhaps −20°C or −25°C, excellent performance is achieved with a system that can rely on a passive heat exchanger rather than needing either a fan or a re-circulating liquid coolant as part of the head. Workers in the areas of deconvolution or confocal microscopy are often concerned about the effect of a fan in the head causing vibration within the microscope assembly, although with good head design this does not seem to be a problem in practice.

38.3.3 Interline architecture CCDs

Full-frame CCDs are attractive because all of the area of the CCD is light sensitive. A mechanical shutter is opened to allow the light to fall on the CCD. It is closed and must remain closed during read-out in order to avoid any image smear in the final picture. When operating a full-frame CCD at high frame rates the shutter may need to operate many times per second, something that does not lead to a long and happy life. In addition, when working at high frame rates you quickly enter a regime where the shutter is opened for a very small amount of time and nearly all the time is used to read out the CCD. This dramatically reduces the system detective quantum efficiency, irrespective of the actual quantum efficiency of the CCD chip employed. The frame transfer CCD gets around this by having two parts, one of which is light sensitive and one of which is covered with an opaque screen. At the end of the exposure the image that has built up in the light-sensitive area is transfered rapidly into the covered area. The next frame builds up in the light-sensitive area while the signal from the previous frame, now held in the storage section, is read out in the usual way. In this way it is possible to achieve essentially 100% exposure time and therefore to achieve the full efficiency of the CCD used.

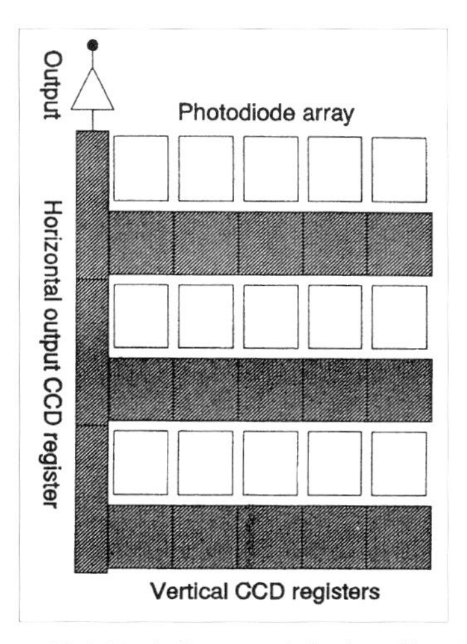

Figure 38.3 Basic layout of the interline CCD.

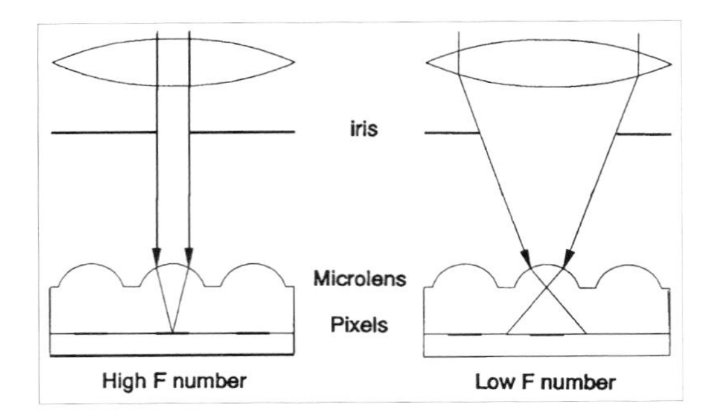

Figure 38.4 The layout of the microlenses on the front of an interline CCD with the rays shown for an input parallel beam.

Another architecture which is popular for TV applications is the interline architecture (Fig. 38.3). In this architecture columns of light-sensitive pixels are interleaved with columns of pixels that are again covered by an opaque screen. At the end of the exposure all the pixels in the array are transferred from the light-sensitive pixels into the screened pixels. The next image is allowed to integrate on the sensitive pixels while the signal held under the opaque stripes is read out in the usual way. This is the interline CCD architecture. In the past it was not popular for scientific applications because only a small fraction of the area of the pixels on the device is available for light detection. For small pixel devices the area available can be as little as 20% of the total. This means that the quantum efficiency in these devices is often as low as 5–10%. New techniques have been developed to allow these devices to have integrated on the surface microlenses so small that one covers each pixel of the device (Fig. 38.4).

The microlenses are arranged so that each can collect the light which would fall on the whole pixel and then focus it onto the light-sensitive part of the array. As expected, this largely restores the overall quantum efficiency of the device to that of the standard full-frame CCD. In order for the microlenses to be focused properly they have to be fed with a beam that is relatively parallel. If you try to use these devices with a microlens array behind a fast camera lens, for example, no benefit is achieved. However, in microscopy the optical arrangement is such that the beam that strikes the CCD is very close to parallel and so in this application users may be confident that good quantum efficiency is obtained.

Interline chips do have some disadvantages. The dark current is somewhat higher than for full-frame or frame transfer devices and the read-out noise is usually somewhat higher. Because part of the pixel is used for light sensitivity and another part is used for read out the overall full well capacity of the interline CCD is significantly lower than that of a full-frame device, possibly only half to one-third of the level. An advantage, however, is that interline CCDs do have good blue sensitivity without coating or thinning and in many applications this can be critical. Interline CCDs are widely used in domestic applications such as video recorders and digital hand-held cameras so these devices are being developed very aggressively. They are now available with higher image resolutions (approaching 2000 × 2000 pixels) and are being used increasingly for scientific imaging. Efforts are being made to improve the performance of the output amplifiers so that they should match that of full-frame devices in this respect before too long.

38.3.4 Small pixel CCDs

The demands of the commercial markets call for CCDs with a high manufacturing yield as well as a low cost. This means that as pixel resolutions improve it is inevitable that pixel sizes will shrink. The latest CCDs that are being produced now have very small pixels indeed, as small as 6.7 μm, and these inevitably lead to a relatively small full well capacity. This means that often the dynamic range that one can get from such a device is much more limited. Full well capacities as low as 20 000 electrons are found in some cases, which makes it very difficult to justify 14-bit conversion in the CCD controller. If the CCD is used in a binned mode, so that charge is added noiseless on the CCD chip itself (see Chapter 37), then the total charge in the binned pixel becomes considerably larger and then perhaps appropriate for a high-resolution A/D converter. From the point of view of microscopists small pixels are helpful in allowing the CCD field of view to fit within the microscope tube and allowing the camera head to be attached via a C-mount adapter. High-resolution CCDs or those with bigger pixels can have an image diagonal that is larger than the field that is properly illuminated by a typical microscope objective. In this case camera head adapters that contain a re-imaging lens to match the large area of the CCD must be used. This sort of unit is normally used for 35 mm camera adapters such as the Nikon F-mount and Pentax K-mount that are widely available.

38.4 SYSTEMS SELECTION

It is seldom wise to choose the CCD camera in isolation from the other parts of the system that will be used with it. Some vendors such as Life Science Resources Ltd, Cambridge, UK, provide complete integrated imaging systems that are carefully tailored for specific applications. The selection of the right camera system does depend critically on good practical experience and knowledge of the application for which it is to be used. The right combination of camera hardware, software, excitation and emission filtering can make a dramatic difference to the performance of the complete system and the effort required to produce results from it. Some examples are given below of how the configuration system needs to be approached.

38.4.1 Digital deconvolution system

At present there are two principal ways of achieving three-dimensional imaging of a sample. Confocal microscopy techniques provide a hardware method of suppressing out-of-focus fluorescence from the sample under investigation. Confocal techniques can often require high levels of illumination which rapidly lead to photobleaching of the sample. An alternative approach is that of digital deconvolution. Here an image is taken of the sample each time that the microscope is stepped in the focus direction. The images taken are then used as a three-dimensional data cube. Special software is able to sort out the true shape of the specimen by locating features that are sharp in one of the planes and then computing what contribution that sharp feature would have in each of the out-of-focus planes. That contribution is then subtracted, increasing resolution substantially. This leads to a corrected three-dimensional image which may be displayed on the screen of the computer with a volume-rendering software package and inspected as a fully three-dimensional image. If the sample has been labelled with multiple fluorophores then images may need to be taken at each of the focus positions in more than one colour. This further increases the amount of data that is being captured. Because the business of deconvolution is essentially one of looking at adjacent image slices, comparing them and looking for slight changes in features as they go in and out of focus, the technique requires excellent signal-to-noise if it is to be successful. This means that a digital imaging system is essential for this work and the volume of data taken implies that it is the essential to use a fast imaging system. As with most fluorescent applications, this is a relatively high-light-level application where the emphasis must be above all on the image quality of the system. Speed is also important because of the number of images that are being taken to provide the full data set, particularly if multiple images need to be taken to provide colour information.

As with all deconvolution methods, the demands placed on computing power are such that it is well worth minimizing the amount of data gathered as far as possible. This makes it important not to take and record in the computer images of a size any greater than is absolutely necessary for the study in hand. This may mean that the user can do with a fairly small area CCD, although the use of a larger CCD to give small captured areas is possible when subarray read-out is used (see Chapter 37). Deconvolution and volume-rendering software is very complex and only available from a very small number of vendors. For this reason it is essential to make sure that the camera hardware you wish to use is fully integrated with the application in a way which makes the entire system easy-to-use and the application software that has been selected really is going to do the job that is required of it.

38.4.2 Neuronal imaging

Considerable interest is now being shown in neuronal imaging. This involves measuring the electrical phenomena that occur within individual nerve cells and networks of nerve cells themselves and examining how these electrical impulses propagate through the cells. Neuronal studies usually call for very fast imaging indeed because the phenomena, particularly in mammalian neurons, are very rapid. In order to achieve very high-speed imaging (in excess of a few hundred frames per second) it is essential to use a small area CCD. Although large area CCDs may be used to produce small data sets by only reading out a sub-array of the device, the overheads implicit in clearing out unwanted charge make it very difficult to achieve a good frame rate. For this reason a small area CCD such as the CCD 39 from EEV, Chelmsford, UK, which has 80 × 80 pixels, is a very attractive device. In addition the CCD is available thinned, something that is especially valuable in neuronal imaging because the high frame rate means that very little light is available for each frame. At high frame rates the read noise is inevitably high (with a CCD 39 and a read rate of 500 frames per second, a typical read-out noise might be 40–60 electrons per pixel), so that the very best quantum efficiency and the lowest possible read-out noise are essential in this application. If, however, absorbance dyes are being used instead of fluorescence dyes, then plenty of light should be available and the emphasis must be on the full well capacity of the CCD since the absorbance that is needed to be measured is often a small fraction of one percent. This requires the use of a CCD with a high full well capacity, and again the CCD 39 meets this requirement well.

38.4.3 Ion ratio imaging

With ion ratio imaging two images are taken at differenct wavelengths in rapid succession so that the fluorescence or absorbance of the dye at two wavelengths may be compared and the ion concentration derived from them. The studies usually involve live cells, and they may well involve live cells that are reacting to an external stimulus of one sort or another. This means that high-speed imaging is almost always required, and the system must of course be closely integrated with the excitation and emission wavelength changer mechanism. Software must provide all the facilities needed for synchronizing the camera and the wavelength changer as well as providing a full range of software facilities for handling and analysing the data. The very nature of ion imaging calls for high-speed operation and a camera should be chosen that is going to give the right combination of speed and noise. In order to maximize the sensitivity to small concentration variations it is important to choose a system with very good dynamic range. The camera should be chosen with a resolution that matches the requirement of the application and runs at high speed with very high dynamic range. Systems using the CCD 37 from EEV, which has 512 × 512 pixels, are attractive. With a camera reading at a 5 MHz pixel rate, 20 frames per second may be recorded (10 pairs of frames per second) and so give the level of performance that is needed.

38.5 CONCLUSIONS

There have been dramatic changes in the technologies available to life scientists to enable them to take images more quickly than ever before without compromising the quality of data they previously obtained at slow speeds. New application areas are opening up rapidly which will lead to great advances in many fields of research. Camera systems rather than camera components are becoming the focus of development so that the user in a specific area of application will be able to select complete packages with a very high degree of integration and sophistication. As is always the case, what can be achieved with these systems will grow and grow as increased computing power allows more and more complex studies to be undertaken with confidence.

PART X

Flow Cytometric Methodologies for Measurement of Optical Probes in Live Cells

CHAPTER THIRTY-NINE

Flow Cytometry: Use of Multiparameter Kinetics to Evaluate Several Activation Parameters Simultaneously in Individual Living Cells

ELIZABETH R. SIMONS
Boston University School of Medicine, Boston, MA, USA

39.1 INTRODUCTION

It has always been important to know whether and how cells function. In that context, it is desirable, wherever possible, to correlate the various expressions of cellular function or stimulus responses temporally with each other and, where applicable, with receptor occupancy. Furthermore, it has become apparent that, in a given sample of cells, all cells do not respond equally or synchronously, especially to subsaturating doses of stimuli, i.e. there is a disproportionate response so that subpopulations respond differently or not at all to that dose, although all cells are capable of responding fully to a saturating dose, thereby lifting the disproportionation.

One of the most important current challenges in cell biochemistry is to delineate the mechanisms and resultant effects of signal transduction, that is, to deduce the mechanisms by which cells transduce information, those by which the resultant signals are transmitted, and those which eventually lead to expressions of the particular cell's function. In order to acquire enough information to resolve these mechanisms, it is essential to correlate these stimulus-induced responses or consequences with the corresponding state of the specific receptors' occupancy in real-time. It is also important to determine whether each of the functions observed depends on another earlier one, and whether it is expressed by all cells, by that fraction of cells which has occupied receptors, or by a fraction of cells which is, *a priori*, identical in terms of receptor occupancy to another which does not respond. Information of this kind can lead to temporal resolution of responses through determination of: Which step comes first? Does it depend upon another which starts earlier? Are more receptors occupied on responding than on non-responding cells, implying receptor cooperativity in the initial response? What step occurs next? How is it controlled? Is the prior event necessary and sufficient?

FLUORESCENT AND LUMINESCENT PROBES, 2ND EDN
ISBN 0-12-447836-0

39.2 FLUORESCENT TECHNIQUES – GENERAL ADVANTAGES AND DISADVANTAGES

These questions, as well as the need to work with fewer cells as both cells and agonists become more expensive and/or difficult to acquire, cannot be resolved by the use of radioactive probes because the latter applications involve separation of cells from their supernatant so that each can be counted separately, a technique which is (1) too slow for the rapid time points needed (often 1–5 s intervals), (2) too wasteful of cells, as each time point requires a new sample of a large number of cells, (3) too difficult to use if more than one parameter is to be followed simultaneously, and (4) incapable of distinguishing subpopulations of cells, so that one cannot determine whether a response which is 50% of the control's reflects a 50% response by each cell or a full response by 50% of the cells.

Routine fluorescence measurements of cell suspensions, or of a monolayer of cells attached to a coverslip, avoid the problems (1) and (2) above; (3) is a disadvantage of fluorometry which is, however, disappearing as instruments capable of rapid automatic scanning of several emissions from separate excitation wavelengths (e.g. four excitation and four emission wavelengths) now exist, and (4) exists for all measurements which reflect the average over large numbers of cells. Conversely, fluorescence microscopy and laser confocal fluorescence microscopy permit evaluation of each individual cell, therefore avoiding (4), but each is laborious to use for large numbers of cells and some microscopes are not equipped to permit more than one excitation wavelength to be used.

Another source of error in fluorescence measurements of cell suspensions is the possibility, indeed the probability, that one is measuring extra- as well as intra-cellular fluorescence. Ideally, one should be measuring 100% of either; any combination of extra- and intra-cellular indications of the same parameter creates problems in the final quantitative evaluation of that parameter. For some distributive probes, such as the thiocyanines used to measure membrane potentials, or 9-aminoacridine used to measure pH_{in}, it is the extracellular probe's fluorescence which gives an indication of how much positively charged probe has penetrated into the negatively charged cell, for the cyanines, or of how much protonated amine is external, for 9-aminoacridine, since the intracellular probe is quenched in each case (Sims *et al.*, 1974; Deamer *et al.*, 1972). That is, one is dealing with fluorescence in supernatant only, and can therefore calculate the overall cell membrane potential or the cytoplasmic pH, respectively. Measurement of extracellular probe is therefore crucial in these cases. Conversely, when measuring intracellular probe, as is true for the *in situ* probes introduced into the cell as esters, one must either assume that there is no de-esterified – and therefore indicating – probe in the extracellular buffer, or one must correct for it. Unfortunately, most investigators assume there is no contribution from any extracellular probe; however, most cells do leak (some more, some less), most have relatively high permeability to small anions, and some cells release esterases capable of hydrolysing any residual esters in the buffer. Furthermore, the esters themselves may be fluorescent at the same wavelengths used for the free acid indicator form (e.g. fura-2-acetoxymethyl ester, Scanlon *et al.*, 1987). The necessary corrections for extra- versus intra-cellular probes have been described in the literature (Davies *et al.*, 1987a) and will be explained briefly in Section 39.6. The problem can also be reduced by using a newer group of *in situ* probes for cytoplasmic pH and $[Ca^{2+}]$ recently offered by Molecular Probes Inc., Eugene, OR, which, after de-esterification by cytoplasmic esterases, combine covalently with glutathione and hence leak markedly less.

In contrast to suspension studies, when one examines intracellular fluorescence on a flow cytometer, the extracellular milieu is so diluted by the sheath buffer that its contribution to the total fluorescence is negligible. Therefore, flow cytometry obviates the necessity for extracellular probe correction and is the method of choice for cells which leak cytoplasmic probes readily, but restricts the probes that can be used to those which report intracellular parameters and whose excitation wavelength matches that of the available lasers.

In addition to the need to eliminate artefacts due to probe leakage, some other precautions are necessary. Among these is the question of ratioing of fluorescences, the basis of some of the most useful fluorometric probes of cell functions. In this context, it must be remembered that the use of the so-called 'ratiometric' fluorometric *in situ* cytoplasmic probes (see below) is really only correct when one uses either the ratio system of a flow cytometer, or that of a microscope, each of which evaluates the ratio directly for each cell. That is, the use of ratios to obviate errors dealing with different sizes of cell and/or different probe concentrations within each cell depends upon not only adherence to Beer's Law but also assumes that the probe concentration (c) and the light path (l) in the ratioed fluorescences are equal, i.e. assumes that the measurement is made in the same cell at two different wavelengths. This means that $R = F_1/F_2 = f_1cl/f_2cl = f_1/f_2$ where f_1 and f_2 are independent molar fluorescences characteristic of the property being measured by the probe. Since mathematically the average of ratios is not equal to the ratio of the averages, the equation holds only when a single cell's c and l (and not the average over all cells, as is true in suspensions) is used. Thus, although ratiometric probes are routinely used in cell suspensions, this procedure involves an implicit assumption that all cells are of equal size and contain equal concentrations of probe. In flow cytometry, no such simplifying assumption needs to be made if one uses a ratio board since flow cytometers are or can be so equipped. A ratio board takes the ratio of selected photomultiplier outputs for each cell, cell by cell; if, however, one takes the ratio of the average fluorescences (mean fluorescence channel) at the two wavelengths (i.e. the ratio of the mean channels in a flow cytometer), this singular advantage of flow cytometry is lost.

39.3 FLOW CYTOMETRY FOR KINETIC STUDIES OF CELLULAR FUNCTIONS

For all of the above reasons, we, as well as a number of other investigators, have adapted the flow cytometer, an instrument originally designed to examine cell surface antigens as cell type markers, to perform multiparameter kinetic studies of stimulus responses of suspendable cells, and to correlate these parameters, for each cell, with the receptor occupancy of that cell (Davies *et al.*, 1988, 1989, 1990; Ryan *et al.*, 1990; Bernardo *et al.*, 1990; Guillemot *et al.*, 1996; Lazzari *et al.*, 1986, 1990; Brunkhorst *et al.*, 1991, 1992; June and Rabinovitch, 1994; Amorino & Fox, 1995; Strohmeier *et al.*, 1995a,b: Model *et al.*, 1997; Bernardo *et al.*, 1997). The time resolution to date is between 1 and 4 s, achievable by using an injection system which adds the stimulus directly to the thermostatted stirred tube containing the already flowing cells; such a system was first described by Kelley (Kelley, 1989) and is now available commercially. The time resolution depends upon the time required to flow from the mixing chamber to release from the nozzle, which in turn depends on the diameter of the tubing and the shear rate which can be tolerated by the cells being examined. Much shorter resolution times are being approached by Sklar and his colleagues with a not-yet-commercially available flow cytometer (Nolan *et al.*, 1995).

We have shown, for platelets, neutrophils and monocytes, that flow cytometry is particularly useful when one is dealing with small cell samples, low subsaturating doses of stimulus, multiple classes of receptors, and/or heterogeneity in the response and in the proportion of cells which is exhibiting that response (Beard *et al.*, 1986; Lazzari *et al.*, 1986, 1990; Sullivan *et al.*, 1987b; Davies *et al.*, 1988, 1989, 1990; Ryan *et al.*, 1990; Bernardo *et al.*, 1990, 1997; Brunkhorst *et al.*, 1991, 1992; Strohmeier *et al.*, 1995a,b).

39.4 TIME OF ONSET OF INITIAL RESPONSE

Many cellular functions are initiated almost instantaneously upon exposure to agonists; some functions are independent of each other while others are interdependent. It is therefore of interest to measure, if possible, the actual time of onset of each of the functions being studied, as well as the response time for each function relative to all the others.

To date, the ability to measure cells very rapidly in flow cytometers has been limited by the length of time required to mix cells and agonist adequately (considerably shortened in a system equipped with a magnetic stirrer), the time necessary to flow from the sample chamber to the drop-forming tip (dependent on the length and diameter of the sample delivery tube), on the flow rate, and on the desired number of cells per time point. For standard flow cytometers with standard mixing chambers, the earliest post-agonist addition time point at which data can be collected is approximately 10 s for cells (e.g. platelets) which cannot be subjected to the high shear rates arising from flow rates greater than 2000 cells s^{-1}. In our hands the 'boost' system designed to push cells through more rapidly caused artefacts as well as broken cells (others may have better luck with this accessory). By moving the stirred thermostatted cell chamber nearer the delivery tip, thereby shortening the delivery tube, and installing an injection system, Kelley (1989) was able to achieve initial time observations of 1–2 s. Sklar and his colleagues built a system capable of observations within approx. 100 ms at Los Alamos, NM (Nolan *et al.*, 1995).

It should be noted that, in this respect, non-flow cytometric measurements have achieved much shorter initial time points. In terms of cell applications, these are continuous flow, stopped flow, and quenched flow systems such as those developed by Gear and by Rink (Gear, 1980; Gear & Burke, 1982; Rink & Sage, 1987; Sage & Rink, 1986, 1987; Jones et al., 1989). However, since these depend on observations of cell suspensions, each yields an average over all the cells, with temporal resolution in the 100 ms range; continuous observation of subpopulations (unless the cells are fixed) is not possible in these rapid kinetics systems, and each time point hence requires a new sample. Except for the systems described by Dr Sklar's group (Nolan *et al.*, 1995), flow cytometers cannot achieve such early initial time points due largely to the need to mix cells and agonists well, then flow the mixture through a tube of some finite length into the drop-forming tip without damaging the cells by excessive shear. Some attempts to place the mixing chamber immediately above the tip have been made, and the time when observations can be made within 100–300 ms after addition of agonist is near.

The discussion above refers to initial times of observation, the times which are critical when the sequence of events in signal transduction, for example, is of interest. Clearly the temporal resolution, i.e. the time interval between recording of observations, depends strongly upon the flow rate which the cells can tolerate – the faster the flow, the higher the number of cells/second, and the shorter the time interval between data points.

39.5 PARAMETERS WHICH CAN BE MEASURED

An important limit on the parameters which can be measured by multiparameter flow kinetics (i.e. multiple simultaneous fluorescences measured on a fluorescence-activated cell sorter, also known as a FACS) is the availability of non-mutually interfering intracellular or cell surface-linkable fluorescent probes. New probes are constantly being developed, in many cases by Molecular Probes, Inc. of Eugene, OR, which has become the world's main supplier of fluorescent probes and whose large catalogue lists hundreds of probes and their properties. Not all probes, however, can be excited by wavelengths corresponding to an accessible laser line, so that probes usable on a laser flow cytometer are more limited than those usable for suspension fluorometry or fluorescence microscopy, where the accessible wavelengths are only limited by the available filters (if microscopes or filter fluorometers are used) or not limited at all (if fluorometers having both excitation and emission monochromators are used).

Although cell types and classes differ in their specific responses to agonists, it is now clear that most (though not all) mammalian cells respond to a specific stimulus (i.e. one acting through a receptor-mediated pathway) by undergoing a series of changes:

(1) A relatively rapid but small (and often missed) hyperpolarization, which is followed by a slower and much larger depolarization. Depending on the cell, the depolarization may be $[K^+]$ and $[Na^+]$ or purely $[Na^+]$ transmembrane-gradient-dependent. While mammalian membranes are considerably more permeable to anions, so that contributions of anion gradients to the transmembrane potential are often ignored as in the Goldman approximation (Hoffman & Laris, 1974). However, it is now clear that neglect of anion gradient contributions can lead to errors in membrane potential calculations (Korchak *et al.*, 1982; Simchowitz, 1988; Grinstein & Foskett, 1990).

(2) A relatively rapid acidification which does not appear to depend upon the $[Na^+]$ gradient, but which is followed by an alkalinization which proceeds via Na^+/H^+ countertransport and can be blocked by amilorides, first described by Grinstein and Sha'afi and their colleagues (Grinstein & Furuya, 1986; Grinstein *et al.*, 1986; Weisman SJ *et al.*, 1987) and recently reviewed by Orlowski & Grinstein (1997).

(3) A transient cytoplasmic $[Ca^{2+}]$ increase (within <1 s in some cells, 30 times slower in others) in which the Ca^{2+} generally comes from intracellular stores such as the endoplasmic reticulum, although some cells count on a Ca^{2+} influx from the extracellular milieu for at least part of the transient, first described by Korchak and her colleagues (Simons *et al.*, 1982).

(4) Activation of at least one phospholipase in the plasma membrane (most are Ca^{2+}-dependent) and release of the appropriate products, e.g. arachidonic acid if phospholipase A_2, inositol trisphosphate and diacylglyceride if phospholipase C is the initially activated enzyme.

(5) For secretory cells, an eventual degranulation though the times and controls of granule content release vary widely.

(6) For phagocytic cells, a phagocytic vacuole (phagovacuole) will form sometime during steps 1–5, and the oxidative burst will be initiated so that oxidizing products, as well as the specific granules' lytic enzymes, are released into the phagovacuole.

(7) If a cell is able to respond and the agonist is a chemotactic agent, chemotaxis will be initiated sometime during the above processes.

For most cells it is not yet known whether any of the above events depends upon the prior occurrence of any of the others, or whether all can be initiated simultaneously by the same event. Almost all of the events except chemotaxis can readily be followed continuously by flow cytometry kinetics.

39.6 CLASSES OF FLUORESCENT PROBES

There are a number of classes of probes, and only examples of each will be given here, rather than the entire and exhaustive list. Many of the older ones are well described in reviews by Waggoner (1977, 1988,

1990), while current ones are described by Haugland (Haugland, 1994), by Johnson (Johnson, 1997) and in the Molecular Probes Handbook (Eugene, OR).

The overall principles which govern the use of these probes differ.

39.6.1 Distributive probes

Membrane potential and some pH probes distribute themselves between the cell and the external milieu according to the property being evaluated. In the case of membrane potentials, this is the net negative charge of the cell with respect to the buffer in which it is suspended. In the case of pH probes, it is the difference between the extra- and the intra-cellular pH. These probes will distribute into every compartment across whose membrane a potential or pH gradient, respectively, exists, and therefore will be distributed within all cellular organelles. Unless one uses an image-enhanced fluorometer, the net membrane potential or pH will therefore be an average over the whole cell, its organelles and interior compartments.

39.6.2 Distributive probe concentration limitations

There are two general precautions with respect to distributive probes: (1) One must use enough probe so that the distribution is dependent upon the property to be evaluated, not upon the quantity of probe available. As Hoffman and Laris (1974) described when these probes were first prepared by Waggoner (Sims *et al.*, 1974), one must therefore evaluate, for each new cell type, the lowest concentration of probe which will yield a result independent of that concentration. (2) One must not use too much cationic probe since it, being of opposite charge, will tend to neutralize the net negative charge of the cell and, within it, tend to concentrate in the most negative compartment. Similarly, one must not use an excess of the amine pH probes since these will become protonated in the cytoplasm and therefore will alter the cytoplasmic pH. The balancing of these two mutually contradictory conditions is critical to the correct use of a distributive probe (Freedman & Novac, 1983; Freedman & Laris, 1988).

It is self-evident that probes which tend to multimerize and self-quench within the cell, such as the dithiocyanine indicators of membrane potentials (Sims *et al.*, 1974) or the aminoacridine indicators of cytoplasmic pH (Deamer *et al.*, 1972), are not suitable for flow cytometry as only the cell and not its surrounding medium register a fluorescence in flow cytometry. There are, however, non-self-associating cyanine probes such as the dioxa- and dicarbo- series (see below) which do fluoresce in the cell and are therefore usable for flow cytometric measurements of membrane potentials (Seligman & Gallin, 1983). Although some other amines such as acridine and methyl orange have been suggested as distributive pH probes (Nuccitelli & Deamer, 1982), these have been found to be difficult to quantitate due to binding, largely to sites on cell membranes, and they are therefore used qualitatively but not quantitatively.

39.6.3 *In situ* cytoplasmic probes

In situ cytoplasmic probes depend upon the principle that uncharged entities can traverse a cell membrane easily, whereas charged ones are trapped on one side or the other (Thomas *et al.*, 1979; Tsien, 1980; Grynkiewicz *et al.*, 1985). This is rather a broad principle, often honoured in the breach since some cell membranes are relatively permeable to anions, but it appears to hold well for many classes of cells. The probes' designers then reasoned that, if an esterified (and preferably non-fluorescent probe) could diffuse into a mammalian cell's cytoplasm, the ubiquitous and relatively unspecific cytoplasmic esterases would hydrolyse the probe, yielding a fluorescent indicator which is trapped in the cytoplasm because of its newly acquired charge. The assumption was that, because most of the esterases are contained in the cytoplasm, the eventual localization of the de-esterified probe would be in the cytoplasm. We have shown this assumption to be correct for at least three such probes, the Ca^{2+} probes indo-1 and fura-2, and the pH probe BCECF (see below).

39.6.4 Intensity of fluorescence as an indicator

As indicated above, there are *in situ* probes whose fluorescence intensity F at a given wavelength is, over the probe concentration region in which Beer's Law is obeyed (i.e. $F = f[X]$), a linear function of the concentration, [X], of the ion for which it is specific, assuming the probe concentration and the light path through the cell (i.e. the cell size) to be constant. A calibration curve can then be used to evaluate the actual concentration from the fluorescence intensity. The calibration curves, for a probe which changes only in the intensity of its fluorescence as a function of [X], will have a slope dependent upon the internal concentration of probe, i.e. will be parallel but not superimposed for different loadings of the same cell with the same probe. To calculate the actual concentration of X, one therefore determines the slope of the calibration curve by measuring the maximal and minimal fluorescences of X under the conditions and for the type and number of cells being studied, to locate the linear correlation. These curves should be used only in the region in which they are approximately linear. It should be remembered, in this connection, that such a calibration curve is always S-shaped with an inflection point at the pK, pCa^{2+} or, in the general case, pX.

39.6.5 Ratio of fluorescences as an indicator

If the *in situ* probe being used undergoes a wavelength shift of the absorbance or the emission maximum upon binding X, one can use the ratio technique described above (Thomas *et al.*, 1979; Tsien, 1980; Grynkiewicz *et al.*, 1985). The ratio R of fluorescence emissions at two wavelengths, elicited at a single excitation wavelength, or of fluorescence emissions at a single wavelength excited by light at the two different excitation wave lengths, is then independent of the cell size (i.e. the length of the light path) and of the probe concentration. In some cases, there is an isobestic point at which the fluorescence is independent of the concentration of X; the ratio technique is then still valid. It is, however, not valid if the two chosen wavelengths for the ratio fall on the same emission or absorbance peak; in such a case, one is taking the ratio of two intensities on essentially the same straight line (the side of the absorbance or emission peak), which will yield a constant and will be independent of the concentration of X.

It should also be noted that, where an appropriate ratiometric probe is not available, the use of two separate *in situ* probes, either both sensitive to the same cation (e.g fluo-3 + Fura Red has been used for Ca^{2+} measurements where only the 488 nm excitation wavelength is available (Novak & Rabinovitch, 1994; Floto *et al.*, 1996; Atsumi *et al.*, 1996)) or one sensitive to the property being measured, the other insensitive to that property as well as to any other parameter which is changing, has been successful; the latter condition is sometimes hard to meet. As discussed below, it is important to use probe ester concentrations low enough to minimize acidification of the cell's cytoplasm as the probe is de-esterified; where two different probes are used the same precautions apply.

39.6.6 *In situ* probe concentration limitations

For any of the *in situ* probes, it is preferable to use the lowest concentration at which the probe's fluorescence changes can be detected. That is, unlike the distributive probes, there is no possible error in using a very low concentration of probe. There are several problems that may occur if the probe concentration chosen is too high:

(1) The probe concentration within the cell may exceed adherence to Beer's Law, and self-quenching of fluorescence may occur. This is particularly true of fluorescein derivatives.
(2) The probe ester may not be fully hydrolysed, leading to a continuous change in intracellular fluorescence and the possibility that extracellular ester may be present and be hydrolysed in the external buffer. Furthermore, excess ester may penetrate into some of the internal organelles or granules; if these contain no esterases, the problem of excess ester leakage remains, while if they do contain esterases one is now not measuring the cytoplasmic pH but rather that of both cytoplasmic and organellar compartments.
(3) Since leakage is concentration gradient-dependent, the probe as well as its ester will leak more readily to the exterior, causing the artefacts already described above.
(4) Cytoplasmic hydrolysis of any ester leads to acidification of the cytoplasm which may or may not have enough protein content to act as buffer. For example, we have found that loading platelets or neutrophils with a 15 μM probe ester for 15 min at 37°C leads to a sustained reduction of the cytoplasmic pH from approx. 7 to 6.85.
(5) It should also be noted that, even if the probe concentration is sufficiently low, the linear relationship between the calculated ratio and the concentration of ion to be measured must be verified. In this connection, it is apparent that equations like the Grynkiewicz equation $[Ca^{2+}] = K_dS\ \{(R_{obs} - R_{min})/(R_{max} - R_{obs})\}$ (Grynkiewicz *et al.*, 1985) which involve differences between maximal (and minimal) and observed ratios, break down at very high R_{obs}, i.e. as the denominator approaches zero. Therefore, at high R_{obs}, one must use an equation which retains linearity at high R_{obs}, but is also linear at the normal R_{obs}. Such an equation, an exponential, has now been developed for indo-1 (Davies *et al.*, 1997); similar curve-fitting will have to be done for other probes and their respective indivated cations.

39.7 SPECIFIC PROBES FOR PARAMETERS OF CELL FUNCTION

(N.B. The following list and brief description of some of the currently most popular and frequently used probes is not meant to be all inclusive; it is merely meant to illustrate the use of certain types of probes, and to present their advantages and disadvantages in specific applications.The examples include those probes we have found most useful for cell function studies by multiparameter flow kinetics.)

39.7.1 Intracellular cations: Ca^{2+} and Mg^{2+}

The initial *in situ* Ca^{2+} probe described by Tsien (1980) was quin-2, a probe whose fluorescence intensity increased upon binding of Ca^{2+}, but which exhibited no wavelength shift. It could be introduced as a membrane-permeable acetoxymethyl ester, making it preferable to previously available Ca^{2+} indicators such as aequorin which is not membrane permeable and must be introduced into the cell cytoplasm by more stressful techniques such as electroporation or 'scrape loading'. Quin-2, however, had several disadvantages, including a relatively low quantum yield, a high Ca^{2+} buffering capacity, and the absence of a wavelength shift upon binding Ca^{2+}. Tsien and his colleagues therefore developed a new series of ultraviolet-excited probes, indo-1 and fura-2 (Grynkiewicz *et al.*, 1985) (later followed by fura-5, a longer wavelength ratiometric analogue). These indicators are introduced into cells as acetoxymethyl esters, and de-esterified in the cytoplasm. However, unlike quin-2, which exhibited no emission or excitation wavelength shift after binding Ca^{2+}, indo-1 undergoes a large emission wavelength change (from 485 nm for Ca^{2+}-free to 405 nm for Ca^{2+}-bound indo-1, both excited at 357 nm), while fura-2 undergoes a somewhat smaller excitation wavelength shift (380 nm Ca^{2+}-free to 340 nm Ca^{2+}-bound, with both emissions at 510 nm). The esters are almost non-fluorescent, but an excess of unhydrolysed fura-2/AM can lead to artefactual ratios as it also emits fluorescence at 510 nm (Scanlon *et al.*, 1987); there is no problem with indo-1/AM, which fluoresces at 455 nm when excited at 357 nm, i.e. at a wavelength far removed from 405 and 485 nm, the two monitored emission wavelengths. Additional UV-excitable Ca^{2+} probes such as bis-fura 1, and comparable Mg^{2+} indicators, such as mag-fura-2, mag-fura-5 and mag-indo-1, also exist. There are now newer and, for some applications, preferable Ca^{2+} and Mg^{2+} indicators, including fluo-3, rhod-2, Fura Red, Calcium Green, Calcium Orange, Calcium Crimson, Oregon green 488 BAPTA 1 and 2, and the comparable Mg^{2+} indicators, Magnesium Green and Magnesium Orange, which can be excited in the visible range, and exhibit large Stokes shifts (some with emissions around 600 nm) but not all exhibit wavelength maximum differences between the ligand-bound and ligand-free states. Other divalent cations of interest to cell biologists, such as Zn^{2+}, Mn^{2+}, Cu^{2+}, Fe^{2+}, Hg^{2+}, Pb^{2+}, Cd^{2+}, Co^{2+} and Ni^{2+} bind to some of the Ca^{2+} and Mg^{2+} indicators as well; the literature on their detection is less extensive but has been reviewed (Haugland, 1996).

The current literature tends to describe experiments using indo-1 for flow cytometric studies and fura-2 for fluorescence microscopy because it is more convenient to use a single excitation wavelength in the former, a single emission in the latter, and ratiometric probes are always preferable to the non-ratiometric (i.e. single wavelength) indicators. We have shown by organellar separation (Borregaard *et al.*, 1983; DelBuono et al., 1989), followed by fluorometry, that the two probes, when used with all the aforementioned corrections, at concentrations below 5 μM, give fully comparable results, and that >95% of the probe resides in the cytoplasm under these conditions.

39.7.2 Intracellular cations: H^+

The earliest fluorescent pH_{in} probes for cells were the distributive probe 9-aminoacridine (Deamer *et al.*, 1972) and other acridine derivatives (Nuccitelli & Deamer, 1982). Because these self-associate and hence self-quench in cells, they are not recommended for flow cytometric measurements. Methyl orange, an unquantitative pH probe, has been used as a qualitative probe of some of the highly negative compartments of mammalian cells.

The first of the *in situ* probes was fluorescein diacetate, soon replaced by 5(6) carboxyfluorescein diacetate, whose de-esterified form was more readily retained within the cell (Thomas *et al.*, 1979), and by dimethyl-5(6) carboxyfluorescein (Simons *et al.*, 1982). A number of fluorescein and rhodamine derivatives have since appeared, the most frequently used one at near-neutral pH, especially for flow cytometry, being BCECF (2',7'-bis-(2–carboxyethyl)-5-(and 6) carboxy-fluorescein)(Grinstein *et al.*, 1986; Siffert *et al.*, 1987b; Davies *et al.*, 1987a,b, 1990; Sullivan *et al.*, 1988; Naccache *et al.*, 1989; Strohmeier *et al.*, 1995a,b; Seetoo *et al.*, 1997), which permeates the cell membrane in the acetoxymethyl ester form. Since all of these pH indicators are weak acids, their p*K* is of prime importance, and the choice of probe must be made on the basis of the range of pH expected to be encountered, and the p*K* of the probes. Thus, while BCECF or its analogues may be the ratiometric probes of choice in the neutral pH region, Oregon Green and Rhodol probes are more suitable for acidic pH evaluations because they have lower p*K*s. Users must also recognize that some pH probes have very short linear regions on the pH calibration curve, and therefore only a short pH span over which they are sensitive enough to be used as pH indicators.

The choice of cytoplasmic pH probe must be predicated not only on the desired pH range to be observed, but also on the excitation wavelengths available and on the desirability of ratiometric versus non-ratiometric probes. For example, BCECF has been such a successful pH probe for flow cytometric measurements, in spite of a relatively low quantum yield, because it has a pH-dependent absorbance maximum at 500 nm, easily excitable by the 488 nm line of an argon laser, and a pH-independent absorbance at 450 nm, excitable by the 457 nm argon line. Emission from both excitations is observable at 500 nm. In general, the emission fluorescence is easy to isolate by appropriate bandpass filters – it is the excitation wavelengths which must be deliverable by a laser if a flow cytometer is to be used. For ordinary fluorometry, the latter restriction is, of course, not applicable. Another advantage of BCECF and other similarly fluorescein-based probes is that they can easily be used in conjunction with indo-1 so that pH and Ca^{2+} changes can be observed simultaneously (Davies *et al.*, 1990).

39.7.3 Intracellular cations: Na^+ and K^+

The measurement of cytoplasmic Na^+ changes has been attempted, both in the flow cytometer and in the fluorescence microscope. Both UV-excitable (SBFI and PBFI, for Na^+ and K^+, respectively) and 488 nm-excitable (Sodium Green, for Na^+), probes exist (Minta & Tsien, 1989; Amorino & Fox, 1995), but the selectivity *in situ* of SBFI for Na^+ over K^+ is limited; for example, the K_d for Na^+ in the absence versus presence of K^+ differs by only a factor of 4, while the comparable ratio for PBFI in the absence versus presence of Na^+ is 2. While each of these indicators may therefore be usable in an artificial medium where the other ion is excludable, it makes them difficult to use in a 120 mM K^+, 25 mM Na^+ environment like the cytoplasm of a cell. A 100-fold higher selectivity for one ion over another (i.e. a difference of two orders of magnitude in the dissociation constant K_d of the selected ion from the probe) is desirable in a 'good' indicator. Such a difference is present, for example, in mag-fura-2, which has a 100-fold lower K_d for Mg^{2+} than for Ca^{2+}. In contrast, the two-fold difference in SBFI affinity for Na^+ and K^+ means that a cytoplasmic change in Na^+ would be difficult to detect in a mammalian cell whose $[K^+]_i$ is approximately 120 nM. Although PBFI could in principle be used to detect a change in that $[K^+]$, the high concentration of probe which would be required is likely, as described above, to lower abnormally the cytoplasmic pH due to the number of protons released as PBFI/AM is lysed by cytoplasmic esterases.

39.7.4 Transmembrane potentials

Any living mammalian cell retains its functional integrity by maintaining an electrical potential, the transmembrane potential, between the interior and the exterior of the cell. If there are organelles, there is generally also a potential across the organellar membranes. The retention of the overall transmembrane potential requires energy, in the form of ATP, since the major portion is attributable to the transmembrane gradients of Na^+ and K^+, usually maintained by an ATP-driven Na^+/K^+ ATPase in the membrane. For that reason, poisoning of the Na^+/K^+ ATPase with compounds such as ouabain, or collapse of the ionic gradients with the appropriate ionophores, profoundly perturbs the transmembrane potential, although, as shown below, the portion of that potential attributable to the K^+ and Na^+ gradients, respectively, appears to differ for each cell.

In most cells, all or part of the membrane potential change observed in response to stimulation is attributable to the opening of either an antiport (e.g. the Na^+/H^+ antiport as described for platelets initially by Horne & Simons (1978) and found to be a countertransport by Davies *et al.* (1987b) or a channel involving Na^+ or K^+ (e.g. the hyperpolarization in neutrophils; Korchak & Weissman, 1978; Lazzari *et al.*, 1986, 1990). Although, in general, anions permeate the plasma membrane freely and therefore establish a gradient based purely upon the existent Donnan potential, that distribution, especially for Cl^- and HCO_3^- ions, also contributes to the overall potential, the extent of contribution being cell and cell environment dependent (Wright & Diamond, 1977; Korchak *et al.*, 1982; Simchowitz, 1988; Grinstein & Foskett 1990; Lazzari *et al.*, 1990). Similarly, each cell type's internal organelles, and the extent to which the potential across their membranes contributes to the overall potential of the cell, varies, but these contributions must be considered. For example, it has been estimated (Sims *et al.*, 1974; Waggoner, 1988) that mitochondria can account for >50% of the overall potential, but, unless the organelles can be isolated (DelBuono *et al.*, 1989) or are absent, as in erythrocytes, the individual organelles' contributions cannot usually be evaluated except in cells large enough so that microelectrodes can reliably be inserted into the desired location.

The thermodynamic parameters contributing to the overall transmembrane potential include the chemical potentials attributable to each individual ion's concentration gradient across internal as well as plasma membranes, to the osmotic pressure across these membranes, and to the Donnan equilibrium (Hoffman & Laris, 1974; Simons, 1988; Simons & Greenberg-Sepersky, 1987; Freedman & Laris, 1988; Simons *et al.*, 1988; Gallin & McKinney, 1988a). It has been shown that, overall, the Goldman constant field equation, also known as the Hodgkin–Katz equation, applies to these cells:

$$V = (RT/F) \ln \{P_{K}K^{+}_{out} + P_{Na}Na^{+}_{out} + P_{Cl}Cl^{-}_{in}\}/\{P_{K}K^{+}_{in} + P_{Na}Na^{+}_{in} + P_{Cl}Cl^{-}_{out}\}$$

where V is the transmembrane potential, F is the Faraday and P represents the membrane permeability of the specific ion named. In general, all other ionic contributions, including those of protons, can be neglected when dealing with the overall potential of a cell.

Since membrane potentials reflect the overall charge difference between the inside and the outside of a cell, all membrane potential probes involve the measurement of their relative distribution, and, of necessity, reflect the overall potential which includes not only the trans-plasma membrane but also trans-organellar membrane potentials.

As discussed earlier in this chapter, two types of distributive fluorescent membrane potential probes exist: those whose contribution to fluorescence is intracellular and those whose fluorescence intracellularly is quenched, so that only the extracellular probe is detectable. Clearly one can use either type in fluorometers, but only the intracellularly fluorescing probe in flow cytometers.

39.7.4.1 Cyanines

The cyanines' dithio-, dioxa- or dicarboxy- series, highly conjugated probes whose properties depend upon the number of carbons in the conjugated chain and upon the length of the *N*-attached C tails, were first described by Waggoner and his colleagues (Sims *et al.*, 1974; Waggoner, 1988). All are lipophilic quaternary ammonium salts which distribute between the external buffer and the cell according to its net negative charge. Some, like the dithio- series, have a tendency to self-associate in the cell membranes, thereby quenching the probe fluorescence. The sole fluorescence which can therefore be observed is that in the extracellular buffer, rendering this probe useless for flow cytometry. In contrast, other cyanines such as the dioxa-series do not exhibit a self-associating tendency; since they tend to have a very low quantum yield in aqueous buffers, the sole fluorescence which can be observed would be that emitted by the intracellular probe. The dioxa-cyanines (diOCx(Y)) with a varying number of carbon atoms (Y) in the conjugated link between the rings, and various numbers of carbon atoms (X) attached to the two heterocyclic rings, have been used successfully in flow cytometry by a number of investigators (Seligman *et al.*, 1982, 1984; Lazzari *et al.*, 1986, 1990; Bernardo *et al.*, 1988; Seligman, 1990; Brunkhorst *et al.*, 1991).

39.7.4.2 Oxonols

Since most oxonols are themselves negatively charged, the reasons why their distribution reflects the overall transmembrane potential remain unclear. They have relatively low Stokes shifts, but have nevertheless been used for membrane potential measurements in cell suspensions (Bashford *et al.*, 1985) and in flow cytometric measurements (Lazzari *et al.*, 1990).

In published cases in which several different probes have been used to evaluate membrane potentials and their changes (Horne *et al.*, 1981; Seligman & Gallin, 1983; Lazzari *et al.*, 1990), the results have been identical. Thus, when used correctly, any of the above probes yields the same information.

39.7.4.3 Phospholipase activation

Although it has long been known that stimulation of secretory cells activates the phospholipid metabolic pathways, it has only been possible to follow the kinetics of activation of phospholipases A_2, C and D by isolation and identification (by HPLC or TLC) of the phospholytic products, at time intervals from milliseconds to seconds and minutes after exposure of the cells to a stimulus.

Ca^{2+} availability is usually required by these enzymes, but the question of whether the Ca^{2+} comes from extra- or intra-cellular stores, and how high a concentration of Ca^{2+} must be available before the phospholipase is active, had not been resolved because simultaneous studies had not been possible.

Fluorometric techniques for following phospholipase activation have been tried, but largely unsuccessfully. Although fluorescently tagged phospholipids, lysophospholipids and fatty acids have been available for a number of years, and can be incorporated into cell membranes, a resultant fluorescence-detectable susceptibility to phospholipase hydrolysis had not been achieved. Recently, however, we reported our success with a new probe, bis-BODIPY™-phosphatidylcholine, developed in collaboration with Molecular Probes, which can be incorporated into neutrophil (and platelet) plasma membranes (Meshulam *et al.*, 1992). Because the phospholipid carries a BODIPY group (a fluorescein derivative) in both the 1 and the 2 positions, there is total quenching in the resting cell membrane and the probe is not fluorescent. Stimulation activates one or more of the aforementioned phospholipases. If the activated enzyme is phospholipase A_1 or A_2 (PLA_1, PLA_2), one of the BODIPY-labelled fatty acids will be lysed off, the quenching will be relieved, and both the lysophospholipid remaining in the membrane and the fatty acid (which usually is released into the exterior buffer or into the cytoplasm, depending on the membrane leaflet in which the phospholipidolysis occurs) will be fluorescent. The appearance of this fluorescence can then be followed continuously. For our experiments, this was done by flow cytometry since we are interested in responding subpopulations. The results permitted us not only to resolve the Ca^{2+} versus activation of PLA question: when neutrophils are stimulated with immune complexes, the maximal Ca^{2+} transient is over and redistribution within the neutrophil is well under way before PLA activation can be detected, at ~20 s. Of perhaps equal importance is the ability to distinguish between different mechanisms of cell activation: in contrast to immune complexes, which activate neutrophil PLA, chemotactic peptides do not activate this enzyme at all within the first 2.5 min after exposure to neutrophils.

More recently we have succeeded, in collaboration with Molecular Probes Inc., in developing a fluorescent probe for phospholipase D (PLD) (Gewirtz & Simons, 1997). BODIPY™lyso-phosphatidylcholine is readily incorporated into mammalian cell (e.g. human neutrophil) membranes and converted to BODIPY™phosphatidylcholine. Lysis by PLD in the presence of a very small amount (<0.5%) ethanol leads to the formation of BODIPY™phosphatidylethanol, which can be easily detected by HPLC in an instrument equipped with a fluorescence detector. Since both the phosphatidylcholine and its ethanol product are fluorescent, flow cytometry is not applicable to analysis of PLD activity with this substrate.

39.7.4.4 Oxidative burst

Among the consequences of phagocyte stimulation is the liberation of oxidative products, via activation of an NADPH-dependent oxidase, some of whose components, in the resting cell, reside in the azurophilic granule membrane (Borregaard *et al.*, 1983), others in the cytoplasm (Curnutte *et al.*, 1987, 1989), and the remainder in the plasma membrane. Activation of the phagocyte leads eventually to degranulation, a process of exocytosis in which granule membranes fuse with the plasma membrane. The consequent assembly of the oxidase, as well as release of the granules' superoxide dismutase and myeloperoxidase, leads to an event called the oxidative burst, the release of O_2^-, O_2^{2-}, O·, OH· and, if Cl^- is present, OCl^-. The release occurs preferentially into the phagocytic vacuole, but also into the extracellular milieu. Quantitation of the oxidative burst, in terms of the rate or extent of release of superoxide, is usually accomplished by following the difference in absorbance when oxidized cytochrome *C* is reduced by the activated neutrophil-released superoxide (Whitin *et al.*, 1980, 1981). Chemiluminescence has also been used. However, neither technique is adaptable to fluorescence spectroscopy. Bass and his colleagues were the first to adapt oxidizable hydrofluorophores to measurements of the rate of appearance of products of activated phagocytes' oxidative burst in a flow cytometer (Bass *et al.*, 1983, 1986). The membrane-permeant non-fluorescent ester of their indicator, dihydrodichlorofluorescein (originally called dichlorofluorescin), remained in the cytoplasm of the resting neutrophils as a non-fluorescent free acid after esterolysis. Upon stimulation of the phagocytes, and consequent initiation of the oxidative burst, the products are released into the extracellular buffer and into the phagocytic vacuoles. A certain proportion of these released products diffuses back into the cytoplasm of the cells and oxidizes the dihydrochlorofluorescein. The resultant increase in fluorescence can be followed on a fluorometer or on a flow cytometer. However, the sensitivity of the assay was limited by the time required for rediffusion from the extracellular buffer or from the phagovacuole into the cell cytoplasm, by the extent of that diffusion, and by the extent to which catalase and peroxidase destroyed the oxidative burst products before they could react with the dihydrochlorofluorescein.

To overcome this problem, the oxidizable fluorophore has now been attached directly to the agonist (Ryan *et al.*, 1990). Dihydrodichloro-fluorescein or dihydrorhodamine-labelled immune complexes and similarly labelled chemotactic peptides are now commercially available and have been used successfully in flow cytometry measurements of neutrophil activation kinetics (Model *et al.*, 1997). We have shown that, for immune complex-stimulated neutrophils, a much larger fluorescence increase is detectable much sooner after exposure of the neutrophils to the stimulus than for the cytoplasmic probe (Ryan *et al.*, 1990; Brunkhorst *et al.*, 1991, 1992; Strohmeier *et al.*, 1995a,b; Guillemot *et al.*, 1996; Bernardo *et al.*, 1997; Seetoo *et al.*, 1997). This is true largely because most of the oxidative products are released directly into the phagocytic vacuole, where they are not as subject to dilution as in extracellular buffer, and the concentration of oxidants is therefore much higher very near the reduced probe. For chemotactic peptides, which elicit a much smaller oxidative burst, detection is possible without resorting to the addition of cytochalasin B, the agent usually added when fMLP-induced oxidative burst is measured by any of the other techniques; addition of cytochalasin B enhances the oxidative burst but the products are diluted in the external medium, making detection much more dependent on the cell number-to-agonist concentration. Furthermore, since there is always a small amount of already oxidized material in the preparations, it is possible to measure agonist binding as well as oxidative product in the same experiment, with the same photomultiplier. These probes, as well as their more recently available dihydro-rhodamine counterparts, are applicable both to fluorescence microscopy and flow cytometry.

The same oxidizable groups have been linked to particles such as zymosan or bacterial fragments which, when opsonized, can stimulate phagocytes via the Fc or the C3b receptors; the particles are then phagocytized, and, in response to the oxidative burst, exhibit an increase in fluorescence proportional to the superoxide and/or peroxide release. These probes have also been used (Guillemot *et al.*, 1996) to evaluate the subpopulation of phagocytic cells in a whole blood preparation or in cultured cells in which only a fraction exhibit oxidative-product generating capabilities.

39.7.4.5 Phagocytosis

For flow cytometry, phagocytosis is most readily demonstrated by the use of fluorophore-labelled particles such as beads, latex particles, zymosan, LPS, bacterial fragments, etc. The same fluorophores described above as antibody and ligand labels can be employed, though coupling is sometimes best achieved by using the succinimidyl ester rather than the thiocyanate or the anhydride of the fluorophore. In order to be ingestible or recognized by phagocytic cells, the labelled particles may need to be opsonized, a technique which generally does not perturb the fluorescence. The fluorescence of agonist which is too well bound to the surface to be washed off, yet is not phagocytosed, can be quenched with an antibody to the fluorophore.

39.8 NON-KINETIC APPLICATIONS OF RELEVANCE TO MULTIPARAMETER FLOW KINETIC CORRELATIONS

It is impossible here to discuss all of the measurements which can be accomplished by flow cytometry. The instrument was originally developed for cell type identification, detection of fluorescently labelled markers on the plasma membrane surface of the cells, and sorting of the cells according to these markers. Similarly, the use of flow cytometers to study cell cycle stages and ploidy will not be covered here. Rather, the only static or equilibrium measurements to be discussed here are those that happen very rapidly and that are associated with kinetic functions, such as receptor occupancy.

39.8.1 *Receptor occupancy*: Surface proteins and their ligands

Since the flow cytometer was originally designed to recognize and, eventually, to sort cells according to their surface 'markers', i.e. the receptors and binding proteins which are exposed on the outer surface of the membrane, the number of possible labels is already too large to be reviewed here *in toto* and increases almost daily.

In some cases the fluorophore is bound directly to the ligand, in others to a primary antibody to the protein being recognized, while in still other situations, especially where magnification of the fluorescent signal is necessary, the fluorophore is bound to the second antibody. For flow cytometry, the usual fluorophores for this purpose are excited at 488 nm and are derivatives of fluorescein, of rhodamine, or of the phycobiliproteins. In some cases the fluorophore is designed for internal energy transfer so that an emission at long wavelengths can be achieved. The lymphocyte literature is particularly rich in usage of cell surface labels.

39.8.2 Nucleic acids

While there are many nucleic acid probes used routinely to determine the ploidy or the phase of a cell, virtually all require fixation and permeabilization of the cells. Since this makes them incompatible with cell function measurements, the propidium and ethidium derivatives will not be discussed here. In contrast, although others are in the planning stages, the only probe which appears to have been used to measure the DNA content of a living and functioning cell is Hoechst 33342, a probe which binds in the minor groove of DNA, which is excited at 357 nm and fluoresces at 485 nm. The AT-selective DAPI probe is also specific for double-stranded DNA, and exhibits similar fluorescence characteristics to the Hoechst probe, albeit with a higher intensity and a somewhat greater photostability. It is, however, less readily inserted into living mammalian cells, and has been used mainly in fungi and prokaryotes. The Hoechst 33342 excitation and emission wavelengths do not interfere with those of fluorescein, rhodamine or phycoerythrin derivatives, so that simultaneous observations of cell cycle and existent surface-exposed proteins can be performed.

39.9 CONCLUSION

It should be clear from the above that this chapter presents only the beginning of the possible applications of flow cytometry to kinetic measurements of cell function. Clearly, as more probes become available and as the ability to design non-interference between the probes increases, the number of possible applications of multiparameter flow kinetic techniques not only to identify cells but to follow their functional responses while monitoring the receptor occupancy, the type of cell, the size, etc., will increase. The limitations arise from our ability as researchers.

ACKNOWLEDGEMENTS

I thank all of the members of my laboratory group who, over the years and now, have done the research and accumulated the data on much of which this review relies. I am also grateful to the National Institutes of Health, whose grants HL15335, HL07501, HL19717, HL33565 and AM31056 supported this laboratory's studies.

REFERENCES

Atsumi T., Sugita K., Kohno M., Takahashi T. & Ueha T. (1996) *Cytometry* **24**, 99–105.
Amorino G.P. & Fox M.H. (1995) *Cytometry* **21**, 248–256.
Bashford C.L., Alder G.M., Gray M.A., Micklem K.J., Taylor C.C., Turek P.J. & Pasternak, C.A. (1985) *J. Cell. Physiol.* **123**, 326–336.
Bass D.A., DeChatelet L.R., Szejda P., Seeds M.C. & Thomas M. (1983) *J. Immunol.* **130**, 1910–1917.
Bass D.A., Olbatantz P., Szejda P., Seeds M.C. & McCall C.E. (1986) *J. Immunol.* **136**, 860–866.
Beard C.J., Key L., Newburger P.E., Ezekowitz R.A.B., Arceci R., Miller B., Proto P., Ryan T., Anast C. & Simons E.R. (1986) *J. Lab. Clin. Med.* **198**, 498–505.
Bernardo J., Brink H.F. & Simons E.R. (1988) *J. Cell. Physiol.* **134**, 131–136.
Bernardo J., Newburger P.E., Brennan L., Brink H.F., Bresnick S.A., Weil G. & Simons E.R. (1990) *J. Leuk. Biol.* **47**, 265–274.
Bernardo J., Billingslea A.M., Ortiz M.F., Seetoo K.F., Macauley J. & Simons E.R. (1997) *J. Immunol. Methods* **209**, 165–175.
Borregaard N., Heiple J.M., Simons E.R. & Clark R.A. (1983) *J. Cell. Biol.* **97**, 52–61.
Brunkhorst B.A., Lazzari K.G., Strohmeier G.R., Weil G. & Simons E.R. (1991) *J. Biol. Chem.* **266**, 13035–13043.
Brunkhorst B.A., Strohmeier G.R., Lazzari K.G., Weil G., Melnick D., Fleit H.B. & Simons E.R. (1992) *J. Biol. Chem.* **267**, 20659–20666.
Curnutte J.T., Kuver R. & Babior B.M. (1987) *J. Biol. Chem.* **262**, 6450–6452.
Curnutte J.T., Scott P.J. & Babior B.M. (1989) *J. Clin. Invest.* **83**, 1236–1240.
Davies T.A., Dunn J.M. & Simons E.R. (1987a) *Anal. Biochem.* **167**, 118–123.
Davies T.A., Katona E., Vasilescu V., Cragoe E.J. Jr & Simons E.R. (1987b) *Biochim. Biophys. Acta.* **903**, 381–387.
Davies T.A., Drotts D., Weil G.J. & Simons E.R. (1988) *Cytometry* **9**, 138–142.
Davies T.A., Drotts D.L., Weil G.J. & Simons E.R. (1989) *J. Biol. Chem.* **264**, 19600–19606.
Davies T.A., Zabe N., Weil G.J., Drotts D. & Simons E.R. (1990) *J. Biol. Chem.* **265**, 11522–11526.
Davies T.A., Billingslea A., Johnson R., Greenberg S., Ortiz M., Long H., Sgro K., Tibbles H., Seetoo K., Rathbun W., Schonhorn J. & Simons E.R. (1997) *J. Lab. Clin. Med.* **130**, 21–32.
Deamer D.W., Princ R.C. & Crofts A.R. (1972) *Biochim. Biophys. Acta.* **274**, 323.
DelBuono B.J., Luscinskas F.W. & Simons E.R. (1989) *J. Cell. Physiol.* **141**, 636–644.
Floto R.A., Mahaut-Smith M.P., Somasundaram B. & Allen J.M. (1996) *Cell Calcium* **18**, 377–389.
Freedman J.C. & Laris P.C. (1988) In *Membrane Probes*, L. Loew (ed). CRC Press, Inc., Boca Raton, 1–49.
Freedman J.C. & Novac T.S. (1983) *J. Membr. Biol.* **72**, 59–74.
Gallin E.K. & McKinney L.C. (1988a) *J. Membr. Biol.* **103**, 55–66.
Gallin E.K. & McKinney L.C. (1988b) In *Cell Physiology of Blood*, R.B. Gunn & J.C. Parker (eds). Rockefeller University Press, 315–332.
Gear A.R.L. (1980) *J. Lab. Clin. Med.* **100**, 866–886.
Gear A.R.L. & Burke D. (1982) *Blood* **60**, 1231–1234.
Gewirtz A.T. & Simons E.R. (1997) *J. Leuk. Biol.* **61**, 522–528.
Grinstein S. & Furuya W. (1986) *Am. J. Physiol.* **250**, C283–C291.
Grinstein S. & Foskett J.K. (1990) *Annu. Rev. Physiol.* **52**, 399–414.
Grinstein S., Furuya W. & Biggar W.D. (1986) *J. Biol. Chem.* **261**, 512–514.
Grynkiewicz G., Poenie M. & Tsien R.Y. (1985) *J. Biol. Chem.* **260**, 3340–3350.
Guillemot J.-C., Kruskal B.A., Adra C.N., Zhu S., Ko J.-L., Burch P., Nocka K., Seetoo K., Simons E.R. & Lim B. (1996) *Blood* **88**, 2722–2731.
Haugland R.P. (1994) *Methods Cell. Biol.* **42B**, 641–663.
Haugland R.P. (1996) *Handbook of Fluorescent Probes and Research Chemicals*, 6th edn. Molecular Probes Inc., Eugene, OR.
Hoffman J.F. & Laris P.C. (1974) *J. Physiol.* **239**, 519–552.
Horne W.C. & Simons E.R. (1978) *Blood* **51**, 741–749.
Horne W.C. & Simons E.R. (1979) *Thromb Res* **13**, 599–607.
Horne W.C., Norman N.E., Schwartz D.B. & Simons E.R. (1981) *Eur. J. Biochem.* **120**, 295–302.
Johnson I.D. (1997) In *Curr. Protocols Cytom*, J.P. Robinson, Z. Darzynkiewicz, P.N. Dean, A. Orfao, P.S. Rabinovitch, C.C. Stewart, H.J. Ranke & L. Wheeless (eds). Wiley, New York, pp. 4.4.1–4.4.17.
Jones G.D., Carty D.J., Freas D.L., Spears J.T. & Gear A.R.L. (1989) *Biochem. J.* **262**, 611.
June C.H. & Rabinovitch P.S. (1994) *Methods Cell. Biol.* **41**, 149–174.
Kelley K. (1989) *Cytometry* **10**, 796–800.
Korchak H.M. & Weissmann G. (1978) *Proc. Natl Acad. Sci. USA* **75**, 3818–3822.

Korchak H.M. & Weissmann G. (1980) *Biochim. Biophys. Acta.* **601**, 180–194.

Korchak H.M., Eisenstat B.A., Smolen J.E., Rutherford L.E., Dunham P.B. & Weissmann G. (1982) *J. Biol. Chem.* **257**, 6919–6922.

Lazzari K., Proto P.J. & Simons E.R. (1986) *J. Biol. Chem.* **261**, 9710–9713.

Lazzari K.G., Proto P. & Simons E.R. (1990) *J. Biol. Chem.* **265**, 10959–10967.

Meshulam T., Herscovitz H., Casavant D., Bernardo J., Roman R., Haugland R.P., Diamond R.D. & Simons E.R. (1992) *J. Biol. Chem.* **267**, 21465–21470.

Minta A. & Tsien R.Y. (1989) *J. Biol. Chem.* **264**, 19449–19457.

Model M.A., KuKuruga M.A. & Todd R.F. (1997) *J. Immunol. Methods* **202**, 105–111.

Naccache P.H., Therrien S., Caon A., Liao N., Gilbert C. & McColl S.R. (1989) *J. Immunol.* **142**, 2438–2444.

Nolan J.P., Posner R.G., Martin J.C., Habbersett R. & Sklar L.A. (1995) *Cytometry* **21**, 223–229.

Novak E.J. & Rabinovitch P.S. (1994) *Cytometry* **17**, 135–141.

Nuccitelli R. & Deamer D. (1982) *Intracellular pH: Its Measurement, Regulation and Utilization in Cellular Functions.* A.R. Liss, New York.

Orlowski S. & Grinstein S. (1997) *J. Biol. Chem.* **272**, 22373–22376.

Rink T.J. & Sage S.D. (1987) *J. Physiol.* **393**, 513–524.

Ryan T.C., Weil G.J., Newburger P.E., Haugland R.P. & Simons E.R. (1990) *J. Immunol. Methods* **130**, 223–233.

Sage S.D. & Rink T.J. (1986) *Biochem. Biophys. Res. Commun.* **139**, 1124–1129.

Sage S.D. & Rink T.J. (1987) *J. Biol. Chem.* **262**, 16364–16360.

Scanlon M., Williams D.A. & Fay F.S. (1987) *J. Biol. Chem.* **262**, 6308.

Seetoo K.F., Schonhorn J.E., Gewirtz A.T., Zhou M.J., McMenamin M.E., Delva L. & Simons E.R. (1997) *J. Leuk. Biol.* **62**, 329–340.

Seligman B.E. & Gallin J.I. (1983) *J. Cell. Physiol.* **115**, 105–115.

Seligman B.E. (1990) *Curr. Topics Membr. Transport* **35**, 103–125.

Seligman B.E., Fletcher M.P. & Gallin J.I. (1982) *J. Biol. Chem.* **257**, 6280–6286.

Seligman B.E., Chused T.M. & Gallin J.I. (1984) *J. Immunol.* **133**, 2641–2646.

Siffert W., Siffert G. & Scheid P. (1987a) *Biochem. J.* **241**, 301–303.

Siffert W., Siffert G., Scheid P., Reimens T. & Ackerman J.W.N. (1987b) *FEBS Lett* **212**, 123–126.

Simchowitz L. (1988) In *Cell Physiology of Blood.*, R.B. Gunn & J.B. Parker (eds). The Rockefeller University Press, New York, pp. 194–208.

Simons E.R., Davies T.A., Greenberg S.M., Dunn J.M. & Horne W.C. (1988) In *Cell Physiology of Blood*, R.B. Gunn & J.C. Parker (eds). The Rockefeller University Press, New York, pp. 266–279.

Simons E.R. & Greenberg-Sepersky S.M. (1987) In *Platelet Responses and Metabolism*, H. Holmsen (ed). Vol. III, CRC Press, pp. 31-49.

Simons E.R., Schwartz D.B., Norman N.E. (1982) In *Intracellular pH: Its Measurement, Regulation and Utilization in Cellular Functions*, R. Nuccitelli & D.M. Deamer (eds). Alan R. Liss, New York, pp. 463–482.

Simons E.R. (1988) In *Energetics of Secretion Responses*, J.W.N. Akkerman (ed). C.R.C. Press, Vol. 1 pp. 105–120.

Sims P.J., Waggoner A.S., Wang C.H. & Hoffman J. (1974) *Biochemistry* **13**, 3315–3329.

Strohmeier G.R., Brunkhorst B.A., Seetoo K.F., Bernardo J., Weil G.J. & Simons E.R. (1995a) *J. Leuk. Biol.* **58**, 403–414.

Strohmeier G.R., Brunkhorst B.A., Seetoo K.F., Meshulam T., Bernardo J. & Simons E.R. (1995b) *J. Leuk. Biol.* **58**, 415–422.

Sullivan R., Griffin J.D., Fredette J.P., Leavitt J.L. & Simons E.R. (1987a) *J. Immunol.* **139**, 3422–3430.

Sullivan R., Melnick D.A., Malech H., Meshulam T., Simons E.R., Lazzari K.G., Proto P., Gadenne A.-S., Leavitt J.L. & Griffin J.D. (1987b) *J. Biol. Chem.* **262**, 1274–1281.

Sullivan R., Griffin J.D., Wright J., Melnick D.A., Leavitt J.J., Fredette J.P., Horne J.H., Lyman C.A., Lazzari K.G. & Simons E.R. (1988) *Blood* **72**, 1665–1673.

Thomas J.A., Buchsbaum R.N., Zimniak A. & Racker E. (1979) *Biochemistry* **18**, 2210–2218.

Tsien R.Y. (1980) *Biochemistry* **19**, 2396–2404.

Waggoner A.S. (1977) *J. Membr. Biol.* **33**, 109–140.

Waggoner A.S. (1988) In *Cell Physiology of Blood*, R.B. Gunn & J.B. Parker (eds). The Rockefeller University Press, New York, pp. 210–215.

Waggoner A.S. (1990) In *Flow Cytometry and Sorting*, M.R. Malamed, T. Lindow & M.L. Mendelsohn (eds). Wiley Liss, New York, pp. 209–225.

Weisman S.J., Punzo A., Ford C. & Sha'afi R.I. (1987) *J. Leukoc. Biol.* **41**, 25–32.

Whitin J.C., Chapman C.E., Simons E.R., Chovaniec M.E. & Cohen J. (1980) *J. Biol. Chem.* **255**, 1874–1878.

Whitin J.C., Clark R.A., Simons E.R. & Cohen H.J. (1981) *J. Biol. Chem.* **256**, 8904–8906.

Wright E.M. & Diamond J.M. (1977) *Physiol. Rev.* **57**, 109–156.

CHAPTER FORTY

Photolabile Caged Compounds

ALISON M. GURNEY
Department of Pharmacology, United Medical & Dental Schools, St Thomas's Hospital, London, UK

40.1 INTRODUCTION

The application of photolabile caged compounds in biology has been the subject of a number of reviews (Lester & Nerbonne, 1982; Lester *et al.*, 1986; Nerbonne, 1986; Kaplan, 1986; Gurney & Lester, 1987; Kaplan & Somlyo, 1989; McCray & Trentham, 1989; Homsher & Millar, 1990; Kaplan, 1990; Somlyo & Somlyo, 1990). Practical aspects in the use of caged compounds have recently been detailed (Gurney, 1991; Walker, 1991), along with methods for their synthesis and purification (Walker, 1991). It is not the aim of this chapter to provide another review on the subject, but to outline the general properties of caged compounds and to point out some of the problems that have been, or might be, encountered when applying them to study biological pathways. Many caged compounds can now be purchased from Molecular Probes, Oregon, USA or Calbiochem, California, USA.

Photolabile 'caged' compounds are inert precursors of biologically active molecules that can be stimulated with near-UV light to release the active species. When applied extracellularly or incorporated into cells, they provide a means of manipulating physiological or pharmacological pathways on a rapid (milliseconds) time-scale. A flash of light triggers the uncaging of the bioactive molecule, so that it is released at its site of action without a diffusional delay. Some molecules are designed for extracellular application, being particularly useful with intact preparations where restricted diffusion slows the response to the bioactive molecule, or where its effects would be modified by, for example, metabolism, uptake or receptor desensitization. Other 'caged' compounds can be loaded directly into cells to release intracellular second messengers upon photolysis.

Early studies with caged compounds were limited by the availability of only a few molecules. Of those, caged ATP was used to study the energetics of muscle contraction (Goldman *et al.*, 1982; reviewed in Homsher & Millar, 1990) and activation of the Na pump (Kaplan *et al.*, 1978; Kaplan, 1986), while molecules that interact with cholinergic receptors were exploited to probe drug-receptor interactions and the kinetics of nicotinic-receptor channels (Lester & Chang, 1977; Lester & Nerbonne, 1982; Lester *et al.*, 1986). Other

FLUORESCENT AND LUMINESCENT PROBES, 2ND EDN
ISBN 0–12–447836–0

Table 40.1 Summary of photolabile cage compounds.

Photochemically caged molecules	*References*
Neurotransmitters	
Glutamate	Messenger *et al.*, 1991
Glycine; γ-Aminobutyric acid (GABA)	McCray & Trentham, 1989
Receptor Ligands	
Carbachol	Milburn *et al.*, 1989
Nicotinic receptor agonists (Bis-Q, QBr); Cholinergic antagonists	Lester & Nerbonne, 1982
Phenylephrine	Walker & Trentham, 1988; Somylo *et al.*, 1988
Nifedipine[a]	
Intracellular second messengers	
inorganic ions: Ca^{2+}	Adams *et al.*, 1988; Kaplan & Ellis-Davis, 1988
Mg^{2+}	Kaplan & Ellis-Davis, 1988
H^+	Nerbonne, 1986; Janko & Reichert, 1987
HPO_4^{2-}	Kaplan *et al.*, 1978; McCray & Trentham, 1989
Zn^{2+}	Blank *et al.*, 1981
Nucleotides: ATP	Kaplan *et al.*, 1978; Walker *et al.*, 1989a
ADP; GTP; GDP	Walker *et al.*, 1989a
Cyclic nucleotides: cAMP; cGMP	Nerbonne *et al.*, 1984; Wootton & Trentham, 1989
Inositol phosphates: IP3	Walker *et al.*, 1987; Walker *et al.*, 1989b
IP2	Walker *et al.*, 1989b
Other intracellular probes	
ATP-γ-S; ATP-β, γ-NH; GTP-γ-S; GTP-β, γ-NH	Walker *et al.*, 1989a
BAPTA	Ferenczi *et al.*, 1989

[a] Nifedipine is inactivated by photolysis.

possible applications of caged compounds became apparent with the development of caged cyclic nucleotides (Engels & Schlaeger, 1977; Engels & Reidys, 1978; Nerbonne *et al.*, 1984), which allowed cyclic AMP-dependent regulation of the calcium conductance and contraction in cardiac muscle to be studied directly (Nargeot *et al.*, 1983; Nerbonne *et al.*, 1984; Richard *et al.*, 1985). More recently, a novel chemical approach to caging compounds has enabled a wider variety of probes to be developed (Walker *et al.*, 1988). Around the same time, photolabile calcium chelators were introduced, which release free calcium upon photolysis (Tsien & Zucker, 1985; Adams *et al.*, 1988; Ellis-Davies & Kaplan, 1988). As a result of this increased availability of suitable probes, caged compounds have now become popular for studies on a wide variety of biological responses. Table 40.1 lists some of the currently available molecules.

Optical probes provide a particularly powerful means for manipulating the concentration of molecules inside or outside cells. This is because photochemical reactions tend to be fast and the intensity, duration and area of activating light can be varied with relative ease. In theory, light could be directed at a whole population of cells, or be focused into a spot small enough to irradiate a single cell or part of a cell. The approach should, therefore, allow the possibility of studying how cells respond to perturbations in the concentration of important signalling molecules imposed throughout the cell, or localized to specialized regions of the cell.

40.2 PROPERTIES OF A PHOTOLABILE PROBE

Light is used to initiate a series of reactions. The first phase occurs when the caged compound absorbs a photon of energy, $E = h\nu$, where h is Planck's constant and ν is the frequency of the light. Absorption of a photon promotes the molecule (M) to an excited state (M^*) with higher energy, represented as $M + h\nu \rightarrow$

M^*. M^* is a new species with distinct chemical and physical properties. Its extra energy results in chemical reactions known as the dark reactions, which then lead to the formation of stable products. One of these products is the physiologically active moiety, while the other by-products should ideally be biologically inert.

Photochemical reactions take place only under the influence of light, and only light that is absorbed by the molecule can trigger the photochemical reaction. The amount of light absorbed depends on the extinction coefficient, the intensity and wavelength of the exciting light and the path length of solution through which the light has to pass. Pigmented cells, such as are found in the nervous system of the marine mollusc *Aplysia*, may absorb some of the exciting light and interfere with photolysis (Tsien & Zucker, 1985; Nerbonne & Gurney, 1987). However, most cells are sufficiently transparent that they contribute little to the absorbance of the exciting light, thereby permitting efficient photolysis of caged compounds in isolated cells and in thin tissue preparations. Most of the available caged compounds have an absorbance peak at around 350 nm, a wavelength that is relatively harmless to cells. Absorbance is usually low above 500 nm, so caged compounds are relatively stable under normal laboratory lighting. Stock solutions of most caged compounds can be sufficiently well-protected during use by keeping them in containers wrapped in aluminium foil. Brief exposures to room light while preparing solutions or applying them to experimental preparations do not usually have any detrimental effect. Photolysis typically requires high intensity light in the near *UV* region.

It is not sufficient that a molecule absorbs light for it to be photolabile. Indeed, many molecules absorb light but remain chemically unchanged, e.g. dyes. With other molecules, the extra energy produced by the absorption of a photon results in fluorescence or phosphorescence. Even with a photolabile molecule, the excited intermediate M^* can lose its extra energy and decay back to the ground state M rather than initiate the dark reactions. Clearly an efficient photochemical reaction would be one in which the absorbed light results in product formation a high percentage of the time. The quantum yield of the reaction, which is defined as the ratio of product molecules formed to photons absorbed, provides a measure of the effectiveness of the absorbed light at triggering the photochemical reaction. Thus less light would be required to trigger the release of active molecules from caged compounds with a high extinction coefficient and a high quantum yield. The quantum yields of commercially available probes are all sufficiently high to permit concentration changes in the physiological range to be produced with flash lamps or lasers. In practice, the amount of photolysis produced by a flash of light also depends on the lifetime of the excited intermediate state, relative to the duration of the flash. If the intermediate is short lived, then multiple excitations may occur during the flash, resulting in a higher percentage conversion than would be predicted from the quantum yield. This phenomenon is exemplified by the caged cyclic nucleotides, which have much lower quantum yields than caged ATP, but are photolysed about 50% as well as caged ATP by a 1 ms flash (Wootton & Trentham, 1989). Multiple excitations can apparently occur even with very brief (50 ns) laser pulses.

The ability of photolabile caged compounds to release active molecules rapidly makes them especially useful to biologists. Millisecond time resolution can be achieved with many caged compounds, which is faster than that provided by most other rapid application techniques. Thus photolabile probes have been used to study the kinetics of such rapid events as the activation of receptors (see, for example, Lester & Chang, 1977; Lester & Nerbonne, 1982; Lester *et al.*, 1986; Gurney & Lester, 1987; Milburn *et al.*, 1989; Somlyo & Somlyo, 1990) and ion channels (Gurney *et al.*, 1987; Karpen *et al.*, 1988; Lando & Zucker, 1988; Ogden *et al.*, 1990), muscle contraction (reviewed in Homsher & Millar, 1990; Somlyo & Somlyo, 1990) and neurotransmitter release (Zucker & Hayden, 1988). Nevertheless, although the speed offered by photolysis is one of the main reasons for developing caged compounds, it should not be assumed that photorelease is always fast. The rate of photorelease varies enormously among different caged compounds (see below), and in some cases it can take 100 ms or so for complete photolysis (e.g. Wootton & Trentham, 1989). Thus, when a caged compound is to be used to measure the kinetics of a biological process, it is important that photorelease from the probe proceeds with sufficient speed that it does not limit the time-course of the response being studied.

40.3 THE CHEMISTRY OF PHOTOLABILE COMPOUNDS

Two types of photochemical reaction have been exploited in the development of photolabile caged compounds. These are the *cis* ⇔ *trans* photoisomerization of azobenzenes (Fig. 40.1(a)) and the photochemical cleavage of *o*-nitrobenzyl groups (Fig. 40.1(b)).

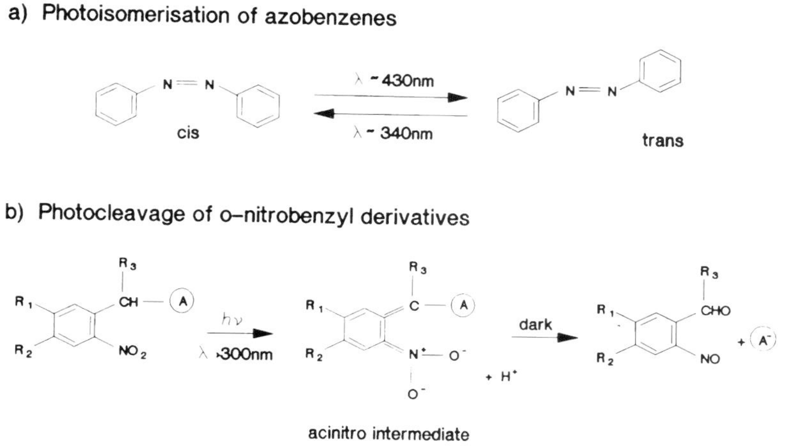

a) Photoisomerisation of azobenzenes

cis ⇌ trans ($\lambda \sim 430$nm / $\lambda \sim 340$nm)

b) Photocleavage of o–nitrobenzyl derivatives

$h\nu$, λ >300nm; dark; acinitro intermediate

Figure 40.1 (a) Photoisomerization of azobenzene is reversible. The *trans* isomer is produced with light of 410–450 nm, while the *cis* isomer is promoted at 300–350 nm. (b) The reaction scheme proposed to account for the photocleavage of *o*-nitrobenzyl derivatives. Substitutions at R_1, R_2 and R_3 alter the speed and efficiency of the reaction, as well as the biological activity of the precursor and photoproducts.

40.3.1 Azobenzenes

Azobenzene derivatives have been used most extensively to probe the kinetics of nicotinic receptors (reviewed in Lester & Nerbonne, 1982; Lester *et al.*, 1986; Gurney & Lester, 1987). These studies were made possible because the *cis* and *trans* stereoisomers differ in their pharmacological properties, and interconversion between the isomeric forms can be controlled by irradiating the molecules at different wavelengths. For example, the best-studied azobenzene, Bis-Q (3,3′bis-(α-(trimethylammonium)methyl) azobenzene), is a potent agonist at nicotinic receptors in the *trans* form, but not in the *cis* form (Lester & Chang, 1977; Lester *et al.*, 1986). Thus preparations bathed in a solution containing mainly the *cis* isomer of Bis-Q show minimal nicotinic receptor activity, but receptors can be activated by exposing the preparation to a flash of light at 420–440 nm, which converts a large proportion of *cis*-Bis-Q to the *trans* form. The reverse reaction is possible with *trans* to *cis* isomerization favoured at wavelengths of around 340 nm. The advantages offered by these azobenzenes (see Lester & Nerbonne, 1982) are the rapidity of photoisomerization, which occurs within 1 μs of absorbing a photon, and the high quantum yield. In addition, the only effect of light is isomerization, with no competing photoreactions or photoproducts to worry about. On the other hand, even complete irradiation does not produce a pure solution of one or other isomer, but a photostationary mixture of the two.

Photoisomerizations are being exploited in other ways which could lead eventually to the development of a general 'cage' for physiologically active molecules. Attempts thus far have involved forming liposomes (Kano *et al.*, 1981) or micelles (Shinkai *et al.*, 1982) from molecules containing azobenzene groups. These approaches (reviewed in Gurney & Lester, 1987) depend on differences in liposome permeability, or in critical micelle concentrations, when the incorporated azobenzene group is in the *cis* or *trans* configuration. Azobenzene chemistry was also exploited in the first approach to caging divalent cations. By synthesizing an azobenzene compound containing two iminodiacetic acid groups, it proved possible to chelate Zn^{2+} in a light-sensitive and reversible manner (Blank *et al.*, 1981). The authors suggested that protons might also be amenable to caging in this manner, but the method has had limited success in caging other divalent cations such as Ca^{2+}.

40.3.2 *o*-Nitrobenzyl derivatives

Currently, the most widely used and available photolabile caged compounds exploit the light sensitivity of the *o*-nitrobenzyl moiety (Patchornik *et al.*, 1970), illustrated in Fig. 40.1(b). By linking the molecule of biological interest to an *o*-nitrobenzyl group, it is hidden from the active site. Irradiation then cleaves the precursor at the benzyl carbon, freeing the active molecule along with a proton and a nitroso by-product, which may be either a ketone or an aldehyde depending on the substitution at the benzyl carbon (R_3 in Fig. 40.1(b)). Proton loss occurs simultaneously with the formation of an *aci*-nitro intermediate (Nerbonne, 1986), and the rate of breakdown of the intermediate determines the rate of photorelease of the active species from the caged compound (McCray *et al.*, 1980; Goldman *et al.*, 1984a; Nerbonne, 1986). The *o*-nitrobenzyl group was first exploited in the development of caged ATP and photolabile cyclic nucleotide analogues. It has since been used successfully to cage a wide variety of intracellular messengers, neurotransmitters, receptor ligands and other useful probes (Table 40.1).

Alternative analogues are available for some caged

compounds, with different substitutions at positions R_1, R_2 and R_3 in Fig. 40.1(b). For example, caged ATP can be purchased as either the *o*-nitrobenzyl (NB) ester where R_1, R_2 and R_3 are all protons, the *o*-nitrophenylethyl (NPE) ester where $R_3 = CH_3$, or the dimethoxy *o*-nitrophenylethyl (DMNPE) ester where R_1 and $R_2 = CH_3O$ and $R_3 = CH_3$. These modifications influence the rate and efficiency of the photochemical reaction, as well as the biological activity (see later) of the precursor and the photoproduct. A number of studies have examined the structural requirements for fast efficient photolysis (McCray *et al.*, 1980; Nerbonne, 1986; Wootton & Trentham, 1989). Unfortunately, there does not seem to be any general rule regarding the effect that each substituent has on photolysis, because the properties of the photolabile molecule are also influenced by the nature of the biologically active molecule being caged. Incorporating methoxy groups at positions R_1 and R_2 has the beneficial effect of causing a red shift of the absorption maximum of caged compounds, thereby improving light absorption in the 300–360 nm range. This modification also improves the speed and efficiency of photorelease from caged cyclic nucleotides (Nerbonne, 1986), but it has the opposite effect on caged ATP and caged phosphate (Wootton & Trentham, 1989). On the other hand, photorelease from caged carbachol (Milburn *et al.*, 1989) and caged ATP (Kaplan *et al.*, 1978) is accelerated when a methyl group is present at R_3, but introducing the same group into dimethoxy *o*-nitrobenzyl cyclic nucleotides makes them unstable in aqueous solution (Wootton & Trentham, 1989).

The rate of photorelease, which is likely to be an important consideration in selecting a photolabile probe, can vary markedly among analogues. The release of ATP from NPE-caged ATP proceeds with a rate constant of 84 s^{-1}, compared with only 18 s^{-1} from the DMNPE analogue at 20°C and pH 7 (Wootton & Trentham, 1989). In contrast, inorganic phosphate is released rapidly (21 000 s^{-1}) from DMNPE-caged phosphate under similar conditions (Wootton & Trentham, 1989). Photorelease of cyclic nucleotides is more rapid from DMNB analogues than from NB analogues (Nerbonne, 1986), although the precise rates are not yet clear. It has proved difficult to measure the rates of photolysis of the DMNB analogues using chemical techniques (Wootton & Trentham, 1989), but a biological assay has shown that DMNB-caged cyclic GMP probably photolyses at >3000 s^{-1} at pH 7 (Karpen *et al.*, 1988). Even the slowest rates of release may be acceptable in some experiments, for example, when the aim is simply to elevate the concentration of an intracellular messenger. Photolabile probes probably provide the simplest way of doing this. However, if the aim is to gain kinetic information from the response, the probes should photolyse more rapidly, preferably by at least an order of magnitude, than the response develops.

40.3.3 Photolabile cation chelators

Caged compounds can be used to control the concentration of cations inside or outside cells. The photolabile cation chelators currently available were reviewed recently (Kaplan, 1990), and practical aspects of their application in biology have been described in detail (Gurney, 1991). Zinc ions were the first to be chelated in a photosensitive manner, using the azobenzene chemistry as indicated above (Blank *et al.*, 1981). Several molecules now exist that can act either as caged cations (nitr-5, DM-nitrophen), which release Ca^{2+} and/or Mg^{2+} upon photolysis, or caged chelators (diazo-2, caged BAPTA), which mop up Ca^{2+} upon photolysis. All of these agents use the photochemistry of the *o*-nitrobenzyl group, the photolysis of which alters the affinity of the molecule toward Ca^{2+} and/or Mg^{2+}. The affinities of some of these agents for Ca^{2+} and Mg^{2+} before and after photolysis are listed in Table 40.2, along with other properties that bear on their use as photolabile cation chelators.

The nitr-5 family of compounds (Fig. 40.2(a)) were designed around BAPTA (Tsien & Zucker, 1985;

Table 40.2 Properties of photolabile Ca^{2+} chelators.

	K_D for Ca^{2+} binding (μm)		*Photolysis rate (s^{-1})*	*Quantum yield*	*KD for Mg^{2+} binding (mM)*	
	Pre-photolysis	*Post photolysis*			*Pre-photolysis*	*Post photolysis*
Nitr-5[a]	0.145	6	4000	0.03–0.1	8.5	8
DM-nitrophen[b]	0.005	2000	3000	0.18	0.005	3
Diazo-2[c]	2.2	0.073	>2000			
Caged BAPTA[d]	160	0.11	>300			

Data from references.
[a] Adams *et al.* (1988); [b] Kaplan & Ellis-Davies (1988); Kaplan (1990); [c] Adams *et al.* (1989); [d] Ferenczi *et al.* (1989).

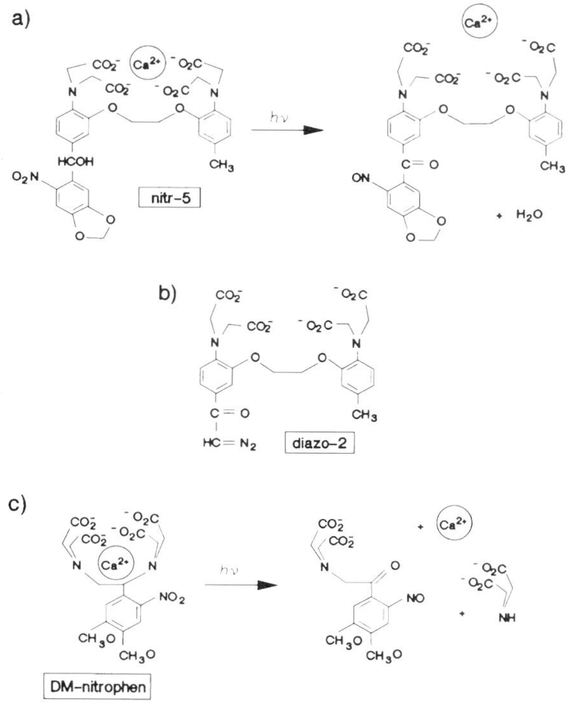

Figure 40.2 Reaction schemes proposed to account for the photolysis of (a) nitr-5 and (c) DM-nitrophen, both of which lose affinity for Ca^{2+} upon photolysis. Diazo-2 (b) is structurally related to nitr-5, but gains affinity for Ca^{2+} on photolysis.

Adams *et al.*, 1988), a Ca^{2+} chelator with high selectivity for Ca^{2+} over Mg^{2+}. This selectivity is retained in the photolabile probes, both before and after photolysis and, like BAPTA, their Ca^{2+} affinity shows little dependence on pH above pH 7. The affinity of BAPTA for Ca^{2+} is known to be increased or decreased depending on the nature of electron-withdrawing or -donating groups present on one of its aromatic rings (Tsien, 1980). With nitr-5 this group becomes more electron withdrawing after photolysis, resulting in reduced Ca^{2+} affinity (Fig. 40.2(a)). In contrast, the opposite change in diazo-2 (Fig. 40.2(b)) results in an increased affinity for Ca^{2+} after photolysis (Adams *et al.*, 1989). Thus nitr-5 or diazo-2 can be used to respectively elevate or reduce the free Ca^{2+} concentration inside cells. Another recently introduced molecule of this type, diazo-3 (Molecular Probes), displays similar photochemical properties to nitr-5 and diazo-2, but it does not bind Ca^{2+} well before or after photolysis. It should therefore prove useful as a control against effects of photolysis produced by the structurally related compounds, that are not due to changes in Ca^{2+} concentration. BAPTA itself has been photochemically caged by derivatizing it with an *o*-nitrobenzyl group (Ferenczi *et al.*, 1989); photolysis results in the release of BAPTA, which binds up free Ca^{2+}.

DM-nitrophen (Fig. 40.2(c)) represents a different approach to caging divalent cations (Ellis-Davies & Kaplan, 1988). This molecule is based around the chelator EDTA, which binds Ca^{2+} ions tightly, but also has a significant affinity for Mg^{2+}. Thus at the millimolar concentrations of Mg^{2+} present inside cells, a significant fraction of DM-nitrophen would be present as the Mg^{2+} complex, and photolysis would release a mixture of Ca^{2+} and Mg^{2+}. This is a clear disadvantage since it would not permit the effects of a rise in intracellular Ca^{2+} to be studied under physiological conditions (i.e. in the presence of Mg^{2+}). Nevertheless, in preparations where Ca^{2+}-dependent effects can be studied in the absence of Mg^{2+}, it is possible to use DM-nitrophen to selectively release Ca^{2+}. It is also possible to exploit DM-nitrophen to selectively release Mg^{2+} as, for example, in the regulation of the sodium pump (Klodos & Forbush, 1988).

Nitr-5 and DM-nitrophen, the most easily available and best-studied of the photolabile chelators, both have distinct advantages and disadvantages for use as Ca^{2+} donors (Table 40.2). Both agents release Ca^{2+} rapidly, with a rate of around 3000 s^{-1}. The clearest advantage of nitr-5 is that it is relatively insensitive to pH and the presence of Mg^{2+} ions. Furthermore, unlike DM-nitrophen, it can be loaded into cells non-invasively (see later), so it permits studies in intact cells. DM-nitrophen, on the other hand, binds Ca^{2+} more tightly. Thus at low resting levels of free Ca^{2+} a greater proportion of the probe is bound with Ca^{2+}, enabling more Ca^{2+} to be released by photolysis. DM-nitrophen also displays a greater quantum yield than nitr-5, and its Ca^{2+} affinity undergoes a greater change upon photolysis, both of which add to the Ca^{2+}-releasing ability of the probe.

The two types of chelator also differ in the nature of the reactions triggered by photolysis. Irradiation of nitr-5 induces a structural rearrangement (Fig. 40.2 (a)) to form a molecule with reduced calcium affinity, and photolysis results in the same end products whether it is bound to Ca^{2+} or not. In contrast, DM-nitrophen is thought to undergo cleavage to yield two photoproducts (Fig. 40.2(c)), each with negligible Ca^{2+} affinity. Moreover, the photolysis of DM-nitrophen may yield different products depending on whether or not it is complexed with Ca^{2+} (Ellis-Davies & Kaplan, 1988).

40.4 APPLICATION OF PHOTOLABILE PROBES TO STUDYING BIOLOGICAL PATHWAYS

40.4.1 Caged receptor ligands

The delay between activation of a receptor and a cell's response can provide insight into the mechanisms coupling the receptor to the response. For example, activation of the nicotinic acetylcholine receptor opens an ion channel that is part of the same protein complex. Events mediated by such direct coupling occur on the millisecond time-scale, so to study their activation kinetics requires a very rapid method for applying agonist. At present photochemistry provides the most rapid means of applying nicotinic agonists, submillisecond resolution having been achieved both with the photoisomerizable azobenzene, Bis-Q (Lester & Nerbonne, 1982; Gurney & Lester, 1987), and with caged carbachol (Milburn *et al.*, 1989). Similar time resolution is available with photoisomerizable azobenzene molecules that block either the nicotinic receptor or its channel (Lester & Nerbonne, 1982; Gurney & Lester, 1987). The advantage of these molecules is that they can bypass the diffusional delays normally associated with drug application. Preparations are bathed in the inactive precursor, with light flashes triggering the release of the active ligand on demand. The speed of application allows receptor activation to be temporally separated from receptor desensitization, and from other processes that inactivate the ligand.

Photolysis studies of nicotinic-receptor activation have mostly been performed with Bis-Q and the related molecule, QBr (Lester & Nerbonne, 1982; Gurney & Lester, 1987), because for a long time they were the only caged agonists available. In studying these receptors, caged carbachol may, however, have some advantages. For example, the *N*-(α-carboxy-2-nitrobenzyl) derivative appears to be more pharmacologically inert than *cis* Bis-Q before photolysis (Milburn *et al.*, 1989), and carbachol is a well-characterized analogue of the natural transmitter, acetylcholine. Furthermore, caged carbachol should also act as a caged agonist at muscarinic receptors, where Bis-Q is an antagonist (Nargeot *et al.*, 1982). Although a caged acetylcholine molecule would be the most useful for probing the physiological activation of cholinergic receptors, such a molecule has not yet been synthesized.

Caged ligands have also been developed for other receptors (see Table 40.1), although they have not been widely tested in biological experiments. The activation of α-adrenergic receptors in vascular smooth muscle has been studied using caged phenylephrine (NPE ester), which has minimal activity before photolysis (Somlyo *et al.*, 1988). The active agonist is released rather slowly from this caged compound, with a time constant of ~300 ms (Walker & Trentham, 1988), reflecting slow dark reactions. This would be too slow to probe the type of direct excitation–response coupling found at nicotinic receptors. However, following photolysis of caged phenylephrine, vascular muscle contracts with a latency of 1.5 s at 30°C, and this appears to be due entirely to the events linking phenylephrine binding to its receptor and the subsequent response (Somlyo *et al.*, 1988). The contractile response is, however, quite far removed from receptor activation. It has yet to be determined if the rate of phenylephrine release from the caged precursor is fast enough to permit studies of the activation of more immediate events, such as membrane conductances.

40.4.2 Caged intracellular second messengers

There are now a wide variety of caged molecules available for activating, or interfering with, second messenger pathways inside cells (see Table 40.1). For light to photolyse such a molecule at its site of action, the caged compound must be present inside the cell. Methods of incorporating caged compounds into cells are dictated by the preparation under study and by the chemical properties of the probe. For example, the caged cyclic nucleotides are lipid-soluble, and will therefore gain access to the cell interior simply by diffusing from the extracellular medium. However, these molecules may accumulate in lipid components of the cell, and this is likely to influence their concentration at the site of action. Similarly, the photolabile calcium buffer nitr-5 can be incorporated into cells in a non-invasive way, using its acetoxymethyl ester (nitr-5-AM) form. This diffuses across the cell membrane and becomes trapped inside the cell by the action of intracellular enzymes, which cleave the ester to release the free buffer. It can therefore be preloaded using methods that were developed earlier for loading fluorescent Ca indicators (Kao *et al.*, 1989; Valdeolmillos *et al.*, 1989). With this approach, the intracellular concentration of nitr-5 achieved is difficult to predict, and changes in Ca^{2+} concentration induced by photolysis are not known with certainty. On the other hand, nitr-5 can be co-loaded with fluorescent Ca^{2+} indicators, which then report the Ca^{2+} jumps induced by flashes (Kao *et al.*, 1989; Gurney, 1991).

Most other caged second messengers are hydrophilic and do not readily cross the cell membrane. A number of compounds have been used with 'skinned' muscle fibres, in which the cell membranes have been removed (Goldman *et al.*, 1982), or in cells permeabilized with staphylococcal α-toxin (Somlyo *et al.*, 1988; Somlyo & Somlyo, 1990), which makes membranes permeable to low-molecular-weight solutes while leaving receptor pathways intact. Most compounds can be incorporated into cells by injection, either through a fine micropipette that impales the cell, or through a fire-polished pipette used in the whole-cell configuration of the tight-seal, patch-clamp technique. Microinjection is the method of choice when second messenger concentrations are to be manipulated while measuring electrical properties of the cell membrane. Flash photolysis is well-suited to electrophysiological studies using the patch-clamp technique, either to study the activation or modulation of ion channels in isolated membrane patches (Chabala *et al.*, 1985; Lester *et al.*, 1986; Karpen *et al.*, 1988), or in whole cells (e.g. Gurney *et al.*, 1985, 1987, 1989; Chabala *et al.*, 1986; Lester *et al.*, 1986; Ogden *et al.*, 1990). The low resistance pathway formed between the recording pipette and the cell interior in the whole-cell configuration, means that the solution filling the pipette equilibrates with, and controls the cell interior. Thus the concentration of the caged compound is known with some degree of certainty. A number of compounds have been successfully incorporated into cells in this way, and millimolar concentrations can be achieved. Provided care is taken to protect the photolabile probe in the pipette from the light, for example by coating the pipette with black Sylgard (Dow Corning, available from BDH), the solution in the pipette provides an essentially unlimited store of unphotolysed chelator. Reproducible responses to photolysis can thus be obtained in a single cell, and it is possible to examine reponses over a wide range of concentrations.

40.4.3 Instrumentation

The only specialized instrumentation required for photolysis experiments with caged compounds is a high-intensity light source, with output in the near *UV.* If time resolution is not important, then an ordinary xenon lamp equipped with a shutter can be used. However, to take advantage of the speed afforded by the photochemistry, the reaction should be started with a very brief flash of light. Flashlamps are available that produce light pulses of less than 1 ms duration with an output of >200 mJ between 300 and 400 nm. This is achieved by discharging a high-voltage capacitor across a short-arc flash tube, which encloses xenon gas. Lamps designed for the photolysis of caged compounds can be obtained from Chadwick Helmuth, El Monte, California (Strobex model 238) or from Hi-Tech Scientific Ltd., Salisbury, the UK distributors for Dr Rapp, Optoelektronik, Hartkrögen 65, D-2000 Hamburg 56, Germany (model JML or JML-E). In each case, the arc light is collected and focused with quartz lenses. The intensity of the flash is determined by the energy discharged through the lamp upon ignition, and is easily varied.

The output of xenon flashlamps has a broad spectrum of 250–1500 nm. The wavelengths most efficient for photolysis (300–400 nm) can be selected by placing filters in the light path. A band-pass filter, such as the Schott UG 11 (Ealing Electro-optics plc, Watford, UK) which shows peak transmission of about 80% at 320 nm, removes short wavelengths that may be damaging to cells and long wavelengths that may warm the cells. We frequently use broader spectrum cut-off filters that only remove wavelengths <300 nm, because this permits greater photolysis. The long wavelengths that remain do not usually present a problem, presumably because they are absorbed as they pass through aqueous solution on their way to the experimental preparation. This can be tested by observing the effects of a flash on the preparation in the absence of any photolabile probe.

It is often desirable to direct flashes through a light guide, which allows the flashlamp to be placed at a distance from the preparation (Fig. 40.3). Light guides are particularly convenient when flash photolysis is to be combined with electrophysiological measurements, because the large trigger pulse (12 kV) and the discharge current (2000 A) are not easily shielded from the recording system. With these kinds of experiments it is also important to isolate mechanically the flashlamp from the recording system, because it thumps its housing when it discharges. Fibre optics do not transmit enough UV light to be useful. Liquid light guides have, however, been used successfully for studies with caged compounds. These light guides

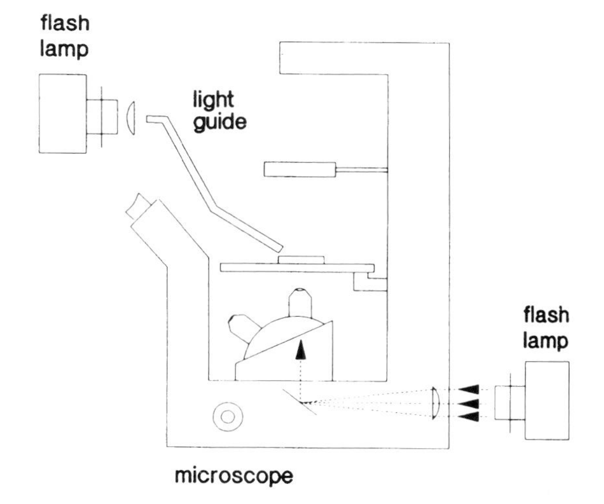

Figure 40.3 Light from a xenon short-arc flashlamp can be directly focused onto a preparation, or directed through a liquid light guide or the optical path of a microscope.

transmit well from 270 to 720 nm, although the efficiency of transmission depends on the length of the light guide. With a 1 m guide, maximum transmittance is about 80%, but it falls off fairly steeply below 400 nm, such that at 300 nm only 40% is transmitted. The loss of light in this region is unacceptably high if the light guide is used in combination with a UG11 filter. We prefer to remove short wavelengths by placing a borosilicate glass coverslip, about 0.1 mm thick (no. 0 thickness; BDH, Poole, Dorset, UK), at the input to the light guide. The coverslip cuts off sharply below 320 nm, but transmits greater than 90% at longer wavelengths.

Flashlamps can also be directed onto cells *via* the optical path of a microscope (Fig. 40.3). We have used the epifluorescence port of an Olympus IMT2 inverted microscope. The flashlamp condenser is placed in front of the port and a 200 mm quartz lens focuses the light onto a dichroic mirror, which is selected to reflect maximally below 500 nm. The dichroic mirror is set at a 45° angle to reflect light through the objective lens, which should have as high a numerical aperture (NA) as possible. With a Nikon 40×, NA 0.85 lens, we can photolyse ~4% of DMNPE-caged ATP with a single flash. Our microscope is limited by a fixed glass lens in the light path. UV-transmitting optics are available with new Olympus microscopes and the comparable Nikon Diaphot. With these it should be possible to achieve greater photolysis, although we have not made any direct comparisons.

Photolysis experiments can also be performed using a pulsed laser (reviewed in McCray & Trentham, 1989), which further improves the time resolution. These lasers have the advantage of emitting high-intensity light at a defined wavelength, with sub-microsecond resolution. Furthermore, the light is emitted as a parallel beam, which can be more easily directed onto small areas. However, lasers capable of producing UV light are expensive. They also tend to be laborious and expensive to maintain and they occupy a lot of bench space. The lasers that have most commonly been used for biological studies with caged compounds include pulsed dye lasers (e.g. Sheridan & Lester, 1982), which are tunable over a range of wavelengths and frequency-doubled ruby lasers (McCray *et al.*, 1980), which produce 50 ns pulses at 347 nm. Others, such as XeF and XeCl excimer lasers, capable of producing near-UV wavelengths can also be used (McCray & Trentham, 1989).

40.4.4 Calibration

For the photosensitive probes to provide useful biological information it is important that the concentration change resulting from photolysis can be determined accurately. With caged calcium, Ca^{2+} release can be simultaneously measured with Ca^{2+} indicators (see Kao *et al.*, 1989), but this is not possible with most other caged compounds. The extent of photolysis can be determined by measuring the concentration of the caged compound and/or its photoproducts in droplets of solution placed at the focal point, before and after flashing. To ensure that all of the drop is irradiated during photolysis, it must be smaller than the light spot that is focused on to it. To prevent evaporation from the droplet, it can be covered with a thin film of mineral oil. When photolysis is directed through a microscope, the area irradiated is very small (~400 μm diameter with 40f objective). To keep the droplet of test solution within this area, small volumes (<0.2 μl) can be microinjected into a blob (~1 μl) of mineral oil. Flashes are then presented after focusing the objective on the centre of the droplet. Since the volume of the droplet in this case is hard to measure accurately, it is useful to include in the test solution a marker molecule, whose concentration is known and does not change with photolysis. For example, when calibrating the photolysis of caged ATP in this way, we include a known concentration of adenosine monophosphate, which has a different retention time on high-performance liquid chromatography (HPLC) to ATP and caged ATP.

The concentration of a caged compound may be measured spectrophotometrically or by HPLC analysis. Most HPLC methods use a C18 reversed-phase column or ion-exchange column. Various conditions have been described for separating different caged compounds from their photolysis products using HPLC (Walker, 1991). These compounds include nucleotides (Walker *et al.*, 1989a), cyclic nucleotides (Nerbonne *et al.*, 1984; Wootton & Trentham, 1989), IP3 (Walker *et al.*, 1989b), calcium (Walker, 1991) and carbachol (Milburn *et al.*, 1989). If one flash does not cause detectable photolysis, then separate droplets can be flashed a varying number of times, and the fraction of molecules photolysed by a single flash extrapolated.

40.5 POTENTIAL PROBLEMS ASSOCIATED WITH THE USE OF CAGED COMPOUNDS

There are three possible artifacts in flash photolysis experiments with caged compounds: (1) the caged compound could itself have effects on the tissue even in the dark, (2) the side products of photolysis could be biologically active, and (3) the flash itself might trigger a biological response. Ideally, the caged compound should be inert, and flashes in the 300–400 nm wavelength range should cause no measurable physiological artifact in the absence or presence of the probe. So far, these conditions appear to have been fulfilled with most preparations that are not normally thought of as light-sensitive, and with many of the available caged compounds. However, a few examples of these artifacts have been reported.

40.5.1 Effects of the precursor

Perhaps the most widely studied caged compound is NPE-caged ATP, which in studies of cross-bridge kinetics in skeletal (e.g. Goldman *et al.*, 1984a,b) and smooth (Somlyo *et al.*, 1988) muscle behaved simply as an inert precursor for ATP. Although this analogue has also been successfully exploited to study the kinetics of activation of the sodium pump by ATP, it was found to bind to the same site on the enzyme as free ATP, albeit with lower affinity (Forbush, 1984). It has since been shown to bind to myosin as well (Dantzig *et al.*, 1989). Unphotolysed caged ATP also causes partial blockade of ATP-sensitive K^+ channels in cardiac muscle, an effect observed with both the NPE and the DMNPE derivatives (Nichols *et al.*, 1990). In pancreatic β-cells, DMNPE-caged ATP was almost as effective at blocking this type of channel as MgATP, the physiological regulator (Ammälä *et al.*, 1991). Similar blockade of an ATP-sensitive K^+ current occurs in vascular muscle cells with NPE-caged ATP in the absence of light, but in this tissue the DMNPE derivative (up to 20 mM) appears to block the current only after photolysis (Clapp & Gurney, 1992).

The first caged cyclic nucleotides (NB esters) to be synthesized (Engels & Schlaeger, 1977; Engels & Reidys, 1978) were employed originally as lipophilic precursors that could spontaneously hydrolyse inside cardiac muscle cells to generate the free cyclic nucleotides (Korth & Engels, 1979). It proved possible to use NB cAMP as a photolabile precursor in heart muscle, because flashes released sufficient cAMP to enhance the slow inward current at concentrations where preflash effects were small (Nargeot *et al.*, 1983). However, the preferred precursor in cardiac tissue is DMNB-caged cAMP, which in the absence of light has no effect on the current or on muscle tension at much higher concentrations than can be tolerated with the NB ester (Nerbonne *et al.*, 1984; Richard *et al.*, 1985). We have similarly compared the effects of the photolabile DMNB and NPE esters of cGMP in vascular muscle (Fig. 40.4), where a rise in intracellular cGMP is thought to mediate relaxation. Without irradiation, concentrations of DMNB-caged cGMP up to 200 μM have no effect on tonic tension developed by exposing the muscle to elevated extracellular K^+, although in the presence of this analogue flashes do produce relaxation (Fig. 40.4(a)). In contrast, NPE-caged cGMP by itself induces large relaxations even at 50 μM (Fig. 40.4(b)), with light flashes having little further effect (not shown). A similar difference in the activities of these caged cGMP analogues has also been noted in isolated cardiac cells (Charnet & Richard, pers. commun.). It is not yet clear whether the effects of NPE-caged cGMP arise as a result of intracellular hydrolysis to release free cGMP, from a direct interaction with cGMP-dependent kinase or from an interaction with some other cGMP-binding site.

In many tissues, a fall in the intracellular pH decreases the conductance of gap junctions between

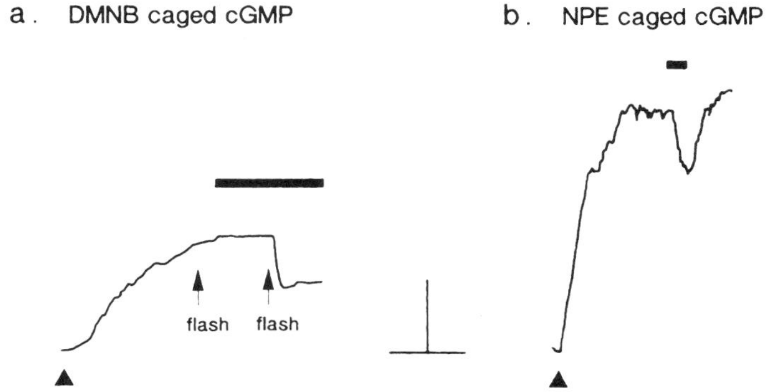

Figure 40.4 Comparison of the effects of DMNB-caged cGMP and NPE-caged cGMP on isometric tension measured in a strip of rabbit main pulmonary artery. Strips were precontracted by changing the perfusing solution to one containing 20 mM K^+ as indicated by the filled triangles. They were then perfused with either DMNB-caged cGMP at 100 μM (a) or NPE-caged cGMP at 10 μM (b), for the times indicated by the horizontal bars. In the presence of the DMNB analogue, the muscle relaxed only when presented with a light flash. In contrast, the NPE analogue caused relaxation on its own. Calibration bars represent 10 min and 50 mg (a) or 40 mg (b). Records a and b from different tissues.

pairs of cells. The photolabile proton donor, *o*-nitrobenzyl acetate, was used successfully to uncouple cells in the salivary gland of the midge *Chironomus*, where protons were released only upon photolysis (Nerbonne *et al.*, 1982). However, in a number of other cell types the same caged compound reversibly decreased junctional conductance at low concentrations, even without photolysis (Spray *et al.*, 1984). The effect was clearly due to intracellular cleavage of the probe, because it was associated with a fall in the intracellular pH.

Effects of other caged compounds have also been noted in the absence of light. For example, NPE-caged carbachol, an early version of the probe, was found to be active at both nicotinic and muscarinic acetylcholine receptors (Walker *et al.*, 1986). Introducing a carboxylate group into the benzyl carbon to make *N*-(α-carboxy-2-nitrobenzyl) carbachol, subsequently eliminated this activity, while flashes presented in its presence produced rapid activation of nicotinic receptors in BC_3H1 cells (Milburn *et al.*, 1989). Bis-Q, the photoisomerizable nicotinic receptor agonist suffers from a similar problem. Although the *cis* isomer is much less potent than the *trans* isomer, it is not completely inactive (Lester & Nerbonne, 1982; Nerbonne *et al.*, 1983; Lester *et al.*, 1986). Both isomers also act as antagonists at muscarinic receptors, although the *cis* isomer is less potent (Nargeot *et al.*, 1982). Some photolabile molecules that are active before photolysis can be rapidly inactivated with light, and this property can also be exploited. For example, the calcium antagonist drug nifedipine is rapidly destroyed with a flash, an approach that has been useful in probing the mechanisms by which the drug interacts with voltage-dependent calcium channels (Gurney *et al.*, 1985; Nerbonne *et al.*, 1985).

40.5.2 Effects of the photolysis side products

Photorelease of the 'protected' molecule from *o*-nitrobenzyl derivatives occurs together with the release of a proton and a nitroso by-product. In most cases the pH change resulting from proton loss can be kept to a minimum by heavily buffering the experimental solution. This may not be possible with lipophilic compounds used as intracellular probes, where access of a buffer to the cell interior may be restricted. Nevertheless, in either case the contribution of protons to a response can be evaluated by carrying out control experiments with caged protons. Although, as indicated above, *o*-nitrobenzyl acetate spontaneously hydrolyses in a number of cell types, there are other caged proton compounds that could be used (Janko & Reichert, 1987; Shimada & Berg, 1987). The nitroso photoproducts present a potentially greater problem; they precipitate out of aqueous solution and are highly reactive toward sulphydryl groups on proteins (McCray *et al.*, 1980). Their reactivity can be reduced by derivatizing the benzyl carbon (McCray *et al.*, 1980) with, for example, a methyl group, as in the NPE derivatives ($R_1 = CH_3$ in Fig. 40.1(b)). Despite this, the nitrosoacetophenone produced alongside photolysis of NPE-caged ATP does have effects on cell proteins, such as inactivating the sodium pump of erythrocyte ghosts (Kaplan *et al.*, 1978). The deleterious effects of the nitroso photoproducts can be avoided by including a hydrophilic thiol, such as glutathione or dithiothreitol (DTT), in the solution, usually at millimolar concentrations (Kaplan *et al.*, 1978; Walker *et al.*, 1988; McCray & Trentham, 1989). Fortunately, most receptor ligands and intracellular messengers work in the micromolar concentration range, so effective concentrations of the biologically interesting molecule can be produced with relatively small amounts of the by-product. The photoproducts are only likely to cause a problem when high concentrations of a caged compound are photolysed, for example, when millimolar concentrations of ATP are needed to activate muscle contraction.

40.5.3 Direct effects of light

Light in the wavelength region of 250 nm is well-known to be damaging to cells. However, there is no evidence that light of 300–400 nm causes any photodynamic damage over the time-course of most physiological experiments, even after exposure to many flashes. Other than in preparations that are expected to respond to light (e.g. photoreceptors) there are only a few reports that light of these wavelengths has any measurable effect in the absence of light-sensitive molecules. Flashes on their own transiently activated voltage-clamp currents in intact strips of frog atrial muscle, apparently through a direct effect on the myocardial tissue (Nargeot *et al.*, 1982). However, these effects were small enough and reproducible enough to be subtracted from the responses of interest (Nargeot *et al.*, 1982, 1983). Furthermore, these non-specific effects of light were absent in later studies on isolated cardiac cells (Gurney *et al.*, 1985, 1989; Naebauer *et al.*, 1989).

Light is known to have pronounced effects on blood vessels. Photo-induced relaxation of vascular muscle was first described by Furchgott *et al.* in 1955, and shown to depend on the same wavelengths as those used for flash photolysis experiments. Most studies of the photo-induced relaxation have employed low-level, maintained illumination. However, as illustrated in Fig. 40.5, brief, intense flashes of similar wavelengths and intensity to those that have been widely used in many experiments to photolyse caged compounds, can

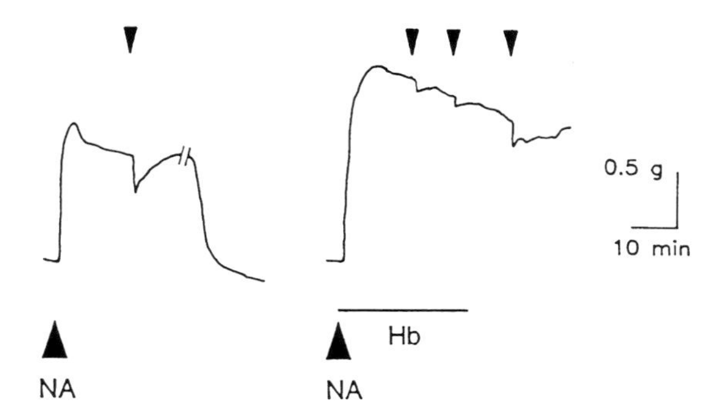

Figure 40.5 Effect of flashes on isometric tension in a strip of rabbit main pulmonary artery. A strip of artery was perfused continuously with 6 μM noradrenaline (NA), and a flash presented during the plateau phase of the contraction. A single flash relaxed the tissue, but this effect was blocked by 5 μM haemoglobin.

also relax arterial muscle. The response is quite variable, with some tissues showing pronounced relaxation (Fig. 40.5), while others show no response (as in Fig. 40.4) or occasionally contraction. The basis of this photo-induced relaxation has been quite widely studied, and evidence is increasing that light stimulates the cytosolic guanylate cyclase either directly (Karlsson *et al.*, 1984) or more likely indirectly (Furchgott *et al.*, 1985; Wigilius *et al.*, 1990; Wolin *et al.*, 1991). Thus it is possible to block the photo-induced relaxation by exposing the tissue to haemoglobin (Fig. 40.5(b)), which inhibits the activation of guanylate cyclase (Matsunaga & Furchgott, 1989; Furchgott & Jothianandan, 1991). Whatever the mechanism, it is clear that light elevates the intracellular cGMP concentration (Karlsson *et al.*, 1984) and mimics the effect of nitrovasodilator drugs as well as the endothelium-derived relaxing factor (EDRF), which plays an important role in the normal regulation of vascular tone. Flash effects independent of caged compounds can also be observed in isolated vascular muscle cells (Fig. 40.6), where changes in membrane potassium and calcium currents can be observed in the absence of photolabile probes (see also Komori & Bolton, 1991). Again the mechanisms underlying these changes are unclear, but the effects of light resemble those of nitrovasodilator drugs (Clapp *et al.*, 1990; Clapp & Gurney, 1991a). For these reasons, the application of photolabile caged compounds in vascular muscle is not as straightforward as it is in other tissues.

Photo-induced relaxation may only be a problem with intact smooth muscle cells, since a number of caged compounds have been applied successfully to study contraction mechanisms in permeabilized vascular muscle, apparently without interference from the activating light (Somlyo *et al.*, 1988; Somlyo,

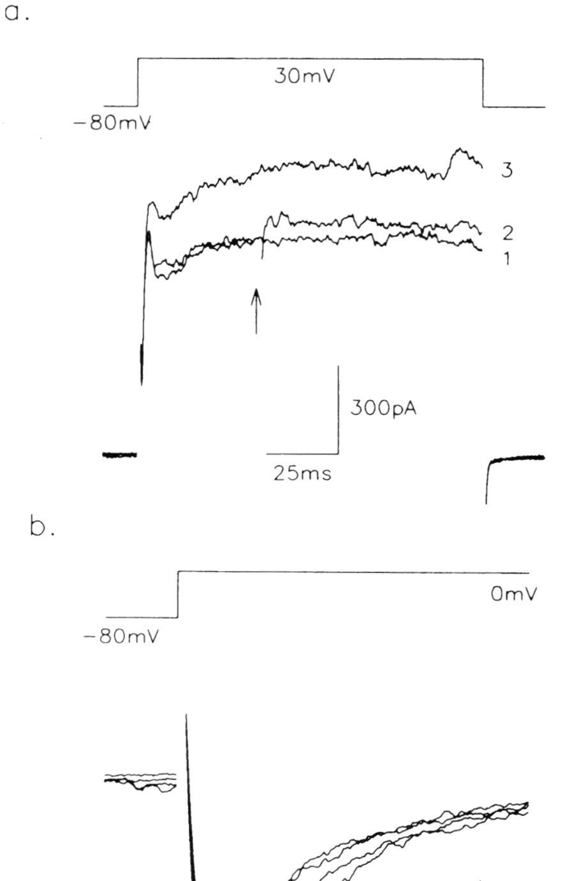

Figure 40.6 Flashes alter outward potassium (a) and inward calcium-channel (b) currents, recorded using the whole-cell patch-clamp technique from isolated smooth muscle cells of rabbit pulmonary artery. Voltage protocols used to activate the currents are illustrated above each set of traces. Currents were activated at 5 s intervals and were recorded in the order indicated by the numbers next to each trace. Experimental details were as described in Clapp and Gurney (1991a, b). The cell in (a) was perfused with physiological solution, and the recording pipette contained an isotonic, HEPES-buffered K^+ solution with 1 mM EGTA. A flash presented to the cell during sweep 2 (arrow) increased current amplitude. The cell in (b) was perfused with a Ba^{2+} (10 mM) containing solution, and the recording pipette contained Cs^+ in place of K^+ to block outward current. A flash presented between sweeps 2 and 3 suppressed the current.

1990). Agents that elevate cytoplasmic cGMP only relax tissues that have been precontracted by agonists or elevated K^+ concentrations, and this is also true of light (e.g. Furchgott *et al.*, 1955). Thus, although flashes may not appear to have an effect under basal conditions, in some tissues they may well trigger biochemical reactions that might alter the response to photoactivation of a caged compound. This could be tested by presenting flashes to preparations in which the pathway of interest has already been stimulated. Direct effects of light could have important consequences for preparations other than vascular smooth muscle which contain guanylate cyclase, because EDRF, which is thought to be nitric oxide, activates this enzyme and is increasingly being shown to have widespread biological effects (Moncada *et al.*, 1991).

40.6 CONCLUSION

Photolabile caged compounds have recently become popular probes for the study of biological functions. They provide a means of rapidly changing the concentration of a biologically active molecule at its site of action, either inside a cell or at its extracellular surface. The technique is, however, still in its infancy, and it is likely that many more useful caged compounds will be developed in the near future. In this chapter I have tried to point out some of the problems that might be encountered in using these probes, and ways in which they might be overcome. Nevertheless, in most cases so far, photolabile caged compounds have provided helpful biological information, with relatively few problems encountered.

ACKNOWLEDGEMENTS

I would like to thank Mr G. Allerton-Ross, Ms S. Morgans and Dr L.H. Clapp for allowing me to use the unpublished data in Figs. 40.4–40.6. I also wish to thank Dr S. Bates and Dr Clapp for reading the manuscript and helpful discussions. The work in my laboratory is funded by grants from the Medical Research Council, the British Heart Foundation, the Wellcome Trust and the Royal Society.

REFERENCES

Adams S.R., Kao J.P.Y., Grynkiewicz G., Minta A. & Tsien R.Y. (1988) *J. Am. Chem. Soc.* **110**, 3212–3220.

Adams S.R., Kao J.P.Y. & Tsien R.Y. (1989) *J. Am. Chem. Soc.* **111**, 7957–7968.

Ammälä C., Bokvist K., Galt S. & Rorsman P. (1991) *Biochim. Biophys. Acta.* **1092**, 347–349.

Blank M., Soo L.M., Wassermann N.H. & Erlanger B.F. (1981) *Science* **214**, 70–72.

Chabala L.D., Gurney A.M. & Lester H.A. (1985) *Biophys. J.* **48**, 241–246.

Chabala L.D., Gurney A.M. & Lester H.A. (1986) *J. Physiol.* **371**, 407–433.

Clapp L.H. & Gurney A.M. (1991a) *Pflügers Arch.* **418**, 462–470.

Clapp L.H. & Gurney A.M. (1991b) *Exp. Physiol.* **76**, 677–693.

Clapp L.H. & Gurney A.M. (1992) *Am. J. Physiol. (Heart)* **262**, H916–H920.

Clapp L.H., Allerton-Ross G. & Gurney A.M. (1990) *Biophys. J.* **57**, 158a (abstract).

Dantzig J.A., Goldman Y.E., Luttman M.L., Trentham D.R. & Woodward S.K.A. (1989) *J. Physiol.* **419**, 64P (abstract).

Ellis-Davies G.C.R. & Kaplan J.H. (1988) *J. Org. Chem.* **53**, 1966–1969.

Engels J. & Reidys R. (1978) *Experentia Basel* **34**, 14–15.

Engels J. & Schlaeger E.-J. (1977) *J. Med. Chem.* **20**, 907–911.

Ferenczi M.A., Goldman Y.E. & Trentham D.R. (1989) *J. Physiol.* **418**, 155P (abstract).

Forbush B. III (1984) *Proc. Natl. Acad. Sci. USA* **81**, 5310–5314.

Furchgott R.F. & Jothianandan D. (1991) *Blood Vessels* **28**, 52–61.

Furchgott R.F., Sleator W., McCaman M.W. & Elchlepp I. (1955) *J. Pharmac. Exp. Ther.* **113**, 22–23.

Furchgott R.F., Martin W., Cherry P.D., Jothianandan D. & Villani G.M. (1985) In *Vascular Neuroeffector Mechanisms*, J.A. Bevan, T. Godfraind, R.A. Maxwell, J.C. Stoclet & M. Worcel (eds). Elsevier, Amsterdam, pp. 105–114.

Goldman Y.E., Hibberd M.G., McCray J.A. & Trentham D.R. (1982) *Nature* **300**, 701–705.

Goldman Y.E., Hibberd M.G. & Trentham D.R. (1984a) *J. Physiol.* **354**, 577–604.

Goldman Y.E., Hibberd M.G. & Trentham D.R. (1984b) *J. Physiol.* **354**, 605–624.

Gurney A.M. (1991) In *Cellular Neurobiology: A Practical Approach*, J. Chad & H. Wheal (eds). IRL Press at Oxford University Press, Oxford, pp. 153–177.

Gurney A.M. & Lester H.A. (1987) *Physiol. Rev.* **67**, 583–617.

Gurney A.M., Nerbonne J.M. & Lester H.A. (1985) *J. Gen. Physiol.* **86**, 353–379.

Gurney A.M., Tsien R.Y. & Lester H.A. (1987) *Proc. Natl. Acad. Sci. USA* **84**, 3496–3500.

Gurney A.M., Charnet P., Pye J.M. & Nargeot J. (1989) *Nature* **341**, 65–68.

Homsher E. & Millar N.C. (1990) *Ann. Rev. Physiol.* **52**, 875–896.

Janko K. & Reichert J. (1987) *Biochim. Biophys. Acta* **905**, 409–416.

Kano K., Tanaka Y., Ogawa T., Shimomura M. & Kunitake T. (1981) *Photochem. Photobiol.* **34**, 323–329.

Kao J.P.Y., Harootunian A.T. & Tsien R.Y. (1989) *J. Biol. Chem.* **264**, 8179–8184.

Kaplan J.H. (1986) In *Optical Methods in Cell Physiology (Soc. Gen. Physiol. Ser.)*, P. De Weer & B. Salzberg (eds). Wiley, New York, pp. 385–396.

Kaplan J.H. (1990) *Ann. Rev. Physiol.* **52**, 897–914.

Kaplan J.H. & Ellis-Davies G.C.R. (1988) *Proc. Natl. Acad. Sci. USA* **85**, 6571–6575.

Kaplan J.H. & Somlyo A.P. (1989) *Trends Neurosci.* **12**, 54–59.

Kaplan J.H., Forbush B. III & Hoffman J.F. (1978) *Biochemistry* **17**, 1929–1935.

Karlsson J.O.G., Axelsson K.L. & Andersson R.G.G. (1984) *Life Sci.* **34**, 1555–1563.

Karpen J.W., Zimmerman A.L., Stryer L. & Baylor D.A. (1988) *Proc. Natl. Acad. Sci. USA* **85**, 1287–1291.

Klodos I. & Forbush B. III (1988) *J. Gen. Physiol.* **92**, 46a (abstract).

Komori S. & Bolton T.B. (1991) *Pflügers Arch.* **418**, 437–441.

Korth M. & Engels J. (1979) *Naunyn-Schmiedeberg's Arch. Pharmacol.* **310**, 103–111.

Lando L. & Zucker R.S. (1988) *J. Gen. Physiol.* **93**, 1017–1060.

Lester H.A. & Chang H.W. (1977) *Nature* **266**, 373–374.

Lester H.A. & Nerbonne J.M. (1982) *Ann. Rev. Biophys. Bioengng* **11**, 151–175.

Lester H.A., Chabala L.D., Gurney A.M. & Sheridan R.E. (1986) In *Optical Methods in Cell Physiology (Soc. Gen. Physiol. Ser.)*, P. De Weer & B. Salzberg (eds). Wiley, New York, pp. 447–462.

McCray J.A. & Trentham D.R. (1989) *Ann. Rev. Biophys. Chem.* **18**, 239–270.

McCray J.A., Herbette L., Kihara T. & Trentham D.R. (1980) *Proc. Natl. Acad. Sci. USA* **77**, 7237–7241.

Matsunaga K. & Furchgott R.F. (1989) *J. Pharmac. Exp. Ther.* **248**, 687–695.

Messenger J.B., Katayama Y., Ogden D.C., Corrie J.E.T. & Trentham D.R. (1991) *J. Physiol.* **438**, 293P.

Milburn T., Matsubara N., Billington A.P., Udgaonkar J.B., Walker J.W., Carpenter B.K., Webb W.W., Marque J., Denk W., McCray J.A. & Hess G.P. (1989) *Biochemistry* **28**, 49–55.

Moncada S., Palmer R.M.J. & Higgs E.A. (1991) *Pharmacol. Rev.* **43**, 109–142.

Nargeot J., Lester H.A., Birdsall N.J.M., Stockton J., Wassermann N.H. & Erlanger B.F. (1982) *J. Gen. Physiol.* **79**, 657–678.

Nargeot J., Nerbonne J.M., Engels J. & Lester H.A. (1983) *Proc. Natl. Acad. Sci. USA* **80**, 2395–2399.

Naebauer M., Ellis-Davies G.C.R., Kaplan J.H. & Morad M. (1989) *Am. J. Physiol.* **256**, H916–H920.

Nerbonne J.M. (1986) In *Optical Methods in Cell Physiology (Soc. Gen. Physiol. Ser.)*, P. De Weer & B. Salzberg (eds). Wiley, New York, pp. 417–445.

Nerbonne J.M. & Gurney A.M. (1987) *J. Neurosci.* **7**, 882–893.

Nerbonne J.M., Lester H.A. & Connor J.A. (1982) *Soc. Neurosci. Abstr.* **8**, 945a.

Nerbonne J.M., Sheridan R.E., Chabala L.D. & Lester H.A. (1983) *Molec. Pharmacol.* **23**, 344–349.

Nerbonne J.M., Richard S., Nargeot J. & Lester H.A. (1984) *Nature* **310**, 74–76.

Nerbonne J.M., Richard S. & Nargeot J. (1985) *J. Molec. Cell. Cardiol.* **17**, 511–515.

Nichols C.G., Niggli E. & Lederer W.J. (1990) *Pflügers Arch.* **415**, 510–512.

Ogden D.C., Capiod T., Walker J.W. & Trentham D.R. (1990) *J. Physiol.* **422**, 585–602.

Patchornik A., Amit B. & Woodward R.B. (1970) *J. Am. Chem. Soc.* **92**, 6333–6335.

Richard S., Nerbonne J.M., Nargeot J., Lester H.A. & Garnier D. (1985) *Pflügers Arch.* **403**, 312–317.

Sheridan R.E. & Lester H.A. (1982) *J. Gen. Physiol.* **80**, 499–515.

Shimada K. & Berg H.C. (1987) *J. Molec. Biol.* **193**, 585–589.

Shinkai S., Matsuo K., Harada A. & Manabe O. (1982) *J. Chem. Soc. Perkin Trans.* **2**, 1261–1265.

Somlyo A.P. & Somlyo A.V. (1990) *Ann. Rev. Physiol.* **52**, 857–874.

Somlyo A.P. Walker J.W., Goldman Y.E., Trentham D.R., Kobayashi S., Kitazawa T. & Somlyo A.V. (1988) *Phil. Trans. R. Soc. Lond. B* **320**, 399–414.

Spray D.C., Nerbonne J.M., Campos De Carvalho A., Harris A.L. & Bennet M.V.L. (1984) *J. Cell Biol.* **99**, 174–179.

Tsien R.Y. (1980) *Biochemistry* **19**, 2396–2404.

Tsien R.Y. & Zucker R.S. (1985) *Biophys. J.* **50**, 843–853.

Valdeolmillos M., O'Neill S.C., Smith G.L. & Eisner D.A. (1989) *Pflügers Arch.* **413**, 676–678.

Walker J.W. (1991) In *Cellular Neurobiology: A Practical Approach* J. Chad & H. Wheal (eds). IRL Press at Oxford University Press, Oxford. pp. 179–203.

Walker J.W. & Trentham D.R. (1988) *Biophys. J.* **53**, 596a.

Walker J.W., McCray J.A. & Hess G.P. (1986) *Biochemistry* **25**, 1799–1805.

Walker J.W., Somlyo A.V., Goldman Y.E., Somlyo A.P. & Trentham D.R. (1987) *Nature* **327**, 249–252.

Walker J.W., Reid G.P., McCray J.A. & Trentham D.R. (1988) *J. Am. Chem. Soc.* **110**, 7170–7177.

Walker J.W., Reid G.P. & Trentham D.R. (1989a) *Methods Enzymol.* **172**, 288–301.

Walker J.W., Feeney J. & Trentham D.R. (1989b) *Biochemistry* **28**, 3272–3280.

Wigilius I.M., Axelsson K.L., Andersson R.G.G., Karlsson J.O.G. & Odman S. (1990) *Biochem. Biophys. Res. Commun.* **169**, 129–135.

Wolin M.S., Omar H.A., Mortelliti M.P. & Cherry P.D. (1991) *Am. J. Physiol.* **261**, H1141–H1147.

Wootton J.F. & Trentham D.R. (1989) In *Photochemical Probes in Biochemistry*, P.E. Nielsen (ed.). NATO ASI series C, Vol. 272. Kluwer, Dordrecht, pp. 277–296.

Zucker R.S. & Hayden P.G. (1988) *Nature* **335**, 360–362.

CHAPTER FORTY-ONE

Fluorescent Analogues: Optical Biosensors of the Chemical and Molecular Dynamics of Macromolecules in Living Cells

K. HAHN, J. KOLEGA, J. MONTIBELLER, R. DeBIASIO, P. POST, J. MYERS & D.L. TAYLOR
Center for Light Microscope Imaging and Biotechnology, Carnegie Mellon University, Pittsburgh, PA, USA

41.1 INTRODUCTION

Fluorescent analogue cytochemistry has grown to produce important insights in a wide variety of fields since 1978, when it was first demonstrated that a protein labelled with a fluorescent dye could function and be investigated within living cells (Taylor & Wang, 1978). The fluorescein-actin used in the first experiments exemplified an approach to analogue production which has been used in the large majority of studies to date. Almost all analogues have been made with dyes whose fluorescence is minimally affected by the intracellular environment, thus focusing on analysis of the analogue's distribution within live cells. Meaningful information has been obtained through adherence to several important principles (Taylor & Wang, 1980; Taylor *et al.*, 1984; Simon & Taylor, 1986; Wang, 1989). Chief among these has been careful *in vitro* characterization of the labelled protein prior to interpretation of live cell data (Wang & Taylor, 1980). Minimal perturbation of the analogues' biological function has been sought during labelling, and the effects of dye attachment have been carefully determined. Furthermore, the smallest possible quantity of analogue has been injected to avoid alteration of normal cell function.

Environmentally insensitive fluorescent analogues of many different proteins have been successfully observed in live cells. Examples include analogues of actin-binding proteins (myoin, vinculin, α-actinin), which have revealed complex changes in the cytoskeleton during a number of physiological processes (Wang *et al.*, 1982; Kreis & Birchmeier, 1982; Simon & Taylor, 1986; Kolega & Taylor, 1991). Labelled tubulin has shown the location and extent of tubulin exchange in the mitotic spindle, thus testing different models of spindle function (Salmon *et al.*, 1984; Wadsworth & Salmon, 1986; Gorbsky *et al.*, 1987). Calmodulin analogues have revealed the changing distribution of calmodulin during mitosis, and indicated transient calmodulin association with various

FLUORESCENT AND LUMINESCENT PROBES, 2ND EDN
ISBN 0-12-447836-0

subcellular organelles (Zavortink *et al.*, 1983; Luby-Phelps *et al.*, 1985). A recent addition to this family of probes have been protein analogues labelled with 'caged' fluorophores, whose fluorescence is activated by irradiation (Ware *et al.*, 1986; Mitchison, 1989; Theriot & Mitchison, 1991). This approach has permitted the marking of a subpopulation of analogues for temporal-spatial tracing at high contrast (Mitchison, 1989; Theriot & Mitchison, 1991).

More recently, the first representatives of a new family of fluorescent protein analogues have been developed using environmentally sensitive fluorophores, as proposed earlier (Taylor & Wang, 1980). These analogues have served as indicators of protein activity, in that the attached dye reflected conformational changes or altered ligand binding by the proteins. An environmentally sensitive dye has been used to make an analogue of calmodulin whose fluorescence reflects calcium-calmodulin binding (Hahn *et al.*, 1990). This analogue has been used in conjunction with a fluorescent calcium indicator to correlate the spatial and temporal dynamics of calcium transients and calium-calmodulin binding in individual, living cells. Details of these studies are described below. Resonance energy transfer between two fluorophores has been applied in an alternate approach to fluorescent sensing of protein activity. Energy transfer is strongly affected by protein changes that alter the distance between donor and acceptor dyes. The approach has been used to monitor actin polymerization *in vitro* (Taylor *et al.*, 1981; Wang & Taylor, 1981). When mixed fluorescein-actin and rhodamine-actin monomers were brought in proximity during polymerization energy transfer increased dramatically. More recently, the catalytic and regulatory subunits of protein kinase A have been labelled with fluorescein and rhodamine (Adams *et al.*, 1991). Dissociation of the subunits by cAMP was observed in live cells as a decrease in energy transfer.

In our laboratory, we are harnessing environmentally sensitive and insensitive protein analogues to decipher the mechanics and regulation of cell motility. We will describe here our efforts to use fluorescent analogue cytochemistry in elucidating the mechanisms by which extracellular stimuli induce cell contraction and initiation of motility. Our aim is to define the often rapid changes in second messengers involved in this process, and their effects on the dynamics of contractile proteins in time and space during cell function.

In the currently established paradigm of stimulus-contraction coupling in non-muscle cells (for reviews see McNeil & Taylor, 1987; Sellers & Adelstein, 1987), binding of ligands to external receptors leads to intracellular production of lipid metabolites and alterations in the concentration and distribution of calcium. We are initially focusing on subsequent regulatory steps involving modulation of the calcium signal through calmodulin and myosin light chain kinase. Calcium binding to calmodulin enables calmodulin to activate myosin light chain kinase, which phosphorylates myosin II regulatory light chains. This phosphorylation activates myosin for contraction.

We will describe protein-based fluorescent indicators (optical biosensors) for observation of individual signalling steps in this process. The production and application of fluorescent analogues used as biosensors of calcium-calmodulin binding will be described, as will progress towards indicators of myosin light chain phosphorylation. The use of fluorescent analogues of actin and myosin to study the dynamics of contractility will also be described.

We study serum stimulation of quiescent fibroblasts (McNeil & Taylor, 1987) and a 'wound healing' model in which a gap is introduced in a monolayer of fibroblasts using a blunt razor blade (DeBiasio *et al.*, 1988; Fisher *et al.*, 1988). Fibroblasts at the edge of this 'wound' migrate into the gap. In these systems, the timing of stimulation can be controlled, and during wound healing, polarized movement occurs with a predetermined orientation. Cells are observed with a multimode microscope imaging system that permits rapid switching between transmitted light microscopy and various forms of fluorescence microscopy using different detectors and excitation and emission filters (Giuliano *et al.*, 1990).

41.2 MEROCAM 1 AND 2: FLUORESCENT INDICATORS OF CALCIUM–CALMODULIN BINDING

Fluorescent calcium indicators have been invaluable in elucidating the kinetics and intracellular distribution of calcium transients during a wide range of cellular processes (Tsien, 1989). However, subsequent signalling steps have to date remained inaccessible to observation *in vivo*. We have designed a fluorescent indicator of calcium–calmodulin binding which we are now using to correlate intracellular calcium changes with calmodulin activation during growth factor stimulation of quiescent fibroblasts (Hahn *et al.*, 1990). We are correlating the spatial and temporal dynamics of calcium changes and calcium–calmodulin binding. Biochemical evidence indicates that calmodulin's calcium response will depend on its intracellular environment. Calmodulin's calcium affinity is affected by target proteins (Cohen & Klee, 1988), and calmodulin binds to different targets at different calcium levels (Andreason *et al.*, 1983; Cohen & Klee,

1988). Changing subcellular distribution of calmodulin seen in previous studies using environmentally insensitive fluorescent analogues supports regulation of calmodulin by factors in addition to calcium (Pardue *et al.*, 1981; Zavortink *et al.*, 1983; Luby-Phelps *et al.*, 1985).

We have designed our indicators of calcium–calmodulin binding on the basis of previous biochemical studies which show that calmodulin–calcium binding produces a hydrophobic site on calmodulin with affinity for specific small molecules (LaPorte *et al.*, 1980, 1981; Malencik *et al.*, 1981; Manalan & Klee, 1984; Cohen & Klee, 1988). Structure–activity studies enabled synthesis of a novel dye with both highly solvent-sensitive fluorescence and strong affinity for calmodulin's calcium-induced hydrophobic binding site. Through affinity labelling, this dye was covalently attached to calmodulin in a position where it would have access to the calcium-induced hydrophobic pocket. Whenever calcium bound calmodulin, the solvent-sensitive dye moved into the hydrophobic pocket, with a consequent shift in fluorescence. The fluorescence change was shown to be fully reversible on removal of calcium. We named the calcium–calmodulin binding indicator meroCaM 1, for merocyanine dye + calmodulin.

Spectral characterization of meroCaM 1 showed that its excitation spectrum changed in both intensity and peak shape in a calium-dependent manner. The spectral changes were suitable for ratio imaging, an important technique used to normalize fluorescence of analogues in the cell for differences in cell thickness, excitation intensity, and other factors (Tanasugarn *et al.*, 1984; Bright *et al.*, 1989). The excitation ratio varied sigmoidally with calcium, and showed an overall 3.4-fold change. Unlike other solvent-sensitive fluorophores, the dye used in meroCaM 1 was designed specifically for use in live cells. Optimum excitation ratios were obtained at 532 and 608 nm (emission 623 nm), long wavelengths which do not cause photodamage of cells and are far from the autofluorescence of mammalian cells. The dyes had high absorbance and quantum yield values, decreasing the amount of intracellular protein analogue required to obtain acceptable fluorescence signal.

In order to apply meroCaM 1 to study regulation of calmodulin activity, it was important to characterize the analogue's target protein binding, and to show that the effects of target proteins on calcium affinity remained intact. MeroCaM 1 was shown to retain calmodulin's ability to activate myosin light chain kinase, a calmodulin target protein. It also showed calcium-dependent mobility on native and SDS gels. Known effects of melittin and cAMP-phosphodiesterase on calcium affinity were also observed in meroCaM 1.

The calcium-induced change in meroCaM 1 fluorescence was half maximal at 0.3–0.4 μM, below the 5–10 μM apparent dissociation constant reported for overall calcium–calmodulin binding (Klee, 1988). This could have resulted from the dye stabilizing calmodulin's calcium-bound form, thus affecting the calcium binding constant. Alternately, the dye could have been positioned on the protein where it would reflect primarily binding to the high-affinity calcium sites. These possibilities were distinguished by comparing the calcium dependence of meroCaM 1 and calmodulin cAMP-phosphodiesterase activation. The similar behaviour of the two proteins in this assay indicated that meroCaM 1 fluorescence reflected binding to high-affinity calcium sites. The effect of cAMP-phosphodiesterase binding on calmodulin's calcium affinity remained intact in meroCaM. The assay did show that meroCaM's maximal activation or affinity for the phosphodiesterase was somewhat reduced from that of calmodulin.

In order to assay the effect of target protein binding on meroCaM 1 fluorescence, the calcium dependence of the excitation ratio was determined in the presence and absence of melittin, a peptide mimic of target proteins' calmodulin binding sites (Comte *et al.*, 1983; Maulet & Cox, 1983; Cox *et al.*, 1985; Seeholzer *et al.*, 1986). These experiments showed that the high and low excitation ratio values, produced at extremes of calcium concentration, were altered by melittin. Although this will complicate quantitation of the extent of calcium-calmodulin binding, quantitative analysis of binding kinetics should not be affected.

We have begun testing meroCaM 1 in live cells. The analogue has shown sufficient brightness and photostability to obtain greater than 10 ratio pairs, and has shown no apparent toxic effects or alteration of cell physiology. In our preliminary studies, meroCaM 1 has been injected in Swiss 3T3 fibroblasts made quiescent by 48 h incubation in low serum medium. Published studies and our own work have shown that these cells respond to whole serum with a rapid and transient upsurge in intracellular calcium (Byron & Villereal, 1989; Tucker & Fay, 1990; Takuwa *et al.*, 1991; McNeil *et al.*, 1985; Taylor, DeBiaiso & Hahn, (pers. commun.)). When the cells containing meroCaM 1 were challenged with whole serum, the intracellular meroCaM 1 excitation ratio increased 10–15% within 1 min and then returned to baseline over a period of several minutes. The filters used to excite and monitor the analogue's fluorescence have not been optimized, so it is anticipated that greater fluorescence responses will ultimately be obtained. Mapping of meroCaM 1 response within individual cells indicated spatial heterogeneity of calmodulin activation.

We have also developed new dyes in an effort to

improve meroCaM 1, and have found that substitution of a benzoxazole moiety for benzothiazole in the meroCaM 1 dye results in a strong enhancement of fluorescence response. The new dye has been attached to calmodulin to produce a protein analogue which we call meroCaM 2. The excitation ratio of this indicator shows a 7-fold calcium-dependent change in excitation ratio (excitation 457, 569 nm; emission 587 nm). In preliminary studies, the indicator has been injected in quiescent Swiss 3T3 fibroblasts together with the fluorescent calcium indicator, Calcium Green (Plate 41.1) (Molecular Probes, Eugene, OR). Stimulation of the cells with whole serum caused a rapid increase in calcium to maximal levels within 1 min, followed by a decrease which sometimes showed smaller additional maxima. The response of MeroCaM 2 to these calcium changes was quite complex. MeroCaM response generally also reached its maximum (up to 80% increase) within the first minute after stimulation, but afterwards sometimes paralleled calcium changes and sometimes underwent independent oscillations.

The correlation of calcium–calmodulin binding with calcium changes in living cells now appears to be within reach. We hope to extend the approach we have developed to produce indicators of calmodulin binding to target proteins. Published studies indicate that sensitivity of fluorescent calmodulin derivatives to calcium vs. protein binding is strongly dependent on dye structure and the position of dye attachment (LaPorte *et al.*, 1981; Mills *et al.*, 1988). We will test new dyes and explore the use of site-specific mutagenesis to introduce cysteines for attachment of dye at precise positions.

41.3 FLUORESCENT ANALOGUE OF MYOSIN II

The simplest form of fluorescent analogue is one whose fluorescence is insensitive to configuration and local environment. Such analogues act as direct reporters of a molecule's distribution and movements. A variety of clever biophysical and optical techniques have been used to parlay this simple information into local chemical concentrations, state of molecular assembly, diffusion constants, and a number of other detailed analyses of molecular dynamics at a subcellular level. As an example of some of these applications, we will discuss fluorescent analogues of myosin II that have been produced to visualize directly the dynamics of myosin II during cell movement in single, living Swiss 3T3 fibroblasts.

Smooth muscle myosin II isolated from chicken gizzard was labelled with tetramethyl rhodamine iodoacetamide (DeBiasio *et al.*, 1988). The dye was covalently bound to both the myosin II heavy chain and the 17 kDa light chain, leading to incorporation of 4–6 mol dye per mol protein. Before performing *in vivo* studies, it was essential to establish the biological activity of the analogue through careful biochemical characterization. The analogue was shown to retain native myosin II's ability to assemble into filaments as measured by right-angle light scattering (Fig. 41.1), and both labelled and unlabelled filaments were depolymerized by 5 mM ATP. Fluorescent labelling did not affect the analogue's K-EDTA ATPase activity. Labelled and native myosin II were phosphorylated to the same extent and at the same rate by chicken gizzard myosin light chain kinase (Fig. 41.2).

The positions of dye attachment on the heavy chain were determined using the papain digestion procedure of Nath *et al.*, (1986). One fluorophore was located in the 70 kDa *N*-terminal portion of myosin II's S1 region, and another in the S2 region. No fluorophore was detected in the 25 kDa *C*-terminal portion of the S1 head. Modification at the latter site has been shown to cause elevation of actin-activated MgATPase activity in unphosphorylated myo-

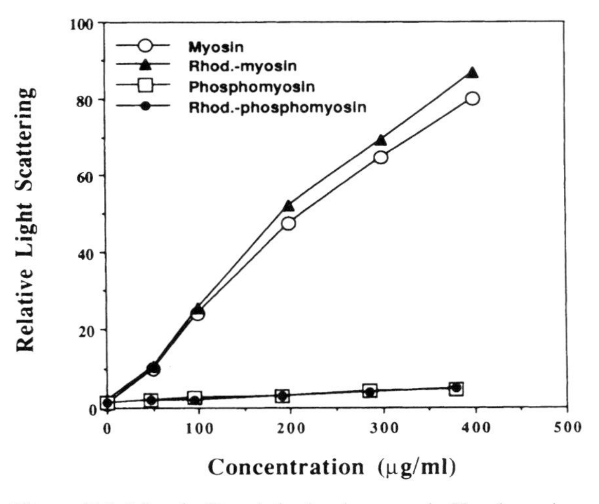

Figure 41.1 Myosin II and rhodamine-myosin II polymerization. The abilities of myosin II and its rhodamine analogue to polymerize were measured by right-angle light scattering using the procedure of McKenna *et al.* (1989). Unlabelled myosin and rhodamine-myosin II were dialysed into 500 mM KCl, 0.1 mM EDTA, 0.1 mM EGTA, 10 mM HEPES, pH 7.5. Dialysates were clarified by centrifugation and the concentrations measured spectrophotometrically using an extinction coefficient of E280 = 0.53 cm^{-1} (0.1%), correcting for 280 nm absorbance due to dye in the case of rhodamine-myosin II. The myosins were diluted 12-fold into an assembly buffer containing 150 mM KCl, 10 mM $MgCl_2$, 1 mM EGTA, 0.1 mM DTT, 10 mM HEPES, pH 7.5 at room temperature. Scattered light intensities were measured at 340 nm using a Perkin-Elmer MPF-3 fluorescence spectrophotometer. To test the sensitivity of the filaments to ATP, the assembly mixtures were made 5 mM in ATP and the measurements were repeated.

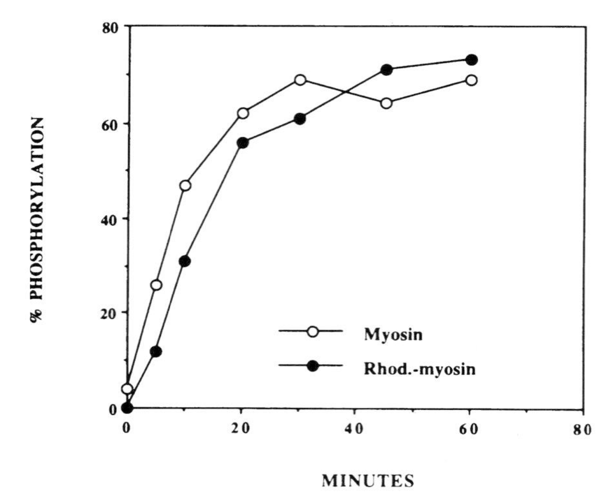

Figure 41.2 Phosphorylation of myosin II and rhodamine-myosin II. Myosin II or rhodamine-myosin II (c. 4 mg ml^{-1}) were phosphorylated at room temperature in a mixture containing 25 mM Tris-HCl, 4 mM $MgCl_2$, 1 mM ATP, 15 μg ml^{-1} bovine brain calmodulin, and $CaCl_2$ 0.2 mM in excess of EGTA, pH 7.5. The reaction was started by the addition of 0.6 μg ml^{-1} smooth muscle MLCK in 1 mg ml^{-1} BSA in Tris. Aliquots of 20 μl were removed at intervals, and the reaction was terminated by dilution. Samples were run on glycerol-urea polyacrylamide gels (Perrie & Perry, 1970). The phosphorylated and dephosphorylated 20 kDa light chain bands were stained with Coomassie blue dye and the relative amounts of each were quantified by densitometry. The rate of phosphorylation was slightly higher for the unlabelled myosin, but this was due to the presence of endogenous kinase which had been eliminated from the rhodamine-myosin II during postlabelling purification.

sin II, and to hinder myosin II's ability to adopt a folded conformation (Chandra *et al.*, 1985). A gel filtration assay (Trybus *et al.*, 1982) indicated that folding was unaffected by labelling (Fig. 41.3). Actin-activated MgATPase activity was decreased by 0–40% (Table 41.1). In the absence of actin, the labelling had no effect on this activity.

After biochemical characterization, the analogue was injected into living Swiss 3T3 fibroblasts (DeBiasio *et al.*, 1988), where it permitted direct visualization of myosin II dynamics. Fluorescent myosin II became distributed throughout the cell within 1 h after injection, incorporating into the same structures and displaying the same periodic, 'pseudosarcomeric' distribution revealed by immunofluorescence staining of endogenous myosin II (Fig. 41.4) (DeBiasio *et al.*, 1988; McKenna *et al.*, 1989). Unlike immunofluorescence experiments, use of the analogue permitted observation of myosin II dynamics over extended periods of time, and time-lapse studies revealed that myosin II was very dynamic. In fibroblasts made quiescent by serum deprivation, myosin II assembled into stress fibres at the cell margins, moved centripetally toward the nucleus, and then disassembled in the perinuclear region, presumably recycling to reassemble at the periphery (Giuliano & Taylor, 1990). The intracellular movements of myosin II have also been examined during amoeboid locomotion of endothelial cells (Kolega, 1997) and during cytokinesis (Sanger *et al.*, 1989; DeBiasio *et al.*, 1996) using similar microinjected fluorescent analogues. Furthermore, by preparing fluorescent analogues using myosin II from different tissues, it is possible to generate isoform-specific analogues (Kolega, 1998). When these analogues are labelled with spectrally distinct fluorophores, the dynamic behaviour of different myosin II isoforms can be monitored simultaneously within the same cell.

Dynamic imaging of fluorescent analogues of myosin II has helped elucidate the nature of actin–myosin contractility in living cells. Stimulation of fibroblasts by serum or growth factors causes shortening of stress fibres with concomitant shortening of the pseudosarcomeric spacing (Giuliano & Taylor, 1990). These observations supported a sliding filament model of fibre shortening, as did quantitative analysis of the fibre shortening induced by cytochalasin, an actin-solating agent (Kolega *et al.*, 1991). The amount of fluorescence associated with stress fibres during cytochalasin-induced shortening was measured by integrating the intensity of digital images. The density of the analogue along fibres increased as the fibres shortened, indicating that myosin II became more concentrated during shortening. The *total* analogue associated with the fibre *decreased* during shortening, indicating a significant loss of myosin II from the shortening fibre. This partial 'self-destruct' process is an important component of the solation–contraction coupling hypothesis (Taylor & Fechheimer, 1982; Kolega *et al.*, 1991). Thus, the use of a myosin II fluorescent analogue enabled demonstration of a basic mechanism of cell contractility *in vivo*.

The ability to follow myosin II dynamics in living cells has also illuminated how contractile structures are regulated and distributed in the cytoplasm. It has been shown *in vitro* and in extracted cell models that myosin II assembly and motor activity can be regulated by phosphorylation of myosin II's 20 kDa regulatory light chain (Sellers & Adelstein, 1987; Lamb *et al.*, 1988). Fluorescent analogues of myosin II revealed changes in myosin II organization *in vivo* when light-chain phosphorylation was increased or decreased by inhibitors of intracellular phosphatases and kinases. When serum-deprived 3T3 fibroblasts were treated with okadaic acid, a phosphatase inhibitor, stress fibres contracted slowly, but incompletely (Fig. 41.5). In contrast, staurosporine, a kinase inhibitor, caused dissolution of fibres without shortening (Fig. 41.6). The dissolution of fibres could be quantified using the fluorescence images of the myosin

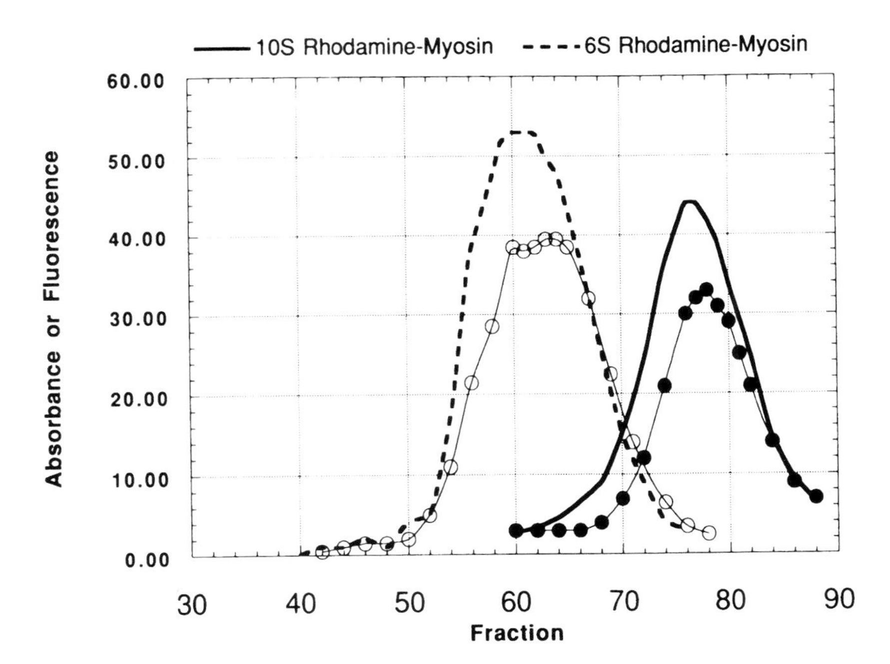

Figure 41.3 Chromatographic assay of ionic strength-dependent myosin II and rhodamine-myosin II conformational change. A 1.5 × 90 cm Sepharose CL-4B column was equilibrated in low salt buffer (150 mM KCl, 5 mM $MgCl_2$, 1 mM EGTA, 10 mM K_nPO4, 1 mM MgATP, pH 7.5) at a flow rate of 6 ml h^{-1} maintained using a peristaltic pump. The column was calibrated with proteins IgG, IgM, and BSA. The void and included volumes were determined using Blue dextran 2000 and bromophenol blue, respectively. After dissolving in buffer, myosins were clarified by centrifugation. Fractions of 27 drops (160 fractions total) were collected. The column was then equilibrated in high salt buffer (600 mM KCl, 5 mM $MgCl_2$, 1 mM EGTA, 10 mM K_nPO4, pH 7.5) and the procedure repeated. Proteins in low salt were detected by fluorescence, and in high salt by absorbance. The elution volumes of the IgG, IgM, BSA, Blue dextran 2000, and bromophenol blue did not change significantly with ionic strength. Thus, both labelled and unlabelled myosins exhibited phosphate-dependent folding and unfolding.

II analogue, and this measurement revealed a dose-dependent loss of fibres that correlated extremely well with dephosphorylation of myosin light chains (Kolega *et al.*, 1993). These observations are consistent with regulation of fibre assembly and contraction through myosin phosphorylation.

One area in which fluorescent analogues provide unique capabilities is in assessing spatial variations in the state of assembly and local concentration of cytoplasmic proteins. A striking example was provided when fluorescent actin and myosin analogues were co-injected into the same cell and the relative distribution of the two proteins followed during early stages of locomotion (DeBiasio *et al.*, 1988). To correct for subcellular differences in accessible volume and optical pathlength, actin and myosin II fluorescence were normalized to inert fluorescence volume indicators (dextran labelled with fluorescein). Newly formed cellular protrusions contained an elevated concentration of actin and very little or no myosin II. No well-defined structures were detected in the protrusions by fluorescence or video-enhanced contrast microscopy. Within 1–2 min, well-defined actin- and myosin II-containing structures were detected in the same protrusion, together with diffuse actin and myosin II. Thus, initial protrusive activity involved a local, relative increase in actin concentration, and did not appear to require structures containing myosin II.

The state of assembly of actin and myosin II in various regions of locomoting cells can be examined by a number of different approaches using fluorescent analogues. Fluorescence recovery after photobleaching (FRAP) is perhaps the most elegant of these. In this method, a laser is used to bleach a small region of the analogue in a cell, then the rate at which fluores-

Table 41.1 Myosin II and rhodamine-myosin II ATPase activity.

	K-EDTA	Mg^{2+} *ATPase* −Actin	*+Actin*
Myosin	381	<0.1	0.4
Myosin-PO4	–	3.1	27
Rhodamine-myosin	372	0.3	0.5
Rhodamine-myosin-PO4	–	3.7	22

Activities were measured at 25°C using the procedure of Sellers *et al.* (1981). K-EDTA activity was measured in a mixture containing 500 mM KCl,1 mM ATP, 2 mM EDTA, 15 mM Tris-HCl, and 45–90 nM myosin, pH 7.5. MgATPase activity was measured in 30 mM KCl, 5 mM $MgCl_2$, 1 mM ATP, 0.1 mM EGTA, and 15 mM Tris, pH 7.5, using 540 nM myosin. For actin activation, actin was added at a concentration of 0.3 mg ml^{-1}. Myosins were phosphorylated using the procedure described in Fig. 41.3, except that ATP was replaced with adenosine 5′-O-(3-thiotriphosphate). ATP hydrolysis rates are given as nmol min^{-1} mg^{-1}.

cence is restored by diffusion of molecules from outside the photobleached spot is measured. In FRAP studies of locomoting fibroblasts, myosin II analogue had essentially no translational mobility in regions near the leading edge, in contrast to a 79% mobile fraction in the perinuclear region (DeBiasio *et al.*, 1988; Kolega & Taylor, 1991). Thus, myosin II was not readily available for diffusion into a protrusion even though it was present at the protrusion's base and in areas within the leading edge. Active transport may be required to mobilize myosin II into established protrusions, possibly by molecular interactions with actin in the protrusion. The increase in myosin II mobility in the perinuclear region is consistent with disassembly of myosin II, and a similar perinuclear increase in myosin II mobility was observed in serum-deprived fibroblasts, where fibres are continuously assembled at the cell periphery, transported centripetally, and disassembled in the perinuclear cytoplasm (Kolega & Taylor, 1991).

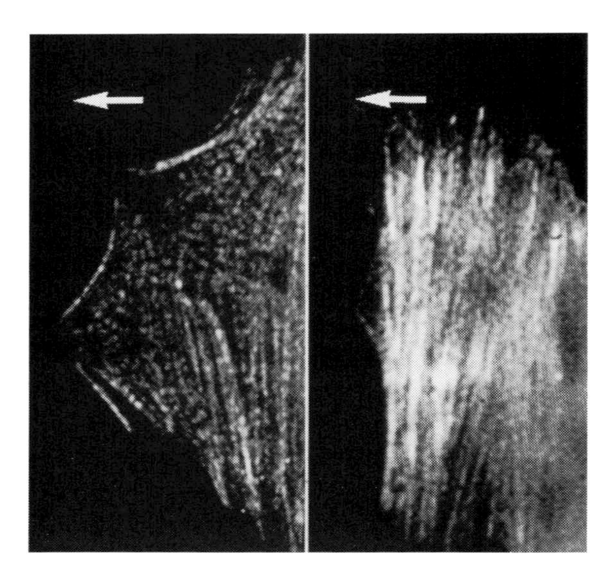

Figure 41.4 Fluorescent images of rhodamine-myosin II and rhodamine-myosin 20 kDa light chain analogues in Swiss 3T3 fibroblasts. Left: light chain analogue. Right: myosin II analogue. These cells are representative of a motile population 3–6 h after wounding. The arrows point in the direction of migration. Both analogues are distributed in a periodic, 'semisarcomeric' pattern.

FRAP measurements are restricted to a small number of small spots within any given cell. Much more extensive maps of assembly can be obtained by following the redistribution of an analogue when the cell is lysed with mild detergent. Freely diffusing material perfuses away from the cell, while material that is associated with structural elements remains in place. Comparison of pre- and post-permeabilization images reveals the proportion of bound material at any point in the cell. This approach has revealed a gradient in the assembly of myosin II across locomoting fribroblasts (Kolega & Taylor, 1993) and endothelial cells (Kolega, 1997). Myosin II was found to be extensively associated with the cytoskeleton at the base of the cells' leading protrusions, even when discrete fibrous structures were not apparent by fluorescence microscopy. Verkhovsky *et al.* (1995) followed the exact distribution of a myosin II analogue at high magnification to verify that the distribution of myosin II-containing structures was preserved during permeabilization and extensive post-permeabilization treatments. This allowed them to perform subsequent electron microscopic studies that have identified heretofore unknown forms of myosin II organization in non-muscle cells.

An additional trick allows one to infer the state of assembly of an analogue without permeabilizing (and stopping the movement of) cells. This entails comparing the distribution of the analogue with an inert, fluorescent tracer of the same size, which serves as an indicator of the volume that is accessible to the analogue. The extent to which the functional analogue becomes locally concentrated relative to the volume indicator can reveal when and where the analogue is binding or assembling in the cytoskeleton. Ratio images comparing the distribution of an assembly-competent analogue of myosin II with an assembly-incompetent counterpart confirmed the spatial variations in myosin II assembly revealed by FRAP and detergent-permeabilization (Kolega & Taylor, 1993). Actin assembly can be visualized in a similar fashion (Giuliano & Taylor, 1994). Thus, time-lapse, ratio imaging can produce dynamic maps of structural

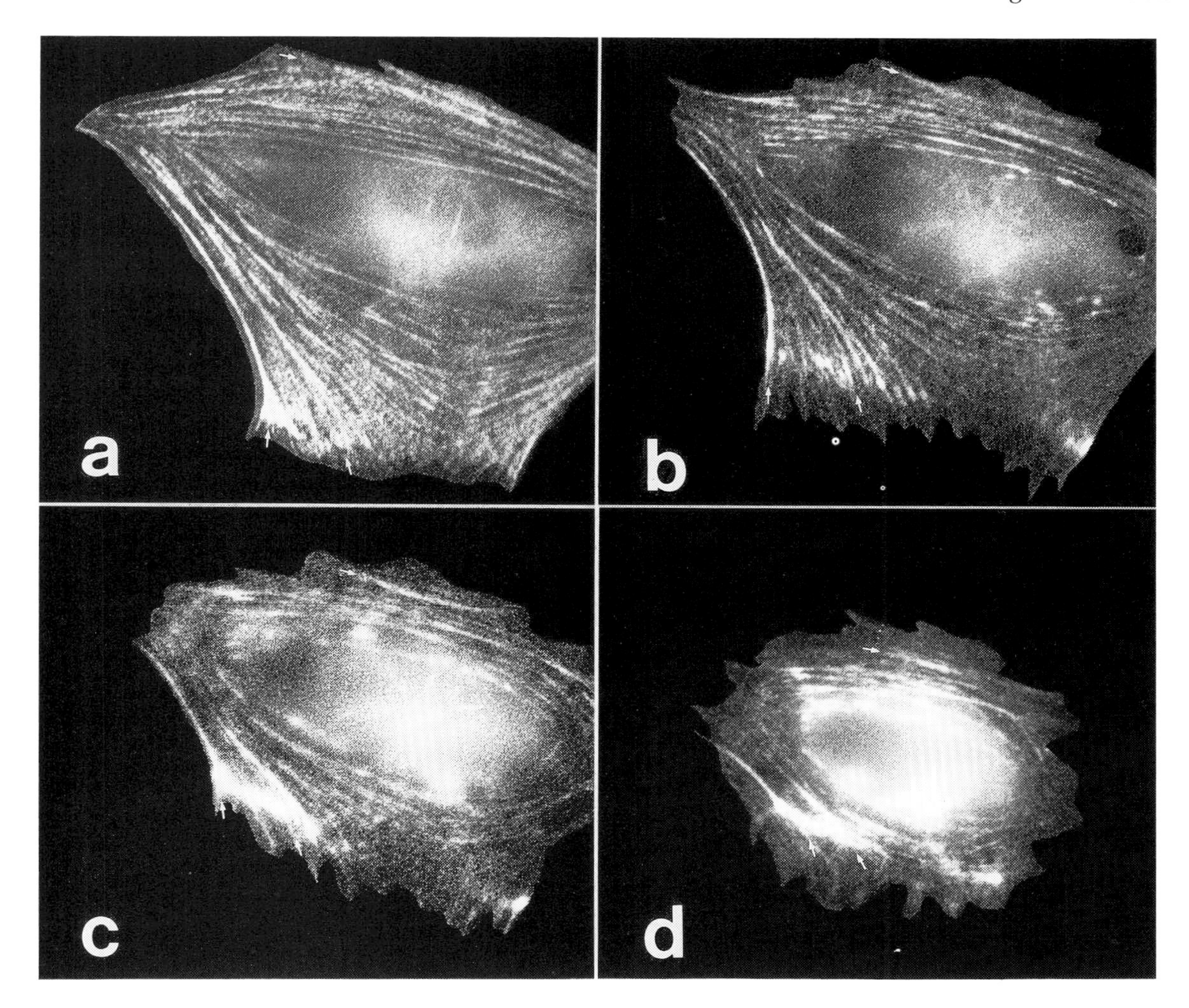

Figure 41.5 Myosin contraction in fibroblasts after treatment with phosphatase inhibitor. Quiescent, serum-deprived Swiss 3T3 fibroblasts were treated with a 400 nM solution of okadaic acid, a phosphatase inhibitor. Images were acquired with a cooled CCD camera at 2.5 min intervals. Small arrows identify selected myosin labelled fibres. (a) Myosin labelled fibres were distributed throughout the quiescent cell, displaying a relatively uniform punctuate distribution, (b) 25 min, (c) 40 min, (d) 55 min after okadaic treatment. Myosin-labelled fibres contracted from the cell periphery towards the mucleus over time. The cell edges detached from the substrate and retracted.

components as they assemble and disassemble in living cells.

Understanding how different myosin motors move within living cells should greatly facilitate our understanding of the mechanisms of cell locomotion. This is particularly true as we attempt to unravel the functions of myosin II and the ever-growing family of unconventional myosins (for reviews, see Mooseker & Cheney, 1995; Sellers *et al.*, 1996). The use of fluorescent analogue cytochemistry in this effort has recently been revolutionized by the cloning of green fluorescent protein (GFP). The portion of the GFP molecule that is required for fluorescence can now be engineered into proteins to give them an integral fluorescent tag. While the large size of the fluorescent GFP piece (20–30 kDa) can potentially disrupt a protein's normal function, if carefully engineered and tested, GFP-analogues can be extremely powerful tools. A GFP-myosin fusion protein has been shown to maintain its ATPase activity (Iwane *et al.*, 1997), and a GFP sequence has been spliced into the heavy chain of myosin II in *Dictyostelium* (Moores *et al.*, 1996), creating an organism in which endogenous myosin II can be monitored by fluorescence microscopy, in any cell at any time, and without microinjection. These new tools, in conjunction with the existing analogues of myosin II and other cytoskeletal proteins, continue to expand our window onto the workings of the cell's locomotive machinery.

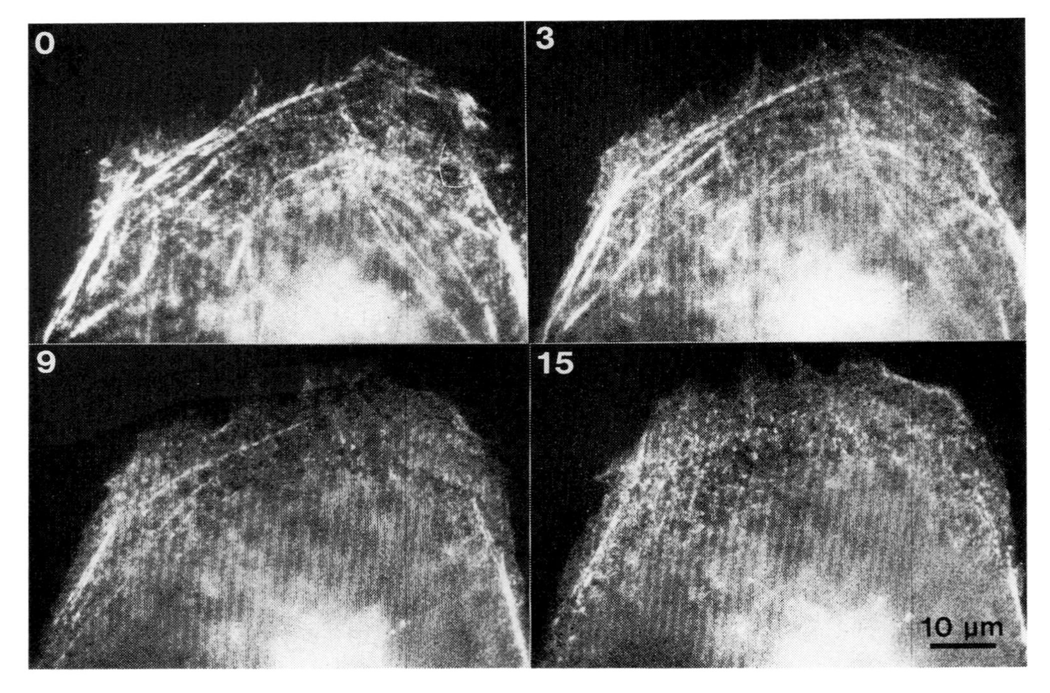

Figure 41.6 Dissolution of stress fibres in the presence of kinase inhibitor. Serum-deprived Swiss 3T3 fibroblasts were microinjected with a rhodamine analogue of myosin II. After 60 min. during which the analogue became distributed through the cytoplasm, the cell was perfused with 50 µM staurosporine while fluorescence images were acquired at 1 min intervals using a cooled CCD camera. Selected images from the sequence are shown with the time in minutes after perfusion indicated in the upper left. Prior to treatment, myosin II was located predominantly in stress fibres. Upon perfusion of inhibitor, stress fibres appeared to dissolve: fluorescence decreased uniformly along the length of the fibres while diffuse cytoplasmic fluorescence increased until little or no fibre structure remained.

41.4 PROTEIN-BASED OPTICAL BIOSENSORS OF MYOSIN II REGULATORY LIGHT CHAIN PHOSPHORYLATION

Smooth muscle and non-muscle cells express a myosin II molecule consisting of a pair of heavy chains that form two globular head domains (which contain the actin- and ATP-binding sites), a neck domain (containing the light chain binding sites), and a coiled-coil tail domain. Non-covalently associated with the neck are two pairs of light chains – the essential light chains and the regulatory light chains. The regulatory light chains undergo covalent, reversible phosphorylation, which helps to regulate the activity of the myosin motor domain (reviewed by Sellers & Adelstein, 1987; Moussavi *et al.*, 1993; Sellers & Goodson, 1995). Serine-19, or analogous serine (and sometimes threonine-18), of the regulatory light chain is phosphorylated by the calcium/calmodulin-dependent enzyme myosin light chain kinase. *In vitro*, phosphorylation at this site exhibits several effects on the entire myosin-II molecule: the actin-activated MgATPase activity is increased by several hundred-fold; the protein shifts from a 10S unphosphorylated, folded conformation to a 6S, phosphorylated extended conformation; and the phosphorylated 6S particle assembles into bipolar thick filaments. In addition, serine-1, serine-2 and threonine-9 of the regulatory light chain can be phosphorylated by phospholipid-dependent protein kinase C and cyclin-p34^{cdc2}. Phosphorylation at these 'inhibitory' sites decreases the actin-activated MgATPase activity of myosin II previously phosphorylated on serine-19 (by myosin light chain kinase), but has little effect on the actin-activated MgATPase of myosin II previously *un*phosphorylated on serine-19. Additionally, phosphorylation of serine-1, serine-2 and threonine-9 decreases the affinity of myosin light chain kinase for the regulatory light chain.

Cellular effects of myosin-II phosphorylation have previously been studied using dilute solution biochemistry or extracted cell models. Recently, phosphorylation-specific myosin-II regulatory light chain antibodies have been used in fixed cells to study regulatory light chain phosphorylation in time and space (Bennett *et al.*, 1988; Matsumura *et al.*, 1998). These measurements of the phosphorylation state of cell populations have provided valuable information about the regulatory role of this event *in vivo*, but these methods can only report the cell population average. Our laboratory has designed novel, phosphorylation-sensitive fluorescent derivatives of myosin II, with the aim of learning more about how the temporal and spatial regulation of myosin II contributes to non-muscle cell motility in individual, living cells. This section describes the design of two fluorescent protein biosensors of myosin II that report myosin II activity *in situ* or *in vivo*. One reagent was used in locomoting and dividing cells to help understand the contribution of myosin II to these events.

Before we produced a phosphorylation-sensitive light chain analogue, it was important to establish that light chains could be fluorescently labelled in such a way that they would retain nearly native biochemical activity and would incorporate into the myosin II of living cells. Model studies were carried out using a turkey gizzard light chain labelled with an environmentally insensitive dye, tetramethylrhodamine-5,(6)iodoacetamide. This analogue could be fully phosphorylated *in vitro* with myosin light chain kinase isolated from smooth muscle; the rate of phosphorylation was 70% that of unlabelled myosin II. When injected into live cells, the analogue exchanged with native myosin II regulatory light chains and incorporated into actin/myosin-based fibres in both non-motile and actively migrating cells. The analogue showed the same punctate, periodic fluorescence pattern that we have observed with the whole myosin II analogue (Fig. 41.4). Similar results have been obtained with light chains from striated muscle (Mittal *et al.*, 1987). This *in vivo* exchange provides a versatile method for the introduction of light chains into myosin II within live cells. Biochemical studies have shown that regulatory light chains are functionally interchangeable among species.

We have designed two phosphorylation-sensitive fluorescent biosensors using two different approaches. Via a molecular genetic approach, a regulatory light chain was engineered to contain a single cysteine (residue 18) adjacent to the phosphorylated serine (residue 19) (Post *et al.*, 1994). This regulatory light chain was labelled with the polarity-sensitive fluorescent dye acrylodan on the newly inserted cysteine. Phosphorylation of this labelled light chain (called AC-cys18·LC_{20}) by MLCK produces a 60% quenching and a 28 nm red shift of fluorescence emission. The labelled, mutant light chain was reconstituted with smooth muscle myosin II heavy chains and essential light chains with an *in vitro* light chain exchange procedure of Katoh & Lowey (1989). Biochemical assays demonstrated that this phosphorylation biosensor retains nearly native levels of rate of phosphorylation, K^+ATPase activity, and *in vitro* motility. The acrylodan-labelled mutant light chain was exchanged into the A-bands of chicken pectoralis myofibrils *in situ* to demonstrate the localization and activity of the biosensor in a highly ordered contractile system. Fluorometry and quantitative fluorescence microscopic imaging experiments demonstrated a phosphorylation-dependent fluorescence change in AC-cys18·LC_{20} exchanged myofibrils. Labelled mutant light chains were also incorporated into stress fibres of living fibroblasts and smooth muscle cells. This reagent has been valuable for *in vitro* and *in situ* studies, but severe intracellular quenching has limited its usefulness for live cell studies.

The second approach employs fluorescence resonance energy transfer between fluorescein-labelled regulatory light chains to rhodamine-labelled essential light chains and heavy chains (Post *et al.*, 1995). Native regulatory light chains were labelled at their single cysteine (cys 108) with fluorescein; these labelled light chains were incorporated into the rhodamine-labelled myosin II analogue (see Section 47.3). The ratio of rhodamine/fluorescein emission increases by up to 26% with regulatory light chain phosphorylation by myosin light chain kinase. The majority of the change in energy transfer is from regulatory light chain phosphorylation by myosin light chain kinase (versus phosphorylation by protein kinase C). Myosin II folding/unfolding, filament assembly, and actin-binding do not significantly affect the energy transfer ratio of this biosensor. Treatment of fibroblasts containing the phosphorylation biosensor with the kinase inhibitor staurosporine produced a lower ratio of rhodamine/fluorescein emission. This corresponds to a lower level of myosin II regulatory light chain phosphorylation and correlates with previously obtained biochemical measurements of the state of phosphorylation in these cells (Giuliano *et al.*, 1992). This biosensor has been microinjected into living cells, where it incorporates into stress fibres. Locomoting fibroblasts containing the biosensor showed a gradient of myosin II phosphorylation that was lowest near the leading edge and highest in the tail region of these cells, which correlates with previously observed gradients of free calcium and activated calmodulin (Plate 47.1) (Brundage *et al.*, 1991; Hahn *et al.*, 1992; Gough & Taylor, 1993). Note that while phosphorylation is highest in the rear of the cell, myosin II concentration is low and actin and myosin II filament assembly is low (Kolega & Taylor, 1993; Guiliano & Taylor, 1994; Post *et al.*, 1995). This suggests that, in the tail region, where myosin II is maximally activated, a solation of the actin-myosin gel occurs to allow for maximum contractility of the activated myosin II motor (Janson & Taylor, 1993).

Fibroblasts (microinjected with this biosensor)

undergoing cell division showed a global increase in serine 19 myosin II regulatory light chain phosphorylation at anaphase onset. The level of phosphorylation remained elevated throughout telophase and into cytokinesis near the equatorial region, while regulatory light chain phosphorylation decreased at the polar regions, creating a bipolar dividing cell (DeBiasio *et al.*, 1996). At the same time, myosin II-based fibres shortened in the cleavage furrow. Elevated levels of phosphorylated myosin II correlated well with elevated concentrations of myosin II in the cleavage furrow (DeBiasio *et al.*, 1996). These results suggest a contractile force at the cell equator may be generated, at least in part, through activation of myosin II (via phosphorylation of serine 19 of the regulatory light chain) at the cleavage furrow in conjunction with deactivation of myosin II at the polar regions.

Future myosin II biosensors will combine both the genetic and the energy transfer approaches. Genetic engineering will be used to move labelling sites to maximize detection of a conformational change. We have demonstrated the feasibility of these approaches and have provided previously unobtainable information about the state of myosin II regulatory light chain phosphorylation in individual, living motile cells.

41.5 FUTURE STUDIES

Recent developments in design of protein-based indicators hold promise for the observation of a wide range of protein activities *in vivo*, including protein conformational changes, post-translational modification, and ligand binding. Proteins have evolved to recognize physiologically relevant molecules with great specificity, and can be modified to become fluorescent indicators of small molecule concentration. Production of protein-based indicators is currently being pursued through two routes, using environmentally sensitive dyes or fluorescence energy transfer. These approaches have complementary strengths and shortcomings which will determine their applicability in particular situations. Genetic engineering of analogues will be a key tool, enabling precise positioning of attached fluorophores.

The ability to use a single environmentally sensitive fluorophore, rather than the two dyes required for energy transfer, will have obvious advantages in some situations. Maintenance of biological activity in a small protein (such as calmodulin) will be much more difficult after attachment of two dyes. Site-specific attachment of even a single dye is difficult. Intact biological activity will be important in indicators where biological regulation of ligand affinity is being studied, as in indicators revealing the regulation of signalling proteins.

When a protein is modified to become an indicator of small molecule concentration, intact biological activity may, in fact, be undesirable. The indicator should strictly reflect the concentration and distribution of the target ligand. The indicator's distribution and target affinity should not be affected by binding of other ligands, post-translational modification, or other regulatory effects. In such cases, energy transfer would be more applicable because disruption of certain protein activities will be desirable. The primary obstacle to designing useful indicators of small signalling molecules is the unavoidable buffering of these molecules by the indicator itself. Buffering must be minimized by loading very small quantities of indicator. This will require development of bright fluorescent proteins which show strong fluorescence changes when binding their targets. Access to practical small molecule indicators may ultimately depend on the development of very bright dyes suitable for use in living cells.

Indicators of protein–protein binding may be more accessible using currently available methodology. The high concentrations of some intracellular proteins permit loading of relatively large indicator concentrations. Indicators of protein-protein binding could be made by modifying one of the two interacting proteins. Either solvent-sensitive dyes or energy transfer pairs could be used to indicate a protein conformational change induced specifically by the targeted ligand. It may be possible to make indicators with great specificity for a single protein-protein interaction by placing a different dye on each of the interacting proteins, and monitoring energy transfer as an indicator of protein binding. This approach will likely require covalent attachment of the two proteins via a flexible tether. Without such a tether, the labelled proteins could bind to unlabelled, endogenous proteins rather than to each other, necessitating the loading of unacceptably large indicator concentrations to produce detectable energy transfer.

Both energy transfer dyes and environmentally sensitive fluorophores may potentially interact with cellular components to produce artifactual fluorescence changes. Solvent-sensitive fluorophores can interact with lipids and hydrophobic proteins, and energy transfer will be sensitive to microviscosity and factors influencing dye orientation. Environmentally sensitive dyes can be designed with a strong affinity for the tagged protein to avoid artifactual binding to intracellular components. Meaningful interpretation of indicator data will require careful standardization of fluorescence response, both *in vitro* and when possible *in vivo*. Fluorescent calcium indicators have been calibrated within living cells using calcium ionophores and external calcium buffers to control intracellular calcium (Spurgeon *et al.*, 1990).

Intracellular indicators of protein activity promise to yield otherwise inaccessible information about transient protein activity and its regulation in individual, living cells. The cell-to-cell heterogeneity observed in calcium signalling studies suggests that observation of single cells will be critical in untangling the complex interactions of signalling proteins (Byron & Villereal, 1989; Tucker & Fay, 1990). In this and other areas, fluorescent protein indicators should elucidate protein function *in vivo*, thus enabling integration of biochemical data into models of global function. The regulation of many proteins has to date been approachable only through observation of interactions with isolated cellular components. It may soon be possible to observe proteins regulated by the full complement of cellular controls, including those not yet characterized or even postulated.

The most exciting experiments will combine multiple reagents in the same cell or tissue to permit correlations between various parameters in time and space. Ultimately, we would like to monitor changes in free calcium, calcium binding to calmodulin, phosphorylation of myosin II regulatory light chains and myosin II dynamics in the same experiment. Therefore, the optimal combination of reagents and instrumentation continues to be a great opportunity and challenge.

REFERENCES

Adams S.R., Harootunian A.T., Buechler Y.J., Taylor S.S. & Tsien R.Y. (1991) *Nature* **349**, 694–697.

Andreason T.J., Luetje C.W., Heideman W. & Storm D.R. (1983) *Biochemistry* **22**, 4615–4618.

Bennett J.P., Cross R.A., Kendrick-Jones J. & Weeds A.G. (1988) *J. Cell Biol.* **107**, 2623–2629.

Bright G.R., Fisher G.W., Rogowska J. & Taylor D.L. (1989) *Methods Cell Biol.* **30**, 157–190.

Brundage R.A., Fogarty K.E., Tuft R.A. & Fay F.S. (1991) *Science* **254**, 703–706.

Byron K.L. & Villereal M.L. (1989) *J. Biol. Chem.* **264**, 18234–18239.

Chandra T.S., Nath N., Suzuki H. & Seidel J.C. (1985) *J. Biol. Chem.* **260**, 202–207.

Cohen P. & Klee C. (eds) (1988) *Calmodulin*. Elsevier, New York.

Comte M., Maulet Y. & Cox J. (1983) *Biochem. J* **209**, 269–272.

Cox J.A., Comte M., Fitton J.E. & DeGrado W.F. (1985) *J. Biol. Chem.* **260**, 2527–2534.

DeBiasio R.L., Wang L., Fisher G.W. & Taylor D.L. (1988) *J. Cell Biol.* **107**, 2631–2645.

DeBiasio R.L., LaRocca G.M., Post P.L. & Taylor, D.L. (1996) *Mol. Biol. Cell* **7**, 1259–1282.

Fisher G.W., Conrad P.A., DeBiasio R.L. & Taylor D.L. (1988) *Cell Motil. Cytoskeleton* **11**, 235–247.

Guilano K.A., Kolega J., DeBiasio R.L. & Taylor D.L. (1992) *Molec. Biol. Cell* **3**, 1037–1048.

Giuliano K.A. & Taylor D.L. (1990) *Cell Motil. Cytoskeleton* **16**, 14–21.

Giuliano K.A. & Taylor D.L. (1994) *J. Cell Biol.* **124**, 971–983.

Giuliano K.A., Nederlof M.A., DeBiasio R., Lanni F., Waggoner A.S. & Taylor D.L. (1990) In *Optical Microscopy for Biology* (Herman B. & Jacobson K. eds). Wiley-Liss, New York, pp. 537–543.

Gorbsky G.J., Sammak P.J. & Borisy G.G. (1987) *J. Cell Biol.* **104**, 9–18.

Gough A.H. & Taylor D.L. (1993) *J. Cell Biol.* **121**, 1095–1107.

Hahn K.M., Waggoner A.W. & Taylor D.L. (1990) *J. Biol. Chem.* **265**, 20335–20345.

Hahn K., DeBiasio R.L. & Taylor, D.L. (1992) *Nature* **359**, 736–738.

Iwane A.H., Funatsu T., Harada Y., Tokunaga M., Ohara O., Morimoto S. & Yanagida T. (1997) *FEBS Lett* **407**, 235–238.

Janson L.W. & Taylor D.L. (1993) *J. Cell Biol.* **123**, 345–356.

Katoh T., & Lowey S. (1989) *J. Cell Biol.* **109**, 1549–1560.

Klee C.B. (1988) In *Calmodulin*, P. Cohen & C. Klee (eds). Elsevier, New York, pp. 35–53.

Kolega J. (1997) *Exp. Cell Res.* **231**, 66–82.

Kolega J. (1988) *J. Cell. Biochem.* **68**, 389–401.

Kolega J. & Taylor D.L. (1991) *Curr. Topics Membr.* **38**, 187–206.

Kolega J. & Taylor D.L. (1993) *Molec. Biol. Cell* **4**, 819–836.

Kolega J., Janson L.W. & Taylor D.L. (1991) *J. Cell Biol.* **114**, 993–1003.

Kolega J., Nederlof M.A. & Taylor D.L. (1993) *Bioimaging* **1**, 136–150.

Kreis, T.E. & Birchmeier W. (1982) *Int. Rev. Cytol.* **75**, 209–227.

Lamb, N.J.C., Fernandez A., Conti M.A., Adelstein R., Glass D.B., Welch W.J. & Feramisco J.R. (1988) *J. Biol. Chem.* **106**, 1955–1971.

LaPorte D.C., Wierman B.M. & Storm D.R. (1980) *Biochemistry* **19**, 3814–3819.

LaPorte D.C., Keller C.H., Olwin B.B. & Storm D.R. (1981) *Biochemistry* **20**, 3965–3972.

Luby-Phelps K., Lanni F., Taylor D.L. (1985) *J. Cell Biol.* **101**, 1245–1256.

McKenna N.M., Wang Y.-l. & Konkel M.E. (1989) *J. Cell Biol.* **109**, 1163–1172.

McNeil P.L. & Taylor D.L. (1987) *Cell Membranes*, vol. 3 (Elson E., Frazier W. & Glaser L., eds) Plenum Publishing, New York, 365–405.

McNeil P.L., McKenna M.P. & Taylor D.L. (1985) *J. Cell Biol.* **101**, 372–379.

Malencik D.A., Anderson S.R., Shalitin Y. & Schimerlik M.I. (1981) *Biochem. Biophys. Res. Commun.* **101**, 390–395.

Manalan A.S. & Klee C.B. (1984) *Adv. Cyclic Nucleotide Protein Phosphorylation Res.* **18**, 227–277.

Matsumura F., Ono S., Yamakita Y., Totsukawa G. & Yamashiro S. (1998) *J. Cell Biol.* **140**, 119–129.

Maulet Y. & Cox J.A. (1983) *Biochemistry* **22**, 5680–5686.

Mills J.S., Walsh M.P., Nemcek K. & Johnson J.D. (1988) *Biochemistry* **27**, 991–996.

Mitchison T.J. (1989) *J. Cell Biol.* **109**, 637–652.

Mittal B., Sanger J.M. & Sanger J.W. (1987) *J. Cell Biol.* **105**, 1753–1760.

Moores S.L., Sabry J.H. & Spudich J.A. (1996) *Proc. Natl Acad. Sci. USA* **93**, 443–446.

Mooseker M.S. & Cheney R.E. (1995) *Annu. Rev. Cell Devel. Biol,* **11**, 633–675.

Moussavi R.S., Kelley C.A. & Adelstein R.S. (1993) *Mol. Cell. Biochem.* **127**, 219–227.

Nath N., Nag S. & Seidel J.C. (1986) *Biochemistry* **25**, 6169–6176.
Pardue R.L., Kaetzel M.A., Hahn S.H., Brinkley B.R. & Dedman J.R. (1981) *Cell* **23**, 533–542.
Perrie W.T. & Perry S.V. (1970) *Biochem. J.* **119**, 31–38.
Post P.L., Trybus K.M. & Taylor D.L. (1994) *J. Biol. Chem.* **269**, 12 880–12 887.
Post P.L., DeBiasio R.L. & Taylor D.L. (1995) *Mol. Biol. Cell* **6**, 1755–1768.
Salmon E.D., Leslie R.J., Saxton W.M., Karow M.L. & McIntosh J.R. (1984) *J. Cell Biol.*, **99**, 2165–2174.
Sanger J.M., Mittal B., Dome J.S. & Sanger J.W. (1989) *Cell Motility Cytoskeleton* **14**, 201–219.
Seeholzer S.H., Cohn M., Putkey J.A., Means A.R. & Crespi H.L. (1986) *Proc. Natl Acad. Sci. USA* **83**, 3634–3683.
Sellers J.R. & Adelstein R.S. (1987) In *The Enzymes* (Boyer P.D. ed.). Academic Press, Orlando, pp. 381–418.
Sellers J.R. & Goodson H.V. (1995) *Prot. Profile* **2**, 1323–1423.
Sellers J.R., Pato M.D. & Adlestein R.S. (1981) *J. Biol. Chem.* **256**, 13137–13142.
Sellers J.R., Goodson H.V. & Wang F. (1996) *J. Muscle Res. Cell Motility* **17**, 7–22.
Simon J.R. & Taylor D.L. (1986) *Methods Enzymol.* **134**, 487–507.
Spurgeon H.A., Stern M.D., Baartz G., Raffaeli S., Hausford R.G., Talo A., Lakatta E.G. & Capogrossi M.C. (1990) *Am. J. Physiol.* **258**, H574–H586.
Takuwa N., Iwamoto A., Kumada M., Yamashita K. & Takuwa Y. (1991) *J. Biol. Chem.* **266**, 1403–1409.
Tanasugarn L., McNeil P., Reynolds G.T. & Taylor D.L. (1984) *J. Cell Biol.* **98**, 717–724.
Taylor D.L. & Fechheimer M. (1982) *Phil. Trans. Roy. Soc. Lond. B* **299**, 185–197.
Taylor D.L. & Wang Y. (1978) *Proc. Natl Acad. Sci. USA* **75**, 857–861.
Taylor D.L. & Wang Y. (1980) *Nature* **284**, 405–410.
Taylor D.L., Reidler J., Spudich J.A. & Stryer L. (1981) *J. Cell Biol.* **89**, 362–367.
Taylor D.L., Amato P.A., Luby-Phelps K. & McNeil P.L. (1984) *Trends Biochem. Sci.* **9**, 88–91.
Theriot J.A. & Mitchison T.J. (1991) *Nature* **352**, 126–131.
Trybus K.M., Huiatt T.W. & Lowey S. (1982) *Proc. Natl Acad. Sci. USA* **79**, 6151–6155.
Tsien R.Y. (1989) *Methods Cell Biol.* **30**, 127–153.
Tucker R.W. & Fay F.S. (1990) *Eur. J. Cell Biol.* **51**, 120–127.
Vekhovsky A.B., Svitkina T.M. & Borisy G.G. (1995) *J. Cell Biol.* **131**, 989–1002.
Wadsworth P. & Salmon E.D. (1986) *J. Cell Biol.* **102**, 1032–1038.
Wang Y. (1989) *Methods Cell Biol.* **29**, 1–12.
Wang Y. & Taylor D.L. (1980) *J. Histochem. Cytochem.* **28**, 1198–1206.
Wang Y. Taylor D.L. (1981) *Cell* **27**, 429–436.
Wang Y., Heiple J.M. & Taylor D.L. (1982) *Methods Cell Biol.* **24B**, 1–11.
Ware B., Brvenik L.J., Cummings R.T., Furukawa R.H. & Krafft G.A. (1986) In *Applications of Fluorescence in the Biomedical Sciences*, D.L. Taylor, A.S. Waggoner, R.F. Murphy, F. Lanni & B.R. Birge (eds). Alan R. Liss, New York, pp. 141–157.
Zavortnik M., Welsh M.J. & McIntosh J.R. (1983) *Exp. Cell Res.* **149**, 375–385.

PART XI

Applications of Optical Probe Imaging to Biological Problems

CHAPTER FORTY-TWO

Fluorescence and Luminescence Techniques to Probe Ion Activities in Living Plant Cells

M.D. FRICKER[1], C. PLIETH[1], H. KNIGHT[1], E. BLANCAFLOR[2], M.R. KNIGHT[1], N.S. WHITE[1] & S. GILROY[2]

[1] Department of Plant Sciences, University of Oxford, Oxford, UK
[2] Biology Department, Pennsylvania State University, University Park, PA, USA

42.1 INTRODUCTION

Fluorescent probes offer almost unparalleled opportunities to visualize and quantify dynamic events within living cells, tissues or even organs with a minimum of perturbation. Although single cells or monolayers maintained in culture can be readily imaged, measurements are often needed from cells within intact tissues. These cells are operating in their correct physiological context of cell–wall and cell–cell interactions. Imaging technology and dye-loading approaches have now progressed to where such *in planta* experiments are a real possibility. A range of different measurement techniques is now available to quantify fluorescence signals from reporter molecules within biological specimens. These include fluorometry, flow cytometry, microscope photometry, camera, confocal and two-photon imaging. In addition, luminescence reporters, such as aequorin, are available to measure calcium using photomultiplier-based luminometer systems or photon-counting camera systems. Each of these approaches performs well for a specific range of measurement conditions and specimens. Thus, several techniques may need to be applied to provide a sufficiently flexible balance between the spatial, temporal and spectral resolution required to understand the physiological questions in the cell(s) of interest.

Although the imaging and detector technology has rapidly advanced, two real problems remain. First, the techniques to introduce ion-selective probes into plant cells are not straightforward. Second, many of the reporter dyes do not behave predictably in the plant cell cytoplasm. Transgenic approaches using aequorin measurements have circumvented many of the problems associated with loading Ca^{2+} indicators. More recently, transgenic fluorescent calcium indicators have become available that exploit fluorescence resonance energy transfer (FRET) between different spectral forms of 'green' fluorescent protein (Miyawaki *et al.*, 1997). These probes offer great potential to combine the advantages of protein engineering and fluorescence techniques. Currently, however, the field of fluorescence and luminescence imaging of plant cell activities remains one fraught with potential artefacts and few generalized protocols. Most studies have required an extensive period of trial and error to determine what will, and will not, work with a particular

FLUORESCENT AND LUMINESCENT PROBES, 2ND EDN
ISBN 0–12–447836–0

cell type. Despite these limitations, perseverance has been rewarded by fascinating glimpses into cellular regulation. Coupling these imaging approaches to techniques such as caged probe technology has begun to allow us both to observe and to manipulate these cellular processes in the ultimate test-tube setting, the cell itself. The following sections provide guidelines on how to apply these approaches to plant specimens and how to identify, and hopefully avoid, many of the unique problems associated with quantitative fluorescence imaging and manipulation of plant cells.

42.2 TISSUE PREPARATION, MOUNTING AND PERFUSION

The experimental systems used by plant biologists are diverse and often unique, and so the direct application of technology developed for preparing animal specimens for ion imaging has not been straightforward. Optical techniques place certain constraints on the type of tissue that can be examined. Some of the best specimens for microscopy are flat and submerged in water. Thus unicellular, filamentous or relatively flat organisms with restricted 3-D growth are readily observed, particularly if they are aquatic. Fluorescence measurements can also be made on cells one to three cell layers deep within intact tissues using confocal optical sectioning and probably twice as deep using two-photon imaging. To expose the cells of interest even further into more bulky higher plant tissue may involve peeling, excision, dissection or the formation of protoplasts. All of these approaches will cause a certain amount of 'wound'-induced artefacts that must be carefully tested for. An alternative for Ca^{2+} measurements is to monitor the luminescence from the Ca^{2+}-dependent photoprotein aequorin. Plants can be transformed with the apoaequorin gene and active aequorin reconstituted *in situ*. Thus, the Ca^{2+}-dependent aequorin signal can be imaged from entire transgenic plants (although most of the signal may well be derived from the surface layers).

The extent of the trauma induced during sample preparation is difficult to define. Plant cells normally alter their metabolic poise continuously in response to changes in their surroundings. For example, the mere act of touching the plant when mounting it in a sample chamber may elicit Ca^{2+}-signals in the stimulated cells (e.g. Knight *et al.*, 1991; Legue *et al.*, 1997) and alter gene expression patterns (e.g. Braam 1992). Depending on the system, the extent of physiological response caused by sample preparation should be assessed from measurement of parameters such as membrane potential, cytoplasmic streaming, progression through division, growth rate or gene expression. The contribution of such perturbation to subsequent measurement may be limited by minimizing the extent of the manipulation and allowing an adequate recovery time (generally 15–60 min) after mounting the sample.

Microscope-based measurements are generally preferred over cuvette-based measurements as they provide additional spatial information. This is particularly important if there are localized responses, gradients or substantial heterogeneity in the population of cells examined. In addition, continuous observation can also provide critical information on the state of the tissue and is essential in many cases to monitor responses such as stomatal closure or cytoplasmic streaming. Observation of the tissue at the cell and subcellular level typically requires magnification 100- to 600-fold. The best clarity or brightness of image is achieved with high numerical aperture (NA) microscope objectives and immersion of the specimen. To maintain viability the immersion medium is usually water. Submersion reduces light scattering from highly reflective surfaces in the sample and may assist in efficient dye loading and application of many stimuli and calibration solutions.

The composition of the bathing medium should ideally mimic the environment around the cells *in vivo*, particularly with respect to ionic composition, water potential and gaseous environment. Continuous perfusion of this medium is preferred to simple immersion as it minimizes boundary layers, reduces anoxic stress and allows rapid solution changes. For aquatic organisms, cultured cells or tissues grown in agarose, continuation of the growth conditions is usually sufficient, though the concentration of certain ions may need careful consideration. For example, Mn^{2+} can enter through Ca^{2+}-channels and quench fluorescence from intracellular Ca^{2+}-sensitive dyes and may need to be omitted from the perfusion medium. Other heavy metals may also interfere with the behaviour of both dyes and luminescent proteins.

Perfusion conditions are more difficult to define for aerial tissues as the turgor relations, apoplastic activities for ions such as H^+, Ca^{2+} and K^+, and the prevailing hormonal status are usually unknown. In addition, the waxy cuticle of aerial tissues may significantly reduce dye uptake. The consequences of immersion of aerial tissues include: (1) build-up of regions depleted in O_2 and enriched in CO_2 next to non-photosynthetic cells and vice versa for actively photosynthesizing tissue; (2) dilution of apoplastic ions and equilibration of local ion gradients; and (3) increased turgor. There are only a limited number of reports examining the significance of these changes (e.g. Mühling *et al.*, 1995).

The temperature is normally set near ambient (20–25°C) through temperature control of the perfusion medium, without additional recourse to a temperature-regulated stage. In a non-perfused system, the sample temperature is likely to increase due to prolonged illumination and this will lead to temperature related artefacts. Good IR-blocking filters are particularly important when a xenon light source is to be used. In addition, the composition of the air above open perfusion systems may need to be regulated and screened to prevent interference from CO_2 exhaled by the microscope operator. It can also not be stressed enough that when designing and using perfusion systems extreme care must be taken to ensure that liquid does not leak into the microscope, where it may permanently damage delicate optical coatings and lens elements.

42.3 SECURING THE SPECIMEN FOR MICROSCOPY

The specimen needs to be securely fixed down to prevent movement during microscope observation, perfusion and especially if microinjection is to be attempted. Various approaches may be appropriate depending on whether an upright or inverted microscope is used and the nature of the tissue, including:

(1) Using silicone grease (e.g. vacuum grease, M494, ICI) as an adhesive leaving a clear window in the grease for observation and measurement.
(2) Silicone contact adhesive (e.g. Corning 355: Blatt, 1991).
(3) Protoplasts, tissue culture cells and pollen tubes often adhere to clean coverslips or more strongly after coating the coverslip with poly-L (or D)-lysine. Poly-lysine can be applied at 0.01–0.1% (w/v) in 10 mM Tris-HCl, pH 8.0 for 5–60 min, followed by washing.
(4) Suction pipettes with 10–20 μm diameter tips can be used to hold protoplasts or single cells. These may be filled with inert solutions (e.g. silicone fluid Dow Corning 200/100 CS).
(5) Immobilization in agarose or Phytagel™ (Gellan gum agar substitute, Sigma, Dorset): Isolated cells and protoplasts can be stabilized on coverslips with a thin film of 1–2% (w/v) low melting point agarose (Hillmer *et al.*, 1993; Parton *et al.*, 1997). A typical protocol for *Arabidopsis* roots is as follows: roots can be grown in place by planting seeds in nutrient media (see Legue *et al.*, 1997 for composition of nutrient medium) with 0.5–1% Phytagel™. The media is autoclaved for 25 min and poured to 1 mm depth onto coverslips contained in sterile Petri dishes. The gel should polymerize at room temperature within 10 min. Seeds are planted by pushing through the gel onto the surface of the coverslip. This ensures that the root grows along the surface of the coverslip and therefore can be imaged. The same planting scheme can also be used for both low pH and AM-ester loading of intact *Arabidopsis* roots (Legue *et al.*, 1997; Wymer *et al.*, 1997).
(6) Immobilization in gelatin: Concentrations of gelatin up to 18% (w/v) made up in nutrient media are useful for embedding *Schizosaccharomyces pombe* or *Saccharomyces cerevisiae*. Samples are mixed with molten gelatin warmed to 40°C and spread thinly on a prewarmed coverslip.
(7) Immobilization in alginate: Alginic acid at concentrations ~1.5% forms a gel at room temperature in the presence of excess (mM) $CaCl_2$ that can be used to trap protoplasts or cells. However, the high Ca^{2+} concentrations may perturb the cell physiology.
(8) Mechanical clips or restraints can be used to secure large cells or tissues (e.g. Plieth, 1995).

It is critical to ensure that the immobilization protocol does not markedly affect cell or tissue responses. For example, poly-lysine may induce K^+-channel activity (Reuveni *et al.*, 1985) and hot agarose or gelatin may heat shock protoplasts before it cools enough to form a gel (S. Gilroy, unpublished). All embedding and immobilization procedures are likely to reduce diffusion to and from the tissue as well as rapid exchange of media, although this is generally less of a problem with mechanical restraints.

Microinjection requires an open perfusion system, although cells can be loaded prior to observation and mounted in a closed perfusion system. Inverted microscopes are an ideal platform for microinjection of thin, relatively transparent tissue. Upright systems with objective rather than stage focusing offer significant advantages for thick, opaque specimens such as roots or leaves, as observation must take place from the same side as loading. The arrival of long working distance, water immersion lenses have reduced the steric problems of microinjection on upright systems, except when a steep angle of penetration is needed (around 45° is currently possible).

42.4 SELECTION AND USE OF FLUORESCENT PROBES

The basic principle of intracellular ion measurement using non-protein fluorescent probes involves introduction of a chelating agent whose fluorescent properties alter with the activity of a particular ion. The change in fluorescence is measured and then converted to an estimate of the change in ion level using some form of calibration. The change may be a simple quantitative alteration in fluorescence intensity (single-wavelength dye) or a shift in either the excitation or emission spectrum (ratiometric dye). A wide range of dyes is now available and their properties are described elsewhere in this book. Applications of many of these dyes in plant cells have been reviewed recently (Gilroy, 1997). Dyes with spectral shifts are preferable as they permit ratio measurements that distinguish fluorescence changes due to ion binding from those due to dye leakage, bleaching or uneven distribution (Grynkiewicz *et al.* 1985).

There are several features that influence selection of a particular dye for measurements in plant cells including:

(1) The ion to be measured. Dyes to monitor Ca^{2+}, H^+, K^+ and Na^+ have all been used successfully in plant cells, but many more are available.
(2) The possibility of interference by changes in other ions in the cell of interest.
(3) The K_d of the dye, i.e. how close is the dissociation constant of the dye to the ion level in the cell compartment to be monitored.
(4) The dynamic range of the dye response.
(5) The ease of loading the dye into a defined compartment, usually the cytosol.
(6) The behaviour of the dye within the cell, including compartmentalization, metabolism and physiological perturbation.
(7) The excitation/emission wavelengths in relation to the spectral sensitivity of the tissue (i.e. will illumination of the dye also trigger a phytochrome or blue light response?).
(8) The level of autofluorescence of the tissue at the measurement wavelengths.
(9) The instrument configuration for dual-excitation or dual-emission dyes.
(10) Compatibility with other optical techniques, such as UV photolysis of caged compounds.

42.4.1 Loading strategies for plant cells

The objective is to introduce the probe into a defined compartment, usually the cytoplasm, of as many cells as possible in sufficient concentration to give good signal-to-noise without causing toxic effects or significantly disturbing the cellular ion buffering. In the latter case it is important to consider not just the magnitude of the increase in buffering capacity, but also whether the increase in the buffer mobility of the dye will disrupt local gradients in ion activity (Neher & Augustine, 1992). Loading the dye into the cytoplasm of plant cells has proved difficult (Cork, 1986; Bush & Jones, 1990; Callaham & Hepler, 1991; Read *et al.*, 1992; Fricker *et al.*, 1994; Gilroy, 1997). A variety of strategies have emerged, but there are no simple rules as to which will be most effective with a particular tissue. There are essentially nine approaches. The least invasive techniques that load a population of cells are preferred but unfortunately do not work with many plant cells.

42.4.1.1 Permeant dyes

The most extensively used permeant dye for Ca^{2+} measurements is chlorotetracycline (also known as aureomycin, chlortetracycline or CTC). CTC is a fluorescent, lipophilic antibiotic isolated from *Streptomyces aureofaciens* that shows enhanced fluorescence upon binding divalent and trivalent cations (excitation 400 nm, emission 530 nm). CTC is readily membrane permeant and cells are simply loaded by incubation with 10–200 μM dye in an acidic environment (pH 4.5 to pH 6.5). CTC has provided a useful first step in identifying cells where changes in 'membrane-associated' Ca^{2+} levels occur (e.g. Caswell, 1979; Reiss & Herth, 1978; Tretyn & Kopcewicz, 1988; Timmers *et al.*, 1989). In most cases the results obtained with CTC have been reproduced with other, more sophisticated methods (see for example Reiss & Nobiling, 1986; Rathore *et al.*, 1991; Timmers *et al.*, 1991, 1996). However, CTC possesses some properties which render its use difficult and which often cause the results of Ca^{2+} measurements being questioned:

- CTC is sensitive to Mg^{2+} ions (Gupta & Berkowitz, 1989).
- CTC fluorescence increases with increasing pH: a fivefold change occurs between pH 5.5 and pH 8.5.
- CTC is a lipophilic dye which predominantly binds to membranes (Schneider *et al.*, 1983). Thus, the Ca^{2+} level monitored by the CTC fluorescence is mainly due to the Ca^{2+} ion concentration near membrane surfaces (e.g. Meindl, 1982; Polito, 1983).
- The CTC K_d for Ca^{2+} depends strongly on the polarity of the environment. For instance, it changes from 440 μM (in pure water) to 9 μM (in 70% methanol) (Caswell & Hutchison, 1971). Thus, as the polarity in living biological systems is unknown, the fluorescence of CTC cannot be calibrated in terms of absolute Ca^{2+} concentration.
- The temperature dependence of fluorescence emissions between 4°C and 35°C is linear. $(\delta F/F)/\delta T = 2.1\%\ °C^{-1}$).
- CTC is an antibiotic which exerts toxic effects especially on prokaryotes (Caswell & Hutchison, 1971). There are also some toxic effects on eukaryotes (e.g. Foissner, 1991).

- CTC stimulates an increased flux of Ca^{2+} into the cell.
- CTC increases the sensitivity of cells to damage. For example, during impalement by microelectrodes a progressive depolarization of about 1 mV min^{-1} was observed which sometimes led to death of the cell (Plieth, 1995).

42.4.1.2 Ester loading

Most of the ion-selective dyes are impermeant due to one or more carboxyl groups in the molecule which are charged in the physiological pH range. Esterification of the carboxyl groups in the molecule with acetate or acetoxymethyl (AM) groups masks their charge and renders the dye membrane permeant. Hydrolysis by intracellular esterases releases the free dye in an active form in the cytoplasm. However, there are numerous problems with applying the ester loading technique to plant cells (see notes).

Protocol

(1) For AM loading, prepare a 1 mM stock solution of the AM-ester indicator dye with DMSO as the solvent (AM-ester forms of indicator dyes specially packaged in dry DMSO are also available from Molecular Probes). Store aliquots of dye at −20°C and avoid exposure to light.
(2) Dilute the indicator stock solution to 1–5 μM with deionized water or media prior to use and incubate cells for 10–120 min.
(3) Wash out excess dye and observe cells under an epifluorescence microscope.

Notes

- Acetoxymethyl ester loading works well with some plant cells (Gehring *et al.*, 1990a,b; Dixon *et al.*, 1989; Williams *et al.*, 1990; Fricker *et al.*, 1997b; Parton *et al.*, 1997), but many mature tissues appear to load poorly (Gilroy *et al.*, 1991; Hodick *et al.*, 1991). For example, fluorescence of BCECF is visible in intact *Arabidopsis* roots within 2–3 min of application in the AM-ester form. However, the AM-ester forms of indo-1 and Calcium Green-1 do not appear to work, possibly because of the larger number of ester groups (*c.* 5) that are clustered together and require cleavage.
- Problems may arise if no hydrolysis or incomplete hydrolysis releases only partially activated fluorescent intermediates with different spectral properties (Highsmith *et al.*, 1986; Scanlon *et al.*, 1987; Elliot & Petkoff, 1990).
- The hydrolysis of AM-esters generates acetic acid and formaldehyde which may exert negative effects on the cells. The DMSO in the stock solution may also cause toxic effects.
- Mild detergents (e.g. 0.02–0.2% Pluronic F-127: Gehring *et al.*, 1990b), temperature, ATP permeabilization, increased external pH (Elliot & Petkoff, 1990) and varying ionic conditions may facilitate AM-ester loading.
- The esterified dye can permeate all cell compartments and the free dye may accumulate in any organelle which has esterases (e.g. mitochondria: Almers & Neher, 1985; Cobbold & Rink, 1987; Roe *et al.*, 1990; Rathore *et al.*, 1991; Brauer *et al.*, 1996). Esterase inhibitors, such as eserine, may also help prevent external hydrolysis of the dye and improve loading (Tretyn *et al.*, 1997).
- After release in the cytoplasm dyes may be sequestered to the vacuole rapidly (sometimes within 5–10 min of application) and hence cannot be used for cytoplasmic measurements (although see Kosegarten *et al.*, 1997). Instead, BCECF loaded as the AM-ester has been used successfully for ratiometric determination of vacuolar pH changes in protoplasts (Swanson & Jones, 1996) and intact roots (Brauer *et al.*, 1995). In these experiments, the dye is used well below its pK_a in a region of the ratio titration curve that is normally relatively insensitive to pH.

42.4.1.3 Low pH loading

Reversible protonation of the carboxyl groups at low external pH can be used to mask their charge and hence allow the dye to cross the plasma membrane. The protons dissociate at the higher pH of the cytoplasm and the dye is effectively trapped in the cell in its anionic (ion sensitive) form.

Protocol

(1) Make a 1 mM stock solution of the salt form of the dye using deionized water. Aliquot into small volumes. Store at −20°C and avoid exposure to light.
(2) Dilute the stock solution prior to use in 25 mM dimethylglutaric acid (DMGA), pH 4.5 to a final dye concentration of 20 to 50 μM.
(3) For plant cells embedded in a gel, hydrate the gel matrix by adding excess nutrient media for 15 min prior to adding the dye.
(4) Add the dye and incubate plants in the dark for 1 to 2 h.
(5) Wash out unloaded dye by rinsing in nutrient media.

Notes

- In some intact cells the dye appears to stick in the wall, either through co-ordination with other charged groups in the apoplast or possibly through precipitation in localized regions of low pH (Gilroy *et al.*, 1991). This is not always the case (Wymer *et al.*, 1997; Legue *et al.*, 1997) but it is not clear why tissues differ. Charge masking with high levels of other ions might reduce this problem.

- Increased temperature and the prescence of saponin may assist dye loading in some tissues (Tretyn *et al.*, 1997).
- Washing out the unloaded dye is difficult for plant roots supported in gels and may result in high background fluorescence. Carefully removing roots from the gel matrix (Cramer & Jones, 1996) or allowing the roots to grow into a gel-free zone prior to incubation in the dye reduces this problem.
- Low pH loading has been successful with some protoplast types (Bush & Jones, 1987), but not others (Gilroy *et al.*, 1986).
- Certain cells do not survive low pH treatment (Elliot & Petkoff, 1990; Hodick *et al.*, 1991) and the physiological consequences of pH stress on pH regulation and signalling need careful scrutiny.
- If dye penetration appears to be a problem, access of the dye to the tissue can be facilitated by a cutinase pretreatment. Tissues are incubated with purified cutinase activities ranging from 0.1 to 10 μmol min^{-1} mg^{-1} protein for 5–30 min at pH 7.6 in 10 mM Tris-HCl and then washed in buffer (Fricker *et al.*, 1994). Cutinase can be prepared according to Kolattukudy *et al.* (1981) and Coleman *et al.* (1994). The cutinase gene has also been cloned and expressed (Soliday *et al.*, 1984).

42.4.1.4 Electroporation

Pores of variable size can be selectively induced in the plasma membrane of protoplasts by short, high-voltage pulses. Resealing is spontaneous, but can be slowed sufficiently at low temperature to allow diffusion of dye or other macromolecules into the cytoplasm. A cocktail of low-molecular-weight factors is normally included to replace cytoplasmic components diffusing out of the permeabilized protoplasts. The precise conditions for successful and reversible electro-permeabilization of the plasma membrane require careful optimization. Many cells do not survive and the remainder are loaded with variable concentrations of dye. A major part of any set of electroporation experiments is optimizing the electrical pulse protocol to maximize permeabilization whilst maintaining viability.

Protocol

(1) Sediment 1 ml of protoplasts at 1–5 ***g*** for 15 min, in a 1.5 ml Eppendorf tube. Remove supernatant and resuspend in 0.8 ml of electroporation buffer (200 mM sorbitol, 300 mM mannitol, 1 mM Mg^{2+}, 100 mM KCl, 10 mM HEPES, pH 7.2). Repeat washing step and resuspend protoplasts to a final concentration of 0.5×10^6 ml^{-1} in buffer pre-chilled to 4°C. Store on ice.

(2) Electroporate with systematically varied electroporation pulse characteristics: field strength (0.1–5 kV cm^{-1}); number of pulses applied (1–5) and capacitor used (1–50 μF). Assess permeabilization efficiency as the percentage of protoplasts staining with ethidium bromide (0.1% w/v, excitation 510 nm, emission 620 nm) within 5 min. Optimize for the lowest voltage, and briefest pulses giving 60–80% permeabilization (Gilroy *et al.*, 1986).

(3) Repeat steps 1–3 using the optimized pulse protocol but excluding the ethidium bromide staining. Incubate electroporated protoplasts on ice for 10 min and then add concentrated medium to restore the composition of buffer to the full strength of the normal protoplast culture medium (e.g. full strength MS medium). Cover cuvette with parafilm and mix by gently inverting the cuvette 5 times.

(4) Incubate for 15 min at room temperature and transfer protoplasts to culture chamber.

(5) After 1 h incubation test for viability with fluoresecein diacetate staining (FDA: Huang *et al.*, 1986). Viable protoplasts will become intensively fluorescent (excitation 480 nm, emission 530 nm) after 2–5 min incubation with 0.05% (w/v) FDA. Optimize the pulse protocol established in step 2 to yield >80% viability.

(6) Repeat steps 1–4 until percentage permeabilization and maintained viability are optimized.

(7) To load with dye repeat steps 1–4, electroporating using the optimized pulse protocol in step 2 in the presence of 20 μl of 1 mM dye.

Notes

- The method is not straightforward and is likely to perturb sensitive signalling systems.
- The electroporated protoplasts need to be carefully analysed with independent cellular assays of function to ensure the electroporation process has not altered cellular responses.
- The osmolarity of the electroporation buffer (mannitol and sorbitol concentrations) will need to be modified to be compatible with each protoplast system used.
- Ethidium bromide slowly penetrates intact cells, therefore it is important to run a non-electroporated control to assess how fast this occurs with each cell type.

42.4.1.5 Ionophoretic microinjection

The cell is impaled with a fine microelectrode containing charged dye and the dye is loaded into the cell by application of a small current or current pulses (see Blatt, 1991; Callaham & Hepler, 1991).

Protocol

(1) Microelectrodes are pulled from filamented borosilicate glass capillaries (e.g. GC150F or GC120F, Clark Electromedical Instruments, Reading, UK) to give a tip diameter of 0.1–0.3 μm and resistances in the range of 10–20 MOhms.
(2) Dilute the stock solution of the dye in free acid form to 0.1–1 mM with deionized water and fill the microelectrode with 2–5 μl using a 5 μl Hamilton syringe or a plastic pipette tip pulled to a long, fine taper. Backfill the rest of the pipette electrode with 1 M KCl.
(3) Connect the impalement electrode to a micro-iontophoresis current generator and place the reference electrode in the medium.
(4) Impale the cell by carefully advancing the microelectrode. A 4-axis manipulator is very useful to help penetrate plant cell walls.
(5) Loading can be achieved by diffusion from the microelectrode tip (*c.* 20–30 min). Iontophoretic injection with continuous current (e.g. 2 nA for 10–20 s) or pulsed current (3–4.5 nA in 2–3 s pulses) is much more rapid.
(6) Carefully withdraw the microelectrode over 3–10 min. In some cases, waiting for a few minutes before withdrawing the electrode can improve cell survival. Allow cells to recover for 30–60 min. Cell mortality most often occurs during pipette removal.

Notes

- Microinjection requires a lot of patience and the number of cells that can be successfully loaded is limited. Aspects such as the angle of impalement, micropipette shape and exactly where to impale the cell all need to be optimized for each cell type. In elongated plant cells, the cytoplasm occupies a very thin region between the vacuole and the cell wall. Between 60 and 80% of injections appear to load the cytoplasm rather than the vacuole and the difference in dye distribution between the two compartments can be readily observed.
- Filtering solutions through a 0.22 μm filter before use may help to reduce tip blockage.
- Care needs to be taken to minimize the disruption caused by passing the iontophoretic current into the cell. Currents of tens of nA into a small plant cell are often highly damaging.
- Mild plasmolysis prior to injection may help reduce blockage of the electrode as less cytoplasm is forced up the electrode by the reduced turgor pressure.
- The impalement pipette can also function as a microelectrode for measurement of the plasma membrane potential to follow the extent of cell disruption and recovery during the injection procedure (Van der Shoot & Lucas, 1995). Additional ions such as KCl or K-acetate may be needed to be added to the dye solution in the pipette to reduce the tip resistance and tip potentials for reliable membrane potential measurements.
- Dye can be loaded from multiple-barrelled electrodes (Blatt, 1991) with the circuit entirely between the electrodes, minimizing the disturbance to the membrane potential (Grabov & Blatt, 1997). This configuration also permits simultaneous voltage-clamping and fluorescence ion measurements.
- For elongating cells, continued growth and maintenance of turgor is one way to assess viability. Unfortunately more than 50% of microinjected cells show loss of turgor, granularity of the cytoplasm and subsequent cessation of growth (Legue *et al.*, 1997).
- It may be convenient to microinject several cells and then select the best cell for measurement after the end of the recovery period. However, this may not be suitable for growth-related studies on whole organ since impaling several cells may cause stress to the plant and affect interpretation of results.
- Iontophoresis is not usually effective for large or uncharged molecules. Electro-osmosis of dyes may also be possible (Erhardt *et al.*, 1996) where the flow of water caused by movement of ions in the iontophoretic current carries with it large and uncharged molecules into the cell.

42.4.1.6 Pressure microinjection

Pressure injection can be used for large or uncharged molecules. Notably, the problem of subsequent dye compartmentation into organelles can be avoided by pressure injection of dextran-conjugated form of the dyes. The cell is impaled with a wider microelectrode and the pipette solution loaded into the cell by application of hydraulic pressure.

Protocol

(1) Pull microelectrodes from borosilicate capillaries (e.g. GC 150F) to give a tip diameter of 0.3–1 μm and backfill with a dextran-conjugated form of the dye.
(2) Insert microelectrode into a holder and connect to a continuous pressure injection system. These systems apply pressure to the pipette using either air pressure (e.g. Plieth & Hansen, 1996; Legue *et al.*, 1997) or an oil-filled displacement system (Oparka *et al.*, 1991).
(3) Prior to penetration increase the holding pressure that is continuously applied to the pipette to 0.1–0.2 MPa to counteract cell turgor. The holding

pressure is usually unnecessary if injecting protoplasts.

(4) Impale the cell and load the dye using a series of 0.14–1.2 MPa pressure pulses or a gradual increase in pressure until the dye exits the pipette. The precise pulse protocol is highly dependent on the cell being injected.

(5) Slowly withdraw the microelectrode and allow the cells to recover for 30–60 min.

Notes

- Pressure injection is the easiest way to load giant cells like *Chara*, *Nitella* or *Eremosphaera* (Plieth & Hansen, 1996; Plieth *et al.*, 1997).
- Although a syringe can be used to apply the injection pressure, control over the duration and intensity of injection pressure pulses afforded by a commercial pressure injection regulator greatly improves the injection success rate.

42.4.1.7 *Loading from patch electrodes*

Fluorescent dyes, dextran-conjugated dyes and even proteins can be loaded into protoplasts in the whole cell patch-clamp configuration by diffusion of the contents of the patch pipette into the cell. The conditions for enzymatic protoplast isolation and successful gigaseal formation have to be determined for the protoplasts to be studied. The following is the protocol optimized for patch-clamp loading of the dye indo-1 into stomatal guard cell protoplasts from *Vicia faba* to monitor simultaneously Ca^{2+} levels and K^+ channel activities (L.A. Romano & S. Assmann, personal communication). Precise bath and pipette solutions will need to be optimized for the cells and channel activities to be studied.

Protocol

(1) The bath solution comprises 10 or 100 mM KCl, 1 mM $CaCl_2$, 1 mM $MgCl_2$, 10 mM MES, pH 5.6, osmolality adjusted to 460 mosmol kg^{-1} with sorbitol. The patch pipette is filled with 80 mM K^+-glutamate, 20 mM KCl, 2 mM Mg-ATP, 10 mM HEPES, pH 7.2, adjusted to 500 mosmol kg^{-1} with sorbitol and 60 μM indo-1 pentapotassium salt. Filter the pipette solution through 0.2 μm syringe filter prior to use and store on ice in the dark.

(2) The resistance of the patch pipettes should be 6–10 MΩ rather than the typical 15–20 MΩ of a normal patch experiment in these protoplasts to increase the rate of dye diffusion from the pipette. Stable dye loading should occur in 5–10 min and initial seal formation is reduced by only 20% under these conditions.

Notes

- The larger pipettes required than normally used for patch-clamping potentially means loss of cell constituents and channel run-down may be accelerated. However, using these conditions robust Ca^{2+} signals and channel activities last for more than 1 h in guard cell protoplasts.
- Care is needed to shield the patch-clamp amplifiers from electrical noise generated by the imaging or photometry system. Also filter wheels and shutters may generate vibrations that cause seals to be lost. This problem is somewhat alleviated by mounting the microscope and patch-clamp manipulators on an independent vibration isolation table and coupling the illumination system via fibreoptics.

42.4.1.8 *Laser ablation and loading via patch electrodes*

An alternative to enzymatic production of protoplasts for patch-clamping is to reveal a small region of naked plasma membrane by laser ablation of the overlying wall. This has the advantage of maintaining the cell in its almost intact setting yet allowing access to a localized region of the plasma membrane. Loading of dyes through the cell-attached patch pipette is identical to that outlined above. The following is a brief outline of the protocol to laser ablate and patch a stomatal guard cell of *Vicia faba*. Details of the laser ablation equipment are reviewed in detail elsewhere (De Boer *et al.*, 1994; Henriksen *et al.*, 1996; Henriksen & Assmann, 1997).

Protocol

(1) Epidermal peels are mounted in a perfusion chamber on an inverted microscope and perfused with 10 mM KCl, 1 mM $CaCl_2$, 1 mM $MgCl_2$, 10 mM MES, pH 5.6.

(2) The guard cells are plasmolysed in the same medium with osmolality adjusted to 460 mosmol kg^{-1} with sorbitol.

(3) The guard cell wall is ablated with 5–10 2 ns pulses of 337 nm light from the ablation laser.

(4) The preparation is de-plasmolysed using a slow decrease in osmolality provided by a linear gradient maker, allowing a small 'bleb' of exposed membrane to protrude through the hole in the wall, taking care not to rupture the membrane.

(5) The bleb is patch-clamped as normal.

Notes

Treating the wall with the UV absorbing dye Calcofluor White may improve absorption of the laser and improve ablation efficiency (Henriksen *et al.*, 1996). Blebs of *Vicia faba* guard cells are relatively easy to patch, but this may not be true for all systems (De Boer *et al.*, 1994).

42.4.1.9 Loading via detergent permeabilization

Somatic embryos have been loaded by incubation in the presence of 100 μM ion indicating dye and then treating with a low concentration of Triton X-100 (0.1% v/v) to permeabilize the plasma membrane (Timmers *et al.*, 1991). The detergent is then washed from the sample to allow the membranes to reseal. It is likely that detergent permeabilization will disrupt many cellular processes and signalling activities, at least in the short term.

It should be stressed at this point that it is important to have accessible markers of cell function to compare in loaded and unloaded cells. These may include parameters such as membrane potential, cytoplasmic streaming, growth rates, elongation rates, cell division rates, and gene expression. Alternative strategies include the monitoring of cell viability using other 'vital' staining techniques, such as fluorescein diacetate (Huang *et al.*, 1986) which probe different aspects of membrane integrity and metabolic activity.

42.4.2 Intracellular dye distribution and concentration

The cytoplasm in a mature plant cell typically occurs as a thin layer less than 1 μm thick sandwiched between the vacuole and the wall, with larger accumulations localized around the nucleus and chloroplasts. In young, rapidly growing cells and some specialized cells, such as tips of root hairs and rhizoids, a greater contiguous volume of cytoplasm may occur, uninterrupted by vacuoles. The distribution of dye follows the distribution of cytoplasm and often appears uneven. The nucleus appears to accumulate high concentrations of many dyes. Confocal microscopy indicates that this signal is genuinely located within the nucleus, not in the perinuclear cytoplasm. The majority of reliable measurements are derived from regions rich in cytoplasm where fluorescence signals are strongest.

Once successfully loaded into the plant cell the fluorescent dye may interfere with the normal function of the cells as the concentration is increased (Wagner & Keizer, 1994). In addition, the interaction of illumination with the fluorescent dye may cause phototoxic damage, particularly if the excited dye reacts with oxygen to give highly reactive free radicals. Thus, the concentration of fluorochrome introduced should be kept low to minimize buffering of the ion to be measured and reduce any potential non-specific chemical or photochemical side-effects. It is also possible to incorporate free radical scavengers such as ascorbic acid at (0.1–1 mg ml^{-1}), Trolox (vitamin E), carotenoids (see Tsien & Waggoner, 1995) or deplete oxygen levels by modification of the chamber atmosphere or adding Oxyrase (Oxyrase Inc., PO Box 1345, Mansfield, OH 44901, USA). However, these treatments can dramatically affect the physiological status of the tissue, particularly the redox equilibrium of the cells, and are not recommended for live preparations.

Usually the best way forward is to ascertain the upper limit of dye loading consistent with minimal disruption of cell physiology. Typically for calcium or pH dyes this is about 3–50 μM intracellular dye. Cytoplasmic dye concentrations have been estimated by comparison with the signal from known concentrations of dye confined to cell-sized volumes, either as droplets in immersion oil or enclosed in microcuvettes (e.g. rectangular glass capilliaries, W5005, Vitro Dynamics Inc., Rockaway, NJ, USA). To generate droplets with a range of sizes, vigorously vortex 100 μl of immersion oil (Fisher Scientific, Type FF) with 5 μl of dye solution in a microfuge tube and transfer to a slide. In confocal measurements, the volume sampled is much better defined and measurement from calibration solutions can be directly related to dye concentrations inside cells, provided the contribution from depth- and sample-dependent attenuation is taken into account (White *et al.*, 1996; Fricker *et al.*, 1997a).

42.4.3 Maintaining dyes in the cytoplasm

Compartmentalization of the dyes into organelles is a major problem with plant tissues (Callaham & Hepler, 1991; Oparka, 1991; Read *et al.*, 1992) and often limits the time window when cytosolic ion activities are faithfully reported. The tissue distribution and phloem mobility of many dyes can be partly explained using structure–activity relationships (SAR) based on parameters such as the log octan-1-ol/water partition coefficient (log$P$), pK_a, conjugated bond number (CBN), charge and molecular weight of the dye (Wright *et al.*, 1996). However, the models are less predictive for subcellular dye distributions, implying either specific transport steps are involved or, in the case of esterified dyes, the esterases are localized in different cells and compartments (Brauer *et al.*, 1996). Dyes are sequestered in the vacuole in many tissues with a variable time constant, from less than 10 min to longer than than 48 h. Several of the dyes are highly charged at physiological pH values, suggesting that this transport step probably involves specific transporters. These may include multi-drug-resistance pumps (e.g. Dudler & Hertig, 1992), glutathione (GS-X) pumps (Ishikawa *et al.*, 1997) or sulfonate transporters (Klein *et al.*, 1997). The signal from the vacuole is complicated by the low pH and ionic conditions, which will alter both the K_d of the dye and its fluorescent properties.

Dye may also accumulate in other organelles such as the endoplasmic reticulum, which has a high (>10 μM) lumenal Ca^{2+} activity (Bush *et al.*, 1989). However, the error introduced into cytosolic measurements is likely to be slight due to the limited amount of dye in the lumen (2% of the total in barley aleurone: Bush *et al.*, 1989; Bush and Jones, 1990), though it may present a problem in the interpretation of 'hot-spots' of Ca^{2+} seen in imaging experiments.

Dye may also be lost across the plasma membrane by an unknown mechanism, particularly during permeabilization treatments for *in situ* calibration. Leakage rates may be 25% h^{-1} or higher, but are highly cell type specific and must be empirically determined.

In some tissues such as pollen tubes the levels of compartmentalization can be estimated after 'selective' permeabilization of the plasma membrane with digitonin (100 μM) to release cytosolic dye (e.g. Fricker *et al.*, 1997b). All intracellular dye should be subsequently released by exposure to 0.1% (v/v) Triton X-100.

42.4.4 Techniques to prevent or minimize compartmentalization

(1) Use of dextran-linked dyes usually (but not always: see Read *et al.*, 1992) reduces compartmentalization. The dye is usually covalently linked to a 10 kDa dextran but requires pressure microinjection to load it into the cytoplasm. This approach remains the method of choice to test for compartmentalization artefacts in plant cells.

(2) Changing the temperature, time and dye concentration may extend the useable window for measurements.

(3) Choosing cells without large vacuoles or analysing regions of large cells where the vacuole is absent minimizes this difficulty, providing the experimenter is confident that small vacuoles are not being overlooked.

(4) Anion channel blockers such as probenecid are known to prevent compartmentalization of some negatively charged dyes such as lucifer yellow and FITC (e.g. Cole *et al.*, 1991), but have not yet been employed during ion measurements. The effect of these blockers on cell physiology is essentially unknown.

(5) Optical sectioning, using confocal or two-photon microscopy, can separate signals from the cytoplasm and vacuole in situations where compartmentalization does occur and may be used to measure cytoplasmic and vacuolar ion activities simultaneously. However, the practical resolution even of these techniques means that signals from ER or small vacuoles are unlikely to be well resolved from a cytosolic background under physiological conditions.

42.5 OBSERVATION AND MEASUREMENT OF DYE FLUORESCENCE

The primary objective of quantitative physiological measurements is to maximize the signal-to-noise ratio (SNR) with minimal disruption to the cell physiology. A variety of measurement systems are currently available, offering a balance between cost, sensitivity, spatial resolution, temporal resolution and sampling rate. The details of these systems are described elsewhere in this book. The main principle for botanical work is to match the hardware to the experiment. Some plant processes occur over long time periods (minutes, hours and days) and involve coordinated interaction of many cells in the tissue, for instance the tropic responses of plant organs. Phenomena in plants that take place in the seconds to minutes or longer range do not require the sophistication of subsecond temporal resolution. Photometry systems have a significantly higher (orders of magnitude) SNR than imaging systems and measurements can be made at second or subsecond rates. Integration over a 1–10 s sampling interval gives respectable signal to noise for camera-based imaging applications, though intermittent sampling and shuttering of the illumination may be required to minimize photobleaching and photodamage during extended periods of data collection. The inherent SNRs for confocal or two-photon systems are much lower than for other systems as the volume of dye sampled and the pixel dwell time are reduced. The available measurement systems are summarized in Table 42.1. Each has its own advantages and disadvantages related to the spatial and temporal sensitivity and cost. Use of several observation techniques provides complementary data characterizing both spatial and temporal components of the processes in the cell of interest.

42.5.1 Controls for physiological measurements

Once the cells are loaded with the appropriate dye, the image collection protocol can be optimized. Keeping cells alive may be at odds with optimal sampling. Measurements of ion concentrations rarely require

Table 42.1 Comparison of measurement techniques used to monitor ion activities in plant cells

Measurement technique	*Comments*
Fluorometry	*Monitors*: Population of (single) cells in suspension. Sampling is rapid (interval *c*. 0.05 s or better). Spectra are easy to measure. Autofluorescence is easy to correct. *Disadvantages*: No spatial resolution in measurement. (Although some spatial information can be collected using dyes for specific compartments or cellular domains). Heterogeneous responses cannot be distinguished. Signals from dead and dying cells also included.
Flow cytometry	*Monitors*: Population of single cells in suspension. Sampling is rapid but of different cells for each data point. Potential for preparative sorting cells by response. *Disadvantages*: No spatial resolution. Needs robust cells. Heterogeneous responses appear as an increase in variance.
Micro-photometry	*Monitors*: Typically single cells. Whole cells or subcellular regions in large cells can be measured. Average measurement defined by a (variable) mechanical aperture. The specimen or aperture may be moved to sample different regions. Sampling is rapid (interval 0.05 s or better minimum, typically 1 s). Requires microscope but allows simultaneous observation of cells. Autofluorescence correction straightforward. Relatively inexpensive. *Disadvantages*: Prone to errors from heterogenous dye distribution and redistribution.
Camera imaging	*Monitors*: Single cells, a population of cells or cells in 'thin' tissues. Subcellular regions typically down to 0.3–0.4 μm in (x,y) however (z) is poorly defined (although it may be possible to remove out-of-focus blur by deconvolution). Dual-excitation is easy to implement with a single camera, but simultaneous dual-emission requires split-view optics or two cameras. Allows mapping of spatial heterogeneity and transients. *Disadvantages*: Sampling interval typically every 1–2 s. More expensive systems can run at video rate. Sampling may need to be intermittent to reduce photobleaching. May require extended integration to increase SNR. Autofluorescence subtraction is difficult. Expensive.
Confocal microscopy	*Monitors*: Single cell, population of cells or cells in intact tissue. The (x,y) and (z) resolution are relatively well defined (maximum $0.2 \times 0.2 \times 0.6$ μm, typically $0.4 \times 0.4 \times 1.2$ μm). The fastest temporal resolution is dependent on the instrument and the volume sampled from milliseconds for a line scan, ms to seconds for 2-D section and seconds to minutes for 3-D data stack. Measurements are possible within intact tissue with subcellular resolution and reduced out-of-focus blur. Simultaneous dual-emission imaging is easy to implement. *Disadvantages*: Excitation wavelengths are limited by available lasers. There are few ratioable visible dyes for Ca^{2+}, conversely many ratio dyes require a UV laser system. Very expensive to very, very expensive for UV systems.
Two-photon imaging	*Monitors*: Single cell, population of cells or cells in intact tissue. Similar spatial and temporal resolution to a confocal system with red illumination. The depth penetration into thick tissues is much better compared with a conventional confocal system. UV dyes can also be excited with long (red) wavelength illumination, minimizing tissue damage. Excitation is restricted to the focal point, minimizing photobleaching. *Disadvantages*: Two-photon excitation spectra are much broader than one-photon spectra and not yet well defined. Very, very, very expensive. May cook the specimen.

spatial resolution greater than ~0.5 μm in (x,y) although the improved (z) resolution in confocal and two-photon systems is an advantage in interpreting the results. It is important to optimize the instrument to maximize SNR and minimize phototoxicity under these conditions. It is also important to minimize the light exposure to the sample, hence even when finding the cells to study this should be done as quickly as possible. The amount of illumination presented at the sample is probably the most critical parameter. For example, values between 76 μW (Tsien & Waggoner, 1995) and 20 μW (Errington *et al.*, 1997) are appropriate for confocal microscopy using high NA lenses to give acceptable SNRs and cell viability whilst maintaining adequate spatial sampling and scan speeds. It is not easy to predict the appropriate intensities for other systems and specimens; however, it is useful to be able to measure the illumination intensity when the imaging conditions have been optimized to act as a guide for other experiments. Four sets of controls should be run:

(1) Sample alone with no dye or fluorescence excitation: to test the effects of the microscope perfusion regime on the physiological response of the cells studied;
(2) Sample plus illumination: to test the biological effects of the excitation illumination and to measure the levels of autofluorescence;
(3) Sample plus dye (but without fluorescence excitation): to test the effects of dye loading on physiological function;
(4) Sample plus dye plus illumination: to test the potential phototoxic effects of illumination levels and dye concentrations.

These steps require good markers of cell function and physiological response.

42.5.2 Data analysis

Measurements from photometry systems intrinsically average the signal from a large area (volume) of the specimen. The key stages in analysing the data are to ensure that the dark current and autofluorescence are correctly measured and subtracted before calculation of the ratio value. In photometry measurements autofluorescence is estimated from the signal measured either prior to loading dye or after quenching the dye at the end of the experiment for the same measurement area. However, errors can be introduced into photometry measurements from uneven dye distribution within the cells. For example, changes in localized regions of the cytoplasm may be swamped by the large signal derived from the nucleus, which may comprise 30–50% of the total. A pragmatic approach to the autofluorescence problem is to calculate a mean autofluorescence value from many cells and ensure that this autofluorescence is less than 10% of the dye signal from the loaded cell.

Extracting useful data from images is somewhat more complex and there is a wide range of different analysis techniques that can be applied. Images collected at two different wavelengths can be ratioed pixel-by-pixel to generate a ratio image that compensates in principle for varying dye levels, dye leakage and bleaching. This method has found wide application in both conventional and confocal imaging (Chapters 12, 13 and 25: see also Fricker *et al.*, 1997a) and provides a good visual indicator of the magnitude of the response and the level of spatial heterogeneity within or between cells.

(1) The SNR in the raw images can be increased at the expense of spatial resolution by an averaging filter (e.g. averaging over a 3×3 box reduces noise by 3) or collecting the data at lower spatial resolution initially.
(2) Images taken at each wavelength can be manually aligned in (x,y) to correct for any minor misregistration between the two wavelength images. Objective criteria are required to perform this alignment based on, for example, imaging a standard fluorescent bead sample with both wavelengths. However, a simple image translation cannot correct for magnificational changes between two wavelengths arising from chromatic aberration.
(3) The instrument background, measured in the absence of the specimen, should be subtracted from all images.
(4) Correction for tissue autofluorescence is more difficult. One approach is to measure autofluorescence from an adjacent region of tissue that is unloaded. An alternative is to record an autofluorescence image at different wavelengths that do not intefere with the loaded dye and subtract the appropriate 'bleed-through' component from the dye images.
(5) Pixels with low values or those outside the object are normally masked and excluded from the ratio image by setting the intensity to zero with a spatially defined mask. Three protocols may be used to define the mask:
    (i) An intensity value at a fixed number of standard deviation (SD) units above the mean background intensity, typically 2 SD units;
    (ii) The 50% threshold between the fluorescence intensity within the object and the background (e.g. Errington *et al.*, 1997);
    (iii) A morphological boundary, such as the edge of the cell, defined from a separate image, such as a bright-field view.
(6) Masking is also required to exclude values in each image approaching saturation of the digitization range. Saturation is related to an assessment of the number of photons contributing to the signal and requires knowledge of the conversion from photons to grey-levels. A pragmatic approach is to measure the distribution of intensities in a fluorescent area at about the concentration of fluorochrome encountered *in vivo* and determine the highest mean value where the distribution is not clipped.
(7) The ratio image is calculated pixel-by-pixel and the mask applied.
(8) Pseudocolour look-up tables are often used to enhance the viewers perception of changes, particularly in publications where grey-scale images are not reproduced well.

Notes

- Graphical presentation of data derived directly from a region of the ratio image should be avoided.

Ratio images are notoriously noisy and it is difficult to interpret the statistics from spatial averaging of the ratio values. Ratioing two normally distributed populations gives a highly skewed distribution of ratio values. Thus it is usually more appropriate to visualize changes using ratio images, but to perform quantitative analysis on the original intensity data from the individual wavelength images directly with an average area analysis as described below.

42.5.3 Measurements on regions of interest

A series of regions of interest (ROI) are defined on one wavelength image and the average intensity and standard deviation measured from that region. The corresponding region is measured on the other wavelength image with appropriate image or area alignment if necessary. In some cases it is useful to segment the object using a mask prior to area measurements to facilitate measurements from regions encompassing irregular structures, such as cytoplasmic strands or subcellular compartments, without recourse to detailed manual delimitation of the area. A useful objective criteria to segment the object is the 50% intensity level between the object and the background.

(1) The average background values are subtracted independently for each ROI at each wavelength.
(2) The average ratio is calculated and displayed graphically.
(3) Error bars can be calculated from the standard deviation of the individual wavelength data from each ROI. If the 90% confidence limits of the individual wavelengths are used, the ratio values will have 81% confidence limits, assuming the variability in both populations arises only from uncorrelated noise.
(4) Alternatively, an estimate of the average ratio and the confidence limits can be made using application of Bayes theorem, where *a priori* information can be incorporated into the analysis (Parton *et al*., 1997).

Notes

- Sample autofluorescence correction is almost impossible with imaging techniques as cytoplasmic streaming constantly moves organelles, potentially causing spatial re-organization of autofluorescence. Partial correction can be made in some cases by sampling autofluorescence in corresponding regions of a similar non-loaded cell (Gilroy *et al*., 1991).

42.6 CALIBRATION *IN VITRO* AND *IN SITU*

The accuracy of absolute measurements depends to a large extent on calibration and becomes increasingly important when small quantitative differences rather than large qualitative changes in ion concentration between cells are thought to be significant. The best method would be to make an *in situ* calibration because all parameters in the cell which would affect the dye fluorescence and the ratio would be automatically considered. For *in situ* calibration the plasma membrane has to be made selectively permeable for the measured ion. Ionophores have been successfully used for this purpose in animal cells. These chemicals lead to an electrochemical equilibrium between bathing medium and cytoplasm (Williams & Fay, 1990; Bright *et al*., 1989). The fluorescence ratio is measured as the external concentration of the ion is varied. Often *in situ* calibration is only performed at two points, usually to determine the maximum and minimum ratio (Groden *et al*., 1991; Rathore *et al*., 1991). After application of the ionophore two different ion concentrations are established in the biological system; at a very low concentration R_{min} and F_{max2} are obtained, and at a saturating concentration R_{max} and F_{min2} are measured. However, in this case the measured ratios can be converted to concentration values only if the dissociation constant of the dye K_d is exactly known inside the cell. This is often not the case in plant cells.

A second major problem in plant cells is that ionophores do not equilibrate ion concentrations to a sufficient or reproducible extent across the plasma membrane (Bush & Jones, 1987; Felle *et al*., 1992). Ionophore concentrations greater than 3 μM often act like detergents and damage the cell, however concentrations up to 100 μM have to be applied to penetrate into the tissues and to shift the internal ion concentration. Thus, in many cases *in situ* calibration is not a reliable method for plant cells, leaving an *in vitro* calibration as the only alternative. There is some confidence if both *in situ* and *in vitro* calibrations match (Brownlee *et al*., 1987; Brownlee & Pulsford, 1988; Gilroy *et al*., 1991; McAinsh *et al*., 1992; Mühling *et al*., 1995). However, this is often not the case (Gilroy *et al*., 1989) and it is not obvious which calibration is more appropriate. One of the most appropriate solutions to this conundrum is to measure independently the calcium level in the same system using a different technique such as a calcium-selective electrode (e.g. Felle & Hepler, 1997).

42.6.1 *In vitro* calibration – calcium dyes

Calcium dyes may be calibrated *in vitro* using Ca^{2+}-EGTA buffers to set $[Ca^{2+}]$ in a solution designed to mimic the plant cytosol. Calibration is simple to perform and calibration kits are now available from companies such as Molecular Probes and World Precision Instruments. *In vitro* calibration of single-wavelength dyes can be used to define the dynamic range expected for the dye response under the collection conditions used for the experiment. It is not possible to use an *in vitro* calibration for a single-wavelength dye to calibrate the Ca^{2+} response *in vivo* as it is only valid for a single concentration and defined pathlength. For ratio dyes, an *in vitro* calibration provides an indication of the $[Ca^{2+}]$ *in vivo*, but the appropriateness of the calibration depends primarily on the degree of correlation with conditions experienced by the dye in the cytosol.

Protocol

(1) A typical buffer contains 100–120 mM KCl, 0–20 mM NaCl, 1 mM $MgSO_4$, 10 mM HEPES pH 7.2, 1–10 μM dye and 10 mM EGTA plus appropriate amounts of $CaCl_2$ to give the required free $[Ca^{2+}]$.
(2) The actual free Ca^{2+} is usually calculated using an iterative computer program that accounts for all the ionic interactions in the calibration buffer (Vivaudou *et al.*, 1991; Fabiato, 1988). These authors distribute their programs free of charge.
(3) Measurements can be made using volumes of calibration solution similar in size to loaded cells in the experimental apparatus.
(4) Experimentally derived values are calibrated using the following equation (Grynkiewicz *et al.*, 1985; Ameloot *et al.*, 1993):

$$[Ca^{2+}]_{cyt} = K_x \cdot \left(\frac{R - R_{min}}{R_{max} - R}\right) = K_d \cdot \left(\frac{R - R_{min}}{R_{max} - R}\right) \cdot \left(\frac{F_{max\lambda 2}}{F_{min\lambda 2}}\right) \quad [42.1]$$

Where R is the measured ratio, R_{min} and R_{max} are the ratio values at zero and saturating calcium levels, respectively, K_d is the (assumed) dissociation constant and $F_{min\lambda 2}$ and $F_{max\lambda 2}$ are the fluorescent intensities at the second wavelength (the denominator of the ratio) at zero and saturated calcium, respectively.
(5) It is more convenient to re-arrange equation [42.1] for use with a non-linear curve fitting package as follows:

$$R = \frac{R_{min} - R_{max}}{(1 + 10^{-(pCa - pK_x)})} + R_{max} \quad [42.2]$$

The measured ratio at each pCa value can be fitted by this function with R_{max}, R_{min} and pK_x, as free running parameters (so called Boltzmann fit) (Plieth *et al.*, 1997). The fitted sigmoidal parameters can be used to estimate the cytoplasmic calcium concentration from the ratios measured in the living system.

Notes

- The *in vitro* calibration solutions should be as similar as possible to the cytoplasmic milieu. For example, when calibration curves obtained with and without Mg^{2+} were compared, the apparent dissociation constants for Fura-dextran ranged from 400–1000 nM measured with Mg^{2+} and 100–250 nM measured without Mg^{2+} (Plieth *et al.*, 1997).
- Ionic strength, viscosity and hydrophobicity of the medium have all been identified as potential factors that influence the response of these dyes (Poenie, 1990; Roe *et al.*, 1990; Uto *et al.*, 1991).
- Viscosity may be increased by addition of 20–60% sucrose (e.g. Zhang *et al.*, 1990) or 500 mM mannitol (e.g. Dixon *et al.*, 1989) and hydrophobicity altered with 25% ethanol (Russ *et al.*, 1991).
- Judicious selection of wavelengths where the dye spectrum is less susceptible to potential interference has been suggested. For example, fura-2 can be measured using the ratio of 340 nm and 365 nm rather than the normal 340 nm and 380 nm (Roe *et al.*, 1990).
- Definition of the appropriate composition of the calibration solution can be best estimated from comparison of the dye spectra *in vivo* with that determined *in vitro* under a variety of hydrophobicity, viscosity and ionic composition regimes (Poenie, 1990; Owen, 1991).
- The spectra for dextran conjugates is different from those of the free dyes and the K_d also varies from batch to batch depending on the length of the dextran molecules and the degree of substitution (Haugland, 1996). For example, Fura-dextran has a dissociation constant around 350 nM compared with 150 nM for Fura-K5 and the peak in the excitation spectrum for the calcium-free form is shifted to shorter wavelengths (from 380 nm to 364 nm).
- Most dyes show a pH dependence of K_d (e.g. Lattanzio, 1990; Roe *et al.*, 1990; Lattanzio & Bartschat, 1991). For example, the apparent K_d of Fura becomes more and more dependent on the proton concentration below pH 7. Calcium concentrations estimated from the measured ratios have to be corrected by a pH-dependent factor X_{corr} that is

detemined from an exponential fit to the K_d values determined by Lattanzio (1990). See Plieth *et al.* (1997) for full details.

$$K_d = K_{d0} \cdot X_{corr} \quad [42.3]$$

$$\text{where } X_{corr} = \left[1 + 9 \cdot \exp - \left(\frac{(pH - 5.45)}{0.5}\right)\right] \quad [42.4]$$

The significance of the pH error will depend on the prevailing cytoplasmic pH value, which is often unknown. However, this correction becomes of increasing importance in experiments where cytoplasmic pH is varied using for example, weak acid loading, or where steep pH gradients have been proposed e.g. at the tip of algal rhizoids.

42.6.2 *In situ* calibration – calcium dyes

Dye-loaded cells are permeabilized with a Ca^{2+}-ionophore or by detergent treatment at the end of the experiment. Cytosolic $[Ca^{2+}]_{cyt}$ is set by extracellular Ca^{2+}-EGTA buffer solutions. In principle, cytosolic conditions experienced by the dye should not have changed significantly between the calibration and *in vivo* measurements of $[Ca^{2+}]_{cyt}$ during an experiment. Results from single-wavelength dyes are difficult to interpret as there is no inherent correction for dye leakage, redistribution or bleaching. However, dyes are available across a wide range of wavelengths and show substantial changes in fluorescence.

Protocol

(1) Add 1–10 μM ionomycin or Br-A23187 (a non-fluorescent analogue of A23187) at the end of the experiment.
(2) Increase external calcium to ~1 mM and allow the signal to stabilize at R_{max} (ratiometric dye) or F_{max} (single-wavelength dye).
(3) Replace the high calcium medium with medium containing 1 mM EGTA to set R_{min} or F_{min}.
(4) Calculate $[Ca^{2+}]_{cyt}$ using published K_d values or K_d values measured *in vitro* using equation [42.2] for ratio dyes or the following equation for single-wavelength dyes (Kao *et al.*, 1989):

$$[Ca^{2+}] = K_d \frac{F - F_{min}}{F_{max} - F} \quad [42.5]$$

Notes

- Ionomycin has been reported to be much less effective than Br-A23187 in plant cells (Bush & Jones, 1987; Gilroy *et al.*, 1991). This may reflect its requirement for alkaline (pH 9) rather than acidic (< pH 7) conditions normally encountered in perfusion solutions (Liu & Hermann, 1978).
- An alternative to determining F_{min} for fluo-3 using EGTA buffers is to use Mn^{2+} to quench the fluorescence from the dye. Conveniently, Mn^{2+} is efficiently transported into cells from external concentrations of 0.1–1 mM $MnCl_2$ during incubation with 10 μM ionophore for 10 min. Fluo-3 fluorescence is quenched to $8 \times F_{min}$. This set point can then be used in a modified form of the above equation (Kao *et al.*, 1989; Minta *et al.*, 1989). Our experience in calibrating fluo-3 in plant cells (Gilroy *et al.*, 1990) is that the Mn^{2+}-quench procedure is more consistent than determining F_{min} with EGTA. Even so, calibration is extremely difficult to perform accurately and in general data from single-wavelength dyes is more qualitatively useful than quantitatively accurate.

42.6.3 *In situ* calibration – ratio pH dyes

The approach for calibration of pH dyes such as BCECF is essentially similar to those outlined for calcium dyes. The K^+/H^+ exchanger, nigericin, has been used as an ionophore to equilibrate internal and external pH in the presence of high K^+ (e.g. Dixon *et al.*, 1989; Thiel *et al.*, 1993; Fricker *et al.* 1997b; Parton *et al.*, 1997).

Protocol

(1) At the end of the experiment the external $[K^+]$ is increased to a value close to the anticipated internal $[K^+]$, typically 100–120 mM.
(2) Nigericin is added to a final concentration of 10 μg ml^{-1} and the pH adjusted to pH <6.0 (for BCECF).
(3) The ratio values are allowed to stabilize (2–15 min) and R_{min} measured.
(4) The external pH is shifted to pH 8.5 to give R_{max} and the ratio allowed to stabilize.
(5) pH values are estimated from a sigmoidal fit to the *in situ* calibration data with an assumed K_d (typically pH 7.0–7.2 for BCECF) or the pH is estimated from equation [42.2] substituting appropriate values for the pH dye used and assuming that nigericin equilibrates pH as follows:

$$\frac{[H^+]_{in}}{[H^+]_{out}} = \frac{[K^+]_{out}}{[K^+]_{in}} \quad [42.6]$$

Notes

- *In situ* calibrations give very poor agreement with *in vitro* measurements as the K_d is rather sensitive to the local environment (Dixon *et al.*, 1989). Additions such as de-proteinized coconut water supplemented with 1% ovalbumin improve the overlap between *in situ* and *in vitro* calibrations (Pheasant & Hepler, 1987).

- pH intervals can be monitored fairly accurately even if the absolute level cannot be determined. The relatively small shift in ratio values means the limits of reliable detection lie between 0.05 and 0.15 pH units (Pheasant & Hepler, 1987; Fricker *et al.*, 1997b; Parton *et al.*, 1997).
- As it is difficult to shift pH to R_{max} and R_{min} *in situ*, an alternative calibration between pH 6.5 and pH 7.8 can be used to cover the near-linear region of the ratio response.
- Ionophore treatments stress cells rapidly (Pheasant & Hepler, 1987). Calibration can also be based on equilibration of permeant weak acids and bases (Pheasant & Hepler, 1987; Gehring *et al.*, 1990a,b; Parton *et al.*, 1997).
- Parallel measurements using pH-sensitive microelectrodes provide an independent check of the fluorescence calibration (e.g. Gibbon & Kropf, 1994).

42.7 ADDITIONAL MEASUREMENT TECHNIQUES

42.7.1 The manganese quench technique

It is possible to extend the basic measurement of cytosolic calcium concentration using fluorescent dyes to identify the possible source of the Ca^{2+} leading to the increase in cytosolic Ca^{2+}. In principle there are three different sources for an increase in Ca^{2+}, namely: (1) influx across the plasma membrane; (2) release from internal stores, such as the vacuole, ER, mitochondria or chloroplasts; or (3) from changes in cytoplasmic Ca^{2+} buffering capacity arising from cytosolic pH changes, for example. Manganese ions (Mn^{2+}) can permeate at least some types of Ca^{2+} channels (Fasolato *et al.*, 1993; Piñeros & Tester, 1995; Striggow & Ehrlich, 1996) and also bind Ca^{2+}-indicator dyes with a very high affinity (K_d Mn^{2+}-Fura ≈ 5 nM whilst K_d Ca^{2+}-Fura ≈ 250 nM) quenching their fluorescence (Thomas & Delaville, 1991; Gilroy & Jones, 1992; Zottini & Zannoni, 1993; Malhó *et al.*, 1995; McAinsh *et al.*, 1995). It is possible to use these properties to discriminate between Ca^{2+} influx and Ca^{2+} mobilization from internal stores by introducing Mn^{2+} into the external medium. Opening of plasma membrane Ca^{2+} allows Mn^{2+} influx and consequent fluorescence quenching. In the case of Fura, this can be readily measured at the iso-excitation wavelength around 360 nm (e.g. Plieth *et al.*, 1998).

A variant of the Mn^{2+}-quench technique can also be used to identify release from intracellular stores. Mn^{2+} is loaded into the intracellular stores by extended incubation in Mn^{2+} followed by washing from the external medium (A. Grabov & M.R. Blatt, unpublished) or by microinjection into the vacuole (Plieth *et al.*, 1998). Under these conditions, fluorescence quenching indicates release of Mn^{2+} from the loaded stores rather than flux across the plasma membrane.

42.7.2 Dissipation of intracellular calcium gradients using buffers with varying pK_d

In certain cell types, most notably tip-growing cells, a steep tip-focused gradient in calcium correlates with tip growth (e.g. Pierson *et al.*, 1996). One technique that has been used to test the importance of the Ca^{2+}-gradient is to disupt it using microinjection of calcium buffers of varying K_d derived from substitutions on the 1,2-*bis*(*o*-aminophenoxy)ethane *N,N,N′,N′*,-tetraacetic acid (BAPTA) moiety. These buffers are thought to increase the mobility of calcium in the cytoplasm, particularly if the K_d falls between the concentrations expected at the high and low points of the gradient (Speksnijder *et al.*, 1989; Pierson *et al.*, 1994).

42.7.3 pH clamping with weak acids

A relatively straightforward method to acidify the cytoplasm of living cells is to apply a weak (HBA) acid to the outer medium (Franchisse *et al.*, 1988). The weak acid (usually butyric acid or acetic acid) has to be membrane permeable in its undissociated (lipophilic) form. Once in the cytosol the weak acid dissociates due to the higher pH in the cytoplasm (pH_c) compared with that in the outer medium (pH_o). Thus the acid anion is trapped in the cytosol (assuming the anion is membrane-impermeant) and a proton is released. Assuming the plasma membrane is permeable only for the undissociated form (HBA) of the acid, an equilibrium will be established between bathing medium and cytoplasm such that the concentration of HBA will be equal in both compartments. The amount of anion that dissociates under these conditions is given by:

$$BA^-_{cyt} = \frac{10^{(pH_c - pH_0)}}{1 + 10^{(pK_d - pH_0)}} \cdot HBA_0 \qquad [42.7]$$

Obviously there occurs a massive accumulation of acid anions in the cytoplasm. With the assumption that the weak acid releases in the cytoplasm as many BA^- ions as protons ($BA^-_{cyt} = \Delta H^+$), equation [42.7] gives an estimate of the amount of protons imported into the cell. Most of these protons equilibrate with cytoplasmic buffer sites. Only a few lead to the measured pH_{cyt} decrease.

The weak acid can be washed out of the cell and, at low concentrations, exhibits no severe side-effects on the cell physiology. Thus, application of weak acids allows pH_{cyt} to be brought under experimental control and can be used to pH-clamp the cytoplasm (e.g. Thiel *et al.*, 1993).

Notes

- Normally there are other cations in the bathing medium (K^+, Na^+, Mg^{2+}, Ca^{2+}) which can also bind to BA^- forming neutral salt molecules (KBA, NaBA, $MgBA_2$, $CaBA_2$). These salt molecules are membrane-permeable to an unknown extent, but will cause an overestimate in the calculated amount of protons imported into the cell.
- The distribution models require that no transport systems exist for the anion and that the molecule is not metabolized.
- Weak acid loading will affect the pH in all compartments of the cell and will lead to massive accumulation of anions in even more alkaline compartments, such as chloroplasts.
- It would be naive to expect that the cell does not respond to such dramatic perturbation of pH, and during extended weak acid-loading treatments, significant changes in ion transport systems have been reported (Reid & Whittington, 1989).
- Iso-butyrate is considerably less smelly than butyrate.

42.8 USING RECOMBINANT AEQUORIN FOR MEASUREMENT OF INTRACELLULAR CALCIUM IN PLANTS

Apoaequorin is a single polypeptide chain of approximately 22 kDa isolated from the coelenterate *Aequorea victoria*. Apoaequorin combines with a low molecular weight luminophore called coelenterazine to form functional aequorin in a process termed reconstitution. Molecular oxygen is also required at this stage. The aequorin molecule is similar in structure to the calcium-binding protein calmodulin and possesses three calcium-binding EF hand domains (analogous to the four EF hands of calmodulin) and a binding site for coelenterazine and oxygen. When calcium is bound, the coelenterazine is oxidized to coelenteramide and the protein undergoes a conformational change accompanied by the release of carbon dioxide and emission of blue (462 nm) light. Aequorin is highly selective for Ca^{2+}; Mg^{2+} and K^+ do not trigger luminescence, though these ions may depress the Ca^{2+}-sensitivity (Thomas, 1982). Aequorin can potentially detect free calcium levels of up to 100 μM, although in practice most measurements are made in the range of 10 nM–10 μM.

Aequorin has several potential advantages over fluorescent dyes as an indicator for $[Ca^{2+}]_{cyt}$. Luminescence measurements usually have an intrinsically high signal-to-background ratio as there is relatively little endogenous luminescence under optimal conditions. However, it is important to ascertain the level of endogenous chemiluminescence as this may comprise up to 30% of the signal in plant cells (Gilroy *et al.*, 1989). As a natural protein, aequorin is expected to be non-toxic and remains in the cytoplasm unless specifically targeted elsewhere. Light emission is unaffected by pH values greater than pH 7. Photodamage associated with excitation illumination for fluorescence is also avoided. Supplies of the apoprotein have increased since the gene for aequorin has been cloned (Prasher *et al.*, 1985; Inouye *et al.*, 1985), although the essential cofactor coelenterazine is still expensive.

42.8.1 Introduction of aequorin into cells

Although aequorin can be microinjected into cells, the method is suitable only for large cell types such as the giant green alga *Chara* (e.g. Williamson & Ashley, 1982). Pressure injection is used in preference to iontophoresis to overcome the low iontophoretic mobility of the 20 kDa apoprotein. Premature discharge of the aequorin in the micropipette by Ca^{2+} in the perfusion medium leads to a reduction in the amount of active probe introduced into the cell and hence a reduction in the signal measured. This can be avoided by either forcing a slow constant stream of aequorin out of the pipette when approaching the cell or by plugging the end of the pipette with a small amount of vegetable oil that can be expelled just prior to injection. It is also possible to inject the apoprotein only and then reconstitute *in planta*.

More recently the problems of introducing aequorin into the cytoplasm have been elegantly solved using recombinant DNA technology to transform plant cells with the cDNA for the apoprotein (Knight *et al.*, 1991; Knight & Knight, 1995). Aequorin can be reconstituted *in vivo* by adding coelenterazine to transgenic plants and it is possible to produce enough luminescence to make reliable measurements without cellular disruption. All cells produce their own Ca^{2+}-indicator and thus calcium changes can be measured in whole intact plants. *Arabidopsis* (Knight *et al.*, 1996), tobacco (Knight *et al.*, 1991) and the moss *Physcomitrella patens* (Russell *et al.*, 1996) have all been stably transformed with an apoaequorin gene. A plasmid containing the cloned gene in a binary vector is available commercially from Molecular Probes under the name of pMAQ2. pMAQ2 encodes a constitutively expressed cytosolic form of apoaequorin, under the control of the cauliflower mosaic virus (CaMV) promoter. This is the most commonly used form in plants. In addition, aequorin has been successfully targeted to a number of subcellular locations in addition to the cytoplasm using targeting sequences consisting of either peptide leader sequences or whole polypeptides encoding proteins which exist naturally in the chosen locale. If aequorin is to be expressed as a fusion protein, then it is necessary to maintain the features of the protein which are required to retain its full activity as a calcium reporter. The *C*-terminal proline residue of apoaequorin is essential for the long-term stability of reconstituted aequorin (Watkins & Campbell, 1993). Even if the proline residue is present but linked to the *N*-terminal amino acid of another peptide, the resulting fusion protein is likely to be very unstable. It is therefore advisable to design aequorin fusions in which the aequorin is the *C*-terminal part of the protein and the other peptide the *N*-terminal portion. This may not always be possible, for instance in the case of endoplasmic reticulum-targeting, for which a *C*-terminal KDEL sequence is required. Targeted forms of apoaequorin have been engineered for use in plants, which express the protein in the chloroplast (Johnson *et al.*, 1995) (pMAQ6, available from Molecular Probes), nucleus (van der Luit *et al.*, submitted) and the cytosolic face of the vacuolar membrane (Knight *et al.*, 1996). All of the plasmids confer kanamycin resistance and therefore the antibiotic is used in the identification of transformants (see below).

42.8.1.1 Stable transformation

Tobacco leaf disc transformation (Draper *et al.*, 1988) and *Arabidopsis* root transformation (Knight *et al.*, 1997) have been successfully used to produce stable transformants. However, if mutants of *Arabidopsis* are to be transformed, it may be worth considering whether or not it would be appropriate to take the mutant through tissue culture, as some mutants are difficult to regenerate from callus. If this is a concern, the worker may choose an alternative method such as vacuum infiltration (Bechtold *et al.*, 1993), which does not include a tissue culture step.

42.8.1.2 Transient expression

Transient expression of aequorin may serve two purposes. First, it may be useful to test whether or not a newly engineered aequorin construct will be transcribed and translated in a plant cell to form functional aequorin. Second, it is often beneficial to be able to perform calcium measurements, for instance in a mutant, as a prelude to stable transformation. For the first purpose, biolistic transformation of whole seedlings provides a very quick method of testing for luminescence from a new construct (Knight *et al.*, 1997). This could alternatively be carried out using onion epidermis (Klein *et al.*, 1987). The transiently transformed tissues can be homogenized and aequorin activity assayed *in vitro* (see later). For *in vivo* Ca^{2+} measurements in plants for which stable transformants are not yet available, PEG transformation of *Arabidopsis* protoplasts (Abel & Theologis, 1994) has proved very useful. Protoplasts from stably transformed *Arabidopsis* and tobacco (Haley *et al.*, 1995) have been used in experiments where a uniform population of cells is required.

42.8.2 Measurement of luminescence

The level of light emission from recombinant aequorin is $4.30–5.16 \times 10^{15}$ photons mg^{-1} of aequorin (Shimomura, 1991). At present the actual levels of aequorin which can be expressed in plant tissues are relatively low (a few pg protein per mg fresh weight of tissue) and therefore detection of the blue light emitted in response to calcium requires the use of very sensitive light-counting equipment. A luminescence detector needs to be able to detect light signals over a wide range varying by several orders of magnitude of intensity; from only a few photons per second to several millions. It is also important that the detector provides appropriate time resolution (Campbell, 1988).

There are two methods routinely used for measuring Ca^{2+}-transients in plants using aequorin luminescence. This first involves luminometry and provides high sensitivity and fast (sub s) time resolution. The second involves low-light-level imaging, which gives additional spatial information.

42.8.2.1 Luminometry

A luminometer is the simpler and by far the cheaper of the two options. Stanley (1992) has published a survey of a great number of luminometers and imaging devices and the reader may wish to refer to this to aid in the selection of a suitable device. The luminometer can be designed to hold whole plants within a sample cuvette (Knight *et al.*, 1991). The luminometer has a light-tight sample housing containing a sample cuvette adjacent to the photomultiplier detector. A cooled photomultiplier (e.g.

EMI 9235B) with a low dark current and a bialkali coating gives a good response in the blue with minimal background noise, especially when used with photon-counting circuitry. It is important to avoid the use of materials such as paints, silicon grease, silica gel and certain glass or plastic tubes, which luminesce themselves. Tubes 12 mm in diameter (Sarstedt, Leicester, UK) are suitable for this purpose. Mirrors or silvering inside the sample housing may help reflect the light from the sample onto the detector (Campbell, 1988).

42.8.2.2 *Low-light-level imaging of aequorin*

Low-light-level imaging cameras can be purchased separately and built into an imaging system; however, there are now a growing number of 'off-the-shelf' complete imaging systems (see Stanley, 1992). The image is sent from the camera to a computer loaded with software for processing and analysis. Copies of the stored image can be printed using a colour video copy processor.

Most imaging systems are based around the charge-coupled device (CCD). CCDs are made up of an array of light sensors in which each acts as an individual light detector (Aikens, 1990). Incoming photons break bonds between the silicon atoms in the semiconductors, and generate electron–hole pairs. The conversion of photon to electron is a linear process, allowing the measurement of light to be quantifiable. The electrons produced in this way are collected in a potential well; each well corresponds to a pixel on the screen of the imaging device. Exposure to light during imaging causes the CCD to acquire a pattern of electronic charge in the wells. This pattern of information is transmitted through a parallel register (of the same layout as the array) onto a serial register which transmits the information row by row to the output circuit. An image is then reconstructed by the computer. 'Blooming' may occur when there is too much charge for one well to hold (i.e. too much light has been allowed to enter the imaging device) and charge spills over to the adjacent wells.

The two types of CCD camera used for luminescence imaging can be classified by the method employed to improve the sensitivity of the camera. They are the cooled CCD and the image-intensified CCD. In the cooled CCD, sensitivity is increased by reducing the background noise by cooling the camera, allowing lower levels of light to be detected. The cooling system may be a Peltier device, circulated water, forced air or others. Improvement of the SNR is thus achieved by lowering the noise component. The disadvantage of this kind of camera is that a slow scan speed is needed and therefore rapid changes in light emission (typical of plant calcium responses) cannot be recorded. In the image-intensified CCD, the signal from the light sensors is intensified (using microchannel plates or a phosphor-cathode sandwich) before detection and so the SNR is improved by increasing the strength of the signal. This is a more sensitive device than the cooled CCD at the low levels of light expected from aequorin luminescence.

Two 'off-the-shelf' intensified systems which have been used successfully to image aequorin are systems from Hamamatsu (Hamamatsu Photonics UK Limited) e.g. the C2400-20 photon-counting camera system with an Argus-50 image processor (Knight *et al.*, 1993), or the Photek (Photek Ltd, Hastings, UK) photon counting camera system (Campbell *et al.*, 1996). In terms of sensitivity and ease of use, both systems are comparable. However, there is one important difference, namely that the Photek system captures luminescence image information at video rate, and this information can be accessed at video rate or above with the software supplied. This is absolutely invaluable for imaging rapid calcium-transients reported by aequorin luminescence. It also means that no *a priori* knowledge of the nature of the kinetics of luminescence is required to start an experiment. We thus favour the Photek system (see below).

42.8.2.3 In vitro *reconstitution of aequorin*

Not all of the kanamycin-resistant plant transformants obtained will yield light, and therefore, although transformants can be screened by Southern and Western blotting, the most direct way of telling whether or not the plants will be of use is to check for the ability to produce blue light after addition of coelenterazine. For this purpose, aequorin should be reconstituted in a plant extract *in vitro*.

Protocol

(1) For primary transformants, select a piece of leaf tissue from individual mature plants. For T1 or T2 generations, select 3–5 green seedlings (7- to 8-day-old tobacco or 6- to 7-day-old *Arabidopsis*) grown on a selection plate containing nutrient media (e.g. Murashige and Skoog) supplemented with kanamycin. Place in an Eppendorf tube and snap freeze in liquid nitrogen.
(2) Grind tissue to a powder using a micropestle or glass homogenizer.
(3) Add 0.1 ml chilled medium (0.5 M NaCl, 5 mM EDTA, 10 mM Tris-HCl, pH 7.4) with 5 mM mercaptoethanol and 0.1% (w/v) gelatin added just prior to use.
(4) Homogenize further using a motorized micropestle.
(5) Add a further 0.4 ml of medium, mix and leave

on ice until all samples have been processed, using a clean micropestle for each sample.

(6) Centrifuge at 13000 ***g*** for 10 min and retain supernatant.

(7) To 100 μl of supernatant, add 1 μl of 100 μM coelenterazine and pipette up and down to mix and to introduce air into the reaction tube. Incubate tubes in the dark at room temperature for 3 h.

(8) Dilute 10- to 50-fold with 0.5 ml Tris-EDTA buffer (200 mM Tris-HCL, 0.5 mM EDTA, pH 7.0) in a plastic luminometer cuvette.

(9) Take up 0.5 ml 50 mM $CaCl_2$ in a 1 ml syringe covered in black electrical tape.

(10) Measure the sample background then quickly inject the $CaCl_2$ and count for a further 10 s. Subtract background to give luminescence counts for each sample.

Notes

- The coelenterazine should be stored dry at −70°C. Aliquot 50 μl samples of 200 μM coelenterazine in MeOH into microcentrifuge tubes, vacuum dry, wrap in foil and store at −70°C. Protect coelenterazine from light during this procedure. Remove aliquots when required and re-dissolve in methanol at the required concentration. Coelenterazine is very light-sensitive and should be protected from light at all times. When adding coelenterazine to reaction tubes, dim room lights are required.
- It is also advisable to test for activity in the debris pellet as well as the supernatant, as this will contain some of the cellular components to which the aequorin may be associated. More sophisticated subcellular fractionation may be required to localize the aequorin activity.

42.8.2.4 In vivo *reconstitution of aequorin*

When the plants producing the highest levels of light have been identified, an *in planta* time course of reconstitution can be carried to ascertain the optimum time to allow reconstitution to maximal levels. Subsequently, the optimum duration of reconstitution should be used for experimental plants. Approximately 6–18 h is satisfactory for cytosolic aequorin plants and levels remain high for at least 24 h (Knight *et al.*, 1991). Other semisynthetic aequorins are stable for shorter periods. To be sure of achieving maximum reconstitution of aequorin in plants of other species or age, it is useful to conduct a preliminary time course study of reconstitution. The amount of aequorin reconstituted increases with the concentration of coelenterazine up to a coelenterazine concentration of at least 10 μM (Knight *et al.*, 1991).

Protocol

(1) Float 7- to 8-day-old tobacco or 6- to 7-day-old *Arabidopsis* seedlings in as low a volume of water as is possible without causing tissue to dry out (typically 1 ml).

(2) Add coelenterazine to give a final concentration of 2.5 μM, mix by gentle swirling and place in the dark at room temperature.

(3) The extent of reconstitution is measured by discharging the reconstituted aequorin. Measure the luminescence counts for 10 s before and for 2 min after injection of 1 ml 2 M $CaCl_2$ with 20% ethanol.

Notes

- Expense may prohibit the use of large amounts of tissue due to the increased amounts of coelenterazine required.

42.8.2.5 Experimental measurements

Once reconstitution of aequorin in the transformed plants has been demonstrated, and the optimum time period for reconstitution determined, intracellular calcium measurements can be made.

Protocol

(1) Dim room lights before switching on luminometer to reduce background counts. Take a background reading either using a wild type seedling in a cuvette or a cuvette containing 1 ml water.

(2) Take a freshly reconstituted seedling (steps 1–2 above) and gently place in the bottom of a luminometer cuvette containing water, if this is to be used.

(3) Allow a minute for seedling to settle. Resting level calcium will be higher just after the seedling has been moved, and luminescence counts will reflect this.

(4) Fill the syringe with substance to be injected through the port.

(5) Start counting luminescence counts and then inject contents of syringe.

(6) If calibrating, discharge the remaining aequorin with an equal volume of 2 M $CaCl_2$, 20% ethanol at the end of the experiment.

Notes

- The syringe should be coated with black PVC tape to reduce light entry via the injection port.
- The time for the injected solution to mix thoroughly in the cuvette must be taken into account.
- Although magnetic stirrers speed up mixing, their use should be avoided as they generate electrical noise in the photon detector (Campbell, 1988).
- The choice of frequency at which to record lumi-

nescence counts depends on the response to be measured. Counting periods of 1–10 s may be used if low levels of luminescence are anticipated. However, for rapid responses, counting every 0.1 s is more appropriate in order to obtain a clear picture of the dynamics of the response.

- Adequate controls must be performed as the act of injection itself may cause an elevation in cytosolic calcium. For instance, the rapid introduction of liquid into a cuvette containing a seedling will effect a touch response and consequent spike of cytosolic calcium. Similarly, introduction of cold liquids may cause a cold shock response and again an elevation in cytosolic calcium.
- The temperature at which the experiment is carried out should be monitored as any calibration formula used will only be accurate if determined at the same temperature as that at which measurements are made.

42.8.2.6 *Imaging aequorin luminescence*

Biological samples (transgenic plants) containing aequorin are imaged in a specially constructed dark box containing a three microchannel plate intensified CCD camera (Photek ICCD325–FTM800 Intensified CCD Camera, Photek Ltd, Hastings, UK) equipped with standard Nikon photographic lenses. The normal background of this camera is approximately 10 counts s^{-1} over the whole pixel array. The camera is controlled by a HRPCS-2 camera control unit (Photek Ltd., Hastings). The camera stores the photon image, as an array of 384×288 pixels, at video rates (50 Hz), or as an array of 768×576 pixels at half video rates (25 Hz). The latter sampling speed has proved the most appropriate for plant measurements, as it offers the highest spatial resolution and the temporal resolution is sufficient for transients detected using aequorin. The software (IFS216 Software, Photek Ltd, Hastings, UK) then analyses the distribution of signals in the image and extracts and stores the (x,y) co-ordinates of the individual photons at (half or full) video rates (25 or 50 Hz). A user-defined number of these images are then integrated and displayed. These two features, in particular, distinguish this imaging system from those widely used for fluorescence imaging.

Regions of interest can be drawn over defined parts of the plants or seedlings and the absolute number of photons detected extracted and plotted. True photon imaging requires at least two microchannel plates to generate a pulse height distribution curve. The high sensitivity of the camera used here is sufficient to detect Ca^{2+} signals in seedlings and plants over integrations of greater or equal to 1 s.

Protocol

(1) Reconstitute aequorin in the sample *in vivo* by treating with coelenterazine as described above.
(2) Place sample inside light-tight box.
(3) Set camera to ‘bright-field mode’. Focus camera on sample using real-time monitor. Take an average of four bright-field images. Save this image.
(4) Close the door of the light-tight box and ensure that the locking mechanism is closed. Set the camera to ‘photon-counting mode’. Set up photon counting integration, with neutral density setting at 100%.
(5) At the end of the experiment, stop the integration and save the sequential time-resolved integration (TRI) file.

Notes

- The light-tight box can be customized for particular types of experiment. A very useful feature is to integrate a Peltier element into the base to perform temperature stress experiments.
- For bright-field images, the door of the light-tight box is left open and the specimen illuminated obliquely with either spectacle torches or a head-torch. The safety cut-out mechanism which will disable the camera when the door is opened needs to be overridden manually (Photek are currently developing a software ‘cure’ for this).
- When photon-counting, with the neutral density setting at 100%, the camera is operating at maximum gain and hence maximum sensitivity. This is fine for aequorin-containing samples as this level of sensitivity is required for small responses and even the largest response will not saturate the camera. In the unlikely event that it did, the camera is equipped with an ‘overbright’ cut-out safety mechanism.
- The TRI sequential file contains all the imaging pixel-array information at full or half video rate. The processing software allows this information to be recalled and displayed for a series of user-defined integrations, a significant advantage of the Photek system.

42.8.3 Calibration of aequorin luminescence

There are several approaches to calibrate the aequorin signal *in planta*. In the first method, cytosolic Ca^{2+} activities are calculated from aequorin luminescence by comparison of the rate of Ca^{2+}-triggered luminescence from the aequorin in the cell to the peak rate of light emission at saturating $[Ca^{2+}]$ (Cobbold & Rink, 1987; Gilroy, *et al.*, 1989). There are significant differences, however, between *in vivo* and *in vitro* calibrations (Gilroy *et al.*, 1989). In the second method, the proportion of the total aequorin consumed at any point in time is related to the cytosolic calcium concentration by an empirically derived formula. The total amount of luminescence is measured as the integral of all luminescence during the experiment and after complete discharge of all the aequorin following permeabilization of the tissue in the presence of high calcium.

Protocol

(1) Record the total luminescence during the experiment.
(2) At the end of each experiment discharge the remaining aequorin with 1 M $CaCl_2$ and 10% ethanol, and record counts for a further 1–2 min, or until they fall to approximately one-thousandth of the maximum value recorded.
(3) Measure the background luminescence from wild-type seedling and subtract the average value from all the experimental data points.
(4) Measure the total number of counts over the course of the whole experiment, including the discharge of aequorin at the end.
(5) The rate constant, k, at each time point in the experiment is given as:

$$k = \frac{\text{luminescence counts s}^{-1}}{\text{total luminescence counts}} \quad [42.8]$$

(6) Conversion to calcium uses an empirically derived (Cobbold & Rink, 1987) calibration formula which is specific for the isoform of aequorin encoded by pMAQ2 (Badminton *et al.*, 1995).

$$pCa = 0.332588(-\log k) + 5.5593 \quad [42.9]$$

Notes

- These calibration coefficients were determined at 25°C. Different values are required at different temperatures. This can cause difficulties when using temperature treatments such as the cold-shock control (see below).

42.8.3.1 The cold-shock control

The cold-shock response occurs when plant tissues experience a rapid drop in temperature, producing an immediate and large increase in $[Ca^{2+}]_{cyt}$. This has proven to be a very reliable response and demonstrates that functional aequorin has been reconstituted in the plant cells and that the luminometer or imaging equipment is capable of measuring Ca^{2+}-induced changes in luminescence. The drop in temperature can be achieved using a Peltier-cooled stage or by placing a small piece of ice next to the seedling on the microscope stage. When the ice melts, cold water reaches and stimulates the seedling. When using the luminometer, the cold-shock response can be caused by injecting ice-cold water into the luminometer cuvette containing the plant.

42.8.4 Additional techniques for aequorin measurements

42.8.4.1 Semisynthetic aequorins

Semisynthetic aequorins are molecules made up of natural apoaequorin coupled with a chemically synthesized analogue of coelenterazine (Shimomura *et al.*, 1988). Semisynthetic aequorins have a very wide range of sensitivities to Ca^{2+} ranging from *c.* 0.01 to 200 times that of natural aequorin (Shimomura, 1991). It should be noted, however, that semisynthetic aequorins have significantly lower stability and a marked reduction in half-life compared with natural aequorins; measurement of luminescence must therefore be carried out within a few hours of reconstitution. Semisynthetic aequorins may be reconstituted *in vitro*, for subsequent loading by traditional methods (Lliñas *et al.*, 1992), but can also be reconstituted in whole plants combining chemical analogues of coelenterazine with the recombinant apoaequorin expressed in the plant tissues (Knight *et al.*, 1993).

Semisynthetic aequorins can be used to measure Ca^{2+} levels over a wider range than normal cytoplasmic Ca^{2+} concentrations. For instance, n-aequorin, with a low sensitivity to calcium, has been used in neuron cells to report very high Ca^{2+} levels (Lliñas *et al.*, 1992), whereas h-aequorin, which is very sensitive to calcium, has been used to demonstrate very small changes in $[Ca^{2+}]_{cyt}$ which occur during the wounding response in plants (Knight *et al.*, 1993).

A variant of coelenterazine known as e-type coelenterazine confers on aequorin a bimodal emission peak. Light emission can be measured at two wavelengths (approx. 405 nm and 465 nm) and the ratio of the amounts of light is directly proportional to the calcium concentration but independent of the concentration of aequorin in the tissues, making this method the most reliable way of quantifying calcium concentrations measured by aequorin bioluminescence (Shimomura *et al.*, 1988).

42.8.4.2 Engineered aequorins

Ten different isoforms of apoaequorin exist naturally and their sensitivities to calcium vary such that the most sensitive is about 10 times more sensitive than the least (Shimomura, 1991). Other modifications in the protein structure can affect the calcium-binding properties of aequorin. Aequorin has three cysteine residues which appear to have a role in the bioluminescence reaction, either in a catalytic function or in the regeneration of active aequorin. Replacing these residues causes a reduction in the amount of luminescence produced by the protein, except when all three are replaced by serine, in which case luminescence increases (Kurose *et al.*, 1989). Single amino acid substitutions in the EF hand regions of apoaequorin produce a protein with a reduced affinity for calcium (Kendall *et al.*, 1992). This type of aequorin can be used for making measurements in cellular locations where Ca^{2+} levels are expected to be orders of magnitude higher than in the cytosol, e.g. the mitochondria, and provides an alternative to the use of n-aequorin, with the advantage of avoiding the use of the less stable n-coelenterazine.

Engineered aequorins can now be produced with different emission spectra (Ohmiya *et al.*, 1992) and these aequorins which emit different colours of light may be of use in future developments (see below).

42.9 MANIPULATION OF INTRACELLULAR EVENTS USING CAGED PROBES

Caged compounds are molecules whose biological activities have been chemically masked by a photolytic 'caging' group. Illuminating the caged compound with UV light causes the cage to dissociate and the biologically active molecule to be released (Adams & Tsien, 1993). Through controlling the timing, intensity, and region of illumination, the dynamics of caged probe release can be tightly regulated, allowing analysis of the spatial and temporal components of signal transduction.

The list of compounds that have been caged is extensive including: ions, ionophores, enzymes, proteins, drugs and hormones (see Chapter 40 for a comprehensive review of the field). Several commercial distributors provide a wide range of these compounds. However, many of the more 'exotic' caged compounds (such as caged plant hormones: see Ward & Beale, 1995) are not commercially available and must be obtained from the laboratory that synthesized them. The alternative is to synthesize the caged compound in-house. The availability of simple nitrophenyl ethyl ester caging technology from Molecular Probes makes this a real possibility. However, as always, chemical synthesis is not for the faint of heart and a clear plan of how to separate and characterize the products of the caging reaction is essential before embarking on this kind of project. There are no good hard and fast rules for such syntheses except that access to a good synthetic chemist is almost essential.

Assuming a ready supply of your caged compound is available storage for several months is possible if the aliquoted stock is stored at −80°C in the dark. The caged compounds are labile under UV light but working in complete darkness is usually not needed and simply handling them under subdued room lighting and storing them in the dark are adequate precautions to prevent their premature discharge.

42.9.1 Experimental protocols

42.9.1.1 Introduction of caged compounds into cells

The successful approaches used to date to incorporate caged probes into plant cells have been microinjection, ester loading and electroporation. The approaches, caveats, controls, need for optimization, and difficulties with these approaches when applied to caged compounds are identical to those outlined above for fluorescent dye experiments.

42.9.1.2 Uncaging equipment

(1) *The epi-illuminator of a standard fluorescence microscope*: 1–10 s illumination through a narrow band UV interference filter around 340 nm releases up to 100% of the caged molecules. Localized uncaging is possible by closing the epifluorescence diaphragm to illuminate only a small spot in the center of the field of view.
(2) *The UV laser of a confocal microscope*: The ability to steer and restrict the scan of such a UV laser provides a very high degree of spatial control on the uncaging process.
(3) *Two (or three)-photon excitation*: Two-photon systems provide a better defined volume of released probe and most of the specimen is only illuminated with longer-wavelength light.
(4) *A photographic flash gun*: Remove the plastic UV absorbing lens and mount on the stage of the microscope or in place of the epi-illuminator. Care needs to be taken with this approach as the electrical surge associated with flash gun discharge can damage sensitive photodetectors. Additionally the flash gun approach makes it much harder to control the intensity and spatial localization of illumination.
(5) *Transilluminator*: Where biochemical analysis is the goal, uncaging can be achieved by placing the loaded cells in a UV-transmissive cuvette on a standard transilluminator. Uncaging efficiency can then be controlled by altering the time of

exposure to UV light or attenuating the illumination with an appropriate neutral density filter.

42.9.1.3 *Artefacts and controls*

Being able to regulate the timing, intensity and localization of illumination become critical factors when designing the caged probe experiment. This is because only when endogenous changes are closely mimicked is true cellular regulation likely to be revealed.

This need to mimic the amplitude and spatial dynamics of endogenous changes highlights the need to monitor the site and efficiency of caged probe release *in situ*. When the caged compound is affecting a detectable cellular parameter, e.g. changes in Ca^{2+} levels caused by caged Ca^{2+}-ionophore, using fluorescent or luminescent probes to monitor this parameter can indicate how well the caged probe is working (Allan *et al.*, 1994; Gilroy, 1996; Malhó & Trewavas, 1996; Franklin-Tong *et al.*, 1996; Bibikova *et al.*, 1997). When the released compound is undetectable, e.g. ABA released from caged-ABA, a useful approach to visualizing the extent of photoactivation is to co-load the cell with caged fluorescein. Caged fluorescein is non-fluorescent until it is uncaged. Thus, the production of fluorescein in the sample is an indication of how efficiently caged compounds are being photoactivated. Although indirect, the caged fluorescein approach is a much preferable alternative to simply assuming that UV irradiation has photoactivated the caged-compound in your cell.

UV irradiation and photolysis by-products produced during the photoactivation of the caging group can be highly damaging and may also have biological effects. Thus extensive controls for the viability of cells used in caged probe experiments are needed.

Minimal controls include repeating experiments:

(1) without the caged compound but with UV irradiation;
(2) with the caged compound but without UV irradiation;
(3) with an inactive caged analogue, if available.

An example of this last control is the use of Diazo-3 as a control for Diazo-2 in experiments to manipulate cellular Ca^{2+} signaling. Both are caged Ca^{2+} chelators with almost identical chemical structures, but photoactivation of Diazo-2 produces a chelator with 5× the affinity for Ca^{2+} of Diazo-3. Similarly, a non-photoactivatable analogue of caged ABA has been synthesized as a control for caged ABA experiments (Allan *et al.*, 1994).

42.9.1.4 *The need for a functional assay*

Perhaps the most significant aspect of caged probe experiments is developing appropriate measures of cellular activities to determine the effect of the regulator released upon caged probe photoactivation. In microscope-based experiments single cell parameters such as growth (Malhó & Trewavas, 1996; Bibikova *et al.*, 1997), cell structure (Fallon *et al.*, 1993; Gilroy *et al.*, 1991; S. Gilroy, 1996), gene expression (Gilroy, 1996), ion channel activity (Blatt *et al.*, 1990; L.A. Romano, S. Gilroy & S. Assmann, unpublished data) and exocytosis (Gilroy, 1996) have all been used to determine whether photoactivation of a caged probe has had an effect on cell regulation.

A few plant studies have used caged probes to manipulate populations of cells where the assay of function is a biochemical change such as protein kinase activity (Fallon *et al.*, 1993). Such biochemical approaches are elegant in that they use the controllability of caged probe technology to manipulate cells but need a large number of loaded cells in order to be able to perform biochemical assays. So far electroporation and ester loading of caged probes have provided the only successful loading strategies for caged probes into large numbers of plant cells.

42.10 PROBES FOR OTHER COMPARTMENTS

A description of events in the cytoplasm gives a limited view point of dynamic metabolism in plant cells. The massive transport events occurring across the plasma membrane and tonoplast and the ionic balance in the apoplast and vacuole are all potential areas needing research. Many of the probes, particularly Ca^{2+}-dyes, have been specifically tailored to respond to cytosolic ion concentrations within a limited range of pH and ionic composition. Alternative probes or modification of the dyes (e.g. higher K_d values) are required if they are to be useful in other compartments.

42.10.1 Apoplast

The ionic composition of the apoplast is poorly defined, mainly as the high ion exchange capacity of the wall creates a host of polarized microenvironments for selective ion binding. Local ion activities are dependent to a large extent on the interaction between the supply of ions from the neighbouring apoplast or bathing medium, the wall polymer composition and the selective ion transport phenomena occurring in adjacent cells (Grignon & Sentenac, 1991). Membrane potentials, activity of extracellular enzymes, wall structure and binding of ligands to receptors are all likely to be affected by the apoplastic environment.

Apoplastic pH has been measured using soluble pH indicators such as umbelliferones (Pfanz & Dietz, 1987) or Cl-NERF (Taylor *et al.*, 1996). In the latter case, the signal was ratioed against the pH-insensitive signal from Texas Red dextran measured in parallel. Direct coupling of ratioable pH dyes to dextrans may also be desirable to prevent permeation or uptake by non-specific transporters in the membrane (Mühling *et al.*, 1995; Hoffmann *et al.*, 1992; Hoffman & Kosegarten, 1995). Typically dyes are introduced into the apoplast via the transpiration stream, cut surfaces and/or vacuum infiltration.

Apoplastic [K^+] has been measured using ratio imaging of benzofuran isophthalate (PBFI) (Mühling & Sattelmacher, 1997) and gives values between 20 and 25 mM on the abaxial leaf surface to 5–8 mM on the adaxial surface, with higher concentrations around stomatal guard cells. These values compare with microelectrode measurements that range from 50 μM (Blatt, 1985) to 3–100 mM (Bowling, 1987) around stomatal guard cells.

42.10.2 The vacuole

Dynamic changes in vacuolar morphology and ion transport are of increasing interest. pH measurements rely on uptake and compartmentalization of appropriate dyes in the vacuole (e.g. esculentin, pyranine and fluorescein derivatives: Yin *et al.*, 1990; CDCF-DA: Yoshida, 1995; BCECF: Swanson & Jones, 1996; Brauer *et al.*, 1995; CF: Davies *et al.*, 1996). Estimates of pH have been based on the ability of intact cells to take up a variety of fluorescent pH indicators with differing K_d values. The changes in fluorescence can be attributed to pH in particular compartments on the basis of the distribution of the dye and the pH range over which the dye is responsive (Yin *et al.*, 1990). Care has to be taken with such measurements from intact tissues as changes in fluorescence may result from changes in other parameters, such as light scattering (Yin *et al.*, 1996). More conventional ratioing approaches may suffer contamination of the vacuolar signal with cytoplasmic signal, and therefore represent a complex average of the pH in the two compartments (Brauer *et al.*, 1995). The calibration response of dyes in the vacuole may also be markedly different from the response in the cytoplasm. Optical sectioning using confocal microscopy allows signals from vacuole and cytoplasm to be distinguished and may permit simultaneous measurement of ion activities in both compartments (Fricker *et al.*, 1994). Certain cells contain autofluorescent compounds in the vacuole that respond to pH, which have been imaged to follow changes in vacuolar morphology during guard cell development (Palevitz *et al.*, 1981), but not yet changes of pH in mature guard cells.

In principle, membrane-permeant weak amines accumulate in acidic compartments in response to the pH gradient across the intervening membrane. Fluorescent acridine derivatives fall into this category and can be readily imaged in plant vacuoles. However, many of these derivatives appear to interact with vacuolar components and their partitioning does not correctly respond to pH (J.L. Wood & M.D. Fricker, unpublished). Acridine derivatives also exhibit complex and diverse staining behaviour and can be highly photo-toxic (Gupta & De, 1988).

42.11 FUTURE DEVELOPMENTS

The number of ions that can be imaged is increasing all the time. Dyes for potassium, sodium, magnesium, nickel, aluminium and chloride exist, but there are still only a limited number of reports on their use in plants (e.g. Lindberg, 1995; Lindberg & Strid, 1997; Mühling and Sattelmacher, 1997; Vitorello & Haug, 1997). Following the trends in recent years, we would expect more multiple parameter measurements combining calcium and/or pH measurements with electrophysiological measurements (e.g. Schroeder & Hagiwara, 1990; Grabov & Blatt, 1997; Bauer *et al.*, 1997; Felle & Hepler, 1997; Thiel *et al.*, 1997) or vibrating probe measurements (e.g. Pierson *et al.*, 1996). Interest is also developing in techniques to measure other components of the signal transduction chain, such as calmodulin distribution (Love *et al.*, 1997).

However, we feel the most significant advances are likely to arise from application of molecular genetics, which offers staggering potential for precisely targeted functional analysis of key signalling and regulatory components, as well as providing optical techniques as *in vivo* assay systems. The aequorin transformation studies have elegantly demonstrated the benefits of a molecular biological solution to calcium measurements. One recent improvement in this area is that the response can now be quantified using dual-wavelength coelenterazines (Knight *et al.*, 1993). The recent development (Miyawaki *et al.*, 1997) of a Ca^{2+}-sensor based on fluorescence energy transfer (FRET) between different-wavelength versions of GFP is reminiscent of the introduction of the first fluorescent Ca^{2+}-indicator quin-2 by Roger Tsien and colleagues in 1985. Quin-2 was the progenitor that heralded the explosive growth in the availability of ion-indicating fluorescent dyes. The chameleon probes can be expressed in plants and can be imaged with dual-emission confocal systems. The approach exemplified by the chameleon probes combine the advantages of fluorescent measurements with genetic engineering to target probes to different cells or subcellular compartments. In addition, it should be possible to engineer different linkers that modify the FRET signal in response to a wide variety of molecules other than just inorganic ions.

ACKNOWLEDGEMENTS

We thank the USDA, NSF, Nuffield Foundation, Royal Society, INTAS and BBSRC for financial support. NSW is a Royal Society Industry Fellow.

REFERENCES

Abel S. & Theologis A. (1994) *Plant J.* **5**, 421–427.
Adams S.R. & Tsien R. (1993) *Annu. Rev. Physiol.* **55**, 755–784.
Aikens R. S. (1990) In *Optical Microscopy for Biology*, B. Herman & K. Jacobson (eds). Wiley-Liss, Inc., New York, pp. 207–218.
Allan A.C., Fricker M.D., Ward J.L., Beale M.H. & Trewavas A.J. (1994) *Plant Cell* **6**, 1319–1328.
Almers W. & Neher E. (1985) *FEBS Lett.* **192**, 13–18.
Ameloot M., van den Bergh V., Boens N., de Schryver F.C. & Steels P. (1993) *J. Fluorescence* **3**, 169–171.
Badminton M.N., Kendall J.M., Scala-Newly G. & Campbell A.K. (1995) *Exp. Cell Res.* **216**, 236–243.
Bauer C.S., Plieth C., Hansen U.-P., Sattelmacher B., Simonis W. and Schönknecht G. (1997) *FEBS Lett.* **405**, 390–393.
Bechtold N., Ellis J., Pelletier G. (1993) *C. R. Acad. Sci. Paris* **316**, 1194–1199.
Bibikova T.N., Zhigilei A. & Gilroy S. (1997) *Planta* **203**, 495–505.
Blatt M.R. (1991) *Methods Plant Biochem.* **6**, 281–321.
Blatt M.R. (1985) *J. Exp. Bot.* **36**, 240–251.
Blatt M.R., Thiel G. & Trentham D.R. (1990) *Nature* **346**, 766–769.
Bowling D.J.F. (1987) *J. Exp. Bot.* **38**, 1351–1355.
Braam J. (1992) *Proc. Natl Acad. Sci. USA* **89**, 3213–3216.
Brauer D., Otto J. & Tu S.-I. (1995) *J Plant Physiol.* **145**, 57–61.
Brauer D., Uknalis J., Triana R. & Tu S.-I. (1996) *Protoplasma* **192**, 70–79.
Bright G.R., Fisher G.W., Rogowska J. & Taylor D.L. (1989) *Methods Cell Biol.* **30**, 157–192.
Brownlee C. & Pulsford A.L. (1988) *J. Cell Sci.* **91**, 249–256.
Brownlee C., Wood J.W. & Briton D. (1987) *Protoplasma* **140**, 118–122.
Bush D.S. & Jones R.L. (1987) *Cell Calcium* **8**, 455–472.
Bush D.S. & Jones R.L. (1990) *Plant Physiol.* **93**, 841–845.
Bush D.S., Biswas A.K. & Jones R.L. (1989) *Planta* **178**, 411–420.
Campbell A.K. (1988) In *Chemiluminescence Principles and Applications in Biology and Medicine*. Ellis Horwood Ltd, Chichester, UK.
Campbell A.K., Trewavas A.J. & Knight M.R. (1996) *Cell Calcium* **19**, 211–218.
Callaham D.A. & Hepler P.K. (1991) In *Cellular Calcium – A Practical Approach*, J.G. McCormack & P.H. Cobbold (eds). Oxford University Press, Oxford.
Caswell A.H. (1979) *Int. Rev. Cytol.* **56**, 145–181.
Caswell A.H. & Hutchinson J.D. (1971) *Biochem. Biophys. Res. Commun.* **43**, 625–630.
Cobbold P.H. & Rink T.J. (1987) *Biochem. J.* **248**, 313–328.
Cole L., Coleman J.O.D., Kearns A., Morgan G. & Hawes C. (1991) *J. Cell Sci.* **99**, 545–555.
Coleman J.O.D., Hiscock S.J. & Dewey F.M. (1994) *Physiol. Mol. Plant Path.* **43**, 391–401.
Cork R.J. (1986) *Plant Cell Environ.* **9**, 157–161.
Cramer G.R. & Jones R.L. (1996) *Plant Cell Environ.* **19**, 1291–1298
Davies T.G.E., Steele S.H., Walker D.J. & Leigh R. (1996) *Planta* **198**, 356–364.
De Boer A.H., Van Duijn B., Giesberg P., Wegner L., Obermeyer G., Kohler K., Linz K.W. (1994) *Protoplasma* **178**, 1–10.
Dixon G.K., Brownlee C. & Merrett M.J. (1989) *Planta* **178**, 443–449.
Draper J., Scott R., Armitage P. & Walden R. (1988) In *Plant Genetic Transformation and Gene Expression: A Laboratory Manual*. Blackwell Science, Oxford.
Dudler R. & Hertig C. (1992) *J. Biol. Chem.* **267**, 5882–2888.
Elliott D.C. & Petkoff H.S. (1990) *Plant Sci.* **67**, 125–131.
Ehrhardt D.W., Wais R. & Long S.R. (1996) *Cell* **85**, 673–681.
Errington R.J., Fricker M.D., Wood J.L., Hall A.C. & White N.S. (1997) *Am. J. Physiol.* **272**, 1040–1051.
Fabiato A. (1988) *Methods Enzymol.* **157**, 378–417.
Fallon K.M., Shacklock P.S. & Trewavas A.J. (1993) *Plant Physiol.* **101**, 1039–1045.
Fasolato C., Hoth M., Matthews G. & Penner R. (1993) *Proc. Natl Acad. Sci. USA* **90**, 3068–3072.
Felle H.H. & Hepler P.K. (1997) *Plant Physiol.* **114**, 39–45.
Felle H.H., Tretyn A. & Wagner G. (1992) *Planta* **188**, 306–313.
Foissner I. (1991) *Plant Cell Environ.* **14**, 907–915.
Franchisse J.-M., Johannes E. & Felle H.H. (1988) *Biochem. Biophys. Acta* **938**, 199–210.
Franklin-Tong V.E., Drøbak B.K., Allan A.C., Watkins P.A.C. & Trewavas A.J. (1996) *Plant Cell* **8**, 1305–1321.
Fricker M.D., Tlalka M., Ermantraut J., Obermeyer G., Dewey M., Gurr S., Patrick J. & White, N.S. (1994) *Scanning Microsc.* **8**, 391–405.
Fricker M.D., Errington R.J., Wood J.L., Tlalka M., May M. & White N.S. (1997a) In *Signal Transduction – Single Cell Research*, B. Van Duijn & A. Wiltnik (eds). Springer-Verlag, Heidelberg, pp 413–445.
Fricker M.D., White N.S. & Obermeyer G. (1997b) *J. Cell Sci.* **110**, 1729–1740.
Gehring C.A., Williams D.A., Cody S.H. & Parish R.W. (1990a) *Nature* **345**, 528–530.
Gehring C.A., Irving H.R. & Parish R.W. (1990b) *Proc. Natl Acad. Sci. USA* **87**, 9645–9649.
Gibbon B.C. & Kropf D.L. (1994) *Science* **263**, 1419–1421.
Gilroy S. (1996) *Plant Cell* **8**, 2193–2209.
Gilroy S. (1997) *Ann. Rev. Plant Physiol. Plant Mol. Biol.* **48**, 165–190.
Gilroy S. & Jones R.L. (1992) *Proc. Natl Acad. Sci. USA* **89**, 3591–3595.
Gilroy S.G., Hughes W.A. & Trewavas A.J. (1986) *FEBS Lett.* **199**, 217–221.
Gilroy S.G., Hughes W.A. & Trewavas A.J. (1989) *Plant Physiol.* **90**, 482–491.
Gilroy S.G., Read N.D. & Trewavas A.J. (1990) *Nature* **346**, 769–771.
Gilroy S.G., Fricker M.D., Read N.D. & Trewavas A.J. (1991) *Plant Cell* **3**, 333–444.
Grignon C. & Sentenac H. (1991) *Annu. Rev. Plant Physiol.* **42**, 103–128.

Grabov A. & Blatt M.R. (1997) *Planta* **201**, 84–95.
Groden D.L., Guan Z. & Stokes B.T. (1991) *Cell Calcium* **12**, 279–287.
Grynkiewicz G., Poenie M. & Tsien R.Y. (1985) *J. Biol. Chem.* **260**, 3440–3450.
Gupta A.S. & Berkowitz G.A. (1989) *Plant Physiol.* **89**, 753–761.
Gupta H.S. & De D.N. (1988) *J. Plant Physiol.* **132**, 254–256.
Haley A., Russell A.J., Wood N., Allan A.C., Knight M., Campbell A.K. & Trewavas A.J. (1995) *Proc. Natl Acad. Sci. USA* **92**, 4124–4128.
Haugland R.P. (1996) In *Handbook of Fluorescent Probes and Research Chemicals*, 6th edn. Molecular Probes, Eugene, Oregon.
Henriksen G.H., Assmann S.M. (1997) *Pflugers Arch. Eur. J. Physiol.* **433**, 832–841.
Henriksen G.H., Taylor A.R., Brownlee C. & Assmann S.M. (1996) *Plant Physiol.* **110**, 1063–1068.
Highsmith S., Bloebaum P., & Snowdown K.W. (1986) *Biochem. Biophys. Res. Commun.* **138**, 1153–1162.
Hillmer S., Gilroy S. & Jones R.L. (1993) *Plant Physiol.* **102**, 279–286.
Hodick D., Gilroy S., Fricker M.D. & Trewavas A.J. (1991) *Bot. Acta* **104**, 221–228.
Hoffmann B. & Kosegarten H. (1995) *Physiol. Plantarum* **95**, 327–335.
Hoffmann B., Planker R. & Mengel K. (1992) *Physiol. Plant.* **84**, 146–153.
Huang C.-N., Cornejo M.J., Bush D.S. & Jones R.L. (1986) *Protoplasma* **135**, 80–87.
Inouye S., Noguchi M., Sakaki Y., Takagi Y., Miyata T., Iwanaga S., Miyata T. & Tsuji F.I. (1985) *Proc. Natl Acad. Sci. USA* **82**, 3154–3158.
Ishikawa T., Li Z.-S., Lu Y.-P. & Rea P.A. (1997) *Biosci. Rep.* **17**, 189–207.
Johnson C.H., Knight M.R., Kondo T., Masson P., Sedbrook J., Haley A. & Trewavas A.J. (1995) *Science* **269**, 1863–1865.
Kao J.P.Y., Harootunian A.C. & Tsien R.Y. (1989) *J. Biol. Chem.* **264**, 8179–8184.
Kendall J.M., Sala-Newby G., Ghalaut V., Dormer R.L. & Campbell A.K. (1992) *Biochem. Biophys. Res. Commun.* **187**, 1091–1097.
Klein M., Martinoia E. & Weissenböck G. (1997) *FEBS Lett.* **420**, 86–92.
Klein T.M., Wolf E.D., Wu R. & Sandford J.C. (1987) *Nature* **327**, 70–73.
Knight H. & Knight M.R. (1995) *Methods Cell Biol.* **49**, 201–216.
Knight H., Trewavas A.J. & Knight M.R. (1996) *Plant Cell* **8**, 489–503.
Knight H., Trewavas A.J. & Knight M.R. (1997) In *Plant Molecular Biology Manual*, S.B. Gelvin & Schilperoort (eds). Kluwer Academic Publishers, Dordrecht, **C4**, 1–22.
Knight M.R., Campbell A.K., Smith S.M. & Trewavas A.J. (1991) *Nature* **352**, 524–526.
Knight M.R., Read N.D., Campbell A.K., & Trewavas A.J. (1993) *J. Cell Biol.* **121**, 83–90.
Kolattukudy P.E., Purdy R.E. & Maitai I.B. (1981) *Methods Enzymol.* **71**, 652–654.
Kosegarten H., Grolig F., Wieneke J., Wilson G. & Hoffman B. (1997) *Plant Physiol.* **113**, 451–461.
Kurose K., Inouye S., Sakaki Y., & Tsuji F.I. (1989) *Proc. Natl Acad. Sci. USA* **86**, 80–84.
Lattanzio F.A. (1990) *Biochem. Biophys. Res. Commun.* **171**, 102–108.
Lattanzio F.A. & Bartschat D.K. (1991) *Biochem. Biophys. Res. Commun.* **177**, 184–191.
Legue V., Blancaflor E., Wymer C., Fantin D., Perbal G. & Gilroy S. (1997) *Plant Physiol.* **114**, 789–800
Lindberg S. (1995) *Planta* **195**, 525–529.
Lindberg S. & Strid H. (1997) *Physiol. Plantarum* **99**, 405–414.
Lliñas R., Sugimori M. & Silver R.B. (1992) *Science* **256**, 677–679.
Liu C. & Hermann T.E. (1978) *J. Biol. Chem.* **253**, 5892–5895.
Love J., Brownlee C. & Trewavas A.J. (1997) *Plant Physiol.* **115**, 249–261.
Malhó R. & Trewavas A.J. (1996) *Plant Cell* **8**, 1935–1949.
Malhó R., Read N.D., Trewavas A.J. & Pais M.S. (1995) *Plant Cell* **7**, 1173–1184.
McAinsh M.R., Brownlee C. & Hetherington A.R. (1992) *Plant Cell* **4**, 1113–1122.
McAinsh M.R., Webb A.A.R., Taylor J.E. & Hetherington A.M. (1995) *Plant Cell* **7**, 1207–1219.
Meindl U. (1982) *Protoplasma* **10**, 143–146.
Minta A., Kao J.P.Y. & Tsien R.Y. (1989) *J. Biol. Chem.* **264**, 8171–8178.
Miyawaki A., Llopis J., Heim R., McCaffery J.M., Adams J.A., Ikura M. & Tsien R.Y. (1997) *Nature* **388**, 882–887.
Mühling K. & Sattelmacher B. (1997) *J. Exp. Bot.* **48**, 1609–1614.
Mühling K.-H., Plieth C., Hansen U.-P. & Sattelmacher B. (1995) *J. Exp. Bot.* **46**, 377–382.
Neher E. & Augustine G.J. (1992) *J. Physiol.* **450**, 273–301.
Ohmiya Y., Ohashi M. & Tsuji F.I. (1992) *FEBS Lett.* **301**, 197–201.
Oparka K.J. (1991) *J. Exp. Bot.* **42**, 565–579.
Oparka K.J., Murphy R., Derrick P.M., Prior D.A.M. & Smith J.A.C. (1991) *J. Cell Sci.* **98**, 539–544.
Owen C.S. (1991) *Cell Calcium* **12**, 385–393.
Palevitz B.A., O'Kane D.J., Kobres R.E. & Raikhel N.V. (1981) *Protoplasma* **109**, 23–55.
Parton R.M., Fischer S., Malhó R., Papasouliotis O., Jelitto T.C., Leonard T. & Read N.D. (1997) *J. Cell Sci.* **110**, 1187–1198.
Pfanz H. & Dietz K.-J. (1987) *J. Plant Physiol.* **129**, 41–48.
Pheasant D.J. & Hepler P.K. (1987) *Eur. J. Cell Biol.* **43**, 10–13.
Pierson E.S., Miller D.D., Callaham D.A., Shipley A.M., Rivers B.A., Cresti M. & Hepler P.K. (1994) *Plant Cell* **6**, 1815–1828.
Pierson E.S., Miller D.D., Callaham D.A., van Aken J., Hackett G. & Hepler P.K. (1996) *Dev. Biol.* **174**, 160–173.
Piñeros M. & Tester M. (1995) *Planta* **195**, 478–488.
Plieth C. (1995) Dissertation am Institut für Angewandte Physik der Christian-Albrechts-Universität zu Kiel.
Plieth C. & Hansen U.-P. (1996) *J. Exp. Bot.* **47**, 1601–1612.
Plieth C., Sattelmacher B. & Hansen U.-P. (1997) *Protoplasma* **198**, 107–124.
Plieth C., Sattelmacher B., Hansen U.-P. & Thiel G. (1998) *Plant J.* **13**, 167–175.
Poenie M. (1990) *Cell Calcium* **11**, 85–91.
Polito V.S. (1983) *Protoplasma* **17**, 226–232.
Prasher D., McCann R.O. & Cormier M.J. (1985) *Biochem. Biophys. Res. Commun.* **126**, 1259–1268.
Rathore K.S., Cork R. J. & Robinson K. R. (1991) *Dev. Biol.* **148**, 612–619.
Read N.D., Allan W.T.G., Knight H, Knight M.R., Malhó R., Russell A., Shacklock P.S. & Trewavas A.J. (1992) *J. Microsc.* **166**, 57–86.
Reid R.J. & Whittington J. (1989) *J. Exp. Bot.* **40**, 883–891.
Reiss H.-D. & Herth W. (1978) *Protoplasma* **97**, 373–377.
Reiss H.-D. & Nobiling R. (1986) *Protoplasma* **131**, 244–246.

Reuveni M., Lerner H.R. & Poljakoff-Mayber A. (1985) *Plant Physiol.* **79**, 406–410.
Roe M.W., Lemesters J.J. & Herman B. (1990) *Cell Calcium* **11**, 63–73.
Russ U., Grolig F. & Wagner G. (1991) *Planta* **184**, 105–112.
Russell A.J., Knight M.R., Cove D.J., Knight C.D., Trewavas A.J. & Wang T.L. (1996) *Transgenic Res.* **5**, 167–170.
Scanlon M., Williams D.A. & Fay F.S. (1987) *J. Biol. Chem.* **262**, 6308–6312.
Schneider A.S., Herz R. & Sonenber M. (1983) *Biochemistry* **22**, 1680–1686.
Schroeder J.I. & Hagiwara S. (1990) *Proc. Natl Acad. USA* **87**, 9305–9309.
Shimomura O. (1991) *Cell Calcium* **12**, 635–643.
Shimomura O., Musicki B. & Kishi Y. (1988) *Biochem. J.* **251**, 405–410.
Soliday C.L., Flurkey W.H., Okita T.W. & Kolattukudy P.E. (1984) *Proc. Natl Acad. Sci. USA* **81**, 3939–3943
Speksnijder J.E., Weisenseel M.H., Chen T.-H. & Jaffe L.F. (1989) *Biol. Bull.* **176**(S), 9–13.
Stanley P.E. (1992) *J. Biolumin. Chemilumin.* **7**, 77–108.
Striggow F. & Ehrlich B.E. (1996) *J. Gen. Physiol.* **108**, 115–124.
Swanson S.J. & Jones R.L. (1996) *Plant Cell* **8**, 2211–2221.
Taylor D.P., Slattery J. & Leopold A.C. (1996) *Physiol. Plant.* **97**, 35–38.
Thiel G., Blatt M.R., Fricker M.D., White I.R. & Millner P. (1993) *Proc. Natl Acad. Sci. USA* **90**, 11493–11497.
Thiel G., Homann U. & Plieth C. (1997) *J. Exp. Bot.* **48**, 609–622.
Thomas M.V. (1982) In *Techniques in Calcium Research*. Academic Press, London.
Thomas P. & Delaville F. (1991) In *Cellular Calcium – A Practical Approach*, McCormack J.G. & Cobbold P.H. (eds). Oxford University Press, Oxford, New York, Tokyo, pp. 1–54.
Timmers A.C.J., de Vries S.C. & Schel J.H.N. (1989) *Protoplasma* **153**, 24–29.
Timmers A.C.J., Reiss H.-D. & Schel J.H.N. (1991) *Cell Calcium* **12**, 515–521.
Timmers A.C.J., Reiss H.-D., Bohsung J., Traxel K. & Schel J.H.N. (1996) *Protoplasma* **190**, 107–118.
Tretyn A. & Kopcewicz J. (1988) *Planta* **175**, 237–240.
Tretyn A., Kado R.T. & Kendrick R.E. (1997) *Folia Hist. Cyto.* **35**, 41–51.
Tsien R.Y. & Waggoner A. (1995) In *Handbook of Confocal Microscopy*, 2nd edn, J.B. Pawley (ed.). Plenum Press, New York, pp. 267–280.
Uto A., Arai H. & Ogawa Y. (1991) *Cell Calcium* **12**, 29–37.
Van der Schoot C. & Lucas W.J. (1995) In *Methods in Plant Molecular Biology. A Laboratory Manual*, P. Maliga, D.F. Klessig, A.R. Cashmore, W. Gruissem, J.E. Varner (eds.). Cold Spring Harbor Laboratory Press, pp. 173–192.
Vivaudou M.B., Arnoult C. & Villaz M. (1991) *J. Membr. Biol.* **122**, 165–175.
Vitorello V.A. & Haug A. (1997) *Plant Sci.* **122**, 35–42.
Wagner J. & Keizer J. (1994) *Biophys. J.* **67**, 447–456.
Ward J.L. & Beale M.H. (1995) *Phytochemistry* **38**, 811–816.
Watkins N.J. & Campbell A.K. (1993) *Biochem. J.* **293**, 181–185.
White N.S., Errington R.J., Fricker M.D. & Wood J.L. (1996) *J. Microsc.* **181**, 99–116.
Williams D.A. & Fay F.S. (1990) *Cell Calcium* **11**, 75–83.
Williams D.A., Cody S.H., Gehring C.A., Parish R.W. & Harris P.J. (1990) *Cell Calcium* **11**, 291–297.
Williamson R.E. & Ashley C.C. (1982) *Nature* **296**, 647–651.
Wright K.M., Horobin R.W. & Oparka K.J. (1996) *J. Exp. Bot.* **47**, 1779–1787.
Wymer C.L., Bibikova T.N. & Gilroy S. (1997) *Plant J.* **12**, 427–439.
Yin Z.-H., Neimanis S., Wagner U. & Heber U. (1990) *Planta* **182**, 244–252.
Yin Z.-H., Hüve K. & Heber U. (1996) *Planta* **199**, 9–17.
Yoshida S. (1995) *Plant Cell Physiol.* **36**, 1075–1079.
Zhang D.H., Callaham D.A. & Hepler P.K. (1990) *J. Cell Biol.* **111**, 171–182.
Zottini M. & Zannoni D. (1993) *Plant Physiol.* **102**, 573–578.

CHAPTER FORTY-THREE

Nuclear Calcium: Concepts and Controversies

GEETHA SHANKAR[1] & MICHAEL A. HORTON[2]

[1] NPS Pharmaceuticals Inc., Salt Lake City, UT, USA
[2] University College London, London, UK

43.1 INTRODUCTION

The transformation of the conventional light microscope from an instrument used to record static events to an instrument that documents dynamic signal transduction events at 'real-time' speeds has opened up a wide range of applications in cell biology. Limiting factors, such as the dynamic range of cameras, computer speed and memory, are undergoing changes at a rapid pace. Thus, the combination of increased sensitivity of signal detection with faster hardware has pushed the frontiers of 'real-time' recording of many biological events. It is evident that several factors have contributed and continue to influence the capabilities of imaging technology. These include advances in computer software and hardware, the evolution of cameras that function under low-light conditions, and the discovery and development of fluorescent dyes that allow us to document changes within a living cell. In essence, imaging technology exemplifies the marriage of several areas that have come together in a simple yet sophisticated harmony.

The importance of recording biological events in 'real-time' goes far beyond the basic quest for academic knowledge. With dramatic improvements in image acquisition and data processing, new methodologies have emerged in cellular imaging that have enhanced biomedical research in the biotechnology and drug industry arena. Faster and better time-resolved cellular events are contributing to the understanding of basic cellular biology as well as speeding up the drug discovery process. The need for greater speed, precision and automation has forced the emergence of novel strategies for labels, better software for image enhancement and analysis, and better integration of imaging with other laboratory functions. 'High-throughput' cellular assays are the driving force in drug discovery screening programmes, with assays utilizing transient end points such as changes in intracellular calcium and membrane potential to screen for compound activity.

The focus of this chapter is to review some of the concepts that have emerged from this technology. One aspect that has received much attention because of the utility of the information that it yields is the ability to resolve signals spatially within a cell. However, this is also an area that is prone to artifactual information

FLUORESCENT AND LUMINESCENT PROBES, 2ND EDN
ISBN 0–12–447836–0

and it has become clear that critical evaluation of the data is necessary before any statements can be made regarding the spatial characteristics of a calcium signal. We will specifically address the concept of nuclear calcium, the tools utilized to resolve the signal and the controversy that surrounds the exact localization of the signal. A review of recent literature in this field makes it abundantly clear that while the concept of nuclear calcium is in itself tremendously exciting, it is still a hotly debated issue, since the implications of such a phenomenon are far reaching. Thus, although the aim here is to discuss the recent advances in nuclear calcium signalling, we have utilized this concept as a basis for the possibilities that lie beyond. In short, nuclear calcium signalling is only a part of a more global, rapidly evolving field of calcium-dependent nuclear events. It should be emphasized that the data referenced here are by no means comprehensive. Recent reviews can be found in Santella & Carafoli (1997) and Perez-Terzic *et al.* (1997a).

Every cell has ionic charge gradients generated by ions like calcium, sodium and potassium. Gradients in intracellular hydrogen ion concentration also are important to many biological processes. Ionic concentrations inside cells change quickly and dramatically and underlie a wide range of cellular processes including development, growth, secretion and reproduction, so it is important to observe and understand them. As we have been able to apply image processing and low-light-level image capture from signals emitted by optical probes, it has also become clear that large standing ionic gradients occur within cells, and may persist for many seconds during stimulus or suppression of cellular activity. The source of ionic changes may also occur in widely different parts of the same cell, and even with small cells of only 10 μm or so, these ionic pools may be detected.

That the nucleus is capable of housing cellular machinery that supports calcium signalling in much the same way that the cytoplasm does, is a controversial issue. Much of the debate arises from data that are prone to artifactual signal contributions and consequent methodological and interpretational errors. The basis for the controversy is the structure of the nucleus itself, and whether it is capable of 'gating' or filtering small molecules and ions that it is exposed to by way of changes in the cytosol (*vide infra*).

43.2 THE NUCLEAR ENVELOPE AND NUCLEAR TRANSPORTERS

The nucleus is surrounded by a nuclear envelope that serves to insulate it from the cytoplasm. The nuclear envelope comprises an inner and outer nuclear membrane, the latter being continuous with the endoplasmic reticulum (ER). The inner and outer nuclear membranes enclose the nuclear cisterna, and fuse with each other periodically to form nuclear pore complexes (NPC). The process of nuclear protein transport has been studied extensively. Proteins that are actively transported into the nucleus possess a nuclear localization sequence (NLS) that is recognized by a transporter protein (importin) (Hicks & Raikhel, 1995) that targets it to the nucleus (Dingwall *et al.*, 1982). The process is energy dependent and is saturable (Hicks & Raikhel, 1995; Gorlich & Mattaj, 1996). There is some evidence to implicate the participation of a small guanosine triphosphatase (GTPase) Ran in providing the driving force for the transport protein importin (Gorlich & Mattaj, 1996), although there is evidence that nuclear transport can also occur in a GTPase-independent manner (Nakielny & Dreyfuss, 1998). The process is believed to be bi-directional (i.e. nuclear import and export of proteins) and there is growing evidence for export of proteins possessing the nuclear export signal (NES), including the regulation of MAP kinase kinase (MEK) activity with and without the NES sequence (Fukuda *et al.*, 1997). Actin, a highly conserved, ubiquitous cytoskeletal protein, was shown to contain two leucine-rich NES sequences; these regulate the spatial control of actin and exclude it from the nucleus (Wada *et al.*, 1998).

In contrast to import and export of proteins from the nucleus, the transport of small molecules and ions across the nuclear membrane is still open to interpretation. In particular, the regulation of calcium in the nucleus is still unresolved. The classical picture is one where the NPC is freely permeable to ions and small molecules, suggesting a rapid equilibration of changes in free calcium concentrations (Ca^{2+}) in the cytoplasm and the nucleus. There are several studies that have challenged this notion and have shown that changes in Ca^{2+} in the nucleus can be independent of changes in the cytoplasm and vice versa. This would further imply that the nucleus is capable of regulating calcium independently, and utilize this mechanism in calcium-dependent nuclear events. The size of the nuclear pore has been determined, based on studies using dextrans of different sizes, to be about 9 nm in diameter and about 15 nm long (Paine *et al.*, 1975). If these dimensions are correct, then an almost instantaneous equilibration of Ca^{2+} in the nucleus and cytoplasm would occur. However, patch-clamp studies using both isolated nuclei and intact cells have shown that pore gating, even to Ca^{2+}, is commonly observed (Mazzanti *et al.*, 1994; Bustamante *et al.*, 1995). Further, depletion of ER lumen Ca^{2+} (and consequently nuclear envelope Ca^{2+}) with thapsigargin and ionophores blocked protein transport as well as small molecule diffusion across the pore (Greber & Gerace, 1995). Ca^{2+} depletion from nuclei was shown to predispose the NPC to become 'plugged' (Perez-Terzic *et al.*, 1997b).

The likelihood that there is no instantaneous equilibrium between cytosolic and nuclear Ca^{2+} is supported by observations using fluorescent indicators such as fura-2 and fluo-3. Gradients have been observed in various cell types, including smooth muscle cells and neurons, among others (Birch *et al.*, 1993; Himpens *et al.*, 1994). In starfish oocytes, microinjection of fura-2 free acid or Calcium Green dextran showed relatively no passage of dye when injected in the nucleus for as long as 60 minutes (Santella & Kyozuka, 1994). In contrast, studies using Calcium Green dextran and chimaeric aequorin with and without the NLS showed that calcium waves spread freely into the nucleus where appropriate nuclear targeting occurred. Thus, the subject of nuclear pore permeability to Ca^{2+} remains a hotly debated one and consequently so does the idea of independent regulation of calcium by the nucleus. It is possible that the explanation lies in the magnitude of the Ca^{2+} signal, with smaller changes equilibrating rapidly, thus creating the notion of a permeable pore, and larger changes remaining confined spatially, supporting observations of pore 'gating' (Al-Mohanna *et al.*, 1994).

43.3 NUCLEAR CALCIUM

The possibility of spatially regulated Ca^{2+} signals, raises the question, in this particular instance, of the location of calcium stores. The nuclear envelope, whose lumen is continuous with the ER, is a likely candidate. In primary cultures of rat hepatocyes, increases in nuclear calcium were found to arise from inositol 1,4,5-trisphosphate (IP_3)-sensitive stores in the nuclear envelope (Lin *et al.*, 1994). Studies have shown that calcium transporters, such as Ca-ATPases, and IP_3-sensitive Ca^{2+} channels, as well as channels regulated by ryanodine and cyclic ADP ribose, are present in the nuclear membrane (Galione *et al.*, 1991; Gerasimenko *et al.*, 1995). IP_3-sensitive channels have been described in the inner membrane of the nuclear envelope, suggesting that IP_3 production occurs within the nucleus (Humbert *et al.*, 1996). There is ample evidence to suggest that phosphatidylinositol (PI) hydrolysis occurs within the nucleus (Divecha *et al.*, 1991), as well as the detection of many of the enzymes of this pathway (Martelli *et al.*, 1992). However, as with most aspects of nuclear calcium signalling, there remains a controversy as to whether nuclear inositides regulate nuclear Ca^{2+}. It has been suggested that nuclear protein kinase activity may be a likely outcome, but the authors also argue that no IP_3 receptors have been demonstrated in the inner membrane of the nuclear envelope (Divecha *et al.*, 1994) while data from Humbert *et al.* (1996) show that IP_3 and IP4 receptors can be localized to the inner membrane. It is therefore likely that, given the scenario where the calcium signalling machinery can exist within the nucleus, and the calcium stores reside in the nuclear envelope, any stimulus that causes an increase in nuclear IP_3 can cause increases in nuclear as well as cytoplasmic Ca^{2+}, depending on activation of one or both ligand-gated channels (i.e. outer or inner nuclear membrane).

The role of calcium in the nucleus gets more intriguing as one considers the complexity of transmitting a cell surface stimulus to a calcium signal in the nucleus with subsequent actions that can alter cellular behaviour and function. We have shown that a cell surface ligand, such as the bone sialoprotein fragment BSP-IIA, that recognizes an integrin, the $\alpha_v\beta_3$ vitronectin receptor on osteoclasts, can cause changes in nuclear calcium (Shankar *et al.*, 1993; Plate 43.1). It has been argued that this observation may be based on compartmentalization of fura-2 when loaded with the acetoxymethylester form of the dye. N. Parkinson and W.T. Mason (personal communication) found that when osteoclasts were loaded with fura-2 dextran by microinjection and compared with fura-2 AM loaded cells, nuclear 340/380 ratio was higher in the latter instance, while osteoclasts with fura-2 dextran showed a uniform increase in calcium in nuclear and cytosolic regions of the cell. While the possibility of artifactual signal measurement with fura-2 AM cannot be ruled out, there are several observations in the original study that do not support this notion. Firstly, the spatial localization of the signal was ligand-dependent (Plate 43.1(a)). While the calciotropic hormone calcitonin gave a largely cytoplasmic response, peptides containing the Arg–Gly–Asp (RGD) sequence, including BSP-IIA, gave a response that was localized to the nuclei of these multinucleated cells. Secondly, similar spatial characteristics were observed when osteoclasts were loaded with either fura-2 AM or fluo-3 AM (Plate 43.1(b)). The latter dye was used in confocal scanning experiments where the signal was observed in the nuclei of osteoclasts prior to an increase in cytosolic calcium. Whether the calcium signal is perinuclear or nuclear is still debatable.

Integrin-dependent cell adhesion to specific extracellular matrix proteins has been shown to trigger changes in gene expression (Akiyama, 1996; Giancotti, 1997). In some cases, such a regulation of gene expression has been shown to be facilitated by a low-abundance phosphoprotein, zyxin, which associates with integrin at focal contacts, and shuttles between the cytosol and nucleus (Nix & Beckerle,

1997). Whether a similar scenario prevails in osteoclasts is unclear.

More recently, elastin peptides were shown to cause increases in nuclear and cytoplasmic free calcium in human endothelial cells (Faury *et al.*, 1998). In trying to establish the origin of nuclear calcium signals, Lipp *et al.* (1997) have demonstrated, by using confocal imaging, that all nuclear calcium signals arise as cytoplasmic events. Interestingly, they found that Ca^{2+} 'puffs', or individual cytoplasmic elementary release events, caused transient increases in nucleoplasmic ionized calcium concentrations. The majority of the Ca^{2+} puffs were found to be in close proximity with the nucleus, suggesting that these elementary release events facilitate nuclear Ca^{2+} signalling.

Thus, in light of emerging data on the role of calcium in gene expression, the significance of a nuclear calcium signal following cell surface receptor activation can be easily envisioned. For instance, the cAMP-response element (CRE) in the c-fos promoter is the site of transcriptional regulation by nuclear calcium signals (Sheng *et al.*, 1990). Thus, the CRE binding protein CREB can function as a nuclear calcium responsive transcription factor. The mechanism of CREB activation is further regulated by phosphorylation catalysed by a number of protein kinases, including cAMP-kinase (Takuma *et al.*, 1997) CaM-kinases (Enslen *et al.*, 1995) and MAPK (Tan *et al.*, 1996).

Other Ca^{2+}-dependent enzyme reactions in the nuclei include the demonstration of myosin light chain kinase (MLCK) in chicken liver nuclear matrix (Simmen *et al.*, 1994) and other tissues. Since MLCK is calmodulin (CaM)-dependent, this implicated the presence of CaM in the nucleus, an issue that was eventually resolved by immunocytochemical studies (Bachs *et al.*, 1991). These findings have opened up a whole array of possibilities of calcium-dependent nuclear functions, many of which continue to emerge. The role of calcium and CaM in the nucleus in apoptosis and gene expression continue to draw much attention. In addition to MLCK, CaM-kinase II and IV are present in nuclei of several cells (Ohta *et al.*, 1990; Matthews *et al.*, 1994). The nuclear substrates of CaM-kinase II are believed to be transcription factors, such as CREB (Dash *et al.*, 1991), which has been implicated in synaptic plasticity and cognition enhancement. Further, several Ca-dependent proteases, including calpain and calcineurin, have been implicated in apoptosis and activation of transcription factors (Squier *et al.*, 1994; Matheos & Cunningham, 1997).

In the aggregate, the data suggest that the nucleus can be powerfully regulated by many Ca^{2+}-dependent processes. Many, such as apoptosis, implicate the elevation of nuclear calcium by apoptotic signals. The removal of extracellular calcium or buffering of intracellular calcium is anti-apoptotic (Nicotera *et al.*, 1993); along a similar vein, the transfection of thymoma cells with the Ca^{2+}-binding protein calbindin protects the cells from glucocorticoid, cAMP or ionophore-induced apoptosis (Dowd *et al.*, 1992).

43.4 SUMMARY

The concept of nuclear calcium signalling is still controversial. The emerging picture on the subject, however, supports the hypothesis of nuclear calcium playing an increasingly important role in cell function and behaviour. What remains to be resolved is whether the cell is capable of regulating nuclear calcium independently of changes in cytoplasmic calcium or whether the latter is always a prerequisite for the former.

REFERENCES

Akiyama, S.K. (1996) *Hum. Cell* **9**, 181–186.

Al-Mohanna F.A., Caddy K.W.T. & Bolsover S.R. (1994) *Nature* **367**, 745–750.

Bachs O., Lanini L., Serratosa J., Coll M.J., Bastos R., Aligue R., Rius E. & Carafoli E. (1991) *J. Biol. Chem.* **265**, 18 595–18 600.

Birch B.D., Erg D.I. & Kocsis J.D. (1993) *Proc. Natl Acad. Sci. USA* **89**, 7978–7982.

Bustamante J.O., Hanover J.A. & Liepins A. (1995) *J. Membr. Biol.* **146**, 239–251.

Dash P.K., Karl K.A., Colicos M.A., Prywes R. & Kandel E.R. (1991) *Proc. Natl Acad. Sci. USA* **88**, 5061–5065.

Dingwall C., Sharnick S.V. & Laskey R.A. (1982) *Cell* **30**, 449–458.

Divecha N., Banfic H. & Irvine R.F. (1991) *EMBO J.* **10**, 3207–3214.

Divecha N., Banfic H. & Irvine R.F. (1994) *Cell Calcium* **16**, 297–300.

Dowd D.R., MacDonald P.N., Komm B.S., Haussler M.R. & Miesfeld R.L. (1992) *Mol. Endocrinol.* **6**, 1843–1848.

Enslen H., Tokumitsu H. & Soderling T.R. (1995) *Biochem. Biophys. Res. Commun.* **207**, 1038–1043.

Faury G., Usson T., Robert-Nicoud M., Robert Z. & Verdetti J. (1998) *Proc. Natl Acad. Sci. USA* **95**, 2967–2972.

Fukuda M., Gotoh I., Adachi M., Gotoh Y. & Nishida E. (1997) *J. Biol. Chem.* **272**, 32 642–32 648.

Galione A., Lee H.C. & Busa W.B. (1991) *Science* **253**, 1143–1146.

Gerasimenko O.V., Gerasimenko J.V., Tepikin A.V. & Peterson O.H. (1995) *Cell* **80**, 439–444.

Giancotti F.G. (1997) *Curr. Opin. Cell Biol.* **9**, 691–700.

Gorlich D. & Mattaj I.W. (1996) *Science* **271**, 1513–1518.

Greber U.F. & Gerace L. (1995) *J. Cell Biol.* **128**, 5–14.

Hicks G.R. & Raikhel N.V. (1995) *Annu. Rev. Cell. Dev. Biol.* **11**, 155–188.

Himpens B., DeSmedt H. & Bollen M. (1994) *FASEB J.* **8**, 879–883.

Humbert J.P., Matter N., Artault J.C., Koppler P. & Malviya A.N. (1996) *J. Biol. Chem.* **271**, 478–475.

Lin C., Hajnoczky G. & Thomas A.P. (1994) *Cell Calcium* **16**, 247–258.
Lipp P., Thomas D., Berridge M.J. & Bootman M.D. (1997) *EMBO J.* **16**, 7166–7173.
Martelli A.M., Gilmour R.S., Bertagnolo V., Neri L.M., Manzoli L. & Cocco L. (1992) *Nature* **358**, 242–245.
Matheos P. & Cunningham K.W. (1997) *10th Int. Symp. on Ca-binding proteins and Ca function in Health and Disease*, Lund, Sweden.
Matthews R.P., Guthrie C.R., Wailes L.M., Zhao X., Means A.R. & McKnight G.S. (1994) *Mol. Cell Biol.* **14**, 6107–6116.
Mazzanti M., Innocenti B. & Rigatelli M. (1994) *FASEB J.* **8**, 231–236.
Nakielny S. & Dreyfuss G. (1998) *J. Biol. Chem.* **8**, 89–95.
Nicotera P., Zhivotovsky B., Bellomo G. & Orrenius S. (1993) In *Apoptosis*, R.T. Schemke & E. Miihlich (eds). Plenum, New York, pp. 97–108.
Nix D.A. & Beckerle M.C. (1997) *J. Cell Biol.* **138**, 1139–1147.
Ohta Y., Ohba T. & Miyamoto E. (1990) *Proc. Natl Acad. Sci. USA* **87**, 5341–5345.
Paine P.L., Moore L.C. & Horowitz S.B. (1975) *Nature* **254**, 109–114.
Perez-Terzic C., Pyle J., Jaconi M., Stehno-Bittel L. & Clapham D.E. (1997a) *Science* **273**, 1875–1877.
Perez-Terzic C., Jaconi M. & Clapham D.E. (1997b) *BioEssays* **19**, 787–792.
Santella L. & Carafoli, E. (1997) *FASEB J.* **11**, 1091–1109.
Santella L. & Kyozuka K. (1994) *Biochem. Biophys. Res. Commun.* **203**, 674–680.
Shankar G., Davison I., Helfrich M.H., Mason W.T. & Horton M.A. (1993) *J. Cell Sci.* **105**, 61–68.
Sheng M., McFadden G. & Greenberg M.E. (1990) *Neuron* **4**, 571–582.
Simmen R.C.M., Dunbar B.S., Guerriero V., Chafouleas J.G., Clark J.H. & Means A.R. (1994) *J. Cell Biol.* **99**, 588–593.
Squier M., Miller A., Malkinson A. & Cohen J. (1994) *J. Cell Phys.* **159**, 229–237.
Takuma T., Tajima Y. & Ichida T. (1997) *Biochem. Mol. Biol. Int.* **43**, 563–570.
Tan Y., Rouse J., Zhang A., Cariati S., Cohen P. & Comb M.J. (1996) *EMBO J.* **15**, 4629–4642.
Wada A., Fukuda M., Mishima M. & Nishida E. (1998) *EMBO J.* **17**, 1635–1641.

CHAPTER FORTY-FOUR

Assessment of Gap Junctional Intercellular Communication in Living Cells Using Fluorescence Techniques

ADRIAAN W. DE FEIJTER,[1] JAMES E. TROSKO[2] & MARGARET H. WADE[1]

[1] Meridian Instruments Inc., Okemos, MI, USA
[2] Department of Pediatrics and Human Development, Michigan State University, MI, USA

44.1 INTRODUCTION

44.1.1 Intercellular communication: its role in maintaining homeostasis

The biological process by which cells influence control over each other in a multicellular organism is referred to as intercellular communication. One could classify intercellular communication as either systemic or local (Potter, 1983). However, with the discovery of a form of intercellular communication between contiguous cells via a membrane protein channel, the gap junction (Loewenstein, 1979, 1981), and the observation that the 'systemic' form of intercellular communication could modulate the gap junctional form of intercellular communication (Larsen & Risinger, 1985; Neyton & Trautman, 1986; Maldonado *et al.*, 1988; Madhukar *et al.*, 1989) via a variety of second messengers (Spray & Bennett, 1985), one needs to re-examine and integrate those different forms of cellular communication.

Three forms of interactive communication (extra-, intra- and inter-) could explain how the secretion of specific molecules by one cell could interact with another target cell, triggering a physiological response in that cell (e.g. stimulation of inhibition of mitogenesis; induction of differentiation; adaptive secretory products in a differentiated cell). The affected cell would, in turn, secrete other products which would signal the original signalling cell, thereby completing the feedback loop (Fig. 44.1).

Extracellular communication would refer to how cells communicate with distal cells via the secretion of a molecular signal (e.g. hormone, peptide, growth regulator, neurotransmitter). Intracellular communication would refer to those transmembrane-signalling mechanisms triggered by the extracellular signals. This would now modulate intercellular communication mediated by gap junctions.

One of the most important implications of this model of extra-, intra- and intercellular communication is that adaptive responses to changes in the environment can be made and that homeostasis of multiple functions of multiple systems can be restored after the challenge. This model provides a mechanistic integration of the various organ systems of the body, e.g. the neuroendocrine–immune system.

FLUORESCENT AND LUMINESCENT PROBES, 2ND EDN
ISBN 0–12–447836–0

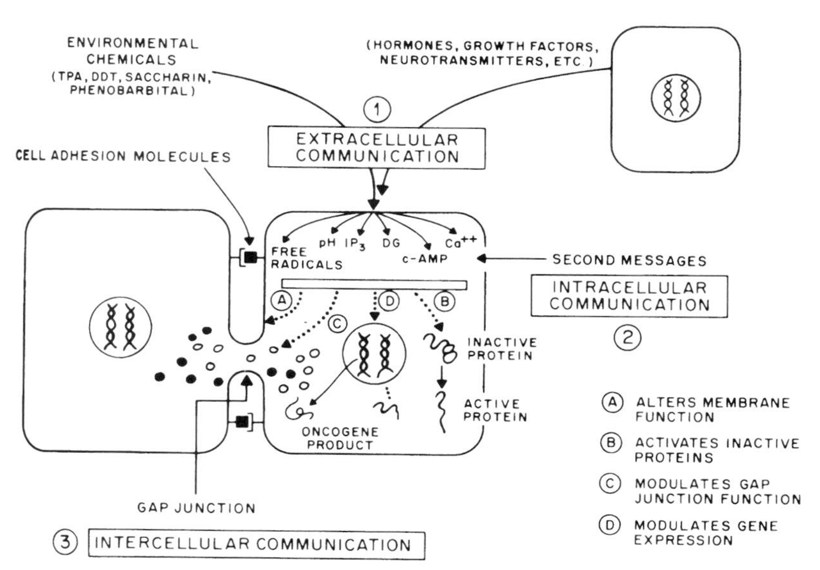

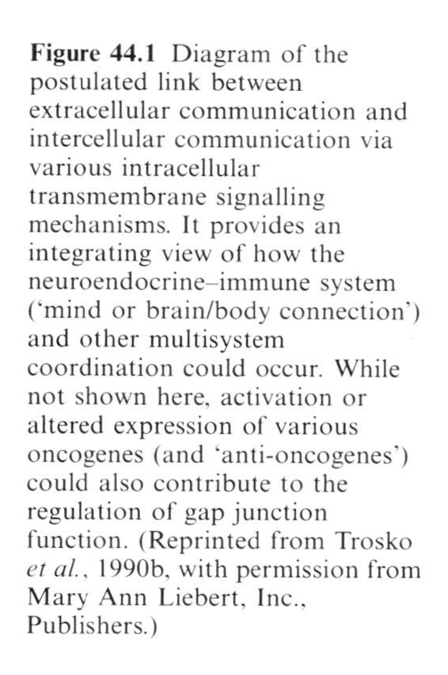
Figure 44.1 Diagram of the postulated link between extracellular communication and intercellular communication via various intracellular transmembrane signalling mechanisms. It provides an integrating view of how the neuroendocrine–immune system ('mind or brain/body connection') and other multisystem coordination could occur. While not shown here, activation or altered expression of various oncogenes (and 'anti-oncogenes') could also contribute to the regulation of gap junction function. (Reprinted from Trosko *et al.*, 1990b, with permission from Mary Ann Liebert, Inc., Publishers.)

Disruption of any of these steps could lead to either adaptive or maladaptive consequences depending on circumstances (Trosko & Chang, 1984). Sustained or permanent blockage of any of the three communication pathways could, in principle, lead to disruption of homeostatic control of cell growth or differentiation. This disruption could be the result of endogenous (e.g. genetic, developmental, sex) or exogenous (e.g. dietary, lifestyle, workplace, pollution, drug) factors (Trosko & Chang, 1988a,b, 1989).

Gap junctions are protein channels found in the membranes of most contiguous normal cells of metazoans (Hertzberg & Johnson, 1988). The subunit of the gap junction found in one membrane is referred to as a connexon and it joins the connexon in the adjoining membrane to form an aqueous channel which allows the passive transfer of ions and small molecules having molecular weights less than 1500 Da (Rammos *et al.*, 1990). Connexon subunits belong to a family of proteins, the connexins, which are related by their sequence but differ in their molecular weight (Nicholson *et al.*, 1987; Beyer *et al.*, 1987). Connexins 26, 30, 31, 32, 33, 36, 37, 38, 43 and 46 have been isolated from several organs in a variety of species (Willecke *et al.*, 1991). Several genes which code for these relatively highly conserved proteins have been cloned (Beyer *et al.*, 1987; Saez *et al.*, 1990).

Gap junction proteins are found in all metazoans and, while different gap junction genes/proteins can be found within one species and within one tissue in one species, the high degree of homology suggests their importance in the evolution of multicellular organisms (Saez *et al.*, 1990). It has been speculated that the variable regions of the gap junction protein might serve as the sites for specific regulation needed for gap junctions in different tissues (e.g. in excitable cells or non-excitable cells). In general, cells coupled by gap junctions are in equilibrium for ions and small regulatory and substrate molecules. The gap junctions can control episodic increases or decreases in these ions or small molecules (Sheridan, 1987). As a result, cells in this communicating network can be synchronized for function (e.g. heart beats – DeMello, 1982; secretions of peptides – Meda *et al.*, 1987), or suppressed for cell division, i.e. contact inhibited (Azarnia *et al.*, 1988).

The biological roles of gap junctions seem to have been designed to be modulated by both endogenous factors (e.g. growth factors – Madhukar *et al.*, 1989; Maldonado *et al.*, 1988; hormones – Larsen & Risinger, 1985; neurotransmitters – Neyton & Trautman, 1986) and a wide variety of exogenous chemicals (pollutants, drugs, food additives, natural plant chemicals, nutrients, heavy metals, toxins, pesticides, herbicides, etc. – Trosko & Chang, 1988a). In other words, gap junctions have evolved to respond to external and internal signals in order to either increase or decrease communication (Trosko & Chang, 1984; Trosko *et al.*, 1990a). The biological consequences of the modulation of communication via gap junctions could be either adaptive or maladaptive (Trosko & Chang, 1984).

The regulation of gap junctional communication could, conceptually, occur by the modulation of cell–cell adhesion (Kanno *et al.*, 1984), by the production, assembly, and function of the gap junction proteins, at the transcriptional, translational or post-translational levels (Spray *et al.*, 1988), by the control of the amount of potential signals through the gap junctions (Loewenstein, 1966), or by the transduction of the gap junction signal passed via gap junctions (Trosko *et al.*, 1990a).

A wide number of intracellular communicating second messengers (intracellular calcium, pH, cAMP, free radicals, protein phosphorylation) have been reported to be factors in the regulation of gap junctions (Spray & Bennett, 1985; Saez *et al.*, 1987, 1990).

To date, a variety of approaches have been used to study functional gap junctional communication. Electrocoupling (Loewenstein, 1979), metabolic cooperation via the transfer of radioactive metabolites (Subak-Sharpe *et al.*, 1966), genetic mutants to metabolic cooperation (Yotti *et al.*, 1979), microinjection and transfer of fluorescent dyes (Enomoto & Yamasaki, 1984), scrape loading and dye transfer of fluorescent dyes (El-Fouly *et al.*, 1987), and fluorescence redistribution after photobleaching ('FRAP') (Wade *et al.*, 1986) have all been reported.

Glycyrrhetinic acid (GA), a folk remedy which has anti-inflammatory activity (Finney & Somers, 1958), was previously demonstrated to suppress 12-*O*-tetradecanoylphorbol-13-acetate (TPA)-induced ornithine decarboxylase activity (Okamoto *et al.*, 1983) and Epstein–Barr virus-associated early antigen (Okamoto *et al.*, 1983) *in vitro*. Furthermore, GA has been shown to possess anti-skin tumour-initiating (Wang *et al.*, 1991) and anti-skin tumour-promoting (Nishino *et al.*, 1984, 1986; Wang *et al.*, 1991) activities *in vivo*. On the other hand, GA and several derivatives have been shown to inhibit gap junctional communication *in vitro*, working by a mechanism other than activation of protein kinase C (PK-C) or mineralocorticoid or glucocorticoid receptors (Davidson *et al.*, 1986; Tsuda & Okamoto, 1986; Davidson & Baumgarten, 1988; Rosen *et al.*, 1988) and to enhance TPA-mediated inhibition of metabolic cooperation (Tsuda & Okamoto, 1986).

Several laboratories have reported that TPA induces a significant reduction in both the number and the area of gap junctions (Yancey *et al.*, 1982; Kalimi & Sirsat, 1984; van der Zandt *et al.*, 1990), and that the mechanism by which TPA exerts its inhibiting effect on gap junctional communication may involve PK-C (Nishizuka, 1984; van der Zandt *et al.*, 1990).

In this chapter, we discuss how fluorescence technology can be used to study the biological and physiological roles of gap junctions and the potential mechanisms of regulation of these gap junctions. In particular, we assess the effect of the number of contacting cells as well as the effect of two chemicals, TPA and 18β-GA, on intercellular gap junctional communication. In addition, an immunofluorescence technique to localize gap junction proteins is described.

44.2 MATERIALS AND METHODS

44.2.1 Instrument description

The ACAS 570 Interactive Laser Cytometer (Meridian Instruments Inc., Okemos, MI) was used in the studies described to monitor fluorescence in living cells. The instrument consists of an inverted phase-contrast microscope, an *x–y* scanning stage, a 5 W argon ion laser for excitation, an acousto-optic modulator to control laser pulse intensity and duration during scanning, and photomultiplier tubes for fluorescence detection, all of which are under microprocessor and computer control. Unless otherwise noted, all data were acquired with a 40× long-working-distance objective. A schematic diagram of the ACAS optical path is depicted in Fig. 44.2. For the immunofluorescent localization of gap junction proteins, an ACAS 570 with confocal accessory was used. In addition to the components shown in Fig. 44.2, the confocal ACAS applies diffraction-limited optics, an adjustable pinhole in front of the detector, and an *x–y* scanning stage with *z*-axis control. Confocal optics reduce out-of-plane fluorescence, thus allowing an 'optical slice' to be scanned in the *z*-plane (see Chapter 21 in this volume for a description of confocal microscopy).

44.2.2 Cell culture

The human teratocarcinoma cell line, PA-1 (ATTC CRL no. 1572) (HT) is near diploid and has functional gap junctional communication (Wade *et al.*, 1986; Schindler *et al.*, 1987). The human kidney epithelial cell line G401.2/6TG.1 (SB-3, a gift from E. Stanbridge, University of California) was originally

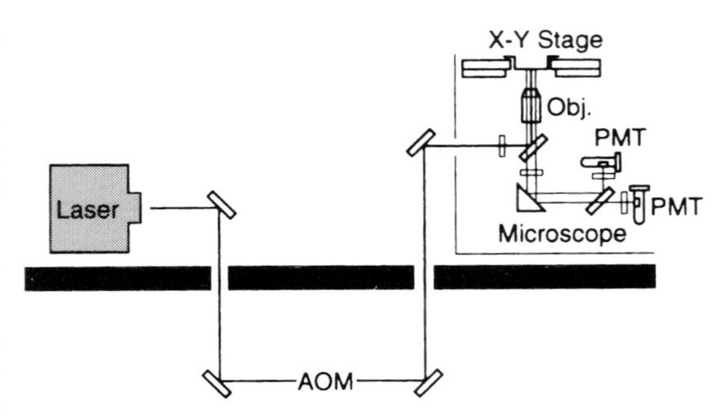

Figure 44.2 Optical path of the ACAS 570 showing the excitation pathway (laser, dichroic mirror, microscope objective) and emission pathway (microscope objective, dichroic mirror, photomultiplier tubes). Data are then digitized, collected and analysed using specific ACAS software.

derived from a Wilm's tumour patient and established as described previously (Weissman *et al.*, 1987). Fischer 344 rat liver epithelial cells (WB344), derived from normal liver (Tsao *et al.*, 1984), were used for the immunofluorescence study of the gap junction protein.

All cell cultures were maintained in D-medium, a modified Eagle's medium containing Earle's balanced salt solution with a 50% increase of vitamins and essential amino acids except glutamine, 100% increase of non-essential amino acids and 1 mM sodium pyruvate, 5.5 mM glucose, 14.3 mM NaCl, and 11.9 mM $NaHCO_3$ (pH 7.3). The medium was supplemented with 3% (SB-3), 5% (WB) or 10% (HT) fetal bovine serum (Gibco Laboratories) and 50 μg ml^{-1} gentamicin. Cells were grown at 37°C and 5% CO_2 in a humidified incubator. Low passage SB-3 cells were treated with the test chemicals 1 day after seeding. In both long- and short-term experiments, SB-3 cells were not trypsinized for the total duration of treatment. HT cells were plated and used within 48 h. All experiments were done at room temperature.

44.2.3 Chemicals

TPA (Sigma Chemical Co.) and 18β-glycyrrhetinic acid (18β-GA) (Aldrich Chemical Co.) were dissolved in ethanol and dimethylsulphoxide (DMSO), respectively. For both solvents, the final concentration in the medium was 0.1%. For the scrape loading experiments, Lucifer yellow CH (Sigma Chemical Co., molecular weight 457.2) and tetramethylrhodamine dextran (Molecular Probes Inc., molecular weight 10 000) were each dissolved in phosphate-buffered saline (PBS) at 0.5 mg ml^{-1}. For the FRAP assays, 5,6-carboxyfluorescein diacetate (molecular weight 376 as carboxyfluorescein, Molecular Probes Inc.) was dissolved in ethanol at 1 mg ml^{-1}.

44.2.4 Scrape loading

The scrape loading technique was performed as described previously by El-Fouly *et al.* (1987) and Oh *et al.* (1988). The cells were plated at high density and allowed to form a monolayer. They were then treated with non-cytotoxic concentrations of test chemicals and rinsed with PBS. Two millilitres of the dye (0.5 mg ml^{-1} Lucifer yellow in PBS) were added and several scrape lines were made in the monolayer with a surgical blade. After 3 min, to allow dye uptake and transfer, the cells were rinsed several times with PBS and examined under a Nikon epifluorescence phase-contrast microscope. Since the purpose of these experiments was to perform a quick assessment of various drug treatments on communication and not an exact quantitation, a fluorescence microscope was sufficient. Results were expressed as the average number of dye-coupled cells at a minimum of 15 random points along either side of the scrape line in each of two dishes. During the assay the cells were maintained at room temperature and care was taken to avoid exposure of the dye mixture to excessive room light. For the long-term incubations, media (with or without test chemicals) were changed every 2 days. In some experiments, tetramethyl rhodamine dextran (0.5 mg ml^{-1}) was included with Lucifer yellow as a control to distinguish the primary dye-loaded cells. Because of its large molecular weight, the dextran will neither diffuse through intact plasma membranes nor cross the junctional channels and can therefore serve to identify the primary loaded cells.

44.2.5 Fluorescence redistribution after photobleaching (FRAP)

For the FRAP assay, originally described by Wade *et al.* (1986), SB-3 or HT cells were plated at low density in 35 mm plates. After exposure to non-cytotoxic concentrations of test chemicals or solvent alone, the cells were rinsed with PBS containing calcium and magnesium (0.1 g $litre^{-1}$ each) (PBS/Ca/Mg) and labelled with 7 μl ml^{-1} 5,6-carboxyfluorescein diacetate. After 15 min, the plates were rinsed several times with PBS/Ca/Mg and examined with the ACAS Interactive Laser Cytometer. An initial scan was made to determine the pre-bleach fluorescence levels. Then, selected cells were photobleached and monitored for the return of fluorescence at various time intervals (1–4 min, total of 5–10 scans).

For the HT experiments, care was taken to choose cells growing as single cells, in doublets, in triplets, or cells nested within clumps of four or more cells to monitor the rate of recovery after bleaching. In addition, the rate of decline in fluorescence of the neighbouring cells was assessed. The cells were examined within 40 min after labelling, as the recovery rates diminished when the cells were left out of culture for longer time periods. Some of the HT experiments were done with cells incubated in Hank's buffer (5.3 mM KCl, 0.34 mM KH_2PO_4, 0.5 mM $MgCl_2$, 1.3 mM $CaCl_2$, 0.8 mM $MgSO_4$, 138 mM NaCl, 4.2 mM $NaHCO_3$, 0.35 mM Na_2HPO_4, and 5.6 mM glucose).

For the experiments with SB-3 cells, 5–8 cells in each of five separate plates were monitored, resulting in at least 25 cells per treatment group. Cells growing in colonies and in direct contact with a minimum of three neighbouring cells were selected for photobleaching. In each plate a single cell was selected, but not photobleached to determine the background decline of fluorescence. Occasionally, a single cell was photobleached as a negative control for fluorescence redistribution.

Results of the FRAP experiments were expressed as the average percentage recovery of fluorescence with a standard error of the mean (SEM). The recovery is calculated as a percentage of the initial pre-bleach fluorescence value for each cell monitored. As in scrape loading, the complete assay was performed at room temperature, and media were changed every 2 days in long-term experiments.

44.2.6 Immunofluorescence labelling for gap junction protein

WB rat liver epithelial cells were grown on coverslips until near confluency and fixed with 5% acetic acid in methanol and processed for immunofluorescence detection of gap junction proteins as described by Dupont *et al.* (1988) and El-Aoumari *et al.* (1990). In brief, fixed cells were incubated with 3% BSA in PBS to block non-specific binding, followed by incubation with a rabbit polyclonal antibody to connexin 43 (2 h). The antigen–antibody complex was localized using biotinylated anti-rabbit IgG and the streptavidin-fluorescein detection method. The coverslips were mounted with anti-fade solution (Aqua-Poly/Mount, Polysciences, Warrington, PA) and scanned with a confocal ACAS 570, using a 100× oil objective.

44.3 RESULTS

44.3.1 Morphological modulation of intercellular gap junction communication measured by FRAP analysis

To investigate a morphological aspect of gap junctional intercellular communication, we examined the effect of the number of contacting cells on the initial rate of fluorescence recovery after photobleaching in HT cells. Plate 44.1 depicts a triplet of cells where the cell in the centre was photobleached and fluorescence levels were monitored in that cell and both adjacent cells. As expected, while the bleached cell (area no. 3) recovered, the fluorescence levels in the neighbouring cells (areas nos 1 and 2) dropped as equilibrium between fluorescent and non-fluorescent carboxyfluorescein was attained. The fluorescence in the entire triplet was also monitored (area no. 4) and shows that the total fluorescence is not changing after photobleaching and that the recovery has nearly reached equilibrium. In some experiments, a single cell in the field was not bleached, but monitored, to correct for any photobleaching caused by repetitive scanning. Non-specific photobleaching was minimal (typically less than 1% per scan), due to the low excitation levels, the small area of the sample being illuminated, and the brief time of illumination. In some cases, the results were corrected for this non-specific bleaching effect.

Table 44.1 FRAP analysis of gap junctional communication between HT cells with varying numbers of cell contacts.

	Recovery rate ± *SEM*	
Single cells	0.65 ± 0.16	(*N* = 23)
Doublets	6.35 ± 0.47	(*N* = 21)
Triplets	9.54 ± 0.95	(*N* = 10)
Multiple contacts	11.59 ± 0.39	(*N* = 71)

Similar experiments were done with single cells, doublets, and cells with more than three cells in contact. The average rate over the first 3 min, expressed as the initial percentage increase per minute, for each of these combinations is shown in Table 44.1. As expected, the single cells did not recover when bleached since there are no contacting cells. Recovery rates in cells growing in groups of two, three or more cells increase in a linear fashion as the number of contacting cells increases (Wade & Schindler, 1988).

44.3.2 Chemical modulation of gap junction communication in SB-3 cells measured by scrape loading and FRAP techniques

Another set of experiments was designed to examine the effects of various chemical treatments on gap junctional communication in SB-3 cells. These cells normally show a high level of intercellular communication as assessed by both scrape loading and FRAP (Madhukar *et al.*, 1993). Figure 44.3 depicts several photomicrographs of scrape-loaded SB-3 cells after treatment with TPA. In control SB-3 cultures, Lucifer yellow migrated to 8–12 rows of adjacent cells on either side of the scrape line (Fig. 44.3(B)). When the cells were exposed to various concentrations of TPA for 1 h, an obvious inhibition of dye transfer was observed at 0.49 nM, resulting in 2–6 rows of dye transfer (Fig. 44.3(D)). At 4.9 nM, communication was completely blocked and the dye was retained only in the initially loaded cells (Fig. 44.3(E)). When the cells were treated with 18β-GA, dye transfer measured by scrape loading was partially inhibited at 25 μM, and virtually completely blocked at 50 μM (Table 44.2). As shown in Table 44.2, treatment with 30, 35, 40 and 45 μM 18β-GA resulted in significant,

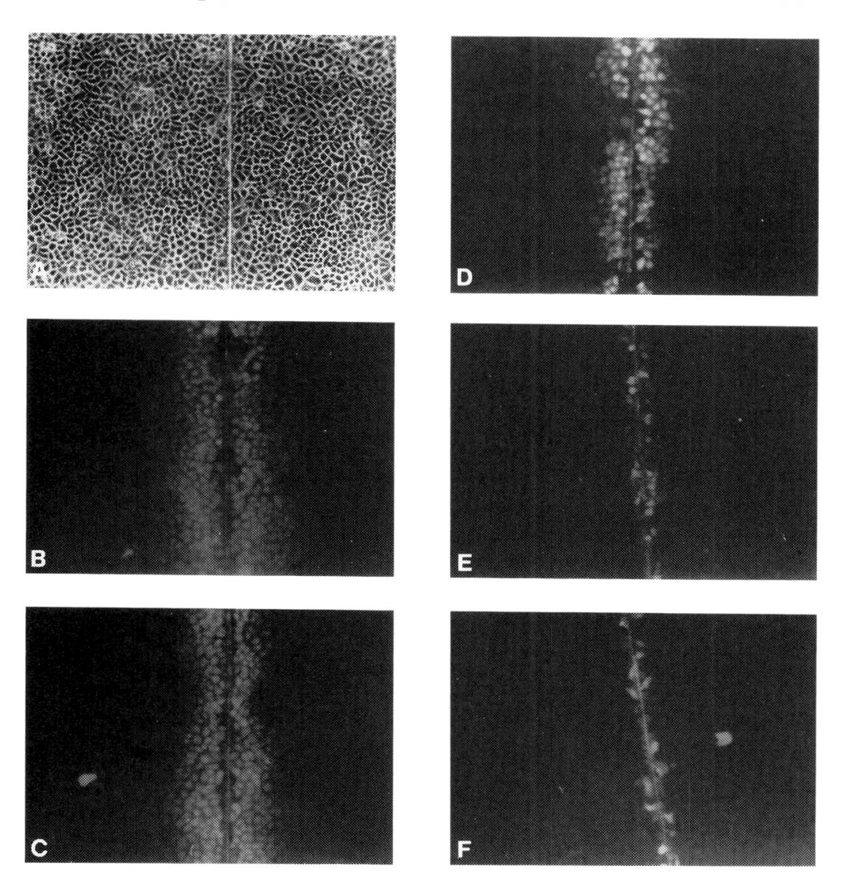

Figure 44.3 Dose–response relationship of TPA effect on gap junctional communication between SB-3 cells. Transfer of Lucifer yellow, 3 min after scrape loading of control cells is shown (B). Panel (A) shows the same field under phase contrast. Cells were treated for 1 h with 0.16 nM (C), 0.49 nM (D), 1.6 nM (E), and 4.9 nM (F) of TPA.

Table 44.2 Scrape loading analysis of dose-dependent inhibition of gap junctional communication in SB-3 cells after a 1 h exposure to 18β-GA.

Treatment	*Transfer of Lucifer yellow (number of rows of cells)*
0.1% DMSO	8–12
20 μM 18β-GA	8–12
25 μM 18β-GA	3–7
30 μM 18β-GA	2–4
35 μM 18β-GA	2–3
40 μM 18β-GA	2–4
45 μM 18β-GA	2–4
50 μM 18β-GA	1–3

but not complete, inhibition of communication to 2–4 rows of cells without showing a clear, reproducible dose–response relationship. Similar results were obtained when the exposure was limited to a 10 min incubation with the compound (data not shown).

Intercellular communication was then quantitated in these cells after identical treatments using the FRAP technique. As expected, the control SB-3 cells show a high level of recovery (Plate 44.2 (A)); after 16 min, an average recovery of 61% was noted. Recovery was inhibited in a dose-dependent fashion by 0.16–1.6 nM TPA (Fig. 44.4). At concentrations over 1.6 nM communication was virtually completely blocked (Plate 44.2(B), Fig. 44.4). When the cells were treated with 18β-GA, only a moderate effect was seen, and concentrations needed to exceed 50 μM before significant reductions in recovery were noted (Fig. 44.5). The maximum inhibition observed under non-cytotoxic conditions (100 μM, 1 h) still resulted in 26% recovery.

The absence of a clear and reproducible dose–response relationship after treatment with 18β-GA in the scrape loading experiments, and the limited effect revealed by FRAP, led us to suppose that removal of 18β-GA immediately prior to the photobleaching experiment permits a fast recovery of communication after initial blockage. After ending exposure to the test chemical, the overall FRAP experiment takes approximately 30–40 min during which staining, rinsing, selection, photobleaching

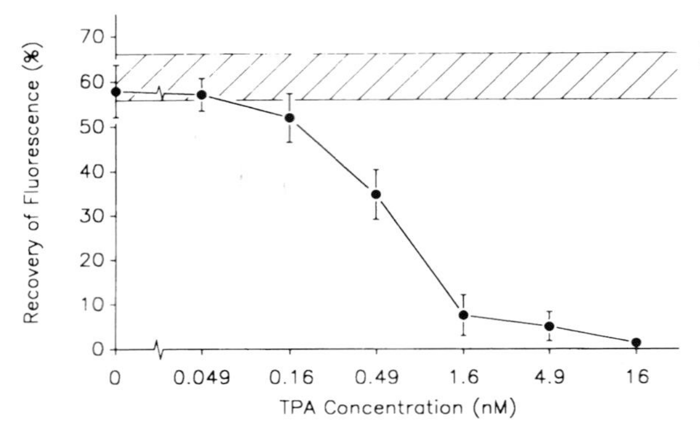

Figure 44.4 Dose–response relationship of TPA effect on gap junctional communication measured by FRAP. SB-3 cells were treated with different concentrations of TPA for 1 h and fluorescence recovery was determined 16 min after photobleaching. TPA concentration is plotted on a logarithmic scale. The shaded area shows the recovery in control cells. At 0 nM, the effect of the solvent (0.1% ethanol) is displayed. Each point in the graph represents the mean fluorescence recovery ± SEM of five experiments.

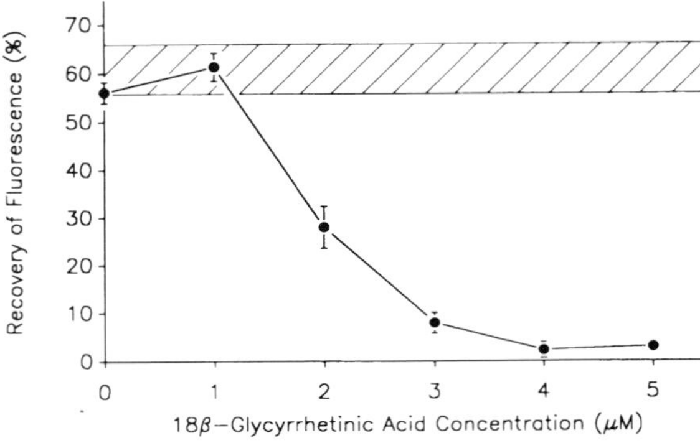

Figure 44.6 Dose–response relationship of 18β-GA effect on gap junctional communication measured by FRAP. SB-3 cells were treated with different concentrations of 18β-GA for 1 h and fluorescence recovery was determined 16 min after photobleaching. In contrast to the original protocol (Fig. 44.5), exposure to the chemical was continued during staining and scanning of the cells. The shaded area shows the recovery in control cells. At 0 μM, the effect of the solvent (0.1% DMSO) is displayed. Each point in the graph represents the mean fluorescence recovery ± SEM of five experiments.

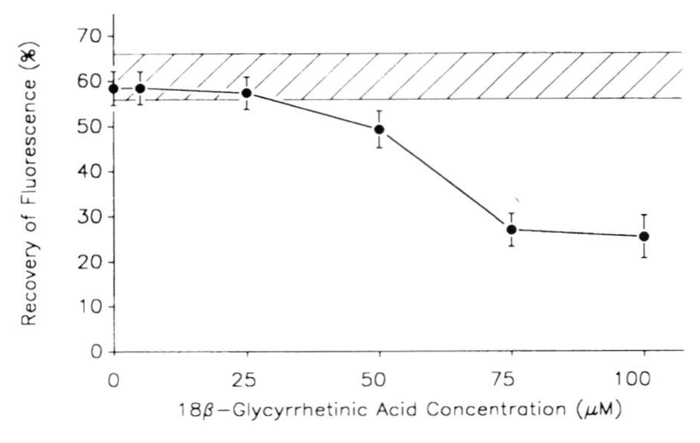

Figure 44.5 Dose–response relationship of 18β-GA effect on gap junctional communication measured by FRAP. SB-3 cells were treated with different concentrations of 18β-GA for 1 h and fluorescence recovery was determined 16 min after photobleaching. The shaded area shows the recovery in control cells. At 0 μM, the effect of the solvent (0.1% DMSO) is displayed. Each point in the graph represents the mean fluorescence recovery ± SEM of five experiments.

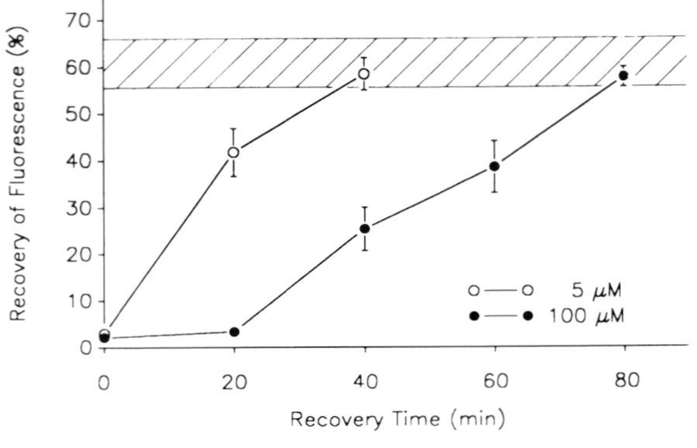

Figure 44.7 FRAP analysis of time-dependent recovery of gap junctional communication between SB-3 cells after complete blockage by a 1 h treatment with 5 or 100 μM 18β-GA. The shaded area shows the recovery in control cells. Each point in the graph represents the mean fluorescence recovery ± SEM of five experiments.

and scanning of the cells take place. Scrape loading, on the other hand, is completed within 4–6 min after drug treatment. To test this hypothesis, the FRAP protocol was slightly modified to allow continuation of 18β-GA treatment during the staining and scanning phases. As shown in Fig. 44.6, this modification had a drastic effect on the dose–response curve of the compound. No significant effect was observed at 1 μM, but at a concentration as low as 3 μM, dye transfer was virtually completely blocked. To establish the rate of return of communication after removal of 18β-GA, a FRAP experiment was performed in which the length of the 'recovery' period was varied. After incubation with the chemical for 1 h, cells were allowed to restore communication in the absence of 18β-GA, for 0, 20, 40, 60 or 80 min prior to measurement. In this experiment two concentrations of 18β-GA were tested to reveal any effect of the dose on the rate of return of communication (Fig. 44.7). The results indicated that the return rate was very fast at the lower concentration of 5 μM 18β-GA; within 20 min communication returned to 69% of the control value while being completely blocked if the chemical was kept on the cells during analysis. After 40 min, no inhibitory effect remained. At 100 μM 18β-GA, more time was required for the recovery process; control levels were reached after 80 min. This fast recovery after 18β-GA removal was confirmed using the scrape loading technique: recovery was evident as soon as 5 min and complete within 40 min after removal of 18β-GA (50 μM, 1 h) (data not shown).

Since inhibition of intercellular communication has been shown to be a transient effect in many *in vitro* studies using different chemicals and cell types (Enomoto & Yamasaki, 1985; Stedman & Welsch, 1985; Jongen *et al.*, 1987; Enomoto *et al.*, 1984; Madhukar *et al.*, 1993), the time–effect relationships of TPA and 18β-GA were examined using the FRAP assay during continuous exposure for up to 4 days. In contrast to the effect exhibited by TPA which was partially reversible, 18β-GA induced blockage of communication remained complete for at least 4 days of incubation (Fig. 44.8). Removal of 18β-GA after a 4 day exposure resulted in a fast return of dye transfer to control levels within 40 min (data not shown). None of the experiments revealed any effects of the solvents, ethanol (for TPA) and DMSO (for 18β-GA), on gap junctional communication between SB-3 cells.

44.3.3 Immunofluorescent localization of gap junction protein

In addition to fluorescence assays to examine functional properties of gap junctions (i.e. intercellular communication), we used an immunofluorescence technique to directly visualize the gap junctions. Fixed WB rat liver epithelial cells were labelled with a rabbit

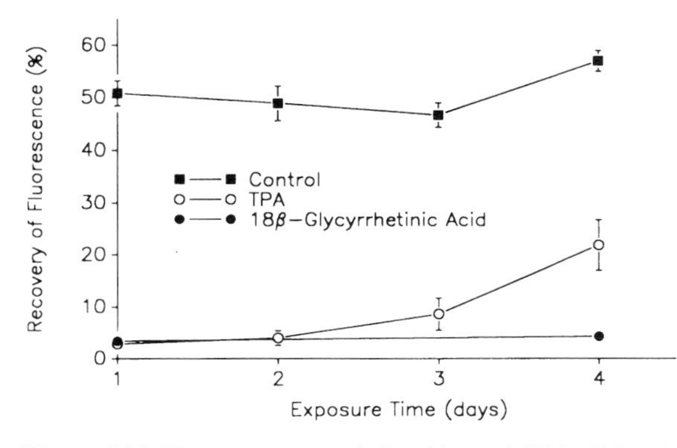

Figure 44.8 Time–response relationships of TPA (16 nM) and 18β-GA (5 μM) effects on gap junctional communication measured by FRAP. SB-3 cells were continuously exposed to the chemicals and their solvents for up to 4 days of culture, and fluorescence recovery was determined 16 min after photobleaching. Exposure to 18β-GA was continued during FRAP. Four days of exposure to 0.1% ethanol and 0.1% DMSO resulted in recoveries of respectively 49.6% ± 2.8 and 50.7 ± 3.1.

polyclonal antibody against connexin 43, the major protein constituent of rat heart junctions (Dupont *et al.*, 1988) followed by a biotinylated anti-rabbit IgG, and streptavidin-fluorescein. The stained cells were examined with the ACAS 570, equipped with a confocal option. Figure 44.9 clearly demonstrates that the gap junction protein is located at areas of cellular contact. In areas where cells are not touching, there is no visible fluorescence.

44.4 DISCUSSION

FRAP and scrape loading have been shown to be very useful techniques for studies on gap junction mediated intercellular communication (Wade *et al.*, 1986; El-Fouly *et al.*, 1987; Schindler *et al.*, 1987; Oh *et al.*, 1988; Evans & Trosko, 1988; Evans *et al.*, 1988; de Feijter *et al.* 1990; Madhukar *et al.*, 1991). Morphological and biochemical aspects of intercellular communication were examined with these techniques.

The studies with the human teratocarcinoma (HT) cells show that the number of cell contacts is directly related to the extent of intercellular communication. As the number of neighbouring cells increased, the initial rate of recovery after photobleaching increased in a linear fashion. It has been shown that the initial rate of recovery is a good measure of total recovery (Bombick & Doolittle, 1991). Similarly, reduced rates of communication have been noted in doublets when compared to cells with multiple contacts using rat glial cells, WB rat liver cells and normal human diploid fibroblasts (Bombick, pers. commun.). It has been shown that one can allow the bleached cell to recover, and bleach it again and obtain a similar rate of recovery (Wade *et al.*, 1986), and that the bleaching process is non-toxic as evidenced by continued cell growth (Wade, data not shown).

If the area of membrane contact is proportional to the number of gap junctions between cells, it should be possible, using antibodies to various connexins, to directly correlate the rate of gap junctional communication with the number of gap junctions between cells in doublets, triplets or groups with more than three contacting cells. Biegon *et al.* (1987) suggest that variations in size of the junction and dye permeance may also be related to cell cycle.

Intercellular communication between two different cell populations can be examined using the FRAP technique by culturing one of the cell types in the presence of fluorescent microspheres before co-culturing (Kalimi *et al.*, 1990, 1992). The microspheres serve to identify a particular cell type, but, unlike carboxyfluorescein molecules, are too large to move through gap junctions.

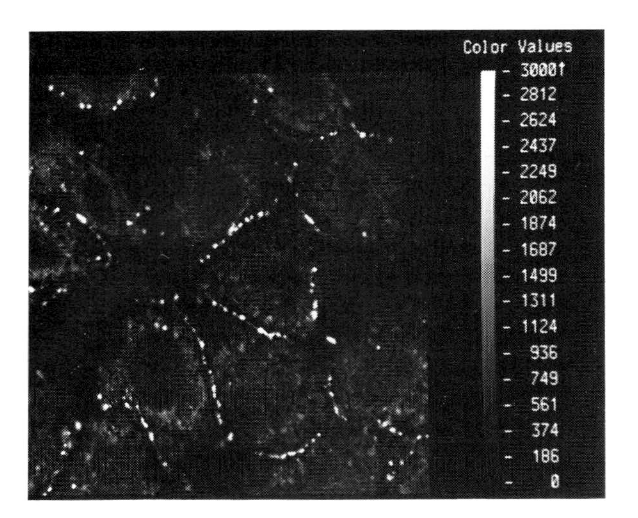

Figure 44.9 Optical section (approximately 0.7 μm thick) through WB rat liver epithelial cells, fluorescently labelled for the gap junction protein, connexin 43. The image (72 × 72 μm) was scanned with a confocal ACAS 570, using the 100× oil objective and displays fluorescence in areas of cell contact.

For the chemical studies on gap junction function, the techniques of scrape loading and FRAP were employed. The scrape loading experiments in this study primarily served to establish appropriate concentrations of test chemicals and confirm FRAP data; exact quantitation of the scrape loading data therefore was not attempted in this study. A more accurate quantitation of scrape loading data is possible using the ACAS (Oh *et al.*, 1988). Because of its low molecular weight, Lucifer yellow can be transferred between adjacent cells via gap junctions. The human kidney epithelial SB-3 cell line exerts a high level of intercellular communication as demonstrated by both the scrape loading and FRAP techniques.

Both TPA and GA proved to be very potent inhibitors of communication in SB-3 cells at non-cytotoxic concentrations. Many authors have observed that the TPA effect on gap junctional intercellular communication was partially or completely reversible when cells were continuously exposed to the chemical for a prolonged period (Enomoto & Yamasaki, 1985; Stedman & Welsch, 1985; Jongen *et al.*, 1987; Enomoto *et al.*, 1984; van der Zandt *et al.*, 1990; Madhukar *et al.*, 1993). Our experiments with TPA on SB-3 cells (Fig. 44.8) confirmed these observations. Exhaustion of TPA in the culture medium could not have accounted for this recovery, since media and test chemicals were changed after 2 days of incubation. This is also supported by the observation that recovery of communication after complete blockage by a 1 h treatment with TPA and subsequent removal of the chemical took at least 2–3 days (data not shown).

The effect of 18β-GA on gap junctional communication has been described (Davidson *et al.*, 1986; Tsuda & Okamoto, 1986; Davidson & Baumgarten, 1988; Rosen *et al.*, 1988). In contrast to TPA, no recovery of communication was observed when cells were continuously exposed to 18β-GA for up to 4 days. Removal of 18β-GA after complete blockage of communication, however, led to full recovery within 40–80 min, depending on the concentration of the compound. This fast recovery was observed regardless of the exposure time (1 h or 4 days). Davidson *et al.* (1986) previously provided strong indications that the working mechanism of GA-induced inhibition of intercellular communication does not involve steroid receptors or the PK-C pathway. Data presented here are in accordance with this assumption. Rate of onset and reversal of the effect suggest that this compound does not exert its inhibitory effect on gap junctional communication via any biochemical route, but rather via a more direct mechanism. The fast recovery rate after chemical removal favours a mechanism in which direct and reversible binding of the molecule to existing gap junctions, as proposed by Davidson and Baumgarten (1988), might play a role. Other mechanisms for inhibition of intercellular communication may involve intracellular calcium ions, pH, decrease in cAMP levels, or free radical production. Further studies may reveal the basis for the different mechanisms by which TPA and 18β-GA inhibit gap junctional communication.

Antibodies to connexin 43 were used to localize this gap junction protein in WB rat liver epithelial cells using immunofluorescence. This method has been used to show that the number of gap junctions in normal human keratinocytes, exposed to specific growth factors, is directly correlated to the extent of intercellular communication, measured by scrape loading (Dupont *et al.*, manuscript in preparation; Madhukar *et al.*, 1989). Jongen *et al.* (1991) demonstrated a correlation between the level of immunocytochemically stained connexin 43 and gap junctional communication, measured by microinjection/dye transfer, in cultured mouse epidermal cells at varying calcium concentrations.

The studies reported here have shown the use of three fluorescent tools, scrape loading, FRAP and immunofluorescent labelling, for studying gap junction function in cultured mammalian cells. Various mechanisms which affect intercellular communication, such as calcium perturbation, pH changes, peroxide formation (Bombick, 1990), lateral diffusion in the membrane, and translocation of PK-C (Rupp *et al.*, 1991) can also be assessed using fluorescent probes in combination with the ACAS 570.

ACKNOWLEDGEMENTS

The research on which this manuscript was based was supported, in part, by grants from the USAFOSR (no. USAFOSR-89-0325) and the NIEHS (1P42ESOY91) to J.E.T.

The authors would like to thank Dr B.V. Madhukar and Dr Mel Schindler for helpful discussions, Dr Emmanuel Dupont for the antibody labelled specimens and Jeanne McHugh for secretarial assistance.

REFERENCES

Azarnia R., Reddy S., Kimiecki T.E., Shalloway D. & Loewenstein W.R. (1988) *Science* **239**, 398–400.

Beyer E., Paul D.L. & Goodenough D.A. (1987) *J. Cell Biol.* **195**, 2621–2623.

Biegon R.P., Atkinson M.M., Liu T-F., Kam E.Y. & Sheridan, J.D. (1987) *J. Membr. Biol.* **96**, 225–233.

Bombick D.W. (1990) *In Vitro Toxicol.* **3**, 27–39.

Bombick D.W. & Doolittle D.G. (1991) *The Toxicologist* **11**, 214.

Davidson J.S. & Baumgarten I.M. (1988) *J. Pharmac. Exp. Ther.* **246**, 1104–1107.

Davidson J.S., Baumgarten I.M. & Harley E.H. (1986) *Biochem. Biophys. Res. Commun.* **1**, 29–36.

de Feijter A.W., Ray J.S., Weghorst C.M., Klaunig J.E., Goodman J.I., Chang C.C., Ruch R.J. & Trosko J.E. (1990) *Molec. Carcin.* **3**, 54–67.

DeMello W.C. (1982) *Prog. Biophys. Molec. Biol.* **39**, 147–182.

Dupont E., El Aoumari A., Roustiau-Severe S., Briand J.P. & Gros D. (1988) *J. Membr. Biol.* **104**, 119–128.

El Aoumari A., Fromaget C., Dupont E., Reggio H., Durbec P., Brian J.-P., Boller K., Kreitman B. & Gros D. (1990) *J. Membr. Biol.* **115**, 229–240.

El-Fouly M.H., Trosko J.E. & Chang C.C. (1987) *Exp. Cell Res.* **168**, 422–430.

Enomoto T. & Yamasaki H. (1984) *Cancer Res.* **44**, 5200–5203.

Enomoto T. & Yamasaki H. (1985) *Cancer Res.* **45**, 2681–2688.

Enomoto T., Martel N., Kanno Y. & Yamasaki H. (1984) *J. Cell. Physiol.* **121**, 323–333.

Evans M.G. & Trosko J.E. (1988) *Cell Biol. Toxicol.* **4**, 163–171.

Evans M.G., El-Fouly M.H., Trosko J.E. & Sleight S.D. (1988) *J. Toxicol. Environ. Hlth* **24**, 261–271.

Finney R.S.H. & Somers G.J. (1958) *J. Pharmac. Pharmacol.* **10**, 613–620.

Hertzberg E.G. & Johnson R.G. (1988) *Gap Junctions.* Alan R. Liss, New York.

Jongen W.M.F., Sijtsma S.R., Zwijsen R.M.L. & Temmink J.H.M. (1987) *Carcinogenesis* **8**, 767–772.

Jongen W.M.F., Fitzgerald D.J., Asamoto M., Piccoli C., Slaga T.J., Gros D., Takeichi M. & Yamasaki H. (1991) *J. Cell Biol.* **114**, 545–555.

Kalimi G.H. & Sirsat S.M. (1984) *Cancer Lett.* **22**, 343–350.

Kalimi G.H., Chang C.C., Edwards P., Dupont E., Madhukar B.B., Stanbridge E. & Trosko J.E. (1990) *Am. Assoc. Cancer Res.* **31**, 319.

Kalimi G.H., Hampton L.L., Trosko J.E., Thorgeirsson S.S. & Huggett A.C. (1992) *Molec. Carcinogenesis* **5**, 301–310.

Kanno Y., Sasaki Y., Shiba Y., Yoshida-Noro C. & Takeichi M. (1984) *Exp. Cell Res.* **152**, 270–274.

Larsen W.J. & Risinger M.A. (1985) *Mod. Cell Biol.* **4**, 151–216.

Loewenstein W.R. (1966) *Ann. NY Acad. Sci.* **137**, 441–472.

Loewenstein W.R. (1979) *Biochim. Biophys. Acta* **560**, 1–65.

Loewenstein W.R. (1981) *Physiol. Rev.* **61**, 829–913.

Madhukar B.V., Oh S.Y., Chang C.C., Wade M.H. & Trosko J.E. (1989) *Carcinogenesis* **10**, 13–20.

Madhukar B.V., de Feijter A.W., Hasler C.M., Lockwood B., Oh S.Y., Chang C.C., Stanbridge E. & Trosko J.E. (1993) *In Vitro Toxicol.* (in press).

Maldonado P.E., Rose B. & Loewenstein W.R. (1988) *J. Membr. Biol.* **106**, 203–210.

Meda P., Bruzzone R., Chanson M., Bosco D. & Orci L. (1987) *Proc. Natl. Acad. Sci. USA* **84**, 4901–4904.

Murray A.W. & Fitzgerald D.J. (1979) *Biochem. Biophys Res. Commun.* **91**, 395–401.

Neyton J. & Trautman A. (1986) *J. Physiol.* **377**, 285–295.

Nicholson B.J., Dermietzel R., Teplov D.B., Traub O., Willecke K. & Revel J.P. (1987) *Nature* **329**, 732–734.

Nishino H., Kitagawa K. & Iwashima A. (1984) *Carcinogenesis* **5**, 1529–1530.

Nishino H., Yoshioka K., Iwashima A., Takizawa H., Konishi S., Okamoto H., Okabe H., Shibata S., Fujiki H. & Sugimura T. (1986) *Japan. J. Cancer Res. (GANN)* **77**, 33–38.

Nishizuka Y. (1984) *Nature* **308**, 693–698.

Oh S.Y., Madhukar B.M. & Trosko J.E. (1988) *Carcinogenesis* **9**, 135–139.

Okamoto H., Yosida D. & Mizusaki S. (1983) *Cancer Lett.* **21**, 29–35.

Pitts J.D. & Finbow M.E. (1986) *J. Cell Sci.* **4**, 239–266.

Potter V.R. (1983) *Prog. Nucleic Acid Res. Molec. Biol.* **29**, 161–173.

Rammos S., Gittenberger-de Groot A.C. & Oppenheimer-Dekker A. (1990) *Int. J. Cardiol.* **29**, 285–295.

Rosen A., Van der Merwe P.A. & Davidson J.S. (1988) *Cancer Res.* **48**, 3485–3489.

Rupp H.L., Trosko J.E. & Madhukar B.V. (1991) *The Toxicologist* **11**, 273 (abstract).

Saez J.D., Bennett M.V.L. & Spray D.C. (1987) *Science* **236**, 967–969.

Saez J.C., Spray D.C. & Hertzberg E.L. (1990) *In Vitro Toxicol.* **3**, 69–86.

Schindler M., Trosko J.E. & Wade M.H. (1987) *Methods Enzymol.* **141**, 439–447.

Sheridan J.D. (1987) In *Cell-to-Cell Communication*, W.C. DeMello (ed.). Plenum Press, New York, pp. 187–222.

Spray D.C. & Bennett M.V.L. (1985) *Ann. Rev. Physiol.* **47**, 281–303.

Spray D.C., Saeiz J.D. & Burt J.M. (1988) In *Gap Junctions*,

E. Hertzberg & R. Johnson (eds). Alan R. Liss, New York, pp. 227–244.

Stedman D.B. & Welsch F. (1985) *Carcinogenesis* **6**, 1599–1605.

Subak-Sharpe H., Burk R.R. & Pitts J.D. (1966) *Hereditary* **21**, 342–343.

Trosko J.E. & Chang C.C. (1984) *Pharmacol. Rev.* **36**, 137–144.

Trosko J.E. & Chang C.C. (1988a) In *Tumor Promoters: Biological Approaches for Mechanistic Studies and Assay Systems*, R. Langenbach, J.C. Barrett & E. Elmore (eds). Raven Press, New York, pp. 97–111.

Trosko J.E. & Chang C.C. (1988b) In *Banbury Report 31: Carcinogen Risk Assessment: New Directions in the Qualitative and Quantitative Aspects*, R.W. Hart & F.G. Hoerger (eds). Cold Spring Harbor Lab, Cold Spring Harbor, NY, pp. 139–170.

Trosko J.E. & Chang C.C. (1989) In *Biologically Based Methods for Cancer Risk Assessment*, C.C. Travis (ed.). Plenum Press, New York, pp. 165–179.

Trosko J.E., Chang C.C., Madhukar B.V. & Klaunig J.E. (1990a) *Pathobiology* **58**, 265–278.

Trosko J.E., Chang C.C., Madhukar B.V. & Oh S.Y. (1990b) *In Vitro Toxicol.* **3**, 9.

Tsao M.S., Smith J.D., Nelson K.G. & Grisham J.W. (1984) *Exp. Cell Res.* **154**, 38–52.

Tsuda H. & Okamoto H. (1986) *Carcinogenesis* **7**, 1805–1807.

van der Zandt P.T.J., de Feijter A.W., Homan E.C., Spaaij C., de Haan L.H.J., van Aelst A.C. & Jongen W.M.F. (1990) *Carcinogenesis* **11**, 883–888.

Wade M.H. & Schindler M. (1988) *FASEB* **2**, A320.

Wade M.H., Trosko J.E. & Schindler M. (1986) *Science* **232**, 525–528.

Wang Z.Y., Agarwal R., Zhou Z.C., Bickers D.R. & Mukhtar H. (1991) *Carcinogenesis* **12**, 187–192.

Weissman B.E., Saxon P.J., Pasquale S.R., Jones G.R., Geiser A.G. & Standbridge E.J. (1987) *Science* **236**, 175–180.

Willecke K., Jungbluth S., Dahl E., Hennemann H., Heynkes R. & Grzeschik K.-H. (1991) *Eur. J. Cell Biol.* **53**, 275–280.

Yamasaki H. (1988) In *Gap Junctions*, E.L. Hertzberg & R.G. Johnson. (eds). Alan R. Liss, New York, pp. 449–465.

Yancey S.B., Edens J.E., Trosko J.E., Chang C.C. & Revel J.P. (1982) *Exp. Cell Res.* **139**, 329–340.

Yotti L.P., Chang C.C. & Trosko J.E. (1979) *Science* **206**, 1089–1091.

CHAPTER FORTY-FIVE

Photoactivation of Fluorescence as a Probe for Cytoskeletal Dynamics in Mitosis and Cell Motility

K.E. SAWIN,[1] J.A. THERIOT[1] & T.J. MITCHISON[2]
[1] Department of Biochemistry and Biophysics
[2] Department of Pharmacology, University of California, San Francisco, CA, USA

45.1 INTRODUCTION

Microtubules (MTs) and actin microfilaments (AFs) are cytoskeletal polymers composed of subunits of the proteins tubulin and actin, respectively, with associated proteins. Tubulin and actin are among the most abundant proteins in cells, and MTs and AFs have important functions in nearly all aspects of the life of the cell – in cell structure and morphogenesis, motility, secretion, and mitosis, as well as in additional processes often specific to specialized cell types (for reviews, see Dustin, 1984; Bershadsky & Vasiliev, 1988). It is our opinion that understanding the role of the cytoskeleton in these and other phenomena depends upon a much deeper understanding of fundamental aspects of the biology of MTs and AFs, which are unusual as biopolymers, yet similar in many respects (for reviews, see Oosawa & Asakura, 1975; Inoue, 1982; Carlier, 1989).

MTs and AFs represent dynamic rather than static networks. This is primarily the result of two factors, both of which have been well-studied *in vitro*: (1) In biophysical terms, MTs and AFs are best described not as self-assembling equilibrium polymers but rather as reversible steady-state polymers. Their assembly and continued persistence involves a classical nucleation/condensation reaction (Oosawa & Asakura, 1975) but normally also requires the presence of monomer-associated nucleotide (ATP for actin, GTP for tubulin) which is hydrolysed at some point after subunit incorporation onto the ends of existing filaments, (for review see Carlier, 1989). Under normal steady-state conditions, MTs and AFs constantly turn over, exchanging with a monomer pool, with a continuous input of energy obtained through nucleotide hydrolysis. Interestingly, under certain conditions nucleotide hydrolysis *per se* is not required for MT or AF polymerization; assembly is essentially entropically driven (Oosawa & Asakura 1975). (2) Both MTs and AFs have a defined polarity that reflects the asymmetric structure of constitutent subunits arranged in a head-to-tail fashion, and as a consequence the assembly properties of the two ends of the polymer can be functionally distinct. In conjunction with reversible, nucleotide-dependent assembly, this confers unusual and unexpected dynamic properties to both MTs and AFs. This was first recognized on

FLUORESCENT AND LUMINESCENT PROBES, 2ND EDN
ISBN 0-12-447836-0

theoretical grounds by Wegner (1976), who showed that a requirement for nucleotide hydrolysis could permit one end of an AF to undergo net assembly (the 'plus', or 'barbed' end) while the other (the 'minus', or 'pointed' end) undergoes a net disassembly, the result being a net flux, or treadmill, of subunits through the filament at steady state. Wegner demonstrated this treadmilling experimentally *in vitro*, and a similar flux was inferred for MT protein by Margolis and Wilson (1978, 1981). It remains controversial what function, if any, this *in vitro* behaviour may serve within cells (Margolis & Wilson, 1981; Kirschner, 1980; Hill & Kirschner, 1982).

At least in the case of MTs, treadmilling appears to depend on the presence of MT-associated proteins (Horio & Hotani, 1986; Hotani & Horio, 1988). By contrast, pure tubulin *in vitro* and most MTs *in vivo* display a different behaviour known as 'dynamic instability' (Mitchison & Kirschner, 1984), in which individual MTs in a steady-state population coexist in growing and shrinking phases and transit stochastically between these two states with characteristic frequencies in interphase and mitosis (Walker *et al.*, 1988; Belmont *et al.*, 1990). The mechanism of dynamic instability is still not well-understood but is probably best ultimately explained through conformational changes in the structure of the MT lattice, induced by the process of nucleotide hydrolysis.

In vivo, the dynamics of MTs and AFs have been studied indirectly for many years. In the 1950s and 1960s, Inoue and collaborators used polarization optics to observe the rapid loss of mitotic spindle birefringence under (what were later found to be) MT-depolymerizing conditions; when such conditions were withdrawn, spindle birefringence quickly recovered. These and other experiments indicated that, *in vivo*, MTs exist in a kind of dynamic equilibrium with the tubulin monomer pool, with rapid turnover rates (reviewed in Inoue & Sato, 1967; Sato *et al.*, 1975; Inoue, 1981). Similar results have also been obtained with inhibitors of the actin cytoskeleton, using phase-contrast and differential-interference contrast microscopy, most recently in the neuronal growth cone (Forscher & Smith, 1988).

The characteristic turnover and instability of AFs and MTs *in vivo* raises questions of fundamental biological importance: What is the function of these unusual dynamics? Why does the cell invest so much energy (in the form of nucleotide hydrolysis) in processes that essentially allow it only to 'run in place'? Are dynamics primarily designed for morphogenetic and organizational purposes, or can they also perform useful work within cells? How are the basic dynamic properties of actin and tubulin regulated *in vivo* to achieve a variety of morphogenetic and motile processes, and what are the components of such regulatory systems? How is the regulation of dynamics coupled to other cytoskeleton-based activities such as force generation and movement through motor proteins such as myosin, dynein, and kinesin?

To begin to answer these and related questions requires new technologies for the quantitative study of cytoskeletal polymer dynamics in a non-perturbing, physiological context. The introduction of vital fluorescent probes for the cytoskeleton has been singularly useful in this regard. Specifically, questions of polymer dynamics have been addressed by microinjection and incorporation of fluorescently labelled subunit proteins (e.g. actin or tubulin) to steady-state, followed by local photobleaching of incorporated fluorescence with a laser microbeam. The disposition and/or recovery of the bleached mark can then be tracked, in both time and space, by fluorescence microscopy. This technique, known as fluorescence redistribution after photobleaching (FRAP) has been used to study subunit turnover of both MTs and AFs in a number of systems and has been of particular value to students of mitosis (Salmon *et al.*, 1984; Saxton *et al.*, 1984; Gorbsky & Borisy, 1989) and cell motility (Wang *et al.*, 1982; Wang, 1985; Lee *et al.*, 1990).

For some years we have been interested in the role of polymer dynamics in the regulation of MT and AF function *in vivo*. The purpose of this paper is to summarize work from our laboratory concerning a novel fluorescence marking technique for studying cytoskeletal polymer dynamics, photoactivation of fluorescence (Mitchison, 1989). In principle, photoactivation is similar to FRAP, but whereas photobleaching involves the generation of a non-fluorescent mark on a bright background, photoactivation generates a fluorescent mark on a dark background. As described below, this is achieved by 'caging' a fluorochrome with a photolabile group that renders it non-fluorescent, whereupon the caged compound is covalently coupled to a specific protein and microinjected in cells. After incorporation of this tagged molecule into structures of interest, fluorescence is locally activated by a brief exposure to a UV microbeam, and followed by time-lapse fluorescence video microscopy. Here we describe the structure and properties of caged fluorochromes synthesized in the laboratory, the design of a computer-controlled microscope to study cytoskeletal dynamics, and some of the results obtained using these technologies. We expect that photoactivation of fluorescence will have wide applicability as a non-perturbing marking technique in a number of areas besides cytoskeletal polymer dynamics (Mitchison, 1989; Sawin & Mitchison, 1991; Theriot & Mitchison, 1991; Reinsch *et al.*, 1991) – for example, in studying membrane dynamics, embryonic fate mapping, intracellular membrane transport, and any other situations where it may be informative to locally mark structures of living cells or organisms under the microscope.

45.2 PHOTOACTIVATABLE FLUORESCENT PROBES

The use of photosensitive compounds in microscopy and biochemistry is well-established. A surprisingly wide variety of biologically active compounds have been made in photoactivatable or photoinactivatable forms, including nucleotides (Walker *et al.*, 1989), phosphatidylinositol metabolites (Walker *et al.*, 1987), neurotransmitters (Marque, 1989), amino acids (Patchornik *et al.*, 1970), calcium chelators (Tsien & Zucker 1986), proton donors (Janko & Reichert, 1987), and microtubule drugs (Hiramoto *et al.*, 1984).

45.2.1 Caged fluorescein

Fluorescein itself exists as two tautomeric forms; a fluorescent carboxylic acid tautomer favoured in neutral or basic aqueous solution, and a non-fluorescent lactone tautomer favoured in acidic or non-aqueous solvents. Fluorescein may be constrained in its lactone tautomer by alkylation of its two phenolic oxygens. Such a non-fluorescent derivative is photoactivatable if one or both of the ether groups are subject to photocleavage. The photolysis of 2-nitrobenzyl compounds to form 2-nitrosobenzaldehyde has been exploited in the caging of compounds containing carboxylic acid, phosphate or hydroxyl functionalities, and this chemistry lends itself well to caging fluorescein (Krafft *et al.*, 1988; Mitchison, 1989; Fig. 45.1(a)). The resultant compound, bis-caged carboxyfluorescein, or C2CF, is colourless and non-fluorescent, and is photoactivated by 360 nm light with simple first-order kinetics to regenerate the fluorescent parent compound (Mitchison, 1989). The side-chain carboxyl group of carboxyfluorescein can be activated for protein labelling, for example, as the *N*-hydroxysulphosuccinimide ester (Mitchison, 1989) (Fig. 45.1(b)). The caged dye can be coupled to specific proteins or dextrans (e.g. as fluid-phase markers) using these and related chemistries.

C2CF has been used to label covalently tubulin purified from bovine brain. C2CF tubulin polymerizes to form fairly short microtubules which appear identical in structure to microtubules formed from unlabelled purified tubulin (Mitchison, 1989) and which can be rapidly photoactivated under the microscope (Mitchison, 1989; Fig. 45.2). C2CF-tubulin injected into cells incorporates into the endogenous microtubule structures including mitotic spindles (Mitchison, 1989) and neuronal axons (Reinsch *et al.*, 1991).

Although C2CF has only limited solubility in water, its solubility can be greatly improved by the addition

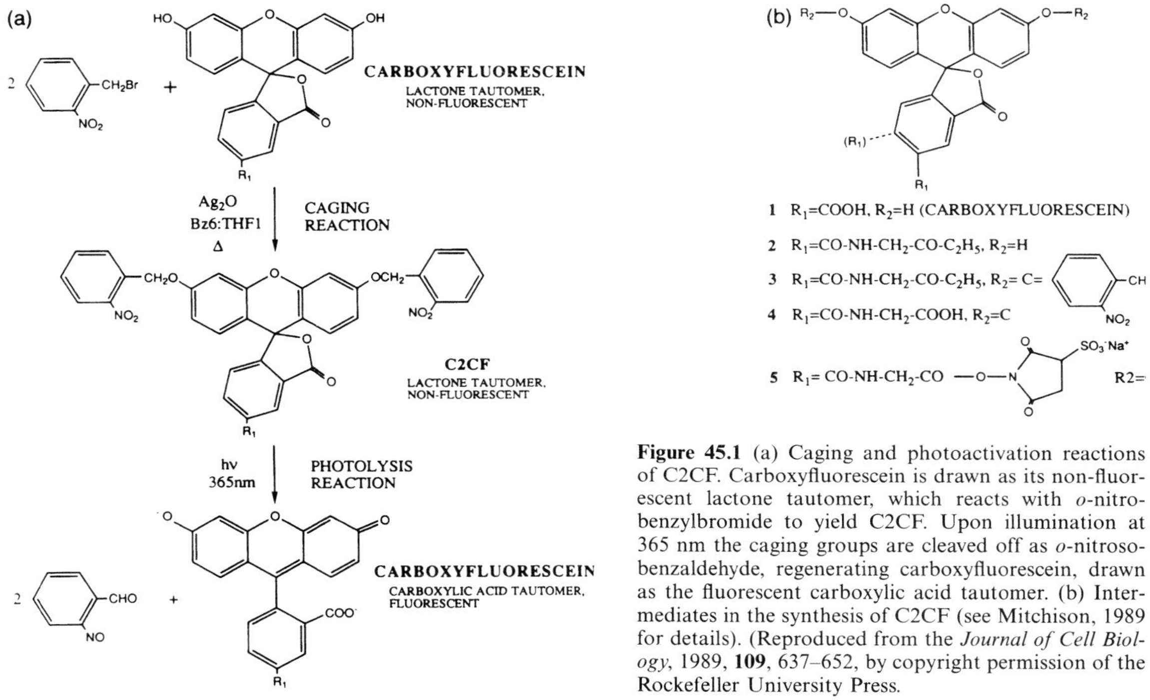

1 R_1=COOH, R_2=H (CARBOXYFLUORESCEIN)
2 R_1=CO-NH-CH_2-CO-C_2H_5, R_2=H
3 R_1=CO-NH-CH_2-CO-C_2H_5, R_2= C=
4 R_1=CO-NH-CH_2-COOH, R_2=C
5 R_1= CO-NH-CH_2-CO —O—N R2=

Figure 45.1 (a) Caging and photoactivation reactions of C2CF. Carboxyfluorescein is drawn as its non-fluorescent lactone tautomer, which reacts with *o*-nitrobenzylbromide to yield C2CF. Upon illumination at 365 nm the caging groups are cleaved off as *o*-nitrosobenzaldehyde, regenerating carboxyfluorescein, drawn as the fluorescent carboxylic acid tautomer. (b) Intermediates in the synthesis of C2CF (see Mitchison, 1989 for details). (Reproduced from the *Journal of Cell Biology*, 1989, **109**, 637–652, by copyright permission of the Rockefeller University Press.

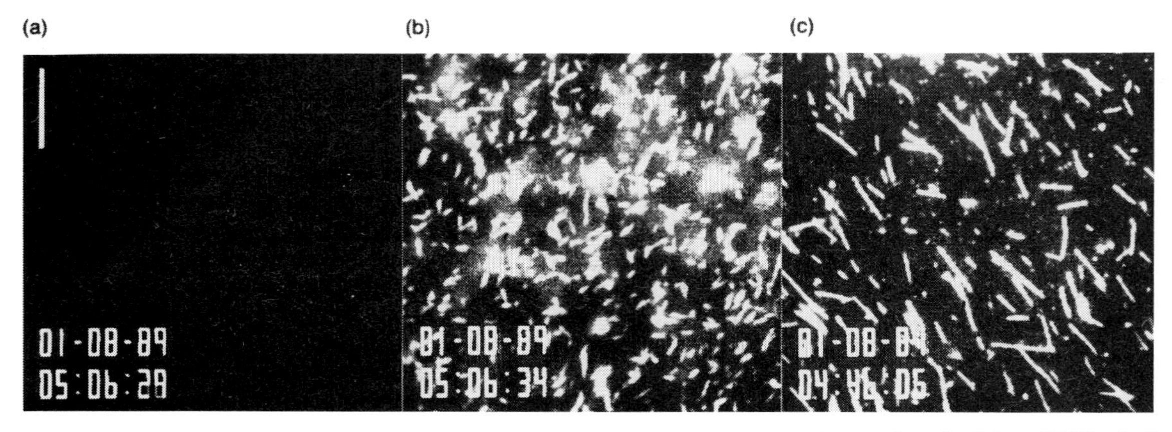

Figure 45.2 C2CF-labelled microtubules before and after photoactivation. (a) Microtubules polymerized from C2CF-tubulin in the dark, before photoactivation. (b) The same microscope field, after photoactivation (1 s exposure under the microscope, using a Hoechst filter set and a mercury arc lamp. (c) Microtubules polymerized from C2CF-labelled tubulin that was photoactivated before polymerization. Bar is 10 μm. (Micrographs from Mitchison, 1989. Reproduced from the *Journal of Cell Biology*, 1989, **109**, 637–652, by copyright permission of the Rockefeller University Press.)

of oxyacetyl sidechains to the caging nitrobenzyl rings (T.J. Mitchison, unpublished observations). The more water-soluble compound is superior to the original C2CF for labelling antibodies, but the efficiency of tubulin labelling is similar for the two compounds (T.J. Mitchison, unpublished observations).

45.2.2 Caged resorufin

The size and relative hydrophobicity of C2CF render it inefficient for labelling some proteins; specifically, we found it difficult to label actin or myosin to high stoichiometry with C2CF (M. Symons, pers. commun.; J.A. Theriot, unpublished observations). We therefore developed a smaller, more hydrophilic caged fluorescent compound based on the fluorochrome resorufin (Fig. 45.3). The non-fluorescent O-alkyl and O-acyl derivatives of resorufin are widely used as fluorogenic enzyme substrates (Haugland, 1989, and references therein), and resorufin modified with convenient protein-labelling groups (either iodoacetamide or *N*-hydroxysuccinimide ester) is available from Boehringer-Mannheim. Both compounds are readily caged on one phenolic oxygen in a single step using the standard protocol for caging nucleotides (Walker *et al.*, 1989), although we have obtained good yields only with the iodoacetamide derivative (Theriot & Mitchison, 1991).

Caged-resorufin (CR) iodoacetamide readily labels actin to high dye-to-protein stoichiometries and does not appreciably inhibit copolymerization of labelled with unlabelled monomer. CR-actin injected into cells incorporates into endogenous actin structures including stress fibres and lamellipodia (Theriot & Mitchison, 1991). Photoactivation with 360 nm light is rapid and efficient, with a time-course somewhat faster than that of C2CF (probably because CR contains only one caging group, as opposed to two for C2CF). However, there are two major drawbacks to the use of CR. First, resorufin bleaches even more rapidly than fluorescein under full mercury arc illumination. Second, the caged compound retains some visible absorption at 450 nm (Fig. 45.3(b)), and can be photoactivated (albeit inefficiently) by visible light. To minimize bleaching, we usually attenuate the mercury arc with neutral-density filters and shutter the illumination (see below). To avoid uncaging with visible light, we use a narrow band-pass 577 nm excitation filter for fluorescence microscopy (see below). Within this band there is a strong mercury line corresponding to the peak of resorufin absorption (about 575 nm); this set-up provides optimal excitation with no undesired uncaging. The uncaging of CR by visible blue light limits its usefulness in double-label applications where fluorescein (or C2CF) is also observed.

Useful protein probes for photoactivation must fulfil the following six requirements (Mitchison, 1989):

(1) They must be stably non-fluorescent inside cells.
(2) They must be readily photoactivated to a stable, highly fluorescent form.
(3) They must be efficiently photoactivated by light at a wavelength and intensity which does not perturb biological structures.
(4) They must be easily coupled to proteins of interest.
(5) They must not perturb the function of the carrier protein.
(6) The side-products of activation must be non-toxic to the cell and non-perturbing to the carrier molecule.

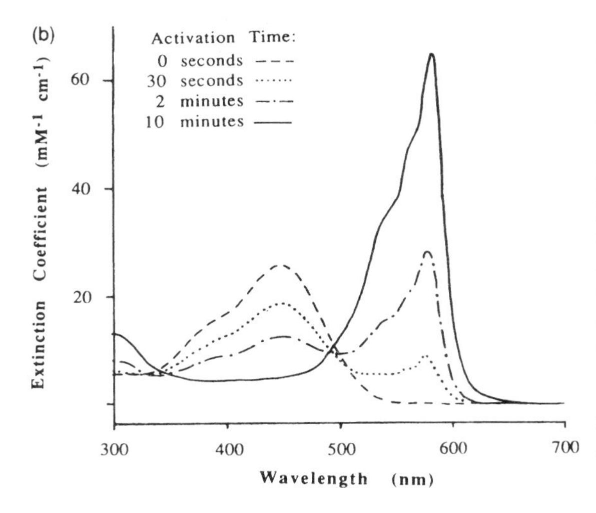

Figure 45.3 (a) Caging and photoactivation reactions of CR. (b) Optical absorption spectra after 0 s, 30 s, 2 min and 10 min of activation in a cuvette with a 360 nm hand-held lamp. (Adapted from Theriot & Mitchison, 1991, with permission.)

C2CF and CR fulfil these requirements to varying degrees: (1) Both C2CF and CR are stably non-fluorescent in the dark inside cells for hours or days (J.A. Theriot, unpublished observations; Reinsch *et al.*, 1991). As mentioned above, CR is slowly activated by visible light, and both compounds particularly must be guarded from the UV components of fluorescent overhead lights and sunlight. (2) C2CF and CR are readily photoactivated in cells, but both fluorescein and resorufin are easily photobleached. This must be kept in mind when determining the rate of turnover of photoactivated subunits (see Fig. 45.9) and when considering possible photodynamic damage to the cell. (3) Both compounds are activated by 360 nm light, which is relatively benign to biological specimens (Hiramoto *et al.*, 1984). Photoactivation has not been observed to have adverse consequences for injected or uninjected cells, and photoactivation does not cause breakage of C2CF-labelled microtubules (Mitchison, 1989). (4) C2CF is capable of labelling tubulin, but only to stoichiometries substantially lower than those achieved with NHS-rhodamine or NHS-biotin (Hyman *et al.*, 1991). It does not efficiently label actin or myosin. CR does efficiently label actin and myosin, but its ability to label tubulin is similar to that of C2CF (J.A. Theriot, unpublished observations). (5) C2CF-labelled tubulin does not polymerize quite as efficiently as unlabelled tubulin, and tends to promote microtubule nucleation (Mitchison, 1989). CR-labelled actin also does not polymerize quite as efficiently as unlabelled actin (Theriot & Mitchison, 1991). However, in most *in vivo* applications, where labelled protein represents only a tiny fraction of the total protein present, the behaviour of labelled structures is presumably dominated by the characteristics of the more abundant unlabelled endogenous protein. (6) Both compounds generate potentially toxic photoactivation products, nitrosobenzaldehyde (from C2CF) or nitrosoacetophenone (from CR). Both are reactive toward sulphydryls and are presumably scavenged by glutathione and related compounds inside cells, but it is possible that they may interfere with the biological activity of some proteins. To date we have not observed any toxicity associated with photoactivation.

45.3 A COMPUTER-CONTROLLED, MULTIPLE-CHANNEL FLUORESCENCE MICROSCOPE FOR PHOTOACTIVATION

45.3.1 The photoactivation beam

A microbeam for local photoactivation of fluorescence requires only (1) a small aperture (typically a pinhole or slit) that can be demagnified by the microscope objective and focused onto a specimen of interest, and (2) a 360 nm light source for illuminating this aperture. An epifluorescence mercury arc lamp is ideal for this purpose, and in principle an existing epifluorescence set-up could also be used for photoactivation. This method, however, suffers from a number of immediate disadvantages. First, under such conditions accurate positioning of a microbeam aperture will be difficult, because the aperture must be placed directly in the epi-illumination light path and therefore must also be easily removed and replaced for illumination of the entire field during image acquisition. Moreover, the most efficient (i.e. intense) configuration for microbeaming requires critical illumination of the sample, while full-field epifluorescence nearly always involves Koehler illumination. For these reasons we use a separate, additional mercury arc lamp (HBO 100) exclusively for photoactivation in our microscope system. This lamp is oriented orthogonal to the epifluorescence light path and is brought into the microscope via a dichroic interference filter (390 nm long-pass) set at a 45° angle in the epifluorescence tube (Fig. 45.4). The photoactivation beam is focused with a secondary lens to critically illuminate the sample. This also focuses the beam as a second, conjugate plane that is close to but outside the body of the microscope. The microbeam aperture or slit is placed at this second plane, first mounted onto x–y positioners and then onto an optical rail bolted to a convenient part of the microscope. (Slits, rails, etc. can be purchased from companies such as Oriel or Newport and modified as needed by most machine shops.) The slit can then be adjusted along x, y and z axes and fixed into position. We usually adjust the slit by looking at a drop of coumarin in glycerol on a microscope slide. Coumarin is excited well by the 360 nm Hg line and can be seen easily using a filter set for fluorescein, yielding a sharp, fluorescent image of the slit through the oculars when the slit is in focus. Care should be taken to avoid exposing one's eyes to too much light when focusing the arc and the slit; in addition to reducing the spectrum of the microbeam at all times with a broad 360 nm band-pass filter (Schott UG11) and a UV/IR absorption filter (Schott KG1), we often add a neutral-density filter to the system when focusing the beam and slit under the microscope.

With nearly all possible geometries the image of the arc at the slit plane is much larger than both the original arc and the slit itself, so a great deal of the available light does not pass through the aperture. (In our system, the mercury arc focuses to a spot of about 1 × 1.5 cm at a distance of about 30 cm from the focusing lens and is used to illuminate a slit of 0.01 × 0.3 cm.) Because the arc is not a true point source, there is nothing one can do about this loss. That is, any attempts to 'shrink' the beam through the use of a telescope (e.g. a laser beam expander placed in reverse orientation) or by repositioning the arc will be countered by an increase in beam divergence, and a corresponding loss of irradiance at the exit pupil of the objective. Because of the high intensity of the source, however, this 'masking effect' is typically not a problem for rapid photoactivation and is therefore of little concern.

We have also examined the possibility of using a UV laser for photoactivation. To date we have not found any suitable lasers operating in the range of 360 nm. Small tunable dye lasers pumped by a pulsed nitrogen laser (337 nm) do not offer much more total average power than a mercury arc, and also may cause photodamage at peak power. The only other potentially reasonable photoactivation source, a continuous-wave helium–cadmium laser (325 nm), requires expensive quartz objectives as well as additional modification to the microscope, since most microscope optics do not transmit below about 340 nm. It is also possible to double or triple a longer wavelength laser, but these lasers must be fairly high power and require considerably more care in operation.

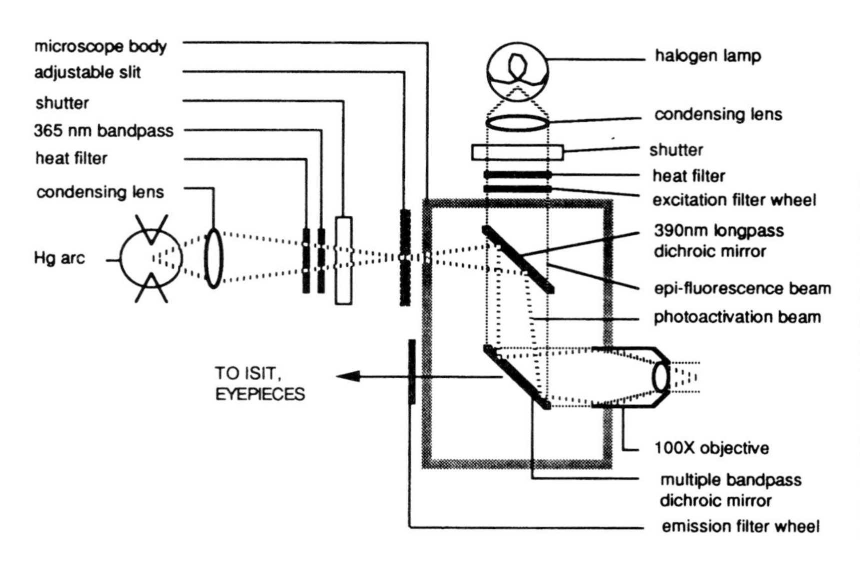

Figure 45.4 Schematic of the optical apparatus used for photoactivation microbeaming. Microscope, Zeiss IM35, with 100 × 1.3 NA Ph3 Neofluar; slit, homemade from Newport, Oriel, and Melles-Griot components; shutters, Vincent associates; 365 nm filter, Schott UG11; heat filter, Schott KG1; Hg arc HBO100, Oriel; halogen lamp, Zeiss 12V; filter wheels, AZI, with Omega filters; dichroic mirrors, Omega Optical. Further details are given in text. (Adapted from Mitchison, 1989.)

45.3.2 Multiple-channel fluorescence microscopy

In most cases one would like to follow the fate of a photoactivated fluorescent mark with respect to some known landmark. In simple cases some fiduciary background fluorescence may be available. In our original work (Mitchison, 1989) we collected alternate fluorescence and phase-contrast images of cells before and after photoactivation by alternating between epi- and trans-illumination and manually sliding back and forth a two-place filter cube containing a fluorescein set in one place and no filters in the other. Where a more complete quantitation of fluorescence is desired, it is essential to be able to normalize locally photoactivated fluorescence to total fluorescence, observed in a separate channel (Sawin & Mitchison, 1991). To avoid the extreme tedium entailed in sliding a two-place fluorescence filter cube back and forth in long-term experiments (such as fate mapping of embryos; see below) we developed a simple computer-controlled multichannel fluorescence microscope that uses filter wheels for fluorescence excitation and emission and a multiple band-pass dichroic mirror in the fluorescence filter cube.

Based on the success of fluorescence ratio indicators, a number of motorized, computer-addressable filter wheels are currently commercially available (e.g. from AZI, Oriel, LEP, Sutter). They differ in many aspects, including: speed; 'intelligence' (e.g. the ability to find the shortest path to a new position, to know what is the current filter position, and to relay this information back to the computer); the number of wheels in a unit (e.g. for combining fluorescence and neutral-density filters); mode of address (serial port vs. TTL, via an additional digital input/output (I/O) board; see below); and price. Unless extreme precision or speed is required, wheels that are relatively simple and slow (0–2 s between positions) are usually sufficient. A low-cost set-up, for example, could probably be put together using bottom-of-the-line filter wheels from Oriel in conjunction with a separately purchased low-cost I/O board (e.g. a Data Translation DT2817), that can also control any other electronics that respond to standard TTL signals.

If one is setting up a microscope for use exclusively with filter wheels, one should avoid purchasing any filters with the microscope itself altogether and instead obtain filters (typically 25 mm diameter for wheels, as opposed to 17–18 mm diameter for most microscopes) from a manufacturer specializing in high-quality interference filters (e.g. Omega Optical or Chroma Technologies). To avoid any signal crossover one should use fairly narrow band-pass interference filters for both fluorescence excitation and emission, and filters should be anti-reflection (AR) coated to prevent any spurious reflection images. Interference-coated fluorescence excitation filters should be 'blocked' with coloured glass, which improves specificity, with an accompanying loss in total transmission. Emission filters are almost always similarly blocked, although in our experience this may not be completely essential.

To avoid sliding the fluorescence cube back and forth also requires having a dichroic mirror that can be used with more than one fluorochrome. Such interference filters ('multiple band-pass dichroic mirrors') can be custom-ordered from Omega Optical or Chroma Technologies. Specifications for such filters should take into consideration: (1) optimizing for the

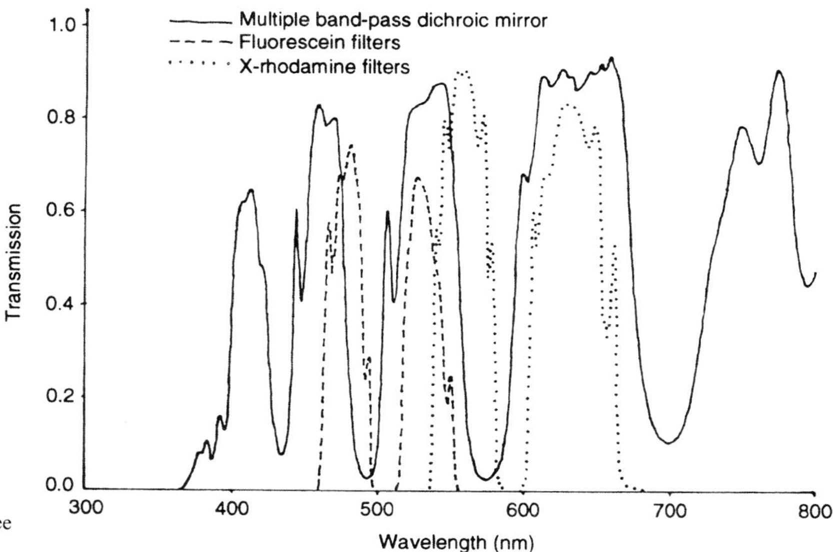

Figure 45.5 Filter sets used for double-label fluorescence imaging. See text for details.

particular fluorochromes to be used, within limits of manufacturing technology; (2) matching the dichroic to excitation and emission filters, and at the same time avoiding any cross-over between fluorochromes; (3) making sure that the dichroic will reflect the 360 nm photoactivation beam onto the specimen. For good spectral separation we typically use C2CF and X-rhodamine or Texas Red as fluorochromes in double-label experiments, using a dichroic and filter set designed for fluorescein and Texas Red, made by Omega Optical (Fig. 45.5). We have also made a triple-label set, for imaging fluorescein, X-rhodamine and the far-red dye CY5 (Ernst *et al.*, 1989).

45.3.3 Putting it all together

Most fluorescence video microscopy requires that the total sample exposure to light be limited, either because of the photosensitivity of the sample and/or the rapid photobleaching rates characteristic of many fluorochromes, including fluorescein. This is achieved by shuttering illumination, usually under the control of an image-processing system (see below). At the same time, a low-light-level camera should be used to minimize the amount of incident light required to collect a decent image. Either a silicon-intensified target camera (SIT) or an intensified SIT (ISIT) may be appropriate, or, in some cases, a much more expensive cooled charge-coupled device (CCD). In our experience, SIT cameras can easily detect anything visible to the dark-adapted eye (which is quite sensitive). The ISIT is more sensitive, but has poor spatial resolution and is much noisier. For example, imaging single fluorescent MTs is quite easy with a SIT camera, but impossible with an ISIT. The ISIT may be preferred, however, where a need for sensitivity outweighs that for resolution. Both cameras are rather sensitive in the red, beyond the range of the human eye, so to avoid unnecessary degradation of signal-to-noise it is advisable to use band-pass rather than long-pass emission filters (as described above). As mentioned above, CCDs are very expensive and require additional hardware support, but they also have the benefit of an enormous linear dynamic range, which may be of great value in quantitating fluorescence. Under normal sampling frequency and duration in our experiments, a cooled CCD has approximately the sensitivity of a SIT.

For low-light, time-lapse experiments, video-frame summing or averaging a digital image processor is useful, if not essential, and in any long-term experiment (e.g. fate-mapping) it is easiest to perform as many operations as possible under computer control. Many types of image processors to perform these functions are now available, and most of these have fairly sophisticated I/O capabilities to control shutters, filter-wheels, and recording devices (see below) in addition to doing standard image acquisition, processing, and analysis. Image processors can vary widely in price, depending on colour capabilities, available software, support, etc. Because of our high use of video microscopy in our laboratory we have found it useful to have a number of relatively low-cost image processors for data acquisition (typically the most time-consuming step in experiments), while more complicated analyses are done on a better-supported instrument. For data collection we use a Datacube MaxVision AT-1 system coupled to an inexpensive PC clone (Fig. 45.6). With this system it is particularly easy to write 'scripts' or macros that string together specific operations (e.g. rotate wheel, open shutter, average *n* frames, record image, wait *n* seconds), and to modify and extend existing scripts as source files. To make some operations run more smoothly and to connect the system to a number of I/O devices we found it necessary to modify the original source code (which is made available by the manufacturer at no extra cost, with some supporting documentation) in a few key places. These alterations, as well as additional tricks for using MaxVision for video microscopy, are available to interested investigators.

Most applications using photoactivation of fluorescence involve time-lapse recording and shuttered illumination rather than recording at 30 frames per second video-rates, so it is important to record images on an instrument that has good single-frame capability, i.e. either an optical memory disk recorder

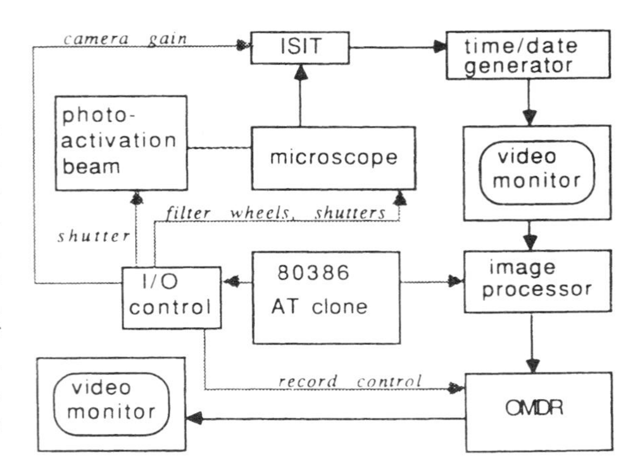

Figure 45.6 Schematic of the imaging system used in photoactivation experiments. The microscope and photoactivation beam are as in Fig. 45.4. Intensified silicon-intensified target camera (ISIT), Cohu; time-date generator, Panasonic; image processor, Datacube (AT-1), 80386-based AT computer, Samsung, optical memory disk recorder (OMDR), Panasonic; digital input/output (I/O) board, Data Translation, (no. 2817).

(OMDR) or direct digital storage. The drawback of the OMDR is its relatively high cost, while any digital storage system requires high-capacity disk storage, as well as memory buffers and a computer system capable of handling all of this. We currently do most recording with an OMDR; however, since most recording is done with an image processor that should be easily interfaced with direct digital image acquisition, and electronics prices are always dropping, it seems likely that in the future direct digital recording will be the method of choice.

45.4 EXPERIMENTS USING PHOTOACTIVATION OF FLUORESCENCE

45.4.1 Poleward MT flux in the mitotic spindle

C2CF was first synthesized in order to study the MT dynamics in the kinetochore fibres of the mitotic spindle (Mitchison, 1989). As described above, MTs are polar structures, and in each half of the spindle, MT 'minus' ends are proximal to the centrosome, while MT 'plus' ends are distal and interact with MTs from the opposing half-spindle and with kinetochores. While many photobleaching studies have shown that most MTs in the spindle are highly dynamic (Salmon *et al.*, 1984; Saxton *et al.*, 1984) MTs of the kinetochore fibre appear to be much more stable, both in terms of their turnover (Gorbsky & Borisy, 1989) and resistance to MT-depolymerizing conditions. We originally demonstrated *in vitro* that isolated kinetochores could 'capture' free MT plus ends and that MT ends thus captured could nevertheless incorporate exogenous, labelled tubulin without disrupting the attachment at the kinetochore (Mitchison & Kirschner, 1985). These studies were followed by *in vivo* microinjection experiments, in which incorporation of a biotin-labelled tubulin into spindle MTs of mitotic tissue culture cells was followed by immunoelectron microscopy (Mitchison *et al.*, 1986). Whereas most spindle MTs incorporated the biotin-tubulin rather quickly, kinetochore MTs incorporated the biotin-tubulin signal more slowly, at the site of kinetochore attachment, and at increasing times after microinjection kinetochore MTs became progressively labelled from their plus ends polewards towards their minus ends at spindle poles. Because the metaphase spindle is at steady state, this result suggested that incorporation of labelled tubulin at kinetochore-attached MT plus ends might be balanced by a net disassembly at MT minus ends, resulting in the steady movement of tubulin subunits in the MT lattice towards spindle poles, as was suggested by Margolis and Wilson on the basis of experiments *in vitro* (Margolis & Wilson, 1981). The notion of a poleward MT flux is of great interest to students of mitosis because of its implications for MT organization and regulation of MT dynamics, and for modes of force generation and chromosome movement. However, pulse-labelling of spindles with biotin-tubulin could not show unequivocally that such a flux was occurring.

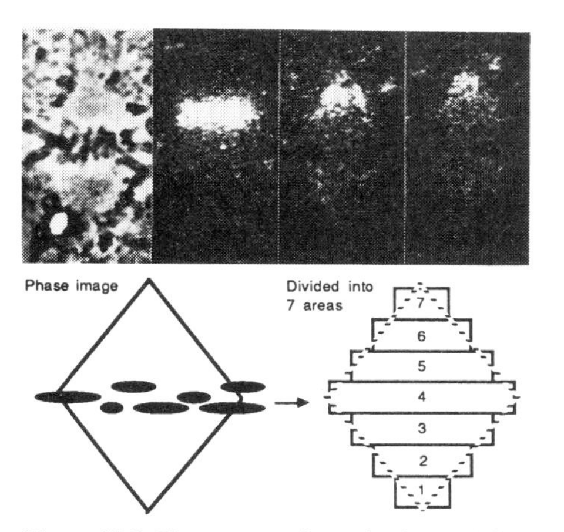

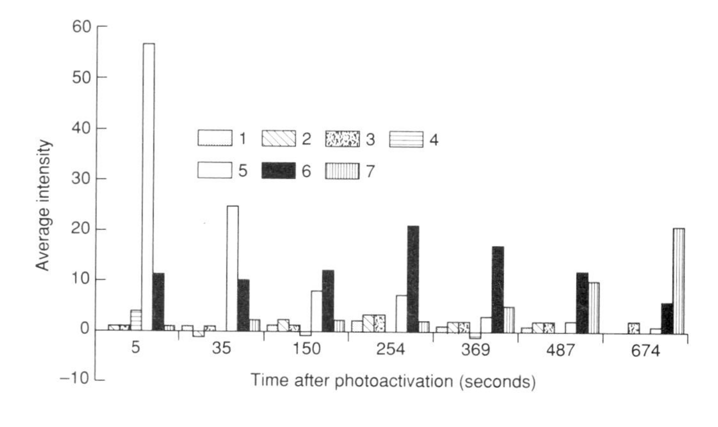

Figure 45.7 Fluorescence intensity in predefined zones of an LLC-PK1 spindle at different times after photoactivation. Left-hand panel shows (above) phase-contrast image of the spindle and three ISIT fluorescence images, at 5, 429, and 674 s after photoactivation, and (below) a diagram of the same spindle, arbitrarily divided into seven zones. Right-hand panel shows the average intensity in each of these zones at various times after photoactivation. Fluorescence intensity in zone 5, the site of photoactivation, decreases with time, while intensity in zone 6 rises, then falls. As intensity in zone 6 falls, intensity in zone 7 rises, indicating the vectorial movement of the microtubule lattice. The spindle remained in metaphase throughout the experiment. (From Mitchison, 1989. Reproduced from the *Journal of Cell Biology*, 1989, **109**, 637–652, by copyright permission of the Rockefeller University Press.)

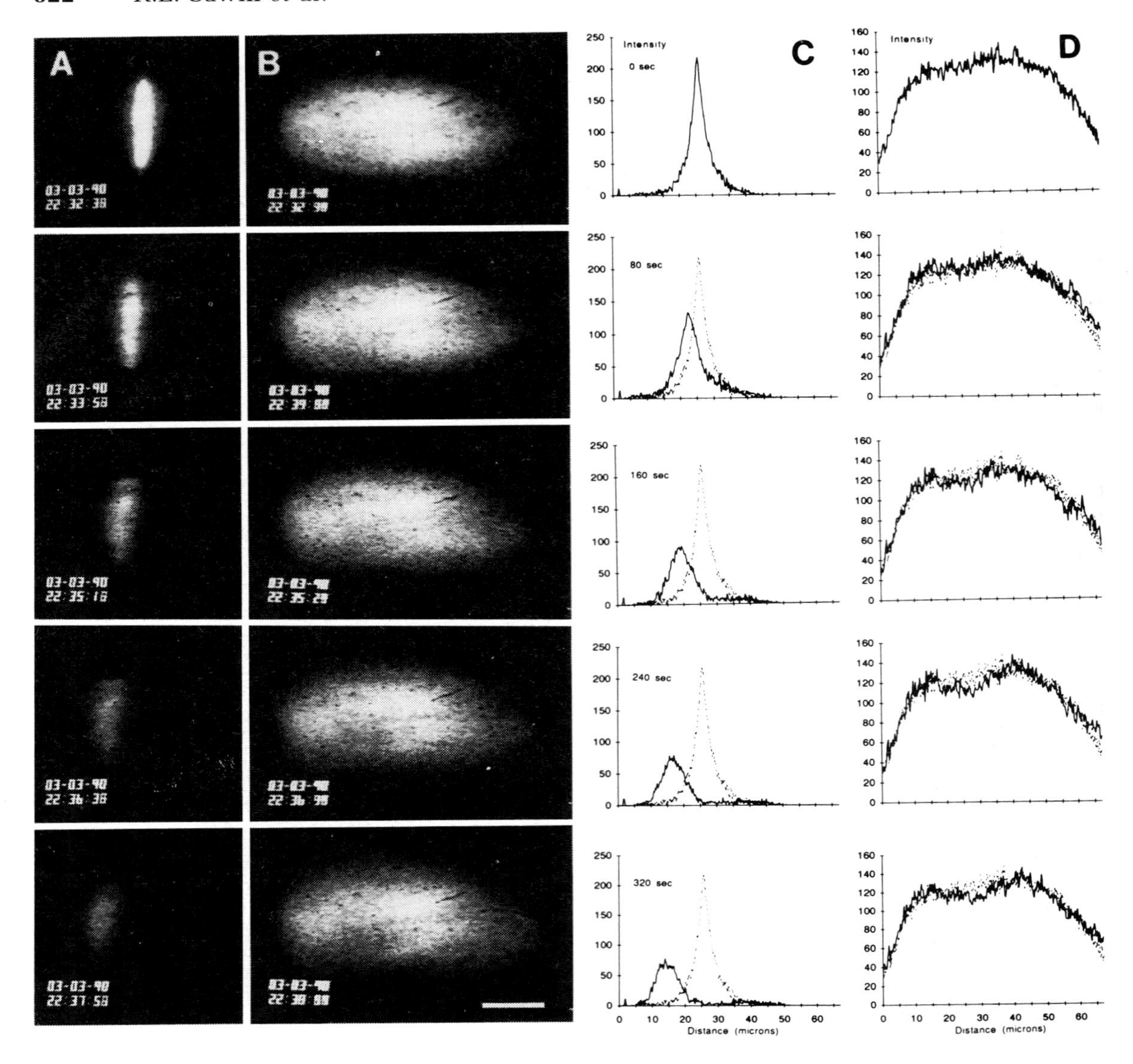

Figure 45.8 Photoactivation and double-label fluorescence imaging in a spindle assembled *in vitro* from a *Xenopus* egg extract. (A and B) ISIT images of photoactivated (fluorescein) and total (X-rhodamine) fluorescence at various times after photoactivation. (C and D) Digitized intensity-tracings of the same images (based on the average intensity of a 15-pixel-wide line, running from the left pole to the right pole), relative to the original photoactivated mark at time zero. Both the vectorial movement of photoactivated fluorescence and the essentially unchanged total fluorescence are apparent. (From Sawin & Mitchison, 1991. Reproduced from the *Journal of Cell Biology*, 1991, **112**, 941–954, by copyright permission of the Rockefeller University Press.)

This inferred poleward movement was demonstrated directly in living tissue culture cells using C2CF covalently coupled to tubulin (Mitchison, 1989). After microinjection of unactivated C2CF-tubulin into metaphase tissue culture cells and incorporation into spindles, fluorescence was locally activated in the spindle by UV microbeam and followed by low-light fluorescence video microscopy. As predicted, activated fluorescence moved polewards, and observed poleward flux in these experiments was apparent only in what appeared to be kinetochore MTs; other spindle MTs lost fluorescence rapidly, presumably because of their high rate of turnover (Fig. 45.7). This movement could be quantitated using digital image analysis. Photoactivated fluorescence increased near spindle poles at later times after photoactivation, while decreasing at the initial site of activation, indicating true movement of fluorescence.

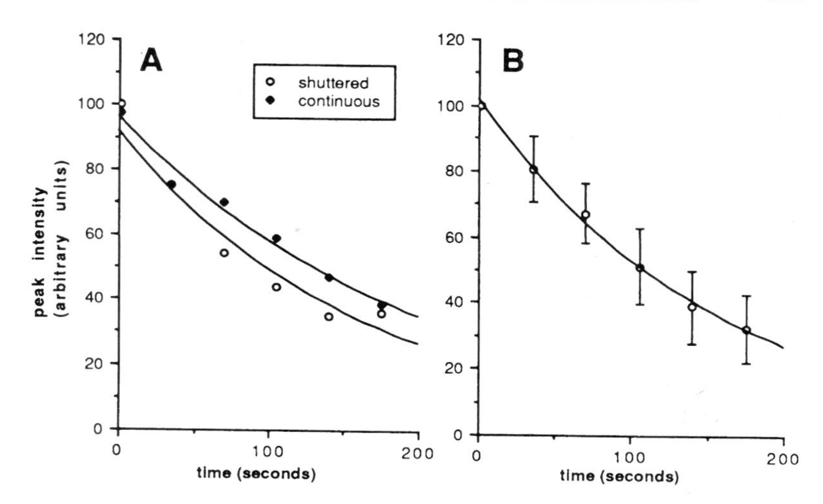

Figure 45.9 Turnover in fluxing microtubules in spindles assembled *in vitro*. (A) Decline of the peak of fluorescence intensity over time, from profiles such as those shown in the preceding figure. The same curve is obtained with continuous and carefully shuttered illumination, indicating that loss of fluorescence is not due to photobleaching. (B) Average rate of loss of fluorescence, from six different experiments. (From Sawin & Mitchison, 1991. Reproduced from the *Journal of Cell Biology*, 1991, **112**, 941–954, by copyright permission of the Rockefeller University Press.)

Quantitation of fluorescence was particularly important in this instance because if MT dynamics were anisotropic in the spindle (so-called 'tempered instability'; Sammak *et al.*, 1987) fluorescence might falsely appear to the eye to move polewards over time, in the absence of flux (Sawin & Mitchison, 1991). By contrast, the measured increase of fluorescence in 'polar' regions of the spindle at increasing times after photoactivation indicates the directed movement of the microtubule lattice (Mitchison, 1989).

More recently we identified a poleward MT flux in spindles reconstituted *in vitro* from extracts of *Xenopus* eggs (Sawin & Mitchison, 1991). In these experiments we were interested in more accurately quantitating fluorescence, to determine rates of flux and of MT turnover. For these reasons we made spindles that contained both C2CF-tubulin and non-caged, X-rhodamine-tubulin, and by collecting alternate images in fluorescein and rhodamine channels using filter wheels we were able to normalize photoactivated fluorescence with respect to total fluorescence during image analysis. As shown in Figs 45.8 and 45.9, this allowed us to align intensity profiles over time to near-perfect register (Fig. 45.8) and to determine approximate rates of MT turnover, based on fluorescence decay after photoactivation (Fig. 45.9).

FRAP experiments have not yet detected any significant poleward flux of MTs in spindles (Gorbsky & Borisy, 1989; Wadsworth & Salmon, 1986, and references therein) with one possible exception (Hamaguchi *et al.*, 1987), even in the same cell types in which flux has been observed by photoactivation (Gorbsky & Borisy, 1989; Mitchison, 1989). The reasons for this apparent negative result are not yet clear. It is possible that photobleaching may not actually be non-perturbing. At very high incident illumination fluorescent MTs will experience significant damage and break up (Vigers *et al.*, 1988). However, when photobleaching experiments are done carefully, spindles remain morphologically normal and progress through mitosis at normal rates after photobleaching, so we consider it unlikely that photodamage is a serious concern. A second possibility is that signal-to-noise problems may make it difficult to detect small changes in the position of a dark spot (representing a stable, bleached kinetochore MT) against a light background (representing dynamic, non-bleached and recovering spindle MTs) as opposed to the converse. One can imagine, for example, that as the more dynamic components of a multi-component system recover quickly after photobleaching, the dark spot of slowly recovering component will be 'obscured' by newly recovered fluorescence, making accurate measurements more difficult (Fig. 45.10(a)). From a theoretical point of view, it would seem immaterial whether a mark is bright against a dark background, or vice versa. However, we note that non-linearities in camera gain may indeed affect the ability to detect movement of bleached fluorescence as compared with photoactivated fluorescence, since identifying movement unambiguously often depends on distinguishing fairly subtle differences in intensity (Fig. 45.10(b,c)). Unfortunately, this explanation may be insufficient to account for the differences seen with photoactivation vs. FRAP, since recent FRAP experiments using CCDs with extremely broad linear dynamic range have also failed to detect poleward flux in the spindle. Further experiments using photoactivation and photobleaching in the same system will help to resolve these discrepancies.

45.4.2 Actin dynamics in the leading edge of cells

The leading edge, or lamellipodium, of motile cells is a dynamic cytoskeletal structure in some ways

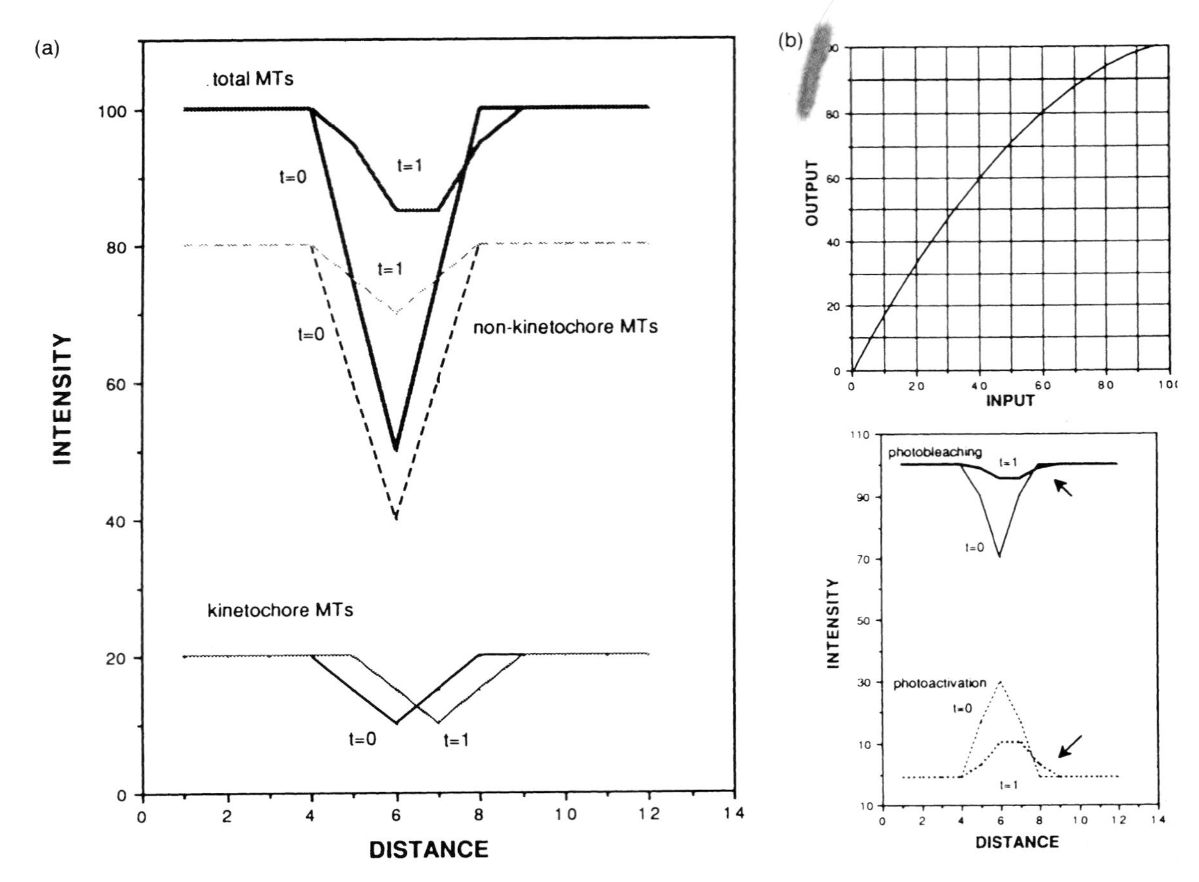

Figure 45.10 Why FRAP and photoactivation of fluorescence may yield different results. (a) Schematic drawing in which microtubules in the spindle are represented as two distinct populations: 80% non-kinetochore MTs, which turn over quickly but are stable in space; and 20% kinetochore MTs, which do not turn over but move from left to right. In this example fluorescence is bleached to 50% of its initial value at $t = 0$, and non-kinetochore MTs recover significantly by $t = 1$. The movement of kinetochore MTs at $t = 1$ is apparent in the profile of total MT fluorescence (kinetochore MTs plus non-kinetochore MTs), and theoretically should be equally obvious using photoactivation or photobleaching tehniques. (b) An example of a typical, slightly non-linear gain function for a video camera. 'Input' denotes actual light levels; 'output' is what the camera 'sees' (and what the investigator measures). (c) How non-linear gain may affect one's ability to judge movement of fluorescence. The original profiles of total MT fluorescence shown in (a) at $t = 0$ and $t = 1$ were transformed using the function shown in (b) (inverting the profile in order to represent photoactivation), such that 'photobleaching' curves have intensities near the flat (top) portion of the gain function, while 'photoactivation' curves (i.e. the photobleaching profiles inverted) have intensities near the steeper (bottom) portion. While after this transformation, the contribution of the movement of the kinetochore MTs is still visible in both curves, it is more obvious in the 'photoactivation' curve, i.e. in the steeper portion of the gain function, where small differences in intensity are exaggerated. This is especially apparent where the profile at $t = 1$ has moved beyond the 'shoulder' of the profile at $t = 0$ (arrows).

analogous to the mitotic spindle. Actin filaments in the lamellipodium are arranged in a loosely organized meshwork and are at least partially polarized with their barbed (dynamic) ends toward the membrane (Small, 1982; Yin & Hartwig, 1988). The most direct observation of actin behaviour in the lamellipodium was the classic photobleaching experiment of Wang (1985) which demonstrated that a spot bleached on the lamellipodium of a fibroblast injected with rhodamine-actin would move slowly backwards toward the cell body. This movement was interpreted as arising due to treadmilling within individual long actin filaments, that is, addition of monomers to the filament barbed end at the membrane coupled with concomitant loss of monomers at the pointed end at the rear of the lamellipodium. This rearward movement of actin polymer may be analogous to the polewards movement of tubulin in the spindle. The continual influx of actin and actin-rich structures from the leading edge was supported by DIC observations of lamellipodia of neuronal growth cones (Forscher & Smith, 1988) and fibroblasts (Fisher *et al.*, 1988). However, the rate of actin filament transport observed by photobleaching was substantially slower than the rates of movement of structures observable by DIC. In addition, these observations of actin filament movements *in vivo* have generally been performed on cells which were immobile during the period of observation. Thus the relationships among rearward flux of actin and other lamellipodial structures, protrusion of the leading edge, and translocation of the bulk cytoplasm and nucleus were unclear.

We addressed these issues by injecting CR-actin into rapidly locomoting goldfish keratocytes (Euteneuer & Schliwa, 1984) and activating fluorescence in a narrow bar in the lamellipodium. We confirmed that actin filaments move backward with respect to the leading edge (Fig. 45.11). In this rapidly moving cell type, however, the rate of forward cell translocation in any given cell was approximately the same as the rate of relative rearward actin movement, so the activated filaments in the lamellipodium remained fixed in space as the cell moved forward over them, regardless

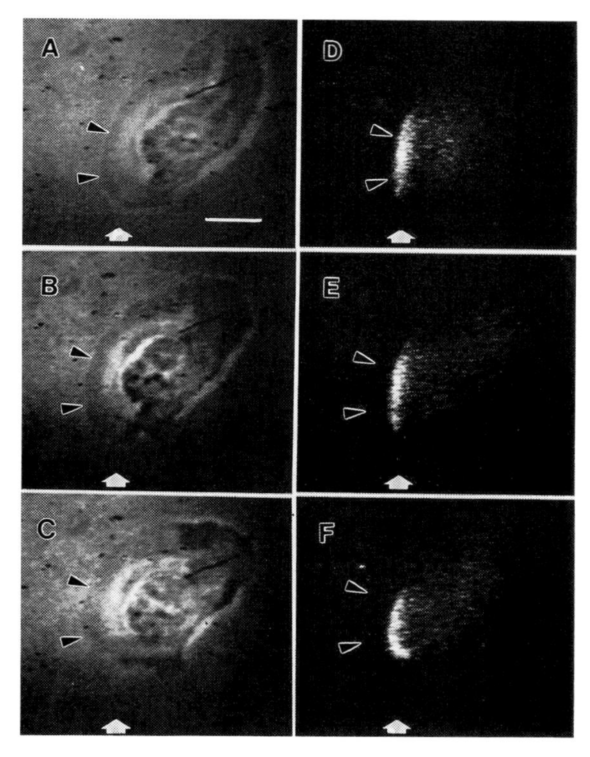

Figure 45.11 Behaviour of actin microfilaments in a moving goldfish keratocyte. Paired phase (A–C) and resorufin fluorescence (D–F) images of a keratocyte injected with CR-actin 4 s (A, D), 48 s (B, E) and 81 s (C, F) after activation of a bar-shaped region in the lamellipodium. Arrowheads indicate the position of the cell's leading edge in each frame; white arrows mark the same position in all panels. Note that the activated region remains approximately fixed in space as the cell moves from right to left. Bar is 10 μm. (Adapted from Theriot & Mitchison, 1991, with permission.)

of the rate of cell speed. We also found that the rate of actin filament turnover in the lamellipodium was remarkably rapid, about 23 s. This rapid turnover rate implies that the filaments must be much shorter than the width of the lamellipodium. This result is not consistent with the treadmilling model proposed by Wang (1985) for fibroblasts. Instead we think that in the keratocyte, short filaments are nucleated continually at the tip of the leading edge, and then move backwards as a cross-linked meshwork (Theriot & Mitchison, 1991).

CR-actin will also be a useful probe for determining the rate of actin filament turnover in lamellipodia in fibroblasts, and for relating the rate of rearward filament movement to the rate of forward translocation in these cells, as in keratocytes. It will also provide a convenient way of observing actin filament movements directly in cells in which actin-based structures can be observed to move by DIC or phase microscopy, to determine whether these rates of movement are in fact different from one another in a given individual cell. In addition to studies of actin-based cell motility, CR-actin may prove itself useful in investigations of other actin-dependent phenomena, such as cytokinesis.

45.4.3 Fate mapping

A promising application for photoactivation of fluorescence involves marking not subcellular components but rather cells themselves. Over the years a number of different general, non-genetic techniques have been developed for fate mapping of embryos, but no single method is without problems. One early method, direct observation (see, for example, Poulson, 1950), is often difficult for a number of reasons, not least among them the opacity and light-scattering qualities of many embryos. A second, common approach involves labelling specific cells or groups of cells with a non-perturbing marker of some sort (often fluorescent or enzymatic), either externally or by microinjection (see, for example, Weisblat *et al.*, 1978). The major problem with this method is that in a multicellular embryo the number of cells accessible by this technique may be limited – for example, it would be impossible to inject cells of an inner germ layer without disrupting outer layers. Moreover, microinjection of single cells becomes increasingly difficult as embryonic cleavages progress. A third method, local killing of cells with a laser microbeam (e.g. Lohs-Schardin *et al.*, 1979), avoids the problems faced by microinjection approaches but only with the clear disadvantage that cell-killing may induce unforeseen abnormalities in development. As a marking technique, photoactivation of fluorescence combines the positive aspects of these latter two methods without their obvious drawbacks, as follows: embryos can be first 'pre-loaded' with caged fluorochromes coupled to soluble carrier proteins or dextrans at times early in development, when microinjection is fairly easy (for example, at the one- or two-cell stage), and then allowed to progress to some predefined point, when single cells or fields of cells can be marked by photoactivation. Preliminary experiments have suggested that this is a reasonable approach to fate mapping (Goetz and T.J. Mitchison, unpublished observations); its success in real experiments appears to depend on coupling photoactivatable fluorochromes to appropriate carriers (J. Minden, pers. commun.; J.-P. Vincent, pers. commun.).

45.5 FUTURE PROSPECTS AND CONCLUSIONS

Photoactivation of fluorescence represents a novel technology for marking structures in living cells and organisms with the precision of a light microbeam. In addition to the experiments described above, future applications of photoactivation of fluorescence may include studying membrane and membrane skeleton dynamics during cell locomotion and adhesion, and perhaps also in following membrane transport and recycling pathways directly under the microscope. In addition to 'standard' embryonic fate mapping, one can imagine that photoactivation of fluorescence could also be used to mark cells or fields of cells in order to study morphogenetic movements in development, nerve growth, and cell–cell coupling with much greater precision than has previously been available *in situ*. More broadly speaking, photoactivation of fluorescence is one of a number of light-directed techniques – for example, photoactivation or photoablation of enzyme activities or enzyme substrates, or even the potential photoactivation of gene expression – which may be applied to complex biological problems. It is exciting to consider how these techniques may be combined in the future in order to uncover the inner workings of cells.

REFERENCES

Belmont L.D., Hyman, A.A., Sawin K.E. & Mitchison T.J. (1990) *Cell* **62**, 579–589.
Bershadsky A.D. & Vasiliev J.M. (1988) *Cytoskeleton.* New York, Plenum Press.
Carlier M.-F. (1989) *Int. Rev. Cytol.* **115**, 139–170.
Dustin P. (1984) *Microtubules.* Berlin, Springer-Verlag.
Ernst L.A., Gupta R.K., Mujumdar R.B. & Waggoner A.S. ÛŠ9:5ÛT(1989) *Cytometry* **10**(1), 3–10.
Euteneuer U. & Schliwa M. (1984) *Nature (Lond.)* **310**, 58–61.
Fisher G.W., Conrad P.A., DeBiasio R.L. & Taylor D.L. (1988) *Cell Motil. Cytoskeleton* **11**, 235–247.
Forscher P. & Smith S.J. (1988) *J. Cell Biol.* **107**, 1505–1516.
Gorbsky G.J. & Borisy G.G. (1989) *J. Cell Biol.* **109**(2), 653–662.
Hamaguchi Y., Toriyama M., Sakai H. & Hiramoto Y. (1987) *Cell Struct. Funct.* **12**, 43–52.
Haugland R.P. (1989) *Handbook of Fluorescent Probes and Research Chemicals.* Molecular Probes Inc., Eugene, OR.
Hill T.L. & Kirschner M.W. (1982) *Proc. Natl. Acad. Sci. USA* **79**, 490–494.
Hiramoto Y., Hamaguchi M.S., Nakano Y. & Shoji Y. (1984) *Zool. Sci.* **1**, 29–34.
Horio T. & Hotani H. (1986) *Nature* **321**, 605–607.
Hotani H. & Horio T. (1988) *Cell Motil. Cytoskeleton* **10**, 229–236.
Hyman A., Drechsel D., Kellogg D., Salser S., Sawin K., Steffen P., Wordeman L. & Mitchison T. (1991) *Methods Enzymol.* **196**, 478–485.
Inoue S. (1981) *J. Cell Biol.* **91**, 131s–147s.
Inoue S. (1982) In *Developmental Order: Its Origin and Regulation (40th Symposium of the Society for Developmental Biology)*, S. Subtelny & P.B. Green (eds). Alan R. Liss, New York, pp. 35–76.
Inoue S. & Sato H. (1967) *J. Gen. Physiol.* **50**, 259–292.
Janko K. & Reichert J. (1987) *Biochim. Biophys. Acta* **905**, 409–416.
Kirschner M.W. (1980) *J. Cell Biol.* **86**, 330–334.
Krafft G.A., Sutton W.R. & Cummings R.T. (1988) *J. Am. Chem. Soc.* **110**, 301–303.
Lee J., Gustafsson M., Magnusson K.-E. & Jacobson K. (1990) *Science (Washington, D.C.)* **247**, 1229–1233.
Lohs-Schardin M., Sander K., Cremer C., Cremer T. & Zorn C. (1979) *Devel. Biol.* **68**, 533–545.
Margolis R.L. & Wilson L. (1978) *Cell* **13**, 1–8.
Margolis R.L. & Wilson L. (1981) *Nature (Lond.)* **293**, 705–711.
Marque J.J. (1989) *Nature* **337**, 583–584.
Mitchison T.J. (1989) *J. Cell Biol.* **109**, 637–652.
Mitchison T.J. & Kirschner M.W. (1984) *Nature* **312**, 237–242.
Mitchison T.J. & Kirschner M.W. (1985) *J. Cell Biol.* **101**, 767–777.
Mitchison T.J., Evans L., Schultze E. & Kirschner M.W. (1986) *Cell* **45**, 515–527.
Oosawa F. & Asakura S. (1975) *Thermodynamics of the Polymerization of Protein.* Academic Press, London.
Patchornik A., Amit B. & Woodward R.B. (1970) *J. Am. Chem. Soc.* **92**, 6333–6335.
Poulson D.F. (1950) In *The Biology of Drosophila.* Wiley-Interscience, New York.
Reinsch S.S., Mitchison T.J. & Kirschner M.W. (1991) *J. Cell Biol.* **115**, 365–379.
Salmon E.D., Leslie R.J., Karow W.M., McIntosh J.R. & Saxton R.J. (1984) *J. Cell Biol.* **99**, 2165–2174.
Sammak P.J., Gorbsky G.J. & Borisy G.G. (1987) *J. Cell Biol.* **104**(3), 395–405.
Sato H., Ellis G.W. & Inoue S. (1975) *J. Cell Biol.* **67**, 501–517.
Sawin K.E. & Mitchison T.J. (1991) *J. Cell Biol.* **112**, 941–954.
Saxton W.M., Stemple D.L., Leslie R.J., Salmon E.D., Zavortink M. & McIntosh J.R. (1984) *J. Cell Biol.* **99**, 2175–2186.
Small J.V. (1982) *Electron Microsc. Rev.* **1**, 155–174.
Theriot J.A. & Mitchison T.J. (1991) *Nature* **352**, 126–131.
Tsien F.Y. & Zucker R.S. (1986) *Biophys. J.* **50**, 843–853.
Wadsworth P. & Salmon E.D. (1986) *J. Cell Biol.* **102**, 1032–1038.

Walker J.W., Somlyo A.V., Goldman Y.E., Somlyo A.P. & Trentham D.R. (1987) *Nature* **327**, 249–252.
Walker J.W., Reid G.P. & Trentham D.R. (1989) *Methods Enzymol.* **172**, 288–301.
Walker R.A., O'Brien E.T., Pryer N.K., Sobeiro M.F., Voter W.A., Erickson H.P. & Salmon E.D. (1988) *J. Cell Biol.* **107**, 1437.
Wang Y.-L. (1985) *J. Cell Biol.* **101**, 597–602.
Wang Y.-L., Lanni F., McNeil P.L., Ware B.R. & Taylor D.L. (1982) *Proc. Natl. Acad. Sci. USA* **79**, 4660–4664.
Wegner A. (1976) *J. Molec. Biol.* **108**, 139–150.
Weisblat D.A., Sawyer R.T. & Stent G.S. (1978) *Science* **202**, 1295–1297.
Vigers G.P.A., Coue M. & McIntosh J.R. (1988) *J. Cell Biol.* **107**, 1011–1024.
Yin H.L. & Hartwig J.H. (1988) *J. Cell Sci. Suppl.* **9**, 169–184.

NOTE ADDED IN PROOF

Since the original submission of this manuscript, a number of papers on photoactivation of fluorescence have been published. They are listed below.

Mitchison T.J. & Salmon E.D. (1992) *J. Cell Biol.* **119**, 569–582.
Okabe S. & Hirokawa N. (1992) *J. Cell Biol.* **117**, 105–120.
Theriot J.A. & Mitchison T.J. (1992) *J. Cell Biol.* **119**, 367–377.
Theriot J.A., Mitchison T.J., Tilney L.G. & Portnoy D.A. (1992) *Nature* **357**, 257–260.
Vincent J.-P. & O'Farrell P.H. (1992) *Cell* **68**, 923–931.

CHAPTER FORTY-SIX

Video Imaging of Lipid Order

KATHRYN FLORINE-CASTEEL,[1] JOHN J. LEMASTERS[2] & BRIAN HERMAN

[1] Department of Pathology, Duke University Medical Center, Durham, NC, USA

[2] Department of Cell Biology and Anatomy, University of North Carolina at Chapel Hill, NC, USA

46.1 INTRODUCTION

Membrane lipid order may be an important parameter in the regulation of certain cellular functions, hence the interest in its measurement at the single cell level. A widely used method of measuring lipid order, usually performed on membrane suspensions, has been fluorescence polarization spectroscopy. In this technique, membranes are labelled with lipophilic fluorophores, and polarized fluorescence is measured in order to determine the average constraint of probe motion in the bilayer. In theory, this is a measure of the degree of lipid order. With the increasing sophistication of microscope/imaging systems as well as the availability of an increasingly diverse selection of suitable fluorescent probes, it is now possible to attempt these measurements microscopically on individual cells. This enables lipid order to be determined with subcellular spatial and temporal resolution.

Microscopic fluorescence polarization measurements on single cells present certain problems not associated with fluorometric measurements. The major difficulties are the depolarizing effect of the microscope optics and the orientation dependence of the observed fluorescence polarization. The first problem is due mainly to the use of a high aperture objective lens for collection of fluorescence, rather than a narrow slit of effectively zero aperture. A theoretical treatment of high aperture observation has been presented (Axelrod, 1979) and will not be discussed in detail here. The problem of oriented samples, such as surface-labelled single cells, requires a knowledge of the probe excited state orientation distribution in order to determine the degree of lipid order from the measured orientation-dependent fluorescence polarization.

In this chapter, we outline a method for obtaining spatially resolved measurements of lipid order using steady-state fluorescence polarization microscopy in combination with digital image processing. We treat in detail the case of spherical membrane surfaces labelled with rod-shaped, diphenylhexatriene- derivative probes. The general applicability of the technique is also discussed.

FLUORESCENT AND LUMINESCENT PROBES, 2ND EDN
ISBN 0–12–447836–0

46.2 THEORY

46.2.1 Description of the experiment

Figure 46.1 depicts a spherical membrane surface on the microscope stage, focused at the X_2–X_3 plane. In the conventional polarization experiment employing an inverted microscope and epi-illumination, incident excitation light polarized in the X_3 direction is reflected from a dichroic mirror below the stage and propagates along the X_1 direction, through the condenser/ objective, to the sample. Fluorescence is collected at 180° to the incident light propagation direction at each of two orthogonal orientations of a polarizer in the emission path, one in the X_2 direction and the other in the X_3 direction (i.e. perpendicular and parallel to the excitation polarization).

The polarized fluorescence observed in the microscope images will depend on four factors: (1) the orientation of probe absorption dipole moments relative to the excitation light polarization direction (X_3); (2) probe rotational diffusion and consequent emission dipole reorientation during the excited state lifetime; (3) objective lens numerical aperture, which determines the amount of mixing in of fluorescence components in the X_1 and X_3 (X_1 and X_2) directions in the recorded $F_{\perp}$ ($F_{\parallel}$) image; and (4) the range of probe orientations on the three-dimensional membrane surface corresponding to each pixel location in the two-dimensional image. The following section outlines how these factors are dealt with in order to infer lipid order from polarized fluorescence images.

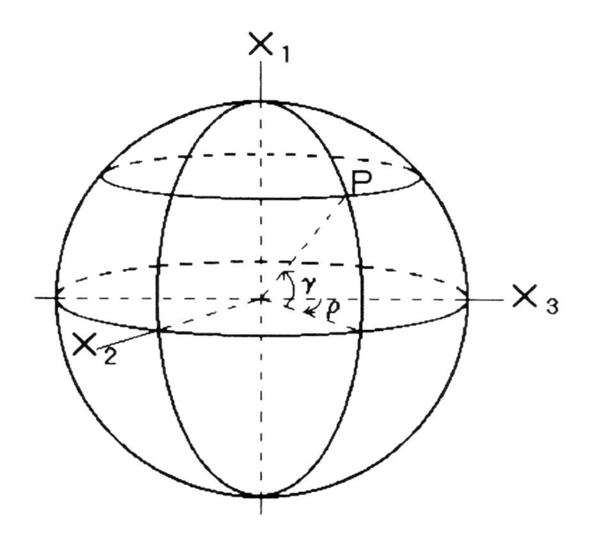

Figure 46.1 Definition of coordinates. The X_1-axis is the optical axis of the microscope, and the X_2- and X_3-axes define the focal plane. An arbitrary point P on the spherical membrane surface is specified by the angles ρ and γ.

46.2.2 Fluorescence polarization as a function of lipid order and membrane surface location

The general procedure for determining the functional dependence of fluorescence polarization on lipid order as well as probe location on the membrane surface is as follows. First, a model of probe rotational diffusion in the membrane is adopted which includes a parameter that indicates the degree of lipid order. For diphenylhexatriene (DPH) probes, we have used the 'wobbling-in-a-cone' model (Kinosita *et al.*, 1977) in which lipid order is expressed in terms of a cone angle, θ_{max}, representing the fluorophore's maximum angular motional freedom in the bilayer. The cone angle θ_{max} is related to the lipid order parameter, S, by $S = 0.5 \cos\theta_{max} (1 + \cos\theta_{max})$ (Lipari & Szabo, 1980). One then solves the appropriate diffusion equation in terms of a Green function to obtain the probability $p(\Omega',t'|\Omega,t)$ that a probe with orientation Ω' (in the membrane reference frame) at time t' will have orientation Ω at time t, where t' and t are the time of absorption and emission, respectively, of a photon. The probe excited state distribution function f can then be found from the relation

$$f = \frac{1}{\tau} \int\int (x_3')^2 \, p(\Omega',t'|\Omega,t) e^{-(t-t')/\tau} \, d\Omega' \, d(t - t') \quad [46.1]$$

where x_3' is the component of a unit magnitude absorption dipole moment along the X_3-axis (the excitation light polarization direction) at the time of absorption ($t - t' = 0$), and τ is the probe excited state lifetime. Finally, the distribution of excited fluorophores, f, is used to calculate the polarized fluorescence collected from the location (ρ,γ) on the membrane surface, for an arbitrary orientation ψ of the emission polarizer, according to

$$F_\psi(\theta_{max},\rho,\gamma) = N \int f \, [K_a x_1^2 + (K_b \cos^2\psi + K_c \sin^2\psi) x_2^2 + (K_b \sin^2\psi + K_c \cos^2\psi) x_3^2 + 2(K_c - K_b) \sin\Psi \cos\psi \, x_2 x_3] d\Omega \quad [46.2]$$

where x_1, x_2, and x_3 are the components of a unit magnitude emission dipole moment along the X_1-, X_2-, and X_3-axes, respectively, ψ is the angle in the X_2–X_3 plane between the X_3-axis and the transmission axis of the emission polarizer, and N represents the combined normalization constants. K_a, K_b and K_c are weighting factors which are functions of the numerical aperture of the objective and the index of refraction of the sample medium (Axelrod, 1979). For ψ = 0° and 90°, which correspond to $F_{\parallel}$ and $F_{\perp}$, respectively, equation [46.2] reduces to

$$F_{\parallel}\ (\theta_{max},\rho,\gamma) = N \int f\,[K_a x_1^2 + K_b x_2^2 + K_c x_3^2]d\Omega \qquad [46.3]$$

$$F_{\perp}\ (\theta_{max},\rho,\gamma) = N \int f\,[K_a x_1^2 + K_c x_2^2 + K_b x_3^2]d\Omega \qquad [46.4]$$

The dependence of F_{ψ} on lipid order, in terms of θ_{max}, arises in the integration of equations [46.1] and [46.2] over Ω' and Ω, where in each case θ_{max} is an integration limit. The dependence on membrane surface location, i.e. the coordinates (ρ,γ), is contained in the absorption and emission dipole moment components x_i' and x_i, respectively, and results from the coordinate transformation from the membrane to the laboratory reference frame.

The fluorescence intensity recorded in each pixel in the two-dimensional fluorescence image is found by integration of F_{ψ} over the appropriate range of ρ and γ. For example, the spherical membrane surface shown in Fig. 46.1 is in focus at $\gamma = 0°$ (the X_2–X_3 plane). However, each image pixel around the in-focus membrane perimeter ($0 \leq \rho \leq 360°$, $\gamma = 0°$) actually contains fluorescence from the region $-\gamma_o$ to $+\gamma_o$, where the angle γ_o depends on pixel size and is typically about 10° (Florine-Casteel, 1990).

46.2.3 Examples

Some examples of theoretical polarized fluorescence patterns for the in-focus membrane perimeter region are illustrated in Figs 46.2–46.4. Fluorescence was calculated according to equations [46.3] and [46.4], with variations in the probe rotational diffusion model (and hence in the distribution function f). The explicit form of these equations for each of the three diffusion models discussed is presented elsewhere (Florine-Casteel, 1990). The first model (Fig. 46.2), appropriate for DPH, is that in which the probe 'wobbles' with a maximum amplitude of angle θ_{max} about an axis normal to the plane of the membrane. In the second model (Fig. 46.3), the symmetry axis is tilted 30° from the bilayer normal, which is the case for probes such as TMA-DPH (a cationic derivative of DPH) that are located at or near the headgroup region of membrane phospholipids (Florine-Casteel, 1990). Finally, we consider the case of probe molecules

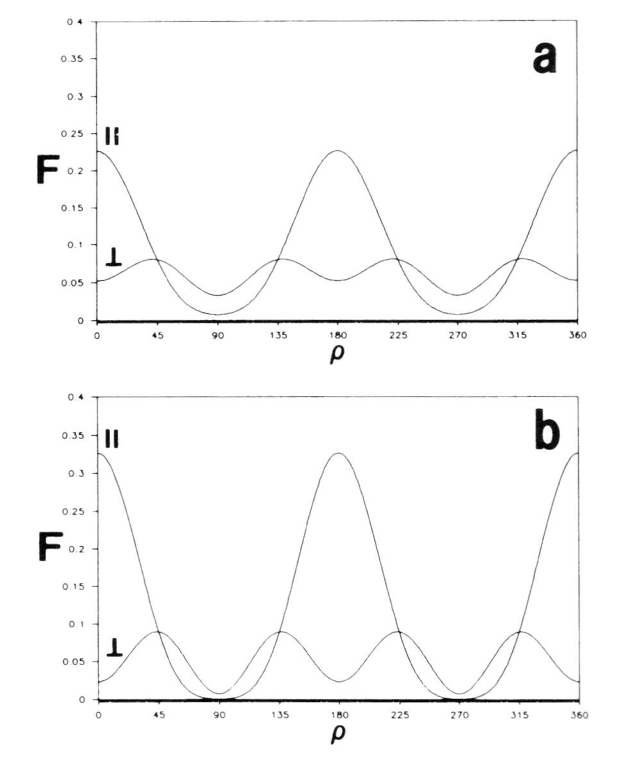

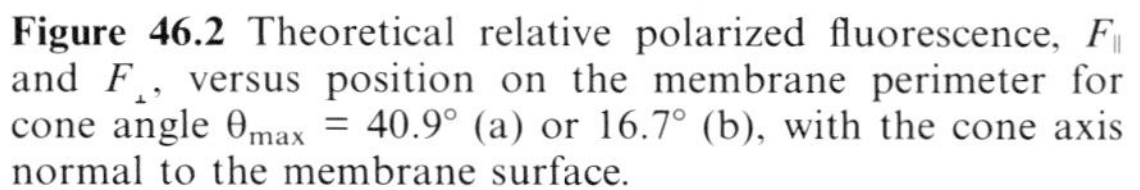
Figure 46.2 Theoretical relative polarized fluorescence, $F_{\parallel}$ and $F_{\perp}$, versus position on the membrane perimeter for cone angle $\theta_{max} = 40.9°$ (a) or 16.7° (b), with the cone axis normal to the membrane surface.

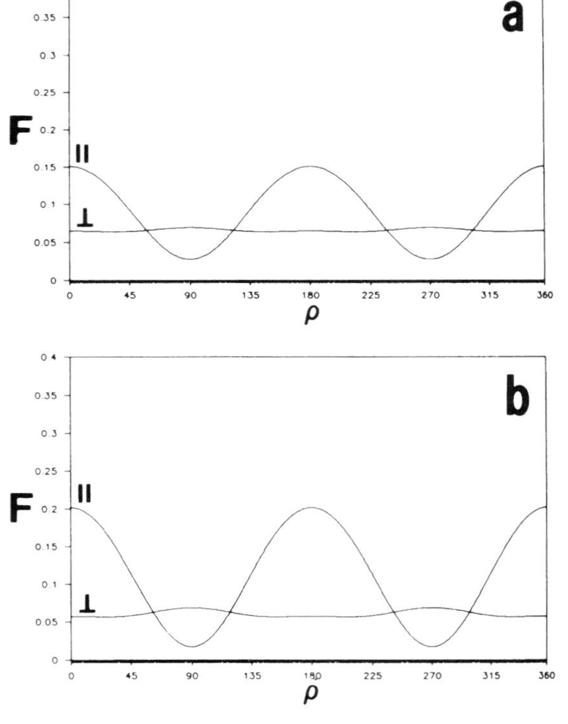

Figure 46.3 Theoretical relative polarized fluorescence, $F_{\parallel}$ and $F_{\perp}$, versus position on the membrane perimeter for cone angle $\theta_{max} = 40.9°$ (a) or 16.7° (b), with the cone axis tilted 30° from the bilayer normal.

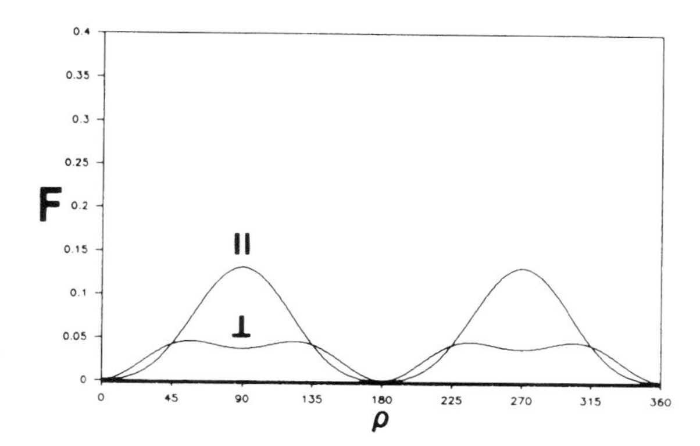

Figure 46.4 Theoretical relative polarized fluorescence, $F_{\parallel}$ and $F_{\perp}$, versus position on the membrane perimeter for the case of probe rotation in the plane of the membrane.

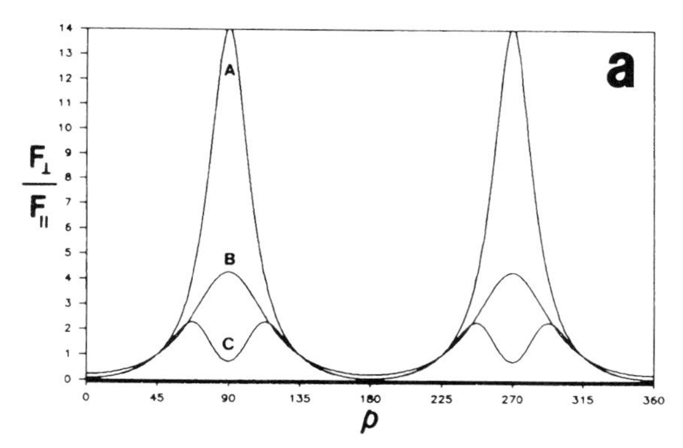

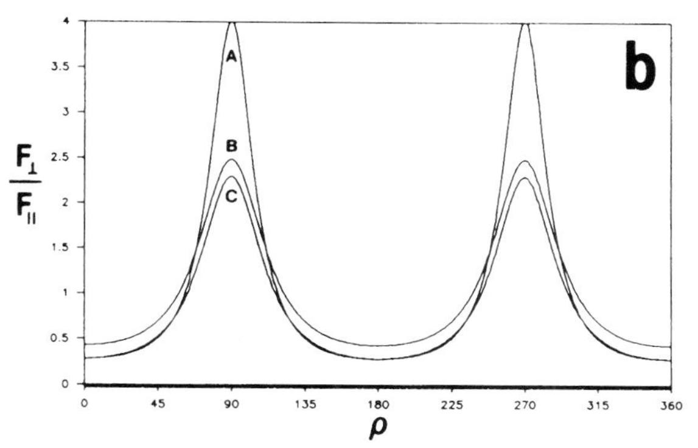

Figure 46.5 Theoretical polarization ratio, $F_{\perp}/F_{\parallel}$, versus position on the membrane perimeter in the absence (a) or presence (b) of a cone tilt of 30°, for θ_{max} = 16.7° (A), θ_{max} = 40.9° (B), or θ_{max} = 16.7° with 10% of probe molecules oriented parallel to the plane of the membrane (C).

trapped in the centre of the bilayer and oriented parallel to the plane of the membrane (Fig. 46.4), which has been reported in the literature for DPH and to a lesser extent, TMA-DPH.

As Figs 46.2 and 46.3 show, an increase in lipid order (i.e. a decrease in θ_{max}) results in a greater change in relative fluorescence between maxima and minima around the membrane perimeter. The presence of a cone axis tilt of 30° reduces the magnitude of this change. Probe molecules aligned parallel to the plane of the membrane (Fig. 46.4) display a different pattern of polarized fluorescence that is not dependent on the degree of lipid order. In all cases, fluorescence is greatest in those regions where the average orientation of probe absorption and emission dipole moments (for DPH, the probe long axis) is parallel to the excitation and emission polarizer orientations, respectively. For example, in the first model, probes in the X_2–X_3 plane (see Fig. 46.1) are, on average aligned radially around the membrane perimeter, along the X_3-axis (the excitation polarization direction) at ρ = 0° and 180° and along the X_2-axis at ρ = 90° and 270°. Therefore fluorescence will be maximum at ρ=0° and 180° with the emission polarizer oriented along the X_3-axis ($F_{\parallel}$).

Figure 46.5 illustrates how the polarization ratio, $F_{\perp}/F_{\parallel}$, changes with lipid order and with position on the membrane perimeter for the various models. The ratio is most sensitive to lipid order (θ_{max}) in the vicinity of ρ = 0°, 90°, 180°, and 270°. Accurate measurement of lipid order over the entire membrane perimeter would require an additional pair of emission polarizer orientations, ψ = 45° and 135°. The polarization ratio is also very sensitive to probe location in the bilayer, with probe molecules located in the hydrophobic core (Fig. 46.5(a)) displaying a greater variation in $F_{\perp}/F_{\parallel}$ than those located in the headgroup region where the symmetry axis is tilted (Fig. 46.5(b)). The presence of a significant fraction of probe molecules aligned parallel to the plane of the membrane results in a dramatic decrease in the polarization ratio in the vicinity of ρ = 90° and 270°, while leaving it virtually unchanged in the vicinity of ρ = 0° and 180°, where these probes are oriented perpendicular to the excitation polarization direction and thus only weakly excited.

46.3 EXPERIMENT

46.3.1 Probe selection

As Fig. 46.5 illustrates, a knowledge of the probe orientation in the membrane is important in interpreting the fluorescence polarization images in terms of lipid order. For example, a significant fraction of probe molecules trapped between the inner and outer leaflets of the bilayer or adsorbed to the outer membrane surface such that the predominant orientation is not along the lipid acyl chains, could lead to ratio values that underestimate the degree of lipid order. Another important consideration is the probe's quantum yield, since light transmission through the emission polarizer can be reduced by as much as 90%. Because a minimum of two fluorescence images is required for a polarization measurement, the probe's susceptibility to photobleaching is also important.

46.3.2 Fluorescence polarization imaging

The optical arrangement for fluorescence polarization microscopy and imaging has been described in detail (Axelrod, 1989; Florine-Casteel, 1990). The essential feature is the addition to the conventional epifluorescence set-up of a pair of polarizers, one in the excitation light path and one in the emission light path. It is important to place the polarizers such that the number of optical elements, i.e. lenses and mirrors, 'downstream' from each polarizer is a minimum. On the excitation side, additional optical elements can affect the polarization purity of the light incident on the sample. Similarly, on the emission side, the transmittivity of fluorescent light to the detector will be polarization-dependent. It is therefore necessary to measure this effective birefringence of the microscope/imaging system before performing experiments and to correct the polarization images accordingly. Since a change in emission polarizer orientation can sometimes cause a slight optical shift of the image, it is also important to make sure that each pair of images is properly aligned before ratioing.

An example of polarization imaging is shown in Plate 46.1. Here, cell-size phosphatidylcholine liposomes have been labelled with TMA-DPH. The polarized fluorescence images were obtained using a 1.3 NA objective lens in a Zeiss IM-35 inverted microscope, with an ISIT video camera as the detector. Each image is an average of 64 frames (2 s illumination time). Fluorescence polarization ratios were computed from digitized image pairs on a pixel-by-pixel basis, after background subtraction and outlining (mapping) of the liposome perimeter, with all pixel intensities outside of the map set to zero for image clarity. Pseudo-colour ratio images were obtained by converting ratios to grey levels (0–255), using a multiplication factor of typically 100, and then assigning colour values to the grey levels. Based on the variation in image intensity between adjacent pixels in the fluorescence images, we estimate that polarization ratios can be accurately determined with a surface spatial resolution of about 1 μm^2.

The polarization ratio images in Plate 46.1 illustrate the difference in the degree of lipid order between fluid-phase POPC and gel-phase DPPC liposomes. Around the membrane perimeter, which is in the focal plane, the ratios are lower at $\rho = 0°$ and 180° and higher at $\rho = 90°$ and 270° for DPPC than for POPC. Moving inward toward the centre of the ratio image, fluorescence becomes increasingly out of focus and ratios contain contributions from both the upper and lower membrane surfaces. In the centre of the image, probe molecules are oriented predominantly along the microscope optical axis, where the probability of excitation is low and fluorescence polarization is only weakly dependent on lipid order. By comparing the polarization ratio around the membrane perimeter with theoretical curves such as those illustrated in Fig. 46.5, we can determine both the degree of lipid order, i.e. θ_{max}, and the fraction of probe molecules, if any, that are aligned parallel to the plane of the membrane. In this example, we find $\theta_{max} = 32 \pm 4°$ for POPC and $17 \pm 2°$ for DPPC, with 10% and 5%, respectively, of probes aligned parallel to the plane of the bilayer (Florine-Casteel, 1990).

46.4 BIOLOGICAL APPLICATIONS

We have used the methodology outlined above to measure lipid order in the blebbing plasma membrane of single, cultured rat hepatocytes during ATP depletion (Florine-Casteel *et al.*, 1990, 1991). Cells on coverslips were incubated in medium containing the metabolic inhibitors cyanide and iodoacetic acid to induce the ATP depletion and bleb formation observed in hypoxic injury (Lemasters *et al.*, 1987). After subsequent transfer to the microscope stage, cells were labelled with the plasma membrane probe TMA-DPH by brief incubation in probe-containing buffer. We chose the membrane probe TMA-DPH because of its significantly slower rate of internalization compared to DPH (Kuhry *et al.*, 1983; Florine-Casteel *et al.*, 1990).

Plate 46.2 illustrates a polarization measurement on a pair of plasma membrane blebs. Experimental conditions were the same as described for Plate 46.1, except that an intensified CCD camera was used as the detector. The blebs are structurally similar to liposomes and give similar polarized fluorescence patterns (compare Plate 46.1 with the left-hand column of Plate 46.2). In this example, an additional pair of fluorescence images, $F_{\perp}$ and $F_{\parallel}$, was acquired at a second orientation of the exitation polarizer, orthogonal to the first (right-hand column of Plate 46.2), in order to more accurately determine lipid order over the entire bleb perimeter. Because of small amounts of probe internalization during the course of the measurements, some probe molecules will be transiently oriented parallel to the plane of the membrane. Lipid order is most accurately measured in those regions of the membrane perimeter where the in-plane component is perpendicular to the excitation polarization direction and thus only weakly excited, i.e. at $\rho = 0°$ and 180° (see Fig. 46.5). Therefore, by combining the information from two ratio images, obtained using orthogonal excitation polarizer orientations, we can measure lipid order over most of the bleb perimeter without regard to probe internalization. As in the case of the liposomes, the central regions of the bleb images contain weak, out-of-focus fluorescence from the upper and lower surfaces, making lipid order measurements difficult in these regions. However, the ratios at $\rho = 0°$ and 180° on the bleb perimeters range from 0.22 to 0.28, corresponding to $\theta_{max} = 10$–20° (Florine-Casteel *et al.*, 1991; Fig. 46.5(b)), indicating a uniformly rigid membrane surface. The plasma membrane of the main cell body displays a fairly uniform polarization ratio of 0.6–0.7, with little variation around the in-focus cell perimeter, due to the randomizing effect of the microvillous structure on probe orientation.

The information on lipid order that is obtainable in a single cell by the method we have described depends on three factors: cell geometry, the properties of the membrane probe, and the particular combination of excitation/emission polarizer orientations used in image acquisition. In the above example, we looked at the simple case of a spherical membrane surface (a plasma membrane bleb) labelled with a rod-shaped probe of known orientation in the membrane (TMA-DPH). We were able to completely describe the plasma membrane lipid order profile around the bleb perimeter with a set of four fluorescence images. A different cell geometry, or a probe with different absorption/emission dipole orientations or diffusion properties, would require the appropriate modifications to equations [46.1]–[46.4] in order to determine the functional dependence of the polarization ratio on lipid order and membrane surface location. Image acquisition at other polarizer orientations might also be necessary.

ACKNOWLEDGEMENTS

This work was supported, in part, by grant AG07218 from the National Institutes of Health and grant J-1433 from the Office of Naval Research.

REFERENCES

Axelrod D. (1979) *Biophys. J.* **26**, 557–574.
Axelrod D. (1989) *Methods Cell Biol.* **30**, 333–352.
Florine-Casteel K. (1990) *Biophys. J.* **57**, 1199–1215.
Florine-Casteel K., Lemasters J.J. & Herman B. (1990) In *Optical Microscopy for Biology*, B. Herman & K. Jacobson (eds). Wiley-Liss, New York, pp. 559–573.
Florine-Casteel K., Lemasters J.J. & Herman B. (1991) *FASEB J.* **5**, 2078–2084.
Kinosita Jr. K., Kawato S. & Ikegami A. (1977) *Biophys. J.* **20**, 289–305.
Kuhry J.-G., Fonteneau P., Duportail G., Maechling C. & Laustriat G. (1983) *Cell Biophys.* **5**, 129–140.
Lemasters J.J., DiGuiseppi J., Niemien A.-L. & Herman B. (1987) *Nature* **325**, 78–81.
Lipari G. & Szabo A. (1980) *Biophys. J.* **30**, 489–506.

Index